Nonlinear Regression

Nonlinear Regression

G. A. F. SEBER and C. J. WILD
Department of Mathematics and Statistics
University of Auckland
Auckland, New Zealand

A JOHN WILEY & SONS, INC., PUBLICATION

Published by John Wiley & Sons, Inc., Hoboken, New Jersey.
Published simultaneously in Canada.

For general information on our other products and services please contact our Customer Care Department within the U.S. at 877-762-2974, outside the U.S. at 317-572-3993 or fax 317-572-4002.

Wiley also publishes its books in a variety of electronic formats. Some content that appears in print, however, may not be available in electronic format.

Library of Congress Cataloging-in-Publication is available.

ISBN 0-471-47135-6

Printed in the United States of America.

10 9 8 7 6 5 4 3 2 1

Preface

Some years ago one of the authors (G.A.F.S.) asked a number of applied statisticians how they got on with fitting nonlinear models. The answers were generally depressing. In many cases the available computing algorithms for estimation had unsatisfactory convergence properties, sometimes not converging at all, and there was some uncertainty about the validity of the linear approximation used for inference. Furthermore, parameter estimates sometimes had undesirable properties. Fortunately the situation has improved over recent years because of two major developments. Firstly, a number of powerful algorithms for fitting models have appeared. These have been designed to handle "difficult" models and to allow for the various contingencies that can arise in iterative optimization. Secondly, there has been a new appreciation of the role of curvature in nonlinear modeling and its effect on inferential procedures. Curvature comes in a two-piece suite: intrinsic curvature, which relates to the geometry of the nonlinear model, and parameter-effects curvature, which depends on the parametrization of the model. The effects of these curvatures have recently been studied in relation to inference and experimental design. Apart from a couple of earlier papers, all the published literature on the subject has appeared since 1980, and it continues to grow steadily. It has also been recognized that the curvatures can be regarded as quantities called connection coefficients which arise in differential geometry, the latter providing a unifying framework for the study of curvature. Although we have not pursued these abstract concepts in great detail, we hope that our book will at least provide an introduction and make the literature, which we have found difficult, more accessible.

As we take most of our cues for nonlinear modeling from linear models, it is essential that the reader be familiar with the general ideas of linear regression. The main results used are summarized in Appendix D. In this respect our book can be regarded as a companion volume to Seber [1977, 1984], which deal with linear regression analysis and multivariate methods.

We originally began writing this book with the intention of covering a wide range of nonlinear topics. However, we found that in spite of a smaller literature than that of linear regression or multivariate analysis, the subject is difficult and diverse, with many applications. We have therefore had to omit a number of important topics, including nonlinear simultaneous equation systems, generalized linear models (and nonlinear extensions), and stochastic approximation. Also, we have been unable to do full justice to the more theoretical econometric literature with its detailed asymptotics (as in Gallant [1987]), and to the wide range of models in the scientific literature at large.

Because of a paucity of books on nonlinear regression when we began this work, we have endeavored to cover both the applied and theoretical ends of the spectrum and appeal to a wide audience. As well as discussing practical examples, we have tried to make the theoretical literature more available to the reader without being too entangled in detail. Unfortunately, most results tend to be asymptotic or approximate in nature, so that asymptotic expansions tend to dominate in some chapters. This has meant some unevenness in the level of difficulty throughout the book. However, although our book is predominantly theoretical, we hope that the balance of theory and practice will make the book useful from both the teaching and the research point of view. It is not intended to be a practical manual on how to do nonlinear fitting; rather, it considers broad aspects of model building and statistical inference.

One of the irritations of fitting nonlinear models is that model fitting generally requires the iterative optimization (minimization or maximization) of functions. Unfortunately, the iterative process often does not converge easily to the desired solution. The optimization algorithms in widespread use are based upon modifications of and approximations to the Newton(–Raphson) and Gauss–Newton algorithms. In unmodified form both algorithms are unreliable. Computational questions are therefore important in nonlinear regression, and we have devoted three chapters to this area. We introduce the basic algorithms early on, and demonstrate their weaknesses. However, rather than break the flow of statistical ideas, we have postponed a detailed discussion of how these algorithms are made robust until near the end of the book. The computational chapters form a largely self-contained introduction to unconstrained optimization.

In Chapter 1, after discussing the notation, we consider the various types of nonlinear model that can arise. Methods of estimating model parameters are discussed in Chapter 2, and some practical problems relating to estimation, like ill-conditioning, are introduced in Chapter 3. Chapter 4 endeavors to summarize some basic ideas about curvature and to bring to notice the growing literature on the subject. In Chapter 5 we consider asymptotic and exact inferences relating to confidence intervals and regions, and hypothesis testing. The role of curvature is again considered, and some aspects of optimal design close the chapter. Autocorrelated errors are the subject of Chapter 6, and Chapters 7, 8, and 9 describe in depth, three broad families of popular models, namely, growth-curve, compartmental, and change-of-phase and spline-regression models. We have not tried to cover every conceivable model, and our coverage thus complements Ratkowsky's [1988] broader description of families of parametric models. Errors-in-variables models are discussed in detail in Chapter 10 for both explicit and implicit nonlinear models, and nonlinear multivariate models are considered briefly in Chapter 11. Almost by way of an appendix, Chapter 12 gives us a glimpse of some of the basic asymptotic theory, and Chapters 13 to 15 provide an introduction to the growing literature on algorithms for optimization and least squares, together with practical advice on the use of such programs. The book closes with five appendices, an author index, an extensive list of references, and a subject index. Appendix A deals with matrix results, Appendix B gives an introduction to some basic concepts of differential geometry and curvature, Appendix C outlines some theory of stochastic differential equations, Appendix D summarizes linear regression theory, and Appendix E discusses a computational method for handling linear equality constraints. A number of topics throughout the book can be omitted at first reading: these are starred in the text and the contents.

We would like to express our sincere thanks to Mrs. Lois Kennedy and her associates in the secretarial office for typing the difficult manuscript. Thanks go also to Betty Fong for the production of many of the figures. We are grateful to our colleagues and visitors to Auckland for many helpful discussions and suggestions, in particular Alastair Scott, Alan Rogers, Jock MacKay, Peter Phillips, and Ray Carroll. Our thanks also go to Miriam Goldberg, David Hamilton, Dennis Cook, and Douglas Bates for help with some queries in Chapter 4.

We wish to thank the authors, editors, and owners of copyright for permission to reproduce the following published tables and figures: Tables 2.1, 6.1, 6.2, 6.4, 6.5, 7.1, 8.3, 8.4, Figs. 2.1, 6.4, 7.4, 7.5, 7.6, 9.6, 9.10, 9.11 (copyright by the Biometric Society); Tables 2.2, 2.3, 4.1, 4.2, 9.4, Figs. 2.2, 2.3, 6.2, 9.1, 9.2, 9.4, 9.12, 9.16 (copyright by the Royal Statistical Society); Tables 2.5, 5.1, 5.3, 8.1, 9.1, 9.2, 9.5, 10.2, 10.3, 10.4, 10.5, 10.6, 11.1, 11.2, 11.4, 11.5, 11.6, Figs. 2.4, 2.5, 5.1, 5.2, 5.6, 5.7, 5.8, 5.9, 9.5, 9.7, 9.9., 9.17, 10.1, 11.1, 11.2, 11.6, 11.7 (copyright by the American Statistical Association); Table 3.5 (copyright by Akademie-Verlag, Berlin, DDR); Tables 9.3, 11.3, Figs. 5.4, 9.8, 9.14, 11.3, 11.4 (copyright by Biometrika Trust); Figs. 4.3, 5.3 (copyright by the Institute of Mathematical Statistics); Table 2.4 (copyright by the American Chemical Society); Table 3.1, Figs. 3.2, 3.3 (copyright by Plenum Press); Tables 8.2, 10.1, Figs. 8.12, 8.13, 8.14 (copyright by North-Holland Physics Publishing Company); Fig. 9.15 (copyright by Marcel Dekker); Table 5.2 (copyright by Oxford University Press); Figs. 7.2, 7.3 (copyright by Alan R. Liss); Table 7.2 and Fig. 7.7 (copyright by Growth Publishing Company); and Table 6.3 and Fig. 6.3 (copyright by John Wiley and Sons).

G. A. F. SEBER
C. J. WILD

Auckland, New Zealand
May, 1988

Contents

1 Model Building **1**

1.1 Notation, 1
1.2 Linear and Nonlinear Models, 4
1.3 Fixed-Regressor Model, 10
1.3.1 Regressors measured without error, 10
1.3.2 Conditional regressor models, 11
1.4 Random-Regressor (Errors-in-Variables) Models, 11
1.4.1 Functional relationships, 11
1.4.2 Structural relationships, 12
1.5 Controlled Regressors with Error, 13
1.6 Generalized Linear Model, 14
1.7 Transforming to Linearity, 15
1.8 Models with Autocorrelated Errors, 18
1.9 Further Econometric Models, 19

2 Estimation Methods **21**

2.1 Least Squares Estimation, 21
2.1.1 Nonlinear least squares, 21
2.1.2 Linear approximation, 23
2.1.3 Numerical methods, 25
2.1.4 Generalized least squares, 27
*2.1.5 Replication and test of fit, 30
2.2 Maximum-Likelihood Estimation, 32
2.2.1 Normal errors, 32

*Starred topics can be omitted at first reading.

2.2.2 Nonnormal data, 34
2.2.3 Concentrated likelihood methods, 37
*2.3 Quasi-likelihood Estimation, 42
*2.4 LAM Estimator, 48
*2.5 L_1-norm Minimization, 50
2.6 Robust Estimation, 50
2.7 Bayesian Estimation, 52
2.7.1 Choice of prior distributions, 53
2.7.2 Posterior distributions, 55
2.7.3 Highest-posterior-density regions, 63
2.7.4 Normal approximation to posterior density, 64
2.7.5 Test for model adequacy using replication, 65
*2.7.6 Polynomial estimators, 66
2.8 Variance Heterogeneity, 68
2.8.1 Transformed models, 70
a. Box–Cox transformations, 71
b. John–Draper transformations, 72
2.8.2 Robust estimation for model A, 73
2.8.3 Inference using transformed data, 74
2.8.4 Transformations to linearity, 75
2.8.5 Further extensions of the Box–Cox method, 76
2.8.6 Weighted least squares: model B, 77
2.8.7 Prediction and transformation bias, 86
2.8.8 Generalized least-squares model, 88

3 Commonly Encountered Problems **91**

3.1 Problem Areas, 91
3.2 Convergence of Iterative Procedures, 91
3.3 Validity of Asymptotic Inference, 97
3.3.1 Confidence regions, 97
3.3.2 Effects of curvature, 98
3.4 Identifiability and Ill-conditioning, 102
3.4.1 Identifiability problems, 102
3.4.2 Ill-conditioning, 103
a. Linear models, 103
b. Nonlinear models, 110
c. Stable parameters, 117
d. Parameter redundancy, 118
e. Some conclusions, 126

4 Measures of Curvature and Nonlinearity **127**

4.1 Introduction, 127
4.2 Relative Curvature, 128
4.2.1 Curvature definitions, 129
4.2.2 Geometrical properties, 131
*4.2.3 Reduced formulae for curvatures, 138
*4.2.4 Summary of formulae, 145
4.2.5 Replication and curvature, 146
*4.2.6 Interpreting the parameter-effects array, 147
*4.2.7 Computing the curvatures, 150
*4.2.8 Secant approximation of second derivatives, 154
4.3 Beale's Measures, 157
*4.4 Connection Coefficients, 159
4.5 Subset Curvatures, 165
4.5.1 Definitions, 165
*4.5.2 Reduced formulae for subset curvatures, 168
*4.5.3 Computations, 170
4.6 Analysis of Residuals, 174
4.6.1 Quadratic approximation, 174
4.6.2 Approximate moments of residuals, 177
4.6.3 Effects of curvature on residuals, 178
4.6.4 Projected residuals, 179
a. Definition and properties, 179
b. Computation of projected residuals, 181
4.7 Nonlinearity and Least-Squares Estimation, 181
4.7.1 Bias, 182
4.7.2 Variance, 182
4.7.3 Simulated sampling distributions, 184
4.7.4 Asymmetry measures, 187

5 Statistical Inference **191**

5.1 Asymptotic Confidence Intervals, 191
5.2 Confidence Regions and Simultaneous Intervals, 194
5.2.1 Simultaneous intervals, 194
5.2.2 Confidence regions, 194
5.2.3 Asymptotic likelihood methods, 196
5.3 Linear Hypotheses, 197
5.4 Confidence Regions for Parameter Subsets, 202
5.5 Lack of Fit, 203

*5.6 Replicated Models, 204
*5.7 Jackknife Methods, 206
*5.8 Effects of Curvature on Linearized Regions, 214
5.8.1 Intrinsic curvature, 214
5.8.2 Parameter-effects curvature, 218
5.8.3 Summary of confidence regions, 220
5.8.4 Reparametrization to reduce curvature effects, 222
5.8.5 Curvature and parameter subsets, 227
5.9 Nonlinear Hypotheses, 228
5.9.1 Three test statistics, 228
5.9.2 Normally distributed errors, 229
5.9.3 Freedom-equation specification, 232
5.9.4 Comparison of test statistics, 234
5.9.5 Confidence regions and intervals, 235
5.9.6 Multiple hypothesis testing, 235
5.10 Exact Inference, 236
5.10.1 Hartley's method, 236
5.10.2 Partially linear models, 240
5.11 Bayesian Inference, 245
*5.12 Inverse Prediction (Discrimination), 245
5.12.1 Single prediction, 245
5.12.2 Multiple predictions, 246
5.12.3 Empirical Bayes interval, 247
*5.13 Optimal Design, 250
5.13.1 Design criteria, 250
5.13.2 Prior estimates, 255
5.13.3 Sequential designs, 257
5.13.4 Multivariate models, 259
5.13.5 Competing models, 260
5.13.6 Designs allowing for curvature, 260
a. Volume approximation, 260
b. An example, 264
c. Conclusions, 269

6 Autocorrelated Errors **271**

6.1 Introduction, 271
6.2 AR(1) Errors, 275
6.2.1 Preliminaries, 275
6.2.2 Maximum-likelihood estimation, 277

6.2.3 Two-stage estimation, 279
6.2.4 Iterated two-stage estimation, 280
6.2.5 Conditional least squares, 281
6.2.6 Choosing between the estimators, 282
6.2.7 Unequally spaced time intervals, 285
6.3 AR(2) Errors, 286
6.4 AR(q_1) Errors, 289
6.4.1 Introduction, 289
6.4.2 Preliminary results, 290
6.4.3 Maximum-likelihood estimation and approximations, 294
a. Ignore the determinant, 295
b. Approximate the derivative of the determinant, 295
c. Asymptotic variances, 296
6.4.4 Two-stage estimation, 301
6.4.5 Choosing a method, 303
6.4.6 Computational considerations, 304
6.5 MA(q_2) Errors, 305
6.5.1 Introduction, 305
6.5.2 Two-stage estimation, 306
6.5.3 Maximum-likelihood estimation, 306
6.6 ARMA(q_1, q_2) Errors, 307
6.6.1 Introduction, 307
6.6.2 Conditional least-squares method, 310
6.6.3 Other estimation procedures, 314
6.7 Fitting and Diagnosis of Error Processes, 318
6.7.1 Choosing an error process, 318
6.7.2 Checking the error process, 321
a. Overfitting, 321
b. Use of noise residuals, 322

7 Growth Models **325**

7.1 Introduction, 325
7.2 Exponential and Monomolecular Growth Curves, 327
7.3 Sigmoidal Growth Models, 328
7.3.1 General description, 328
7.3.2 Logistic (autocatalytic) model, 329
7.3.3 Gompertz growth curve, 330
7.3.4 Von Bertalanffy model, 331

7.3.5 Richards curve, 332
7.3.6 Starting values for fitting Richards models, 335
7.3.7 General procedure for sigmoidal curves, 337
7.3.8 Weibull model, 338
7.3.9 Generalized logistic model, 339
7.3.10 Fletcher family, 339
7.3.11 Morgan–Mercer–Flodin (MMF) family, 340

7.4 Fitting Growth Models: Deterministic Approach, 342

7.5 Stochastic Growth-Curve Analysis, 344
7.5.1 Introduction, 344
7.5.2 Rates as functions of time, 346
a. Uncorrelated error process, 347
b. Autocorrelated error processes, 348
7.5.3 Rates as functions of size, 353
a. A tractable differential equation, 354
b. Approximating the error process, 356

7.6 Yield–Density Curves, 360
7.6.1 Preliminaries, 360
7.6.2 Bleasdale–Nelder model, 363
7.6.3 Holliday model, 364
7.6.4 Choice of model, 364
7.6.5 Starting values, 365

8 Compartmental Models **367**

8.1 Introduction, 367

8.2 Deterministic Linear Models, 370
8.2.1 Linear and nonlinear models, 370
8.2.2 Tracer exchange in steady-state systems, 372
8.2.3 Tracers in nonsteady-state linear systems, 375

8.3 Solving the Linear Differential Equations, 376
8.3.1 The solution and its computation, 376
8.3.2 Some compartmental structures, 383
8.3.3 Properties of compartmental systems, 384

*8.4 Identifiability in Deterministic Linear Models, 386

8.5 Linear Compartmental Models with Random Error, 393
8.5.1 Introduction, 393
8.5.2 Use of the chain rule, 395
8.5.3 Computer-generated exact derivatives, 396
a. Method of Bates et al., 396

b. Method of Jennrich and Bright, 400
c. General algorithm, 401
8.5.4 The use of constraints, 402
8.5.5 Fitting compartmental models without using derivatives, 404
8.5.6 Obtaining initial parameter estimates, 406
a. All compartments observed with zero or linear inputs, 406
b. Some compartments unobserved, 407
c. Exponential peeling, 407
8.5.7 Brunhilda example revisited, 410
8.5.8 Miscellaneous topics, 412
8.6 More Complicated Error Structures, 413
*8.7 Markov-Process Models, 415
8.7.1 Background theory, 415
a. No environmental input, 416
b. Input from the environment, 420
8.7.2 Computational methods, 423
a. Unconditional generalized least squares, 423
b. Conditional generalized least squares, 424
8.7.3 Discussion, 429
8.8 Further Stochastic Approaches, 431

9 Multiphase and Spline Regressions 433

9.1 Introduction, 433
9.2 Noncontinuous Change of Phase for Linear Regimes, 438
9.2.1 Two linear regimes, 438
9.2.2 Testing for a two-phase linear regression, 440
9.2.3 Parameter inference, 445
9.2.4 Further extensions, 446
9.3 Continuous Case, 447
9.3.1 Models and inference, 447
9.3.2 Computation, 455
9.3.3 Segmented polynomials, 457
a. Inference, 460
b. Computation, 463
9.3.4 Exact tests for no change of phase in polynomials, 463
9.4 Smooth Transitions between Linear Regimes, 465
9.4.1 The sgn formulation, 465
9.4.2 The max–min formulation, 471

a. Smoothing max$(0, z)$, 472
b. Limiting form for max$\{z_i\}$, 474
c. Extending sgn to a vector of regressors, 476
9.4.3 Examples, 476
9.4.4 Discussion, 480
9.5 Spline Regression, 481
9.5.1 Fixed and variable knots, 481
a. Fixed knots, 484
b. Variable knots, 484
9.5.2 Smoothing splines, 486

***10 Errors-In-Variables Models 491**

10.1 Introduction, 491
10.2 Functional Relationships: Without Replication, 492
10.3 Functional Relationships: With Replication, 496
10.4 Implicit Functional Relationships: Without Replication, 501
10.5 Implicit Functional Relationships: With Replication, 508
10.5.1 Maximum-likelihood estimation, 508
10.5.2 Bayesian estimation, 510
10.6 Implicit Relationships with Some Unobservable Responses, 516
10.6.1 Introduction, 516
10.6.2 Least-squares estimation, 516
10.6.3 The algorithm, 519
10.7 Structural and Ultrastructural Models, 523
10.8 Controlled Variables, 525

11 Multiresponse Nonlinear Models 529

11.1 General Model, 529
11.2 Generalized Least Squares, 531
11.3 Maximum-Likelihood Inference, 536
11.3.1 Estimation, 536
11.3.2 Hypothesis testing, 538
11.4 Bayesian Inference, 539
11.4.1 Estimates from posterior densities, 539
11.4.2 H.P.D. regions, 542
11.4.3 Missing observations, 544
*11.5 Linear Dependencies, 545
11.5.1 Dependencies in expected values, 545
11.5.2 Dependencies in the data, 546

11.5.3 Eigenvalue analysis, 547
11.5.4 Estimation procedures, 549
*11.6 Functional Relationships, 557

***12 Asymptotic Theory** **563**

12.1 Introduction, 563
12.2 Least-Squares Estimation, 563
12.2.1 Existence of least-squares estimate, 563
12.2.2 Consistency, 564
12.2.3 Asymptotic normality, 568
12.2.4 Effects of misspecification, 572
12.2.5 Some extensions, 574
12.2.6 Asymptotics with vanishingly small errors, 575
12.3 Maximum-Likelihood Estimation, 576
12.4 Hypothesis Testing, 576
12.5 Multivariate Estimation, 581

13 Unconstrained Optimization **587**

13.1 Introduction, 587
13.2 Terminology, 588
13.2.1 Local and global minimization, 588
13.2.2 Quadratic functions, 590
13.2.3 Iterative algorithms, 593
a. Convergence rates, 593
b. Descent methods, 594
c. Line searches, 597
13.3 Second-Derivative (Modified Newton) Methods, 599
13.3.1 Step-length methods, 599
a. Directional discrimination, 600
b. Adding to the Hessian, 602
13.3.2 Restricted step methods, 603
13.4 First-Derivative Methods, 605
13.4.1 Discrete Newton methods, 605
13.4.2 Quasi-Newton methods, 605
*13.4.3 Conjugate-gradient methods, 609
13.5 Methods without Derivatives, 611
13.5.1 Nonderivative quasi-Newton methods, 611
*13.5.2 Direction-set (conjugate-direction) methods, 612
13.5.3 Direct search methods, 615

13.6 Methods for Nonsmooth Functions, 616
13.7 Summary, 616

14 Computational Methods for Nonlinear Least Squares **619**

14.1 Gauss–Newton Algorithm, 619
14.2 Methods for Small-Residual Problems, 623
14.2.1 Hartley's method, 623
14.2.2 Levenberg–Marquardt methods, 624
14.2.3 Powell's hybrid method, 627
14.3 Large-Residual Problems, 627
14.3.1 Preliminaries, 627
14.3.2 Quasi-Newton approximation of $\mathbf{A}(\boldsymbol{\theta})$, 628
14.3.3 The Gill–Murray method, 633
a. Explicit second derivatives, 636
b. Finite-difference approximation, 636
c. Quasi-Newton approximation, 637
14.3.4 Summary, 639
14.4 Stopping Rules, 640
14.4.1 Convergence criteria, 640
14.4.2 Relative offset, 641
14.4.3 Comparison of criteria, 644
14.5 Derivative-Free Methods, 646
14.6 Related Problems, 650
14.6.1 Robust loss functions, 650
14.6.2 L_1-minimization, 653
14.7 Separable Least-Squares Problems, 654
14.7.1 Introduction, 654
14.7.2 Gauss–Newton for the concentrated sum of squares, 655
14.7.3 Intermediate method, 657
14.7.4 The NIPALS method, 659
14.7.5 Discussion, 660

15 Software Considerations **661**

15.1 Selecting Software, 661
15.1.1 Choosing a method, 661
15.1.2 Some sources of software, 663
a. Commercial libraries, 663
b. Noncommercial libraries, 663
c. ACM algorithms, 663

15.2 User-Supplied Constants, 664
15.2.1 Starting values, 665
a. Initial parameter value, 665
b. Initial approximation to the Hessian, 666
15.2.2 Control constants, 666
a. Step-length accuracy, 666
b. Maximum step length, 666
c. Maximum number of function evaluations, 667
15.2.3 Accuracy, 667
a. Precision of function values, 667
b. Magnitudes, 668
15.2.4 Termination criteria (stopping rules), 668
a. Convergence of function estimates, 669
b. Convergence of gradient estimates, 670
c. Convergence of parameter estimates, 670
d. Discussion, 671
e. Checking for a minimum, 672
15.3 Some Causes and Modes of Failure, 672
15.3.1 Programming error, 672
15.3.2 Overflow and underflow, 673
15.3.3 Difficulties caused by bad scaling (parametrization), 673
15.4 Checking the Solution, 675

APPENDIXES

A. Vectors and Matrices **677**

A1 Rank, 677
A2 Eigenvalues, 677
A3 Patterned matrices, 678
A4 Positive definite and semidefinite matrices, 678
A5 Generalized inverses, 679
A6 Ellipsoids, 679
A7 Optimization, 679
A8 Matrix factorizations, 680
A9 Multivariate *t*-distribution, 681
A10 Vector differentiation, 682
A11 Projection matrices, 683
A12 Quadratic forms, 684

A13 Matrix operators, 684
A14 Method of scoring, 685

B. Differential Geometry **687**

B1 Derivatives for curves, 687
B2 Curvature, 688
B3 Tangent planes, 690
B4 Multiplication of 3-dimensional arrays, 691
B5 Invariance of intrinsic curvature, 692

C. Stochastic Differential Equations **695**

C1 Rate equations, 695
C2 Proportional rate equations, 697
C3 First-order equations, 698

D. Multiple Linear Regression **701**

D1 Estimation, 701
D2 Inference, 702
D3 Parameter subsets, 703

E. Minimization Subject to Linear Constraints **707**

References **711**

Author Index **745**

Subject Index **753**

Nonlinear Regression

CHAPTER 1

Model Building

1.1 NOTATION

Matrices and vectors are denoted by boldface letters **A** and **a**, respectively, and scalars by italics. If **a** is an $n \times 1$ column vector with elements $a_1, a_2, \ldots, a_n$, we write $\mathbf{a} = [(a_j)]$, and the *length* or *norm* of **a** is denoted by $\|\mathbf{a}\|$. Thus

$$\|\mathbf{a}\| = (\mathbf{a}'\mathbf{a})^{1/2} = (a_1^2 + a_2^2 + \cdots + a_n^2)^{1/2}.$$

The $n \times 1$ vector with all its elements equal to unity is represented by $\mathbf{1}_n$.

If the $m \times n$ matrix **A** has elements a_{ij}, we write $\mathbf{A} = [(a_{ij})]$, and the sum of the diagonal elements, called the *trace* of **A**, is denoted by $\operatorname{tr} \mathbf{A}$ $(= a_{11} + a_{22} + \cdots + a_{kk})$, where k is the smaller of m and n). The transpose of **A** is represented by $\mathbf{A}' = [(a'_{ij})]$, where $a'_{ij} = a_{ji}$. A matrix $\mathbf{A}^-$ that satisfies $\mathbf{A}\mathbf{A}^-\mathbf{A} = \mathbf{A}$ is called a generalized *inverse* of **A**. If **A** is square, its determinant is written $|\mathbf{A}|$, or $\det \mathbf{A}$ and if **A** is nonsingular ($|\mathbf{A}| \neq 0$), its inverse is denoted by $\mathbf{A}^{-1}$. The $n \times n$ matrix with diagonal elements $d_1, d_2, \ldots, d_n$ and zero elsewhere is represented by $\operatorname{diag}(d_1, d_2, \ldots, d_n)$, and when all the d_i's are unity, we have the identity matrix $\mathbf{I}_n$.

For any $m \times n$ matrix **A** the space spanned by the columns of **A**, called the *range space* or column space of **A**, is denoted by $\mathscr{R}[\mathbf{A}]$. The *null space* or kernel of **A** $(= \{\mathbf{x} : \mathbf{A}\mathbf{x} = \mathbf{0}\})$ is denoted by $\mathscr{N}[\mathbf{A}]$. If V is a vector space, then $V^\perp$ represents the *orthogonal complement* of V, i.e.

$$V^\perp = \{\mathbf{x} : \mathbf{x}'\mathbf{v} = 0 \quad \text{for all } \mathbf{v} \text{ in } V\}$$

The symbol $\mathbb{R}^d$ will represent d-dimensional Euclidean space. An orthogonal projection onto a vector subspace can be achieved using a symmetric idempotent matrix (called a *projection matrix*). Properties of projection matrices are described in Appendix A11.

It is common practice to distinguish between random variables and their values by using upper- and lowercase letters, for example Y and y. This is particularly helpful in discussing errors-in-variables models. However, it causes difficulty with multivariate models, so that, on balance, we have decided not to observe this convention. Instead we shall use boldface upper- and lowercase to denote matrices and vectors respectively of both constants and random variables, for example $\mathbf{Y}$ and $\mathbf{y}$.

If x and y are random variables, then the symbols $E[x]$, $\text{var}[x]$, $\text{cov}[x, y]$, $\text{corr}[x, y]$ and $E[x|y]$ represent the expectation, variance, covariance, correlation, and conditional expectation, respectively. Multivariate analogues are $\mathscr{E}[\mathbf{x}]$ for the expected value of random $\mathbf{x}$,

$$\mathscr{C}[\mathbf{x}, \mathbf{y}] = [(\text{cov}[x_i, y_j])] = \mathscr{E}[(\mathbf{x} - \boldsymbol{\mu}_x)(\mathbf{y} - \boldsymbol{\mu}_y)'] \tag{1.1}$$

for the covariance of two random vectors $\mathbf{x}$ and $\mathbf{y}$ with expected values (means) $\boldsymbol{\mu}_x$ and $\boldsymbol{\mu}_y$ respectively, and

$$\mathscr{D}[\mathbf{x}] = \mathscr{C}[\mathbf{x}, \mathbf{x}],$$

the *dispersion* or *variance–covariance* matrix of $\mathbf{x}$. The following will be used frequently:

$$\mathscr{C}[\mathbf{Ax}, \mathbf{By}] = \mathbf{A}\mathscr{C}[\mathbf{x}, \mathbf{y}]\mathbf{B}', \tag{1.2}$$

from which we deduce

$$\mathscr{D}[\mathbf{Ax}] = \mathbf{A}\mathscr{D}[\mathbf{x}]\mathbf{A}'. \tag{1.3}$$

We say that $y \sim N(\theta, \sigma^2)$ if y is normally distributed with mean θ and variance σ^2. If $\mathbf{y}$ has a d-dimensional multivariate normal distribution with mean $\boldsymbol{\theta}$ and dispersion matrix $\boldsymbol{\Sigma}$, we write $\mathbf{y} \sim N_d(\boldsymbol{\theta}, \boldsymbol{\Sigma})$. The t- and chi-square distributions with k degrees of freedom are denoted by t_k and χ_k^2, respectively, and the F-distribution with m and n degrees of freedom is denoted by $F_{m,n}$. An important distribution used in Bayesian methods is the inverted gamma distribution. This has density function

$$p(x|a, b) = \frac{2}{\Gamma(b/2)}\left(\frac{ab}{2}\right)^{a/2} \frac{1}{x^{b+1}} e^{-ab/(2x^2)}, \qquad a, b > 0 \text{ and } x > 0, \tag{1.4}$$

and provides the useful integral

$$\int_0^\infty x^{-b-1} \exp\left(-\frac{ab}{2x^2}\right) dx = \frac{\Gamma(b/2)}{2(ab/2)^{a/2}}. \tag{1.5}$$

If $\{\mathbf{z}_n\}$ is a sequence of random vectors, we write

$$\text{plim}\, \mathbf{z}_n = \mathbf{z} \tag{1.6}$$

if, for every $\delta > 0$,

$$\lim_{n\to\infty} \text{pr}(\|\mathbf{z}_n - \mathbf{z}\| \leq \delta) = 1.$$

Also, if $g(\cdot)$ is a positive function of n, we write

$$\mathbf{z}_n = \mathbf{o}_p[g(n)] \tag{1.7}$$

if plim $\mathbf{z}_n/g(n) = \mathbf{0}$, and

$$\mathbf{z}_n = \mathbf{O}_p[g(n)] \tag{1.8}$$

if, for each ε, there exists $A(\varepsilon) > 0$ such that

$$\text{pr}[\|\mathbf{z}_n\| \leq A(\varepsilon)g(n)] \geq 1 - \varepsilon$$

for all n.

Differentiation with respect to vectors is used frequently in this book, and as a variety of notations are used in practice, we now state our own. If $a(\cdot)$ and $\mathbf{b}(\cdot)$ are, respectively, a differentiable function and a differentiable $n \times 1$ vector function of the p-dimensional vector $\boldsymbol{\theta}$, we define

$$\frac{\partial a(\boldsymbol{\theta})}{\partial \boldsymbol{\theta}} = \left(\frac{\partial a(\boldsymbol{\theta})}{\partial \theta_1}, \frac{\partial a(\boldsymbol{\theta})}{\partial \theta_2}, \ldots, \frac{\partial a(\boldsymbol{\theta})}{\partial \theta_p}\right)' = \left(\frac{\partial a(\boldsymbol{\theta})}{\partial \boldsymbol{\theta}'}\right)', \tag{1.9}$$

$$\frac{\partial \mathbf{b}(\boldsymbol{\theta})}{\partial \theta_r} = \left(\frac{\partial b_1(\boldsymbol{\theta})}{\partial \theta_r}, \frac{\partial b_2(\boldsymbol{\theta})}{\partial \theta_r}, \ldots, \frac{\partial b_n(\boldsymbol{\theta})}{\partial \theta_r}\right)',$$

and

$$\frac{\partial \mathbf{b}(\boldsymbol{\theta})}{\partial \boldsymbol{\theta}'} = \left[\left(\frac{\partial b_r(\boldsymbol{\theta})}{\partial \theta_s}\right)\right] = \left(\frac{\partial \mathbf{b}'(\boldsymbol{\theta})}{\partial \boldsymbol{\theta}}\right)'. \tag{1.10}$$

If $a(\cdot)$ is twice differentiable, then we define

$$\frac{\partial^2 a(\boldsymbol{\theta})}{\partial \boldsymbol{\theta}\, \partial \boldsymbol{\theta}'} = \left[\left(\frac{\partial^2 a(\boldsymbol{\theta})}{\partial \theta_r \partial \theta_s}\right)\right], \tag{1.11}$$

where

$$\frac{\partial^2 a(\boldsymbol{\theta})}{\partial \theta_r \partial \theta_s} = \frac{\partial}{\partial \theta_r}\left(\frac{\partial a(\boldsymbol{\theta})}{\partial \theta_s}\right).$$

In this book we will frequently use a dot to denote differentiation, e.g. $\dot{f}(x) = df(x)/dx$.

Finally we note that a number of results are included in the appendices, and references to these are given, for example, as A4.2.

1.2 LINEAR AND NONLINEAR MODELS

An important task in statistics is to find the relationships, if any, that exist in a set of variables when at least one is random, being subject to random fluctuations and possibly measurement error. In regression problems typically one of the variables, often called the *response* or dependent variable, is of particular interest and is denoted by y. The other variables $x_1, x_2, \ldots, x_k$, usually called *explanatory*, *regressor*, or independent variables, are primarily used to predict or explain the behavior of y. If plots of the data suggest some relationship between y and the x_i's, then we would hope to express this relationship via some function f, namely

$$y \approx f(x_1, x_2, \ldots, x_k). \tag{1.12}$$

Using f, we might then predict y for a given set of x's. For example, y could be the price of a used car of a certain make, x_1 the number of previous owners, x_2 the age of the car, and x_3 the mileage. As in (1.12), relationships can never hold exactly, as data will invariably contain unexplained fluctuations or noise, and some degree of measurement error is usually present.

Explanatory variables can be random or fixed (e.g. controlled). Consider an experiment conducted to measure the yield (y) of wheat at several different specified levels of density of planting (x_1) and fertilizer application (x_2). Then both x_1 and x_2 are fixed. If, at the time of planting, soil pH (x_3) was also measured on each plot, then x_3 would be random.

Sometimes the appropriate mathematical form for the relationship (1.12) is known, except for some unknown constants or coefficients (called *parameters*), the relationship being determined by a known underlying physical process or governed by accepted scientific laws. Mathematically (1.12) can be written as

$$y \approx f(x_1, x_2, \ldots, x_k; \boldsymbol{\theta}), \tag{1.13}$$

where f is entirely known except for the parameter vector $\boldsymbol{\theta} = (\theta_1, \theta_2, \ldots, \theta_p)'$, which needs to be estimated. For example

$$y \approx \alpha e^{\beta x},$$

and $\boldsymbol{\theta} = (\alpha, \beta)'$. Since these parameters often have physical interpretations, a major aim of the investigation is to estimate the parameters as precisely as possible. Frequently the relationship (1.13) is tentatively suggested by theoretical investigations and a further aim is to test the fit of the data to the model. With two variables (x, y) an obvious first step is to graph the estimated functional relationship and see how well it fits a scatterplot of the data points (x_i, y_i), $i = 1, 2, \ldots, n$. Models with a known f are the main subject of this book.

However, in much of statistical practice, particularly in the biological as opposed to the physical sciences, the underlying processes are generally complex and not well understood. This means that we have little or no idea about the form of the relationship and our aim is to simply find some function f for which (1.13) holds as closely as possible. Again one usually assumes that the "true" relationship belongs to a parametric family (1.13), so that finding a model often reduces to estimating $\boldsymbol{\theta}$. In this case, it is important to use a family that is large and flexible enough to approximate a sufficiently wide variety of functional forms. If several models fit the data equally well, we would usually choose the simplest one. Even in this situation, where the unknown parameters do not have special significance, a model which fits the data well will be useful. For example it could be used for predicting y, or even controlling y, by adjusting the x-variables. In some situations we may also use the model for calibration (inverse prediction), in which we predict x for a new y-value. If we are interested in simply fitting the best "black box" model to use, for example, in prediction, or in finding the maximum and minimum values of the curve or the slope at particular points, then splines (segmented polynomials) may be appropriate. These are discussed in Section 9.5.

In linear regression analysis (see Seber [1977]) models of the form $y \approx \beta_0 + \beta_1 x_1 + \cdots + \beta_{p-1} x_{p-1}$ with additive errors are used. We thus have

$$y = \beta_0 + \beta_1 x_1 + \cdots + \beta_{p-1} x_{p-1} + \varepsilon,$$

where $E[\varepsilon] = 0$. Here $\boldsymbol{\theta} = (\beta_0, \beta_1, \ldots, \beta_{p-1})'$ and the x_i's can include squares, cross products, higher powers, and even transformations (e.g. logarithms) of the original measurements. The important requirement is that the expression should be linear in the parameters. For example

$$y \approx \beta_0 + \beta_1 x_1 + \beta_2 x_2 + \beta_3 x_1^2 + \beta_4 x_2^2 + \beta_5 x_1 x_2$$

and

$$y \approx \beta_0 + \beta_1 \sin x_1 + \beta_2 \sin x_2$$

are both linear models, whereas

$$y \approx \beta_0 + \beta_1 e^{\beta_2 x}$$

is a nonlinear model, being nonlinear in β_2. We see from the above examples that the family of linear models is very flexible, and so it is often used in the absence of a theoretical model for f. As results from linear regression theory are used throughout this book, the relevant theory is summarized briefly in Appendix D.

Nonlinear models tend to be used either when they are suggested by theoretical considerations or to build known nonlinear behavior into a model. Even when a linear approximation works well, a nonlinear model may still be used to retain a clear interpretation of the parameters. Nonlinear models have been applied to a wide range of situations, even to finite populations (Valiant

[1985]), and we now give a selection of examples. The approximate equality sign of (1.13) will be replaced by an equality, through it should be understood that the given relationships are not expected to hold exactly in a set of data. We postpone considerations of random error until Section 1.3. For a review of some of the issues involved in nonlinear modeling see Cox [1977].

Example 1.1 Theoretical chemistry predicts that, for a given sample of gas kept at a constant temperature, the volume v and pressure p of the gas satisfy the relationship $pv^{\gamma} = c$. Writing $y = p$ and $x = v^{-1}$, we have

$$y = cx^{\gamma} = f(x; c, \gamma). \tag{1.14}$$

Here γ is a constant for each gas, which is to be estimated from the data. This particular model is said to be nonlinear, as f is nonlinear in one of the parameters, γ. In fact the model is nonlinear in γ but linear in c: a number of models have this property. We can, of course, take a log transformation so that setting $y = \log p$ and $x = \log v$, we have the re-expression

$$y = \alpha + \beta x,$$

where $\alpha = \log c$ and $\beta = -\gamma$. This model is now linear in its unknown parameters α and β. The decision whether or not to make linearizing transformation depends very much on the nature of the errors or fluctuations in the variables. This point is discussed later, in Section 1.7. ■

Example 1.2 In studying the relationships between migrant streams of people and the size of urban areas, Olsson [1965] (see also King [1969: 126] for a summary) proposed the following "gravity"-type model:

$$I = \frac{kP_1P_2}{D^b},$$

where k and b are constants, and I is the level of migration (interaction) between two areas at distance D apart with respective population sizes P_1 and P_2. This model can be constructed by arguing that migration will be inversely related to distance but proportional to the size of each population. Olsson, in fact, linearized the model using the transformation

$$\log \frac{I}{P_1P_2} = \log k - b \log D$$

or

$$y = \alpha + \beta x,$$

and showed that the absolute values of b for different pairs of areas were consistent with the hypothesis that "migrants from small places move shorter distances than migrants from large places." ■

Example 1.3 In an irreversible chemical reaction, substance A changes into substance B, which in turn changes into substance C. Denote by $A(t)$ the quantity of substance A at time t. The differential equations governing the changes are

$$\frac{dA(t)}{dt} = -\theta_1 A(t),$$

$$\frac{dB(t)}{dt} = \theta_1 A(t) - \theta_2 B(t),$$

$$\frac{dC(t)}{dt} = \theta_2 B(t),$$

where θ_1 and θ_2 are unknown constants. Assuming $A(0) = 1$, $B(t)$ has the solution

$$B(t) = \frac{\theta_1}{\theta_1 - \theta_2}\{\exp(-\theta_2 t) - \exp(-\theta_1 t)\}. \tag{1.15}$$

Here the relationship between B and t is nonlinear in θ_1 and θ_2. Since it is not possible to linearize this model by means of a suitable transformation, this model can be described as *intrinsically nonlinear*. This is an example of a "compartmental" model. Such models are discussed in Chapter 8. ■

Example 1.4 The relative rate of growth of an organism or plant of size $w(t)$ at time t is often found to be well approximated by a function g of the size, i.e.,

$$\frac{d \log w(t)}{dt} = \frac{1}{w}\frac{dw}{dt} = g(w). \tag{1.16}$$

Richards [1959], for example, proposed the function

$$g(w) = k\left[1 - \left(\frac{w}{\alpha}\right)^{1/\delta}\right], \qquad k > 0. \tag{1.17}$$

Setting $w = \alpha v^{\delta}$, (1.16) becomes

$$\delta \frac{dv}{dt} = kv(1 - v).$$

This has solution

$$v^{-1} = 1 + \exp\left(-\frac{\lambda + kt}{\delta}\right),$$

where λ is a constant of integration. Thus

$$w = \frac{\alpha}{\{1 + \exp[-(\lambda + kt)/\delta]\}^{\delta}}. \tag{1.18}$$

Again we have an intrinsically nonlinear model relating w and t, with unknown parameters α, k, λ and δ. The special case $\delta = 1$ gives the logistic curve

$$w = \frac{\alpha}{1 + \exp[-(\lambda + kt)]}, \tag{1.19}$$

often seen in the form (with $t = x$)

$$\frac{1}{w} = \frac{1}{\alpha} + \frac{1}{\alpha} e^{-\lambda} e^{-kx}$$

or

$$y = \alpha_0 + \beta e^{-kx} \tag{1.20}$$

$$= \alpha_0 + \beta \rho^x, \qquad 0 < \rho < 1, \tag{1.21}$$

where $\rho = e^{-k}$. This model has also been used as a yield–fertilizer model in agriculture, and as a learning curve in psychology. Models like (1.18) and (1.21) are discussed in greater detail in Chapter 7. ■.

Example 1.5 In the study of allometry, attention is focused on differences in shape associated with size. For example, instead of relating two size measurements x and y (e.g. lengths of two bones) to time, we may be interested in relating them to each other. Suppose the two relative growth rates are given by

$$\frac{1}{x}\frac{dx}{dt} = k_x \quad \text{and} \quad \frac{1}{y}\frac{dy}{dt} = k_y.$$

Defining $\beta = k_y/k_x$ we have, canceling out dt,

$$\frac{dy}{y} = \beta \frac{dx}{x}.$$

Making the strong assumption that β is constant, we can integrate the above equation and obtain

$$\log y = \log \alpha + \beta \log x$$

or

$$y = \alpha x^{\beta}. \tag{1.22}$$

This model has been studied for over 50 years and, as a first approximation, seems to fit a number of growth processes quite well. An intrinsically nonlinear extension of the above model is also used, namely

$$y = \gamma + \alpha x^{\beta}. \tag{1.23}$$

Griffiths and Sandland [1984] discuss allometric models with more than two variables and provide references to the literature of allometry. ■

Example 1.6 The average yield y per plant for crop planting tends to decrease as the density of planting increases, due to competition between plants for resources. However, y is not only related to density, or equivalently the area available to each plant, but also to the shape of the area available to each plant. A square area tends to be more advantageous than an elongated rectangle. Berry [1967] proposed a model for y as a function of the two explanatory variables x_1 (the distance between plants within a row) and x_2 (the distance between rows), namely

$$y^{-\delta} = \alpha + \beta\left(\frac{1}{x_1} + \frac{1}{x_2}\right) + \frac{\gamma}{x_1 x_2}. \tag{1.24}$$

Yield–density models are discussed in Section 7.6. ■

Example 1.7 The forced expiratory volume y, a measure of how much air a subject can breathe, is widely used in screening for chronic respiratory disease. Cole [1975] discusses a nonlinear model relating y to height (x_1) and age (x_2) of the form

$$y = x_1^{\gamma}(\alpha + \beta x_2). \tag{1.25}$$

■

Example 1.8 The freezing point y of a solution of two substances is related to the concentration x of component 1 by the equation

$$\log x = \alpha\left(\frac{1}{T_0} - \frac{1}{y}\right) + \beta \log\left(\frac{y}{T_0}\right), \tag{1.26}$$

where T_0 is the freezing point of pure component 1. In this model we cannot express y explicitly in terms of x, which would be the natural way of presenting the relationship. One would vary x (the controlled explanatory variable) and measure y (the response). ■

In the above discussion we have considered models of the form $y = f(\mathbf{x}; \boldsymbol{\theta})$, or $g(\mathbf{x}, y; \boldsymbol{\theta}) = 0$ as in Example 1.8, relating y and $\mathbf{x}$, with $\boldsymbol{\theta} = (\theta_1, \theta_2, \ldots, \theta_p)'$ being a vector of unknown parameters. This deterministic approach does not allow for the fact that y is random and x may be random also. In addition, y and x may be measured with error. The above relationships therefore need to be modeled stochastically, and we now consider various types of model.

1.3 FIXED-REGRESSOR MODEL

1.3.1 Regressors Measured without Error

Suppose μ is the expected size of an organism at time x, and it is assumed that $\mu = \alpha x^{\beta}$. However, due to random fluctuations and possibly errors of measurement, the actual size is y such that $E[y] = \mu$. Then the model becomes

$$y = \alpha x^{\beta} + \varepsilon \tag{1.27}$$

or, in general,

$$y = f(x; \boldsymbol{\theta}) + \varepsilon, \tag{1.28}$$

where $E[\varepsilon] = E[y - \mu] = 0$. If the time is measured accurately so that its variance is essentially zero, or negligible compared with $\mathrm{var}[y]$, then one can treat x as fixed rather than random.

Example 1.9 A model which has wide application in industrial, electrical, and chemical engineering (Shah and Khatri [1965]) is

$$y = \alpha + \delta x + \beta \rho^{x} + \varepsilon, \qquad 0 < \rho < 1. \tag{1.29}$$

This model is a simple extension of (1.21) and, in the context of demography, is known as Makeham's second modification (Spergeon [1949]). It has also been used in pharmacology, where, for example, y is the amount of unchanged drug in the serum and x is the dose or time. ■

In the above examples we have seen that theoretical considerations often give the form of the trend relationship between y and $\mathbf{x}$ but throw no light on an appropriate form for an error structure to explain the scatter. Using the deterministic approach, models are typically fitted by least squares. As in linear regression, this is appropriate only when the random errors are uncorrelated with constant variance, and such an assumption may be unreasonable. For example, in the fitting of a growth curve such as (1.21) it is often found that the scatter about the growth curve increases with the size of the individual(s). However, the variability of the scatter in a plot of the logarithm of y versus x may appear fairly constant. Thus rather than using the model

$$y = \alpha_0 + \beta \rho^{x} + \varepsilon,$$

it may be preferable to use

$$\log y = \log(\alpha_0 + \beta\rho^x) + \varepsilon. \tag{1.30}$$

More generally, a relationship $\mu = f(x; \boldsymbol{\theta})$ may have to be transformed before constructing a model of the form (1.28). A more formal discussion of the effect of transformations on the error structure follows in Section 1.7, and transformations and weighting are discussed in Section 2.8.

1.3.2 Conditional Regressor Models

There are two variations on the above fixed-regressor approach in which x is in fact random but can be treated as though it were fixed. The first occurs when f is a theoretical relationship and x is random but measured accurately, i.e. $x = x_0$, say. For example, y could be the length of a small animal and x its weight. Although for an animal randomly selected from the population, x and y are both random, x could be measured accurately but y would not have the same accuracy. A suitable model would then be

$$E[y|x = x_0] = f(x_0; \boldsymbol{\theta}), \tag{1.31}$$

and (1.28) is to be interpreted as a conditional regression model, conditional on the observed value of x.

The second variation occurs when f is chosen empirically to model the relationship between y and the measured (rather than the true) value of x, even when x is measured with error. In this case (1.28) is used to model the conditional distribution of y, given the measured value of x. We shall still call the above two models with random x fixed-regressor models, as, once f is specified, they are analyzed the same way, namely conditionally on x.

There is a third possible situation, namely, f is a theoretical model connecting y to the *true* value of x, where x is measured with error. In this case the true value of x is unknown and a new approach is needed, as described below.

1.4 RANDOM-REGRESSOR (ERRORS-IN-VARIABLES) MODEL

1.4.1 Functional Relationships

Suppose there is an exact functional relationship

$$\mu = f(\xi; \boldsymbol{\theta}) \tag{1.32}$$

between the realizations ξ and μ of two variables. However, both variables are measured with error, so that what is observed is

$$x = \xi + \delta \quad \text{and} \quad y = \mu + \varepsilon,$$

where $E[\delta] = E[\varepsilon] = 0$. Then

$$\begin{aligned} y &= f(\xi; \boldsymbol{\theta}) + \varepsilon \\ &= f(x - \delta; \boldsymbol{\theta}) + \varepsilon. \end{aligned} \tag{1.33}$$

Using the mean-value theorem,

$$f(x - \delta; \boldsymbol{\theta}) = f(x; \boldsymbol{\theta}) - \delta \dot{f}(\tilde{x}; \boldsymbol{\theta}), \tag{1.34}$$

where $\dot{f}$ is the derivative of f, and $\tilde{x}$ lies between x and $x - \delta$. Substituting (1.34) in (1.33) gives us

$$\begin{aligned} y &= f(x; \boldsymbol{\theta}) - \delta \dot{f}(\tilde{x}; \boldsymbol{\theta}) + \varepsilon \\ &= f(x; \boldsymbol{\theta}) + \varepsilon^*, \end{aligned} \tag{1.35}$$

say. This model is not the same as (1.28), as x is now regarded as random and, in general, $E[\varepsilon^*] \neq 0$. If it is analyzed in the same manner as (1.28) using least squares, biases result (cf. Seber [1977: Section 6.4] for a discussion of the linear case). For an example of this model see Example 1.1 [Equation (1.14)]. The theory of such models is discussed in Section 10.2.

Sometimes a functional relationship cannot be expressed explicitly but rather is given implicitly in the form $g(\xi, \mu; \boldsymbol{\theta}) = 0$ [or $g(\boldsymbol{\zeta}; \boldsymbol{\theta}) = 0$] with

$$\begin{pmatrix} x \\ y \end{pmatrix} = \begin{pmatrix} \xi \\ \mu \end{pmatrix} + \begin{pmatrix} \delta \\ \varepsilon \end{pmatrix},$$

or

$$\mathbf{z} = \boldsymbol{\zeta} + \mathbf{v},$$

say. Here both variables are on the same footing, so that we do not distinguish between response and explanatory variables.

1.4.2 Structural Relationships

A different type of model is obtained if the relationship (1.32) is a relationship between random variables (u and v, say) rather than their realizations. We then have the so-called *structural* relationship

$$v = f(u; \boldsymbol{\theta}) \tag{1.36}$$

with $y = v + \varepsilon$ and $x = u + \delta$. Arguing as in (1.34) leads to a model of the same form as (1.35) but with a different structure for ε^*. The linear case

$$v = \alpha + \beta u \tag{1.37}$$

has been investigated in some detail (Moran [1971]), and it transpires that, even for this simple case, there are identifiability problems when it comes to the estimation of unknown parameters. For further details and references see Section 10.7.

Example 1.10 In a sample of organisms, the allometric model (1.22) of Example 1.5 relating two size measurements is usually formulated as a structural relationship

$$v = \alpha u^{\beta}, \tag{1.38}$$

or

$$\log v = \beta_0 + \beta \log u, \tag{1.39}$$

where $\beta_0 = \log \alpha$. Instead of assuming additive errors for v and u, another approach (Kuhry and Marcus [1977]) is to assume additive errors for $\log v$ and $\log u$, namely $y = \log v + \varepsilon$ and $x = \log u + \delta$, so that (1.39) becomes

$$y = \beta_0 + \beta x + (\varepsilon - \beta\delta). \tag{1.40}$$

∎

1.5 CONTROLLED REGRESSORS WITH ERROR

A third type of model, common to many laboratory experiments, is due to Berkson [1950]. We begin with the structural relationship $v = f(u; \boldsymbol{\theta})$ and try to set u at the target value x_0. However, x_0 is not attained exactly, but instead we have $u = x_0 + \delta$, where $E[\delta] = 0$ and u is unknown. For example, Ohm's law states that $v = ru$, where u is the current in amperes through a resistor of r ohms, and v volts is the voltage across the resistor. We could then choose various readings on the ammeter, say 1.1, 1.2, 1.3,... amperes, and measure the corresponding voltages. In each case the true current will be different from the ammeter reading. Thus, for a general model,

$$\begin{aligned} y &= v + \varepsilon \\ &= f(u; \boldsymbol{\theta}) + \varepsilon \\ &= f(x_0 + \delta; \boldsymbol{\theta}) + \varepsilon, \end{aligned}$$

which can be expanded once again using the mean-value theorem to give

$$\begin{aligned} y &= f(x_0; \boldsymbol{\theta}) + \delta \dot{f}(\tilde{x}; \boldsymbol{\theta}) + \varepsilon \\ &= f(x_0; \boldsymbol{\theta}) + \tilde{\varepsilon}, \end{aligned}$$

say, where $\tilde{x}$ lies between x_0 and $x_0 + \delta$. In general $E[\tilde{\varepsilon}] \neq 0$, but if δ is small so that $\tilde{x} \approx x_0$, then $E[\tilde{\varepsilon}] \approx 0$. In this case we can treat the model as a fixed-regressor model of the form (1.31). If f is linear, so that

$$f(u; \boldsymbol{\theta}) = \theta_1 + \theta_2 u,$$

then $\tilde{\varepsilon} = \theta_2 \delta + \varepsilon$ and $E[\tilde{\varepsilon}] = 0$.

Although the fixed-regressor model is the main model considered in this book, the other models are also considered briefly in Chapter 10.

1.6 GENERALIZED LINEAR MODEL

Example 1.11 In bioassay, the probability p_x of a response to a quantity x of a drug can sometimes be modeled by the logistic curve [cf. (1.19)]

$$p_x = \frac{\exp(\alpha + \beta x)}{\exp(\alpha + \beta x) + 1} = \frac{1}{1 + \exp[-(\alpha + \beta x)]}. \tag{1.41}$$

If y is an indicator (binary) variable taking the values 1 for a cure and 0 otherwise, then

$$E[y|x] = 1 \cdot p_x + 0 \cdot q_x = p_x \qquad (q_x = 1 - p_x).$$

Hence

$$\begin{aligned} y &= p_x + \varepsilon \\ &= \{1 + \exp[-(\alpha + \beta x)]\}^{-1} + \varepsilon, \end{aligned}$$

where $E[\varepsilon] = 0$. We note that

$$\log(p_x/q_x) = \alpha + \beta x,$$

where the left-hand side is called the logistic transform of p_x. This model is therefore called a *linear logistic* model (Cox [1970]), and $\log(p_x/q_x)$ is commonly called the *log odds*. The model is a special case of the so-called *generalized linear model*

$$E[y|\mathbf{x}] = \mu_{\mathbf{x}}, \qquad g(\mu_{\mathbf{x}}) = \alpha + \boldsymbol{\beta}'\mathbf{x}, \tag{1.42}$$

with the distribution of y belonging to the exponential family. This generalized linear model includes as special cases the linear models $[g(\mu) = \mu]$, models which can be transformed to linearity, and models which are useful for data with a discrete or ordinal (ordered multistate) response variable y. For a full discussion of this extremely useful class of models see McCullagh and Nelder [1983]. In their Chapter 10 they show how such models can be adapted to allow for additional nonlinear parameters. The statistical package GLIM (see Baker and Nelder [1978]) was developed for fitting this class of models. ■

1.7 TRANSFORMING TO LINEARITY

A number of examples have been given in Section 1.2 (Examples 1.1, 1.2, and 1.5) in which a nonlinear trend relationship between y and $\mathbf{x}$ can be transformed to give a linear relationship. Because of the relative simplicity of using linear regression methods, working with the linearized model is very attractive.

Suppose we have the model

$$E[y] = e^{\alpha+\beta x},$$

where the error in y is proportional to the expected magnitude of y but is otherwise independent of x (in some fields this situation is termed "constant relative error"). Then we can write

$$\begin{aligned} y &= e^{\alpha+\beta x}(1+\varepsilon_0) \\ &= e^{\alpha+\beta x}+\varepsilon, \end{aligned} \tag{1.43}$$

where $E[\varepsilon_0]=0$ and $\text{var}[\varepsilon_0]=\sigma_0^2$ independently of x. However, the variance of $\varepsilon\,(=\sigma_0^2\{E[y]\}^2)$ varies with x. If instead we take logarithms, then

$$\begin{aligned} \log y &= \alpha+\beta x+\log(1+\varepsilon_0) \\ &= \alpha+\beta x+\varepsilon_0^*, \end{aligned}$$

where $E[\varepsilon_0^*]\approx E[\varepsilon_0]=0$ (for small ε_0), and $\text{var}[\varepsilon_0^*]$ is independent of x. We note that if ε_0 is normally distributed then ε_0^* is not, and vice versa.

On the other hand, if the error in y is additive and independent of x ("constant absolute error"), we have

$$\begin{aligned} y &= e^{\alpha+\beta x}+\varepsilon_0 \\ &= e^{\alpha+\beta x}\left(1+\frac{\varepsilon_0}{E[y]}\right). \end{aligned} \tag{1.44}$$

Then

$$\begin{aligned} \log y &= \alpha+\beta x+\log\left(1+\frac{\varepsilon_0}{E[y]}\right) \\ &= \alpha+\beta x+\nu_0, \end{aligned}$$

say, where $E[\nu_0]\approx E\{\varepsilon_0/E[y]\}=0$ for ε_0 small compared with $E[y]$, and $\text{var}[\nu_0]$ varies with $E[y]$. Thus in the first case [the model (1.43)], taking logarithms stabilizes the variance of the error, while in the second case [the model (1.44)] the variance of the error becomes dependent on x through $E[y] = \exp(\alpha + \beta x)$. The decision to transform or not depends very much on the nature of the error.

There are three main reasons for wanting to transform: firstly to achieve linearity, secondly to obtain errors that are approximately normally distributed, and thirdly to achieve a constant error variance. Sometimes a transformation to linearity fortuitously achieves the remaining objectives as well. The problem of nonconstant error variance can be approached by weighting. With linear models, having an error distribution that is closely approximated by the normal is important because the normal assumption allows us to obtain exact inferences in small samples. With nonlinear models, however, we rely almost entirely on obtaining approximate inferences by applying asymptotic results to finite samples. In this respect the asymptotic theory for, say, least squares does not require normally distributed errors. Unfortunately the relative importance of an approximately normal error distribution and of constant error variance does not appear to have been explored in any detail. However, finding a linearizing transformation is not an overriding factor, as we shall see in Chapter 2. For example, Nelder [1961] used the logarithm of the model (1.18), namely

$$\log w = \alpha_0 - \delta \log\left[1 + \exp\left(-\frac{\lambda + kt}{\delta}\right)\right] + \varepsilon \qquad (1.45)$$

on the grounds that ε appears to be approximately normally distributed with constant variance.

Example 1.12 A family of models used in economics, for example in the study of demand and production functions, is the Cobb–Douglas family

$$y = \theta_0 x_1^{\theta_1} x_2^{\theta_2} \cdots x_k^{\theta_k}.$$

Various models for the error component have been proposed. For example, there is the additive model

$$y = \theta_0 x_1^{\theta_1} x_2^{\theta_2} \cdots x_k^{\theta_k} + \varepsilon, \qquad (1.46)$$

or the multiplicative model

$$y = \theta_0 x_1^{\theta_1} x_2^{\theta_2} \cdots x_k^{\theta_k} e^{\varepsilon}, \qquad (1.47)$$

where ε is assumed to be $N(0, \sigma^2)$. Since

$$E[e^{\varepsilon}] = E[e^{\varepsilon t}]_{t=1},$$

it follows from the moment generating function of the normal distribution that

$$E[e^{\varepsilon}] = e^{\sigma^2/2}$$

and hence for (1.47),

$$E[y|x_1, x_2, \ldots, x_k] = \theta_0 x_1^{\theta_1} x_2^{\theta_2} \cdots x_k^{\theta_k} e^{\sigma^2/2}. \tag{1.48}$$

Taking logarithms will transform (1.47) into a linear model. A mixed model for the error structure has also been proposed (Goldfeld and Quandt [1972: 140ff.]), namely

$$y = \theta_0 x_1^{\theta_1} x_2^{\theta_2} \cdots x_k^{\theta_k} e^{\varepsilon} + \nu. \tag{1.49}$$

■

Example 1.13 The Holliday model relating the average yield (y) per plant to the density of crop planting (x) is given by

$$y \approx (\alpha + \beta x + \gamma x^2)^{-1}. \tag{1.50}$$

The linearizing transformation is clearly to use y^{-1}. However, in practice it is often found that the transformation that stabilizes the error variance is the logarithmic transformation

$$\log y = -\log(\alpha + \beta x + \gamma x^2) + \varepsilon. \tag{1.51}$$

Models for yields versus densities are discussed further in Section 7.6. ■

We see from our three cases of additive, proportional, and multiplicative error that the error structure must be considered before making linearizing transformations. Although a residual plot is usually the most convenient method of checking an error structure, occasionally physical considerations will point towards the appropriate model (e.g. Just and Pope [1978]). The general questions of transformations and weighting are discussed in Section 2.8, and the dangers of ignoring variance heterogeneity are demonstrated there.

It should also be noted that the parameters of a linearized model are often not as interesting or as important as the original parameters. In physical and chemical models the original parameters usually have a physical meaning, e.g. rate constants, so that estimates and confidence intervals for these parameters are still required. Therefore, given the availability of efficient nonlinear algorithms, the usefulness of linearization is somewhat diminished.

It is interesting to note that all of the models discussed in this section (Section 1.7) fit into the generalized linear-model framework of (1.42) when they are expressed in the form

$$y = f(x;\theta) + \varepsilon$$

with normally distributed errors ε having constant variance.

1.8 MODELS WITH AUTOCORRELATED ERRORS

In many examples in which nonlinear regression models have been fitted to data collected sequentially over time, plots of the data reveal long runs of positive residuals and long runs of negative residuals. This may be due to the inadequacy of the model postulated for $E[y|x]$, or it may be caused by a high degree of correlation between successive error terms ε_i. A simple autocorrelation structure which is sometimes applicable to data collected at equally spaced time intervals is given by an autoregressive process of order 1 [AR(1)], namely

$$\varepsilon_i = \rho\varepsilon_{i-1} + a_i, \tag{1.52}$$

where the a_i are uncorrelated, $E[a_i] = 0$, $\mathrm{var}[a_i] = \sigma_a^2$, and $|\rho| < 1$. Under such a structure (see Section 6.2.1)

$$\mathrm{corr}[\varepsilon_i, \varepsilon_j] = \rho^{|i-j|},$$

so that the correlation between ε_i and ε_j decreases exponentially as the distance increases between the times at which y_i and y_j were recorded.

Example 1.14 Consider the simple linear model

$$y_i = \beta_0 + \beta_1 x_i + \varepsilon_i \tag{1.53}$$

for $i = 1, 2, \ldots, n$, in which the ε_i have an AR(1) autocorrelation structure. From (1.52) $a_i = \varepsilon_i - \rho\varepsilon_{i-1}$, so that for $i = 2, 3, \ldots, n$ we have

$$y_i = \beta_0(1-\rho) + \rho y_{i-1} + \beta_1 x_i - \beta_1\rho x_{i-1} + a_i. \tag{1.54}$$

This is of the form

$$y_i = \beta_0^* + \rho x_{i1}^* + \beta_1 x_{i2}^* + \beta_1\rho x_{i3}^* + a_i.$$

Thus a simple linear model with an autocorrelated error structure can be converted into a nonlinear model with an uncorrelated constant-variance error structure. The resulting nonlinear model can then be fitted by ordinary least-squares methods. We should note that there can be problems with this approach caused by the fact that it does not make full use of the first observation pair (x_1, y_1). ■

Methods for fitting linear and nonlinear models with autocorrelated (ARMA) error structures are discussed in Chapter 6.

1.9 FURTHER ECONOMETRIC MODELS

There are two types of model which are widely used in econometrics. The first is a model which includes "lagged" y-variables in with the x-variables, and the second is the so-called simultaneous-equations model. These are illustrated by the following examples.

Example 1.15 Suppose that the consumption y_i at time i depends on income variables $\mathbf{x}_i$ and the consumption y_{i-1} in the previous year through the model

$$y_i = f(y_{i-1}, \mathbf{x}_i; \boldsymbol{\theta}) + \varepsilon_i \qquad (i = 1, 2, \ldots, n), \tag{1.55}$$

where the ε_i are usually autocorrelated. This is called a *lagged* or *dynamic* model because of its dependence on y_{i-1}. It differs from previous models in that f and ε_i in (1.55) are usually correlated through y_{i-1}. ■

Example 1.16 Consider the model (Amemiya [1983: p. 374])

$$y_{i1} = \theta_1 \log y_{i2} + \theta_2 x_{i1} + \theta_3 + \varepsilon_{i1},$$
$$y_{i2} = (x_{i3})^{\theta_4} y_{i1} + \theta_5 x_{i2} + \varepsilon_{i2}, \qquad (i = 1, 2, \ldots, n),$$

where ε_{i1} and ε_{i2} are correlated. This model is called a simultaneous-equations model and has general representation

$$y_{ij} = f_j(y_{i1}, y_{i2}, \ldots, y_{im}, \mathbf{x}_i; \boldsymbol{\theta}) + \varepsilon_{ij}$$
$$(i = 1, 2, \ldots, n, \quad j = 1, 2, \ldots, m)$$

or, in vector form,

$$\mathbf{y}_i = \mathbf{f}(\mathbf{y}_i, \mathbf{x}_i; \boldsymbol{\theta}) + \boldsymbol{\varepsilon}_i \qquad (i = 1, 2, \ldots, n). \tag{1.56}$$

This model is frequently expressed in the more general form

$$q_j(\mathbf{y}_i, \mathbf{x}_i; \boldsymbol{\theta}) = \varepsilon_{ij},$$

or

$$\mathbf{q}(\mathbf{y}_i, \mathbf{x}_i; \boldsymbol{\theta}) = \boldsymbol{\varepsilon}_i.$$

If i refers to time, the above model can be generalized further by allowing $\mathbf{x}_i$ to include lagged values $\mathbf{y}_{i-1}, \mathbf{y}_{i-2}$, etc. ■

The above models are not considered further in this book, and the interested reader is referred to Amemiya [1983] and Maddala [1977] for further details. The asymptotic theory for the above models is developed in detail by Gallant [1987: Chapters 6, 7].

CHAPTER 2

Estimation Methods

2.1 LEAST-SQUARES ESTIMATION

2.1.1 Nonlinear Least Squares

Suppose that we have n observations $(\mathbf{x}_i, y_i)$, $i = 1, 2, \ldots, n$, from a fixed-regressor nonlinear model with a known functional relationship f. Thus

$$y_i = f(\mathbf{x}_i; \boldsymbol{\theta}^*) + \varepsilon_i \qquad (i = 1, 2, \ldots, n), \tag{2.1}$$

where $E[\varepsilon_i] = 0$, $\mathbf{x}_i$ is a $k \times 1$ vector, and the true value $\boldsymbol{\theta}^*$ of $\boldsymbol{\theta}$ is known to belong to Θ, a subset of $\mathbb{R}^p$. The least-squares estimate of $\boldsymbol{\theta}^*$, denoted by $\hat{\boldsymbol{\theta}}$, minimizes the error sum of squares

$$S(\boldsymbol{\theta}) = \sum_{i=1}^{n} [y_i - f(\mathbf{x}_i; \boldsymbol{\theta})]^2 \tag{2.2}$$

over $\boldsymbol{\theta} \in \Theta$. It should be noted that, unlike the linear least-squares situation, $S(\boldsymbol{\theta})$ may have several relative minima in addition to the absolute minimum $\hat{\boldsymbol{\theta}}$. Assuming the ε_i to be independently and identically distributed with variance σ^2, it is shown in Chapter 12 that, under certain regularity assumptions, $\hat{\boldsymbol{\theta}}$ and $s^2 = S(\hat{\boldsymbol{\theta}})/(n-p)$ are consistent estimates of $\boldsymbol{\theta}^*$ and σ^2 respectively. With further regularity conditions, $\hat{\boldsymbol{\theta}}$ is also asymptotically normally distributed as $n \to \infty$. These results are proved using a linear approximation discussed in the next section. If, in addition, we assume that the ε_i are normally distributed, then $\hat{\boldsymbol{\theta}}$ is also the maximum-likelihood estimator (see Section 2.2.1).

When each $f(\mathbf{x}_i; \boldsymbol{\theta})$ is differentiable with respect to $\boldsymbol{\theta}$, and $\hat{\boldsymbol{\theta}}$ is in the interior of Θ, $\hat{\boldsymbol{\theta}}$ will satisfy

$$\left.\frac{\partial S(\boldsymbol{\theta})}{\partial \theta_r}\right|_{\hat{\boldsymbol{\theta}}} = 0 \qquad (r = 1, 2, \ldots, p). \tag{2.3}$$

We shall use the notation $f_i(\boldsymbol{\theta}) = f(\mathbf{x}_i; \boldsymbol{\theta})$,

$$\mathbf{f}(\boldsymbol{\theta}) = \Big(f_1(\boldsymbol{\theta}), f_2(\boldsymbol{\theta}), \ldots, f_n(\boldsymbol{\theta}) \Big)', \tag{2.4}$$

and

$$\mathbf{F}_{\bullet}(\boldsymbol{\theta}) = \frac{\partial \mathbf{f}(\boldsymbol{\theta})}{\partial \boldsymbol{\theta}'} = \left[\left(\frac{\partial f_i(\boldsymbol{\theta})}{\partial \theta_j} \right) \right]. \tag{2.5}$$

Also, for brevity, let

$$\mathbf{F}_{\bullet} = \mathbf{F}_{\bullet}(\boldsymbol{\theta}^*) \quad \text{and} \quad \hat{\mathbf{F}}_{\bullet} = \mathbf{F}_{\bullet}(\hat{\boldsymbol{\theta}}). \tag{2.6}$$

Thus a dot subscript will denote first derivatives; two dots represent second derivatives (see Chapter 4). Using the above notation we can write

$$\begin{aligned} S(\boldsymbol{\theta}) &= [\mathbf{y} - \mathbf{f}(\boldsymbol{\theta})]'[\mathbf{y} - \mathbf{f}(\boldsymbol{\theta})] \\ &= \|\mathbf{y} - \mathbf{f}(\boldsymbol{\theta})\|^2. \end{aligned} \tag{2.7}$$

Equation (2.3) then leads to

$$\sum_i \{y_i - f_i(\boldsymbol{\theta})\} \left.\frac{\partial f_i(\boldsymbol{\theta})}{\partial \theta_r}\right|_{\boldsymbol{\theta}=\hat{\boldsymbol{\theta}}} = 0 \qquad (r = 1, 2, \ldots, p), \tag{2.8}$$

or

$$\begin{aligned} \mathbf{0} &= \hat{\mathbf{F}}_{\bullet}'\{\mathbf{y} - \mathbf{f}(\hat{\boldsymbol{\theta}})\} \\ &= \hat{\mathbf{F}}_{\bullet}'\hat{\boldsymbol{\varepsilon}}, \end{aligned} \tag{2.9}$$

say. If $\hat{\mathbf{P}}_F = \hat{\mathbf{F}}_{\bullet}(\hat{\mathbf{F}}_{\bullet}'\hat{\mathbf{F}}_{\bullet})^{-1}\hat{\mathbf{F}}_{\bullet}'$, the idempotent matrix projecting $\mathbb{R}^n$ orthogonally onto $\mathscr{R}[\hat{\mathbf{F}}_{\bullet}]$ (see A11.4), then (2.9) can also be written as

$$\hat{\mathbf{P}}_F \hat{\boldsymbol{\varepsilon}} = \mathbf{0}. \tag{2.10}$$

The equations (2.9) are called the *normal equations* for the nonlinear model. For most nonlinear models they cannot be solved analytically, so that iterative methods are necessary (see Section 2.1.3 and later chapters).

Example 2.1 Consider the allometric model

$$y_i = \alpha x_i^{\beta} + \varepsilon_i \qquad (i = 1, 2, \ldots, n). \tag{2.11}$$

The normal equations are

$$\sum_i (y_i - \alpha x_i^{\beta}) x_i^{\beta} = 0$$

and

$$\sum_i (y_i - \alpha x_i^\beta)\alpha x_i^\beta \log x_i = 0.$$

These equations do not admit analytic solutions for α and β. ∎

2.1.2 Linear Approximation

We now introduce a number of asymptotic results heuristically. Our aim is to give the flavor of the asymptotic theory without worrying about rigor at this stage. Further details and a discussion of the regularity conditions needed for the validity of the asymptotics are discussed in Chapter 12.

We begin by first noting that in a small neighborhood of $\boldsymbol{\theta}^*$, the true value of $\boldsymbol{\theta}$, we have the linear Taylor expansion

$$f_i(\boldsymbol{\theta}) \approx f_i(\boldsymbol{\theta}^*) + \sum_{r=1}^{p} \left.\frac{\partial f_i}{\partial \theta_r}\right|_{\boldsymbol{\theta}^*} (\theta_r - \theta_r^*), \tag{2.12}$$

or

$$\mathbf{f}(\boldsymbol{\theta}) \approx \mathbf{f}(\boldsymbol{\theta}^*) + \mathbf{F}_{\bullet}(\boldsymbol{\theta} - \boldsymbol{\theta}^*), \tag{2.13}$$

where, from (2.6), $\mathbf{F}_{\bullet} = \mathbf{F}_{\bullet}(\boldsymbol{\theta}^*)$. Hence

$$\begin{aligned} S(\boldsymbol{\theta}) &= \|\mathbf{y} - \mathbf{f}(\boldsymbol{\theta})\|^2 \\ &\approx \|\mathbf{y} - \mathbf{f}(\boldsymbol{\theta}^*) - \mathbf{F}_{\bullet}(\boldsymbol{\theta} - \boldsymbol{\theta}^*)\|^2 \\ &= \|\mathbf{z} - \mathbf{F}_{\bullet}\boldsymbol{\beta}\|^2, \end{aligned} \tag{2.14}$$

say, where $\mathbf{z} = \mathbf{y} - \mathbf{f}(\boldsymbol{\theta}^*) = \boldsymbol{\varepsilon}$ and $\boldsymbol{\beta} = \boldsymbol{\theta} - \boldsymbol{\theta}^*$. From the properties of the linear model (Appendix D1.1), (2.14) is minimized when $\boldsymbol{\beta}$ is given by

$$\hat{\boldsymbol{\beta}} = (\mathbf{F}_{\bullet}'\mathbf{F}_{\bullet})^{-1}\mathbf{F}_{\bullet}'\mathbf{z}.$$

When n is large we find, in Chapter 12, that under certain regularity conditions $\hat{\boldsymbol{\theta}}$ is almost certain to be within a small neighborhood of $\boldsymbol{\theta}^*$. Hence $\hat{\boldsymbol{\theta}} - \boldsymbol{\theta}^* \approx \hat{\boldsymbol{\beta}}$ and

$$\hat{\boldsymbol{\theta}} - \boldsymbol{\theta}^* \approx (\mathbf{F}_{\bullet}'\mathbf{F}_{\bullet})^{-1}\mathbf{F}_{\bullet}'\boldsymbol{\varepsilon}. \tag{2.15}$$

Furthermore, from (2.13) with $\boldsymbol{\theta} = \hat{\boldsymbol{\theta}}$,

$$\begin{aligned} \mathbf{f}(\hat{\boldsymbol{\theta}}) - \mathbf{f}(\boldsymbol{\theta}^*) &\approx \mathbf{F}_{\bullet}(\hat{\boldsymbol{\theta}} - \boldsymbol{\theta}^*) \\ &\approx \mathbf{F}_{\bullet}(\mathbf{F}_{\bullet}'\mathbf{F}_{\bullet})^{-1}\mathbf{F}_{\bullet}'\boldsymbol{\varepsilon} \\ &= \mathbf{P}_F\boldsymbol{\varepsilon} \end{aligned} \tag{2.16}$$

and

$$\begin{aligned}\mathbf{y}-\mathbf{f}(\hat{\boldsymbol{\theta}}) &\approx \mathbf{y}-\mathbf{f}(\boldsymbol{\theta}^*)-\mathbf{F}_{\bullet}(\hat{\boldsymbol{\theta}}-\boldsymbol{\theta}^*)\\ &\approx \boldsymbol{\varepsilon}-\mathbf{P}_F\boldsymbol{\varepsilon}\\ &=(\mathbf{I}_n-\mathbf{P}_F)\boldsymbol{\varepsilon},\end{aligned} \tag{2.17}$$

where $\mathbf{P}_F=\mathbf{F}_{\bullet}(\mathbf{F}_{\bullet}'\mathbf{F}_{\bullet})^{-1}\mathbf{F}_{\bullet}'$ and $\mathbf{I}_n-\mathbf{P}_F$ are symmetric and idempotent (Appendix A11). Hence, from (2.17) and (2.16), we have

$$\begin{aligned}(n-p)s^2 &= S(\hat{\boldsymbol{\theta}})\\ &=\|\mathbf{y}-\mathbf{f}(\hat{\boldsymbol{\theta}})\|^2\\ &\approx \|(\mathbf{I}_n-\mathbf{P}_F)\boldsymbol{\varepsilon}\|^2\\ &=\boldsymbol{\varepsilon}'(\mathbf{I}_n-\mathbf{P}_F)\boldsymbol{\varepsilon},\end{aligned} \tag{2.18}$$

and

$$\begin{aligned}\|\mathbf{f}(\hat{\boldsymbol{\theta}})-\mathbf{f}(\boldsymbol{\theta}^*)\|^2 &\approx \|\mathbf{F}_{\bullet}(\hat{\boldsymbol{\theta}}-\boldsymbol{\theta}^*)\|^2\\ &=(\hat{\boldsymbol{\theta}}-\boldsymbol{\theta}^*)'\mathbf{F}_{\bullet}'\mathbf{F}_{\bullet}(\hat{\boldsymbol{\theta}}-\boldsymbol{\theta}^*)\\ &\approx \|\mathbf{P}_F\boldsymbol{\varepsilon}\|^2\\ &=\boldsymbol{\varepsilon}'\mathbf{P}_F\boldsymbol{\varepsilon}.\end{aligned} \tag{2.19}$$

Therefore, using (2.18) and (2.19), we get

$$\begin{aligned}S(\boldsymbol{\theta}^*)-S(\hat{\boldsymbol{\theta}}) &\approx \boldsymbol{\varepsilon}'\boldsymbol{\varepsilon}-\boldsymbol{\varepsilon}'(\mathbf{I}_n-\mathbf{P}_F)\boldsymbol{\varepsilon}\\ &=\boldsymbol{\varepsilon}'\mathbf{P}_F\boldsymbol{\varepsilon}\\ &\approx(\hat{\boldsymbol{\theta}}-\boldsymbol{\theta}^*)'\mathbf{F}_{\bullet}'\mathbf{F}_{\bullet}(\hat{\boldsymbol{\theta}}-\boldsymbol{\theta}^*).\end{aligned} \tag{2.20}$$

Within the order of the linear approximation used we can replace $\mathbf{F}_{\bullet}$ by $\hat{\mathbf{F}}_{\bullet}=\mathbf{F}_{\bullet}(\hat{\boldsymbol{\theta}})$ in the above expressions, when necessary. Also (2.15) and (2.18) hold to $\mathbf{o}_p(n^{-1/2})$ and $o_p(1)$, respectively (e.g. Gallant [1987: 259–260]). We now have the following theorem.

Theorem 2.1 Given $\boldsymbol{\varepsilon}\sim N(\mathbf{0},\sigma^2\mathbf{I}_n)$ and appropriate regularity conditions (Section 12.2) then, for large n, we have approximately:

(i) $\hat{\boldsymbol{\theta}}-\boldsymbol{\theta}^*\sim N_p(\mathbf{0},\sigma^2\mathbf{C}^{-1})$, where $\mathbf{C}=\mathbf{F}_{\bullet}'\mathbf{F}_{\bullet}=\mathbf{F}_{\bullet}'(\boldsymbol{\theta}^*)\mathbf{F}_{\bullet}(\boldsymbol{\theta}^*)$; (2.21)

(ii) $(n-p)s^2/\sigma^2\approx\boldsymbol{\varepsilon}'(\mathbf{I}_n-\mathbf{P}_F)\boldsymbol{\varepsilon}/\sigma^2\sim\chi^2_{n-p}$; (2.22)

(iii) $\hat{\boldsymbol{\theta}}$ is statistically independent of s^2; and (2.23)

(iv)

$$\begin{aligned}\frac{[S(\boldsymbol{\theta}^*)-S(\hat{\boldsymbol{\theta}})]/p}{S(\hat{\boldsymbol{\theta}})/(n-p)} &\approx \frac{\boldsymbol{\varepsilon}'\mathbf{P}_F\boldsymbol{\varepsilon}}{\boldsymbol{\varepsilon}'(\mathbf{I}_n-\mathbf{P}_F)\boldsymbol{\varepsilon}}\cdot\frac{n-p}{p}\\ &\sim F_{p,n-p}.\end{aligned} \tag{2.24}$$

The normality of $\boldsymbol{\varepsilon}$ is not required for the proof of (i).

Proof: Parts (i) to (iii) follow from the exact linear theory of Appendix D1 with $\mathbf{X} = \mathbf{F}_{\bullet}$, while (iv) follows from Appendix A11.6. The normality of the ε_i is not needed to prove (i), as (2.15) implies that $\hat{\boldsymbol{\theta}} - \boldsymbol{\theta}^*$ is asymptotically a linear combination of the i.i.d. ε_i. An appropriate version of the central limit theorem then gives us (i). ■

Finally, using (iv) and (2.20) we have, approximately,

$$\frac{(\hat{\boldsymbol{\theta}} - \boldsymbol{\theta}^*)'\mathbf{F}_{\bullet}'\mathbf{F}_{\bullet}(\hat{\boldsymbol{\theta}} - \boldsymbol{\theta}^*)}{ps^2} \sim F_{p,n-p}, \tag{2.25}$$

so that estimating $\mathbf{F}_{\bullet}$ by $\hat{\mathbf{F}}_{\bullet}$, an approximate $100(1-\alpha)\%$ confidence region for $\boldsymbol{\theta}^*$ is given by

$$\{\boldsymbol{\theta}^*: (\boldsymbol{\theta}^* - \hat{\boldsymbol{\theta}})'\hat{\mathbf{F}}_{\bullet}'\hat{\mathbf{F}}_{\bullet}(\boldsymbol{\theta}^* - \hat{\boldsymbol{\theta}}) \le ps^2 F^{\alpha}_{p,n-p}\}. \tag{2.26}$$

Thus $\mathbf{F}_{\bullet}$ (or $\hat{\mathbf{F}}_{\bullet}$) plays the same role as the $\mathbf{X}$-matrix in linear regression. This idea is taken further in Chapter 5, where we shall develop approximate confidence intervals and hypothesis tests for $\boldsymbol{\theta}^*$ or subsets of $\boldsymbol{\theta}^*$.

2.1.3 Numerical Methods

Suppose $\boldsymbol{\theta}^{(a)}$ is an approximation to the least-squares estimate $\hat{\boldsymbol{\theta}}$ of a nonlinear model. For $\boldsymbol{\theta}$ close to $\boldsymbol{\theta}^{(a)}$, we again use a linear Taylor expansion

$$\mathbf{f}(\boldsymbol{\theta}) \approx \mathbf{f}(\boldsymbol{\theta}^{(a)}) + \mathbf{F}_{\bullet}^{(a)}(\boldsymbol{\theta} - \boldsymbol{\theta}^{(a)}), \tag{2.27}$$

where $\mathbf{F}_{\bullet}^{(a)} = \mathbf{F}_{\bullet}(\boldsymbol{\theta}^{(a)})$. Applying this to the residual vector $\mathbf{r}(\boldsymbol{\theta})$, we have

$$\begin{aligned}\mathbf{r}(\boldsymbol{\theta}) &= \mathbf{y} - \mathbf{f}(\boldsymbol{\theta}) \\ &\approx \mathbf{r}(\boldsymbol{\theta}^{(a)}) - \mathbf{F}_{\bullet}^{(a)}(\boldsymbol{\theta} - \boldsymbol{\theta}^{(a)}).\end{aligned}$$

Substituting in $S(\boldsymbol{\theta}) = \mathbf{r}'(\boldsymbol{\theta})\mathbf{r}(\boldsymbol{\theta})$ leads to

$$S(\boldsymbol{\theta}) \approx \mathbf{r}'(\boldsymbol{\theta}^{(a)})\mathbf{r}(\boldsymbol{\theta}^{(a)}) - 2\mathbf{r}'(\boldsymbol{\theta}^{(a)})\mathbf{F}_{\bullet}^{(a)}(\boldsymbol{\theta} - \boldsymbol{\theta}^{(a)}) + (\boldsymbol{\theta} - \boldsymbol{\theta}^{(a)})'\mathbf{F}_{\bullet}^{(a)\prime}\mathbf{F}_{\bullet}^{(a)}(\boldsymbol{\theta} - \boldsymbol{\theta}^{(a)}). \tag{2.28}$$

The right-hand side is minimized with respect to $\boldsymbol{\theta}$ when

$$\begin{aligned}\boldsymbol{\theta} - \boldsymbol{\theta}^{(a)} &= (\mathbf{F}_{\bullet}^{(a)\prime}\mathbf{F}_{\bullet}^{(a)})^{-1}\mathbf{F}_{\bullet}^{(a)\prime}\mathbf{r}(\boldsymbol{\theta}^{(a)}) \\ &= \boldsymbol{\delta}^{(a)}, \qquad \text{say.}\end{aligned} \tag{2.29}$$

This suggests that, given a current approximation $\boldsymbol{\theta}^{(a)}$, the next approximation should be

$$\boldsymbol{\theta}^{(a+1)} = \boldsymbol{\theta}^{(a)} + \boldsymbol{\delta}^{(a)}. \tag{2.30}$$

This provides an iterative scheme for obtaining $\hat{\boldsymbol{\theta}}$. The approximation of $S(\boldsymbol{\theta})$ by the quadratic (2.28), and the resulting updating formulae (2.29) and (2.30), are usually referred to as the Gauss–Newton method. It forms the basis of a number of least-squares algorithms used throughout this book. The Gauss–Newton algorithm is convergent, i.e. $\boldsymbol{\theta}^{(a)} \to \hat{\boldsymbol{\theta}}$ as $a \to \infty$, provided that $\boldsymbol{\theta}^{(1)}$ is close enough to $\boldsymbol{\theta}^*$ and n is large enough.

A more general approach, which applies to any function satisfying appropriate regularity conditions, is the Newton method, in which $S(\boldsymbol{\theta})$ is expanded directly using a quadratic Taylor expansion. Using the notation of (1.9) and (1.11), let

$$\mathbf{g}(\boldsymbol{\theta}) = \frac{\partial S(\boldsymbol{\theta})}{\partial \boldsymbol{\theta}} \tag{2.31}$$

and

$$\mathbf{H}(\boldsymbol{\theta}) = \frac{\partial^2 S(\boldsymbol{\theta})}{\partial \boldsymbol{\theta}\, \partial \boldsymbol{\theta}'} \tag{2.32}$$

denote, respectively, the so-called gradient vector and Hessian matrix of $S(\boldsymbol{\theta})$. Then we have the quadratic approximation

$$\begin{aligned} S(\boldsymbol{\theta}) &\approx q_S^{(a)}(\boldsymbol{\theta}) \\ &= S(\boldsymbol{\theta}^{(a)}) + \mathbf{g}'(\boldsymbol{\theta}^{(a)})(\boldsymbol{\theta} - \boldsymbol{\theta}^{(a)}) + \tfrac{1}{2}(\boldsymbol{\theta} - \boldsymbol{\theta}^{(a)})'\mathbf{H}(\boldsymbol{\theta}^{(a)})(\boldsymbol{\theta} - \boldsymbol{\theta}^{(a)}), \end{aligned} \tag{2.33}$$

which differs from (2.28) only in that $\mathbf{H}(\boldsymbol{\theta}^{(a)})$ is approximated there by $2\mathbf{F}_{\bullet}'(\boldsymbol{\theta}^{(a)})\mathbf{F}_{\bullet}(\boldsymbol{\theta}^{(a)})$. However, since

$$\frac{\partial^2 S(\boldsymbol{\theta})}{\partial \theta_r\, \partial \theta_s} = 2 \sum_{i=1}^{n} \left\{ \frac{\partial f_i(\boldsymbol{\theta})}{\partial \theta_r} \cdot \frac{\partial f_i(\boldsymbol{\theta})}{\partial \theta_s} - [y_i - f_i(\boldsymbol{\theta})] \frac{\partial^2 f_i(\boldsymbol{\theta})}{\partial \theta_r\, \partial \theta_s} \right\}, \tag{2.34}$$

then

$$\mathscr{E}\left[\frac{\partial^2 S(\boldsymbol{\theta})}{\partial \boldsymbol{\theta}\, \partial \boldsymbol{\theta}'}\right] = 2\mathbf{F}_{\bullet}'(\boldsymbol{\theta})\mathbf{F}_{\bullet}(\boldsymbol{\theta}) \tag{2.35}$$

and $\mathbf{H}(\boldsymbol{\theta}^{(a)})$ is approximated by its expected value evaluated at $\boldsymbol{\theta}^{(a)}$ in (2.28).

The minimum of the quadratic (2.33) with respect to $\boldsymbol{\theta}$ occurs when

$$\begin{aligned} \boldsymbol{\theta} - \boldsymbol{\theta}^{(a)} &= -[\mathbf{H}(\boldsymbol{\theta}^{(a)})]^{-1}\mathbf{g}(\boldsymbol{\theta}^{(a)}) \\ &= -[\mathbf{H}^{-1}\mathbf{g}]_{\boldsymbol{\theta} = \boldsymbol{\theta}^{(a)}}. \end{aligned} \tag{2.36}$$

This is the so-called Newton method, and the "correction" $\boldsymbol{\delta}^{(a)}$ in (2.30) now given by (2.36) is called the Newton step. Various modifications to this method have been developed, based on (for example) scaling $\mathbf{g}$, modifying or approximating $\mathbf{H}$, and estimating the various first and second derivatives in $\mathbf{g}$ and $\mathbf{H}$ using finite-difference techniques. Further details of these methods are given in Chapters 13 and 14.

A feature of the quadratic approximation $q_S^{(a)}(\boldsymbol{\theta})$ of (2.33) is that its contours are concentric ellipsoids with center $\boldsymbol{\theta}^{(a+1)}$. To see this we write

$$\begin{aligned} q_S^{(a)}(\boldsymbol{\theta}) &= \boldsymbol{\theta}'\mathbf{A}\boldsymbol{\theta} - 2\boldsymbol{\theta}'\mathbf{b} + c \\ &= (\boldsymbol{\theta} - \mathbf{c}_0)'\mathbf{A}(\boldsymbol{\theta} - \mathbf{c}_0) + d_0, \end{aligned}$$

where, for well-behaved functions, $\mathbf{A} = \frac{1}{2}\mathbf{H}(\boldsymbol{\theta}^{(a)})$ is positive definite (and therefore nonsingular), $\mathbf{c}_0 = \mathbf{A}^{-1}\mathbf{b}$, and $d_0 = c - \mathbf{c}_0'\mathbf{A}\mathbf{c}_0$. Then setting

$$\frac{\partial q_S^{(a)}(\boldsymbol{\theta})}{\partial \boldsymbol{\theta}} (= 2\mathbf{A}\boldsymbol{\theta} - 2\mathbf{b}) \qquad \text{[cf. (Appendix A10.5)]}$$

equal to zero gives us $\boldsymbol{\theta}^{(a+1)} = \mathbf{A}^{-1}\mathbf{b} = \mathbf{c}_0$, where $\mathbf{c}_0$ is the center of the ellipsoidal contours of $(\boldsymbol{\theta} - \mathbf{c}_0)'\mathbf{A}(\boldsymbol{\theta} - \mathbf{c}_0)$. This result also follows directly from $\mathbf{A}$ being positive definite. We note that, for linear models,

$$\begin{aligned} S(\boldsymbol{\beta}) &= (\mathbf{y} - \mathbf{X}\boldsymbol{\beta})'(\mathbf{y} - \mathbf{X}\boldsymbol{\beta}) \\ &= \mathbf{y}'\mathbf{y} - 2\boldsymbol{\beta}'\mathbf{X}'\mathbf{y} + \boldsymbol{\beta}'\mathbf{X}'\mathbf{X}\boldsymbol{\beta}, \end{aligned}$$

a quadratic in $\boldsymbol{\beta}$. Then both the Gauss–Newton and Newton approximations are exact, and both find $\hat{\boldsymbol{\beta}}$ in a single iteration.

Finally it should be noted that we do not advocate the unmodified Gauss–Newton and Newton (Newton–Raphson) algorithms as practical algorithms for computing least-squares estimates. Some of the problems that can occur when using the basic unmodified algorithms are illustrated in Sections 3.2 and 14.1. A range of techniques for modifying the algorithms to make them more reliable are described in Chapter 14. Nevertheless, throughout this book we describe, for simplicity, just the bare Gauss–Newton scheme for a particular application. Such iterative techniques should not be implemented without employing some of the safeguards described in Chapter 14.

2.1.4 Generalized Least Squares

Before leaving numerical methods, we mention a generalization of the least-squares procedure called weighted or generalized least squares (GLS). The function to be minimized is now

$$S(\boldsymbol{\theta}) = [\mathbf{y} - \mathbf{f}(\boldsymbol{\theta})]'\mathbf{V}^{-1}[\mathbf{y} - \mathbf{f}(\boldsymbol{\theta})],$$

where $\mathbf{V}$ is a known positive definite matrix. This minimization criterion usually arises from the so-called generalized least-squares model $\mathbf{y} = \mathbf{f}(\boldsymbol{\theta}) + \boldsymbol{\varepsilon}$, where $\mathscr{E}[\boldsymbol{\varepsilon}] = \mathbf{0}$ and $\mathscr{D}[\boldsymbol{\varepsilon}] = \sigma^2\mathbf{V}$. Thus the ordinary least squares (OLS) previously discussed is a special case in which $\mathbf{V} = \mathbf{I}_n$. Denote by $\hat{\boldsymbol{\theta}}_G$ the generalized least-squares estimate which minimizes $S(\boldsymbol{\theta})$ above.

Let $\mathbf{V} = \mathbf{U}'\mathbf{U}$ be the Cholesky decomposition of $\mathbf{V}$, where $\mathbf{U}$ is an upper triangular matrix (Appendix A8.2). Multiplying the nonlinear model through by $\mathbf{R} = (\mathbf{U}')^{-1}$, we obtain

$$\mathbf{z} = \mathbf{k}(\boldsymbol{\theta}) + \boldsymbol{\eta},$$

where $\mathbf{z} = \mathbf{R}\mathbf{y}$, $\mathbf{k}(\boldsymbol{\theta}) = \mathbf{R}\mathbf{f}(\boldsymbol{\theta})$ and $\boldsymbol{\eta} = \mathbf{R}\boldsymbol{\varepsilon}$. Then $\mathscr{E}[\boldsymbol{\eta}] = \mathbf{0}$ and, from (1.3), $\mathscr{D}[\boldsymbol{\eta}] = \sigma^2\mathbf{R}\mathbf{V}\mathbf{R}' = \sigma^2\mathbf{I}_n$. Thus our original GLS model has now been transformed to an OLS model. Furthermore,

$$\begin{aligned} S(\boldsymbol{\theta}) &= [\mathbf{y} - \mathbf{f}(\boldsymbol{\theta})]'\mathbf{V}^{-1}[\mathbf{y} - \mathbf{f}(\boldsymbol{\theta})] \\ &= [\mathbf{y} - \mathbf{f}(\boldsymbol{\theta})]'\mathbf{R}'\mathbf{R}[\mathbf{y} - \mathbf{f}(\boldsymbol{\theta})] \\ &= [\mathbf{z} - \mathbf{k}(\boldsymbol{\theta})]'[\mathbf{z} - \mathbf{k}(\boldsymbol{\theta})]. \end{aligned}$$

Hence the GLS sum of squares in the same as the OLS sum of squares for the transformed model, and $\hat{\boldsymbol{\theta}}_G$ is the OLS estimate from the transformed model.

Let $\mathbf{K}_{\boldsymbol{\cdot}}(\boldsymbol{\theta}) = \partial\mathbf{k}(\boldsymbol{\theta})/\partial\boldsymbol{\theta}'$ and $\hat{\mathbf{K}}_{\boldsymbol{\cdot}} = \mathbf{K}_{\boldsymbol{\cdot}}(\hat{\boldsymbol{\theta}})$. Then

$$\mathbf{K}_{\boldsymbol{\cdot}}(\boldsymbol{\theta}) = \mathbf{R}\frac{\partial\mathbf{f}(\boldsymbol{\theta})}{\partial\boldsymbol{\theta}'} = \mathbf{R}\mathbf{F}_{\boldsymbol{\cdot}}(\boldsymbol{\theta}).$$

Since $\hat{\boldsymbol{\theta}}_G$ is the OLS estimate in the transformed model, it has for large n a variance–covariance matrix given by [cf. (2.21)]

$$\begin{aligned} \mathscr{D}[\hat{\boldsymbol{\theta}}_G] &\approx \sigma^2[\mathbf{K}_{\boldsymbol{\cdot}}'(\boldsymbol{\theta}^*)\mathbf{K}_{\boldsymbol{\cdot}}(\boldsymbol{\theta}^*)]^{-1} \\ &= \sigma^2[\mathbf{F}_{\boldsymbol{\cdot}}'(\boldsymbol{\theta}^*)\mathbf{R}'\mathbf{R}\mathbf{F}_{\boldsymbol{\cdot}}(\boldsymbol{\theta}^*)]^{-1} \\ &= \sigma^2[\mathbf{F}_{\boldsymbol{\cdot}}'(\boldsymbol{\theta}^*)\mathbf{V}^{-1}\mathbf{F}_{\boldsymbol{\cdot}}(\boldsymbol{\theta}^*)]^{-1}. \end{aligned}$$

This matrix is estimated by

$$\hat{\mathscr{D}}[\hat{\boldsymbol{\theta}}_G] = \hat{\sigma}^2(\hat{\mathbf{K}}_{\boldsymbol{\cdot}}'\hat{\mathbf{K}}_{\boldsymbol{\cdot}})^{-1} = \hat{\sigma}^2(\hat{\mathbf{F}}_{\boldsymbol{\cdot}}'\mathbf{V}^{-1}\hat{\mathbf{F}}_{\boldsymbol{\cdot}})^{-1},$$

where

$$\begin{aligned} \hat{\sigma}^2 &= \frac{1}{n-p}[\mathbf{z} - \mathbf{k}(\hat{\boldsymbol{\theta}}_G)]'[\mathbf{z} - \mathbf{k}(\hat{\boldsymbol{\theta}}_G)] \\ &= \frac{1}{n-p}[\mathbf{y} - \mathbf{f}(\hat{\boldsymbol{\theta}}_G)]'\mathbf{V}^{-1}[\mathbf{y} - \mathbf{f}(\hat{\boldsymbol{\theta}}_G)]. \end{aligned}$$

Repeatedly throughout this book we shall require the computation of a GLS estimate. In each case the computational strategy is to transform the model to an OLS model using the Cholesky decomposition $\mathbf{V} = \mathbf{U}'\mathbf{U}$. The important point of the above analysis is that treating the transformed problem as an ordinary nonlinear least-squares problem produces the correct results for the generalized least-squares problem. In particular, ordinary least-squares programs produce the correct variance–covariance matrix estimates.

To illustrate, let us apply the linearization method to the original form of $S(\boldsymbol{\theta})$. Using (2.27) we have

$$S(\boldsymbol{\theta}) \approx [\mathbf{y} - \mathbf{f}(\boldsymbol{\theta}^{(a)}) - \mathbf{F}_{\bullet}^{(a)}(\boldsymbol{\theta} - \boldsymbol{\theta}^{(a)})]'\mathbf{V}^{-1}[\mathbf{y} - \mathbf{f}(\boldsymbol{\theta}^{(a)}) - \mathbf{F}_{\bullet}^{(a)}(\boldsymbol{\theta} - \boldsymbol{\theta}^{(a)})].$$

This approximation produces a linear generalized least-squares problem and is thus minimized by

$$\boldsymbol{\theta} - \boldsymbol{\theta}^{(a)} = (\mathbf{F}_{\bullet}^{(a)\prime}\mathbf{V}^{-1}\mathbf{F}_{\bullet}^{(a)})^{-1}\mathbf{F}_{\bullet}^{(a)\prime}\mathbf{V}^{-1}[\mathbf{y} - \mathbf{f}(\boldsymbol{\theta}^{(a)})].$$

The above leads to the iterative scheme $\boldsymbol{\theta}^{(a+1)} = \boldsymbol{\theta}^{(a)} + \boldsymbol{\delta}^{(a)}$, where

$$\begin{aligned}\boldsymbol{\delta}^{(a)} &= (\mathbf{F}_{\bullet}^{(a)\prime}\mathbf{V}^{-1}\mathbf{F}_{\bullet}^{(a)})^{-1}\mathbf{F}_{\bullet}^{(a)\prime}\mathbf{V}^{-1}[\mathbf{y} - \mathbf{f}(\boldsymbol{\theta}^{(a)})] \\ &= (\mathbf{F}_{\bullet}^{(a)\prime}\mathbf{R}'\mathbf{R}\mathbf{F}_{\bullet}^{(a)})^{-1}\mathbf{F}_{\bullet}^{(a)\prime}\mathbf{R}'\mathbf{R}[\mathbf{y} - \mathbf{f}(\boldsymbol{\theta}^{(a)})] \\ &= (\mathbf{K}_{\bullet}^{(a)\prime}\mathbf{K}_{\bullet}^{(a)})^{-1}\mathbf{K}_{\bullet}^{(a)\prime}[\mathbf{z} - \mathbf{k}(\boldsymbol{\theta}^{(a)})]. \end{aligned} \tag{2.37}$$

The last expression is the Gauss–Newton step for the transformed model and minimizes the ordinary linear least-squares function

$$[\mathbf{z} - \mathbf{k}(\boldsymbol{\theta}^{(a)}) - \mathbf{K}_{\bullet}^{(a)}\boldsymbol{\delta}]'[\mathbf{z} - \mathbf{k}(\boldsymbol{\theta}^{(a)}) - \mathbf{K}_{\bullet}^{(a)}\boldsymbol{\delta}]$$

with respect to $\boldsymbol{\delta}$. Thus $\boldsymbol{\delta}^{(a)}$ is obtained by an ordinary linear regression of $\mathbf{z} - \mathbf{k}(\boldsymbol{\theta}^{(a)})$ on $\mathbf{K}_{\bullet}^{(a)}$. Linear least squares has, as virtual by-products, $(\mathbf{X}'\mathbf{X})^{-1}$ and the residual sum of squares. Therefore, at the final iteration where $\boldsymbol{\theta}^{(a)} \approx \hat{\boldsymbol{\theta}}_G$,

$$(\mathbf{X}'\mathbf{X})^{-1} = (\hat{\mathbf{K}}_{\bullet}'\hat{\mathbf{K}}_{\bullet})^{-1} = (\hat{\mathbf{F}}_{\bullet}'\mathbf{V}^{-1}\hat{\mathbf{F}}_{\bullet})^{-1},$$

and the residual sum of squares is

$$\begin{aligned}[\mathbf{z} - \mathbf{k}(\hat{\boldsymbol{\theta}}_G)]'[\mathbf{z} - \mathbf{k}(\hat{\boldsymbol{\theta}}_G)] &= [\mathbf{y} - \mathbf{f}(\hat{\boldsymbol{\theta}}_G)]'\mathbf{V}^{-1}[\mathbf{y} - \mathbf{f}(\hat{\boldsymbol{\theta}}_G)] \\ &= (n-p)\hat{\sigma}^2.\end{aligned}$$

If a linear least-squares program is used to perform the computations, then the estimated variance–covariance matrix from the final least-squares regression is $\hat{\mathscr{D}}[\hat{\boldsymbol{\theta}}_G]$.

In practice we would not compute $\mathbf{R} = (\mathbf{U}')^{-1}$ and multiply out $\mathbf{R}\mathbf{y}$ $(=\mathbf{z})$. Instead it is better to solve the lower triangular system $\mathbf{U}'\mathbf{z} = \mathbf{y}$ for $\mathbf{z}$ directly by forward substitution. We similarly solve $\mathbf{U}'\mathbf{k}(\boldsymbol{\theta}^{(a)}) = \mathbf{f}(\boldsymbol{\theta}^{(a)})$ for $\mathbf{k}(\boldsymbol{\theta}^{(a)})$ and, in the case of the Gauss–Newton method, $\mathbf{U}'\mathbf{K}_{\bullet}^{(a)} = \mathbf{F}_{\bullet}^{(a)}$ for $\mathbf{K}_{\bullet}^{(a)}$.

Many applications of the above method arise from weighted least-squares problems in which $\mathbf{V}$ is diagonal. Since $\mathbf{U}$ is then diagonal with diagonal elements equal to the square roots of the corresponding elements of $\mathbf{V}$, the computation of $\mathbf{z}$ etc. becomes particularly simple.

It is now convenient, for easy reference, to summarize the main steps leading to the computation of $\hat{\boldsymbol{\theta}}_G$, the value of $\boldsymbol{\theta}$ minimizing

$$S(\boldsymbol{\theta}) = [\mathbf{y} - \mathbf{f}(\boldsymbol{\theta})]'\mathbf{V}^{-1}[\mathbf{y} - \mathbf{f}(\boldsymbol{\theta})],$$

and $\hat{\mathscr{D}}[\hat{\boldsymbol{\theta}}_G] = \hat{\sigma}^2(\hat{\mathbf{F}}_{\bullet}'\mathbf{V}^{-1}\hat{\mathbf{F}}_{\bullet})^{-1}$, an estimate of the asymptotic covariance matrix of $\hat{\boldsymbol{\theta}}_G$:

1. Perform a Cholesky decomposition $\mathbf{V} = \mathbf{U}'\mathbf{U}$ of $\mathbf{V}$.
2. Solve $\mathbf{U}'\mathbf{z} = \mathbf{y}$ and $\mathbf{U}'\mathbf{k}(\boldsymbol{\theta}) = \mathbf{f}(\boldsymbol{\theta})$ for $\mathbf{z}$ and $\mathbf{k}(\boldsymbol{\theta})$.
3. Apply an ordinary nonlinear least-squares technique to

$$[\mathbf{z} - \mathbf{k}(\boldsymbol{\theta})]'[\mathbf{z} - \mathbf{k}(\boldsymbol{\theta})].$$

4. If a Gauss–Newton method is used for (3), then $\boldsymbol{\theta}^{(a+1)} = \boldsymbol{\theta}^{(a)} + \boldsymbol{\delta}^{(a)}$, where

$$\boldsymbol{\delta}^{(a)} = (\mathbf{K}_{\bullet}^{(a)\prime}\mathbf{K}_{\bullet}^{(a)})^{-1}\mathbf{K}_{\bullet}^{(a)\prime}[\mathbf{z} - \mathbf{k}(\boldsymbol{\theta}^{(a)})]$$

   is obtained from a linear regression of $\mathbf{z} - \mathbf{k}(\boldsymbol{\theta}^{(a)})$ on $\mathbf{K}_{\bullet}^{(a)}$. The matrix $\mathbf{K}_{\bullet}^{(a)}$ is found by solving $\mathbf{U}'\mathbf{K}_{\bullet}(\boldsymbol{\theta}^{(a)}) = \mathbf{F}_{\bullet}(\boldsymbol{\theta}^{(a)})$.
5. $\hat{\mathscr{D}}[\hat{\boldsymbol{\theta}}_G] = \hat{\sigma}^2(\hat{\mathbf{K}}_{\bullet}'\hat{\mathbf{K}}_{\bullet})^{-1}$ is the estimated dispersion matrix obtained in (3). If the Gauss–Newton method is used, $(\hat{\mathbf{K}}_{\bullet}'\hat{\mathbf{K}}_{\bullet})^{-1}$ and $\hat{\sigma}^2$ can be obtained from the linear regression in (4) at the final iteration.

A more general version of generalized least squares in which $\mathbf{V}$ may depend on $\boldsymbol{\theta}$ is described in Section 2.8.8. If $\mathbf{V} = \mathbf{V}(\boldsymbol{\theta})$, a known function of $\boldsymbol{\theta}$, then $\mathbf{V}$ can, for example, be replaced by $\mathbf{V}(\boldsymbol{\theta}^{(a)})$ in (2.37). A new Cholesky factorisation of $\mathbf{V}$ is then required at each iteration.

*2.1.5 Replication and Test of Fit

Suppose a design is replicated, say J_i times for point $\mathbf{x}_i$, so that our nonlinear model (2.1) now becomes

$$\begin{aligned} y_{ij} &= \mu_i + \varepsilon_{ij} \\ &= f(\mathbf{x}_i; \boldsymbol{\theta}^*) + \varepsilon_{ij} \qquad (i = 1, 2, \ldots, n \quad j = 1, 2, \ldots, J_i) \end{aligned} \tag{2.38}$$

where the fluctuations (or "errors") ε_{ij} are assumed to be i.i.d. $N(0, \sigma^2)$. As stressed by Draper and Smith [1981: 35], it is important to distinguish between a genuine repeated point in which the response y is measured for two different experiments with the same value of $\mathbf{x}_i$, and a "reconfirmed" point in which y is measured twice for the same experiment. Although the latter method would supply information on that part of σ^2 relating to the accuracy of the measuring method, it is the former which forms the basis for (2.38). This model includes the genuine variation between different experiments with the same $\mathbf{x}$ observation as well as error of measurements.

Ignoring the structure on μ_i, we have the usual decomposition

$$\begin{aligned}\sum_i\sum_j(y_{ij}-\mu_i)^2 &= \sum_i\sum_j(y_{ij}-\bar{y}_{i\cdot}+\bar{y}_{i\cdot}-\mu_i)^2\\ &= \sum_i\sum_j(y_{ij}-\bar{y}_{i\cdot})^2+\sum_i J_i(\bar{y}_{i\cdot}-\mu_i)^2,\end{aligned} \tag{2.39}$$

where $\sum\sum(y_{ij}-\bar{y}_{i\cdot})^2$ is usually referred to as the "pure error" sum of squares. Defining $N=\sum_i J_i$, an unbiased estimate of σ^2 is

$$s_e^2=\frac{1}{N-n}\sum_i\sum_j(y_{ij}-\bar{y}_{i\cdot})^2, \tag{2.40}$$

and, under the normality assumptions of the ε_{ij},

$$\frac{(N-n)s_e^2}{\sigma^2}\sim\chi^2_{N-n}. \tag{2.41}$$

To find the least-squares estimate $\hat{\boldsymbol{\theta}}$ of $\boldsymbol{\theta}$ we minimize $\sum\sum(y_{ij}-\mu_i)^2$ with respect to $\boldsymbol{\theta}$, which is equivalent to minimizing (cf. (2.39)] $\sum_i J_i[\bar{y}_{i\cdot}-f(\mathbf{x}_i;\boldsymbol{\theta})]^2$, i.e. a weighted least-squares analysis with weights J_i. The normal equations for $\hat{\boldsymbol{\theta}}$ are therefore

$$-2\sum_i J_i(\bar{y}_i)-f(\mathbf{x}_i;\boldsymbol{\theta})\frac{\partial f(\mathbf{x}_i;\boldsymbol{\theta})}{\partial\theta_r}\bigg|_{\boldsymbol{\theta}=\hat{\boldsymbol{\theta}}}=0 \qquad (r=1,2,\ldots,p). \tag{2.42}$$

If μ_i is replaced by $\hat{\mu}_i=f(\mathbf{x}_i;\hat{\boldsymbol{\theta}})$, the identity (2.39) still applies, so that

$$\sum\sum(y_{ij}-\hat{\mu}_i)^2=\sum\sum(y_{ij}-\bar{y}_{i\cdot})^2+\sum J_i(\bar{y}_{i\cdot}-\hat{\mu}_i)^2,$$

or

$$Q_H=Q+(Q_H-Q),$$

say. The left-hand side, called the residual sum of squares, is therefore split into a pure-error sum of squares Q and a lack-of-fit sum of squares Q_H-Q. For the case when μ_i is a linear function of $\boldsymbol{\theta}$ (i.e. $\boldsymbol{\mu}=\mathbf{X}\boldsymbol{\theta}$, say), a test for the validity of this linear model is given by (Seber [1977: 181])

$$F = \frac{(Q_H - Q)/(n-p)}{Q/(N-n)}$$

$$= \frac{\sum J_i(\bar{y}_{i\cdot} - \hat{\mu}_i)^2}{\sum\sum(y_{ij} - \bar{y}_{i\cdot})^2} \cdot \frac{N-n}{n-p}, \tag{2.43}$$

where $F \sim F_{n-p,N-n}$ when the model is valid. Because of asymptotic linearity, we find that the above F-test is approximately valid for large N when the model is nonlinear. Confidence intervals are discussed in Section 5.6.

2.2 MAXIMUM-LIKELIHOOD ESTIMATION

If the joint distribution of the ε_i in the model (2.1) is assumed known, then the maximum-likelihood estimate of $\boldsymbol{\theta}$ is obtaind by maximizing the likelihood function. Suppose the ε_i's are i.i.d. with density function $\sigma^{-1}g(\varepsilon/\sigma)$, so that g is the error distribution for errors standardized to have unit variance. Then the likelihood function is

$$p(\mathbf{y}|\boldsymbol{\theta}, \sigma^2) = \prod_{i=1}^{n}\left[\sigma^{-1}g\left(\frac{y_i - f(\mathbf{x}_i;\boldsymbol{\theta})}{\sigma}\right)\right]. \tag{2.44}$$

We shall discuss both normally and nonnormally distributed errors below. In both cases we find that the maximum-likelihood estimator of $\boldsymbol{\theta}$ can be found using least-squares methods.

2.2.1 Normal Errors

If the ε_i are i.i.d $N(0, \sigma^2)$, then (2.44) becomes

$$p(\mathbf{y}|\boldsymbol{\theta}, \sigma^2) = (2\pi\sigma^2)^{-n/2}\exp\left(-\frac{1}{2}\sum_{i=1}^{n}\frac{[y_i - f(\mathbf{x}_i;\boldsymbol{\theta})]^2}{\sigma^2}\right). \tag{2.45}$$

Ignoring constants, we denote the logarithm of the above likelihood by $L(\boldsymbol{\theta}, \sigma^2)$ and obtain

$$L(\boldsymbol{\theta}, \sigma^2) = -\frac{n}{2}\log\sigma^2 - \frac{1}{2\sigma^2}\sum_{i=1}^{n}[y_i - f(\mathbf{x}_i;\boldsymbol{\theta})]^2$$

$$= -\frac{n}{2}\log\sigma^2 - \frac{1}{2\sigma^2}S(\boldsymbol{\theta}). \tag{2.46}$$

Given σ^2, (2.46) is maximized with respect to $\boldsymbol{\theta}$ when $S(\boldsymbol{\theta})$ is minimized, that is, when $\boldsymbol{\theta} = \hat{\boldsymbol{\theta}}$ (the least-squares estimate). Furthermore, $\partial L/\partial \sigma^2 = 0$ has solution $\sigma^2 = S(\boldsymbol{\theta})/n$, which gives a maximum (for given $\boldsymbol{\theta}$) as the second derivative is negative. This suggests that $\hat{\boldsymbol{\theta}}$ and $\hat{\sigma}^2 = S(\hat{\boldsymbol{\theta}})/n$ are the maximum-likelihood estimates, and we now verify this directly. Since $S(\boldsymbol{\theta}) \geq S(\hat{\boldsymbol{\theta}})$,

$$\begin{aligned} L(\hat{\boldsymbol{\theta}}, \hat{\sigma}^2) - L(\boldsymbol{\theta}, \sigma^2) &= -\frac{n}{2}\log \hat{\sigma}^2 - \frac{n}{2} - L(\boldsymbol{\theta}, \sigma^2) \\ &\geq -\frac{n}{2}\log\frac{\hat{\sigma}^2}{\sigma^2} - \frac{n}{2} + \frac{1}{2}\frac{S(\hat{\boldsymbol{\theta}})}{\sigma^2} \\ &= -\frac{n}{2}\left(\log\frac{\hat{\sigma}^2}{\sigma^2} + 1 - \frac{\hat{\sigma}^2}{\sigma^2}\right) \\ &\geq 0, \end{aligned}$$

as $\log x \leq x - 1$ for $x \geq 0$. Hence $\hat{\boldsymbol{\theta}}$ and $\hat{\sigma}^2$ maximize $L(\boldsymbol{\theta}, \sigma^2)$. The maximum value of (2.45) is

$$p(\mathbf{y} \mid \hat{\boldsymbol{\theta}}, \hat{\sigma}^2) = (2\pi\hat{\sigma}^2)^{-n/2} e^{-n/2}. \tag{2.47}$$

Jennrich [1969: p. 640] notes that the least-squares estimate is now not only the maximum-likelihood estimate but, under appropriate regularity conditions (see Section 12.2), is also asymptotically efficient (i.e., $\mathbf{a}'\hat{\boldsymbol{\theta}}$ is asymptotically minimum-variance for every $\mathbf{a}$). The usual asymptotic maximum-likelihood theory does not apply directly, but needs modification, as the y_i are not identically distributed, having different means. Thus if $\boldsymbol{\delta} = (\boldsymbol{\theta}', v)'$, where $v = \sigma^2$, then from (2.46) and (2.35) the (expected) information matrix is given by

$$\begin{aligned} -\mathscr{E}\left[\frac{\partial^2 L}{\partial \boldsymbol{\delta}\, \partial \boldsymbol{\delta}'}\right] &= \begin{bmatrix} \dfrac{1}{2\sigma^2}\mathscr{E}\left[\dfrac{\partial^2 S}{\partial \boldsymbol{\theta}\, \partial \boldsymbol{\theta}'}\right] & -\mathscr{E}\left[\dfrac{\partial^2 L}{\partial \boldsymbol{\theta}\, \partial v}\right] \\ -\mathscr{E}\left[\dfrac{\partial^2 L}{\partial v\, \partial \boldsymbol{\theta}'}\right] & -E\left[\dfrac{\partial^2 L}{\partial v^2}\right] \end{bmatrix} \\ &= \begin{bmatrix} \dfrac{1}{\sigma^2}\mathbf{F}_{\cdot}'(\boldsymbol{\theta}^*)\mathbf{F}_{\cdot}(\boldsymbol{\theta}^*) & \mathbf{0} \\ \mathbf{0}' & \dfrac{n}{2\sigma^4} \end{bmatrix}. \end{aligned} \tag{2.48}$$

Hence, under appropriate regularity conditions, we have from (12.20) that

$$\begin{aligned} \lim_{n \to \infty} \mathscr{D}[\sqrt{n}\hat{\boldsymbol{\theta}}] &= \sigma^2 \boldsymbol{\Omega}^{-1} \\ &= \sigma^2 \lim_{n \to \infty} n[\mathbf{F}_{\cdot}'(\boldsymbol{\theta}^*)\mathbf{F}_{\cdot}(\boldsymbol{\theta}^*)]^{-1}. \end{aligned}$$

As with the i.i.d. case, we see that the variance–covariance matrix of the maximum-likelihood estimator $\hat{\boldsymbol{\theta}}$ is given asymptotically by the inverse of the (expected) information matrix in (2.48).

Let $\hat{\boldsymbol{\delta}} = (\hat{\boldsymbol{\theta}}', \hat{\sigma}^2)'$. Then, since $\partial S(\boldsymbol{\theta})/\partial\boldsymbol{\theta} = \mathbf{0}$ at $\boldsymbol{\theta} = \hat{\boldsymbol{\theta}}$, we also have

$$\left\{-\frac{\partial^2 L}{\partial\boldsymbol{\delta}\,\partial\boldsymbol{\delta}'}\right\}^{-1}_{\hat{\boldsymbol{\delta}}} = \begin{bmatrix} \dfrac{1}{2\hat{\sigma}^2}\left\{\dfrac{\partial^2 S}{\partial\boldsymbol{\delta}\,\partial\boldsymbol{\delta}'}\right\}_{\hat{\boldsymbol{\theta}}} & \mathbf{0} \\ \mathbf{0}' & \dfrac{n}{2\hat{\sigma}^4} \end{bmatrix}^{-1} = \begin{bmatrix} 2\hat{\sigma}^2\left\{\dfrac{\partial^2 S}{\partial\boldsymbol{\theta}\,\partial\boldsymbol{\theta}'}\right\}^{-1}_{\hat{\boldsymbol{\theta}}} & \mathbf{0} \\ \mathbf{0}' & \dfrac{2\hat{\sigma}^4}{n} \end{bmatrix}. \tag{2.49}$$

2.2.2 Nonnormal Data

For completeness we shall now digress from our main theme of nonlinear regression and consider the case where the log-likelihood function is a general function of $\mathscr{E}[\mathbf{y}] = \mathbf{f}(\boldsymbol{\theta})$, say $L([\mathbf{f}(\boldsymbol{\theta})])$. Our discussion is drawn from the extensive paper by Green [1984]. It is assumed that the model is sufficiently regular so that the maximum-likelihood estimate $\hat{\boldsymbol{\theta}}$ is the solution of the likelihood equations

$$\begin{aligned}\mathbf{0} &= \frac{\partial L}{\partial\boldsymbol{\theta}} \\ &= \left(\frac{\partial\mathbf{f}}{\partial\boldsymbol{\theta}'}\right)'\frac{\partial L}{\partial\mathbf{f}} \qquad \text{(by Appendix A10.1).}\end{aligned} \tag{2.50}$$

Let $\boldsymbol{\theta}^{(a)}$ be the ath approximation for $\hat{\boldsymbol{\theta}}$ and consider the Taylor expansion

$$\begin{aligned}\mathbf{0} &= \left.\frac{\partial L}{\partial\boldsymbol{\theta}}\right|_{\hat{\boldsymbol{\theta}}} \\ &\approx \left.\frac{\partial L}{\partial\boldsymbol{\theta}}\right|_{\boldsymbol{\theta}^{(a)}} + \left.\frac{\partial^2 L}{\partial\boldsymbol{\theta}\partial\boldsymbol{\theta}'}\right|_{\boldsymbol{\theta}^{(a)}}(\hat{\boldsymbol{\theta}} - \boldsymbol{\theta}^{(a)}).\end{aligned}$$

Then

$$\begin{aligned}\hat{\boldsymbol{\theta}} - \boldsymbol{\theta}^{(a)} &\approx \left[\left(-\frac{\partial^2 L}{\partial\boldsymbol{\theta}\,\partial\boldsymbol{\theta}'}\right)^{-1}\frac{\partial L}{\partial\boldsymbol{\theta}}\right]_{\boldsymbol{\theta}^{(a)}} \\ &= \boldsymbol{\delta}^{(a)},\end{aligned} \tag{2.51}$$

so that an updated estimate is $\boldsymbol{\theta}^{(a+1)} = \boldsymbol{\theta}^{(a)} + \boldsymbol{\delta}^{(a)}$. This is Newton's method of solving the equations, though in the above context statisticians often refer to it as the Newton–Raphson method. In general, the negative second-derivative matrix contains random variables, and it is often recommended that it be replaced by its expected value, the so-called (expected) information matrix. This technique is known as Fisher's "scoring" algorithm. (cf. Appendix A14). We now find the above expected matrix.

From Appendix A10.2 we have

$$\frac{\partial^2 L}{\partial \boldsymbol{\theta}\, \partial \boldsymbol{\theta}'} = \left(\frac{\partial \mathbf{f}}{\partial \boldsymbol{\theta}'}\right)' \frac{\partial^2 L}{\partial \mathbf{f}\, \partial \mathbf{f}'} \left(\frac{\partial \mathbf{f}}{\partial \boldsymbol{\theta}'}\right) + \sum_i \frac{\partial L}{\partial f_i} \cdot \frac{\partial^2 f_i}{\partial \boldsymbol{\theta}\, \partial \boldsymbol{\theta}'}. \tag{2.52}$$

Assuming that the order of differentiation with respect to $\boldsymbol{\theta}$ and integration (or summation) with respect to $\mathbf{y}$ can be interchanged, we have [with $p(\cdot)$ the density function of $\mathbf{y}$, and $L = \log p$]

$$\begin{aligned}
\mathscr{E}\left[\frac{\partial L}{\partial \mathbf{f}}\right] &= \int \frac{1}{p} \frac{\partial p}{\partial \mathbf{f}} \cdot p\, d\mathbf{y} \\
&= \frac{\partial}{\partial \mathbf{f}} \int p\, d\mathbf{y} \\
&= \mathbf{0},
\end{aligned} \tag{2.53}$$

and, using a similar argument,

$$\mathscr{E}\left[\frac{1}{p} \frac{\partial^2 p}{\partial \mathbf{f}\, \partial \mathbf{f}'}\right] = \mathbf{0}. \tag{2.54}$$

Hence

$$\begin{aligned}
\mathscr{E}\left[\frac{\partial^2 L}{\partial \mathbf{f}\, \partial \mathbf{f}'}\right] &= \mathscr{E}\left[\frac{\partial}{\partial \mathbf{f}}\left(\frac{1}{p} \frac{\partial p}{\partial \mathbf{f}'}\right)\right] \\
&= \mathscr{E}\left[-\frac{1}{p^2} \frac{\partial p}{\partial \mathbf{f}} \cdot \frac{\partial p}{\partial \mathbf{f}'} + \frac{1}{p} \frac{\partial^2 p}{\partial \mathbf{f}\, \partial \mathbf{f}'}\right] \\
&= -\mathscr{E}\left[\frac{\partial L}{\partial \mathbf{f}} \cdot \frac{\partial L}{\partial \mathbf{f}'}\right] \\
&= -\mathbf{G}.
\end{aligned} \tag{2.55}$$

We note that $\mathbf{G}$ is positive definite, as

$$\mathbf{a}'\mathbf{G}\mathbf{a} = E\left[\left(\mathbf{a}' \frac{\partial L}{\partial \mathbf{f}}\right)^2\right] \geq 0,$$

with equality only if $\mathbf{a} = \mathbf{0}$ (under general conditions on L). Applying (2.55) and (2.53) to (2.52) gives us

$$\mathscr{E}\left[-\frac{\partial^2 L}{\partial \boldsymbol{\theta}\, \partial \boldsymbol{\theta}'}\right] = \mathbf{F}_.'\mathbf{G}\mathbf{F}_., \tag{2.56}$$

which is positive definite (Appendix A4.5) for $\mathbf{F}_{\bullet}$ of full rank, and is therefore nonsingular. Thus Fisher's scoring algorithm becomes

$$\boldsymbol{\theta}^{(a+1)} = \boldsymbol{\theta}^{(a)} + \left((\mathbf{F}_{\bullet}'\mathbf{G}\mathbf{F}_{\bullet})^{-1} \frac{\partial L}{\partial \boldsymbol{\theta}} \right)_{\boldsymbol{\theta}^{(a)}}. \tag{2.57}$$

The negative second-derivative matrix may not be positive definite at every $\boldsymbol{\theta}^{(a)}$, and this can cause the Newton method of (2.51) to fail (see Section 13.3.1). The Fisher scoring method therefore uses an approximation of the negative second-derivative matrix which is always positive definite, so that the step taken at the ath iteration, $\boldsymbol{\delta}^{(a)} = \boldsymbol{\theta}^{(a+1)} - \boldsymbol{\theta}^{(a)}$, leads off in an uphill direction (see Section 13.2.3b with downhill replaced by uphill, as here we are referring to maximization rather than minimization). It has a further advantage in that only first derivatives of L are required, so that the approximation can often be calculated more quickly than the second-derivative matrix. The scoring algorithm pays for these advantages by converging more slowly than the Newton method as $\boldsymbol{\theta}^{(a)}$ approaches $\hat{\boldsymbol{\theta}}$. More widely applicable methods of protecting against matrices which may not be positive definite are described in Section 13.3.1. With a number of well-known models such as the linear logistic regression model (Section 1.6), the scoring algorithm and Newton's method coincide.

If we now set $\mathbf{g} = \partial L/\partial \mathbf{f}$, we obtain from (2.51) and (2.57)

$$\begin{aligned} \boldsymbol{\delta}^{(a)} &= [(\mathbf{F}_{\bullet}'\mathbf{G}\mathbf{F}_{\bullet})^{-1}\mathbf{F}_{\bullet}'\mathbf{g}]_{\boldsymbol{\theta}^{(a)}} \\ &= [(\mathbf{F}_{\bullet}'\mathbf{V}^{-1}\mathbf{F}_{\bullet})^{-1}\mathbf{F}_{\bullet}'\mathbf{V}^{-1}\mathbf{v}]_{\boldsymbol{\theta}^{(a)}}, \end{aligned} \tag{2.58}$$

where $\mathbf{V} = \mathbf{G}^{-1}$ and $\mathbf{v} = \mathbf{V}\mathbf{g}$. We note that the equation (2.58) represents the generalized least-squares estimate $\boldsymbol{\delta}^{(a)}$ of $\boldsymbol{\delta}$ for the model (evaluated at $\boldsymbol{\theta}^{(a)}$)

$$\mathbf{v} = \mathbf{F}_{\bullet}^{(a)}\boldsymbol{\delta} + \boldsymbol{\nu}, \tag{2.59}$$

where $\mathscr{D}[\boldsymbol{\nu}] = \mathbf{V}^{(a)}$. Using the same idea as in Section 2.1.4, let $\mathbf{V}^{(a)} = \mathbf{U}'\mathbf{U}$ be the upper triangular Cholesky decomposition of $\mathbf{V}^{(a)}$ (Appendix A8.2), and let $\mathbf{R} = (\mathbf{U}')^{-1}$. Then multiplying (2.59) by $\mathbf{R}$, we obtain

$$\mathbf{z} = \mathbf{X}\boldsymbol{\delta} + \boldsymbol{\eta}, \tag{2.60}$$

where $\mathbf{z} = \mathbf{R}\mathbf{v}$, $\mathbf{X} = \mathbf{R}\mathbf{F}_{\bullet}^{(a)}$, and $\mathscr{D}[\boldsymbol{\eta}] = \sigma^2\mathbf{I}_n$. Thus $\boldsymbol{\delta}^{(a)}$ can be computed by using an ordinary linear least-squares program and regressing $\mathbf{z}$ on $\mathbf{X}$. A standard output from a linear-regression program is $(\mathbf{X}'\mathbf{X})^{-1}$ [or s^2 and $s^2(\mathbf{X}'\mathbf{X})^{-1}$]: in (2.60)

$$(\mathbf{X}'\mathbf{X})^{-1} = (\mathbf{F}_{\bullet}'\mathbf{V}^{-1}\mathbf{F}_{\bullet})^{-1}_{\boldsymbol{\theta}^{(a)}}.$$

Thus, by (2.56), the regression output at the final iteration provides

$$\left(\mathscr{E}\left[-\frac{\partial^2 L}{\partial \boldsymbol{\theta} \partial \boldsymbol{\theta}'}\right]_{\hat{\boldsymbol{\theta}}}\right)^{-1}$$

which, under certain regularity conditions, is the asymptotic variance–covariance matrix of $\hat{\boldsymbol{\theta}}$.

The above method is a very general one and is commonly referred to as iteratively reweighted least squares (IRLS). It can be applied to a wide range of models, and one such application, quasi-likelihood models, is considered in Section 2.3. The method enables a large range of models to be grafted onto existing statistical packages with minimal additional programming by making use of existing least-squares regression procedures. However, as IRLS is not central to our main theme, we refer the interested reader to the paper of Green [1984] and the ensuing discussion for further details and examples. In commenting on Green's paper Ross claimed that it was preferable to use the optimization facilities of good statistical packages rather than rely on the IRLS method, especially when the problem is unfamiliar. He noted that IRLS needs the correct algebraic specification and good initial estimates, and may still be slow or divergent. However, the optimization applications require only the model formulation and the distribution of errors. There is also protection from divergence, and parameter transformations are easily incorporated.

The special case of a linear model $f(\mathbf{x}_i;\boldsymbol{\theta}) = \mathbf{x}_i'\boldsymbol{\theta}$ is considered by Stirling [1984]. An application to regression diagnostics is given by Moolgavkar et al. [1984]. Green's method has been extended by Jørgensen [1984].

2.2.3 Concentrated Likelihood Methods

In finding maximum-likelihood estimates from general likelihoods it is sometimes convenient to use the following stepwise method of maximization. Suppose that we have a general log-likelihood function $L(\boldsymbol{\theta}, \boldsymbol{\tau}|\mathbf{z})$ to be maximized with respect to $\boldsymbol{\theta}$ and $\boldsymbol{\tau}$, where $\mathbf{z}$ represents the data and $\boldsymbol{\tau}$ is $q \times 1$. It is assumed that L is uniquely maximized with respect to $\boldsymbol{\theta}$ and $\boldsymbol{\tau}$ for every $\mathbf{z}$. Then the first step of the maximization is to find $\boldsymbol{\gamma}(\boldsymbol{\theta}, \mathbf{z})$, the unique value of $\boldsymbol{\tau}$ which maximizes L with respect to $\boldsymbol{\tau}$, $\boldsymbol{\theta}$ being regarded as a constant. The second step consists of finding $\hat{\boldsymbol{\theta}} = \hat{\boldsymbol{\theta}}(\mathbf{z})$, the value of $\boldsymbol{\theta}$ which maximizes

$$M(\boldsymbol{\theta}|\mathbf{z}) \equiv L\{\boldsymbol{\theta}, \boldsymbol{\gamma}(\boldsymbol{\theta}, \mathbf{z})|\mathbf{z}\}. \tag{2.61}$$

Since

$$\begin{aligned}
L\{\hat{\boldsymbol{\theta}}(\mathbf{z}), \boldsymbol{\gamma}(\hat{\boldsymbol{\theta}}, \mathbf{z})|\mathbf{z}\} &= M(\hat{\boldsymbol{\theta}}|\mathbf{z}) \\
&\geq M(\boldsymbol{\theta}|\mathbf{z}) \\
&= L\{\boldsymbol{\theta}, \boldsymbol{\gamma}(\boldsymbol{\theta}, \mathbf{z})|\mathbf{z}\} \\
&\geq L\{\boldsymbol{\theta}, \boldsymbol{\tau}|\mathbf{z}\},
\end{aligned}$$

the maximum-likelihood estimates are $\hat{\boldsymbol{\theta}}$ and $\hat{\boldsymbol{\tau}} = \boldsymbol{\gamma}(\hat{\boldsymbol{\theta}}, \mathbf{z})$. If there are any relative maxima, the above argument holds in a neighborhood of the absolute maximum which does not contain any relative maxima. The key part of the argument is that $\boldsymbol{\gamma}(\boldsymbol{\theta}, \mathbf{z})$ must be unique for a given $\boldsymbol{\theta}$. The function $M(\boldsymbol{\theta}|\mathbf{z})$ is called the concentrated log-likelihood function (Hood and Koopmans [1953: pp. 156–157]) because it is concentrated on the set of parameters $\boldsymbol{\theta}$. In what follows we drop the functional dependence on the data $\mathbf{z}$ for notational convenience.

We now consider some advantages of using $M(\boldsymbol{\theta})$ instead of $L(\boldsymbol{\theta}, \boldsymbol{\tau})$. Firstly, we shall see in Theorem 2.2 below that $\hat{\boldsymbol{\theta}}$ can be obtained directly as the solution of $\partial M(\boldsymbol{\theta})/\partial \boldsymbol{\theta} = \mathbf{0}$, so that the nuisance parameter $\boldsymbol{\tau}$ is effectively eliminated. Secondly, under certain conditions, an estimate of the asymptotic variance–covariance matrix of $\hat{\boldsymbol{\delta}} = (\hat{\boldsymbol{\theta}}', \hat{\boldsymbol{\tau}}')'$ is given by

$$\left\{ -\frac{\partial^2 L(\boldsymbol{\delta})}{\partial \boldsymbol{\delta}\, \partial \boldsymbol{\delta}'} \right\}_{\hat{\boldsymbol{\delta}}}^{-1},$$

a $(p+q) \times (p+q)$ matrix which can be partitioned to give an estimate of $\mathscr{D}[\hat{\boldsymbol{\theta}}]$. However, if we are interested in just $\hat{\boldsymbol{\theta}}$, then from Theorem 2.2 below we find that $\mathscr{D}[\hat{\boldsymbol{\theta}}]$ can be estimated by

$$\left\{ -\frac{\partial^2 M(\boldsymbol{\theta})}{\partial \boldsymbol{\theta}\, \partial \boldsymbol{\theta}'} \right\}_{\hat{\boldsymbol{\theta}}}^{-1}.$$

Example 2.2 We now demonstrate the above methodology by rederiving the maximum-likelihood estimates for the log-likelihood (2.46) with $\tau = \sigma^2$. Dropping the dependence on the data for notational convenience, we have

$$L(\boldsymbol{\theta}, \sigma^2) = -\frac{n}{2} \log \sigma^2 - \frac{1}{2\sigma^2} S(\boldsymbol{\theta}).$$

For fixed $\boldsymbol{\theta}$ the above expression is maximized when $\sigma^2 = S(\boldsymbol{\theta})/n$, so that the concentrated log-likelihood function is

$$\begin{aligned} M(\boldsymbol{\theta}) &= L\left\{ \boldsymbol{\theta}, \frac{S(\boldsymbol{\theta})}{n} \right\} \\ &= -\frac{n}{2} \log S(\boldsymbol{\theta}) + \frac{n}{2}(\log n - 1). \end{aligned} \tag{2.62}$$

This is maximized when $S(\boldsymbol{\theta})$ is minimized, that is, at the least-squares estimate $\hat{\boldsymbol{\theta}}$. Then $S(\hat{\boldsymbol{\theta}})/n$ is the maximum-likelihood estimator $\hat{\sigma}^2$ of σ^2. Furthermore, since

$[\partial S(\boldsymbol{\theta})/\partial \boldsymbol{\theta}]_{\hat{\boldsymbol{\theta}}} = \mathbf{0}$,

$$\left\{-\frac{\partial^2 M(\boldsymbol{\theta})}{\partial \boldsymbol{\theta}\, \partial \boldsymbol{\theta}'}\right\}_{\hat{\boldsymbol{\theta}}} = \left\{-\frac{n}{2[S(\boldsymbol{\theta})]^2}\frac{\partial S(\boldsymbol{\theta})}{\partial \boldsymbol{\theta}}\cdot\frac{\partial S(\boldsymbol{\theta})}{\partial \boldsymbol{\theta}'} + \frac{n}{2S(\boldsymbol{\theta})}\frac{\partial^2 S(\boldsymbol{\theta})}{\partial \boldsymbol{\theta}\, \partial \boldsymbol{\theta}'}\right\}_{\hat{\boldsymbol{\theta}}}$$
$$= \frac{1}{2\hat{\sigma}^2}\left\{\frac{\partial^2 S(\boldsymbol{\theta})}{\partial \boldsymbol{\theta}\, \partial \boldsymbol{\theta}'}\right\}_{\hat{\boldsymbol{\theta}}} \tag{2.63}$$
$$\approx \frac{1}{2\hat{\sigma}^2}\cdot 2\mathbf{F}_{.}'(\hat{\boldsymbol{\theta}})\mathbf{F}_{.}(\hat{\boldsymbol{\theta}}) \qquad \text{[by (2.35)]}$$
$$= \frac{\hat{\mathbf{F}}_{.}'\hat{\mathbf{F}}_{.}}{\hat{\sigma}^2},$$

so that $\mathscr{D}[\hat{\boldsymbol{\theta}}]$ is estimated by $\hat{\sigma}^2(\hat{\mathbf{F}}_{.}'\hat{\mathbf{F}}_{.})^{-1}$. This is our usual estimator but with s^2 replaced by $\hat{\sigma}^2$ [cf. (2.21)]. ■

We note that the inverse of (2.63) is the upper diagonal block of (2.49). This result is generalized in the following theorem (Richards [1961], Kale [1963]).

Theorem 2.2 Let $L(\boldsymbol{\theta}, \boldsymbol{\tau})$ be the log-likelihood function defined above, where $\boldsymbol{\theta}$ is the $p \times 1$ vector parameter of interest and $\boldsymbol{\tau}$ is a $q \times 1$ vector of "nuisance" parameters. We assume that L is twice differentiable. Define $\boldsymbol{\delta} = (\boldsymbol{\theta}', \boldsymbol{\tau}')'$, and let $\hat{\boldsymbol{\delta}} = (\hat{\boldsymbol{\theta}}', \hat{\boldsymbol{\tau}}')'$ solve $\partial L/\partial \boldsymbol{\delta} = \mathbf{0}$, that is, solve

$$\left.\frac{\partial L(\boldsymbol{\theta}, \boldsymbol{\tau})}{\partial \boldsymbol{\theta}}\right|_{\hat{\boldsymbol{\delta}}} = \mathbf{0} \quad \text{and} \quad \left.\frac{\partial L(\boldsymbol{\theta}, \boldsymbol{\tau})}{\partial \boldsymbol{\tau}}\right|_{\hat{\boldsymbol{\delta}}} = \mathbf{0}. \tag{2.64}$$

Define

$$\mathbf{I}(\boldsymbol{\delta}) = -\frac{\partial^2 L}{\partial \boldsymbol{\delta}\, \partial \boldsymbol{\delta}'}$$
$$= \begin{bmatrix} -\dfrac{\partial^2 L}{\partial \boldsymbol{\theta}\, \partial \boldsymbol{\theta}'} & -\dfrac{\partial^2 L}{\partial \boldsymbol{\theta}\, \partial \boldsymbol{\tau}'} \\ -\dfrac{\partial^2 L}{\partial \boldsymbol{\tau}\, \partial \boldsymbol{\theta}'} & -\dfrac{\partial^2 L}{\partial \boldsymbol{\tau}\, \partial \boldsymbol{\tau}'} \end{bmatrix}$$
$$= \begin{bmatrix} \mathbf{I}_{\theta\theta} & \mathbf{I}_{\theta\tau} \\ \mathbf{I}_{\tau\theta} & \mathbf{I}_{\tau\tau} \end{bmatrix}, \tag{2.65}$$

say, and assume that it is positive definite at $\boldsymbol{\delta} = \hat{\boldsymbol{\delta}}$. Also define

$$\mathbf{I}^{-1}(\boldsymbol{\delta}) = \begin{pmatrix} \mathbf{J}_{\theta\theta} & \mathbf{J}_{\theta\tau} \\ \mathbf{J}_{\tau\theta} & \mathbf{J}_{\tau\tau} \end{pmatrix}. \tag{2.66}$$

Suppose, for any fixed $\boldsymbol{\theta}$, that $\boldsymbol{\tau} = \boldsymbol{\gamma}(\boldsymbol{\theta})$ solves

$$\frac{\partial L(\boldsymbol{\theta}, \boldsymbol{\tau})}{\partial \boldsymbol{\tau}} = \mathbf{0}, \tag{2.67}$$

and let $M(\boldsymbol{\theta}) = L[\boldsymbol{\theta}, \boldsymbol{\gamma}(\boldsymbol{\theta})]$. Then

(i) $\partial M(\boldsymbol{\theta})/\partial \boldsymbol{\theta}|_{\hat{\boldsymbol{\theta}}} = \mathbf{0}$, and
(ii) we have

$$\left\{ -\frac{\partial^2 M(\boldsymbol{\theta})}{\partial \boldsymbol{\theta}\, \partial \boldsymbol{\theta}'} \right\}_{\hat{\boldsymbol{\theta}}} = \mathbf{J}_{\theta\theta}^{-1} \Big|_{\hat{\boldsymbol{\delta}}} = (\mathbf{I}_{\theta\theta} - \mathbf{I}_{\theta\tau}\mathbf{I}_{\tau\tau}^{-1}\mathbf{I}_{\theta\tau}')_{\hat{\boldsymbol{\delta}}}. \tag{2.68}$$

[It should be noted that this theorem is true for any function L and not just log-likelihoods.]

Proof: The assumption that $\mathbf{I}(\hat{\boldsymbol{\delta}})$ is positive definite means that $\mathbf{I}(\boldsymbol{\delta})$ is positive definite in a neighborhood $\mathcal{N}$ of $\hat{\boldsymbol{\delta}}$, and both $\mathbf{I}_{\theta\theta}$ and $\mathbf{I}_{\tau\tau}$ are nonsingular in $\mathcal{N}$. As pointed out by Kale [1963], the existence of $\hat{\boldsymbol{\delta}}$ and the nonsingularity of $\mathbf{I}_{\tau\tau}$ in $\mathcal{N}$ imply that the implicit-function theorem can be applied to the set of equations (2.67). Hence in $\mathcal{N}$ we have (a) $\boldsymbol{\gamma}(\boldsymbol{\theta})$ is uniquely defined, (b) $\hat{\boldsymbol{\tau}} = \boldsymbol{\gamma}(\hat{\boldsymbol{\theta}})$ and (c) $\boldsymbol{\gamma}(\boldsymbol{\theta})$ has continuous first-order derivatives. In the following we use the differential notation of Section 1.1 and assume that expressions are valid for $\boldsymbol{\delta}$ in $\mathcal{N}$.

(i): We note that, for $\boldsymbol{\tau} = \boldsymbol{\gamma}(\boldsymbol{\theta})$,

$$\frac{\partial M(\boldsymbol{\theta})}{\partial \theta_r} = \sum_j \frac{\partial L(\boldsymbol{\theta}, \boldsymbol{\tau})}{\partial \theta_j} \cdot \frac{\partial \theta_j}{\partial \theta_r} + \sum_k \frac{\partial L(\boldsymbol{\theta}, \boldsymbol{\tau})}{\partial \tau_k} \cdot \frac{\partial \tau_k}{\partial \theta_r},$$

that is,

$$\begin{aligned} \frac{\partial M(\boldsymbol{\theta})}{\partial \boldsymbol{\theta}} &= \frac{\partial L(\boldsymbol{\theta}, \boldsymbol{\tau})}{\partial \boldsymbol{\theta}} + \left(\frac{\partial \boldsymbol{\gamma}}{\partial \boldsymbol{\theta}'} \right)' \frac{\partial L(\boldsymbol{\theta}, \boldsymbol{\tau})}{\partial \boldsymbol{\tau}} \\ &= \frac{\partial L(\boldsymbol{\theta}, \boldsymbol{\tau})}{\partial \boldsymbol{\theta}} \Big|_{\boldsymbol{\tau} = \boldsymbol{\gamma}(\boldsymbol{\theta})}, \end{aligned} \tag{2.69}$$

since the second term is zero by (2.67). Hence using $\hat{\boldsymbol{\tau}} = \boldsymbol{\gamma}(\hat{\boldsymbol{\theta}})$, we have

$$\begin{aligned} \frac{\partial M(\boldsymbol{\theta})}{\partial \boldsymbol{\theta}} \Big|_{\hat{\boldsymbol{\theta}}} &= \frac{\partial L(\boldsymbol{\theta}, \boldsymbol{\gamma})}{\partial \boldsymbol{\theta}} \Big|_{\hat{\boldsymbol{\theta}}} \\ &= \frac{\partial L(\boldsymbol{\theta}, \boldsymbol{\tau})}{\partial \boldsymbol{\theta}} \Big|_{\hat{\boldsymbol{\theta}}, \hat{\boldsymbol{\tau}}} \\ &= \mathbf{0}, \end{aligned}$$

by (2.64). Thus (i) is proved.

(ii): Now $\boldsymbol{\tau} = \boldsymbol{\gamma}(\boldsymbol{\theta})$ is the solution of (2.67), that is, of

$$\frac{\partial L(\boldsymbol{\theta}, \boldsymbol{\tau})}{\partial \tau_s} = 0 \qquad (s = 1, 2, \ldots, q).$$

In $\mathcal{N}$, the above represents an identity in $\boldsymbol{\theta}$, so that differentiating with respect to θ_r, we get

$$\begin{aligned} 0 &= \frac{\partial}{\partial \theta_r} \left\{ \left. \frac{\partial L(\boldsymbol{\theta}, \boldsymbol{\tau})}{\partial \tau_s} \right|_{\boldsymbol{\tau} = \boldsymbol{\gamma}(\boldsymbol{\theta})} \right\} \\ &= \left\{ \frac{\partial^2 L(\boldsymbol{\theta}, \boldsymbol{\tau})}{\partial \theta_r \, \partial \tau_s} + \sum_u \frac{\partial^2 L(\boldsymbol{\theta}, \boldsymbol{\tau})}{\partial \tau_u \, \partial \tau_s} \cdot \frac{\partial \gamma_u}{\partial \theta_r} \right\}_{\boldsymbol{\tau} = \boldsymbol{\gamma}(\boldsymbol{\theta})} \end{aligned}$$

that is

$$\mathbf{0} = \left\{ \frac{\partial^2 L(\boldsymbol{\theta}, \boldsymbol{\tau})}{\partial \boldsymbol{\theta} \, \partial \boldsymbol{\tau}'} + \left(\frac{\partial \boldsymbol{\gamma}}{\partial \boldsymbol{\theta}'} \right)' \frac{\partial^2 L(\boldsymbol{\theta}, \boldsymbol{\tau})}{\partial \boldsymbol{\tau} \, \partial \boldsymbol{\tau}'} \right\}_{\boldsymbol{\tau} = \boldsymbol{\gamma}(\boldsymbol{\theta})}. \tag{2.70}$$

Now $-\partial^2 L(\boldsymbol{\theta}, \boldsymbol{\tau}) / \partial \boldsymbol{\tau} \, \partial \boldsymbol{\tau}'$ evaluated at $\boldsymbol{\tau} = \boldsymbol{\gamma}(\boldsymbol{\theta})$ cannot be identified with $\mathbf{I}_{\tau\tau}$, as the former is a function of $\boldsymbol{\theta}$ ($\boldsymbol{\gamma}$ being a particular function of $\boldsymbol{\theta}$), whereas the latter is a function of both $\boldsymbol{\theta}$ and $\boldsymbol{\tau}$ with $\boldsymbol{\tau}$ unconstrained. However, when $\boldsymbol{\theta} = \hat{\boldsymbol{\theta}}$ we have $\boldsymbol{\gamma}(\hat{\boldsymbol{\theta}}) = \hat{\boldsymbol{\tau}}$ and the two matrices have the same value. The same argument applies to the first matrix in (2.70). Hence from (2.70) we have

$$\left(\frac{\partial \boldsymbol{\gamma}}{\partial \boldsymbol{\theta}'} \right)'_{\hat{\boldsymbol{\theta}}} = (-\mathbf{I}_{\theta\tau} \mathbf{I}_{\tau\tau}^{-1})_{\hat{\boldsymbol{\theta}}, \hat{\boldsymbol{\tau}}}. \tag{2.71}$$

Using a similar argument leads to

$$-\frac{\partial^2 M(\boldsymbol{\theta})}{\partial \boldsymbol{\theta} \, \partial \boldsymbol{\theta}'} = \left\{ -\frac{\partial^2 L(\boldsymbol{\theta}, \boldsymbol{\tau})}{\partial \boldsymbol{\theta} \, \partial \boldsymbol{\theta}'} + \left(\frac{\partial \boldsymbol{\gamma}}{\partial \boldsymbol{\theta}'} \right)' \left[-\frac{\partial^2 L(\boldsymbol{\theta}, \boldsymbol{\tau})}{\partial \boldsymbol{\tau} \, \partial \boldsymbol{\theta}'} \right] \right\}_{\boldsymbol{\tau} = \boldsymbol{\gamma}(\boldsymbol{\theta})}.$$

Setting $\boldsymbol{\theta} = \hat{\boldsymbol{\theta}}$ and using (2.71) gives us

$$\left(-\frac{\partial^2 M(\boldsymbol{\theta})}{\partial \boldsymbol{\theta} \, \partial \boldsymbol{\theta}'} \right)_{\hat{\boldsymbol{\theta}}} = (\mathbf{I}_{\theta\theta} - \mathbf{I}_{\theta\tau} \mathbf{I}_{\tau\tau}^{-1} \mathbf{I}_{\tau\theta})_{\hat{\boldsymbol{\delta}}}.$$

Now if $\mathbf{A}$ and $\mathbf{C}$ are symmetric and all inverses exist,

$$\begin{pmatrix} \mathbf{A} & \mathbf{B} \\ \mathbf{B} & \mathbf{C} \end{pmatrix}^{-1} = \begin{pmatrix} \mathbf{F}^{-1} & -\mathbf{F}^{-1}\mathbf{G}' \\ -\mathbf{G}\mathbf{F}^{-1} & \mathbf{C}^{-1} + \mathbf{G}\mathbf{F}^{-1}\mathbf{G}' \end{pmatrix}, \tag{2.72}$$

where $\mathbf{F} = \mathbf{A} - \mathbf{B}\mathbf{C}^{-1}\mathbf{B}'$ and $\mathbf{G} = \mathbf{C}^{-1}\mathbf{B}'$. Applying this result to $\mathbf{I}^{-1}(\hat{\boldsymbol{\delta}})$ of (2.66) gives us

$$\mathbf{J}_{\theta\theta}|_{\hat{\theta}} = \mathbf{F}^{-1} = (\mathbf{I}_{\theta\theta} - \mathbf{I}_{\theta\tau}\mathbf{I}_{\tau\tau}^{-1}\mathbf{I}_{\tau\theta})_{\hat{\delta}}^{-1},$$

and (ii) is proved. ■

The methodology implied by the above lemma for finding the maximum-likelihood estimates $\hat{\theta}$ and $\hat{\tau}$, and their asymptotic variance–covariance matrices, can be summarized as follows:

1. Differentiate the log-likelihood function with respect to τ, and solve the resulting equations for $\tau = \gamma(\theta)$ as a function of θ.
2. Replace τ by $\gamma(\theta)$ in the log-likelihood, and obtain the concentrated log-likelihood function $M(\theta)$.
3. Use $M(\theta)$ as though it were the true log-likelihood function for θ, and go through the usual maximum-likelihood process, namely (a) differentiate $M(\theta)$ with respect to θ and solve for $\hat{\theta}$, and (b) find the estimated information matrix (2.68). Under general regularity conditions the latter matrix is an estimate of the asymptotic dispersion matrix of $\hat{\theta}$.
4. $\hat{\tau}$ is given by $\gamma(\hat{\theta})$.
5. The remainder of $\mathbf{I}^{-1}(\hat{\delta})$ can be obtained using (2.72).

As already noted, a nice feature of the above method is that through using the concentrated log-likelihood we can effectively eliminate the nuisance-parameter vector τ and deal with θ directly. A second feature is that instead of having to invert the $(p+q) \times (p+q)$ matrix $\mathbf{I}(\hat{\delta})$, we need only invert the $q \times q$ matrix $\mathbf{I}_{\tau\tau}$ in (2.68). Finally, as mentioned in Section 5.9.1, various large-sample tests for hypotheses about θ can also be expressed in terms of the concentrated log-likelihood rather than the log-likelihood.

*2.3 QUASI-LIKELIHOOD ESTIMATION

We now introduce an estimator of θ which has properties akin to a maximum-likelihood estimator but with the difference that the distribution of $\mathbf{y}$ does not need to be known explicitly, only its first two moments. The following discussion is based on McCullagh [1983], but with some notational changes.

Let $\mathscr{E}[\mathbf{y}] = \mathbf{f}(\theta)$ ($= \boldsymbol{\mu}$, say) and $\mathscr{D}[\mathbf{y}] = \sigma^2 \mathbf{V}(\boldsymbol{\mu})$. The log quasi-likelihood function, considered as a function $\ell(\boldsymbol{\mu})$ of $\boldsymbol{\mu}$, is defined by the system of partial differential equations

$$\frac{\partial \ell(\boldsymbol{\mu})}{\partial \boldsymbol{\mu}} = \mathbf{V}^{-1}(\boldsymbol{\mu})(\mathbf{y} - \boldsymbol{\mu}). \tag{2.73}$$

This definition is given by McCullagh [1983], and it extends Wedderburn's [1974] definition. For notational convenience we shall write $\ell[\mathbf{f}(\boldsymbol{\theta})]$ and $\mathbf{V}[\mathbf{f}(\boldsymbol{\theta})]$ as simply $\ell(\boldsymbol{\theta})$ and $\mathbf{V}(\boldsymbol{\theta})$, respectively. The maximum quasi-likelihood estimate $\breve{\boldsymbol{\theta}}$ of $\boldsymbol{\theta}$ is obtained by solving the equations

$$\begin{aligned} \mathbf{0} &= \frac{\partial \ell}{\partial \boldsymbol{\theta}} \\ &= \left(\frac{\partial \mathbf{f}}{\partial \boldsymbol{\theta}'}\right)' \frac{\partial \ell}{\partial \mathbf{f}} \qquad \text{[by Appendix A10.1]} \\ &= \mathbf{F}_{\bullet}'(\boldsymbol{\theta})\mathbf{V}^{-1}(\boldsymbol{\theta})[\mathbf{y} - \mathbf{f}(\boldsymbol{\theta})] \qquad \text{[by (2.73)]}. \end{aligned} \tag{2.74}$$

To do this we can use a Gauss–Newton method based on the linear Taylor expansion

$$\mathbf{f}(\breve{\boldsymbol{\theta}}) \approx \mathbf{f}(\boldsymbol{\theta}^{(a)}) + \mathbf{F}_{\bullet}(\boldsymbol{\theta}^{(a)})(\breve{\boldsymbol{\theta}} - \boldsymbol{\theta}^{(a)}) \tag{2.75}$$

for $\breve{\boldsymbol{\theta}}$ about an approximation $\boldsymbol{\theta}^{(a)}$. Substituting (2.75) into (2.74), and approximating $\mathbf{F}_{\bullet}(\boldsymbol{\theta})$ and $\mathbf{V}(\boldsymbol{\theta})$ by $\mathbf{F}_{\bullet}^{(a)} = \mathbf{F}_{\bullet}(\boldsymbol{\theta}^{(a)})$ and $\mathbf{V}^{(a)} = \mathbf{V}(\boldsymbol{\theta}^{(a)})$, we have

$$\mathbf{F}_{\bullet}^{(a)\prime}\mathbf{V}^{(a)-1}[\mathbf{y} - \mathbf{f}(\boldsymbol{\theta}^{(a)}) - \mathbf{F}_{\bullet}^{(a)}(\breve{\boldsymbol{\theta}} - \boldsymbol{\theta}^{(a)})] \approx \mathbf{0},$$

or

$$\begin{aligned} \breve{\boldsymbol{\theta}} - \boldsymbol{\theta}^{(a)} &\approx (\mathbf{F}_{\bullet}^{(a)\prime}\mathbf{V}^{(a)-1}\mathbf{F}_{\bullet}^{(a)})^{-1}\mathbf{F}_{\bullet}^{(a)\prime}\mathbf{V}^{(a)-1}[\mathbf{y} - \mathbf{f}(\boldsymbol{\theta}^{(a)})] \\ &= \boldsymbol{\delta}^{(a)}, \end{aligned} \tag{2.76}$$

say, where the superscript a denotes evaluation at $\boldsymbol{\theta}^{(a)}$. The next approximation for $\breve{\boldsymbol{\theta}}$ is then $\boldsymbol{\theta}^{(a+1)} = \boldsymbol{\theta}^{(a)} + \boldsymbol{\delta}^{(a)}$. As in the maximum-likelihood method [cf. (2.58)] we have, once again, a generalized least-squares technique in which $\boldsymbol{\delta}^{(a)}$ is obtained by minimizing

$$[\mathbf{y} - \mathbf{f}(\boldsymbol{\theta}^{(a)}) - \mathbf{F}_{\bullet}^{(a)}\boldsymbol{\delta}]'\mathbf{V}^{(a)-1}[\mathbf{y} - \mathbf{f}(\boldsymbol{\theta}^{(a)}) - \mathbf{F}_{\bullet}^{(a)}\boldsymbol{\delta}] \tag{2.77}$$

with respect to $\boldsymbol{\delta}$ (see Section 2.1.4 for computational details).

Given certain conditions, $\breve{\boldsymbol{\theta}}$ has a number of useful properties. These conditions are: (a) $\partial^3 f_i(\boldsymbol{\theta})/\partial\theta_r\,\partial\theta_s\,\partial\theta_t$ is bounded ($i = 1, 2, \ldots, n$; $r, s, t = 1, 2, \ldots, p$), (b) the third moments of $\mathbf{y}$ exist, and (c) $n^{-1}\mathbf{I}_n(\boldsymbol{\theta}^*)$ has a positive definite limit as $n \to \infty$, where $\boldsymbol{\theta}^*$ is the true value of $\boldsymbol{\theta}$ and

$$\mathbf{I}_n(\boldsymbol{\theta}^*) = -\mathscr{E}\left[\frac{\partial^2 \ell}{\partial \boldsymbol{\theta}\,\partial \boldsymbol{\theta}'}\right] \tag{2.78}$$

$$= \mathbf{F}_{\bullet}'(\boldsymbol{\theta}^*)\mathbf{V}^{-1}(\boldsymbol{\theta}^*)\mathbf{F}_{\bullet}(\boldsymbol{\theta}^*). \tag{2.79}$$

Assuming these conditions, McCullagh [1983] derived the following properties.

Firstly, using (2.79) and (2.76) (with $\boldsymbol{\theta}^{(a)}$ replaced by $\boldsymbol{\theta}^*$), we have

$$\sqrt{n}(\breve{\boldsymbol{\theta}} - \boldsymbol{\theta}^*) \sim N_p(\mathbf{0}, \sigma^2[n^{-1}\mathbf{I}_n(\boldsymbol{\theta}^*)]^{-1}) + \mathbf{O}_p(n^{-1/2}) \tag{2.80}$$

[see (1.8) for the notation $\mathbf{O}_p$]. Secondly, if the third moment is infinite then the error term in (2.80) is $\mathbf{o}_p(1)$. Thirdly, without altering the order of approximation, (2.78) can be replaced by $-\partial^2\ell/\partial\boldsymbol{\theta}\,\partial\boldsymbol{\theta}'$. Also σ^2 can be estimated by

$$\breve{\sigma}^2 = \frac{[\mathbf{y} - \mathbf{f}(\breve{\boldsymbol{\theta}})]'\mathbf{V}^{-1}(\breve{\boldsymbol{\theta}})[\mathbf{y} - \mathbf{f}(\breve{\boldsymbol{\theta}})]}{n-p}. \tag{2.81}$$

Fourthly, among all the estimators of $\boldsymbol{\theta}^*$ for which the influence function is linear (i.e. estimators $\breve{\boldsymbol{\theta}}_L$ satisfying

$$\breve{\boldsymbol{\theta}}_L - \boldsymbol{\theta}^* = \mathbf{L}[\mathbf{y} - \mathbf{f}(\boldsymbol{\theta}^*)] + \mathbf{o}_p(n^{-1/2}),$$

where $\mathbf{L}$ is a $p \times n$ matrix with elements which are functions of $\boldsymbol{\theta}^*$), $\breve{\boldsymbol{\theta}}$ has the "minimum" dispersion matrix. Finally, suppose we wish to test $H_0{:}\boldsymbol{\theta}\in\omega$ versus $H_1{:}\boldsymbol{\theta}\in\Omega$, where ω and Ω have respective dimensions q and p, and $\omega \subset \Omega\,(q < p)$. If $\ell(\breve{\boldsymbol{\theta}}_0)$ and $\ell(\breve{\boldsymbol{\theta}}_1)$ are the maximum values of $\ell(\boldsymbol{\theta})$ under H_0 and H_1 respectively, then

$$2\frac{\ell(\breve{\boldsymbol{\theta}}_1) - \ell(\breve{\boldsymbol{\theta}}_0)}{\sigma^2} \sim \chi^2_{p-q} + O_p(n^{-1/2}) \tag{2.82}$$

when H_0 is true. If $\ell(\boldsymbol{\theta})$ is the log-likelihood function as opposed to an artificially constructed log quasi-likelihood, and $\mathbf{y}$ has a continuous distribution, then the error term in (2.82) turns out to be $O_p(n^{-1})$. If $\mathbf{V}$ is singular, then $\mathbf{V}^{-1}$ may be replaced by an appropriate generalized inverse $\mathbf{V}^-$ (Scott et al [1989]).

A nice feature of the above theory is that it does not require the existence of a family of distributions for which $\ell(\boldsymbol{\mu})$ is the log-likelihood. Furthermore, $\ell(\boldsymbol{\mu})$ is not required explicitly for the computation of $\breve{\boldsymbol{\theta}}$, though it is needed for some forms of hypothesis testing. A solution for $\ell(\boldsymbol{\mu})$ can sometimes be found by constructing functions $\boldsymbol{\beta}(\boldsymbol{\mu})$ and $c(\boldsymbol{\beta})$ satisfying $\boldsymbol{\mu} = \partial c/\partial\boldsymbol{\beta}$ and $\mathbf{V}(\boldsymbol{\mu}) = \partial^2 c/\partial\boldsymbol{\beta}\,\partial\boldsymbol{\beta}'$, and writing

$$\ell(\boldsymbol{\mu}) = \boldsymbol{\beta}'\mathbf{y} - c(\boldsymbol{\beta}) - d(\mathbf{y}, \sigma),$$

where $d(\mathbf{y}, \sigma)$ is arbitrary. To see that $\ell(\boldsymbol{\mu})$ is a solution of (2.73) we note that

$$\frac{\partial\boldsymbol{\beta}}{\partial\boldsymbol{\mu}'} = \left(\frac{\partial\boldsymbol{\mu}}{\partial\boldsymbol{\beta}'}\right)^{-1} = \left(\frac{\partial^2 c}{\partial\boldsymbol{\beta}\,\partial\boldsymbol{\beta}'}\right)^{-1} = \mathbf{V}^{-1}(\boldsymbol{\mu}).$$

An important special case of the above theory is when the y_i are uncorrelated

and

$$\mathbf{V}(\boldsymbol{\mu}) = \mathrm{diag}\,[v(\mu_1), v(\mu_2), \ldots, v(\mu_n)],$$

where $v(\mu_i) = v[f(\mathbf{x}_i; \boldsymbol{\theta})] = v_i(\boldsymbol{\theta})$, say. Then (2.73) reduces to

$$\frac{\partial \ell(\boldsymbol{\mu})}{\partial \mu_i} = \frac{y_i - \mu_i}{v(\mu_i)}. \tag{2.83}$$

Also, the quadratic described by (2.81) now takes the form

$$S_v(\boldsymbol{\theta}) = \sum_{i=1}^{n} v_i^{-1}(\boldsymbol{\theta})[y_i - f(\mathbf{x}_i; \boldsymbol{\theta})]^2, \tag{2.84}$$

the familiar weighted sum of squares. Because of some misleading statements in the literature it should be noted that $\check{\boldsymbol{\theta}}$ is not the minimum of (2.84): it is the "minimum" obtained by treating $v_i(\boldsymbol{\theta})$ as though it were functionally independent of $\boldsymbol{\theta}$ (cf. Section 2.8.8).

We now consider a special family of density functions

$$\begin{aligned} p(y|\mu, \sigma) &= \exp\{\sigma^{-2}[a(\mu)y - b(\mu) + c(\sigma, y)]\}, \qquad y \geq 0, \\ &= e^A, \end{aligned} \tag{2.85}$$

say, where $\mu = E[y]$, and $a(\mu)$ and $b(\mu)$ are at least twice differentiable. When $\sigma = 1$, (2.85) reduces to the so-called (univariate) regular exponential family, though expressed somewhat differently here in terms of μ rather than $a(\mu) = \phi$, say. Following Bradley [1973], we differentiate both sides of $1 = \int e^A\,dy$ with respect to μ and obtain

$$\begin{aligned} 0 &= \int e^A \sigma^{-2}[\dot{a}(\mu)y - \dot{b}(\mu)]\,dy \\ &= \sigma^{-2} E[\dot{a}(\mu)y - \dot{b}(\mu)] \end{aligned} \tag{2.86}$$

so that

$$\dot{b}(\mu) = \mu \dot{a}(\mu). \tag{2.87}$$

Using (2.87) and differentiating (2.86) again under the integral sign leads to

$$0 = \sigma^{-4} E[\{\dot{a}(\mu)(y - \mu)\}^2] + E[\sigma^{-2}\{\ddot{a}(\mu)(y - \mu) - \dot{a}(\mu)\}],$$

or

$$\mathrm{var}\,[y] = \sigma^2[\dot{a}(\mu)]^{-1} = \sigma^2 v(\mu). \tag{2.88}$$

Hence if $y_1, y_2, \ldots, y_n$ are independently distributed with $E[y] = \mu_i = f(\mathbf{x}_i; \boldsymbol{\theta})$, then the log-likelihood function is [cf. (2.85)]

$$L = \sigma^{-2} \sum_i [a(\mu_i) y_i - b(\mu_i) + c(y_i, \sigma)],$$

and, using (2.87),

$$\begin{aligned} \frac{\partial L}{\partial \mu_i} &= \sigma^{-2} \dot{a}(\mu_i)(y_i - \mu_i) \\ &= \frac{1}{\sigma^2} \frac{y_i - \mu_i}{v(\mu_i)}. \end{aligned} \tag{2.89}$$

Apart from σ^2, which does not affect the estimation of $\boldsymbol{\theta}$, we see that (2.89) is of the same form as (2.83). Hence the maximum-likelihood estimate $\check{\boldsymbol{\theta}}$ is the same as the maximum quasi-likelihood estimate $\tilde{\boldsymbol{\theta}}$ and can be found using the method of iteratively reweighted least squares. This result was proved by Charnes et al. [1976] by a different route. They established that the maximum-likelihood estimate was a solution of (2.74) and gave sufficient conditions for a solution of (2.74) to maximize $L(\boldsymbol{\theta})$ at a unique value of $\boldsymbol{\theta}$. The above theory can be applied to members of the family (2.85) and thus includes the Poisson, binomial, exponential, multinomial, normal, inverse normal, and gamma distributions. Applications are discussed by Jennrich and Moore [1975], McCullagh [1983], and McCullagh and Nelder [1983].

The above method of computing maximum-likelihood estimates for regular exponential families using iteratively reweighted least squares was originally demonstrated by a number of authors for the linear model $\mu_i = \mathbf{x}_i'\boldsymbol{\theta}$ (Bradley [1973]) and the generalized linear models $g(\mu_i) = \mathbf{x}_i'\boldsymbol{\theta}$ (for some "link" function g) introduced by Nelder and Wedderburn [1972] (see also McCullagh and Nelder [1983], Fahrmeir and Kaufmann [1985]). Further extensions to more general distributions are given by Jørgenson [1983], Cox [1984], and Stirling [1984]. The reader is also referred to Nelder and Pregibon [1987], Firth [1987], and Morton [1987a].

An attractive aspect of quasi-likelihood theory in this context is the following. When a data analyst is fairly confident that the mean function and the relationship between the mean and variance has been modeled fairly well, but is unsure of other aspects of the parametric distribution used, quasi-likelihood theory assures him or her of the asymptotic correctness of the resulting inferences. In this way it is a generalization of the asymptotic applicability of least-squares theory beyond the restrictive assumption of normally distributed errors. However, there can be problems (Crowder [1987]).

Example 2.3 Suppose the y_i are independent Poisson random variables with means $\mu_i = \theta \exp(-\theta x_i) = \mu_i(\theta)$. Then the probability function of y is

$$\begin{aligned} p(y|\mu) &= e^{-\mu}\mu^y/y! \\ &= \exp\{y \log \mu - \mu - \log y!\}, \end{aligned}$$

which is of the form (2.85) with $\sigma = 1$. Hence the maximum-likelihood estimate is obtained by IRLS using $v(\mu_i) = \mu_i$. ■

The above quasi-likelihood approach is a special case of a general technique usually referred to as the method of (unbiased) estimating equations (Durbin [1960b], Godambe [1960]). This method consists of finding a suitable vector function $\mathbf{g}(\mathbf{y}, \boldsymbol{\theta})$ such that

$$\mathscr{E}[\mathbf{g}(\mathbf{y}, \boldsymbol{\theta})] = \mathbf{0}$$

and choosing an estimate $\hat{\boldsymbol{\theta}}$ such that

$$\mathbf{g}(\mathbf{y}, \hat{\boldsymbol{\theta}}) = \mathbf{0}.$$

For example, if $\ell(\boldsymbol{\theta}|\mathbf{y})$ is the likelihood function and

$$\mathbf{g}(\mathbf{y}, \boldsymbol{\theta}) = \frac{\partial \ell(\boldsymbol{\theta}|\mathbf{y})}{\partial \boldsymbol{\theta}},$$

then $\mathscr{E}[\mathbf{g}(\mathbf{y}, \boldsymbol{\theta})] = \mathbf{0}$ and $\hat{\boldsymbol{\theta}}$ is, under fairly general conditions, the maximum likelihood estimate of $\boldsymbol{\theta}$. In the quasi-likelihood case we assume a specific form for $\partial \ell / \partial \boldsymbol{\theta}$ [cf. (2.73) and (2.74)] and set

$$\mathbf{g}(\mathbf{y}, \boldsymbol{\theta}) = \mathbf{F}_{\cdot}'(\boldsymbol{\theta}) \mathbf{V}^{-1}(\boldsymbol{\theta})[\mathbf{y} - \mathbf{f}(\boldsymbol{\theta})].$$

This has zero expectation, as $\mathscr{E}[\mathbf{y}] = \mathbf{f}(\boldsymbol{\theta})$, and $\hat{\boldsymbol{\theta}} = \check{\boldsymbol{\theta}}$, the maximum quasi-likelihood estimator of $\boldsymbol{\theta}$.

If suitable conditions are imposed on the distribution of $\mathbf{y}$ and the function $\mathbf{g}$, then $\hat{\boldsymbol{\theta}}$ will be consistent and close to the true value $\boldsymbol{\theta}$ for n sufficiently large. Hence using a Taylor expansion,

$$\begin{aligned} \mathbf{0} &= \mathbf{g}(\mathbf{y}, \hat{\boldsymbol{\theta}}) \\ &\approx \mathbf{g}(\mathbf{y}, \boldsymbol{\theta}) + \mathbf{G}_{\cdot}(\hat{\boldsymbol{\theta}} - \boldsymbol{\theta}), \end{aligned}$$

where $\mathbf{G}_{\cdot} = \partial \mathbf{g} / \partial \boldsymbol{\theta}'$ evaluated at $\boldsymbol{\theta}$. If the rows of $\mathbf{G}_{\cdot}$ are linearly independent, then

$$\begin{aligned} \sqrt{n}(\hat{\boldsymbol{\theta}} - \boldsymbol{\theta}) &\approx \sqrt{n}(\mathbf{G}_{\cdot}'\mathbf{G}_{\cdot})^{-1}\mathbf{G}_{\cdot}'\mathbf{g}(\mathbf{y}, \boldsymbol{\theta}) \\ &= \mathbf{H}_n \mathbf{g}(\mathbf{y}, \boldsymbol{\theta}), \end{aligned}$$

say. Given that $\operatorname{plim} \mathbf{H}_n$ exists and is nonsingular, and $\mathbf{g}(\mathbf{y}, \boldsymbol{\theta})$ (suitably scaled) tends to normality as $n \to \infty$, then $\hat{\boldsymbol{\theta}}$ will be approximately normal.

The above estimating-equation method has been used in a wide range of applications, e.g. distribution-free methods (Maritz [1981: 7]), nonlinear instrumental variables estimation (Hansen and Singleton [1982]), robust estimation with i.i.d. random variables (Godambe and Thompson [1984]), and generalized linear models with clustered data (Liang and Zeger [1986]). Hansen [1982] discusses the theory under the title of generalized method of moment estimators.

How is $\mathbf{g}$ chosen? At present this seems to depend very much on the nature of the problem considered. For the one-parameter i.i.d. case, Godambe and Thompson [1978] suggest choosing g to minimize

$$\frac{E[g^2]}{(E[\partial g/\partial\theta])^2}.$$

*2.4 LAM ESTIMATOR

Byron and Bera [1983] introduced a new method of estimation called the linear approximation method (LAM). Using a second-order Taylor expansion about the k-dimensional vector $\mathbf{x}_0$, we have

$$\begin{aligned} y_i &= f(\mathbf{x}_i;\boldsymbol{\theta}) + \varepsilon_i \\ &= f(\mathbf{x}_0;\boldsymbol{\theta}) + \left(\frac{\partial f(\mathbf{x};\boldsymbol{\theta})}{\partial \mathbf{x}}\right)'_{\mathbf{x}_0}(\mathbf{x}_i - \mathbf{x}_0) \\ &\quad + \tfrac{1}{2}(\mathbf{x}_i - \mathbf{x}_0)'\left(\frac{\partial^2 f(\mathbf{x};\boldsymbol{\theta})}{\partial \mathbf{x}\,\partial \mathbf{x}'}\right)_{\mathbf{x}_0}(\mathbf{x}_i - \mathbf{x}_0) + r_i + \varepsilon_i, \end{aligned} \tag{2.90}$$

where r_i is the remainder. For illustrative purposes we assume $k = 1$, so that the above model takes the form

$$y_i = \mathbf{w}_i'\boldsymbol{\beta} + r_i + \varepsilon_i \qquad (i = 1, 2, \ldots, n)$$

where $\mathbf{w}_i' = [1, x_i - x_0, \frac{1}{2}(x_i - x_0)^2]$ and

$$\boldsymbol{\beta} = \left(f(x;\boldsymbol{\theta}), \frac{\partial f(x;\boldsymbol{\theta})}{\partial x}, \frac{\partial^2 f(x;\boldsymbol{\theta})}{\partial x^2} \right)'_{x_0}.$$

Then combining the n equations we have

$$\begin{aligned} \mathbf{y} &= \mathbf{f}(\boldsymbol{\theta}) + \boldsymbol{\varepsilon} \\ &= \mathbf{W}\boldsymbol{\beta} + \mathbf{r} + \boldsymbol{\varepsilon}, \end{aligned}$$

where $\mathbf{W}$ has ith row $\mathbf{w}_i'$. For k-dimensional $\mathbf{x}_0$, the above model still holds with $\mathbf{W}$ an appropriate $n \times m$ matrix $[m = 1 + k + \frac{1}{2}k(k+1)]$, and $\boldsymbol{\beta}$ an $m \times 1$ vector function of $\boldsymbol{\theta}$. Depending on f, p, and k, the vector $\boldsymbol{\theta}$ will be uniquely under- or overdetermined in terms of $\boldsymbol{\beta}$. Byron and Bera [1983] introduced the following four assumptions:

(i) There exists at least one way of determining $\boldsymbol{\theta}$ in terms of $\boldsymbol{\beta}$, namely $\boldsymbol{\theta} = \mathbf{h}(\boldsymbol{\beta})$, where $\mathbf{h}$ is differentiable.

(ii) $\mathbf{W}$ has rank m, so that $\mathbf{W}'\mathbf{W}$ is positive definite.

(iii) $\mathbf{W}'\mathbf{W} + \mathbf{W}'(\partial \mathbf{r}/\partial \boldsymbol{\beta}')$ is positive definite in some neighborhood of $\boldsymbol{\beta}^*$, the true value of $\boldsymbol{\beta}$. [In other words, since $\mathbf{W}'\mathbf{W}$ is positive definite, $\mathbf{W}'(\partial \mathbf{r}/\partial \boldsymbol{\beta}')$ does not dominate $\mathbf{W}'\mathbf{W}$. A second interpretation of this assumption is that $\mathbf{f}$ is not excessively nonlinear in terms of $\boldsymbol{\beta}$, since $\partial \mathbf{r}/\partial \boldsymbol{\beta}' = \partial \mathbf{f}/\partial \boldsymbol{\beta}' - \mathbf{W}$.]

(iv) $\lim n^{-1}\mathbf{W}'\mathbf{W} = \mathbf{Q}_1$ where $\mathbf{Q}_1$ is a finite matrix, $\operatorname{plim} n^{-1}\mathbf{W}'[\mathbf{y} - \mathbf{f}(\boldsymbol{\theta})] = \mathbf{0}$, $\operatorname{plim}[-n^{-1}\mathbf{W}'(\partial \mathbf{r}/\partial \boldsymbol{\beta}')] = \mathbf{P}$ exists in a neighborhood of $\boldsymbol{\beta}^*$, and $\mathbf{I} - \mathbf{Q}_1^{-1}\mathbf{P}$ is nonsingular.

They then suggested the following method for estimating $\boldsymbol{\theta}$. Firstly, ignore $\mathbf{r}$ and obtain the least-squares estimate $\boldsymbol{\beta}^{(1)} = (\mathbf{W}'\mathbf{W})^{-1}\mathbf{W}'\mathbf{y}$. Secondly, estimate $\boldsymbol{\theta}$ by $\boldsymbol{\theta}^{(1)} = \mathbf{h}(\boldsymbol{\beta}^{(1)})$ and then estimate $\mathbf{r}$ by $\mathbf{r}^{(1)} = \mathbf{f}(\boldsymbol{\theta}^{(1)}) - \mathbf{W}\boldsymbol{\beta}^{(1)}$. Thirdly, re-estimate $\boldsymbol{\beta}$ using $\boldsymbol{\beta}^{(2)} = (\mathbf{W}'\mathbf{W})^{-1}\mathbf{W}'(\mathbf{y} - \mathbf{r}^{(1)})$. This process can be continued so that, in general,

$$\boldsymbol{\theta}^{(a)} = \mathbf{h}(\boldsymbol{\beta}^{(a)}),$$

$$\mathbf{r}^{(a)} = \mathbf{f}(\boldsymbol{\theta}^{(a)}) - \mathbf{W}\boldsymbol{\beta}^{(a)},$$

and

$$\begin{aligned} \boldsymbol{\beta}^{(a+1)} &= (\mathbf{W}'\mathbf{W})^{-1}\mathbf{W}'(\mathbf{y} - \mathbf{r}^{(a)}) \\ &= (\mathbf{W}'\mathbf{W})^{-1}\mathbf{W}'[\mathbf{y} - \mathbf{f}(\boldsymbol{\theta}^{(a)}) + \mathbf{W}\boldsymbol{\beta}^{(a)}] \\ &= \boldsymbol{\beta}^{(a)} + (\mathbf{W}'\mathbf{W})^{-1}\mathbf{W}'[\mathbf{y} - \mathbf{f}(\boldsymbol{\theta}^{(a)})], \end{aligned} \tag{2.91}$$

which is similar in form to the Gauss–Newton method (2.29). The estimator resulting from this iteration process is called the LAM estimator.

Byron and Bera [1983] showed that assumptions (i)–(iii) imply that the iteration process (2.91) converges. Furthermore, with the addition of assumption (iv), the convergent solution of (2.91) yields a consistent estimator of $\boldsymbol{\beta}^*$, provided the process is started sufficiently close to $\boldsymbol{\beta}^*$. They also proved that the LAM estimate is, asymptotically, not as efficient as the Gauss–Newton (least-squares) estimate of $\boldsymbol{\beta}$, the difference in their asymptotic variance–covariance matrices being positive semidefinite. In applying the method, $\mathbf{x}_0$ must be chosen and various starting estimates $\boldsymbol{\beta}^{(1)}$ can be used. Convergence can be improved by using such standard devices as changing the step length of the correction $\boldsymbol{\beta}^{(a+1)} - \boldsymbol{\beta}^{(a)}$ in (2.91). The method seems to have more robust convergence properties than the Gauss–Newton method, and, at the very least, the LAM estimator is a source of consistent starting values for the Gauss–Newton method. The reader is referred to Byron and Bera [1983] for further details and three examples.

*2.5 L_1-NORM MINIMIZATION

We note that $S(\boldsymbol{\theta})$ of (2.2) is based on a Euclidean norm. An alternative approach is to use an L_1-norm and minimize $\sum_i |y_i - f(\mathbf{x}_i; \boldsymbol{\theta})|$ with respect to $\boldsymbol{\theta}$. Sufficient conditions for the weak consistency of the solution are given by Oberhofer [1982]. This L_1-norm approach gives the maximum-likelihood estimate for the case where the ε_i have a double exponential (Laplace) distribution

$$g(\varepsilon) = \frac{1}{4\sigma} e^{-|\varepsilon|/2\sigma}, \qquad -\infty < \varepsilon < \infty.$$

This has heavier tails than the normal distribution. As the L_1-criterion gives lesser weight to large deviations than least squares, it is more resistant to the effects of outliers and is often used as a technique for robust regression. For further comments and references about L_p-norm estimation in general, see Gonin and Money [1985a, b].

2.6 ROBUST ESTIMATION

For the nonlinear model (2.1), the normal equations for the least-squares estimate $\hat{\boldsymbol{\theta}}$ are, from (2.8),

$$\sum_{i=1}^{n} \frac{\partial f(\mathbf{x}_i; \hat{\boldsymbol{\theta}})}{\partial \theta_r} [y_i - f(\mathbf{x}_i; \hat{\boldsymbol{\theta}})] = 0 \qquad (r = 1, 2, \ldots, p). \tag{2.92}$$

However, if outliers are a problem, a more robust method of estimation is needed. By analogy with robust linear regression, an M-estimate $\tilde{\boldsymbol{\theta}}$ of $\boldsymbol{\theta}$ is the solution of (Huber [1972, 1977])

$$\sum_{i=1}^{n} \frac{\partial f(\mathbf{x}_i; \tilde{\boldsymbol{\theta}})}{\partial \theta_r} \psi\left(\frac{y_i - f(\mathbf{x}_i; \tilde{\boldsymbol{\theta}})}{\tilde{\sigma}} \right) = 0, \tag{2.93}$$

where ψ is a suitable function which downweights or omits extreme values, and $\tilde{\sigma}$ is a robust estimate of dispersion (see Seber [1984: Section 4.4] for a brief summary of robust estimation for simpler models). Further computational details are given in Section 14.6.1.

Example 2.4 The choice of ψ is a matter of taste; Tiede and Pagano [1979] used the sine function

$$\psi(z) = \begin{cases} \sin(z/a), & |z| \le \pi a, \\ 0, & |z| > \pi a, \end{cases}$$

in fitting their calibration model

$$y_i = \alpha + \beta(1 + \gamma x_i^\delta)^{-1} + \varepsilon_i. \tag{2.94}$$

They found that of the two recommended values 1.5 and 2.1 for a, the latter seemed to perform better. Experimentation with various scale parameters led them to choose $\tilde{\sigma}$ as the median of the largest $n - p + 1$ absolute residuals $|y_i - f(\mathbf{x}_i; \tilde{\boldsymbol{\theta}})|$.

The authors applied their method to 124 data sets, and their algorithm failed to converge in only 9 very poor-quality sets with large numbers of outliers. One of their data sets is given in Table 2.1, and it has a very obvious outlier (see Fig. 2.1) that one would normally omit before carrying out a standard least-squares analysis. However, the advantage of using a robust method which automatically rejects extreme observations is that it does not require a subjective decision on the part of the experimenter. Tiede and Pagano [1979] note that it is unlikely for a typical clinical laboratory to employ a full-time statistician, so that automatic procedures for routine analyses seem appropriate here. ■

As bias can be a problem with least-squares estimates, the jackknife method has been proposed as a possible alternative. In the examples considered by several authors it seems that the jackknife is better in some cases and worse in others, and generally cannot be relied on as a method of bias reduction in nonlinear models (Simonoff and Tsai [1986: p. 106]). Its role in the construction of confidence intervals and regions is discussed in Section 5.7. A number of useful modifications of the jackknife are also presented there.

Table 2.1 Parameter Estimates for Data from the Model of Equation (2.94)[a]

x	y	x	y
0	7720	20	4478[b]
	8113		2396
2	6664	50	1302
	6801		1377
5	4994	100	1025
	4948		1096
10	3410		
	3208		

	α	β	γ	δ
Robust	919.90	7022.29	0.086	1.330
Least squares	443.53	7550.86	0.133	0.958

[a]Reproduced from: J. J Tiede and M. Pagano, "The Application of Robust Calibration to Radioimmunoassay", Biometrics 35, 567–574, 1979. With permission from the Biometric Society.
[b]Outlier.

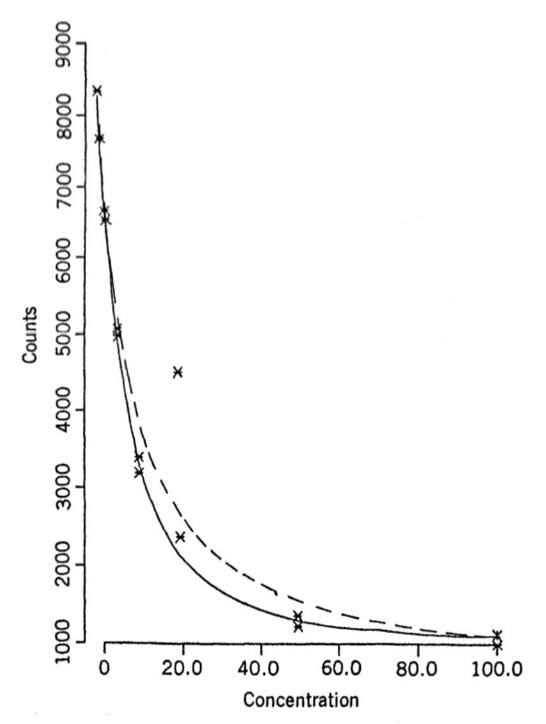

Figure 2.1 Plot of the *TSH* data reproduced from: J. J. Tiede and M. Pagano, "The Application of Robust Calibration to Radioimmunoassay". Biometrics 35, 567–574, 1979. With permission from the Biometric Society. The observation (20, 4478) is an outlier. The solid curve is a robust fit, and the dashed curve is a standard least-squares fit.

2.7 BAYESIAN ESTIMATION

Consider the nonlinear model (2.1) in vector form, namely

$$\mathbf{y} = \mathbf{f}(\boldsymbol{\theta}) + \boldsymbol{\varepsilon}, \tag{2.95}$$

where $\boldsymbol{\varepsilon} \sim N_n(\mathbf{0}, \sigma^2 \mathbf{I}_n)$. If $p(\boldsymbol{\theta}, \sigma)$ is the prior distribution of the unknown parameters, then the posterior density function is given by Bayes' theorem, namely

$$\begin{aligned} p(\boldsymbol{\theta}, \sigma \mid \mathbf{y}) &= \frac{p(\boldsymbol{\theta}, \sigma, \mathbf{y})}{p(\mathbf{y})} \\ &= \frac{p(\mathbf{y} \mid \boldsymbol{\theta}, \sigma) p(\boldsymbol{\theta}, \sigma)}{p(\mathbf{y})} \\ &\propto p(\mathbf{y} \mid \boldsymbol{\theta}, \sigma) p(\boldsymbol{\theta}, \sigma). \end{aligned} \tag{2.96}$$

Here $p(\mathbf{y}|\boldsymbol{\theta},\sigma)$ is the likelihood function, which is frequently written as $\ell(\boldsymbol{\theta},\sigma|\mathbf{y})$, with any factor not involving $\boldsymbol{\theta}$ or σ omitted, to emphasize that it is treated as a function of $\boldsymbol{\theta}$ and σ^2. The posterior distribution for $\boldsymbol{\theta}$ alone is then given by

$$p(\boldsymbol{\theta}|\mathbf{y}) = \int p(\boldsymbol{\theta},\sigma|\mathbf{y})\, d\sigma. \tag{2.97}$$

If the prior distributions of $\boldsymbol{\theta}$ and σ are assumed to be independent, then we can find $p(\boldsymbol{\theta}|\mathbf{y})$ as follows. Firstly,

$$\begin{aligned} p(\mathbf{y}|\boldsymbol{\theta}) &= \frac{p(\mathbf{y},\boldsymbol{\theta})}{p(\boldsymbol{\theta})} \\ &= \frac{1}{p(\boldsymbol{\theta})}\int p(\mathbf{y},\boldsymbol{\theta},\sigma)\, d\sigma \\ &= \frac{1}{p(\boldsymbol{\theta})}\int p(\mathbf{y}|\boldsymbol{\theta},\sigma)p(\boldsymbol{\theta},\sigma)\, d\sigma \\ &= \int p(\mathbf{y}|\boldsymbol{\theta},\sigma)p(\sigma)\, d\sigma, \end{aligned} \tag{2.98}$$

since $p(\boldsymbol{\theta},\sigma) = p(\boldsymbol{\theta})p(\sigma)$. Secondly, from (2.96) with σ omitted,

$$p(\boldsymbol{\theta}|\mathbf{y}) \propto p(\mathbf{y}|\boldsymbol{\theta})p(\boldsymbol{\theta}). \tag{2.99}$$

2.7.1 Choice of Prior Distributions

A number of prior distributions have been proposed. An important class of priors is the so-called noninformative priors discussed by Box and Tiao [1973]. They are intended to approximate prior information or opinion about the model parameters which is very vague in comparison with the information supplied by the data; the information from the data being contained in the likelihood function. According to Box and Tiao [1973: 58], noninformative priors should, ideally, function as reference priors to be "employed as a point of *reference* against which to judge the kind of unprejudiced inference that can be drawn from the data." Their criteria led to a rule proposed by Jeffreys [1961] to construct priors which are invariant under parameter transformations. This rule is described below.

Let $L(\boldsymbol{\gamma})$ be the log-likelihood function for the parameters $\boldsymbol{\gamma}$ of a model; then Jeffreys' prior for $\boldsymbol{\gamma}$ is

$$p(\boldsymbol{\gamma}) \propto |\mathbf{I}(\boldsymbol{\gamma})|^{1/2}, \tag{2.100}$$

where $\mathbf{I}(\boldsymbol{\gamma})$ is the expected (Fisher) information matrix with (r,s)th element

$$E\left[-\frac{\partial^2 L(\boldsymbol{\gamma})}{\partial\gamma_r\,\partial\gamma_s}\right]. \tag{2.101}$$

When $\gamma = (\boldsymbol{\alpha}', \boldsymbol{\beta}')'$ and the components are taken to be a priori independent, i.e. $p(\boldsymbol{\alpha}, \boldsymbol{\beta}) = p_1(\boldsymbol{\alpha}) p_2(\boldsymbol{\beta})$, Jeffreys' rule is to obtain $p_1(\boldsymbol{\alpha})$ and $p_2(\boldsymbol{\beta})$ separately, treating the other parameters as known. For example,

$$\mathbf{p}_1(\boldsymbol{\alpha}) \propto |\mathbf{I}_1(\boldsymbol{\alpha})|^{1/2}$$

is treated as a function of $\boldsymbol{\alpha}$ only, where

$$[\mathbf{I}_1(\boldsymbol{\alpha})]_{r,s} = E\left[-\frac{\partial^2 L(\boldsymbol{\alpha}, \boldsymbol{\beta})}{\partial \alpha_r \, \partial \alpha_s} \right].$$

Applying the above ideas to (2.95) we have, from (2.46),

$$L(\boldsymbol{\theta}, \sigma) = -n \log \sigma - \frac{1}{2\sigma^2} S(\boldsymbol{\theta}),$$

where $S(\boldsymbol{\theta}) = \| \mathbf{y} - \mathbf{f}(\boldsymbol{\theta}) \|^2$. Thus by (2.35)

$$\begin{aligned} \mathbf{I}_1(\boldsymbol{\theta}) &= \frac{1}{2\sigma^2} \mathscr{E}\left(\frac{\partial^2 S(\boldsymbol{\theta})}{\partial \boldsymbol{\theta}\, \partial \boldsymbol{\theta}'} \right), \\ &= \frac{1}{\sigma^2} \mathbf{F}'_{\bullet}(\boldsymbol{\theta}) \mathbf{F}_{\bullet}(\boldsymbol{\theta}) \\ &\propto \mathbf{F}'_{\bullet}(\boldsymbol{\theta}) \mathbf{F}_{\bullet}(\boldsymbol{\theta}). \end{aligned} \tag{2.102}$$

Also

$$I_2(\sigma) = -\frac{n}{\sigma^2} + \frac{3}{\sigma^4} E[S(\boldsymbol{\theta})] = \frac{2n}{\sigma^2}.$$

Here the noninformative prior for $(\boldsymbol{\theta}, \sigma)$, where $\boldsymbol{\theta}$ and σ are a priori independent, is

$$p(\boldsymbol{\theta}, \sigma) \propto |\mathbf{F}'_{\bullet}(\boldsymbol{\theta}) \mathbf{F}_{\bullet}(\boldsymbol{\theta})|^{1/2} \sigma^{-1}, \tag{2.103}$$

as proposed by Eaves [1983]. These priors are improper in that $p(\boldsymbol{\theta}, \sigma)$ is not a proper distribution. However, they give the same results as any proper prior which is proportional to $p(\boldsymbol{\theta}, \sigma)$ over the support (domain) of the likelihood function.

In the linear model case (see Appendix D1.1) $\mathbf{F}'_{\bullet}(\boldsymbol{\theta}) \mathbf{F}_{\bullet}(\boldsymbol{\theta}) = \mathbf{X}'\mathbf{X}$, which is independent of $\boldsymbol{\theta}$, so that $p_1(\boldsymbol{\theta})$ is proportional to a constant and

$$p(\boldsymbol{\theta}, \sigma) \propto \sigma^{-1}. \tag{2.104}$$

By analogy with the linear case, Box and Tiao [1973: 431] argue that if there is a parametrization of a nonlinear model for which the model acts in a close to linear manner (i.e. a small parameter-effects curvature: see Chapter 4), (2.104) should be a reasonable approximation to the noninformative prior (2.103). The prior (2.104) has been discussed, for example, by Zellner [1971]. It has often been used indiscriminantly in that for problems with high intrinsic nonlinearity or parametrizations with high parameter-effects nonlinearity (as discussed in Chapter 4) it may be a bad approximation to the noninformative prior.

2.7.2 Posterior Distributions

Using the prior distribution (2.104), we now obtain joint and marginal posterior distributions for the parameters. Since the likelihood function for our normal nonlinear model (2.95) is

$$\begin{aligned} p(\mathbf{y}|\boldsymbol{\theta},\sigma) &= \frac{1}{(2\pi\sigma^2)^{n/2}} \exp\left(-\frac{1}{2\sigma^2} [\mathbf{y}-\mathbf{f}(\boldsymbol{\theta})]'[\mathbf{y}-\mathbf{f}(\boldsymbol{\theta})] \right) \\ &= \frac{1}{(2\pi\sigma^2)^{n/2}} \exp\left(-\frac{1}{2\sigma^2} S(\boldsymbol{\theta}) \right), \end{aligned} \tag{2.105}$$

it follows from (2.96) and (2.105) that

$$p(\boldsymbol{\theta},\sigma|\mathbf{y}) \propto \sigma^{-(n+1)} \exp\left(-\frac{1}{2\sigma^2} S(\boldsymbol{\theta}) \right). \tag{2.106}$$

Integrating out σ using the inverted gamma integral [cf. (1.5)], we have

$$p(\boldsymbol{\theta}|\mathbf{y}) \propto S(\boldsymbol{\theta})^{-n/2}, \tag{2.107}$$

or

$$p(\boldsymbol{\theta}|\mathbf{y}) = k^{-1} S(\boldsymbol{\theta})^{-n/2}, \qquad \boldsymbol{\theta} \in \Theta. \tag{2.108}$$

Usually $\Theta = \mathbb{R}^n$, and

$$k = \int_{\Theta} S(\boldsymbol{\theta})^{-n/2} \, d\boldsymbol{\theta} \tag{2.109}$$

is found by numerical integration. The above theory also holds for discrete $\boldsymbol{\theta}$ but with the integral in (2.109) replaced by a sum.

If (2.104) is replaced by $p(\boldsymbol{\theta},\sigma) \propto \sigma^{-1} p(\boldsymbol{\theta})$, then it follows from (2.98) that

$$\begin{aligned} p(\mathbf{y}|\boldsymbol{\theta}) &\propto \int \sigma^{-(n+1)} \exp\left(-\frac{1}{2\sigma^2} S(\boldsymbol{\theta}) \right) d\sigma \\ &\propto S(\boldsymbol{\theta})^{-n/2}, \end{aligned}$$

by the argument leading to (2.107). Hence, from (2.99),

$$p(\boldsymbol{\theta}|\mathbf{y}) \propto p(\boldsymbol{\theta})S(\boldsymbol{\theta})^{-n/2}, \tag{2.110}$$

which, as expected, reduces to (2.107) when $p(\boldsymbol{\theta})$ is constant.

The posterior distribution (2.108) can be used for estimation (e.g., the posterior mean $\mathscr{E}[\boldsymbol{\theta}|\mathbf{y}]$ can be used as an estimate of $\boldsymbol{\theta}$), and for posterior intervals and regions (Katz et al. [1981]). We note that the mode of (2.108) is the value of $\boldsymbol{\theta}$ which minimizes $S(\boldsymbol{\theta})$, namely the least-squares (and maximum-likelihood) estimate $\hat{\boldsymbol{\theta}}$.

If the model is linear with $\mathbf{f}(\boldsymbol{\theta}) = \mathbf{X}\boldsymbol{\theta}$ then, by Appendix D1.4,

$$\begin{aligned} S(\boldsymbol{\theta}) &= (\mathbf{y} - \mathbf{X}\boldsymbol{\theta})'(\mathbf{y} - \mathbf{X}\boldsymbol{\theta}) \\ &= (\mathbf{y} - \mathbf{X}\hat{\boldsymbol{\theta}})'(\mathbf{y} - \mathbf{X}\hat{\boldsymbol{\theta}}) + (\boldsymbol{\theta} - \hat{\boldsymbol{\theta}})'\mathbf{X}'\mathbf{X}(\boldsymbol{\theta} - \hat{\boldsymbol{\theta}}) \\ &= \nu s^2 + (\boldsymbol{\theta} - \hat{\boldsymbol{\theta}})'\mathbf{X}'\mathbf{X}(\boldsymbol{\theta} - \hat{\boldsymbol{\theta}}), \end{aligned} \tag{2.111}$$

where $\nu = n - p$ and $\hat{\boldsymbol{\theta}} = (\mathbf{X}'\mathbf{X})^{-1}\mathbf{X}'\mathbf{y}$. From (2.106) it follows that

$$p(\boldsymbol{\theta}, \sigma|\mathbf{y}) \propto \sigma^{-(n+1)} \exp\left(-\frac{1}{2\sigma^2} [\nu s^2 + (\boldsymbol{\theta} - \hat{\boldsymbol{\theta}})'\mathbf{X}'\mathbf{X}(\boldsymbol{\theta} - \hat{\boldsymbol{\theta}})] \right), \tag{2.112}$$

while (2.107) leads to

$$p(\boldsymbol{\theta}|\mathbf{y}) \propto \left(1 + \frac{(\boldsymbol{\theta} - \hat{\boldsymbol{\theta}})'\mathbf{X}'\mathbf{X}(\boldsymbol{\theta} - \hat{\boldsymbol{\theta}})}{\nu s^2} \right)^{-n/2}. \tag{2.113}$$

This distribution can be recognized as the multivariate t-distribution, namely $t_p(\nu, \hat{\boldsymbol{\theta}}, s^2(\mathbf{X}'\mathbf{X})^{-1})$ (cf. Appendix A9), so that the constant of proportionality is known. The marginal posterior distribution of θ_r is a t-distribution given by

$$\frac{\theta_r - \hat{\theta}_r}{s\sqrt{c^{rr}}} \sim t_{n-p},$$

where $[(c^{rs})] = (\mathbf{X}'\mathbf{X})^{-1}$. The highest posterior density (h.p.d.) interval (see Section 2.7.3) for θ_r is the same as the usual confidence interval based on $\hat{\theta}_r$. Furthermore, using the properties of the multivariate normal integral, $\boldsymbol{\theta}$ can be integrated out of (2.112) to give

$$p(\sigma|\mathbf{y}) \propto \sigma^{-(\nu+1)} e^{-\nu s^2/\sigma^2}, \tag{2.114}$$

the so-called inverted gamma distribution [see (1.4)]. Since s^2 and $\hat{\boldsymbol{\theta}}$ are jointly sufficient for σ^2 and $\boldsymbol{\theta}$, we can write $p(\sigma|s)$ instead of $p(\sigma|\mathbf{y})$.

If σ is known, it follows from (2.112) that

$$p(\boldsymbol{\theta}|\mathbf{y}) \propto \exp\left(-\frac{1}{2\sigma^2}(\boldsymbol{\theta} - \hat{\boldsymbol{\theta}})'\mathbf{X}'\mathbf{X}(\boldsymbol{\theta} - \hat{\boldsymbol{\theta}}) \right), \tag{2.115}$$

so that the posterior distribution of $\boldsymbol{\theta}$ is now $N_p(\hat{\boldsymbol{\theta}}, \sigma^2(\mathbf{X}'\mathbf{X})^{-1})$. The case of a mixture of linear and nonlinear parameters is considered by Katz et al. [1982].

The above discussion assumes that the model is linear. However, if the model is nonlinear, but has a reasonable linear approximation [with $\mathbf{X}$ replaced by $\mathbf{F}_{\bullet}(\boldsymbol{\theta})$], then the above analysis will be approximately true. Also, as $n \to \infty$ the multivariate t-distribution tends to the multivariate normal, so that the posterior density will be asymptotically normal. Part of the value of the above analysis is that it gives the results of the usual least-squares analysis a Bayesian posterior probability interpretation in addition to the usual sampling-theory interpretation. However, there are difficulties associated with noninformative and improper priors in multiparameter situations, and some controversy surrounds their use (e.g. Stein [1962, 1965]; Stone [1976] and discussants).

When subjective opinion or further information about the parameters can be expressed in the form of a suitable prior, then the posterior distribution can be obtained using (2.96). This distribution can also be used as a prior for any new data set. A natural estimate of the parameter vector is the posterior mode, as it is the most probable value of the parameter vector, given the data. Furthermore, the mode can be found by simply maximizing (2.96), without the need for finding normalizing constants like k of (2.109). However, most of the usual Bayesian summary statistics such as the posterior moments $E[\theta_r|\mathbf{y}]$ and $E[\theta_r\theta_s|\mathbf{y}]$ will require $(p+1)$-dimensional integrals. For the linear model we find that if we use (2.112) as our prior, then the posterior density has the same functional form as the prior. Priors with this property are called conjugate priors (cf. Maddala [1977: 412]). Usually integration can then be done analytically so that marginal posterior distributions and posterior moments can be found using analytic rather than numerical integration. However, in nonlinear models, analytic integration is rarely feasible and numerical methods must be used. For fairly straightforward likelihood functions like that arising from (2.85) we can evaluate the function at a large number of grid points and approximate integrals by sums (Reilly [1976]). Assuming $\boldsymbol{\theta}$ to be approximated by a discrete random variable over a p-dimensional rectangular grid of N points $\boldsymbol{\theta}_g$ $(g = 1, 2, \ldots, N)$ then, using a noninformative prior, we have from (2.108)

$$\begin{aligned} \operatorname{pr}(\boldsymbol{\theta} = \boldsymbol{\theta}_g|\mathbf{y}) &= k^{-1}S(\boldsymbol{\theta}_g)^{-n/2} \\ &= P(\boldsymbol{\theta}_g), \end{aligned} \tag{2.116}$$

say, and

$$k = \sum_{g=1}^{N} P(\boldsymbol{\theta}_g). \tag{2.117}$$

Once the p-dimensional array, $P(\boldsymbol{\theta}_g)$, associated with the grid is set up, marginal posterior distributions can be obtained by summing out the unwanted elements of $\boldsymbol{\theta}_g$. If only posterior modes are required, then the scale factor k is not needed.

Reilly [1976], in his examples, initially used a very coarse adjustable grid to determine the appropriate range of $\boldsymbol{\theta}$ and to shift most of the posterior probability to the central part of the array. He then used a much finer grid, $\mathscr{G}$ say, to find the $P(\boldsymbol{\theta}_g)$. One advantage of this discrete method is that posterior densities for functions of $\boldsymbol{\theta}$ can be readily found. For example, suppose $p=2$, i.e. $\boldsymbol{\theta}=(\theta_1,\theta_2)'$, and we wish to find the posterior density of $\phi=\theta_1/\theta_2$. We first determine a range for ϕ and then divide this range into cells as in constructing a histogram. For each element of $\mathscr{G}$ we then calculate a ratio ϕ_g. If ϕ_g lies in a particular cell, we add $P(\boldsymbol{\theta}_g)$ into that cell and thus build up a histogram for ϕ.

Other discrete priors can, of course, be used. If $\boldsymbol{\theta}$ and σ have independent priors, then we can use (2.110). The grid method can also be applied to other likelihood functions which do not involve the nuisance parameter σ, as seen by the following example. In such cases we use the general relationship $p(\boldsymbol{\theta}|\mathbf{y}) \propto p(\mathbf{y}|\boldsymbol{\theta})p(\boldsymbol{\theta})$.

Example 2.5 Reilly [1976] gave the set of data in Table 2.2 representing the number of megacycles y to failure on a fatigue trial on eight specimens of a joint between two pieces of rubber, all made under the same conditions. There was good reason to believe that y had a Weibull distribution with probability density function

$$p(y|\boldsymbol{\theta})=\frac{c}{b^c}(y-a)^{c-1}\exp\left[-\left(\frac{y-a}{b}\right)^c\right] \tag{2.118}$$

with $\boldsymbol{\theta}=(a,b,c)'$, where a, b, and c can be regarded as location, scale, and shape parameters respectively. On mathematical grounds Reilly decided to work with the logarithms of b and c, so that uniform priors were assumed for a, $\log b$, and $\log c$. Using equally spaced values in the ranges of these transformed parameters led to the following grid of $N=21\times 51\times 51$ points for the untransformed parameters:

$$\begin{aligned} a_u &= 0.005u-0.005 & (u&=1,2,\ldots,21),\\ b_v &= \exp(0.144v-2.644) & (v&=1,2,\ldots,51),\\ c_w &= \exp(0.072w-2.072) & (w&=1,2,\ldots,51). \end{aligned}$$

As mentioned above, the ranges of the parameters were initially determined by a coarse grid (see Table 2.3). The likelihood function for n observations $\mathbf{y}=(y_1,y_2,\ldots,y_n)'$ and grid point $\boldsymbol{\theta}_g$ ($g=1,2,\ldots,N$) is given by

$$p(\mathbf{y}|\boldsymbol{\theta}=\boldsymbol{\theta}_g)=\prod_{i=1}^{n} p(y_i|\boldsymbol{\theta}_g), \tag{2.119}$$

Table 2.2 Megacycles to Failure in a Fatigue Test[a]

y	0.12	1.28	1.72	2.36	2.80	3.16	3.84	4.68

[a]From Reilly [1976].

Table 2.3 Probability Distribution of Weibull Parameters[a]

	Probability				
C	B = 0.08	0.50	3.00	18.17	109.95
			$A = 0.000$		
0.14	0.0000001	0.0000002	0.0000002	0.0000001	0.0000001
0.33	0.0000002	0.0000594	0.0001652	0.0000334	0.0000016
0.82	0.0000000	0.0000010	0.0435437	0.0000565	0.0000000
2.01	0.0000000	0.0000000	0.2310455	0.0000000	0.0000000
4.95	0.0000000	0.0000000	0.0000007	0.0000000	0.0000000
			$A = 0.025$		
0.14	0.0000001	0.0000002	0.0000002	0.0000002	0.0000001
0.33	0.0000003	0.0000796	0.0002127	0.0000421	0.0000020
0.82	0.0000000	0.0000014	0.0489642	0.0000605	0.0000000
2.01	0.0000000	0.0000000	0.1889979	0.0000000	0.0000000
4.95	0.0000000	0.0000000	0.0000003	0.0000000	0.0000000
			$A = 0.050$		
0.14	0.0000002	0.0000003	0.0000004	0.0000003	0.0000001
0.33	0.0000004	0.0001130	0.0002890	0.0000558	0.0000026
0.82	0.0000000	0.0000020	0.0558285	0.0000658	0.0000000
2.01	0.0000000	0.0000000	0.1434556	0.0000000	0.0000000
4.95	0.0000000	0.0000000	0.0000001	0.0000000	0.0000000
			$A = 0.075$		
0.14	0.0000003	0.0000005	0.0000006	0.0000004	0.0000002
0.33	0.0000007	0.0001786	0.0004342	0.0000815	0.0000037
0.82	0.0000000	0.0000029	0.0653258	0.0000733	0.0000000
2.01	0.0000000	0.0000000	0.0946208	0.0000000	0.0000000
4.95	0.0000000	0.0000000	0.0000000	0.0000000	0.0000000
			$A = 0.100$		
0.14	0.0000007	0.0000012	0.0000013	0.0000009	0.0000005
0.33	0.0000016	0.0003743	0.0008514	0.0001541	0.0000069
0.82	0.0000000	0.0000045	0.0818809	0.0000874	0.0000000
2.01	0.0000000	0.0000000	0.0428379	0.0000000	0.0000000
4.95	0.0000000	0.0000000	0.0000000	0.0000000	0.0000000

[a]From Reilly [1976].

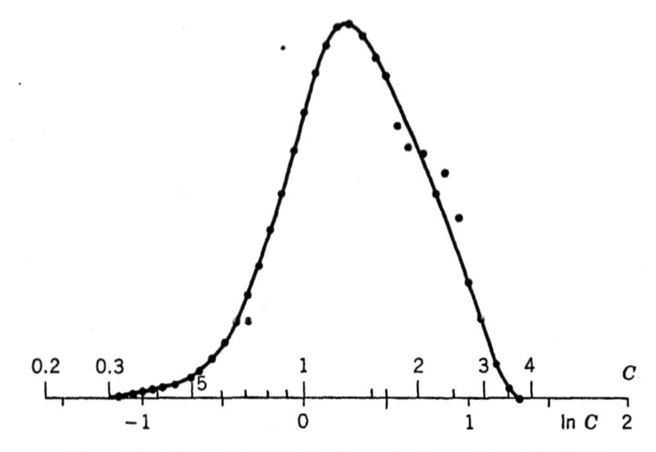

Figure 2.2 Marginal distribution of c (from Reilly [1976]).

where $p(y_i|\boldsymbol{\theta}_g)$ is obtained from (2.118). By way of illustration, Reilly obtained the marginal prior distribution of c and plotted the density function (Fig. 2.2). He also considered the posterior distribution of the function of the parameter p

$$t_p = a + b[-\log(1-p)]^{1/c},$$

the pth fractile of y. The plot for $p = 0.1$, corresponding to the bottom decile $t_{0.1}$, is given in Fig. 2.3. We note that the scale factor was not computed, as only the shape of this distribution was of interest. ■

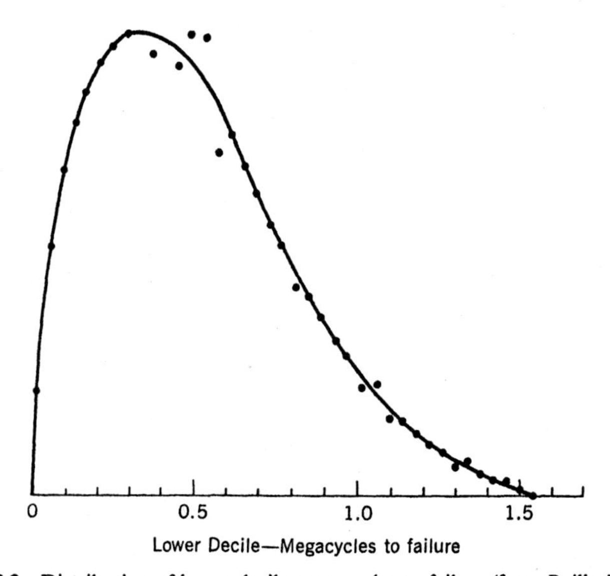

Figure 2.3 Distribution of lower-decile megacycles to failure (from Reilly [1976]).

The above method of Reilly is suitable only for straightforward and well-behaved likelihoods. Naylor and Smith [1982] (see also Smith et al. [1985]) point out that there are important classes of problems for which large-scale function evaluation of the above kind would be prohibitively expensive in computer time. They give a more general method which is much faster for Reilly's examples, in fact by a factor of 4 in one example and a factor of at least 25 in the other. Firstly they argue that, at least for moderate samples, it is reasonable to assume $p(\boldsymbol{\theta}|\mathbf{y})$ [or equivalently $p(\mathbf{y}|\boldsymbol{\theta})p(\boldsymbol{\theta})$] to be adequately approximated by the product of a p-dimensional multivariate normal probability density and a polynomial in the elements of $\boldsymbol{\theta}$. Secondly, for a function $g(t)$ satisfying

$$g(t) = h(t)(2\pi\sigma^2)^{-1/2}\exp\left[-\frac{1}{2}\left(\frac{t-\mu}{\sigma}\right)^2\right],$$

where μ and σ^2 (>0) are given and $h(t)$ is a suitably regular function, they develop the approximation

$$\int_{-\infty}^{\infty} g(t)\,dt \approx \sum_{j=1}^{N} m_j g(z_j). \tag{2.120}$$

Here N is the number of grid points,

$$m_j = w_j \exp(t_j^2)\sigma\sqrt{2}, \qquad z_j = \mu + \sigma t_j\sqrt{2}, \qquad w_j = \frac{2^{N-1}N!\sqrt{\pi}}{N^2[H_{N-1}(t_j)]^2}, \tag{2.121}$$

and t_j is the jth zero of the Hermite polynomial $H_N(t)$ (see, for example, Davis and Rabinowitz [1967]). Tables of t_j, w_j, and $w_j\exp(t_j^2)$ are available for $N = 1(1)20$ (Salzer et al. [1952]), and the relationship (2.120) is exact if $h(t)$ is a polynomial of degree at most $2N-1$. For a multivariate function

$$g(\boldsymbol{\theta}) = h(\boldsymbol{\theta})(2\pi)^{-p/2}|\boldsymbol{\Sigma}|^{-1/2}\exp[-\tfrac{1}{2}(\boldsymbol{\theta}-\boldsymbol{\mu})'\boldsymbol{\Sigma}^{-1}(\boldsymbol{\theta}-\boldsymbol{\mu})],$$

an "orthogonalizing" transformation of $\boldsymbol{\theta}$ can be used so that the multivariate normal component splits into a product of univariate normal factors. The following approximation can then be obtained:

$$\int\cdots\int g(\boldsymbol{\theta})\,d\theta_1\,d\theta_2\cdots d\theta_q \approx \sum_{j_q} m_{j_q}^{(q)}\cdots\sum_{j_2} m_{j_2}^{(2)}\sum_{j_1} m_{j_1}^{(1)} g(z_{j_1}^{(1)}, z_{j_2}^{(2)}, \ldots, z_{j_q}^{(q)}), \tag{2.122}$$

where the $m_{j_r}^{(r)}$ and $z_{j_r}^{(r)}$ are found using (2.121) but with μ and σ^2 replaced by estimates of the marginal posterior mean and variance of θ_r, and $\theta_1, \theta_2, \ldots, \theta_q$ are any q elements of the p-dimensional vector $\boldsymbol{\theta}$ $(q \leq p)$.

There are three applications of the above numerical integration. Firstly, if

$g(\boldsymbol{\theta}) \propto p(\boldsymbol{\theta}|\mathbf{y})$, then (2.122), with $q = p$, gives the scale factor which converts $g(\boldsymbol{\theta})$ into $p(\boldsymbol{\theta}|\mathbf{y})$. Secondly, if $q < p$ and $g(\boldsymbol{\theta}) = p(\boldsymbol{\theta}|\mathbf{y})$, then (2.122) gives the posterior density of a q-dimensional subset of $\boldsymbol{\theta}$. Finally, if $g(\boldsymbol{\theta})$ equals $\theta_i p(\boldsymbol{\theta}|\mathbf{y})$ or $\theta_i \theta_j p(\boldsymbol{\theta}|\mathbf{y})$, then we can use (2.122) to find posterior moments. Further details of the method, together with examples, are given by Naylor and Smith [1982], Smith et al. [1985], and Smith and Naylor [1987]. The latter paper also discusses briefly the computation of the marginal univariate and bivariate posterior densities using splines.

A third, analytic method of approximating posterior moments and marginal posterior densities from a unimodal joint posterior density is given by Tierney and Kadane [1986]. The posterior mean of a smooth positive function $r(\boldsymbol{\theta})$ is given by [cf. (2.99)]

$$\begin{aligned} E[r] &= E[r(\boldsymbol{\theta})|\mathbf{y}] \\ &= \int r(\boldsymbol{\theta}) p(\boldsymbol{\theta}|\mathbf{y})\, d\boldsymbol{\theta} \\ &= k^{-1} \int r(\boldsymbol{\theta}) p(\mathbf{y}|\boldsymbol{\theta}) p(\boldsymbol{\theta})\, d\boldsymbol{\theta} \\ &= \frac{\int r(\boldsymbol{\theta}) p(\mathbf{y}|\boldsymbol{\theta}) p(\boldsymbol{\theta})\, d\boldsymbol{\theta}}{\int p(\mathbf{y}|\boldsymbol{\theta}) p(\boldsymbol{\theta})\, d\boldsymbol{\theta}}, \end{aligned} \tag{2.123}$$

and, using Laplace's method, the above authors approximate this ratio by

$$\hat{E}[r] = \frac{|\boldsymbol{\Sigma}_N|^{1/2}}{|\boldsymbol{\Sigma}_M|^{1/2}} \exp[nN(\hat{\boldsymbol{\theta}}_N) - nM(\hat{\boldsymbol{\theta}}_M)].$$

Here $\hat{\boldsymbol{\theta}}_N$ and $\hat{\boldsymbol{\theta}}_M$ respectively minimize

$$N(\boldsymbol{\theta}) = n^{-1} \log[r(\boldsymbol{\theta}) p(\mathbf{y}|\boldsymbol{\theta}) p(\boldsymbol{\theta})]$$

and (2.124)

$$M(\boldsymbol{\theta}) = n^{-1} \log[p(\mathbf{y}|\boldsymbol{\theta}) p(\boldsymbol{\theta})].$$

Also

$$\boldsymbol{\Sigma}_N = -\left.\frac{\partial^2 N(\boldsymbol{\theta})}{\partial \boldsymbol{\theta}\, \partial \boldsymbol{\theta}'}\right|_{\hat{\boldsymbol{\theta}}_N} \quad \text{and} \quad \boldsymbol{\Sigma}_M = -\left.\frac{\partial^2 M(\boldsymbol{\theta})}{\partial \boldsymbol{\theta}\, \partial \boldsymbol{\theta}'}\right|_{\hat{\boldsymbol{\theta}}_M}.$$

Furthermore, if $V[r]$ is the posterior variance of $r(\boldsymbol{\theta})$, and $C[r, s]$ is the covariance of two positive functions $r(\boldsymbol{\theta})$ and $s(\boldsymbol{\theta})$, then the following approximations can be used:

$$\hat{V}[r] = \hat{E}[r^2] - (\hat{E}[r])^2 \tag{2.125}$$

and

$$\hat{C}[r,s] = \hat{E}[rs] - \hat{E}[r]\hat{E}[s]. \tag{2.126}$$

These are found by replacing r by r^2 and rs, respectively, in (2.123) and (2.124).

Tierney and Kadane [1986] show that when $\boldsymbol{\theta}$ is one-dimensional, some of the error terms in the approximations for the numerator and denominator of (2.123) cancel out, so that

$$E[r] = \hat{E}[r][1 + O(n^{-2})],$$
$$V[r] = \hat{V}[r][1 + O(n^{-2})],$$
$$C[r,s] = \hat{C}[r,s] + O(n^{-3}).$$

They note that when n is very large, (2.125) should be used with caution, as it involves computing a small number as the difference of two large numbers. Obviously computations should be carried out with sufficient precision. When n is very small, the variance approximation (2.125) could possibly be negative, and a covariance matrix computed from (2.126) may not be positive semidefinite. In computing $\hat{\boldsymbol{\theta}}_N$, the posterior mode $\hat{\boldsymbol{\theta}}_M$ can be used as a starting value. It should be stressed that $r(\boldsymbol{\theta})$ must be a positive function, which can be a serious limitation of the method. For example, if $r(\boldsymbol{\theta}) = \theta_j$, we require $\theta_j \geq 0$. However, the authors show that Laplace's method can also be used to give a promising approximation for the marginal density of any θ_j. They also give an example which demonstrates that their methods compare very favorably with those of Naylor and Smith [1982].

In conclusion we note that Smith et al. [1985] give a brief review of various methods for handling posterior densities. They mention the above method of Tierney and Kadane [1986] very briefly, as well as a method of successive transformations to approximate normality due to Dagenais and Liem [1981].

2.7.3 Highest-Posterior-Density Regions

Once the posterior density $p(\boldsymbol{\gamma}|\mathbf{y})$ has been found for a vector $\boldsymbol{\gamma}$ of parameters, we can make statements about the most likely values of $\boldsymbol{\gamma}$. For example, we may wish to find a set Γ_α such that $\text{pr}(\boldsymbol{\gamma} \in \Gamma_\alpha) = 1 - \alpha$ for given α. In contrast to confidence regions, $\boldsymbol{\gamma}$ is now random rather than the set Γ_α. When constructing a confidence interval with a given confidence level we would endeavor to choose an interval with minimum length. Similarly, for confidence regions we would try to choose a region with minimum volume. The same idea applies to Bayesian regions: we choose Γ_α to be as small as possible. This will occur when Γ_α is associated with those values which have the highest probability of occurrence. Such a region is called a highest-posterior-density (h.p.d.) region. We now give a formal definition (Box and Tiao [1973]). A region Γ_α in the parameter space of γ is called an h.p.d. region of content $1-\alpha$ if

(i) $\text{pr}[\gamma \in \Gamma_\alpha | \mathbf{y}] = 1-\alpha$, and
(ii) for $\gamma_1 \in \Gamma_\alpha$ and $\gamma_2 \notin \Gamma_\alpha$, $p(\gamma_1|\mathbf{y}) \geq p(\gamma_2|\mathbf{y})$.

A marginal h.p.d. interval for a single parameter γ_r is similarly derived from the marginal posterior density $p(\gamma_r|\mathbf{y})$. This marginal distribution is obtained, usually numerically, by integrating out the other γ's in $p(\gamma|\mathbf{y})$. The h.p.d. interval, say $[c, d]$ is that interval with minimum width $d-c$. Thus for a unimodal distribution, c and d must satisfy

$$\int_c^d p(\gamma_r|\mathbf{y})\, d\gamma_r = 1-\alpha$$

and $p(c|\mathbf{y}) = p(d|\mathbf{y})$. To see that this second condition is needed to minimise $d-c$, suppose that $p(c|\mathbf{y}) < p(d|\mathbf{y})$. Then we can shift the interval to the right and reduce its width, as the left-hand end will move further than the right-hand end in order to encompass the same probability. Methods of solving for c and d are described by Wright [1986].

2.7.4 Normal Approximation to Posterior Density

Suppose the y_i are mutually independent with probability distribution given by $p_i(y_i|\gamma)$, where γ is a $q \times 1$ vector of unknown parameters with prior distribution $p(\gamma)$. Then, from (2.96), the posterior distribution of γ is

$$p(\gamma|\mathbf{y}) = k^{-1}\left(\prod_{i=1}^{n} p_i(y_i|\gamma)\right)p(\gamma),$$

where k does not depend on γ. The posterior mode $\tilde{\gamma}$ is found by maximizing

$$\begin{aligned} A(\gamma) &= \sum_{i=1}^{n} \log p_i(y_i|\gamma) + \log p(\gamma) \\ &= L(\gamma) + \log p(\gamma). \end{aligned}$$

Since the sum of the logarithms is of order n, it is clear that the first term, the log-likelihood, will dominate as $n \to \infty$. Hence $\tilde{\gamma}$ will converge to the maximum-likelihood estimate $\hat{\gamma}$ as $n \to \infty$ provided $p(\gamma)$ is continuous and does not vanish at $\hat{\gamma}$. Under general regularity conditions $\hat{\gamma}$ will be a consistent estimator of γ^*, the true value of γ, so that for large enough n, $\tilde{\gamma}$ will be close to γ^*.

Now in a neighborhood of $\tilde{\gamma}$ we have the Taylor expansion (Bard [1974: p. 191])

$$\log p(\gamma|\mathbf{y}) \approx \log p(\tilde{\gamma}|\mathbf{y}) - \tfrac{1}{2}(\gamma - \tilde{\gamma})'\tilde{\mathbf{V}}^{-1}(\gamma - \tilde{\gamma}), \tag{2.127}$$

where

$$\tilde{\mathbf{V}}^{-1} = \left[\frac{-\partial^2 \log p(\boldsymbol{\gamma}|\mathbf{y})}{\partial \boldsymbol{\gamma}\, \partial \boldsymbol{\gamma}'}\right]_{\tilde{\boldsymbol{\gamma}}}. \tag{2.128}$$

Hence

$$p(\boldsymbol{\gamma}|\mathbf{y}) \approx p_{\max} \exp\left[-\tfrac{1}{2}(\boldsymbol{\gamma} - \tilde{\boldsymbol{\gamma}})'\tilde{\mathbf{V}}^{-1}(\boldsymbol{\gamma} - \tilde{\boldsymbol{\gamma}})\right],$$

where $p_{\max} = p(\tilde{\boldsymbol{\gamma}}|\mathbf{y})$, so that close to $\tilde{\boldsymbol{\gamma}}$ we see that $\boldsymbol{\gamma}$ is approximately $N_q(\tilde{\boldsymbol{\gamma}}, \tilde{\mathbf{V}})$. We can therefore write

$$\begin{aligned} 2[\log p(\tilde{\boldsymbol{\gamma}}|\mathbf{y}) - \log p(\boldsymbol{\gamma}|\mathbf{y})] &\approx (\boldsymbol{\gamma} - \tilde{\boldsymbol{\gamma}})'\tilde{\mathbf{V}}^{-1}(\boldsymbol{\gamma} - \tilde{\boldsymbol{\gamma}}) \\ &\sim \chi^2_q, \end{aligned} \tag{2.129}$$

approximately. If $\boldsymbol{\gamma} = \boldsymbol{\theta}$ then an h.p.d. region for $\boldsymbol{\theta}$ is given by (Box and Draper [1972])

$$\{\boldsymbol{\theta} : p(\boldsymbol{\theta}|\mathbf{y}) \geq p(\tilde{\boldsymbol{\theta}}|\mathbf{y}) \exp[-\tfrac{1}{2}\chi^2_q(\alpha)]\}, \tag{2.130}$$

where $\text{pr}[\chi^2_q \geq \chi^2_q(\alpha)] = \alpha$. This region has a content of approximately $1 - \alpha$. Equivalently we can use

$$\{\boldsymbol{\theta} : (\boldsymbol{\theta} - \tilde{\boldsymbol{\theta}})'\tilde{\mathbf{V}}^{-1}(\boldsymbol{\theta} - \tilde{\boldsymbol{\theta}}) \leq \chi^2_q(\alpha)\}. \tag{2.131}$$

2.7.5 Test for Model Adequacy Using Replication

Consider the replicated model

$$y_{ij} = f(\mathbf{x}_i; \boldsymbol{\theta}) + \phi_i + \varepsilon_{ij} \qquad (i = 1, 2, \ldots, n, \quad j = 1, 2, \ldots, J_i),$$

where the ε_{ij} are i.i.d. $N(0, \sigma^2)$ and ϕ_i represents a possible departure from the postulated nonlinear model. Let

$$S(\boldsymbol{\theta}, \boldsymbol{\phi}) = \sum_i \sum_j [y_{ij} - f(\mathbf{x}_i; \boldsymbol{\theta}) - \phi_i]^2,$$

where $\boldsymbol{\phi} = (\phi_1, \phi_2, \ldots, \phi_n)'$. Using the prior distribution

$$p(\boldsymbol{\theta}, \boldsymbol{\phi}, \sigma) \propto \sigma^{-1},$$

Steece [1979] argued as in Section 2.7.2 and obtained the posterior distribution [cf. (2.108)]

$$p(\boldsymbol{\theta}, \boldsymbol{\phi}|\mathbf{y}) = k_1^{-1}[S(\boldsymbol{\theta}, \boldsymbol{\phi})]^{-n/2}, \qquad \boldsymbol{\theta} \in \Theta, \quad \boldsymbol{\phi} \in \Phi.$$

This leads to

$$p(\boldsymbol{\phi}|\mathbf{y}) = k_1^{-1} \int_{\Theta} [S(\boldsymbol{\theta}, \boldsymbol{\phi})]^{-n/2} \, d\boldsymbol{\theta}.$$

To test model adequacy, i.e. $\boldsymbol{\phi} = \mathbf{0}$, Steece [1979] suggested constructing a $100(1-\alpha)\%$ h.p.d. region for $\boldsymbol{\phi}$ and seeing if it contained $\boldsymbol{\phi} = \mathbf{0}$. Steece [1979] also gave a linear approximation method similar in spirit to the argument leading to (2.115).

*2.7.6 Polynomial Estimators

Consider, by way of illustration, the compartmental model (1.15), namely $E[y_i] = f(t_i; \boldsymbol{\theta})$, where

$$f(t; \boldsymbol{\theta}) = \frac{\theta_1(e^{-t\theta_2} - e^{-t\theta_1})}{\theta_1 - \theta_2} \qquad (\theta_1 \neq \theta_2), \tag{2.132}$$

with parameter space $\Theta = \{(\theta_1, \theta_2)' : \theta_1 > \theta_2\}$. In addition to estimating the rates θ_1 and θ_2, the parametric function

$$\phi = g(\boldsymbol{\theta}) = (\theta_1 - \theta_2)^{-1} \log(\theta_1/\theta_2) \tag{2.133}$$

is also of interest, being the time t at which the maximum expected yield, $E[y]$, occurs. If $\hat{\boldsymbol{\theta}}$ is the maximum-likelihood estimate of $\boldsymbol{\theta}$ (also least squares under normality), then a natural estimator of ϕ is $\hat{\phi} = g(\hat{\boldsymbol{\theta}})$. This estimator will be asymptotically normal if $\hat{\boldsymbol{\theta}}$ is asymptotically normal. If we set $\phi_1 = g(\boldsymbol{\theta})$ and $\phi_2 = \theta_2$, then $\hat{\phi}_1$ $[= g(\hat{\boldsymbol{\theta}})]$ and $\hat{\phi}_2$ $(= \hat{\theta}_2)$ are the maximum-likelihood estimates of ϕ_1 and ϕ_2, as the transformation is one-to-one. For some functions it may be preferable to transform from $\boldsymbol{\theta}$ to $\boldsymbol{\phi}$ right from the beginning, particularly if the parameter-effects curvature for $\boldsymbol{\phi}$ is much smaller (see Chapter 4).

We now consider a general nonlinear model and a general function $\phi = g(\boldsymbol{\gamma})$ of $\boldsymbol{\gamma} = (\boldsymbol{\theta}', \sigma)'$, where $\boldsymbol{\gamma} \in \Gamma$. Instead of using $\hat{\phi} = g(\hat{\boldsymbol{\gamma}})$, where $\hat{\boldsymbol{\gamma}}$ is the maximum-likelihood estimate of $\boldsymbol{\gamma}$, Gallant [1980] proposed a different method which is particularly useful for the situation where ϕ is to be estimated repetitively and a premium is placed on the simplicity of the computations, e.g. the daily calibration of an industrial process. He suggested selecting an "explicit" estimator from a class of polynomial functions of the data according to the criterion of best average mean squared error (AMSE). Such an estimator takes the form of $\mathbf{a}'\mathbf{z}$, where the elements of $\mathbf{z}$ are polynomial functions of the y_i, for example

$$\mathbf{z} = (1, y_1, y_2, \ldots, y_n)' \tag{2.134}$$

or

$$\mathbf{z} = (1, y_1, y_2, \ldots, y_n, y_1^2, y_1 y_2, y_2^2, \ldots, y_n^2)'. \tag{2.135}$$

Since the estimation is repetitive, previous knowledge may be incorporated in a prior density function $p(\gamma)$. Using the AMSE criterion, Gallant showed that $g(\gamma)$ is estimated by $\tilde{\phi} = \mathbf{a}'\mathbf{z}$, where

$$\mathbf{a} = \mathbf{A}^{-}\mathbf{b}. \tag{2.136}$$

Here $\mathbf{A}^{-}$ is a generalized inverse of

$$\mathbf{A} = \int_{\Gamma} \mathscr{E}[\mathbf{z}\mathbf{z}'|\gamma]p(\gamma)\,d\gamma \tag{2.137}$$

and

$$\mathbf{b} = \int_{\Gamma} g(\gamma)\mathscr{E}[\mathbf{z}|\gamma]p(\gamma)\,d\gamma. \tag{2.138}$$

For example, suppose that $\mathscr{D}[\mathbf{y}|\gamma] = \sigma^2\mathbf{I}_n$. Then applying this to, say, (2.134) gives us

$$\mathscr{E}[\mathbf{z}|\gamma] = \begin{pmatrix} 1 \\ \mathbf{f}(\boldsymbol{\theta}) \end{pmatrix} = \boldsymbol{\mu}_z$$

and

$$\mathscr{E}[\mathbf{z}\mathbf{z}'|\gamma] = \mathscr{D}[\mathbf{z}|\gamma] + \boldsymbol{\mu}_z\boldsymbol{\mu}_z',$$

where

$$\mathscr{D}[\mathbf{z}|\gamma] = \begin{pmatrix} 0 & \mathbf{0}' \\ \mathbf{0} & \sigma^2\mathbf{I}_n \end{pmatrix}.$$

Gallant showed that, under fairly general conditions, $\tilde{\phi}$ has good properties relative to the Bayes estimator, which has the smallest AMSE attainable for any estimator, polynomial or otherwise.

Various choices of the prior $p(\gamma)$ are available. For example, one could choose a smooth approximation $P(\gamma)$ of the empirical distribution function of previous maximum-likelihood estimates, so that $p(\gamma)\,d\gamma$ is replaced by $dP(\gamma)$. In the case of the above compartmental model (2.132), Gallant used prior knowledge about the sampling distributions of $\hat{\boldsymbol{\theta}}$ and s^2 to propose a uniform prior for $\boldsymbol{\theta}$ on the ellipsoid

$$\mathscr{D} = \{\boldsymbol{\theta}:(\boldsymbol{\theta} - \mathbf{c})'\mathbf{C}^{-1}(\boldsymbol{\theta} - \mathbf{c}) \le r^2\},$$

where $\mathbf{C} = [\mathbf{F}_.'(\mathbf{c})\mathbf{F}_.(\mathbf{c})]^{-1}$ and $\mathbf{F}_.(\boldsymbol{\theta}) = [(\partial f_i/\partial\theta_j)]$, and an independent inverted gamma prior for σ [see (1.4)]. The values of $\mathbf{c}$ and r are chosen by the analyst—in Gallant's case, from prior experiments. If v is the volume of $\mathscr{D}$ $(= \pi r^2|\mathbf{C}|$ for

$p=2$, from Appendix A6), then

$$p(\gamma)=p(\boldsymbol{\theta})p(\sigma), \tag{2.139}$$

where

$$p(\boldsymbol{\theta})=v^{-1}, \qquad \boldsymbol{\theta}\in\mathscr{D},$$

and

$$p(\sigma)=\frac{2}{\Gamma(\nu/2)}\left(\frac{\nu s^2}{2}\right)^{\nu/2}\frac{1}{\sigma^{\nu+1}}\exp\left(-\frac{\nu s^2}{2\sigma^2}\right)$$

for suitable ν. In order to use (2.136) we need to be able to evaluate integrals of the form $\int k(\boldsymbol{\theta},\sigma)p(\gamma)\,d\gamma$. Using the transformations $\mathbf{w}=(\mathbf{R}')^{-1}(\boldsymbol{\theta}-\mathbf{c})$ and $u=\nu s^2/2\sigma^2$, where $\mathbf{C}=\mathbf{R}'\mathbf{R}$ is the Cholesky decomposition of $\mathbf{C}$, Gallant obtained the expression

$$\int k(\boldsymbol{\theta},\sigma)p(\gamma)\,d\gamma=\frac{1}{\pi r\Gamma(\nu/2)}\int_0^\infty u^{\nu/2-1}\int_{\mathbf{w}'\mathbf{w}\le r^2}k(\mathbf{R}'\mathbf{w}+\mathbf{c},[\nu s^2/2]^{1/2})\,d\mathbf{w}\,e^{-u}\,du \tag{2.140}$$

and evaluated the above integral using suitable quadrature formulae with specified $\mathbf{c}$ and $r^2=8s^2$. He found that the linear estimate $\tilde{\phi}$ had smaller average mean squared error than the maximum-likelihood estimate $\hat{\phi}$, and that it compared very favorably with the Bayes estimator.

2.8 VARIANCE HETEROGENEITY

In Section 1.7 we discussed the possibility of linearizing a nonlinear model using an appropriate transformation. We saw that the errors in the original model could be additive, proportional, or multiplicative, so that the appropriateness of the transformation depends on this error structure. It was also noted in Chapter 1 that theoretical considerations frequently provided a functional form for the trend part of the model, so that $y\approx f(x;\boldsymbol{\theta})$ with f known. Although it is then usually assumed that

$$y_i=f(\mathbf{x}_i;\boldsymbol{\theta})+\varepsilon_i \qquad (i=1,2,\ldots,n), \tag{2.141}$$

where the ε_i are i.i.d. $N(0,\sigma^2)$, there is often little understanding of the stochastic nature of the model, particularly with regard to the distribution and variance structure of the errors ε_i. If such a model is fitted by, for example, least squares, the residuals

$$\begin{aligned}\hat{\varepsilon}_i&=y_i-\hat{y}_i\\&=y_i-f(\mathbf{x}_i;\hat{\boldsymbol{\theta}})\end{aligned} \tag{2.142}$$

may provide some insight into the validity of the model and error assumptions. However, such residuals can be misleading, particularly if the model is highly nonlinear (as indicated by a high intrinsic curvature: see Chapter 4), and modified residuals are discussed in Section 4.6.4. These so-called "projected" residuals have similar properties to the ordinary residuals (2.142) for linear models and can be used to diagnose nonnormality and variance heterogeneity. When assumptions are clearly violated, a new or more complex model must be built and this in turn must be scrutinized. This process is continued until a model is arrived at that appears adequate, and this final model is then used to make inferences from the data.

The method of least squares does not require the assumption of normality. However, we shall see later that this method may give completely misleading estimates if there is variance heterogeneity. This heterogeneity can take two important forms: (a) var $[y]$ is proportional to some function of $E[y]$, and (b) var $[y]$ is proportional to some function of $\mathbf{x}$ (and possibly $\boldsymbol{\theta}$). If var $[y]$ is a known function of $E[y]$, then the quasi-likelihood method of Section 2.3 can be used. Otherwise, two other methods of modeling (a) are available, and these are described as models A and B below. Case (b) is considered in Section 2.8.8.

Box and Cox [1964] discussed the family of transformations

$$y^{(\lambda)} = \begin{cases} \dfrac{y^{\lambda}-1}{\lambda}, & \lambda \neq 0, \\ \log y, & \lambda = 0 \end{cases} \tag{2.143}$$

for use when $y > 0$. Using a Taylor expansion, we have the first-order approximations

$$E[y^{(\lambda)}] \approx \{E[y]\}^{(\lambda)}$$

and

$$\begin{aligned} \operatorname{var}[y^{(\lambda)}] &\approx \operatorname{var}[y] \left[\frac{dy^{(\lambda)}}{dy} \right]^2_{y=E[y]} \\ &= \operatorname{var}[y] (E[y])^{2\lambda-2}. \end{aligned} \tag{2.144}$$

We can then choose λ to make (2.144) constant. This means we can either work with $y^{(\lambda)}$ and assume homogeneity, or else work with y and allow for heterogeneity. The variance of y is then proportional to $(E[y])^{2-2\lambda}$. Therefore, when the variability in the appropriate residuals appears to increase or decrease with the predicted value $\hat{y}$, we have the following two approaches for extending (2.141). We adopt the terminology of Carroll and Ruppert [1984].

Model A The PTBS, or "power-transform both sides," model is (Leech [1975])

$$y_i^{(\lambda_1)} = f(\mathbf{x}_i;\boldsymbol{\theta})^{(\lambda_1)} + \varepsilon_i \tag{2.145}$$

for some λ_1, where the ε_i are i.i.d. $N(0, \sigma^2)$. This model is only approximate, as, apart from the log-normal distribution ($\lambda_1 = 0$), $y^{(\lambda_1)}$ cannot be normal for positive y. In fact, from (2.143) we must have $y^{(\lambda_1)} > -\lambda_1^{-1}$ when $\lambda_1 > 0$ and $y^{(\lambda_1)} < -\lambda_1^{-1}$ when $\lambda_1 < 0$. Some consequences of this normal approximation are discussed by Hernandez and Johnson [1980] and Amemiya and Powell [1981].

Model B The PTWLS, or "power-transformed weighted least squares," model is

$$y_i = f(\mathbf{x}_i; \boldsymbol{\theta}) + \varepsilon_i, \tag{2.146}$$

where the ε_i are independent $N(0, \sigma^2 f(\mathbf{x}_i; \boldsymbol{\theta})^{\lambda_2})$. From (2.144) we have $\lambda_2 \approx 2 - 2\lambda_1$.

Here model B is a *weighted least squares* model in which the weights depend on the parameters to be estimated; it has often been preferred to model A when $f(\mathbf{x}; \boldsymbol{\theta})$ is a linear function of $\boldsymbol{\theta}$. In contrast, *transformed models* like A have been preferred in nonlinear work. Models A and B differ mainly in that A transforms so that $y^{(\lambda)}$ has a different distribution from y as well as having a homogeneous variance structure, while B models the variance heterogeneity but leaves the distribution of y unchanged. However, the close similarity between the approaches is emphasized by using the linear Taylor approximation

$$y^{\lambda} \approx f(\mathbf{x}; \boldsymbol{\theta})^{\lambda} + \lambda f(\mathbf{x}; \boldsymbol{\theta})^{\lambda-1}[\mathbf{y} - \mathbf{f}(\mathbf{x}; \boldsymbol{\theta})].$$

Then the sum of squares for model A is

$$\begin{aligned} S_A(\boldsymbol{\theta}) &= \sum_i [y_i^{(\lambda)} - f(\mathbf{x}_i; \boldsymbol{\theta})^{(\lambda)}]^2 \\ &= \sum_i \left(\frac{y_i^{\lambda} - f(\mathbf{x}_i; \boldsymbol{\theta})^{\lambda}}{\lambda} \right)^2 \\ &\approx \sum_i f(\mathbf{x}_i; \boldsymbol{\theta})^{2\lambda-2} [y_i - f(\mathbf{x}_i; \boldsymbol{\theta})]^2 \\ &= S_B(\boldsymbol{\theta}), \end{aligned}$$

the weighted sum of squares for model B. In conclusion we note that transformations other than the power transformation (2.143) could be used in the two models.

2.8.1 Transformed Models

Here we assume that a transformation function h exists under which the usual model assumptions hold, i.e.

$$h(y_i) = h[f(\mathbf{x}_i; \boldsymbol{\theta})] + \varepsilon_i, \tag{2.147}$$

where the ε_i are i.i.d. $N(0, \sigma^2)$. In practice the transformation chosen may not achieve the desired objective, and the fitted model should itself be subject to scrutiny. We will now consider the families of transformations discussed by Box and Cox [1964] and John and Draper [1980].

a Box–Cox Transformations

Suppose $h(y_i) = y_i^{(\lambda)}$, where $y^{(\lambda)}$ is defined by (2.143); then (2.147) becomes model A. The joint distribution of the original y_i is approximately

$$\frac{1}{(2\pi\sigma^2)^n} \prod_{i=1}^{n} \left[\exp\left(-\frac{1}{2\sigma^2} [y_i^{(\lambda)} - f(\mathbf{x}_i; \boldsymbol{\theta})^{(\lambda)}]^2 \right) \left| \frac{dy_i^{(\lambda)}}{dy_i} \right| \right],$$

so that, excluding constants, the log-likelihood is

$$L(\boldsymbol{\theta}, \sigma^2, \lambda) = -\frac{n}{2} \log \sigma^2 - \frac{1}{2\sigma^2} \sum_{i=1}^{n} [y_i^{(\lambda)} - f(\mathbf{x}_i; \boldsymbol{\theta})^{(\lambda)}]^2 + (\lambda - 1) \sum_{i=1}^{n} \log y_i. \tag{2.148}$$

For *fixed* λ, maximizing L with respect to $\boldsymbol{\theta}$ and σ^2 is equivalent to minimizing

$$\sum_{i=1}^{n} [y_i^{(\lambda)} - f(\mathbf{x}_i; \boldsymbol{\theta})^{(\lambda)}]^2$$

to obtain $\tilde{\boldsymbol{\theta}}_\lambda$, and then setting

$$\tilde{\sigma}_\lambda^2 = \frac{1}{n} \sum_{i=1}^{n} [y_i^{(\lambda)} - f(\mathbf{x}_i; \tilde{\boldsymbol{\theta}}_\lambda)^{(\lambda)}]^2. \tag{2.149}$$

Hence the maximum-likelihood (Box–Cox) estimate of λ maximizes the concentrated log-likelihood function (Section 2.2.3)

$$M(\lambda) = -\frac{n}{2} \log \tilde{\sigma}_\lambda^2 + (\lambda - 1) \sum_{i=1}^{n} \log y_i. \tag{2.150}$$

In conjunction with a nonlinear least-squares program, the above can be maximized using a one-dimensional maximization program, or by plotting $M(\lambda)$ over a suitable range of λ-values. From Section 5.2.3, an approximate $100(1-\alpha)\%$ confidence interval for λ is given by

$$\{\lambda : M(\lambda) \geq c_\alpha = M(\hat{\lambda}) - \tfrac{1}{2}\chi_1^2(\alpha)\}, \tag{2.151}$$

where $P[\chi_1^2 \geq \chi_1^2(\alpha)] = \alpha$. Since a nonlinear least-squares problem must be solved to obtain $M(\lambda)$ for each new value of λ, maximization of the full function $L(\boldsymbol{\theta}, \sigma^2, \lambda)$ using a quasi-Newton method (Section 13.4.2), for example, may be faster than using $M(\lambda)$. However, very few iterations tend to be required to obtain $\tilde{\boldsymbol{\theta}}_\lambda$ from its value at a neighboring λ, and a plot of $M(\lambda)$ versus λ is useful for several reasons. Firstly, $\hat{\lambda}$ can be read from it to an accuracy that will usually be sufficient for practical purposes. For example, if $\hat{\lambda}=0.1$ and the interval for λ is wide, one would probably use $\lambda=0$ and the corresponding log transformation. Secondly, the asymptotic $100(1-\alpha)\%$ confidence interval (2.151) can be read from the graph (see Fig. 8.12 of Section 8.5.7). Thirdly, the general shape of $M(\lambda)$ gives an indication of how well the data support a particular λ-value. Finally we note that Bayesian estimation of λ is also possible along the lines of Box and Cox [1964], Pericchi [1981], and Sweeting [1984]. An expression for $\text{var}[\hat{\lambda}]$ has recently been given by Lawrance [1987].

The family (2.143) is clearly of no value if any of the y_i's are negative, so that Box and Cox [1964] suggested the more general family

$$y^{(\lambda)}=\begin{cases}\dfrac{(y+\lambda_2)^{\lambda_1}-1}{\lambda_1}, & \lambda_1\neq 0,\\ \log(y+\lambda_2), & \lambda_1=0,\end{cases} \tag{2.152}$$

for $y>-\lambda_2$. Even for samples with positive y_i, this enlarged family gives a more flexible range of transformations. The log-likelihood function is still given by (2.148), but with $y^{(\lambda)}$ replaced by $y^{(\boldsymbol{\lambda})}$ and $(\lambda-1)\log y$ replaced by $(\lambda_1-1)\log(y+\lambda_2)$; and similarly with $M(\boldsymbol{\lambda})$. However, there are problems. The maximum-likelihood estimate of λ_2 is $-y_{\min}$, and the maximum value of $M(\boldsymbol{\lambda})$ is ∞. A satisfactory estimation procedure is needed. For further discussion, see Carroll and Ruppert [1988].

Transformations such as those in the Box–Cox family affect the shape of the distribution of $y^{(\lambda)}$ as well as affecting the relationship between the variance of $y^{(\lambda)}$ and its expected value. Indeed, the family (2.143) is often used to transform a distribution to symmetry (see Hinkley [1975]). Values of $\lambda>1$ stretch out the upper tail of the distribution of y, while values of $\lambda<1$ draw out the lower tail.

It should be stressed, once again, that the above theory is only approximate in that, for $\lambda\neq 0$, $y^{(\lambda)}$ cannot be normal, as it is bounded either above or below. Moreover, the model may be incorrect. The residuals from the fit of the transformed data should therefore be checked for normality, as the transformation may not be successful.

b John–Draper Transformations

John and Draper [1980] presented an example in which the "best" choice of λ in (2.143) led to a model which was still unsatisfactory. The residuals for the untransformed data were symmetric but heavy-tailed and had heterogeneity of variance. However, they obtained a successful transformation using the modulus family

$$y^{(\lambda)}=\begin{cases}(\text{sign}\, y)\dfrac{(|y|+1)^{\lambda}-1}{\lambda}, & \lambda\neq 0,\\ (\text{sign}\, y)\log(|y|+1), & \lambda=0.\end{cases} \tag{2.153}$$

When the y_i's are distributed about zero, this transformation tends to influence the tails of the distribution but has little effect on the skewness. If all the observations are positive, the modulus and power transformations are equivalent. In conclusion it should be stressed that once the best choice from a family of transformations is made, the transformed model should still be checked for adequacy.

2.8.2 Robust Estimation for Model A

In the linear-model case, estimation of λ by maximum likelihood, and likelihood-ratio tests to assess the need for a transformation, can be heavily influenced by the presence of outliers. Carroll [1980] noted that since the power transformations are "intended to achieve only approximate normality, the methods mentioned above are deficient in not considering the possibility that the errors ε might be only close to normality with perhaps heavier tail behaviour." Cook and Wang [1983] further noted that an observation which is outlying or influential in the untransformed data may not be noticed after transformation because the transformation was largely determined to accommodate that observation.

The method of Carroll [1980, 1982a] is readily adapted to model A in (2.145). Consider a modification of the log-likelihood function (2.148) in which

$$L(\boldsymbol{\theta}, \sigma^2, \lambda) = -\frac{n}{2}\log \sigma^2 - \sum_{i=1}^{n} \rho\left(\frac{y_i^{(\lambda)} - f(\mathbf{x}_i; \boldsymbol{\theta})^{(\lambda)}}{\sigma}\right) + (\lambda - 1)\sum_{i=1}^{n} \log y_i. \tag{2.154}$$

Here ρ is a robust loss function, for example Huber's function given by

$$\rho(z) = \begin{cases} \frac{1}{2}z^2, & |z| \le k, \\ k(|z| - k/2), & |z| > k, \end{cases}$$

for some k (e.g. $k = 1.5$ or 2.0). For fixed λ and an initial estimate σ_0 of σ, $\hat{\boldsymbol{\theta}}(\lambda, \sigma_0)$ is chosen to maximize (2.154). Then the estimate of σ is updated (Carroll [1982a]) as the solution to

$$\sum_{i=1}^{n} [r_i \psi(r_i) - 1] = 0,$$

where $r_i = [y_i^{(\lambda)} - f\{\mathbf{x}_i; \hat{\boldsymbol{\theta}}(\lambda, \sigma_0)\}^{(\lambda)}]/\sigma$ and $\psi(z) = d\rho(z)/dz$. These steps of maximizing with respect to $\boldsymbol{\theta}$ and updating σ are iterated until convergence to give a robust concentrated log-likelihood for λ, $M(\lambda)$. The transformation parameter $\hat{\lambda}$ is then chosen to maximize $M(\lambda)$.

The above approach does not prevent observations with extreme **x**-values (or "outliers in **x**") having a large influence on the choice of $\hat{\lambda}$. The question of influence has been discussed in the case of Box–Cox transformations to linearity (see Section 2.8.4 below) by Cook and Wang [1983] and by Carroll and Ruppert

[1985]. Cook and Wang concentrate on diagnosing influential cases, while Carroll and Ruppert seek to lessen the influence of "outliers in x" on $\hat{\lambda}$. An excellent discussion of the very important problem of robust estimation and influence diagnostics in the "transform both sides" model (2.147) is given in the recent paper by Carroll and Ruppert [1987].

2.8.3 Inference using Transformed Data

Having estimated λ by $\hat{\lambda}$, the approach most often used in practice is to treat $\hat{\lambda}$ as if it were the true value and then fit the model to $y_i^{(\hat{\lambda})}$, namely

$$y_i^{(\hat{\lambda})} \doteq f(\mathbf{x}_i;\boldsymbol{\theta})^{(\hat{\lambda})} + \varepsilon_i, \tag{2.155}$$

where the ε_i are i.i.d. $N(0,\sigma^2)$. This approach is attractive in that a standard nonlinear regression program can be used. However, by ignoring the variability of $\hat{\lambda}$, such an approach must lead to inferences that are too precise, for example, a variance–covariance matrix for $\hat{\boldsymbol{\theta}}$ that is too "small."

Carroll and Ruppert [1984] have investigated the seriousness of this problem for fairly general transformations indexed by a parameter λ. Suppose $\hat{\lambda}$ is obtained by maximum likelihood and we fit the model (2.155) above by least squares to obtain $\hat{\boldsymbol{\theta}}$. Let $\boldsymbol{\Sigma}(\hat{\lambda})$ be the asymptotic variance–covariance matrix for $\hat{\boldsymbol{\theta}}$ obtained from least-squares theory on the assumption that $\hat{\lambda}$ is not a random variable but is the true value of λ. Let $\boldsymbol{\Sigma}(\lambda)$ be the true asymptotic variance–covariance matrix for $\hat{\boldsymbol{\theta}}$ obtained from maximum-likelihood theory for $L(\boldsymbol{\theta},\sigma^2,\lambda)$ (i.e., λ is treated as another unknown parameter). Then Carroll and Ruppert [1984] proved under fairly general conditions that as $n\rightarrow\infty$:

(i). $\boldsymbol{\Sigma}(\hat{\lambda}) \le \boldsymbol{\Sigma}(\lambda) \le (\pi/2)\boldsymbol{\Sigma}(\hat{\lambda})$ (where $\mathbf{A}\le\mathbf{B}$ means that $\mathbf{B}-\mathbf{A}$ is positive semidefinite).

(ii). As $\sigma^2\rightarrow 0$, the limiting values of $\boldsymbol{\Sigma}(\hat{\lambda})$ and $\boldsymbol{\Sigma}(\lambda)$ are the same, but the limiting distribution of $\hat{\sigma}^2$ depends on whether or not λ is known.

The first result tells us that the true asymptotic standard errors of the elements of $\hat{\boldsymbol{\theta}}$ could be at most $\sqrt{\pi/2}$ (≈ 1.25) times larger than the values given in the output of a nonlinear regression program. The second tells us that the standard errors given by a nonlinear regression program will be essentially correct, in large samples, if σ^2 is small enough. The above results were supported by a small simulation study. For the model simulated, the bound of $\pi/2$ for the variance ratios was too conservative, and the mean squared errors for the elements of $\hat{\boldsymbol{\theta}}$ when λ is unknown were always less than 1.2 times the mean squared errors for λ known. The errors incurred by treating $\hat{\lambda}$ as known and using the "usual" method of analysis are therefore, at worst, moderate. This is very different from the situation where Box–Cox transformations are used to transform to linearity (see Section 2.8.4 below).

In applications, rather than use $\hat{\lambda}$ itself as above, $\hat{\lambda}$ is frequently approximated by some "preferred" value (e.g. $-1, -\frac{1}{2}, 0, \frac{1}{2}, 1$) for Box–Cox transformations, which also gives a good fit to the data. Such simple approximations often provide an aid to interpretation; e.g., $\lambda = 0$ corresponds to constant relative errors [cf. (1.43)]. Examples of the use of PTBS (model A) follow in Example 2.6 of Section 2.8.6 and Example 8.8 of Section 8.5.7.

2.8.4 Transformations to Linearity

Box and Cox [1964] introduced their transformation methods in the context of the model

$$y_i^{(\lambda)} = \mathbf{x}_i'\boldsymbol{\theta} + \varepsilon_i, \tag{2.156}$$

where the ε_i are i.i.d. $N(0, \sigma^2)$. By using this model one hopes to find a power transformation that simultaneously achieves linearity, constant variance, and normality. This is in contrast to the situation we have described above (PTBS), where the trend relationship $y \approx f(\mathbf{x}; \boldsymbol{\theta})$ is known and the transformation is then undertaken only to produce constant variance and normality. Box and Cox's method provides an estimate $\hat{\lambda}$ of λ using either a maximum-likelihood estimate or a posterior mode. Applying the transformation $y^{(\hat{\lambda})}$, a linear regression model can then be fitted and inferences made using standard linear theory, on the assumption that $\hat{\lambda}$ is the correct value of λ.

The above procedure, which we shall call the Box–Cox procedure, has recently aroused some controversy. In particular, Bickel and Doksum [1981] showed that when the model (2.156) is true for some unknown triple $(\boldsymbol{\theta}, \sigma^2, \lambda)$, the estimates $\hat{\boldsymbol{\theta}}$ and $\hat{\lambda}$ are highly correlated and the standard errors for $\hat{\boldsymbol{\theta}}$ produced by the Box–Cox procedure are far too small. However, Box and Cox [1982] questioned the scientific relevance of these results and introduced the following illustrative example.

Consider two groups of observations modelled by

$$y_{ij}^{(\lambda)} = \mu_i + \varepsilon_{ij} \qquad (i = 1, 2).$$

The parameter $\theta = \mu_2 - \mu_1$ measures the "difference" between the two groups. Suppose one group of observations is concentrated around 995 and the other around 1005. If $\lambda = 1$, θ represents the difference between groups for the untransformed data and is estimated by 10 y-units. If $\lambda = -1$ we work with units of $z = y^{-1}$, so that θ is estimated approximately by $\frac{1}{995} - \frac{1}{1005}$, or about 10^{-5} z-units. Changing the value of λ makes huge differences to the estimate. Only with large data sets will the data be able to distinguish sharply between these two values -1 and 1 of λ. The effect will be that the estimates of λ and θ will be highly correlated, and the estimate of θ in the true unknown transformation scale $y^{(\lambda)}$ will have a much larger standard error than it would have if λ were known. They argue that θ, representing a difference in location under an unknown transformation (and therefore in unknown units), has no physical interepretation and is not, therefore, a useful parameter for estimation. (Note that in contrast to this estimation problem, the hypothesis $H_0:\theta=0$ always has meaning whether or not we know λ.) According to Box and Cox [1982], the goal of such transformations is to find a transformation for which a linear model provides a reasonable approximation. For an elaboration on the debate, see Hinkley and Runger [1984] and the discussants.

The debate about the Box–Cox procedure is important because it provides a framework for discussing, to quote Hinkley and Runger [1984], "the general issue of whether and how to adjust a 'model-given' analysis to allow for model selection." Even in the above fairly simple situation, the question has not yet been satisfactorily resolved. There remains the problem that inferences made by treating $\hat{\lambda}$ as known must still be overly precise. One area in which it is possible to asses this effect is in prediction. Carroll and Ruppert [1981a] assumed (2.156) is true for some λ and compared the mean squared error of prediction (MSEP) for a predicted value $\hat{y}_{BC}$ of y at a new $\mathbf{x}_0$, obtained by the Box–Cox procedure, with the MSEP of $\hat{y}$ obtained when the true value of λ is known. The estimate $\hat{y}_{BC}$ is given by

$$\hat{y}_{BC}^{(\hat{\lambda})} = \mathbf{x}_0'\hat{\boldsymbol{\theta}}(\hat{\lambda}), \tag{2.157}$$

where $\hat{\boldsymbol{\theta}}(\hat{\lambda})$ is found by regressing $y^{(\hat{\lambda})}$ on $\mathbf{x}$. Carroll and Ruppert [1981a] found that MSEP($\hat{y}_{BC}$) is, on average, not much larger than MSEP($\hat{y}$).

Doksum and Wong [1983] showed that using the Box–Cox procedure for hypothesis testing gives tests with the correct asymptotic levels and good local power in two-sample t-tests, linear-regression t-tests, and a number of important analysis-of-variance tests. Carroll [1982b] produced similar results for linear-regression t-tests that some of the regression coefficients may be zero. Doksum and Wong [1983] also found that if the model (2.156) is incorrect, transformed t-tests can perform worse than untransformed ones. This emphasizes the importance of model checking even after choosing the "best" transformation.

Even though the Box–Cox transformations to linearity were intended to "massage" the data so that linear models could be used, we have discussed them in some detail for the following reasons: (a) the model (2.156) is strictly a nonlinear model in y, (b) the controversies focus on important aspects of model building and inference from models, and (c) the controversy surrounding the Box–Cox procedure in linear models does not carry over to the "transform both sides" method of (2.155).

2.8.5 Further Extensions of the Box–Cox Method

In empirical model building, a single transformation does not always achieve both linearity and homogeneity of variance. Wood [1974] allowed for one transformation to achieve linearity, $g_1(y)$ say, and a separate transformation, $g_2(y)$, to achieve constant variance. The resulting model is then

$$g_2(y_i) = g_2(g_1^{-1}(\mathbf{x}_i'\boldsymbol{\beta})) + \varepsilon_i, \tag{2.158}$$

where the ε_i are i.i.d. $N(0,\sigma^2)$. When Box–Cox transformation are used, $g_1(y) = y^{(\gamma)}$, $g_2(y) = y^{(\phi)}$, and

$$y_i^{(\phi)} = [(1 + \mathbf{x}'\boldsymbol{\beta}\gamma)^{1/\gamma}]^{(\phi)} + \varepsilon_i. \tag{2.159}$$

When $\gamma \neq 0$ we can incorporate one of the γ's in with $\boldsymbol{\beta}$, giving the model

$$y_i^{(\phi)} = [(1 + \mathbf{x}'\boldsymbol{\beta})^{1/\gamma}]^{(\phi)} + \varepsilon_i.$$

The model $y \approx (1 + x\beta)^{1/\gamma}$ has been used in agriculture, under the name of Bleasdale's simplified model, in which x is the crop density and y is the average yield per plant (see Section 7.6.1).

Lahiri and Egy [1981] have proposed a further generalization which allows for a transformation to linearity as in (2.156) but also models possible heterogeneity of variance using, for example, the PTWLS model of (2.146).

2.8.6 Weighted Least Squares: Model B

Up till now we have dealt with variance heterogeneity using the model (2.145), whereby we transform both y and $E[y]$ to stabilize the error variance. Another way of tackling this is to leave the model unchanged, as in the PTWLS model (2.146), but allow for variance heterogeneity in the analysis by using a weighted least-squares method. This approach is particularly relevant when the model that we have to hand has already been linearized. As we saw in Section 1.7, such linearization can lead to variance heterogeneity. To set the scene we consider the following example.

Example 2.6 Carr [1960] studied the catalytic isometrization of n-pentane to i-pentane in the presence of hydrone. One model proposed for the rate of this reaction, based on a single-site mechanism, was

$$r \approx \frac{\theta_1\theta_3(x_2 - x_3/1.632)}{1 + \theta_2 x_1 + \theta_3 x_2 + \theta_4 x_3}, \tag{2.160}$$

where r is the rate of disappearance of n-pentane [in g/(g catalyst)-hr]; x_1, x_2, and x_3 are the partial pressures of hydrogen, n-pentane, and i-pentane, respectively (in psia); θ_1 is a constant depending on the catalyst, and θ_2, θ_3, and θ_4 are equilibrium adsorption constants (in psia^{-1}). Carr [1960] used the following transformation to linearize the model:

$$y \approx \frac{x_2 - x_3/1.632}{r}$$

$$= \frac{1}{\theta_1\theta_3} + \frac{\theta_2}{\theta_1\theta_3} x_1 + \frac{1}{\theta_1} x_2 + \frac{\theta_4}{\theta_1\theta_3} x_3, \tag{2.161}$$

or

$$y \approx \beta_0 + \beta_1 x_1 + \beta_2 x_2 + \beta_3 x_3. \tag{2.162}$$

If a standard linear least-squares procedure is carried out using the data of Table 2.4, we obtain negative estimates of θ_2, θ_3, and θ_4, contrary to the usual physical interpretation. Box and Hill [1974] studied this model and argued,

Table 2.4 Reaction Rate for Catalytic Isomerization of *n*-Pentane to Isopentane[a]

Run No.	Partial Pressures (psia)			Rate
	x_1	x_2	x_3	r
1	205.8	90.9	37.1	3.541
2	404.8	92.9	36.3	2.397
3	209.7	174.9	49.4	6.694
4	401.6	187.2	44.9	4.722
5	224.9	92.7	116.3	.593
6	402.6	102.2	128.9	.268
7	212.7	186.9	134.4	2.797
8	406.2	192.6	134.9	2.451
9	133.3	140.8	87.6	3.196
10	470.9	144.2	86.9	2.021
11	300.0	68.3	81.7	.896
12	301.6	214.6	101.7	5.084
13	297.3	142.2	10.5	5.686
14	314.0	146.7	157.1	1.193
15	305.7	142.0	86.0	2.648
16	300.1	143.7	90.2	3.303
17	305.4	141.1	87.4	3.054
18	305.2	141.5	87.0	3.302
19	300.1	83.0	66.4	1.271
20	106.6	209.6	33.0	11.648
21	417.2	83.9	32.9	2.002
22	251.0	294.4	41.5	9.604
23	250.3	148.0	14.7	7.754
24	145.1	291.0	50.2	11.590

[a]From Carr [1960] with permission of the American Chemical Society.

from the wedge-shaped plot of the residual $y-\hat{y}$ versus $\hat{y}$ $(=\hat{\beta}_0+\cdots+\hat{\beta}_3x_3)$, that an unweighted analysis is inappropriate, as the variance of y increases with the mean. ■

Consider now the general nonlinear model

$$y_i = f(\mathbf{x}_i;\boldsymbol{\theta}) + \varepsilon_i,$$

where the ε_i are independently distributed as $N(0,\sigma_i^2)$. Box and Hill [1974] assumed that there is a set of transformed responses $y_i^{(\phi)}$ having constant variance σ^2. Using the Box–Cox power transformation, with $\lambda=\phi$, we see that (2.144) leads to

$$\sigma^2 \approx \sigma_i^2 (E[y_i])^{2\phi-2}$$

or

$$\sigma_i^2 \approx \sigma^2/w_i, \tag{2.163}$$

where

$$w_i = [f(\mathbf{x}_i;\boldsymbol{\theta})]^{2\phi-2}. \tag{2.164}$$

The likelihood function for the parameters $\boldsymbol{\theta},\phi$, and σ (or σ^2) is then given by

$$\begin{aligned} p(\mathbf{y}|\boldsymbol{\theta},\phi,\sigma) &= \frac{\left(\prod_i w_i\right)^{1/2}}{(2\pi\sigma^2)^{n/2}} \exp\left(-\frac{1}{2\sigma^2}\sum_{i=1}^n w_i[y_i - f(\mathbf{x}_i;\boldsymbol{\theta})]^2\right) \\ &\propto \left(\prod_i w_i\right)^{1/2} \sigma^{-n} \exp\left(-\frac{1}{2\sigma^2} S(\boldsymbol{\theta},\phi)\right), \end{aligned} \tag{2.165}$$

say. Three methods of estimation are available. The first is the usual method of maximum likelihood, in which we maximize the logarithm of (2.165) with respect to $\boldsymbol{\theta},\phi$, and σ^2. The second method is to eliminate σ by using the concentrated log-likelihood function (cf. Section 2.2.3). Setting $\sigma^2 = S(\boldsymbol{\theta},\phi)/n$, and ignoring constants, we obtain

$$M(\boldsymbol{\theta},\phi) = \frac{1}{2}\sum_i \log w_i - \frac{n}{2}\log S(\boldsymbol{\theta},\phi), \tag{2.166}$$

which we maximize with respect to $\boldsymbol{\theta}$, and ϕ. This again leads to the maximum-likelihood estimates $\hat{\boldsymbol{\theta}},\hat{\phi}$, and $\hat{\sigma}^2 = S(\hat{\boldsymbol{\theta}},\hat{\phi})/n$. The third method is to use a Bayesian analysis and derive a joint posterior density for the parameters.

This approach was utilized by Box and Hill [1974] and Pritchard et al. [1977], and we now discuss their methods.

Box and Hill used the prior distribution (2.104), for $\boldsymbol{\theta}$ and σ, and an independent uniform prior for ϕ so that

$$p(\boldsymbol{\theta}, \phi, \sigma) \propto \sigma^{-1}.$$

The joint posterior distribution is then given by (c.f. (2.96)]

$$\begin{aligned} p(\boldsymbol{\theta}, \phi, \sigma \mid \mathbf{y}) &\propto p(\mathbf{y} \mid \boldsymbol{\theta}, \phi, \sigma) p(\boldsymbol{\theta}, \phi, \sigma) \\ &\propto \left(\prod_i w_i \right)^{1/2} \sigma^{-(n+1)} \exp\left(-\frac{1}{2\sigma^2} S(\boldsymbol{\theta}, \phi) \right), \end{aligned} \tag{2.167}$$

and σ can be integrated out [see (2.107)] to get

$$p(\boldsymbol{\theta}, \phi \mid \mathbf{y}) \propto \left(\prod_i w_i \right)^{1/2} S(\boldsymbol{\theta}, \phi)^{-n/2}. \tag{2.168}$$

We recall [see (2.164)] that w_i is a function of $\boldsymbol{\theta}$ and ϕ. Pritchard et al. [1977] suggested using the modal parameter values of the above posterior distribution. They proposed finding these values by a $(p+1)$-dimensional search on the parameter space, which they call the *direct procedure* using derivative-free or gradient methods. However, as (2.166) is the logarithm of (2.168), we see that these modal estimates are, in fact, the maximum-likelihood estimates. Therefore all three methods of estimation give identical point estimates. Also, if the normal approximation for the posterior density of Section 2.7.4 is used, the approximate Bayesian h.p.d. region for $\boldsymbol{\theta}$ coincides with the usual asymptotic confidence region obtained from maximum-likelihood theory.

For a linear model with $\boldsymbol{\beta}$ replaced by $\boldsymbol{\theta}$, we have $f(\mathbf{x}_i; \boldsymbol{\theta}) = \mathbf{x}_i' \boldsymbol{\theta}$, or $\mathbf{f}(\boldsymbol{\theta}) = \mathbf{X}\boldsymbol{\theta}$. This leads to

$$S(\boldsymbol{\theta}, \boldsymbol{\phi}) = (\mathbf{y} - \mathbf{X}\boldsymbol{\theta})' \mathbf{W} (\mathbf{y} - \mathbf{X}\boldsymbol{\theta}), \tag{2.169}$$

where $\mathbf{W} = \operatorname{diag}(w_1, w_2, \ldots, w_n)$ and $w_i = (\mathbf{x}_i'\boldsymbol{\theta})^{2\phi - 2}$. Suppose, then, that we have a current estimate $\check{\boldsymbol{\theta}}$ of $\boldsymbol{\theta}$ and we estimate $\mathbf{x}_i'\boldsymbol{\theta}$ by $\mathbf{x}_i'\check{\boldsymbol{\theta}}$ $(= \check{y}_i)$ in w_i. With this estimate of $\mathbf{W}$ we can use the theory of weighted least squares (Seber [1977: 61]) to obtain

$$\begin{aligned} S(\boldsymbol{\theta}, \phi) &= (\mathbf{y} - \mathbf{X}\boldsymbol{\theta}^+)' \mathbf{W} (\mathbf{y} - \mathbf{X}\boldsymbol{\theta}^+) + (\boldsymbol{\theta}^+ - \boldsymbol{\theta})' \mathbf{X}' \mathbf{W} \mathbf{X} (\boldsymbol{\theta}^+ - \boldsymbol{\theta}) \\ &= c + d, \end{aligned}$$

where

$$\boldsymbol{\theta}^+ = (\mathbf{X}'\mathbf{W}\mathbf{X})^{-1} \mathbf{X}' \mathbf{W} \mathbf{y}. \tag{2.170}$$

Hence from (2.167) we have, approximately,

$$p(\boldsymbol{\theta}, \phi, \sigma \mid \mathbf{y}) \propto |\mathbf{W}|^{1/2} \sigma^{-(n+1)} \exp\left(-\frac{1}{2\sigma^2}(c+d)\right).$$

Since $\mathbf{W}$ is now treated as a function of ϕ only, we can integrate out $\boldsymbol{\theta}$ using the properties of the multivariate normal integral, and then integrate out σ using (1.5) to obtain

$$\begin{aligned} p(\phi \mid \mathbf{y}) &\propto \frac{|\mathbf{W}|^{1/2}}{|\mathbf{X}'\mathbf{W}\mathbf{X}|^{1/2}} c^{-(n-p)/2} \\ &= \frac{|\mathbf{W}|^{1/2}}{|\mathbf{X}'\mathbf{W}\mathbf{X}|^{1/2}} [(\mathbf{y} - \mathbf{X}\boldsymbol{\theta}^+)'\mathbf{W}(\mathbf{y} - \mathbf{X}\boldsymbol{\theta}^+)]^{-(n-p)/2}. \end{aligned} \tag{2.171}$$

A suitable estimate of ϕ is then the mode, ϕ^+, of the above distribution. We then re-estimate w_i and repeat the cycle, thus giving the following iterative procedure of Box and Hill [1974]:

(i) Let $w_i = (\breve{y}_i)^{2\phi - 2}$.
(ii) Choose a starting value of ϕ and calculate the w_i.
(iii) Compute $\boldsymbol{\theta}^+$ from (2.170), and evaluate $p(\phi \mid \mathbf{y})$ of (2.171).
(iv) Return to step (ii) and continue until a ϕ^+ is found maximizing $p(\phi \mid \mathbf{y})$. (In w_i the same initial value of $\breve{y}_i$ is used: only ϕ changes.)
(v) With ϕ^+ as the value of ϕ, calculate w_i and then $\boldsymbol{\theta}^+$. Using this value of $\boldsymbol{\theta}^+$ as the new value of $\breve{\boldsymbol{\theta}}$, return to step (i).
(vi) Start a new iteration to obtain a new value of ϕ using the previous value as a starting value. Continue the whole cycle until the iterations converge on a common value of $\boldsymbol{\theta}$. Use $\boldsymbol{\theta}^+$, the current estimate of $\boldsymbol{\theta}$.

The above procedure, called the *staged procedure* by Pritchard et al. [1977], can be started by using y_i instead of $\breve{y}_i$ initially in w_i. Steps (ii)–(iv) may be replaced by a standard algorithm for maximizing $p(\phi \mid \mathbf{y})$.

The staged procedure can also be used for fitting nonlinear models using an approximate linearization with $\mathbf{X} = \mathbf{F}_{\bullet}(\boldsymbol{\theta})$ [cf. (2.14)]. However, this requires further iterations in finding $\boldsymbol{\theta}^+$, this now being the value of $\boldsymbol{\theta}$ which minimizes $\sum_i w_i \{y_i - f(\mathbf{x}_i; \boldsymbol{\theta})\}^2$. The direct procedure (i.e. maximum likelihood) is therefore preferred in this case.

The adequacy of the weights can be checked using residual plots. However, for this purpose we use weighted observations $\sqrt{w_i} y_i$, the fitted values $\sqrt{w_i} f(\mathbf{x}_i; \hat{\boldsymbol{\theta}})$ $(= \sqrt{w_i}\, \hat{y}_i)$, and the weighted residuals $\sqrt{w_i}(y_i - \hat{y}_i)$, where $w_i = (\hat{y}_i)^{2\hat{\phi} - 2}$. If the model is highly nonlinear, then projected residuals may be needed (cf. Section 4.6.4).

If the data are replicated so that

$$y_{ij} = f(\mathbf{x}_i; \boldsymbol{\theta}) + \varepsilon_{ij} \qquad (i = 1, 2, \ldots, n \quad j = 1, 2, \ldots, J_i),$$

where the ε_{ij}, $j = 1, 2, \ldots, J_i$, are i.i.d. $N(0, \sigma^2/w_i)$, then a rough test of the adequacy of the model can be made using [cf. (2.43)]

$$F = \frac{Q_H - Q}{Q} \cdot \frac{N - n}{n - p}, \tag{2.172}$$

where $N = \sum_i J_i$,

$$Q_H = \sum_i \sum_j w_i [y_{ij} - f(\mathbf{x}_i; \boldsymbol{\theta}_e)]^2,$$

and

$$Q = \sum_i \sum_j w_i (y_{ij} - \bar{y}_{i \cdot})^2.$$

Here $\boldsymbol{\theta}_e$ is $\hat{\boldsymbol{\theta}}$ or $\boldsymbol{\theta}^+$, depending on which procedure is used, and the weights w_i are appropriately estimated. If the model is linear, or there exists a parametrization for which it can be adequately approximated by a linear model, then F will be roughly distributed as $F_{n-p, N-n}$ when the model is valid. Pritchard et al. [1977] suggested reducing the usual number of degrees of freedom, $N - n$, by 1 to allow for the estimation of ϕ.

Example 2.6 **(*Continued*)** Pritchard et al. [1977] considered fitting three models to the data of Table 2.4—the nonlinear model (2.160), the linearized model (2.161), and the following linearized model:

$$z = \frac{1}{r} = \beta_0 u_0 + \beta_1 u_1 + \beta_2 u_2 + \beta_3 u_3, \tag{2.173}$$

where $u_0 = (x_2 - x_3/1.632)^{-1}$ and $u_i = u_0 x_i$ $(i = 1, 2, 3)$. In the nonlinear model both unweighted and weighted (direct search) methods were used for fitting $\boldsymbol{\theta}$, while in the linearized models $\boldsymbol{\beta}$ was estimated using weighted and unweighted methods and then transformed to give an estimate of $\boldsymbol{\theta}$. The results are given in Table 2.5: the estimates of θ_2, θ_3, and θ_4 have been multiplied by 14.7 to convert the units from psia^{-1} to atm^{-1}. The original units, psia, were used in Pritchard and Bacon [1977: p. 112] for the model (2.173).

The first thing we notice from Table 2.5 is the inappropriateness of unweighted least squares for the linearized models. The estimates of θ_2, θ_3, and θ_4 not only are negative, but they are also very different for the two linearized models described by columns (i) and (ii). The difference is not unexpected, since compared with (2.173), (2.161) is a weighted least-squares model with weights $(x_2 - x_3/1.632)^2$. As noted by Pritchard et al. [1977], this emphasizes the danger of using a linearized model to find an initial approximation of $\boldsymbol{\theta}$ for the nonlinear

Table 2.5 Least-Squares Estimates of Parameters from Fitting Linearized and Nonlinear Models to Carr's Data

	Linearized Models[a]				Nonlinear Model[a]		PTBS Model[b]
	(2.161)	(2.173)			(2.160)		(2.145)
	Unweighted	Unweighted	Staged	Direct	Unweighted	Direct	
Parameter	(i)	(ii)	(iii)	(iv)	(v)	(vi)	(vii)
θ_1	193.0	16.3	40.0	39.5	35.9($\pm$16.1)[c]	38.6($\pm$21.3)[c]	39.2
θ_2 (atm^{-1})	−0.40	−0.043	0.73	0.76	1.04($\pm$5.15)	0.65($\pm$2.43)	0.043
θ_3 (atm^{-1})	−0.03	−0.014	0.35	0.36	0.55($\pm$2.88)	0.32($\pm$1.33)	0.021
θ_4 (atm^{-1})	−1.09	−0.098	1.82	1.86	2.46($\pm$11.99)	1.57($\pm$5.73)	0.104
ϕ	1.0	1.0	0.81	−0.91	1.0	0.72($\pm$0.38)	0.71 (λ_1)
F-test[d]	10.6	75.8	2.61	2.61	1.82	1.82	

[a] From Pritchard et al. [1977].
[b] From Caroll and Ruppert [1984].
[c] Approximate 95% h.p.d. intervals obtained from marginal posterior densities.
[d] 5% point of the reference F-distribution is 8.7.

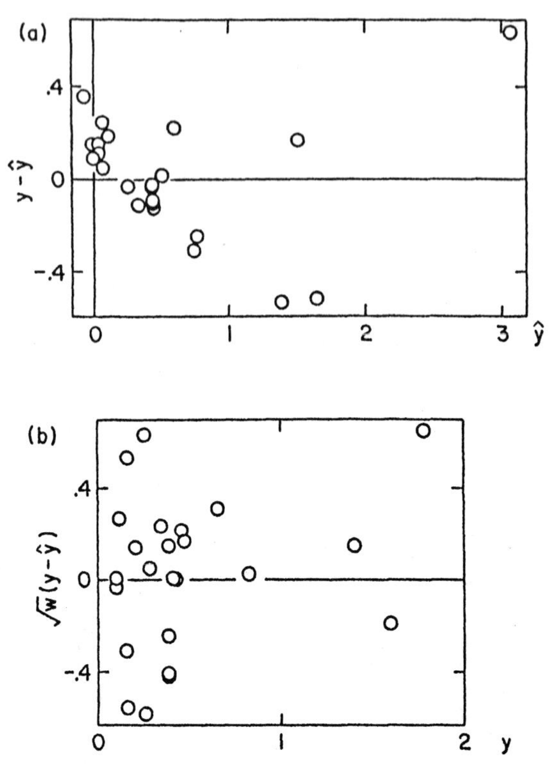

Figure 2.4 (*a*) Residual plot from unweighted fitting of the linearized model (2.173). (*b*) Residual plot from weighted fitting of the model (2.173) by the direct procedure. From Pritchard et al. [1977].

model. However, a good minimization algorithm will often achieve convergence with poor starting values. The lack of homogeneity for the linearized model (2.173) is demonstrated clearly by the unweighted residual plot in Fig. 2.4*a*. This problem seems to be corrected by the weighted procedure (Fig. 2.4*b*).

The second feature of Table 2.5 is the wide h.p.d. intervals given in parentheses in columns (v) and (vi). Those in column (vi) are based on the approximation (cf. Section 2.7.4) $\boldsymbol{\gamma} = (\boldsymbol{\theta}', \phi)' \sim N_{p+1}(\hat{\boldsymbol{\gamma}}, \hat{\mathbf{V}})$, where

$$\hat{\mathbf{V}} = \left[\frac{-\partial^2 \log p(\boldsymbol{\gamma}|\mathbf{y})}{\partial\boldsymbol{\gamma}\,\partial\boldsymbol{\gamma}'}\right]_{\hat{\boldsymbol{\gamma}}}^{-1}, \tag{2.174}$$

and $p(\boldsymbol{\gamma}|\mathbf{y})$, with mode $\hat{\boldsymbol{\gamma}}$, is given by (2.168). The matrix $\hat{\mathbf{V}}$ can be estimated numerically or else computed using analytic expressions given by Pritchard et al. [1977] in their appendix. With the above normal approximation, the marginal posterior distributions of the θ_r and ϕ are univariate normal, and h.p.d. intervals are then readily calculated. The intervals in column (v) were obtained in a similar manner, but with $\phi = 1$ in (2.168) and $\boldsymbol{\gamma} = \boldsymbol{\theta}$. Although it is comforting to see that the two sets of weighted estimates for the linearized model are close together [column (iii) and (iv)], h.p.d. intervals for the *original* parameters may not be readily calculated. This is a serious shortcoming of linearization.

We have seen that linearization leads to heterogeneity of variance. However, we might ask whether there is any intrinsic heterogeneity in the nonlinear model itself. According to Pritchard and Bacon [1977] the unweighted residual plot does not suggest this, and the estimates from the unweighted and weighted analyses [columns (v) and (vi)] do not appear to be very different. Yet the value $\phi = 1$ is only just contained in the h.p.d. interval 0.72 ± 0.38 for ϕ, and the h.p.d. intervals for θ_2, θ_3, and θ_4 are about half as wide for the weighted analysis. This suggests the presence of some heterogeneity.

Of the 24 observations there is just one set of four approximate replicates (runs 15 to 18 inclusive in Table 2.4), so that the approximate F-statistic of (2.172) can be calculated for each model. These F-ratios are given in Table 2.5 and (with $N = 24$, $n = 21$, and $p = 4$) are to be compared with $F_{17,3}^{0.05} = 8.7$ (or $F_{17,2}^{0.05} = 9.4$ if one degree of freedom is allowed for the estimation of ϕ). Although this test is only approximate and the denominator has few degrees of freedom, it is interesting to note that all the weighted analyses lead to low values of F, thus suggesting adequate fits. This supports the evidence from the weighted-residual plot given by Fig. 2.5.

So far our discussion has been in terms of the PTWLS model (2.146) with $\lambda_2 = 2 - 2\phi$. Since ϕ is estimated to be -0.72 in column (vi) using a linearized model based on the inverse transformation, we would expect $\lambda_1 = -\phi \approx 0.72$ for the PTBS model (2.145). Column (viii) from Carroll and Ruppert [1984] gives the estimate $\lambda_1 = .71$. However, it should be noted that Carroll and Ruppert [1984] could not reproduce column (v). Their estimates for unweighted nonlinear least squares using several programs were very close to their results in column (vii). They suspected that the differences in the estimates of θ_2, θ_3, and θ_4 between columns (vi) and (vii) could be due to the presence of several local maxima. ■

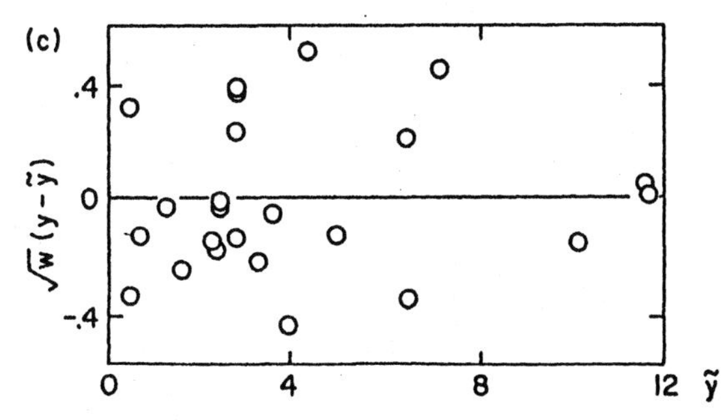

Figure 2.5 Residual plot from weighted fitting of the nonlinear model (2.160) by the direct procedure. From Pritchard et al. [1977].

The problem of heterogeneity in linear models has received considerable attention in the literature. Various ways of modeling the heterogeneity as a function of the mean, in addition to the method of Box and Hill [1974] considered above, have been suggested (see Carroll and Ruppert [1982a] for references and a robust estimation procedure). Various estimation procedures have been proposed, such as ordinary, weighted, and generalized least squares, as well as maximum likelihood (Jobson and Fuller [1980], Carroll and Ruppert [1982b]). However, in the case of nonlinear models there seems little point in using a linearizing transformation if it leads to heterogeneity. Also, as already noted in the above example, the computational simplicity that comes with linear models is often lost when one is faced with obtaining interval estimates for the original rather than the transformed parameters. Currie [1982], for example, studied the Michaelis–Menten model

$$E[y_i] = \frac{vx_i}{k + x_i},$$

and found that the best linearizing inverse transformation

$$E\left[\frac{x_i}{y_i}\right] = \frac{k}{v} + \frac{1}{v}x_i$$

produced unreliable estimates. He came to the conclusion that transformations should not be used except in cases where they stabilize the variance.

We note that the above iterative procedure of Box and Hill [1974] is the iteratively reweighted least squares (IRLS) method of Sections 2.2.2 and 2.3. The latter technique can be applied to more general models with heterogeneity, as described in Section 2.8.8. But first we discuss further the effect of linearizing transformations on estimation.

2.8.7 Prediction and Transformation Bias

It was noted above that transforming to linearity is not appropriate if the variance is not stabilized. Even then one is often more interested in the original parameters, so that the point and interval estimators of the transformed parameters need to be "detransformed." Where the emphasis is on estimating $E[y]$ for a given $\mathbf{x}$, then simply detransforming the usual linear least-squares prediction equation can lead to severe bias, as we shall see below.

Suppose that

$$z_i = h(y_i) = \mathbf{x}_i'\boldsymbol{\beta} + \varepsilon_i \qquad (i = 1, 2, \ldots, n),$$

where the ε_i are i.i.d. $N(0, \sigma^2)$, holds for some known monotonic function h. We can assume, without loss of generality, that h is monotonically increasing; otherwise we can replace $h(y)$ by $-h(y)$. Writing $\mathbf{z} = \mathbf{X}\boldsymbol{\beta} + \boldsymbol{\varepsilon}$, then $\hat{\boldsymbol{\beta}} = (\mathbf{X}'\mathbf{X})^{-1}\mathbf{X}'\mathbf{z}$

is the least-squares estimate of $\boldsymbol{\beta}$. If y_0 is a new and independent response at $\mathbf{x} = \mathbf{x}_0$ and $z_0 = h(y_0)$, then $\hat{z}_0 = \mathbf{x}_0'\hat{\boldsymbol{\beta}}$ is an unbiased estimator of $E[z_0]$. However, we shall now see that $\hat{y}_0 = h^{-1}(\hat{z}_0)$ is a biased estimator of $E[y_0]$. Since $z_0 \sim N(\mathbf{x}_0'\boldsymbol{\beta}, \sigma^2 v_0)$, where $v_0 = \mathbf{x}_0'(\mathbf{X}'\mathbf{X})^{-1}x_0$, z_0 has median $\mathbf{x}_0'\boldsymbol{\beta}$. Hence, from

$$\begin{aligned}\tfrac{1}{2} &= \text{pr}\,[h(y_0) \leq \mathbf{x}_0'\boldsymbol{\beta}] \\ &= \text{pr}\,[y_0 \leq h^{-1}(\mathbf{x}_0'\boldsymbol{\beta})],\end{aligned}$$

it follows that $h^{-1}(\mathbf{x}_0'\boldsymbol{\beta})$ is the median of y_0, where the distribution of y_0 is generally skewed. Thus $\hat{y}_0$ tends to estimate the median rather than the mean of the distribution of y_0. For example, if $h(y) = \log y$, then

$$\begin{aligned}E[\hat{y}_0] &= E[e^{\hat{z}_0}] \\ &= E[e^{\hat{z}_0 t}]_{t=1} \\ &= \exp(\mathbf{x}_0'\boldsymbol{\beta} + \tfrac{1}{2}\sigma^2 v_0) && (2.175) \\ &\to \exp(\mathbf{x}_0'\boldsymbol{\beta}) \qquad (\text{as} \quad n \to \infty) \\ &= h^{-1}(\mathbf{x}_0'\boldsymbol{\beta}), && (2.176)\end{aligned}$$

and

$$\begin{aligned}E[y_0] &= E[e^{z_0}] \\ &= \exp(\mathbf{x}_0'\boldsymbol{\beta} + \tfrac{1}{2}\sigma^2). && (2.177)\end{aligned}$$

From (2.176) we see that $\hat{y}_0$ consistently estimates the median of y_0 but underestimates $E[y_0]$ of (2.177). In highlighting the above problem, Miller [1984] recommended using

$$\begin{aligned}\hat{E}[y_0] &= \exp(\mathbf{x}_0'\hat{\boldsymbol{\beta}} + \tfrac{1}{2}s^2) \\ &= \hat{y}_0 e^{s^2/2},\end{aligned}$$

where $s^2 = \|\mathbf{z} - \mathbf{X}\hat{\boldsymbol{\beta}}\|^2/(n-p)$, as an estimator of $E[y_0]$. Although this estimator is still biased, the adjustment factor $e^{s^2/2}$ substantially reduces the bias. Miller produced similar results for $h(y) = y^k$, where $k = \pm N^{-1}$ and N is a positive integer. He showed that $E[y_0] = g(\mathbf{x}_0'\boldsymbol{\beta}, \sigma^2)$, where g is a straightforward function, and suggested the estimator $\hat{E}[y_0] = g(\mathbf{x}_0'\hat{\boldsymbol{\beta}}, s^2)$. For the square-root, cube-root, and logarithmic transformations, Miller found that his low-bias estimators were numerically very similar to the very complex minimum-variance unbiased estimators.

A nonparametric approach to the problem is given by Duan [1983]. He proposed the "smearing" estimate

$$\hat{E}[y_0] = \frac{1}{n}\sum_{i=1}^{n} h^{-1}(\mathbf{x}_0'\hat{\boldsymbol{\beta}} + \hat{\varepsilon}_i).$$

2.8.8 Generalized Least-Squares Model

Consider the general nonlinear model

$$\mathbf{y} = \mathbf{f}(\boldsymbol{\theta}) + \boldsymbol{\varepsilon},$$

where $\mathscr{E}[\boldsymbol{\varepsilon}] = \mathbf{0}$ and $\mathscr{D}[\boldsymbol{\varepsilon}] = \sigma^2 \mathbf{V}(\boldsymbol{\theta})$. It is assumed that the functional form of $\mathbf{V}$ is known and that $\mathbf{V}(\boldsymbol{\theta})$ is positive definite for $\boldsymbol{\theta}$ sufficiently close to $\boldsymbol{\theta}^*$, the true value of $\boldsymbol{\theta}$. Define

$$S[\boldsymbol{\theta}, \mathbf{V}(\boldsymbol{\theta})] = [\mathbf{y} - \mathbf{f}(\boldsymbol{\theta})]' \mathbf{V}^{-1}(\boldsymbol{\theta}) [\mathbf{y} - \mathbf{f}(\boldsymbol{\theta})]. \tag{2.178}$$

Then one method of estimating $\boldsymbol{\theta}$ is to use iteratively reweighted least squares. If $\boldsymbol{\theta}^{(a)}$ is the current estimate of $\boldsymbol{\theta}$, then the next estimate $\boldsymbol{\theta}^{(a+1)}$ is obtained by minimizing

$$S_a(\boldsymbol{\theta}) = S[\boldsymbol{\theta}, \mathbf{V}(\boldsymbol{\theta}^{(a)})]. \tag{2.179}$$

This can be done several ways. Using the generalized least-squares method of Section 2.1.4, we have the "nested" updating scheme

$$\begin{aligned} \boldsymbol{\delta}^{(a,b)} &= \boldsymbol{\theta}^{(a,b+1)} - \boldsymbol{\theta}^{(a,b)} \\ &= [\mathbf{F}_{\bullet}^{(a,b)\prime} \mathbf{V}^{-1}(\boldsymbol{\theta}^{(a)}) \mathbf{F}_{\bullet}^{(a,b)}]^{-1} \mathbf{F}_{\bullet}^{(a,b)\prime} \mathbf{V}^{-1}(\boldsymbol{\theta}^{(a)}) [\mathbf{y} - \mathbf{f}(\boldsymbol{\theta}^{(a,b)})], \end{aligned} \tag{2.180}$$

where $\mathbf{F}_{\bullet}(\boldsymbol{\theta}) = \partial \mathbf{f}(\boldsymbol{\theta})/\partial \boldsymbol{\theta}'$, $\mathbf{F}_{\bullet}^{(a,b)} = \mathbf{F}_{\bullet}(\boldsymbol{\theta}^{(a,b)})$, and $\boldsymbol{\theta}^{(a,1)} = \boldsymbol{\theta}^{(a)}$. The converged value of $\boldsymbol{\theta}^{(a,b)}$ as $b \to \infty$ becomes the new $\boldsymbol{\theta}^{(a)}$ namely $\boldsymbol{\theta}^{(a+1)}$, to be substituted into $\mathbf{V}(\boldsymbol{\theta})$. An alternative algorithm is to use the updating formula

$$\begin{aligned} \boldsymbol{\delta}^{(a)} &= \boldsymbol{\theta}^{(a+1)} - \boldsymbol{\theta}^{(a)} \\ &= [\mathbf{F}_{\bullet}^{(a)\prime} \mathbf{V}^{-1}(\boldsymbol{\theta}^{(a)}) \mathbf{F}_{\bullet}^{(a)}]^{-1} \mathbf{F}_{\bullet}^{(a)\prime} \mathbf{V}^{-1}(\boldsymbol{\theta}^{(a)}) [\mathbf{y} - \mathbf{f}(\boldsymbol{\theta}^{(a)})]. \end{aligned} \tag{2.181}$$

The difference between (2.181) and (2.180) is that (2.181) takes a single Gauss–Newton step towards the minimum of $S_a(\boldsymbol{\theta})$ before updating $\mathbf{V}(\boldsymbol{\theta})$ and going on to $S_{a+1}(\boldsymbol{\theta})$, whereas (2.180) actually minimizes $S_a(\boldsymbol{\theta})$ (or, in practice, gets closer to the minimum). We have seen no guidance in the literature about whether it is better to update $\mathbf{V}^{-1}(\boldsymbol{\theta})$ at every iteration or to go for several iterations with the same value of $\mathbf{V}^{-1}(\boldsymbol{\theta})$.

If the algorithms converge, then $\boldsymbol{\delta} \to \mathbf{0}$ and $\boldsymbol{\theta}^{(a,b)}$ and $\boldsymbol{\theta}^{(a)}$ converge to the solution, $\tilde{\boldsymbol{\theta}}$ say, of [cf. (2.181)]

$$\mathbf{F}_{\bullet}'(\boldsymbol{\theta}) \mathbf{V}^{-1}(\boldsymbol{\theta}) [\mathbf{y} - \mathbf{f}(\boldsymbol{\theta})] = \mathbf{0},$$

as in (2.74). Under fairly general conditions (e.g. Fedorov [1974] for the case when $\mathbf{V}(\boldsymbol{\theta})$ is diagonal) it can be shown that $\sqrt{n}(\tilde{\boldsymbol{\theta}} - \boldsymbol{\theta}^*)$ is asymptotically

$N_p(\boldsymbol{\theta}, \sigma^2 \boldsymbol{\Omega})$, where $\boldsymbol{\Omega}$ is consistently estimated by

$$\boldsymbol{\Omega} = \frac{1}{n}[\mathbf{F}'_{\cdot}(\tilde{\boldsymbol{\theta}})\mathbf{V}^{-1}(\tilde{\boldsymbol{\theta}})\mathbf{F}_{\cdot}(\tilde{\boldsymbol{\theta}})]^{-1}.$$

A consistent estimate of σ^2 is then given by

$$\tilde{\sigma}^2 = \frac{1}{n-p} S[\tilde{\boldsymbol{\theta}}, \mathbf{V}(\tilde{\boldsymbol{\theta}})]. \tag{2.182}$$

Writing

$$S[\boldsymbol{\theta}, \mathbf{V}(\boldsymbol{\theta})] = \sum_i \sum_j [y_i - f(\mathbf{x}_i; \boldsymbol{\theta})][y_j - f(\mathbf{x}_j; \boldsymbol{\theta})] v^{ij},$$

we see that

$$\frac{\partial S[\boldsymbol{\theta}, \mathbf{V}(\boldsymbol{\theta})]}{\partial \boldsymbol{\theta}} = 2\mathbf{F}'_{\cdot}(\boldsymbol{\theta})\mathbf{V}^{-1}(\boldsymbol{\theta})[\mathbf{y} - \mathbf{f}(\boldsymbol{\theta})]$$

$$+ \Sigma \Sigma \{y_i - f(\mathbf{x}_i; \boldsymbol{\theta})][y_j - f(\mathbf{x}_j; \boldsymbol{\theta})] \frac{\partial v^{ij}}{\partial \boldsymbol{\theta}}, \tag{2.183}$$

so that $\tilde{\boldsymbol{\theta}}$ does not minimize $S[\boldsymbol{\theta}, \mathbf{V}(\boldsymbol{\theta})]$. Only the first term of the above equation vanishes at $\boldsymbol{\theta} = \tilde{\boldsymbol{\theta}}$. Interestingly enough, $\hat{\boldsymbol{\theta}}$, the value of $\boldsymbol{\theta}$ minimizing $S[\boldsymbol{\theta}, \mathbf{V}(\boldsymbol{\theta})]$, is generally not a consistent estimator of $\boldsymbol{\theta}$ (Fedorov [1972]: p. 45] with $d_{ii} = 0$ and $\mathbf{V}(\boldsymbol{\theta})$ diagonal).

In conclusion we note that the use of a Cholesky decomposition of $\mathbf{V}(\boldsymbol{\theta}^{(a)})$ to convert (2.180) or (2.181) to a usual linear-regression computation is discussed in Section 2.1.4. A by-product of the linear regression is

$$\tilde{\mathscr{D}}[\tilde{\boldsymbol{\theta}}] = \tilde{\sigma}^2 (\mathbf{F}'_{\cdot}\mathbf{V}^{-1}\mathbf{F}_{\cdot})^{-1}_{\tilde{\boldsymbol{\theta}}},$$

an estimate of the asymptotic dispersion matrix of $\tilde{\boldsymbol{\theta}}$.

CHAPTER 3

Commonly Encountered Problems

3.1 PROBLEM AREAS

At this point we briefly consider some problem areas commonly encountered when fitting nonlinear regression models. We shall discuss problems associated with (a) the convergence of the iterative methods of computation, (b) the finite application of asymptotic inference theory, and (c) ill-conditioning in parameter estimation. The difficulties arising in (c) are akin to multicollinearity in linear regression and therefore may be due to the choice of the data points $\mathbf{x}_i$ in the model (2.1). However, unlike the linear case, they may also be due to the nature of the model. Each of these closely interconnected topics will be expanded on as they arise in later chapters.

3.2 CONVERGENCE OF ITERATIVE PROCEDURES

Most of the methods of estimation discussed in the previous chapter, namely least-squares and maximum-likelihood estimation, robust estimation, and Bayesian estimation using posterior modes, require an estimate ($\hat{\boldsymbol{\theta}}$ say) obtained by maximizing or minimizing an appropriate function of $\boldsymbol{\theta}$, say $h(\boldsymbol{\theta})$. For example, in least squares we minimize $h(\boldsymbol{\theta}) = \|\mathbf{y} - \mathbf{f}(\boldsymbol{\theta})\|^2$. We shall refer only to minimization, as maximizing $h(\boldsymbol{\theta})$ is equivalent to minimizing $-h(\boldsymbol{\theta})$.

The function $h(\boldsymbol{\theta})$ can rarely be minimized analytically, and an iterative procedure for finding $\hat{\boldsymbol{\theta}}$ is usually required. We start with an initial guess $\boldsymbol{\theta}^{(1)}$ and sequentially find $\boldsymbol{\theta}^{(2)}, \boldsymbol{\theta}^{(3)}, \ldots$ in such a way that the sequence converges to $\hat{\boldsymbol{\theta}}$. Obtaining $\boldsymbol{\theta}^{(a+1)}$ using $\boldsymbol{\theta}^{(a)}$ (and possibly previous iterations) is called the ath iteration. In practice, at some finite value of a we decide that $\boldsymbol{\theta}^{(a)}$ is "close enough" to $\hat{\boldsymbol{\theta}}$ and use $\boldsymbol{\theta}^{(a)}$ as an approximation for $\hat{\boldsymbol{\theta}}$. We discuss minimization algorithms in Chapters 13 to 15.

Most of the algorithms we use can be considered as adaptations of, or approximations to, the classical Newton method described in Section 2.1.3

[see (2.33) with $S(\boldsymbol{\theta})$ replaced by $h(\boldsymbol{\theta})$]. The essence of the method is to approximate $h(\boldsymbol{\theta})$ by a quadratic $q_h^{(a)}(\boldsymbol{\theta})$ in the neighborhood of the ath iterate $\boldsymbol{\theta}^{(a)}$. The next iterate $\boldsymbol{\theta}^{(a+1)}$ is found by minimizing $q_h^{(a)}(\boldsymbol{\theta})$ with respect to $\boldsymbol{\theta}$. Many adaptations and protective strategies are used to transform this basic algorithm into algorithms which are useful for general smooth functions of $\boldsymbol{\theta}$. Because of the underlying Newton method, it is not surprising that the resulting algorithms work best on functions that are close to quadratic. Since any smooth function is approximately quadratic in a small enough neighborhood of $\hat{\boldsymbol{\theta}}$ (e.g., take a Taylor expansion about $\hat{\boldsymbol{\theta}}$ rather than $\boldsymbol{\theta}^{(a)}$), we find that in theory the Newton method will converge provided the initial value $\boldsymbol{\theta}^{(1)}$ is close enough to $\hat{\boldsymbol{\theta}}$.

It is helpful to think of function minimization geometrically. For two-dimensional $\boldsymbol{\theta}$ we can take the θ_1 and θ_2 axes as horizontal and the function axis h as vertical, so that $h(\boldsymbol{\theta})$ represents a three-dimensional surface. Then, if the absolute minimum $\hat{\boldsymbol{\theta}}$ is in the interior of the parameter space, the function-minimization problem is equivalent to finding the bottom of the deepest valley in the surface. However, as we see from the following example, our function may have several valleys, so that we end up at a local rather than a global minimum.

Example 3.1 Consider the function

$$h(\boldsymbol{\theta}) = S(\alpha, \beta) = \sum_{i=1}^{4} (y_i - \alpha e^{-\beta x_i})^2, \tag{3.1}$$

where the (x_i, y_i) are $(-2, 0)$, $(-1, 1)$, $(1, -0.9)$, (2.0) (suggested by S. P. Fitzpatrick). The surface contours for S, given in Fig. 3.1, indicate two valleys representing two relative minima. ■

As function-minimization algorithms tend to head off in a downhill direction from $\boldsymbol{\theta}^{(1)}$, there is a tendency to end up at the bottom of the valley nearest to $\boldsymbol{\theta}^{(1)}$. As noted above, the iteration process is based upon a quadratic approximation $q_S^{(a)}(\boldsymbol{\theta})$ of $S(\boldsymbol{\theta})$. Therefore the rate of convergence will depend on how well the contours through $\boldsymbol{\theta}^{(a)}$ of the two functions match up for each a. We saw, in Section 2.1.3, that the contours of $q_S^{(a)}(\boldsymbol{\theta})$ are concentric ellipsoids with center $\boldsymbol{\theta}^{(a+1)}$.

Example 3.2 Watts [1981] fitted the Michaelis–Menten model

$$y = \frac{\theta_1 x}{\theta_2 + x} + \varepsilon \tag{3.2}$$

to the data in Table 3.1 by least squares. This model relates the rate of formation of product (y = enzyme velocity) in an enzyme-catalyzed chemical reaction to

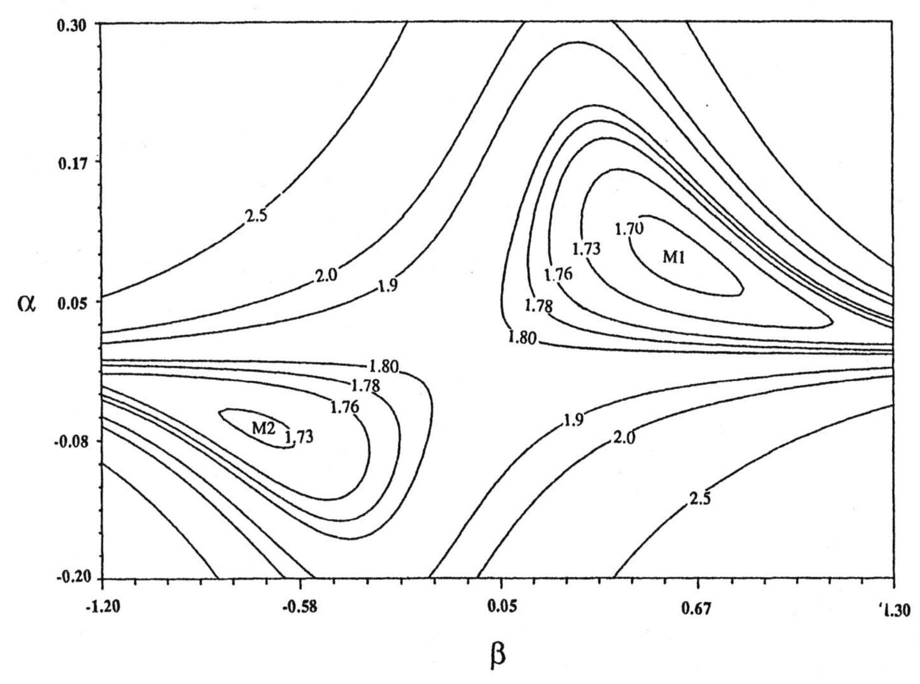

Figure 3.1 Surface contours for $S(\alpha, \beta)$ of Equation (3.1) showing two relative minima, $M1$ ($S = 1.69$, $\alpha = 0.087$, $\beta = 0.620$) and $M2$ ($S = 1.727, \alpha = -0.063, \beta = -0.699$).

Table 3.1 Enzyme Velocity (y) and Substrate Concentration (x)[a]

x	y	x	y
2.0	0.0615	0.286	0.0129
2.0	.0527	.286	.0183
0.667	.0334	.222	.0083
0.667	.0258	.222	.0169
0.400	.0138	.200	.0129
0.400	.0258	.200	.0087

[a]From Watts [1981]

the concentration of substrate, x. Figure 3.2 shows the contours of

$$S(\boldsymbol{\theta}) = \sum_{i=1}^{12} \left(y_i - \frac{\theta_1 x_i}{\theta_2 + x_i} \right)^2 \tag{3.3}$$

together with $\hat{\boldsymbol{\theta}}$, the least-squares estimate minimizing $S(\boldsymbol{\theta})$. A blown-up version of these contours is given in Fig. 3.3 along with the approximating ellipsoids for the first iterations $\boldsymbol{\theta}^{(1)}$ and $\boldsymbol{\theta}^{(2)}$ [denoted by A and B respectively] of a

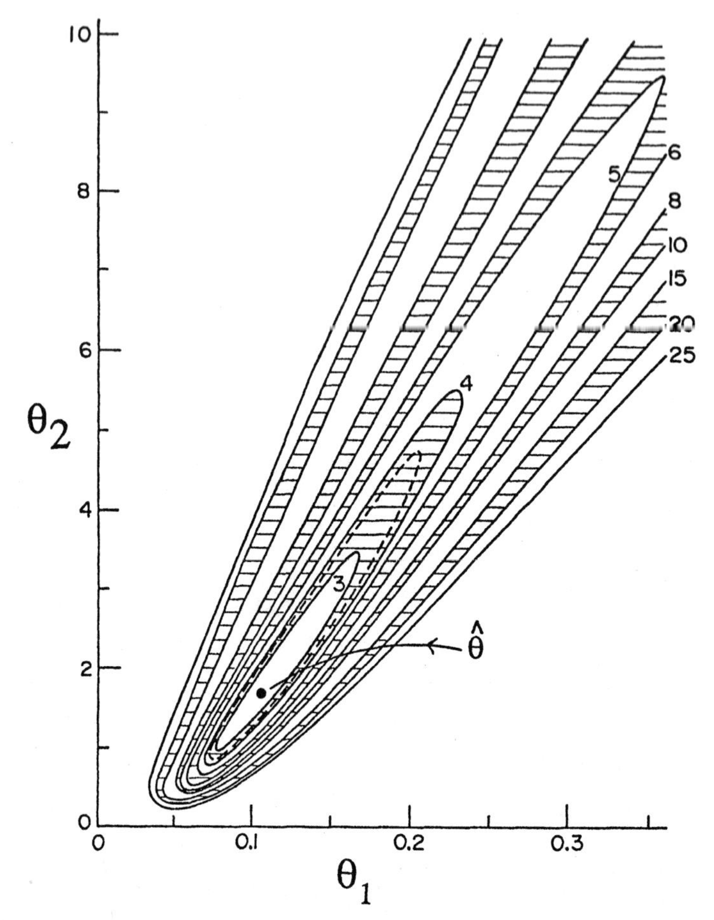

Figure 3.2 Sum-of-squares contours for Equation (3.3) using the enzyme data of Table 3.1. From Watts [1981].

Gauss–Newton process [see (2.29)]. The initial approximation $\boldsymbol{\theta}^{(1)} = (0.057, 0.6)'$ is well away from $\hat{\boldsymbol{\theta}} = (0.10579, 1.7077)'$. However, the approximating ellipse through $\boldsymbol{\theta}^{(1)}$ is close to the true contour of $S(\boldsymbol{\theta})$, and its center $\boldsymbol{\theta}^{(2)} = (0.088, 1.24)'$ is much nearer to $\hat{\boldsymbol{\theta}}$. The next approximating ellipse, through $\boldsymbol{\theta}^{(2)}$, fits even better, so that its center C is close to $\hat{\boldsymbol{\theta}}$. As expected from such a well-behaved function $S(\boldsymbol{\theta})$, convergence to $\hat{\boldsymbol{\theta}}$ (point D) is rapid.

Example 3.3 Consider fitting the model

$$y = \beta(1 - e^{-\gamma x}) + \varepsilon \tag{3.4}$$

to the six data points $(x, y) = (0.1, 0.095)$, $(0.5, 0.393)$, $(1.0, 0.632)$, $(1.4, 1.0)$, $(1.5, 0.777)$, $(2.0, 0.865)$. Figure 3.4a shows the contours of the sum of squares

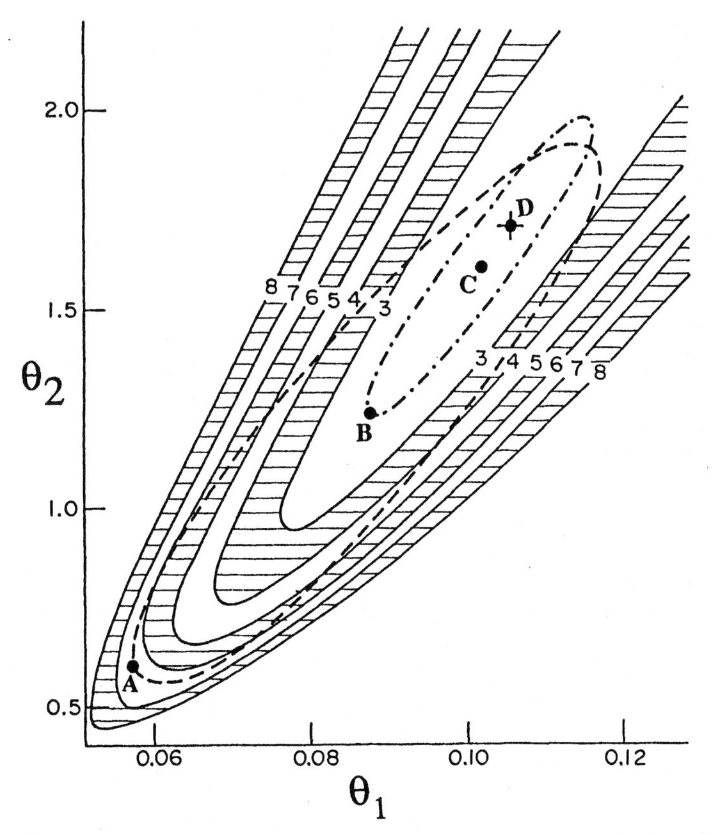

Figure 3.3 Sum-of-squares contours and approximating ellipsoidal contours for data of Table 3.1: A, first approximation $\boldsymbol{\theta}^{(1)} = (0.057, 0.6)'$ (dashed curve); B, second approximation $\boldsymbol{\theta}^{(2)} = (0.088, 1.24)'$ (dot-dash curve); C, third approximation $\boldsymbol{\theta}^{(3)}$; D, least squares estimate $\hat{\boldsymbol{\theta}} = (0.10579, 1.7077)'$. From Watts [1981].

$S(\boldsymbol{\theta})$, where $\boldsymbol{\theta} = (\beta, \gamma)'$. This function has a long narrow curved valley with a fairly flat valley floor. The Gauss–Newton algorithm diverges from the initial approximation $(\beta^{(1)}, \gamma^{(1)}) = (8, 0.5)$. (However, a trivial modification of the Gauss–Newton algorithm, namely step halving if $S(\boldsymbol{\theta}^{(a+1)}) > S(\boldsymbol{\theta}^{(a)})$, will prevent the divergence.)

Changing the parametrization can have a marked effect on the performance of an algorithm. For example, Fig. 3.4*b* shows the contours of S using parameters $\beta\gamma$ and γ: these contours are much closer to the elliptical shape. From the same starting point, i.e. $\beta\gamma = 4$, $\gamma = 0.5$, the Gauss–Newton algorithm converges in four iterations to the point M.

In both Fig. 3.4*a* and *b* the contours of $S(\boldsymbol{\theta})$ are elongated, and in 3.4*a* they are also banana-shaped. We shall see in Section 3.3 that elongated contours usually suggest some difficulty with parameter identification or "ill-conditioning" of the data in that some function of the parameters (e.g. $\theta_1 + \theta_2$) can be estimated

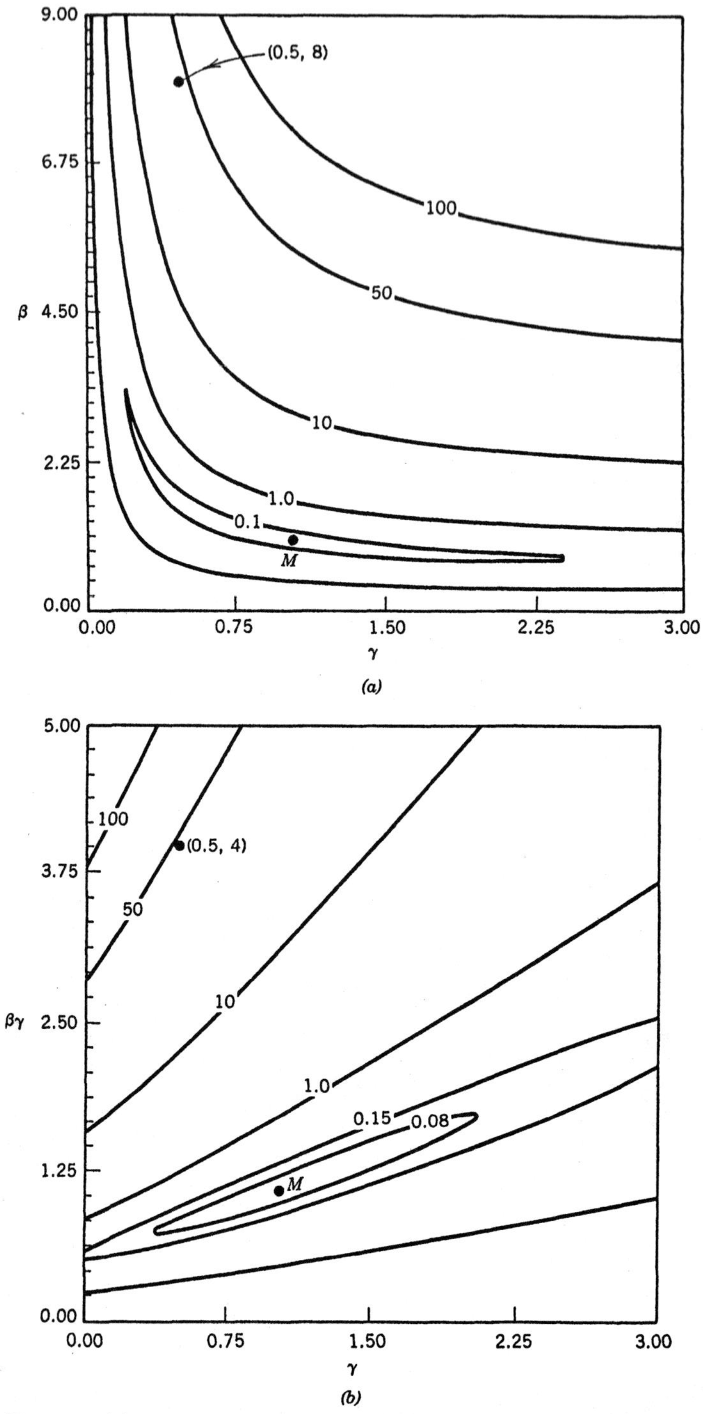

Figure 3.4 Contours of the sum-of-squares function $S(\boldsymbol{\theta})$ for two different parametrizations of the model (3.4): (*a*) β versus γ, (*b*) $\beta\gamma$ versus γ. The position of the minimum is labeled by M.

precisely, but not the individual parameters. A bending of the contours as in Fig. 3.4*a* sufficiently close to $\hat{\boldsymbol{\theta}}$ indicates the inadequacy of any quadratic approximation, particularly that obtained by the linearization method (2.27). In the next section we shall see that curvature of the contours can indicate problems with asymptotic inference.

3.3 VALIDITY OF ASYMPTOTIC INFERENCE

3.3.1 Confidence Regions

Although asymptotic inference is discussed in detail in Chapter 5, we shall introduce some results here. Let us consider the usual nonlinear regression model

$$y_i = f(\mathbf{x}_i; \boldsymbol{\theta}) + \varepsilon_i \qquad (i = 1, 2, \ldots, n),$$

where the ε_i are i.i.d. $N(0, \sigma^2)$. We noted earlier in Theorem 2.1 of Section 2.1.2 that, as $n \to \infty$, the least-squares estimate $\hat{\boldsymbol{\theta}}$ is asymptotically $N_p(\boldsymbol{\theta}^*, \sigma^2(\mathbf{F}_{\bullet}'\mathbf{F}_{\bullet})^{-1})$, where $\boldsymbol{\theta}^*$ is the true value of $\boldsymbol{\theta}$. Also, the matrix $\mathbf{F}_{\bullet} = [(\partial f(\mathbf{x}_i; \boldsymbol{\theta})/\partial \theta_j]_{\boldsymbol{\theta}^*}$ plays the same role as $\mathbf{X}$ in the linear theory of Appendix D. In particular, by analogy with the linear confidence region (cf. Appendix D2.2)

$$(\boldsymbol{\beta} - \hat{\boldsymbol{\beta}})'\mathbf{X}'\mathbf{X}(\boldsymbol{\beta} - \hat{\boldsymbol{\beta}}) \leq ps^2 F^{\alpha}_{p,n-p}, \tag{3.5}$$

we saw, from (2.26), that

$$\{\boldsymbol{\theta} : (\boldsymbol{\theta} - \hat{\boldsymbol{\theta}})'\hat{\mathbf{F}}_{\bullet}'\hat{\mathbf{F}}_{\bullet}(\boldsymbol{\theta} - \hat{\boldsymbol{\theta}}) \leq ps^2 F^{\alpha}_{p,n-p}\} \tag{3.6}$$

is an approximate $100(1-\alpha)\%$ confidence region for $\boldsymbol{\theta}$. Here $s^2 = S(\hat{\boldsymbol{\theta}})/(n-p)$, $\hat{\mathbf{F}}_{\bullet} = \mathbf{F}_{\bullet}(\hat{\boldsymbol{\theta}})$, and $F^{\alpha}_{p,n-p}$ is the upper α critical value of the $F_{p,n-p}$ distribution. As the linear approximation is valid asymptotically, (3.6) will have the correct confidence level of $1-\alpha$ asymptotically. For varying α the regions (3.6) are enclosed by ellipsoids which are also contours of the approximate multivariate normal density function of $\hat{\boldsymbol{\theta}}$ (with $\mathbf{F}_{\bullet}$ replaced by $\hat{\mathbf{F}}_{\bullet}$). Since $S(\boldsymbol{\theta})$ measures the "closeness" of the observations to the fitted equation for any $\boldsymbol{\theta}$, it would seem appropriate to also base confidence regions for $\boldsymbol{\theta}$ on the contours of $S(\boldsymbol{\theta})$. Such a region could take the form

$$\{\boldsymbol{\theta} : S(\boldsymbol{\theta}) \leq cS(\hat{\boldsymbol{\theta}})\} \tag{3.7}$$

for some c, $c > 1$. Regions of this type are often called "exact" confidence regions, as they are not based on any approximations. However, the confidence levels, or coverage probabilities, of such regions are generally unknown, though approximate levels can be obtained from asymptotic theory as follows.

For large enough n, the consistent estimator $\hat{\boldsymbol{\theta}}$ will be sufficiently close to $\boldsymbol{\theta}^*$ for the approximation (2.20) to hold with $\mathbf{F}_.$ replaced by $\hat{\mathbf{F}}_.$ Therefore, substituting

$$S(\boldsymbol{\theta}^*) - S(\hat{\boldsymbol{\theta}}) \approx (\hat{\boldsymbol{\theta}} - \boldsymbol{\theta}^*)' \hat{\mathbf{F}}_.' \hat{\mathbf{F}}_. (\hat{\boldsymbol{\theta}} - \boldsymbol{\theta}^*) \tag{3.8}$$

from (2.20) in (3.6), or using (2.24), we obtain the confidence region (Beale [1960])

$$\left\{ \boldsymbol{\theta} : S(\boldsymbol{\theta}) \le S(\hat{\boldsymbol{\theta}}) \left(1 + \frac{p}{n-p} F^{\alpha}_{p,n-p} \right) \right\}. \tag{3.9}$$

This will have the required asymptotic confidence level of $100(1-\alpha)\%$, as asymptotically the regions (3.6) and (3.9) are the same (they are identical for linear models). However, for finite n, these regions can be very different (see Fig. 3.13*c* in Section 3.4.2b), thus indicating inadequacy of the linear approximation (2.20). They are discussed further in Section 5.2.2. As we shall see in Chapter 5, (3.9) is more reliable, though it is more difficult to compute and to display.

Under the assumption of normal errors we see that the above clash between the two confidence regions embodies a wider principle. We have a choice between confidence regions based on the asymptotic normality of the maximum-likelihood estimator $\hat{\boldsymbol{\theta}}$ (which in this case is also the least-squares estimator) and those based on the contours of the likelihood function via the likelihood-ratio criterion.

3.3.2 Effects of Curvature

The set $\Omega = \{\mathbf{f}(\boldsymbol{\theta}) : \boldsymbol{\theta} \in \Theta\}$ gives a surface consisting of the set of all possible values of $\mathscr{E}[\mathbf{y}]$ and is called the expectation surface (or solution locus). Then the essential ingredient of the above asymptotics is the linear approximation (cf. Section 2.1.2)

$$\mathbf{f}(\boldsymbol{\theta}) \approx \mathbf{f}(\hat{\boldsymbol{\theta}}) + \hat{\mathbf{F}}_.(\boldsymbol{\theta} - \hat{\boldsymbol{\theta}}) \tag{3.10}$$

to the expectation surface. We shall see in Chapter 4 that this approximation states that the secant joining $\mathbf{f}(\boldsymbol{\theta})$ and $\mathbf{f}(\hat{\boldsymbol{\theta}})$, namely $\mathbf{f}(\boldsymbol{\theta}) - \mathbf{f}(\hat{\boldsymbol{\theta}})$, lies in the tangent plane to this surface at the point $\hat{\boldsymbol{\theta}}$, the tangent plane being the space spanned by the columns of $\hat{\mathbf{F}}_.$. This local approximation of $\mathbf{f}(\boldsymbol{\theta})$ in a neighborhood of $\hat{\boldsymbol{\theta}}$ by the tangent plane will be appropriate only if $\mathbf{f}(\boldsymbol{\theta})$ is fairly flat in that neighborhood. If f and θ are one-dimensional, then the first derivative of $f(\theta)$ gives the slope of the curve, while the second derivative gives the rate of change of slope, which is related to the curvature. More generally, the (local) curvature of $\mathbf{f}(\boldsymbol{\theta})$ can be described in terms of two orthogonal components involving partial second derivatives—one being normal to the expectation surface $\mathbf{f}(\boldsymbol{\theta})$, and the other tangential. The first (normal) component is called the intrinsic curvature. This simply measures how much the expectation surface is bending as $\boldsymbol{\theta}$ changes near $\hat{\boldsymbol{\theta}}$. The second (tangential) component is called the parameter-effects

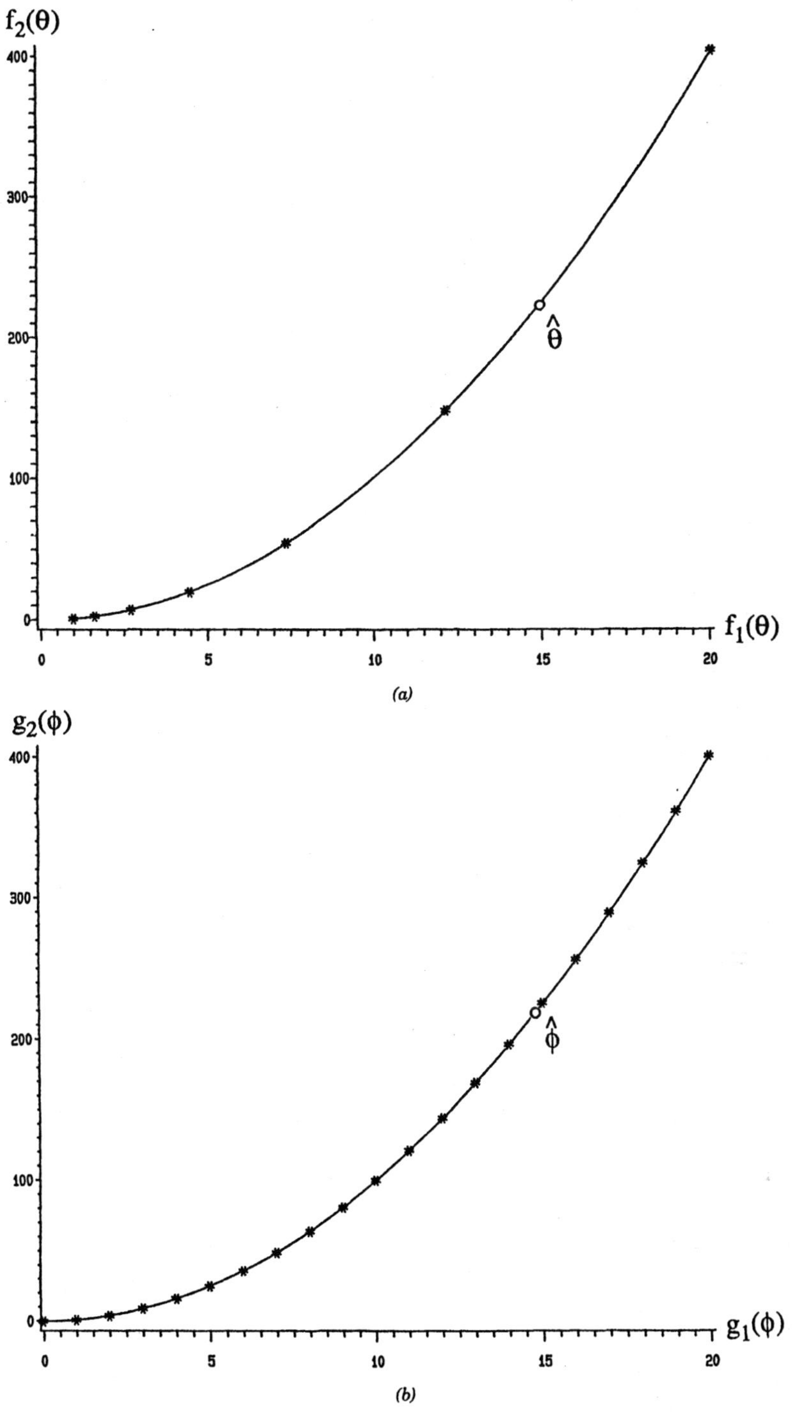

Figure 3.5 Plot of the expectation surface for $x_1 = 1$ and $x_2 = 2$: (*a*) from the model (3.12), $E[y] = e^{\theta x}$; (*b*) from the model (3.13), $E[y] = \phi^x$. The points in (*a*) correspond to equally spaced θ-values, $\theta = 0, 0.5, 1.0, \ldots, 3.0$. The points in (*b*) correspond to $\phi = 0, 1, 2, \ldots, 20$.

curvature. Its role is more subtle, as it reflects how closely a local rectangular grid of $\boldsymbol{\theta}$-values centered at $\hat{\boldsymbol{\theta}}$ remains rectangular when translated into points on $\mathbf{f}(\boldsymbol{\theta})$. Basically it measures the curvature of the projections on the tangent plane of the lines of the expectation surface corresponding to a rectangular grid of $\boldsymbol{\theta}$-values. Therefore a judicious change from $\boldsymbol{\theta}$ to some other parameter $\boldsymbol{\phi}$ may reduce or even eliminate this component of curvature. For example, consider the following simple model (inspired by Fig. 1.2 of Ratkowsky [1983]):

$$E[y_i] = f(x_i;\theta) = e^{\theta x_i} \qquad (i = 1, 2), \tag{3.11}$$

where $x_1 = 1$ and $x_2 = 2$. Then the expectation surface

$$\mathbf{f}(\theta) = \begin{pmatrix} e^{\theta} \\ e^{2\theta} \end{pmatrix} \tag{3.12}$$

represents a curve in two dimensions (Fig. 3.5*a*). If $y_1 = 12$ and $y_2 = 221.5$, we find that $\hat{\theta} = 2.7$ and we see from Fig. 3.5*a* that $\mathbf{f}(\theta)$ is fairly linear near $\hat{\theta}$, i.e., the intrinsic curvature is small. However, for an equally spaced grid of θ-values, we see that the corresponding points on the curve are not equally spaced but rapidly get further apart as θ increases, i.e., the parameter-effects curvature is substantial. Since we want to linearize the θ-scale on $\mathbf{f}(\theta)$ as much as possible, an obvious transformation is $\phi = e^{\theta}$, so that out model now becomes

$$\begin{aligned} E[y_i] &= g(x_i;\phi) \\ &= f(x_i;\log\phi) \\ &= \phi^{x_i}. \end{aligned}$$

Then the expectation surface has representation

$$\mathbf{g}(\phi) = \begin{pmatrix} \phi \\ \phi^2 \end{pmatrix}, \tag{3.13}$$

and this is graphed in Fig. 3.5*b*. We obtain the same curve as before, but the ϕ-spacing is much more uniform, indicating a reduction in the parameter-effects curvature. However, it should be noted that these curvatures depend very much on the choice of the design points x_i as well as on the model. For example, if $x_1 = -0.1$ and $x_2 = -0.5$, then Fig. 3.6 shows that the θ-parametrization now has a more uniform spacing near $\hat{\theta}$ than the ϕ-parametrization near $\hat{\phi} = e^{\hat{\theta}}$.

How important are these curvature effects? If the intrinsic curvature is high, then the model is highly nonlinear and the linear tangent-plane approximation will not be appropriate. Therefore, no matter what parametrization is used, approximate inferences based on the asymptotic normality of $\hat{\boldsymbol{\theta}}$ and the confidence regions (3.6) and (3.9) will be misleading. However, in Chapter 5 we shall see that it is possible to correct for intrinsic curvature to some extent.

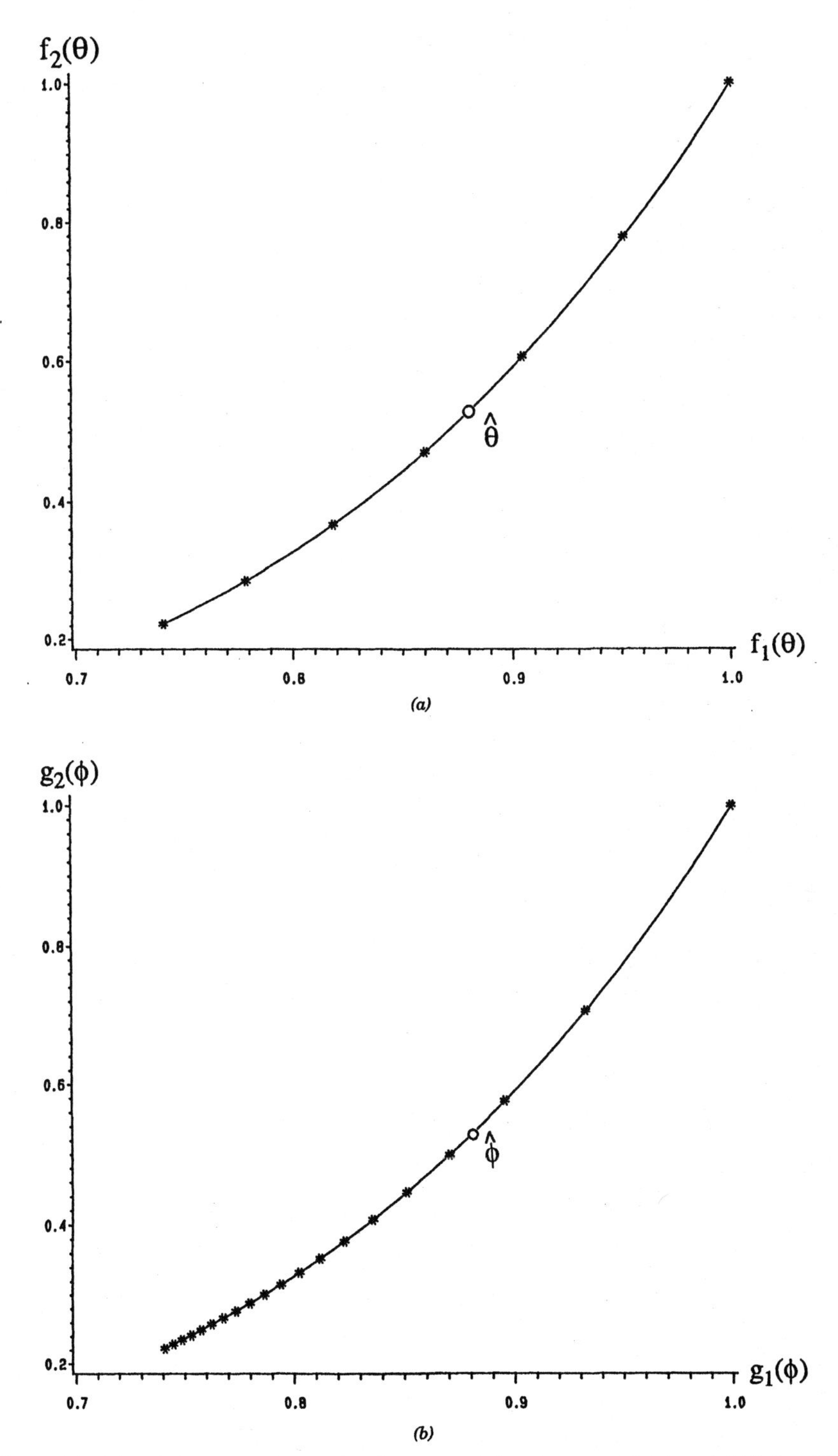

Figure 3.6 Plot of the expectation surface for $x_1 = -0.1$ and $x_2 = -0.5$: (*a*) model $E[y] = e^{\theta x}$, (*b*) model $E[y] = \phi^x$. The points in (*a*) correspond to $\theta = 0, 0.5, 1.0, \ldots, 3.0$, and in (*b*) to $\phi = 1, 2, 3, \ldots, 20$.

Parameter-effects curvature can also invalidate inference procedures, and a method of measuring its effect on (3.6) [(3.9) is unaffected] is given in Section 5.8.2. However, as we saw above, a careful choice of parameters can ameliorate the problem.

The importance of carefully selecting the design points $\mathbf{x}_i$ has long been recognized and is usually discussed under the heading of optimal designs. The emphasis has been on choosing design points which minimize some aspect of the design such as the volume of the confidence ellipsoid (3.6) or the asymptotic variance of a parameter, or which discriminate between competing models in some optimal fashion. The effects of curvature also need to be allowed for in any design criterion, and we need methods for finding design points which will minimize these effects. Optimal designs are discussed in Section 5.13.

3.4 IDENTIFIABILITY AND ILL-CONDITIONING

3.4.1 Identifiability Problems

The linear regression model (Appendix D1.1) can be written in the form $\mathbf{y}=\boldsymbol{\mu}+\boldsymbol{\varepsilon}$, where $\boldsymbol{\mu}\in\mathscr{R}[\mathbf{X}]$ (the range space of $\mathbf{X}$), i.e. $\boldsymbol{\mu}=\mathbf{X}\boldsymbol{\beta}$ for some $\boldsymbol{\beta}\in\mathbb{R}^p$. If $\mathbf{X}$ is not of full rank, then $\boldsymbol{\beta}$ is not identifiable, because there are an infinite number of $\boldsymbol{\beta}$-values which satisfy $\mathbf{X}\boldsymbol{\beta}=\boldsymbol{\mu}$ for a given $\boldsymbol{\mu}$ (see Seber [1977: Section 3.8]). Therefore, even if we knew the true value of $\boldsymbol{\mu}$, we would not know which of the $\boldsymbol{\beta}$-values we wanted. In linear modeling, such rank-deficient design matrices $\mathbf{X}$ arise in analysis-of-variance models such as

$$y_{ij}=\mu+\alpha_i+\tau_j+\varepsilon_{ij}.$$

Here we usually impose identifiability constraints, e.g. $\sum_i\alpha_i=\sum_j\tau_j=0$, that identify which solution to $\mathbf{X}\boldsymbol{\beta}=\boldsymbol{\mu}$ is required. Such constraints also define the meaning of the parameters. Another approach to the problem of unidentifiability is to look for a parametric function of the form $\mathbf{a}'\boldsymbol{\beta}$ which is estimable. The least-squares estimate $\mathbf{a}'\hat{\boldsymbol{\beta}}$ of such a function, where $\hat{\boldsymbol{\beta}}$ is a solution of the (singular) normal equations $\mathbf{X}'\mathbf{X}\hat{\boldsymbol{\beta}}=\mathbf{X}'\mathbf{y}$, then turns out to be invariant with respect to $\hat{\boldsymbol{\beta}}$.

Parameters can also be unidentifiable in nonlinear models. We define lack of identifiability in the model $E[y]=f(\mathbf{x};\boldsymbol{\theta})$ as occurring when there exist two values $\boldsymbol{\theta}_1$ and $\boldsymbol{\theta}_2$ such that $f(\mathbf{x};\boldsymbol{\theta}_1)=f(\mathbf{x};\boldsymbol{\theta}_2)$ for all $\mathbf{x}$. For example, consider the model (Bird and Milliken [1976])

$$f(\mathbf{x};\boldsymbol{\theta})=\exp(-\beta\tau x_1)+\frac{\alpha}{\beta}[1-\exp(-\beta\tau x_1)]x_2, \tag{3.14}$$

with $\boldsymbol{\theta}=(\alpha,\beta,\tau)'$. Here β and τ only occur as the product $\beta\tau$, so that $\boldsymbol{\theta}_c=(c\alpha,c\beta,\tau/c)$ gives that same value of $f(\mathbf{x};\boldsymbol{\theta})$ for any nonzero c. A way out of this problem would be to reparametrize using $\phi_1=\beta\tau$ and $\phi_2=\alpha/\beta$. Although a careful inspection of the parameters will usually reveal any lack of identifiability in a simple model, a more formal approach using the above invariance property of linear estimable functions is given by Bird and Milliken [1976]. If $\hat{\boldsymbol{\theta}}$ is any solution of the normal equations (2.8), then $\boldsymbol{\phi} = \mathbf{a}(\boldsymbol{\theta})$ is said to be estimable if $\hat{\boldsymbol{\phi}} = \mathbf{a}(\hat{\boldsymbol{\theta}})$ is invariant for all solutions $\hat{\boldsymbol{\theta}}$. Methods for checking identifiability for compartmental models are discussed in Section 8.4.

The restrictions placed on the parameters by a hypothesis test can sometimes induce a loss of identifiability. For example, we cannot readily test $H_0: \gamma = 0$ in the model

$$f(x;\boldsymbol{\theta}) = \alpha_1 + \alpha_2 e^{\gamma x}. \tag{3.15}$$

Under H_0,

$$f(x;\boldsymbol{\theta}) = \alpha_1 + \alpha_2, \tag{3.16}$$

and α_1 and α_2 are not identifiable. There is, however, another type of problem associated with nonlinear models, which might be described as approximate nonidentifiability. Consider the model

$$f(x;\boldsymbol{\theta}) = \alpha_1 + \alpha_2 x^{1/\gamma}. \tag{3.17}$$

When γ is large, $x^{1/\gamma} \approx x^0 = 1$ and (3.16) is approximately true. A similar situation arises when $\gamma \approx 0$ in (3.15). Here it can be shown that as $\gamma \to 0$, the least-squares estimates of α_1 and α_2 become infinite. However, this problem can be avoided if we reparametrize the model (Stirling [1985]) and use

$$f(x;\boldsymbol{\theta}) = \beta_1 + \beta_2 \left\{ \frac{e^{\gamma x} - 1}{\gamma} \right\}.$$

As $\gamma \to 0$ the term in braces tends to x, and β_1 and β_2 become the ordinary linear-regression coefficients. However, problems still remain when $\beta_2 \approx 0$, as the model is now insensitive to changes in γ. A third, less obvious example is given below (Example 3.5 in Section 3.4.2d).

3.4.2 Ill-conditioning

The above difficulties associated with identifiability stem from the structure of the model and the method of parametrization rather than from inappropriate design points $\mathbf{x}_i$. The problems encountered in the above nonlinear examples remain no matter how well the $\mathbf{x}_i$ are placed. Algebraically, a lack of identifiability is signaled by $\mathbf{F}'(\boldsymbol{\theta})\mathbf{F}(\boldsymbol{\theta})$ from $S(\boldsymbol{\theta})$ [or $\partial^2 h(\boldsymbol{\theta})/\partial\boldsymbol{\theta}\,\partial\boldsymbol{\theta}'$ for more general functions h] being singular, or nearly so.

a Linear Models

In the case of linear models, the parameters are unidentifiable if the $n \times p$ matrix $\mathbf{X}$ has rank less than p and $\mathbf{X}'\mathbf{X}$ is singular. As we noted above, the most common examples of this occur in analysis of variance. Otherwise, in the usual multiple linear-regression model with continuous explanatory x-variables, it is unlikely that $\mathbf{X}$ is rank deficient. However, a more important problem, usually referred to as (approximate) multicollinearity, occurs when some of the columns of $\mathbf{X}$ are highly correlated, rather than being exactly linearly related. Then $\mathbf{X}$ has full rank p, but its columns are close to being linearly dependent and $\mathbf{X}'\mathbf{X}$ is nearly singular. When this happens, $\mathbf{X}'\mathbf{X}$ is said to be ill-conditioned. An example of this is given in Fig. 3.7, which represents a plot of the data in Table 3.2. This data consists of 50 observations on monozygote twins which were analyzed by Lee and Scott [1986]. Here y is the birthweight of a baby, x_1 ($=$AC) is the abdominal circumference, and x_2 ($=$BPD) is the biparietal (head) diameter. We shall now see what happens when we fit the centered model (c.f. Seber [1977: p. 330])

$$y_i = \beta_0 + \beta_1(x_{i1} - \bar{x}_1) + \beta_2(x_{i2} - \bar{x}_2) + \varepsilon_i. \tag{3.18}$$

Since $\hat{\beta}_0$ ($= \bar{y}$) is statistically independent of $\hat{\beta}_1$ and $\hat{\beta}_2$ under normality of the

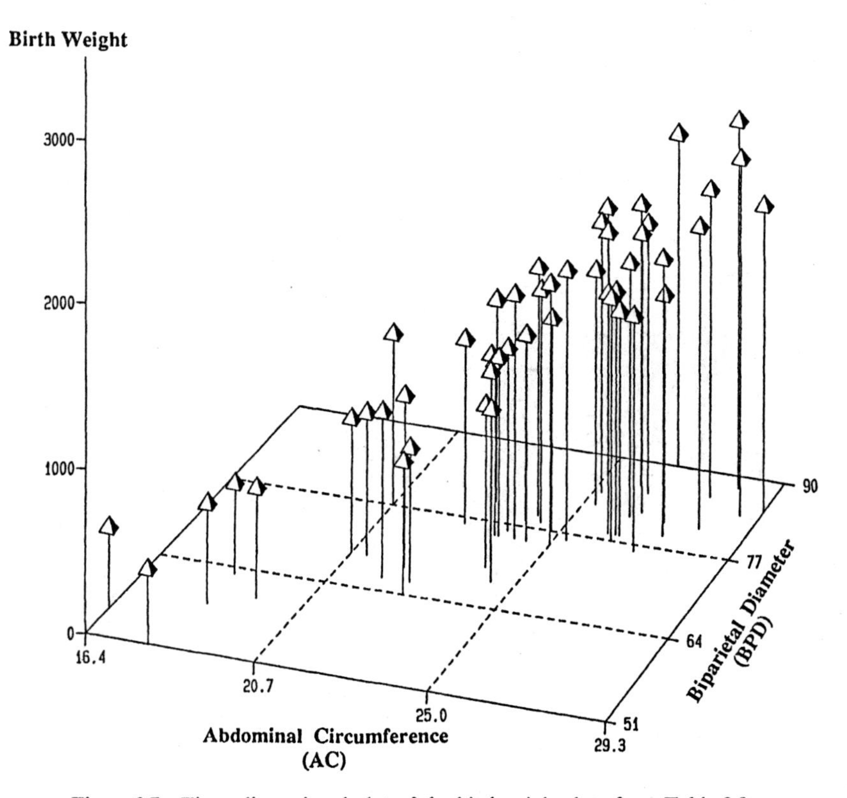

Figure 3.7 Three-dimensional plot of the birthweight data from Table 3.2.

Table 3.2 Birthweight Data for Monozygote Twins[a]

Observation Number	Birth-weight BW (g)	Biparietal Diameter BPD (mm)	Abdominal Circumference AC (cm)
1	1350	77	24.8
2	1500	82	25.3
3	2170	90	26.5
4	1780	82	26.5
5	1820	86	26.0
6	525	55	16.4
7	1300	80	23.0
8	1730	87	25.0
9	1580	79	24.2
10	1620	81	26.3
11	850	69	20.7
12	1915	81	28.1
13	1740	84	25.2
14	1640	75	25.0
15	1520	78	26.1
16	700	60	19.5
17	1040	66	21.9
18	640	58	18.5
19	900	69	21.1
20	580	63	18.5
21	1080	68	22.2
22	1480	78	24.4
23	1720	83	25.5
24	1180	78	24.0
25	1140	75	23.5
26	1020	70	24.0
27	1960	85	29.3
28	1160	76	23.8
29	1100	77	20.7
30	1175	76	22.7
31	1450	81	27.2
32	1480	76	27.0
33	1740	85	26.3
34	840	64	22.7
35	850	66	22.6
36	1080	68	24.4
37	1130	75	23.7
38	1100	75	23.6
39	1300	75	24.4
40	1540	78	26.3
41	2380	88	28.3
42	1980	86	27.8
43	1540	75	24.1
44	2280	84	28.8

Table 3.2 (*Continued*).

Observation Number	Birth-weight BW (g)	Biparietal Diameter BPD (mm)	Abdominal Circumference AC (cm)
45	1700	76	25.3
46	1420	78	26.4
47	1750	79	27.4
48	1190	71	24.0
49	1530	77	26.3
50	480	51	18.0

[a]Data courtesy of A. J. Lee and A. J. Scott.

ε_i, we can obtain a separate confidence region for $\boldsymbol{\beta} = (\beta_1, \beta_2)'$, namely the ellipse

$$(\boldsymbol{\beta} - \hat{\boldsymbol{\beta}})'\tilde{\mathbf{X}}'\tilde{\mathbf{X}}(\boldsymbol{\beta} - \hat{\boldsymbol{\beta}}) \leq 2s^2 F^{\alpha}_{2,n-3} = c^2 \tag{3.19}$$

with center $\hat{\boldsymbol{\beta}} = (\hat{\beta}_1, \hat{\beta}_2)'$ and

$$\tilde{\mathbf{X}}'\tilde{\mathbf{X}} = \begin{pmatrix} \sum(x_{i1} - \bar{x}_1)^2 & \sum(x_{i1} - \bar{x}_1)(x_{i2} - \bar{x}_2) \\ \sum(x_{i1} - \bar{x}_1)(x_{i2} - \bar{x}_2) & \sum(x_{i2} - \bar{x}_2)^2 \end{pmatrix}.$$

We note that

$$|\tilde{\mathbf{X}}'\tilde{\mathbf{X}}| = \sum(x_{i1} - \bar{x}_1)^2 \sum(x_{i2} - \bar{x}_2)^2 (1 - r^2),$$

where r is the correlation coefficient of the pairs (x_{i1}, x_{i2}), $i = 1, 2, \ldots, n$. From the plot in Fig. 3.8 it is clear that the variables x_1 and x_2 are approximately linearly related; in fact $r = 0.85$. The model is therefore moderately ill-conditioned. Furthermore, since $\mathscr{D}[\hat{\boldsymbol{\beta}}] = \sigma^2(\tilde{\mathbf{X}}'\tilde{\mathbf{X}})^{-1}$, we find that $\hat{\beta}_1$ and $\hat{\beta}_2$ are highly correlated with correlation $-r$.

Suppose, in the parameter space, we now shift the origin to $\hat{\boldsymbol{\beta}}$ and make a rotation using an orthogonal matrix $\mathbf{T}$ which diagonalizes $\tilde{\mathbf{X}}'\tilde{\mathbf{X}}$. Then if $\mathbf{z} = \mathbf{T}'(\boldsymbol{\beta} - \hat{\boldsymbol{\beta}})$,

$$\begin{aligned}(\boldsymbol{\beta} - \hat{\boldsymbol{\beta}})'\tilde{\mathbf{X}}'\tilde{\mathbf{X}}(\boldsymbol{\beta} - \hat{\boldsymbol{\beta}}) &= \mathbf{z}'\mathbf{T}'\tilde{\mathbf{X}}'\tilde{\mathbf{X}}\mathbf{T}\mathbf{z} \\ &= \lambda_1 z_1^2 + \lambda_2 z_2^2.\end{aligned}$$

Here λ_1 and λ_2 are the (positive) eigenvalues of $\tilde{\mathbf{X}}'\tilde{\mathbf{X}}$ with corresponding unit eigenvectors $\mathbf{t}_1$ and $\mathbf{t}_2$, the columns of $\mathbf{T}$, representing the directions of the new axes. In the new z-coordinate system the ellipse (3.19) is given by the boundary

$$\lambda_1 z_1^2 + \lambda_2 z_2^2 = c^2 \tag{3.20}$$

with principal axes of lengths $\sqrt{c^2/\lambda_i}$ $(i = 1, 2)$ (c.f. Appendix A6). In our example

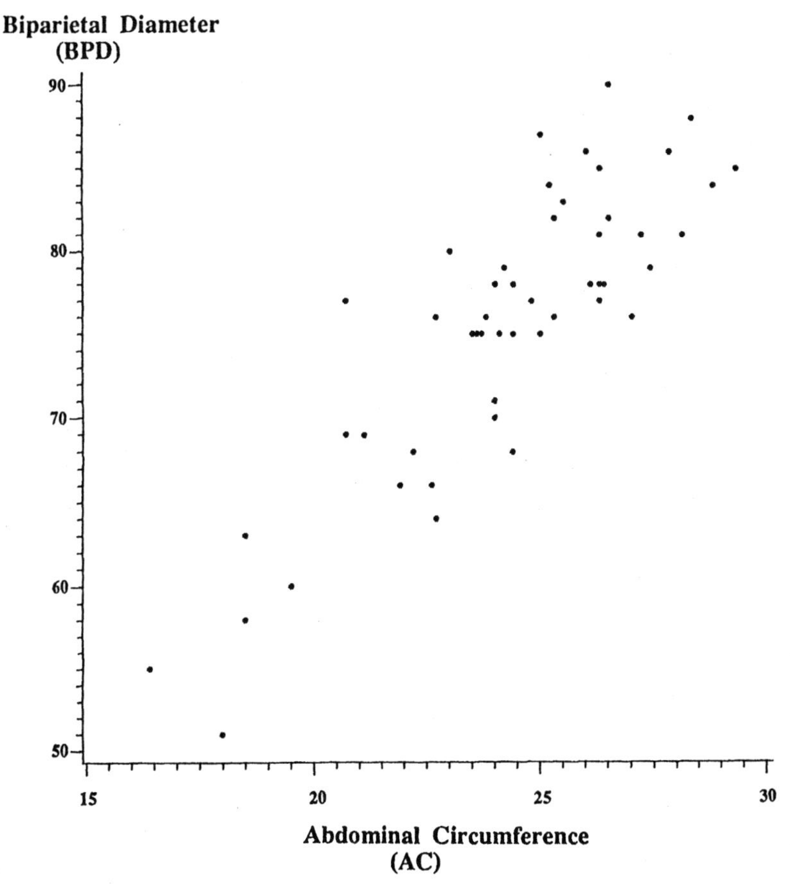

Figure 3.8 Scatterplot of abdominal circumference versus biparietal diameter for the birthweight data of Table 3.2.

$\lambda_1 = 3948.4$, $\lambda_2 = 101.5$, and the 95% and 99% confidence ellipses are drawn in Fig. 3.9: the z-coordinate axes are also indicated.

Having described the geometry of the confidence ellipse, we now consider the statistical ramifications. Suppose we wish to find a confidence interval for $\mathbf{a}'\boldsymbol{\beta}$, where $\mathbf{a}$ is any 2×1 vector. According to Scheffé's simultaneous confidence-interval procedure (Scheffé [1959: 68]), $\boldsymbol{\beta}$ lies in the confidence ellipse if and only if it lies between every pair of parallel tangent planes. Statistically this means that we can translate a confidence region into an equivalent set of confidence intervals, one for each $\mathbf{a}'\boldsymbol{\beta}$ where $\mathbf{a}$ is perpendicular to the pair of tangent planes. The perpendicular distance between the tangent planes, namely the projection of the ellipse in direction $\mathbf{a}$, therefore describes how precisely $\mathbf{a}'\boldsymbol{\beta}$ is estimated. The greatest precision is thus obtained in the direction of the shorter principal axis, and the least precision in the direction of the longer principal axis. Other directions will have an intermediate precision. For example

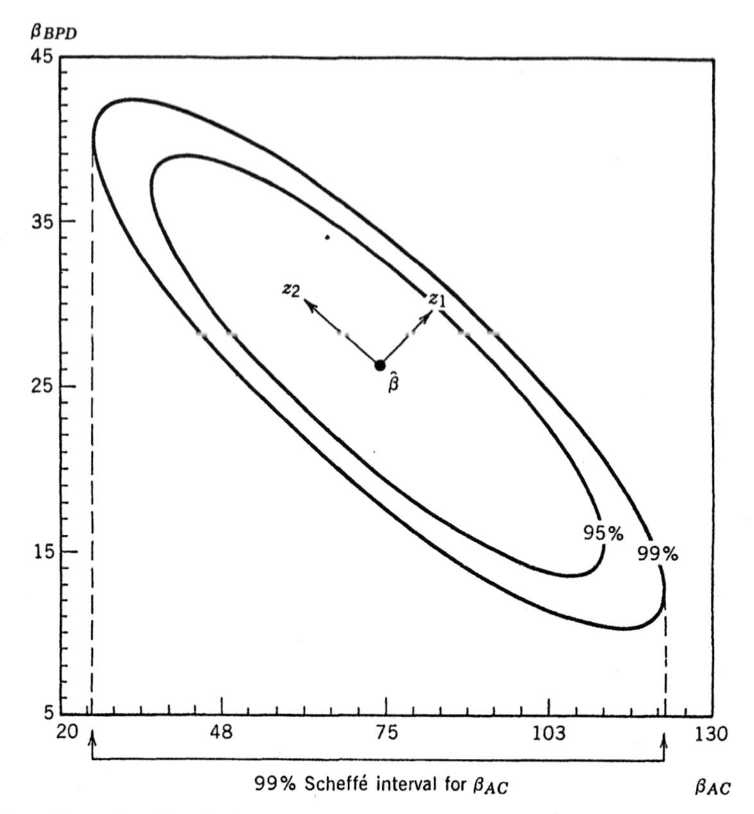

Figure 3.9 Plots of 95% and 99% confidence regions for $\boldsymbol{\beta}$, given by Equation (3.19), for the birthweight data of Table 3.2.

the confidence interval for β_2 is simply the projection of the ellipse onto the β_2-axis, and the 99% confidence interval for $\beta_2 = \beta_{AC}$ is shown in Fig. 3.9.

We can also discuss precision in terms of the original data and principal components. The eigenvalues are an index of precision: the larger the eigenvalue, the better the precision in the direction of the corresponding eigenvector (principal component). If $\bar{\mathbf{x}} = (\bar{x}_1, \bar{x}_2)'$ is the sample mean of the data points $\mathbf{x}_i$ in the (x_1, x_2) plane, then the first principal component of $\mathbf{x}_i$ is $\mathbf{t}_1'(\mathbf{x}_i - \bar{\mathbf{x}})$, where $\mathbf{t}_1$ is the direction of the line through $\bar{\mathbf{x}}$ for which the sum of squared distances of the points from the line is minimized (Seber [1984: 185]). It is therefore not surprising that $\mathbf{t}_1$ is the direction best estimated from the data and $\mathbf{t}_2$ is the worst.

When $\tilde{\mathbf{X}}'\tilde{\mathbf{X}}$ is diagonal, $\mathscr{D}[\hat{\boldsymbol{\beta}}]$ $[= \sigma^2(\tilde{\mathbf{X}}'\tilde{\mathbf{X}})^{-1}]$ is also diagonal and $\hat{\beta}_1$ and $\hat{\beta}_2$ are uncorrelated (or independent with normal data). Then $\mathbf{T} = \mathbf{I}_2$ and the axes of the ellipse are parallel to the β-axes. For example, in Fig. 3.10*a* $\hat{\beta}_1$ and $\hat{\beta}_2$ are uncorrelated, and β_2 is much more precisely estimated than β_1, as the projection of the ellipse onto the β_2-axis is much smaller than the projection

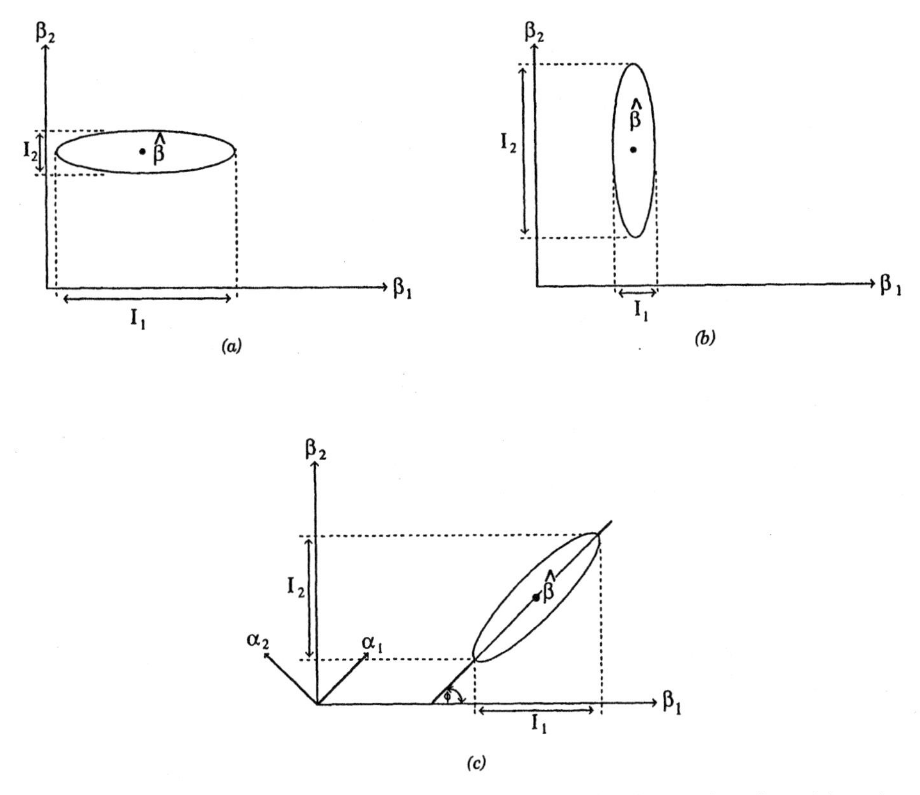

Figure 3.10 Various 95% confidence regions for $\boldsymbol{\beta} = (\beta_1, \beta_2)'$ with different orientations of the major axes: $\alpha_1 = \beta_1 \cos\phi + \beta_2 \sin\phi$ and $\alpha_2 = \beta_2 \cos\phi - \beta_1 \sin\phi$. The intervals I_1 and I_2 are the Scheffé confidence intervals for β_1 and β_2 respectively.

onto the β_1-axis. In Fig. 3.10*b* $\hat{\beta}_1$ and $\hat{\beta}_2$ are again uncorrelated, but β_1 is now more precisely estimated. Finally, in Fig. 3.10*c* $\hat{\beta}_1$ and $\hat{\beta}_2$ are highly correlated. As noted by Draper and Smith [1981:499], projecting in the directions of the major axes of the ellipse, we see that $\alpha_2 = \beta_2 \cos\phi - \beta_1 \sin\phi$ is comparatively precisely estimated, whereas $\alpha_1 = \beta_1 \cos\phi + \beta_2 \sin\phi$ is imprecisely estimated. A consequence of this is that both β_1 and β_2 are themselves imprecisely estimated. However, such comparisons of precision are not appropriate unless the explanatory variables x_1 and x_2 are measured on comparable scales. Transforming $x_1 \to 10x_1$ transforms $\hat{\beta}_1 \to 0.1\hat{\beta}_1$ and has the effect of reducing the projection onto the β_1 axis to $\frac{1}{10}$ of its original size.

Although the above discussion has been in terms of two dimensions, the same problems arise in general. Ill-conditioning firstly leads to small eigenvalues with the consequent stretching out of one or more axes of the confidence ellipsoids. Precision is poor in these directions. Secondly, as the ellipsoid is generally rotated (because of correlated x-variables), the precision of each individual estimate $\hat{\beta}_i$ is affected, usually for the worse. Thirdly, the parameter estimates can be highly correlated so that there is some "instability" in the estimates. Finally, as the columns of $\mathbf{X}$ are almost linearly dependent, a range of values of $\boldsymbol{\beta}$ will give almost equally good fits to the data.

b Nonlinear Models

From the above discussion we saw that multicollinearity in linear models can lead to highly correlated estimates and ill-conditioned matrices $\tilde{\mathbf{X}}'\tilde{\mathbf{X}}$ (and $\mathbf{X}'\mathbf{X}$). Unfortunately, all these problems are inherited by nonlinear-regression models with the added complication that the contours of $S(\boldsymbol{\theta})$, and therefore the confidence contours of (3.9), are often curved as in Fig. 3.11. Here the "projection" of the contour onto $g(\theta_1, \theta_2) = c$ has substantial length, so that $g(\theta_1, \theta_2)$ will be estimated with poor precision. The individual parameters θ_1 and θ_2 do not fare much better. Unfortunately, the problems of approximate nonidentifiability, correlated estimates, and poor precision of estimation in certain directions are not so clearly distinguished in nonlinear models. Nor do we have a good standard terminology representing these phenomena at present. However, we shall use the term "ill-conditioning" to describe all these problems in a nonlinear setting

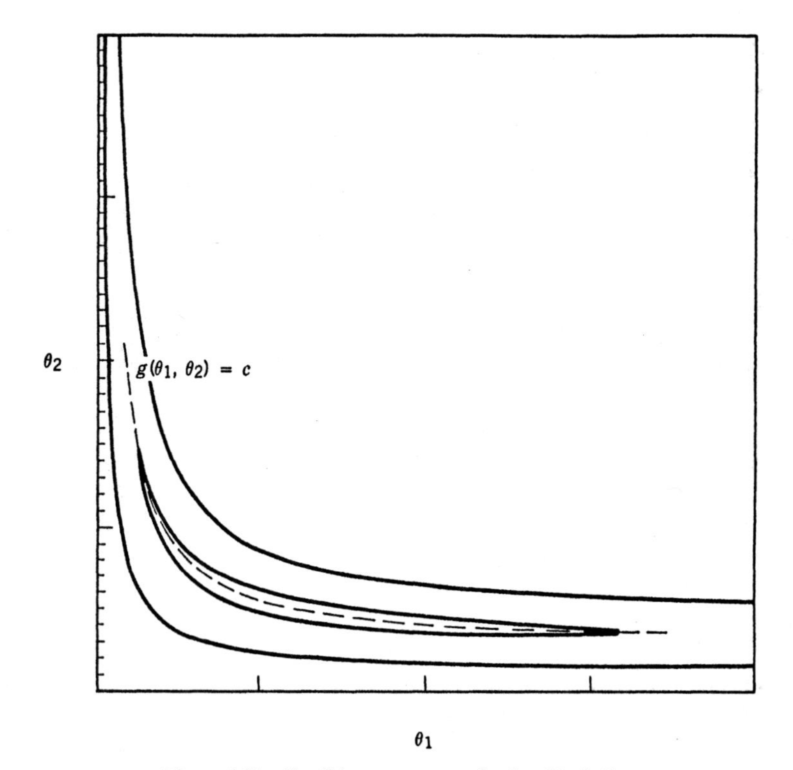

Figure 3.11 Confidence contours for $\boldsymbol{\theta} = (\theta_1, \theta_2)'$.

on the gounds that a major symptom of the problems is generally an ill-conditioned matrix $\mathbf{F}_{\bullet}'(\boldsymbol{\theta})\mathbf{F}_{\bullet}(\boldsymbol{\theta})$.

Example 3.4 The three sets of data given in Table 3.3 were simulated from the model

$$y_i = \beta(1 - e^{-\gamma x_i}) + \varepsilon_i \qquad (i = 1, 2, \ldots, 10), \tag{3.21}$$

where $\beta = 1$, $\gamma = 1$, and the ε_i are i.i.d. $N(0, 0.05^2)$. The same set of random errors $\{\varepsilon_i\}$ was used for each data set. (This example was prompted by Ross [1970].) In Fig. 3.12, y_i is plotted against x_i for data data set I, and superimposed are graphs A–F of

$$\mu = E[y] = \beta(1 - e^{-\gamma x}) \tag{3.22}$$

for different values of the parameter pair (β, γ) as specified in Table 3.4. The curves B, C, D, and E appear to fit the data well, whereas A and F clearly do

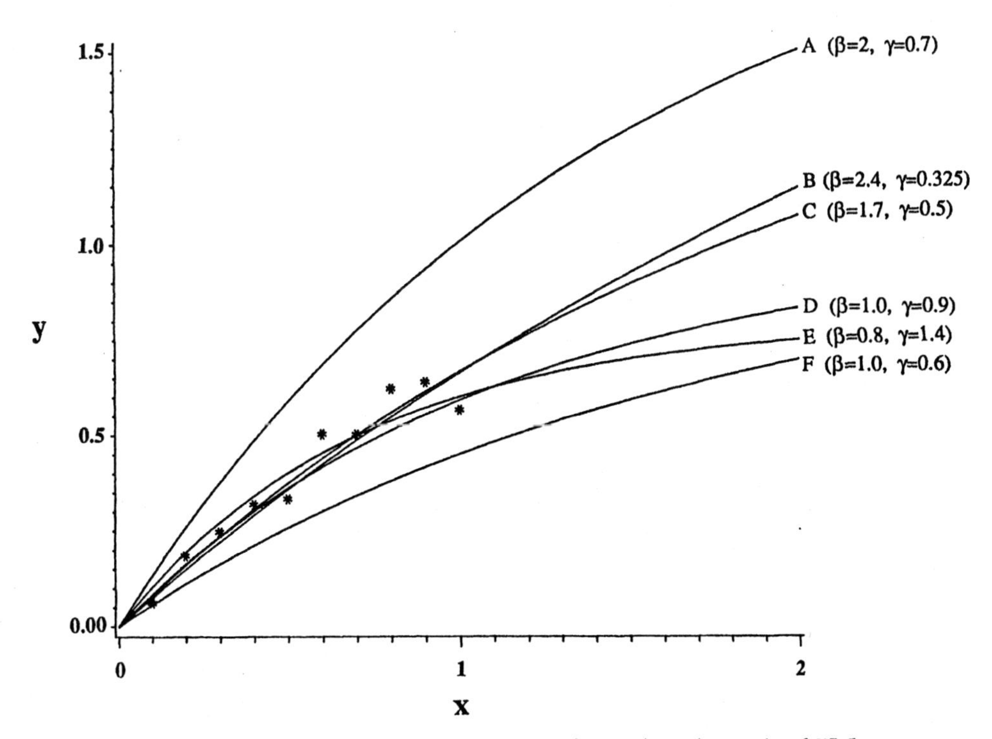

Figure 3.12 Plot of the data set I from Table 3.3. Superimposed are the graphs of $E[y]$ versus x, given by Equation (3.22), for various values of (β, γ).

Table 3.3 Three Sets of Simulated Data from the Model (3.21) Using the Same Errors ε_i for Each Set

	I		II		III		
i	x_i	y_i	x_i	y_i	x_i	y_i	ε_i
1	0.1	0.0638	0.1	0.0638	0.954	0.5833	−.03135
2	0.2	0.1870	0.6	0.4569	0.954	0.6203	.00574
3	0.3	0.2495	1.1	0.6574	0.954	0.6049	−.00969
4	0.4	0.3207	1.6	0.7891	0.954	0.6056	−.00898
5	0.5	0.3356	2.1	0.8197	0.954	0.5567	−.05787
6	0.6	0.5040	2.6	0.9786	4.605	1.0428	.05285
7	0.7	0.5030	3.1	0.9545	4.605	0.9896	−.00040
8	0.8	0.6421	3.6	1.0461	4.605	1.0634	.07344
9	0.9	0.6412	4.1	1.0312	4.605	1.0377	.04773
10	1.0	0.5678	4.6	0.9256	4.605	0.9257	−.06432

Table 3.4 Curves *A–F* and Their Error Sum of Squares $S(\beta, \gamma)$ from Equation (3.23) for Data Set I in Table 3.3

Curve	β	γ	$S(\beta,\gamma)$
A	2.0	0.7	0.560912
B	2.4	0.325	0.02624
C	1.7	0.5	0.02188
D	1.0	0.9	0.03136
E	0.8	1.4	0.022904
F	1.0	0.6	0.21908

not. For each of the curves *A–F*, the error sum of squares

$$S(\beta,\gamma) = \Sigma\,[y_i - \beta(1 - e^{-\gamma x_i})]^2 \tag{3.23}$$

is given in Table 3.4. Using (3.23) as the criterion, we see that curve *C* fits the best. In Fig. 3.13*a*, using Set I, we have also plotted the contours of $S(\beta,\gamma)$ at levels corresponding, respectively, to the boundaries of the "exact" 99% and 95% confidence regions for (β,γ) defined by (3.9).

The curve μ versus x of (3.22) is graphed in Fig. 3.14. As $x \to \infty$, $\mu \to \beta$. Also

$$\frac{d\mu}{dx} = \beta\gamma e^{-\gamma x}$$

so that for small x, $d\mu/dx \approx \beta\gamma = \lambda$, say. If we consider μ as a proportion of its limiting value β, then γ is a parameter governing the rate at which μ approaches this limit. Looking at data set I plotted in Fig. 3.12, we see that the range of x

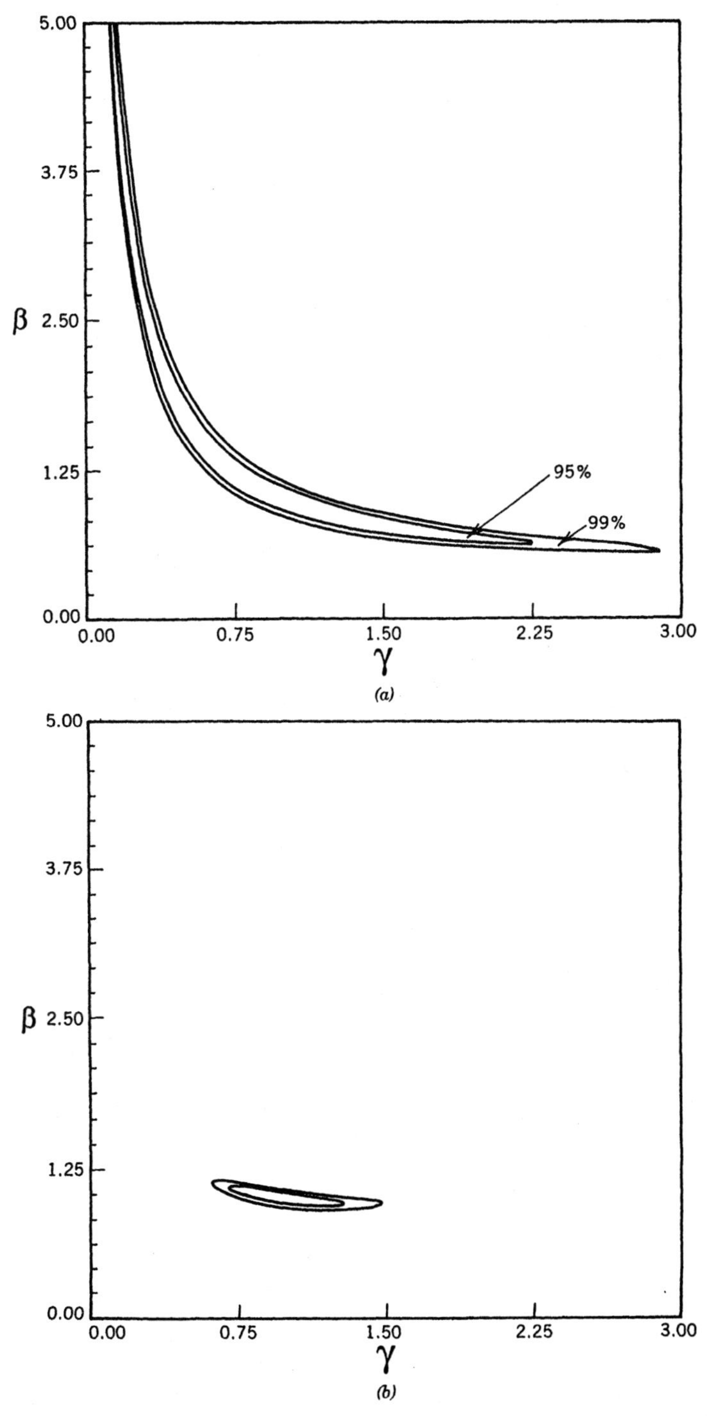

Figure 3.13 Plots of the 99% and 95% "exact" confidence regions (solid curves) and their linearized approximations (dashed curves) for (β, γ), based on Equations (3.9) and (3.6) respectively, for the three data sets in Table 3.3: (*a*) set I, (*b*) set II, (*c*) set II (closer view), (*d*) set III.

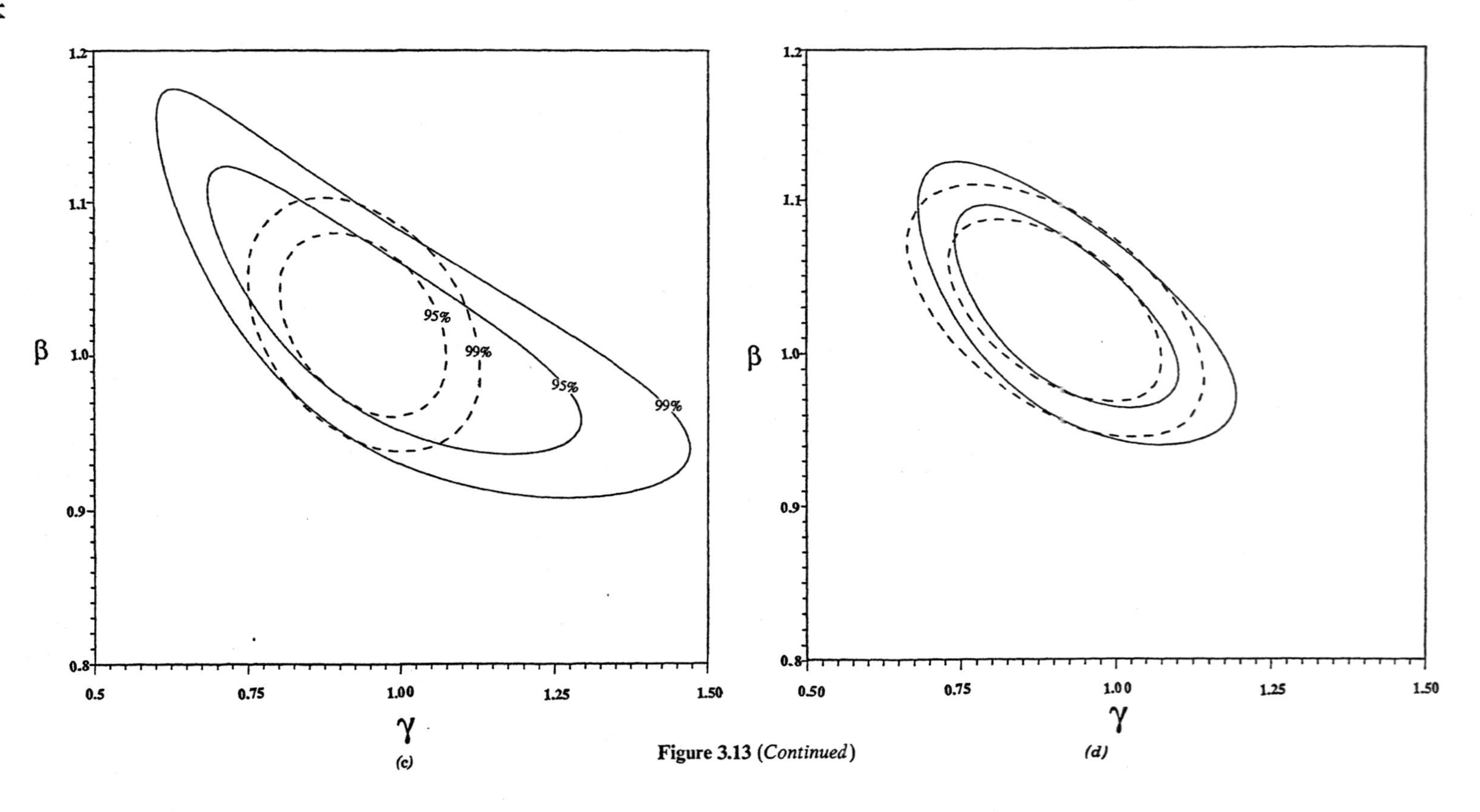

Figure 3.13 *(Continued)*

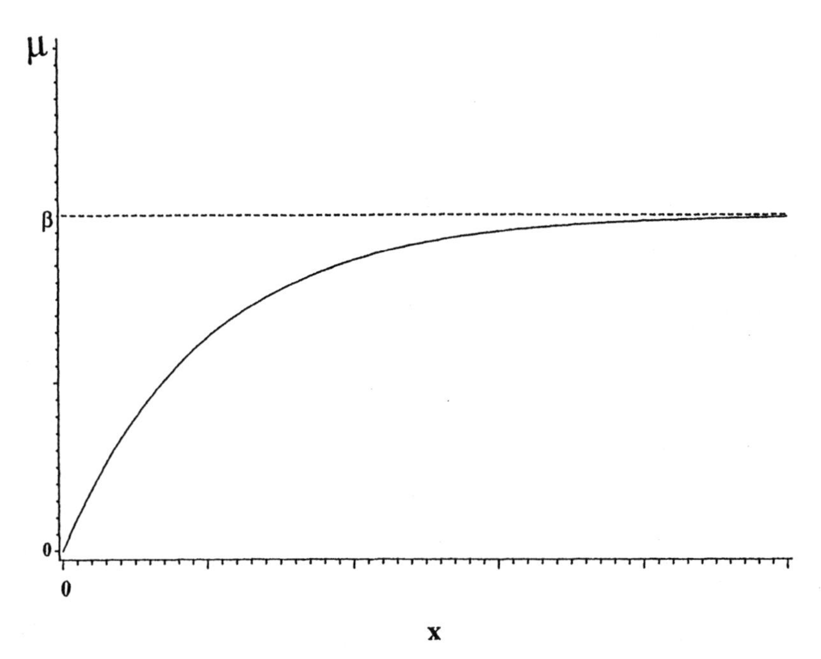

Figure 3.14 Plot of $\mu = E[y]$ versus x, where $\mu = \beta(1 - e^{-\gamma x})$.

is too small for the level of the asymptote $\mu = \beta$ in Fig. 3.14 to be clearly established. Thus much of the information about the parameters comes from the early rate of increase where $\mu \approx \beta\gamma x$, and we tend to have information about the product $\beta\gamma$ rather than the individual parameters. Furthermore, in estimating β we are essentially "predicting" outside the range of our data.

Inferences made from the model (3.22), albeit imprecise for data set I, depend critically upon the assumed function for $\mu = f(\mathbf{x}; \boldsymbol{\theta})$. Even though this function fits the data well over the observed range of x-values, predictions using the fitted model could be grossly misleading. If we do not have complete confidence in the model, it may be more realistic to look for a simpler model, such as a quadratic in x, which is only intended to fit the data well over the observed range of x-values and not explain asymptotic behavior. [For a discussion of (3.22) in the context of growth curves see Section 7.2. The parameter β is of particular interest there.]

The situation can be greatly improved from that of data set I if we have the freedom to design the experiment so that the x-values can be placed further along the curve. For example, for the model with $\beta = \gamma = 1$, we have $\mu = 0.99\beta$ at $x = 4.6$. Thus in data set II in Table 3.3 the x-values are equally spaced between 0.1 and 4.6. The random errors added in are the same as for data set I, but the effect on the confidence regions for (β,γ) is dramatic (see Fig. 3.13*b*). The parameters are now much more precisely estimated, there is less ill-conditioning, and the bending of the contours (banana shape) has lessened. For comparison, the linearized confidence regions (3.6) are also shown as dashed curves on Fig. 3.13*c*.

The situation can be improved even further by using an optimal experimental design [see Equation (5.148)] in which we take half our observations at $x = 0.954$ (where $\mu = 0.615\beta$) and the other half at 4.605 ($\mu = 0.99\beta$). This gives data set III in Table 3.3. The corresponding 99% and 95% "exact" confidence regions for (β, γ), together with the linearized regions, are plotted in Fig. 3.13*d*. We see that there is a further improvement in precision and conditioning, and again less bending of the contours. Clearly the linearized regions are a better approximation to the exact regions for data set III than those for data set II. ■

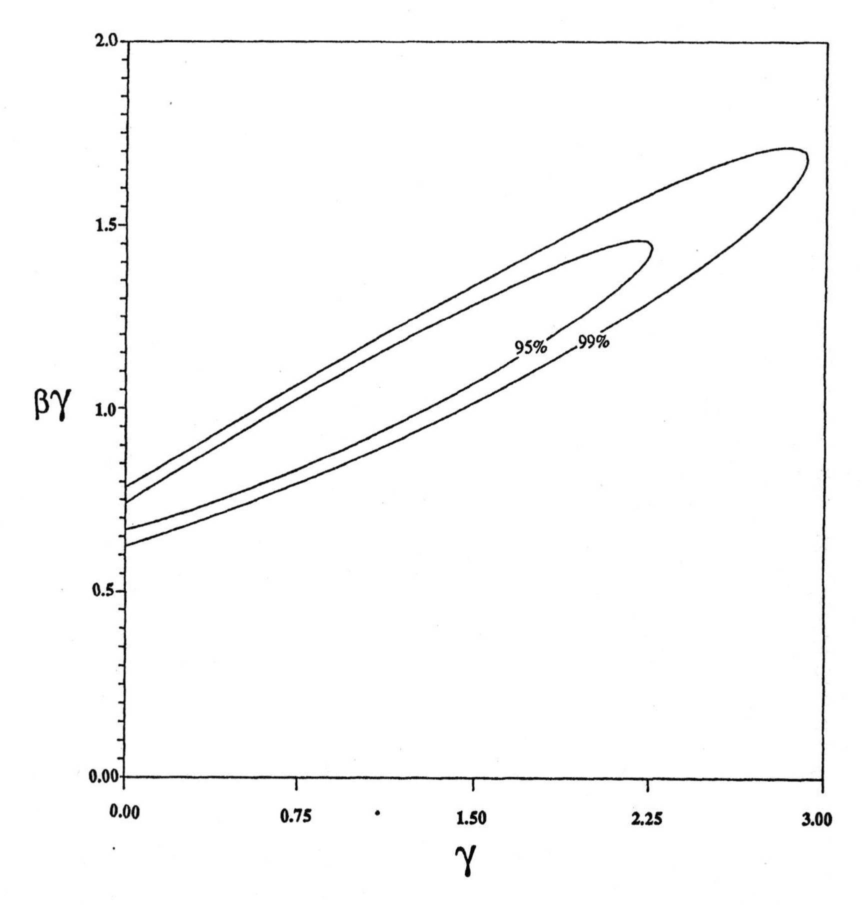

Figure 3.15 Plots of the 95% and 99% "exact" confidence regions for $(\beta\gamma, \gamma)$ for Example 3.4, data set I (given in Table 3.3).

c Stable Parameters

Ross [1970] discussed the idea of *stable parameters* using an example similar to Example 3.4. These parameters constitute a parameter vector $\boldsymbol{\theta}$ for which the elements of $\hat{\boldsymbol{\theta}}$ are as close as possible to being uncorrelated. In particular, the contours of $S(\boldsymbol{\theta})$ should not be banana-shaped and in the two-dimensional case should be oriented in the form of Fig. 3.10*a* or *b*. The parameters are called stable because a change in the value of one parameter has little effect on the estimates of the others (when the first parameter is assumed known and held fixed while the rest are estimated). For example, using data set I of Table 3.3, a contour plot of $S(\beta, \gamma)$ using parameters $\beta\gamma$ and γ gives approximately elliptical contours (Fig. 3.15). To remove the correlation and preserve the physically interpretable parameter $\beta\gamma$, we take

$$\theta_1 = \beta\gamma, \qquad \theta_2 = \beta\gamma - 0.35\gamma. \tag{3.24}$$

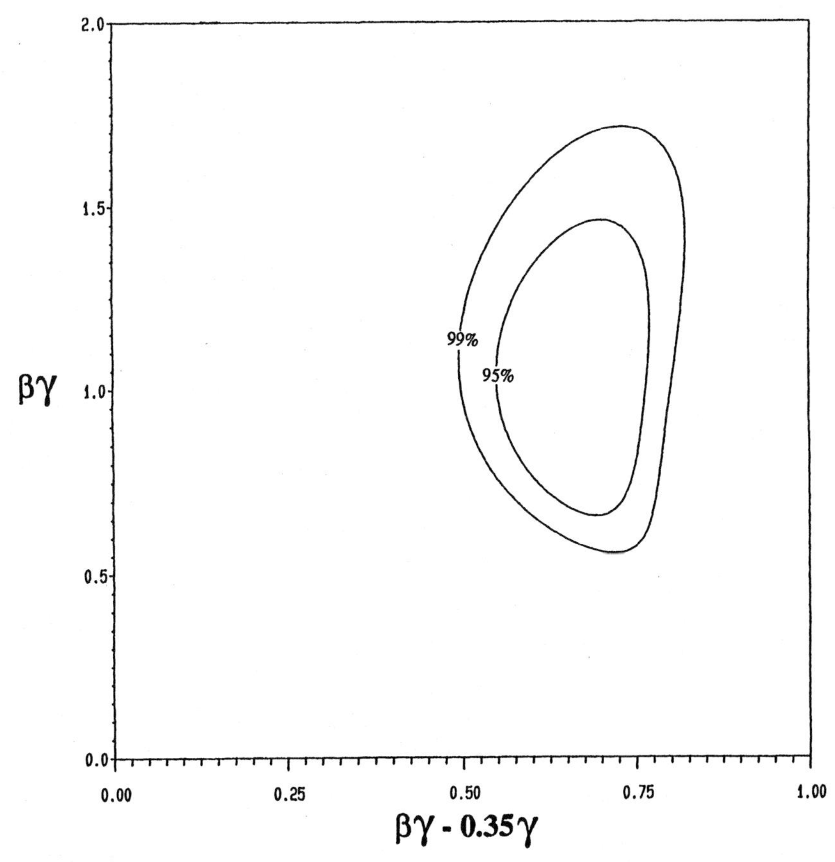

Figure 3.16 Plots of the 95% and 99% "exact" confidence regions for $(\beta\gamma, \beta\gamma - 0.35\gamma)$ for Example 3.4, data set I.

A contour plot for these parameters is given in Fig. 3.16, and we see that these contours have approximately the correct form for stability. One advantage of finding stable parameters lies in forcing us to think about those aspects of the model for which the data provide good information and those aspects for which there is little information.

Approximate ill-conditioning in linear regression (Fig. 3.7) is a problem with the data and not the model. If the experimenter has freedom to place his or her x-values (or design points) and does so in such a way that the variables are uncorrelated (or nearly so), there will be no ill-conditioning. However, with nonlinear models ill-conditioning can be a feature of the model itself. As we have seen in the previous example, good experimental design can reduce the problem, but it may not be able to eliminate it.

d Parameter Redundancy

Models of the form

$$y = \sum_{i=1}^{d} \alpha_i e^{-\beta_i x} + \varepsilon \tag{3.25}$$

arise frequently in many subject areas (see Chapter 8 on compartmental models). We now discuss one such model below, which introduces a new problem relating to ill-conditioning.

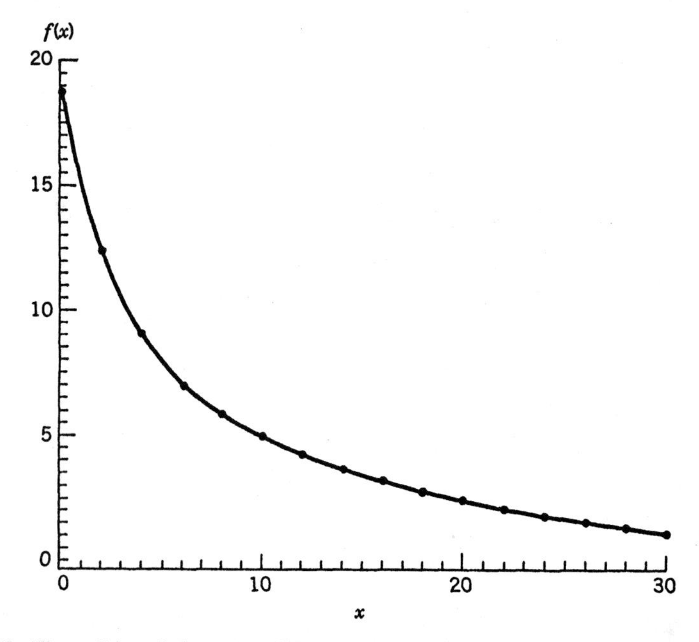

Figure 3.17 Ill-conditioned character of exponential functions. The points were calculated from Equation (3.26a) and the solid curve from (3.26b).

Example 3.5 In Fig. 3.17, inspired by Cornish-Bowden [1976], we have superimposed plots of

$$f(x;\boldsymbol{\theta}) = 5e^{-x/1.5} + 5e^{-x/4} + 9e^{-x/14}, \tag{3.26a}$$

indicated by the dots, and

$$f(x;\boldsymbol{\theta}) = 8.54e^{-x/2.1} + 10.40e^{-x/12.85}, \tag{3.26b}$$

indicated by a solid curve. Even though these functions are so different, the curves are visually indistinguishable. This approximate lack of identifiability is called *parameter redundancy* by Reich [1981]. Models such as this give rise to bad ill-conditioning, no matter where the x's are placed—a common problem with linear combinations of exponentials. Figure 3.18 displays plots of two curves, each based on sums of two exponentials. Even though they have very different parameter values, the curves almost coincide. ■

Example 3.6 Consider the modified Adair equation

$$f(x;\boldsymbol{\theta}) = V_{\max}\frac{k_1x + 2k_1k_2x^2 + 3k_1k_2k_3x^3 + 4k_1k_2k_3k_4x^4}{1 + k_1x + k_1k_2x^2 + k_1k_2k_3x^3 + k_1k_2k_3k_4x^4},$$

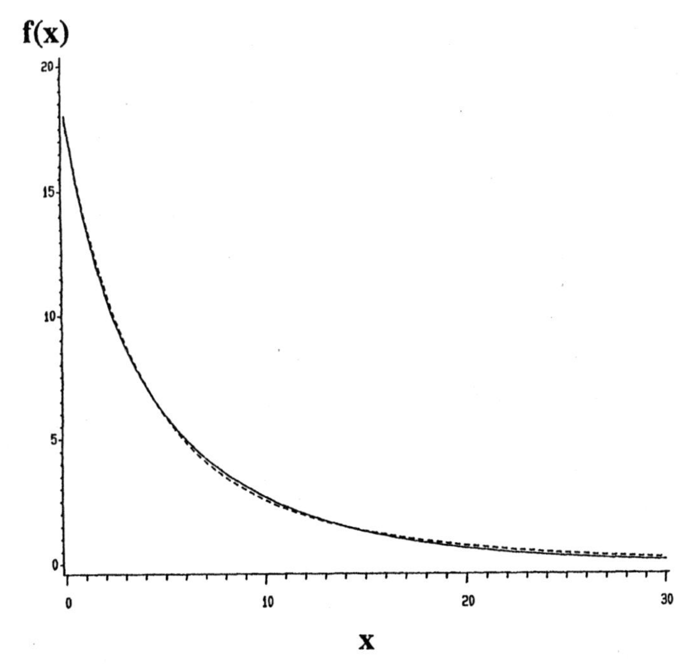

Figure 3.18 Ill-conditioned character of exponential functions. The solid curve is $f(x) = 7e^{-x/2} + 11e^{-x/7}$. The dashed curve is $11.78e^{-x/3.1} + 6.06e^{-x/9.4}$.

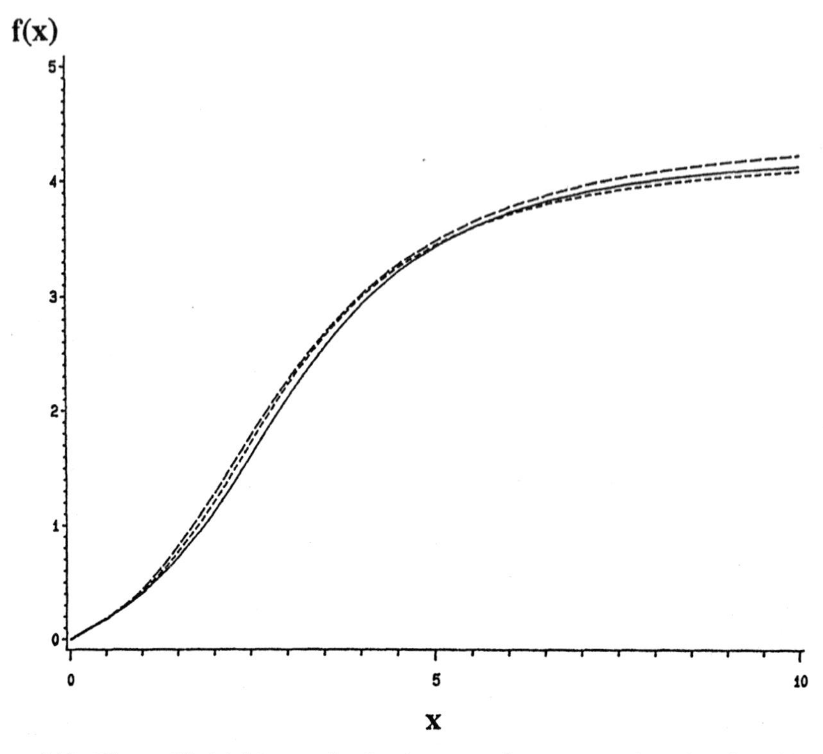

Figure 3.19 The modified Adair equation for three sets of parameter values given in Table 3.5.

where $\theta = (k_1, k_2, k_3, k_4, V_{max})'$. Using the values of the parameters given in Table 3.5 (Reich [1974]), we again find that the three curves (Fig. 3.19) are very similar in spite of the big differences in parameter values. ■

Example 3.7 The so-called Hill equation is given by the family of curves

$$f(x; \theta) = V_{max} \frac{x^n}{k^n + x^n}$$

where $\theta = (V_{max}, k, n)'$. In Fig. 3.20, adapted from Reich [1974], we have two curves from the family corresponding to parameter values (a) $\theta = (1.108, 0.3, 0.8),'$ the solid curve, and (b) $\theta = (1, 1, 3.1)$, the dashed curve. We see that the two curves fit the observed data points equally well, as they are very similar over the range of the observed x-values. Clearly observations at smaller values of x are needed to discriminate between these competing curves. ■

For a single continuous x-variable on $[0, x_{max}]$, Reich [1981] advocates checking for parameter redundancy *prior* to data collection using methods based

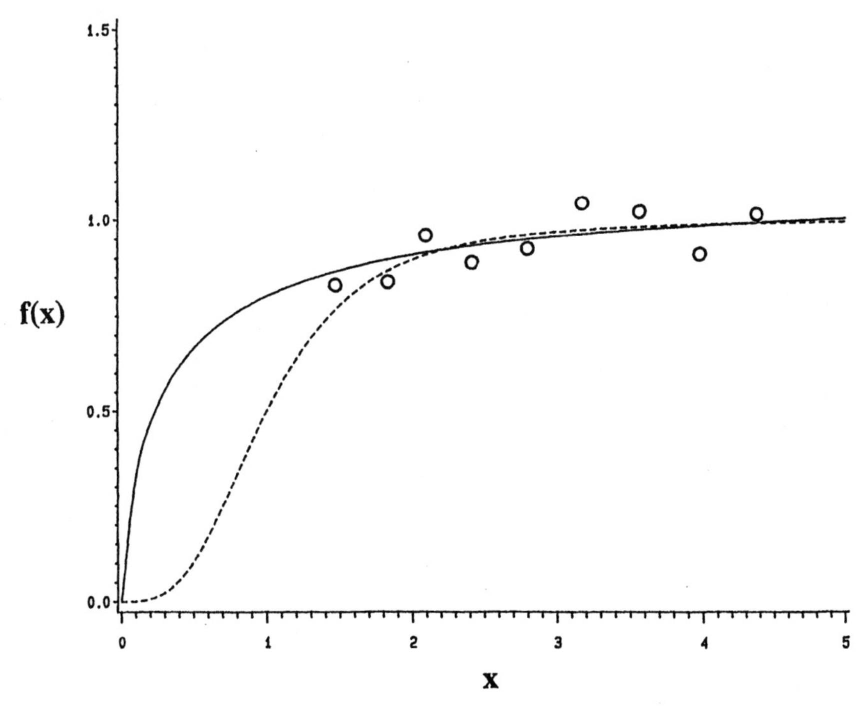

Figure 3.20 The Hill equation for two sets of parameter values. A set of observed data points is also shown.

on the matrix

$$\mathbf{M}_\theta = \frac{1}{x_{max}} \int_0^{x_{max}} \frac{\partial f(x;\boldsymbol{\theta})}{\partial \boldsymbol{\theta}} \cdot \frac{\partial f(x;\boldsymbol{\theta})}{\partial \boldsymbol{\theta}'} dx. \tag{3.27}$$

He motivates the choice of this matrix as follows. For a small change $d\boldsymbol{\theta}$ in $\boldsymbol{\theta}$ we have the following change in the model structure:

$$\mathbf{f}(\boldsymbol{\theta} + d\boldsymbol{\theta}) - \mathbf{f}(\boldsymbol{\theta}) \approx \mathbf{F}.(\boldsymbol{\theta})\, d\boldsymbol{\theta}.$$

For a "sensitive" model we would like the absolute value of this change to be

Table 3.5 Parameter Values for Three Curves Satisfying the Modified Adair Equation[a]

Model	$\log k_1$	$\log k_2$	$\log k_3$	$\log k_4$	V_{max}
I	−1.01	−1.87	−4.0	2.87	1.066
II	−1.0	−2.33	−0.69	0.0	1.07
III	−1.4	−0.90	−2.18	0.30	1.12

[a]From Reich [1974].

as large as possible, so that a measure of sensitivity is the average

$$
\begin{aligned}
T_n &= \frac{1}{n}\|\mathbf{f}(\boldsymbol{\theta}+d\boldsymbol{\theta})-\mathbf{f}(\boldsymbol{\theta})\|^2 \\
&\approx \frac{1}{n}(d\boldsymbol{\theta})'\mathbf{F}'_.(\boldsymbol{\theta})\mathbf{F}_.(\boldsymbol{\theta})\,d\boldsymbol{\theta} \\
&= d\boldsymbol{\theta}'\left(\frac{1}{n}\sum_{i=1}^{n}\dot{\mathbf{f}}_i\dot{\mathbf{f}}_i'\right)d\boldsymbol{\theta},
\end{aligned}
$$

where $\dot{\mathbf{f}}_i = \partial f(x_i;\boldsymbol{\theta})/\partial\boldsymbol{\theta}$. If we take equally spaced x_i's over the interval $[0, x_{\max}]$ and set $\delta x = x_{\max}/n$, then

$$
\begin{aligned}
\frac{1}{n}\sum_{i=1}^{n}\dot{\mathbf{f}}_i\dot{\mathbf{f}}_i' &= \frac{1}{x_{\max}}\sum_{i=1}^{n}\dot{\mathbf{f}}_i\dot{\mathbf{f}}_i'\frac{x_{\max}}{n} \\
&= \frac{1\cdot}{x_{\max}}\sum_{x}\frac{\partial f(x;\boldsymbol{\theta})}{\partial\boldsymbol{\theta}}\cdot\frac{\partial f(x;\boldsymbol{\theta})}{\partial\boldsymbol{\theta}'}\,\delta x \\
&\to \mathbf{M}_{\boldsymbol{\theta}},
\end{aligned}
\tag{3.28}
$$

as $n\to\infty$. Thus a sensitivity measure for change $d\boldsymbol{\theta}$ can be defined as

$$T_{\boldsymbol{\theta}} = d\boldsymbol{\theta}'\,\mathbf{M}_{\boldsymbol{\theta}}\,d\boldsymbol{\theta}.$$

Reich and Zinke [1974] and Reich [1981] proposed separating the diagonal and off-diagonal elements m_{ij} of $\mathbf{M}_{\boldsymbol{\theta}}$ by using the correlation-like matrix

$$\mathbf{R}_{\boldsymbol{\theta}} = [(r_{ij})] = \mathbf{D}^{-1}\mathbf{M}_{\boldsymbol{\theta}}\mathbf{D}^{-1}, \tag{3.29}$$

where $\mathbf{D} = \operatorname{diag}(\sqrt{m_{11}}, \sqrt{m_{22}}, \ldots, \sqrt{m_{pp}})$ and $r_{ij} = m_{ij}/(m_{ii}m_{jj})^{1/2}$. Let $\lambda_1 \geq \lambda_2 \geq \cdots \geq \lambda_p$ be the eigenvalues of $\mathbf{R}_{\boldsymbol{\theta}}$, and let $\mathbf{T} = [\mathbf{t}_1, \mathbf{t}_2, \ldots, \mathbf{t}_p]$ be the orthogonal matrix such that $\mathbf{T}'\mathbf{R}_{\boldsymbol{\theta}}\mathbf{T} = \operatorname{diag}(\lambda_1, \lambda_2, \ldots, \lambda_p)$, i.e. $\mathbf{R}_{\boldsymbol{\theta}}\mathbf{t}_k = \lambda_k\mathbf{t}_k$ $(k = 1, 2, \ldots, p)$ (c.f. Appendix A2.2). Then we can describe those directions $\mathbf{u}_k = \mathbf{D}^{-1}\mathbf{t}_k$ for which $\lambda_k \approx 0$ as redundant directions, since $d\boldsymbol{\theta} \propto \mathbf{u}_k$ implies

$$
\begin{aligned}
T_{\boldsymbol{\theta}} &\propto \mathbf{u}_k'\mathbf{M}_{\boldsymbol{\theta}}\mathbf{u}_k \\
&= \mathbf{t}_k'\mathbf{D}^{-1}\mathbf{M}_{\boldsymbol{\theta}}\mathbf{D}^{-1}\mathbf{t}_k \\
&= \mathbf{t}_k'\mathbf{R}_{\boldsymbol{\theta}}\mathbf{t}_k \qquad [\text{by (3.29)}] \\
&= \mathbf{t}_k'\lambda_k\mathbf{t}_k \\
&= \lambda_k \\
&\approx 0.
\end{aligned}
$$

Reich [1981] describes the diagonal elements of $\mathbf{D}$ as giving the "sensitivities" of the individual parameters and $\mathbf{R}_\theta$ as representing the effect of redundancy. Reich and Zinke [1974] proposed using $|\mathbf{R}_\theta|^{-1} = (\Pi_k \lambda_k)^{-1}$ as a measure of redundancy, large values indicating severe redundancy. These will arise when one or more of the λ_k are small. We note that, in general, the λ_k are positive because $\mathbf{R}_\theta$ is positive definite, and $p = \operatorname{tr} \mathbf{R}_\theta = \Sigma_k \lambda_k$. Hence some of the eigenvalues will be greater than 1 and some will be less than 1, with an average of 1. An "acceptable" threshold level for $|\mathbf{R}_\theta|^{-1}$ will depend on σ^2 and p: Reich and Zinke [1974] suggest levels of 10^3 or 10^4.

From Theorem 2.1 in Section 2.1.2 with $\boldsymbol{\theta}^*$ replaced by $\boldsymbol{\theta}$ for notational convenience, the variance–covariance matrix of $\sqrt{n}\hat{\boldsymbol{\theta}}$ is approximately

$$
\begin{aligned}
\mathbf{V}_n(\boldsymbol{\theta}) &= n\sigma^2 [\mathbf{F}_{.}'(\boldsymbol{\theta})\mathbf{F}_{.}(\boldsymbol{\theta})]^{-1} \\
&= \sigma^2 \left[\frac{1}{n} \sum_{i=1}^{n} \dot{\mathbf{f}}_i \dot{\mathbf{f}}_i' \right]^{-1} \qquad (3.30) \\
&\to \sigma^2 \mathbf{M}_\theta^{-1} \quad \text{as } n \to \infty \qquad [\text{by } (3.28)].
\end{aligned}
$$

Now if $\mathbf{v}_k = \mathbf{D}\mathbf{t}_k$, then

$$
\begin{aligned}
\operatorname{var}[\sqrt{n}\mathbf{v}_k'\hat{\boldsymbol{\theta}}] &\approx \sigma^2 \mathbf{v}_k' \mathbf{M}_\theta^{-1} \mathbf{v}_k \\
&= \sigma^2 \mathbf{t}_k' \mathbf{D}\mathbf{D}^{-1} \mathbf{R}_\theta^{-1} \mathbf{D}^{-1} \mathbf{D} \mathbf{t}_k \qquad [\text{by } (3.29)] \\
&= \sigma^2 \mathbf{t}_k' \mathbf{R}_\theta^{-1} \mathbf{t}_k \\
&= \sigma^2 \mathbf{t}_k' \lambda_k^{-1} \mathbf{t}_k \\
&= \sigma^2 \lambda_k^{-1}. \qquad (3.31)
\end{aligned}
$$

When λ_k is small, there is a direction $\mathbf{v}_k$ which leads to a parametric function $\mathbf{v}_k'\hat{\boldsymbol{\theta}}$ with large variance. Clearly the problems of large variances and redundancy are related. If we worked with $\mathbf{M}_\theta$ instead of $\mathbf{R}_\theta$, we would get a different set of eigenvalues and eigenvectors. Redefining λ_k and $\mathbf{t}_k$ to be these quantities and setting $\mathbf{D} = \mathbf{I}_d$ above, we then have $\mathbf{u}_k = \mathbf{v}_k = \mathbf{t}_k$ so that redundant directions correspond directly to parametric functions with large variances. However, the use of $\mathbf{R}_\theta$ here is often sensible, as we would want our measure of parameter redundancy to be independent of the measurement scale used for the x-variable. The criterion $|\mathbf{R}_\theta|^{-1}$ may therefore be a useful addition to the predicted asymptotic variances of the $\sqrt{n}\hat{\theta}_r$, the diagonal elements of $\sigma^2 \mathbf{M}_\theta^{-1}$.

One interpretation of the criterion $|\mathbf{R}_\theta|^{-1}$ is given by Reich [1981]. Suppose the parameters are scaled for "equal sensitivity," so that $\mathbf{D} \approx \mathbf{I}_p$ and $\mathbf{R}_\theta \approx \mathbf{M}_\theta$. Then, for $\|\mathbf{a}\| = 1$,

$$
\operatorname{var}[\sqrt{n}\mathbf{a}'\hat{\boldsymbol{\theta}}] \approx \sigma^2 \mathbf{a}' \mathbf{R}_\theta^{-1} \mathbf{a},
$$

which has maximum and minimum values of $\sigma^2 \lambda_{\min}^{-1}$ and $\sigma^2 \lambda_{\max}^{-1}$ respectively (by

Appendix A7.3). Hence the standard deviations of some parametric functions will be of order $\sigma\lambda_{\min}^{-1/2}$, which is $\lambda_{\min}^{-1/2}$ times σ. For example, if $\lambda_{\min} = 0.01$, then the scale factor for converting the standard deviation of the error into the standard deviation of $\sqrt{n}$ times the parametric estimate is $\lambda_{\min}^{-1/2} = 10$. Reich [1981] suggests that $|\mathbf{R}_\theta|^{-1/2} = (\prod_k \lambda_k)^{-1/2}$ is a "global" measure of this scale factor.

Example 3.6 Reich [1981] considered the model

$$f(x;\boldsymbol{\theta}) = \alpha_1 e^{-\beta_1 x} + \alpha_2 e^{-\beta_2 x},$$

where $\boldsymbol{\theta} = (\alpha_1, \alpha_2, \beta_1, \beta_2)'$. He assumed that $x_{\max}$ is large enough so that the exponential terms have decayed virtually to zero by the time $x = x_{\max}$. Then integrals from 0 to $x_{\max}$ can be replaced by integrals from 0 to ∞, and

$$m_{11} = \int_0^\infty \left(\frac{\partial f}{\partial \alpha_1}\right)^2 dx = \int_0^\infty e^{-2\beta_1 x} dx = \frac{1}{2\beta_1} \qquad \text{etc.}$$

Hence, from Reich [1981],

$$\mathbf{R}_\theta = \begin{bmatrix} 1 & b & -1/\sqrt{2} & -b(1-c) \\ b & 1 & bc & -1/\sqrt{2} \\ -1/\sqrt{2} & bc & 1 & b^3 \\ -b(1-c) & -1/\sqrt{2} & b^3 & 1 \end{bmatrix}$$

Table 3.6 Variation of the Redundancy Measure with the Ratio of Decay Rates[a]

Ratio r	$1/\lvert\mathbf{R}_\theta\rvert$
1	∞
0.9	6.8×10^{10}
0.8	1.7×10^{8}
0.7	4.3×10^{6}
0.6	2.6×10^{5}
0.5	2.6×10^{4}
0.4	3515
0.3	566
0.2	103
0.1	20
0.01	4.7
0.001	4.06
0	4

[a] Adapted from Reich [1981: Table 1].

where

$$b = \frac{2\sqrt{\beta_1\beta_2}}{\beta_1 + \beta_2} \quad \text{and} \quad c = \frac{\beta_1}{\beta_1 + \beta_2}.$$

We note that $\mathbf{R}_\theta$ depends only on the decay rates β_1 and β_2 and not on the amplitudes α_1 and α_2. Then, defining $r = \beta_2/\beta_1 < 1$, it transpires that

$$|\mathbf{R}_\theta| = \frac{1}{4}\left(\frac{1-r}{1+r}\right)^8.$$

A table of $|\mathbf{R}_\theta|^{-1}$ for different values of r is given in Table 3.6. Using a threshold value of $|\mathbf{R}_\theta|^{-1} = 100$, we see that parameters are redundant if $r > 0.2$, i.e. if the decay rates differ by less than a factor of 5. ■

Example 3.7 Reich and Zinke [1974] investigated the Monod–Wyman–Changeux kinetic equation

$$f(x;\boldsymbol{\theta}) = V_{\max}\frac{\dfrac{k_1x}{1+k_1x} + L_1\dfrac{k_2x}{1+k_2x}}{L_1+1}, \qquad L_1 = \left(L\frac{1+k_1x}{1+k_2x}\right)^q, \tag{3.32}$$

with $q = 4$ and $\boldsymbol{\theta} = (V_{\max}, k_1, k_2, L)'$. They found that $|\mathbf{R}_\theta|^{-1}$ varied considerably, depending on which of the parameters were known from other sources. Without going into numerical details regarding the range of x etc., we summarize some of their results briefly. When the full model is fitted, $|\mathbf{R}_\theta|^{-1}$ is 8300, whereas if one of the parameters is known, the criterion is 1000 ($V_{\max}$ known), 1.9 (k_1 known), 1400 (k_2 known), or 2.1 (L known). The authors proposed a threshold level of 100 for $|\mathbf{R}_\theta|^{-1}$, so that there is strong evidence of redundancy. There are too many parameters for the specification of the sigmoidal curve (3.32). However, a knowledge of $V_{\max}$ or k_2 does not help much, whereas knowledge of k_1 or L leads to a drastic improvement. ■

If we are interested in assessing the degree of redundancy in a model for chosen experimental design points $x_1, x_2, \ldots, x_n$, then we use $n^{-1}\sum_i \dot{\mathbf{f}}_i \dot{\mathbf{f}}_i' = n^{-1}\mathbf{F}'_\cdot(\boldsymbol{\theta})\mathbf{F}_\cdot(\boldsymbol{\theta})$. If the columns of $\mathbf{F}_\cdot(\boldsymbol{\theta})$ are close to being linearly dependent, then $\mathbf{F}'_\cdot(\boldsymbol{\theta})\mathbf{F}_\cdot(\boldsymbol{\theta})$—and therefore $\mathbf{R}_\theta$—will be nearly singular, and $|\mathbf{R}_\theta|^{-1}$ will be large. Also the asymptotic variance–covariance matrix (3.30) will be ill-conditioned. Reich and Zinke [1974] allowed for unequal weighting in the model. The reader is referred to their paper for details and further examples.

Reich's work is really part of experimental design. In Section 5.13 we discuss the placing of the **x**-values to maximize the "precision" of $\hat{\boldsymbol{\theta}}$. Typically we minimize the determinant of $\mathscr{D}[\hat{\boldsymbol{\theta}}]$, though experimental designing should not stop here. We should investigate $\mathscr{D}[\hat{\boldsymbol{\theta}}]$ [for example $\sigma^2\{\mathbf{F}_\cdot(\boldsymbol{\theta})'\mathbf{F}_\cdot(\boldsymbol{\theta})\}^{-1}$]

for the optimal design and intended sample size to ensure that the experiment is worth doing. If ill-conditioning will prevent a satisfactory separate estimation of physically important parameters, then another source of information may be required to separate them. It may not be worth while proceeding without additional resources. As asymptotic theory forms the basis of design considerations, it is probably a good idea to simulate the model and check the applicability of the asymptotics. Unfortunately, a feature of nonlinear models is that we usually need to have a fairly good idea about the value of $\boldsymbol{\theta}$ before useful design work can be done. Sequential designs in which the estimate of $\boldsymbol{\theta}$ is sequentially updated are appropriate here.

e Some Conclusions

It is helpful to now give a brief summary of some of the problems described above:

1. Ill-conditioning problems of various kinds can be inherent in the model itself.
2. Bad experimental designs with x-values in the wrong places can make matters worse, but even with good designs and lots of data it may not be possible to eliminate severe ill-conditioning.
3. Very accurate experimental data can still give rise to extremely imprecise and highly correlated estimates.
4. Problem 3 may occur even though the model being used is the correct one and fits the data well.
5. Scientists are sometimes reluctant to recognize the above problems.

Apart from a careful choice of design points, some of the above problems can be overcome by using appropriate transformations. Ross [1980] lists five types of transformations, each achieving a different purpose. They include transformations of the parameters to achieve approximately zero correlations among the estimates of the new parameters, transformations to improve the computational processes either applied initially or throughout the optimization, transformations to straighten curved valleys, and transformations to minimize the curvature effects described in Chapter 4. This latter aspect is discussed in Section 5.8.4.

CHAPTER 4

Measures of Curvature and Nonlinearity

4.1 INTRODUCTION

In Section 2.1.3 we indicated briefly how the least-squares estimate $\hat{\boldsymbol{\theta}}$ for the nonlinear model $\mathscr{E}[\mathbf{y}] = \mathbf{f}(\boldsymbol{\theta})$ can be calculated iteratively using a linear approximation. We found that $\mathbf{F}_{\bullet}(\boldsymbol{\theta})$ plays the same role as the matrix $\mathbf{X}$ in the approximating linear model, and that asymptotically $\hat{\boldsymbol{\theta}} \sim N_p(\boldsymbol{\theta}, \sigma^2[\mathbf{F}_{\bullet}'(\boldsymbol{\theta})\mathbf{F}_{\bullet}(\boldsymbol{\theta})]^{-1})$. Furthermore, in Chapter 5 we shall see how this approximate linearization can be exploited for making inferences about $\boldsymbol{\theta}$ and its elements. However, the linear approximation is asymptotic, and we would like to know what happens when n, the dimension of $\mathbf{y}$, is moderate or small. We then find that $\hat{\boldsymbol{\theta}}$ can have an appreciable bias with a variance–covariance matrix $\mathscr{D}[\hat{\boldsymbol{\theta}}]$ that exceeds $\sigma^2[\mathbf{F}_{\bullet}'(\boldsymbol{\theta})\mathbf{F}_{\bullet}(\boldsymbol{\theta})]^{-1}$ in the sense that the difference is generally positive definite. Later, in Section 4.7, we shall show that, to a first order of approximation, the bias and additional variance–covariance terms are simple functions of certain arrays of projected second derivatives called curvature arrays. These arrays are fundamental to the study of nonlinearity and come in two varieties: the first, called the intrinsic array, measures the degree of bending and twisting of the surface $\mathbf{f}(\boldsymbol{\theta})$; and the second, called the parameter-effects array, describes the degree of curvature induced by the choice of parameters $\boldsymbol{\theta}$.

Based on intuitive ideas from differential geometry, the above arrays can be used to develop the concepts of intrinsic and parameter-effects curvature at a point $\boldsymbol{\theta}$. These curvatures are introduced in Section 4.2 along with their interpretation and "reduced" formulae for their computation. They can be useful in assessing the appropriateness of the linear approximation mentioned above. Several other curvature measures have also been introduced, and these are discussed briefly in Section 4.3; related quantities called statistical curvatures and connection coefficients are introduced in Section 4.4. In Section 4.5 the curvatures developed in Section 4.2 are generalized to deal with subsets of $\boldsymbol{\theta}$ rather than the full parameter set.

As might be expected from the properties of $\hat{\boldsymbol{\theta}}$, the above intrinsic array also plays a role in the interpretation of residual plots, as we see in Section 4.6. The presence of moderate intrinsic-array values can invalidate residual plots, and new residuals called projected residuals are introduced. Similar problems arise in the construction of confidence regions and confidence intervals, but the discussion of these is postponed to Chapter 5.

Finally, in Section 4.7, several theoretical and empirical methods based on simulation are developed for assessing the effects of nonlinearity on the distributional properties of $\hat{\boldsymbol{\theta}}$.

4.2 RELATIVE CURVATURE

Using concepts from differential geometry (see Appendix B for some background), Bates and Watts [1980, 1981a] and Goldberg et al. [1983] extended the pioneering work of Beale [1960] and developed a number of useful measures of nonlinearity based on the notion of curvature. These measures are independent of scale changes in both the data and the parameters, so that they can be used for comparing different data sets as well as different parametrizations of the same data set. The discussion in much of this section is based on the papers of Bates and Watts. We begin by distinguishing between the concepts of a model and a model function. This distinction is akin to the difference between a linear vector space and its matrix representation in the form of, say, the range space or null space (kernel) of a matrix. The model takes the form

$$\mathbf{y} = \boldsymbol{\mu} + \boldsymbol{\varepsilon}, \tag{4.1}$$

where $\boldsymbol{\mu} \in \Omega$, a subset of the sample space of $\mathbf{y}$. Usually the sample space is $\mathbb{R}^n$. For the usual nonlinear regression model [c.f. (2.1) and (2.4)], Ω can be described in terms of a p-dimensional parameter $\boldsymbol{\theta} \in \Theta$ and a model function $\mathbf{f}(\boldsymbol{\theta})$; thus

$$\Omega = \{\boldsymbol{\mu} : \boldsymbol{\mu} = \mathbf{f}(\boldsymbol{\theta}), \boldsymbol{\theta} \in \Theta\}, \tag{4.2}$$

where

$$[\mathbf{f}(\boldsymbol{\theta})]_i = f_i(\boldsymbol{\theta}) = f(\mathbf{x}_i; \boldsymbol{\theta}).$$

In this case Ω is a p-dimensional surface in the n-dimensional sample space. Since Ω contains all possible values of $\mathscr{E}[\mathbf{y}]$, it is called the *expectation surface* (Goldberg et al. [1983]). Furthermore, as each $\boldsymbol{\mu} \in \Omega$ satisfies $\boldsymbol{\mu} = \mathbf{f}(\boldsymbol{\theta})$ for some $\boldsymbol{\theta}$, Ω is also known as the *solution locus* (Beale [1960]).

It is always possible to represent Ω by a different parametrization and a different model function. For example, for any bijective (one-to-one) function defined by $\boldsymbol{\theta} = \mathbf{a}(\boldsymbol{\phi})$ we have

$$\boldsymbol{\mu} = \mathbf{f}(\boldsymbol{\theta}) = \mathbf{f}\{\mathbf{a}(\boldsymbol{\phi})\} = \mathbf{g}(\boldsymbol{\phi}), \tag{4.3}$$

say. We shall conveniently abuse the function notation and write $\boldsymbol{\mu}(\boldsymbol{\theta}) = \mathbf{f}(\boldsymbol{\theta})$ and $\boldsymbol{\mu}(\boldsymbol{\phi}) = \mathbf{g}(\boldsymbol{\phi})$.

For a linear model we have that

(i) Ω is a linear subspace of the sample space; and
(ii) for any parametric description of Ω in the form

$$\Omega = \{\boldsymbol{\mu} : \boldsymbol{\mu} = \mathbf{X}\boldsymbol{\theta}, \boldsymbol{\theta} \in \mathbb{R}^p\}, \tag{4.4}$$

equally spaced $\boldsymbol{\theta}$-values are mapped onto equally spaced $\boldsymbol{\mu}$-values in Ω.

We shall be interested in the extent of departures from these properties in the nonlinear model.

The least-squares estimate $\hat{\boldsymbol{\theta}}$ of $\boldsymbol{\theta}$ for any linear or nonlinear model is obtained by choosing $\hat{\boldsymbol{\theta}}$ to minimize $S(\boldsymbol{\theta}) = \|\mathbf{y} - \boldsymbol{\mu}(\boldsymbol{\theta})\|^2$, the squared distance between $\mathbf{y}$ and a point $\boldsymbol{\mu}(\boldsymbol{\theta})$ in Ω. For a linear model $S(\boldsymbol{\theta})$ is minimized when $\mathbf{y} - \boldsymbol{\mu}(\boldsymbol{\theta})$ is perpendicular to Ω. For a nonlinear model we shall see below that $S(\boldsymbol{\theta})$ is minimized when $\mathbf{y} - \boldsymbol{\mu}(\boldsymbol{\theta})$ is perpendicular to the tangent plane to the expectation surface at $\boldsymbol{\mu}(\boldsymbol{\theta})$.

4.2.1 Curvature Definitions

Before reading this section it may be helpful for the reader to look again at the two-dimensional examples in Section 3.3.2, where we saw that there were two aspects of curvature in studying nonlinear models. The first is related to the bending of the expectation surface, and the second is related to the method of parametrization.

Having discussed ideas of curvature informally, we now develop formal definitions. Suppose $\boldsymbol{\theta}$ is close to $\hat{\boldsymbol{\theta}}$; then we have the following quadratic Taylor approximation:

$$\begin{aligned}\boldsymbol{\mu} - \hat{\boldsymbol{\mu}} &= \mathbf{f}(\boldsymbol{\theta}) - \mathbf{f}(\hat{\boldsymbol{\theta}}) \\ &\approx \hat{\mathbf{F}}_{\bullet}(\boldsymbol{\theta} - \hat{\boldsymbol{\theta}}) + \tfrac{1}{2}(\boldsymbol{\theta} - \hat{\boldsymbol{\theta}})' \hat{\boldsymbol{F}}_{\bullet\bullet}(\boldsymbol{\theta} - \hat{\boldsymbol{\theta}}) \\ &= \hat{\mathbf{F}}_{\bullet}\boldsymbol{\delta} + \tfrac{1}{2}\boldsymbol{\delta}' \hat{\boldsymbol{F}}_{\bullet\bullet}\boldsymbol{\delta},\end{aligned} \tag{4.5}$$

where $\boldsymbol{\delta} = \boldsymbol{\theta} - \hat{\boldsymbol{\theta}}$. The arrays of derivatives are

$$\underset{n \times p}{\hat{\mathbf{F}}_{\bullet}} = \left[\left(\frac{\partial f_i(\boldsymbol{\theta})}{\partial \theta_j}\right)\right]_{\boldsymbol{\theta} = \hat{\boldsymbol{\theta}}}, \tag{4.6}$$

$$\underset{n \times p \times p}{\hat{\boldsymbol{F}}_{\bullet\bullet}} = \left[\left(\frac{\partial^2 \mathbf{f}(\boldsymbol{\theta})}{\partial \theta_r \partial \theta_s}\right)\right]_{\boldsymbol{\theta} = \hat{\boldsymbol{\theta}}} = [(\hat{\mathbf{f}}_{rs})], \tag{4.7}$$

and we have the symbolic representation in (4.5) of

$$\sum_r \sum_s \hat{\mathbf{f}}_{rs}\delta_r\delta_s = \boldsymbol{\delta}'\hat{\boldsymbol{F}}_{..}\boldsymbol{\delta}. \tag{4.8}$$

Here it is convenient to represent the $p \times p$ array of n-dimensional vectors $\hat{\mathbf{f}}_{rs}$ $(=\hat{\mathbf{f}}_{sr})$ by a three-dimensional $n \times p \times p$ array denoted by $\hat{\boldsymbol{F}}_{..}$ with typical element

$$\hat{f}_{irs} = \left[\frac{\partial^2 f_i(\boldsymbol{\theta})}{\partial\theta_r\,\partial\theta_s}\right]_{\boldsymbol{\theta}=\hat{\boldsymbol{\theta}}}. \tag{4.9}$$

The first dimension can be regarded as being perpendicular to this page, and the $p \times p$ array consisting of the ith slice or "face"—one for each observation—is denoted by $\hat{\mathbf{F}}_{i..} = [(\hat{f}_{irs})]$. Thus (4.8) is a vector with ith element $\boldsymbol{\delta}'\mathbf{F}_{i..}\boldsymbol{\delta}$, and we use boldface italic for $\hat{\boldsymbol{F}}_{..}$ to indicate we are using the form of mutliplication described by (4.8) [see Appendix B4].

If we ignore the quadratic term in (4.5), we have the linear approximation for $\boldsymbol{\theta}$ in the vicinity of $\hat{\boldsymbol{\theta}}$

$$\boldsymbol{\mu} - \hat{\boldsymbol{\mu}} \approx \hat{\mathbf{F}}_.(\boldsymbol{\theta} - \hat{\boldsymbol{\theta}}). \tag{4.10}$$

We shall see below from (4.18) that the range (column space) of the matrix $\hat{\mathbf{F}}_.$ is the tangent plane to the expectation surface at the point $\hat{\boldsymbol{\theta}}$, and (4.10) shows that, under the approximation, $\boldsymbol{\mu}$ lies in this tangent plane. Therefore the linear approximation (4.10) amounts to approximating the expectation surface in the neighborhood of $\hat{\boldsymbol{\theta}}$ by the tangent plane at $\hat{\boldsymbol{\theta}}$. Using this linear approximation, we see from (2.26) that a $100(1-\alpha)\%$ confidence region for $\boldsymbol{\theta}$ is the set of $\boldsymbol{\theta}$ in the tangent plane such that

$$\|\boldsymbol{\mu} - \hat{\boldsymbol{\mu}}\|^2 \approx \|\hat{\mathbf{F}}_.(\boldsymbol{\theta} - \hat{\boldsymbol{\theta}})\|^2 \le ps^2 F_\alpha, \tag{4.11}$$

where $F_\alpha = F^\alpha_{p,n-p}$, the upper α quantile value of the $F_{p,n-p}$ distribution, and $s^2 = \|\mathbf{y} - \hat{\boldsymbol{\mu}}\|^2/(n-p)$. From (4.11) we see that $\boldsymbol{\mu}$ lies approximately on a sphere with center $\hat{\boldsymbol{\mu}}$ and radius $(ps^2 F_\alpha)^{1/2}$. It is convenient at this stage to introduce the parameter

$$\rho = s\sqrt{p}, \tag{4.12}$$

called the standard radius, so that $\rho\sqrt{F_\alpha}$ is the radius of the above sphere. We note that

$$(\boldsymbol{\theta} - \hat{\boldsymbol{\theta}})'\hat{\mathbf{F}}_.'\hat{\mathbf{F}}_.(\boldsymbol{\theta} - \hat{\boldsymbol{\theta}}) \le \rho^2 F_\alpha \tag{4.13}$$

is an ellipsoid with center $\hat{\boldsymbol{\theta}}$.

The validity of the tangent-plane approximation (4.10) will depend on the magnitude of the quadratic term $\boldsymbol{\delta}'\hat{F}_{..}\boldsymbol{\delta}$ in (4.5) relative to the linear term $\hat{\mathbf{F}}_{.}\boldsymbol{\delta}$. In making this comparison we find that it is helpful to split the quadratic term, an $n \times 1$ vector, into two orthogonal components, the respective projections onto the tangent plane and normal to the tangent plane. This decomposition can be achieved using the projection matrix (Appendix A11.4)

$$\hat{\mathbf{P}}_F = \hat{\mathbf{F}}_{.}(\hat{\mathbf{F}}_{.}'\hat{\mathbf{F}}_{.})^{-1}\hat{\mathbf{F}}_{.}'. \tag{4.14}$$

These two components are then given by

$$\hat{F}_{..} = \hat{F}_{..}^T + \hat{F}_{..}^N,$$

where $\hat{F}_{..}^T = [(\hat{\mathbf{f}}_{rs}^T)]$, $\hat{F}_{..}^N = [(\hat{\mathbf{f}}_{rs}^N)]$, $\hat{\mathbf{f}}_{rs}^T = \hat{\mathbf{P}}_F\hat{\mathbf{f}}_{rs}$, and $\hat{\mathbf{f}}_{rs}^N = (\mathbf{I}_n - \hat{\mathbf{P}}_F)\hat{\mathbf{f}}_{rs}$. Here T and N stand for the tangential and normal components respectively. Bates and Watts [1980] now define two measures for comparing each quadratic component with the linear term, namely

$$K_{\delta}^T = \frac{\| \boldsymbol{\delta}'\hat{F}_{..}^T\boldsymbol{\delta} \|}{\| \hat{\mathbf{F}}_{.}\boldsymbol{\delta} \|^2} \tag{4.15}$$

and

$$K_{\delta}^N = \frac{\| \boldsymbol{\delta}'\hat{F}_{..}^N\boldsymbol{\delta} \|}{\| \hat{\mathbf{F}}_{.}\boldsymbol{\delta} \|^2}. \tag{4.16}$$

For reasons described in the next section, the above ratios are called the *parameter-effects curvature* and the *intrinsic curvature*, respectively, in direction $\boldsymbol{\delta}$. Since $\mathbf{f}_{rs} = \mathbf{f}_{rs}^T + \mathbf{f}_{rs}^N$, we have

$$\| \boldsymbol{\delta}'F_{..}\boldsymbol{\delta} \|^2 = \| \boldsymbol{\delta}'F_{..}^T\boldsymbol{\delta} \|^2 + \| \boldsymbol{\delta}'F_{..}^N\boldsymbol{\delta} \|^2,$$

and both of the curvatures need to be small if the tangent-plane approximation is to be valid.

4.2.2 Geometrical Properties

To justify calling (4.15) and (4.16) curvatures, Bates and Watts [1980] used the following ideas from differential geometry (see Appendix B). An arbitrary straight line through $\hat{\boldsymbol{\theta}}$ with direction $\mathbf{h}$ in the parameter set Θ can be expressed in the form

$$\boldsymbol{\theta}(b) = \hat{\boldsymbol{\theta}} + b\mathbf{h} = \hat{\boldsymbol{\theta}} + \boldsymbol{\delta}, \tag{4.17}$$

where b is arbitrary. As b varies, this line generates in the expectation surface Ω the curve

$$\boldsymbol{\mu}_{\mathbf{h}}(\boldsymbol{\theta}) = \boldsymbol{\mu}(\hat{\boldsymbol{\theta}} + b\mathbf{h}),$$

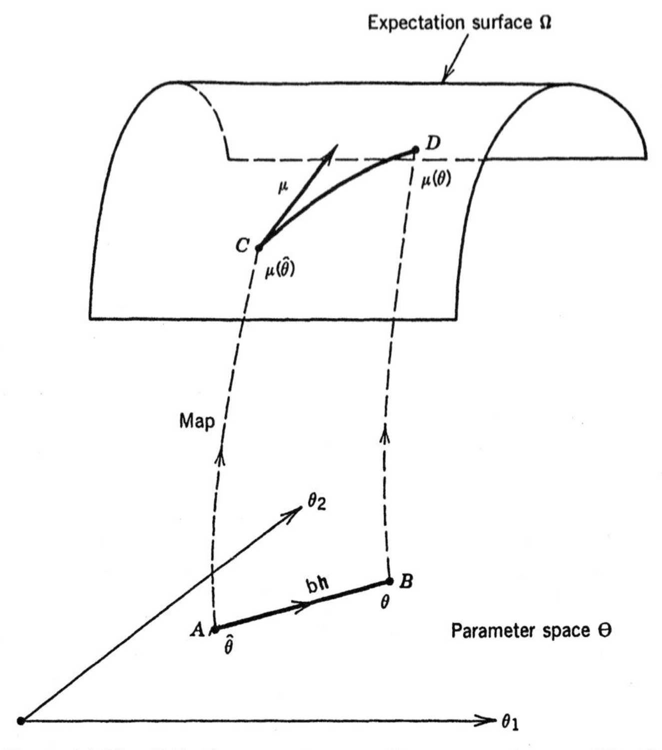

Figure 4.1 The straight line AB in the parameter space Θ maps onto the curve (lifted line) CD in the expectation surface Ω. The tangent vector at C to the curve CD is given by $\dot{\boldsymbol{\mu}}$.

called a lifted line (see Fig. 4.1). Since $\boldsymbol{\mu}(\boldsymbol{\theta}) = \mathbf{f}(\boldsymbol{\theta})$, the tangent to this curve at $b = 0$ is

$$\begin{aligned}
\dot{\boldsymbol{\mu}}_{\mathbf{h}} &= \left[\frac{d\boldsymbol{\mu}_{\mathbf{h}}(\boldsymbol{\theta})}{db}\right]_{\boldsymbol{\theta}=\hat{\boldsymbol{\theta}},\, b=0} \\
&= \left[\sum_{r=1}^{p} \frac{\partial \mathbf{f}(\boldsymbol{\theta})}{\partial \theta_r} \frac{\partial \theta_r(b)}{\partial b}\right]_{\boldsymbol{\theta}=\hat{\boldsymbol{\theta}},\, b=0} \\
&= \sum_{r=1}^{p} \hat{\mathbf{f}}_r h_r \\
&= \hat{\mathbf{F}}_{\bullet}\mathbf{h}.
\end{aligned} \tag{4.18}$$

Since $\dot{\boldsymbol{\mu}}_{\mathbf{h}}$, the tangent vector to the lifted line (Appendix (B1.3)), is a linear combination of the columns of $\hat{\mathbf{F}}_{\bullet}$ for every $\mathbf{h}$, the range of the matrix $\hat{\mathbf{F}}_{\bullet}$ is the tangent plane at $\boldsymbol{\mu}(\hat{\boldsymbol{\theta}})$.

The second directional derivative of $\boldsymbol{\mu}_{\mathbf{h}}(\boldsymbol{\theta})$ at $b=0$ is

$$
\begin{aligned}
\ddot{\boldsymbol{\mu}}_{\mathbf{h}} &= \left[\sum_s \frac{\partial}{\partial \theta_s}\left\{\sum_r \frac{\partial \mathbf{f}(\boldsymbol{\theta})}{\partial \theta_r} h_r\right\}\frac{d\theta_s(b)}{db}\right]_{\boldsymbol{\theta}=\hat{\boldsymbol{\theta}},\, b=0} \\
&= \left[\sum_r \sum_s \frac{\partial^2 \mathbf{f}(\boldsymbol{\theta})}{\partial \theta_r \partial \theta_s} h_r h_s\right]_{\boldsymbol{\theta}=\hat{\boldsymbol{\theta}}} \\
&= \sum_r \sum_s \hat{\mathbf{f}}_{rs} h_r h_s \\
&= \mathbf{h}'\hat{F}_{..}\mathbf{h},
\end{aligned}
\tag{4.19}
$$

as in (4.8). We then have the orthogonal decomposition

$$\ddot{\boldsymbol{\mu}}_{\mathbf{h}} = \ddot{\boldsymbol{\mu}}_{\mathbf{h}}^T + \ddot{\boldsymbol{\mu}}_{\mathbf{h}}^N,$$

where $\ddot{\boldsymbol{\mu}}_{\mathbf{h}}^T = \hat{\mathbf{P}}_F \ddot{\boldsymbol{\mu}}_{\mathbf{h}}$. From the discussion at the end of Appendix B3.3 we see that we can then interpret

$$K_{\mathbf{h}}^T = \frac{\|\ddot{\boldsymbol{\mu}}_{\mathbf{h}}^T\|}{\|\dot{\boldsymbol{\mu}}_{\mathbf{h}}\|^2} = \frac{\|\mathbf{h}'\hat{F}_{..}^T\mathbf{h}\|}{\|\hat{\mathbf{F}}_{.}\mathbf{h}\|^2} \tag{4.20}$$

and

$$K_{\mathbf{h}}^N = \frac{\|\ddot{\boldsymbol{\mu}}_{\mathbf{h}}^N\|}{\|\dot{\boldsymbol{\mu}}_{\mathbf{h}}\|^2} = \frac{\|\mathbf{h}'\hat{F}_{..}^N\mathbf{h}\|}{\|\hat{\mathbf{F}}_{.}\mathbf{h}\|^2} \tag{4.21}$$

as "curvatures" at $\hat{\boldsymbol{\theta}}$. Now $K_{\mathbf{h}}^T$ depends on the particular parametrization used, so that the designation "parameter-effects curvature in direction $\mathbf{h}$" at $\hat{\boldsymbol{\theta}}$ seems appropriate. However, $K_{\mathbf{h}}^N$, the intrinsic curvature in direction $\mathbf{h}$, is an "intrinsic" property of the expectation surface, as it does not depend on the parametric system used. A proof of this invariance is given in Appendix B5.

To make the curvatures scale-free, Bates and Watts [1980] suggested standardizing the model and associated quantities by the scale factor ρ of (4.12): ρ^2 was also used by Beale [1960] as a normalizing factor. If we now divide $\mathbf{y}, \boldsymbol{\mu}, \hat{\boldsymbol{\mu}}, \hat{\mathbf{F}}_{.}$, and $\hat{F}_{..}$ by ρ and let $F_\alpha = F_{p,n-p}^\alpha$, then our two curvatures and the curvature (inverse radius) of the sphere in (4.11) are

$$\gamma_{\mathbf{h}}^T (= K_{\mathbf{h}}^T \rho), \qquad \gamma_{\mathbf{h}}^N (= K_{\mathbf{h}}^N \rho), \quad \text{and} \quad 1/\sqrt{F_\alpha} \tag{4.22}$$

respectively. As a final step Bates and Watts suggested finding the maximum curvatures $\gamma_{\max}^T$ and $\gamma_{\max}^N$ by respectively maximizing $\gamma_{\mathbf{h}}^T$ and $\gamma_{\mathbf{h}}^N$ with respect to $\mathbf{h}$. An algorithm for doing this is described below in Section 4.2.7. A good parameter choice would be one for which $\gamma_{\max}^T = 0$, or approximately so, as the smaller the curvature, the fewer problems there are likely to be with finding a suitable confidence region for $\boldsymbol{\theta}$.

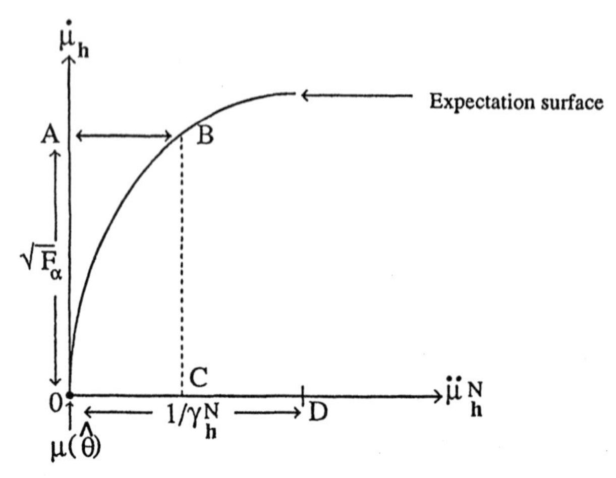

Figure 4.2 Deviation of the circular approximation from the tangent plane. Based on Bates and Watts [1980: Fig. 1].

In using the tangent-plane approximation (4.10) for inference we assume two things about the local behavior of the expectation surface. Firstly there is the *planar assumption*: the expectation surface is replaced locally by the tangent plane. For example, in Fig. 4.2 we show the various "radii of curvature," namely the reciprocals of (4.22) (technically the Bates–Watts curvatures do not have this interpretation, as emphasized by Hougaard [1985]). Using scaled units, one measure of planarity in terms of confidence regions is the radius BC $(=\sqrt{F_\alpha})$ of the sphere (4.11) determined by $\boldsymbol{\mu}$, relative to the intrinsic "radius of curvature" OD $(=1/\gamma_{\text{h}}^N)$, namely

$$BC/OD = \sqrt{F_\alpha}\gamma_{\text{h}}^N. \tag{4.23}$$

We would like this ratio to be small. Another measure is the distance AB of the expectation surface from the boundary of the sphere, relative to OD. If AB is not too big, we have $BD \approx OD$, so that

$$\begin{aligned} \frac{AB}{OD} &= \frac{OD - CD}{OD} \\ &= 1 - CD/OD \\ &= 1 - \frac{(BD^2 - BC^2)^{1/2}}{OD} \\ &\approx 1 - \{1 - (\gamma_{\text{h}}^N)^2 F_\alpha\}^{1/2}, \end{aligned} \tag{4.24}$$

and this will be large if γ_{h}^N is large. We conclude that the tangent-plane approximation will be untenable if the maximum intrinsic curvature is too large at $\hat{\boldsymbol{\theta}}$.

A second feature of the tangent-plane approximation is the *uniform-coordinate assumption*: straight parallel equispaced lines in the parameter space Θ map into straight parallel equispaced lines in the expectation surface (as they do in the tangent plane). This approximation will not be tenable if the maximum parameter-effects curvature is large at $\boldsymbol{\mu}(\hat{\boldsymbol{\theta}})$. Recalling from Appendix B2.3 that the parameter-effects curvature for a given curve is of the form $|\ddot{s}|/|\dot{s}|^2$, where s is the length of curve, then the uniform-coordinate assumption implies that $\dot{s}$ $(=ds/db)$ is constant, so that $\ddot{s}=0$, or $\gamma_{\max}^T=0$. We note that the planar assumption must be satisfied to a sufficient degree before it is sensible to discuss whether the uniform-coordinate assumption is acceptable.

On the basis of (4.23) and (4.24) Bates and Watts [1980] suggested comparing $\gamma_{\max}^N$ and $\gamma_{\max}^T$ with $1/\sqrt{F_\alpha}$, where $F_\alpha = F_{p,n-p}^\alpha$. They did this for 24 published data sets, of which the 16 unreplicated sets are used in Table 4.1. If

$$\gamma_{\max}^N < \frac{1}{2\sqrt{F_\alpha}}, \tag{4.25}$$

then

$$1-\{1-(\gamma_{\max}^N)^2 F_\alpha\}^{1/2} < 1-(\tfrac{3}{4})^{1/2} = 0.134$$

and the maximum departure described by (4.24) is less than 14%. In this case the planar assumption would seem acceptable, and in fact 13 out of the 16

Table 4.1 Relative Curvatures for 16 Published Unreplicated Data Sets[a]

Data Set no.	n	p	$\gamma_{\max}^N$	γ_{RMS}^N [b]	$\gamma_{\max}^T$	γ_{RMS}^T [b]	$1/\sqrt{F_{.05}}$
1	7	2	0.032	0.020	0.079	0.048	0.416
6	4	2	0.065	0.045	0.114	0.081	0.229
7	6	2	0.068	0.046	0.101	0.068	0.379
10	4	2	0.171	0.102	2.669	1.646	0.229
11	6	2	0.177	0.107	2.818	1.735	0.379
13	11	2	0.013	0.008	0.659	0.402	0.485
14	15	2	0.150	0.016	2.242	1.375	0.513
15	5	3	0.037	0.017	12.814	5.717	0.228
16	10	3	0.038	0.017	7.565	3.378	0.480
18	16	3	0.000	0.000	39.680	17.745	0.541
19	12	3	0.014	0.009	2.487	1.086	0.509
20	57	4	0.201	0.070	1.031	0.426	0.627
21	57	4	0.897	0.330	3.063	1.337	0.627
22	20	4	0.078	0.028	4.197	1.487	0.577
23	24	4	0.088	0.048	140.140	48.388	0.591
24	54	9	1.245	0.218	2.629	0.591	0.691

[a]From Bates and Watts [1980].
[b]These curvatures are defined in Section 4.3.

examples in Table 4.1 satisfy the bound (4.25). By contrast, the parameter-effects curvature $\gamma^T_{\max}$ is unacceptable in 13 of the 16 examples if the same criterion is used. It is indeed fortunate that in most cases $\gamma^N_{\max}$ is acceptably small, as little can be done about intrinsic curvature, though some conservative modifications can be made to confidence regions (see Section 5.8.3). However, a large $\gamma^T_{\max}$ can be substantially reduced by an appropriate reparametrization. Care is needed in choosing a suitable parameter transformation, as it will generally be data-dependent. For example Bates and Watts found that using $\exp(xe^{\beta})$ instead of $e^{\theta x}$ reduced $\gamma^T_{\max}$ by a factor of 2.3 for one data set, and increased it by a factor of 2.4 for another data set. This phenomenon was demonstrated geometrically in Section 3.3.2. Bates and Watts also found that using additional design points need not reduce nonlinearity, although the inferences may be improved because the F-value will be decreased.

The usefulness of the rule (4.25) will be discussed further in Section 5.8.3 when the construction of confidence regions has been considered. However, it can be mentioned at this point that Ratkowsky (personal communication) has found examples where $\gamma^T_{\max}$ seems spuriously high when the model appears satisfactory in all other respects. This unreliability of $\gamma^T_{\max}$ has also been demonstrated by Cook and Witmer [1985], who give examples of $\gamma^T_{\max}$ below $1/\sqrt{F_\alpha}$ but having an unsatisfactory tangent-plane approximation. Clearly further research is needed here.

A major disadvantage of the above curvature measures is that they measure the worst possible curvature in any given direction. If we are interested in confidence regions for a subset of $\boldsymbol{\theta}$, then we may be interested in the curvatures for certain directions only. This problem is discussed further in Section 4.5.

Example 4.1 We now demonstrate the above theory using the following nonlinear model taken from Cook and Witmer [1985], namely

$$f(x_i;\boldsymbol{\theta}) = \theta_1 x_i + \theta_1\theta_2(1-x_i), \qquad i=1,2,\ldots,n, \tag{4.26}$$

where $n=2m$, $x_i=1$ for $i=1,2,\ldots,m$, and $x_i=0$ for $i=m+1,m+2,\ldots,2m$. This model represents the usual two-sample problem in which the ratio, θ_2, of the means is of interest. By reparametrizing using the one-to-one transformation $\phi_1=\theta_1$ and $\phi_2=\theta_1\theta_2$ the model becomes linear. Since the intrinsic curvature is invariant under one-to-one transformations of the parameters (Appendix B5), it follows that $\gamma^N_{\mathbf{h}}=0$ for all $\mathbf{h}$. To find the parameter-effects curvature we require

$$\hat{\mathbf{F}}_{\bullet} = \left[\left(\frac{\partial f_i}{\partial\theta_r}\right)\right]_{\boldsymbol{\theta}=\hat{\boldsymbol{\theta}}} = \begin{pmatrix} \mathbf{1}_m & \mathbf{0} \\ \hat{\theta}_2\mathbf{1}_m & \hat{\theta}_1\mathbf{1}_m \end{pmatrix}, \tag{4.27}$$

and the ith face of $\hat{F}_{\bullet\bullet} = [(\hat{\mathbf{f}}_{rs})]$, namely

$$\hat{\mathbf{F}}_{i\bullet\bullet} = \left[\left(\frac{\partial^2 f_i}{\partial\theta_r\,\partial\theta_s}\right)\right]_{\boldsymbol{\theta}=\hat{\boldsymbol{\theta}}} = \begin{pmatrix} 0 & 1-x_i \\ 1-x_i & 0 \end{pmatrix}.$$

Thus $\hat{\mathbf{f}}_{11} = \mathbf{0}$, $\hat{\mathbf{f}}_{22} = \mathbf{0}$, and

$$\hat{\mathbf{f}}_{12} = \hat{\mathbf{f}}_{21} = \begin{pmatrix} \mathbf{0} \\ \mathbf{1}_m \end{pmatrix}. \tag{4.28}$$

Although not required in this example (as $\gamma_{\mathbf{h}}^N = 0$), we can find $\hat{\mathbf{P}}_{\mathbf{F}}$ of (4.14) readily by noting that the following range spaces are equal:

$$\mathscr{R}[\hat{\mathbf{F}}.] = \mathscr{R}\begin{bmatrix} \mathbf{1}_m & \mathbf{0} \\ \mathbf{0} & \mathbf{1}_m \end{bmatrix}.$$

As the columns of the right-hand matrix are orthogonal, the projection matrix onto $\mathscr{R}[\hat{\mathbf{F}}.]$, namely $\hat{\mathbf{P}}_{\mathbf{F}} = \hat{\mathbf{F}}.(\hat{\mathbf{F}}'_{\cdot}\hat{\mathbf{F}}.)^{-1}\hat{\mathbf{F}}'_{\cdot}$ (c.f. Appendix A11.4), is

$$\hat{\mathbf{P}}_{\mathbf{F}} = \begin{bmatrix} \frac{1}{m}\mathbf{J}_m & \mathbf{0} \\ \mathbf{0} & \frac{1}{m}\mathbf{J}_m \end{bmatrix},$$

where $\mathbf{J}_m$ is an $m \times m$ matrix of ones. Then, as expected, $\hat{\mathbf{f}}_{rs}^T = \hat{\mathbf{P}}_{\mathbf{F}}\hat{\mathbf{f}}_{rs} = \hat{\mathbf{f}}_{rs}$ and $\mathbf{f}_{rs}^N = (\mathbf{I}_n - \hat{\mathbf{P}}_{\mathbf{F}})\hat{\mathbf{f}}_{rs} = \mathbf{0}$. Hence, from (4.20) and (4.21),

$$\begin{aligned} \gamma_{\mathbf{h}}^T &= \rho \frac{\left\| \sum_r \sum_s \hat{\mathbf{f}}_{rs} h_r h_s \right\|}{\| \hat{\mathbf{F}}.\mathbf{h} \|^2} \\ &= \rho \frac{\| 2\hat{\mathbf{f}}_{12} h_1 h_2 \|}{\| \hat{\mathbf{F}}.\mathbf{h} \|^2} \\ &= \rho \frac{2|h_1 h_2|}{\sqrt{m}[h_1^2 + (\hat{\theta}_2 h_1 + \hat{\theta}_1 h_2)^2]}. \end{aligned} \tag{4.29}$$

By setting $h_2 = ch_1$, and considering positive and negative c separately, we find that $\gamma_{\mathbf{h}}^T$ is maximized when

$$c = \frac{-(\text{sign of } \hat{\theta}_2)[1 + \hat{\theta}_2^2]^{1/2}}{\hat{\theta}_1}.$$

Substituting in (4.29) leads to

$$\gamma_{\max}^T = \rho \frac{(\theta_2^2 + 1)^{1/2} + |\theta_2|}{\sqrt{m}|\hat{\theta}_1|}. \tag{4.30}$$

For ease of exposition Cook and Witmer [1985] assume that σ^2 is known, so that the tangent-plane approximation (4.13) is replaced by

$$\begin{aligned}\sigma^2\chi_p^2 &\geq (\boldsymbol{\theta}-\hat{\boldsymbol{\theta}})'\hat{\mathbf{F}}_{.}'\hat{\mathbf{F}}_{.}(\boldsymbol{\theta}-\hat{\boldsymbol{\theta}})\\ &=(\theta_1-\hat{\theta}_1)^2 m(1+\theta_2^2)+2(\theta_1-\hat{\theta}_1)(\theta_2-\hat{\theta}_2)m\hat{\theta}_1\hat{\theta}_2+(\theta_2-\hat{\theta}_2)^2 m\theta_1^2,\end{aligned}$$

where $\mathrm{pr}[\chi_p^2 \geq \chi_p^2(\alpha)] = \alpha$. The above expression then reduces to

$$(\theta_1-\hat{\theta}_1)^2+[\hat{\theta}_2(\theta_1-\hat{\theta}_1)+\hat{\theta}_1(\theta_2-\hat{\theta}_2)]^2 \leq \frac{\sigma^2\chi_p^2(\alpha)}{m}. \tag{4.31}$$

Also, in (4.29) and (4.30) ρ is replaced by σ. ■

The reader who wishes to avoid computational details can skip to Section 4.2.5.

*4.2.3 Reduced Formulae for Curvatures

In order to compute the scaled curvatures (4.22) and, in particular, the maximum curvatures

$$\gamma_{\max}^T = \max_{\mathbf{h}} \rho \frac{\|\ddot{\boldsymbol{\mu}}_{\mathbf{h}}^T\|}{\|\dot{\boldsymbol{\mu}}_{\mathbf{h}}\|^2} \quad \text{and} \quad \gamma_{\max}^N = \max_{\mathbf{h}} \rho \frac{\|\ddot{\boldsymbol{\mu}}_{\mathbf{h}}^N\|}{\|\dot{\boldsymbol{\mu}}_{\mathbf{h}}\|^2}, \tag{4.32}$$

it is convenient to reparametrize the model and then rotate the axes of the sample space using an orthogonal matrix, $\hat{\mathbf{Q}}'$ say. Here $\hat{\mathbf{Q}}$ is chosen so that the first p coordinate vectors $\mathbf{e}_1, \mathbf{e}_2, \ldots, \mathbf{e}_p$ [with $\mathbf{e}_1' = (1,0,\ldots,0)$ etc.] are parallel to, and form the basis for, the tangent plane; and the last $n-p$ $(\mathbf{e}_{p+1}, \mathbf{e}_{p+2}, \ldots, \mathbf{e}_n)$ are orthogonal to the tangent plane. As we shall see below, this has the effect of reducing the vectors $(\mathbf{h}'\hat{\boldsymbol{F}}_{..}\mathbf{h})^T$ and $(\mathbf{h}'\hat{\boldsymbol{F}}_{..}\mathbf{h})^N$ to simply the first p and last $n-p$ elements, respectively, of the new transformed $\mathbf{h}'\hat{\boldsymbol{F}}_{..}\mathbf{h}$. Then, with reparametrization, we shall find that $\hat{\mathbf{F}}_{.}$ is replaced by a matrix with orthonormal columns, and $\|\hat{\mathbf{F}}_{.}\mathbf{h}\|^2$ becomes equal to 1 when $\mathbf{h}$ is suitably scaled.

To find $\hat{\mathbf{Q}}$, we consider the *QR* decomposition (see Appendix A8.3) of the $n \times p$ matrix $\hat{\mathbf{F}}_{.}$ (assumed to have rank p), namely

$$\hat{\mathbf{F}}_{.} = \hat{\mathbf{Q}}\hat{\mathbf{R}}_1 = (\hat{\mathbf{Q}}_p | \hat{\mathbf{Q}}_{n-p}) \begin{pmatrix} \mathbf{R}_{11} \\ \mathbf{O} \end{pmatrix} \begin{matrix} \}p\times p \\ \}(n-p)\times p \end{matrix} = \hat{\mathbf{Q}}_p\hat{\mathbf{R}}_{11}, \tag{4.33}$$

where $\hat{\mathbf{R}}_{11}$ is a nonsingular upper triangular matrix. Since $\hat{\mathbf{R}}_{11}'\hat{\mathbf{R}}_{11}$ is the Cholesky decomposition of $\hat{\mathbf{F}}_{.}'\hat{\mathbf{F}}_{.}$ (c.f. Appendix A8.2), $\hat{\mathbf{R}}_{11}$ is unique if its diagonal elements are all positive or all negative (Bates and Watts [1980] use the latter). If $\hat{\mathbf{R}}_{11}$ is unique, then so is $\hat{\mathbf{Q}}_p$, as

$$\hat{\mathbf{Q}}_p = \hat{\mathbf{F}}_{.}\hat{\mathbf{R}}_{11}^{-1} = \hat{\mathbf{F}}_{.}\hat{\mathbf{K}}, \tag{4.34}$$

say. However, $\hat{\mathbf{Q}}_{n-p}$ (and therefore $\hat{\mathbf{Q}}$) is not unique, as, for example, any permutation of the columns of $\hat{\mathbf{Q}}_{n-p}$ will do. It then follows from (4.18) that

$$\dot{\boldsymbol{\mu}}_{\mathbf{h}} = \hat{\mathbf{F}}_{.}\mathbf{h} = \hat{\mathbf{Q}}_p\hat{\mathbf{R}}_{11}\mathbf{h} = \hat{\mathbf{Q}}_p\mathbf{d}, \tag{4.35}$$

say, so that the columns of $\hat{\mathbf{Q}}_p$ form a basis for the tangent plane. As a bonus, the columns of $\hat{\mathbf{Q}}_{n-p}$ form a basis for vectors perpendicular to the tangent plane. If we then rotate all the vectors in the sample space by premultiplying by $\hat{\mathbf{Q}}'$, then the basis vectors, given by the columns of $\hat{\mathbf{Q}}$, rotate to $\mathbf{e}_1, \mathbf{e}_2, \ldots, \mathbf{e}_n$ as

$$\hat{\mathbf{Q}}'\hat{\mathbf{Q}} = \mathbf{I}_n = (\mathbf{e}_1, \mathbf{e}_2, \ldots, \mathbf{e}_n).$$

Here $\mathbf{e}_1, \mathbf{e}_2, \ldots, \mathbf{e}_p$ now form a basis for the tangent plane.

We will now try to reparametrize the model so that $\hat{\mathbf{Q}}_p$ for the new coordinate system plays the same role as the matrix of first derivatives $\hat{\mathbf{F}}_{.}$ for the original system. Since $\mathbf{d} = \hat{\mathbf{R}}_{11}\mathbf{h}$ in (4.35), we are led to consider the transformation

$$\boldsymbol{\phi} = \hat{\mathbf{R}}_{11}\boldsymbol{\theta}, \qquad \text{or} \quad \boldsymbol{\theta} = \hat{\mathbf{K}}\boldsymbol{\phi}, \tag{4.36}$$

where

$$\hat{\mathbf{K}} = \hat{\mathbf{R}}_{11}^{-1} \; (= [(\hat{k}_{rs})], \text{ say}). \tag{4.37}$$

(Bates and Watts [1980] actually use $\boldsymbol{\phi} = \hat{\mathbf{R}}_{11}(\boldsymbol{\theta} - \hat{\boldsymbol{\theta}})$, but we prefer (4.36) for ease of exposition later.) Then

$$f\{\boldsymbol{\theta}(\boldsymbol{\phi})\} = \mathbf{g}(\boldsymbol{\phi}) \; (= \boldsymbol{\mu}(\boldsymbol{\phi}), \text{ say}), \tag{4.38}$$

and $\hat{\boldsymbol{\phi}} = \hat{\mathbf{R}}_{11}\hat{\boldsymbol{\theta}}$ is the least-squares estimate for the model $\mathbf{g}(\boldsymbol{\phi})$.

Also a line in the θ-system with direction $\mathbf{h}$ now becomes

$$\begin{aligned}
\boldsymbol{\phi} &= \hat{\boldsymbol{\phi}} + \boldsymbol{\phi} - \hat{\boldsymbol{\phi}} \\
&= \hat{\boldsymbol{\phi}} + \hat{\mathbf{R}}_{11}(\boldsymbol{\theta} - \hat{\boldsymbol{\theta}}) \\
&= \hat{\boldsymbol{\phi}} + b\hat{\mathbf{R}}_{11}\mathbf{h} \qquad [\text{by } (4.17)] \\
&= \hat{\boldsymbol{\phi}} + b\mathbf{d},
\end{aligned} \tag{4.39}$$

which is a line through $\hat{\boldsymbol{\phi}}$ with direction $\mathbf{d}$. We can now proceed as in (4.18) but with the ϕ-parametrization and $\mathbf{g}$ instead of $\mathbf{f}$. Thus

$$\begin{aligned}
\dot{\boldsymbol{\mu}}_{\mathbf{d}} &= \left[\frac{d\mu_{\mathbf{h}}(\boldsymbol{\phi})}{db}\right]_{\boldsymbol{\phi}=\hat{\boldsymbol{\phi}},b=0} \\
&= \left[\sum_{r=1}^{p} \frac{\partial \mathbf{g}(\boldsymbol{\phi})}{\partial \phi_r}\frac{\partial \phi_r}{\partial b}\right]_{\boldsymbol{\phi}=\hat{\boldsymbol{\phi}},b=0} \\
&= \sum_{r=1}^{p} \hat{\mathbf{g}}_r d_r \qquad [\text{by } (4.39)] \\
&= \mathbf{G}_{.}\mathbf{d},
\end{aligned} \tag{4.40}$$

say, where $\hat{\mathbf{G}}_{\cdot} = (\hat{\mathbf{g}}_1, \hat{\mathbf{g}}_2, \ldots, \hat{\mathbf{g}}_p)$. However, from (4.38) and (4.36),

$$\hat{\mathbf{g}}_r = \left[\sum_u \frac{\partial \mathbf{f}(\boldsymbol{\theta})}{\partial \theta_u} \frac{\partial \theta_u}{\partial \phi_r} \right]_{\boldsymbol{\theta} = \hat{\boldsymbol{\theta}}}$$

$$= \sum_u \hat{\mathbf{f}}_u k_{ur}, \tag{4.41}$$

say, where $(\hat{\mathbf{f}}_1, \hat{\mathbf{f}}_2, \ldots, \hat{\mathbf{f}}_p) = \hat{\mathbf{F}}_{\cdot}$. Hence, from (4.34)

$$\hat{\mathbf{G}}_{\cdot} = \hat{\mathbf{F}}_{\cdot} \hat{\mathbf{K}} = \hat{\mathbf{Q}}_p. \tag{4.42}$$

Also,

$$\dot{\boldsymbol{\mu}}_{\mathbf{d}} = \hat{\mathbf{G}}_{\cdot} \mathbf{d} = \hat{\mathbf{F}}_{\cdot} \hat{\mathbf{K}} \mathbf{d} = \hat{\mathbf{F}}_{\cdot} \mathbf{h} = \dot{\boldsymbol{\mu}}_{\mathbf{h}}. \tag{4.43}$$

We have now achieved our aim of making $\hat{\mathbf{Q}}_p$ the matrix of derivatives for the new parametric system $\boldsymbol{\phi}$. We now show that the curvatures are the same for both systems.

Working with this new parametrization, we proceed as in (4.19) and form

$$\ddot{\boldsymbol{\mu}}_{\mathbf{d}} = \left[\sum_r \sum_s \frac{\partial^2 \mathbf{g}(\boldsymbol{\phi})}{\partial \phi_r \partial \phi_s} d_r d_s \right]_{\boldsymbol{\phi} = \hat{\boldsymbol{\phi}}}$$

$$= \sum_r \sum_s \hat{\mathbf{g}}_{rs} d_r d_s$$

$$= \mathbf{d}' \hat{G}_{\cdot\cdot} \mathbf{d}, \tag{4.44}$$

say, as in (4.8), where $\hat{G}_{\cdot\cdot} = [(\hat{\mathbf{g}}_{rs})]$. We note that

$$\hat{\mathbf{g}}_{rs} = \left[\frac{\partial g_r(\boldsymbol{\phi})}{\partial \phi_s} \right]_{\boldsymbol{\phi} = \hat{\boldsymbol{\phi}}}$$

$$= \left[\sum_u \frac{\partial}{\partial \phi_s} (\mathbf{f}_u k_{ur}) \right]_{\boldsymbol{\phi} = \hat{\boldsymbol{\phi}}} \qquad \text{[by (4.41)]}$$

$$= \left[\sum_u \sum_v \frac{\partial \mathbf{f}_u}{\partial \theta_v} \frac{\partial \theta_v}{\partial \phi_s} k_{ur} \right]_{\boldsymbol{\phi} = \hat{\boldsymbol{\phi}}}$$

$$= \left[\sum_u \sum_v \frac{\partial^2 \mathbf{f}}{\partial \theta_v \partial \theta_u} \frac{\partial \theta_v}{\partial \phi_s} k_{ur} \right]_{\boldsymbol{\phi} = \hat{\boldsymbol{\phi}}}$$

$$= \sum_u \sum_v \hat{\mathbf{f}}_{uv} k_{vs} k_{ur} \qquad \text{[by (4.7)]}$$

$$= \sum_u \sum_v k_{ru} \hat{\mathbf{f}}_{uv} k_{vs}. \tag{4.45}$$

Comparing the ith elements of both sides, we have the following relationship between the ith faces of the arrays $\hat{G}_{\cdot\cdot}$ and $\hat{F}_{\cdot\cdot}$:

$$\hat{\mathbf{G}}_{i\cdot\cdot} = \hat{\mathbf{K}}' \hat{\mathbf{F}}_{i\cdot\cdot} \hat{\mathbf{K}}, \tag{4.46}$$

or, symbolically (see Appendix B4),

$$\hat{G}_{..} = \hat{\mathbf{K}}'\hat{F}_{..}\hat{\mathbf{K}}. \tag{4.47}$$

Furthermore, from (4.46),

$$\begin{aligned}(\ddot{\boldsymbol{\mu}}_{\mathbf{d}})_i &= \mathbf{d}'\hat{\mathbf{G}}_{i..}\mathbf{d} \\ &= \mathbf{d}'\hat{\mathbf{K}}'\hat{\mathbf{F}}_{i..}\hat{\mathbf{K}}\mathbf{d} \\ &= \mathbf{h}'\hat{\mathbf{F}}_{i..}\mathbf{h} \\ &= (\ddot{\boldsymbol{\mu}}_{\mathbf{h}})_i,\end{aligned}$$

so that

$$\ddot{\boldsymbol{\mu}}_{\mathbf{d}} = \ddot{\boldsymbol{\mu}}_{\mathbf{h}}. \tag{4.48}$$

From (4.43) the tangent plane for the ϕ-parameters is the same as that for the θ-parameters, both being spanned by the columns of $\hat{\mathbf{Q}}_p$ [see also (4.35) and (4.42)]. Since $\ddot{\boldsymbol{\mu}}_{\mathbf{h}}^T$ is the projection of $\ddot{\boldsymbol{\mu}}_{\mathbf{h}}$ onto the tangent plane, we have

$$\ddot{\boldsymbol{\mu}}_{\mathbf{d}}^T = \ddot{\boldsymbol{\mu}}_{\mathbf{h}}^T \quad \text{and} \quad \ddot{\boldsymbol{\mu}}_{\mathbf{d}}^N = \ddot{\boldsymbol{\mu}}_{\mathbf{h}}^N. \tag{4.49}$$

If $\Gamma_{\mathbf{d}}^T$ is the scaled parameter-effects curvature for the ϕ-parameters, defined in a similar fashion to $\gamma_{\mathbf{h}}^T$ of (4.22), then from (4.43) and (4.49) (or from the more general result of Appendix B5),

$$\begin{aligned}\Gamma_{\mathbf{d}}^T &= \rho\frac{\|\mathbf{d}'\hat{G}_{..}^T\mathbf{d}\|}{\|\hat{\mathbf{G}}_{.}\mathbf{d}\|^2} \\ &= \rho\frac{\|\ddot{\boldsymbol{\mu}}_{\mathbf{d}}^N\|}{\|\dot{\boldsymbol{\mu}}_{\mathbf{d}}\|^2} \\ &= \rho\frac{\|\ddot{\boldsymbol{\mu}}_{\mathbf{h}}^N\|}{\|\dot{\boldsymbol{\mu}}_{\mathbf{h}}\|^2} \\ &= \rho\frac{\|\mathbf{h}'\hat{F}_{..}^T\mathbf{h}\|}{\|\hat{\mathbf{F}}_{.}\mathbf{h}\|^2} \\ &= \gamma_{\mathbf{h}}^T,\end{aligned}$$

and, similarly, $\Gamma_{\mathbf{d}}^N = \gamma_{\boldsymbol{\gamma}}^N$. Since the above ratios depend on the directions but not the magnitudes of $\mathbf{d}$ and $\mathbf{h}$, we can assume $\|\mathbf{d}\| = 1$. This leads to [c.f. (4.40)]

$$\begin{aligned}\|\dot{\boldsymbol{\mu}}_{\mathbf{d}}\| &= \mathbf{d}'\hat{\mathbf{G}}_{.}'\hat{\mathbf{G}}_{.}\mathbf{d} \\ &= \mathbf{d}'\hat{\mathbf{Q}}_p'\hat{\mathbf{Q}}_p\mathbf{d} \\ &= \mathbf{d}'\mathbf{d} \\ &= 1.\end{aligned} \tag{4.50}$$

If we now apply the rotation based on $\hat{\mathbf{Q}}'$ $[=(\hat{\mathbf{Q}}_p, \hat{\mathbf{Q}}_{n-p})']$ to the above vectors, we do not change their lengths, as, for all $\mathbf{x}$,

$$\|\hat{\mathbf{Q}}'\mathbf{x}\| = \mathbf{x}'\hat{\mathbf{Q}}\hat{\mathbf{Q}}'\mathbf{x} = \mathbf{x}'\mathbf{x} = \|\mathbf{x}\|^2. \tag{4.51}$$

The columns of $\hat{\mathbf{Q}}_p$ form a basis for the tangent plane, so that $\ddot{\boldsymbol{\mu}}_{\mathbf{d}}^T$ is a linear combination of the columns of $\hat{\mathbf{Q}}_p$. However, we saw above that the rotation using $\hat{\mathbf{Q}}'$ rotates the columns of $\hat{\mathbf{Q}}_p$ to the vectors $\mathbf{e}_1, \mathbf{e}_2, \ldots, \mathbf{e}_p$. Hence we have the representation

$$\hat{\mathbf{Q}}'\ddot{\boldsymbol{\mu}}_{\mathbf{d}}^T = \begin{pmatrix} \boldsymbol{\xi}_1 \\ \mathbf{0} \end{pmatrix}, \tag{4.52}$$

where $\boldsymbol{\xi}_1$ is $p \times 1$. A similar argument applies to the columns of $\hat{\mathbf{Q}}_{n-p}$, so that

$$\hat{\mathbf{Q}}'\ddot{\boldsymbol{\mu}}_{\mathbf{d}}^N = \begin{pmatrix} \mathbf{0} \\ \boldsymbol{\xi}_2 \end{pmatrix}, \tag{4.53}$$

where $\boldsymbol{\xi}_2$ is $(n-p) \times 1$. Adding (4.52) and (4.53) gives us

$$\begin{pmatrix} \boldsymbol{\xi}_1 \\ \boldsymbol{\xi}_2 \end{pmatrix} = \hat{\mathbf{Q}}'\ddot{\boldsymbol{\mu}}_{\mathbf{d}} = \begin{pmatrix} \hat{\mathbf{Q}}_p'\ddot{\boldsymbol{\mu}}_{\mathbf{d}} \\ \hat{\mathbf{Q}}_{n-p}'\ddot{\boldsymbol{\mu}}_{\mathbf{d}} \end{pmatrix}.$$

Now

$$\begin{aligned} \ddot{\boldsymbol{\mu}}_{\mathbf{d}} &= \mathbf{d}'\hat{G}_{..}\mathbf{d} \\ &= \sum_r \sum_s \hat{\mathbf{g}}_{rs} d_r d_s \end{aligned}$$

so that

$$\begin{aligned} \hat{\mathbf{Q}}'\ddot{\boldsymbol{\mu}}_{\mathbf{d}} &= \sum_r \sum_s \hat{\mathbf{Q}}'\hat{\mathbf{g}}_{rs} d_r d_s \\ &= \begin{bmatrix} \sum_r \sum_s \hat{\mathbf{Q}}_p'\hat{\mathbf{g}}_{rs} d_r d_s \\ \sum_r \sum_s \hat{\mathbf{Q}}_{n-p}'\hat{\mathbf{g}}_{rs} d_r d_s \end{bmatrix}. \end{aligned} \tag{4.54}$$

This leads us to consider the array

$$\begin{aligned} \hat{A}_{..} &= \{\hat{a}_{irs}\} \\ &= [(\hat{\mathbf{a}}_{rs})] \\ &= [(\hat{\mathbf{Q}}'\hat{\mathbf{g}}_{rs})] \\ &= [\hat{\mathbf{Q}}'][\hat{G}_{..}], \end{aligned} \tag{4.55}$$

say, where square-bracket multiplication represents summation over $i = 1, 2, \ldots, n$

(see Appendix B4). Let

$$\hat{A}^T_{\cdot\cdot} = [(\hat{\mathbf{Q}}'_p \hat{\mathbf{g}}_{rs})] = [(\hat{\mathbf{a}}^T_{rs})] \tag{4.56}$$

and

$$\hat{A}^N_{\cdot\cdot} = [(\hat{\mathbf{Q}}'_{n-p} \hat{\mathbf{g}}_{rs})] = [(\hat{\mathbf{a}}^N_{rs})] \tag{4.57}$$

represent the first p and last $n-p$ faces, respectively, of $\hat{A}_{\cdot\cdot}$, so that

$$\hat{a}_{irs} = \begin{cases} \hat{a}^T_{irs} & (i = 1, 2, \ldots, p) \\ \hat{a}^N_{i-p,rs} & (i = p+1, p+2, \ldots, n). \end{cases}$$

Then if $\|\mathbf{d}\| = 1$, we have

$$\begin{aligned} \Gamma^T_{\mathbf{d}} &= \rho \|\ddot{\boldsymbol{\mu}}^T_{\mathbf{d}}\| \qquad [\text{by } (4.50)] \\ &= \rho \|\hat{\mathbf{Q}}' \ddot{\boldsymbol{\mu}}^T_{\mathbf{d}}\| \qquad [\text{by } (4.51)] \\ &= \rho \|\boldsymbol{\xi}_1\| \qquad [\text{by } (4.52)] \\ &= \rho \left\| \sum_r \sum_s \hat{\mathbf{Q}}'_p \hat{\mathbf{g}}_{rs} d_r d_s \right\| \qquad [\text{by } (4.54)] \\ &= \rho \left\| \sum_r \sum_s \hat{\mathbf{a}}^T_{rs} d_r d_s \right\| \\ &= \rho \|\mathbf{d}' \hat{A}^T_{\cdot\cdot} \mathbf{d}\| \\ &= \rho \left(\sum_{i=1}^{p} |\mathbf{d}' \hat{\mathbf{A}}_{i\cdot\cdot} \mathbf{d}|^2 \right)^{1/2} \end{aligned} \tag{4.58}$$

and

$$\begin{aligned} \Gamma^N_{\mathbf{d}} &= \rho \|\mathbf{d}' A^N_{\cdot\cdot} \mathbf{d}\| \\ &= \rho \left(\sum_{i=p+1}^{n} |\mathbf{d}' \hat{\mathbf{A}}_{i\cdot\cdot} \mathbf{d}|^2 \right)^{1/2}. \end{aligned} \tag{4.59}$$

Since there is a one-to-one correspondence between $\mathbf{d}$ and $\mathbf{h} = \hat{\mathbf{K}}\mathbf{d}$, $\hat{\mathbf{K}}$ being nonsingular, maximizing with respect to $\mathbf{h}$ is equivalent to maximizing with respect to $\mathbf{d}$. We therefore finally require

$$\begin{aligned} \gamma^T_{\max} &= \max_{\mathbf{h}} \gamma^T_{\mathbf{h}} \\ &= \max_{\|\mathbf{d}\|=1} \Gamma^T_{\mathbf{d}} \\ &= \max_{\|\mathbf{d}\|=1} \rho \left(\sum_{i=1}^{p} |\mathbf{d}' \hat{\mathbf{A}}_{i\cdot\cdot} \mathbf{d}|^2 \right)^{1/2} \end{aligned} \tag{4.60}$$

and

$$\gamma_{\max}^{N} = \max_{\|\mathbf{d}\|=1} \rho \left(\sum_{i=p+1}^{n} |\mathbf{d}'\hat{\mathbf{A}}_{i..}\mathbf{d}|^2 \right)^{1/2}. \tag{4.61}$$

An algorithm for doing this is described in the next section.

As ρ $(=s\sqrt{p})$ is a scale factor, Bates and Watts [1980] assume that $\mathbf{y}, \boldsymbol{\mu}, \hat{\boldsymbol{\varepsilon}} = \mathbf{y} - \hat{\boldsymbol{\mu}}, \hat{\mathbf{F}}_{.}$, and $\hat{F}_{..}$ have all been divided by ρ right from the beginning. Then the new $\hat{\mathbf{K}}, \hat{G}_{...}$, and $\hat{A}_{..}$ become $\rho\hat{\mathbf{K}}, \rho\hat{G}_{...}$, and $\rho\hat{A}_{..}$ respectively, and ρ disappears from (4.60) and (4.61).

In the above mathematical development we have been concerned with finding curvatures and curvature arrays at $\boldsymbol{\theta} = \hat{\boldsymbol{\theta}}$. Such curvatures can, of course, be evaluated at any other $\boldsymbol{\theta}$, and in Section 4.6 we are concerned with curvatures at $\boldsymbol{\theta} = \boldsymbol{\theta}^*$, where $\boldsymbol{\theta}^*$ is the true value of $\boldsymbol{\theta}$. In this case we simply omit the caret and write, for example,

$$\mathbf{F}_{.} = \mathbf{F}_{.}(\boldsymbol{\theta}^*) = \left[\left(\frac{\partial f_i(\boldsymbol{\theta})}{\partial \theta_j} \right) \right]_{\boldsymbol{\theta}=\boldsymbol{\theta}^*} \tag{4.62}$$

with $\mathbf{G}_{..}, \mathbf{Q}_p, \mathbf{R}_{11}, A_{...}$, etc. being similarly defined.

Example 4.2 (continued) From (4.27) we have

$$\hat{\mathbf{F}}_{.} = \begin{pmatrix} \mathbf{1}_m & \mathbf{0} \\ \hat{\theta}_2 \mathbf{1}_m & \hat{\theta}_1 \mathbf{1}_m \end{pmatrix} = \hat{\mathbf{Q}}_2 \hat{\mathbf{R}}_{11},$$

where

$$\hat{\mathbf{Q}}_2 = a \begin{pmatrix} \mathbf{1}_m & -\hat{\theta}_2 \mathbf{1}_m \\ \hat{\theta}_2 \mathbf{1}_m & \mathbf{1}_m \end{pmatrix}.$$

$$\hat{\mathbf{R}}_{11} = \frac{1}{a} \begin{pmatrix} 1 & \hat{\theta}_1 \hat{\theta}_2/(1+\hat{\theta}_2^2) \\ 0 & \hat{\theta}_1/(1+\hat{\theta}_2^2) \end{pmatrix}, \tag{4.63}$$

and, with $m = n/2$,

$$a = [m(1+\hat{\theta}_2^2)]^{-1/2}.$$

Hence

$$\begin{aligned} \hat{\mathbf{K}} &= \hat{\mathbf{R}}_{11}^{-1} \\ &= a \begin{pmatrix} 1 & -\hat{\theta}_2 \\ 0 & (1+\hat{\theta}_2^2)/\hat{\theta}_1 \end{pmatrix} \\ &= \begin{pmatrix} \hat{k}_{11} & \hat{k}_{12} \\ \hat{k}_{21} & \hat{k}_{22} \end{pmatrix}, \end{aligned}$$

say. From (4.28) and the following equations, $\hat{\mathbf{f}}_{11} = \hat{\mathbf{f}}_{22} = \mathbf{0}$, $\hat{\mathbf{f}}_{12} = \hat{\mathbf{f}}_{21} = (\mathbf{0}', \mathbf{1}'_m)'$, and $\hat{\mathbf{f}}_{uv} = \hat{\mathbf{f}}^T_{uv}$ $(u, v = 1, 2)$. Then using (4.56) and (4.45), we have

$$\begin{aligned}\hat{\mathbf{a}}^T_{rs} &= \sum_{u=1}^{2} \sum_{v=1}^{2} \hat{\mathbf{Q}}'_2 \hat{\mathbf{f}}_{uv} \hat{k}_{ur} \hat{k}_{vs} \\ &= \hat{\mathbf{Q}}'_2 \hat{\mathbf{f}}_{12} (\hat{k}_{1r} \hat{k}_{2s} + \hat{k}_{2r} \hat{k}_{1s}),\end{aligned}$$

which, on simplification, leads to the following faces of $\hat{A}^T_{..}$:

$$\begin{aligned}\hat{\mathbf{A}}^T_{1..} &= \frac{\hat{\theta}_2 a}{\hat{\theta}_1}\begin{pmatrix} 0 & 1 \\ 1 & -2\hat{\theta}_2 \end{pmatrix}, \\ \hat{\mathbf{A}}^T_{2..} &= \hat{\mathbf{A}}^T_{1..}/\hat{\theta}_2,\end{aligned} \tag{4.64}$$

a result quoted by Cook and Witmer [1985]. ■

*4.2.4 Summary of Formulae

At this stage it is helpful to list some of the formulae derived thus far, with all expressions evaluated at $\boldsymbol{\theta}^*$, the true value of $\boldsymbol{\theta}$ [c.f. (4.62)]. To evaluate them at $\hat{\boldsymbol{\theta}}$ we simply add a caret to each symbol. These formulae are:

$$\mathbf{F}_. = (\mathbf{Q}_p, \mathbf{Q}_{n-p})\begin{pmatrix} \mathbf{R}_{11} \\ \mathbf{0} \end{pmatrix} = \mathbf{Q}_p \mathbf{R}_{11}, \qquad \mathbf{K} = \mathbf{R}_{11}^{-1}, \qquad \mathbf{G}_. = \mathbf{Q}_p, \qquad G_{..} = \mathbf{K}' F_{..} \mathbf{K},$$

$$\boldsymbol{\theta} = \mathbf{K}\boldsymbol{\phi}, \qquad A_{..} = [\mathbf{Q}'][G_{..}], \qquad A^T_{..} = [\mathbf{Q}'_p][G_{..}], \qquad A^N_{..} = [\mathbf{Q}'_{n-p}][G_{..}],$$

$\Gamma^T_{\mathbf{d}} = \rho \, \| \mathbf{d}' A^T_{..} \mathbf{d} \|$, and $\Gamma^N_{\mathbf{d}} = \rho \, \| \mathbf{d}' A^N_{..} \mathbf{d} \|$.

We also give some further results for future reference, though these may be omitted at first reading. Firstly, from (4.45),

$$\begin{aligned}\mathbf{a}_{rs} &= \mathbf{Q}' \mathbf{g}_{rs} \\ &= \sum_u \sum_v \mathbf{Q}' k'_{ru} \mathbf{f}_{uv} k_{vs} \\ &= \sum_u \sum_v k'_{ru} \mathbf{Q}' \mathbf{f}_{uv} k_{vs},\end{aligned} \tag{4.65}$$

so that using the square-bracket multiplication of (4.55) we have, from (4.65),

$$\begin{aligned}A_{..} &= [\mathbf{Q}'][G_{..}] \\ &= \mathbf{K}'[\mathbf{Q}'][F_{..}]\mathbf{K}.\end{aligned} \tag{4.66}$$

Hence

$$[\mathbf{Q}'][\mathbf{K}' F_{..} \mathbf{K}] = \mathbf{K}'[\mathbf{Q}'][F_{..}]\mathbf{K}. \tag{4.67}$$

This relationship demonstrates that square-bracket and ordinary multiplication commute (see Appendix B4.1).

Secondly, we note from (4.56) that

$$\begin{aligned} A^T_{..} &= [\mathbf{Q}'_p][G_{..}] \\ &= [\mathbf{Q}'_p][\mathbf{K}'F_{..}\mathbf{K}] \\ &= [\mathbf{K}'\mathbf{F}'_{.}][\mathbf{K}'F_{..}\mathbf{K}] \qquad [\text{by (4.34)}], \end{aligned} \tag{4.68}$$

or

$$\begin{aligned} a^T_{irs} &= \sum_{h=1}^{n} \sum_{a,b,c=1}^{p} k'_{ia} \frac{\partial f_h}{\partial \theta_a} k'_{rb} \frac{\partial^2 f_h}{\partial \theta_b \partial \theta_c} k_{cs} \\ &= \sum_{h=1}^{n} \sum_{a,b,c=1}^{p} \frac{\partial f_h}{\partial \theta_a} \frac{\partial^2 f_h}{\partial \theta_b \partial \theta_c} k_{ai} k_{br} k_{cs} \qquad (i,r,s = 1,2,\ldots,p). \end{aligned} \tag{4.69}$$

Similarly

$$A^N_{..} = [\mathbf{Q}'_{n-p}][\mathbf{K}'F_{..}\mathbf{K}],$$

or

$$a^N_{irs} = \sum_{h=1}^{n} \sum_{b,c=1}^{p} q_{hi} \frac{\partial^2 f_h}{\partial \theta_b \partial \theta_c} k_{br} k_{cs}$$

$$(i = 1,2,\ldots,n-p; \quad r,s = 1,2,\ldots,p). \tag{4.70}$$

Here $f_h = f(\mathbf{x}_h;\boldsymbol{\theta})$, $[(k_{ab})] = \mathbf{K}$, and $[(q_{ab})] = \mathbf{Q}_{n-p}$. For later use we add one further array, namely

$$c_{irst} = \sum_{h=1}^{n} \sum_{m,a,b,c=1}^{p} \frac{\partial f_h}{\partial \theta_m} \frac{\partial^3 f_h}{\partial \theta_a \partial \theta_b \partial \theta_c} k_{mi} k_{ar} k_{bs} k_{ct} \qquad (i,r,s,t = 1,2,\ldots,p). \tag{4.71}$$

All the above derivatives are evaluated at $\boldsymbol{\theta} = \boldsymbol{\theta}^*$, and (4.69) and (4.70) are symmetric in (r,s), i.e., r and s can be interchanged without changing the value of the array element.

4.2.5 Replication and Curvature

As noted by Bates and Watts [1980], completely replicating a design J times [see (2.38)] has the beneficial effect of reducing the curvature at any point in any direction by a factor of $J^{-1/2}$. This is seen by noting that, for the replicated design, the corresponding derivative vectors rep $\dot{\boldsymbol{\mu}}_h$ and rep $\ddot{\boldsymbol{\mu}}_h$ are, respectively, J copies of $\dot{\boldsymbol{\mu}}_h$ and $\ddot{\boldsymbol{\mu}}_h$. Thus

$$\begin{aligned} \|\text{rep}\,\dot{\boldsymbol{\mu}}_h\| &= \|(\dot{\boldsymbol{\mu}}'_h, \dot{\boldsymbol{\mu}}'_h, \ldots, \dot{\boldsymbol{\mu}}'_h)\| = \sqrt{J}\,\|\dot{\boldsymbol{\mu}}_h\|, \\ \|\text{rep}\,\ddot{\boldsymbol{\mu}}_h\| &= \sqrt{J}\,\|\ddot{\boldsymbol{\mu}}_h\|, \end{aligned}$$

and the curvatures take the form

$$\text{rep}\,\gamma_{\mathbf{h}} = \frac{\|\text{rep}\,\ddot{\boldsymbol{\mu}}_{\mathbf{h}}\|}{\|\text{rep}\,\dot{\boldsymbol{\mu}}_{\mathbf{h}}\|^2} = \frac{\gamma_{\mathbf{h}}}{\sqrt{J}}.$$

In calculating the standard radius we now use $\rho = s_e\sqrt{p}$, where s_e^2 is given by (2.40). The reader who wishes to avoid some of the details should now skip to Section 4.3.

*4.2.6 Interpreting the Parameter-Effects Array

In the discussion following (4.35) we noted that the orthogonal transformation $\hat{\mathbf{Q}}'$ $[=(\hat{\mathbf{Q}}_p, \hat{\mathbf{Q}}_{n-p})']$ made it easy to set up a coordinate system on the tangent plane. Thus

$$\hat{\mathbf{Q}}'[\boldsymbol{\mu}(\boldsymbol{\theta}) - \hat{\boldsymbol{\mu}}] = \begin{pmatrix} \boldsymbol{\tau} \\ \boldsymbol{\xi} \end{pmatrix} \begin{matrix} \}\, p \\ \}\, n-p \end{matrix}$$

represents the new coordinates of $\boldsymbol{\mu}(\boldsymbol{\theta})$ relative to the origin $\hat{\boldsymbol{\mu}}$ $[=\boldsymbol{\mu}(\hat{\boldsymbol{\theta}})]$, and

$$\begin{pmatrix} \boldsymbol{\tau} \\ \mathbf{0} \end{pmatrix} = \tau_1\mathbf{e}_1 + \tau_2\mathbf{e}_2 + \cdots + \tau_p\mathbf{e}_p$$

is the component in the tangent plane. In terms of the ϕ-coordinates we have the transformation

$$\boldsymbol{\tau} = \mathbf{Q}_p'[\boldsymbol{\mu}(\mathbf{K}\boldsymbol{\phi}) - \hat{\boldsymbol{\mu}}] = \boldsymbol{\psi}(\boldsymbol{\phi}), \tag{4.72}$$

say, which takes a vector $\boldsymbol{\phi}$ into a vector $\boldsymbol{\tau}$ in the tangent plane. Just as we can trace parameter curves (called ϕ_i parameter curves), on the expectation surface by varying one coordinate of $\boldsymbol{\phi}$ (ϕ_i say) while holding the others fixed, we can also trace parameter curves on the tangent plane. If the function $\boldsymbol{\psi}$ is in fact linear, then the parameter curves on the tangent plane will be straight, equispaced parallel lines, and by (4.56) all the elements $\hat{a}_{irs}^T$ in $\hat{A}_{\cdot\cdot}^T$ will be zero, as $\hat{\mathbf{g}}_{rs} = \mathbf{0}$. Nonzero entries in $\hat{A}_{\cdot\cdot}^T$ then indicate a failure of the parameter curves to be straight, or equidistant, or parallel. In this case Bates and Watts [1981a] give the following geometrical interpretation of the elements $\hat{a}_{ijk}^T$ with respect to parameter lines on the tangent plane:

(i) $\hat{a}_{iii}^T$ produces a change in length only. It is then called a *compansion* term, since it causes a compression or expansion of scale along a ϕ_i parameter line.

(ii) $\hat{a}_{ijj}^T$ $(i \neq j)$ is called an *arcing* term, as it causes a change in the coordinate τ_i as we move along the ϕ_j parameter line, that is, produces a curved ϕ_j parameter line.

(iii) $\hat{a}^T_{iij}$ ($=\hat{a}^T_{iji}$, $i \neq j$) is called a *fanning* term, since the ϕ_j parameter lines will appear to fan out from a common point on the τ_j axis.

(iv) $\hat{a}^T_{ijk}$ (i, j, k all different) is called a *torsion* curve, since it causes a twisting of the (ϕ_j, ϕ_k) parameter surface, this surface being obtained by varying ϕ_j and ϕ_k but keeping the other ϕ's fixed.

Example 4.3 Bates and Watts [1980] simulated the twelve observations given in Table 4.2 for the Michaelis–Menten model

$$f(x;\boldsymbol{\theta}) = \frac{\theta_1 x}{\theta_2 + x}.$$

This was a replicated design with two replicates for each x-value. Bates and Watts [1980, 1981a] computed the following values (we use the *scaled* values of $\hat{A}^T_{..}$ from their 1981 paper):

$$\hat{\boldsymbol{\theta}}' = (0.10579, 1.7007), \qquad p = 2,$$

$$s_e^2 = 1.998 \times 10^{-4}/6, \qquad \text{[cf. (2.40)]}$$

$$\rho = \sqrt{p s_e^2} = 0.008165,$$

and

$$\hat{\mathbf{K}} = \begin{pmatrix} -0.00860 & -0.03099 \\ 0 & -0.87016 \end{pmatrix}.$$

(Here Bates and Watts define the Cholesky decomposition in terms of an upper triangular matrix with *negative* diagonal elements.) The two faces of $A^T_{..}$ are given by

$$\hat{\mathbf{A}}^T_{1..} = \begin{pmatrix} 0 & -0.292 \\ -0.292 & -0.163 \end{pmatrix}$$

Table 4.2 Simulated Data from the Michaelis–Menten Model[a]

x	y	x	y
2.000	0.0615	0.286	0.0129
2.000	0.0527	0.286	0.0183
0.667	0.0334	0.222	0.0083
0.667	0.0258	0.222	0.0169
0.400	0.0138	0.200	0.0129
0.400	0.0258	0.200	0.0087

[a]From Bates and Watts [1980].

and

$$\hat{\mathbf{A}}_{2\cdot\cdot}^{T} = \begin{pmatrix} 0 & -0.081 \\ -0.081 & -0.716 \end{pmatrix},$$

and the curvatures are

$$\gamma_{\max}^{N} = 0.0836$$

and

$$\gamma_{\max}^{T} = 0.7710.$$

Bates and Watts compare these curvatures with $1/\sqrt{F_\alpha} = 1/\sqrt{F_{2,6}^{.05}} = 0.441$. From (4.23) we have

$$\gamma_{\max}^{N}\sqrt{F_\alpha} = \frac{0.0836}{0.441} = 0.19,$$

and from (4.24)

$$1 - \{1 - (\gamma_{\max}^{N})^2 F_\alpha\}^{1/2} = 0.018,$$

which give measures of the worst deviation (since we are using $\max \gamma_{\mathbf{h}}^{N}$). From the values 0.19 and 0.018 of the above two criteria it would seem that we can ignore the intrinsic curvature. However,

$$\gamma_{\max}^{T}\sqrt{F_\alpha} = 1.75,$$

which indicates an appreciable parameter-effects curvature.

Using the ϕ-coordinate system (but with a shift of origin to $\hat{\boldsymbol{\phi}}$ so that $\boldsymbol{\phi} - \hat{\boldsymbol{\phi}}$ becomes "$\boldsymbol{\phi}$"), we obtain the ϕ_i parameter curves on the tangent plane as shown in Fig. 4.3. In interpreting this figure it should be noted that the lines labeled $\phi_1 = 0, \phi_1 = 1$, and so on are actually ϕ_2 parameter curves, as ϕ_2 is the parameter that is varying. An examination of $\hat{A}_{\cdot\cdot}^{T}$ reveals the following curvature effects. The ϕ_1 compansion term at $\hat{a}_{111}^{T}$ is zero, so that the ϕ_2 parameter curves (i.e. $\phi_1 = 0$, $\phi_1 = 1$, etc.) are perfectly uniformly spaced (but curved in this case). However, $\hat{a}_{222}^{T} = 0.716$, which indicates that the ϕ_1 curves are markedly compressed together as ϕ_2 increases. This lack of scale uniformity seems to be the main reason for the large $\gamma_{\max}^{T}$. The comparatively small arcing term $\hat{a}_{122}^{T} = -0.163$ will cause little bending of the ϕ_2 parameter curves, and the ϕ_1 parameter curves will be straight, as $\hat{a}_{211}^{T} = 0$. The small ϕ_1 fanning term $(\hat{a}_{221}^{T} = \hat{a}_{212}^{T} = -0.081)$ causes a slight convergence of the ϕ_1 parameter curves, but the larger value for the ϕ_2 fanning term $(\hat{a}_{121}^{T} = \hat{a}_{112}^{T} = -0.292)$ causes considerable convergence of the ϕ_2 parameter curves. These conclusions are supported by Fig. 4.3.

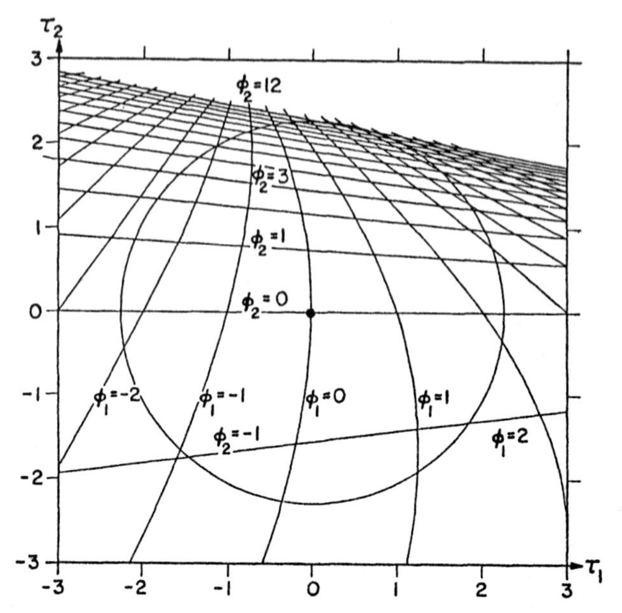

Figure 4.3 ϕ-parameter curves on the tangent plane. From Bates and Watts [1981a], with permission from the Institute of Mathematical Statistics.

*4.2.7 Computing the Curvatures

The basic steps in calculating the curvature arrays can be summarized as follows (Bates and Watts [1980], Bates et al. [1983]):

1. Calculate $\hat{\mathbf{F}}_{\bullet}$ and $\hat{F}_{\bullet\bullet} = [(\hat{\mathbf{f}}_{rs})]$, the derivative arrays evaluated at $\boldsymbol{\theta} = \hat{\boldsymbol{\theta}}$. The array $\hat{F}_{\bullet\bullet}$ is arranged in symmetric storage mode (ssm) using the $n \times q$ matrix, with $q = p(p+1)/2$,

$$\hat{\mathbf{F}}_{\text{ssm}} = (\hat{\mathbf{f}}_{11}, \hat{\mathbf{f}}_{12}, \hat{\mathbf{f}}_{22}, \hat{\mathbf{f}}_{13}, \hat{\mathbf{f}}_{23}, \hat{\mathbf{f}}_{33}, \ldots, \hat{\mathbf{f}}_{pp}). \tag{4.73}$$

(We note that $\hat{\mathbf{f}}_{rs} = \hat{\mathbf{f}}_{sr}$.)

2. Form the $n \times (p+q)$ matrix $\hat{\mathbf{D}} = (\hat{\mathbf{F}}_{\bullet}, \hat{\mathbf{F}}_{\text{ssm}})$, and rescale this matrix by dividing each element by $\rho = s\sqrt{p}$.

3. Use a QR decomposition with orthogonal matrix $\mathbf{Q}_1$, and pivoting in $\hat{\mathbf{F}}_{\text{ssm}}$, to reduce $\hat{\mathbf{D}}$ to an upper triangular form. Thus

$$\hat{\mathbf{Q}}_1' \hat{\mathbf{D}} \begin{pmatrix} \mathbf{I}_p & \mathbf{0} \\ \mathbf{0} & \boldsymbol{\Pi}_1 \end{pmatrix} = \hat{\mathbf{Q}}_1' (\hat{\mathbf{F}}_{\bullet}, \hat{\mathbf{F}}_{\text{ssm}} \boldsymbol{\Pi}_1)$$

$$= \begin{pmatrix} \hat{\mathbf{U}} \\ \mathbf{0} \end{pmatrix}, \tag{4.74}$$

where

$$\hat{\mathbf{U}} = \begin{pmatrix} \hat{\mathbf{R}}_{11} & \hat{\mathbf{R}}_{12} \\ \mathbf{0} & \hat{\mathbf{R}}_{22} \end{pmatrix} \begin{matrix} \}p \\ \}q \end{matrix}$$

is an upper triangular matrix, and $\mathbf{\Pi}_1$ is a $q \times q$ permutation matrix (and therefore orthogonal). The matrix $\mathbf{\Pi}_1$, which is introduced by pivoting, consists of a product of matrices each representing a column interchange. Although $\hat{\mathbf{F}}.$ is assumed to have linearly independent columns, $\hat{\mathbf{D}}$ need not have linearly independent columns, so that some of the diagonal elements of $\mathbf{R}_{22}$ (and the remaining elements in the corresponding rows) may be zero. Pivoting using a permutation of the columns of $\hat{\mathbf{F}}_{\text{ssm}}$ moves the zero rows to the bottom of $\hat{\mathbf{R}}_{22}$ so that $\hat{\mathbf{U}}$ is upper trapezoidal.

4. Find $\hat{\mathbf{K}} = \hat{\mathbf{R}}_{11}^{-1}$: $\hat{\mathbf{R}}_{11}$ is the upper triangular matrix defined in (4.33). This follows from (4.74), which gives

$$\hat{\mathbf{Q}}_1' \hat{\mathbf{F}}. = \begin{pmatrix} \hat{\mathbf{R}}_{11} \\ \mathbf{0} \\ \mathbf{0} \end{pmatrix}, \tag{4.75}$$

and this implies that $\hat{\mathbf{R}}_{11}' \hat{\mathbf{R}}_{11}$ is the (unique) Cholesky decomposition of $\hat{\mathbf{F}}_.'\hat{\mathbf{F}}_.$. Also from (4.74) we have

$$\begin{pmatrix} \hat{\mathbf{R}}_{12} \\ \hat{\mathbf{R}}_{22} \\ \mathbf{0} \end{pmatrix} = \hat{\mathbf{Q}}_1' \hat{\mathbf{F}}_{\text{ssm}} \mathbf{\Pi}_1. \tag{4.76}$$

We noted after (4.34) that $\hat{\mathbf{Q}}$ is not unique, being any orthogonal matrix whose first p columns are $\hat{\mathbf{Q}}_p$. As $\hat{\mathbf{Q}}_1$ has this property [by (4.75) and the uniqueness of $\hat{\mathbf{R}}_{11}$], it can be used wherever $\hat{\mathbf{Q}}$ occurs in the remaining steps.

5. Calculate the $\hat{\mathbf{a}}_{rs}$ using (4.65), namely

$$\begin{aligned} \hat{\mathbf{a}}_{rs} &= \sum_u \sum_v \hat{\mathbf{Q}}_1' \hat{\mathbf{f}}_{uv} \hat{k}_{ur} \hat{k}_{vs} \\ &= \hat{\mathbf{Q}}_1'(\hat{\mathbf{f}}_{11}, \hat{\mathbf{f}}_{12}, \ldots, \hat{\mathbf{f}}_{pp}) \begin{pmatrix} \hat{k}_{1r}\hat{k}_{1s} \\ \hat{k}_{1r}\hat{k}_{2s} + \hat{k}_{2r}\hat{k}_{1s} \\ \vdots \\ \hat{k}_{pr}\hat{k}_{ps} \end{pmatrix} \\ &= \hat{\mathbf{Q}}_1' \hat{\mathbf{F}}_{\text{ssm}} \mathbf{m}_{rs} \qquad \text{(say)} \\ &= \hat{\mathbf{Q}}_1' \hat{\mathbf{F}}_{\text{ssm}} \mathbf{\Pi}_1 \mathbf{\Pi}_1' \mathbf{m}_{rs}, \end{aligned}$$

since $\mathbf{\Pi}_1$ is orthogonal. Using symmetric storage model for the $\hat{\mathbf{a}}_{rs}$ and $\mathbf{m}_{rs}$, we have

$$\begin{aligned}\hat{\mathbf{A}}_{\text{ssm}} &= (\hat{\mathbf{a}}_{11}, \hat{\mathbf{a}}_{12}, \ldots, \hat{\mathbf{a}}_{pp}) \\ &= \mathbf{Q}_1' \hat{\mathbf{F}}_{\text{ssm}} \mathbf{\Pi}_1 (\mathbf{\Pi}_1' \mathbf{M}_{\text{ssm}}) \\ &= \begin{pmatrix} \hat{\mathbf{R}}_{12} \\ \hat{\mathbf{R}}_{22} \\ \mathbf{0} \end{pmatrix} (\mathbf{\Pi}_1' \mathbf{M}_{\text{ssm}}) \end{aligned} \tag{4.77}$$

by (4.76), where $\mathbf{M}_{\text{ssm}}$ is a $q \times q$ matrix composed of products of elements in $\hat{\mathbf{K}}$.

6. The two component arrays are then given by

$$\hat{\mathbf{A}}_{\text{ssm}}^T = \hat{\mathbf{R}}_{12} \mathbf{\Pi}_1' \mathbf{M}_{\text{ssm}} \tag{4.78}$$

and

$$\hat{\mathbf{A}}_{\text{ssm}}^N = \begin{pmatrix} \hat{\mathbf{R}}_{22} \mathbf{\Pi}_1' \mathbf{M}_{\text{ssm}} \\ \mathbf{0} \end{pmatrix}. \tag{4.79}$$

An advantage in using the symmetric storage mode is the presence of the zero matrix in $\hat{\mathbf{A}}_{\text{ssm}}^N$. This implies a reduction in the dimension of the intrinsic-curvature array, with a consequent saving in storage. Furthermore, once $\mathbf{M}_{\text{ssm}}$ has been computed we can obtain $\mathbf{\Pi}_1' \mathbf{M}_{\text{ssm}}$ $[= (\mathbf{M}_{\text{ssm}}' \mathbf{\Pi}_1)']$ by simply applying the same column interchanges to $\mathbf{M}_{\text{ssm}}'$ that were applied to $\hat{\mathbf{F}}_{\text{ssm}}$.

The next stage is maximize (4.58) and (4.59), or their squares, with respect to the unit vector $\mathbf{d}$. To do this for $\Gamma_{\mathbf{d}}^T$, say, we introduce a Lagrange multiplier λ and differentiate

$$\sum_{i=1}^{p} (\mathbf{d}' \hat{\mathbf{A}}_{i\cdot\cdot} \mathbf{d})^2 + \lambda(1 - \mathbf{d}'\mathbf{d})$$

with respect to $\mathbf{d}$. This leads to (Appendix A10.5)

$$4 \sum_{i=1}^{p} (\mathbf{d}' \hat{\mathbf{A}}_{i\cdot\cdot} \mathbf{d}) \hat{\mathbf{A}}_{i\cdot\cdot} \mathbf{d} - 2\lambda \mathbf{d} = \mathbf{0},$$

or

$$\mathbf{r}(\mathbf{d}) = \tfrac{1}{2} \lambda \mathbf{d},$$

where

$$\mathbf{r}(\mathbf{d}) = \sum_{i=1}^{p} (\mathbf{d}' \hat{\mathbf{A}}_{i\cdot\cdot} \mathbf{d}) \hat{\mathbf{A}}_{i\cdot\cdot} \mathbf{d}. \tag{4.80}$$

Multiplying on the left by $\mathbf{d}'$, we have

$$\lambda = 2 \sum_{i=1}^{p} (\mathbf{d}'\hat{\mathbf{A}}_{i..}\mathbf{d})^2 > 0,$$

so that a stationary value occurs at $\mathbf{d} = \mathbf{d}_0$, say, when $\mathbf{r}_0\ [= \mathbf{r}(\mathbf{d}_0)]$ is in the same direction as $\mathbf{d}_0$. The algorithm can be summarized as follows:

(i) Choose an initial direction $\mathbf{d}^{(a)}$.
(ii) Using (4.80), calculate $\mathbf{r}^{(a)} = \mathbf{r}(\mathbf{d}^{(a)})$ and $\tilde{\mathbf{r}}^{(a)} = \mathbf{r}^{(a)}/\|\mathbf{r}^{(a)}\|$.
(iii) If $\tilde{\mathbf{r}}^{(a)\prime}\mathbf{d}^{(a)} > 1 - \delta$ then set $\mathbf{d}^{(a+1)} = \tilde{\mathbf{r}}^{(a)}$ and repeat from step (ii); otherwise set $\mathbf{d}_0 = \mathbf{d}^{(a)}$ and calculate

$$\begin{aligned} \gamma_{\max}^T &= \| \mathbf{d}'\hat{\mathcal{A}}_{..}^T\mathbf{d} \|_{\mathbf{d}=\mathbf{d}_0} \\ &= \left\| \sum_r \sum_s \hat{\mathbf{a}}_{rs}^T d_r d_s \right\|_{\mathbf{d}=\mathbf{d}_0} \\ &= \| \hat{\mathbf{A}}_{\text{ssm}}^T \mathbf{c}_{\mathbf{d}} \|_{\mathbf{d}=\mathbf{d}_0}, \end{aligned} \tag{4.81}$$

where

$$\mathbf{c}_{\mathbf{d}} = (d_1^2, 2d_1d_2, d_2^2, 2d_1d_3, 2d_2d_3, d_3^2, \ldots, d_p^2)'. \tag{4.82}$$

In step (iii) the convergence criterion, δ, determines how close $\tilde{\mathbf{r}}^{(a)}$ and $\mathbf{d}^{(a)}$ are to being in the same direction. As they are unit vectors, they will have an inner product of unity if they are in the same direction. Bates and Watts [1980] note that the algorithm tends to oscillate about the optimum as it gets close to convergence, so they suggest modifying step (iii) and setting

$$\mathbf{d}^{(a+1)} = \frac{3\tilde{\mathbf{r}}^{(a)} + \mathbf{d}^{(a)}}{\| 3\tilde{\mathbf{r}}^{(a)} + \mathbf{d}^{(a)} \|}. \tag{4.83}$$

They also noticed that in using a QR decomposition with pivoting, $\mathbf{d}_0$ tended to lie close to $\mathbf{e}_p = (0, 0, \ldots, 1)'$, so that the initial value $\mathbf{d}^{(1)} = \mathbf{e}_p$ was used to start the algorithm. Finally, the maximum intrinsic curvature $\gamma_{\max}^N$ is computed in the same manner.

Bates and Watts [1980] found that with the above starting value and convergence criterion, their modified algorithm using (4.83) usually converged in about four iterations and never required more than ten iterations in any of their 24 examples. To check convergence to the global rather than a local maximum, they also calculated bounds on the curvatures. Details of the bounds are given in their paper. If the calculated maximum fell within the bounds, they accepted that they had found the global maximum. Alternatively, one can find the maximum using a standard program for optimization subject to constraints (Section 13.1).

As noted by Goldberg et al. [1983], the above method has several problems. Firstly, the second derivatives in the array $\hat{F}_{..}$ can be very complex, especially for multiparameter models. This makes their routine use difficult, particularly when considering different methods of reparametrization. Secondly, analytic derivatives represent only the local behavior of the expectation surface, and may not represent the shape of the surface over the region of interest. Goldberg et al. [1983] therefore propose an alternative formulation which effectively uses a secant approximation to second derivatives. Their method is described below.

*4.2.8 Secant Approximation of Second Derivatives

We now consider a method of estimating the second derivative vectors $\hat{\mathbf{f}}_{rs}$ in $\hat{\mathbf{F}}_{..}$ evaluated at $\hat{\boldsymbol{\theta}}$ using a number of secants emanating out from $\hat{\boldsymbol{\theta}}$. Let $\boldsymbol{\eta}(\boldsymbol{\delta})$ be the secant joining the points $\boldsymbol{\mu}(\hat{\boldsymbol{\theta}})$ and $\boldsymbol{\mu}(\hat{\boldsymbol{\theta}}+\boldsymbol{\delta})$ on the expectation surface (see Fig. 4.4). Then, from (4.5),

$$\begin{aligned}\boldsymbol{\eta}(\boldsymbol{\delta}) &\approx \hat{\mathbf{F}}_{.}\boldsymbol{\delta} + \tfrac{1}{2}\boldsymbol{\delta}'\hat{\boldsymbol{F}}_{..}\boldsymbol{\delta} \\ &= \mathbf{t}_{\delta} + \tfrac{1}{2}\mathbf{a}_{\delta},\end{aligned} \tag{4.84}$$

say, where $\mathbf{t}_{\delta}\,(=\mathbf{F}_{.}\boldsymbol{\delta})$ is in the tangent plane: $\mathbf{a}_{\delta}\,(=\boldsymbol{\delta}'\hat{\boldsymbol{F}}_{..}\boldsymbol{\delta})$ is usually called the

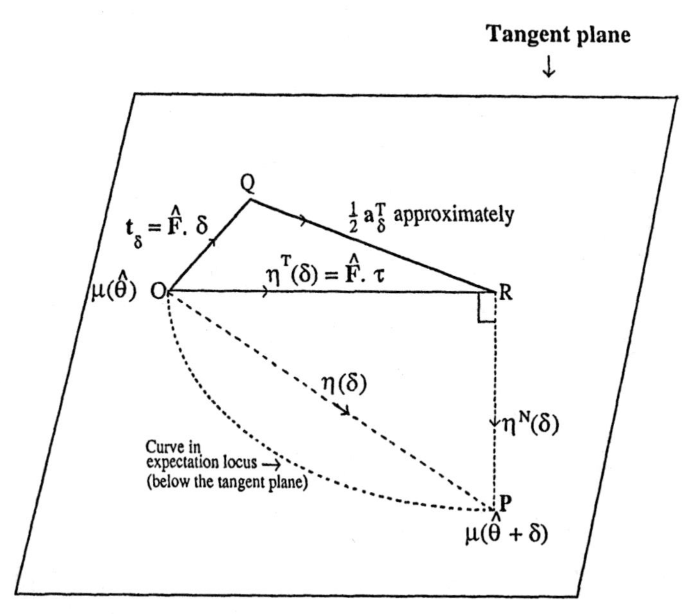

Figure 4.4 Decomposition of the secant OP into tangential and orthogonal components OR and RP respectively.

"acceleration" vector. From (4.84) we have

$$\mathbf{a}_\delta \approx 2\{\boldsymbol{\eta}(\boldsymbol{\delta}) - \mathbf{t}_\delta\}, \tag{4.85}$$

so that we can compute the components of $\mathbf{a}_\delta$ using the approximations

$$\mathbf{a}_\delta^T \approx 2\{\boldsymbol{\eta}^T(\boldsymbol{\delta}) - \mathbf{t}_\delta\}$$

and

$$\mathbf{a}_\delta^N = 2\boldsymbol{\eta}^N(\boldsymbol{\delta}).$$

Also, using the symmetric storage mode defined by (4.73), we have

$$\begin{aligned}
\mathbf{a}_\delta^T &= (\boldsymbol{\delta}'\hat{F}_{..}\boldsymbol{\delta})^T \\
&= \left(\sum_r \sum_s \hat{\mathbf{f}}_{rs}\delta_r\delta_s\right)^T \qquad \text{[by (4.8)]} \\
&= \sum_r \sum_s \hat{\mathbf{f}}_{rs}^T\delta_r\delta_s \\
&= \hat{\mathbf{F}}_{\text{ssm}}^T \mathbf{c}_\delta,
\end{aligned} \tag{4.86}$$

where

$$\mathbf{c}_\delta = (\delta_1^2, 2\delta_1\delta_2, \delta_2^2, \ldots, \delta_p^2)'.$$

We shall now show that by choosing several $\boldsymbol{\delta}$-directions, a system of equations like (4.86) can be set up and solved for $\hat{\mathbf{F}}_{\text{ssm}}^T$. A similar method will be used below to find $\hat{\mathbf{F}}_{\text{ssm}}^N$.

To compute the components of $\mathbf{a}_\delta$ we need $\boldsymbol{\eta}^T(\boldsymbol{\delta})$ and $\boldsymbol{\eta}^N(\boldsymbol{\delta})$, which can be found by orthogonal projection. For example, $\boldsymbol{\eta}^T(\boldsymbol{\delta})$ is the orthogonal projection of $\boldsymbol{\eta}(\boldsymbol{\delta})$ onto the tangent plane, the range space $\mathscr{R}[\hat{\mathbf{F}}_.]$ of $\hat{\mathbf{F}}_.$. Hence (Appendix A11.4)

$$\begin{aligned}
\boldsymbol{\eta}^T(\boldsymbol{\delta}) &= \hat{\mathbf{P}}_F\boldsymbol{\eta}(\boldsymbol{\delta}) \\
&= \hat{\mathbf{F}}_.(\hat{\mathbf{F}}_.'\hat{\mathbf{F}}_.)^{-1}\mathbf{F}_.'\boldsymbol{\eta}(\boldsymbol{\delta}) \\
&= \hat{\mathbf{F}}_.\boldsymbol{\tau} \\
&= \mathbf{t}_\tau,
\end{aligned} \tag{4.87}$$

say, where $\boldsymbol{\tau} = (\hat{\mathbf{F}}_.'\hat{\mathbf{F}}_.)^{-1}\hat{\mathbf{F}}_.'\boldsymbol{\eta}(\boldsymbol{\delta})$. Also

$$\boldsymbol{\eta}^N(\boldsymbol{\delta}) = (\mathbf{I}_n - \hat{\mathbf{P}}_F)\boldsymbol{\eta}(\boldsymbol{\delta}). \tag{4.88}$$

By analogy with the regression model $\mathbf{y} \approx \hat{\mathbf{F}}_.\boldsymbol{\beta}$, with $\mathbf{y}$ replaced by $\boldsymbol{\eta}(\boldsymbol{\delta})$, we see that $\boldsymbol{\tau}$ is the "least-squares estimate" of $\boldsymbol{\beta}$ and that $\boldsymbol{\eta}^N(\boldsymbol{\delta})$ is the residual vector for the fitted model (c.f. Appendix D1).

We now turn our attention to the estimation of $\hat{\mathbf{F}}^N_{\text{ssm}}$. As $\boldsymbol{\eta}^T(\boldsymbol{\delta}) = \mathbf{t}_\tau$ [by (4.87)], Goldberg et al. [1983] suggest that $\boldsymbol{\tau}$, rather than $\boldsymbol{\delta}$, is the appropriate direction in which to consider the bending of the surface. This is because PR, the length of $\boldsymbol{\eta}^N(\boldsymbol{\delta})$ in Fig. 4.4, is the orthogonal offset of P from the tangent plane in the direction of $\mathbf{t}_\tau$ (and not $\mathbf{t}_\delta$). The authors then approximate this offset by $\frac{1}{2}\mathbf{a}^N_\tau$, so that

$$\mathbf{a}^N_\tau \approx 2\boldsymbol{\eta}^N(\boldsymbol{\delta}),$$

where [by (4.86)]

$$\begin{aligned}\mathbf{a}^N_\tau &= \boldsymbol{\tau}'\hat{F}^N_{\cdot\cdot}\boldsymbol{\tau}\\ &= \hat{\mathbf{F}}^N_{\text{ssm}}\mathbf{c}_\tau.\end{aligned} \tag{4.89}$$

To estimate the arrays $\hat{\mathbf{F}}^T_{\text{ssm}}$ and $\hat{\mathbf{F}}^N_{\text{ssm}}$ we use K directions $\boldsymbol{\delta}_k$ (and $\boldsymbol{\tau}_k$) $(k = 1, 2, \ldots, K)$ and obtain, by transposing (4.86), a set of equations of the form $\mathbf{Y} \approx \mathbf{XB}$. Here the K rows of $\mathbf{Y}$ are estimated acceleration vectors, $\mathbf{B}$ is the transpose of the relevant $\mathbf{F}_{\text{ssm}}$-matrix, and the rows of $\mathbf{X}$ are $p(p+1)/2$-dimensional c-vectors. Therefore, provided $K > p(p+1)/2$, we can solve for $\mathbf{B}$, namely $\tilde{\mathbf{B}} = (\mathbf{X}'\mathbf{X})^{-1}\mathbf{X}'\mathbf{Y}$. This means that any column of $\mathbf{B}$ can be found by regressing the corresponding column of $\mathbf{Y}$ on $\mathbf{X}$, and $\tilde{\mathbf{B}}$ can be regarded as the least-squares estimate of $\mathbf{B}$. We now summarize the above discussion in the following algorithm.

1. Choose evaluation points $\hat{\boldsymbol{\theta}} + \boldsymbol{\delta}_k$ (this is discussed below), and calculate

$$\boldsymbol{\eta}(\boldsymbol{\delta}_k) = \boldsymbol{\mu}(\hat{\boldsymbol{\theta}} + \boldsymbol{\delta}_k) - \boldsymbol{\mu}(\hat{\boldsymbol{\theta}}), \qquad k = 1, 2, \ldots, K. \tag{4.90}$$

2. Regress each $\boldsymbol{\eta}(\boldsymbol{\delta}_k)$ on $\hat{\mathbf{F}}_\cdot$ to get

$$\begin{aligned}\boldsymbol{\tau}_k &= (\hat{\mathbf{F}}_\cdot'\hat{\mathbf{F}}_\cdot)^{-1}\hat{\mathbf{F}}_\cdot'\boldsymbol{\eta}(\boldsymbol{\delta}_k),\\ \boldsymbol{\eta}^T(\boldsymbol{\delta}_k) &= \hat{\mathbf{F}}_\cdot\boldsymbol{\tau}_k,\\ \boldsymbol{\eta}^N(\boldsymbol{\delta}_k) &= \boldsymbol{\eta}(\boldsymbol{\delta}_k) - \hat{\mathbf{F}}_\cdot\boldsymbol{\tau}_k.\end{aligned}$$

3. Form the following $\mathbf{Z}$ and $\mathbf{C}$ matrices:

$$\begin{aligned}\mathbf{Z}^T &= [2\boldsymbol{\eta}^T(\boldsymbol{\delta}_1), \ldots, 2\boldsymbol{\eta}^T(\boldsymbol{\delta}_K)]',\\ \mathbf{C}_\delta &= [\mathbf{c}_{\delta_1}, \mathbf{c}_{\delta_2}, \ldots, \mathbf{c}_{\delta_K}]'.\end{aligned}$$

$\mathbf{Z}^N = [2\boldsymbol{\eta}^N(\boldsymbol{\delta}_1), \ldots, 2\boldsymbol{\eta}^N(\boldsymbol{\delta}_K)]'$, and $\mathbf{C}_\tau$.

4. Since the rows of $\mathbf{Z}^T$ and $\mathbf{Z}^N$ estimate the acceleration vectors, we have from (4.86) and (4.89) the equations $(\mathbf{Z}^T)' \approx \hat{\mathbf{F}}^T_{\text{ssm}}\mathbf{C}_\delta'$ or $\mathbf{Z}^T \approx \mathbf{C}_\delta(\hat{\mathbf{F}}^T_{\text{ssm}})'$, and $\mathbf{Z}^N \approx \mathbf{C}_\tau(\hat{\mathbf{F}}^N_{\text{ssm}})'$. Obtain the "least-squares" solutions of these equations,

namely

$$(\tilde{\mathbf{F}}^T_{\text{ssm}})' = (\mathbf{C}'_\delta \mathbf{C}_\delta)^{-1}\mathbf{C}'_\delta \mathbf{Z}^T$$

and

$$(\tilde{\mathbf{F}}^N_{\text{ssm}})' = (\mathbf{C}'_\tau \mathbf{C}_\tau)^{-1}\mathbf{C}'_\tau \mathbf{Z}^N,$$

using a regression package.

5. The closeness of the secant approximations can be measured by the residuals from step 4, namely

$$\mathbf{Z}^T - \mathbf{C}_\delta(\tilde{\mathbf{F}}^T_{\text{ssm}})' = \mathbf{Z}^T - \tilde{\mathbf{Z}}^T \qquad \text{etc.}$$

Various measures of fit are possible. One method is to convert the multivariate model into the univariate regression model

$$\operatorname{vec}\mathbf{Z}^T \approx \operatorname{diag}(\mathbf{C}_\delta, \mathbf{C}_\delta, \ldots, \mathbf{C}_\delta)\operatorname{vec}(\hat{\mathbf{F}}^T_{\text{ssm}})'$$

i.e.

$$\mathbf{z} \approx \mathbf{X}\boldsymbol{\beta},$$

where vec denotes the columns of a matrix "rolled out" into a single vector (Appendix A13.2). The least-squares estimate of $\boldsymbol{\beta}$ is $\operatorname{vec}(\tilde{\mathbf{F}}^T_{\text{ssm}})'$, and we can, for example, use the coefficient of multiple correlation (R^2) as a measure of fit.

All that remains now is to choose the directions $\boldsymbol{\delta}_k$. Goldberg et al. [1983] work with the transformed parameters $\boldsymbol{\phi}$ for which $\hat{\mathbf{F}}_{.}$ becomes $\hat{\mathbf{Q}}_p$, and $\boldsymbol{\mu}(\boldsymbol{\phi}) = \boldsymbol{\mu}(\hat{\boldsymbol{\phi}} + \boldsymbol{\delta})$. They choose a total of $2p + 4p(p-1)/2$ [$= 2p^2$] directions $\pm\mathbf{e}_i$ and $(\pm\mathbf{e}_i \pm \mathbf{e}_j)/\sqrt{2}$ ($i < j$; $i, j = 1, 2, \ldots, p$), where $\mathbf{e}_i$ is the usual unit vector with one in the ith dimension and zeros elsewhere, but in the ϕ-parameter space. The vectors $\boldsymbol{\delta}_k$ are obtained by multiplying each of the $2p^2$ unit vectors by $\rho(F^\alpha_{p,n-p})^{1/2}$, as this gives points on the boundary of the approximate linearized confidence region [see (4.13)].

4.3 BEALE'S MEASURES

Beale [1960] proposed four measures of nonlinearity. The first measure was defined to be

$$\hat{N}_\theta = \rho^2 \frac{\sum_{k=1}^{K} \|\boldsymbol{\mu}(\boldsymbol{\theta}_k) - \mathbf{T}(\boldsymbol{\theta}_k)\|^2}{\sum_{k=1}^{K} \|\boldsymbol{\mu}(\boldsymbol{\theta}_k) - \boldsymbol{\mu}(\hat{\boldsymbol{\theta}})\|^4}, \tag{4.91}$$

where

$$\mathbf{T}(\boldsymbol{\theta}_k) = \boldsymbol{\mu}(\hat{\boldsymbol{\theta}}) + \hat{\mathbf{F}}.(\boldsymbol{\theta}_k - \hat{\boldsymbol{\theta}})$$

and $\boldsymbol{\theta}_k$ $(k = 1, 2, \ldots, K)$ are K points near $\boldsymbol{\theta}$. Since the choice of the number and location of the points is arbitrary, and the linear approximation $\mathbf{T}(\boldsymbol{\theta})$ to $\boldsymbol{\mu}(\boldsymbol{\theta})$ depends on the particular parametrization chosen, $\hat{N}_\theta$ was called an empirical measure of (total) nonlinearity for the parameter vector $\boldsymbol{\theta}$. Guttman and Meeter [1965] tested $\hat{N}_\theta$ on two models, each with three experimental-design combinations (involving 4, 8, and 12 observations respectively) and found, by looking at the contours of $S(\boldsymbol{\theta})$, that $\hat{N}_\theta$ tended to underestimate severe nonlinearity. Linssen [1975] also gave an example, based on 54 observations, which showed that $\hat{N}_\theta$ could be small for a severely nonlinear model. He proposed a modification

$$M^* = \left\{ \rho^2 \frac{\sum_{k=1}^{K} \|\boldsymbol{\mu}(\boldsymbol{\theta}_k) - \mathbf{T}(\boldsymbol{\theta}_k)\|^2}{\sum_{k=1}^{K} \|\mathbf{T}(\boldsymbol{\theta}_k) - \boldsymbol{\mu}(\hat{\boldsymbol{\theta}})\|^2} \right\}^{1/2}$$

in the hope of partly compensating for this defect in $\hat{N}_\theta$. Similar shortcomings in $\hat{N}_\theta$ were found by Gillis and Ratkowsky [1978] in their yield-density models, and by Bates and Watts [1980] in several of their data sets.

Beale's second measure, $\hat{N}_\phi$, is the minimum value of $\hat{N}_\theta$ for any parametrization and therefore measures the nonlinearity intrinsic to the model itself. Thus $\hat{N}_\phi$ was called the empirical measure of intrinsic nonlinearity.

Beale's third measure, N_θ, is the limit of $\hat{N}_\theta$ as $K \to \infty$ with the test points θ_k having a particular configuration. His fourth measure is N_ϕ, the corresponding limit of $\hat{N}_\phi$. Bates and Watts [1980] give an elegant interpretation of N_ϕ and N_θ in terms of mean square curvature. For example, they showed that

$$\begin{aligned} N_\phi &= \tfrac{1}{4}(\gamma_{\mathrm{RMS}}^N)^2 \\ &= \frac{\rho^2}{p(p+2)} \left(\frac{1}{4} \left\| \sum_{r=1}^{p} \mathbf{a}_{rr}^N \right\|^2 + \frac{1}{2} \sum_{r=1}^{p} \sum_{s=1}^{p} \|\mathbf{a}_{rs}^N\|^2 \right), \end{aligned} \tag{4.92}$$

where

$$(\gamma_{\mathrm{RMS}}^N)^2 = \frac{\int_{\|\mathbf{d}\|=1} (\gamma_{\mathbf{d}}^N)^2 \, dS}{\int_{\|\mathbf{d}\|=1} dS},$$

dS is an element of the surface area on a p-dimensional unit sphere, and $\gamma_{\mathbf{d}}^N$ is the scaled normal curvature in direction $\mathbf{d}$. Similarly

$$N_\theta - N_\phi = \tfrac{1}{4}(\gamma_{\text{RMS}}^T)^2$$
$$= \frac{\rho^2}{p(p+2)}\left(\frac{1}{4}\left\|\sum_{r=1}^{p} \mathbf{a}_{rr}^T\right\|^2 + \frac{1}{2}\sum_{r=1}^{p}\sum_{s=1}^{p} \|\mathbf{a}_{rs}^T\|^2\right). \tag{4.93}$$

Guttman and Meeter [1965] concluded from their small-sample experiment that N_θ and N_ϕ seem to reflect fairly well the overall picture of nonlinearity for their examples. Furthermore, $\hat{N}_\theta$ and $\hat{N}_\phi$ were reasonable estimates for N_θ and N_ϕ provided the degree of nonlinearity was not too "high." Beale also gave an estimate of N_ϕ which effectively uses a secant-type approximation to the second-derivative vector. This point is discussed further by Goldberg et al. [1983].

The difference between the measures of Bates and Watts [1980] and those of Beale [1960] is now clear. The former measure maximum curvature, while the latter measure average curvature. It is therefore possible for a model with a high degree of nonlinearity in just one direction on the expectation surface to score high for the Bates–Watts intrinsic curvature but score low on Beale's criterion because of the averaging effect. A comparison of the maximum curvatures with the root-mean-square measures is given in Table 4.1 (Section 4.2.2). For the parameter-effects curvatures, the maximum curvature is often more than double the root-mean-square curvature.

*4.4 CONNECTION COEFFICIENTS

The measures of curvature described above focus on the geometrical properties of the expectation surface (solution locus) $\mathscr{E}[\mathbf{y}] = \boldsymbol{\mu}(\boldsymbol{\theta})$. A different approach to the problem was earlier suggested by Efron [1975: 1196], who considered one-parameter ($p = 1$) models with a general (not necessarily normal) probability density function $p(\mathbf{y}|\theta)$ belonging to a family $\mathscr{F}$ with appropriate regularity properties. Let $L(\theta) = \log p(\mathbf{y}|\theta)$, and let $\dot{L}(\theta)$ and $\ddot{L}(\theta)$ represent first and second derivatives with respect to θ. It is assumed that the operations of differentiation with respect to θ and integration over the sample space can be interchanged. For example

$$\begin{aligned} E[\dot{L}(\theta)] &= \int \dot{L}(\theta) p(\mathbf{y}|\theta)\, d\mathbf{y} \\ &= \int \frac{\partial p(\mathbf{y}|\theta)}{\partial \theta} d\mathbf{y} \\ &= \frac{\partial}{\partial \theta} \int p(\mathbf{y}|\theta)\, d\mathbf{y} \\ &= \frac{\partial}{\partial \theta}(1) \\ &= 0. \end{aligned} \tag{4.94}$$

Then Efron defined the *statistical curvature* for $\mathscr{F}$ as

$$\gamma_\theta = \left\{ \frac{\nu_{02}(\theta)}{I^2(\theta)} - \frac{\nu_{11}^2(\theta)}{I^3(\theta)} \right\}^{1/2}, \tag{4.95}$$

where

$$\begin{aligned} I(\theta) &= E[\dot{L}(\theta)^2] = -E[\ddot{L}(\theta)], \\ \nu_{02}(\theta) &= E[\ddot{L}(\theta)^2] - I^2(\theta) = \text{var}[\ddot{L}(\theta)], \end{aligned} \tag{4.96}$$

and

$$\nu_{11}(\theta) = E[\dot{L}(\theta)\ddot{L}(\theta)] = \text{cov}[\dot{L}(\theta), \ddot{L}(\theta)]. \tag{4.97}$$

When the underlying distribution belongs to the one-parameter exponential family

$$p(\mathbf{y}|\theta) = k(\mathbf{y}) \exp\{\theta \mathbf{b}'\mathbf{y} - \psi(\theta)\},$$

with θ the so-called "natural parameter," then $\ddot{L}(\theta)$ is a constant and $\gamma_\theta = 0$. Efron therefore regards γ_θ as a measure of how close $\mathscr{F}$ is to being the exponential family. It is zero for the exponential family and positive, for at least some θ-values, otherwise. He shows that families with small curvatures enjoy, nearly, some of the good statistical properties of the exponential family, such as the applicability of standard linear methods and the existence of locally most powerful tests and efficient estimators.

How does Efron's statistical curvature γ_θ compare with the intrinsic curvature of Bates and Watts? It transpires that in this one-parameter case they are essentially the same when the ε_i $[= y_i - \mu_i(\theta)]$ are i.i.d. $N(0, \sigma^2)$ with known σ^2. This fact was noted, without proof, by Reid and Hinkley in the discussion of the paper by Bates and Watts [1980]. A proof follows.

Since

$$L(\theta) = \log p(\mathbf{y}|\theta) = \text{constant} - \frac{1}{2\sigma^2} \|\mathbf{y} - \boldsymbol{\mu}(\boldsymbol{\theta})\|^2,$$

we have (suppressing the dependence on θ for notational convenience, and using Appendix A10.5)

$$\sigma^2 \dot{L} = \dot{\boldsymbol{\mu}}'(\mathbf{y} - \boldsymbol{\mu}) \quad \text{and} \quad \sigma^2 \ddot{L} = \ddot{\boldsymbol{\mu}}'(\mathbf{y} - \boldsymbol{\mu}) - \dot{\boldsymbol{\mu}}'\dot{\boldsymbol{\mu}}.$$

Taking expected values, we have from (4.96) and (4.97)

$$\frac{\nu_{02}(\theta)}{I^2(\theta)} = \frac{\sigma^2 \ddot{\boldsymbol{\mu}}'\ddot{\boldsymbol{\mu}}}{(\dot{\boldsymbol{\mu}}'\dot{\boldsymbol{\mu}})^2}, \tag{4.98}$$

$$\frac{\nu_{11}^2(\theta)}{I^3(\theta)} = \frac{\sigma^2(\ddot{\boldsymbol{\mu}}'\dot{\boldsymbol{\mu}})^2}{(\dot{\boldsymbol{\mu}}'\dot{\boldsymbol{\mu}})^3}$$

$$= \frac{\sigma^2\ddot{\boldsymbol{\mu}}'\dot{\boldsymbol{\mu}}(\dot{\boldsymbol{\mu}}'\dot{\boldsymbol{\mu}})^{-1}\dot{\boldsymbol{\mu}}'\ddot{\boldsymbol{\mu}}}{(\dot{\boldsymbol{\mu}}'\dot{\boldsymbol{\mu}})^2}$$

$$= \frac{\sigma^2\ddot{\boldsymbol{\mu}}'\mathbf{P}_{\dot{\mu}}\ddot{\boldsymbol{\mu}}}{(\dot{\boldsymbol{\mu}}'\dot{\boldsymbol{\mu}})^2}, \tag{4.99}$$

where (by Appendix A11.4) $\mathbf{P}_{\dot{\mu}}$ represents an orthogonal projection in the direction of the tangent vector $\dot{\boldsymbol{\mu}}$ to the curve $\boldsymbol{\mu}(\boldsymbol{\theta})$. Substituting (4.98) and (4.99) into (4.95) leads to

$$\gamma_\theta = \sigma\left(\frac{\ddot{\boldsymbol{\mu}}'(\mathbf{I}-\mathbf{P}_{\dot{\mu}})\ddot{\boldsymbol{\mu}}}{(\dot{\boldsymbol{\mu}}'\dot{\boldsymbol{\mu}})^2}\right)^{1/2}$$

$$= \sigma\frac{\|(\mathbf{I}-\mathbf{P}_{\dot{\mu}})\ddot{\boldsymbol{\mu}}\|}{\|\dot{\boldsymbol{\mu}}\|^2}$$

$$= \sigma^2\frac{\|\ddot{\boldsymbol{\mu}}^N\|}{\|\dot{\boldsymbol{\mu}}\|^2}, \tag{4.100}$$

since $\mathbf{I}-\mathbf{P}_{\dot{\mu}}$ represents the projection perpendicular to the tangent, that is, in the direction of the normal. Equation (4.100) may be compared with (4.21) when $p=1$.

Efron [1975] also considered the so-called "curved exponential family" (i.e. curved subsets of a larger multiparameter family indexed by a single parameter θ) with density function

$$p(\mathbf{y}|\theta) = k(\mathbf{y})\exp\{\boldsymbol{\lambda}'\mathbf{y} - \psi(\boldsymbol{\lambda})\},$$

where $\boldsymbol{\lambda} = \boldsymbol{\lambda}(\theta)$, $\theta \in \Theta$. He showed that the bias and asymptotic variance of the maximum-likelihood estimate contain terms involving γ_θ and a "naming curvature." The latter is essentially the parameter-effects curvature of Bates and Watts.

When we move away from the normal-distribution model, the approach of Efron and that of Bates and Watts are quite different. Statistical curvature takes into account the change in the form of $p(\mathbf{y}|\theta)$, whereas the intrinsic curvature apparently does not. Although the Bates–Watts curvatures are geometrical, depending only on $\boldsymbol{\mu}(\theta)$, Bates and Watts in the discussion of their paper point out that the normal-error assumption imposes a Euclidean metric to which standard differential geometry can be applied. Changing the error distribution should then lead to a different metric, thus changing the geometry and concepts of curvature. Therefore the distribution of $\mathbf{y}$ can, in fact, be regarded as having an indirect role in the Bates–Watts approach to curvature.

The next questions of interest are: (i) can Efron's statistical and naming curvatures be extended to deal with multiparameter models, and (ii) can the Bates–Watts curvatures be extended to nonnormal families such as, for example, the generalized linear model? The answer to (i) is yes, and Reeds [1975] and Madsen [1979] (see Amari [1982: 358]) give such extensions, though details are not published. However, a more general geometrical framework which endeavors to deal with these problems is provided by Dawid [1975], Amari [1982, 1985], and Kass [1984]. This framework makes use of the so-called information metric and α-connections, instead of the Euclidean inner product and the ordinary derivative operators. As the theory is beyond the scope of this book, we shall just highlight the main features.

The change in $L(\boldsymbol{\theta}) = \log p(\mathbf{y}|\boldsymbol{\theta})$ in going from $\boldsymbol{\theta}$ to $\boldsymbol{\theta} + d\boldsymbol{\theta}$ is given by

$$dL = \sum_{r=1}^{p} \frac{\partial L(\boldsymbol{\theta})}{\partial \theta_r} d\theta_r = \sum_{r=1}^{p} u_r(\boldsymbol{\theta})\, d\theta_r. \tag{4.101}$$

As $d\boldsymbol{\theta}$ changes, dL changes, but the coefficients $u_r(\boldsymbol{\theta}) = \partial L(\boldsymbol{\theta})/\partial \theta_r$ remain the same and can be regarded as basis elements (for a tangent space $T_{\boldsymbol{\theta}}$, say). To define the inner product between two such dL increments we first define an inner product for the basis elements. In a statistical context a natural inner product is

$$\begin{aligned} \langle u_r, u_s \rangle &= E[u_r(\boldsymbol{\theta}) u_s(\boldsymbol{\theta})] \\ &= u_{rs}, \end{aligned}$$

say. The "length" of $d\boldsymbol{\theta}$ can now be defined in terms of the "distributional distance" using (4.101), namely

$$\begin{aligned} \langle dL, dL \rangle &= \sum_r \sum_s \langle u_r, u_s \rangle\, d\theta_r\, d\theta_s \\ &= \sum_r \sum_s u_{rs}\, d\theta_r\, d\theta_s. \end{aligned}$$

Here $\{u_{rs}\}$ is called the information metric, and it gives the appropriate coefficients or "weights" in defining $\| d\boldsymbol{\theta} \|^2$. By comparing the tangent spaces $T_{\boldsymbol{\theta}}$ and $T_{\boldsymbol{\theta}+d\boldsymbol{\theta}}$ and establishing a relationship between their bases using the information metric, one can establish a set of "connecting" coefficients called affine connections. In particular one can define a one-parameter family of connections called α-connections (Amari [1982]), namely

$$\Gamma^{\alpha}_{rst}(\boldsymbol{\theta}) = E\left[\frac{\partial^2 L}{\partial \theta_r \partial \theta_s} \frac{\partial L}{\partial \theta_t}\right] + \frac{1-\alpha}{2} E\left[\frac{\partial L}{\partial \theta_r} \frac{\partial L}{\partial \theta_s} \frac{\partial L}{\partial \theta_t}\right], \qquad -1 \le \alpha \le 1. \tag{4.102}$$

When we have a multivariate exponential family

$$p(\mathbf{y}|\boldsymbol{\theta}) = k(\mathbf{y}) \exp\{\boldsymbol{\theta}'\mathbf{y} - \psi(\boldsymbol{\theta})\},$$

we find that $\partial^2 L(\boldsymbol{\theta})/\partial\theta_r\,\partial\theta_s$ is a constant, so that, by (4.94), the first term of (4.102) is zero. The second term also vanishes when $\alpha = 1$. We therefore call the 1-connection in (4.102), namely

$$\Gamma^{1}_{rst}(\boldsymbol{\theta}) = E\left[\frac{\partial^2 L}{\partial\theta_r\,\partial\theta_s}\frac{\partial L}{\partial\theta_t}\right], \tag{4.103}$$

the exponential connection, as it vanishes for the exponential family.

For a similar reason we call the -1-connection

$$\Gamma^{-1}_{rst}(\boldsymbol{\theta}) = \Gamma^{1}_{rst}(\boldsymbol{\theta}) + E\left[\frac{\partial L}{\partial\theta_r}\frac{\partial L}{\partial\theta_s}\frac{\partial L}{\partial\theta_t}\right] \tag{4.104}$$

the mixture connection, as it vanishes when $p(\mathbf{y}|\boldsymbol{\theta})$ is a certain mixture distribution. We express (4.102) in the form

$$\begin{aligned}\Gamma^{\alpha}_{rst}(\boldsymbol{\theta}) &= \frac{1+\alpha}{2}\Gamma^{1}_{rst}(\boldsymbol{\theta}) + \frac{1-\alpha}{2}\Gamma^{-1}_{rst}(\boldsymbol{\theta}) \\ &= \delta\Gamma^{1}_{rst}(\boldsymbol{\theta}) + (1-\delta)\Gamma^{-1}_{rst}(\boldsymbol{\theta}), \qquad 0 \le \delta \le 1.\end{aligned}$$

The above connections are akin to curvature coefficients and can be related to the work of Efron [1975] and Bates and Watts [1980]. For example, suppose the ε_i are i.i.d. $N(0,\sigma^2)$ with σ^2 known; then

$$L(\boldsymbol{\theta}) = \text{const} - \frac{1}{2\sigma^2}\sum_i [y_i - f_i(\boldsymbol{\theta})]^2.$$

Applying the theory following (4.33), but with matrices evaluated at the true value of $\boldsymbol{\theta}$ rather than $\hat{\boldsymbol{\theta}}$, we define $\boldsymbol{\theta} = \mathbf{K}\boldsymbol{\phi}$ with $\mathbf{K} = [(k_{rs})]$. Then

$$\frac{\partial L}{\partial\phi_i} = \frac{1}{\sigma^2}\sum_{h=1}^{n}\sum_{a=1}^{p}[y_h - f_h(\boldsymbol{\theta})]\frac{\partial f_h}{\partial\theta_a}k_{ai} \tag{4.105}$$

and

$$\frac{\partial^2 L}{\partial\phi_r\,\partial\phi_s} = \frac{1}{\sigma^2}\sum_{h=1}^{n}\sum_{b,c=1}^{p}[y_h - f_h(\boldsymbol{\theta})]\frac{\partial^2 f_h}{\partial\theta_a\,\partial\theta_b}k_{br}k_{cs} + g_{rs},$$

where g_{rs} contains no random variables. Hence

$$E\left[\frac{\partial L}{\partial\phi_i}\right] = 0,$$

$$\begin{aligned}E\left[\frac{\partial L}{\partial\phi_i}\frac{\partial^2 L}{\partial\phi_r\,\partial\phi_s}\right] &= \frac{1}{\sigma^2}\sum_{h=1}^{n}\sum_{a,b,c=1}^{p}\frac{\partial f_h}{\partial\theta_a}\frac{\partial^2 f_h}{\partial\theta_b\,\partial\theta_c}k_{ai}k_{br}k_{cs} \\ &= \sigma^{-2}a^{T}_{irs} \qquad [\text{by } (4.69)],\end{aligned}$$

and

$$\Gamma^1_{rsi}(\phi) = a^T_{irs} (= a^T_{isr}). \tag{4.106}$$

We can therefore interpret the exponential connection as a parameter-effects array. Furthermore, for any parametrization, we have from (4.105)

$$E\left[\frac{\partial L}{\partial \theta_r}\frac{\partial L}{\partial \theta_s}\frac{\partial L}{\partial \theta_t}\right] = 0,$$

since $E[\varepsilon_i \varepsilon_j \varepsilon_k] = 0$, which is also true for any symmetric ε_i and not just normal ε_i (Kass [1984]). Hence, from (4.104),

$$\Gamma^1_{rst}(\phi) = \Gamma^{-1}_{rst}(\phi),$$

and the family of α-connections reduces to just one connection, namely (4.106).

To tidy up loose ends we now link up the above with a result due to Kass [1984]. We first note that in the ϕ-system the metric matrix is now the identity matrix. This follows from

$$\begin{aligned} u_{rs} &= E\left[\frac{\partial L}{\partial \phi_r}\frac{\partial L}{\partial \phi_s}\right] \\ &= \sum_{a=1}^{p}\sum_{b=1}^{p}\sum_{h=1}^{n}\frac{\partial f_h}{\partial \theta_a}\frac{\partial f_h}{\partial \theta_b}k_{ar}k_{bs} \qquad \text{[by (4.105)]}, \end{aligned}$$

which, by (4.37) evaluated at $\boldsymbol{\theta}$ and the Cholesky decomposition $\mathbf{F}'_{\bullet}(\boldsymbol{\theta})\mathbf{F}_{\bullet}(\boldsymbol{\theta}) = \mathbf{R}'_{11}\mathbf{R}_{11}$, leads to

$$\begin{aligned} \mathbf{U} &= [(u_{rs})] \\ &= \mathbf{K}'\mathbf{F}'_{\bullet}\mathbf{F}_{\bullet}\mathbf{K} \\ &= \mathbf{K}'\mathbf{R}'_{11}\mathbf{R}_{11}\mathbf{K} \\ &= \mathbf{I}_p. \end{aligned}$$

The related covariant derivatives Γ^i_{rs} are then given by

$$\begin{aligned} \Gamma^i_{rs} &= \sum_a \Gamma^1_{rsa} u^{ai} \\ &= \sum_a \Gamma^1_{rsa}\delta_{ai} \\ &= \Gamma^1_{rsi} \\ &= a^T_{irs}, \end{aligned} \tag{4.107}$$

which is Kass's equation (5) (see also Ross [1987: 93]).

In conclusion, we see that α-connections can provide possible curvature arrays for general nonnormal nonlinear models. When the data are normal these arrays reduce to the Bates–Watts parameter-effects curvature array.

4.5 SUBSET CURVATURES

4.5.1 Definitions

A major disadvantage of the Bates–Watts curvatures is that they measure the worst possible curvature in any direction from $\hat{\boldsymbol{\theta}}$. If we are interested in particular subsets of $\boldsymbol{\theta}$, then certain directions may not be relevant. Curvatures for subsets are developed by Cook and Goldberg [1986] on the assumption of negligible intrinsic curvature, and the following discussion is based on their work. A recent, related reference is Clarke [1987b].

We begin, once again, with the nonlinear model

$$\mathbf{y} = \mathbf{f}(\boldsymbol{\theta}) + \boldsymbol{\varepsilon}, \tag{4.108}$$

where $\boldsymbol{\varepsilon} \sim N_n(\mathbf{0}, \sigma^2\mathbf{I}_n)$ and $\boldsymbol{\theta} \in \Theta$. Consider the partition $\boldsymbol{\theta} = (\boldsymbol{\theta}_1', \boldsymbol{\theta}_2')'$, where $\boldsymbol{\theta}_i$ is $p_i \times 1$ $(1 = 1, 2)$ and $p_1 + p_2 = p$. Suppose that $\boldsymbol{\theta}_2$, our parameter subset of interest, belongs to Θ_2, so that $\Theta_2 = \{\boldsymbol{\theta}_2 : \boldsymbol{\theta} \in \Theta\}$. We shall also use the notation $f(\boldsymbol{\theta}) = f(\boldsymbol{\theta}_1, \boldsymbol{\theta}_2)$ for convenience. Let $L(\boldsymbol{\theta}_1, \boldsymbol{\theta}_2, \sigma^2)$ be the log-likelihood for (4.108), and let $\hat{\boldsymbol{\theta}}_1, \hat{\boldsymbol{\theta}}_2$, and $\hat{\sigma}^2$ be the maximum-likelihood (and least-squares) estimates of the parameters. Then the standard likelihood confidence region for $\boldsymbol{\theta}_2$ [c.f. (5.22)] can be written in the form

$$\begin{aligned} \{\boldsymbol{\theta}_2 : S[\tilde{\boldsymbol{\theta}}_1(\boldsymbol{\theta}_2), \boldsymbol{\theta}_2] \le k_\alpha\} &= \{\boldsymbol{\theta}_2 : \| \mathbf{y} - \mathbf{f}[\mathbf{a}_1(\boldsymbol{\theta}_2), \boldsymbol{\theta}_2] \|^2 \le k_\alpha\} \\ &= \{\boldsymbol{\theta}_2 : \| \mathbf{y} - \mathbf{b}(\boldsymbol{\theta}_2) \|^2 \le k_\alpha\}, \end{aligned} \tag{4.109}$$

say. Here k_α is determined by the significance level, and $\mathbf{a}_1(\boldsymbol{\theta}_2) = \tilde{\boldsymbol{\theta}}_1(\boldsymbol{\theta}_2)$ represents the value of $\boldsymbol{\theta}_1$ which minimizes $S(\boldsymbol{\theta}_1, \boldsymbol{\theta}_2)$ for each value of $\boldsymbol{\theta}_2$. Now define

$$\mathbf{a}(\boldsymbol{\theta}_2) = \begin{pmatrix} \mathbf{a}_1(\boldsymbol{\theta}_2) \\ \boldsymbol{\theta}_2 \end{pmatrix} = \begin{pmatrix} \mathbf{a}_1(\boldsymbol{\theta}_2) \\ \mathbf{a}_2(\boldsymbol{\theta}_2) \end{pmatrix} \begin{matrix} \}\, p_1 \\ \}\, p_2 \end{matrix} \tag{4.110}$$

say. Then

$$\begin{aligned} \mathbf{f}(\boldsymbol{\theta}) &= \mathbf{f}(\mathbf{a}_1(\boldsymbol{\theta}_2), \boldsymbol{\theta}_2) \\ &= \mathbf{f}\{\mathbf{a}(\boldsymbol{\theta}_2)\} \\ &= \mathbf{b}(\boldsymbol{\theta}_2), \end{aligned} \tag{4.111}$$

as mentioned in (4.109). As $\boldsymbol{\theta}_2$ varies over Θ_2, $\mathbf{a}(\boldsymbol{\theta}_2)$ will vary over a subset of Θ and $\mathbf{b}(\boldsymbol{\theta}_2)$ will trace out a subset (submanifold) of the expectation surface,

Ω_2 say. The form of $\mathbf{b}(\boldsymbol{\theta}_2)$ will determine the behavior of the confidence region (4.109). If $\mathbf{b}$ is essentially linear over a sufficiently large neighborhood of $\hat{\boldsymbol{\theta}}_2$, we can then use a submanifold tangent-plane approximation. We also note that Ω_2 is invariant under one-to-one transformations $\boldsymbol{\theta}_2 = \mathbf{k}(\boldsymbol{\phi}_2)$, as maximum-likelihood estimates remain maximum-likelihood under such transformations: all we have is simply a reparametrization.

As a first step in developing subset curvatures we proceed in an analogous fashion to (4.5) and develop a quadratic Taylor expansion for $\mathbf{b}(\boldsymbol{\theta}_2)$. In order to do this we require subset analogues of $\hat{\mathbf{F}}_.$ and $\hat{F}_{..}$ for the function $\mathbf{a}(\boldsymbol{\theta}_2) = [a_1(\boldsymbol{\theta}_2), a_2(\boldsymbol{\theta}_2), \ldots, a_p(\boldsymbol{\theta}_2)]'$ in (4.110). If $\boldsymbol{\theta}_2 = (\theta_{21}, \theta_{22}, \ldots, \theta_{2p_2})'$, let

$$\Delta_. = \left[\left(\frac{\partial a_i}{\partial \theta_{2j}}\right)\right] \qquad (i = 1, 2, \ldots, p; j = 1, 2, \ldots, p_2)$$

and

$$\Delta_{..} = \left[\left(\frac{\partial^2 \mathbf{a}(\boldsymbol{\theta}_2)}{\partial \theta_{2j}\partial \theta_{2k}}\right)\right] = [(\boldsymbol{\alpha}_{jk})] \qquad (j, k = 1, 2, \ldots, p_2).$$

Assuming that $\mathbf{b}(\boldsymbol{\theta}_2)$ has continuous first and second derivatives, and noting that $\mathbf{a}_1(\hat{\boldsymbol{\theta}}_2) = \hat{\boldsymbol{\theta}}_1$ [i.e. $\mathbf{b}(\hat{\boldsymbol{\theta}}_2) = \mathbf{f}(\hat{\boldsymbol{\theta}})$], we have for $i = 1, 2, \ldots, p$,

$$\begin{aligned} b_i(\boldsymbol{\theta}_2) \approx f_i(\hat{\boldsymbol{\theta}}) &+ \left[\sum_{j=1}^{p_2} \frac{\partial f_i(\boldsymbol{\theta})}{\partial \theta_{2j}}\right]_{\hat{\boldsymbol{\theta}}} (\theta_{2j} - \hat{\theta}_{2j}) \\ &+ \frac{1}{2}\left[\sum_{j=1}^{p_2}\sum_{k=1}^{p_2} \frac{\partial^2 f_i(\boldsymbol{\theta})}{\partial \theta_{2j}\partial \theta_{2k}}\right]_{\hat{\boldsymbol{\theta}}} (\theta_{2j} - \hat{\theta}_{2j})(\theta_{2k} - \hat{\theta}_{2k}). \end{aligned} \tag{4.112}$$

Now

$$\left[\left(\frac{\partial f_i}{\partial \theta_{2j}}\right)\right] = \left[\left(\sum_{r=1}^{p_2} \frac{\partial f_i}{\partial a_r}\frac{\partial a_r}{\partial \theta_{2j}}\right)\right] = \mathbf{F}_.\boldsymbol{\Delta}_. \tag{4.113}$$

and

$$\left[\left(\frac{\partial^2 f_i}{\partial \theta_{2j}\partial \theta_{2k}}\right)\right] = \left[\left(\sum_{r=1}^{p}\sum_{s=1}^{p} \frac{\partial^2 f_i}{\partial a_r \partial a_s}\frac{\partial a_r}{\partial \theta_{2j}}\frac{\partial a_s}{\partial \theta_{2k}} + \sum_{r=1}^{p}\frac{\partial f_i}{\partial a_r}\frac{\partial^2 a_r}{\partial \theta_{2j}\partial \theta_{2k}}\right)\right]. \tag{4.114}$$

Hence substituting (4.113) and (4.114) into (4.112), and setting $\boldsymbol{\delta} = \boldsymbol{\theta}_2 - \hat{\boldsymbol{\theta}}_2$, we have

$$\mathbf{b}(\boldsymbol{\theta}_2) \approx \mathbf{f}(\hat{\boldsymbol{\theta}}) + \hat{\mathbf{F}}_.\hat{\boldsymbol{\Delta}}_.\boldsymbol{\delta} + \tfrac{1}{2}\boldsymbol{\delta}'\hat{\boldsymbol{\Delta}}_.'\hat{F}_{..}\hat{\boldsymbol{\Delta}}_.\boldsymbol{\delta} + \tfrac{1}{2}\hat{\mathbf{F}}_.(\boldsymbol{\delta}'\hat{\Delta}_{..}\boldsymbol{\delta}). \tag{4.115}$$

Cook and Goldberg [1986] prove that the linear part of the above equation,

namely

$$\mathbf{b}(\boldsymbol{\theta}_2) \approx \mathbf{f}(\hat{\boldsymbol{\theta}}) + \hat{\mathbf{F}}_{.}\hat{\boldsymbol{\Delta}}_{.}\boldsymbol{\delta}, \tag{4.116}$$

gives the appropriate tangent-plane approximation for $\mathbf{b}(\boldsymbol{\theta}_2)$ at $\hat{\boldsymbol{\theta}}_2$, leading to the correct linearized confidence region for $\boldsymbol{\theta}_2$. They do this by showing that, with negligible intrinsic curvature,

$$\begin{aligned}\hat{\mathbf{F}}_{.}\hat{\boldsymbol{\Delta}}_{.} &= (\hat{\mathbf{F}}_1, \hat{\mathbf{F}}_2)\hat{\boldsymbol{\Delta}}_{.} \\ &= (\mathbf{I}_n - \hat{\mathbf{P}}_{\mathbf{F}_1})\hat{\mathbf{F}}_2,\end{aligned} \tag{4.117}$$

where $\hat{\mathbf{F}}_i$ is $n \times p_i$ $(i = 1, 2)$ and $\hat{\mathbf{P}}_{\mathbf{F}_i} = \hat{\mathbf{F}}_i(\hat{\mathbf{F}}_i'\hat{\mathbf{F}}_i)^{-1}\hat{\mathbf{F}}_i'$. If we use the linear approximation for the expectation surface Ω, then $(\hat{\mathbf{F}}_1, \hat{\mathbf{F}}_2)$ behaves like $(\mathbf{X}_1, \mathbf{X}_2)$ in the linear regression model of Appendix D3. The confidence region for $\boldsymbol{\beta}_2$ given there (D3.3) is then interpreted as the following region for $\boldsymbol{\theta}_2$:

$$\{\boldsymbol{\theta}_2 : \|(\mathbf{I}_n - \hat{\mathbf{P}}_{\mathbf{F}_1})\hat{\mathbf{F}}_2(\boldsymbol{\theta}_2 - \hat{\boldsymbol{\theta}}_2)\|^2 \leq p_2 s^2 F^{\alpha}_{p_2, n-p}\}. \tag{4.118}$$

The matrix part of this expression is the same as (4.117), so that (4.116) leads to the correct linearized region.

In Section 4.2.2 we developed curvature definitions by comparing the tangential and normal components of the quadratic term of the Taylor expansion (4.5) with the linear term. We now do the same thing for (4.115), except that we now find that three mutually orthogonal components are appropriate. If $\frac{1}{2}\mathbf{q}$ is the quadratic part of (4.115), then

$$\begin{aligned}\mathbf{q} &= \boldsymbol{\delta}'\hat{\boldsymbol{\Delta}}_{.}'\hat{F}_{..}\hat{\boldsymbol{\Delta}}_{.}\boldsymbol{\delta} + \hat{\mathbf{F}}_{.}\boldsymbol{\delta}'\Delta_{..}\boldsymbol{\delta} \\ &= \boldsymbol{\delta}'\hat{\boldsymbol{\Delta}}_{.}'\hat{F}_{..}^N\hat{\boldsymbol{\Delta}}_{.}\boldsymbol{\delta} + (\boldsymbol{\delta}'\hat{\boldsymbol{\Delta}}_{.}'\hat{F}_{..}^T\hat{\boldsymbol{\Delta}}_{.}\boldsymbol{\delta} + \hat{\mathbf{F}}_{.}\boldsymbol{\delta}'\hat{\Delta}_{..}\boldsymbol{\delta}) \\ &= \mathbf{q}_1 + \mathbf{q}_2,\end{aligned} \tag{4.119}$$

say. Here $\mathbf{q}_1$ takes the form

$$\mathbf{q}_1 = \mathbf{u}'\hat{F}_{..}^N\mathbf{u} = \sum_r \sum_s \hat{\mathbf{f}}_{rs}^N u_r u_s,$$

which is normal to the tangent plane to Ω, while $\mathbf{q}_2$ takes the form

$$\mathbf{q}_2 = \mathbf{u}'\hat{F}_{..}^T\mathbf{u} + \hat{\mathbf{F}}_{.}\mathbf{w},$$

which is in the tangent plane, $\mathscr{R}[\hat{\mathbf{F}}_{.}]$. Hence $\mathbf{q}_1$ and $\mathbf{q}_2$ are the normal and tangential components of $\mathbf{q}$ with respect to the expectation surface Ω.

We now go a step further and split $\mathbf{q}_2$ into two orthogonal components, both still in the tangent plane to Ω: $\mathbf{q}_2^T$ in the tangent plane to the submanifold Ω_2, and $\mathbf{q}_2^N$ normal to this tangent plane. If $\ell = \hat{\mathbf{F}}_{.}\hat{\boldsymbol{\Delta}}_{.}\boldsymbol{\delta}$, the linear term, then by analogy with (4.15) we can define the parameter-effects and intrinsic "subset" curvatures as $\|\mathbf{q}_2^T\|/\|\ell\|^2$ and $\|\mathbf{q}_2^N\|/\|\ell\|^2$ respectively. In order to obtain simple "reduced" forms for these curvatures as in Section 4.2.3, Cook and Goldberg [1986] made the assumption that the intrinsic curvature array $F_{..}^N$ has zero elements. From the examples given in the literature (e.g. Table 4.1 in Section 4.2.2) this assumption seems a reasonable approximation for many nonlinear models. Although $\mathbf{q}_1$ is not the intrinsic vector $\boldsymbol{\delta}' F_{..}^N \boldsymbol{\delta}$ [see (4.119)], $\mathbf{q}_1 = \mathbf{0}$ when $\hat{F}_{..} = \mathbf{0}$, and $\mathbf{q}_2 = \mathbf{q}_2^T + \mathbf{q}_2^N$. It may seem strange to the reader, as it did at first to the authors, that we can have a zero intrinsic curvature and a nonzero intrinsic subset curvature. However, these curvatures are measured in two orthogonal directions. The intrinsic curvature measures the degree of bending of the whole expectation surface, and not just $\mathbf{b}(\boldsymbol{\theta}_2)$, in the direction perpendicular to the tangent plane. On the other hand, the intrinsic subset curvature measures the degree of bending of the "curve" $\mathbf{b}(\boldsymbol{\theta}_2)$ in the direction perpendicular to the tangent $\mathbf{q}_2^T$. If the intrinsic curvature is zero, then the expectation surface and the tangent plane are the same, with $\mathbf{b}(\boldsymbol{\theta}_2)$ a curve in this plane. However, unless $\mathbf{b}(\boldsymbol{\theta}_2)$ is a straight line, the intrinsic subset curvature will not be zero.

Readers who are not interested in technical details may skip to Section 4.6. Rules of thumb for subset curvatures are given in the paragraph following Equation (4.130).

*4.5.2 Reduced Formulae for Subset Curvatures

The first step in obtaining reduced forms for the subset curvatures is to begin, once again, with the QR decomposition of $\hat{\mathbf{F}}_.$, namely [cf. (4.33)]

$$\hat{\mathbf{F}}_. = \hat{\mathbf{Q}}_p \hat{\mathbf{R}}_{11},$$

where the columns of $\hat{\mathbf{Q}}_p$ are orthonormal. We then make the transformation $\boldsymbol{\phi} = \hat{\mathbf{R}}_{11}\boldsymbol{\theta}$, or $\boldsymbol{\theta} = \hat{\mathbf{R}}_{11}^{-1}\boldsymbol{\phi} = \hat{\mathbf{K}}\boldsymbol{\phi}$, and we define $\hat{\boldsymbol{\phi}} = \hat{\mathbf{R}}_{11}\hat{\boldsymbol{\theta}}$. Setting $\mathbf{f}(\boldsymbol{\theta}) = \mathbf{f}(\mathbf{K}\boldsymbol{\phi}) = \mathbf{g}(\boldsymbol{\phi})$, we obtain the ϕ-coordinate derivatives [c.f. (4.42) and (4.47)] $\hat{\mathbf{G}}_. = \hat{\mathbf{Q}}_p$ and $\hat{G}_{..} = \hat{\mathbf{K}}'\hat{F}_{..}\hat{\mathbf{K}}$. As with $\boldsymbol{\theta}' = (\boldsymbol{\theta}_1', \boldsymbol{\theta}_2')$, consider the partition

$$\hat{\mathbf{Q}}_p = (\hat{\mathbf{Q}}_{p_1}, \hat{\mathbf{Q}}_{p_2}), \tag{4.120}$$

where $\hat{\mathbf{Q}}_{p_i}$ is $n \times p_i$ with orthonormal columns ($i = 1, 2$). We also partition the $n \times p \times p$ array $\hat{G}_{..}$ in a similar fashion, namely

$$\begin{aligned}
\hat{G}_{..} &= [(\hat{\mathbf{g}}_{rs})] \qquad (r, s = 1, 2, \ldots, p) \\
&= \begin{bmatrix} (\hat{\mathbf{g}}_{r_1 s_1}) & (\hat{\mathbf{g}}_{r_1 s_2}) \\ (\hat{\mathbf{g}}_{r_2 s_1}) & (\hat{\mathbf{g}}_{r_2 s_2}) \end{bmatrix} \qquad (r_1, s_1 = 1, 2, \ldots, p_1, \quad r_2, s_2 = p_1 + 1, \ldots, p_1 + p_2) \\
&= \begin{bmatrix} \hat{G}_{..(11)} & \hat{G}_{..(12)} \\ \hat{G}_{..(21)} & \hat{G}_{..(22)} \end{bmatrix} \begin{matrix} \} p_1 \\ \} p_2 \end{matrix}
\end{aligned} \tag{4.121}$$

say, where $\hat{G}_{..(ij)}$ is the $n \times p_i \times p_j$ array $[(\hat{\mathbf{g}}_{r_i s_j})]$. Then if $\boldsymbol{\phi}$ is partitioned in the same way as $\boldsymbol{\theta}$ and $\mathbf{d} = \boldsymbol{\phi}_2 - \hat{\boldsymbol{\phi}}_2$, a $p_2 \times 1$ vector, Cook and Goldberg [1986] show that the linear and quadratic terms of (4.115) [see also (4.119)] take the form

$$\ell = \hat{\mathbf{F}}.\hat{\Delta}.\delta = \hat{\mathbf{Q}}_{p_2}\mathbf{d},$$

$$\tfrac{1}{2}\mathbf{q}_2^T = \tfrac{1}{2}\mathbf{d}'[\hat{\mathbf{Q}}_{p_2}\hat{\mathbf{Q}}'_{p_2}][\hat{G}_{..(22)}]\mathbf{d} \tag{4.122}$$

$$= \tfrac{1}{2}\mathbf{Q}_{p_2}\mathbf{d}'[(\hat{\mathbf{Q}}'_{p_2}\hat{\mathbf{g}}_{r_2 s_2})]\mathbf{d} \tag{4.123}$$

and

$$\tfrac{1}{2}\mathbf{q}_2^N = -\hat{\mathbf{Q}}_{p_1}[\mathbf{d}'\hat{\mathbf{Q}}'_{p_2}][\hat{G}_{..(12)}]\mathbf{d}$$

$$= -\hat{\mathbf{Q}}_{p_1}[\mathbf{d}'][(\hat{\mathbf{Q}}'_{p_2}\hat{\mathbf{g}}_{r_1 s_2})]\mathbf{d}, \tag{4.124}$$

where the rules of square-bracket multiplication are described in Appendix B4. We note that the columns of $\hat{\mathbf{Q}}_p$ form a basis for the tangent plane, while the columns of $\hat{\mathbf{Q}}_{p_2}$ form a basis for the submanifold tangent plane.

The next stage of the reduction in Section 4.2.3 was to go from the array $\hat{G}_{..}$ to the arrays $\hat{A}_{..}^N$ and $\hat{A}_{..}^T$. Here $\hat{A}_{..}^N$ is assumed to be zero, while $\hat{A}_{..}^T$ can be partitioned in a similar fashion to $\hat{G}_{..}$ in (4.121), namely,

$$\hat{A}_{..}^T = [(\hat{\mathbf{Q}}'_p\hat{\mathbf{q}}_{rs})]$$

$$= \begin{pmatrix} (\hat{\mathbf{Q}}'_p\hat{\mathbf{g}}_{r_1 s_1}) & (\hat{\mathbf{Q}}'_p\hat{\mathbf{g}}_{r_1 s_2}) \\ (\hat{\mathbf{Q}}'_p\hat{\mathbf{g}}_{r_2 s_1}) & (\hat{\mathbf{Q}}'_p\hat{\mathbf{g}}_{r_2 s_2}) \end{pmatrix}$$

$$= \begin{pmatrix} (\hat{\mathbf{a}}_{r_1 s_1}^T) & (\hat{\mathbf{a}}_{r_1 s_2}^T) \\ (\hat{\mathbf{a}}_{r_2 s_1}^T) & (\hat{\mathbf{a}}_{r_2 s_2}^T) \end{pmatrix},$$

$$= \begin{pmatrix} \hat{A}_{..(11)}^T & \hat{A}_{..(12)}^T \\ \hat{A}_{..(21)}^T & \hat{A}_{..(22)}^T \end{pmatrix}, \tag{4.125}$$

say. Let the $p_2 \times p_2 \times p_2$ subarray $\hat{A}_{..(22)}^{T(2)}$ denote the last p_2 faces of $\hat{A}_{..(22)}^T$, and let the $p_2 \times p_1 \times p_2$ subarray $\hat{A}_{..(12)}^{T(2)}$ denote the last p_2 faces of $\hat{A}_{..(12)}^T$. Since

$$\hat{\mathbf{Q}}'_p\hat{\mathbf{g}}_{rs} = \begin{pmatrix} \hat{\mathbf{Q}}'_{p_1}\hat{\mathbf{g}}_{rs} \\ \hat{\mathbf{Q}}'_{p_2}\hat{\mathbf{g}}_{rs} \end{pmatrix},$$

it follows that

$$\hat{A}_{..(22)}^{T(2)} = [(\hat{\mathbf{Q}}'_{p_2}\hat{\mathbf{g}}_{r_2 s_2})]$$

and

$$\hat{A}_{..(12)}^{T(2)} = [(\hat{\mathbf{Q}}'_{p_2}\hat{\mathbf{g}}_{r_1 s_2})].$$

Now $\hat{\mathbf{Q}}'_{p_t}\hat{\mathbf{Q}}_{p_t} = \mathbf{I}_{p_t}$, so that $\|\hat{\mathbf{Q}}_{p_t}\mathbf{x}\| = \|\mathbf{x}\|$. Hence from (4.122), (4.123), and (4.124)

$$\|\ell\| = \|\hat{\mathbf{Q}}_{p_2}\mathbf{d}\| = \|\mathbf{d}\|, \tag{4.126}$$

$$\begin{aligned}\|\mathbf{q}_2^T\| &= \|\mathbf{d}'[(\hat{\mathbf{Q}}'_{p_2}\hat{\mathbf{g}}_{r_2 s_2})]\mathbf{d}\| \\ &= \|\mathbf{d}'A^{T(2)}_{\cdot\cdot(22)}\mathbf{d}\|\end{aligned} \tag{4.127}$$

and

$$\begin{aligned}\|\mathbf{q}_2^N\| &= 2\|[\mathbf{d}'][(\hat{\mathbf{Q}}'_{p_2}\hat{\mathbf{g}}_{r_1 s_2})]\mathbf{d}\| \\ &= 2\|[\mathbf{d}'][\hat{A}^{T(2)}_{\cdot\cdot(12)}]\mathbf{d}\| \\ &= 2\left\|\sum_{i=p_1+1}^{p} d_{i-p_1}\hat{\mathbf{A}}^T_{i12}\mathbf{d}\right\|,\end{aligned} \tag{4.128}$$

where $\hat{\mathbf{A}}^T_{i12}$ is the ith face of $\hat{A}^T_{\cdot\cdot(12)}$. Since we can assume $\|\mathbf{d}\| = 1$ in (4.126), we can now define the maximum parameter-effects and intrinsic subset curvatures for $\boldsymbol{\theta}_2$ at $\hat{\boldsymbol{\theta}}_2$ as

$$\gamma^T_{\max,s} = \rho_2 \max_{\|\mathbf{d}\|=1} \|\mathbf{d}'A^{T(2)}_{\cdot\cdot(22)}\mathbf{d}\| \tag{4.129}$$

and

$$\gamma^N_{\max,s} = \rho_2 \max_{\|d\|=1} 2\left\|\sum_{i=p_1+1}^{p} d_{i-p_1}\hat{\mathbf{A}}^T_{i12}\mathbf{d}\right\|, \tag{4.130}$$

respectively, where $\rho_2 = \sqrt{p_2}s$ [see (4.118)] and $s^2 = \|\mathbf{y} - \mathbf{f}(\hat{\boldsymbol{\theta}})\|^2/(n-p)$.

Cook and Goldberg [1986] note that the intrinsic subset curvature $\gamma^N_{\max,s}$ depends only on the fanning and torsion components of $\hat{A}^T_{\cdot\cdot}$ and not on the compansion and arcing components (c.f. Section 4.2.6). As the last p_2 columns of $\hat{\mathbf{Q}}_p$ are assumed to correspond to $\boldsymbol{\theta}_2$, it is, however, the fanning and torsion with respect to this ordering which are relevant. If both subset curvatures are small, the likelihood and tangent-plane confidence regions for $\boldsymbol{\theta}_2$ [c.f. (4.109) and (4.118)] will be similar. A numerical example is given in Section 3.4.2, Fig. 3.13. Although, following (4.25), $(F^{\alpha}_{p_2,n-p})^{-1/2}$ can be used as a rough cutoff value, Cook and Goldberg [1986] found from their experience that subset curvatures must be much smaller than this to get close agreement between tangent-plane and likelihood confidence regions. They also recommend combining the two subset curvatures into a total subset curvature

$$\gamma_{\max,s} = \{(\gamma^T_{\max,s})^2 + (\gamma^N_{\max,s})^2\}^{1/2}, \tag{4.131}$$

as it may be more relevant than the individual curvatures.

*4.5.3 Computations

With regard to the computation of the subset curvatures it should be stressed that the ordering of the parameters is important. As $\boldsymbol{\theta}_2$ denotes the last p_2

elements of $\boldsymbol{\theta}$, it is the last p_2 columns of $\hat{\mathbf{F}}_{\bullet} = [(\partial f_i/\partial \theta_j)]_{\hat{\theta}}$ that correspond to $\boldsymbol{\theta}_2$, and the subset arrays $\hat{A}^T_{\cdot\cdot(22)}$ and $\hat{A}^T_{\cdot\cdot(12)}$ follow directly from $\hat{A}^T_{\cdot\cdot}$. If, however, $\boldsymbol{\theta}_2$ does not refer to the last p_2 elements, then $\boldsymbol{\theta}$ must be reordered with a corresponding reordering of the columns of $\hat{\mathbf{F}}_{\bullet}$ and the calculation of a new $\hat{A}^T_{\cdot\cdot}$ array. To avoid a lot of additional calculation Cook and Goldberg [1986] suggest a more effective method for obtaining the new $\hat{A}^T_{\cdot\cdot}$ array, $\hat{A}^T_{\cdot\cdot\pi}$ say, using permutation matrices. Suppose $\boldsymbol{\Pi}$ is the $p \times p$ permutation matrix which permutes $\boldsymbol{\theta}$ so that $\boldsymbol{\theta}_2$ is now given by the last p_2 elements of $\boldsymbol{\theta}$. Let the subscript π denote evaluation in the permuted coordinate system. Thus

$$\begin{aligned} \hat{\mathbf{F}}_{\bullet\pi} &= \hat{\mathbf{F}}_{\bullet}\boldsymbol{\Pi}' \\ &= \hat{\mathbf{Q}}_p \hat{\mathbf{R}}_{11} \boldsymbol{\Pi}' \end{aligned} \tag{4.132}$$

represents the same permutation applied to the columns of $\hat{\mathbf{F}}_{\bullet}$. Let $\mathbf{Q}_* \hat{\mathbf{R}}_\pi$ be the QR decomposition of the $p \times p$ matrix $\hat{\mathbf{R}}_{11}\boldsymbol{\Pi}'$, where $\mathbf{Q}_*$ is orthogonal and $\hat{\mathbf{R}}_\pi$ is upper triangular. Cook and Goldberg [1986] then give the following useful relationship:

$$\hat{A}^T_{\cdot\cdot\pi} = [\mathbf{Q}'_*][\mathbf{Q}'_* \hat{A}^T_{\cdot\cdot} \mathbf{Q}_*] \tag{4.133}$$

relating the arrays in the old and permuted systems.

To prove (4.133) we note first of all that

$$\begin{aligned} \hat{\mathbf{F}}_{\bullet\pi} &= \hat{\mathbf{Q}}_p \hat{\mathbf{R}}_{11} \boldsymbol{\Pi}' \\ &= \hat{\mathbf{Q}}_p \mathbf{Q}_* \hat{\mathbf{R}}_\pi \\ &= \hat{\mathbf{Q}}_\pi \hat{\mathbf{R}}_\pi, \end{aligned} \tag{4.134}$$

say, where $\hat{\mathbf{Q}}_\pi$ is an $n \times p$ matrix with orthonormal columns as

$$\hat{\mathbf{Q}}'_\pi \hat{\mathbf{Q}}_\pi = \mathbf{Q}'_* \hat{\mathbf{Q}}'_p \hat{\mathbf{Q}}_p \mathbf{Q}_* = \mathbf{Q}'_* \mathbf{I}_p \mathbf{Q}_* = \mathbf{I}_p.$$

Since the QR decomposition is unique (Appendix A8.3). $\hat{\mathbf{Q}}_\pi \hat{\mathbf{R}}_\pi$ is the QR decomposition of $\hat{\mathbf{F}}_{\bullet\pi}$. Hence using the theory of Section 4.2.3 with the matrix $\hat{\mathbf{F}}_{\bullet\pi}$ instead of $\hat{\mathbf{F}}_{\bullet}$ and $\hat{\mathbf{K}}_\pi = \hat{\mathbf{R}}_\pi^{-1}$ instead of $\hat{\mathbf{K}} = \hat{\mathbf{R}}_{11}^{-1}$, we have from (4.66) and (4.67) [see also Appendix B4.1]

$$\begin{aligned} \hat{A}^T_{\cdot\cdot\pi} &= [\hat{\mathbf{Q}}'_\pi][\hat{\mathbf{K}}'_\pi \hat{F}^T_{\cdot\cdot\pi} \hat{\mathbf{K}}_\pi] \\ &= \hat{\mathbf{K}}'_\pi [\hat{\mathbf{Q}}'_\pi][\hat{F}^T_{\cdot\cdot\pi}] \hat{\mathbf{K}}_\pi \\ &= \hat{\mathbf{K}}'_\pi [\hat{\mathbf{Q}}'_\pi][\boldsymbol{\Pi} \hat{F}^T_{\cdot\cdot} \boldsymbol{\Pi}'] \hat{\mathbf{K}}_\pi \\ &= \hat{\mathbf{K}}'_\pi \boldsymbol{\Pi} [\hat{\mathbf{Q}}'_\pi][\hat{F}^T_{\cdot\cdot}] \boldsymbol{\Pi}' \hat{\mathbf{K}}_\pi. \end{aligned} \tag{4.135}$$

Now $\hat{\mathbf{R}}_{11}\boldsymbol{\Pi}' = \mathbf{Q}_* \hat{\mathbf{R}}_\pi$ so that $\mathbf{Q}'_* \hat{\mathbf{R}}_{11} = \hat{\mathbf{R}}_\pi \boldsymbol{\Pi}$. Taking inverses, $\hat{\mathbf{K}}\mathbf{Q}_* = \boldsymbol{\Pi}'\hat{\mathbf{K}}_\pi$.

Finally, substituting in (4.135), we have

$$\begin{aligned}\hat{A}^T_{..\pi} &= \mathbf{Q}'_*[\hat{\mathbf{Q}}'_\pi][\hat{\mathbf{K}}'\hat{F}^T_{..}\hat{\mathbf{K}}]\mathbf{Q}_* &&\text{(by Appendix B4.1)}\\ &= \mathbf{Q}'_*[\mathbf{Q}'_*\hat{\mathbf{Q}}'_p][\hat{G}_{..}]\mathbf{Q}_* \\ &= \mathbf{Q}'_*[\mathbf{Q}'_*][(\hat{\mathbf{Q}}'_p\hat{\mathbf{g}}_{rs})]\mathbf{Q}_* &&\text{(by B4.2)}\\ &= \mathbf{Q}'_*[\mathbf{Q}'_*][\hat{A}^T_{..}]\mathbf{Q}_* \\ &= [\mathbf{Q}'_*][\mathbf{Q}'_*\hat{A}^T_{..}\mathbf{Q}_*], &&\text{(by B4.1).}\end{aligned} \tag{4.136}$$

Using the algorithm described in Section 4.2.7, $\hat{A}^T_{..}$ will be in symmetric storage mode, $\hat{\mathbf{A}}^T_{\text{ssm}}$ [c.f. (4.73)]. To obtain the corresponding matrix, $\hat{\mathbf{A}}^T_{\text{ssm},\pi}$ say, for the permuted coordinates, we note from (4.136) that with $\mathbf{Q}_* = [(q_{rs})]$,

$$\begin{aligned}\hat{\mathbf{a}}^T_{rs,\pi} &= \mathbf{Q}'_*(\mathbf{Q}'_*\hat{A}^T_{..}\mathbf{Q}_*)_{rs} \\ &= \mathbf{Q}'_*\sum_u\sum_v \hat{\mathbf{a}}^T_{uv}q_{ur}q_{vs} \\ &= \mathbf{Q}'_*(\hat{\mathbf{a}}^T_{11},\hat{\mathbf{a}}^T_{12},\ldots,\hat{\mathbf{a}}^T_{pp})\begin{bmatrix} q_{1r}q_{1s} \\ q_{1r}q_{2s}+q_{2r}q_{1s} \\ \vdots \\ q_{pr}q_{ps}\end{bmatrix} \\ &= \mathbf{Q}'_*\hat{\mathbf{A}}^T_{\text{ssm}}\mathbf{w}_{rs},\end{aligned}$$

say. Thus

$$\hat{\mathbf{A}}^T_{\text{ssm},\pi} = \mathbf{Q}'_*\hat{\mathbf{A}}^T_{\text{ssm}}\mathbf{W}_{\text{ssm}}, \tag{4.137}$$

where $\mathbf{W}_{\text{ssm}}$ contains the vectors $\mathbf{w}_{rs}$ in symmetric storage mode.

Example 4.2 (continued) With $a = [m(1+\hat{\theta}_2^2)]^{-1/2}$, we have from (4.64) the faces

$$\hat{\mathbf{A}}^T_{1..} = \frac{a\hat{\theta}_2}{\hat{\theta}_1}\begin{pmatrix} 0 & 1 \\ 1 & -2\hat{\theta}_2\end{pmatrix}$$

and

$$\hat{\mathbf{A}}^T_{2..} = \hat{\mathbf{A}}^T_{1..}/\hat{\theta}_2.$$

Then, using (4.125), we have the subarrays

$$\hat{A}^T_{..(12)} = \frac{a\hat{\theta}_2}{\hat{\theta}_1}\begin{pmatrix} 1 \\ \hat{\theta}_2^{-1}\end{pmatrix} \tag{4.138}$$

and

$$\hat{A}^T_{..(22)} = \frac{a\hat{\theta}_2}{\hat{\theta}_1}\begin{pmatrix} -2\hat{\theta}_2 \\ -2 \end{pmatrix}. \tag{4.139}$$

To calculate curvatures we require the second face of these arrays, in this case the second vector elements. Since θ_2 is one-dimensional we do not need to carry out any minimization with respect to **d**, which is also one-dimensional, i.e. $\mathbf{d} = \pm 1$. Thus, from (4.129) to (4.131) we have the subset curvatures for θ_2, namely

$$\gamma^T_{\max,s} = \rho_2 |\hat{a}_{222}| = \frac{2\rho_2 |\hat{\theta}_2| a}{|\hat{\theta}_1|}, \tag{4.140}$$

$$\gamma^N_{\max,s} = 2\rho_2 |\hat{a}_{212}| = \frac{2\rho_2 a}{|\hat{\theta}_1|}, \tag{4.141}$$

and

$$\gamma_{\max,s} = \frac{2\rho_2}{\sqrt{m}|\hat{\theta}_1|}, \tag{4.142}$$

as quoted by Cook and Goldberg [1986].

It is instructive to use the above permutation method to find the subset curvatures for θ_1. The first step is to find the permutation matrix which interchanges θ_1 and θ_2, namely

$$\begin{pmatrix} \theta_2 \\ \theta_1 \end{pmatrix} = \begin{pmatrix} 0 & 1 \\ 1 & 0 \end{pmatrix}\begin{pmatrix} \theta_1 \\ \theta_2 \end{pmatrix} = \mathbf{\Pi\theta}.$$

From (4.63)

$$\begin{aligned} \hat{\mathbf{R}}_{11}\mathbf{\Pi}' &= \frac{1}{a}\begin{pmatrix} \hat{\theta}_1\hat{\theta}_2/(1+\hat{\theta}_2^2) & 1 \\ \hat{\theta}_1/(1+\hat{\theta}_2^2) & 0 \end{pmatrix} \\ &= \mathbf{Q}_*\hat{\mathbf{R}}_\pi, \end{aligned}$$

where $\hat{\mathbf{R}}_\pi$ is upper triangular. We can find the 2×2 orthogonal matrix $\mathbf{Q}_*$ by noting that the (2, 1) element of $\mathbf{Q}'_*\hat{\mathbf{R}}_{11}\mathbf{\Pi}'$ $(= \hat{\mathbf{R}}_\pi)$ is zero and the diagonal elements are positive. This gives us

$$\mathbf{Q}'_* = \frac{1}{(1+\hat{\theta}_2^2)^{1/2}}\begin{pmatrix} c\hat{\theta}_2 & c \\ 1 & -\hat{\theta}_2 \end{pmatrix},$$

where c is the sign of $\hat{\theta}_1$. To apply (4.136) we note that the $\hat{\mathbf{a}}^T_{rs}$ in $A^T_{..}$ are all multiples of the vector $\mathbf{u} = (\hat{\theta}_2, 1)'$. Hence each vector in the array $\mathbf{Q}'_*\hat{A}^T_{..}\mathbf{Q}_*$, being a linear combination of the $\hat{\mathbf{a}}_{rs}$ [c.f. (4.45) and (4.47)], is also a multiple of $\mathbf{u}$. Now $\mathbf{Q}'_*\mathbf{u}$ is a 2-dimensional vector with second element zero. Therefore, from (4.136), $\hat{a}_{222,\pi} = \hat{a}_{212,\pi} = 0$ and the subset curvatures for θ_1 [c.f. (4.140) and (4.141) with subscript π] are zero. This is to be expected, as θ_1 enters linearly into the original model (4.26) when θ_2 is regarded as a constant. ■

4.6 ANALYSIS OF RESIDUALS

4.6.1 Quadratic Approximation

Let $\hat{\boldsymbol{\theta}}$ be the least-squares estimate for the model

$$y_i = f(\mathbf{x}_i; \boldsymbol{\theta}) + \varepsilon_i, \qquad i = 1, 2, \ldots, n,$$

where the ε_i are i.i.d. $N(0, \sigma^2)$. Then the ith residual is defined to be $\hat{\varepsilon}_i = y_i - f(\mathbf{x}_i; \hat{\boldsymbol{\theta}})$, and the vector of residuals is given by

$$\hat{\boldsymbol{\varepsilon}} = \mathbf{y} - \hat{\mathbf{y}} = \mathbf{y} - \mathbf{f}(\hat{\boldsymbol{\theta}}). \tag{4.143}$$

Cook and Tsai [1985] show that these residuals, which are so useful as a diagnostic aid for linear regression models, can produce misleading results in nonlinear models when there is intrinsic curvature.

In this section we discuss the effects of curvature on the usual residual plots and recommend some transformed residuals which are sometimes more appropriate for diagnostics. The reader who wishes to skip the algebraic derivations may proceed directly to Section 4.6.3, where the main results are summarized.

Before discussing the results of Cook and Tsai [1985], we note the relationship between their notation and our own. Firstly, their treatment follows that of Section 4.1 except that all matrices and arrays are evaluated at $\boldsymbol{\theta} = \boldsymbol{\theta}^*$, the true value of $\boldsymbol{\theta}$, rather than $\hat{\boldsymbol{\theta}}$: we simply omit the caret from all our symbols. Thus following (4.62) we have the QR decomposition

$$\begin{aligned}
\mathbf{F}_{.} &= \mathbf{F}_{.}(\boldsymbol{\theta}^*) \\
&= \mathbf{Q}\begin{pmatrix} \mathbf{R}_{11} \\ \mathbf{0} \end{pmatrix} \\
&= (\mathbf{Q}_p, \mathbf{Q}_{n-p})\begin{pmatrix} \mathbf{R}_{11} \\ \mathbf{0} \end{pmatrix} \\
&= \mathbf{Q}_p \mathbf{R}_{11} \\
&= \mathbf{Q}_p \mathbf{K}^{-1}.
\end{aligned} \tag{4.144}$$

Secondly, they use the transformed parameters $\boldsymbol{\phi} = \mathbf{R}_{11}\boldsymbol{\theta}$ [c.f. (4.36)], so that $\hat{\boldsymbol{\phi}} = \mathbf{R}_{11}\hat{\boldsymbol{\theta}}$, or $\hat{\boldsymbol{\theta}} = \mathbf{K}\hat{\boldsymbol{\phi}}$, and

$$\begin{aligned}
\hat{\boldsymbol{\varepsilon}} &= \mathbf{y} - \mathbf{f}(\mathbf{K}\hat{\boldsymbol{\phi}}) \\
&= \mathbf{y} - \mathbf{g}(\hat{\boldsymbol{\phi}}).
\end{aligned} \tag{4.145}$$

Under this transformation we find, using the arguments that led to (4.42), that $\mathbf{G}_{.} = \mathbf{Q}_p$. We note from $\mathbf{Q}_p'\mathbf{Q}_p = \mathbf{I}_p$ that

$$\begin{aligned} \mathbf{P}_1 &= \mathbf{Q}_p(\mathbf{Q}_p'\mathbf{Q}_p)^{-1}\mathbf{Q}_p' = \mathbf{Q}_p\mathbf{Q}_p', \\ \mathbf{I}_n &= \mathbf{Q}\mathbf{Q}' = \mathbf{Q}_p\mathbf{Q}_p' + \mathbf{Q}_{n-p}\mathbf{Q}_{n-p}', \end{aligned} \tag{4.146}$$

and

$$\mathbf{I}_n - \mathbf{P}_1 = \mathbf{I}_n - \mathbf{Q}_p\mathbf{Q}_p' = \mathbf{Q}_{n-p}\mathbf{Q}_{n-p}'. \tag{4.147}$$

Here $\mathbf{P}_1$ and $\mathbf{I}_n - \mathbf{P}_1$ represent the orthogonal projections of $\mathbb{R}^n$ onto the tangent plane V_1 $(= \mathscr{R}[\mathbf{Q}_p])$ at $\boldsymbol{\theta}^*$, and onto $V_1^{\perp}(= \mathscr{R}[\mathbf{Q}_{n-p}])$ respectively. Finally, they use $\boldsymbol{\theta}$ instead of $\boldsymbol{\phi}$ and $\boldsymbol{\phi}$ instead of $\hat{\boldsymbol{\phi}} - \boldsymbol{\phi}^*$.

Assuming that $\hat{\boldsymbol{\phi}}$ is sufficiently close to $\boldsymbol{\phi}^*$ $(= \mathbf{R}_{11}\boldsymbol{\theta}^*)$, we can use a quadratic Taylor expansion for $\mathbf{g}$ of (4.145) about $\boldsymbol{\phi}^*$ and get [c.f. (4.5)]

$$\begin{aligned} \hat{\boldsymbol{\varepsilon}} &\approx \mathbf{y} - [\mathbf{g}(\boldsymbol{\phi}^*) + \mathbf{G}_{.}(\hat{\boldsymbol{\phi}} - \boldsymbol{\phi}^*) + \tfrac{1}{2}(\hat{\boldsymbol{\phi}} - \boldsymbol{\phi}^*)' G_{..}(\hat{\boldsymbol{\phi}} - \boldsymbol{\phi}^*)] \\ &= \boldsymbol{\varepsilon} - \mathbf{G}_{.}\hat{\boldsymbol{\gamma}} - \tfrac{1}{2}\hat{\boldsymbol{\gamma}}' G_{..}\hat{\boldsymbol{\gamma}}. \end{aligned} \tag{4.148}$$

Here $\hat{\boldsymbol{\gamma}} = \hat{\boldsymbol{\phi}} - \boldsymbol{\phi}^*$, and the last term on the right of (4.148) is a vector with ith element $\frac{1}{2}\hat{\boldsymbol{\gamma}}'\mathbf{G}_{i..}\hat{\boldsymbol{\gamma}}$, where $\mathbf{G}_{i..}$ is the ith face of the $n \times p \times p$ array $G_{...}$. We can also use the expression

$$\hat{\boldsymbol{\gamma}}' G_{..}\hat{\boldsymbol{\gamma}} = \sum_r \sum_s \mathbf{g}_{rs}\hat{\gamma}_r\hat{\gamma}_s. \tag{4.149}$$

In order to use (4.148) for finding means and variances of the $\hat{\varepsilon}_i$ we need to express $\hat{\boldsymbol{\gamma}}$ in terms of $\boldsymbol{\varepsilon}$. Such an expression can be found by expanding the least-square normal equations for $\hat{\boldsymbol{\phi}}$ [c.f. (2.9)], namely

$$\hat{\mathbf{G}}_{.}'\hat{\boldsymbol{\varepsilon}} = \mathbf{G}_{.}'(\hat{\boldsymbol{\phi}})[\mathbf{y} - \mathbf{g}(\hat{\boldsymbol{\phi}})] = \mathbf{0},$$

about $\boldsymbol{\phi}^*$. Thus

$$\begin{aligned} 0 &= \sum_{i=1}^{n} \frac{\partial g_i(\hat{\boldsymbol{\phi}})}{\partial \phi_r}\hat{\varepsilon}_i \\ &\approx \sum_{i=1}^{n} \left\{ \frac{\partial g_i(\boldsymbol{\phi}^*)}{\partial \phi_r} + \sum_{s=1}^{p} \frac{\partial^2 g_i(\boldsymbol{\phi}^*)}{\partial \phi_s \partial \phi_r}(\hat{\phi}_s - \phi_s^*) \right\} \hat{\varepsilon}_i. \end{aligned}$$

Rewriting the above equation in vector form, and using (4.148), we obtain

$$\begin{aligned} \mathbf{0} &\approx \mathbf{G}_{.}'\hat{\boldsymbol{\varepsilon}} + \sum_{i=1}^{n} \mathbf{G}_{i..}\hat{\varepsilon}_i\hat{\boldsymbol{\gamma}} \\ &\approx \mathbf{G}_{.}'(\boldsymbol{\varepsilon} - \mathbf{G}_{.}\hat{\boldsymbol{\gamma}} - \tfrac{1}{2}\hat{\boldsymbol{\gamma}}' G_{..}\hat{\boldsymbol{\gamma}}) + \sum_i \mathbf{G}_{i..}\hat{\varepsilon}_i\hat{\boldsymbol{\gamma}}. \end{aligned}$$

Since $\mathbf{G} = \mathbf{Q}_p$, we have $\mathbf{G}'_{.}\mathbf{G}_{.} = \mathbf{I}_p$ and, from the above equation,

$$\hat{\boldsymbol{\gamma}} \approx \mathbf{G}'_{.}\boldsymbol{\varepsilon} + [(\hat{\boldsymbol{\varepsilon}}'\mathbf{g}_{rs})]\hat{\boldsymbol{\gamma}} - \tfrac{1}{2}\mathbf{G}'_{.}(\hat{\boldsymbol{\gamma}}'G_{..}\hat{\boldsymbol{\gamma}}), \tag{4.150}$$

as

$$\sum_{i=1}^{n} \hat{\varepsilon}_i \mathbf{G}_{i..} = \left[\left(\sum_{i=1}^{n} \hat{\varepsilon}_i g_{irs}\right)\right] = [(\hat{\boldsymbol{\varepsilon}}'\mathbf{g}_{rs})]. \tag{4.151}$$

By repeated substitution for $\hat{\boldsymbol{\gamma}}$ and $\hat{\boldsymbol{\varepsilon}}$ in the right-hand side of (4.150) we can obtain an expression for $\hat{\boldsymbol{\varepsilon}}$ from (4.148) that involves $\boldsymbol{\varepsilon}$ up to any desired power. We now find a quadratic expression.

To begin with, linear approximations for $\hat{\boldsymbol{\gamma}}$ and $\hat{\boldsymbol{\varepsilon}}$ are [c.f. (4.150)]

$$\hat{\boldsymbol{\gamma}} \approx \mathbf{G}'_{.}\boldsymbol{\varepsilon} = \mathbf{Q}'_p\boldsymbol{\varepsilon} = \mathbf{z}, \tag{4.152}$$

say, and [using the linear part of (4.148) with (4.152)]

$$\begin{aligned}
\hat{\boldsymbol{\varepsilon}} &\approx \boldsymbol{\varepsilon} - \mathbf{G}_{.}(\mathbf{G}'_{.}\boldsymbol{\varepsilon}) \\
&= (\mathbf{I}_n - \mathbf{Q}_p\mathbf{Q}'_p)\boldsymbol{\varepsilon} \\
&= \mathbf{Q}_{n-p}\mathbf{Q}'_{n-p}\boldsymbol{\varepsilon} \qquad \text{[by (4.147)]} \\
&= \mathbf{Q}_{n-p}\mathbf{w},
\end{aligned} \tag{4.153}$$

where

$$\mathbf{w} = \mathbf{Q}'_{n-p}\boldsymbol{\varepsilon}. \tag{4.154}$$

Substituting (4.152) and (4.153) in the right-hand side of (4.150) gives the following quadratic approximation for $\hat{\boldsymbol{\gamma}}$:

$$\hat{\boldsymbol{\gamma}} \approx \mathbf{z} + [(\mathbf{w}'\mathbf{Q}'_{n-p}\mathbf{g}_{rs})]\mathbf{z} - \tfrac{1}{2}\mathbf{Q}'_p(\mathbf{z}'G_{..}\mathbf{z}). \tag{4.155}$$

To obtain a quadratic approximation for $\hat{\boldsymbol{\varepsilon}}$ we replace $\hat{\boldsymbol{\gamma}}$ by the quadratic approximation (4.155) in the second term of (4.148), and by the linear approximation (4.152) in the third term of (4.148). Recalling (4.57), we define

$$\mathbf{B}_{\mathbf{w}} = [(\mathbf{w}'\mathbf{Q}'_{n-p}\mathbf{g}_{rs})] = [(\boldsymbol{\varepsilon}'\mathbf{Q}_{n-p}\mathbf{a}^N_{rs})]. \tag{4.156}$$

We thus have from (4.148) and (4.147)

$$\begin{aligned}
\hat{\boldsymbol{\varepsilon}} &\approx \boldsymbol{\varepsilon} - \mathbf{Q}_p\mathbf{z} - \mathbf{Q}_p[(\mathbf{w}'\mathbf{Q}_{n-p}\mathbf{g}_{rs})]\mathbf{z} - \tfrac{1}{2}(\mathbf{I}_n - \mathbf{Q}_p\mathbf{Q}'_p)(\mathbf{z}'G_{..}\mathbf{z}) \\
&= (\mathbf{I}_n - \mathbf{Q}_p\mathbf{Q}'_p)\boldsymbol{\varepsilon} - \mathbf{Q}_p\mathbf{B}_{\mathbf{w}}\mathbf{z} - \tfrac{1}{2}(\mathbf{Q}_{n-p}\mathbf{Q}'_{n-p})(\mathbf{z}'G_{..}\mathbf{z}) \qquad (4.157) \\
&= \mathbf{Q}_{n-p}\mathbf{w} - \mathbf{Q}_p\mathbf{B}_{\mathbf{w}}\mathbf{z} - \tfrac{1}{2}\mathbf{Q}_{n-p}\mathbf{z}'A^N_{..}\mathbf{z}, \qquad (4.158)
\end{aligned}$$

as

$$\begin{aligned}
\mathbf{Q}'_{n-p}(\mathbf{z}'G_{..}\mathbf{z}) &= \sum_r \sum_s z_r z_s \mathbf{Q}'_{n-p}\mathbf{g}_{rs} \\
&= \sum_r \sum_s z_r z_s \mathbf{a}^N_{rs}.
\end{aligned} \tag{4.159}$$

Cook and Tsai [1985] note that when $\boldsymbol{\theta}^*$ is replaced by $\hat{\boldsymbol{\theta}}$, we have $\boldsymbol{\varepsilon} \to \hat{\boldsymbol{\varepsilon}}$, $\mathbf{a}_{rs}^N \to \hat{\mathbf{a}}_{rs}^N$, and $\mathbf{B}_w \to \hat{\mathbf{B}}$, where $\hat{\mathbf{B}}$ is the special matrix introduced by Hamilton et al. [1982] [c.f. (5.62)].

4.6.2 Approximate Moments of Residuals

Referring to (4.144), we note that, since $\mathbf{Q}$ is orthogonal,

$$\begin{pmatrix} \mathbf{z} \\ \mathbf{w} \end{pmatrix} = \begin{pmatrix} \mathbf{Q}_p' \\ \mathbf{Q}_{n-p}' \end{pmatrix} \boldsymbol{\varepsilon} = \mathbf{Q}'\boldsymbol{\varepsilon} \sim N(\mathbf{0}, \sigma^2 \mathbf{I}_n). \tag{4.160}$$

Hence using (4.158), (4.159), and the statistical independence of $\mathbf{z}$ and $\mathbf{w}$,

$$\begin{aligned} \mathscr{E}[\hat{\boldsymbol{\varepsilon}}] &\approx -\tfrac{1}{2}\mathscr{E}\left[\sum_r \sum_s \mathbf{Q}_{n-p} \mathbf{a}_{rs}^N z_r z_s\right] \\ &= -\tfrac{1}{2}\sigma^2 \mathbf{Q}_{n-p} \sum_{r=1}^{n-p} \mathbf{a}_{rr}^N \\ &= -\tfrac{1}{2}\sigma^2 \mathbf{Q}_{n-p} \sum_{r=p+1}^{n} \mathbf{a}_{rr}. \end{aligned} \tag{4.161}$$

Now if $\mathbf{u}$ and $\mathbf{v}$ are any random vectors, and at least one has mean zero, then $\mathscr{C}[\mathbf{u}, \mathbf{v}] = \mathscr{E}[\mathbf{u}\mathbf{v}']$ [c.f. (1.1)]. Applying this to the three terms in (4.158), and using $E[z_r z_s z_t] = 0$ and the statistical independence of $\mathbf{z}$ and $\mathbf{w}$ (or any functions of them), we find that the three terms are uncorrelated. We use this result to find an approximation for $\mathscr{D}[\hat{\boldsymbol{\varepsilon}}]$; but first a lemma.

Lemma: Let $\mathbf{u} = \mathbf{B}_w \mathbf{z}$ and $\mathbf{v} = \mathbf{z}' A_{\cdot\cdot}^N \mathbf{z}$. Then

$$\begin{aligned} \mathscr{D}[\mathbf{u}] &= \sigma^4 \sum_{i=1}^{n-p} (\mathbf{A}_{i\cdot\cdot}^N)^2 \qquad (= \sigma^4 \mathbf{L}^N, \text{ say}), \\ \mathscr{D}[\mathbf{v}] &= 2\sigma^4 [(\operatorname{tr}\{\mathbf{A}_{i\cdot\cdot}^N \mathbf{A}_{j\cdot\cdot}^N\})] \qquad (= 2\sigma^4 \mathbf{M}^N, \text{ say}). \end{aligned} \tag{4.162}$$

■

Proof: Noting that $A_{\cdot\cdot}^N$ represents the last $n - p$ faces of $A_{\cdot\cdot}$, and letting $\mathbf{A}_{i\cdot\cdot}^N$ represent the ith face of $A_{\cdot\cdot}^N$ [also the $(p + i)$th face of $A_{\cdot\cdot}$], we have from (4.156) and an analogous equation to (4.151)

$$\begin{aligned} \mathbf{B}_w &= [(\boldsymbol{\varepsilon}' \mathbf{Q}_{n-p} \mathbf{a}_{rs}^N)] \\ &= [(\mathbf{w}' \mathbf{a}_{rs}^N)] \\ &= \sum_{i=1}^{n-p} w_i \mathbf{A}_{i\cdot\cdot}^N. \end{aligned} \tag{4.163a}$$

Since $\mathbf{w}$ and $\mathbf{z}$ are statistically independent, and $\mathbf{A}_{i..}^N$ is symmetric,

$$\begin{aligned}\mathscr{D}[\mathbf{u}] &= \mathop{\mathscr{E}}_{\mathbf{w}}\{\mathscr{D}[\mathbf{B}_{\mathbf{w}}\mathbf{z}|\mathbf{w}]\} + \mathop{\mathscr{D}}_{\mathbf{w}}\{\mathscr{E}[\mathbf{B}_{\mathbf{w}}\mathbf{z}|\mathbf{w}]\} \\ &= \sigma^2\mathscr{E}[\mathbf{B}_{\mathbf{w}}\mathbf{B}_{\mathbf{w}}'], \qquad \text{[by (1.3)]} \qquad (4.163\text{b}) \\ &= \sigma^2\mathscr{E}\left[\left(\sum_i w_i \mathbf{A}_{i..}^N\right)^2\right] \qquad \text{[by (4.163a)]} \\ &= \sigma^4 \sum_{i=1}^{n-p} (\mathbf{A}_{i..}^N)^2 \qquad \text{[by (4.160)]}.\end{aligned}$$

Also, by Appendix A12.2,

$$\begin{aligned}\operatorname{cov}[v_i, v_j] &= \operatorname{cov}[\mathbf{z}'\mathbf{A}_{i..}^N\mathbf{z}, \mathbf{z}'\mathbf{A}_{j..}^N\mathbf{z}] \\ &= 2\sigma^4 \operatorname{tr}\{\mathbf{A}_{i..}^N\mathbf{A}_{j..}^N\}.\end{aligned}$$

■

We can now use the above lemma and (1.3) to get an approximation for $\mathscr{D}[\hat{\boldsymbol{\varepsilon}}]$ from (4.158), namely

$$\begin{aligned}\mathscr{D}[\hat{\boldsymbol{\varepsilon}}] &\approx \mathscr{D}[\mathbf{Q}_{n-p}\mathbf{z}] + \mathscr{D}[\mathbf{Q}_p\mathbf{B}_{\mathbf{w}}\mathbf{z}] + \tfrac{1}{4}\mathscr{D}\{\mathbf{Q}_{n-p}\mathbf{z}'A_{..}^N\mathbf{z}\} \\ &= \sigma^2\mathbf{Q}_{n-p}\mathbf{Q}_{n-p}' + \sigma^4\mathbf{Q}_p\mathbf{L}^N\mathbf{Q}_p' + \tfrac{1}{2}\sigma^4\mathbf{Q}_{n-p}\mathbf{M}^N\mathbf{Q}_{n-p}'. \qquad (4.164)\end{aligned}$$

Also

$$\begin{aligned}\mathscr{C}[\hat{\boldsymbol{\varepsilon}}, \hat{\mathbf{y}}] &= \mathscr{C}[\hat{\boldsymbol{\varepsilon}}, \mathbf{y} - \hat{\boldsymbol{\varepsilon}}] \\ &= \mathscr{C}[\hat{\boldsymbol{\varepsilon}}, \mathbf{y}] - \mathscr{D}[\hat{\boldsymbol{\varepsilon}}] \\ &= \mathscr{C}[\mathbf{Q}_{n-p}\mathbf{Q}_{n-p}'\boldsymbol{\varepsilon}, \boldsymbol{\varepsilon}] - \mathscr{D}[\hat{\boldsymbol{\varepsilon}}], \qquad (4.165)\end{aligned}$$

since $E[\varepsilon_r\varepsilon_s\varepsilon_t] = 0$ for all r, s, t implies that the covariance of $\boldsymbol{\varepsilon}$ with all but the first term of (4.158) is zero. Thus from (4.164)

$$\begin{aligned}\mathscr{C}[\hat{\boldsymbol{\varepsilon}}, \hat{\mathbf{y}}] &\approx \sigma^2\mathbf{Q}_{n-p}\mathbf{Q}_{n-p}' - \mathscr{D}[\hat{\boldsymbol{\varepsilon}}] \\ &\approx -\sigma^4\mathbf{Q}_p\mathbf{L}^M\mathbf{Q}_p' - \tfrac{1}{2}\sigma^4\mathbf{Q}_{n-p}\mathbf{M}^N\mathbf{Q}_{n-p}', \qquad (4.166)\end{aligned}$$

which is negative semidefinite, as both terms are negative semidefinite.

4.6.3 Effects of Curvature on Residuals

For the linear model with $\mathbf{X} = \mathbf{F}_{\boldsymbol{\cdot}}$ (c.f. Appendix D1) we have

$$\mathscr{E}[\mathbf{y}] = \mathbf{X}\boldsymbol{\theta}^* = \mathbf{Q}_p\mathbf{R}_{11}\boldsymbol{\theta}^* = \mathbf{Q}_p\boldsymbol{\phi}^* \qquad \text{[by (4.144)]}$$

and $\hat{\boldsymbol{\varepsilon}} = (\mathbf{I}_n - \mathbf{P}_1)\boldsymbol{\varepsilon}$, where $\mathbf{P}_1$ represents the orthogonal projection onto

$\Omega_1 = \mathscr{R}[\mathbf{Q}_p]$. Furthermore, $\mathscr{E}[\hat{\boldsymbol{\varepsilon}}] = \mathbf{0}$, $\mathscr{D}[\hat{\boldsymbol{\varepsilon}}] = \sigma^2(\mathbf{I}_n - \mathbf{P}_1) = \sigma^2 \mathbf{Q}_{n-p}\mathbf{Q}'_{n-p}$ [by (4.147)], and $\mathscr{C}[\hat{\boldsymbol{\varepsilon}}, \hat{\mathbf{y}}] = \mathbf{0}$, where $\hat{\mathbf{y}} = \mathbf{X}\hat{\boldsymbol{\theta}} = \mathbf{Q}_p\hat{\boldsymbol{\phi}}$. An important residual plot is that of $\tilde{\varepsilon}_i = \hat{\varepsilon}_i/\{(1 - r_{ii})^{1/2}s\}$ versus $\hat{y}_i$, where $s^2 = \|\mathbf{y} - \hat{\mathbf{y}}\|^2/(n-p)$ and r_{ii} is the ith diagonal element of $\mathbf{Q}_p\mathbf{Q}'_p = \mathbf{X}(\mathbf{X}'\mathbf{X})^{-1}\mathbf{X}'$. Under the normality assumptions for $\boldsymbol{\varepsilon}$, the $\tilde{\varepsilon}_i$ may be treated, for plotting purposes, as though they were approximately i.i.d. $N(0,1)$. Also, $\tilde{\boldsymbol{\varepsilon}}$ and $\hat{\mathbf{y}}$ are approximately uncorrelated, so that there is no linear trend in the plot. However, for nonlinear models we have additional terms in the moment expressions which are functions of the intrinsic curvature array $A^N_{..}$ and of $\boldsymbol{\phi}^*$. Hence residuals plots can be seriously affected by these additional terms if there is substantial intrinsic curvature. In particular the $\tilde{\varepsilon}_i$ will have nonzero means [by (4.161)] and different variances greater than one [by (4.164)], and the plot of $\tilde{\varepsilon}_i$ versus $\hat{y}_i$ will tend to slope downwards [by (4.166)]. An ordinary residual may therefore appear to be unusually large simply because it has a large positive expectation.

Various other types of residuals and diagnostic measures have been proposed for linear models. However, Cook and Tsai [1985] suggest, by way of example, that for nonlinear models these alternative types will suffer from the same shortcomings as ordinary residuals. For example, the measures proposed by Moolgavkar et al. [1984] will be affected by intrinsic curvature and need to be modified in a similar fashion. The influence methods of Ross [1987], however, look promising.

In conclusion we note that the parameter-effects array $A^T_{..}$ does not feature in the above mathematics. This is to be expected, as $\hat{\boldsymbol{\varepsilon}}$ is invariant under one-to-one transformations, i.e., $\mathbf{f}(\hat{\boldsymbol{\theta}}) = \mathbf{g}(\hat{\boldsymbol{\phi}})$.

4.6.4 Projected Residuals

a Definition and Properties

The intrinsic curvature affects the residuals through the second and third terms of (4.158). The second term, $\mathbf{Q}_p\mathbf{B}_w\mathbf{z}$, belongs to $\mathscr{R}[\mathbf{Q}_p]$ $(= \Omega_1)$; and the third term, $\sum_r\sum_s \mathbf{Q}_{n-p}\mathbf{a}^N_{rs}z_r z_s$, belongs to Ω_2, the space spanned by the vectors $\mathbf{Q}_{n-p}\mathbf{a}^N_{rs}$ $(r, s = 1, 2, \ldots, p)$. We note that $\Omega_2 \subset \mathscr{R}[\mathbf{Q}_{n-p}] = \Omega_1^\perp$. Then the contribution of these terms can be removed by projecting $\hat{\boldsymbol{\varepsilon}}$ onto the orthogonal complement of $\Omega_{12} = \Omega_1 \oplus \Omega_2$, the direct sum of the two orthogonal spaces Ω_1 and Ω_2. This projection can be carried out using the projection matrix $\mathbf{I}_n - \mathbf{P}_{12}$, where $\mathbf{P}_{12}$ represents the orthogonal projection onto Ω_{12}, and we obtain the "projected" residuals $(\mathbf{I}_n - \mathbf{P}_{12})\hat{\boldsymbol{\varepsilon}}$ of Cook and Tsai [1985]. To study the properties of these residuals we note, from (4.157) with the second and third terms dropping out under projection, that

$$\begin{aligned}(\mathbf{I}_n - \mathbf{P}_{12})\hat{\boldsymbol{\varepsilon}} &\approx (\mathbf{I}_n - \mathbf{P}_{12})(\mathbf{I}_n - \mathbf{Q}_p\mathbf{Q}'_p)\boldsymbol{\varepsilon} \\ &= (\mathbf{I}_n - \mathbf{P}_{12})\boldsymbol{\varepsilon}, \end{aligned} \tag{4.167}$$

as $\Omega_1 \subset \Omega_{12}$ (by considering $\Omega_1 + 0$), and this implies

$$(\mathbf{I}_n - \mathbf{P}_{12})\mathbf{Q}_p = \mathbf{0}. \tag{4.168}$$

The vector $(\mathbf{I}_n - \mathbf{P}_{12})\boldsymbol{\varepsilon}$ has similar moment properties to those of the ordinary residual vector $(\mathbf{I}_n - \mathbf{P}_1)\mathbf{y}$ $[=(\mathbf{I}_n - \mathbf{P}_1)\boldsymbol{\varepsilon}]$ from the linear model (c.f. Appendix D1.3), namely

$$\begin{aligned} \mathscr{E}[(\mathbf{I}_n - \mathbf{P}_{12})\boldsymbol{\varepsilon}] &= \mathbf{0}, \\ \mathscr{D}[(\mathbf{I}_n - \mathbf{P}_{12})\boldsymbol{\varepsilon}] &= (\mathbf{I}_n - \mathbf{P}_{12})\sigma^2, \end{aligned} \tag{4.169}$$

and, since $\mathbf{I}_n - \mathbf{P}_{12}$ is symmetric and idempotent,

$$\begin{aligned} E[\hat{\boldsymbol{\varepsilon}}'(\mathbf{I}_n - \mathbf{P}_{12})\hat{\boldsymbol{\varepsilon}}] &\approx E[\boldsymbol{\varepsilon}'(\mathbf{I}_n - \mathbf{P}_{12})\boldsymbol{\varepsilon}] \qquad \text{[by (4.167)]} \\ &= \sigma^2(n - \operatorname{tr}\mathbf{P}_{12}) \qquad \text{[by Appendix A12.1]}. \end{aligned} \tag{4.170}$$

Equation (4.169) can be used to standardize the projected residuals to have approximately constant variance, while (4.170) provides an estimate $\hat{\boldsymbol{\varepsilon}}'(\mathbf{I}_n - \mathbf{P}_{12})\hat{\boldsymbol{\varepsilon}}/(n - \operatorname{tr}\mathbf{P}_{12})$ of σ^2.

A further property from the linear model which also holds is that the projected residuals are uncorrelated with the fitted values, as

$$\begin{aligned} &4\mathscr{C}[(\mathbf{I}_n - \mathbf{P}_{12})\hat{\boldsymbol{\varepsilon}}, \mathbf{g}(\hat{\boldsymbol{\phi}})] \\ &= 4(\mathbf{I}_n - \mathbf{P}_{12})\mathscr{C}[\hat{\boldsymbol{\varepsilon}}, \hat{\mathbf{y}}] \\ &= 4\sigma^2(\mathbf{I}_n - \mathbf{P}_{12})(\mathbf{Q}_{n-p}\mathbf{Q}'_{n-p} - \mathscr{D}[\hat{\boldsymbol{\varepsilon}}]) \qquad \text{[by (4.165)]} \\ &= -2\sigma^4(\mathbf{I}_n - \mathbf{P}_{12})\mathbf{Q}_{n-p}\mathbf{M}^N\mathbf{Q}'_{n-p} \qquad \text{[by (4.166) and (4.168)]} \\ &= -(\mathbf{I}_n - \mathbf{P}_{12})\mathscr{D}[\mathbf{Q}_{n-p}\mathbf{z}'A^N_{..}\mathbf{z}] \qquad \text{[by (4.162)]} \\ &= -\mathscr{C}[(\mathbf{I}_n - \mathbf{P}_{12})\mathbf{Q}_{n-p}\mathbf{z}'A^N_{..}\mathbf{z}, \mathbf{Q}_{n-p}\mathbf{z}'A^N_{..}\mathbf{z}] \\ &= \mathbf{0}. \end{aligned}$$

The last line follows from the fact that, by definition, $\mathbf{Q}_{n-p}\mathbf{z}'A^N_{..}\mathbf{z}$ belongs to Ω_{12}, so that its projection $(\mathbf{I}_n - \mathbf{P}_{12})\mathbf{Q}_{n-p}\mathbf{z}'A^N_{..}\mathbf{z}$ onto $\Omega_{12}^{\perp}$ is zero.

We see then that many of the problems associated with the usual residuals $\hat{\boldsymbol{\varepsilon}}$ are largely overcome by the projected residuals. These latter residuals have properties similar to those of the residuals from linear models and can be plotted and interpreted in much the same manner. As, however, least-squares (and maximum-likelihood) estimates have certain asymptotic properties, there will be some loss of information in using the projected residuals. Although this loss is difficult to quantify, being essentially the difference between (4.164) and (4.169),

Cook and Tsai [1985] suggest considering the loss in rank in going from $\mathbf{I}_n - \mathbf{P}_1$ to $\mathbf{I}_n - \mathbf{P}_{12}$. Let p_2 be the dimension of Ω_2. Then $\Omega_{12}\,(=\Omega_1 \oplus \Omega_2)$ has dimension $p + p_2$, and the loss is $(n-p)-(n-p-p_2) = p_2$. If p_2 is small relative to $n-p$, then this rank loss will be negligible.

b Computation of Projected Residuals

In the formula (4.167) for the projected residuals, the projection matrix $\mathbf{P}_{12}$ is a function of $\boldsymbol{\theta}^*$, the true value of $\boldsymbol{\theta}$, and can be estimated by replacing $\boldsymbol{\theta}^*$ by $\hat{\boldsymbol{\theta}}$. Using a "hat" (caret) to represent this estimation process, we note from $\hat{\Omega}_2 \perp \hat{\Omega}_1$ that

$$\begin{aligned} \hat{\Omega}_2 &= \hat{\Omega}_1^{\perp} \cap (\hat{\Omega}_1 \oplus \hat{\Omega}_2) \\ &= \hat{\Omega}_1^{\perp} \cap \hat{\Omega}_{12}, \end{aligned} \tag{4.171}$$

and the corresponding relationship between projection matrices is (Appendix A11.7)

$$\hat{\mathbf{P}}_2 = \hat{\mathbf{P}}_{12} - \hat{\mathbf{P}}_1. \tag{4.172}$$

However, from the normal equations (2.10), $\hat{\mathbf{P}}_1 \hat{\boldsymbol{\varepsilon}} = \mathbf{0}$, so that $\hat{\mathbf{P}}_{12}\hat{\boldsymbol{\varepsilon}} = \hat{\mathbf{P}}_2 \hat{\boldsymbol{\varepsilon}}$. Cook and Tsai [1985] obtained the last equation and noted that $\hat{\mathbf{P}}_2\hat{\boldsymbol{\varepsilon}}$ is found by regressing $\hat{\boldsymbol{\varepsilon}}$ on the matrix with columns

$$\hat{\mathbf{Q}}_{n-p}\hat{\mathbf{a}}_{rs}^N = \hat{\mathbf{Q}}_{n-p}\hat{\mathbf{Q}}_{n-p}'\hat{\mathbf{g}}_{rs} = (\mathbf{I}_n - \hat{\mathbf{P}}_1)\hat{\mathbf{g}}_{rs}.$$

4.7 NONLINEARITY AND LEAST-SQUARES ESTIMATION

In Chapter 12 we see that, under fairly general conditions, the least-squares estimate θ_r of $\hat{\theta}_r$ is asymptotically unbiased, has asymptotic minimum variance, and is asymptotically normal [c.f. (2.21)]. The main basis for these properties is, once again, the tangent-plane approximation (4.10) for the expectation surface in the neighborhood of $\hat{\boldsymbol{\theta}}$. If this approximation is unsatisfactory, then $\hat{\theta}_r$ will not have the three large-sample properties listed above. Although the measures of curvature provide useful insights into the geometrical effects of nonlinearity, they are apparently not always good indicators of the statistical properties of the estimators (Lowry and Morton [1983]). Therefore, studying the sampling distribution of $\hat{\theta}_r$ for each $r = 1, 2, \ldots, p$ will also shed some light on the effects of nonlinearity. This approach, which we discuss below, has been hinted at by various people, but it is Ratkowsky [1983] who successfully exploits this idea in a practical manner. It should be stressed that the following methods depend on normally distributed "errors" ε_i.

4.7.1 Bias

If $\hat{\gamma} = \hat{\phi} - \phi^*$, then we have from (4.155) and (4.156)

$$\begin{aligned}\hat{\gamma} &\approx \mathbf{z} + \mathbf{B}_w \mathbf{z} - \tfrac{1}{2}\mathbf{Q}_p'(\mathbf{z}' G_{..} \mathbf{z}) \\ &= (\mathbf{I}_p + \mathbf{B}_w)\mathbf{z} - \tfrac{1}{2}\mathbf{z}' A_{..}^T \mathbf{z}.\end{aligned} \tag{4.173}$$

Since the elements of $\mathbf{z}$ and $\mathbf{w}$ are all i.i.d. $N(0,\sigma^2)$, and $\hat{\theta} - \theta^* = \mathbf{K}(\hat{\phi} - \phi^*)$, it follows that

$$\begin{aligned}\mathscr{E}[\hat{\theta} - \theta^*] &= \mathbf{K}\mathscr{E}[\hat{\gamma}] \\ &\approx \mathbf{K}\mathscr{E}\{-\tfrac{1}{2}\mathbf{z}' A_{..}^T \mathbf{z}\} \\ &= \mathbf{K}\left\{-\tfrac{1}{2}\sum_r \sum_s \mathbf{a}_{rs}^T E[z_r z_s]\right\} \\ &= -\tfrac{1}{2}\sigma^2 \mathbf{K} \sum_{r=1}^{p} \mathbf{a}_{rr}^T.\end{aligned} \tag{4.174}$$

[We shall see below that this result also holds for a cubic expansion instead of the quadratic (4.173).] M.J. Box [1971a] discussed bias in nonlinear estimation and showed that the approximate bias can be related to Beale's measure of nonlinearity (see also Beale's discussion of that paper). Bates and Watts [1980] then showed that Box's expression for the bias was the same as (4.174): their term is scaled by ρ. This approximate bias can then be estimated by

$$\hat{\boldsymbol{\beta}} = -\tfrac{1}{2}s^2 \hat{\mathbf{K}} \sum_{r=1}^{p} \hat{\mathbf{a}}_{rr}^T. \tag{4.175}$$

To the extent of the quadratic approximation (4.173), the bias depends on the parameter-effects array and can therefore be reduced or virtually eliminated with an appropriate reparametrization. Since $\hat{\mathbf{K}}$ and $A_{..}^T$ will already be available from previous computations, $\hat{\boldsymbol{\beta}}$ is readily calculated. It is convenient to express the bias as a percentage, namely $100\hat{\beta}_r/\hat{\theta}_r$, and Ratkowsky [1983] suggests that an absolute value in excess of 1% is a good rule for indicating nonlinear behavior in θ_r.

How good is Box's approximation? For normal data, the empirical studies of Box [1971a], Gillis and Ratkowsky [1978], and Ratkowsky [1983] indicate that the approximation works very well, particularly when the bias is large. The bias can be used for indicating which are the problem parameters in the vector $\boldsymbol{\theta}$. For a further study of the properties and computation of the bias approximation see Cook et al. [1986].

4.7.2 Variance

Using the same arguments that led to (4.164), and applying (4.162), we have from the quadratic approximation (4.173) and $\mathscr{E}[\mathbf{B}_w]=\mathbf{0}$

$$
\begin{aligned}
\mathscr{D}[\hat{\boldsymbol{\theta}}] &= \mathbf{K}\mathscr{D}[\hat{\boldsymbol{\gamma}}]\mathbf{K}' \\
&\approx \mathbf{K}\{\mathscr{D}[(\mathbf{I}_p+\mathbf{B}_w)\mathbf{z}]+\tfrac{1}{4}\mathscr{D}(\mathbf{z}'A^T_{..}\mathbf{z})\}\mathbf{K}' \\
&= \mathbf{K}\{\sigma^2\mathscr{E}[(\mathbf{I}_p+\mathbf{B}_w)(\mathbf{I}_p+\mathbf{B}_w)']+\tfrac{1}{4}\mathscr{D}(\mathbf{z}'A^T_{..}\mathbf{z})\}\mathbf{K}' \\
&= \mathbf{K}\{\sigma^2(\mathbf{I}_p+\mathscr{E}[\mathbf{B}_w\mathbf{B}_w'])+\tfrac{1}{4}\mathscr{D}(\mathbf{z}'A^T_{..}\mathbf{z})\}\mathbf{K}' \\
&= \mathbf{K}\{\sigma^2\mathbf{I}_p+\sigma^4\mathbf{L}^N+\tfrac{1}{2}\sigma^4\mathbf{M}^T\}\mathbf{K}' \qquad \text{[by (4.163b)]} \\
&= \sigma^2\mathbf{K}\mathbf{K}'+\sigma^4\mathbf{K}(\mathbf{L}^N+\tfrac{1}{2}\mathbf{M}^T)\mathbf{K}', \qquad (4.176)
\end{aligned}
$$

where $\mathbf{L}^N$ and $\mathbf{M}^T$ are defined analogously to (4.162). We note from (4.144) that

$$
\begin{aligned}
\sigma^2\mathbf{K}\mathbf{K}' &= \sigma^2(\mathbf{F}_.'\mathbf{F}_.)^{-1} \\
&= \sigma^2\mathbf{C}^{-1}, \qquad (4.177)
\end{aligned}
$$

say. From (2.21) we see that the first term of (4.176), namely (4.177), is the variance–covariance matrix of $\hat{\boldsymbol{\theta}}$ when the linear approximation is valid. The second and third terms give some indication as to the effect of the nonlinearity on the variance and involve both the intrinsic and parameter-effects arrays. These can be readily estimated.

Since $\mathbf{z}$ and $\mathbf{w}$ are linear functions of the ε_i, Equation (4.173) represents the approximation of $\hat{\boldsymbol{\gamma}}=\hat{\boldsymbol{\phi}}-\boldsymbol{\phi}^*$ by a quadratic in the elements of $\boldsymbol{\varepsilon}$. However, Clarke [1980] obtained a third-degree expansion for $\hat{\boldsymbol{\gamma}}$. In his notation $\mathbf{D}=\mathbf{F}_.$, $\mathbf{DK}=\mathbf{Q}_p$, $\mathbf{H}=\mathbf{Q}_{n-p}'$, $\mathbf{A}_i=\mathbf{A}^N_{i..}$, $\mathbf{z}$ and $\mathbf{w}$ are unchanged, and, using (4.69) and (4.70), $A_{irs}=a^N_{irs}$ and $B_{irs}=a^T_{irs}$. Furthermore he used

$$
\begin{aligned}
\mathbf{a} &= \mathbf{z}'A^N_{..}\mathbf{z}\left(=\sum_r\sum_s \mathbf{a}^N_{rs}z_rz_s\right), \\
\mathbf{b} &= \mathbf{z}'A^T_{..}\mathbf{z}, \\
\mathbf{c} &= \sum_r\sum_s\sum_t \mathbf{c}_{rst}z_rz_sz_t,
\end{aligned}
$$

where the ith element of $\mathbf{c}_{rst}$ is given by (4.71). We recall, from (4.163a), that

$$
\mathbf{B}_w=\sum_{i=1}^{n-p} w_i\mathbf{A}^N_{i..}.
$$

Thus from Clarke [1980: Equation (2)] we have

$$
\hat{\boldsymbol{\gamma}}\approx\mathbf{z}+\mathbf{B}_w\mathbf{z}-\tfrac{1}{2}\mathbf{b}+\{-\tfrac{1}{2}\mathbf{B}_a\mathbf{z}+\tfrac{1}{2}\mathbf{z}'A^T_{..}\mathbf{b}-\tfrac{1}{6}\mathbf{c}+\mathbf{B}_w^2\mathbf{z}\}, \qquad (4.178)
$$

where the third-degree terms are given in braces. The first thing we note is that the expected value of the terms in braces is zero because $\mathbf{z}$ and $\mathbf{w}$ are independent and $E[z_r z_s z_t] = 0$. Hence $\mathscr{E}[\hat{\gamma}] \approx -\frac{1}{2}\mathscr{E}[\mathbf{b}]$ and we have (4.174) again. Next, if we include fourth-order moments, then, in comparison with (4.176), $\mathscr{D}[\hat{\gamma}]$ now has additional terms due to the covariance of the bracketed third order terms in (4.178) with the linear term $\mathbf{z}$. Clarke, in fact, applied

$$\begin{aligned}\mathscr{D}[\hat{\gamma}] &= \mathscr{E}[\hat{\gamma}\hat{\gamma}'] - (\mathscr{E}[\hat{\gamma}])(\mathscr{E}[\hat{\gamma}])' \\ &\approx \mathscr{E}[\hat{\gamma}\hat{\gamma}'] - \tfrac{1}{4}(\mathscr{E}[\mathbf{b}])(\mathscr{E}[\mathbf{b}])'\end{aligned}$$

and ignored fifth-degree terms and higher in the expansion of $\hat{\gamma}\hat{\gamma}'$. His final result is

$$\mathscr{D}[\hat{\boldsymbol{\theta}}] \approx \sigma^2 \mathbf{K}\mathbf{K}' + \sigma^4 \mathbf{K}(\mathbf{V}_1 + \mathbf{V}_2 + \mathbf{V}_3)\mathbf{K}', \tag{4.179}$$

where the matrices $\mathbf{V}_m$ $(m = 1, 2, 3)$ have the following respective (r, s)th elements:

$$\sum_{i=1}^{n-p} \sum_{u=1}^{p} \{a_{iru}^N a_{ius}^N - a_{irs}^N a_{iuu}^N\}, \tag{4.180}$$

$$\sum_{u=1}^{p} \sum_{v=1}^{p} \{\tfrac{1}{2} a_{ruv}^T a_{suv}^T + \tfrac{1}{2} a_{rsu}^T a_{uvv}^T + \tfrac{1}{2} a_{sru}^T a_{uvv}^T + a_{ruv}^T a_{vsu}^T + a_{suv}^T a_{vru}^T\}, \tag{4.181}$$

and [see (4.71)]

$$-\frac{1}{2} \sum_{u=1}^{p} \{c_{rsuu} + c_{sruu}\}.$$

Unfortunately the c_{rsuu} involve third derivatives of $\mathbf{f}(\boldsymbol{\theta})$ which are often messy to find algebraically. Hence (4.179) may have only a limited appeal in the routine analysis of multiparameter problems. Perhaps $\mathbf{V}_3$ is not always necessary and (4.176) is adequate for some models. However, this point remains to be investigated. We note that the first terms of (4.180) and (4.181) lead to $\mathbf{L}^N$ and $\frac{1}{2}\mathbf{M}^T$, respectively, in (4.176).

In conclusion we note that the asymptotic third moments of $\hat{\boldsymbol{\theta}}$ and related parameters are given by Hougaard [1985].

4.7.3 Simulated Sampling Distributions

We recall that we have a sample of n observations y_i satisfying

$$y_i = f(\mathbf{x}_i; \boldsymbol{\theta}) + \varepsilon_i \qquad (i = 1, 2, \ldots, n), \tag{4.182}$$

where the ε_i are i.i.d $N(0, \sigma^2)$. Also $\hat{\boldsymbol{\theta}}$ and s^2 are the least-squares estimates of $\boldsymbol{\theta}$ and σ^2. Under certain regularity conditions (Section 12.2) we find that, asymptotically, $\sqrt{n}(\hat{\boldsymbol{\theta}} - \boldsymbol{\theta}^*) \sim N_p(\mathbf{0}, \sigma^2 \mathbf{V}^{-1})$, where

$$\mathbf{V} = \lim_{n \to \infty} \frac{1}{n}(\mathbf{F}_{\bullet}'\mathbf{F}_{\bullet})$$

has a strongly consistent estimator $\hat{\mathbf{V}} = \hat{\mathbf{F}}_{\bullet}'\hat{\mathbf{F}}_{\bullet}/n$. This asymptotic theory follows from the fact that, in the limit, the nonlinear model can be replaced by the linear tangent-plane approximation. For moderate samples we would hope that this linear approximation still holds, so that $\hat{\theta}_r$ is approximately normally distributed and unbiased with a variance estimated by

$$v_r = s^2 \hat{c}^{rr}, \tag{4.183}$$

where $\hat{\mathbf{C}} = \hat{\mathbf{F}}_{\bullet}'\hat{\mathbf{F}}_{\bullet}$ [c.f. (4.177)]. However, if n is not large enough, then the bias of $\hat{\theta}_r$ [approximately estimated by $\hat{\boldsymbol{\beta}}$ of (4.175)] will be appreciable, and v_r will underestimate the variance [as indicated by (4.176) and (4.179)]. What we need is tests for normality, bias, and variance excess. These are given by Ratkowsky [1983], who proposes the following methods based on simulating the n-observation experiment M times, say.

Suppose that $\hat{\boldsymbol{\theta}}$ and s^2 are treated as the "true" values of $\boldsymbol{\theta}$ and σ^2, and M sets of size-n samples are simulated from

$$Y_i = f(\mathbf{x}_i; \hat{\boldsymbol{\theta}}) + \delta_i \qquad (i = 1, 2, \ldots, n),$$

where δ_i are i.i.d. $N(0, s^2)$. For each set we can then obtain the least-squares estimate $\tilde{\boldsymbol{\theta}}_m$ $(m = 1, 2, \ldots, M)$ for the model

$$Y_i = f(\mathbf{x}_i; \boldsymbol{\theta}) + \varepsilon_i.$$

These M values of $\tilde{\boldsymbol{\theta}}_m$ will then provide information on the sampling distribution of least-squares estimators.

Let us now consider just the rth element θ_r of $\boldsymbol{\theta}$, and let u_{rm} $(m = 1, 2, \ldots, M)$ be the rth element of $\tilde{\boldsymbol{\theta}}_m$. The simulation sample mean, variance, skewness, and kurtosis are then computed as follows:

$$\text{mean}_r = \bar{u}_{r\cdot}, \tag{4.184}$$

$$\text{var}_r = \frac{1}{M} \sum_{m=1}^{M} (u_{rm} - \bar{u}_{r\cdot})^2, \tag{4.185}$$

$$g_{1r} = \frac{M^{1/2} \sum_m (u_{rm} - \bar{u}_{r\cdot})^2}{[\sum_m (u_{rm} - \bar{u}_{r\cdot})^2]^{3/2}} \quad (= \sqrt{b_1}), \tag{4.186}$$

$$g_{2r} = \frac{M \sum_m (u_{rm} - \bar{u}_{r\cdot})^4}{[\sum_m (u_{rm} - \bar{u}_{r\cdot})^2]^2} - 3 \quad (= b_2 - 3). \tag{4.187}$$

Ratkowsky recommended that M be large and suggested $M = 1000$. Using $\hat{\theta}_r$ and s^2 as the true parameter values, we have

$$\% \text{ bias} = 100\left(\frac{\text{mean}_r - \hat{\theta}_r}{\hat{\theta}_r}\right) \tag{4.188}$$

and, using (4.183),

$$\% \text{ excess variance} = 100\left(\frac{\text{var}_r - v_r}{v_r}\right). \tag{4.189}$$

Approximate formal tests are also available. To test whether the bias is significant we calculate

$$Z = \frac{\text{mean}_r - \hat{\theta}_r}{(v_r/M)^{1/2}} \tag{4.190}$$

and compare it with the usual $N(0,1)$ tables. In a similar manner we can carry out a chi-square test for an increase in variance using the approximate χ^2_{M-1} variable

$$W = \frac{M\,\text{var}_r}{v_r}. \tag{4.191}$$

Since M is large, the chi-square tables cannot be used. Instead we have the normal approximation

$$Z = \sqrt{2W} - \sqrt{2(M-1) - 1}, \tag{4.192}$$

which is approximately $N(0,1)$ under the null hypothesis.

To test for normality we note that under the null hypothesis the u_{rm} are normally distributed,

$$g_{1r} \sim N(0, 6/M) \quad \text{and} \quad g_{2r} \sim N(0, 24/M),$$

approximately. For $M = 1000$ the two-tailed critical values are therefore as follows:

$$\begin{aligned} g_{1r}&: \pm 0.152 \quad (5\%), \quad \pm 0.200 \quad (1\%), \\ g_{2r}&: \pm 0.304 \quad (5\%), \quad \pm 0.399 \quad (1\%). \end{aligned}$$

Finally, we note that $\hat{\boldsymbol{\theta}}$ and s^2 are not the true parameter values, so that, as a precaution, the above procedures could be carried out for a range of values bracketing $\hat{\boldsymbol{\theta}}$ and s^2.

4.7.4 Asymmetry Measures

Lowry and Morton [1983] found in their examples that asymmetry (bias, skewness, etc.) appeared to be the most important factor in assessing the effect of non-linearity on the statistical properties of the estimators. A nearly symmetric distribution always seemed to have a close fit to normality. R.K. Lowry (see Ratkowsky [1983: 34]) introduced a new simulation method of estimating bias, while Morton proposed a variance measure of nonlinearity. We now discuss these methods. (A more detailed exposition is given by Morton [1987b].)

In the previous section we saw that we could get some idea of the properties of the least-squares estimator $\hat{\boldsymbol{\theta}}$ by simulating (4.182) with $\boldsymbol{\theta}$ and σ^2 equal to $\hat{\boldsymbol{\theta}}$ and s^2, respectively. Since ε_i is symmetric, there are in fact two models we can simulate at the same time with the above parameter values, namely

$$y_i^+ = f(\mathbf{x}_i;\boldsymbol{\theta}) + \varepsilon_i$$

and

$$y_i^- = f(\mathbf{x}_i;\boldsymbol{\theta}) - \varepsilon_i.$$

The least-squares estimates for these two models are denoted by $\tilde{\boldsymbol{\theta}}^+$ and $\tilde{\boldsymbol{\theta}}^-$. These estimates will have the same distribution (as ε_i and $-\varepsilon_i$ have the same distribution), and we can compare the simulated mean of $\frac{1}{2}(\tilde{\boldsymbol{\theta}}^+ + \tilde{\boldsymbol{\theta}}^-)$ with the true value of $\boldsymbol{\theta}$ $(=\hat{\boldsymbol{\theta}})$. Lowry and Morton therefore proposed the measure

$$\begin{aligned}\tilde{\boldsymbol{\psi}} &= [(\tilde{\psi}_r)] \\ &= \tfrac{1}{2}(\tilde{\boldsymbol{\theta}}^+ + \tilde{\boldsymbol{\theta}}^-) - \boldsymbol{\theta} \\ &= \mathbf{K}\{\tfrac{1}{2}(\tilde{\boldsymbol{\phi}}^+ + \tilde{\boldsymbol{\phi}}^-) - \boldsymbol{\phi}\}. \end{aligned} \tag{4.193}$$

We can now describe $\hat{\boldsymbol{\gamma}}$ $(=\hat{\boldsymbol{\phi}} - \boldsymbol{\phi}^*)$ of (4.178) as $\boldsymbol{\gamma}^+$ and, by changing the signs of $\mathbf{z}$ and $\mathbf{w}$, obtain $\boldsymbol{\gamma}^-$. Adding these $\boldsymbol{\gamma}$'s together, we find that the linear and cubic terms disappear and

$$\tilde{\boldsymbol{\psi}} \approx \mathbf{K}\{\mathbf{B}_w\mathbf{z} - \tfrac{1}{2}\mathbf{z}'[A_{..}^T]\mathbf{z}\}. \tag{4.194}$$

Thus, from (4.174) and the independence of $\mathbf{w}$ and $\mathbf{z}$,

$$\mathscr{E}[\tilde{\boldsymbol{\psi}}] \approx -\tfrac{1}{2}\sigma^2\mathbf{K}\sum_{r=1}^{p}\mathbf{a}_{rr}^T, \tag{4.195}$$

so that $\tilde{\boldsymbol{\psi}}$ has the same approximate bias as $\tilde{\boldsymbol{\theta}}$. Also, from (4.176),

$$\mathscr{D}[\tilde{\boldsymbol{\psi}}] \approx \sigma^4\mathbf{K}(\mathbf{L}^N + \tfrac{1}{2}\mathbf{M}^T)\mathbf{K}'. \tag{4.196}$$

With the notational correspondence $\sqrt{n}c_{lm,j}^* = a_{jlm}^T$, $\mathbf{R} = \sqrt{n}\mathbf{K}$ and $\sqrt{n}c_{ji,kl}^* =$

$\sum_l a_{ijl} a_{ikl}$, it is readily shown that (4.196) is the same as the expression obtained by Lowry and Morton [1983]. They proposed using the ratio

$$\lambda_r = \frac{\operatorname{var}[\tilde{\psi}_r]}{\sigma^2 c^{rr}} \tag{4.197}$$

evaluated at $\boldsymbol{\theta} = \hat{\boldsymbol{\theta}}$ and $\sigma^2 = s^2$, where c^{rr} is the rth diagonal element of $\mathbf{KK}'$ (c.f. (4.177)], as a measure of asymmetry. Using the quadratic approximation (4.173), it follows from (4.176) that

$$\mathscr{D}[\hat{\boldsymbol{\theta}}] \approx \sigma^2 \mathbf{KK}' + \mathscr{D}[\tilde{\boldsymbol{\psi}}],$$

which has diagonal elements

$$\operatorname{var}[\hat{\theta}_r] \approx \sigma^2 c^{rr} + \operatorname{var}[\tilde{\psi}_r].$$

Hence $\operatorname{var}[\tilde{\psi}_r]$ also represents an "excess variance" due to the curvature arrays.

We note that $\lambda_r \geq 0$, and we now show that $\lambda_r = 0$ when the model is linear, say $\mathbf{y} = \mathbf{X}\boldsymbol{\theta} + \boldsymbol{\varepsilon}$, where $\mathbf{X}$ is $n \times p$ of rank p. Then

$$\begin{aligned}
\tilde{\boldsymbol{\theta}}^+ &= (\mathbf{X}'\mathbf{X})^{-1}\mathbf{X}'\mathbf{y}^+ \\
&= (\mathbf{X}'\mathbf{X})^{-1}\mathbf{X}'(\mathbf{X}\boldsymbol{\theta} + \boldsymbol{\varepsilon}) \\
&= \boldsymbol{\theta} + (\mathbf{X}'\mathbf{X})^{-1}\mathbf{X}'\boldsymbol{\varepsilon}
\end{aligned}$$

and

$$\tilde{\boldsymbol{\theta}}^- = \boldsymbol{\theta} + (\mathbf{X}'\mathbf{X})^{-1}\mathbf{X}'(-\boldsymbol{\varepsilon}).$$

We find that $\tilde{\boldsymbol{\psi}}$ is now identically zero, so that $\operatorname{var}[\tilde{\psi}_r] = 0$ and $\lambda_r = 0$. Since $\tilde{\theta}_r^+ - \theta_r = -(\tilde{\theta}_r^- - \theta_r) + 2\psi_r$, and $\tilde{\theta}_r^+ - \theta$ and $\tilde{\theta}_r^- - \theta$ have the same distribution, it follows that $\tilde{\theta}_r^+ - \theta_r$ has a symmetric distribution when $\psi_r = 0$: otherwise the distribution of $\tilde{\theta}_r^+ - \theta_r$ would be asymmetric. For this reason Lowry and Morton refer to λ_r as a measure of asymmetry.

D. Ratkowsky (personal communication, 1984), who has studied a wide range of nonlinear models, found λ_r to be the most useful single tool available for studying the behavior of nonlinear models. He suggested the rule of thumb that if $\lambda_r < 0.01$, the behavior of $\hat{\theta}_r$ is close to linear, whereas if $0.01 \leq \lambda_r \leq 0.05$, the nonnormality of $\hat{\theta}_r$ is usually small enough to be ignored for most practical purposes. If $\lambda_r > 0.05$, then perceptible skewness will be observed in the simulation distribution associated with $\hat{\theta}_r$.

The computational procedure may be described as follows. We simulate M sets of size-n samples from each of the models

$$y_i^+ = f(\mathbf{x}_i; \hat{\boldsymbol{\theta}}) + \delta_i \qquad (i = 1, 2, \ldots, n)$$

Table 4.3 A Comparison of the Simultation Bias and Standard Deviation of $\tilde{\psi}_r$, and the Simulated $\hat{\lambda}_r$, with the Corresponding Theoretical Expressions[a]

Data[b] Set	Curvature[c]	Parameter Estimate	Theor. Bias	$\bar{\psi}_r$	Simulated s.d. ψ_r	Theor. s.d. ψ_r	Simulated $\hat{\lambda}_r$	Theor. λ_r
(1)	IN = .053	$\alpha = 539$	.628	.645	.990	.927	.0105	.0097
	PE = .571	$\beta = 308$	.641	.657	.988	.928	.0101	.0091
	$n = 5$	$\rho = .537$	$.0^{3}2$	$.0^{3}2$	.001	.001	.0017	.0016
(2)	IN = .246	$\alpha = 41.7$	1.64	1.67	26.5	2.38	.494	.419
	PE = 6.99	$\beta = 15.8$	1.66	1.66	26.5	2.39	.495	.465
	$n = 5$	$\gamma = .956$	−.001	−.001	.005	.004	.051	.039
(3)	IN = $.0^{3}2$[d]	$\theta_1 = .0056$	$.0^{5}18$	$.0^{5}20$	$.0^{5}28$	$.0^{5}27$	$.0^{3}29$	$.0^{3}29$
	PE = 39.7	$\theta_2 = 6181$	.0597	.0701	.0955	.0927	$.0^{4}16$	$.0^{4}16$
	$n = 16$	$\theta_3 = 345$	.0013	.0016	.0022	.0021	$.0^{5}70$	$.0^{5}70$
(4)	IN = .151	$\alpha = .0019$	$.0^{3}62$	$.0^{3}69$	.0012	$.0^{3}88$	.233	.299
	PE = 23.5	$\beta = .0^{3}11$	$.0^{5}88$	$.0^{5}87$	$.0^{4}13$	$.0^{4}13$	.073	.079
	$n = 20$	$\vartheta = 1.00$	−.0013	−.0017	.0122	.0119	.013	.012

[a] From Lowry and Morton [1983].

[b] Sources of the data sets and parameter definitions are as follows: (1) Pimentel–Gomes [1953]; (2) Snedecor and Cochran [1967: Example 15.8.1, p. 471]; (3) Meyer and Roth [1972: Example 8]; (4) Frappell [1973: 1969/70 data set]. For each example 1000 simulations were used, except that in (2) one simulation failed to get estimates to converge.

[c] IN = γ_{max}^N, the intrinsic curvature, and PE = γ_{max}^T, the parameter-effects curvature.

[d] To be read as 0.0002

and

$$y_i^- = f(\mathbf{x}_i; \hat{\boldsymbol{\theta}}) - \delta_i,$$

where the δ_i are i.i.d. $N(0, s^2)$, and calculate $\tilde{\boldsymbol{\theta}}_m^+$, $\tilde{\boldsymbol{\theta}}_m^-$, and $\tilde{\boldsymbol{\psi}}_m = \frac{1}{2}(\tilde{\boldsymbol{\theta}}_m^+ + \tilde{\boldsymbol{\theta}}_m^-) - \hat{\boldsymbol{\theta}}$ $(m = 1, 2, \ldots, M)$. If $\tilde{\psi}_{rm}$ is the rth element of $\tilde{\boldsymbol{\psi}}_m$, then $E[\tilde{\psi}_{rm}]$ and $\text{var}[\tilde{\psi}_{rm}]$ are estimated by the simulation mean and variance $\tilde{\psi}_{r\cdot} = \sum_m \tilde{\psi}_{rm}/M$ and $\sum_m (\tilde{\psi}_{rm} - \tilde{\psi}_{r\cdot})^2/M$, respectively; λ_r can then be estimated by

$$\hat{\lambda}_r = \sum_m \frac{(\tilde{\psi}_{rm} - \tilde{\psi}_{r\cdot})^2}{M v_r}, \tag{4.198}$$

where v_r is given by (4.183). In Table 4.3 from Lowry and Morton [1983] the simulated mean and standard deviation of $\tilde{\psi}_r$, and $\hat{\lambda}_r$, are each compared with the theoretical bias from (4.195), the theoretical standard deviation from (4.196), and λ_r from (4.197), respectively (with σ^2 replaced by s^2). Data set (3) had a large parameter-effects curvature ($\gamma_{\max}^T$), yet λ was very small and the distributions close to normal. On the other hand, data set (2) was exceptionally badly behaved, but did not have such extreme curvatures. Judging by the closeness of $\hat{\lambda}_r$ and λ_r, it would appear that λ_r, based on the approximation (4.194), is a reasonable approximation to the true value of the variance ratio.

CHAPTER 5

Statistical Inference

5.1 ASYMPTOTIC CONFIDENCE INTERVALS

Let

$$y_i = f(\mathbf{x}_i; \boldsymbol{\theta}) + \varepsilon_i \qquad (i = 1, 2, \ldots, n), \tag{5.1}$$

where the ε_i are i.i.d. $N(0, \sigma^2)$. For notational convenience we now omit the star from $\boldsymbol{\theta}^*$ and denote the true value of this parameter vector by $\boldsymbol{\theta}$. If we are interested in constructing a confidence interval for a given linear combination $\mathbf{a}'\boldsymbol{\theta}$, then we can apply Theorem 2.1 in Section 2.1.2. In particular we have the asymptotic (linearization) result

$$\hat{\boldsymbol{\theta}} \sim N_p(\boldsymbol{\theta}, \sigma^2 \mathbf{C}^{-1}), \qquad \mathbf{C} = \mathbf{F}'_{\cdot}\mathbf{F}_{\cdot}, \tag{5.2}$$

which holds under appropriate regularity conditions. From (5.2) we have, asymptotically, $\mathbf{a}'\hat{\boldsymbol{\theta}} \sim N(\mathbf{a}'\boldsymbol{\theta}, \sigma^2\mathbf{a}'\mathbf{C}^{-1}\mathbf{a})$ independently of $s^2 = \| \mathbf{y} - \mathbf{f}(\hat{\boldsymbol{\theta}}) \|^2/(n-p)$, the latter being an unbiased estimate of σ^2 to order n^{-1}. Hence for large n we have, approximately,

$$T = \frac{\mathbf{a}'\hat{\boldsymbol{\theta}} - \mathbf{a}'\boldsymbol{\theta}}{s(\mathbf{a}'\mathbf{C}^{-1}\mathbf{a})^{1/2}} \sim t_{n-p}, \tag{5.3}$$

where t_{n-p} is the t-distribution with $n-p$ degrees of freedom. An approximate $100(1-\alpha)\%$ confidence interval for $\mathbf{a}'\boldsymbol{\theta}$ is then

$$\mathbf{a}'\hat{\boldsymbol{\theta}} \pm t_{n-p}^{\alpha/2} s(\mathbf{a}'\mathbf{C}^{-1}\mathbf{a})^{1/2}. \tag{5.4}$$

Here $\mathbf{C}$ can be estimated by $\hat{\mathbf{C}} = \hat{\mathbf{F}}'_{\cdot}\hat{\mathbf{F}}_{\cdot}$ (Donaldson and Schnabel [1987]). Setting $\mathbf{a}' = (0, 0, \ldots, 0, 1, 0, \ldots, 0)$, where the rth element of $\mathbf{a}$ is one and the remaining elements are zero, and defining $[(\hat{c}^{rs})] = \hat{\mathbf{C}}^{-1}$, a confidence interval for the rth element of $\boldsymbol{\theta}$, θ_r, is

$$\hat{\theta}_r \pm t_{n-p}^{\alpha/2} s \sqrt{\hat{c}^{rr}}. \tag{5.5}$$

Gallant [1975b] simulated the true distribution of $(\hat{\theta}_r - \theta_r)/(s^2 \hat{c}^{rr})^{1/2}$ $(r = 1, 2, 3, 4)$ for the model

$$y_i = \theta_1 x_{i1} + \theta_2 x_{i2} + \theta_4 \exp(\theta_3 x_{i3}) + \varepsilon_i, \tag{5.6}$$

$i = 1, 2, \ldots, 30$, where (x_{i1}, x_{i2}) is $(1, 1)$ for $i = 1, 2, \ldots, 15$ and $(0, 1)$ for $i = 16, 17, \ldots, 30$, and the x_{3i} are a random sample from $[0, 10]$. He showed that, for each r, the critical values of the distribution are reasonably well approximated by the t_{n-p}-distribution. (Note that our **C**-matrix is the inverse of Gallant's.) From the above experiment and general experience he suggests that "the asymptotic theory gives results which appear reasonable in applications." However, we shall see later that the usefulness of (5.5) depends on the curvature properties of $\mathscr{E}[\mathbf{y}] = \mathbf{f}(\boldsymbol{\theta})$. Without adequate checks (5.5) could be completely misleading. For example, Donaldson and Schnabel [1987] give examples where (5.5) has poor coverage properties: in two cases, 44% and 10.8% for a nominally 95% confidence interval.

If we are interested in a set of intervals, one for each θ_r $(r = 1, 2, \ldots, p)$, then we can use, for example, the Bonferroni method and calculate

$$\hat{\theta}_r \pm t_{n-p}^{\alpha/(2p)} s \sqrt{\hat{c}^{rr}} \qquad (r = 1, 2, \ldots, p). \tag{5.7}$$

These p intervals will have an overall confidence level of at least $100(1 - \alpha)\%$. In the same way we can construct confidence intervals for m prechosen linear combinations $\mathbf{a}_j'\boldsymbol{\theta}$ $(j = 1, 2, \ldots, m)$. We use (5.4) but with a t-value of $t_{n-p}^{\alpha/(2m)}$. Tables for finding these t-values for $\alpha = .05$ and $\alpha = .01$ are given by Bailey [1977]. Further methods of constructing confidence intervals are discussed later as they are related to confidence regions and hypothesis tests.

Confidence intervals for nonlinear functions, $a(\boldsymbol{\theta})$ say, of $\boldsymbol{\theta}$ can be handled in a similar fashion using the linear Taylor expansion

$$a(\hat{\boldsymbol{\theta}}) \approx a(\boldsymbol{\theta}) + \mathbf{a}'(\hat{\boldsymbol{\theta}} - \boldsymbol{\theta}),$$

where $\mathbf{a}' = (\partial a/\partial\theta_1, \partial a/\partial\theta_2, \ldots, \partial a/\partial\theta_p)$. Then

$$\begin{aligned} \mathscr{D}[a(\hat{\boldsymbol{\theta}})] &\approx \mathscr{D}[\mathbf{a}'(\hat{\boldsymbol{\theta}} - \boldsymbol{\theta})] \\ &= \mathbf{a}'\mathscr{D}[\hat{\boldsymbol{\theta}}]\mathbf{a} \qquad \text{[by (1.3)]} \\ &\approx \sigma^2 \mathbf{a}'\mathbf{C}^{-1}\mathbf{a}, \end{aligned}$$

and, from (5.2), we have the asymptotic result

$$a(\hat{\boldsymbol{\theta}}) \sim N(a(\boldsymbol{\theta}), \sigma^2 \mathbf{a}'\mathbf{C}^{-1}\mathbf{a}).$$

Therefore an approximate $100(1-\alpha)\%$ confidence interval for $a(\boldsymbol{\theta})$ is

$$a(\hat{\boldsymbol{\theta}}) \pm t_{n-p}^{\alpha/2} s(\mathbf{a}'\mathbf{C}^{-1}\mathbf{a})_{\hat{\boldsymbol{\theta}}}^{1/2}. \tag{5.8}$$

If several such intervals are required, then the Bonferroni method can be used as above. However, such intervals will be affected by curvature in much the same way as (5.5).

Using the asymptotic linearization of (5.1), we can apply existing linear methods to finding a prediction interval for y at $\mathbf{x}=\mathbf{x}_0$. Let

$$y_0 = f(\mathbf{x}_0;\boldsymbol{\theta}) + \varepsilon_0,$$

where $\varepsilon_0 \sim N(0,\sigma^2)$ and is independent of $\boldsymbol{\varepsilon}$. An obvious estimate of y_0 is that provided by the fitted model, namely the "prediction" $\hat{y}_0 = f(\mathbf{x}_0;\hat{\boldsymbol{\theta}})$. Since for large n, $\hat{\boldsymbol{\theta}}$ is close to the true value $\boldsymbol{\theta}$, we have the usual Taylor expansion

$$f(\mathbf{x}_0;\hat{\boldsymbol{\theta}}) \approx f(\mathbf{x}_0;\boldsymbol{\theta}) + \mathbf{f}_0'(\hat{\boldsymbol{\theta}}-\boldsymbol{\theta}),$$

where

$$\mathbf{f}_0' = \left(\frac{\partial f(\mathbf{x}_0;\boldsymbol{\theta})}{\partial \theta_1}, \frac{\partial f(\mathbf{x}_0;\boldsymbol{\theta})}{\partial \theta_2}, \dots, \frac{\partial f(\mathbf{x}_0;\boldsymbol{\theta})}{\partial \theta_p}\right).$$

Hence

$$y_0 - \hat{y}_0 \approx y_0 - f(\mathbf{x}_0;\boldsymbol{\theta}) - \mathbf{f}_0'(\hat{\boldsymbol{\theta}}-\boldsymbol{\theta}) = \varepsilon_0 - \mathbf{f}_0'(\hat{\boldsymbol{\theta}}-\boldsymbol{\theta}).$$

From (5.2) and the statistical independence of $\hat{\boldsymbol{\theta}}$ and ε_0,

$$\begin{aligned} E[y_0 - \hat{y}_0] &\approx E[\varepsilon_0] - \mathbf{f}_0'\mathscr{E}[\hat{\boldsymbol{\theta}}-\boldsymbol{\theta}] \approx 0, \\ \operatorname{var}[y_0 - \hat{y}_0] &\approx \operatorname{var}[\varepsilon_0] + \operatorname{var}[\mathbf{f}_0'(\hat{\boldsymbol{\theta}}-\boldsymbol{\theta})] \\ &\approx \sigma^2 + \sigma^2\mathbf{f}_0'(\mathbf{F}_.'\mathbf{F}_.)^{-1}\mathbf{f}_0 \\ &= \sigma^2(1+v_0), \end{aligned}$$

say, and $y_0 - \hat{y}_0$ is asymptotically $N(0,\sigma^2[1+v_0])$. Now s^2 is independent of y_0 and is asymptotically independent of $\hat{\boldsymbol{\theta}}$, so that s^2 is asymptotically independent of $y_0 - \hat{y}_0$. Hence, asymptotically,

$$\frac{y_0 - \hat{y}_0}{s\sqrt{1+v_0}} \sim t_{n-p},$$

and an approximate $100(1-\alpha)\%$ confidence interval for y_0 is therefore given by

$$\hat{y}_0 \pm t_{n-p}^{\alpha/2} s[1 + \mathbf{f}_0'(\mathbf{F}_.'\mathbf{F}_.)^{-1}\mathbf{f}_0]^{1/2}.$$

Since $\mathbf{f}_0$ and $\mathbf{F}_.$ are functions of $\boldsymbol{\theta}$, v_0 is a function of $\boldsymbol{\theta}$ and can be estimated by replacing $\boldsymbol{\theta}$ by $\hat{\boldsymbol{\theta}}$.

The construction of simultaneous confidence bands for $f(\mathbf{x};\boldsymbol{\theta})$ is discussed by Khorasani and Milliken [1982]. The problem of inverse prediction, which arises for example in the use of calibration curves, is discussed in Section 5.12.

5.2 CONFIDENCE REGIONS AND SIMULTANEOUS INTERVALS

5.2.1 Simultaneous Intervals

As a first step in studying simultaneous confidence intervals for *all* linear combinations $\mathbf{b}'\boldsymbol{\theta}$ ($\mathbf{b} \neq \mathbf{0}$), we now consider the approximate $100(1-\alpha)\%$ confidence region for $\boldsymbol{\theta}$ given by (2.26), namely

$$\{\boldsymbol{\theta}:(\boldsymbol{\theta}-\hat{\boldsymbol{\theta}})'\hat{\mathbf{C}}(\boldsymbol{\theta}-\hat{\boldsymbol{\theta}}) \leq ps^2 F^{\alpha}_{p,n-p}\}, \tag{5.9}$$

sometimes called the linearization region. Following the usual linear theory for Scheffé's *S*-method (e.g. Seber [1977: Chapter 5]), we have

$$\begin{aligned}
1-\alpha &= \mathrm{pr}\left[F_{p,n-p} \leq F^{\alpha}_{p,n-p}\right] \\
&\approx \mathrm{pr}\left[(\hat{\boldsymbol{\theta}}-\boldsymbol{\theta})'\hat{\mathbf{C}}(\hat{\boldsymbol{\theta}}-\boldsymbol{\theta}) \leq ps^2 F^{\alpha}_{p,n-p} = c\right] \\
&= \mathrm{pr}\left[\sup_{\mathbf{b}} \frac{\{\mathbf{b}'(\hat{\boldsymbol{\theta}}-\boldsymbol{\theta})\}^2}{\mathbf{b}'\hat{\mathbf{C}}^{-1}\mathbf{b}} \leq c\right] \qquad \text{(by Appendix A7.2)} \\
&= \mathrm{pr}\left[|\mathbf{b}'(\hat{\boldsymbol{\theta}}-\boldsymbol{\theta})| \leq (c\mathbf{b}'\hat{\mathbf{C}}^{-1}\mathbf{b})^{1/2}, \text{ all } \mathbf{b} \neq \mathbf{0}\right].
\end{aligned} \tag{5.10}$$

Hence, with approximate probability $1-\alpha$,

$$\mathbf{b}'\hat{\boldsymbol{\theta}} \pm (pF^{\alpha}_{p,n-p})^{1/2} s(\mathbf{b}'\hat{\mathbf{C}}^{-1}\mathbf{b})^{1/2} \tag{5.11}$$

contains $\mathbf{b}'\boldsymbol{\theta}$, for all $\mathbf{b}$. Included in this set are the intervals for individual θ_r, though these will be much wider than the Bonferroni intervals of (5.7). Intervals like those of (5.11) are used for looking at linear combinations suggested by the data, whereas the Bonferroni intervals are used for prechosen linear combinations. The latter situation, however, is usually the appropriate one in nonlinear models. The intervals (5.11) tend to be very wide so that they are of limited usefulness.

5.2.2 Confidence Regions

Confidence regions were discussed in Section 3.3. Setting $S(\boldsymbol{\theta}) = \|\boldsymbol{\varepsilon}(\boldsymbol{\theta})\|^2 = \|\mathbf{y}-\mathbf{f}(\boldsymbol{\theta})\|^2$, a competing region is [c.f. (3.9)]

$$\left\{\boldsymbol{\theta}:\frac{S(\boldsymbol{\theta})-S(\hat{\boldsymbol{\theta}})}{S(\hat{\boldsymbol{\theta}})} \leq \frac{p}{n-p} F^{\alpha}_{p,n-p}\right\}, \tag{5.12}$$

or, writing $\hat{\boldsymbol{\varepsilon}} = \boldsymbol{\varepsilon}(\hat{\boldsymbol{\theta}})$,

$$\{\boldsymbol{\theta} : \|\boldsymbol{\varepsilon}(\boldsymbol{\theta})\|^2 \leq (1+f)\|\hat{\boldsymbol{\varepsilon}}\|^2\}, \qquad f = \frac{p}{n-p} F^{\alpha}_{p,n-p}. \tag{5.13}$$

Although (5.12) and (5.9) are asymptotically the same for nonlinear models and exactly the same for linear models, they can be quite different for small samples: (5.12) seems to be superior, as it has better coverage (i.e. closer to the nominal value) and is less affected by curvature (Donaldson and Schnabel [1987]). In fact, as $S(\boldsymbol{\theta})$ is invariant to reparametrizations, we can assume that a transformation has been found to remove the parameter-effects curvature. Thus (5.13) is only affected by intrinsic curvature, which is often negligible. Before discussing the regions further, we mention an approximate version of (5.12) proposed by Box and Coutie [1956]. Expanding $S(\boldsymbol{\theta})$ about $\hat{\boldsymbol{\theta}}$, as in (2.33), we have

$$\begin{aligned} S(\boldsymbol{\theta}) &\approx S(\hat{\boldsymbol{\theta}}) + \left(\frac{\partial S}{\partial \boldsymbol{\theta}'}\right)_{\hat{\boldsymbol{\theta}}} (\boldsymbol{\theta} - \hat{\boldsymbol{\theta}}) + \tfrac{1}{2}(\boldsymbol{\theta} - \hat{\boldsymbol{\theta}})' \left(\frac{\partial^2 S}{\partial \boldsymbol{\theta}\, \partial \boldsymbol{\theta}'}\right)_{\hat{\boldsymbol{\theta}}} (\boldsymbol{\theta} - \hat{\boldsymbol{\theta}}) \\ &= S(\hat{\boldsymbol{\theta}}) + \tfrac{1}{2}(\boldsymbol{\theta} - \hat{\boldsymbol{\theta}})' \hat{\mathbf{H}} (\boldsymbol{\theta} - \hat{\boldsymbol{\theta}}), \end{aligned} \tag{5.14}$$

say, since $\partial S/\partial \boldsymbol{\theta} = \mathbf{0}$ at $\hat{\boldsymbol{\theta}}$. Then (5.12) becomes, approximately,

$$\{\boldsymbol{\theta} : (\boldsymbol{\theta} - \hat{\boldsymbol{\theta}})' \hat{\mathbf{H}} (\boldsymbol{\theta} - \hat{\boldsymbol{\theta}}) \leq 2s^2 p F^{\alpha}_{p,n-p}\}, \tag{5.15}$$

where $s^2 = S(\hat{\boldsymbol{\theta}})/(n-p)$. Also, from (2.34) and the square-bracket multiplication of Appendix B4 we have

$$\begin{aligned} \tfrac{1}{2}\hat{\mathbf{H}} &= \hat{\mathbf{C}} - \sum_{i=1}^{n} \{y_i - f_i(\hat{\boldsymbol{\theta}})\} \left\{\frac{\partial^2 f_i}{\partial \boldsymbol{\theta}\, \partial \boldsymbol{\theta}'}\right\}_{\hat{\boldsymbol{\theta}}} \\ &= \hat{\mathbf{C}} - [\hat{\boldsymbol{\varepsilon}}'][\hat{\boldsymbol{F}}_{\cdot\cdot}] \\ &= \hat{\mathbf{F}}_{\cdot}'\hat{\mathbf{F}}_{\cdot} - [\hat{\boldsymbol{\varepsilon}}'][\hat{\boldsymbol{F}}_{\cdot\cdot}]. \end{aligned} \tag{5.16}$$

If we replace $\hat{\mathbf{H}}$ by $(\mathscr{E}[\mathbf{H}])_{\hat{\boldsymbol{\theta}}} = 2\hat{\mathbf{F}}_{\cdot}'\hat{\mathbf{F}}_{\cdot}$ [c.f. (2.35)], we see that we are back to (5.9). We can expect (5.16) to approximate (5.12) better than (5.9), as it includes second-order terms in the Taylor expansion, whereas (5.9) has first-order terms only. However, we note that regions like (5.12) may be awkward to compute and difficult to represent graphically, particularly if p, the dimension of $\boldsymbol{\theta}$, is greater than 3 (or even 2). Usually we would be more interested in confidence intervals for the individual elements θ_r of $\boldsymbol{\theta}$ (see later).

We shall see in Section 5.8 that the validity of the above asymptotic theory depends very much on both the degree of nonlinearity in the model and the choice of parameters $\boldsymbol{\theta}$, as reflected in the intrinsic and parameter-effects curvatures, respectively. If the linear approximation behind the theory is satisfactory, then the wealth of results from linear theory can be applied with

$\mathbf{X} = \hat{\mathbf{F}}_.$. The linear methods for confidence regions and simultaneous confidence intervals are discussed by Seber [1977], Miller [1981], and Savin [1984]. If the curvature effects cannot be ignored, then some modifications to the above theory are possible, as we shall see later.

As there is a close connection between confidence regions and hypothesis tests, the latter are discussed below. This connection is exploited to give confidence regions for subsets of $\boldsymbol{\theta}$ in Section 5.4. In particular, by using a subset of just a single element we can obtain confidence intervals for the individual θ_r.

5.2.3 Asymptotic Likelihood Methods

Before completing our discussion on confidence regions it is appropriate to highlight a very general technique based on likelihood functions. Suppose $L(\boldsymbol{\gamma})$ is the log-likelihood function for a general model with unknown r-dimensional vector parameter $\boldsymbol{\gamma}$. Then, under certain regularity conditions, the hypothesis $H_0: \boldsymbol{\gamma} = \boldsymbol{\gamma}_0$ can be tested using the statistic

$$\begin{aligned} \text{LR} &= -2 \log \ell \\ &= 2[L(\hat{\boldsymbol{\gamma}}) - L(\boldsymbol{\gamma}_0)], \end{aligned}$$

where ℓ is the likelihood-ratio test statistic and $\hat{\boldsymbol{\gamma}}$, the maximum likelihood estimate of $\boldsymbol{\gamma}$, maximizes $L(\boldsymbol{\gamma})$. The statistic LR is approximately distributed as χ_r^2 when H_0 is true. An approximate $100(1-\alpha)\%$ confidence region for $\boldsymbol{\gamma}$ is therefore given by "inverting" the test to obtain

$$\{\boldsymbol{\gamma} : 2[L(\hat{\boldsymbol{\gamma}}) - L(\boldsymbol{\gamma})] \leq \chi_r^2(\alpha)\},$$

where $\text{pr}[\chi_r^2 \geq \chi_r^2(\alpha)] = \alpha$. Using a similar argument, if $\boldsymbol{\gamma} = (\boldsymbol{\gamma}_1', \boldsymbol{\gamma}_2')'$ and $\tilde{\boldsymbol{\gamma}}_1(\boldsymbol{\gamma}_2)$ is the maximum-likelihood estimate of $\boldsymbol{\gamma}_1$ given the value of r_2-dimensional $\boldsymbol{\gamma}_2$, then we have the confidence region

$$\{\boldsymbol{\gamma}_2 : 2[L(\hat{\boldsymbol{\gamma}}) - L(\tilde{\boldsymbol{\gamma}}_1(\boldsymbol{\gamma}_2), \boldsymbol{\gamma}_2)] \leq \chi_{r_2}^2(\alpha)\}.$$

Under certain conditions (c.f. Amemiya [1983]), the above theory also holds when $L(\boldsymbol{\gamma})$ is replaced by a concentrated log-likelihood function $M(\boldsymbol{\theta})$ (c.f. Section 2.2.3). The corresponding confidence region for $\boldsymbol{\theta}$ is

$$\{\boldsymbol{\theta} : 2[M(\hat{\boldsymbol{\theta}}) - M(\boldsymbol{\theta})] \leq \chi_p^2(\alpha)\}.$$

In the case of the nonlinear regression model with $\boldsymbol{\gamma} = (\boldsymbol{\theta}', \sigma^2)'$ we find that $M(\hat{\boldsymbol{\theta}}) = L(\hat{\boldsymbol{\theta}}, \hat{\sigma}^2)$ and $M(\boldsymbol{\theta}) = L(\boldsymbol{\theta}, \tilde{\sigma}^2(\boldsymbol{\theta}))$, where $\tilde{\sigma}^2(\boldsymbol{\theta})$ is the maximum-likelihood estimate of σ^2 for fixed $\boldsymbol{\theta}$. Under the assumption of normal errors, $M(\boldsymbol{\theta})$ is given by (2.62) and the confidence region becomes

$$\{\boldsymbol{\theta}: n[\log S(\boldsymbol{\theta}) - \log S(\hat{\boldsymbol{\theta}})] \le \chi_p^2(\alpha)\}.$$

The above test procedures are discussed further in Section 5.9.1.

5.3 LINEAR HYPOTHESES

Suppose we wish to test the hypothesis $H_0:\boldsymbol{\theta}=\boldsymbol{\theta}_0$ versus $H_1:\boldsymbol{\theta}\neq\boldsymbol{\theta}_0$. Two asymptotically equivalent methods are available to us. The first method uses (5.9): we simply calculate

$$F_1 = (\hat{\boldsymbol{\theta}} - \boldsymbol{\theta}_0)'\hat{\mathbf{C}}(\hat{\boldsymbol{\theta}} - \boldsymbol{\theta}_0)/ps^2, \tag{5.17}$$

which is approximately distributed as $F_{p,n-p}$ when H_0 is true. We reject H_0 at the α level of significance if $F_1 > F^{\alpha}_{p,n-p}$. The second method uses the likelihood-ratio statistic, which, from (2.47) and (2.46) [leading to $\tilde{\sigma}_0^2 = S(\boldsymbol{\theta}_0)/n$], is given by

$$\begin{aligned}
\ell &= \frac{\sup_{\sigma^2} p(\mathbf{y};\boldsymbol{\theta}_0,\sigma^2)}{\sup_{\sigma^2,\theta} p(\mathbf{y};\boldsymbol{\theta},\sigma^2)} \\
&= \frac{(2\pi\tilde{\sigma}_0^2)^{-n/2}e^{-n/2}}{(2\pi\hat{\sigma}^2)^{-n/2}e^{-n/2}} \\
&= (\hat{\sigma}^2/\tilde{\sigma}_0^2)^{n/2} \\
&= [S(\hat{\boldsymbol{\theta}})/S(\boldsymbol{\theta}_0)]^{n/2}.
\end{aligned} \tag{5.18}$$

We reject H_0 if ℓ is too small or, equivalently, if

$$F_2 = (\ell^{-2/n} - 1)\frac{n-p}{p} = \frac{S(\boldsymbol{\theta}_0) - S(\hat{\boldsymbol{\theta}})}{S(\hat{\boldsymbol{\theta}})}\frac{n-p}{p} \tag{5.19}$$

is too large. From (2.24) we see that F_2 is approximately $F_{p,n-p}$ when H_0 is true. We therefore reject H_0 if $\boldsymbol{\theta}_0$ lies outside the region (5.12). The power of this test is discussed by Gallant [1975a, b].

Suppose that we now partition $\boldsymbol{\theta} = (\boldsymbol{\theta}_1', \boldsymbol{\theta}_2')'$, where $\boldsymbol{\theta}_2$ is $p_2 \times 1$, and we wish to test $H_{02}:\boldsymbol{\theta}_2 = \boldsymbol{\theta}_{02}$. Again both procedures can be used. Partitioning $\hat{\boldsymbol{\theta}} = (\hat{\boldsymbol{\theta}}_1', \hat{\boldsymbol{\theta}}_2')'$ and

$$\mathbf{C}^{-1} = \begin{pmatrix} \hat{\mathbf{C}}_{11} & \hat{\mathbf{C}}_{12} \\ \hat{\mathbf{C}}_{21} & \hat{\mathbf{C}}_{22} \end{pmatrix}^{-1} = \begin{pmatrix} \hat{\mathbf{C}}^{11} & \hat{\mathbf{C}}^{12} \\ \hat{\mathbf{C}}^{21} & \hat{\mathbf{C}}^{22} \end{pmatrix},$$

we note from (5.2) that $\hat{\boldsymbol{\theta}}_2 \sim N_{p_2}(\boldsymbol{\theta}_2, \sigma^2 \hat{\mathbf{C}}^{22})$, approximately. Partitioning $\hat{\mathbf{F}}_{\cdot} = (\hat{\mathbf{F}}_1, \hat{\mathbf{F}}_2)$, where $\hat{\mathbf{F}}_2$ is $n \times p_2$, we have (c.f. Appendix D3.1)

$$\begin{aligned}(\hat{\mathbf{C}}^{22})^{-1} &= \hat{\mathbf{C}}_{22} - \hat{\mathbf{C}}_{21}\hat{\mathbf{C}}_{11}^{-1}\hat{\mathbf{C}}_{12} \\ &= \hat{\mathbf{F}}_2'\hat{\mathbf{F}}_2 - \hat{\mathbf{F}}_2'\hat{\mathbf{F}}_1(\hat{\mathbf{F}}_1'\hat{\mathbf{F}}_1)^{-1}\hat{\mathbf{F}}_1'\hat{\mathbf{F}}_2 \\ &= \hat{\mathbf{F}}_2'(\mathbf{I}_n - \hat{\mathbf{P}}_{\mathbf{F}_1})\hat{\mathbf{F}}_2. \end{aligned} \tag{5.20}$$

Then, analogously to (5.17), we have the test statistic (cf. Appendix D3.2)

$$\tilde{F}_1 = (\hat{\boldsymbol{\theta}}_2 - \boldsymbol{\theta}_{02})'(\hat{\mathbf{C}}^{22})^{-1}(\hat{\boldsymbol{\theta}}_2 - \boldsymbol{\theta}_{02})/p_2 s^2. \tag{5.21}$$

This statistic is approximately distributed as $F_{p_2, n-p}$ when H_0 is true.

To apply the likelihood-ratio procedure we need to minimize $S(\boldsymbol{\theta})$ [$= S(\boldsymbol{\theta}_1, \boldsymbol{\theta}_2)$, say] subject to $\boldsymbol{\theta}_2 = \boldsymbol{\theta}_{02}$. Let $S(\tilde{\boldsymbol{\theta}}) = S[\tilde{\boldsymbol{\theta}}_1(\boldsymbol{\theta}_{02}), \boldsymbol{\theta}_{02}]$ be the minimum value of $S(\boldsymbol{\theta})$ under H_{02}. Then, arguing as in (5.18), the likelihood-ratio statistic is

$$\ell = [S(\hat{\boldsymbol{\theta}})/S(\tilde{\boldsymbol{\theta}})]^{n/2}. \tag{5.22}$$

Using (5.19), we arrive at the test statistic

$$\tilde{F}_2 = \frac{S(\tilde{\boldsymbol{\theta}}) - S(\hat{\boldsymbol{\theta}})}{S(\hat{\boldsymbol{\theta}})} \frac{n-p}{p_2}, \tag{5.23}$$

which is also approximately distributed as $F_{p_2, n-p}$ when H_0 is true.

A third method of constructing test statistics is also available, called the Lagrange-multiplier method (cf. Section 5.9.1). It is available in two versions, giving the test statistics

$$\tilde{F}_3 = \frac{\{\mathbf{y} - \mathbf{f}(\tilde{\boldsymbol{\theta}})\}'[\mathbf{F}_{\cdot}(\mathbf{F}_{\cdot}'\mathbf{F}_{\cdot})^{-1}\mathbf{F}_{\cdot}']_{\tilde{\boldsymbol{\theta}}}\{\mathbf{y} - \mathbf{f}(\tilde{\boldsymbol{\theta}})\}/p_2}{S(\hat{\boldsymbol{\theta}})/(n-p)}$$

and

$$\begin{aligned}\mathrm{LM} &= \frac{\{\mathbf{y} - \mathbf{f}(\tilde{\boldsymbol{\theta}})\}'[\mathbf{F}_{\cdot}(\mathbf{F}_{\cdot}'\mathbf{F}_{\cdot})^{-1}\mathbf{F}_{\cdot}']_{\tilde{\boldsymbol{\theta}}}\{\mathbf{y} - \mathbf{f}(\tilde{\boldsymbol{\theta}})\}}{S(\tilde{\boldsymbol{\theta}})/n} \\ &= \frac{\tilde{\boldsymbol{\varepsilon}}'\tilde{\mathbf{P}}_{\mathbf{F}}\tilde{\boldsymbol{\varepsilon}}}{\tilde{\boldsymbol{\varepsilon}}'\tilde{\boldsymbol{\varepsilon}}/n},\end{aligned}$$

where $\tilde{\boldsymbol{\varepsilon}} = \mathbf{y} - \mathbf{f}(\tilde{\boldsymbol{\theta}})$. Here $\tilde{F}_3$ is asymptotically equivalent to $\tilde{F}_2$, so that $\tilde{F}_3$ is approximately distributed as $F_{p_2, n-p}$ when H_0 is true. The statistic LM is the usual form of the Lagrange-multiplier test and is asymptotically $\chi^2_{p_2}$ when H_0 is true. It has the advantage that only the restricted estimate $\tilde{\boldsymbol{\theta}}$ is needed for its computation. Using projection matrices (cf. Appendix A11), it can be shown (e.g. Gallant [1987: p. 87]) that

$$R = \frac{n^{-1}\mathrm{LM}}{1 - n^{-1}\mathrm{LM}} \frac{n-p}{p_2} \sim F_{p_2, n-p} \tag{5.24}$$

approximately when H_0 is true. Thus we can use LM with a critical value of

$$d_\alpha = \frac{np_2 F_\alpha}{n - p + p_2 F_\alpha}, \qquad F_\alpha = F^\alpha_{p_2, n-p}. \tag{5.25}$$

Although the statistic $\tilde{F}_3$ requires both the restricted and unrestricted estimators $\tilde{\boldsymbol{\theta}}$ and $\hat{\boldsymbol{\theta}}$, Gallant shows that it has a simpler distribution theory and slightly better power than LM. He also notes that the numerator of $\tilde{F}_3$ is quite a good approximation for the numerator of $\tilde{F}_2$, so that the two tests are very similar. However, $\tilde{F}_2$ is simpler to compute. Gallant [1987: Chapter 1, Section 5] derived asymptotic expressions for the power functions of all four test statistics LM, $\tilde{F}_k$ $(k = 1, 2, 3)$ and showed that $\tilde{F}_2$ has better power than $\tilde{F}_3$, though in many cases there is little difference between $\tilde{F}_2$, $\tilde{F}_3$, and LM. The statistic $\tilde{F}_1$ is a special case of the so-called Wald test (5.93), and Gallant [1987: p. 83] demonstrated that its F-approximation can be seriously affected by severe curvature, whether it be intrinsic or, more commonly, parameter-effects curvature. However, the likelihood-ratio statistic $\tilde{F}_2$ holds up well in this regard, as we noted in Section 5.2.2 when discussing the full parameter set $\boldsymbol{\theta}$. This statistic is only affected by intrinsic curvature.

Hamilton [1986] considered a further "efficient score" statistic

$$\tilde{F}_4 = \frac{\tilde{\boldsymbol{\varepsilon}}'(\tilde{\mathbf{P}}_{\mathbf{F}} - \tilde{\mathbf{P}}_{\mathbf{F}_1})\tilde{\boldsymbol{\varepsilon}}/p_2}{\tilde{\boldsymbol{\varepsilon}}'(\mathbf{I}_n - \tilde{\mathbf{P}}_{\mathbf{F}})\tilde{\boldsymbol{\varepsilon}}/(n-p)},$$

where $\tilde{\mathbf{P}}_{\mathbf{F}_1} = [\mathbf{F}_1(\mathbf{F}_1'\mathbf{F}_1)^{-1}\mathbf{F}_1']_{\tilde{\boldsymbol{\theta}}}$ and $\mathbf{F}_1$ consists of the first p_1 $(= p - p_2)$ columns of $\mathbf{F}_{\boldsymbol{\cdot}}$ (corresponding to $\partial \mathbf{f}/\partial \boldsymbol{\theta}_1'$). When H_0 is true, it can be shown that $\tilde{F}_4 \to R$ of (5.24) as $n \to \infty$. Hence the null distribution of $\tilde{F}_4$ is approximately $F_{p_2, n-p}$. We note, in passing, that the numerator of $\tilde{F}_4$ can be simplified. As $\tilde{\boldsymbol{\theta}}_1(\boldsymbol{\theta}_{02})$ satisfies the normal (subset) equations

$$\begin{aligned} \mathbf{0} &= \frac{\partial S(\boldsymbol{\theta}_1, \boldsymbol{\theta}_{02})}{\partial \boldsymbol{\theta}_1} \\ &= -2\mathbf{F}_1'\{\mathbf{y} - \mathbf{f}(\boldsymbol{\theta}_1, \boldsymbol{\theta}_{02})\}, \end{aligned}$$

we have that $\tilde{\mathbf{P}}_{\mathbf{F}_1}\tilde{\boldsymbol{\varepsilon}} = \mathbf{0}$, and the numerator of $\tilde{F}_4$ reduces to $\tilde{\boldsymbol{\varepsilon}}'\tilde{\mathbf{P}}_{\mathbf{F}}\tilde{\boldsymbol{\varepsilon}}/p_2$.

Which of the above five statistics should be used for testing H_{02}? The statistic $\tilde{F}_1$ only requires the unconstrained least-squares estimate $\hat{\boldsymbol{\theta}}$, and the inverses of $\hat{\mathbf{C}}$ and $\hat{\mathbf{C}}^{22}$. Usually $\hat{\mathbf{C}}^{-1}$ will be a by-product of the least-squares program, and if p_2 is small (say 1 or 2), then $\hat{\mathbf{C}}^{22}$ will be trivial to invert. However, although $\tilde{F}_1$ has these desirable computational features, it can be seriously affected by any parameter-effects curvature, and its indiscriminate use is not recommended.

The other four statistics are fairly similar, though, as already observed, $\tilde{F}_2$ is slightly more powerful. Since $\mathbf{f}(\boldsymbol{\theta})$, $S(\boldsymbol{\theta})$, and $\mathbf{P}_F$ are invariant under one-to-one transformations of $\boldsymbol{\theta}$ [cf. Appendix B5, Equation (1)], we can expect these statistics to be unaffected by any parameter-effects curvature: only the effect of intrinsic curvature needs to be considered. Recently Hamilton and Wiens [1987] studied second-order approximations for the distributions of $\tilde{F}_2$ and $\tilde{F}_4$ and developed correction terms which are functions of the intrinsic-curvature arrays. Assuming normality of the ε_i, they showed that as $\sigma^2 \to 0$ (rather than $n \to \infty$), the limiting null distribution of $\tilde{F}_k$ $(k=2,4)$ is, to order σ^3, $c_k F_{p_2, n-p}$, where

$$c_k = (1 - \gamma_k \sigma^2) \qquad (k = 2,4).$$

They gave the following formula for γ_k (with the notational change that their $\boldsymbol{\theta}_1$ becomes our $\boldsymbol{\theta}_2$):

$$\gamma_k = \frac{\alpha_{k2}}{n-p} - \frac{\alpha_{k1}}{p_2}, \tag{5.26}$$

where

$$\begin{aligned}
\alpha_{21} &= \alpha_{41} + \alpha_{42} - \alpha_{22} \\
\alpha_{22} &= -\tfrac{1}{2}\sum_{i=1}^{n-p} \operatorname{tr}(\mathbf{A}_{i..}^N)^2 + \tfrac{1}{4}\sum_{i=1}^{n-p}\{\operatorname{tr}\mathbf{A}_{i..}^N\}^2, \\
\alpha_{41} &= -\tfrac{1}{2}\sum_{i=1}^{p_2} \operatorname{tr}(\mathbf{A}_{0i..}^N)^2 + \tfrac{1}{4}\sum_{i=1}^{p_2}\{\operatorname{tr}\mathbf{A}_{0i..}^N\}^2, \\
\alpha_{42} &= -\tfrac{1}{2}\sum_{i=p_2+1}^{n-p_1} \operatorname{tr}(\mathbf{A}_{0i..}^N)^2 + \tfrac{1}{4}\sum_{i=p_2+1}^{n-p_1}\{\operatorname{tr}\mathbf{A}_{0i..}^N\}^2.
\end{aligned} \tag{5.27}$$

Here the $p \times p$ matrix $\mathbf{A}_{i..}^N$ $(i = 1,2,\dots,n-p)$ is the ith face of the intrinsic curvature array $A_{..}^N$ for the full model [cf. (4.57)], and the $p_1 \times p_1$ matrix $\mathbf{A}_{0i..}^N$ $(i = 1,2,\dots,n-p_1)$ is the ith face of the intrinsic curvature array for the restricted model obtained by fixing $\boldsymbol{\theta}_2 = \boldsymbol{\theta}_{02}$ and working with $\boldsymbol{\theta}_1$. To evaluate γ_k we use $\boldsymbol{\theta}_2 = \boldsymbol{\theta}_{02}$ and $\boldsymbol{\theta}_1 = \tilde{\boldsymbol{\theta}}_1(\boldsymbol{\theta}_{02})$, the restricted least-squares estimator of $\boldsymbol{\theta}$. In c_k, σ^2 is estimated by $s^2 = S(\hat{\boldsymbol{\theta}})/(n-p)$. The computations can be simplified by an appropriate choice of basis for the sample space (Hamilton and Wiens [1986]).

Hamilton and Wiens [1987] summarized the results of computing the correction factors for 89 data sets corresponding to 37 different models with up to 5 parameters. The likelihood-ratio factors c_2 were all between 0.6499 and 1.0612, while the c_4's were all between 0.6499 and 1.0021. This suggests that for σ^2 small enough, the standard procedure of comparing $\tilde{\mathbf{F}}_2$ with F_α may usually be conservative. The study also showed that c_2 was typically further from 1 than c_4, which was generally less than 1. Furthermore, the smallest subsets tended to have the most extreme factors for a given data set.

The authors noted that when the full parameter set is considered, i.e. $\boldsymbol{\theta}_2 = \boldsymbol{\theta}$ and $p_2 = p$, then α_{41} and α_{42} disappear and

$$\gamma_2 = \frac{n\alpha_{22}}{p(n-p)},$$

a result obtained by Johansen [1983]. For the 89 data sets they found that c_2 was usually closer to 1 for the full set of $\boldsymbol{\theta}$ than for subsets.

Unfortunately the above methods for testing H_{02} are not appropriate in every case. In Section 3.4 we saw that some models cause problems because of ill-conditioning. As pointed out by Gallant [1977b], similar problems can arise in hypothesis testing. For example, consider testing $H_{02}: \tau = 0$ for the model (5.1) with

$$f(\mathbf{x}; \boldsymbol{\theta}) = g(\mathbf{x}; \boldsymbol{\phi}) + \tau h(\mathbf{x}; \boldsymbol{\gamma}) \tag{5.28}$$

and $\boldsymbol{\theta} = (\boldsymbol{\phi}', \boldsymbol{\gamma}', \tau)'$. We note that the two test statistics $\tilde{F}_1$ and $\tilde{F}_2$ given above for testing H_{02} require the computation of the unconstrained least-squares estimates $\hat{\boldsymbol{\theta}}$. Also certain regularity assumptions (see Chapter 12) are needed for $\hat{\boldsymbol{\theta}}$ to have appropriate asymptotic properties so that the asymptotic null distributions of the test statistics described above are valid. However, if H_{02} is true or nearly so, $\boldsymbol{\gamma}$ can vary without having much effect on the model (5.28). This will cause a convergence problem in using iterative procedures to find $\hat{\boldsymbol{\theta}}$. Gallant [1977b] demonstrated this problem by simulating the model

$$f(\mathbf{x}; \boldsymbol{\theta}) = \theta_1 x_1 + \theta_2 x_2 + \theta_4 e^{\theta_3 x_3}$$

with $g(\mathbf{x}; \boldsymbol{\phi}) = \theta_1 x_1 + \theta_2 x_2$, $h(\mathbf{x}; \gamma) = \exp(\theta_3 x_3)$, and $\tau = \theta_4$. Using simulated data he showed that as θ_4 became smaller in absolute value, the sum-of-squares surface $\sum_i [y_i - f(\mathbf{x}_i; \boldsymbol{\theta})]^2$ became very flat with respect to θ_3. He also noted that two of the standard regularity assumptions needed for the convergence in probability and the asymptotic normality of $\hat{\boldsymbol{\theta}}$ were violated. Therefore, as $\tilde{F}_1$ and $\tilde{F}_2$ are no longer appropriate test statistics, he proposed using an extension of the "exact" methods of Section 5.10. This consists of testing $H_{02}: \boldsymbol{\delta} = \mathbf{0}$ for the model

$$y_i = g(\mathbf{x}_i; \boldsymbol{\phi}) + \mathbf{z}_i' \boldsymbol{\delta} + \varepsilon_i,$$

where $\mathbf{z}_i$ is a $k \times 1$ vector of regressor variables which do not depend on any unknown parameters. This model does not have the same problems as (5.28), so that, provided certain regularity conditions are satisfied (c.f. the Appendix of Gallant [1977b]), $\tilde{F}_2$ can be used for testing $\boldsymbol{\delta} = \mathbf{0}$. Some methods for finding a suitable number of additional regressors $\mathbf{z}$ are given by Gallant [1977b; 1987:

139], and the reader is referred to his work for further details and worked examples. An application is described in Section 9.3.4.

In hypothesis testing or model building, one frequently wishes to fit submodels of a general "umbrella" model. These submodels are often obtained by assigning values to some of the elements of $\boldsymbol{\theta}$ (e.g. $\boldsymbol{\theta}_2 = \boldsymbol{\theta}_{02}$, as discussed above), equating elements of $\boldsymbol{\theta}$ (e.g. several models with some common rate constants combined into a single model), and imposing general linear equality constraints. To test such a submodel we consider the general linear hypothesis $H:\mathbf{A}\boldsymbol{\theta} = \mathbf{c}$, where $\mathbf{A}$ is $q \times p$ of rank q. In theory, the above test procedures are readily adapted to handle H_0. We simply choose a $(p-q) \times p$ matrix $\mathbf{K}$ such that the $p \times p$ matrix

$$\mathbf{B} = \begin{pmatrix} \mathbf{K} \\ \mathbf{A} \end{pmatrix}$$

is nonsingular. If $\boldsymbol{\beta} = \mathbf{B}\boldsymbol{\theta}$, then $\mathbf{f}(\boldsymbol{\theta}) = \mathbf{f}(\mathbf{B}^{-1}\boldsymbol{\beta})$ [$= \mathbf{g}(\boldsymbol{\beta})$, say], and testing H_0 is equivalent to testing $\boldsymbol{\beta}_2 = \mathbf{A}\boldsymbol{\theta} = \mathbf{c}$. Thus, in theory, H_0 can be tested by reparametrizing the model so that $E[\mathbf{y}] = \mathbf{g}(\boldsymbol{\beta})$ and testing $H_{02}:\boldsymbol{\beta}_2 = \mathbf{c}$, where $\boldsymbol{\beta}_2$ consists of the last q elements of $\boldsymbol{\beta}$. However, in practice, more appropriate computational methods are available particularly when several such hypotheses (submodels) are to be considered. Although each submodel can be fitted separately, this would require modifying the computer subroutines which calculate function values and first and second derivatives every time a new model is fitted. We describe in Appendix E a standard method for handling linear equality constraints in which a single piece of code is written to calculate function and derivative values for the umbrella model. Different submodels are then fitted simply by specifying the relevant linear constraint matrix. Variance-covariance matrices for constrained parameter estimates can also be obtained this way.

5.4 CONFIDENCE REGIONS FOR PARAMETER SUBSETS

If $\boldsymbol{\theta}$ is partitioned in the form $\boldsymbol{\theta} = (\boldsymbol{\theta}_1', \boldsymbol{\theta}_2')'$, where $\boldsymbol{\theta}_2$ is $p_2 \times 1$, then a number of approximate confidence regions are available for $\boldsymbol{\theta}_2$. We can use the fact that a $100(1-\alpha)\%$ confidence region for $\boldsymbol{\theta}_2$ is the set of all $\boldsymbol{\theta}_{02}$ such that $H_0:\boldsymbol{\theta}_2 = \boldsymbol{\theta}_{02}$ is not rejected at the α level of significance. For example, from (5.20) and (5.21) we have the approximate $100(1-\alpha)\%$ region

$$\{\boldsymbol{\theta}_2 : (\boldsymbol{\theta}_2 - \hat{\boldsymbol{\theta}}_2)'\hat{\mathbf{F}}_2'(\mathbf{I}_n - \hat{\mathbf{P}}_{\mathbf{F}_1})\hat{\mathbf{F}}_2(\boldsymbol{\theta}_2 - \hat{\boldsymbol{\theta}}_2) \le p_2 s^2 F^{\alpha}_{p_2, n-p}\}, \tag{5.29}$$

where $s^2 = S(\hat{\boldsymbol{\theta}})/(n-p)$ and $\hat{\mathbf{P}}_{\mathbf{F}_1} = \hat{\mathbf{F}}_1(\hat{\mathbf{F}}_1'\hat{\mathbf{F}}_1)^{-1}\hat{\mathbf{F}}_1'$. This region is based on an appropriate subset tangent-plane approximation [see Equations (4.115) to (4.118)] and, as mentioned in the previous section, can be affected by the usual intrinsic and parameter-effects curvatures (see also Section 5.8.5). However, as noted by Hamilton [1986: 59] it has the advantage of having a simple shape which can be readily conceptualized and displayed. Unfortunately, this is not the case with the confidence region based on the corrected version of the likelihood-ratio statistic (5.23), namely

$$\{\boldsymbol{\theta}_2 : S[\hat{\boldsymbol{\theta}}_1(\boldsymbol{\theta}_2), \boldsymbol{\theta}_2] - S(\hat{\boldsymbol{\theta}}) \leq p_2 s^2 (1 - \hat{\gamma}_2 s^2) F^{\alpha}_{p_2, n-p}\}, \tag{5.30}$$

where $\tilde{\boldsymbol{\theta}}_1(\boldsymbol{\theta}_2)$, which minimizes $S(\boldsymbol{\theta}_1, \boldsymbol{\theta}_2)$ for specified $\boldsymbol{\theta}_2$, has to be computed for each value of $\boldsymbol{\theta}_2$. The scale factor $\hat{\gamma}_2$ is γ_2 of (5.26) evaluated at the unconstrained least-squares estimate $\hat{\boldsymbol{\theta}}$ (Hamilton and Wiens [1987]). When the intrinsic curvature is small, which is usually the case, then $\hat{\gamma}_2$ can be ignored. Although it is straightforward to see if a particular $\boldsymbol{\theta}_2$ lies in the region (5.30), it can be expensive to compute the entire region, though the methods of Appendix E can be used here. Also the region can be difficult to display if $p_2 > 2$. Similar comments apply to the LM method of (5.24) or $\tilde{F}_4$ when they are adapted to constructing confidence regions along the lines of (5.30).

If $\boldsymbol{\theta}_2$ is a single element, say θ_r, then (5.30) takes the form

$$\{\theta_r : L(\theta_r) \leq F_\alpha = F^{\alpha}_{1, n-p}\}.$$

This set of points can be unbounded or, if L has local minima, can consist of several disjoint intervals. However, such occurrences are infrequent, though, as Gallant [1987: 112] notes, they are more likely with LM regions. Gallant [1987: 107] describes a simple quadratic interpolation method for solving $L(z) = F_\alpha$. Here L is approximated by a quadratic q which has the same value as L at three trial values of z. Solving $q(z) = F_\alpha$ gives a value of z, which can be used for locating three more trial values if necessary, and so on.

If we are interested in confidence intervals for all the θ_r, then we can use the Bonferroni method [c.f. (5.7)] and replace α by α/p in the above discussion.

In conclusion we note that there is another method for constructing a confidence region for $\boldsymbol{\theta}$ which, given normal errors ε_i, has an exact confidence level of $100(1-\alpha)\%$. This method is discussed later (Section 5.10). Unfortunately it cannot be used for handling subsets of $\boldsymbol{\theta}$ except in one special case (Section 5.10.2).

5.5 LACK OF FIT

Once a nonlinear model has been chosen and the unknown parameters estimated, the question of goodness of fit of the model to the data arises. Usually one would look at the residuals $y_i - f(\mathbf{x}_i; \hat{\boldsymbol{\theta}})$ and carry out various plots as suggested in Section 4.6. If the intrinsic curvature is negligible, as is often the case, then the usual residual plots carried out for linear models are also appropriate here. However, if there is substantial intrinsic curvature, then the projected residuals of Section 4.6.4 can be used.

When the model is replicated, the adequacy of the model can be tested directly as in Section 2.1.5. If there are no replicates, then there are a number of test procedures available, generally based on the idea of near-replicates. A number of these, which are also suitable for nonlinear models, are given by Neill and Johnson [1984].

A problem which sometimes arises is that of determining whether the absolute error or the relative error is homogeneous. An approximate test, utilizing an idea of Goldfeld and Quandt [1965], was proposed by Endrenyi and Kwong [1981] in the context of kinetic models. They first fit the model by (unweighted) least squares and order the residuals, say $\hat{\varepsilon}_{(1)}, \hat{\varepsilon}_{(2)}, \ldots, \hat{\varepsilon}_{(n)}$, according to the rank order of the corresponding fitted values $\hat{y}_i = f(\mathbf{x}_i; \hat{\boldsymbol{\theta}})$, say $\hat{y}_{(1)} > \hat{y}_{(2)} > \cdots > \hat{y}_{(n)}$. Then they compute

$$F_{(k)} = \frac{\hat{\varepsilon}^2_{(n)} + \hat{\varepsilon}^2_{(n-1)} + \cdots + \hat{\varepsilon}^2_{(n-k+1)}}{\hat{\varepsilon}^2_{(1)} + \hat{\varepsilon}^2_{(2)} + \cdots + \hat{\varepsilon}^2_{(k)}},$$

the ratio of the last k squared residuals to the first k squared residuals. Under the null hypothesis H_0 of homogeneous absolute error, namely $\text{var}[\varepsilon_i] = \sigma^2$, the ordering of the residuals will be random. Therefore, regarding the residuals as being approximately i.i.d. with mean zero and a normal distribution, $F_{(k)}$ is approximately distributed as $F_{k,k}$ where H_0 is true. If however $\text{var}[\varepsilon_i] \propto [f(\mathbf{x}_i; \boldsymbol{\theta})]^{-2}$, as under the alternative hypothesis of homogeneous relative error, then the first k squared residuals will tend to be smaller than the last k, and $F_{(k)}$ will be much greater than one. We therefore reject H_0 if $F_{(k)}$ is too large. Endrenyi and Kwong [1981] suggested choosing k equal to the integer part of $(n+3)/4$. Their examples seem to indicate that the method can be applied when n is quite small, say $n = 5$. The test, however, will be unsatisfactory if the intrinsic curvature is too large, as the residuals will be biased away from zero (Section 4.6.3).

*5.6 REPLICATED MODELS

The replicated model of Section 2.1.5 in which there are J_i replications for each design point $\mathbf{x}_i$ is just a special case of the usual model discussed above. The only difference is that groups of observations have the same value of $\mathbf{x}$, and the general theory applies irrespective of whether some of the $\mathbf{x}_i$'s are equal or not. However, with replication we can use the approximate goodness-of-fit test (2.43) to test the fit of the nonlinear model to the data. This test is based on the "pure error" sum of squares (2.40), which provides an estimate of σ^2 that is valid irrespective of whether the model is true or not. We shall see below that we can take advantage of the replication structure and compute estimators and confidence intervals more efficiently.

As we have already noted, the equality of any x's does not affect the distribution theory for the least-squares estimate $\hat{\boldsymbol{\theta}}$ obtained by minimizing

$$S(\boldsymbol{\theta}) = \sum_i \sum_j [y_{ij} - f(\mathbf{x}_i;\boldsymbol{\theta})]^2$$
$$= \sum_i \sum_j (y_{ij} - \bar{y}_{i\cdot})^2 + \sum_i J_i [\bar{y}_{i\cdot} - f(\mathbf{x}_i;\boldsymbol{\theta})]^2. \quad (5.31)$$

If

$$(N-p)s_R^2 = S(\hat{\boldsymbol{\theta}}),$$

where $N = \sum_i J_i$ is the total number of observations, then from Theorem 2.1 in Section 2.1.2 we have asymptotically

a. $(N-p)s_R^2/\sigma^2 \sim \chi^2_{N-p}$, and
b. $\hat{\boldsymbol{\theta}}$ is statistically independent of s_R^2.

By differentiating (5.31) we see that the normal equations for $\hat{\boldsymbol{\theta}}$ [c.f. (2.42)] can be expressed in matrix form

$$\hat{\mathbf{F}}_{\cdot}'\boldsymbol{\Delta}[\bar{\mathbf{y}} - \mathbf{f}(\hat{\boldsymbol{\theta}})] = \mathbf{0}, \quad (5.32)$$

where $\bar{\mathbf{y}} = (\bar{y}_{1\cdot}, \bar{y}_{2\cdot}, \ldots, \bar{y}_{n\cdot})' \sim N_n(\mathbf{f}(\boldsymbol{\theta}), \sigma^2\boldsymbol{\Delta}^{-1})$, and $\boldsymbol{\Delta} = \mathrm{diag}(J_1, J_2, \ldots, J_n)$. Using a Taylor expression for $\boldsymbol{\theta}$ about $\hat{\boldsymbol{\theta}}$, (5.32) becomes

$$\hat{\mathbf{F}}_{\cdot}'\boldsymbol{\Delta}[\bar{\mathbf{y}} - \mathbf{f}(\boldsymbol{\theta}) - \hat{\mathbf{F}}_{\cdot}(\hat{\boldsymbol{\theta}} - \boldsymbol{\theta})] \approx \mathbf{0},$$

or

$$\hat{\boldsymbol{\theta}} - \boldsymbol{\theta} \approx (\hat{\mathbf{F}}_{\cdot}'\boldsymbol{\Delta}\hat{\mathbf{F}}_{\cdot})^{-1}\hat{\mathbf{F}}_{\cdot}'\,\boldsymbol{\Delta}[\bar{\mathbf{y}} - \mathbf{f}(\boldsymbol{\theta})].$$

This means that, asymptotically,

$$\hat{\boldsymbol{\theta}} - \boldsymbol{\theta} \sim N_p(\mathbf{0}, \sigma^2(\mathbf{F}_{\cdot}'\boldsymbol{\Delta}\mathbf{F}_{\cdot})^{-1}),$$

and the above minimization procedure is equivalent to weighted least squares on the $\bar{y}_{i\cdot}$.

We note that with the replicated model formulation, we need only store the mean response $\bar{y}_{i\cdot}$ and $\sum_j (y_{ij} - \bar{y}_{i\cdot})^2$ (or equivalently the sample variance) for the set of responses at each design point $\mathbf{x}_i$. The least-squares estimate $\hat{\boldsymbol{\theta}}$ and the matrix $\mathbf{C}$ of (5.2) [the $N \times p$ matrix $\mathbf{F}_{\cdot}(\boldsymbol{\theta})$ described there contains n blocks with the ith block consisting of J_i identical rows] can be obtained using a weighted least-squares program on the vector $\bar{\mathbf{y}}$. Also $S(\hat{\boldsymbol{\theta}})$ is obtained by simply adding $\sum\sum(y_{ij} - \bar{y}_{i\cdot})^2$ to the weighted residual sum of squares produced by this program [c.f. (5.31)]. Thus instead of working with an N-dimensional response vector $\mathbf{y}$ and an $N \times p$ derivative matrix $\mathbf{F}_{\cdot}$, computations are performed using an n-dimensional response and an $n \times p$ matrix $\mathbf{F}_{\cdot}$.

We now proceed, as in Section 5.1 above, to construct confidence intervals. For example, using s_R^2 as an estimate of σ^2, an approximate $100(1-\alpha)\%$ confidence interval for $\mathbf{a}'\boldsymbol{\theta}$ is

$$\mathbf{a}'\hat{\boldsymbol{\theta}} \pm t_{N-p}^{\alpha/2} s_R [\mathbf{a}'(\hat{\mathbf{F}}_{\bullet}'\boldsymbol{\Delta}\hat{\mathbf{F}}_{\bullet})^{-1}\mathbf{a}]^{1/2}.$$

Alternatively, we can use the unbiased estimate

$$s_e^2 = \frac{1}{N-n}\sum_i \sum_j (y_{ij} - \bar{y}_{i\cdot})^2$$

of (2.40), which does not depend on the validity of the model $E[y_{ij}] = f(\mathbf{x}_i; \boldsymbol{\theta})$. Under the normality assumptions, s_e^2 is independent of $\bar{\mathbf{y}}$ and therefore of $\hat{\boldsymbol{\theta}}$, as the solution $\hat{\boldsymbol{\theta}}$ of (5.32) is a function of $\bar{\mathbf{y}}$. Hence, from (2.41), an approximate $100(1-\alpha)\%$ confidence interval for $\mathbf{a}'\boldsymbol{\theta}$ is

$$\mathbf{a}'\hat{\boldsymbol{\theta}} \pm t_{N-n}^{\alpha/2} s_e [\mathbf{a}'(\hat{\mathbf{F}}_{\bullet}'\boldsymbol{\Delta}\hat{\mathbf{F}}_{\bullet})^{-1}\mathbf{a}]^{1/2}.$$

Unfortunately, as the replicated data model is essentially a special case of the unreplicated data model, we can expect the above region based on the asymptotic distribution of $\hat{\boldsymbol{\theta}}$ to be affected by both parameter-effects and intrinsic curvature. However, these curvatures are smaller with replicated data (c.f. Section 4.2.5).

An alternative approach is to use likelihood-ratio methods, which are only affected by intrinsic curvature. Thus, analogous to (5.12), we have

$$\{\boldsymbol{\theta} : S(\boldsymbol{\theta}) - S(\hat{\boldsymbol{\theta}}) \le p s_e^2 F_{p,N-n}^{\alpha}\}.$$

If we are interested in just the subset $\boldsymbol{\theta}_2$, then we can readily extend the theory in Section 5.4 to the case of replication. Corresponding to (5.30) we have

$$\{\boldsymbol{\theta}_2 : S(\tilde{\boldsymbol{\theta}}_1(\boldsymbol{\theta}_2), \boldsymbol{\theta}_2) - S(\hat{\boldsymbol{\theta}}) \le p_2 s_e^2 F_{p_2,N-n}^{\alpha}\}.$$

Correction factors for the above two regions could no doubt be derived using the method of Hamilton and Wiens [1986].

*5.7 JACKKNIFE METHODS

Up till now we have focused our attention on the least-squares estimate $\hat{\boldsymbol{\theta}}$ for the nonlinear model (5.1), and the theory has been heavily dependent on the asymptotic normality of $\hat{\boldsymbol{\theta}}$. We now ask whether a jackknife technique may be more appropriate, as it is less dependent on underlying distribution theory. Suppose that $n = Gh$, and let $\mathbf{y}' = (y_1, y_2, \ldots, y_n)$ be partitioned in the form $\mathbf{y}' = (\mathbf{y}_1', \mathbf{y}_2', \ldots, \mathbf{y}_G')$, where each $\mathbf{y}_g$ $(g = 1, 2, \ldots, G)$ is a $h \times 1$ vector. Let $\hat{\boldsymbol{\theta}}_{(g)}$ be the least-squares estimate of $\boldsymbol{\theta}$ for the model with the data group $\mathbf{y}_g$ omitted.

Usually $\hat{\boldsymbol{\theta}}_{(g)}$ is computed iteratively using $\hat{\boldsymbol{\theta}}$ as a starting value. We now define G pseudovalues

$$\tilde{\boldsymbol{\theta}}_g = G\hat{\boldsymbol{\theta}} - (G-1)\hat{\boldsymbol{\theta}}_{(g)} \qquad (g = 1, 2, \ldots, G), \tag{5.33}$$

with sample mean

$$\bar{\boldsymbol{\theta}}_J = \frac{1}{G}\sum_{g=1}^{G} \tilde{\boldsymbol{\theta}}_g$$

and sample variance–covariance (dispersion) matrix

$$\begin{aligned} \mathbf{S}_J &= [(s_{ij})] \\ &= \frac{1}{G-1}\sum_{g=1}^{G} (\tilde{\boldsymbol{\theta}}_g - \bar{\boldsymbol{\theta}}_J)(\tilde{\boldsymbol{\theta}}_g - \bar{\boldsymbol{\theta}}_J)'. \end{aligned} \tag{5.34}$$

Then $\bar{\boldsymbol{\theta}}_J$ is the jackknife estimate of $\boldsymbol{\theta}$, and $\mathbf{S}_J/G$ is an estimate of $\mathscr{D}[\bar{\boldsymbol{\theta}}_J]$. Treating the $\tilde{\boldsymbol{\theta}}_g$ $(g = 1, 2, \ldots, G)$ as being approximately i.i.d. $N_p(\mathbf{0}, \boldsymbol{\Sigma})$, we can construct a $100(1-\alpha)\%$ confidence region for $\boldsymbol{\theta}$ using Hotelling's T^2 distribution (e.g. Seber [1984: 63]), namely

$$\left\{ \boldsymbol{\theta} : (\boldsymbol{\theta} - \bar{\boldsymbol{\theta}}_J)'\mathbf{S}_J^{-1}(\boldsymbol{\theta} - \bar{\boldsymbol{\theta}}_J) \le \frac{p}{G-p}\,\frac{G-1}{G}\,F^{\alpha}_{p,G-p} \right\}. \tag{5.35}$$

Also, a $100(1-\alpha)\%$ interval for a prechosen θ_r is given by

$$\bar{\theta}_{Jr} \pm t^{\alpha/2}_{G-p}\sqrt{\frac{s_{rr}}{G}}, \tag{5.36}$$

where $\bar{\theta}_{Jr}$ is the rth element of $\bar{\boldsymbol{\theta}}_J$.

The properties of (5.35) and (5.36) were studied by Duncan [1978] for $n = 24$ (four replicates at six points) and different values of h using simulation. [There is an omission of $(G-1)/G$ in his formula for (5.35).] Although having h as large as possible is computationally appealing, the evidence from his simulations clearly favors one-at-a-time omissions ($h = 1$). He also included the "conventional" region (5.12) and the t-intervals (5.5) in his study, which encompassed three functions:

$$f(x; \boldsymbol{\theta}) = \frac{\theta_1}{\theta_1 - \theta_2}[\exp(-\theta_2 x) - \exp(-\theta_1 x)], \tag{I}$$

$$f(x; \boldsymbol{\theta}) = 1 - \frac{1}{\theta_1 - \theta_2}[\theta_1 \exp(-\theta_2 x) - \theta_2 \exp(-\theta_1 x)], \tag{II}$$

Table 5.1 Coverages of Nominal 95% Confidence Intervals and Regions for Data Simulated from Three Models Using Two Error Structures[a,b]

Distribution	Parameters	Errors: Additive		Errors: Multiplicative	
		Conventional	Jack-knife[c]	Conventional	Jack-knife[c]
		Model I			
Normal	(θ_1, θ_2)	93.2	90.8	92.2	93.6
	$\theta_1\ (=0.2)$[d]	92.4	94.0	95.6	94.8
	$\theta_2\ (=0.5)$	89.4	91.4	93.8	93.8
Contaminated normal	(θ_1, θ_2)	94.0	94.0	88.0	93.6
	$\theta_1\ (=0.2)$	93.6	94.4	93.6	94.2
	$\theta_2\ (=0.5)$	92.4	92.8	93.6	91.8
		Model II			
Normal	(θ_1, θ_2)	94.8	67.6	80.4	60.6
	$\theta_1\ (=0.2)$	91.8	87.8	83.2	81.4
	$\theta_2\ (=0.5)$	98.6	94.0	96.0	86.0
Normal	(θ_1, θ_2)	94.6	80.0	94.4	76.8
	$\theta_1\ (=1.4)$	98.4	97.2	98.2	97.0
	$\theta_2\ (=0.4)$	95.4	94.2	93.2	93.0

Contaminated normal	(θ_1, θ_2)	96.2	72.4	83.4	65.6
	$\theta_1\ (=0.2)$	94.4	88.2	84.6	84.2
	$\theta_2\ (=0.5)$	98.8	93.6	96.2	87.8
Double exponential	(θ_1, θ_2)	96.2	72.2	83.4	66.2
	$\theta_1\ (=0.2)$	94.2	90.8	83.0	85.0
	$\theta_2\ (=0.5)$	98.4	94.8	96.8	89.4
		Model III			
Normal	(θ_1, θ_2)	95.2	93.4	68.8	89.2
	$\theta_1\ (=0.016)$	94.0	95.8	88.4	94.2
	$\theta_2\ (=0.85)$	93.0	94.2	65.6	88.8
Contaminated normal	(θ_1, θ_2)	96.4	95.4	95.4	95.2
	$\theta_1\ (=0.016)$	96.0	97.2	95.2	96.8
	$\theta_2\ (=0.85)$	94.4	95.2	95.4	95.2
Double exponential	(θ_1, θ_2)	95.4	95.2	67.0	91.6
	$\theta_1\ (=0.016)$	95.2	96.8	90.8	94.2
	$\theta_2\ (=0.85)$	95.4	95.2	66.2	93.0

[a] From Duncan [1978].
[b] The standard error of the coverages does not exceed 2.24%.
[c] One-at-a-time method ($h = 1$).
[d] These values, in brackets, are used for the simulations.

and

$$f(x;\boldsymbol{\theta}) = 1000\theta_1^{\theta_2^x}. \tag{III}$$

He also considered two models for the error structure, the additive form (5.1) and the multiplicative form

$$y_i = f(x_i;\boldsymbol{\theta})\varepsilon_i, \qquad E[\varepsilon_i] = 1,$$

and simulated three distributions for ε_i—the normal, the scale-contaminated normal, and the double exponential. The coverages for nominal 95% confidence regions and intervals are summarized in Table 5.1, where the standard error of any entry does not exceed 2.24%. For the additive case, the bivariate region (5.12) and the t-intervals gave adequate coverage for all three models and all error distributions. The one-at-a-time jackknife performed in a similar fashion for the confidence intervals but performed badly for the confidence region from model II. For the multiplicative model the conventional procedures were often inadequate, particularly in model III, where the jackknife showed some superiority. Generally one would hope to distinguish between an additive or multiplicative model by examining plots of residuals or pseudovalues (c.f. Hinkley [1978: 19]) for the multiplicative model the variance of y_i would vary with the mean. From Duncan's small experiment it would appear that the conventional methods are safer. However, the relative performance of the conventional and jackknife methods is clearly model-dependent, and further research is needed to provide adequate guidelines. One can at least conclude that, for moderate sample sizes, the jackknife method cannot be trusted for establishing joint confidence regions.

One problem with the above one-at-a-time jackknife procedure is that it requires $n+1$ nonlinear fits. Fox et al. [1980], however, proposed a linear jackknife which requires only one nonlinear fit and some subsidiary calculations associated with a linear fit. Since $h=1$, we can now write $\hat{\boldsymbol{\theta}}_{(i)}$ instead of $\hat{\boldsymbol{\theta}}_{(g)}$, and it is obtained by minimizing

$$Q_{(i)} = \sum_{j:j\neq i} [y_j - f(\mathbf{x}_j;\boldsymbol{\theta})]^2. \tag{5.37}$$

Since $\hat{\boldsymbol{\theta}}_{(i)}$ will be close to $\hat{\boldsymbol{\theta}}$, we have, in the neighborhood of $\hat{\boldsymbol{\theta}}$,

$$\begin{aligned} f(\mathbf{x}_j;\boldsymbol{\theta}) &\approx f(\mathbf{x}_j;\hat{\boldsymbol{\theta}}) + \left(\frac{\partial f(\mathbf{x}_j;\boldsymbol{\theta})}{\partial \boldsymbol{\theta}}\right)'_{\hat{\boldsymbol{\theta}}} (\boldsymbol{\theta}-\hat{\boldsymbol{\theta}}) \\ &= f(\mathbf{x}_j;\hat{\boldsymbol{\theta}}) + \hat{\mathbf{f}}_j'(\boldsymbol{\theta}-\hat{\boldsymbol{\theta}}), \end{aligned} \tag{5.38}$$

say. Hence setting $\hat{\varepsilon}_j = y_j - f(\mathbf{x}_j;\hat{\boldsymbol{\theta}})$, $\boldsymbol{\beta} = \boldsymbol{\theta} - \hat{\boldsymbol{\theta}}$, and substituting (5.38) into (5.37), gives us

$$Q_{(i)} \approx \sum_{j:j\neq i} (\hat{\varepsilon}_j - \hat{\mathbf{f}}_j'\boldsymbol{\beta})^2, \tag{5.39}$$

which may be interpreted as the sum of squares for a regression model. Let $\hat{\mathbf{F}}_{\bullet(i)}$ and $\hat{\boldsymbol{\varepsilon}}_{(i)}$ be the $(n-1) \times p$ matrix and the $(n-1) \times 1$ vector with the ith row and element, respectively, removed from $\hat{\mathbf{F}}_{\bullet}$ and $\hat{\boldsymbol{\varepsilon}}$. Then the above quadratic (5.39) is minimized at

$$\begin{aligned}\hat{\boldsymbol{\beta}}_{(i)} &= (\hat{\mathbf{F}}'_{\bullet(i)}\hat{\mathbf{F}}_{\bullet(i)})^{-1}\hat{\mathbf{F}}'_{\bullet(i)}\hat{\boldsymbol{\varepsilon}}_{(i)} \\ &= (\hat{\mathbf{F}}'_{\bullet}\hat{\mathbf{F}}_{\bullet} - \hat{\mathbf{f}}_i\hat{\mathbf{f}}'_i)^{-1}(\hat{\mathbf{F}}'_{\bullet}\hat{\boldsymbol{\varepsilon}} - \hat{\mathbf{f}}_i\hat{\varepsilon}_i) \\ &= -(\hat{\mathbf{F}}'_{\bullet}\hat{\mathbf{F}}_{\bullet} - \hat{\mathbf{f}}_i\hat{\mathbf{f}}'_i)^{-1}\hat{\mathbf{f}}_i\hat{\varepsilon}_i, \end{aligned} \tag{5.40}$$

since $\hat{\mathbf{F}}_{\bullet} = (\hat{\mathbf{f}}_1, \hat{\mathbf{f}}_2, \dots, \hat{\mathbf{f}}_n)'$ and $\hat{\mathbf{F}}'_{\bullet}\,\hat{\boldsymbol{\varepsilon}} = \mathbf{0}$, these being the normal equations for $\hat{\boldsymbol{\theta}}$ [see (2.9)]. Using the identity

$$(\mathbf{A} - \mathbf{v}\mathbf{v}')^{-1}\mathbf{v} = \frac{\mathbf{A}^{-1}\mathbf{v}}{1 - \mathbf{v}'\mathbf{A}^{-1}\mathbf{v}}, \tag{5.41}$$

we see that

$$\begin{aligned}\hat{\boldsymbol{\theta}}_{(i)} - \hat{\boldsymbol{\theta}} &\approx \hat{\boldsymbol{\beta}}_{(i)} \\ &= \frac{(\hat{\mathbf{F}}'_{\bullet}\hat{\mathbf{F}}_{\bullet})^{-1}\hat{\mathbf{f}}_i\hat{\varepsilon}_i}{1 - \hat{h}_{ii}} \\ &= \mathbf{d}_i\hat{\varepsilon}_i, \end{aligned} \tag{5.42}$$

say, where $\hat{h}_{ii} = \hat{\mathbf{f}}'_i(\hat{\mathbf{F}}'_{\bullet}\hat{\mathbf{F}}_{\bullet})^{-1}\hat{\mathbf{f}}_i$ is the ith diagonal element of the so-called "hat" matrix $\hat{\mathbf{P}}_{\mathbf{F}} = \hat{\mathbf{F}}_{\bullet}(\hat{\mathbf{F}}'_{\bullet}\hat{\mathbf{F}}_{\bullet})^{-1}\hat{\mathbf{F}}'_{\bullet}$. We can therefore use the approximation (5.42), as in (5.33) but with $G = n$, to obtain the pseudovalue

$$\begin{aligned}\tilde{\boldsymbol{\theta}}_{i(\mathrm{LP})} &= n\hat{\boldsymbol{\theta}} - (n-1)\hat{\boldsymbol{\theta}}_{(i)} \\ &= \hat{\boldsymbol{\theta}} + (n-1)\mathbf{d}_i\hat{\varepsilon}_i \\ &\approx \hat{\boldsymbol{\theta}} + \frac{n(\hat{\mathbf{F}}'_{\bullet}\hat{\mathbf{F}}_{\bullet})^{-1}\hat{\mathbf{f}}_i\hat{\varepsilon}_i}{1 - \hat{h}_{ii}}, \end{aligned} \tag{5.43}$$

which is linear in $\hat{\varepsilon}_i$. The sample mean and sample variance–covariance matrix of these pseudovalues are defined to be

$$\bar{\boldsymbol{\theta}}_{\mathrm{LP}} = \frac{1}{n}\sum_{i=1}^{n} \tilde{\boldsymbol{\theta}}_{i(\mathrm{LP})} \tag{5.44}$$

and

$$\mathbf{S}_{\mathrm{LP}} = \frac{1}{n-1}\sum_{i=1}^{n} (\tilde{\boldsymbol{\theta}}_{i(\mathrm{LP})} - \bar{\boldsymbol{\theta}}_{\mathrm{LP}})(\tilde{\boldsymbol{\theta}}_{i(\mathrm{LP})} - \bar{\boldsymbol{\theta}}_{\mathrm{LP}})'. \tag{5.45}$$

Confidence regions and intervals can be obtained as before. For example, (5.35) becomes

$$\left\{ \theta : (\theta - \bar{\theta}_{LP})' \mathbf{S}_{LP}^{-1} (\theta - \bar{\theta}_{LP}) \leq \frac{p}{n-p} \frac{n-1}{n} F_{p,n-p}^{\alpha} \right\}. \tag{5.46}$$

Fox et al. [1980] used a chi-square rather than an F-approximation and obtained the region

$$\{ \theta : (\theta - \bar{\theta}_{LP})' \mathbf{S}_{LP}^{-1} (\theta - \bar{\theta}_{LP}) \leq n^{-1} \chi_p^2(\alpha) \}. \tag{5.47}$$

A small simulation study of Fox et al. [1980] compared the conventional methods and the two jackknife methods given above and came to similar conclusions to those of Duncan [1978]. Although the linear jackknife appears to be better than the standard jackknife, the region (5.12) still seems to be the most reliable of the three when it comes to confidence regions. For this reason modifications of the linear jackknife were sought. Fox et al. [1980] suggested multiplying the second term of (5.43) by $1 - \hat{h}_{ii}$ (i.e., high leverage points are downweighted). This leads to the weighted pseudovalues

$$\tilde{\theta}_{i(LQ)} = \hat{\theta} + n(\hat{\mathbf{F}}'_{\cdot}\hat{\mathbf{F}}_{\cdot})^{-1} \hat{\mathbf{f}}_i \hat{\varepsilon}_i.$$

However, the mean of these values is simply $\hat{\theta}$, as the normal equations (2.9) imply that $\sum_i \hat{\mathbf{f}}_i \hat{\varepsilon}_i = \mathbf{0}$. The above method is therefore useful for variance estimation but not for bias reduction. Simonoff and Tsai [1986] proposed using the weights $(1 - \hat{h}_{ii})^2$, giving the reweighted pseudovalues

$$\tilde{\theta}_{i(RLQ)} = \hat{\theta} + n(\hat{\mathbf{F}}'_{\cdot}\hat{\mathbf{F}}_{\cdot})^{-1} \hat{\mathbf{f}}_i \hat{\varepsilon}_i (1 - \hat{h}_{ii}).$$

The corresponding estimates $\bar{\theta}_{RLQ}$ and $\mathbf{S}_{RLQ}$ are analogous to (5.44) and (5.45). In a simulation study, however, $n^{-1}\mathbf{S}_{RLQ}$ consistently underestimated $\mathscr{D}[\bar{\theta}_{RLQ}]$, and the authors considered several alternatives, one of which is given below.

We recall that $\hat{\theta}_{(i)}$, which minimizes $Q_{(i)}$ of (5.39), will be close to $\hat{\theta}$. Hence

$$\begin{aligned} \mathbf{0} &= \left. \frac{\partial Q_{(i)}}{\partial \theta} \right|_{\hat{\theta}_{(i)}} \\ &\approx \left. \frac{\partial Q_{(i)}}{\partial \theta} \right|_{\hat{\theta}} + \left. \frac{\partial^2 Q_{(i)}}{\partial \theta \, \partial \theta'} \right|_{\hat{\theta}} (\hat{\theta}_{(i)} - \hat{\theta}) \\ &= -2\hat{\mathbf{F}}'_{\cdot(i)} \hat{\varepsilon}_{(i)} + \hat{\mathbf{H}}_{(i)} (\hat{\theta}_{(i)} - \hat{\theta}), \end{aligned} \tag{5.48}$$

where, from (5.16),

$$\begin{aligned} \hat{\mathbf{H}}_{(i)} &= 2(\hat{\mathbf{F}}'_{\cdot(i)} \hat{\mathbf{F}}_{\cdot(i)} - [\hat{\varepsilon}'_{(i)}][\hat{F}_{(i)\cdot\cdot}]) \\ &= 2(\hat{\mathbf{F}}'_{\cdot}\hat{\mathbf{F}}_{\cdot} - [\hat{\varepsilon}'_{(i)}][\hat{F}_{(i)\cdot\cdot}] - \hat{\mathbf{f}}_i \hat{\mathbf{f}}'_i). \end{aligned}$$

Here $\hat{F}_{(i)\cdot\cdot}$ is the $(n-1) \times p \times p$ array with the ith component (face) removed from $\hat{F}_{\cdot\cdot} = [(\partial^2 \mathbf{f}/\partial\theta_r \partial\theta_s)]_{\hat{\theta}}$. Using the same approach which produced (5.42) from (5.40), we see that (5.48) leads to

$$\begin{aligned}\hat{\boldsymbol{\theta}}_{(i)} &\approx \hat{\boldsymbol{\theta}} - 2\hat{\mathbf{H}}_{(i)}^{-1}\hat{\mathbf{F}}'_{\cdot(i)}\hat{\boldsymbol{\varepsilon}}_{(i)} \\ &= \hat{\boldsymbol{\theta}} - \frac{\hat{\mathbf{T}}_i^{-1}\hat{\mathbf{f}}_i\hat{\varepsilon}_i}{1-\hat{h}_{ii}^*},\end{aligned}$$

where $\hat{\mathbf{T}}_i = \hat{\mathbf{F}}'_{\cdot}\hat{\mathbf{F}}_{\cdot} - [\hat{\boldsymbol{\varepsilon}}'_{(i)}][\hat{F}_{(i)\cdot\cdot}]$ and $\hat{h}_{ii}^* = \hat{\mathbf{f}}_i'\hat{\mathbf{T}}_i^{-1}\hat{\mathbf{f}}_i$. The above equation leads to the modified pseudovalues (with n instead of $n-1$ in the second term)

$$\tilde{\boldsymbol{\theta}}_{i(\mathrm{MLP})} = \hat{\boldsymbol{\theta}} + \frac{n\hat{\mathbf{T}}_i^{-1}\hat{\mathbf{f}}_i\hat{\varepsilon}_i}{1-\hat{h}_{ii}^*}, \tag{5.49}$$

with corresponding estimates $\bar{\boldsymbol{\theta}}_{\mathrm{MLP}}$ and $\mathbf{S}_{\mathrm{MLP}}$. Simonoff and Tsai [1986] found that the best procedure (called RLQM) is to use $\bar{\boldsymbol{\theta}}_{\mathrm{RLQ}}$ and $\mathbf{S}_{\mathrm{MLP}}$. Thus an approximate $100(1-\alpha)\%$ confidence region for $\boldsymbol{\theta}$ and a confidence interval for θ_r are given by

$$\left\{\boldsymbol{\theta} : (\boldsymbol{\theta} - \bar{\boldsymbol{\theta}}_{\mathrm{RLQ}})'\mathbf{S}_{\mathrm{MLP}}^{-1}(\boldsymbol{\theta} - \bar{\boldsymbol{\theta}}_{\mathrm{RLQ}}) \le \frac{p}{n-p}\,\frac{n-1}{n}\,F_{p,n-p}^{\alpha}\right\} \tag{5.50}$$

and

$$\bar{\theta}_{r(\mathrm{RLQ})} \pm t_{n-p}^{\alpha/2}\sqrt{\frac{s_{rr(\mathrm{MLP})}}{n}}$$

respectively, where $[(s_{ij(\mathrm{MLP})})] = \mathbf{S}_{\mathrm{MLP}}$. In a study of real and simulated data sets, Simonoff and Tsai [1986] came to the following conclusions:

1. For well-behaved data (no outliers or leverage points, and little curvature), there is little difference among the various procedures.
2. Both least-squares and jackknife procedures can be heavily affected by outliers, leverage points, and curvature effects. The RLQM procedure is reasonably robust to these problems and is preferred over least squares, the ordinary jackknife, and the linear jackknife ($\bar{\boldsymbol{\theta}}_{\mathrm{LP}}$). The use of the latter three procedures can lead to poor inferences in many cases. The RLQM joint confidence region is also far more robust to the presence of outliers and leverage points than is the likelihood region (5.12).
3. The RLQM coverage properties are quite good, even for joint regions, unless curvature effects are quite pronounced. In that case (5.12) is preferred, though its unusual shape would make interpretation difficult (or impossible).

In concluding this section it should be noted that the leaving-one-out estimates $\hat{\boldsymbol{\theta}}_{(i)}$, obtained by minimizing (5.37), are useful in their own right. They can be used diagnostically for assessing the influence of individual observations on the least squares estimates (see Cook [1986] for references).

*5.8 EFFECTS OF CURVATURE ON LINEARIZED REGIONS

5.8.1 Intrinsic Curvature

In Section 5.2 we considered the confidence region (5.12) and the linearized region (5.9). The latter was obtained from (5.12) via (3.8) using the linear tangent-plane approximation

$$f(\boldsymbol{\theta}) - f(\hat{\boldsymbol{\theta}}) \approx \hat{\mathbf{F}}_{\bullet}(\boldsymbol{\theta} - \hat{\boldsymbol{\theta}}) \tag{5.51}$$

of the expectation surface (solution locus) in the neighborhood of $\hat{\boldsymbol{\theta}}$. Drawing heavily on the theory of Chapter 4, we shall now examine the effect of intrinsic curvature on the validity of (5.9). This will be done by using a quadratic rather than a linear approximation around $\hat{\boldsymbol{\theta}}$, with the second-order terms depending on the intrinsic curvature array. Measures are derived which can be used to check whether the usual linearized region (5.9) differs appreciably from a region developed using second-order corrections. We shall see that the second-order corrected region involves a parameter transformation obtained by solving a certain set of nonlinear equations (5.54) described below. As much of what follows is a theoretical digression, the reader who is not interested in technical details may wish to skip to Section 5.9.

We first recall that $\boldsymbol{\mu} = \mathscr{E}[\mathbf{y}]$ and, for the model (5.1), $\boldsymbol{\mu} = \mathbf{f}(\boldsymbol{\theta}) = \boldsymbol{\mu}(\boldsymbol{\theta})$, say. In order to compute curvatures we found it convenient in Section 4.2.3 to use an orthogonal transformation $\hat{\mathbf{Q}}'$ $[=(\hat{\mathbf{Q}}_p, \hat{\mathbf{Q}}_{n-p})']$ on the sample space, where the columns of $\hat{\mathbf{Q}}_p$ form a basis for the tangent plane at $\hat{\boldsymbol{\theta}}$ and, from (4.34),

$$\hat{\mathbf{Q}}_p = \hat{\mathbf{F}}_{\bullet}\hat{\mathbf{R}}_{11}^{-1} = \hat{\mathbf{F}}_{\bullet}\hat{\mathbf{K}}. \tag{5.52}$$

The basis vectors $\mathbf{e}_1, \mathbf{e}_2, \ldots, \mathbf{e}_p$ then provided a coordinate system for the tangent plane at $\hat{\boldsymbol{\mu}} = \boldsymbol{\mu}(\hat{\boldsymbol{\theta}})$. For example, if

$$\hat{\mathbf{Q}}'[\boldsymbol{\mu}(\boldsymbol{\theta}) - \hat{\boldsymbol{\mu}}] = \begin{pmatrix} \boldsymbol{\tau} \\ \boldsymbol{\xi} \end{pmatrix} \begin{matrix} \}p \\ \}n-p \end{matrix} \tag{5.53}$$

represents the new coordinates of the secant $\boldsymbol{\mu}(\boldsymbol{\theta}) - \boldsymbol{\mu}(\hat{\boldsymbol{\theta}})$ relative to an origin at $\boldsymbol{\mu}(\hat{\boldsymbol{\theta}})$, then

$$\begin{pmatrix} \boldsymbol{\tau} \\ \mathbf{0} \end{pmatrix} = \tau_1\mathbf{e}_1 + \tau_2\mathbf{e}_2 + \cdots + \tau_p\mathbf{e}_p$$

is the component in the tangent plane. Since intrinsic curvature is independent of the method of parametrization (by Appendix B5), any suitable one-to-one transformation of $\boldsymbol{\theta}$ will do. Hamilton et al. [1982] therefore proposed a transformation based on (5.53), namely

$$\boldsymbol{\tau} = \hat{\mathbf{Q}}_p'[\boldsymbol{\mu}(\boldsymbol{\theta}) - \hat{\boldsymbol{\mu}}] = \boldsymbol{\tau}(\boldsymbol{\theta}). \tag{5.54}$$

This is one-to-one (bijective) in a neighborhood of $\hat{\boldsymbol{\theta}}$, as

$$\begin{aligned} \left.\frac{\partial\boldsymbol{\tau}}{\partial\boldsymbol{\theta}'}\right|_{\hat{\boldsymbol{\theta}}} &= \left.\frac{\partial\boldsymbol{\tau}}{\partial\boldsymbol{\mu}'}\frac{\partial\boldsymbol{\mu}}{\partial\boldsymbol{\theta}'}\right|_{\hat{\boldsymbol{\theta}}} \qquad \text{(by Appendix A10.3)} \\ &= \hat{\mathbf{Q}}_p'\hat{\mathbf{F}}_{\bullet} \\ &= \hat{\mathbf{Q}}_p'\hat{\mathbf{Q}}_p\hat{\mathbf{R}}_{11} \qquad \text{[by (5.52)]} \\ &= \hat{\mathbf{R}}_{11} \end{aligned} \tag{5.55}$$

is nonsingular. It is a convenient transformation for studying intrinsic curvature in that, as we show below, the parameter-effects curvature is zero. Therefore, reparametrizing the underlying model in terms of $\boldsymbol{\tau}$ instead of $\boldsymbol{\theta}$, we have $\mathbf{f}[\boldsymbol{\theta}(\boldsymbol{\tau})]$, or $\boldsymbol{\mu}(\boldsymbol{\tau})$ for short. A Taylor expansion of $\boldsymbol{\mu}(\boldsymbol{\tau})$ at $\boldsymbol{\tau} = \mathbf{0}$ (which corresponds to $\boldsymbol{\theta} = \hat{\boldsymbol{\theta}}$) using the same format as (4.5) then gives us

$$\begin{aligned} \boldsymbol{\mu} - \hat{\boldsymbol{\mu}} &= \boldsymbol{\mu}(\boldsymbol{\tau}) - \boldsymbol{\mu}(\mathbf{0}) \\ &\approx \hat{\mathbf{T}}_{\bullet}\boldsymbol{\tau} + \tfrac{1}{2}\boldsymbol{\tau}'\hat{\boldsymbol{T}}_{\bullet\bullet}\boldsymbol{\tau}. \end{aligned} \tag{5.56}$$

Here

$$\begin{aligned} \hat{\mathbf{T}}_{\bullet} &= \left.\frac{\partial\boldsymbol{\mu}}{\partial\boldsymbol{\tau}'}\right|_{\boldsymbol{\tau}=\mathbf{0}} \\ &= \left.\frac{\partial\boldsymbol{\mu}}{\partial\boldsymbol{\theta}'}\frac{\partial\boldsymbol{\theta}}{\partial\boldsymbol{\tau}'}\right|_{\boldsymbol{\tau}=\mathbf{0}} \qquad \text{(by Appendix A10.3).} \\ &= \left[\frac{\partial\boldsymbol{\mu}}{\partial\boldsymbol{\theta}'}\left(\frac{\partial\boldsymbol{\tau}}{\partial\boldsymbol{\theta}'}\right)^{-1}\right]_{\hat{\boldsymbol{\theta}}} \\ &= \hat{\mathbf{F}}_{\bullet}\hat{\mathbf{R}}_{11}^{-1} \qquad \text{[by (5.55)]} \\ &= \hat{\mathbf{Q}}_p \qquad \text{[by (5.52)],} \end{aligned} \tag{5.57}$$

and, from the Appendix of Hamilton et al. [1982],

$$\begin{aligned} \hat{\boldsymbol{T}}_{\bullet\bullet} &= \left[\left(\frac{\partial^2\boldsymbol{\mu}(\boldsymbol{\tau})}{\partial\tau_r\,\partial\tau_s}\right)\right]_{\boldsymbol{\tau}=\mathbf{0}} \\ &= [(\hat{\mathbf{Q}}_{n-p}\hat{\mathbf{a}}_{rs}^N)] \\ &= [\hat{\mathbf{Q}}_{n-p}][\hat{\boldsymbol{A}}_{\bullet\bullet}^N], \end{aligned} \tag{5.58}$$

where $\hat{A}^N_{..}$ is defined by (4.57), and square-bracket multiplication is described in Appendix B4. We note that if $\hat{T}_{..} = [(\hat{\mathbf{t}}_{rs})]$, then from the discussion following (4.14), and Equations (5.57) and (5.58), we see that

$$\begin{aligned}\hat{\mathbf{t}}^T_{rs} &= \hat{\mathbf{P}}_T \hat{\mathbf{t}}_{rs} \\ &= \hat{\mathbf{T}}_.(\hat{\mathbf{T}}'_.\hat{\mathbf{T}}_.)^{-1}\hat{\mathbf{T}}'_.\hat{\mathbf{t}}_{rs} \\ &= \hat{\mathbf{Q}}_p\hat{\mathbf{Q}}'_p\hat{\mathbf{Q}}_{n-p}\mathbf{a}^N_{rs} \\ &= \mathbf{0},\end{aligned}$$

as $\hat{\mathbf{Q}}'_p\hat{\mathbf{Q}}_{n-p} = \mathbf{0}$. Hence $\hat{T}^T_{..} = \mathbf{0}$ and the parameter-effects curvature at $\hat{\boldsymbol{\theta}}$ for the $\boldsymbol{\tau}$-system is zero.

Having established the properties of our transformation we now derive a confidence region for $\boldsymbol{\theta}$ based on an approximation of $\boldsymbol{\varepsilon}'\boldsymbol{\varepsilon}$. Substituting (5.57) and (5.58) into (5.56), we can move $\hat{\mathbf{Q}}_{n-p}$ to the left (by Appendix B4.1) to get

$$\begin{aligned}\boldsymbol{\mu} &\approx \hat{\boldsymbol{\mu}} + \hat{\mathbf{Q}}_p\boldsymbol{\tau} + \tfrac{1}{2}[\hat{\mathbf{Q}}_{n-p}][\boldsymbol{\tau}'\hat{A}^N_{..}\boldsymbol{\tau}] \\ &= \hat{\boldsymbol{\mu}} + \hat{\mathbf{Q}}_p\boldsymbol{\tau} + \tfrac{1}{2}\sum_r\sum_s \hat{\mathbf{Q}}_{n-p}\hat{\mathbf{a}}^N_{rs}\tau_r\tau_s\end{aligned} \tag{5.59}$$

and

$$\begin{aligned}\boldsymbol{\varepsilon} &= \mathbf{y} - \boldsymbol{\mu} \\ &\approx \hat{\boldsymbol{\varepsilon}} - \hat{\mathbf{Q}}_p\boldsymbol{\tau} - \tfrac{1}{2}[\hat{\mathbf{Q}}_{n-p}][\boldsymbol{\tau}'\hat{A}^N_{..}\boldsymbol{\tau}],\end{aligned} \tag{5.60}$$

where $\hat{\boldsymbol{\varepsilon}} = \mathbf{y} - \hat{\boldsymbol{\mu}}$. Recalling that $\hat{\mathbf{Q}}'_p\hat{\mathbf{Q}}_p = \mathbf{I}_p$, $\hat{\mathbf{Q}}'_{n-p}\hat{\mathbf{Q}}_{n-p} = \mathbf{I}_{n-p}$, and $\hat{\mathbf{Q}}'_{n-p}\hat{\mathbf{Q}}_p = \mathbf{0}$, we have from (5.60)

$$\boldsymbol{\varepsilon}'\boldsymbol{\varepsilon} \approx \hat{\boldsymbol{\varepsilon}}'\hat{\boldsymbol{\varepsilon}} + \boldsymbol{\tau}'\boldsymbol{\tau} + \tfrac{1}{4}(\boldsymbol{\tau}'\hat{A}^N_{..}\boldsymbol{\tau})'(\boldsymbol{\tau}'\hat{A}^N_{..}\boldsymbol{\tau}) - 2\hat{\boldsymbol{\varepsilon}}'\hat{\mathbf{Q}}_p\boldsymbol{\tau} - \tfrac{1}{2}[\hat{\boldsymbol{\varepsilon}}'\hat{\mathbf{Q}}_{n-p}][\boldsymbol{\tau}'\hat{A}^N_{..}\boldsymbol{\tau}]. \tag{5.61}$$

[This expansion of $\boldsymbol{\varepsilon}'\boldsymbol{\varepsilon}$ does not contain all the third and fourth powers of $\boldsymbol{\tau}$, as third powers are ignored in (5.60).] We define

$$\begin{aligned}\hat{\mathbf{B}} &= [(\hat{\boldsymbol{\varepsilon}}'\hat{\mathbf{Q}}_{n-p}\hat{\mathbf{a}}^N_{rs})] \\ &= [(\hat{\boldsymbol{\varepsilon}}'\hat{\mathbf{Q}}_{n-p})][\hat{A}^N_{..}],\end{aligned} \tag{5.62}$$

and note from (5.52) and the normal equations (2.9) that

$$\hat{\mathbf{Q}}'_p\hat{\boldsymbol{\varepsilon}} = (\hat{\mathbf{R}}^{-1}_{11})'\hat{\mathbf{F}}'_.\hat{\boldsymbol{\varepsilon}} = \mathbf{0}. \tag{5.63}$$

If we ignore fourth powers of $\boldsymbol{\tau}$ in (5.61), we obtain the approximation

$$\begin{aligned}\boldsymbol{\varepsilon}'\boldsymbol{\varepsilon} &\approx \hat{\boldsymbol{\varepsilon}}'\hat{\boldsymbol{\varepsilon}} + \boldsymbol{\tau}'\boldsymbol{\tau} - \tfrac{1}{2}\sum_r\sum_s \hat{\boldsymbol{\varepsilon}}'\hat{\mathbf{Q}}_{n-p}\hat{\mathbf{a}}_{rs}\tau_r\tau_s \\ &= \hat{\boldsymbol{\varepsilon}}'\hat{\boldsymbol{\varepsilon}} + \boldsymbol{\tau}'(\mathbf{I}_p - \hat{\mathbf{B}})\boldsymbol{\tau}.\end{aligned} \tag{5.64}$$

Now the confidence region (5.13) is given by

$$\left\{\boldsymbol{\theta}:\boldsymbol{\varepsilon}'\boldsymbol{\varepsilon}-\hat{\boldsymbol{\varepsilon}}'\hat{\boldsymbol{\varepsilon}}\le\frac{pF_\alpha\hat{\boldsymbol{\varepsilon}}'\hat{\boldsymbol{\varepsilon}}}{n-p}\right\}, \tag{5.65}$$

where $F_\alpha = F^\alpha_{p,n-p}$. Substituting the approximation (5.64) into (5.65) leads to

$$\begin{aligned}\boldsymbol{\tau}'(\mathbf{I}_p-\hat{\mathbf{B}})\boldsymbol{\tau}&\le\frac{pF_\alpha\hat{\boldsymbol{\varepsilon}}'\hat{\boldsymbol{\varepsilon}}}{n-p}\\&=F_\alpha\rho^2,\end{aligned} \tag{5.66}$$

where $\rho = s\sqrt{p} = [p\hat{\boldsymbol{\varepsilon}}'\hat{\boldsymbol{\varepsilon}}/(n-p)]^{1/2}$ is the standard radius [c.f. (4.12)].

The region (5.66) will represent an ellipsoid in the $\boldsymbol{\tau}$-coordinates if $\mathbf{I}_p-\hat{\mathbf{B}}$ is positive definite. To show this, Hamilton et al. [1982] establish the following:

$$\begin{aligned}\hat{\mathbf{B}}&=[(\hat{b}_{rs})]\\&=[(\hat{\boldsymbol{\varepsilon}}'\hat{\mathbf{Q}}_{n-p}\hat{\mathbf{a}}^N_{rs})]\\&=[(\hat{\boldsymbol{\varepsilon}}'\hat{\mathbf{Q}}_{n-p}\hat{\mathbf{Q}}'_{n-p}\hat{\mathbf{g}}_{rs})]\qquad[\text{by (4.57)}]\\&=[(\hat{\boldsymbol{\varepsilon}}'\hat{\mathbf{g}}_{rs})],\end{aligned} \tag{5.67}$$

since (5.63) implies that $\hat{\boldsymbol{\varepsilon}}$ is orthogonal to the tangent plane, thus giving $\hat{\boldsymbol{\varepsilon}}=\hat{\mathbf{Q}}_{n-p}\boldsymbol{\alpha}$ for some $\boldsymbol{\alpha}$ and

$$\hat{\mathbf{Q}}_{n-p}\hat{\mathbf{Q}}'_{n-p}\hat{\boldsymbol{\varepsilon}}=\hat{\mathbf{Q}}_{n-p}(\hat{\mathbf{Q}}'_{n-p}\hat{\mathbf{Q}}_{n-p})\boldsymbol{\alpha}=\hat{\boldsymbol{\varepsilon}}.$$

Therefore, using (4.47) and the properties of three-dimensional arrays (see Appendix B4.1), we have, from (5.67),

$$\begin{aligned}\mathbf{I}_p-\hat{\mathbf{B}}&=\mathbf{I}_p-[\hat{\boldsymbol{\varepsilon}}'][\hat{G}_{..}]\\&=\hat{\mathbf{Q}}'_p\hat{\mathbf{Q}}_p-[\hat{\boldsymbol{\varepsilon}}'][\hat{\mathbf{K}}'\hat{F}_{..}\hat{\mathbf{K}}]\qquad[\text{by (4.47)}]\\&=\hat{\mathbf{K}}'\hat{\mathbf{F}}'_{.}\hat{\mathbf{F}}_{.}\hat{\mathbf{K}}-\hat{\mathbf{K}}'[\hat{\boldsymbol{\varepsilon}}'][\hat{F}_{..}]\hat{\mathbf{K}}\qquad[\text{by (5.52)}]\\&=\tfrac{1}{2}\hat{\mathbf{K}}'\hat{\mathbf{H}}\hat{\mathbf{K}},\end{aligned} \tag{5.68}$$

where $\hat{\mathbf{H}}=[\partial^2 S(\boldsymbol{\theta})/\partial\boldsymbol{\theta}\,\partial\boldsymbol{\theta}']_{\hat{\boldsymbol{\theta}}}$, as shown in (5.16). As $S(\boldsymbol{\theta})$ is minimized at $\hat{\boldsymbol{\theta}}$, $\hat{\mathbf{H}}$ is positive definite, and so is $\mathbf{I}_p-\hat{\mathbf{B}}$. Hence the region (5.66) is an ellipsoid with center $\hat{\boldsymbol{\theta}}$ ($\boldsymbol{\tau}=\mathbf{0}$) in the tangent plane. This may be compared with the corresponding region based on the linear approximation in the $\boldsymbol{\tau}$-parameters [c.f. (5.59)]

$$\boldsymbol{\mu}\approx\hat{\boldsymbol{\mu}}+\hat{\mathbf{Q}}_p\boldsymbol{\tau}, \tag{5.69}$$

obtained by setting $\hat{A}^N_{..}$ (and therefore $\hat{\mathbf{B}}$) equal to zero in (5.66), namely the sphere

$$\|\boldsymbol{\mu} - \hat{\boldsymbol{\mu}}\|^2 \approx \boldsymbol{\tau}'\boldsymbol{\tau} \leq F_\alpha \rho^2. \tag{5.70}$$

Since the $\boldsymbol{\tau}$-parameters have a zero parameter-effects curvature, the effect of intrinsic curvature on (5.66) can be gauged by comparing the lengths of the axes of (5.66) with the radius of the sphere (5.70). If $\lambda_{max} = \lambda_1 > \lambda_2 > \cdots > \lambda_p = \lambda_{min}$ are the eigenvalues of $\hat{\mathbf{B}}$, then the ith smallest eigenvalue of $\mathbf{I}_p - \hat{\mathbf{B}}$ is $1 - \lambda_i$, and the length of the corresponding major axis is proportional to $(1 - \lambda_i)^{-1/2}$ (by Appendix A6). To compare (5.66) with (5.70), Hamilton et al. [1982] suggested comparing $(1 - \lambda_{max})^{-1/2}$ and $(1 - \lambda_{min})^{-1/2}$ with unity. A conservative allowance for the intrinsic curvature embodied in $\hat{\mathbf{B}}$ can be made by replacing each major axis of $\boldsymbol{\tau}'(\mathbf{I}_p - \hat{\mathbf{B}})\boldsymbol{\tau} = 1$ by the longest axis $(1 - \lambda_{max})^{-1/2}$. The ellipsoid (5.66) will then be contained in the conservative spherical confidence region (Bates and Watts [1981a])

$$\boldsymbol{\tau}'\boldsymbol{\tau} \leq \frac{F_\alpha \rho^2}{1 - \lambda_{max}} = r_\alpha^2. \tag{5.71}$$

In deriving (5.66), terms involving fourth powers of $\boldsymbol{\tau}$ were ignored in (5.61). A crude upper bound was given by the authors for the proportion of the ignored term relative to the quadratic term. However, Hamilton (personal communication, 1986) does not currently favor the use of this bound, because, as already mentioned after (5.61), there are other ignored terms and these may be more important.

5.8.2 Parameter-Effects Curvature

We saw in the discussion following (4.25) that the intrinsic curvature is often negligible in nonlinear models and it is usually the parameter-effects curvature that causes the problems. However, if the intrinsic curvature is not negligible, then we can make a conservative adjustment to the linearized region (5.70) by extending the radius of this spherical region as in (5.71). Unfortunately, this region is defined in terms of the new parameters $\boldsymbol{\tau}$ which are related nonlinearly to the original parameters $\boldsymbol{\theta}$ by (5.54). In general it will not be possible to actually solve (5.54) for $\boldsymbol{\theta}$ in order to apply (5.71) exactly. To assess the impact of the parameter-effects curvature on this region it is therefore more convenient to express (5.71) in terms of the parameters $\boldsymbol{\phi}$. These parameters form the basis of our curvature calculations in Chapter 4 and have the same curvature effects as those of $\boldsymbol{\theta}$ [c.f. the discussion following (4.49)]. Since $\boldsymbol{\theta} = \hat{\mathbf{R}}_{11}^{-1}\boldsymbol{\phi}$, we have from (5.54)

$$\begin{aligned}\boldsymbol{\tau} &= \hat{\mathbf{Q}}_p'[\boldsymbol{\mu}(\hat{\mathbf{R}}_{11}^{-1}\boldsymbol{\phi}) - \hat{\boldsymbol{\mu}}] \\ &= \boldsymbol{\tau}(\boldsymbol{\phi}),\end{aligned} \tag{5.72}$$

say, and (5.71) becomes

$$\|\boldsymbol{\tau}(\boldsymbol{\phi})\|^2 \leq r_\alpha^2. \tag{5.73}$$

To obtain an approximation for this region we shall use a second order Taylor expansion of $\boldsymbol{\tau}$ in terms of $\boldsymbol{\phi}$. We will find that if a first-order expansion is in fact adequate, then the region (5.71) translates into the usual linearized region for $\boldsymbol{\theta}$, namely (5.9), but modified by an inflation factor. Otherwise the linearized region will differ appreciably from (5.71) and will therefore be inadequate. A different parametrization from $\boldsymbol{\theta}$ should then be used, and some practical guidelines are outlined in Section 5.8.4.

Now

$$\begin{aligned}\left.\frac{\partial \boldsymbol{\tau}}{\partial \boldsymbol{\phi}'}\right|_{\hat{\theta}} &= \left.\frac{\partial \boldsymbol{\tau}}{\partial \boldsymbol{\theta}'}\frac{\partial \boldsymbol{\theta}}{\partial \boldsymbol{\phi}'}\right|_{\hat{\theta}} \\ &= \hat{\mathbf{R}}_{11}\hat{\mathbf{R}}_{11}^{-1} \qquad \text{[by (5.55)]} \\ &= \mathbf{I}_p, \end{aligned} \tag{5.74}$$

and, using a similar argument (Bates and Watts [1981a]),

$$\left.\left[\left(\frac{\partial^2 \boldsymbol{\tau}}{\partial \phi_r \partial \phi_s}\right)\right]\right|_{\hat{\phi}} = \hat{A}^T_{..}. \tag{5.75}$$

If we now carry out another Taylor expansion along the lines of (5.56) and incorporate the first- and second-derivative arrays given by (5.74) and (5.75), we obtain

$$\begin{aligned}\boldsymbol{\tau}(\boldsymbol{\phi}) &= \boldsymbol{\tau}(\boldsymbol{\phi}) - \boldsymbol{\tau}(\hat{\boldsymbol{\phi}}) \\ &\approx (\boldsymbol{\phi}-\hat{\boldsymbol{\phi}}) + \tfrac{1}{2}(\boldsymbol{\phi}-\hat{\boldsymbol{\phi}})'\hat{A}^T_{..}(\boldsymbol{\phi}-\hat{\boldsymbol{\phi}}) \\ &= \boldsymbol{\gamma} + \tfrac{1}{2}\boldsymbol{\gamma}'\hat{A}^T_{..}\boldsymbol{\gamma}, \end{aligned} \tag{5.76}$$

where $\boldsymbol{\gamma} = \boldsymbol{\phi} - \hat{\boldsymbol{\phi}}$. The confidence region (5.73) based on the above quadratic approximation (5.76) may be compared with the corresponding linearized region obtained by setting $\hat{A}^T_{..}$ equal to zero, namely

$$\begin{aligned} r_\alpha^2 &\geq \|\boldsymbol{\gamma}\|^2 \\ &= (\boldsymbol{\phi}-\hat{\boldsymbol{\phi}})'(\boldsymbol{\phi}-\hat{\boldsymbol{\phi}}) \\ &= (\boldsymbol{\theta}-\hat{\boldsymbol{\theta}})'\hat{\mathbf{R}}_{11}'\hat{\mathbf{R}}_{11}(\boldsymbol{\theta}-\hat{\boldsymbol{\theta}}) \\ &= (\boldsymbol{\theta}-\hat{\boldsymbol{\theta}})'\hat{\mathbf{R}}_{11}'\hat{\mathbf{Q}}_p'\hat{\mathbf{Q}}_p\hat{\mathbf{R}}_{11}(\boldsymbol{\theta}-\hat{\boldsymbol{\theta}}) \\ &= (\boldsymbol{\theta}-\hat{\boldsymbol{\theta}})'\hat{\mathbf{F}}_{.}'\hat{\mathbf{F}}_{.}(\boldsymbol{\theta}-\hat{\boldsymbol{\theta}}), \qquad \text{[by (5.52)]}, \end{aligned} \tag{5.77}$$

which is our familiar linearized confidence region (5.9) but with the conservative inflation factor of $(1-\lambda_{\max})^{-1}$. The above linear approximation will be satisfactory if the quadratic term is small compared with the length of the linear term, that is, if [c.f. (5.76)]

$$\max_{\gamma} \frac{\frac{1}{2}\|\gamma' \hat{A}^T_{..}\gamma\|}{\|\gamma\|} = \max_{\gamma} \frac{\frac{1}{2}\|\gamma' \hat{A}^T_{..}\gamma\|}{\|\gamma\|^2} \|\gamma\|$$

$$\leq \max_{\|\mathbf{d}\|=1} \{\tfrac{1}{2}\rho \|\mathbf{d}' \hat{A}^T_{..}\mathbf{d}\|\} \max_{\gamma} \frac{\|\gamma\|}{\rho}$$

$$= \tfrac{1}{2}\gamma^T_{\max} r_\alpha/\rho \qquad \text{[by (4.58), (4.60), and (5.71)]}$$

$$= \tfrac{1}{2}\gamma^T_{\max} R_\alpha \tag{5.78}$$

is small (say less than $\frac{1}{4}$), where r_α is given by (5.71). To assess, therefore, the effect of $\gamma^T_{\max}$ or particular terms of $\rho\hat{A}^T_{..}$ on the linearized confidence region (5.77), we simply compare them with $\frac{1}{2}R_\alpha^{-1} = \frac{1}{2}[(1-\lambda_{\max})/F_\alpha]^{1/2}$. Terms exceeding $\frac{1}{2}R_\alpha^{-1}$ in magnitude indicate that the linear approximation is inadequate. If scaling by ρ is used, then ρ disappears from all the above expressions.

When the intrinsic curvature is negligible, $\lambda_{\max} \approx 0$ and we now compare $\gamma^T_{\max}$ with $1/(2\sqrt{F_\alpha})$ [c.f. (4.25)].

5.8.3 Summary of Confidence Regions

We now summarize the various confidence regions described above. Let $\boldsymbol{\tau}$ be defined as in (5.54). If third and fourth powers of $\boldsymbol{\tau}$ can be ignored in a Taylor series expansion of $\boldsymbol{\varepsilon}'\boldsymbol{\varepsilon}$, then we can use the confidence region

$$\{\boldsymbol{\theta} : \boldsymbol{\tau}'(\mathbf{I}_p - \hat{\mathbf{B}})\boldsymbol{\tau} \leq \rho^2 F_\alpha, \boldsymbol{\tau} = \boldsymbol{\tau}(\boldsymbol{\theta})\}, \tag{5.79}$$

or the simpler but conservative region

$$\{\boldsymbol{\theta} : \boldsymbol{\tau}'\boldsymbol{\tau} \leq r_\alpha^2, \boldsymbol{\tau} = \boldsymbol{\tau}(\boldsymbol{\theta})\}, \tag{5.80}$$

where $\rho^2 = ps^2$, $F_\alpha = F^\alpha_{p,n-p}$,

$$r_\alpha^2 = \frac{\rho^2 F_\alpha}{1-\lambda_{\max}} = \rho^2 R_\alpha^2, \tag{5.81}$$

and $\lambda_{\max}$ is the maximum eigenvalue of $\hat{\mathbf{B}} = [(\hat{\boldsymbol{\varepsilon}}'\hat{\mathbf{g}}_{rs})]$. These regions are corrected, at least partially, for the effects of intrinsic curvature. For use with the parameter vector $\boldsymbol{\theta}$ they require the application of the inverse of $\boldsymbol{\tau} = \boldsymbol{\tau}(\boldsymbol{\theta})$ [c.f. (5.54)] to the ellipsoid (5.79) or the sphere (5.80). Alternatively, an explicit expression for $\boldsymbol{\theta}$ can be obtained using the quadratic approximation (5.76), namely

$$\{\boldsymbol{\theta} : \|\hat{\mathbf{R}}_{11}(\boldsymbol{\theta}-\hat{\boldsymbol{\theta}}) + \tfrac{1}{2}(\boldsymbol{\theta}-\hat{\boldsymbol{\theta}})'\hat{\mathbf{R}}'_{11}\hat{A}^T_{..}\hat{\mathbf{R}}_{11}(\boldsymbol{\theta}-\hat{\boldsymbol{\theta}})\|^2 \leq r_\alpha^2\}. \tag{5.82}$$

When (5.78) is small, e.g. $\gamma_{\max}^{T} < 1/(2\sqrt{F_\alpha})$, then the parameter-effects curvature can be ignored and (5.82) reduces to

$$\{\boldsymbol{\theta}:(\boldsymbol{\theta}-\hat{\boldsymbol{\theta}})'\hat{\mathbf{F}}_{.}'\hat{\mathbf{F}}_{.}(\boldsymbol{\theta}-\hat{\boldsymbol{\theta}}) \le r_\alpha^2\}. \tag{5.83}$$

If scaling by ρ is used, r_α is replaced by R_α in the above.

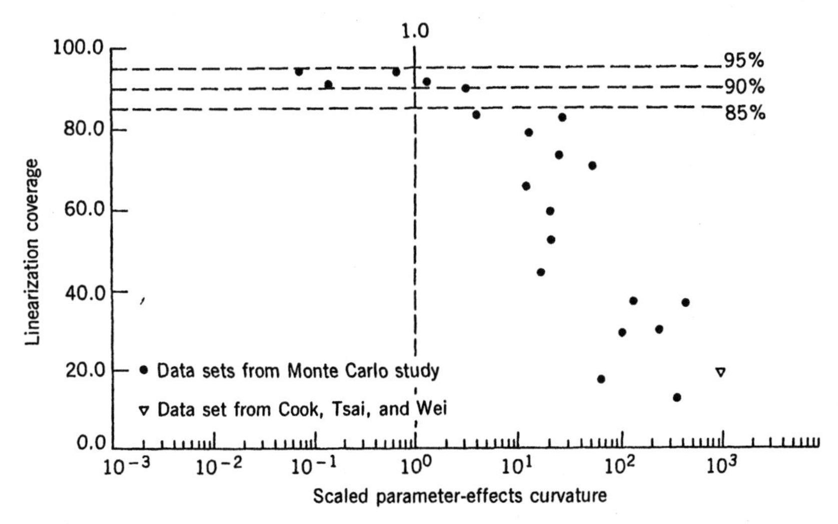

Figure 5.1 Coverage of the linearized confidence region versus scaled parameter-effects curvature (c). From Donaldson and Schnabel [1987].

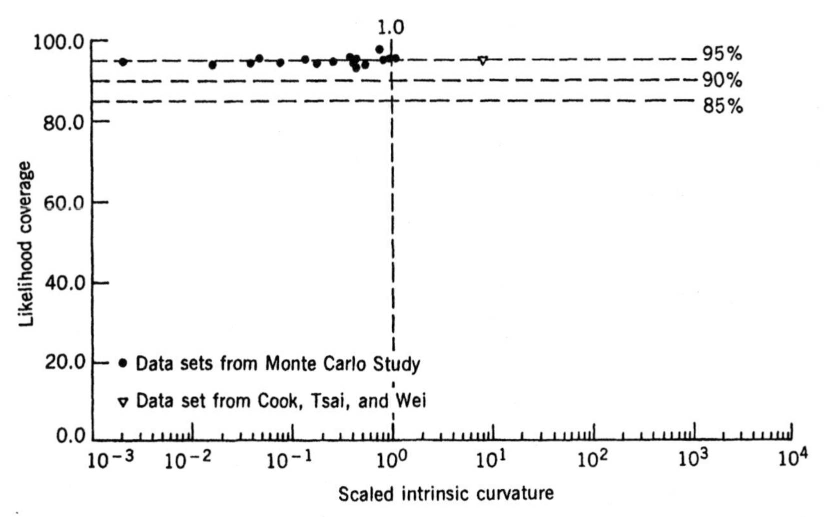

Figure 5.2 Coverage of likelihood confidence region versus scaled intrinsic curvature. From Donaldson and Schnabel [1987].

Donaldson and Schnabel [1987] applied (5.83), with $r_\alpha = \rho\sqrt{F_\alpha}$ and $\alpha = 0.05$, to 20 data sets with negligible intrinsic curvature ($\lambda_{\text{max}} \approx 0$) but varying amounts of parameter-effects curvature. They plotted the true (linearization) coverage of (5.83) (nominally 95%) versus the scaled version of the parameter-effects curvature $c = \gamma_{\text{max}}^T R_\alpha = \gamma_{\text{max}}^T \sqrt{F_\alpha}$ of (5.78) in Fig. 5.1. We see that when c increases above unity (shown dotted), the true coverage begins to fall off dramatically. It is clear, therefore, that the popular linearization region (5.83) should not be used unless the parameter-effects curvature is small enough. However, since the 20 data sets have negligible intrinsic curvature, the confidence region (5.12) can be used when (5.83) is inappropriate, since, under normality, it is based on contours of $S(\boldsymbol{\theta})$. As previously mentioned in Section 5.2 these are not affected by the method of parametrization, so that the observed coverages are close to the nominal 95% as seen in Fig. 5.2.

5.8.4 Reparametrization to Reduce Curvature Effects

If the parameter-effects curvature or some of the elements of $\hat{A}_{..}^T$ are too large, so that the linear approximation of the confidence region (5.83) is unsatisfactory, then we would like to choose another parametrization. At present little guidance is available, though occasionally certain transformations are recommended from practical experience or are suggested by the model itself. However, such transformations may not work for all data sets, as we saw in Section 3.3.2, so that a certain amount of trial and error seems unavoidable. In fact the book by Ratkowsky [1983] is largely a search for parametrizations that are suitable for models and types of data often found in agricultural applications. A similar search in the four-parameter logistic model is made by Ratkowsky and Reedy [1986]. A useful class of transformations is the class of "expected value" transformations (Ross [1970], Clarke [1987a]). Their use is demonstrated in Example 5.2 below.

Suppose we choose another one-to-one parametrization

$$\boldsymbol{\delta} = \boldsymbol{\delta}(\boldsymbol{\theta}) \tag{5.84}$$

with corresponding first-derivative matrix

$$\hat{\mathbf{D}}_{.} = \left.\frac{\partial \boldsymbol{\delta}}{\partial \boldsymbol{\theta}'}\right|_{\hat{\boldsymbol{\theta}}}$$

and second-derivative array

$$\hat{D}_{..} = \left.\frac{\partial \boldsymbol{\delta}}{\partial \boldsymbol{\theta}\, \partial \boldsymbol{\theta}'}\right|_{\hat{\boldsymbol{\theta}}}.$$

Then, for the new parameter system $\boldsymbol{\delta}$, $\hat{A}_{..}^T$ now becomes (Bates and Watts

[1981a: Equation (3.14)])

$$\hat{A}^T_{..\delta} = \hat{A}^T_{..} - [\hat{\mathbf{K}}^{-1}][\hat{\mathbf{K}}'\hat{C}^T_{..}\hat{\mathbf{K}}], \tag{5.85}$$

where

$$\hat{C}^T_{..} = [\hat{\mathbf{D}}_.^{-1}][\hat{D}_{..}], \tag{5.86}$$

and $\hat{\mathbf{K}}$ is defined by (5.52). A similar expression for $\hat{C}^T_{..}$ was obtained by Clarke [1980: Equation (6)], but expressed less conveniently in terms of derivatives with respect to $\boldsymbol{\delta}$ rather than with respect to $\boldsymbol{\theta}$. If the linear approximation is now satisfactory, and the intrinsic curvature (which is independent of the parametrization used) is zero, then the linearized confidence region for $\boldsymbol{\delta}$ is

$$\{\boldsymbol{\delta} : (\boldsymbol{\delta} - \hat{\boldsymbol{\delta}})'\hat{\boldsymbol{\Delta}}'\hat{\boldsymbol{\Delta}}(\boldsymbol{\delta} - \hat{\boldsymbol{\delta}}) \leq r_\alpha^2\}, \tag{5.87}$$

where $\hat{\boldsymbol{\delta}} = \boldsymbol{\delta}(\hat{\boldsymbol{\theta}})$ and

$$\boldsymbol{\Delta} = \frac{\partial \mathbf{f}}{\partial \boldsymbol{\delta}'} = \frac{\partial \mathbf{f}}{\partial \boldsymbol{\theta}'}\frac{\partial \boldsymbol{\theta}}{\partial \boldsymbol{\delta}'} = \mathbf{F}_.\mathbf{D}_.^{-1}.$$

The region (5.87) can be inverted, using the inverse of (5.84), to obtain a confidence region for $\boldsymbol{\theta}$.

Ideally we would like to choose the transformation for which $\hat{A}^T_{..\delta}$ of (5.85) is zero, that is,

$$\hat{A}^T_{..} = [\hat{\mathbf{K}}^{-1}][\hat{\mathbf{K}}'\hat{C}^T_{..}\hat{\mathbf{K}}]. \tag{5.88}$$

Taking $\hat{\mathbf{K}}^{-1}$ to the left-hand side and comparing ith faces, we see that (5.88) has solution

$$C^* = [\hat{\mathbf{K}}][(\hat{\mathbf{K}}')^{-1}\hat{A}^T_{..}\hat{\mathbf{K}}^{-1}],$$

which Bates and Watts [1981a] call the "target" transformation curvature array. Using (4.68),

$$\begin{aligned}\hat{A}^T_{..} &= [\hat{\mathbf{Q}}'_p][\hat{\mathbf{K}}'\hat{F}_{..}\hat{\mathbf{K}}] \\ &= \hat{\mathbf{K}}'[\hat{\mathbf{Q}}'_p][\hat{F}_{..}]\hat{\mathbf{K}} \qquad \text{(by Appendix B4.1)}\end{aligned}$$

and

$$\begin{aligned}C^* &= [\hat{\mathbf{K}}][[\hat{\mathbf{Q}}'_p][\hat{F}_{..}]] \\ &= [\hat{\mathbf{K}}\hat{\mathbf{Q}}'_p][\hat{F}_{..}] \qquad \text{(by Appendix B4.2)} \\ &= [(\hat{\mathbf{F}}'_.\hat{\mathbf{F}}_.)^{-1}\hat{\mathbf{F}}'_.][\hat{F}_{..}] \qquad \text{[by (5.52)]} \\ &= [(\hat{\mathbf{F}}'_.\hat{\mathbf{F}}_.)^{-1}\hat{\mathbf{F}}'_.\hat{\mathbf{f}}_{rs})],\end{aligned} \tag{5.89}$$

where the (r, s)th vector can be calculated by regressing $\hat{\mathbf{f}}_{rs}$ on $\hat{\mathbf{F}}_.$.

Example 5.1 **(*Bates and Watts* [*1981a: p. 1161*])** The model

$$y = \frac{\theta_1 \theta_3 x_1}{1 + \theta_1 x_1 + \theta_2 x_2} + \varepsilon \tag{5.90}$$

was fitted to the data in Table 5.2 taken from Meyer and Roth [1972]. Here $\hat{\boldsymbol{\theta}} = (3.1320, 15.160, 0.77998)'$, $\gamma_{max}^N = 0.037$, and $\gamma_{max}^T = 12.8$. Although the intrinsic curvature is small, the latter curvature is unacceptable by the criterion (5.78), as it exceeds $R_{.05}^{-\frac{1}{2}} = 0.23$. To try and improve the situation, an obvious transformation is $\delta_1 = \theta_1$, $\delta_2 = \theta_2$, and $\delta_3 = \theta_1 \theta_3$ as this reduces some of the nonlinearity in (5.90). Then

$$\mathbf{D}_{\cdot} = \frac{\partial \boldsymbol{\delta}}{\partial \boldsymbol{\theta}'} = \begin{bmatrix} 1 & 0 & 0 \\ 0 & 1 & 0 \\ \theta_3 & 0 & \theta_1 \end{bmatrix},$$

and

$$\begin{aligned} \boldsymbol{D}_{\cdot\cdot} &= \{\delta_{irs}\} \\ &= \left[\left(\frac{\partial \boldsymbol{\delta}}{\partial \theta_r \partial \theta_s} \right) \right] \end{aligned}$$

is zero except for $\delta_{313} = \delta_{331} = 1$. With the $\boldsymbol{\delta}$-parametrization, $\gamma_{max}^T = 0.264$, which is a substantial improvement. We can also compare the two second-derivative arrays. Their three symmetric "faces" are given below [c.f. (5.85)]:

$$\rho \hat{A}_{\cdot\cdot}^T \qquad\qquad\qquad \rho \hat{A}_{\cdot\cdot\delta}^T$$

$$\begin{bmatrix} 0.00 & -0.04 & -0.33 \\ \cdot & -0.05 & 0.69 \\ \cdot & \cdot & 12.77 \end{bmatrix} \qquad \begin{bmatrix} 0.00 & -0.04 & 0.11 \\ \cdot & -0.05 & 0.00 \\ \cdot & \cdot & -0.05 \end{bmatrix}$$

$$\begin{bmatrix} 0.00 & -0.02 & 0.00 \\ \cdot & -0.03 & 0.10 \\ \cdot & \cdot & -0.03 \end{bmatrix} \qquad \begin{bmatrix} 0.00 & -0.02 & -0.00 \\ \cdot & -0.03 & 0.10 \\ \cdot & \cdot & -0.03 \end{bmatrix}$$

$$\begin{bmatrix} 0.00 & 0.00 & -0.02 \\ \cdot & -0.02 & -0.07 \\ \cdot & \cdot & 0.22 \end{bmatrix} \qquad \begin{bmatrix} 0.00 & 0.00 & -0.02 \\ \cdot & -0.02 & -0.07 \\ \cdot & \cdot & 0.21 \end{bmatrix}$$

We see that the three elements with the largest absolute values are substantially reduced. The linear approximation in the $\boldsymbol{\delta}$-parameters will provide a confidence region much closer to the actual one than that given by a linear approximation in the original parameters. Using the inverse of (5.84), the region in $\boldsymbol{\delta}$ can then be mapped into a region for $\boldsymbol{\theta}$. ■

Table 5.2 Data [a] Fitted to the Model given by Equation (5.90)

y	x_1	x_2
0.126	1.0	1.0
0.219	2.0	1.0
0.076	1.0	2.0
0.126	2.0	2.0
0.186	0.1	0.0

[a]From Meyer and Roth [1972].

Unfortunately the appropriate choice of $\boldsymbol{\delta}$ is not always as straightforward as in the above example, and systematic methods for finding an appropriate transformation are needed. Three methods for attempting to find a $\boldsymbol{\delta}$ which satisfies $\hat{\boldsymbol{C}}^T_{..} = \boldsymbol{C}^*$ are given by Bates and Watts [1981a]. However, as noted by the authors, two of the methods seem to be of limited practical use, as they involve the solution of a complex system of differential equations. Their third procedure consists of selecting a transformation from a restricted class and "tuning" it to produce small curvatures for the particular data set at hand. They demonstrated their method using the following example taken from Bates and Watts [1980: Table 1] and used as Example 4.3 in Section 4.2.6.

Example 5.2 The data in Table 4.2 were fitted with the model

$$f(x;\boldsymbol{\theta}) = \frac{\theta_1 x}{\theta_2 + x}. \tag{5.91}$$

We saw in Example 4.3 that the intrinsic curvature could be ignored but the parameter-effects curvature was substantial. A transformation is therefore needed. Bates and Watts [1981a] chose the class of expected-value transformations suggested by Ross [1970, 1980]. In this approach the expected values of the model at particular design points are chosen as the parameters. For example, consider the transformation

$$\delta_1 = f(u;\boldsymbol{\theta}) = \frac{\theta_1 u}{\theta_2 + u}$$

and

$$\delta_2 = f(v;\boldsymbol{\theta}) = \frac{\theta_1 v}{\theta_2 + v},$$

where u and v are appropriate values of x. Then the array (5.86) with its two faces is given by

$$\hat{C}^T_{..} = \begin{pmatrix} \begin{pmatrix} 0 & 0 \\ 0 & -2\hat{\theta}_1/[(\hat{\theta}_2+u)(\hat{\theta}_2+v)] \end{pmatrix} \\ \begin{pmatrix} 0 & 1/\hat{\theta}_1 \\ 1/\hat{\theta}_1 & -2(2\hat{\theta}_2+u+v)/[(\hat{\theta}_2+u)(\hat{\theta}_2+v)] \end{pmatrix} \end{pmatrix}.$$

The target array (5.89) is calculated from the data and this is given by

$$C^* = \begin{pmatrix} \begin{pmatrix} 0 & 0 \\ 0 & a \end{pmatrix} \\ \begin{pmatrix} 0 & \hat{\theta}_1^{-1} \\ \hat{\theta}_1^{-1} & b \end{pmatrix} \end{pmatrix},$$

where $(a, b)' = (\hat{\mathbf{F}}'_{.}\hat{\mathbf{F}}_{.})^{-1}\hat{\mathbf{F}}'_{.}\mathbf{f}_{22}$ can be obtained from a linear regression program. Equating $C^T_{..} = C^*$ and solving for u and v gives

$$u = -\hat{\theta}_2 + \frac{b - [b^2 + (8a/\hat{\theta}_1)]^{1/2}}{2a/\hat{\theta}_1} = 0.396,$$

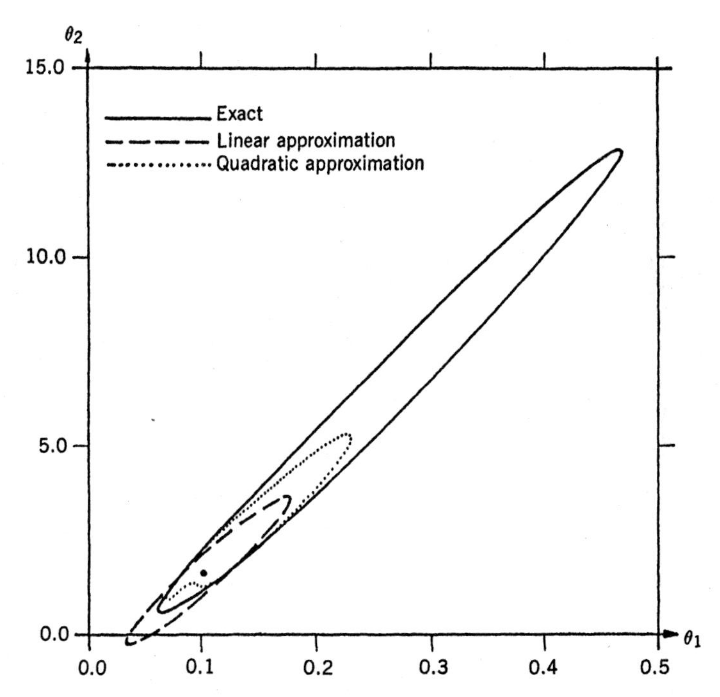

Figure 5.3 Exact and approximate 95% confidence regions for the Michaelis–Menten example. From Bates and Watts [1981a] with permission from the Institute of Mathematical Statistics.

$$v = -\hat{\theta}_2 + \frac{b + [b^2 + (8a/\hat{\theta}_1)]^{1/2}}{2a/\hat{\theta}_1} = 1.945.$$

In order to compare the linearized region (5.83) with that obtained by inverting the linearized region (5.87) for $\boldsymbol{\delta}$, an exact confidence region was constructed using the method described later in Section 5.10.1 [see (5.107) with $\mathbf{P} = \mathbf{F}_{.}(\boldsymbol{\theta})[\mathbf{F}'_{.}(\boldsymbol{\theta})\mathbf{F}_{.}(\boldsymbol{\theta})]^{-1}\mathbf{F}'_{.}(\boldsymbol{\theta})$]. Figure 5.3 compares the three regions, and we see that (5.87) provides an accurate linearized region. ■

5.8.5 Curvature and Parameter Subsets

In Section 5.4 we considered a number of confidence regions for the parameter subset $\boldsymbol{\theta}_2$ of $\boldsymbol{\theta} = (\boldsymbol{\theta}'_1, \boldsymbol{\theta}'_2)'$. The region based on the likelihood ratio was given by [c.f. (5.30) without the correction factor]

$$\{\boldsymbol{\theta}_2 : S[\tilde{\boldsymbol{\theta}}_1(\boldsymbol{\theta}_2), \boldsymbol{\theta}_2] - S(\hat{\boldsymbol{\theta}}) \leq p_2 s^2 F^{\alpha}_{p_2, n-p}\},$$

where $\tilde{\boldsymbol{\theta}}_1$ minimizes $S(\boldsymbol{\theta}_1, \boldsymbol{\theta}_2) = \|\mathbf{y} - \mathbf{f}(\boldsymbol{\theta})\|^2$ for specified $\boldsymbol{\theta}_2$. From (5.54) and (5.64) we have the approximation

$$S(\boldsymbol{\theta}_1, \boldsymbol{\theta}_2) - S(\hat{\boldsymbol{\theta}}) \approx \boldsymbol{\tau}'(\mathbf{I}_p - \hat{\mathbf{B}})\boldsymbol{\tau},$$

where $\hat{\mathbf{B}}$ is given by (5.62). Now $\tilde{\boldsymbol{\theta}}_1$ and $\boldsymbol{\theta}_2$ satisfy the first p_1 normal equations

$$\left[\frac{\partial \mathbf{f}(\boldsymbol{\theta})}{\partial \boldsymbol{\theta}'_1}\right]'_{\tilde{\boldsymbol{\theta}}_1} [\mathbf{y} - \mathbf{f}(\tilde{\boldsymbol{\theta}}_1, \boldsymbol{\theta}_2)] = \mathbf{0}.$$

Let $\boldsymbol{\tau} = (\boldsymbol{\tau}'_1, \boldsymbol{\tau}'_2)'$, where $\boldsymbol{\tau}_i$ is $p_i \times 1$, and define

$$\hat{\mathbf{B}} = \begin{bmatrix} \hat{\mathbf{B}}_{11} & \hat{\mathbf{B}}_{12} \\ \hat{\mathbf{B}}_{21} & \hat{\mathbf{B}}_{22} \end{bmatrix} \begin{matrix} \}p_1 \\ \}p_2 \end{matrix} \qquad (\hat{\mathbf{B}}_{21} = \hat{\mathbf{B}}'_{12}).$$

Then, using a Taylor expansion of the above normal equations and a suitable choice of $\mathbf{Q}_p$ in (5.54), Hamilton [1986] showed that (in our notation)

$$\tilde{\boldsymbol{\tau}}_1 \approx (\mathbf{I}_{p_1} - \hat{\mathbf{B}}_{11})^{-1}\hat{\mathbf{B}}_{12}\boldsymbol{\tau}_2,$$

so that

$$\begin{aligned} \boldsymbol{\tau}'(\mathbf{I}_p - \hat{\mathbf{B}})\boldsymbol{\tau} &= \tilde{\boldsymbol{\tau}}'_1(\mathbf{I}_{p_1} - \hat{\mathbf{B}}_{11})\tilde{\boldsymbol{\tau}}_1 - 2\tilde{\boldsymbol{\tau}}'_1\hat{\mathbf{B}}'_{12}\boldsymbol{\tau}_2 + \boldsymbol{\tau}'_2(\mathbf{I}_{p_2} - \hat{\mathbf{B}}_{22})\boldsymbol{\tau}_2 \\ &= \boldsymbol{\tau}'_2[\mathbf{I}_{p_2} - \hat{\mathbf{B}}_{22} - \hat{\mathbf{B}}_{21}(\mathbf{I} - \hat{\mathbf{B}}_{11})^{-1}\hat{\mathbf{B}}_{12}]\boldsymbol{\tau}_2. \end{aligned}$$

Therefore

$$S(\tilde{\boldsymbol{\theta}}_1, \boldsymbol{\theta}_2) - S(\hat{\boldsymbol{\theta}}) \approx \boldsymbol{\tau}'_2(\mathbf{I}_{p_2} - \hat{\mathbf{T}})\boldsymbol{\tau}_2,$$

where $\hat{\mathbf{T}} = \hat{\mathbf{B}}_{22} + \hat{\mathbf{B}}_{21}(\mathbf{I}_{p_1} - \hat{\mathbf{B}}_{11})^{-1}\hat{\mathbf{B}}_{12}$. Since $\mathbf{I}_p - \hat{\mathbf{B}}$ is positive definite [by (5.68)], its inverse is also positive definite. Hence the lower diagonal block of $(\mathbf{I}_p - \hat{\mathbf{B}})^{-1}$, namely $(\mathbf{I}_{p_2} - \hat{\mathbf{T}})^{-1}$ [by Appendix A3.1], is positive definite, and $\mathbf{I}_{p_2} - \hat{\mathbf{T}}$ is positive definite. Thus corresponding to the full parameter-set approximation (5.79) we have the subset approximation

$$\left\{\boldsymbol{\theta}_2 : \boldsymbol{\tau}_2'(\mathbf{I}_{p_2} - \hat{\mathbf{T}})\boldsymbol{\tau}_2 \le \frac{p_2 s^2 F^{\alpha}_{p_2, n-p}}{1 - \lambda_{\max}}\right\},$$

where $\lambda_{\max}$ is now the maximum eigenvalue of $\hat{\mathbf{T}}$. The reader is referred to Hamilton [1986] for further details.

5.9 NONLINEAR HYPOTHESES

5.9.1 Three Test Statistics

We consider, once again, the nonlinear model (5.1) but without the assumption of normally distributed errors. Suppose we wish to test a nonlinear hypothesis consisting of q $(< p)$ independent nonlinear constraints, that is, $H_0 : \mathbf{a}(\boldsymbol{\theta}) = [a_1(\boldsymbol{\theta}), a_2(\boldsymbol{\theta}), \ldots, a_q(\boldsymbol{\theta})]' = \mathbf{0}$. Three asymptotically equivalent test methods for H_0 have been developed, namely the likelihood-ratio (LR) test, Wald's [1943] (W) test (extended further by Stroud [1971]), and the Lagrange multiplier (LM) test due originally to Rao [1947] and developed further by Silvey [1959] (see also Breusch and Pagan [1980]). If $M(\boldsymbol{\theta})$ is the concentrated log-likelihood function for $\boldsymbol{\theta}$ [c.f. Section 2.2.3], or the log-likelihood function $[= L(\boldsymbol{\theta})]$ from a more general model for $\mathbf{y}$ not involving additional nuisance parameters like σ, then the three test statistics are defined as follows:

$$\text{LR} = 2[M(\hat{\boldsymbol{\theta}}) - M(\tilde{\boldsymbol{\theta}})], \tag{5.92}$$

$$\text{W} = \mathbf{a}'(\hat{\boldsymbol{\theta}})(\hat{\mathbf{A}}\hat{\mathbf{D}}^{-1}\hat{\mathbf{A}}')^{-1}\mathbf{a}(\hat{\boldsymbol{\theta}}) \tag{5.93}$$

and

$$\text{LM} = \left(\frac{\partial M}{\partial \boldsymbol{\theta}'}\mathbf{D}^{-1}\frac{\partial M}{\partial \boldsymbol{\theta}}\right)_{\tilde{\boldsymbol{\theta}}}, \tag{5.94}$$

where

$$\mathbf{A} = \left[\left(\frac{\partial a_i(\boldsymbol{\theta})}{\partial \theta_j}\right)\right] \quad \text{and} \quad \mathbf{D} = \frac{-\partial^2 M(\boldsymbol{\theta})}{\partial \boldsymbol{\theta}\, \partial \boldsymbol{\theta}'}. \tag{5.95}$$

Here $\hat{\mathbf{D}}$ and $\tilde{\mathbf{D}}$ [implicit in (5.94)] represent $\mathbf{D}$ evaluated at $\hat{\boldsymbol{\theta}}$ and $\tilde{\boldsymbol{\theta}}$, the respective unrestricted and restricted (by H_0) maximum-likelihood estimates of $\boldsymbol{\theta}$. Under fairly general conditions (c.f. Amemiya [1983: 351], Engle [1984]), the above three statistics are asymptotically equivalent and asymptotically distributed as

χ_q^2 when H_0 is true. The statistic LR comes from $-2\log\ell$, where ℓ is the likelihood ratio, and this is asymptotically chi-square under H_0 (Section 5.2.3). The Wald statistic W is based on the asymptotic linearization (under H_0)

$$\begin{aligned}\mathbf{a}(\hat{\boldsymbol{\theta}}) &\approx \mathbf{a}(\boldsymbol{\theta}^*) + \mathbf{A}_*(\hat{\boldsymbol{\theta}} - \boldsymbol{\theta}^*)\\ &= \mathbf{A}_*(\hat{\boldsymbol{\theta}} - \boldsymbol{\theta}^*),\end{aligned}$$

where $\mathbf{A}_*$ is $\mathbf{A}$ evaluated at $\boldsymbol{\theta}^*$, the true value of $\boldsymbol{\theta}$. Since $\hat{\boldsymbol{\theta}} - \boldsymbol{\theta}^*$ is asymptotically $N_p(\mathbf{0}, \mathbf{D}_*^{-1})$ it follows that $\mathbf{A}_*(\hat{\boldsymbol{\theta}} - \boldsymbol{\theta}^*)$ is asymptotically $N_q(\mathbf{0}, \mathbf{A}_*\mathbf{D}_*^{-1}\mathbf{A}_*')$. Hence when H_0 is true we have, from (5.93),

$$\begin{aligned}\mathrm{W} &\approx \mathbf{a}'(\hat{\boldsymbol{\theta}})(\mathbf{A}\mathbf{D}^{-1}\mathbf{A}')_*^{-1}\mathbf{a}(\hat{\boldsymbol{\theta}})\\ &\approx [\mathbf{A}_*(\hat{\boldsymbol{\theta}} - \boldsymbol{\theta}^*)]'(\mathbf{A}\mathbf{D}^{-1}\mathbf{A}')_*^{-1}\mathbf{A}_*(\hat{\boldsymbol{\theta}} - \boldsymbol{\theta}^*),\end{aligned}$$

which is asymptotically χ_q^2. In a similar fashion the asymptotic distribution of LM follows from the fact that $\partial M/\partial\boldsymbol{\theta}$ is asymptotically normal when H_0 is true. Furthermore, for an appropriate limiting sequence of alternatives [technically $\boldsymbol{\theta}^* - \boldsymbol{\theta}^+ = O(n^{-1/2})$, where $\tilde{\boldsymbol{\theta}} \to \boldsymbol{\theta}^+$ and $\hat{\boldsymbol{\theta}} \to \boldsymbol{\theta}^*$ as $n \to \infty$], the three statistics are asymptotically distributed as noncentral chi-square when H_0 is false. The equivalence of the statistics follows from the fact that the model and hypothesis are asymptotically equivalent to a linear model with known variance–covariance matrix and a linear hypothesis. The three test statistics are asymptotically equivalent to three corresponding statistics which are indentical for the linear model. Further details are given in Section 12.4.

There is one subtle point about the use of the matrix $\mathbf{D}$, defined by (5.95), that needs highlighting. When the three test statistics were first introduced, the theory was derived for the case of i.i.d. random variables and no nuisance parameters like σ^2. The extensions to the cases where the y_i have different expected values, as in our nonlinear model (5.1), or where concentrated likelihoods are used to avoid problems with nuisance parameters, are more recent developments. Originally the matrix $\mathscr{E}[\mathbf{D}]$ was used instead of $\mathbf{D}$, where $n^{-1}\mathscr{E}[\mathbf{D}]$ is the so-called information matrix, though it makes no difference asymptotically whether we use $\hat{\mathbf{D}}$ or $(\mathscr{E}[\mathbf{D}])_{\hat{\boldsymbol{\theta}}}$: they are asymptotically the same as $\mathscr{E}[\mathbf{D}]$. However, the situation is different with concentrated likelihoods. There the matrix $\hat{\mathbf{D}}$ must be used, because the relationship (2.68), which undergirds the theory, holds for derivative matrices but not for their expected values.

5.9.2 Normally Distributed Errors

If the "errors" ε_i are normally distributed, then the concentrated likelihood function is given by (2.62), which, ignoring constants, is $-(n/2)\log S(\boldsymbol{\theta})$. The maximum-likelihood estimates are now also the least-squares estimates. Using (2.63), and ignoring some smaller-order terms in (5.94), the three statistics take the form (Amemiya [1983:351])

$$\text{LR} = n[\log S(\tilde{\boldsymbol{\theta}}) - \log S(\hat{\boldsymbol{\theta}})], \tag{5.96}$$

$$\text{W} = \frac{\mathbf{a}'(\hat{\boldsymbol{\theta}})[\hat{\mathbf{A}}(\hat{\mathbf{F}}'_{.}\hat{\mathbf{F}}_{.})^{-1}\hat{\mathbf{A}}']^{-1}\mathbf{a}(\hat{\boldsymbol{\theta}})}{S(\hat{\boldsymbol{\theta}})/n}, \tag{5.97}$$

$$\text{LM} = \frac{[\mathbf{y} - \mathbf{f}(\tilde{\boldsymbol{\theta}})]'\tilde{\mathbf{F}}_{.}(\tilde{\mathbf{F}}'_{.}\tilde{\mathbf{F}}_{.})^{-1}\tilde{\mathbf{F}}'_{.}[\mathbf{y} - \mathbf{f}(\tilde{\boldsymbol{\theta}})]}{S(\tilde{\boldsymbol{\theta}})/n}. \tag{5.98}$$

As special cases of the above general theory they are asymptotically χ_q^2 when H_0 is true. Since the asymptotic normality of the least-squares estimates provides the normality part of the asymptotic linear model, we find that the assumption of normal errors is not needed. The above three statistics are still asymptotically χ_q^2 without the normality of the ε_i (Amemiya [1983]). We note that LR is also mentioned briefly in Section 5.2.3.

Using the approximation $\log(1 + x) \approx x$ for small x, we have

$$\begin{aligned} \text{LR} &\approx (n-p)\log\frac{S(\tilde{\boldsymbol{\theta}})}{S(\hat{\boldsymbol{\theta}})} \\ &= (n-p)\log\left(1 + \frac{S(\tilde{\boldsymbol{\theta}}) - S(\hat{\boldsymbol{\theta}})}{S(\hat{\boldsymbol{\theta}})}\right) \\ &\approx (n-p)\frac{S(\tilde{\boldsymbol{\theta}}) - S(\hat{\boldsymbol{\theta}})}{S(\hat{\boldsymbol{\theta}})} \\ &= q\tilde{F}_2, \end{aligned} \tag{5.99}$$

where $\tilde{F}_2$ is given by (5.23). Milliken and DeBruin [1978] demonstrated that, under general regularity conditions, $q\tilde{F}_2$ is asymptotically χ_q^2 when H_0 is true. This is not surprising in view of (5.99). They also presented some evidence, which is supported by Gallant [1987], that the null distribution of $\tilde{F}_2$ is generally well approximated by the $F_{q,n-p}$ distribution. Milliken et al. [1977] carried out a simulation study for the Michaelis–Menten model

$$y_{ij} = \frac{\theta_i x_{ij}}{\mu_i + x_{ij}} + \varepsilon_{ij}, \qquad i = 1, 2, 3, \quad j = 1, 2, \ldots, n,$$

and found that the F-approximation for $\tilde{F}_2$ worked satisfactorily with smaller sample sizes than the chi-squared approximation for $q\tilde{F}_2$ ($n = 6$ as opposed to $n = 24$).

Following Section 5.3, where we considered the special case of testing

$H_{02}:\mathbf{a}(\boldsymbol{\theta}) = \boldsymbol{\theta}_2 - \boldsymbol{\theta}_{02} = \mathbf{0}$, we can define a modified version of W, namely

$$\begin{aligned}\tilde{F}_1 &= \frac{n-p}{nq} W \\ &= \mathbf{a}'(\hat{\boldsymbol{\theta}})[\hat{\mathbf{A}}(\hat{\mathbf{F}}_{\bullet}'\hat{\mathbf{F}}_{\bullet})^{-1}\hat{\mathbf{A}}']^{-1}\mathbf{a}(\hat{\boldsymbol{\theta}})/qs^2. \end{aligned} \tag{5.100}$$

This reduces to (5.21) when $H_0 = H_{02}$. The statistics $\tilde{F}_3$ and $\tilde{F}_4$ in Section 5.3 also apply here and, along with $\tilde{F}_1$, are approximately distributed as $F_{q,n-p}$ when H_0 is true. Also LM can, once again, be related to $F_{q,n-p}$ via the relationship (5.24).

It is of interest to apply the above theory to the case of a linear model, as in the following example.

Example 5.3 Suppose we wish to test $H_0:\boldsymbol{\beta} = \boldsymbol{\beta}_0$ for the linear model $\mathbf{y} = \mathbf{X}\boldsymbol{\beta} + \boldsymbol{\varepsilon}$, where $\mathbf{X}$ is $n \times p$ of rank p and $\boldsymbol{\varepsilon} \sim N_n(\mathbf{0}, \sigma^2\mathbf{I}_n)$. Then using $\boldsymbol{\beta} = \boldsymbol{\theta}$, $\mathbf{A} = \mathbf{I}_p$, $\mathbf{F}_{\bullet} = \mathbf{X}$, and $\hat{\boldsymbol{\beta}} = (\mathbf{X}'\mathbf{X})^{-1}\mathbf{X}'\mathbf{y}$ in the general theory, (5.96) to (5.98) become

$$\begin{aligned}
\mathrm{LR} &= n \log \{S(\boldsymbol{\beta}_0)/S(\hat{\boldsymbol{\beta}})\}, \\
\mathrm{W} &= n(\hat{\boldsymbol{\beta}} - \boldsymbol{\beta}_0)'\mathbf{X}'\mathbf{X}(\hat{\boldsymbol{\beta}} - \boldsymbol{\beta}_0)/S(\hat{\boldsymbol{\beta}}) \\
&= n\frac{S(\boldsymbol{\beta}_0) - S(\hat{\boldsymbol{\beta}})}{S(\hat{\boldsymbol{\beta}})} \qquad \text{(by Appendix D1.4)}, \\
\mathrm{LM} &= n(\mathbf{y} - \mathbf{X}\boldsymbol{\beta}_0)'\mathbf{X}(\mathbf{X}'\mathbf{X})^{-1}\mathbf{X}'(\mathbf{y} - \mathbf{X}\boldsymbol{\beta}_0)/S(\boldsymbol{\beta}_0) \\
&= n(\mathbf{X}'\mathbf{y} - \mathbf{X}'\mathbf{X}\boldsymbol{\beta}_0)'(\mathbf{X}'\mathbf{X})^{-1}(\mathbf{X}'\mathbf{y} - \mathbf{X}'\mathbf{X}\boldsymbol{\beta}_0)/S(\boldsymbol{\beta}_0) \\
&= n(\mathbf{X}'\mathbf{X}\hat{\boldsymbol{\beta}} - \mathbf{X}'\mathbf{X}\boldsymbol{\beta}_0)'(\mathbf{X}'\mathbf{X})^{-1}(\mathbf{X}'\mathbf{X}\hat{\boldsymbol{\beta}} - \mathbf{X}'\mathbf{X}\boldsymbol{\beta}_0)/S(\boldsymbol{\beta}_0) \\
&= n(\hat{\boldsymbol{\beta}} - \boldsymbol{\beta}_0)'\mathbf{X}'\mathbf{X}(\hat{\boldsymbol{\beta}} - \boldsymbol{\beta}_0)/S(\boldsymbol{\beta}_0) \\
&= n\frac{S(\boldsymbol{\beta}_0) - S(\hat{\boldsymbol{\beta}})}{S(\boldsymbol{\beta}_0)}.
\end{aligned}$$

If $b = S(\boldsymbol{\beta}_0)/S(\hat{\boldsymbol{\beta}})$, then $\mathrm{LR} = n\log b$, $\mathrm{W} = n(b-1)$, and $\mathrm{LM} = n(1 - b^{-1})$. Now $e^{x-1} \geq 1 + (x-1)$, so $x - 1 \geq \log x$ for $x > 0$. Therefore, setting $x = b$ and $x = b^{-1}$, we obtain $b - 1 \geq \log b \geq 1 - b^{-1}$. Hence

$$\mathrm{W} \geq \mathrm{LR} \geq \mathrm{LM}. \tag{5.101}$$

We note from Appendix D2.1 that the standard F-test for H_0 is given by $F = (n-p)\mathrm{W}/(np)$, so that W, LR, and LM are all monotonic functions of F. Therefore exact tests based on these statistics are equivalent to the standard F-test. We also find that $\tilde{F}_1, \tilde{F}_2, \tilde{F}_3, R$ [or LM with d_α of (5.25)] and $\tilde{F}_4$ of Section 5.3 are all the same as F. ■

Although LR, W, and LM are asymptotically equal in the general case, we see that the ordering (5.101) holds for the above linear model. This ordering was established by Breusch [1979] for a general linear hypothesis and a more general error covariance matrix. He used an elegant conditional-likelihood argument, thereby extending previous cases considered by Savin [1976: autoregressive errors] and Berndt and Savin [1977: multivariate linear regression]. Hence, using (5.101) and the chi-square distribution, we find that whenever the LM test rejects the others also reject, and whenever W fails to reject the others fail also. However, as noted by Engle [1984], the inequality has nothing to say about the relative powers of the tests, because it applies under the null hypothesis as well. If the W test has a size of 5%, then the LR and LM will have sizes less than 5%. A reduction in size means a reduction in the power as well. Evidence is given by Rothenberg [1980] and Evans and Savin [1982] that the powers are approximately the same when the sizes are approximately corrected.

In the case of nonlinear models, the inequalities no longer hold, though Mizon [1977] found that W $\geq$ LR most of the time in his examples. However, this problem does not seem to arise with the modified statistics $\tilde{F}_k$ $(k = 1, 2, 3, 4)$ and LM (with d_α).

5.9.3 Freedom-Equation Specification

A major problem in testing H_0 using LR or LM with normal data is the computation of the restricted least-squares estimate $\tilde{\boldsymbol{\theta}}$. This is obtained by minimizing $S(\boldsymbol{\theta})$ subject to $\mathbf{a}(\boldsymbol{\theta}) = \mathbf{0}$, and general methods are available for doing this (e.g. Gill et al. [1981], Fletcher [1981]). However, as emphasized by Gallant [1987], H_0 can generally be rewritten in a "freedom equation" format $\boldsymbol{\theta} = \boldsymbol{\theta}(\boldsymbol{\alpha})$, where $\boldsymbol{\alpha}$ has dimension r $(= p - q)$.

Example 5.4 Suppose $\boldsymbol{\theta} = (\theta_1, \theta_2, \theta_3, \theta_4)'$,

$$a_1(\boldsymbol{\theta}) = \theta_1 e^{\theta_2} - 1 = 0$$

and

$$a_2(\boldsymbol{\theta}) = \theta_3 e^{-\theta_2} - 1 = 0.$$

Then $r = 3 - 2 = 1$ and we can, for example, choose $\alpha = \theta_3$. Since $\theta_1 = 1/\theta_3$ and $\theta_2 = \log \theta_3$, we have

$$\boldsymbol{\theta}(\alpha) = \begin{bmatrix} \theta_1(\alpha) \\ \theta_2(\alpha) \\ \theta_3(\alpha) \end{bmatrix} = \begin{bmatrix} \alpha^{-1} \\ \log \alpha \\ \alpha \end{bmatrix}.$$

■

Given the specification $\boldsymbol{\theta} = \boldsymbol{\theta}(\boldsymbol{\alpha})$, then $\mathbf{f}(\boldsymbol{\theta}) = \mathbf{f}[\boldsymbol{\theta}(\boldsymbol{\alpha})] = \mathbf{g}(\boldsymbol{\alpha})$, say. Thus, minimizing $S(\boldsymbol{\theta})$ subject to H_0 is equivalent to the *unconstrained* minimization of $S(\boldsymbol{\alpha})$ with respect to $\boldsymbol{\alpha}$. This can be done using a standard algorithm. All we require is the Jacobian matrix (c.f. Appendix A10.3)

$$\frac{\partial \mathbf{g}}{\partial \boldsymbol{\alpha}'} = \frac{\partial \mathbf{f}}{\partial \boldsymbol{\theta}'} \frac{\partial \boldsymbol{\theta}}{\partial \boldsymbol{\alpha}'} = \mathbf{F}_{.}[\boldsymbol{\theta}(\boldsymbol{\alpha})]\boldsymbol{\Theta}(\boldsymbol{\alpha}), \qquad \text{say.} \tag{5.102}$$

In some situations it may not be possible to find the freedom-equation specification analytically, as in the following example.

Example 5.5 Suppose we have the model

$$E[y] = (\theta_1 + \theta_2)e^{-x(\theta_2 - \theta_1)} \qquad (\theta_2 > \theta_1),$$

and we wish to test whether $E[y] = 1$ when $x = 1$, i.e. test $H_0: (\theta_1 + \theta_2) = \exp(\theta_2 - \theta_1)$. Then, setting $\theta_1 = \alpha$, we have

$$\begin{pmatrix} \theta_1(\alpha) \\ \theta_2(\alpha) \end{pmatrix} = \begin{pmatrix} \alpha \\ a_2(\alpha) \end{pmatrix},$$

where $a_2(\alpha)$ is the value of θ_2 satisfying $\theta_2 - \alpha = \log(\theta_2 + \alpha)$. Unfortunately the latter equation does not have an analytic solution. However, given α, this equation can be solved iteratively for θ_2 using the iterative process $\theta_2^{(a+1)} = \log(\theta_2^{(a)} + \alpha) + \alpha$. Also

$$\frac{\partial \theta_2}{\partial \alpha} = \frac{1}{\theta_2 + \alpha}\left(\frac{\partial \theta_2}{\partial \alpha} + 1\right) + 1,$$

so that

$$\frac{\partial \boldsymbol{\theta}}{\partial \alpha} = \begin{pmatrix} 1 \\ (\theta_2 + \alpha + 1)/(\theta_2 + \alpha - 1) \end{pmatrix}.$$

■

We see from the above example that it is not necessary to derive the freedom specification analytically. All that is required are the values of $\boldsymbol{\theta} = \boldsymbol{\theta}(\boldsymbol{\alpha})$ and $\partial \mathbf{g}/\partial \boldsymbol{\alpha}'$ for a given $\boldsymbol{\alpha}$, and these can be found numerically. More generally, we only need $\boldsymbol{\theta}(\boldsymbol{\alpha})$, as $\partial \boldsymbol{\theta}/\partial \boldsymbol{\alpha}'$ in (5.102) can be approximated using first differences, e.g.

$$\frac{\partial \boldsymbol{\theta}(\boldsymbol{\alpha})}{\partial \alpha_j} \approx \frac{\boldsymbol{\theta}(\boldsymbol{\alpha} + h\mathbf{e}_j) - \boldsymbol{\theta}(\boldsymbol{\alpha})}{h},$$

where $\mathbf{e}_j$ has 1 for the jth component and zeros elsewhere.

5.9.4 Comparison of Test Statistics

The modified statistics $\tilde{F}_k$ $(k=1,2,3)$ and LM (with d_α) were compared by Gallant [1987] for the general nonlinear hypothesis and not just for $H_{02}:\boldsymbol{\theta}_2=\boldsymbol{\theta}_{02}$. Hence the comments of Section 5.3 about the relative merits of the statistics apply in general. The Wald statistic $\tilde{F}_1$ of (5.100) has the advantage of not requiring the restricted estimator $\tilde{\boldsymbol{\theta}}$. This is particularly useful if several hypotheses are to be checked. One simply evaluates $\mathbf{a}(\hat{\boldsymbol{\theta}})$ for the set of constraints $\mathbf{a}(\boldsymbol{\theta})$ imposed by each null hypothesis using the same value of $\hat{\boldsymbol{\theta}}$. Also, given $\hat{\boldsymbol{\theta}}, \tilde{F}_1$ can be readily calculated using a standard program for solving a linear least-squares problem subject to linear constraints (Gallant [1987: 50]). The linear model used is $\hat{\boldsymbol{\varepsilon}}=\hat{\mathbf{F}}_{.}\boldsymbol{\beta}+\mathbf{u}$, and the constraints are $\hat{\mathbf{A}}\boldsymbol{\beta}=\mathbf{a}(\hat{\boldsymbol{\theta}})$. This leads to $\hat{\boldsymbol{\beta}}=(\hat{\mathbf{F}}_{.}'\hat{\mathbf{F}}_{.})^{-1}\hat{\mathbf{F}}_{.}'\hat{\boldsymbol{\varepsilon}}=\mathbf{0}$ [as $\hat{\mathbf{F}}_{.}'\hat{\boldsymbol{\varepsilon}}=\mathbf{0}$, by the normal equations (2.9)] and a test statistic (by Appendix D2.4)

$$\begin{aligned} F &= \frac{\{\hat{\mathbf{A}}\hat{\boldsymbol{\beta}}-\mathbf{a}(\hat{\boldsymbol{\theta}})\}'[\hat{\mathbf{A}}(\hat{\mathbf{F}}_{.}'\hat{\mathbf{F}}_{.})^{-1}\hat{\mathbf{A}}']^{-1}\{\hat{\mathbf{A}}\hat{\boldsymbol{\beta}}-\mathbf{a}(\hat{\boldsymbol{\theta}})\}/q}{\hat{\boldsymbol{\varepsilon}}'\{\mathbf{I}_n-\hat{\mathbf{F}}_{.}(\hat{\mathbf{F}}_{.}'\hat{\mathbf{F}}_{.})^{-1}\hat{\mathbf{F}}_{.}'\}\hat{\boldsymbol{\varepsilon}}/(n-p)} \\ &= \mathbf{a}'(\hat{\boldsymbol{\theta}})[\hat{\mathbf{A}}(\hat{\mathbf{F}}_{.}'\hat{\mathbf{F}}_{.})^{-1}\hat{\mathbf{A}}']^{-1}\mathbf{a}(\hat{\boldsymbol{\theta}})/qs^2 \\ &= \tilde{F}_1, \end{aligned}$$

since $s^2=\hat{\boldsymbol{\varepsilon}}'\hat{\boldsymbol{\varepsilon}}/(n-p)$. Unfortunately, $\tilde{F}_1$ is not invariant to reparametrization, so that a change in the parameters defining the model can lead to a different value of $\tilde{F}_1$. It is perhaps not surprising, therefore, that $\tilde{F}_1$ can be seriously affected by any parameter-effects curvature. Another problem associated with $\tilde{F}_1$ is that it is not invariant to the method of expressing the nonlinear hypothesis $\mathbf{a}(\boldsymbol{\theta})=\mathbf{0}$. For example, $\theta_1-\theta_2^{-1}=0$ and $\theta_1\theta_2-1=0$ will lead to different statistics (Gregory and Veall [1985]) which, although asymptotically equivalent, can be appreciably different for small samples.

The likelihood-ratio statistic $\tilde{F}_2$ tends to be the most powerful of the test statistics. Along with R and $\tilde{F}_4$, it is invariant under reparametrization, so that it is not affected by any parameter-effects curvature. However, it can be affected by intrinsic curvature, and the corrections for this obtained from (5.26) still apply to general H_0, but with $(\tilde{\boldsymbol{\theta}}_1'(\boldsymbol{\theta}_{02}),\boldsymbol{\theta}_{02}')'$ corresponding to H_{02} replaced by the general restricted estimate $\tilde{\boldsymbol{\theta}}$ corresponding to H_0 (Hamilton and Wiens [1986]).

If one has a particular null hypothesis in mind and wishes to consider several possible generalizations of the hypothesis, then an LM statistic has the advantage that only one estimate of $\boldsymbol{\theta}$, $\tilde{\boldsymbol{\theta}}$, is required. Such generalizations are useful in diagnostic testing: this role of the LM method is discussed by Engle [1984].

In conclusion we note that further extensions of the above test procedures are given by a number of authors, including Gallant and Holly [1980; implicit simultaneous-equations model], Gourieroux et al. [1980, 1982: inequality constraints], Burguete et al. [1982: a general model encompassing robust, multivariate, and simultaneous-equations models], Amemiya [1983: a general survey], Engle [1984: LM tests], Rogers [1986: one-sided alternatives], and Gallant [1987: general theory].

5.9.5 Confidence Regions and Intervals

Confidence regions and intervals for a set of q nonlinear functions $\mathbf{b}(\boldsymbol{\theta})$ can be obtained, as in Section 5.4 for the special case $\mathbf{b}(\boldsymbol{\theta}) = \boldsymbol{\theta}_2$, by "inverting" any of the test procedures for testing $H_0{:}\mathbf{a}(\boldsymbol{\theta}) = \mathbf{b}(\boldsymbol{\theta}) - \mathbf{b}_0 = \mathbf{0}$. For example, using a statistic $\widetilde{F}_k(\mathbf{b}_0)$, say, we have an approximate $100(1-\alpha)\%$ confidence region

$$\{\mathbf{b}_0 : \widetilde{F}_k(\mathbf{b}_0) \le F^{\alpha}_{q,n-p}\}$$

for $\mathbf{b}(\boldsymbol{\theta})$. Thus if $\widetilde{\boldsymbol{\theta}}(\mathbf{b}_0)$ is the restricted estimate of $\boldsymbol{\theta}$ under H_0, the corrected likelihood-ratio region corresponding to (5.30) is

$$\{\mathbf{b}_0 : S[\widetilde{\boldsymbol{\theta}}(\mathbf{b}_0)] - S(\hat{\boldsymbol{\theta}}) \le qs^2(1 - \hat{\gamma}_2 s^2)F^{\alpha}_{q,n-p}\}.$$

Of special interest is the problem of finding a confidence interval for a single function ($q = 1$). The above region then reduces to the form

$$\{b_0 : L(b_0) \le F^{\alpha}_{q,n-p}\},$$

which can be found by standard interpolation methods. If k such intervals are required, then we can use the Bonferroni method, i.e., α is replaced by α/k. In the case of the Wald test the confidence interval for a single function $a(\boldsymbol{\theta})$ reduces to (5.8). However, as already noted, this interval can be affected by any parameter-effects curvature. A recent method of Clarke [1987a] looks promising.

5.9.6 Multiple Hypothesis Testing

An important problem is that of choosing between alternative models. Sometimes one model is simply a special case of another and can be tested using a nested scheme (c.f. Mizon [1977]). When nesting is not possible, the hypotheses are sometimes called "separate" or non-nested. In the simplest situation we have just two models and the problem is to test a specific null model against a non-nested alternative, that is one not specified by the same set of parameters as the null model. For example consider the non-nested competing models

$$\begin{aligned} H_0{:}y &= \theta_{10} + \theta_{11}x + \theta_{12}x^2 + \varepsilon \\ &= f^{(1)}(x; \boldsymbol{\theta}^{(1)}) + \varepsilon \end{aligned} \tag{5.103}$$

and

$$\begin{aligned} H_1{:}y &= \theta_{21}\exp(\theta_{22}x) + \varepsilon \\ &= f^{(2)}(x; \boldsymbol{\theta}^{(2)}) + \varepsilon. \end{aligned} \tag{5.104}$$

To date, the approach which has been most widely used is that of Cox [1961, 1962], who proposed comparing the value of the usual likelihood-ratio statistic with its expected value under the null hypothesis (see also Fisher and McAleer [1979] for useful comments on the issues involved). For nonlinear models this approach has been extended in a number of directions, and the reader is referred to the articles in the Journal of Econometrics, Volume 21 (1983) and to Kent [1986] for further details and references. In particular, Aguirre-Torres and Gallant [1983] considered the case of comparing multivariate nonlinear models and introduced a new nonparametric test based on the bootstrap method.

A second method of testing alternative models is to embed both models in a more general model and then test each alternative against the general model. For example

$$y = \theta_0 + \theta_1 x + \theta_2 x^2 + \theta_3 \exp(\theta_4 x) + \varepsilon \tag{5.105}$$

combines (5.103) and (5.104) into one model. However, it may happen that both models are acceptable and we are still undecided. Another variation on the same theme is to use the artificial model

$$y = (1-\alpha) f^{(1)}(x; \boldsymbol{\theta}^{(1)}) + \alpha f^{(2)}(x; \boldsymbol{\theta}^{(2)}) + \varepsilon$$

and test $H_0: \alpha = 0$ versus $H_1: \alpha = 1$. Unfortunately, by itself the above model is not very useful, as α, $\boldsymbol{\theta}^{(1)}$, and $\boldsymbol{\theta}^{(2)}$ will generally be unidentifiable. One possibility is to replace $\boldsymbol{\theta}^{(2)}$ by its least-squares estimate: for further details see MacKinnon et al. [1983]. The density functions for the two models rather than expected values can also be combined in an appropriate manner (e.g. exponential weighting in Atkinson [1970], Breusch and Pagan [1980]; linear weighting in Quandt [1974]).

A third method, for handling either nested or non-nested hypotheses, based on the concept of "pseudo-true" values, is given by Gourieroux et al. [1983].

In conclusion we note that the problem of competing models is considered further in Section 5.13.5 under the general title of optimal designs. There we discuss the question of choosing data points to facilitate model selection.

5.10 EXACT INFERENCE

5.10.1 Hartley's method

Hartley [1964] suggested the following method for providing exact tests and confidence regions for $\boldsymbol{\theta}$ in the model $\mathbf{y} = \mathbf{f}(\boldsymbol{\theta}) + \boldsymbol{\varepsilon}$, where $\boldsymbol{\varepsilon} \sim N_n(\mathbf{0}, \sigma^2 \mathbf{I}_n)$. Let $\mathbf{P}$ be any $n \times n$ symmetric idempotent (projection) matrix of rank p. Since

$\mathbf{P}(\mathbf{I}_n - \mathbf{P}) = \mathbf{0}$, it follows from Appendix A11.6 that

$$\begin{aligned} F_{\boldsymbol{\theta}} &= \frac{\boldsymbol{\varepsilon}'\mathbf{P}\boldsymbol{\varepsilon}/p}{\boldsymbol{\varepsilon}'(\mathbf{I}_n - \mathbf{P})\boldsymbol{\varepsilon}/(n-p)} \\ &= \frac{[\mathbf{y} - \mathbf{f}(\boldsymbol{\theta})]'\mathbf{P}[\mathbf{y} - \mathbf{f}(\boldsymbol{\theta})]}{[\mathbf{y} - \mathbf{f}(\boldsymbol{\theta})]'(\mathbf{I}_n - \mathbf{P})[\mathbf{y} - \mathbf{f}(\boldsymbol{\theta})]} \frac{n-p}{p} \\ &\sim F_{p,n-p}. \end{aligned} \tag{5.106}$$

To test $H_0: \boldsymbol{\theta} = \boldsymbol{\theta}_0$ we evaluate $F_{\boldsymbol{\theta}_0}$ and reject H_0 at the α level of significance if $F_{\boldsymbol{\theta}_0} > F^{\alpha}_{p,n-p}$. Also

$$\{\boldsymbol{\theta} : F_{\boldsymbol{\theta}} \leq F^{\alpha}_{p,n-p}\}$$

or

$$\left\{\boldsymbol{\theta} : \boldsymbol{\varepsilon}'(\boldsymbol{\theta})\mathbf{P}\boldsymbol{\varepsilon}(\boldsymbol{\theta}) \leq \boldsymbol{\varepsilon}'(\boldsymbol{\theta})\boldsymbol{\varepsilon}(\boldsymbol{\theta}) \frac{f}{1+f}\right\} \tag{5.107}$$

is a $100(1-\alpha)\%$ confidence region for $\boldsymbol{\theta}$, where $\boldsymbol{\varepsilon}(\boldsymbol{\theta}) = \mathbf{y} - \mathbf{f}(\boldsymbol{\theta})$ and

$$f = pF^{\alpha}_{p,n-p}/(n-p).$$

Given that the normality assumption for $\boldsymbol{\varepsilon}$ is reasonable, the use of (5.106) not only has the advantage of providing exact tests and confidence regions, but it also avoids the regularity conditions and the large samples needed for the applicability of asymptotic least-squares theory. However, the power of the exact test will depend very much on the choice of $\mathbf{P}$. Gallant [1971] has given examples where the test can have either very poor power or good power, so that the choice of $\mathbf{P}$ is critical. An exact test with poor power is no substitute for an approximate test with good power.

The choice of $\mathbf{P}$ is also important in "unraveling" the confidence region for $\boldsymbol{\theta}$. Even for $p = 1$ or 2, the presence of $\boldsymbol{\theta}$ in the denominator of $F_{\boldsymbol{\theta}}$ can make the computation difficult, and may result in a confidence region made up of disjoint subregions. By approximating $\mathbf{f}(\boldsymbol{\theta})$, Hartley [1964] proposed choosing a $\mathbf{P}$ which does not depend on $\boldsymbol{\theta}$ such that $(\mathbf{I}_n - \mathbf{P})\mathbf{f}(\boldsymbol{\theta}) \approx \mathbf{0}$. The denominator of $F_{\boldsymbol{\theta}}$ is now approximately independent of $\boldsymbol{\theta}$ and we have, from (5.106),

$$F_{\boldsymbol{\theta}} \approx \frac{n-p}{p} \frac{[\mathbf{y} - \mathbf{f}(\boldsymbol{\theta})]'[\mathbf{y} - \mathbf{f}(\boldsymbol{\theta})] - \mathbf{y}(\mathbf{I}_n - \mathbf{P})\mathbf{y}}{\mathbf{y}'(\mathbf{I}_n - \mathbf{P})\mathbf{y}}. \tag{5.108}$$

We now have the desirable feature that contours of $F_{\boldsymbol{\theta}} = a$ with respect to values of $\boldsymbol{\theta}$ correspond to contours $\boldsymbol{\varepsilon}'(\boldsymbol{\theta})\boldsymbol{\varepsilon}(\boldsymbol{\theta}) = b$ of constant likelihood. Hartley [1964] suggested several possibilities for $\mathbf{P}$, including one based on Lagrangian interpolation.

The test statistic F_{θ_0} can also be derived in the following manner. Set $\mathbf{z} = \mathbf{y} - \mathbf{f}(\boldsymbol{\theta}_0)$. Then, under H_0, $\mathbf{z} = \boldsymbol{\varepsilon}$ and $\boldsymbol{\delta} = \mathbf{0}$ in the linear model

$$\mathbf{z} = \mathbf{D}_0\boldsymbol{\delta} + \boldsymbol{\varepsilon}, \tag{5.109}$$

where $\mathbf{D}_0$ is an arbitrary $n \times p$ matrix of rank p. Hence H_0 can be tested by testing $\boldsymbol{\delta} = \mathbf{0}$, and from Appendix D2.3 the appropriate F-statistic is

$$\frac{n-p}{p} \frac{\mathbf{z}'\mathbf{D}_0(\mathbf{D}_0'\mathbf{D}_0)^{-1}\mathbf{D}_0'\mathbf{z}}{\mathbf{z}'[\mathbf{I}_n - \mathbf{D}_0(\mathbf{D}_0'\mathbf{D}_0)^{-1}\mathbf{D}_0']\mathbf{z}}, \tag{5.110}$$

namely, F_{θ_0} of (5.105) with $\mathbf{P} = \mathbf{D}_0(\mathbf{D}_0'\mathbf{D}_0)^{-1}\mathbf{D}_0'$ $(= \mathbf{P}_{\mathbf{D}_0}$, say). This statistic can be computed using a standard regression program. Since the rows of $\mathbf{D}_0$ may be regarded as observations on p regressors, the above method might be described as the method of "redundant additional regressors."

A natural choice of $\mathbf{D}_0$ can be obtained from linearization. For $\boldsymbol{\theta}$ close to $\boldsymbol{\theta}_0$,

$$\mathbf{y} - \mathbf{f}(\boldsymbol{\theta}) \approx \mathbf{y} - \mathbf{f}(\boldsymbol{\theta}_0) - \mathbf{F}_{.}(\boldsymbol{\theta}_0)(\boldsymbol{\theta} - \boldsymbol{\theta}_0).$$

Thus we might hope to test $\boldsymbol{\theta} = \boldsymbol{\theta}_0$ by fitting the linear model

$$\mathbf{y} - \mathbf{f}(\boldsymbol{\theta}_0) = \mathbf{F}_{.}(\boldsymbol{\theta}_0)\boldsymbol{\delta} + \boldsymbol{\varepsilon}, \tag{5.111}$$

where $\boldsymbol{\delta} = \boldsymbol{\theta} - \boldsymbol{\theta}_0$, and testing $\boldsymbol{\delta} = \mathbf{0}$. This corresponds to using $\mathbf{D}_0 = \mathbf{F}_{.}(\boldsymbol{\theta}_0)$ in (5.109) and (5.110). It should be noted that the precision of the approximation used to get (5.111) in no way affects the exactness of the statement (5.106). The Taylor expansion is simply used to find a $\mathbf{D}$-matrix with good local power. In this case $\mathbf{P}$ represents the projection onto the tangent plane to the expectation surface at $\boldsymbol{\theta}_0$, namely $\mathscr{R}[\mathbf{F}_{.}(\boldsymbol{\theta}_0)]$. If this linear tangent-plane approximation works, it can be shown [e.g. by using (5.108)] that the test based on $\mathbf{D}_0 = \mathbf{F}_{.}(\boldsymbol{\theta}_0)$ is approximately equal to the likelihood-ratio test (5.19) and will therefore have good power.

Halperin [1963] arrived at $\mathbf{D} = \mathbf{F}_{.}(\boldsymbol{\theta})$ by requiring $F_{\hat{\theta}} = 0$, or equivalently $\hat{\mathbf{D}}'[\mathbf{y} - \mathbf{f}(\hat{\boldsymbol{\theta}})] = \mathbf{0}$ [c.f. Equation (2.9)]. Sundararaj [1978] also arrived at the same method by looking at the distribution of $\mathbf{F}_{.}'(\boldsymbol{\theta})[\mathbf{y} - \mathbf{f}(\boldsymbol{\theta})]$.

Example 5.6 Consider the model

$$\begin{aligned} y_i &= f(x_i; \boldsymbol{\theta}) + \varepsilon_i \\ &= \alpha + \beta g(x_i; \gamma) + \varepsilon_i, \end{aligned} \tag{5.112}$$

with $\boldsymbol{\theta}' = (\alpha, \beta, \gamma)$. To test $\boldsymbol{\theta} = \boldsymbol{\theta}_0$ we carry out the test based on $\mathbf{D}_0 = \mathbf{F}_{.}(\boldsymbol{\theta}_0)$,

where

$$\begin{aligned}\mathbf{F}_{\bullet}(\boldsymbol{\theta}) &= \left[\frac{\partial \mathbf{f}(\boldsymbol{\theta})}{\partial \alpha}, \frac{\partial \mathbf{f}(\boldsymbol{\theta})}{\partial \beta}, \frac{\partial \mathbf{f}(\boldsymbol{\theta})}{\partial \gamma}\right] \\ &= \left[\mathbf{1}_n, \mathbf{g}(\gamma), \beta \frac{\partial \mathbf{g}}{\partial \gamma}\right] \\ &= [\mathbf{1}_n, \mathbf{g}(\gamma), \beta \dot{\mathbf{g}}(\gamma)], \\ \mathbf{g}(\gamma) &= [g(x_1;\gamma), g(x_2;\gamma), \ldots, g(x_n;\gamma)]'. \end{aligned} \tag{5.113}$$

However, Williams [1962] showed that it is possible to test $\gamma = \gamma_0$ by using a Taylor expansion on just γ. Assuming γ is close to γ_0, we have the approximate model

$$\begin{aligned} y_i &= \alpha + \beta[g(x_i;\gamma_0) + (\gamma - \gamma_0)\dot{g}(x_i;\gamma_0)] + \varepsilon_i \\ &= \alpha + \beta g(x_i;\gamma_0) + \delta \dot{g}(x_i;\gamma_0) + \varepsilon_i, \end{aligned} \tag{5.114}$$

where $\delta = \beta(\gamma - \gamma_0)$. To test $\gamma = \gamma_0$ we simply test $\delta = 0$ in the linear model (5.114) using a standard linear regression package. If we wished to test $(\alpha, \beta, \gamma) = (\alpha_0, \beta_0, \gamma_0)$, then we could use the general method based on (5.111). However, this general approach is equivalent to testing $(\alpha, \beta, \delta) = (\alpha_0, \beta_0, 0)$ in (5.114).

We note that (5.114) is a special case of

$$y_i = \alpha + \beta g(x_i;\gamma_0) + \delta d_i + \varepsilon_i, \tag{5.115}$$

where d_i is the ith value of a redundant additional regressor—redundant in the sense that if γ_0 is the true value of γ then $\delta = 0$. A generalization of this procedure for handling just the nonlinear parameters in a model, like γ above, is given in the next section. ■

Having discussed exact hypothesis tests, we now turn our attention to the exact confidence region (5.107). A natural question to ask is under what conditions this region is similar to the asymptotic region (5.13). Suppose that n is large enough for the true parameter value $\boldsymbol{\theta}$ to be close to $\hat{\boldsymbol{\theta}}$. Then using the transformation (5.54) from $\boldsymbol{\theta}$ to the $\boldsymbol{\tau}$-parameters, and developing expressions similar to those in Section 5.8.1, Hamilton et al. [1982] showed that when $\mathbf{P}$ projects onto the tangent plane, i.e. $\mathbf{P} = \mathbf{F}_{\bullet}(\boldsymbol{\theta})[\mathbf{F}_{\bullet}'(\boldsymbol{\theta})\mathbf{F}_{\bullet}(\boldsymbol{\theta})]^{-1}\mathbf{F}_{\bullet}'(\boldsymbol{\theta})$, then (5.107) is approximately equivalent to

$$\{\boldsymbol{\tau} : \boldsymbol{\tau}'(\mathbf{I}_p - \hat{\mathbf{B}})[\mathbf{I}_p - (1 + f)\hat{\mathbf{B}}]\boldsymbol{\tau} \le F_\alpha \rho^2\}. \tag{5.116}$$

This may be compared with the corresponding approximation to (5.13), namely [c.f. (5.66)]

$$\{\boldsymbol{\tau} : \boldsymbol{\tau}'(\mathbf{I}_p - \hat{\mathbf{B}})\boldsymbol{\tau} \leq F_\alpha \rho^2\}, \tag{5.117}$$

where $\hat{\mathbf{B}}$ is given by (5.67), namely $\hat{\mathbf{B}} = [(\hat{\boldsymbol{\varepsilon}}'\hat{\mathbf{Q}}_{n-p})][\hat{A}^N_{..}]$. We saw, from (5.68), that $\mathbf{I}_p - \hat{\mathbf{B}}$ is positive definite, so that (5.116) will be an ellipsoid also if $\mathbf{I}_p - (1+f)\hat{\mathbf{B}}$ is also positive definite. The difference between (5.116) and (5.117) depends on $\hat{\mathbf{B}}$, which is a function of the intrinsic curvature array $\hat{A}^N_{..}$. If these effects are sufficiently small, then the two regions will be similar. Defining $\lambda_{max} = \lambda_1 > \lambda_2 > \cdots > \lambda_p = \lambda_{min}$ to be the eigenvalues of $\hat{\mathbf{B}}$, then $(\mathbf{I}_p - \hat{\mathbf{B}})[\mathbf{I}_p - (1+f)\hat{\mathbf{B}}]$ has eigenvalues $(1-\lambda_i)[1-(1+f)\lambda_i]$. If these are all positive, so that (5.116) is an ellipsoid, then the lengths of the major axes are $\{(1-\lambda_i)[1-(1+f)\lambda_i]\}^{-1/2}$, which can be used, as in Section 5.8.1, to assess the effect of intrinsic curvature. Also, a conservative confidence region for $\boldsymbol{\tau}$ is given by [c.f. (5.71)]

$$\boldsymbol{\tau}'\boldsymbol{\tau} \leq \frac{F_\alpha \rho^2}{(1-\lambda_{max})[1-(1+f)\lambda_{max}]}. \tag{5.118}$$

This can be converted, as in Section 5.8.1, to give a conservative region for $\boldsymbol{\theta}$.

5.10.2 Partially Linear Models

For $i = 1, 2, \ldots, n$ consider the following models:

$$y_i = \alpha + \beta e^{\gamma x_i} + \varepsilon_i,$$

$$y_i = \alpha + \beta_1(x_i - \gamma) + \beta_2(x_i - \gamma)^2 + \varepsilon_i,$$

and

$$y_i = \alpha + \beta_1 x_i + \beta_2 x_i^\gamma + \varepsilon_i. \tag{5.119}$$

They are all special cases of

$$\mathbf{y} = \mathbf{X}_\gamma \boldsymbol{\beta} + \boldsymbol{\varepsilon}, \tag{5.120}$$

where the elements of $\mathbf{X}_\gamma$ are functions of a $g \times 1$ vector of unknown parameters $\boldsymbol{\gamma}$, $\boldsymbol{\beta}$ is a $b \times 1$ vector of unknown parameters, and $\mathbf{X}_\gamma$ is an $n \times b$ matrix of rank b for all $\boldsymbol{\gamma}$ in a suitable region. For example, in the model (5.119)

$$\mathbf{X}_\gamma \boldsymbol{\beta} = \begin{bmatrix} 1 & x_1 & x_1^\gamma \\ 1 & x_2 & x_2^\gamma \\ \vdots & \vdots & \vdots \\ 1 & x_n & x_n^\gamma \end{bmatrix} \begin{bmatrix} \alpha \\ \beta_1 \\ \beta_2 \end{bmatrix}. \tag{5.121}$$

Now for fixed $\boldsymbol{\gamma}$, (5.120) is a standard linear model with linear parameters $\boldsymbol{\beta}$, so that such models are called partially linear or separable nonlinear models. Here $\boldsymbol{\theta} = (\boldsymbol{\beta}', \boldsymbol{\gamma}')'$ is $p \times 1$ with $p = g + b$: the elements of $\boldsymbol{\beta}$ are called the linear parameters, and those of $\boldsymbol{\gamma}$ the nonlinear parameters. We shall consider tests of the following two hypotheses:

1. $\boldsymbol{\theta} = \boldsymbol{\theta}_0$, i.e. $(\boldsymbol{\beta}', \boldsymbol{\gamma}') = (\boldsymbol{\beta}_0', \boldsymbol{\gamma}_0')$; and
2. $\boldsymbol{\gamma} = \boldsymbol{\gamma}_0$.

No exact methods have yet been developed for just $\boldsymbol{\beta}$ or proper subsets of $\boldsymbol{\gamma}$.

Generalizing (5.120), we introduce an additional dummy regressor for each nonlinear parameter, as in (5.109), and consider (Halperin [1963])

$$\begin{aligned} \mathbf{y} &= \mathbf{X}_\gamma \boldsymbol{\beta} + \mathbf{D}\boldsymbol{\delta} + \boldsymbol{\varepsilon} \\ &= (\mathbf{X}_\gamma, \mathbf{D}) \begin{pmatrix} \boldsymbol{\beta} \\ \boldsymbol{\delta} \end{pmatrix} + \boldsymbol{\varepsilon} \\ &= \mathbf{Z}\boldsymbol{\lambda} + \boldsymbol{\varepsilon}, \end{aligned} \tag{5.122}$$

say, where, for all $\boldsymbol{\theta}$ in a suitable region Θ, $\mathbf{D}$ is any $n \times g$ matrix of rank g whose elements depend on $\boldsymbol{\gamma}$ and possibly $\boldsymbol{\beta}$, and the $n \times p$ matrix $\mathbf{Z}$ has rank p. If $\boldsymbol{\theta}$ takes its true value, then $\boldsymbol{\delta} = \mathbf{0}$. With this format we can now use general linear regression theory and establish F-statistics below for testing $\boldsymbol{\theta} = \boldsymbol{\theta}_0$ and $\boldsymbol{\gamma} = \boldsymbol{\gamma}_0$, respectively.

Let $\tilde{\boldsymbol{\lambda}} = (\mathbf{Z}'\mathbf{Z})^{-1}\mathbf{Z}'\mathbf{y}$ $[= (\tilde{\boldsymbol{\beta}}', \tilde{\boldsymbol{\delta}}')']$ be the "least squares" estimate of $\boldsymbol{\lambda}$. Then, from Appendix D2.1,

$$F_{1,\boldsymbol{\theta}} = \frac{n-p}{p} \frac{(\tilde{\boldsymbol{\lambda}} - \boldsymbol{\lambda})'\mathbf{Z}'\mathbf{Z}(\tilde{\boldsymbol{\lambda}} - \boldsymbol{\lambda})}{\mathbf{y}'\mathbf{y} - \tilde{\boldsymbol{\lambda}}'\mathbf{Z}'\mathbf{Z}\tilde{\boldsymbol{\lambda}}} \tag{5.123}$$

is distributed as $F_{p,n-p}$ when $\boldsymbol{\theta}$ takes its true value. In (5.123) $\boldsymbol{\delta} = \mathbf{0}$ and $F_{1,\boldsymbol{\theta}}$ is a function of $\boldsymbol{\beta}$ and $\boldsymbol{\gamma}$. A $100(1-\alpha)\%$ confidence region for $\boldsymbol{\theta}$ is therefore given by

$$\{\boldsymbol{\theta} : F_{1,\boldsymbol{\theta}} \le F^{\alpha}_{p,n-p}\}. \tag{5.124}$$

We can also test the hypothesis $\boldsymbol{\theta} = \boldsymbol{\theta}_0$ by calculating $F_{1,\boldsymbol{\theta}_0}$ and rejecting H_0 if

$$F_{1,\boldsymbol{\theta}_0} > F^{\alpha}_{p,n-p}. \tag{5.125}$$

The test statistic for $\boldsymbol{\delta} = \mathbf{0}$ in (5.122) takes the form [c.f. Appendix D3.4 with $(\mathbf{X}_\gamma, \mathbf{D}) = (\mathbf{X}_1, \mathbf{X}_2)$]

$$\begin{aligned} F_2 &= \frac{n-p}{g} \frac{\tilde{\boldsymbol{\delta}}'\mathbf{D}'\mathbf{R}\mathbf{D}\tilde{\boldsymbol{\delta}}}{\mathbf{y}'\mathbf{y} - \tilde{\boldsymbol{\lambda}}'\mathbf{Z}'\mathbf{Z}\tilde{\boldsymbol{\lambda}}} \\ &= \frac{n-p}{g} \frac{\tilde{\boldsymbol{\delta}}'\mathbf{D}'\mathbf{R}\mathbf{D}\tilde{\boldsymbol{\delta}}}{\mathbf{y}'\mathbf{R}\mathbf{y} - \tilde{\boldsymbol{\delta}}'\mathbf{D}'\mathbf{R}\mathbf{D}\tilde{\boldsymbol{\delta}}}, \end{aligned} \tag{5.126}$$

where

$$\mathbf{R} = \mathbf{I}_n - \mathbf{X}_\gamma(\mathbf{X}_\gamma'\mathbf{X}_\gamma)^{-1}\mathbf{X}_\gamma'$$

and

$$\tilde{\boldsymbol{\delta}} = (\mathbf{D}'\mathbf{R}\mathbf{D})^{-1}\mathbf{D}'\mathbf{R}\mathbf{y}.$$

Since $\boldsymbol{\delta} \equiv \mathbf{0}$, $F_2 \sim F_{g,n-p}$. If $\mathbf{D}$ does not depend on $\boldsymbol{\beta}$ then F_2 is a function of just γ, say $F_{2,\gamma}$. We can then construct a confidence region for γ and a test for $\gamma = \gamma_0$ similar to (5.124) and (5.125), respectively. If $\mathbf{D}$ is also a function of $\boldsymbol{\beta}$, we can use the method of El-Shaarawi and Shah [1980] to generalize (5.126). They simply replace $\boldsymbol{\beta}$ by $\hat{\boldsymbol{\beta}} = (\mathbf{X}_\gamma'\mathbf{X}_\gamma)^{-1}\mathbf{X}_\gamma'\mathbf{y}$ in $\mathbf{D}$ to get $\hat{\mathbf{D}}$ and then use the latter in (5.126). The test statistic becomes

$$\hat{F}_{2,\gamma} = \frac{n-p}{g}\,\frac{\hat{\boldsymbol{\delta}}'\hat{\mathbf{D}}'\mathbf{R}\hat{\mathbf{D}}\hat{\boldsymbol{\delta}}}{\mathbf{y}'\mathbf{R}\mathbf{y} - \hat{\boldsymbol{\delta}}'\hat{\mathbf{D}}'\mathbf{R}\hat{\mathbf{D}}\hat{\boldsymbol{\delta}}}, \tag{5.127}$$

where $\hat{\boldsymbol{\delta}} = (\hat{\mathbf{D}}'\mathbf{R}\hat{\mathbf{D}})^{-1}\hat{\mathbf{D}}'\mathbf{R}\mathbf{y}$. Now from the model $\mathscr{E}[\mathbf{y}] = \mathbf{X}_\gamma\boldsymbol{\beta}$ we have that $\hat{\boldsymbol{\beta}}$ is statistically independent of $\mathbf{y}'\mathbf{R}\mathbf{y}$ (by Appendix D1.2). If we condition on $\hat{\boldsymbol{\beta}}$, the quadratic forms $\hat{\boldsymbol{\delta}}'\hat{\mathbf{D}}'\mathbf{R}\hat{\mathbf{D}}\hat{\boldsymbol{\delta}}$ and $\mathbf{y}'\mathbf{R}\mathbf{y}$ in (5.127) will have the same distributional properties as the corresponding quadratics in (5.126). Therefore with $\boldsymbol{\delta} = \mathbf{0}$, the conditional distribution of (5.127) given $\hat{\boldsymbol{\beta}}$ is the same as the distribution of F_2, namely $F_{g,n-p}$. As this F-distribution does not depend on $\hat{\boldsymbol{\beta}}$, it is also the unconditional distribution, so that

$$\hat{F}_{2,\gamma} \sim F_{g,n-p}.$$

We can use this statistic in the same way as F_2 to construct a confidence region for γ and to test $\gamma = \gamma_0$.

How do we find a suitable $\mathbf{D}$-matrix? One way is to proceed as in Example 5.6 above and use a Taylor expansion of $\mathbf{X}_\gamma\boldsymbol{\beta}$ about $\gamma = \gamma_0$. This leads to $\mathbf{D} = [(d_{ij})]$, where

$$d_{ij} = \sum_{k=1}^{b} \frac{\partial x_{ik}}{\partial \gamma_j}\beta_k. \tag{5.128}$$

Halperin [1963] obtained this $\mathbf{D}$ and noted that in one important instance it is possible to define $\mathbf{D}$ free of $\boldsymbol{\beta}$. Suppose that each γ_j of γ occurs in one and only one column of $\mathbf{X}_\gamma$ (as in Example 5.7 below); then the β_k's will appear simply as multiplicative factors which can be removed by incorporating them in $\boldsymbol{\delta}$ when defining $\mathbf{D}\boldsymbol{\delta}$ in (5.122).

Example 5.7 Consider the model

$$y_i = \beta_1 e^{-\gamma_1 t_i} + \beta_2 e^{-\gamma_2 t_i} + \varepsilon_i, \tag{5.129}$$

that is,

$$\begin{bmatrix} y_1 \\ y_2 \\ \vdots \\ y_n \end{bmatrix} = \begin{bmatrix} e^{-\gamma_1 t_1} & e^{-\gamma_2 t_1} \\ e^{-\gamma_1 t_2} & e^{-\gamma_2 t_2} \\ \vdots & \vdots \\ e^{-\gamma_1 t_n} & e^{-\gamma_2 t_n} \end{bmatrix} \begin{pmatrix} \beta_1 \\ \beta_2 \end{pmatrix} + \begin{bmatrix} \varepsilon_1 \\ \varepsilon_2 \\ \vdots \\ \varepsilon_n \end{bmatrix},$$

or

$$\mathbf{y} = \mathbf{X}_\gamma \boldsymbol{\beta} + \boldsymbol{\varepsilon}.$$

Then (5.128) leads to $\beta_j \partial x_{ij}/\partial \gamma_j$, and, ignoring β_j, we can choose

$$d_{ij} = \frac{\partial x_{ij}}{\partial \gamma_j} = -t_i x_{ij},$$

or

$$\mathbf{D} = -\operatorname{diag}(t_1, t_2, \ldots, t_n)\,\mathbf{X}_\gamma.$$

∎

Before concluding this section we mention three extensions. Firstly, if any of the matrices are rank-deficient, e.g. $\mathbf{X}_\gamma$ or $(\mathbf{X}_\gamma, \mathbf{D})$ is less than full rank, then we simply replace matrix inverses by generalized inverses, replace b by rank$[\mathbf{X}_\gamma]$, and replace g by rank$[\mathbf{D}'\mathbf{R}\mathbf{D}]$. Secondly, El-Shaarawi and Shah [1980] considered the model

$$\mathbf{y}_\gamma = \mathbf{X}_\gamma \boldsymbol{\beta} + \boldsymbol{\varepsilon}, \tag{5.130}$$

where the elements of $\mathbf{y}_\gamma$ now depend on $\boldsymbol{\gamma}$. Expanding $\mathbf{y}_\gamma - \mathbf{X}_\gamma \boldsymbol{\beta}$ about $\boldsymbol{\gamma}_0$ leads to

$$d_{ij} = \sum_{k=1}^{b} \frac{\partial x_{ik}}{\partial \gamma_j} \beta_k - \frac{\partial y_i}{\partial \gamma_j}, \tag{5.131}$$

which now depends on $\mathbf{y}$ as well as $\boldsymbol{\beta}$ and $\boldsymbol{\gamma}$. Fortunately it transpires that we can adapt the above method by simply replacing $\mathbf{y}$ in $\mathbf{D}$ by $\hat{\mathbf{y}} = \mathbf{X}\hat{\boldsymbol{\beta}}$.

Finally, Halperin [1964] gave a multivariate extension

$$\mathbf{y}_i = \mathbf{X}_\gamma \boldsymbol{\beta} + \boldsymbol{\varepsilon}_i \qquad (i = 1, 2, \ldots, N),$$

where the $\boldsymbol{\varepsilon}_i$ are i.i.d. $N_n(\mathbf{0}, \boldsymbol{\Sigma})$. We now have a random sample of size N from a multivariate normal distribution with mean $\boldsymbol{\mu}$ given by $\boldsymbol{\mu} = \mathbf{X}_\gamma \boldsymbol{\beta}$. Halperin [1964] once again used the augmented model (5.122), namely

$$\mathbf{y}_i = \mathbf{Z}\boldsymbol{\lambda} + \boldsymbol{\varepsilon}_i,$$

where $\mathbf{D}$ is given by (5.128), and obtained multivariate analogues of (5.123) and (5.126), which we now describe. If

$$\mathbf{S}=\frac{1}{N-1}\sum_{i=1}^{N}(\mathbf{y}_i-\bar{\mathbf{y}})(\mathbf{y}_i-\bar{\mathbf{y}})' \quad \text{and} \quad \mathbf{C}=\mathbf{Z}'\mathbf{S}^{-1}\mathbf{Z},$$

then the maximum-likelihood estimate of $\boldsymbol{\lambda}$, when $\boldsymbol{\gamma}$ is assumed known, is (Seber [1984: 495])

$$\tilde{\boldsymbol{\lambda}}=\mathbf{C}^{-1}\mathbf{Z}'\mathbf{S}^{-1}\bar{\mathbf{y}}.$$

(Note that N or $N-1$ can be used in the definition of $\mathbf{S}$, as the scale factor cancels out of $\tilde{\boldsymbol{\lambda}}$.) Let

$$T_n-T_{n-p}=\frac{N}{N-1}(\tilde{\boldsymbol{\lambda}}-\boldsymbol{\lambda})'\mathbf{C}(\tilde{\boldsymbol{\lambda}}-\boldsymbol{\lambda})$$

and

$$\begin{aligned}T_{n-p}&=\frac{N}{N-1}(\bar{\mathbf{y}}-\mathbf{Z}\tilde{\boldsymbol{\lambda}})'\mathbf{S}^{-1}(\bar{\mathbf{y}}-\mathbf{Z}\tilde{\boldsymbol{\lambda}})\\&=\frac{N}{N-1}\bar{\mathbf{y}}'(\mathbf{S}^{-1}-\mathbf{S}^{-1}\mathbf{Z}\mathbf{C}^{-1}\mathbf{Z}'\mathbf{S}^{-1})\bar{\mathbf{y}},\end{aligned}$$

where $\boldsymbol{\lambda}=(\boldsymbol{\beta}',\boldsymbol{\delta}')'$ and $\boldsymbol{\delta}=\mathbf{0}$. Then, with $p=b+g$, it can be shown that

$$\begin{aligned}F_{3,\theta}&=\frac{N-n}{p}\,\frac{T_n-T_{n-p}}{1+T_{n-p}}\\&\sim F_{p,N-n},\end{aligned} \tag{5.132}$$

and we can use $F_{3,\theta}$ to construct hypothesis tests and confidence regions for $\boldsymbol{\theta}\,[=(\boldsymbol{\beta}',\boldsymbol{\gamma}')']$. Furthermore, if $\mathbf{U}=(\mathbf{0},\mathbf{I}_g)$, where $\mathbf{U}$ is $g\times p$, $\tilde{\boldsymbol{\delta}}=\mathbf{U}\tilde{\boldsymbol{\lambda}}$, and

$$T_{n-b}-T_{n-p}=\frac{N}{N-1}\tilde{\boldsymbol{\delta}}'(\mathbf{U}\mathbf{C}^{-1}\mathbf{U}')^{-1}\tilde{\boldsymbol{\delta}},$$

then

$$\begin{aligned}F_4&=\frac{N-n+b}{g}\,\frac{T_{n-b}-T_{n-p}}{1+T_{n-p}}\\&\sim F_{g,N-n+b}.\end{aligned} \tag{5.133}$$

This can be used for making joint inferences about $\boldsymbol{\beta}$ and $\boldsymbol{\gamma}$. Halperin [1963] noted that (5.132) and (5.133) follow from Rao [1959: 51–52]. They can also be derived from the analysis-of-covariance method applied to a growth-curve model (Seber [1984: 484–485, with the notational changes $n \to N$, $K \to Z$, $D \to I$ or U]). When $\mathbf{D}$ does not depend on $\boldsymbol{\beta}$, F_4 is a function of $\boldsymbol{\gamma}$ only and can be used for inferences about $\boldsymbol{\gamma}$.

In conclusion it should be noted that the various confidence regions discussed above are difficult to compute and difficult to represent graphically for $p > 3$. Also, a bad choice of $\mathbf{D}$ can lead to unrealistic regions which are disjoint.

5.11 BAYESIAN INFERENCE

Up till now we have discussed methods of inference based on the asymptotic properties of the least-squares estimate $\hat{\boldsymbol{\theta}}$ of $\boldsymbol{\theta}$, and on the contours of the sum-of-squares function $S(\boldsymbol{\theta}) = \| \mathbf{y} - \mathbf{f}(\boldsymbol{\theta}) \|^2$. Under normality of the data, $\hat{\boldsymbol{\theta}}$ is also the maximum-likelihood estimate, and the sum-of-squares contours become the contours of the likelihood function. However, as noted for example by Smith et al. [1985], some likelihood and sum-of-squares functions have peculiar properties which cause problems with both inference and computation. Also, as discussed in Section 5.8, intrinsic and parameter-effects curvatures can have serious effects on the asymptotic methods of inference proposed thus far. Therefore, with the development of efficient methods for finding posterior density functions and associated moments (c.f. Section 2.7.2), it is clear that Bayesian methods of inference based on the joint posterior density of $\boldsymbol{\theta}$ and its marginal densities can have considerable advantages. Parameters can be estimated by the mode or the mean of the posterior density, and inferences can be based on the h.p.d. (highest posterior density) intervals described in Section 2.7.3. The hypothesis $H_0: \boldsymbol{\theta} = \boldsymbol{\theta}_0$ can be tested by seeing if $\boldsymbol{\theta}_0$ lies in the h.p.d. region for $\boldsymbol{\theta}$.

*5.12 INVERSE PREDICTION (DISCRIMINATION)

5.12.1 Single Prediction

Given the nonlinear model

$$y_i = f(x_i; \boldsymbol{\theta}) + \varepsilon_i, \qquad i = 1, 2, \ldots, n,$$

where the ε_i are i.i.d. $N(0, \sigma^2)$, we wish to predict the value of x_0 given an observation y_0, say. A natural estimate of x_0 is given by solving $y_0 = f(x; \hat{\boldsymbol{\theta}})$ for x, where f will generally be monotonic in the region of interest. As in the straight-line case (Seber [1977: 187]), this estimate of x_0 will usually be biased. To obtain a confidence interval for x_0 we assume that the linear approximation for f is valid and argue as in Section 5.1. Thus if $\hat{y}_x = f(x; \hat{\boldsymbol{\theta}})$ then

$$\mathrm{var}[\hat{y}_x] \approx \sigma^2 \mathbf{f}_x' (\mathbf{F}_.' \mathbf{F}_.)^{-1} \mathbf{f}_x = \sigma^2 v_x,$$

where

$$\mathbf{f}'_x = \left[\frac{\partial f(x;\boldsymbol{\theta})}{\partial \theta_1}, \frac{\partial f(x;\boldsymbol{\theta})}{\partial \theta_2}, \ldots, \frac{\partial f(x;\boldsymbol{\theta})}{\partial \theta_p}\right].$$

We therefore have, approximately,

$$t = \frac{y_0 - \hat{y}_{x_0}}{s(1 + \hat{v}_{x_0})^{1/2}} \sim t_{n-p},$$

where $\hat{\mathbf{f}}_x$ and $\hat{v}_x$ denote evaluation at $\boldsymbol{\theta} = \hat{\boldsymbol{\theta}}$. Hence

$$\begin{aligned} 1 - \alpha &= \mathrm{pr}[|t| \le t_{n-p}^{\alpha/2}] \\ &= \mathrm{pr}[|t|^2 \le (t_{n-p}^{\alpha/2})^2 = F_{1,n-p}^{\alpha}], \end{aligned}$$

and the set of all values of x satisfying the inequality

$$(y_0 - \hat{y}_x)^2 \le F_{1,n-p}^{\alpha} s^2 [1 + \hat{\mathbf{f}}'_x (\hat{\mathbf{F}}'_{\cdot} \hat{\mathbf{F}}_{\cdot})^{-1} \hat{\mathbf{f}}_x] \tag{5.134}$$

will provide an approximate $100(1-\alpha)\%$ confidence interval for x_0. We note that x occurs nonlinearly in $\hat{y}_x$ and $\hat{\mathbf{f}}_x$, so that the interval may be awkward to compute. As with linear regression, we may get a finite interval, two semi-infinite lines, or the entire real line (c.f. Seber [1977: 188]). If the linear approximation underlying the above theory is not adequate and curvature effects need to be taken into consideration, we could replace $(\mathbf{F}'_{\cdot}\mathbf{F}_{\cdot})^{-1}\sigma^2$ by $\mathscr{D}[\hat{\boldsymbol{\theta}}]$ of (4.179).

5.12.2 Multiple Predictions

If we are interested in making k inverse predictions, then we can use the Bonferroni method and replace α by α/k in each of the intervals given by (5.134). These intervals will simultaneously contain their values of x_0 with an overall probability of at least $1-\alpha$ (c.f. Seber [1977: 126]).

If a large (unknown) number of inverse predictions are required, as is often the case with nonlinear calibration curves, then a method of Lieberman et al. [1967] can be generalized to nonlinear models (c.f. Tiede and Pagano [1979]). Using the linear approximation again, we first note that a $100(1-\alpha/2)\%$ band for the calibration curve $f(x;\boldsymbol{\theta})$ is given approximately by (Miller [1981: 111])

$$f(x;\boldsymbol{\theta}) \in f(x;\hat{\boldsymbol{\theta}}) \pm (pF_{p,n-p}^{\alpha/2})^{1/2}(s^2\hat{v}_x)^{1/2}. \tag{5.135}$$

Secondly, if σ was known, $P(Z > z_{\gamma/2}) = \gamma/2$ and $E[y_x] = f(x;\boldsymbol{\theta})$, then each interval

$$y_x \pm z_{\gamma/2}\sigma \tag{5.136}$$

would have probability $1-\gamma$ of containing $f(x;\boldsymbol{\theta})$. However, if σ is unknown, then it can be bounded above using the approximate inequality

$$\text{pr}\left[\sigma \le \left(\frac{n-p}{c}\right)^{1/2} s\right] = 1 - \frac{\alpha}{2}, \tag{5.137}$$

where $\text{pr}[\chi^2_{n-p} \le c] = 1 - \alpha/2$. Hence all intervals (5.136) must be contained in

$$y_x \pm z_{\gamma/2}\left(\frac{n-p}{c}\right)^{1/2} s \tag{5.138}$$

with probability $1 - \alpha/2$. Finally, the inverse prediction interval for x_0 is obtained by intersecting (5.138) with (5.135) and projecting the intersection onto the x-axis (Miller [1981: 126–127]). This is done by equating the lower end of (5.135) with the upper end of (5.138), and vice versa. Hence the lower bound $\hat{x}_L$ is the solution x to

$$f(x;\hat{\boldsymbol{\theta}}) - (pF^{\alpha/2}_{p,n-p})^{1/2} s[\hat{\mathbf{f}}'_x(\hat{\mathbf{F}}'_.\hat{\mathbf{F}}_.)^{-1}\hat{\mathbf{f}}_x]^{1/2} = y_x + z_{\gamma/2}\left(\frac{n-p}{c}\right)^{1/2} s,$$

and the upper bound $\hat{x}_U$ is the solution to

$$f(x;\hat{\boldsymbol{\theta}}) + (pF^{\alpha/2}_{p,n-p})^{1/2} s[\hat{\mathbf{f}}'_x(\hat{\mathbf{F}}'_.\hat{\mathbf{F}}_.)^{-1}\hat{\mathbf{f}}_x]^{1/2} = y_x - z_{\gamma/2}\left(\frac{n-p}{c}\right)^{1/2} s.$$

Then, for at least $100(1-\gamma)\%$ of the inverse prediction intervals,

$$\text{pr}[\hat{x}_L \le x_0 \le \hat{x}_U] \ge 1 - \alpha.$$

5.12.3 Empirical Bayes Interval

We now consider a Bayesian technique for finding a single inverse prediction interval which accommodates known nonnormal distribution families for ε. The regressor x is assumed to be random but measured accurately, so that the regression of y on x can be considered as conditional on the observed value of x (see Section 1.3.2). Since x and y are both random, a prediction of x for given y should clearly be based on the conditional distribution of x given y, if this can be obtained or estimated. Following Lwin and Maritz [1980], we assume that the density function $p(y|x)$ of y given x comes from a location and scale family so that

$$p(y|x) = \sigma^{-1} h\left[\frac{y - f(x;\boldsymbol{\theta})}{\sigma}\right],$$

where h is the density function for a distribution with mean 0 and variance 1.

If σ and $\boldsymbol{\theta}$ are known, then it is readily shown that the prediction of x_0, $x^*(y_0)$ say, with minimum mean squared error is given by

$$x^*(y_0) = E[x \mid y = y_0].$$

Now by Bayes' theorem

$$\begin{aligned} p(x \mid y) &= \frac{p(y \mid x) p_1(x)}{p_2(y)} \\ &= \frac{p(y \mid x) p_1(x)}{\int p(y \mid x)\, dP_1(x)}, \end{aligned}$$

so that

$$x^*(y_0) = \frac{\int x h[\{y_0 - f(x; \boldsymbol{\theta})\}/\sigma]\, dP_1(x)}{\int h[\{y_0 - f(x; \boldsymbol{\theta})\}/\sigma]\, dP_1(x)}, \tag{5.139}$$

where $P_1(x)$ is the distribution function of x. If the calibration observations $x_1, x_2, \ldots, x_n$ can be regarded as independent observations from the *same* random process which generates the future samples to be measured (a somewhat unrealistic assumption in many calibration experiments), then $P_1(x)$ can be estimated by the empirical distribution function

$$\hat{P}_1(x) = \frac{\text{number of } x_i \leq x}{n}.$$

Using any reasonable estimates of $\boldsymbol{\theta}$ and σ^2 such as the least-squares estimates $\hat{\boldsymbol{\theta}}$ and $\hat{\sigma}^2$, x_0 can now be estimated by

$$\hat{x}(y_0) = \frac{\sum_{i=1}^{n} x_i h[\{y_0 - f(x_i; \hat{\boldsymbol{\theta}})\}/\hat{\sigma}]}{\sum_{i=1}^{n} h[\{y_0 - f(x_i; \hat{\boldsymbol{\theta}})\}/\hat{\sigma}]} = \sum_{i=1}^{n} x_i w_i, \tag{5.140}$$

say. The performance of this nonlinear predictor can be measured by comparing each x_i with its predicted value $\hat{x}(y_i)$ calculated from the data with (x_i, y_i) omitted. The predictor can also be used for straight lines and, under the rather restrictive assumption above, has several advantages over the usual predictors based on regressing x on y or y on x (Seber [1977: 187]; Lwin and Maritz [1982]). Firstly, if $E[x^2] < \infty$, then it can be shown from Stone [1977: Corollary 1, p. 598] that $\hat{x}(y_0)$ tends to $x^*(y_0)$ in the mean square: this implies $\hat{x}(y_0)$ is consistent. Secondly, the estimator can be used for nonnormal distributions $p(y|x)$. However, if normality is assumed, then $h(z)$ is simply the density function of the standard normal distribution $N(0,1)$. This assumption of normality can be checked using a normal probability plot of the scaled residuals $\{y_i - f(x_i;\hat{\boldsymbol{\theta}})\}/\hat{\sigma}$. Thirdly, an approximate confidence interval for x_0 can be constructed as follows.

Lwin and Maritz [1980] regarded $p(y|x)$ as a "data" density, $P_1(x)$ as a "prior" distribution function, and the optimal predictor $x^*(y_0)$ as the posterior mean with "posterior" distribution given by [c.f. (2.99)]

$$dG(x|y_0) = \frac{h[\{y_0 - f(x;\boldsymbol{\theta})\}/\sigma]\,dP_1(x)}{\int h[\{y_0 - f(x;\boldsymbol{\theta})\}/\sigma]\,dP_1(x)}. \tag{5.141}$$

Then two-sided $100(1-\alpha)\%$ "posterior" probability limits for x_0 may be obtained as

$$[q(\tfrac{1}{2}\alpha|y_0), q(1-\tfrac{1}{2}\alpha|y_0)],$$

where $q(k|y_0)$ is the kth quantile of the distribution function $G(x|y_0)$. Because G may be a step function (which is certainly the case with its estimate), Lwin and Maritz [1980] follow Stone [1977: 603] and define $q(k|y_0)$ as follows. Let

$$L(k|y_0) = \inf\{x : G(x|y_0) \geq k\}$$

and

$$U(k|y_0) = \sup\{x : G(x|y_0) \leq k\};$$

then we define

$$q(k|y_0) = \tfrac{1}{2}[L(k|y_0) + U(k|y_0)].$$

An empirical Bayes-type estimator of $q(k|y_0)$ can now be obtained by applying these formulae to the empirical distribution used in (5.140). Let

$$\hat{L}(k|y_0) = \inf\left\{x : \frac{\sum_{x_i \leq x} h\left[\dfrac{y_0 - f(x_i;\hat{\boldsymbol{\theta}})}{\hat{\sigma}}\right]}{\sum_{i=1}^{n} h\left[\dfrac{y_0 - f(x_i;\hat{\boldsymbol{\theta}})}{\hat{\sigma}}\right]} \geq k\right\}$$

and

$$\hat{U}(k|y_0) = \sup\left\{x : \frac{\sum_{x_i \leq x} h\left[\dfrac{y_0 - f(x_i;\hat{\boldsymbol{\theta}})}{\hat{\sigma}}\right]}{\sum_{i=1}^{n} h\left[\dfrac{y_0 - f(x_i;\hat{\boldsymbol{\theta}})}{\hat{\sigma}}\right]} \leq k\right\};$$

then $q(k|y_0)$ is estimated by

$$\hat{q}(k|y_0) = \tfrac{1}{2}\{\hat{L}(k|y_0) + \hat{U}(k|y_0)\}. \tag{5.142}$$

Assuming $E[x^2] < \infty$ and applying Corollary 6 of Stone [1977: 604], it transpires that $\hat{q}(k|y_0) \to q(k|y_0)$ in the mean square. Lwin and Maritz [1980] stress that the nominal $100(1-\alpha)\%$ confidence interval

$$[\hat{q}(\tfrac{1}{2}\alpha|y_0), \hat{q}(1-\tfrac{1}{2}\alpha|y_0)] \tag{5.143}$$

is asymptotic and may have a coverage probability very different from $1-\alpha$ for small samples.

*5.13 OPTIMAL DESIGN

5.13.1 Design Criteria

If we are confident that a given nonlinear model $E[y_i] = f(\mathbf{x}_i; \boldsymbol{\theta})$ is appropriate for a certain experimental situation, then there is the question of designing the experiment, i.e. choosing the $\mathbf{x}_i$ so that the unknown parameters are estimated in an optimal fashion. The theory of optimal designs for linear models has been extensively developed (see Atkinson [1982] and Steinberg and Hunter [1984] for general reviews, and St. John and Draper [1975] for a review of D-optimality), but the adaption to nonlinear models is more difficult, as the criteria involve the unknown parameters $\boldsymbol{\theta}$ as well as the design points $\mathbf{x}_i$. References to the design of nonlinear models are given by Cochran [1973] and Bates and Hunter [1984], as well as by the above-mentioned reviews. The first real advance in design methodology was made in the pioneering paper of Box and Lucas [1959]. They suggested the criterion of minimizing the determinant of the asymptotic variance–covariance matrix of $\hat{\boldsymbol{\theta}}$, namely maximizing [c.f. (2.21)]

$$\Delta = |\mathbf{F}'_{\bullet}(\boldsymbol{\theta})\mathbf{F}_{\bullet}(\boldsymbol{\theta})| \tag{5.144}$$

with respect to $\mathbf{x}_1, \mathbf{x}_2, \ldots, \mathbf{x}_n$ in the region of "operability", $\mathscr{X}$, for values of $\boldsymbol{\theta}$ close to the true value $\boldsymbol{\theta}^*$. This criterion is equivalent to using an approximate linearization of the nonlinear model [with $\mathbf{X} = \mathbf{F}_{\bullet}(\boldsymbol{\theta})$] and then, with $\boldsymbol{\theta}^*$ appropriately estimated, looking for a D-optimal design for the linear model. This criterion also has other attractive features (M. J. Box and Draper [1971: 732]). Firstly, it minimizes the volume of the asymptotic confidence region (2.26) for $\boldsymbol{\theta}$, the volume being proportional to $\Delta^{-1/2}$ (by Appendix A6). Secondly, it maximizes the peak of the posterior density function of $\boldsymbol{\theta}$ for the linearized model, given the noninformative prior for $\boldsymbol{\theta}$. Thirdly, it is invariant under changes of scale of the θ_r.

If var $[y_i] = \sigma^2/w_i$, where $w_i = w_i(\mathbf{x}_i; \boldsymbol{\theta})$, then weighted least-squares estimation is appropriate. The criterion (5.144) now becomes

$$\Delta_W = |\mathbf{F}'_{\bullet}(\boldsymbol{\theta})\mathbf{W}(\boldsymbol{\theta})\mathbf{F}_{\bullet}(\boldsymbol{\theta})|, \tag{5.145}$$

where $\mathbf{W}(\boldsymbol{\theta}) = \text{diag}(w_1, w_2, \ldots, w_n)$. For example, in pharmacokinetic models a common method of weighting is (Endrenyi [1981b]) $w_i = [f(\mathbf{x}_i; \boldsymbol{\theta})]^{-2}$, which corresponds to y_i having a standard deviation proportional to its mean. This arises if the model has a "constant relative error," namely a constant coefficient of variation for the response y_i. However, in this case an alternative approach is to use a log transformation, i.e. $\lambda_1 = 0$, in (2.145). Then (5.145) for the original model is approximately equal to (5.144) for the transformed model.

As the above designs depend on $\boldsymbol{\theta}^*$, a sequential scheme which allows for updating the current estimate of $\boldsymbol{\theta}^*$ is a sensible strategy: sequential designs are discussed briefly below in Section 5.13.3. If a sequential approach is not possible, then one strategy is to design the experiment with a prior guess for $\boldsymbol{\theta}^*$. Another strategy is *maximin*, in which the best design is sought for the "worst" value of $\boldsymbol{\theta}^*$ (Silvey [1980: 59]). Unfortunately, the worst value of $\boldsymbol{\theta}^*$ may lead to an extreme design which is impractical. If prior information is available, then Atkinson [1982] notes that the prior distribution of $\boldsymbol{\theta}$ can be discretized to provide weights for the components of a so-called S-optimal design (Laüter [1976]). Alternatively, the design criterion can be integrated over the prior distribution of $\boldsymbol{\theta}^*$ (Laüter [1974]).

Box and Lucas [1959] restricted their investigation to the case where the number of design points (n) equals the number of parameters (p). The experiment could then be replicated any number of times to achieve the required precision. When $n = p$, $\mathbf{F}_{\bullet}(\boldsymbol{\theta})$ is square and $\Delta = |\mathbf{F}_{\bullet}(\boldsymbol{\theta})|^2$, so that the criterion reduces to maximizing the absolute value, mod$|\mathbf{F}_{\bullet}(\boldsymbol{\theta})|$, of $|\mathbf{F}_{\bullet}(\boldsymbol{\theta})|$ over $\mathscr{X}$. Maximization is usually carried out numerically using a preliminary guess at $\boldsymbol{\theta}^*$, though analytical solutions are possible in simple cases, as demonstrated by Box and Lucas [1959] and Endrenyi [1981b]—the latter author including the case of constant relative error. Corresponding to an experimental point $\mathbf{x}$ in $\mathscr{X}$, there is a column vector $\mathbf{a} = \partial f(\mathbf{x}; \boldsymbol{\theta}^*)/\partial\boldsymbol{\theta}$ in $\mathbb{R}^p$. As $\mathbf{x}$ varies in $\mathscr{X}$, $\mathbf{a}$ will range over a subset $\mathscr{A} \subset \mathbb{R}^p$. Given the p observations $\mathbf{x}_1, \mathbf{x}_2, \ldots, \mathbf{x}_p$, $|\mathbf{F}_{\bullet}(\boldsymbol{\theta}^*)|$ is then proportional to the volume of the simplex formed by the origin and the corresponding p values of $\mathbf{a}$. It can then be shown that the values of $\mathbf{a}$ which maximize mod$|\mathbf{F}_{\bullet}(\boldsymbol{\theta})|$ must lie on the boundary of $\mathscr{A}$: this is also true for Δ of (5.144) when $n \geq p$ (Atkinson and Hunter [1968: 288] with $\mathbf{F}_{\bullet}(\boldsymbol{\theta}^*) \to \mathbf{X}$).

Example 5.8 Box and Lucas [1959] considered the model [see also (3.21)]

$$y = \beta(1 - e^{-\gamma x}) + \varepsilon, \qquad \gamma > 0. \tag{5.146}$$

To find an optimal design with $n = p = 2$ we have to find two values of x, x_1 and x_2 say, which we can assume satisfy $0 \leq x_1 < x_2 \leq b$. Then $\boldsymbol{\theta}' = (\beta, \gamma)$

and

$$\mathbf{F}_{\bullet}(\boldsymbol{\theta}) = \begin{bmatrix} \dfrac{\partial f_1}{\partial \beta} & \dfrac{\partial f_1}{\partial \gamma} \\ \dfrac{\partial f_2}{\partial \beta} & \dfrac{\partial f_2}{\partial \gamma} \end{bmatrix} = \begin{bmatrix} 1 - e^{-\gamma x_1} & -\beta x_1 e^{-\gamma x_1} \\ 1 - e^{-\gamma x_2} & -\beta x_2 e^{-\gamma x_2} \end{bmatrix},$$

where $f_i = E[y \mid x = x_i]$. Hence

$$\begin{aligned} |\mathbf{F}_{\bullet}(\boldsymbol{\theta})| &= (1 - e^{-\gamma x_2})\beta x_1 e^{-\gamma x_1} - (1 - e^{-\gamma x_1})\beta x_2 e^{-\gamma x_2} \\ &= \beta(1 - e^{-\gamma x_1}) x_2 e^{-\gamma x_2} \left[\frac{e^{\gamma x_2} - 1}{\gamma x_2} \frac{\gamma x_1}{e^{\gamma x_1} - 1} - 1 \right] \\ &\geq 0, \end{aligned} \tag{5.147}$$

since $(e^y - 1)/y$ is a monotonically increasing function for $y > 0$. We now wish to maximize $|\mathbf{F}_{\bullet}(\boldsymbol{\theta}^*)|$. To do this we first observe that, as x_2 increases, the first term of (5.147) increases and the second term decreases. Hence at the maximum we have $x_2 = b$, and differentiating the resulting equation with respect to x_1 leads to the optimal solution

$$x_1 = \frac{1}{\gamma} - \frac{b e^{-\gamma b}}{1 - e^{-\gamma b}}. \tag{5.148}$$

Suppose we guess that $\beta = \gamma = 1$ and we choose b so that $\mu = E[y]$ is close to its asymptotic value of β [c.f. (5.146)]. For example, when $\mu = 0.99\beta$ we have $x_2 = b = 4.6$ and

$$x_1 = 1 - \frac{(4.6)(0.01)}{0.99} = 0.95.$$

For the case of constant relative error (Endrenyi [1981b]), the criterion of (5.145) is to maximize

$$\begin{aligned} |\mathbf{F}_{\bullet}'\mathbf{W}\mathbf{F}_{\bullet}| &= |\mathbf{F}_{\bullet}|^2 |\mathbf{W}| \\ &= \left(\frac{|\mathbf{F}_{\bullet}|}{f(x_1;\boldsymbol{\theta}) f(x_2;\boldsymbol{\theta})} \right)^2 \\ &= \left(\frac{x_1 e^{-\gamma x_1}}{1 - e^{-\gamma x_1}} - \frac{x_2 e^{-\gamma x_2}}{1 - e^{-\gamma x_2}} \right)^2 \\ &= \frac{1}{\gamma^2} \left(\frac{\gamma x_1}{e^{\gamma x_1} - 1} - \frac{\gamma x_2}{e^{\gamma x_2} - 1} \right)^2. \end{aligned} \tag{5.149}$$

As $y/(e^y - 1)$ is monotonically decreasing, the above expression is minimized by making x_1 as small as possible and x_2 as large as possible. If $a \le x_1 < x_2 \le b$, then our optimal design points are $x_1 = a$ and $x_2 = b$. Endrenyi [1981b] gives graphs to show how the efficiency, as measured by the Δ or Δ_W criterion, falls off as less than optimal designs are used. When the number of observations is increased, the efficiencies can be quite low. ∎

There is a further question relating to the case of $n = p$ with replication, and that is whether unequal replication is preferable. If r_i replicates are made at $\mathbf{x}_i$ for $i = 1, 2, \ldots, p$ ($\sum_i r_i = N$), M. J. Box [1968] showed that Δ of (5.144) for the N observations is maximized when the r_i do not differ by more than unity. Unfortunately, the design with r equal replicates of the optimal design with $n = p$ need not be the optimal design for a total of $N = pr$ observations (see Atkinson and Hunter [1968: 280] for a counterexample). However, the latter authors and M J. Box [1968, 1970a] give empirical and theoretical evidence that, for many experimental situations, near-equal replication leads to the optimal or a near-optimal design for all N-observation designs. A sufficient condition for optimal designs to have the replication property, namely that $\mathscr{X}$ is contained in a certain ellipsoid, is given by Atkinson and Hunter [1968]. Box [1968] also discussed some methods for constructing designs with this replication property. Such designs have a number of advantages, which Box [1970a] lists. These include the nondependence of N on the number of design points, and the reduction in the number of variables from kN to kp in optimizing the design criterion.

If interest is focused on just a $p_1 \times 1$ subset $\boldsymbol{\theta}_1$ of $\boldsymbol{\theta}$, then the criterion (5.144) can be modified accordingly. Let $\mathbf{F}_{\bullet}(\boldsymbol{\theta}^*) = (\mathbf{F}_1, \mathbf{F}_2)$, where $\mathbf{F}_1$ is $n \times p_1$, and let $\mathbf{F}_{\bullet}'(\boldsymbol{\theta}^*)\mathbf{F}_{\bullet}(\boldsymbol{\theta}^*)$ be partitioned as follows:

$$\mathbf{F}_{\bullet}'(\boldsymbol{\theta}^*)\mathbf{F}_{\bullet}(\boldsymbol{\theta}^*) = \begin{pmatrix} \mathbf{F}_1'\mathbf{F}_1 & \mathbf{F}_1'\mathbf{F}_2 \\ \mathbf{F}_2'\mathbf{F}_1 & \mathbf{F}_2'\mathbf{F}_2 \end{pmatrix} = \begin{pmatrix} \mathbf{F}_{11} & \mathbf{F}_{12} \\ \mathbf{F}_{21} & \mathbf{F}_{22} \end{pmatrix}, \qquad \text{say.}$$

Then using arguments similar to those used for the full parameter set, i.e. looking at the asymptotic variance–covariance matrix of $\hat{\boldsymbol{\theta}}_1$, the corresponding quantity to be maximized with respect to the $\mathbf{x}_i$ is

$$\begin{aligned} \Delta_s &= |\mathbf{F}_{11} - \mathbf{F}_{12}\mathbf{F}_{22}^{-1}\mathbf{F}_{21}| \\ &= \frac{\left|\begin{pmatrix} \mathbf{F}_{11} & \mathbf{F}_{12} \\ \mathbf{F}_{21} & \mathbf{F}_{22} \end{pmatrix}\right|}{|\mathbf{F}_{22}|} \qquad \text{(by Appendix A3.2).} \end{aligned} \tag{5.150}$$

As with Δ, this criterion (called D_s-optimality) is equivalent to maximizing the peak of a linearized posterior density for $\boldsymbol{\theta}_1$ based on the noninformative prior (2.104) for $\boldsymbol{\theta}$ and σ (Hunter et al. [1969], Hill and Hunter [1974]), or to minimizing the volume of the linearized confidence region for $\boldsymbol{\theta}_1$.

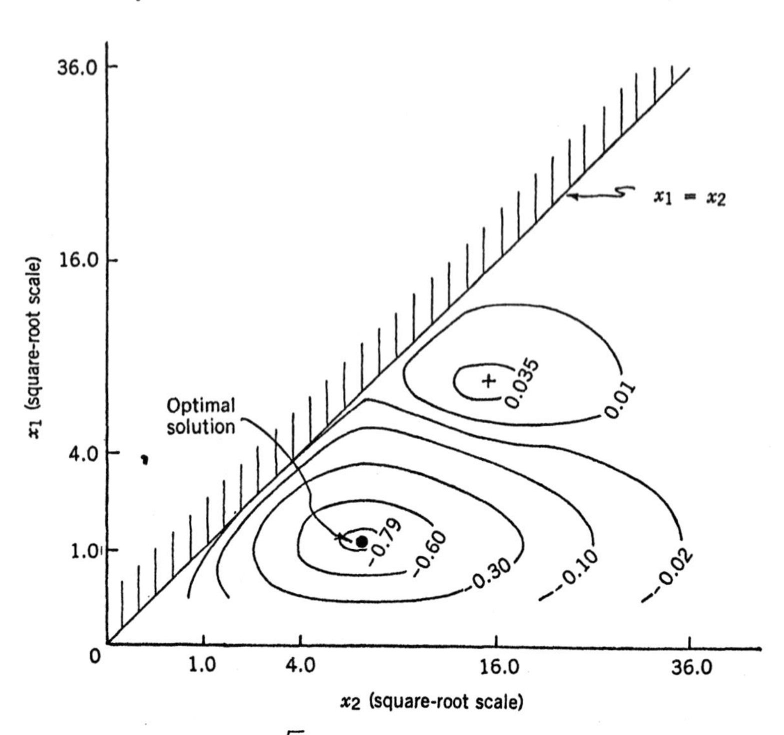

Figure 5.4 Contours of criterion $\sqrt{\Delta}$ for the model (5.151) in the space of x_1, x_2 for $0 \le x_1 \le x_2$. From Box and Lucas [1959] with permission from the Biometrika Trustees.

Example 5.9 Box and Lucas [1959] considered the model

$$y = \frac{\theta_1}{\theta_1 - \theta_2}(e^{-\theta_2 x} - e^{-\theta_1 x}) + \varepsilon. \tag{5.151}$$

If we are interested in the full parameter set $\boldsymbol{\theta} = (\theta_1, \theta_2)'$, then the $n = p = 2$ optimal design requires finding x_1 and x_2 $(0 \le x_1 < x_2)$ which maximize the modulus of $|(\partial f_i/\partial \theta_j)|$, where $f_i = E[y|x = x_i]$. If preliminary guesses are $\theta_1^* = 0.7$ and $\theta_2^* = 0.2$, we find that $(x_1, x_2) = (1.23, 6.86)$. It is interesting to note that $\text{mod}|\mathbf{F}_.(\boldsymbol{\theta}^*)|$ $(= \sqrt{\Delta})$ has two local maxima (see Fig. 5.4), and in cases such as this, care is needed to ensure that the right maximum is chosen.

Hill and Hunter [1974] considered the above model further and found optimal designs for just the single parameters θ_1 and θ_2 respectively. The criterion to be maximized is denoted by $\Delta_s(\theta_i)$, where $\Delta_s(\theta_1)$ is given by (5.150) with $p_1 = 1$, and $\Delta_s(\theta_2)$ follows from (5.150) by interchanging the subscripts 1 and 2. The authors also considered a second case in which the values of θ_1^* and θ_2^* are swapped around. The results for both cases and all three optimal designs are given in Table 5.3 and indicate that the different criteria can lead to quite different solutions. ■

Table 5.3 Optimal Parameter-Estimation Designs Using the Model (5.151)[a]

Criterion Employed	Optimal Settings for x_1, x_2	
	Case 1 $(\theta_1^*, \theta_2^*) = (0.7, 0.2)$	Case 2 $(\theta_1^*, \theta_2^*) = (0.2, 0.7)$
Δ	1.23, 6.86	1.59, 5.89
$\Delta_s(\theta_1)$	1.17, 7.44	1.27, 9.11
$\Delta_s(\theta_2)$	3.52, 5.64	6.66, 6.87

[a]From Hill and Hunter [1974].

A major disadvantage of having $n = p$ is that the adequacy of the model cannot be investigated. If model adequacy is an important issue, then the design should take this into account. Atkinson [1972] proposed checking a model by embedding it in a more general model and testing whether the extra terms are necessary. The experiment can then be designed to detect specific departures in an optimal fashion. For linear models, Atkinson [1972] discussed various optimality criteria and applied them to a number of examples. He also touched briefly on nonlinear models and suggested three methods of extending these, namely:

1. adding a low-order polynomial in the x-variables;
2. adding squared and cross-product terms in the partial derivatives $\partial f(\mathbf{x}; \boldsymbol{\theta})/\partial \theta_r$; and
3. embedding the nonlinear model in a more general model which reduces to the original form for particular values of some nonlinear parameters.

If the experimenter has reasonable faith in the model, then Atkinson recommends method 2.

For linear models $\mathscr{E}[\mathbf{y}] = \mathbf{X}\boldsymbol{\theta}$, (5.144) reduces to $|\mathbf{X}'\mathbf{X}|$, which does not depend on $\boldsymbol{\theta}$. In the case of partially linear models $\mathscr{E}[\mathbf{y}] = \mathbf{X}_\gamma \boldsymbol{\beta}$, where $\boldsymbol{\theta} = (\boldsymbol{\beta}', \boldsymbol{\gamma}')'$ [c.f. Section 5.10.2], Hill [1980] showed that the D-optimal design for $\boldsymbol{\theta}$ depends only on the values of the nonlinear parameters $\boldsymbol{\gamma}$. Conditions for extending this property to D_s-optimality for subsets of $\boldsymbol{\theta}$ are given by Khuri [1984].

5.13.2 Prior Estimates

We have seen that an imporant feature of optimal-design theory for nonlinear models is the need for a prior guess at $\boldsymbol{\theta}^*$. In many situations this will not be a problem if the right information is solicited from the experimenter. Pike [1984] demonstrates this using the model

$$y = \frac{x + \alpha}{\beta_0 + \beta_1(x + \alpha) + \beta_2(x + \alpha)^2}, \tag{5.152}$$

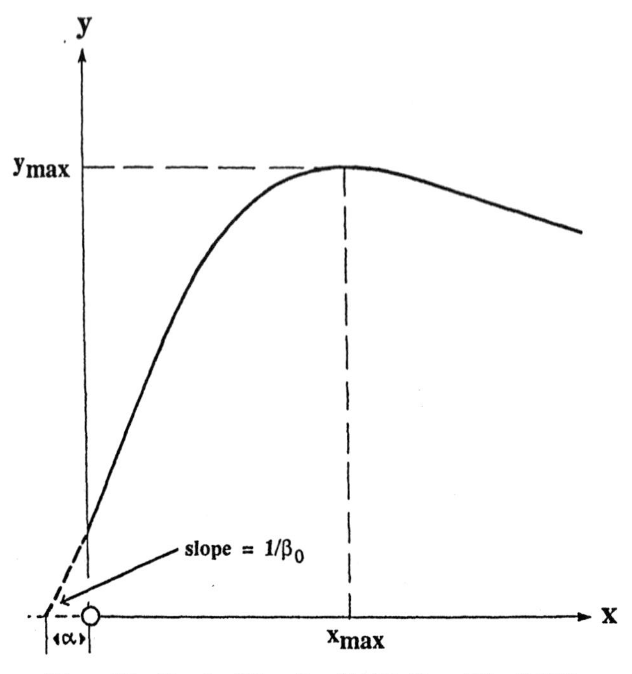

Figure 5.5 Graph of Equation (5.151). From Pike [1984].

first discussed by Nelder [1966]. This model is graphed in Fig. 5.5 and the challenge to the statistician is to ask the right questions so as to obtain information about the four parameters α, β_0, β_1, and β_2. What is known about the function is that

(i) $y = 0$ at $x = -\alpha$, so that α is an "origin" parameter;
(ii) y is a maximum at $x + \alpha = \sqrt{\beta_0/\beta_2}$ $(= x_{\text{max}})$;
(iii) the rate of change of the response, dy/dx, at $x = -\alpha$ is $1/\beta_0$; and
(iv) the optimum response is $y = (\beta_1 + 2\sqrt{\beta_0\beta_2})^{-1}$ $(= y_{\text{max}})$.

For example, suppose that y is the response of a crop to the addition of x units of fertilizer, so that α is the initial level of fertilizer and $x + \alpha$ the level after x has been added. Pike [1984] notes that the prior information can be obtained from the answers to the following four questions, which correspond to the above four formulae:

(i) What is the basal level of fertilizer in the soil?
(ii) Where do you believe the maximum response might lie?
(iii) When small amounts of fertilizers are added, how much does the response change?

(iv) What is the likely value of the maximum response?

Answers to these questions will often be available to the accuracy required by the design criterion. Clearly some imagination may be needed by the statistician to come up with the right characteristics of the model, especially when there are several x-variables instead of one as in (5.152). Plots of the postulated model will always be helpful, as in the above example. If the parameter-effects curvature is a problem, then a parameter transformation like that suggested above, namely

$$(\alpha, \beta_0, \beta_1, \beta_2) \to (\alpha, \beta_0^{-1}, \sqrt{\beta_0/\beta_2}, y_{\max}),$$

may be appropriate.

5.13.3 Sequential Designs

A natural way of designing an experiment is to use a sequential method. After observations have been made at n design points $\mathbf{x}_1, \mathbf{x}_2, \ldots, \mathbf{x}_n$, we wish to choose the next n_0 points $\mathbf{x}_{n+1}, \mathbf{x}_{n+2}, \ldots, \mathbf{x}_{n+n_0}$ in some optimal fashion. Box and Hunter [1965] proposed choosing the n_0 points to maximize the peak of the posterior distribution of $\boldsymbol{\theta}$ based on the $n + n_0$ points. Using a uniform prior for $\boldsymbol{\theta}$ and various approximations, they showed that this criterion leads once again to maximizing (5.144) constructed from the $n + n_0$ points, but with respect to just $\mathbf{x}_{n+1}, \ldots, \mathbf{x}_{n+n_0}$. Their result was extended to the case of a multivariate normal prior for $\boldsymbol{\theta}$ by Draper and Hunter [1967a]. Pritchard and Bacon [1977] discussed further extensions for handling variance heterogeneity. We note that the theory for nonsequential designs readily follows by setting $n = 0$. An optimal nonsequential design can be used for the first part of a sequential experiment.

A case of particular interest is $n_0 = 1$ when additional design points are introduced one at a time. If $\mathbf{f}_{n+1} = \partial f(\mathbf{x}_{n+1}; \boldsymbol{\theta})/\partial \boldsymbol{\theta}$ then, for $n + 1$ observations, Δ of (5.144) becomes

$$\Delta_{(n+1)} = |\mathbf{C}_{(n+1)}|,$$

where

$$\begin{aligned} \mathbf{C}_{(n+1)} &= \left| \begin{pmatrix} \mathbf{F}_{.}(\boldsymbol{\theta}) \\ \mathbf{f}'_{n+1} \end{pmatrix}' \begin{pmatrix} \mathbf{F}_{.}(\boldsymbol{\theta}) \\ \mathbf{f}'_{n+1} \end{pmatrix} \right| \\ &= |\mathbf{F}'_{.}(\boldsymbol{\theta}) \mathbf{F}_{.}(\boldsymbol{\theta}) + \mathbf{f}_{n+1} \mathbf{f}'_{n+1}| \\ &= |\mathbf{C}_{(n)} + \mathbf{f}_{n+1} \mathbf{f}'_{n+1}| \\ &= |\mathbf{C}_{(n)}| (1 + \mathbf{f}'_{n+1} \mathbf{C}_{(n)}^{-1} \mathbf{f}_{n+1}) \qquad \text{(by Appendix A3.3).} \end{aligned}$$

Since $|\mathbf{C}_{(n)}|$ does not involve $\mathbf{x}_{n+1}$, the above criterion of maximizing Δ_{n+1} with respect to $\mathbf{x}_{n+1}$ reduces to maximizing $\mathbf{f}'_{n+1} \mathbf{C}_{(n)}^{-1} \mathbf{f}_{n+1}$ with respect to $\mathbf{x}_{n+1}$. This criterion is given by Fedorov [1972: 188–189], who shows that it has certain limiting optimal properties as $n \to \infty$. The parameter vector $\boldsymbol{\theta}^*$ can be estimated by the least-squares estimate $\hat{\boldsymbol{\theta}}_{(n)}$ based on the n observations, and updated to give $\hat{\boldsymbol{\theta}}_{(n+1)}$ after a further observation is made. As $\mathbf{C}_{(n+1)}^{-1}$, the asymptotic variance–covariance matrix of $\hat{\boldsymbol{\theta}}_{(n+1)}$, is needed for the next step, we can obtain it using Appendix A3.4, namely

$$\mathbf{C}_{(n+1)}^{-1} = \mathbf{C}_{(n)}^{-1} - \frac{\mathbf{C}_{(n)}^{-1}\mathbf{f}_{n+1}\mathbf{f}'_{n+1}\mathbf{C}_{(n)}^{-1}}{1 + \mathbf{f}'_{n+1}\mathbf{C}_{(n)}^{-1}\mathbf{f}_{n+1}}. \tag{5.153}$$

However, this expression requires evaluation at $\hat{\boldsymbol{\theta}}_{(n+1)}$ and is only useful if $\hat{\boldsymbol{\theta}}_{(n+1)} \approx \hat{\boldsymbol{\theta}}_{(n)}$, as $\mathbf{C}_{(n)}^{-1}$ is previously evaluated at $\hat{\boldsymbol{\theta}}_{(n)}$.

We note that for $\hat{\boldsymbol{\theta}}_{(n)}$ close to $\boldsymbol{\theta}^*$,

$$f(\mathbf{x}_{n+1};\hat{\boldsymbol{\theta}}_{(n)}) \approx f(\mathbf{x}_{n+1};\boldsymbol{\theta}^*) + \mathbf{f}'_{n+1}(\hat{\boldsymbol{\theta}}_{(n)} - \boldsymbol{\theta}^*),$$

so that

$$\begin{aligned} \operatorname{var}[f(\mathbf{x}_{n+1};\hat{\boldsymbol{\theta}}_{(n+1)})] &\approx \mathbf{f}'_{n+1}\mathscr{D}[\hat{\boldsymbol{\theta}}_{(n)}]\mathbf{f}_{n+1} \\ &\approx \sigma^2\mathbf{f}'_{n+1}\mathbf{C}_{(n)}^{-1}\mathbf{f}_{n+1}. \end{aligned} \tag{5.154}$$

Hence when the above approximate linearization is valid, the design criterion is approximately equivalent to finding $\mathbf{x}_{n+1}$ which maximizes the asymptotic variance of the "prediction" $f(\mathbf{x}_{n+1};\hat{\boldsymbol{\theta}}_{(n+1)})$.

In concluding this discussion we emphasize that an appropriate stopping rule for stopping the sequential experiment is needed. Some function of $\mathscr{D}[\hat{\boldsymbol{\theta}}_{(n)}]$ would seem appropriate. For example, the process could stop when $|\hat{\mathbf{C}}_{(n)}|^{-1}$ becomes sufficiently small (Fedorov [1972: 196]).

Table 5.4 Least-Squares Estimates Using a Sequential Experiment[a]

Experiment	x_1	x_2	y	θ_1	θ_2	θ_3
1	1.0	1.0	0.126			
2	2.0	1.0	0.129			
3	1.0	2.0	0.076			
4	2.0	2.0	0.126	10.39	48.83	0.74
5	0.1	0.0	0.186	3.11	15.19	0.79
6	3.0	0.0	0.606	3.96	15.32	0.66
7	0.2	0.0	0.268	3.61	14.00	0.66
8	3.0	0.0	0.614	3.56	13.96	0.67
9	0.3	0.0	0.318	3.32	13.04	0.67
10	3.0	0.8	0.298	3.33	13.48	0.67
11	3.0	0.0	0.509	3.74	13.71	0.63
12	0.2	0.0	0.247	3.58	13.15	0.63
13	3.0	0.8	0.319	3.57	12.77	0.63

[a]From Fedorov [1972].

Example 5.10 Fedorov [1972: 193] simulated the model

$$y = f(\mathbf{x}; \boldsymbol{\theta}) + \varepsilon$$
$$= \frac{\theta_3 \theta_1 x_1}{1 + \theta_1 x_1 + \theta_2 x_2} + \varepsilon, \tag{5.155}$$

where x_1 and x_2 lie in $[0, 3]$, and $\varepsilon \sim N(0, 0.01)$. He obtained the values of y given in Table 5.4 for four values of $\mathbf{x} = (x_1, x_2)'$ and calculated the least-squares estimate

$$\hat{\boldsymbol{\theta}}_{(4)} = (10.38, 48.83, 0.74)'.$$

Setting $\mathbf{f}_i = \partial f(\mathbf{x}_i; \boldsymbol{\theta})/\partial \boldsymbol{\theta}$ we have

$$\mathbf{C}_{(4)} = \sum_{i=1}^{4} \mathbf{f}_i \mathbf{f}_i'.$$

We next find $\mathbf{x}_5$ such that $\mathbf{f}_5' \mathbf{C}_{(4)}^{-1} \mathbf{f}_5$, evaluated at $\boldsymbol{\theta} = \hat{\boldsymbol{\theta}}_{(4)}$, is maximized, and perform the experiment at this value of $\mathbf{x}$ to get y_5. Then $\hat{\boldsymbol{\theta}}_{(5)} = (3.11, 15.19, 0.79)'$ and

$$\mathbf{C}_{(5)} = \mathbf{C}_{(4)} + \mathbf{f}_5 \mathbf{f}_5'.$$

The sequence for a total of 13 observations is given in Table 5.4. At each stage $\mathbf{x}_{n+1}$ was found by evaluating $\mathbf{f}_{n+1}$ over a grid of $\mathbf{x}$-values. ■

5.13.4 Multivariate Models

The above sequential design theory for $N = n + n_0$ observations was extended to multiresponse (multivariate) models, with a known variance–covariance matrix $\boldsymbol{\Sigma}$ for the response vector, by Draper and Hunter [1966, 1967b], using uniform and multivariate normal priors, respectively, for $\boldsymbol{\theta}$. For a normal prior with variance–covariance matrix $\mathbf{V}$, the design criterion is maximize

$$|\mathbf{F}_{\cdot}'(\boldsymbol{\theta})[\boldsymbol{\Sigma}^{-1} \otimes \mathbf{I}_N]\mathbf{F}_{\cdot}(\boldsymbol{\theta}) + \mathbf{V}^{-1}| \tag{5.156}$$

with respect to $\mathbf{x}_{n+1}, \mathbf{x}_{n+2}, \ldots, \mathbf{x}_N$. Here $\mathbf{F}_{\cdot}(\boldsymbol{\theta})$ and $\boldsymbol{\Sigma}^{-1} \otimes \mathbf{I}_N$ are defined by (11.6) and Appendix A13.1. For a uniform prior, $\mathbf{V}^{-1}$ is replaced by zero in (5.156), and for the univariate case $\boldsymbol{\Sigma}$ reduces to σ^2. One likely use of a prior like the multivariate normal is to indicate how the design is affected by variations in the prior information, in this case changes in $\mathbf{V}$.

Box [1970a] discussed the nonsequential case ($n = 0$) for the multivariate model with uniform prior. He found that in a number of models, near-replication led once again to designs which were optimum, or close to optimum. In a further paper, Box [1971b] developed a similar criterion for the case when one is interested in a subset of the parameters, the criterion being a multivariate extension of (5.150). Again Σ was regarded as known, but possibly nonconstant. The cases when Σ is unknown, or nonconstant with some variance–covariance matrices known and the remainder unknown, were investigated by Box and Draper [1972].

5.13.5 Competing Models

Frequently there is more than one possible model for a given experimental situation and the experiment needs to be designed to select the most appropriate model. This problem has been studied by a number of authors, usually in a linear-model framework (though their theory is applicable to nonlinear models with an adequate linear approximation). References to this literature are given by Atkinson and Cox [1974], Hill [1978], and Atkinson [1981]. If it is known that one of the models is true, and the models are separate (i.e. not nested), then the general methods of Atkinson and Fedorov [1975a, b] and Fedorov [1975] can be used. These authors introduced a new class of designs, called T-optimal designs, and discussed the construction of both sequential and nonsequential designs. Unfortunately, even if the models are linear, the T-optimal designs depend on the unknown parameter vector so that sequential designs are more appropriate. The rationale behind the latter is to choose, at each stage, the next observation $\mathbf{x}$ such that the least-squares predictions for the two best-fitting models are furthest apart (Atkinson and Fedorov [1975b: 293]). The design given by this intuitively attractive procedure converges to the T-optimal design and, with its emphasis on squared prediction differences, is similar in spirit to the Bayesian method of Box and Hill [1967] (see also Box [1969]). However, it differs in that the variances of the predicted responses are ignored. The two approaches are compared by Atkinson [1981], who found no difference in the behavior of the two criteria for his examples. A criterion based on the sum of the absolute prediction differences is proposed by Mannervik [1981].

5.13.6 Design Allowing for Curvature

a *Volume Approximation*

In Chapter 4 we saw that the asymptotic distributional theory for $\hat{\boldsymbol{\theta}}$, the least-squares estimate of $\boldsymbol{\theta}^*$, needed to be modified for moderate samples to allow for curvature effects. The linear approximation which led to the result that $\hat{\boldsymbol{\theta}}$ is approximately $N_p(\boldsymbol{\theta}^*, [\mathbf{F}_.'\mathbf{F}_.]^{-1})$, where $\mathbf{F}_. = \mathbf{F}_.(\boldsymbol{\theta}^*)$, was not always adequate, and second-order terms were then needed. Box [1971a] obtained an approximation for the bias of $\hat{\boldsymbol{\theta}}$ [c.f. (4.174)] in terms of the parameter-effects curvature array and suggested designing experiments to minimize this bias. Clarke [1980] calculated additional terms for $\mathscr{D}[\hat{\boldsymbol{\theta}}]$, as given in (4.179), which involved both intrinsic and parameter-effects curvature arrays, and recommended choosing the design to minimize the mean squared error of $\hat{\boldsymbol{\theta}}$.

Bates and Watts [1981a] suggested choosing the design points to minimize the parameter-effects curvature in order to simplify inference procedures. We noted in Section 5.13.1 that the criterion based on Δ of (5.144) could be interpreted in terms of the volume of the linearized region. This volume, being proportional to $|\mathbf{F}_{\cdot}'\mathbf{F}_{\cdot}|^{-1/2} = \Delta^{-1/2}$, can be regarded as a first-order approximation for the volume of the exact confidence region. However, the approximation may be inadequate because of curvature effects, and further terms should then be considered. Hamilton and Watts [1985] introduced a quadratic design criterion based on a second-order approximation for the volume in question, and the following discussion is from their paper.

We begin with the one-to-one transformation (5.54), namely

$$\boldsymbol{\tau} = \hat{\mathbf{Q}}_p'[\boldsymbol{\mu}(\boldsymbol{\theta}) - \hat{\boldsymbol{\mu}}], \tag{5.157}$$

from $\boldsymbol{\theta}$ to a vector $\boldsymbol{\tau}$ in the tangent plane at $\boldsymbol{\mu}(\hat{\boldsymbol{\theta}})$. In Section 5.8.1 we saw that an approximate $100(1-\alpha)\%$ confidence region for $\boldsymbol{\theta}$ is given by the second-order approximation [Equation (5.66) with $\hat{\mathbf{B}}$ defined in (5.62)]

$$\{\boldsymbol{\theta} : \boldsymbol{\tau}'(\mathbf{I}_p - \hat{\mathbf{B}})\boldsymbol{\tau} \leq F_\alpha \rho^2 = \rho_\alpha^2\},$$

or

$$\{\boldsymbol{\theta} : \boldsymbol{\tau}'\hat{\mathbf{D}}\boldsymbol{\tau} \leq \rho_\alpha^2\}, \tag{5.158}$$

say, where

$$\hat{\mathbf{D}} = \mathbf{I}_p - \hat{\mathbf{B}}, \tag{5.159}$$

$$\begin{aligned} \hat{\mathbf{B}} &= [\hat{\boldsymbol{\varepsilon}}'\hat{\mathbf{Q}}_{n-p}][\hat{\mathcal{A}}_{\cdot\cdot}^N] \\ &= \hat{\mathbf{K}}'[\hat{\boldsymbol{\varepsilon}}'][\hat{F}_{\cdot\cdot}]\hat{\mathbf{K}} \qquad \text{[by (5.67) and (4.47)]} \end{aligned} \tag{5.160}$$

and $\rho_\alpha = (ps^2 F_{p,n-p}^\alpha)^{1/2}$. When σ^2 is known, $\rho_\alpha = [\sigma^2 \chi_p^2(\alpha)]^{1/2}$. The volume of (5.158) can be obtained by integrating the absolute value of the Jacobian, $\text{mod}|\partial\boldsymbol{\theta}/\partial\boldsymbol{\tau}'|$, over the region $\boldsymbol{\tau}'\hat{\mathbf{D}}\boldsymbol{\tau} \leq \rho_\alpha^2$. Hamilton and Watts [1985] showed that a quadratic approximation to the Jacobian expanded about $\boldsymbol{\theta} = \hat{\boldsymbol{\theta}}$ ($\boldsymbol{\tau} = \mathbf{0}$) is given by

$$J = \left|\frac{\partial\boldsymbol{\theta}}{\partial\boldsymbol{\tau}'}\right| \approx |\hat{\mathbf{K}}|(1 - \hat{\mathbf{m}}'\boldsymbol{\tau} + \tfrac{1}{2}\boldsymbol{\tau}'\hat{\mathbf{M}}\boldsymbol{\tau}), \tag{5.161}$$

where

$$\begin{aligned} \hat{\mathbf{K}} &= \hat{\mathbf{R}}_{11}^{-1} \qquad \text{[by (5.52)]}, \\ \hat{\mathbf{m}}' &= (\text{tr}\,\hat{\mathbf{A}}_{(1)}^T, \text{tr}\,\hat{\mathbf{A}}_{(2)}^T, \ldots, \text{tr}\,\hat{\mathbf{A}}_{(p)}^T), \\ \hat{\mathbf{M}} &= \hat{\mathbf{m}}\hat{\mathbf{m}}' + \hat{\mathbf{W}} + [\hat{\mathbf{m}}'][\hat{\mathcal{A}}_{\cdot\cdot}^T], \\ \hat{\mathbf{W}} &= [(\hat{w}_{\alpha\beta})], \qquad \hat{w}_{\alpha\beta} = \text{tr}[\hat{\mathbf{A}}_{(\alpha)}^T \hat{\mathbf{A}}_{(\beta)}^T], \end{aligned} \tag{5.162}$$

and $\hat{\mathbf{A}}_{(s)}^T$ is the sth *column* slice of the three-dimensional parameter-effects curvature array $\hat{A}_{..}^T$. Thus $\hat{\mathbf{A}}_{(s)}$ is a $p \times p$ matrix with (i,r)th element $\hat{a}_{irs}$, the (r,s)th element in the ith face of $\hat{A}_{..}^T$. As expected, when $\boldsymbol{\tau} = \mathbf{0}$ (5.161) reduces to $|\hat{\mathbf{K}}|$ [c.f. (5.55)]. We note that

$$\begin{aligned} \mathrm{mod}|\hat{\mathbf{K}}| &= \mathrm{mod}|\hat{\mathbf{R}}_{11}|^{-1} \\ &= |\hat{\mathbf{R}}_{11}'\hat{\mathbf{R}}_{11}|^{-1/2} \\ &= |\hat{\mathbf{F}}_{.}'\hat{\mathbf{F}}_{.}|^{-1/2}. \end{aligned}$$

If the Cholesky decomposition of $\hat{\mathbf{F}}_{.}'\hat{\mathbf{F}}_{.}$ is carried out using positive diagonal elements for $\hat{\mathbf{R}}_{11}$ (see Appendix A8.2), then $\mathrm{mod}|\hat{\mathbf{K}}| = |\hat{\mathbf{K}}|$. Assuming that the quadratic expression in (5.161) is positive, so that there is no problem with integrating its modulus, Hamilton and Watts showed that the volume of (5.158) based on the approximation (5.161) is

$$v = \frac{\pi^{p/2}\rho_\alpha^p}{\Gamma[\frac{1}{2}(p+2)]}|\hat{\mathbf{F}}_{.}'\hat{\mathbf{F}}_{.}|^{-1/2}|\hat{\mathbf{D}}|^{1/2}\left(1 + \frac{\rho_\alpha^2 \operatorname{tr}[\hat{\mathbf{D}}^{-1}\hat{\mathbf{M}}]}{2(p+2)}\right) \tag{5.163}$$

$$= c|\hat{\mathbf{F}}_{.}'\hat{\mathbf{F}}_{.}|^{-1/2}|\hat{\mathbf{D}}|^{1/2}(1 + k^2 \operatorname{tr}[\hat{\mathbf{D}}^{-1}\hat{\mathbf{M}}]), \tag{5.164}$$

where

$$c = \frac{\pi^{p/2}\rho_\alpha^p}{\Gamma[\frac{1}{2}(p+2)]}, \tag{5.165}$$

$\rho_\alpha = \rho\sqrt{F_\alpha}$, and $k = \rho_\alpha/\sqrt{2(p+2)}$ is called the "effective noise level" by the authors. Hence this second-order approximation represents the first-order approximation $c|\hat{\mathbf{F}}_{.}'\hat{\mathbf{F}}_{.}|^{-1/2}$ multiplied by a term involving the intrinsic-curvature array through $\hat{\mathbf{D}}$ and the parameter-effects curvature array through $\hat{\mathbf{M}}$.

The expression (5.164) cannot be used as a criterion for optimal design as it stands, since k depends on s^2, $\hat{\mathbf{D}}$ depends on the residuals $\hat{\boldsymbol{\varepsilon}}$, and the derivatives in $\hat{\mathbf{F}}_{.}$, $\hat{\mathbf{D}}$, and $\hat{\mathbf{M}}$ are all evaluated at $\boldsymbol{\theta} = \hat{\boldsymbol{\theta}}$. To generate an anticipated volume for a proposed experimental design, Hamilton and Watts [1985] suggested choosing initial estimates $\boldsymbol{\theta}_0$ and σ_0^2; setting $\rho_\alpha = \rho_0 = \sigma_0[\chi_p^2(\alpha)]^{1/2}$, $k = k_0 = \rho_0/[2(p+2)]^{1/2}$, and $c = c_0 = \pi^{p/2}\rho_0^p/\Gamma[\frac{1}{2}(p+2)]$; and setting $\hat{\boldsymbol{\varepsilon}} = \mathbf{0}$ in $\hat{\mathbf{D}}$, so that $\hat{\mathbf{D}} = \mathbf{I}_p$. The latter choice for $\hat{\mathbf{D}}$ was justified on the grounds that, in general, $\mathscr{E}[\hat{\mathbf{B}}] \approx \mathbf{0}$ and for $n = p$ designs we have $\hat{\mathbf{B}} = \mathbf{0}$ (see below). The remaining matrices are evaluated at $\boldsymbol{\theta}_0$ to give $\mathbf{F}_0 = \mathbf{F}_{.}(\boldsymbol{\theta}_0)$ and $\mathbf{M}_0$, so that (5.164) becomes

$$v_0 = c_0|\mathbf{F}_0'\mathbf{F}_0|^{-1/2}\{1 + k_0^2 \operatorname{tr}\mathbf{M}_0\}. \tag{5.166}$$

This is a function of the design points $\mathbf{x}_i$ $(i = 1, 2, \ldots, n)$ which can be chosen to minimize (5.166). When $n = p$, $\mathbf{F}_{.}(\boldsymbol{\theta})$ is square and therefore generally nonsingular in a neighborhood of $\hat{\boldsymbol{\theta}}$. Hence $\mathbf{f}^{-1}$ exists, so that using the transformation

$\boldsymbol{\theta} = \mathbf{f}^{-1}(\boldsymbol{\beta})$, we have the linear reparametrization

$$\boldsymbol{\eta}(\boldsymbol{\beta}) = \mathbf{f}[\mathbf{f}^{-1}(\boldsymbol{\beta})] = \boldsymbol{\beta}.$$

Thus the intrinsic curvature is zero and $\mathbf{B} = \mathbf{0}$, as mentioned above.

For sequential designs with an initial n-observation experiment and n_0 further observations anticipated, we can use the current least-squares estimates $\hat{\boldsymbol{\theta}}_n$ and s^2 to estimate $\boldsymbol{\theta}, \sigma^2$, and the residuals $\hat{\varepsilon}_i$ $(i = 1, 2, \ldots, n)$. The formula (5.164) is expressed in terms of $n + n_0$ observations but with the residuals in $\hat{\mathbf{D}}$ for $i = n+1, n+2, \ldots, n+n_0$ set equal to zero. Since $\hat{\mathbf{B}} = \hat{\mathbf{K}}'[(\hat{\boldsymbol{\varepsilon}}'\hat{\mathbf{f}}_{rs})]\hat{\mathbf{K}}$, where $\hat{\boldsymbol{\varepsilon}}$ and $\hat{\mathbf{f}}_{rs}$ are $(n + n_0) \times 1$ vectors, we see that the last n_0 elements in $\hat{\mathbf{f}}_{rs}$ do not appear in $\hat{\mathbf{B}}$, as they are multiplied by the last n_0 zero elements in $\hat{\boldsymbol{\varepsilon}}$. Hence $\hat{\mathbf{B}}$ depends on the new design points $\mathbf{x}_{n+1}, \mathbf{x}_{n+2}, \ldots, \mathbf{x}_{n+n0}$ only through the elements of $\hat{\mathbf{K}}$. A suitable design criterion would be to minimize v of (5.164) with respect to the new design points.

In the above theory it was assumed that $1 - \hat{\mathbf{m}}'\boldsymbol{\tau} + \frac{1}{2}\boldsymbol{\tau}'\hat{\mathbf{M}}\boldsymbol{\tau}$ is positive throughout the region of integration [see (5.161)]. Hamilton and Watts [1985: Appendix B] derived the following method for checking this assumption. Writing $\boldsymbol{\tau} = \|\boldsymbol{\tau}\|(\boldsymbol{\tau}/\|\boldsymbol{\tau}\|) = r\mathbf{u}$, say, we have from (5.161) and $\text{mod}|\hat{\mathbf{K}}| = |\hat{\mathbf{K}}|$ that

$$J = \text{mod}|\hat{\mathbf{K}}| \left(1 - \hat{\mathbf{m}}'\mathbf{u}r + \frac{\mathbf{u}'\hat{\mathbf{M}}\mathbf{u}r^2}{2}\right),$$

where $\mathbf{u}'\mathbf{u} = 1$. In the direction $\mathbf{u}$, J is positive for r less than the smaller positive real root of the above quadratic, which is

$$r_{\mathbf{u}} = \frac{\hat{\mathbf{m}}'\mathbf{u} - [(\hat{\mathbf{m}}'\mathbf{u})^2 - 2\mathbf{u}'\hat{\mathbf{M}}\mathbf{u}]^{1/2}}{\mathbf{u}'\hat{\mathbf{M}}\mathbf{u}},$$

if such a root exists. An approximation to the region of positive J is the sphere with radius equal to the smallest value of $r_{\mathbf{u}}$ for any direction $\mathbf{u}$. By looking at derivatives we see that $r_{\mathbf{u}}$ is a decreasing function of $\hat{\mathbf{m}}'\mathbf{u}$ and an increasing function of $\mathbf{u}'\hat{\mathbf{M}}\mathbf{u}$. Hence a lower bound of $r_{\min}$ for $r_{\mathbf{u}}$ is obtained by taking the maximum value of $\hat{\mathbf{m}}'\mathbf{u}$, namely $\|\hat{\mathbf{m}}\|$, since $|\hat{\mathbf{m}}'\mathbf{u}| \leq \|\hat{\mathbf{m}}\| \; \|\mathbf{u}\| = \|\hat{\mathbf{m}}\|$, and the minimum value of $\mathbf{u}'\hat{\mathbf{M}}\mathbf{u}$, namely γ_1, the smallest eigenvalue of $\hat{\mathbf{M}}$. Thus

$$r_{\min} = \frac{\|\hat{\mathbf{m}}\| - [\|\hat{\mathbf{m}}\|^2 - 2\gamma_1]^{1/2}}{\gamma_1}. \tag{5.167}$$

It can be shown that an upper bound, $r_{\max}$, is given by the smaller of the values obtained with $\mathbf{u} = \hat{\mathbf{m}}/\|\hat{\mathbf{m}}\|$ and $\mathbf{u} = \mathbf{v}_1$, where $\mathbf{v}_1$ is the eigenvector associated with γ_1.

The next stage is to see where the tangent-plane confidence ellipsoid [c.f. (5.158)] $\boldsymbol{\tau}'(\mathbf{I}_p - \hat{\mathbf{B}})\boldsymbol{\tau} \leq \rho_\alpha^2$ lies with respect to the region of positive J. From (5.71) we see that the confidence ellipsoid is contained in a sphere $\|\boldsymbol{\tau}\| \le r_\alpha = \rho_\alpha/(1-\lambda_{\max})$, where $\lambda_{\max}$ is the maximum eigenvalue of $\hat{\mathbf{B}}$. If $r_\alpha < r_{\min}$, then J is positive within the confidence region, while if $r_{\min} \le r_\alpha \le r_{\max}$, J may be negative and the volume approximation should be used with care. However, if $r_\alpha > r_{\max}$, then J is likely to be negative and the approximation should not be used. We note that the above quantities are all evaluated at $\hat{\boldsymbol{\theta}}$, the center of the ellipsoids and spheres. However, in practice, various other estimates of $\boldsymbol{\theta}$, like the initial guess $\boldsymbol{\theta}_0$, will be used.

We mentioned above that for $n = p$ designs the expectation surface is flat, $\hat{\mathbf{B}} = \mathbf{0}$, and the confidence region (5.158) becomes $\{\boldsymbol{\theta}: \|\boldsymbol{\tau}\| \le \rho_\alpha\}$. As the tangent plane is the same as the expectation surface, we have $\mathbf{Q}_p = \mathbf{I}_p$. It then follows from (5.157) that $\boldsymbol{\tau} = \boldsymbol{\mu}(\boldsymbol{\theta}) - \hat{\boldsymbol{\mu}}$ $[= \mathbf{f}(\boldsymbol{\theta}) - \mathbf{f}(\hat{\boldsymbol{\theta}})]$, and the confidence region is

$$\{\boldsymbol{\theta}: \|\boldsymbol{\mu}(\boldsymbol{\theta}) - \hat{\boldsymbol{\mu}}\| \le \rho_\alpha\}. \tag{5.168}$$

The linear approximation to this region is

$$\{\boldsymbol{\theta}: (\boldsymbol{\theta} - \hat{\boldsymbol{\theta}})'\hat{\mathbf{F}}_{.}'\hat{\mathbf{F}}_{.}(\boldsymbol{\theta} - \hat{\boldsymbol{\theta}}) \le \rho_\alpha^2\}. \tag{5.169}$$

b An Example

A numerical search routine is used to find the experimental design that minimizes the anticipated volume. Initial estimates of the parameters and k, together with starting values for the design points $\mathbf{x}_i$, are required. The volume criterion is computed and the various quantities re-estimated. The process is repeated until a minimum is obtained. Hamilton and Watts [1985] demonstrated the procedure using the following example (with $n = p = 2$).

Example 5.11 Consider the model

$$f(x;\boldsymbol{\theta}) = \frac{\theta_1}{\theta_1 - \theta_2}(e^{-\theta_2 x} - e^{-\theta_1 x}) \qquad (\theta_1, \theta_2, x > 0). \tag{5.170}$$

Box and Lucas [1959] used initial parameter values $\boldsymbol{\theta}_0 = (0.7, 0.2)'$ and found, by minimizing $|\mathbf{F}_0'\mathbf{F}_0|^{-1/2}$, that the D-optimal two-point design is [c.f. Table 5.3] $\mathbf{x}_D = (1.23, 6.86)'$. Hamilton and Watts [1985] used these values of x as starting values for their search. They decided that $\sigma_0 = 0.1$ was reasonable for this situation, where the maximum expected response $E[y]$ is about 0.6. Since $\chi_2^2(.05) = 6.0$, we have $\rho_0 = 0.1\sqrt{6} = 0.25$, $k_0 = 0.25\sqrt{8} = 0.0884$, and $c_0 = \pi\rho_0^2$. For the above model (5.170)

$$\begin{aligned} \mathbf{F}_0 &= \left|\left(\frac{\partial f(x_i;\boldsymbol{\theta})}{\partial \theta_j}\right)\right|_{\boldsymbol{\theta}_0} \\ &= \begin{pmatrix} -0.4406 & -0.3408 \\ -0.1174 & -1.7485 \end{pmatrix}, \end{aligned}$$

and a matrix $\mathbf{K}_0$ which satisfies $\mathbf{K}_0 = (\mathbf{R}_{11}^{-1})_0$ and $(\mathbf{R}'_{11}\mathbf{R}_{11})_0 = \mathbf{F}'_0\mathbf{F}_0$ is

$$\mathbf{K}_0 = \begin{pmatrix} -2.1930 & 0.1493 \\ 0 & -0.5627 \end{pmatrix}.$$

[We note that the authors use a Cholesky decomposition in which the upper triangular matrix $\mathbf{R}_{11}$, and therefore $\mathbf{K}$ has negative diagonal elements (see Appendix A8.2).] Then $|\mathbf{K}_0| = |\mathbf{F}'_0\mathbf{F}_0|^{-1/2} = 1.23$. The $2 \times 2 \times 2$ array of second derivatives $[(\partial^2 \mathbf{f}(\boldsymbol{\theta})/\partial\theta_r\,\partial\theta_s)]_{\boldsymbol{\theta}_0}$ is given by

$$(F_{..})_0 = \begin{matrix} \begin{pmatrix} -0.5780 & -0.3608 \\ -0.3608 & -0.2931 \end{pmatrix} \\ \begin{pmatrix} 0.1540 & -0.2173 \\ -0.2173 & 9.7140 \end{pmatrix} \end{matrix}.$$

Using (4.69), the parameter-effects curvature array is

$$(A_{..}^T)_0 = \begin{matrix} \begin{pmatrix} 2.8765 & 0.1653 \\ 0.1653 & 0.6665 \end{pmatrix} \\ \begin{pmatrix} -0.0003 & -0.3737 \\ -0.3737 & 3.0464 \end{pmatrix} \end{matrix},$$

while the intrinsic curvature array has zero elements, as $n = p$ implies that the expectation surface is planar. From (5.162), namely

$$(m_{ij})_0 = \left\{ \sum_{r=1}^{p} \sum_{s=1}^{p} (a_{rri}^T a_{ssj}^T + a_{srj}^T a_{ris}^T + a_{ssr}^T a_{rij}^T) \right\}_0,$$

we obtain

$$\mathbf{M}_0 = \begin{pmatrix} 21.8768 & 6.5270 \\ 6.5270 & 30.5770 \end{pmatrix}$$

with $\operatorname{tr} \mathbf{M}_0 = 52.45$. Hence, from (5.166),

$$v_0 = \pi(0.25)^2(1.23)[1 + (0.0884)^2 52.45] = 0.34,$$

which is substantially different from the linear approximation $\pi(0.25)^2(1.23) = 0.24$, and much closer to the "volume" (actually area in this case) $v_D = 0.37$ of the exact region [c.f. (5.168)]

$$\{\boldsymbol{\theta} : \|\boldsymbol{\mu}(\boldsymbol{\theta}) - \boldsymbol{\mu}(\boldsymbol{\theta}_0)\| \leq \rho_0\}$$

found using Monte Carlo integration. An iterative minimization of the quadratic

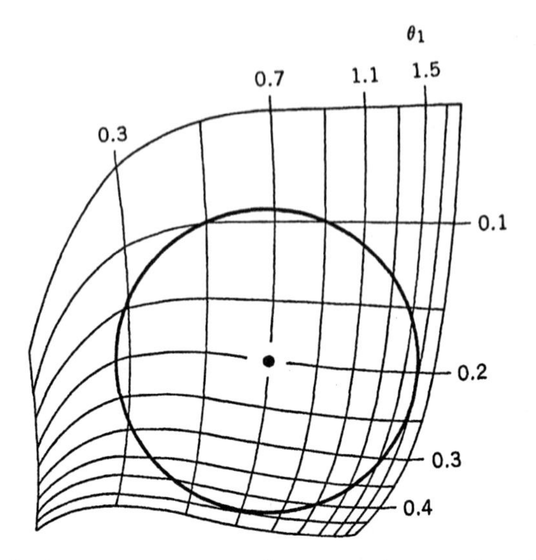

Figure 5.6 Plot of the θ-parameter curves on the tangent plane together with a circle of radius $\rho_0 = 0.25$. From Hamilton and Watts [1985].

criterion v_0 of (5.166) with respect to x_1 and x_2 led to the choice of design points $\mathbf{x}_Q = (1.04, 5.56)'$, giving an exact confidence region with area $v_Q = 0.32$, 86% of v_D. As a measure of efficiency the authors proposed using $E = 100(v_Q/v_D)^{1/p}$, which is 93% in this case.

The nonlinear aspects of the design are illustrated in two diagrams. Figure 5.6 shows, for the D-optimal design $\mathbf{x}_D$, the $\boldsymbol{\theta}$-parameter curves on the tangent plane (in this case the expectation surface also), together with a circle of radius $\rho_0 = 0.25$ [c.f. (5.158) with $\hat{\mathbf{D}} = \mathbf{I}_p$]. Since there is no intrinsic curvature, all the curvature comes from the parameter-effects array and is reflected in the irregularly shaped areas enclosed by pairs of contiguous parameter lines. This demonstrates that the Jacobian changes markedly with changing parameter values. Figure 5.7 shows the exact and linear-approximation ellipsoidal regions, (5.168) and (5.169) respectively, with $\rho_\alpha = \rho_0$.

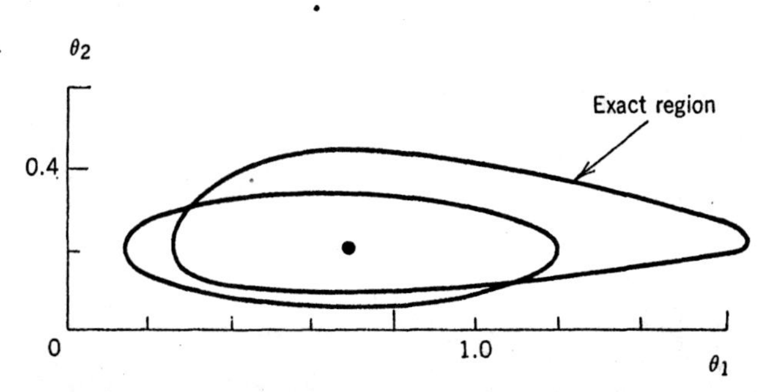

Figure 5.7 Exact and linear-approximation confidence regions. From Hamilton and Watts [1985].

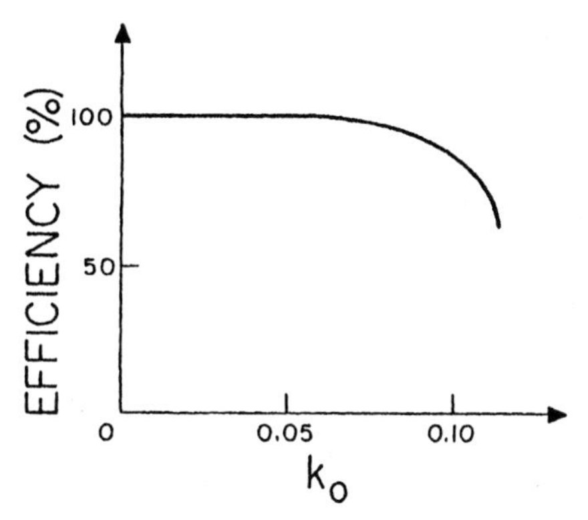

Figure 5.8 Efficiency of the D-optimal design relative to the quadratic optimal design. From Hamilton and Watts [1985].

The effects of reparametrization and of changing k_0, n, and the initial parameter values $\boldsymbol{\theta}_0$ were investigated in detail by the authors. Changing k_0 affects the quadratic design criterion by altering the design points and the area of the exact region. To explore this, optimal designs were obtained using $\boldsymbol{\theta}_0 = (0.7, 0.2)'$ and k_0 in the range from 0 to 0.2. The appropriate Jacobian J was checked in each case to see whether it remained positive over the region of integration. For $k_0 < 0.15$ there were no problems, but for $k_0 = 0.15$ the Jacobian

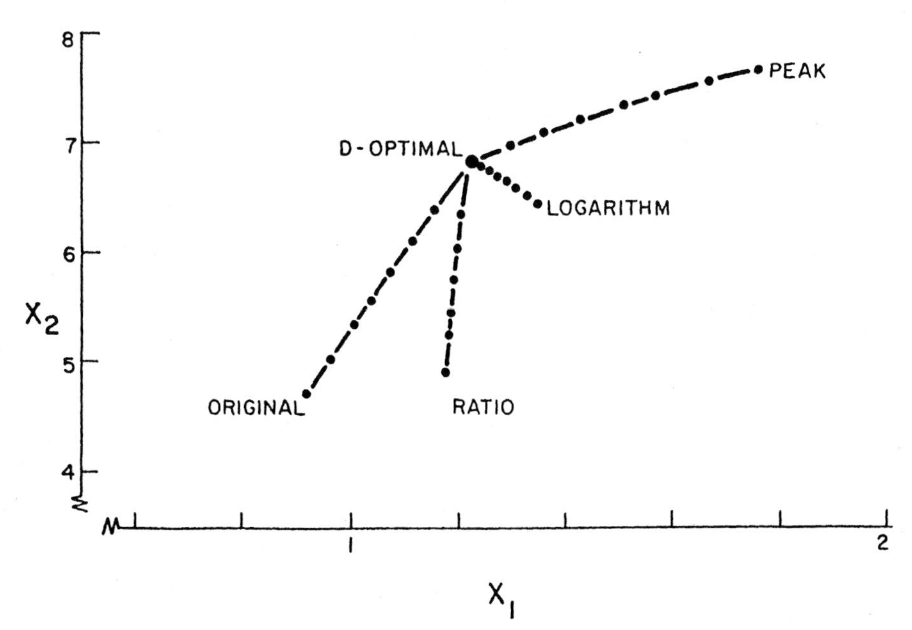

Figure 5.9 Changes in design variables due to parametrization and noise level k_0. From Hamilton and Watts [1985].

was possibly negative, and for $k = 0.2$ it was definitely negative. For $k_0 = 0$ the two criteria are identical, but as k_0 increase both x_1 and x_2 decrease for the quadratic design. For small k_0, the efficiency E is close to 100%, but it drops off rapidly for $k_0 > 0.088$ as seen in Fig. 5.8. However, for large values of k_0 the efficiency E is zero, as the D-optimal design gives an open exact confidence region, with infinite area. On the other hand, the quadratic design criterion produces a closed region over the entire range of k_0 allowed by the Jacobian check.

In contrast to D-optimal designs, quadratic designs depend on the parametrization used, as the parameter-effects curvature array and $\mathbf{M}_0$ change with the parametrization. For example, Hamilton and Watts [1985] considered three parametrizations:

a. $\phi_1 = \theta_1$, $\phi_2 = \theta_2/\theta_1$;
b. $\phi_1 = \log\theta_1$, $\phi_2 = \log\theta_2$; and
c. $\phi_1 = [\log(\theta_1/\theta_2)]/(\theta_1 - \theta_2)$, $\phi_2 = \exp(-\theta_2\phi_1)$,

where the parameters in case c correspond to the maximum $E[y]$ $(=\phi_2)$ and to the time $x = \phi_1$ when it occurs. Graphs of optimal design points x_1 and x_2 for several values of k_0 were plotted in Fig. 5.9 for the original parameters and the three cases. When $k_0 = 0$ all the optimal designs are the same as the D-optimal design $\mathbf{x}_D$. Increasing k_0 causes the optimal designs to move away from $\mathbf{x}_D$ in different directions, depending on the reparametrization. The dependence on k_0 is smallest for case b. For case a, J was possibly negative for the largest value of k_0, so that the corresponding design point was not included in the plot.

The effects of changing $\boldsymbol{\theta}_0$ were studied for $k_0 = 0.088$, and the authors concluded that the quadratic design is slightly less sensitive to such changes. By the same token the quadratic design is less sensitive to incorrect initial parameter estimates.

In addition to the $n = p = 2$ designs discussed above, optimal n-point designs were obtained for $n = 3, 4, \ldots, 7$ using the same initial values $\boldsymbol{\theta}_0 = (0.7, 0.2)'$ and $k_0 = 0.088$. Two random starting designs were used, and in every case the search procedure terminated at a replicated two-point design. However, local minima were found in some cases, so that a smaller anticipated area was sometimes found by starting the search with a different replicated design. For instance, when $n = 3$ the design with the three points $(x_1, x_2, x_3) = (1.09, 1.09, 5.73)$ gives an anticipated area of 0.22, whereas a smaller anticipated area of 0.21 corresponding to $(1.05, 5.80, 5.80)$ is obtained by starting from $(1.09, 5.73, 5.73)$. It was found for n even, the two replicated points correspond to the optimal two-point design with $k_0 = 0.088/\sqrt{(n/2)}$. Hence additional experiments are equivalent to a two-point design with a smaller value of k_0. These results are not unexpected in the light of the comments about replication in Section 5.13.1. We can exploit the fact that replications of the optimal p-point designs for general models are optimal or near-optimal, and choose the design to minimize

$$c|\mathbf{W}|^{-1/2}|\mathbf{F}_0'\mathbf{F}_0|^{-1/2}(1 + k_0^2 \operatorname{tr}[\mathbf{W}^{-1}\mathbf{M}_0]),$$

the anticipated volume for a replicated p-point design. The diagonal matrix $\mathbf{W}$ has its ith diagonal element equal to the number of replicates to be run at the ith point. ■

c Conclusions

Summarizing the results of several applications of their theory, Hamilton and Watts [1985] came to the following conclusions. Firstly, the quadratic design criterion is sensitive to the "effective noise level" (k), the method of parametrization, and the results from previous experiments (when used), whereas the D-optimal design is unaffected by any of these factors, as the linear approximation fails to allow for curvature effects. Secondly, when k is small or moderate, the gain in efficiency through using the quadratic rather than the D-optimal design is slight but consistent. However, for large k the D-optimal design can be disastrous, giving a confidence region of infinite size. Unfortunately there does not seem to be a single way of determining whether a given k is too large for the D-optimal criterion to be used safely. In similar situations the quadratic criterion will produce reasonable results for a wide range of k-values, and it provides a warning when the nonlinearity is too great. Finally, the quadratic criterion is less sensitive to changes in the initial parameter estimates.

CHAPTER 6

Autocorrelated Errors

6.1 INTRODUCTION

Situations in which data are collected sequentially over time may give rise to substantial serial correlation in the errors. This often happens with economic data in which the response y and the explanatory variables $\mathbf{x}$ measure the state of a market at a particular time. Both $\{y_i\}$ and $\{\mathbf{x}_i\}$ are time series. Another well-known example where serially correlated errors usually arise is in the fitting of growth curves to data on a single animal as it ages.

Suppose we wish to fit a sigmoidal growth curve such as the Richards curve [c.f. (1.18)] to data collected on the size of an animal taken at various ages. Our model is of the form

$$y = f(x; \boldsymbol{\theta}) + \varepsilon, \tag{6.1}$$

in which y is the size at time x, $f(x; \boldsymbol{\theta})$ is the theoretical trend curve, and ε is some form of random error. If we could monitor size continuously through time, we would expect the resulting empirical curve of y versus x to be smooth and cross the theoretical curve $f(x; \boldsymbol{\theta})$ a number of times as illustrated in Fig. 6.1. If the size is measured at n equally spaced times $x_1, x_2, \ldots, x_n$, then

$$y_i = f(x_i; \boldsymbol{\theta}) + \varepsilon_i, \tag{6.2}$$

or

$$\mathbf{y} = \mathbf{f}(\boldsymbol{\theta}) + \boldsymbol{\varepsilon}. \tag{6.3}$$

If n is large, so that x_i and x_{i+1} are sufficiently close together, models such as (6.1) but with uncorrelated errors become untenable because of long runs of residuals with the same sign. Thus any reasonable model must allow for serial correlations in the errors. However, if the time points are widely separated, the errors may be approximately uncorrelated. The above types of behavior have been noted in

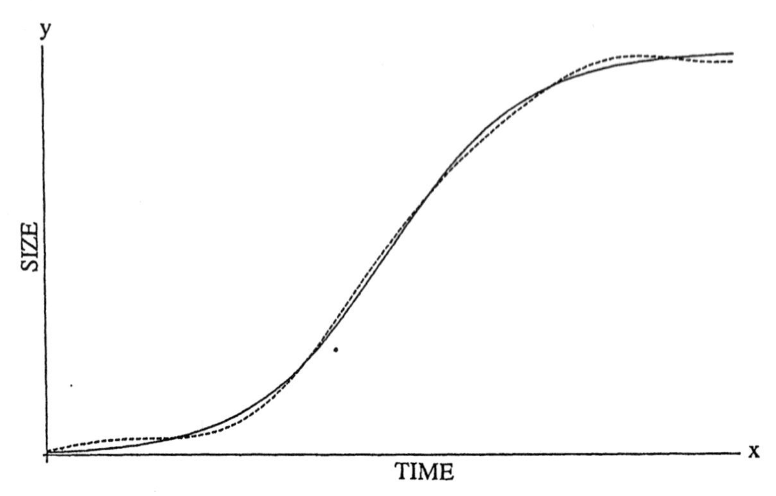

Figure 6.1 Hypothetical growth of animal over time. Solid line is theoretical curve. Dashed line is actual size continuously monitored.

growth data by researchers such as Grassia and De Boer [1980] and White and Brisbin [1980]. This leads to the consideration of models in which the correlation between errors ε_i and ε_{i+j} decreases as the time between them increases. Almost all the work in this area has been done with equally spaced x_i's, usually $x_i = i$ (with a suitable time origin).

In addition to the above statistical effects, physical factors can also lead to the consideration of serially correlated errors. Glasbey [1979] discussed factors influencing growth curves for cows. He noted that contributions to the ε_i's such as measurement error and how full the animal's stomach is at the time of weighing may often be approximately uncorrelated. However, contributions from uncontrolled environmental factors such as illness and variation in the nutritional quality of the diet could be expected to have a lingering effect on size. Figure 6.1 illustrates another problem. In purely empirical model building, there is a conflict between using more complicated models for the trend, and using a simple model for the trend and a more complicated error structure.

Historically, nonlinear models such as growth curves have often been fitted by ordinary least squares (OLS), i.e. by minimizing $\| \mathbf{y} - \mathbf{f}(\boldsymbol{\theta}) \|^2$, thus ignoring the possibility of serial correlation. However, with fairly substantial autocorrelation [e.g. $\rho_1 = \text{corr}(\varepsilon_i, \varepsilon_{i+1}) = 0.9$] OLS estimates can be extremely inefficient. Gallant and Goebel [1976] reported a simulation in which the mean squared error (MSE) for an OLS estimate was 6 times greater than the MSE of more appropriate estimators. Ordinary least squares can also lead to estimated variances of the parameter estimators that are too small, for instance up to a factor of 3 in the example of Glasbey [1979].

In this chapter we assume that $\boldsymbol{\varepsilon} = (\varepsilon_1, \varepsilon_2, \ldots, \varepsilon_n)'$ forms part of a weakly stationary time series $\{\varepsilon_i : i = 0, \pm 1, \ldots\}$. Thus for all i, $E[\varepsilon_i] = 0$, $\text{var}[\varepsilon_i] = \sigma^2$

$< \infty$, and the correlation between ε_i and $\varepsilon_{i \pm k}$ depends only upon k, the "distance" between them. For *lag k* we define the *autocovariance function* as

$$\gamma_k = \text{cov}[\varepsilon_i, \varepsilon_{i \pm k}], \qquad k = 0, 1, \ldots, \tag{6.4}$$

and the *autocorrelation function* as

$$\rho_k = \text{corr}[\varepsilon_i, \varepsilon_{i \pm k}] = \gamma_k / \sigma^2, \qquad k = 0, 1, \ldots. \tag{6.5}$$

We note that

$$\gamma_0 = \sigma^2 \quad \text{and} \quad \rho_k = \gamma_k / \gamma_0. \tag{6.6}$$

For most of this chapter we shall go further and assume stationary autoregressive moving-average models (ARMA) for the ε_i. Such models have representation

$$\begin{aligned} \varepsilon_i &= a_i + \psi_1 a_{i-1} + \psi_2 a_{i-2} + \cdots \\ &= \sum_{r=0}^{\infty} \psi_r a_{i-r}, \qquad \psi_0 = 1, \end{aligned} \tag{6.7}$$

where the number of nonzero coefficients ψ_r may be finite or infinite, and the a_i are i.i.d. with mean 0 and variance σ_a^2. It is assumed that

$$\sum_{r=0}^{\infty} |\psi_r| < \infty, \tag{6.8}$$

which is equivalent to the stationarity assumption and guarantees that all the moments exist and are independent of i. Then

$$\begin{aligned} \sigma^2 &= \text{var}[\varepsilon_i] \\ &= \text{var}\left[\sum_r \psi_r a_{i-r}\right] \\ &= \sum_r \psi_r^2 \, \text{var}[a_{i-r}] \\ &= \sigma_a^2 \sum_r \psi_r^2 \end{aligned} \tag{6.9}$$

and

$$\begin{aligned} \gamma_k &= E[\varepsilon_i \varepsilon_{i+k}] \\ &= E\left[\sum_r \psi_r a_{i-r} \sum_s \psi_s a_{i+k-s}\right] \\ &= \sum_r \sum_s \psi_r \psi_s \, \text{cov}[a_{i-r}, a_{i+k-s}] \\ &= \sum_r \sum_s \psi_r \psi_s \delta_{i-r, i+k-s} \sigma_a^2 \\ &= \sigma_a^2 \sum_r \psi_r \psi_{r+k}. \end{aligned}$$

Then

$$\rho_k = \frac{\sum_r \psi_r \psi_{r+k}}{\sum_r \psi_r^2} \tag{6.10}$$

for the general model (6.7). Special cases of this model will be discussed in detail throughout the chapter.

Let $\mathscr{D}[\boldsymbol{\varepsilon}] = \boldsymbol{\Sigma} = \sigma^2 \mathbf{V}$, so that $\mathbf{V}$ is the correlation matrix of the ε_i's. Hannan [1971] (see also Amemiya [1983]) proved that if $\boldsymbol{\varepsilon}$ comes from a stationary process with continuous spectral density, then, under appropriate regularity conditions, the OLS estimator $\hat{\boldsymbol{\theta}}_{\text{OLS}}$ is a consistent estimator of $\boldsymbol{\theta}^*$ (the true value of $\boldsymbol{\theta}$) and is asymptotically normally distributed with asymptotic variance–covariance matrix

$$\mathscr{D}(\hat{\boldsymbol{\theta}}_{\text{OLS}}) = \sigma^2 (\mathbf{F}_{.}'\mathbf{F}_{.})^{-1} \mathbf{F}_{.}'\mathbf{V}\mathbf{F}_{.}(\mathbf{F}_{.}'\mathbf{F}_{.})^{-1}.$$

Gallant [1987: 137] gives an estimated asymptotic variance–covariance matrix for the OLS estimate $\hat{\boldsymbol{\theta}}_{\text{OLS}}$ that can be used even where the time series $\{\varepsilon_i\}$ is not stationary. This estimate, formed using the OLS residuals $\hat{\varepsilon}_i$, is

$$\hat{\mathscr{D}}(\hat{\boldsymbol{\theta}}_{\text{OLS}}) = (\hat{\mathbf{F}}_{.}'\hat{\mathbf{F}}_{.})^{-1} \hat{\mathbf{B}} (\hat{\mathbf{F}}_{.}'\hat{\mathbf{F}}_{.})^{-1}.$$

If $\langle n \rangle$ is the integer nearest $n^{1/5}$, then

$$\hat{\mathbf{B}} = \sum_{r=-\langle n \rangle}^{\langle n \rangle} w\left(\frac{r}{\langle n \rangle}\right) \hat{\mathbf{B}}_{nr},$$

$$w(t) = \begin{cases} 1 - 6|t|^2 + 6|t|^3, & 0 \le |t| \le \frac{1}{2}, \\ 2(1-|t|)^3, & \frac{1}{2} \le |t| \le 1, \end{cases}$$

and

$$\hat{\mathbf{B}}_{nr} = \begin{cases} \displaystyle\sum_{s=1+r}^{n} \hat{\varepsilon}_s \hat{\varepsilon}_{s-r} \left[\frac{\partial f_s}{\partial \boldsymbol{\theta}} \frac{\partial f_s}{\partial \boldsymbol{\theta}'}\right]_{\boldsymbol{\theta} = \hat{\boldsymbol{\theta}}_{\text{OLS}}}, & r \ge 0, \\ \hat{\mathbf{B}}_{n,-r}', & r < 0. \end{cases}$$

This asymptotic result is justified by Gallant [1987: Section 7.4].

However, if $\mathbf{V}$ were known, a more appropriate estimate of $\boldsymbol{\theta}$ would be the generalized least-squares (GLS) estimate obtained by minimizing

$$\{\mathbf{y} - \mathbf{f}(\boldsymbol{\theta})\}' \mathbf{V}^{-1} \{\mathbf{y} - \mathbf{f}(\boldsymbol{\theta})\}.$$

Throughout this chapter we shall use the GLS method frequently and usually in an iterative mode, with $\mathbf{V}$ being estimated from a previous iteration. The reader is referred to Section 2.1.4 for computational details relating to GLS.

Finally we note note that (6.2) can be quite general. Here x_i need not refer just to time: $\{x_i\}$ may also be a time series. In this latter case x_i could be replaced by a vector of regressors.

6.2 AR(1) ERRORS

6.2.1 Preliminaries

In many applications, the autocorrelation ρ_k might reasonably be expected to decrease steadily as the "distance" k between ε_i and $\varepsilon_{i\pm k}$ increases. The simplest correlation structure of this form is

$$\rho_k = \rho_1^k,$$

so that from (6.5),

$$\text{corr}[\varepsilon_i, \varepsilon_j] = \rho_1^{|i-j|} = v_{ij}, \tag{6.11}$$

say, and $\mathscr{D}[\boldsymbol{\varepsilon}] = \boldsymbol{\Sigma} = \sigma^2\mathbf{V} = \sigma^2[(v_{ij})]$. This is the correlation structure obtained if we assume a first-order autoregressive model AR(1) for the ε_i's, namely

$$\varepsilon_i = \phi\varepsilon_{i-1} + a_i, \qquad |\phi| < 1, \tag{6.12}$$

where $\{a_i : i = 0, \pm 1, \pm 2, \ldots\}$ are independent with $E[a_i] = 0$ and $\text{var}[a_i] = \sigma_a^2$. To see that (6.12) leads to the appropriate structure (6.7), we repeatedly substitute into the right-hand side of (6.12) to give

$$\varepsilon_i = \sum_{r=0}^{\infty} \phi^r a_{i-r}.$$

Alternatively, we can use the backward-shift operator B, where $Ba_i = a_{i-1}$. Then (6.12) becomes

$$(I - \phi B)\varepsilon_i = a_i,$$

and

$$\begin{aligned} \varepsilon_i &= (I - \phi B)^{-1} a_i \\ &= (I + \phi B + \phi^2 B^2 + \cdots) a_i \\ &= \sum_{r=0}^{\infty} \phi^r a_{i-r}, \end{aligned}$$

where $|\phi| < 1$ is needed for convergence. This model is a special case of (6.7) with $\psi_r = \phi^r$, and (6.8) is satisfied. Hence from (6.10) we have $\rho_k = \phi^k$ and, also applying (6.9) to the model (6.12), we identify

$$\phi = \rho_1, \qquad \sigma_a^2 = (1 - \phi^2)\sigma^2. \tag{6.13}$$

Defining $\mathscr{D}[\boldsymbol{\varepsilon}] = \sigma^2 \mathbf{V}_\phi$ for the above model, we can verify directly by multiplication that

$$\mathbf{V}_\phi^{-1} = \frac{1}{1-\phi^2} \begin{bmatrix} 1 & -\phi & 0 & 0 & \cdots & 0 & 0 \\ -\phi & 1+\phi^2 & -\phi & 0 & \cdots & 0 & 0 \\ 0 & -\phi & 1+\phi^2 & -\phi & \cdots & 0 & 0 \\ \vdots & \vdots & \vdots & \vdots & & \vdots & \vdots \\ 0 & 0 & 0 & 0 & \cdots & -\phi & 1 \end{bmatrix} \tag{6.14}$$

$$= \frac{1}{1-\phi^2} \mathbf{R}_\phi' \mathbf{R}_\phi,$$

where

$$\mathbf{R}_\phi = \begin{bmatrix} \sqrt{1-\phi^2} & 0 & 0 & \cdots & 0 & 0 \\ -\phi & 1 & 0 & \cdots & 0 & 0 \\ 0 & -\phi & 1 & \cdots & 0 & 0 \\ \vdots & \vdots & \vdots & & \vdots & \vdots \\ 0 & 0 & 0 & \cdots & -\phi & 1 \end{bmatrix}. \tag{6.15}$$

Hence

$$|\mathbf{V}_\phi^{-1}| = (1-\phi^2)^{-n} |\mathbf{R}_\phi|^2 = (1-\phi^2)^{1-n}, \tag{6.16}$$

since the determinant of the triangular matrix $\mathbf{R}_\phi$ is the product of its diagonal elements. Also

$$\begin{aligned} \boldsymbol{\varepsilon}' \mathbf{V}_\phi^{-1} \boldsymbol{\varepsilon} &= \frac{1}{1-\phi^2} (\mathbf{R}_\phi \boldsymbol{\varepsilon})' (\mathbf{R}_\phi \boldsymbol{\varepsilon}) \\ &= \frac{1}{1-\phi^2} \left((1-\phi^2)\varepsilon_1^2 + \sum_{i=2}^n (\varepsilon_i - \phi \varepsilon_{i-1})^2 \right). \end{aligned} \tag{6.17}$$

Setting $\varepsilon_i = y_i - f(\mathbf{x}_i; \boldsymbol{\theta})$ we have, from (6.3),

$$\{\mathbf{y} - \mathbf{f}(\boldsymbol{\theta})\}' \mathbf{V}_\phi^{-1} \{\mathbf{y} - \mathbf{f}(\boldsymbol{\theta})\} = \frac{1}{1-\phi^2} S_1(\boldsymbol{\theta}, \phi), \tag{6.18}$$

where

$$S_1(\boldsymbol{\theta}, \phi) = (1-\phi^2)\{y_1 - f(\mathbf{x}_1; \boldsymbol{\theta})\}^2 + S_2(\boldsymbol{\theta}, \phi) \tag{6.19}$$

and

$$S_2(\boldsymbol{\theta}, \phi) = \sum_{i=2}^n \{y_i - \phi y_{i-1} - f(\mathbf{x}_i; \boldsymbol{\theta}) + \phi f(\mathbf{x}_{i-1}; \boldsymbol{\theta})\}^2. \tag{6.20}$$

We note in passing that, when ϕ is known, the generalized least-squares criterion requires the minimization with respect to $\boldsymbol{\theta}$ of (6.18), that is, of $S_1(\boldsymbol{\theta}, \phi)$.

Under normality assumptions we have $\boldsymbol{\varepsilon} \sim N_n(\mathbf{0}, \sigma^2 \mathbf{V}_\phi)$, and—using (6.16), (6.17), and (6.13)—the joint probability density function of $\boldsymbol{\varepsilon}$ is

$$\begin{aligned} p(\boldsymbol{\varepsilon}) &= (2\pi\sigma^2)^{-n/2} |\mathbf{V}_\phi|^{-1/2} \exp\left\{ -\frac{1}{2\sigma^2} \boldsymbol{\varepsilon}' \mathbf{V}_\phi^{-1} \boldsymbol{\varepsilon} \right\} \\ &= (2\pi\sigma_a^2)^{-n/2} (1-\phi^2)^{1/2} \exp\left\{ -\frac{1}{2\sigma_a^2} \left[(1-\phi)^2 \varepsilon_1^2 + \sum_{i=2}^{n} (\varepsilon_i - \phi\varepsilon_{i-1})^2 \right] \right\}. \end{aligned} \tag{6.21}$$

Alternatively, this density can be obtained by noting that the elements of $\mathbf{a}^* = (\varepsilon_1, a_2, \ldots, a_n)'$ are mutually independent with $\varepsilon_1 \sim N(0, \sigma_a^2/[1-\phi^2])$ and $a_i \sim N(0, \sigma_a^2)$: the determinant of the transformation from $\mathbf{a}^*$ to $\boldsymbol{\varepsilon}$ is unity. Finally, from (6.21), the joint density function of $\mathbf{y}$ is

$$p(\mathbf{y}|\boldsymbol{\theta}, \phi, \sigma_a^2) = (2\pi\sigma_a^2)^{-n/2} (1-\phi^2)^{1/2} \exp\left\{ -\frac{1}{2\sigma_a^2} S_1(\boldsymbol{\theta}, \phi) \right\}. \tag{6.22}$$

We shall find below that Equations (6.18) and (6.22) suggest three approaches to the estimation problem, namely maximum-likelihood, two-stage, and iterated two-stage estimation. These methods and a conditional least-squares approach are discussed in the next section.

6.2.2 Maximum-Likelihood Estimation

Taking logarithms of (6.22), the log-likelihood function for $\boldsymbol{\gamma} = (\boldsymbol{\theta}', \phi, \sigma_a^2)'$ under the normality assumption is

$$L(\boldsymbol{\theta}, \phi, \sigma_a^2) = \text{const} - \frac{n}{2} \log \sigma_a^2 + \tfrac{1}{2} \log(1-\phi^2) - \frac{1}{2\sigma_a^2} S_1(\boldsymbol{\theta}, \phi). \tag{6.23}$$

We obtain an asymptotically efficient estimator $\hat{\boldsymbol{\gamma}}_{\text{ML}}$ of $\boldsymbol{\gamma}$ by maximizing (6.23). Frydman [1980] established the strong consistency of $\hat{\boldsymbol{\gamma}}_{\text{ML}}$ under the following conditions:

1. $\mathbf{x}_i$ lies in a compact (i.e. closed and bounded) space $\mathscr{X}$ for all i.
2. The admissible parameter space for $\boldsymbol{\gamma}$ is compact, and $\hat{\boldsymbol{\gamma}}_{\text{ML}}$ always lies in this space. Moreover, $\phi \in [-1+\delta, 1-\delta]$ for some small $\delta > 0$ [c.f. (6.12)].
3. $f(\mathbf{x}; \boldsymbol{\theta})$ is continuous, one-to-one, and twice continuously differentiable in $\mathbf{x}$ and $\boldsymbol{\theta}$.
4. $\lim_{n\to\infty} n^{-1} \{\mathbf{f}(\boldsymbol{\theta}^*) - \mathbf{f}(\boldsymbol{\theta})\}' \mathbf{V}_\phi^{-1} \{\mathbf{f}(\boldsymbol{\theta}^*) - \mathbf{f}(\boldsymbol{\theta})\}$ exists and equals $c(\boldsymbol{\theta}, \phi)$, say, for all $(\boldsymbol{\theta}, \phi)$ in the parameter space, where $0 \le c(\boldsymbol{\theta}, \phi) < \infty$ and $\boldsymbol{\theta}^*$ is the true value of $\boldsymbol{\theta}$. Also $\boldsymbol{\theta} \ne \boldsymbol{\theta}^*$ implies $c(\boldsymbol{\theta}, \phi) > 0$.

With additional assumptions, the asymptotic normality and independence of $\hat{\boldsymbol{\theta}}_{\text{ML}}$, $\hat{\phi}_{\text{ML}}$, and $\hat{\sigma}^2_{a,\text{ML}}$ are established by Frydman [1978], namely (dropping the star from the true values for notational convenience),

$$\sqrt{n}(\hat{\boldsymbol{\theta}}_{\text{ML}} - \boldsymbol{\theta}) \sim N_p(\mathbf{0}, \sigma^2[n^{-1}\mathbf{F}_{\boldsymbol{\cdot}}'\mathbf{V}_{\phi}^{-1}\mathbf{F}_{\boldsymbol{\cdot}}]^{-1}),$$

$$\sqrt{n}(\hat{\phi}_{\text{ML}} - \phi) \sim N(0, 1 - \phi^2), \tag{6.24}$$

and

$$\sqrt{n}(\hat{\sigma}^2_{a,\text{ML}} - \sigma_a^2) \sim N(0, 2\sigma_a^4).$$

These results were established for linear models by Hildreth [1969]. The estimated asymptotic dispersion matrix for $\hat{\boldsymbol{\theta}}_{\text{ML}}$ can be computed either as $\{\hat{\sigma}^2(\hat{\mathbf{F}}_{\boldsymbol{\cdot}}'\mathbf{V}_{\hat{\phi}}^{-1}\hat{\mathbf{F}}_{\boldsymbol{\cdot}})^{-1}\}_{\text{ML}}$ or by using the inverse of the observed information matrix. By comparing (6.18) (c.f. Section 2.1.4) and (6.24) we see that for unknown ϕ, $\hat{\boldsymbol{\theta}}_{\text{ML}}$ has the same asymptotic distribution as that obtained using the generalized least-squares estimator with known ϕ.

Example 6.1 Glasbey [1979] fitted the Richards growth model [c.f. (1.18)]

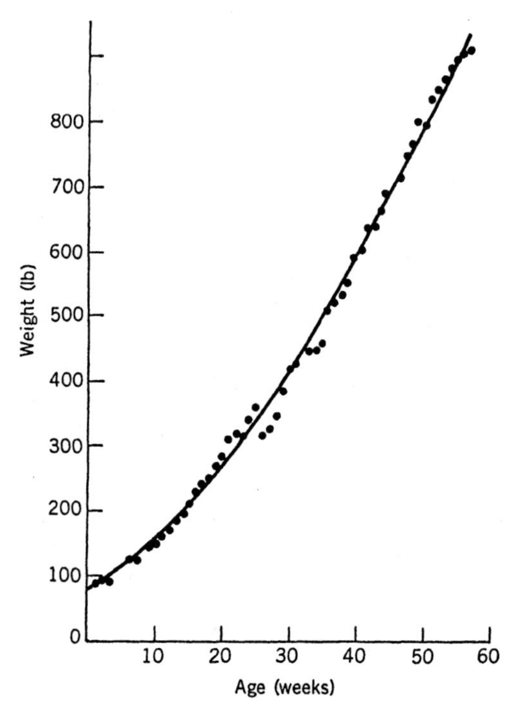

Figure 6.2 Animal-4 growth data together with generalized logistic curve (6.25) estimated after a log transformation. From Glasbey [1979].

$y = f(x;\boldsymbol{\theta}) + \varepsilon$, where x is time,

$$f(x;\boldsymbol{\theta}) = \alpha\left[1 + \exp\left\{-\frac{\lambda + kx}{\delta}\right\}\right]^{\delta}, \tag{6.25}$$

and ε has an AR(1) error structure, to growth data on Holstein cows. He calls the Richards model the generalized logistic curve. Because of variance heterogeneity, two transformations—$\log y$ [c.f. (1.45)] and $\sqrt{y}$—of the weight y were considered. Any improvement in the fit of the model to the data can be measured by the increase in the maximum value of the log-likelihood function. Under this criterion the square-root transformation is preferred. In Fig. 6.2, 58 observations on animal number 4 are plotted along with the fitted model based on $\log y$. The effect of serial correlation is very apparent from the graph, as the observed values tend to linger either above or below the fitted curve. The reader is referred to Glasbey's paper for further details. ■

6.2.3 Two-Stage Estimation

Instead of using maximum-likelihood estimators, the following two stage method can be used.

1. (i) Fit the model by ordinary least squares to obtain $\hat{\boldsymbol{\theta}}_{\text{OLS}}$.
   (ii) Compute the residuals $\hat{\varepsilon}_i = y_i - f(\mathbf{x}_i; \hat{\boldsymbol{\theta}}_{\text{OLS}})$.
   (iii) Obtain an estimate $\hat{\phi}$ of ϕ from $\{\hat{\varepsilon}_i\}$ and hence calculate $\mathbf{V}_{\hat{\phi}}^{-1}$ from (6.14).
2. Choose the two-stage estimator $\hat{\boldsymbol{\theta}}_{\text{TS}}$ to minimize [c.f. (6.18)]

$$\{\mathbf{y} - \mathbf{f}(\boldsymbol{\theta})\}'\mathbf{V}_{\hat{\phi}}^{-1}\{\mathbf{y} - \mathbf{f}(\boldsymbol{\theta})\} = \frac{S_1(\boldsymbol{\theta}, \hat{\phi})}{1 - \hat{\phi}^2}.$$

Two-stage estimators of this type have been used extensively with linear regression models since the work of Prais and Winsten [1954]. From simulation studies Park and Mitchell [1980] recommend the following estimate of ϕ $(=\rho_1)$:

$$\hat{\phi} = \frac{\sum_{i=2}^{n} \hat{\varepsilon}_i \hat{\varepsilon}_{i-1}}{\sum_{i=2}^{n-1} \hat{\varepsilon}_i^2}. \tag{6.26}$$

This estimate minimizes $S_1(\hat{\boldsymbol{\theta}}_{\text{OLS}}, \phi)$ of (6.19) with respect to ϕ. Other minor variants which are asymptotically equivalent have been used. The usual lag-1 sample autocorrelation

$$\hat{\rho}_1 = \frac{\sum_{i=2}^{n} \hat{\varepsilon}_i \hat{\varepsilon}_{i-1}}{\sum_{i=1}^{n} \hat{\varepsilon}_i^2} \tag{6.27}$$

is also widely used to estimate ϕ.

The two-stage estimator is a special case of an estimator proposed by Gallant and Goebel [1976] for AR(q_1) errors, which they actually call a one-stage estimator. They showed that under suitable regularity conditions the two-stage estimator $\hat{\boldsymbol{\theta}}_{TS}$ has the same asymptotic distribution as that given for $\hat{\boldsymbol{\theta}}_{ML}$ in (6.24). For further discussion see Section 6.4.

The two-stage estimator can be computed by fitting a new nonlinear regression model with i.i.d. $N(0, \sigma_a^2)$ errors using ordinary least squares. The new model is

$$z_i = g_i(\boldsymbol{\theta}) + a_i \qquad (i = 1, 2, \ldots, n), \tag{6.28}$$

in which

$$z_1 = (1 - \hat{\phi}^2)^{1/2} y_1, \qquad g(\mathbf{x}_1; \boldsymbol{\theta}) = (1 - \hat{\phi}^2)^{1/2} f(\mathbf{x}_1; \boldsymbol{\theta})$$

and, for $i = 2, 3, \ldots, n$,

$$z_i = y_i - \hat{\phi} y_{i-1}, \qquad g(\mathbf{x}_i; \boldsymbol{\theta}) = f(\mathbf{x}_i; \boldsymbol{\theta}) - \hat{\phi} f(\mathbf{x}_{i-1}; \boldsymbol{\theta}).$$

The error sum of squares for this model is $S_1(\boldsymbol{\theta}, \hat{\phi})$ [c.f. (6.19)]. Also, the matrix of first derivatives $\hat{\mathbf{G}}_{\cdot} = [(\partial g_i(\boldsymbol{\theta})/\partial \theta_j)]_{\hat{\boldsymbol{\theta}}}$ satisfies $\hat{\mathbf{G}}_{\cdot} = \hat{\mathbf{R}}\hat{\mathbf{F}}_{\cdot}$, where $\hat{\mathbf{R}} = \mathbf{R}_{\hat{\phi}}$ [c.f. (6.15)]. Thus

$$\begin{aligned} \sigma_a^2(\hat{\mathbf{G}}_{\cdot}'\hat{\mathbf{G}}_{\cdot})^{-1} &= \sigma_a^2(\hat{\mathbf{F}}_{\cdot}'\hat{\mathbf{R}}'\hat{\mathbf{R}}\hat{\mathbf{F}}_{\cdot})^{-1} \\ &= \sigma^2(\hat{\mathbf{F}}_{\cdot}'\mathbf{V}_{\hat{\phi}}^{-1}\hat{\mathbf{F}}_{\cdot})^{-1}, \end{aligned}$$

and this method of computation produces the correct estimated dispersion matrix.

6.2.4 Iterated Two-Stage Estimation

The steps of the above two-stage algorithm can be repeated using the "improved" estimator $\hat{\boldsymbol{\theta}}_{TS}$ in place of $\hat{\boldsymbol{\theta}}_{OLS}$ for the first stage. This produces a third-stage estimator. Gallant and Goebel [1976] do this to obtain their "two-stage" estimator. The process can be iterated further, or even to convergence. The resulting estimators at any subsequent stage of the iterative process have the same asymptotic distribution as the two-stage estimator (Gallant and Goebel [1976]) but may differ with respect to their performance in finite samples. The iterated two-stage estimator differs from the maximum-likelihood estimator in finite samples with respect to the calculation of $\hat{\phi}$. The former chooses ϕ to minimize $S_1(\boldsymbol{\theta}, \phi)$, while the latter chooses ϕ to minimize [c.f. (6.23)]

$$-\log(1 - \phi^2) + \frac{1}{\sigma_a^2} S_1(\boldsymbol{\theta}, \phi).$$

The effect of the $\log(1 - \phi^2)$ term is to keep the estimate of ϕ in the interval

$(-1,1)$ and bound it away from the end points. The two-stage and iterated two-stage estimators can have problems caused by inadmissible estimates of ϕ.

6.2.5 Conditional Least Squares

In Section 1.8 we noted that by considering the equations

$$y_i = f(\mathbf{x}_i;\boldsymbol{\theta}) + \varepsilon_i$$

and

$$\varepsilon_i = \phi\varepsilon_{i-1} + a_i, \quad |\phi| < 1 \qquad (i = 1, 2, \ldots, n),$$

we could form a more complicated nonlinear regression model with uncorrelated errors. This model, namely

$$y_i = \phi y_{i-1} + f(\mathbf{x}_i;\boldsymbol{\theta}) - \phi f(\mathbf{x}_{i-1};\boldsymbol{\theta}) + a_i \qquad (i = 2, 3, \ldots, n), \tag{6.29}$$

could then be fitted by least squares. This leads to choosing values of $\boldsymbol{\theta}$ and ϕ to minimize $S_2(\boldsymbol{\theta}, \phi)$ of (6.20). Alternatively, under normality assumptions the joint density function of $\mathbf{y}$ is given by (6.22), from which the conditional density of $y_2, y_3, \ldots, y_n$ given y_1 is found to be

$$p(\mathbf{y}|y_1;\boldsymbol{\theta},\phi,\sigma_a^2) = (2\pi\sigma_a^2)^{-(n-1)/2}\exp\left\{-\frac{1}{2\sigma_a^2}S_2(\boldsymbol{\theta},\phi)\right\}. \tag{6.30}$$

Taking logarithms, the conditional log-likelihood is thus

$$L_c(\boldsymbol{\theta},\phi,\sigma_a^2) = \text{const} - \frac{n-1}{2}\log\sigma_a^2 - \frac{1}{2\sigma_a^2}S_2(\boldsymbol{\theta},\phi). \tag{6.31}$$

For any σ_a^2, L_c is maximized with respect to $\boldsymbol{\theta}$ and ϕ by once again minimizing $S_2(\boldsymbol{\theta},\phi)$. Thus we will call the resulting estimator, $\hat{\boldsymbol{\theta}}_{\text{CLS}}$ say, the conditional least-squares estimator. When this method has been used with linear regression models, the resulting estimator has sometimes been called the "nonlinear estimator" because of (6.29).

From (6.23) we see that

$$\begin{aligned} L(\boldsymbol{\theta},\phi,\sigma_a^2) = {} & L_c(\boldsymbol{\theta},\phi,\sigma_a^2) - \tfrac{1}{2}\log\sigma_a^2 + \tfrac{1}{2}\log(1-\phi^2) \\ & - \frac{1}{2\sigma_a^2}(1-\phi^2)\{y_1 - f(\mathbf{x}_1;\boldsymbol{\theta})\}^2. \end{aligned} \tag{6.32}$$

However, under reasonable asymptotic conditions to ensure consistency, $S_2(\boldsymbol{\theta},\phi)$ of (6.20) dominates $(1-\phi^2)\{y_1 - f(\mathbf{x}_1;\boldsymbol{\theta})\}^2/2\sigma_a^2$, and $\frac{1}{2}(n-1)\log\sigma_a^2$ dominates $-\frac{1}{2}\log\sigma_a^2 + \frac{1}{2}\log(1-\phi^2)$ as $n\to\infty$. Thus maximizing L_c is, in the limit, equivalent to maximizing L, so that $\hat{\boldsymbol{\theta}}_{CLS}$ has the same asymptotic behavior as the maximum-likelihood estimator $\hat{\boldsymbol{\theta}}_{ML}$. Since, however, $\hat{\boldsymbol{\theta}}_{CLS}$ ignores to some extent the first observation, we would expect $\hat{\boldsymbol{\theta}}_{ML}$ to be more efficient in small samples. Furthermore, the loss of the $\log(1-\phi^2)$ term makes it possible for $\hat{\phi}$ to lie outside $(-1,1)$.

We can also use the two-stage or iterated two-stage estimators with the conditional least-squares approach. In the two-stage approach, ϕ is estimated from the ordinary least-squares residuals and then $S_2(\boldsymbol{\theta},\hat{\phi})$ is minimized with respect to $\boldsymbol{\theta}$ alone. Park and Mitchell [1980] recommended the estimate

$$\hat{\phi} = \frac{\sum_{i=2}^{n}\hat{\varepsilon}_i\hat{\varepsilon}_{i-1}}{\sum_{i=1}^{n-1}\hat{\varepsilon}_i^2}, \tag{6.33}$$

which minimizes $S_2(\hat{\boldsymbol{\theta}}_{OLS}, \phi)$ with respect to ϕ. The estimate and the corresponding estimated dispersion matrix are computed as in Section 6.2.3, but omitting the first pair $[z_1, g(\mathbf{x}_1;\boldsymbol{\theta})]$.

With a linear regression model $y_i = \boldsymbol{\beta}'\mathbf{x}_i + \varepsilon_i$ and AR(1) errors, the procedure is particularly simple, since the two-stage estimator and its dispersion matrix are obtained by using ordinary linear regression on the "pseudo-data" $(z_i, \mathbf{w}_i)$, $i=2,3,\dots,n$, where $z_i = y_i - \hat{\phi}y_{i-1}$ and $\mathbf{w}_i = \mathbf{x}_i - \hat{\phi}\mathbf{x}_{i-1}$. This procedure was developed by Cochrane and Orcutt [1949], and the resulting estimator is generally known as the Cochrane–Orcutt estimator. However, the procedure of Prais and Winsten [1954] (see Section 6.2.3) is only slightly more complicated, since it involves just adding the pair $(z_1, \mathbf{w}_1)$ with a slightly different definition, namely $z_1 = (1-\hat{\phi}^2)^{1/2}y_1$ and $\mathbf{w}_1 = (1-\hat{\phi}^2)^{1/2}\mathbf{x}_1$.

6.2.6 Choosing between the Estimators

The estimators we have described in Sections 6.2.2 to 6.2.5 are all asymptotically equivalent. Of them, the two-stage estimators are probably the easiest to use with the computer software that is widely available. In the special case of linear models, the small-sample behavior of the various estimators has been investigated using simulation studies by Rao and Griliches [1969], Hong and L'Esperance [1973], Spitzer [1979], and Park and Mitchell [1980]: further studies are referenced by these authors. These studies have concentrated on the relative efficiencies of the estimates of the regression coefficients. Kramer [1980] looked at the finite-sample efficiency of ordinary least squares estimators using second-order asymptotic approximations. Before discussing efficiencies we outline some of the models simulated.

Spitzer [1979] repeated the simulation experiments of Rao and Griliches [1969] using more replications and adding the true maximum-likelihood estimator for comparative purposes. The model used was

$$y_i = \beta x_i + \varepsilon_i, \qquad \varepsilon_i = \rho_1\varepsilon_{i-1} + a_i, \quad \text{and} \quad x_i = \lambda x_{i-1} + v_i. \tag{6.34}$$

The $\{a_i\}$ are i.i.d. $N(0, \sigma_a^2)$, the v_i are i.i.d. $N(0, \sigma_v^2)$, and σ_a^2 and σ_v^2 are chosen so that the squared correlation of $\{x_i\}$ with $\{y_i\}$ is 0.9. Varying λ allows varying degrees of autocorrelation in the x_i-sequence. A sample size of $n = 20$ was used throughout. Some of the qualitative effects of Rao and Griliches [1969] could not be reproduced by Spitzer.

Park and Mitchell [1980] simulated the model

$$y_i = \beta_0 + \beta_1 x_i + \varepsilon_i, \qquad \varepsilon_i = \rho_1 \varepsilon_{i-1} + a_i \tag{6.35}$$

using three x_i sequences:

(i) $x_i = i$;
(ii) a strongly trended x_i sequence obtained from the U.S. gross national product (GNP) from 1950–1970; and
(iii) an untrended U.S. economic time series consisting of the manufacturing-capacity utilization rate, 1950–1970.

They used samples of size 20 and 50. In simulations where $|\rho_1|$ is large, inadmissable values of $\hat{\rho}_1$ can occur, so that they restricted $|\hat{\rho}_1| \leq .99999$. The method of handling such extreme cases can affect comparisons of the relative efficiencies of the estimators. Such effects can be very serious at extreme ρ_1-values, as we shall see below.

Several strong patterns emerge from this work. All of the methods discussed for estimating ρ_1 produce negatively biased estimates. However, in the context of nonlinear modeling, we regard $\phi = \rho_1$ as a nuisance parameter and are chiefly interested in the beta regression parameters [c.f. (6.34) and (6.35)]. In this respect none of the estimation techniques is uniformly better than any other for estimating the betas, though it would appear that methods based upon conditional least squares (CLS) should be avoided. For instance, no CLS method discussed in Section 6.2.5 does substantially better than its counterpart in Sections 6.2.2 to 6.2.4. Indeed, with a trended x_i-sequence, CLS methods can do substantially worse. For example, with ρ_1-values 0.4, 0.8, 0.9 and 0.99, CLS methods tend to be much less efficient than ordinary least squares (OLS), whereas the ML and two-stage methods of Sections 6.2.2 to 6.2.4 are more efficient. The problem with conditional least-squares-based methods is not caused by poor estimators of $\hat{\rho}_1$. In fact, even when the true value of $\phi = \rho_1$ is used in the two-stage CLS estimator (TRUECLS), we find that with a trended x_i-series, TRUECLS is less efficient than the two-stage estimator itself in all cases tabulated. For example, when $\rho_1 = 0.98$, $n = 20$, $x_i = i$, and efficiency is measured in terms of the ratio of the (theoretical) standard deviations of the estimators, TRUECLS had a relative efficiency of only 4% relative to OLS when estimating β_0, and 11% when estimating β_1. In contrast, the methods of Sections 6.2.2 to 6.2.4 were more efficient than OLS. Kramer [1982] investigated this problem theoretically for the linearly trended model $x_i = i$ of Park and Mitchell [1980] and showed that, for finite samples, the efficiency of TRUECLS relative to OLS tends to zero as $\rho_1 \to 1$.

Park and Mitchell explained this behavior with a trended x_i series as follows. When x_i is linear in i, say $x_i = a_1 i + a_2$ with known a_j, then for $\rho_1 \approx 1$ we have $x_i - \rho x_{i-1} \approx x_i - x_{i-1} = a_1$. Hence, from (6.35)

$$\begin{aligned} E[y_i - \rho_1 y_{i-1}] &= \beta_0(1-\rho_1) + \beta_1(x_i - \rho x_{i-1}) \\ &\approx \beta_0(1-\rho_1) + \beta_1 a_1, \end{aligned} \tag{6.36}$$

and β_0 and β_1 are almost unidentifiable. This results in large standard errors for the parameter estimators. In such situations, the first observation supplies a disproportionately large amount of information about the beta parameters. An asymptotic analysis of this issue is given by Taylor [1981].

To put the above results in perspective we should note that an analysis based on an AR(1) error process really assumes that the process generating the errors has been going on for some time and has settled down into some sort of equilibrium before data collection begins. For data in which y_1 is measured following some cataclysmic event such as the birth of an animal or a war, so that initial errors do not behave in the same way as subsequent ones, the efficiency of conditional least squares may not be so bad.

Spitzer [1979] measured the relative efficiency of two estimators as the ratio of their mean squared errors. His estimators of β in (6.34) included the maximum-likelihood (ML) and two-stage (TS) estimators, the latter being called PW for Prais–Winsten. With his model, ML and TS are less efficient than OLS for an interval of values about $\rho_1 = 0$, although the relative efficiency is never worse than 89% for ML and 93% for TS when $\rho_1 = 0$. OLS was very inefficient when $|\rho_1|$ was large (e.g. 16% compared to ML when $\rho_1 = 0.9$). TS was slightly more efficient than ML for $-.5 \le \rho_1 \le .4$, while ML was more efficient than TS outside this range. The differences in efficiency were not serious except for very large ρ_1-values ($\rho_1 > .9$). Park and Mitchell [1980] included the iterated two-stage method (ITTS), which was slightly more efficient than ML. However, when $n = 20$, the efficiency gains over OLS for any of the methods were modest (less than 15% reduction in standard deviation) except for the untrended x_i series when $\rho_1 \ge 0.8$. For $n = 50$, substantial gains in efficiency over OLS were also achievable in the trended series for $\rho_1 \ge 0.9$. Park and Mitchell [1980] advocated the use of iterated two-stage estimators.

Hong and L'Esperance [1973] simulated a linear model with two correlated x-variables and a model with distributed lags (i.e., y_i depends on previous y-values). From their results it appears that the interval about zero of ρ_1-values in which OLS beats the other estimators shrinks as the sample size increases. Also the size of the interval depends on the model used.

In summary, all three methods—ML, TS, and iterated TS—provide good estimators which guard against possibly serious losses in efficiency by OLS in the presence of AR(1) autocorrelated errors. In terms of efficiency they appear to be comparable for $\rho_1 \le .8$, but the iterative methods appear preferable for higher degrees of autocorrelation.

Unfortunately, little attention has been paid to the asymptotic standard errors and their usefulness for approximate confidence intervals and hypothesis testing. Park and Mitchell [1980] looked at error rates for asymptotic t-tests of $H_0:\beta_0=0$ and $H_0:\beta_1=0$ in their model (6.35) using the trended GNP x_t-series with $n=50$ and using 1000 trials. For this series, OLS is reasonably efficient, but the observed significance levels for a test nominally at the 5% significance level gave a range of values of approximately 20% ($\rho_1=0.4$), 65% ($\rho_1=0.9$), and 80% ($\rho_1=0.98$). The observed levels for the ML, TS, and iterated TS methods were better, giving a range of values of approximately 9% ($\rho_1=0.4$), 20% ($\rho_1=0.9$), and 35% ($\rho_1=0.98$). However, they are still much greater than the nominal value of 5%. The differences between the estimators were relatively small for $\rho_1 \le 0.9$, with the iterated TS method performing slightly better than ML, which was itself slightly better than TS. More importantly, however, for Park and Mitchell's simple linear regression model, samples of size $n=50$ were still far too small for the asymptotic theory to provide a good approximation to the true distribution of the parameter estimators.

Recently Kobayashi [1985] investigated the comparative efficiencies of the iterated two-stage, ML, and iterated CLS estimators using second-order expansions for the asymptotic variances. His results are consistent with the empirical results we have just discussed. To this order to approximation, the iterated two-stage and ML methods are equivalent in asymptotic efficiency. Both are more efficient than conditional least squares, the loss in efficiency being due to the effect of the initial observation. He also investigated the downward bias in the estimators of ρ_1 and found this to be the least severe with the iterated two-stage method. However, the bias in ML and iterated CLS is substantially the same when $\rho_1>0$, and the serial correlation in x is greater than that in ε.

In conclusion we mention a two-stage estimator for linear models due to Durbin [1960a]. Although it compares favorably with any of the other methods, it does not generalize to nonlinear models.

6.2.7 Unequally Spaced Time Intervals

Frequently data are collected at irregular time intervals. Suppose that the ith data point was obtained at time t_i and $\mathscr{D}[\boldsymbol{\varepsilon}]=\boldsymbol{\Sigma}=[(\sigma_{ij})]$. By taking

$$\sigma_{ij}=\operatorname{cov}[\varepsilon_i,\varepsilon_j]=\sigma^2\phi^{|t_i-t_j|}, \tag{6.37}$$

Glasbey [1979] extended the AR(1) model in a natural way. Let us write $\Delta t_i = t_i - t_{i-1}$ and consider

$$a_i=\varepsilon_i-\phi^{\Delta t_i}\varepsilon_{i-1}. \tag{6.38}$$

If we assume that $\boldsymbol{\varepsilon}\sim N_n(\mathbf{0},\boldsymbol{\Sigma})$, then $\varepsilon_i\sim N(0,\sigma^2)$, $a_i\sim N(0,\sigma^2[1-\phi^{2\Delta t_i}])$, and a straightforward evaluation of the covariances shows that the elements of $\mathbf{a}^*=(\varepsilon_1,a_2,\ldots,a_n)'$ are uncorrelated and hence independent under normality.

Since $\mathbf{a}^* = \mathbf{K}\boldsymbol{\varepsilon}$, where $\mathbf{K}$ is upper triangular with unit diagonal elements [by (6.38)], the Jacobian of this transformation $|\mathbf{K}|$ is unity. Hence $\boldsymbol{\varepsilon}$ has joint density function

$$p(\boldsymbol{\varepsilon}) = (2\pi\sigma^2)^{-n/2} \prod_{i=2}^{n} (1-\phi^{2\Delta t_i})^{-1/2} \exp\left[-\frac{1}{2\sigma^2}\left\{\varepsilon_1^2 + \sum_{i=2}^{n} \frac{(\varepsilon_i - \phi^{\Delta t_i}\varepsilon_{i-1})^2}{(1-\phi^{2\Delta t_i})}\right\}\right]. \tag{6.39}$$

The joint density function for $\mathbf{y}$, and hence the likelihood function, is obtained by replacing ε_i by $y_i - f(\mathbf{x}_i;\boldsymbol{\theta})$ in (6.39). Maximum-likelihood estimates can then be obtained in the usual way.

The two-step method could be generalized to this model using an estimate $\hat{\phi}$ of ϕ from the least-squares residuals. The procedure again involves fitting an artificial model to pseudo-data by ordinary least squares. The model is [c.f. (6.28)]

$$z_i = g_i(\boldsymbol{\theta}) + a_i \qquad (i = 1, 2, \ldots, n),$$

where $z_1 = y_1$, $g_1(\boldsymbol{\theta}) = f(\mathbf{x}_1;\boldsymbol{\theta})$, and for $i = 2, 3, \ldots, n$,

$$z_i = \frac{y_i - \hat{\phi}^{\Delta t_i} y_{i-1}}{(1-\hat{\phi}^{2\Delta t_i})^{1/2}}$$

and

$$g_i(\boldsymbol{\theta}) = \frac{f(\mathbf{x}_i;\boldsymbol{\theta}) - \hat{\phi}^{\Delta t_i} f(\mathbf{x}_{i-1};\boldsymbol{\theta})}{(1-\hat{\phi}^{2\Delta t_i})^{1/2}}. \tag{6.40}$$

The main difficulty is to find a suitable estimate $\hat{\phi}$. One possibility would be to choose $\hat{\phi}$ so that $(\hat{\phi}, \hat{\sigma}^2)$ maximized $L(\hat{\boldsymbol{\theta}}_{OLS}, \phi, \sigma^2)$, where L is the log-likelihood.

6.3 AR(2) ERRORS

An autoregressive structure or order 2, denoted by AR(2), is of the form

$$\varepsilon_i = \phi_1 \varepsilon_{i-1} + \phi_2 \varepsilon_{i-2} + a_i, \tag{6.41}$$

where the a_i are uncorrelated with $E[a_i] = 0$, $\mathrm{var}[a_i] = \sigma_a^2$, and $\mathrm{var}[\varepsilon_i] = \sigma^2$. Multiplying (6.41) by ε_{i-k} and taking expected values gives us

$$E[\varepsilon_i \varepsilon_{i-k}] = \phi_1 E[\varepsilon_{i-1}\varepsilon_{i-k}] + \phi_2 E[\varepsilon_{i-2}\varepsilon_{i-k}] + E[a_i \varepsilon_{i-k}],$$

or, using (6.4),

$$\gamma_k = \phi_1 \gamma_{k-1} + \phi_2 \gamma_{k-2} + E[a_i \varepsilon_{i-k}].$$

Since ε_{i-k} depends only on $a_{i-k}, a_{i-k-1}, \ldots,$

$$E[a_i \varepsilon_{i-k}] = \begin{cases} \sigma_a^2, & k=0, \\ 0, & k=1,2,\ldots. \end{cases}$$

Hence

$$\begin{aligned} \gamma_0 &= \phi_1 \gamma_{-1} + \phi_2 \gamma_{-2} + \sigma_a^2 \\ &= \phi_1 \gamma_1 + \phi_2 \gamma_2 + \sigma_a^2 \qquad (\text{since} \quad \gamma_{-k} = \gamma_k) \end{aligned} \tag{6.42}$$

and

$$\gamma_k = \phi_1 \gamma_{k-1} + \phi_2 \gamma_{k-2}, \qquad k > 0. \tag{6.43}$$

Dividing by γ_0 $(=\sigma^2)$ gives us

$$\rho_k = \phi_1 \rho_{k-1} + \phi_2 \rho_{k-2}, \qquad k > 0. \tag{6.44}$$

Setting $k=1$ and noting that $\rho_0 = 1$ leads to

$$\rho_1 = \phi_1 \rho_0 + \phi_2 \rho_{-1} = \phi_1 + \phi_2 \rho_1$$

and hence

$$\rho_1 = \frac{\phi_1}{1-\phi_2}. \tag{6.45}$$

For $k=2$,

$$\rho_2 = \phi_1 \rho_1 + \phi_2 = \frac{\phi_1^2}{1-\phi_2} + \phi_2. \tag{6.46}$$

The remaining ρ_k can be obtained recursively from (6.44). From (6.42), $\gamma_0(1 - \phi_1 \rho_1 - \phi_2 \rho_2) = \sigma_a^2$, and substituting for ρ_1 and ρ_2 gives us

$$\sigma^2 = \frac{1-\phi_2}{(1+\phi_2)} \frac{\sigma_a^2}{(1-\phi_2)^2 - \phi_1^2}. \tag{6.47}$$

Properties of the AR(2) model are discussed further by Box and Jenkins [1976] and Abraham and Ledolter [1983: Chapter 5]. They note that (6.41) takes the form

$$(I - \phi_1 B - \phi_2 B^2)\varepsilon_i = a_i$$

or

$$\varepsilon_i = (I - \phi_1 B - \phi_2 B^2)^{-1} a_i, \tag{6.48}$$

where B is the usual backward-shift operator. The expression (6.48) takes the general form [c.f. (6.7)]

$$\begin{aligned}\varepsilon_i &= \psi(B)a_i \\ &= \sum_{r=0}^{\infty} \psi_r a_{i-r},\end{aligned}$$

which is stationary provided $\sum_r |\psi_r| < \infty$. Let G_1^{-1} and G_2^{-1} be the roots of the characteristic equation

$$1 - \phi_1 B - \phi_2 B^2 = (1 - G_1 B)(1 - G_2 B) = 0.$$

Then the condition for stationarity is that both G_1^{-1} and G_2^{-1} lie outside the unit circle in the complex plane (i.e. $|G_i| < 1$, $i = 1, 2$) or equivalently,

$$\phi_1 + \phi_2 < 1, \qquad \phi_2 - \phi_1 < 1, \quad \text{and} \quad -1 < \phi_2 < 1. \tag{6.49}$$

If the roots are distinct, then

$$\rho_k = A_1 G_1^k + A_2 G_2^k, \qquad k = 0, 1, 2, \ldots, \tag{6.50}$$

where A_1 and A_2 are determined by $\rho_0 = 1$ and $\rho_1 = \rho_{-1}$. If the roots are real, then the autocorrelation function ρ_k corresponds to a (possibly oscillating) mixture of damped exponentials (Fig. 6.3*a* and *c*). If they are complex, the autocorrelation function is a damped sine wave (Fig. 6.3*b* and *d*). When the roots are equal,

$$\rho_k = (A_1 + A_2 k) G^k.$$

Using the same method as in Section 6.2.7, we can obtain the joint probability density function of $\boldsymbol{\varepsilon}$ under normality assumptions by considering the set of (independent) transformed variables $(\varepsilon_1, \varepsilon_2 - \rho_1 \varepsilon_1, a_3, \ldots, a_n)$ where a_i is related to ε_i via (6.41). The density is

$$p(\boldsymbol{\varepsilon}) = (2\pi)^{-n/2} \sigma^{-2} (1 - \rho_1^2)^{-1/2} (\sigma_a^2)^{-(n-2)/2} \exp\left\{ -\frac{1}{2\sigma^2} S(\boldsymbol{\varepsilon}: \phi_1, \phi_2) \right\}, \tag{6.51}$$

where

$$S(\boldsymbol{\varepsilon}; \phi_1, \phi_2) = \varepsilon_1^2 + \frac{(\varepsilon_2 - \rho_1 \varepsilon_1)^2}{1 - \rho_1^2} + \frac{\sigma^2}{\sigma_a^2} \sum_{i=3}^{n} (\varepsilon_i - \phi_1 \varepsilon_{i-1} - \phi_2 \varepsilon_{i-2})^2,$$

and ρ_1 and σ^2/σ_a^2 are given by (6.45) and (6.47). The likelihood function for $(\boldsymbol{\theta}, \phi_1, \phi_2, \sigma_a^2)$ is obtained by replacing ε_i by $y_i - f(\mathbf{x}_i; \boldsymbol{\theta})$. We postpone further discussion of the above model until the following section, which deals with general autoregressive error processes.

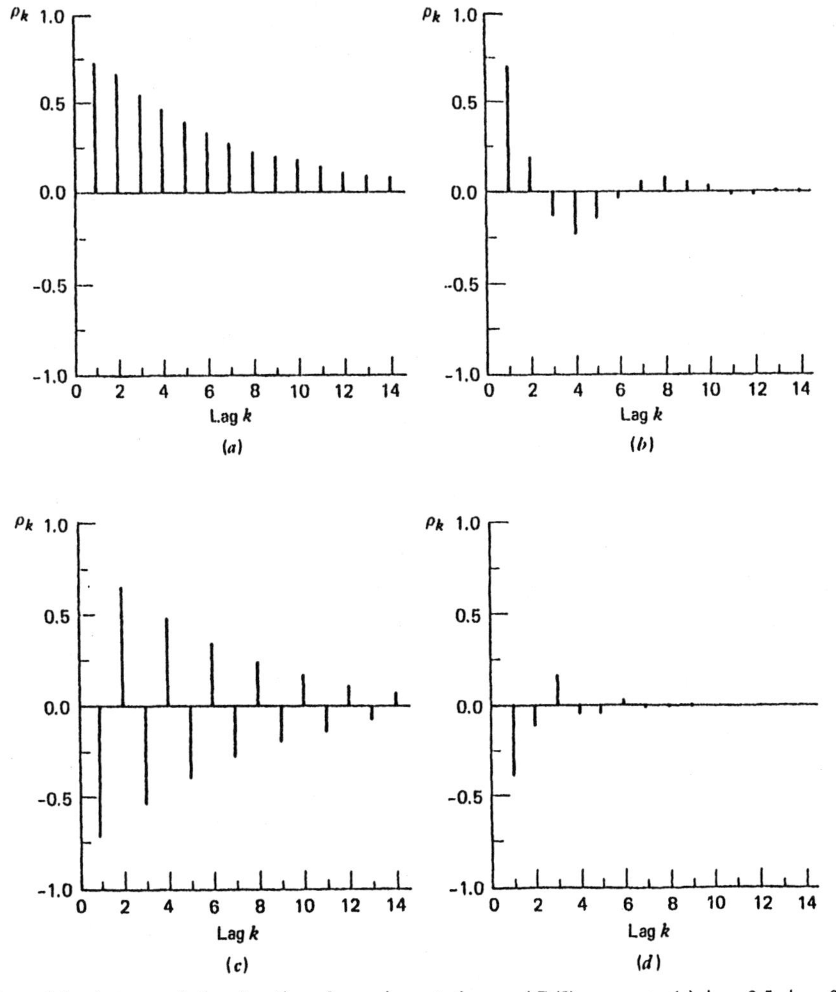

Figure 6.3 Autocorrelation functions for various stationary AR(2) processes: (*a*) $\phi_1 = 0.5$, $\phi_2 = 0.3$; (*b*) $\phi = 1.0$, $\phi_2 = -0.5$; (*c*) $\phi_1 = -0.5$, $\phi_2 = 0.3$; (*d*) $\phi_1 = -0.5$, $\phi_2 = -0.3$. From Abraham and Ledolter [1983] with permission from John Wiley and Sons.

6.4 AR(q_1) ERRORS

6.4.1 Introduction

An autoregressive process of order q_1 is of the form

$$\varepsilon_i = \phi_1 \varepsilon_{i-1} + \phi_2 \varepsilon_{i-2} + \cdots + \phi_{q_1} \varepsilon_{i-q_1} + a_i, \tag{6.52}$$

where the a_i are uncorrelated with mean 0 and variance σ_a^2. Multiplying (6.52) by

ε_{i-k} for $k = 0, 1, \ldots$ and taking expectations, we obtain, as in Section 6.3 for $q_1 = 2$,

$$\sigma^2 = \frac{\sigma_a^2}{1 - \phi_1\rho_1 - \cdots - \phi_{q_1}\rho_{q_1}}, \tag{6.53}$$

$$\gamma_k = \phi_1\gamma_{k-1} + \phi_2\gamma_{k-2} + \cdots + \phi_{q_1}\gamma_{k-q_1} \qquad (k > 0), \tag{6.54}$$

$$\rho_k = \phi_1\rho_{k-1} + \phi_2\rho_{k-2} + \cdots + \phi_{q_1}\rho_{k-q_1} \qquad (k > 0). \tag{6.55}$$

Noting that $\rho_{-j} = \rho_j$, the first k of the equations (6.55) are called the *Yule–Walker equations*. If the roots of the characteristic equation

$$1 - \phi_1 B - \phi_2 B^2 - \cdots - \phi_{q_1} B^{q_1} = 0$$

are distinct, it can be shown (Box and Jenkins [1976: Chapter 3]) that

$$\rho_k = A_1 G_1^k + A_2 G_2^k + \cdots + A_{q_1} G_{q_1}^k,$$

where $|G_i| < 1$, $i = 1, 2, \ldots, q_1$, for stationarity. This implies that the autocorrelation function consists of mixtures of damped exponentials and sine waves.

Since with all autoregressive processes ρ_k dies out as $k \to \infty$, it is often difficult to distinguish between models of different orders. This is to be expected, since it can be hard to distinguish between one or two exponential decays and the sum of several [see Fig. 3.17 for models (3.26a) and (3.26b)]. We would expect this to be an advantage in nonlinear model fitting when the AR(q_1) model is introduced merely to produce a reasonable approximation to the correlation structure of the ε_i. This existence of low-order models with very similar autocorrelation structures to higher-order models means that low-order AR models for the ε_i's should often be adequate in practice. We now discuss maximum-likelihood and two-stage estimation procedures.

6.4.2 Preliminary Results

As in Section 6.2, it is more convenient, for theoretical purposes, to work with the parameter $\sigma_a^2 = \text{var}[a_i]$ than $\sigma^2 = \text{var}[\varepsilon_i]$. We shall use the following notation. For $1 \leq m \leq n$ let

$$\boldsymbol{\varepsilon}_m = (\varepsilon_1, \varepsilon_2, \ldots, \varepsilon_m)'$$

and

$$\mathbf{M}_{m(q_1)} = \left(\frac{1}{\sigma_a^2}\mathscr{D}[\boldsymbol{\varepsilon}_m]\right)^{-1}. \tag{6.56}$$

Since $\mathscr{D}[\boldsymbol{\varepsilon}] = \sigma_a^2 \mathbf{M}_{n(q_1)}^{-1}$, where $\boldsymbol{\varepsilon} = \boldsymbol{\varepsilon}_n$, we shall be particularly interested in $\mathbf{M}_{n(q_1)}$ and $\mathbf{M}_{q_1(q_1)}$: expressions for these are derived below using the method of Box and Jenkins [1976: Appendix A7.5].

Under normality, the a_i are i.i.d. $N(0, \sigma_a^2)$ for $i = 0, \pm 1, \ldots$ and

$$\mathbf{a}_{q_1+} = (a_{q_1+1}, a_{q_1+2}, \ldots, a_n)' \sim N_{n-q_1}(\mathbf{0}, \sigma_a^2 \mathbf{I}_{n-q_1}).$$

Since

$$a_i = \varepsilon_i - \phi_1 \varepsilon_{i-1} - \cdots - \phi_{q_1} \varepsilon_{i-q_1} \qquad (i = q_1 + 1, \ldots, n),$$

we have

$$\mathbf{a}_{q_1+} = \mathbf{K}\boldsymbol{\varepsilon}_{q_1+} + \mathbf{k},$$

where $\boldsymbol{\varepsilon}_{q_1+} = (\varepsilon_{q_1+1}, \varepsilon_{q_1+2}, \ldots, \varepsilon_n)'$, $\mathbf{K}$ is a lower triangular matrix with unit diagonal elements, and $\mathbf{k}$ is a function of $\boldsymbol{\varepsilon}_{q_1} = (\varepsilon_1, \varepsilon_2, \ldots, \varepsilon_{q_1})'$. As $|\mathbf{K}| = 1$, the transformation has a unit Jacobian. The joint distribution of $\boldsymbol{\varepsilon}_{q_1+}$, conditional on $\boldsymbol{\varepsilon}_{q_1}$, is therefore

$$p(\boldsymbol{\varepsilon}_{q_1+} | \boldsymbol{\varepsilon}_{q_1}; \boldsymbol{\phi}, \sigma_a^2) = (2\pi\sigma_a^2)^{-(n-q_1)/2} \exp\left\{ -\frac{1}{2\sigma_a^2} S_2(\boldsymbol{\phi}) \right\}^2, \tag{6.57}$$

where

$$S_2(\boldsymbol{\phi}) = \sum_{i=q_1+1}^{n} (\varepsilon_i - \phi_1 \varepsilon_{i-1} - \cdots - \phi_{q_1} \varepsilon_{i-q_1})^2. \tag{6.58}$$

However, $\boldsymbol{\varepsilon}_{q_1} \sim N_{q_1}(\mathbf{0}, \sigma_a^2 \mathbf{M}_{q_1(q_1)}^{-1})$ with density

$$p(\boldsymbol{\varepsilon}_{q_1}) = (2\pi\sigma_a^2)^{-q_1/2} |\mathbf{M}_{q_1(q_1)}|^{1/2} \exp\left\{ -\frac{1}{2\sigma_a^2} \boldsymbol{\varepsilon}_{q_1}' \mathbf{M}_{q_1(q_1)} \boldsymbol{\varepsilon}_{q_1} \right\}. \tag{6.59}$$

Therefore multiplying (6.57) and (6.59) together gives us

$$p(\boldsymbol{\varepsilon}; \boldsymbol{\phi}, \sigma_a^2) = (2\pi\sigma_a^2)^{-n/2} |\mathbf{M}_{q_1(q_1)}|^{1/2} \exp\left\{ -\frac{1}{2\sigma_a^2} S_1(\boldsymbol{\phi}) \right\}, \tag{6.60}$$

where

$$S_1(\boldsymbol{\phi}) = \boldsymbol{\varepsilon}_{q_1}' \mathbf{M}_{q_1(q_1)} \boldsymbol{\varepsilon}_{q_1} + S_2(\boldsymbol{\phi}). \tag{6.61}$$

However, $\boldsymbol{\varepsilon} \sim N_n(\mathbf{0}, \sigma_a^2 \mathbf{M}_{n(q_1)}^{-1})$, so that from (6.60)

$$|\mathbf{M}_{n(q_1)}| = |\mathbf{M}_{q_1(q_1)}|$$

and

$$\boldsymbol{\varepsilon}' \mathbf{M}_{n(q_1)} \boldsymbol{\varepsilon} = \boldsymbol{\varepsilon}_{q_1}' \mathbf{M}_{q_1(q_1)} \boldsymbol{\varepsilon}_{q_1} + S_2(\boldsymbol{\phi}). \tag{6.62}$$

In particular, taking $n = q_1 + 1$, we obtain

$$\boldsymbol{\varepsilon}_{q_1+1}' \mathbf{M}_{q_1+1(q_1)} \boldsymbol{\varepsilon}_{q_1+1} = \boldsymbol{\varepsilon}_{q_1}' \mathbf{M}_{q_1(q_1)} \boldsymbol{\varepsilon}_{q_1} + (\varepsilon_{q_1+1} - \phi_1 \varepsilon_{q_1} - \cdots - \phi_{q_1} \varepsilon_1)^2 \tag{6.63}$$

for all $\boldsymbol{\varepsilon}_{q_1+1}$. Since for symmetric matrices $\mathbf{A}$ and $\mathbf{B}$ the result $\mathbf{x}'\mathbf{A}\mathbf{x} = \mathbf{x}'\mathbf{B}\mathbf{x}$ for all $\mathbf{x}$ implies $\mathbf{A} = \mathbf{B}$, we have

$$\mathbf{M}_{q_1+1(q_1)} = \left[\begin{array}{c|c} \mathbf{M}_{q_1(q_1)} & \begin{matrix} 0 \\ 0 \\ \vdots \\ 0 \end{matrix} \\ \hline 0 \cdots 0 & 0 \end{array}\right] + \left[\begin{array}{cccc|c} \phi_{q_1}^2 & \phi_{q_1}\phi_{q_1-1} & \cdots & \phi_{q_1}\phi_1 & -\phi_{q_1} \\ \phi_{q_1-1}\phi_{q_1} & \phi_{q_1-1}^2 & \cdots & \phi_{q_1-1}\phi_1 & -\phi_{q_1-1} \\ \vdots & \vdots & & \vdots & \vdots \\ \phi_1\phi_{q_1} & \phi_1\phi_{q_1-1} & \cdots & \phi_1^2 & -\phi_1 \\ \hline -\phi_{q_1} & -\phi_{q_1-1} & \cdots & -\phi_1 & 1 \end{array}\right]. \tag{6.64}$$

Now for a stationary time series, $\boldsymbol{\Sigma}_m = \mathscr{D}[\boldsymbol{\varepsilon}_m]$ not only is symmetric about the main diagonal but also has transverse symmetry, i.e. symmetry about the opposite diagonal. Matrices which are symmetric about both diagonals are called *doubly symmetric*. This occurs in $\boldsymbol{\Sigma}_m$, as $\operatorname{cov}[\varepsilon_i, \varepsilon_j]$ depends only on $|i-j|$. It can be shown that matrix inversion preserves double symmetry, so that from (6.56) $\mathbf{M}_{m(q_1)}$ is doubly symmetric for each m. In particular $\mathbf{M}_{q_1+1(q_1)}$ and $\mathbf{M}_{q_1(q_1)}$ are both doubly symmetric. For given q_1, these two facts together with Equation (6.64) are sufficient to obtain the elements of $\mathbf{M}_{q_1(q_1)}$ explicitly as follows.

If $\mathbf{M}_{q_1(q_1)} = [(m_{rs})]$, then transverse symmetry gives $m_{rs} = m_{q_1+1-s,q_1+1-r}$. For convenience we define

$$\phi_0 = -1. \tag{6.65}$$

Then applying the transverse symmetry of $\mathbf{M}_{q_1+1(q_1)}$ to the first row of the matrix gives us

$$m_{1s} + \phi_{q_1}\phi_{q_1+1-s} = -\phi_{s-1} \qquad (s = 1, 2, \ldots, q_1). \tag{6.66}$$

We can then obtain each succeeding row as

$$\begin{aligned} m_{rs} + \phi_{q_1+1-r}\phi_{q_1+1-s} &= m_{q_1+2-s,q_1+2-r} + \phi_{r-1}\phi_{s-1} \\ &= m_{r-1,s-1} + \phi_{r-1}\phi_{s-1} \qquad (r, s = 2, 3, \ldots, q_1) \end{aligned} \tag{6.67}$$

from the transverse symmetry of $\mathbf{M}_{q_1(q_1)}$. For computational purposes we can now apply Equation (6.66) for $s = 1, 2, \ldots, q_1$ and Equation (6.67) for $r = 2, 3, \ldots, [(q_1+1)/2]$ and $s = r, r+1, \ldots, q_1+1-r$, where $[a]$ is the integer part of a. The rest of the matrix can be filled in using double symmetry. For theoretical, rather than computational use, we now obtain explicit expressions for the elements of $\mathbf{M}_{q_1(q_1)}$. It is natural to apply (6.66) and (6.67) sequentially, working down the diagonal and minor diagonals, from which we obtain

$$m_{rr} = \sum_{j=0}^{r-1} \phi_j^2 - \sum_{j=q_1+1-r}^{q_1} \phi_j^2 \qquad (r = 1, 2, \ldots, q_1) \tag{6.68}$$

and

$$m_{r,r+k} = \sum_{j=0}^{r-1} \phi_j \phi_{j+k} - \sum_{j=q_1+1-r-k}^{q_1-k} \phi_j \phi_{j+k},$$
$$(r = 1,2,\ldots,q_1 - 1, \quad k = 1,2,\ldots,q_1 - r). \quad (6.69)$$

We now turn our attention to $\mathbf{M}_{n(q_1)}$. Since $\boldsymbol{\varepsilon} \sim N_n(\mathbf{0}, \sigma_a^2 \mathbf{M}_{n(q_1)}^{-1})$, we have, from (6.62),

$$[\mathbf{M}_{n(q_1)}]_{rs} = -\sigma_a^2 \frac{\partial^2 \log p(\boldsymbol{\varepsilon})}{\partial \varepsilon_r \, \partial \varepsilon_s}$$
$$= \frac{1}{2} \frac{\partial^2}{\partial \varepsilon_r \, \partial \varepsilon_s} \left(\sum_{a=1}^{q_1} \sum_{b=1}^{q_1} m_{ab} \varepsilon_a \varepsilon_b + \sum_{c=q_1+1}^{n} (\varepsilon_c - \phi_1 \varepsilon_{c-1} - \cdots - \phi_{q_1} \varepsilon_{c-q_1})^2 \right).$$

From this it can be established that (Verbyla [1985]; there are several misprints in this paper)

$$[\mathbf{M}_{n(q_1)}]_{rs} = \begin{cases} 0, & |s-r| > q_1, \\ 1 + \sum_{j=1}^{u_r} \phi_j^2, & r = s, \\ -\phi_{s-r} + \sum_{j=1}^{u_{rs}} \phi_j \phi_{j+s-r}, & r < s, \quad |s-r| \le q_1. \end{cases}$$

where the upper summation limits are

$$u_r = \begin{cases} r-1, & 1 \le r \le q_1, \\ q_1, & q_1 < r \le n - q_1, \\ n-r, & n - q_1 < r \le n, \end{cases}$$

and

$$u_{rs} = \begin{cases} r-1, & 1 \le r \le q_1, \\ q_1 - (s-r), & q_1 < r \le n - q_1, \\ n-s, & n - q_1 < r \le n. \end{cases}$$

We can now write $\mathbf{M}_{n(q_1)}$ compactly as

$$\mathbf{M}_{n(q_1)} = \mathbf{I}_n + \sum_{j=1}^{q_1} \phi_j^2 \mathbf{E}_j - \sum_{j=1}^{q_1} \phi_j \mathbf{F}_j + \sum_{j=1}^{q_1-1} \sum_{i=1}^{q_1-j} \phi_i \phi_{i+j} \mathbf{G}_{i,i+j}, \quad (6.70)$$

where $\mathbf{E}_j$ differs from the identity matrix $\mathbf{I}_n$ only in that the first and last j ones are set to zero, $\mathbf{F}_j$ is zero except for ones on the upper and lower jth minor diagonals, and $\mathbf{G}_{i,i+j}$ is the same as $\mathbf{F}_j$ except that the first and last i ones on both minor diagonals are set to zero.

In order to consider various methods of estimation, we now wish to express (6.62) as a polynomial in the ϕ_j. Let $\mathbf{D}$ be the $(q_1+1)\times(q_1+1)$ matrix with elements

$$d_{ij} = \varepsilon_i\varepsilon_j + \varepsilon_{i+1}\varepsilon_{j+1} + \cdots + \varepsilon_{n+1-j}\varepsilon_{n+1-i}, \tag{6.71}$$

a sum of $n-(i-1)-(j-1)$ terms. From (6.70) it can be shown that

$$\begin{aligned} S_1(\boldsymbol{\phi}) &= \boldsymbol{\varepsilon}'\mathbf{M}_{n(q_1)}\boldsymbol{\varepsilon} \\ &= d_{11} - 2\boldsymbol{\phi}'\mathbf{b} + \boldsymbol{\phi}'\mathbf{C}\boldsymbol{\phi}, \end{aligned} \tag{6.72}$$

where $\mathbf{b}$ is a $q_1 \times 1$ vector with elements $b_i = d_{1,i+1}$ and $\mathbf{C}$ is a $q_1 \times q_1$ matrix with elements $c_{ij} = d_{i+1,j+1}$, and $\boldsymbol{\phi} = (\phi_1, \phi_2, \ldots, \phi_{q_1})'$. This result is stated without proof by Box and Jenkins [1976: 276].

For known $\boldsymbol{\phi}$, the generalized least-squares criterion is to choose $\boldsymbol{\theta}$ to minimize $\{\mathbf{y}-\mathbf{f}(\boldsymbol{\theta})\}'(\mathscr{D}[\boldsymbol{\varepsilon}])^{-1}\{\mathbf{y}-\mathbf{f}(\boldsymbol{\theta})\}$, or equivalently to minimize

$$S(\boldsymbol{\theta},\boldsymbol{\phi}) = \{\mathbf{y}-\mathbf{f}(\boldsymbol{\theta})\}'\mathbf{M}_{n(q_1)}\{\mathbf{y}-\mathbf{f}(\boldsymbol{\theta})\} \tag{6.73}$$

with respect to $\boldsymbol{\theta}$. If we substitute $\varepsilon_i = y_i - f(\mathbf{x}_i;\boldsymbol{\theta})$ into the definition of (6.71), and hence into $\mathbf{b}$ and $\mathbf{C}$, the preceding theory gives an explicit expression for $S(\boldsymbol{\theta},\boldsymbol{\phi})$, namely

$$\begin{aligned} S(\boldsymbol{\theta},\boldsymbol{\phi}) &= \boldsymbol{\varepsilon}'_{q_1}\mathbf{M}_{q_1(q_1)}\boldsymbol{\varepsilon}_{q_1} + \sum_{i=q_1+1}^{n}(\varepsilon_i - \phi_1\varepsilon_{i-1} - \cdots - \phi_{q_1}\varepsilon_{i-q_1})^2 \\ &= d_{11} - 2\boldsymbol{\phi}'\mathbf{b} + \boldsymbol{\phi}'\mathbf{C}\boldsymbol{\phi}. \end{aligned} \tag{6.74}$$

The computation of generalized least-squares estimates under autocorrelated error processes is discussed in Section 6.4.6.

6.4.3 Maximum-Likelihood Estimation and Approximations

The joint density function of $\mathbf{y}$, and hence the likelihood function, is obtained from (6.60) by substituting $\varepsilon_i - y_i - f(\mathbf{x}_i;\boldsymbol{\theta})$ to form $S(\boldsymbol{\theta},\boldsymbol{\phi})$ in place of $S_1(\boldsymbol{\phi})$. Thus the log-likelihood is

$$L(\boldsymbol{\theta},\boldsymbol{\phi},\sigma_a) = \text{const} - \frac{n}{2}\log\sigma_a^2 + \tfrac{1}{2}\log|\mathbf{M}_{q_1(q_1)}| - \frac{1}{2\sigma_a^2}S(\boldsymbol{\theta},\boldsymbol{\phi}). \tag{6.75}$$

Maximization of L and the finding of asymptotic variance estimates require us to be able to obtain first and second partial derivatives of this function. This is straightforward except for the term in $|\mathbf{M}_{q_1(q_1)}|$, which is a function of $\boldsymbol{\phi}$ alone. Unfortunately, as q_1 increases the derivative of this determinant quickly becomes very complicated, and it is desirable to use a simple approximation to maximum likelihood. Box and Jenkins [1976: Appendix A7.5] describe three possibilities.

a Ignore the Determinant

If we ignore the determinant we have the approximation

$$L(\boldsymbol{\theta}, \boldsymbol{\phi}, \sigma_a^2) \sim \text{const} - \frac{n}{2}\log \sigma_a^2 - \frac{1}{2\sigma_a^2} S(\boldsymbol{\theta}, \boldsymbol{\phi}), \tag{6.76}$$

which can be justified heuristically. As $n \to \infty$ we would expect both terms in (6.76) to dominate $|\log \mathbf{M}_{q_1(q_1)}|$, since it does not increase with n. Given an estimate $\hat{\boldsymbol{\theta}}$, the (approximate) maximum-likelihood estimator of $\boldsymbol{\phi}$ from (6.76) satisfies $\partial S(\hat{\boldsymbol{\theta}}, \boldsymbol{\phi})/\partial \boldsymbol{\phi} = \mathbf{0}$. From (6.74), with $\boldsymbol{\theta}$ replaced by $\hat{\boldsymbol{\theta}}$ in $\mathbf{b}$ and $\mathbf{C}$, we have (Appendix A10.5)

$$\frac{\partial S(\hat{\boldsymbol{\theta}}, \boldsymbol{\phi})}{\partial \boldsymbol{\phi}} = -2\hat{\mathbf{b}} + 2\hat{\mathbf{C}}\boldsymbol{\phi} \tag{6.77}$$

so that

$$\hat{\boldsymbol{\phi}} = \hat{\mathbf{C}}^{-1}\hat{\mathbf{b}}. \tag{6.78}$$

The estimation of $\boldsymbol{\theta}$ is discussed later.

b Approximate the Derivative of the Determinant

Using the actual log-likelihood (6.75), and (6.77), we have [c.f. (6.71)]

$$\frac{\partial L}{\partial \phi_j} = m_j + \frac{1}{\sigma_a^2}(d_{1,j+1} - \phi_1 d_{2,j+1} - \cdots - \phi_{q_1} d_{q_1+1,j+1}), \tag{6.79}$$

where $m_j = \frac{1}{2}\partial \log|\mathbf{M}_{q_1(q_1)}|/\partial \phi_j$, $j = 1, 2, \ldots, q_1$. Each of the $n-(i-1)-(j-1)$ terms in d_{ij} has the same expectation, namely $\text{cov}[\varepsilon_i, \varepsilon_j] = \gamma_{|i-j|}$, since $\{\varepsilon_i\}$ forms a stationary series. Using this fact and $E[\partial L/\partial \phi_j] = 0$ [c.f. (2.53)], we obtain

$$m_j\sigma_a^2 + (n-j)\gamma_j - (n-j-1)\phi_1\gamma_{j-1} - \cdots - (n-j-q_1)\phi_{q_1}\gamma_{j-q_1} = 0. \tag{6.80}$$

If we multiply (6.54) by n, set $k = j$, and subtract (6.80), we obtain

$$m_j\sigma_a^2 = j\gamma_j - (j+1)\phi_1\gamma_{j-1} - \cdots - (j+q_1)\phi_{q_1}\gamma_{j-q_1}. \tag{6.81}$$

Given a $\hat{\boldsymbol{\theta}}$, we can follow Box and Jenkins [1976: 278] and use [c.f. (6.71)] $\hat{d}_{i+1,j+1}/(n-i-j)$ as a "sample" estimate of $\gamma_{|i-j|}$ and substitute into (6.81). This estimate contains fewer terms than the usual estimate $\hat{\gamma}_k = \sum_{i=1}^{n-k}\hat{\varepsilon}_i\hat{\varepsilon}_{i+k}/n$, but its use results in a simple formula. Upon substitution we obtain

$$j\frac{\hat{d}_{1,j+1}}{n-j} - (j+1)\phi_1\frac{\hat{d}_{2,j+1}}{n-j-1} - \cdots - (j+q_1)\phi_{q_1}\frac{\hat{d}_{q_1+1,j+1}}{n-j-q_1}$$

as an estimate of $\sigma_a^2 m_j$. By substituting this into (6.79) and estimating each $d_{r,j+1}$ we obtain

$$\frac{\partial L}{\partial \phi_j} \approx \frac{n}{\sigma_a^2}\left(\frac{\hat{d}_{1,j+1}}{n-j} - \phi_1 \frac{\hat{d}_{2,j+1}}{n-j-1} - \cdots - \phi_{q_1} \frac{\hat{d}_{q_1+1,j+1}}{n-j-q_1}\right).$$

With this approximation, $\partial L/\partial \boldsymbol{\phi} = \mathbf{0}$ leads to $\mathbf{C}^* \boldsymbol{\phi} = \mathbf{b}^*$, where, noting $d_{ij} = d_{ji}$, we have

$$d_{ij}^* = \frac{n\hat{d}_{ij}}{n-(i-1)-(j-1)}, \quad c_{ij}^* = d_{i+1,j+1}^*, \quad \text{and} \quad b_i^* = d_{1,i+1}^*.$$

We thus obtain

$$\hat{\boldsymbol{\phi}} = \mathbf{C}^{*-1}\mathbf{b}^*, \tag{6.82}$$

a slight modification of the previous estimate (6.78).

The third method for obtaining $\hat{\boldsymbol{\phi}}$ is described in Section 6.4.4.

c Asymptotic Variances

Before discussing methods of approximating $\hat{\boldsymbol{\theta}}$, the maximum-likelihood estimator of $\boldsymbol{\theta}$, we shall first consider its asymptotic dispersion matrix. From (6.75), the second derivatives of the log-likelihood $\partial^2 L/\partial \boldsymbol{\theta}\, \partial \boldsymbol{\phi}'$ and $\partial^2 L/\partial \boldsymbol{\theta}\, \partial \sigma_a^2$ both consist of linear combinations of terms of the form

$$\frac{\partial \varepsilon_i \varepsilon_j}{\partial \boldsymbol{\theta}} = -\varepsilon_i \frac{\partial f(\mathbf{x}_j; \boldsymbol{\theta})}{\partial \boldsymbol{\theta}} - \varepsilon_j \frac{\partial f(\mathbf{x}_i; \boldsymbol{\theta})}{\partial \boldsymbol{\theta}},$$

which has expectation zero. Thus both $\mathscr{E}[\partial^2 L/\partial \boldsymbol{\theta}\, \partial \boldsymbol{\phi}'] = \mathbf{0}$ and $\mathscr{E}[\partial^2 L/\partial \boldsymbol{\theta}\, \partial \sigma_a^2] = \mathbf{0}$. Assuming the applicability of the usual asymptotic maximum-likelihood theory, the asymptotic variance–covariance (dispersion) matrix for all the estimates is given by the inverse of the expected information matrix. As this matrix consists of two diagonal blocks, $\hat{\boldsymbol{\theta}}$ is asymptotically independent of the maximum-likelihood estimates of $\boldsymbol{\phi}$ and σ_a^2, and has asymptotic dispersion matrix $\{-\mathscr{E}[\partial^2 L/\partial \boldsymbol{\theta}\, \partial \boldsymbol{\theta}']\}^{-1}$. Now from (6.75) and (6.73)

$$-\frac{\partial^2 L}{\partial \boldsymbol{\theta}\, \partial \boldsymbol{\theta}'} = \frac{1}{2\sigma_a^2} \frac{\partial^2 S(\boldsymbol{\theta}, \boldsymbol{\phi})}{\partial \boldsymbol{\theta}\, \partial \boldsymbol{\theta}'}$$

$$= \frac{1}{2\sigma_a^2} \frac{\partial^2 \boldsymbol{\varepsilon}' \mathbf{M}_{n(q_1)} \boldsymbol{\varepsilon}}{\partial \boldsymbol{\theta}\, \partial \boldsymbol{\theta}'},$$

so that

$$-\frac{\partial^2 L}{\partial \theta_r\, \partial \theta_s} = \frac{1}{2\sigma_a^2} \frac{\partial^2}{\partial \theta_r\, \partial \theta_s} \sum_i \sum_j a_{ij} \varepsilon_i \varepsilon_j$$

$$= \frac{1}{2\sigma_a^2}\left\{\sum_i \sum_j a_{ij}\left(\frac{\partial f_i}{\partial \theta_r}\frac{\partial f_j}{\partial \theta_s} + \frac{\partial f_i}{\partial \theta_s}\frac{\partial f_j}{\partial \theta_r}\right) - \sum_i \sum_j a_{ij}\left(\varepsilon_i \frac{\partial^2 f_j}{\partial \theta_r\, \partial \theta_s} + \varepsilon_j \frac{\partial^2 f_i}{\partial \theta_r\, \partial \theta_s}\right)\right\},$$

where $f_i = f(\mathbf{x}_i; \boldsymbol{\theta})$ and $\mathbf{M}_{n(q_1)} = [(a_{ij})]$. The first term is constant, and the second has expectation zero. Therefore, since $\mathbf{M}_{n(q_1)}$ is symmetric,

$$\mathscr{E}\left[-\frac{1}{n}\frac{\partial^2 L}{\partial\boldsymbol{\theta}\,\partial\boldsymbol{\theta}'}\right] = \frac{1}{n\sigma_a^2}\mathbf{F}_{\bullet}'\mathbf{M}_{n(q_1)}\mathbf{F}_{\bullet} \tag{6.83}$$

$$= \mathbf{O}(1),$$

where $\mathbf{F}_{\bullet} = [(\partial f_i/\partial\theta_j)]$. We thus find that the asymptotic dispersion matrix of $\sqrt{n}\hat{\boldsymbol{\theta}}$ is the same as that of the generalized least squares estimate, namely the inverse of (6.83). Hence

$$\mathscr{D}[\sqrt{n}\hat{\boldsymbol{\theta}}] \approx \sigma_a^2(n^{-1}\mathbf{F}_{\bullet}'\mathbf{M}_{n(q_1)}\mathbf{F}_{\bullet})^{-1}. \tag{6.84}$$

The factor n^{-1} in (6.83) is introduced here for later comparisons. This equivalence to generalized least squares suggests the following algorithm for approximate maximum-likelihood estimation.

We begin with the ordinary least-squares estimator $\boldsymbol{\theta}^{(1)} = \hat{\boldsymbol{\theta}}_{\text{OLS}}$ and then iterate to convergence the following two steps:

1. From $\boldsymbol{\theta}^{(t)}$ calculate

$$\boldsymbol{\phi}^{(t)} = \begin{cases} \hat{\mathbf{C}}^{-1}\hat{\mathbf{b}}, & \text{method (a)}, \\ \mathbf{C}^{*-1}\mathbf{b}^{*}, & \text{method (b)}. \end{cases}$$

Use $\boldsymbol{\phi}^{(t)}$ in $\mathbf{M}_{n(q_1)}$ to calculate $\mathbf{M}_{n(q_1)}^{(t)}$ using the method following (6.67).

2. Obtain $\boldsymbol{\theta}^{(t+1)}$ by minimizing

$$S(\boldsymbol{\theta}, \boldsymbol{\phi}^{(t)}) = \{\mathbf{y} - \mathbf{f}(\boldsymbol{\theta})\}'\mathbf{M}_{n(q_1)}^{(t)}\{\mathbf{y} - \mathbf{f}(\boldsymbol{\theta})\}.$$

Using the converged estimates, we then compute $\hat{\sigma}_a^2 = S(\hat{\boldsymbol{\theta}}, \hat{\boldsymbol{\phi}})/n$ and estimate $\mathscr{D}[\hat{\boldsymbol{\theta}}]$ by $\hat{\sigma}_a^2(\hat{\mathbf{F}}_{\bullet}'\hat{\mathbf{M}}_{n(q_1)}\hat{\mathbf{F}}_{\bullet})^{-1}$. The above method is simply an iterated two-stage method, and we discuss computational considerations in Section 6.4.6.

Glasbey [1980] used both methods of approximating the maximum likelihood estimates using real data. He found method (a) unsatisfactory in moderate to small samples, as it sometimes resulted in an estimated AR process which was not stationary. He did not complete step 2 above, which itself involves a sequence of iterations, but simply updated $\hat{\boldsymbol{\phi}}$ and $\hat{\boldsymbol{\sigma}}_a$ after just the first step in minimizing $S(\boldsymbol{\theta}, \boldsymbol{\phi}^{(t)})$. This method converged satisfactorily with his data.

We shall now continue the above analysis to obtain asymptotic variance estimates for $\hat{\boldsymbol{\phi}}$ and $\hat{\sigma}_a^2$. Following experience with linear models, much larger samples are likely to be needed for these estimates to be adequate than are needed for $\hat{\boldsymbol{\theta}}$.

We have from (6.79)

$$-\frac{\partial^2 L}{\partial \sigma_a^2 \partial \phi_j} = \frac{1}{\sigma_a^4}(d_{1,j+1} - \phi_1 d_{2,j+1} - \cdots - \phi_{q_1} d_{q_1+1,j+1}),$$

and using the same arguments that led to (6.81), we obtain

$$E\left[-\frac{1}{n}\frac{\partial^2 L}{\partial \sigma_a^2 \partial \phi_j}\right] = \frac{1}{n\sigma_a^4}[j\gamma_j - (j+1)\phi_1\gamma_{j-1} - \cdots - (j+q_1)\phi_{q_1}\gamma_{j-q_1}]$$

$$= O(n^{-1}). \tag{6.85}$$

This is negligible compared with (6.83) for large n, so that $\sqrt{n}\hat{\boldsymbol{\phi}}$ and $\sqrt{n}\hat{\sigma}_a^2$ are asymptotically independent. Furthermore, from (6.79),

$$\frac{\partial^2 L}{\partial \phi_i \partial \phi_j} = \frac{\partial m_j}{\partial \phi_i} - \frac{1}{\sigma_a^2} d_{i+1,j+1},$$

so that

$$\begin{aligned} E\left[-\frac{1}{n}\frac{\partial^2 L}{\partial \phi_i \partial \phi_j}\right] &= -\frac{1}{n}\frac{\partial m_j}{\partial \phi_i} + \frac{1}{\sigma_a^2}\frac{n-i-j}{n}\gamma_{|i-j|} \\ &\approx \frac{1}{\sigma_a^2}\gamma_{|i-j|} \\ &= \frac{1}{\sigma_a^2}\operatorname{cov}[\varepsilon_i, \varepsilon_j] \\ &= \frac{1}{\sigma_a^2}(\mathcal{D}[\boldsymbol{\varepsilon}_{q_1}])_{ij}. \end{aligned}$$

Hence, from (6.56)

$$\begin{aligned} \mathcal{D}[\sqrt{n}\hat{\boldsymbol{\phi}}] &\approx \left(\mathcal{E}\left[-\frac{1}{n}\frac{\partial^2 L}{\partial \boldsymbol{\phi}\, \partial \boldsymbol{\phi}'}\right]\right)^{-1} \\ &\approx \mathbf{M}_{q_1(q_1)}. \end{aligned} \tag{6.86}$$

Finally, $E[-\partial^2 L/\partial(\sigma_a^2)^2] = n/(2\sigma_a^4)$, so that

$$\operatorname{var}[\sqrt{n}\hat{\sigma}_a^2] \approx 2\sigma_a^4. \tag{6.87}$$

Example 6.2 Glasbey [1980] fitted the exponential-plus-linear trend model

$$y_i = \theta_1 e^{-x_i/\theta_2} + \theta_3 + \theta_4 x_i + \varepsilon_i \tag{6.88}$$

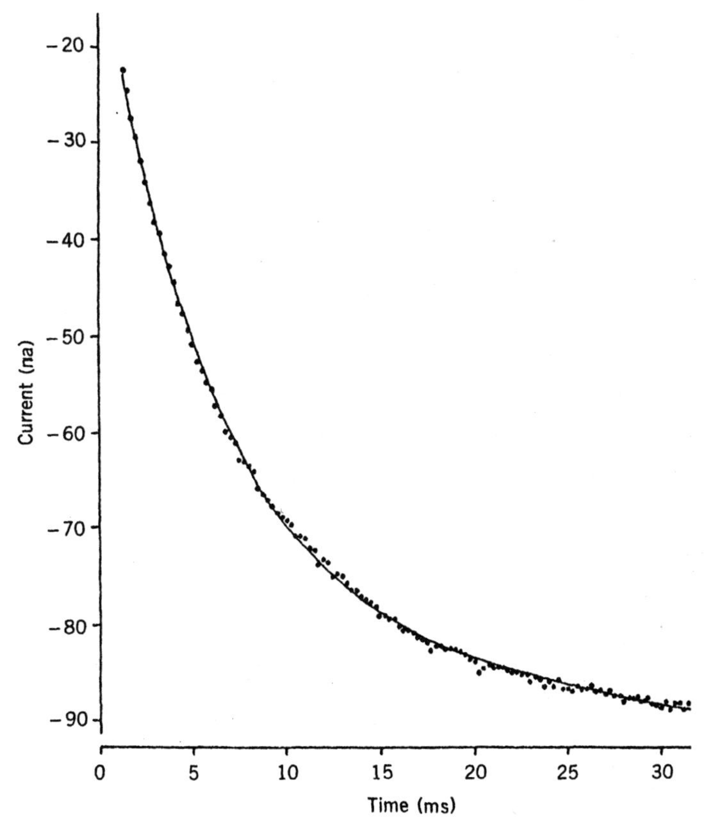

Figure 6.4 Exponential plus linear trend model (6.88) fitted by ordinary least squares to observed data. Reproduced from C. A. Glasbey, "Nonlinear Regression with Autoregressive Time Series Errors". Biometrics 36:135–140, 1980. With permission from The Biometric Society.

to the data depicted in Fig. 6.4. Here x_i is time and the responses y_i are measurements taken every 0.25 milliseconds on the relaxation in electrical current (following a voltage jump) flowing through the end plate membrane of a muscle fibre. The fitted curve is obtained from the ordinary least-squares fit. The model clearly fits well apart from the tendency of runs of points to lie above or below the fitted line, this tendency indicating positive autocorrelation in the residuals. Parameter estimates and estimated standard errors from OLS are given in Table 6.1. Glasbey decided to use an AR(3) error structure using model selection methods described later in Section 6.7.1 based on sample partial autocorrelations of the least-squares residuals. Parameter estimates for an AR(3) fit and their standard errors are presented in Table 6.2. The changes in the parameter estimates $\hat{\boldsymbol{\theta}}$ are fairly small, but the increases in the standard errors of these estimates are quite substantial, with a doubling for $\hat{\theta}_4$. ■

The argument used under method (a) [c.f. (6.76)] can be extended: omitting a

Table 6.1 Ordinary Least-Squares Estimates (± Standard Errors)[a]

$\hat{\theta}_1$(nA):	72.2 ± 0.4
$\hat{\theta}_2$(ms):	6.15 ± 0.09
$\hat{\theta}_3$(nA):	-81.3 ± 0.6
$\hat{\theta}_4$(nA/ms):	-0.25 ± 0.02
$\hat{\sigma}^2$	0.161 ± 0.020

[a] Reproduced from C. A. Glasbey, "Nonlinear Regression with Autoregressive Time Series Errors". Biometrics 36: 135–140, 1980. With permission from The Biometric Society.

finite number of terms from $S(\boldsymbol{\theta}, \boldsymbol{\phi})$ will have no effect asymptotically. Thus we can use

$$-\tfrac{1}{2}(n-q_1)\log\sigma_a^2 - \frac{1}{2\sigma_a^2} S_2(\boldsymbol{\theta}, \boldsymbol{\phi}), \tag{6.89}$$

where

$$S_2(\boldsymbol{\theta}, \boldsymbol{\phi}) = \sum_{i=q_1+1}^{n} (\varepsilon_i - \phi_1\varepsilon_{i-1} - \cdots - \phi_{q_1}\varepsilon_{i-q_1})^2,$$

as an approximation to the log-likelihood. From (6.57), however, (6.89) is also the log-likelihood for $(\boldsymbol{\theta}, \boldsymbol{\phi}, \sigma_a^2)$ based upon the conditional distribution of $\mathbf{y}$ given $y_1, y_2, \ldots, y_{q_1}$. This approach results in conditional least-squares estimation, thus generalizing Section 6.2.5. However, because of the inadequacies of conditional least squares in the AR(1) case we will not explore this method further.

A more accurate approximation to maximum likelihood than any considered above would be to use analytic derivatives of L with respect to all terms except $|\mathbf{M}_{q_1(q_1)}|$, whose derivatives could be approximated by finite differences. This should be feasible if q_1 is not too large.

Table 6.2 Parameter Estimates (± Standard Errors) for $q_1 = 3$[a]

$\hat{\theta}_1$(nA):	72.6 ± 1.0
$\hat{\theta}_2$(ms):	6.08 ± 0.18
$\hat{\theta}_3$(nA):	-81.2 ± 1.2
$\hat{\theta}_4$(nA/ms):	-0.25 ± 0.04
$\hat{\sigma}^2$:	0.117 ± 0.015
$\hat{\phi}_1$:	0.27 ± 0.09
$\hat{\phi}_2$:	0.23 ± 0.09
$\hat{\phi}_3$:	0.18 ± 0.09

[a] Reproduced from C. A. Glasbey, "Nonlinear Regression with Autoregressive Time Series Errors". Biometrics 36: 135–140, 1980. With permission from The Biometric Society.

6.4.4 Two-Stage Estimation

The two-stage method described after (6.84) is the same in essence as that given in Section 6.2.3, and is recalled below.

1. (i) Obtain $\hat{\boldsymbol{\theta}}_{OLS}$.
   (ii) Compute an estimate $\hat{\boldsymbol{\phi}}$ from the OLS residuals.
   (iii) Use $\hat{\boldsymbol{\phi}}$ and the method following (6.67) to obtain an estimate $\hat{\mathbf{M}}_{n(q_1)}$ of $\mathbf{M}_{n(q_1)}$. (Recall that $\mathscr{D}[\boldsymbol{\varepsilon}] = \sigma_a^2 \mathbf{M}_{n(q_1)}^{-1}$.)
2. Choose $\hat{\boldsymbol{\theta}}_{TS}$ to minimize $\{\mathbf{y} - \mathbf{f}(\boldsymbol{\theta})\}' \hat{\mathbf{M}}_{n(q_1)} \{\mathbf{y} - \mathbf{f}(\boldsymbol{\theta})\}$.

Either of methods (a) and (b) of the previous section can be used in 1 (ii). Method (a) generalizes the estimator (6.26) proposed by Park and Mitchell [1980]. The method described below, which was introduced by Gallant and Goebel [1975, 1976], generalizes the use of $\hat{\phi}_1 = \hat{\rho}_1$, the usual lag-1 first-order sample autocorrelation (6.27).

Recalling that $\gamma_k = \gamma_{-k}$, where γ_k is the lag-k autocovariance (6.4), the (Yule–Walker) equations (6.54) for $k = 1, 2, \ldots, q_1$ can be written in matrix form as

$$\boldsymbol{\gamma}_{q_1} = \boldsymbol{\Gamma}_{(q_1)} \boldsymbol{\phi}, \tag{6.90}$$

where $\boldsymbol{\gamma}_{q_1} = (\gamma_1, \gamma_2, \ldots, \gamma_{q_1})'$ and

$$\boldsymbol{\Gamma}_{(q_1)} = \mathscr{D}[\boldsymbol{\varepsilon}_{q_1}] = \begin{bmatrix} \gamma_0 & \gamma_1 & \cdots & \gamma_{q_1-1} \\ \gamma_1 & \gamma_0 & \cdots & \gamma_{q_1-2} \\ \vdots & \vdots & & \vdots \\ \gamma_{q_1-1} & \gamma_{q_1-2} & \cdots & \gamma_0 \end{bmatrix}.$$

Gallant and Goebel [1976] suggested estimating γ_k by the sample autocovariance

$$\tilde{\gamma}_k = n^{-1} \sum_{i=1}^{n-k} \hat{\varepsilon}_i \hat{\varepsilon}_{i+k} \qquad (k = 0, 1, \ldots). \tag{6.91}$$

They substitute these estimates into $\boldsymbol{\Gamma}_{(q_1)}$ and $\boldsymbol{\gamma}_{q_1}$ to give

$$\tilde{\boldsymbol{\phi}} = \tilde{\boldsymbol{\Gamma}}_{(q_1)}^{-1} \tilde{\boldsymbol{\gamma}}_{q_1}, \tag{6.92}$$

and corresponding estimators $\tilde{\mathbf{M}}_{n(q_1)}$ and

$$\tilde{\sigma}_a^2 = \frac{\{\mathbf{y} - \mathbf{f}(\hat{\boldsymbol{\theta}}_{TS})\}' \tilde{\mathbf{M}}_{n(q_1)} \{\mathbf{y} - \mathbf{f}(\hat{\boldsymbol{\theta}}_{TS})\}}{n - p}.$$

In a technical report (Gallant and Goebel [1975]) they prove that $\sqrt{n}(\hat{\boldsymbol{\theta}}_{TS} - \boldsymbol{\theta}^*)$

is asymptotically normally distributed with mean zero and variance–covariance matrix consistently estimated by

$$\tilde{\sigma}_a^2(n^{-1}\tilde{\mathbf{F}}_{\bullet}'\tilde{\mathbf{M}}_{n(q_1)}\tilde{\mathbf{F}}_{\bullet})^{-1} = n\tilde{\sigma}_a^2\tilde{\mathbf{K}}, \tag{6.93}$$

say, under appropriate regularity conditions. The proof only requires conditions on the moments, and not normality of the errors. It involves showing that, with an AR(q_1) error structure, the asymptotic distribution of the generalized least-squares estimator of $\boldsymbol{\theta}$ is the same irrespective of whether $\boldsymbol{\phi}$ is known or suitably estimated. Their estimator is therefore asymptotically fully efficient for normally distributed errors.

Gallant and Goebel [1976] also use a three-stage estimator (which they call two-stage), which is simply one more iteration of the process. The residuals from the two-stage fit are used to produce an updated estimate of $\boldsymbol{\phi}$ and hence of $\mathbf{M}_{n(q_1)}$. Then their estimator $\hat{\boldsymbol{\theta}}_{3S}$ is obtained by generalized least squares as before. Iterating the process one or more times does not change the asymptotic behavior of the estimate of $\boldsymbol{\theta}$, but intuitively we might expect it to be more efficient in small samples. We recall that the approximate maximum likelihood methods discussed in the previous section were iterated two-stage methods.

Example 6.3 Gallant and Goebel [1976] carried out simulations using the model

$$y_i = \theta_1 e^{\theta_2 x_i} + \varepsilon_i \tag{6.94}$$

with 60 data points and 2000 simulations. They used normally distributed errors with the a_i i.i.d. $N(0, 0.25)$ and the following four correlation structures:

(i) i.i.d. errors: $\varepsilon_i = a_i$ $(\mathrm{var}[\varepsilon_i] = 0.25,\ \rho_1 = 0)$;

(ii) MA(4): $\varepsilon_i = 1.5a_i + 1.0a_{i-1} + 0.85a_{i-2} + 0.33a_{i-3} + 0.5a_{i-4}$ $(\mathrm{var}[\varepsilon_i] = 1.08,\ \rho_1 = 0.645)$, MA(4) is defined in Section 6.5;

(iii) AR(1): $\varepsilon_i = 0.735\varepsilon_{i-1} + a_i$ $(\mathrm{var}[\varepsilon_i] = 0.54,\ \rho_1 = 0.735)$;

(iv) AR(2): $\varepsilon_i = 1.048\varepsilon_{i-1} - 0.1287\varepsilon_{i-2} + a_i$ $(\mathrm{var}[\varepsilon_i] = 1.84,\ \rho_1 = 0.93)$.

For each error structure $\boldsymbol{\theta}$ was estimated using ordinary least squares, and two-stage and three-stage estimators appropriate for AR(2) errors. In terms of mean squared error, the AR(2) estimators were only slightly less efficient than ordinary least squares with i.i.d. errors (96%), but showed substantial gains for any of the other error structures, ranging from about 140% with MA(4) errors to 500% or 600% when the errors were really AR(2). The three-stage estimator was more efficient than the two-stage estimator for AR(2) errors (125% for both θ_1 and θ_2) and essentially of equivalent efficiency for the other error structures.

The simulation also studied the behavior of the "t-statistics"

$$\tilde{t}_r = \frac{\tilde{\theta}_r - \theta_r}{(\tilde{\sigma}_a^2 \tilde{k}_{rr})^{1/2}}, \tag{6.95}$$

where $\tilde{k}_{rr}$ is the rth diagonal element of $\tilde{\mathbf{K}}$ [c.f. (6.93)]. For asymptotic inference, for example setting confidence intervals for θ_r, we would be relying on $\tilde{t}_r$ having approximately a t_{n-p}-distribution. With AR(2) errors, an analysis based on OLS led to 90% confidence intervals which were one-third the width that they should have been, and significance tests nominally at the 5% level had a true size of about 50%. The results in other situations were less extreme. However, in all cases the use of OLS when autocorrelated errors were present led to confidence intervals for the θ_r which were too narrow, and to tests whose nominal size was considerably smaller than their true size. Under an i.i.d. error structure, the empirical distribution of the $\tilde{t}$'s from OLS conformed very well to the t_{58}-distribution, as expected, while the t_{58}-approximation was reasonable for both the two-stage and three-stage estimators. With the other error structures the asymptotic approximations were not as good. This is predictable to some extent, because the variance of the ε_i's is much smaller for the i.i.d. simulation and we would expect this to have an effect similar to a larger sample size.

Two features are particularly notable. Firstly, the distributions of the $\tilde{t}_r$'s for the two- and three-stage estimators under the MA(4) error structure were surprisingly close to t_{58}. We recall that these estimators are based on the assumption of an AR(2) error structure. Secondly, in general, the $\tilde{t}_r$'s had distributions closer to t_{58} for the three-stage estimator than they do for the two-stage estimator. ■

6.4.5 Choosing a Method

Although the asymptotic theory has only been worked out in detail by Gallant and Goebel [1975, 1976] for their own procedure, it is clear that all the methods discussed in Sections 6.4.3 and 6.4.4 are asymptotically equivalent. This can be seen heuristically by noting that for each of the methods, the estimating equations for $\boldsymbol{\phi}$ are asymptotically equivalent, so that the estimates of $\boldsymbol{\phi}$ are also asymptotically equivalent. For example, using method (a) of Section 6.4.3 we have $\hat{\boldsymbol{\phi}} = \hat{\mathbf{C}}^{-1}\hat{\mathbf{b}}$, whereas Gallant and Goebel use $\tilde{\boldsymbol{\phi}} = \tilde{\boldsymbol{\Gamma}}_{(q_1)}^{-1}\tilde{\boldsymbol{\gamma}}_{q_1}$. For $i<j$, as $c_{ij} = d_{i+1,j+1}$, we have from (6.71)

$$\hat{c}_{ij} = \hat{\varepsilon}_{i+1}\hat{\varepsilon}_{j+1} + \hat{\varepsilon}_{i+2}\hat{\varepsilon}_{j+2} + \cdots + \hat{\varepsilon}_{n-j}\hat{\varepsilon}_{n-i}$$

and

$$[n\tilde{\boldsymbol{\Gamma}}_{(q_1)}]_{ij} = \sum_{r=1}^{n-j+i} \hat{\varepsilon}_r\hat{\varepsilon}_{r+j-i}. \tag{6.96}$$

Now $\hat{c}_{ij}$ is obtained from (6.96) by dropping $i < q_1$ terms off both ends of (6.96). The expressions are therefore asymptotically equivalent, since q_1 is fixed as $n \to \infty$. The elements of $\hat{\mathbf{b}}$ and $n\tilde{\boldsymbol{\gamma}}_{q_1}$ bear a similar relationship.

Unfortunately there appear to have been no empirical comparisons of the methods for finite samples. Experience from the AR(1) case together with Gallant and Goebel's simulation would tend to favor an iterative method. It also seems that rather large samples will be required for asymptotic theory to provide a good approximation to the distribution of $\hat{\boldsymbol{\theta}}$, even for models that are linear. Estimators using the method of Section 6.4.3a or Equation (6.92) from Gallant and Goebel to obtain an estimate of $\boldsymbol{\phi}$ may lead to estimated error processes that are not stationary. The good performance of the estimators based upon AR(2) errors, when the real error process was MA(4), indicates that the estimators may have desirable robustness properties. This question needs further investigation.

6.4.6 Computational Considerations

We now describe a computational method of Gallant and Goebel [1976] which can be adapted to any of the other methods. It makes use of an ordinary nonlinear least-squares program and requires minimal additional programming, provided a routine is available for computing a Cholesky decomposition (Appendix A8.2).

We recall from (6.62)

$$\boldsymbol{\varepsilon}'\mathbf{M}_{n(q_1)}\boldsymbol{\varepsilon} = \boldsymbol{\varepsilon}'_{q_1}\mathbf{M}_{q_1(q_1)}\boldsymbol{\varepsilon}_{q_1} + \sum_{r=q_1+1}^{n} (\varepsilon_r - \phi_1\varepsilon_{r-1} - \cdots - \phi_{q_1}\varepsilon_{r-q_1})^2. \qquad (6.97)$$

Let $\mathbf{L}_{(q_1)}$ be the lower triangular Cholesky factor of $\mathbf{M}_{q_1(q_1)}$ with positive diagonal elements (c.f. Appendix A8.2), so that $\mathbf{M}_{q_1(q_1)} = \mathbf{L}'_{(q_1)}\mathbf{L}_{(q_1)}$. Adapting the method of Gallant and Goebel [1976], define the $n \times n$ lower triangular matrix

$$\mathbf{L} = \left[\begin{array}{cccccc|cccc} \multicolumn{6}{c|}{\overbrace{\qquad\qquad \mathbf{L}_{(q_1)} \qquad\qquad}^{q_1}} & \multicolumn{4}{c}{\overbrace{\qquad \mathbf{0} \qquad}^{n-q_1}} \\ \hline -\phi_{q_1} & -\phi_{q_1-1} & \cdots & \cdots & -\phi_1 & 1 & & & & \\ & -\phi_{q_1} & -\phi_{q_1-1} & \cdots & \cdots & -\phi_1 & 1 & & & \\ & & & \vdots & & & \vdots & \vdots & & \\ & & & -\phi_{q_1} & -\phi_{q_1-1} & \cdots & \cdots & -\phi_1 & 1 & \end{array}\right] \begin{array}{l} \}\, q_1 \\ \\ \left.\begin{array}{l} \\ \\ \\ \end{array}\right\} n-q_1 \end{array} \qquad (6.98)$$

Then the right-hand side of (6.97) is $(\mathbf{L}\boldsymbol{\varepsilon})'\mathbf{L}\boldsymbol{\varepsilon}$, and we have $\boldsymbol{\varepsilon}'\mathbf{M}_{n(q_1)}\boldsymbol{\varepsilon} = \boldsymbol{\varepsilon}'\mathbf{L}'\mathbf{L}\boldsymbol{\varepsilon}$ for all $\boldsymbol{\varepsilon}$. This establishes that $\mathbf{M}_{n(q_1)} = \mathbf{L}'\mathbf{L}$ and $\mathbf{L}$ is the lower triangular Cholesky factor of $\mathbf{M}_{n(q_1)}$. For *fixed* $\boldsymbol{\phi}$, the generalized least-squares criterion is to minimize with respect to $\boldsymbol{\theta}$

$$\begin{aligned} S(\boldsymbol{\theta}, \boldsymbol{\phi}) &= \{\mathbf{y} - \mathbf{f}(\boldsymbol{\theta})\}'\mathbf{M}_{n(q_1)}\{\mathbf{y} - \mathbf{f}(\boldsymbol{\theta})\} \\ &= \{\mathbf{z} - \mathbf{g}(\boldsymbol{\theta})\}'\{\mathbf{z} - \mathbf{g}(\boldsymbol{\theta})\}, \end{aligned} \qquad (6.99)$$

where $\mathbf{z} = \mathbf{L}\mathbf{y}$ and $\mathbf{g}(\boldsymbol{\theta}) = \mathbf{L}\mathbf{f}(\boldsymbol{\theta})$. The latter expression can be minimized using an ordinary least-squares program. The derivative matrix $\mathbf{G}_{.}(\boldsymbol{\theta}) = \partial\mathbf{g}(\boldsymbol{\theta})/\partial\boldsymbol{\theta}'$ required by such a program is readily obtained as $\mathbf{L}\mathbf{F}_{.}(\boldsymbol{\theta})$. The program also gives the asymptotic estimated variance–covariance matrix of $\hat{\boldsymbol{\theta}}$ as

$$\hat{\sigma}_a^2(\hat{\mathbf{G}}_{\cdot}'\hat{\mathbf{G}}_{\cdot})^{-1} = \hat{\sigma}_a^2(\hat{\mathbf{F}}_{\cdot}'\mathbf{L}'\mathbf{L}\hat{\mathbf{F}}_{\cdot})^{-1} = \hat{\sigma}_a^2(\hat{\mathbf{F}}_{\cdot}'\mathbf{M}_{n(q_1)}\hat{\mathbf{F}}_{\cdot})^{-1},$$

as required. The matrix multiplication by $\mathbf{L}$ does not require the storage of $\mathbf{L}$, only the lower triangle of $\mathbf{L}_{(q_1)}$ and the vector $\boldsymbol{\phi}$. To apply this method for generalized least squares within a two-step or iterated method, current estimates of $\mathbf{L}_{(q_1)}$ and $\boldsymbol{\phi}$ are required. Methods (a) and (b) give a current estimate $\hat{\boldsymbol{\phi}}$ using (6.78) and (6.82) respectively, and $\hat{\mathbf{M}}_{q_1(q_1)}$ is computed from (6.66) and (6.67): $\hat{\mathbf{L}}_{(q_1)}$ and $\mathbf{L}$ then follow.

Gallant and Goebel's method begins with the estimates $\tilde{\gamma}_k$ from (6.91) for $k = 0, 1, \ldots, q_1$ which leads to an estimator $\tilde{\boldsymbol{\Gamma}}_{(q_1)}$ of $\mathscr{D}[\boldsymbol{\varepsilon}_{q_1}]$. The estimator $\tilde{\boldsymbol{\phi}}$ is then computed from (6.92), and from (6.53) we have

$$\begin{aligned}\tilde{\sigma}_a^2 &= \tilde{\sigma}^2(1 - \tilde{\phi}_1\tilde{\rho}_1 - \tilde{\phi}_2\tilde{\rho}_2 - \cdots - \tilde{\phi}_{q_1}\tilde{\rho}_{q_1}) \\ &= \tilde{\gamma}_0 - \tilde{\phi}_1\tilde{\gamma}_1 - \tilde{\phi}_2\tilde{\gamma}_2 - \cdots - \tilde{\phi}_{q_1}\tilde{\gamma}_{q_1}.\end{aligned}$$

As $\mathbf{M}_{q_1(q_1)} = (\sigma_a^{-2}\boldsymbol{\Gamma}_{(q_1)})^{-1}$ [from (6.56)], the lower Cholesky factor of $\tilde{\sigma}_a^{-2}\tilde{\boldsymbol{\Gamma}}_{(q_1)}$ is $\tilde{\mathbf{L}}_{(q_1)}^{-1}$.

6.5 MA (q_2) ERRORS

6.5.1 Introduction

Moving average models of order q_2, denoted by MA (q_2), are of the form

$$\varepsilon_i = a_i - \xi_1 a_{i-1} - \xi_2 a_{i-2} - \cdots - \xi_{q_2} a_{i-q_2}, \tag{6.100}$$

where the a_i are i.i.d. with mean 0 and variance σ_a^2. Then

$$\gamma_0 = E[\varepsilon_i^2] = (1 + \xi_1^2 + \xi_2^2 + \cdots + \xi_{q_2}^2)\sigma_a^2, \tag{6.101}$$

$$\gamma_k = E[\varepsilon_i\varepsilon_{i-k}] = (-\xi_k + \xi_1\xi_{k+1} + \xi_2\xi_{k+2} + \cdots + \xi_{q_2-k}\xi_{q_2})\sigma_a^2 \qquad (k = 1, 2, \ldots, q_2), \tag{6.102}$$

and

$$\gamma_k = 0 \qquad (k > q_2).$$

Thus there are no autocorrelations further apart than the order of the process. Provided the roots of the characteristic equation

$$1 - \xi_1 B - \xi_2 B^2 - \cdots - \xi_{q_2} B^{q_2} = 0$$

lie outside the unit circle, an MA (q_2) process can be represented as an infinite AR process, $\varepsilon_i = \sum_r \phi_r \varepsilon_{i-r} + a_i$[c.f. (6.52)].

6.5.2 Two-Stage Estimation

Two-stage estimation with MA(q_2) errors is the same in principle as with an AR error structure. The time-series parameters $\boldsymbol{\xi}$ and σ_a^2 are estimated from the OLS residuals, and this gives an estimate of $\boldsymbol{\Sigma} = \mathscr{D}[\boldsymbol{\varepsilon}]$ which can be used for estimating $\boldsymbol{\theta}$ by generalized least squares. Unfortunately, estimates of $\boldsymbol{\xi}$ cannot be obtained explicitly. Algorithms for estimating the time-series parameters for general ARMA(q_1, q_2) models, of which MA(q_2) [= ARMA$(0, q_2)$] models are a special case, are referenced in Section 6.6. In addition, specialized methods for MA models are also available, but their use is probably unwarranted unless the two-stage method is to be iterated repeatedly.

Methods for estimating $\boldsymbol{\xi}$ which are exactly maximum-likelihood have been described by Pagan and Nicholls [1976] (in the context of linear models with MA errors), Phadke and Kedem [1978], and Box and Jenkins [1976: Appendix A7.4]. The last two references describe methods for the evaluation of the log-likelihood function $L(\boldsymbol{\xi}, \sigma_a^2)$ for an MA(q_2) time series. For fixed $\boldsymbol{\xi}$ the maximum-likelihood estimate of σ_a^2, $\hat{\sigma}_a^2(\boldsymbol{\xi})$, can be readily obtained analytically and the concentrated log-likelihood function $L[\boldsymbol{\xi}, \hat{\sigma}_a^2(\boldsymbol{\xi})]$ then computed using any of the above methods. The concentrated log-likelihood then has to be maximized using an algorithm which does not use derivatives (see Section 13.5). The method of Pagan and Nicholls [1976] also produces most of the derivatives and is easily extended to nonlinear models. We describe this method in Section 6.5.3. Pesaran [1973] gave an explicit formula for $L[\boldsymbol{\xi}, \hat{\sigma}_a(\boldsymbol{\xi})]$ and its first derivative in the special case of $q_2 = 1$. Approximate maximum-likelihood methods are described by Box and Jenkins [1976: Chapter 7]. Godolphin and cowriters used alternative approximations which result in a very simple iterative scheme. A computer program implementing the method is given by Angell and Godolphin [1978] (see also Godolphin and de Gooijer [1982]).

Since $\boldsymbol{\Sigma} = \mathscr{D}[\boldsymbol{\varepsilon}]$ is a band matrix, the lower Cholesky factor $\mathbf{W}$ of its estimate is also a lower triangular matrix with the same bandwidth (i.e. q_2 nonzero subdiagonals). This results in savings in both the computation and the storage of $\mathbf{W}$. Using the methodology of Section 2.1.4, we can perform generalized least-squares estimation using an OLS algorithm to minimize $\|\mathbf{z} - \mathbf{g}(\boldsymbol{\theta})\|^2$, where $\mathbf{z} = \mathbf{W}^{-1}\mathbf{y}$ and $\mathbf{g}(\boldsymbol{\theta}) = \mathbf{W}^{-1}\mathbf{f}(\boldsymbol{\theta})$.

6.5.3 Maximum-Likelihood Estimation

As in the autoregressive case, we find that iterated two-stage estimation again gives maximum-likelihood or approximate maximum-likelihood estimates under normality assumptions. However, the following method due to Pagan and Nicholls [1976] allows for the calculation or the approximation of the first and second derivatives of the likelihood function. Our model is

$$y_i = f(\mathbf{x}_i; \boldsymbol{\theta}) + \varepsilon_i \qquad (i = 1, 2, \ldots, n),$$

where ε_i is given by (6.100). Defining $\boldsymbol{\varepsilon} = (\varepsilon_1, \varepsilon_2, \ldots, \varepsilon_n)'$, $\mathbf{a} = (a_1, a_2, \ldots, a_n)'$, and

$\mathbf{a}^* = (a_{-q_2+1},\ldots,a_{-1}, a_0)'$, we can express (6.100) in the form

$$\boldsymbol{\varepsilon} = \mathbf{M}\mathbf{a} + \mathbf{N}\mathbf{a}^*.$$

Here $\mathbf{M}$ is an $n \times n$ lower diagonal band matrix with unit main diagonal elements and q_2 nonzero subdiagonals, with the jth subdiagonal away from the main diagonal having all its elements $-\xi_j$. The matrix $\mathbf{N}$ is zero apart from the first q_2 rows, with row i of the form $(0,\ldots,0,\ -\xi_{q_2},\ldots,\ -\xi_i)$. Then, since $\mathbf{N}'\mathbf{M} = \mathbf{0}$, we have

$$\begin{aligned}\boldsymbol{\Sigma} &= \mathscr{D}[\boldsymbol{\varepsilon}]\\ &= \sigma_a^2(\mathbf{M}'\mathbf{M} + \mathbf{N}'\mathbf{N})\\ &= \sigma_a^2\mathbf{P},\end{aligned}$$

say. Under normality, the log-likelihood function is

$$L(\boldsymbol{\theta}, \boldsymbol{\xi}, \sigma_a^2) = -\frac{n}{2}\log(2\pi\sigma_a^2) - \tfrac{1}{2}\log|\mathbf{P}| - \frac{1}{2\sigma_a^2}\boldsymbol{\varepsilon}'\mathbf{P}^{-1}\boldsymbol{\varepsilon}, \qquad (6.103)$$

where $\boldsymbol{\varepsilon} = \mathbf{y} - \mathbf{f}(\boldsymbol{\theta})$. Pagan and Nicholls [1976] showed that the maximization of (6.103) is equivalent to the maximization of

$$\hat{L}(\boldsymbol{\theta}, \boldsymbol{\xi}, \sigma_a^2, \mathbf{a}^*) = -\frac{n}{2}\log(2\pi\sigma_a^2) - \tfrac{1}{2}\log|P| - \frac{1}{2\sigma_a^2}(\mathbf{a}'\mathbf{a} + \mathbf{a}^{*\prime}\mathbf{a}^*). \qquad (6.104)$$

The proof involves showing that when $\mathbf{a}^*$ is concentrated out of $\hat{L}$, the resulting function is L of (6.103). Thus from Section 2.2.3 we can get the correct maximum-likelihood estimates and elements of the inverse matrix of second derivatives by maximizing $\hat{L}$. The reader is referred to the original paper for details.

6.6 ARMA(q_1, q_2) ERRORS

6.6.1 Introduction

Combining the autoregressive and the moving-average models leads us to the autoregressive moving-average model of orders q_1 and q_2, denoted by ARMA(q_1, q_2), of the form

$$\varepsilon_i - \sum_{r=1}^{q_1} \phi_r \varepsilon_{i-r} = a_i - \sum_{s=1}^{q_2} \xi_s a_{i-s}. \qquad (6.105)$$

Data which require many parameters using a pure AR or MA model can often be modeled with only a few parameters using an ARMA model. To derive two-stage methods for such error processes it is again desirable to be able to form $\mathscr{D}[\boldsymbol{\varepsilon}]$ from the model parameters. Let

$$\gamma_{\varepsilon a}(k) = E[\varepsilon_{i-k} a_i]. \tag{6.106}$$

Then $\gamma_{\varepsilon a}(k) = 0$ for $k > 0$, as ε_{i-k} depends only on the previous and current a's. From (6.105),

$$\begin{aligned}\gamma_k &= E[\varepsilon_i \varepsilon_{i-k}] \\ &= \sum_{r=1}^{q_1} \phi_r \gamma_{k-r} + \gamma_{\varepsilon a}(k) - \sum_{s=1}^{q_2} \xi_s \gamma_{\varepsilon a}(k-s)\end{aligned} \tag{6.107}$$

and

$$\gamma_k = \sum_{r=1}^{q_1} \phi_r \gamma_{k-r} \qquad (k \geq q_2 + 1). \tag{6.108}$$

This is of the same form as for an AR(q_1) process. It is not surprising, therefore, that for ARMA(q_1, q_2) processes the pattern of the autocorrelation (or autocovariance) function is the same as for an AR(q_1) process except for the first $q_2 - q_1 + 1$ values, which do not obey the pattern if $q_2 \geq q_1$ (see Box and Jenkins [1976: Chapter 3]). Once we have obtained $\gamma_0, \gamma_1, \ldots, \gamma_Q$, where $Q = \max(q_1, q_2)$, all the other autocovariances can be obtained recursively using (6.108). To find these initial autocovariances we require $\gamma_{\varepsilon a}(0)$, $\gamma_{\varepsilon a}(-1), \ldots, \gamma_{\varepsilon a}(-Q)$, which are found recursively by multiplying (6.105) by a_i, $a_{i-1}, \ldots$ in turn and taking expectations. Thus

$$\begin{aligned}\gamma_{\varepsilon a}(0) &= \sigma_a^2 \\ \gamma_{\varepsilon a}(-1) &= \phi_1 \gamma_{\varepsilon a}(0) - \xi_1 \sigma_a^2 \\ \gamma_{\varepsilon a}(-2) &= \phi_1 \gamma_{\varepsilon a}(-1) + \phi_2 \gamma_{\varepsilon a}(0) - \xi_2 \sigma_a^2 \\ &\ \vdots \\ \gamma_{\varepsilon a}(-k) &= \phi_1 \gamma_{\varepsilon a}(-k+1) + \cdots + \phi_{q_1} \gamma_{\varepsilon a}(-k+q_1) - [\xi_k]\sigma_a^2,\end{aligned} \tag{6.109}$$

where $\gamma_{\varepsilon a}(m) = 0$ for $m > 0$ and

$$[\xi_k] = \begin{cases} -1, & k = 0, \\ \xi_k, & k = 1, 2, \ldots, q_2, \\ 0 & \text{otherwise.} \end{cases} \tag{6.110}$$

If we now write out (6.107) for $k = 0, 1, \ldots, Q$, using the fact that $\gamma_{-k} = \gamma_k$, we obtain

$$\gamma_0 = \phi_1 \gamma_1 + \phi_2 \gamma_2 + \cdots + \phi_{q_1} \gamma_{q_1} + c_0,$$

$$\begin{aligned}
\gamma_1 &= \phi_1\gamma_0 + \phi_2\gamma_1 + \cdots + \phi_{q_1}\gamma_{q_1-1} + c_1, \\
&\vdots \\
\gamma_{q_1} &= \phi_1\gamma_{q_1-1} + \phi_2\gamma_{q_1-2} + \cdots + \phi_{q_1}\gamma_0 + c_{q_1}, \\
&\vdots \\
\gamma_Q &= \phi_1\gamma_{Q-1} + \phi_2\gamma_{Q-2} + \cdots + \phi_{q_1}\gamma_{Q-q_1} + c_Q,
\end{aligned} \tag{6.111}$$

where

$$c_k = \begin{cases} \gamma_{\varepsilon a}(0) - \sum_{s=1}^{q_2} \xi_s \gamma_{\varepsilon a}(-s), & k=0, \\ -\sum_{s=k}^{q_2} \xi_s \gamma_{\varepsilon a}(k-s), & 1 \le k \le q_2, \\ 0, & k > q_2. \end{cases}$$

This gives a set of $Q+1$ linear equations in $Q+1$ unknowns which can be written in the form $\mathbf{B}\boldsymbol{\gamma} = \mathbf{c}$ and hence solved for $\boldsymbol{\gamma} = (\gamma_0, \gamma_1, \ldots, \gamma_Q)'$. A computer algorithm implementing this scheme was given by McCleod [1975] (who used slightly different notation) and is as follows:

1. Set $\phi_0 = \xi_0 = -1$, $d_0 = 1$.
   For $k = 1, 2, \ldots, q_2$ set $d_k = -\xi_k + \sum_{r=1}^{\min(q_1,k)} \phi_r d_{k-r}$.
   [Note: $d_k \sigma_a^2 = \gamma_{\varepsilon a}(-k)$.]
2. For $k = 0, 1, \ldots, q_2$ set $b_k = -\sum_{s=k}^{q_2} \xi_s d_{s-k}$.
   For $k > q_2$ set $b_k = 0$.
   [Note: $b_k \sigma_a^2 = c_k$.]
3. If $q_1 = 0$, $\gamma_k = b_k \sigma_a^2$ for $k = 0, 1, \ldots, q_2$.
   If $q_1 > 0$, solve the linear equations $\mathbf{A}\mathbf{x} = \mathbf{y}$, where

   $$a_{ij} = \begin{cases} [\phi_{i-1}], & j=1, \quad i=1,2,\ldots,Q+1, \\ [\phi_{i-j}] + [\phi_{i+j-2}], & j=2,\ldots,Q+1, \quad i=1,2,\ldots,Q+1, \end{cases}$$

   where

   $$[\phi_k] = \begin{cases} \phi_k, & k=0,1,\ldots,q_1, \\ 0 & \text{otherwise,} \end{cases}$$

   and

   $$y_i = -b_{i-1}\sigma_a^2 \qquad \text{for} \quad i=1,2,\ldots,Q+1.$$

   [Note: $\mathbf{A} = -\mathbf{B}$ above and $\mathbf{y} = -\mathbf{c}$; thus $\mathbf{x} = \boldsymbol{\gamma}$.]
4. Set $\gamma_k = x_{k+1}$ for $k = 0, 1, \ldots, Q$.

6.6.2 Conditional Least-Squares Method

White and Ratti [1977] and White and Brisbin [1980] apply to a nonlinear growth model the methods developed by Pierce [1971a, b] for linear models with an ARMA error structure. We will be discussing Pierce's work in some detail below. But first, for later reference, we rewrite (6.105) as

$$a_i = \sum_{s=1}^{q_2} \xi_s a_{i-s} + \varepsilon_i - \sum_{r=1}^{q_1} \phi_r \varepsilon_{i-r}. \tag{6.112}$$

We now obtain the conditional distribution of $\boldsymbol{\varepsilon} = (\varepsilon_1, \varepsilon_2, \ldots, \varepsilon_n)'$ given $\mathbf{a}^* = (a_{-q_2+1}, \ldots, a_{-1}, a_0)'$ and $\boldsymbol{\varepsilon}^* = (\varepsilon_{-q_1+1}, \ldots, \varepsilon_{-1}, \varepsilon_0)$.

For $i = 1, 2, \ldots, n$, (6.105) can be expressed in the form

$$\mathbf{L}_1 \boldsymbol{\varepsilon} = \mathbf{L}_2 \mathbf{a} + \mathbf{h} \tag{6.113}$$

or

$$\mathbf{a} = \mathbf{L}_2^{-1} \mathbf{L}_1 \boldsymbol{\varepsilon} - \mathbf{L}_2^{-1} \mathbf{h}, \tag{6.114}$$

where $\mathbf{L}_1$ and $\mathbf{L}_2$ are $n \times n$ lower triangular matrices with unit diagonals, and $\mathbf{h}$ has elements which are functions of $\mathbf{a}^*$ and $\boldsymbol{\varepsilon}^*$. Assuming that the a_i are i.i.d. $N(0, \sigma_a^2)$ then, conditional on $\mathbf{a}^*$, the elements of $\mathbf{a} = (a_1, a_2, \ldots, a_n)'$ are still i.i.d. $N(0, \sigma_a^2)$. As the Jacobian of the transformation from $\mathbf{a}$ to $\boldsymbol{\varepsilon}$ defined by (6.114) is $|\mathbf{L}_1|\,|\mathbf{L}_2|^{-1} = 1$, the joint conditional density of $\boldsymbol{\varepsilon}$ can be written as

$$p(\boldsymbol{\varepsilon} | \boldsymbol{\varepsilon}^*, \mathbf{a}^*) = (2\pi\sigma_a^2)^{-n/2} \exp\left\{ -\frac{1}{2\sigma_a^2} \sum_{i=1}^{n} a_i^2 \right\}, \tag{6.115}$$

where $\mathbf{a}$ is given by (6.114). It is convenient to have a notation which distinguishes between the a_i as i.i.d. $N(0, \sigma_a^2)$ random variables and the a_i which have representation (6.114), i.e. are functions of $\boldsymbol{\varepsilon}$ and the parameters $\boldsymbol{\xi}$ and $\boldsymbol{\phi}$. We denote the latter by $\breve{a}_i$ and note that they are i.i.d. $N(0, \sigma_a^2)$ only when $\boldsymbol{\xi}$ and $\boldsymbol{\phi}$ take their true values. Then substituting $\varepsilon_i = y_i - f(\mathbf{x}_i; \boldsymbol{\theta})$ and taking logarithms of (6.115) gives us the conditional log-likelihood function

$$L_c(\boldsymbol{\theta}, \boldsymbol{\phi}, \boldsymbol{\xi}, \sigma_a^2) = \text{const} - \frac{n}{2} \log \sigma_a^2 - \frac{1}{2\sigma_a^2} \sum_{i=1}^{n} \breve{a}_i^2. \tag{6.116}$$

This likelihood is based on the distribution of $y_1, y_2, \ldots, y_n$ conditional upon $a_{-q_2+1}, \ldots, a_{-1}, a_0, y_{-q_1+1}, \ldots, y_{-1}, y_0$ and $\mathbf{x}_{-q_1+1}, \ldots, \mathbf{x}_{-1}, \mathbf{x}_0$. The conditional maximum-likelihood estimators $\hat{\boldsymbol{\theta}}$, $\hat{\boldsymbol{\phi}}$, and $\hat{\boldsymbol{\xi}}$ are conditional least-squares estimators in that they minimize $\sum_i \breve{a}_i^2$. If $\hat{a}_i = \breve{a}_i(\hat{\boldsymbol{\theta}}, \hat{\boldsymbol{\phi}}, \hat{\boldsymbol{\xi}})$, the maximum (conditional) likelihood estimator of σ_a^2 is simply $\sum_i \hat{a}_i^2 / n$.

To make the above theory the basis for a practical method, starting values for the a_{-r}, y_{-r}, and $\mathbf{x}_{-r}$ terms must be specified. Pierce [1971a] suggested that the simplest method is to set $\mathbf{a}^* = \mathbf{0}$ and $\boldsymbol{\varepsilon}^* = \mathbf{0}$, but noted that more sophisticated procedures such as backforecasting should have better small-sample properties. Backforecasting means that starting values can be forecasted using the backward process (see Box and Jenkins [1976: Section 6.4]) applied to the OLS residuals $\hat{\varepsilon}_1, \hat{\varepsilon}_2, \ldots, \hat{\varepsilon}_n$. If, instead, we minimized $\breve{a}_{q_1+1}^2 + \cdots + \breve{a}_n^2$ using starting values for the q_2 a_i-terms up to a_{q_1}, and initial values for the parameters, we would not have to specify starting values for the y_i's and $\mathbf{x}_i$'s, or alternatively the ε_i's. These would be supplied by the first q_1 observations $(\mathbf{x}_i, y_i)$, $i = 1, 2, \ldots, q_1$. This modification also emphasizes that Pierce's approach is a direct extension of the conditional least-squares approaches discussed in Sections 6.4.3 and 6.2.5.

Pierce [1971a] showed that the effect of different starting values is asymptotically negligible. Also, his asymptotic arguments, which he applied to linear models, do not appear to be affected in any important way by moving from linear to nonlinear models. The changes appear to be essentially confined to imposing regularity conditions on the nonlinear model similar to those described in Chapter 12, to achieve the desired convergence properties of sequences. However, we have not worked this through in detail.

We can minimize $\sum_i \breve{a}_i^2$ using a nonlinear least-squares algorithm (see Chapter 14). The derivatives required by such algorithms, namely $\partial \breve{a}_i / \partial \theta_r$, $\partial \breve{a}_i / \partial \phi_r$, and $\partial \breve{a}_i / \partial \xi_r$, can be obtained recursively from the starting values and the formulae derived below. To simplify notation we adopt the backward shift-operator notation so that (6.105) becomes

$$(1 - \phi_1 B - \phi_2 B^2 - \cdots - \phi_{q_1} B^{q_1})\varepsilon_i = (1 - \xi_1 B - \xi_2 B^2 - \cdots - \xi_{q_2} B^{q_2}) a_i,$$

or

$$\phi(B)\varepsilon_i = \xi(B) a_i. \tag{6.117}$$

We also assume invertibility conditions on the ARMA(q_1, q_2) model to allow for manipulations such as

$$\varepsilon_i = \phi^{-1}(B)\xi(B) a_i. \tag{6.118}$$

Here the inverse is obtained by treating the inverse of a polynomial in B in the usual manner, e.g. $(1 - \phi B)^{-1} = 1 + \phi B + \phi^2 B^2 + \cdots$. If we define $b_{ir} = \partial \breve{a}_i / \partial \theta_r$, then, from (6.117) with a_i replaced by $\breve{a}_i$, we have

$$\xi(B) b_{ir} = \phi(B) \frac{\partial \varepsilon_i}{\partial \theta_r}, \tag{6.119}$$

where

$$\frac{\partial \varepsilon_i}{\partial \theta_r} = \begin{cases} 0, & i \leq 0, \\ -\dfrac{\partial f(\mathbf{x}_i; \boldsymbol{\theta})}{\partial \theta_r}, & i > 0. \end{cases}$$

With this definition, $b_{ir}=0$ for $i \le 0$ and the subsequent values can be obtained recursively using (6.119). Differentiating (6.112) with respect to ϕ_r, we obtain

$$\xi(B)\frac{\partial \breve{a}_i}{\partial \phi_r} = -\varepsilon_{i-r}$$

$$= -\phi^{-1}(B)\xi(B)\breve{a}_{i-r} \qquad \text{[by (6.118)]}.$$

Hence

$$\phi(B)\frac{\partial \breve{a}_i}{\partial \phi_r} = -\phi(B)\xi^{-1}(B)\phi^{-1}(B)\xi(B)\breve{a}_{i-r}$$

$$= -\phi(B)\phi^{-1}(B)\xi^{-1}(B)\xi(B)\breve{a}_{i-r}$$

$$= -\breve{a}_{i-r}, \tag{6.120}$$

as the operators commute. This equation allows recursive calculation of $\partial \breve{a}_i/\partial \phi_r$ for $i = 1, 2, \ldots, n$. Finally, differentiating (6.112) with respect to ξ_s gives us

$$\xi(B)\frac{\partial \breve{a}_i}{\partial \xi_s} = \breve{a}_{i-s}. \tag{6.121}$$

As with the b_{ir}, $\partial \breve{a}_i/\partial \phi_r = 0$ and $\partial \breve{a}_i/\partial \xi_s = 0$ for $i \le 0$, so that using Equations (6.120) and (6.121) for $i = 1, 2, \ldots, n$ allows recursive calculation of these derivatives.

Under certain regularity conditions, Pierce [1971a] found that the conditional estimates $\hat{\boldsymbol{\theta}}$, $\hat{\boldsymbol{\phi}}$, $\hat{\boldsymbol{\xi}}$, and $\hat{\sigma}_a^2$ were consistent and asymptotically normally distributed. The normality assumption for the a_i's could also be relaxed. The conditions involved are listed as follows:

(i) The a_i are i.i.d. with mean zero, variance σ_a^2, and finite kurtosis.
(ii) The roots of $\phi(B) = 0$ and $\xi(B) = 0$ all lie outside the unit circle, ensuring stationarity of $\{\varepsilon_i\}$ and invertibility, and they have no common roots.
(iii) $\lim_{n\to\infty} n^{-1}\mathbf{F}_{\cdot}'\mathbf{F}_{\cdot}$ is positive definite, and for all fixed h and k,

$$\lim_{n\to\infty} n^{-1} \sum_{i=1}^{n} \frac{\partial f(\mathbf{x}_{i-h};\boldsymbol{\theta})}{\partial \boldsymbol{\theta}} \frac{\partial f(\mathbf{x}_{i-k};\boldsymbol{\theta})}{\partial \boldsymbol{\theta}'} \text{ exists.}$$

Pierce found the asymptotic dispersion matrix of $\sqrt{n}\hat{\boldsymbol{\theta}}$ to be

$$\mathscr{D}[\sqrt{n}\hat{\boldsymbol{\theta}}] \approx \sigma_a^2 \mathbf{D}^{-1},$$

where the (r,s)th element of $\mathbf{D}$ is $\lim_{n\to\infty} n^{-1}\sum_i b_{ir}b_{is}$. For both linear and nonlinear models

$$\sigma_a^2 \mathbf{D}^{-1} = \left(\frac{1}{n}\mathscr{E}\left[-\frac{\partial^2 L_c}{\partial \boldsymbol{\theta}\,\partial \boldsymbol{\theta}'}\right]\right)^{-1},$$

where L_c is given by (6.116). This is the same asymptotic dispersion matrix that $\sqrt{n}\hat{\boldsymbol{\theta}}$ would have if $\boldsymbol{\phi}$ and $\boldsymbol{\xi}$ were known. It should be noted that in Pierce's notation $\boldsymbol{\theta}$ is $\boldsymbol{\beta}$, $f(\mathbf{x}_i; \boldsymbol{\theta}) = \mathbf{x}_i'\boldsymbol{\theta}$, and $\boldsymbol{\xi}$ is $\boldsymbol{\theta}$.

The asymptotic dispersion matrix of $\sqrt{n}\hat{\boldsymbol{\zeta}} = \sqrt{n}(\hat{\boldsymbol{\phi}}', \hat{\boldsymbol{\xi}}')'$ is of the form

$$\sigma_a^2 \begin{bmatrix} \mathbf{U} & \mathbf{W} \\ \mathbf{W}' & \mathbf{V} \end{bmatrix}^{-1}, \tag{6.122}$$

where $\mathbf{U}$ and $\mathbf{V}$ are the autocovariance matrices of the respective autoregressive processes $\phi(B)u_t = a_t$ and $\xi(B)v_t = a_t$. Also $\mathbf{W} = [(w_{r-s})]$, where $w_k = E[u_i v_{i+k}]$ is the lag-k "cross covariance" between the two processes. The matrix (6.122) is the same asymptotic dispersion matrix that $\sqrt{n}\hat{\boldsymbol{\zeta}}$ would have had if there had been no associated regression model.

Finally, $\sqrt{n}\hat{\sigma}_a^2$ has asymptotic variance $2\sigma_a^4(1 + \frac{1}{2}c_2)$, where c_2 is the (excess in) kurtosis: for normal errors $c_2 = 0$. If the skewness and kurtosis of the distribution of the a_i's are both zero, Pierce [1971a] also showed that the three sets of parameter estimates $\hat{\boldsymbol{\theta}}$, $\hat{\boldsymbol{\zeta}}$, and $\hat{\sigma}_a^2$ are asymptotically independent. Otherwise $\hat{\sigma}_a^2$ and $\hat{\boldsymbol{\theta}}$ are correlated.

Pierce [1971a] carried out simulations with the model

$$y_i = \alpha + \beta x_i + \varepsilon_i \tag{6.123}$$

with $\alpha = -5$, $\beta = 2$, and the x_i i.i.d. $N(15, 9)$. He used three error processes: AR(1) with $\phi = 0.5$, MA(1) with $\xi = 0.5$, and ARMA(1, 1) with $\phi = 0.5$ and $\xi = -0.5$. In each case $\sigma_a^2 = 1$. Sample sizes of 15 (small), 50 (moderate), and 200 (large) were taken, and each experiment was replicated 100 times. For small samples, parameter estimates $\hat{\alpha}$, $\hat{\beta}$, and $\hat{\sigma}_a^2$ showed little bias and the variances were reasonable, giving for the worst case the empirical simulation variance of $\hat{\alpha}$ as 2.6 compared with 1.9 from asymptotic theory. However, there were severe biases in $\hat{\phi}$ and $\hat{\xi}$, and their asymptotic variances were too small, possibly by a factor of 2 to 3. There was strong evidence that $\hat{\sigma}_a^2$ and $\hat{\xi}$ were not independent for error processes involving MA terms. For moderate and large samples there were negligible biases in $\hat{\alpha}$, $\hat{\beta}$, and $\hat{\sigma}_a^2$, and there was excellent agreement between empirical and theoretical variances. The $\hat{\phi}$ and $\hat{\xi}$ estimates still showed some bias for moderate samples, although the variance estimates were good. For large samples with AR errors, $\hat{\phi}$ had variances of 0.0027 (empirical) and 0.0037 (theoretical). Although this was the worst case, the empirical variances of both $\hat{\phi}$ and $\hat{\xi}$ tended to be further away from the theoretical variances for large samples ($n = 200$) than for moderate samples ($n = 50$). No indication of the sampling error of the empirical variances was given. An overall impression remains that large-sample theory begins to work faster for the parameters α, β, and σ_a^2 than it does for the time-series parameters. Pierce noted that increasing the number of time-series parameters or allowing their values to approach the boundary of the stationary region would slow the rate at which large-sample theory gave reasonable approximations to the true distributions of the parameter estimates.

Although the above discussion relates to the straight-line model (6.123), it is clear that similar considerations will apply to simple nonlinear models. Pierce [1972] also extended the work of this section to allow for models with lagged values of the y- and x-variables.

6.6.3 Other Estimation Procedures

Conditional least-squares procedures for ordinary time-series data are known to be unreliable when the process is close to nonstationarity (Box and Jenkins [1976: Chapter 7]). We have already seen situations in which conditional least squares performs badly with an AR(1) error process (Section 6.2.6). In time-series analysis there is now much more emphasis upon exact maximum-likelihood estimation of model parameters, particularly following the algorithm of Ansley [1979]. It would clearly be desirable to generalize to the whole ARMA family of error processes the methods which have proved successful for AR errors, namely two-stage, iterated two-stage, and exact maximum-likelihood estimation. From (6.115) the full log-likelihood function can be written in the form

$$L(\boldsymbol{\theta}, \boldsymbol{\phi}, \boldsymbol{\xi}, \sigma_a^2) = L_c(\boldsymbol{\theta}, \boldsymbol{\phi}, \boldsymbol{\xi}, \sigma_a^2) + \log p(\mathbf{a}^*, \boldsymbol{\varepsilon}^*), \tag{6.124}$$

where $p(\mathbf{a}^*, \boldsymbol{\varepsilon}^*)$ is the joint density function of $\mathbf{a}^*$ and $\boldsymbol{\varepsilon}^*$. From the usual heuristic arguments of Box and Jenkins [1976: Chapter 7] we would expect L_c to dominate (6.124) as $n \to \infty$, as $\log p(\mathbf{a}^*, \boldsymbol{\varepsilon}^*)$ involves only a finite number of terms. Therefore, under suitable conditions we would expect the asymptotic theory of the previous section for conditional estimation to apply to the exact maximum-likelihood estimators as well.

Extension of the two-stage approach is conceptually simple. After fitting the model by (ordinary) least squares, the time-series parameters $\boldsymbol{\phi}$, $\boldsymbol{\xi}$, and σ_a^2 can be estimated from the OLS residuals. Equations (6.109) and (6.111) allow us to calculate a corresponding estimate of $\mathscr{D}[\boldsymbol{\varepsilon}] = \boldsymbol{\Sigma} = \sigma^2\mathbf{V}$, and hence of the correlation matrix $\mathbf{V}$. The second-stage estimator of $\boldsymbol{\theta}$ may then be obtained by generalized least squares. This procedure can then be iterated in the usual way. When iterated to convergence, the method gives maximum-likelihood estimates provided that a maximum-likelihood procedure is used to estimate $\boldsymbol{\phi}$ and $\boldsymbol{\xi}$ from the residuals of the previous fit.

An obvious drawback of this two-stage approach is the need to store the $n \times n$ symmetric matrix $\boldsymbol{\Sigma}$ (or $\mathbf{V}$, or its Cholesky factor). Galbraith and Galbraith [1974] gave a clever factorization of $\mathbf{M}_n = \sigma_a^2\boldsymbol{\Sigma}^{-1}$ that can be used if computer storage is a problem, namely

$$\mathbf{M}_n = \mathbf{C}_n'\mathbf{C}_n - (\mathbf{H}'\mathbf{C}_n)'(\mathbf{D} + \mathbf{H}'\mathbf{H})^{-1}\mathbf{H}'\mathbf{C}_n. \tag{6.125}$$

Here $\mathbf{C}_n$ is an $n \times n$ lower triangular matrix with (r, s)th element c_{r-s} given by the recursive relationship

$$\begin{aligned} c_m &= 0, && m < 0, \\ c_m &= \xi_1 c_{m-1} + \cdots + \xi_{q_2} c_{m-q_2} - \phi_m, && m \geq 0, \end{aligned}$$

where the convention $\phi_0 = -1$ and $\phi_m = 0$ for $m > q_1$ is used. Thus $\mathbf{C}_n$ requires only n storage positions. The matrix $\mathbf{D}$ is such that $\sigma_a^2 \mathbf{D}^{-1}$ is the $(q_1 + q_2) \times (q_1 + q_2)$ dispersion matrix of

$$(\varepsilon_{1-q_1}, \varepsilon_{2-q_1}, \ldots, \varepsilon_0, a_{1-q_2}, a_{2-q_2}, \ldots, a_0)'.$$

Finally, $\mathbf{H}$ is $n \times (q_1 + q_2)$ with $h_{ij} = k_{i+q_1+q_2,j}$, where

$$k_{ij} = \begin{cases} \delta_{ij}, & 1 \leq i \leq q_1 + q_2, \quad 1 \leq j \leq q_1 + q_2, \\ \displaystyle\sum_{s=1}^{q_2} \xi_s k_{i-s,j} - \phi_{i-j-q_2}, & q_1 + q_2 + 1 \leq i \leq n + q_2 + q_2, \\ & \quad 1 \leq j \leq q_1, \\ \displaystyle\sum_{s=1}^{q_2} \xi_s k_{i-s,j}, & q_1 + q_2 + 1 \leq i \leq q_1 + q_2 + n, \\ & \quad q_1 + 1 \leq j \leq q_1 + q_2. \end{cases}$$

and

$$\delta_{ij} = \begin{cases} 1, & i = j, \\ 0, & i \neq j. \end{cases}$$

Computer programs for estimating the ARMA parameters $\boldsymbol{\phi}$, $\boldsymbol{\xi}$, and σ_a^2 from time-series data are available in major statistical packages and as published programs. A number of algorithms are available to calculate the exact value of the log-likelihood function for a given set of parameter values by efficient methods that do not involve the computation of $\boldsymbol{\Sigma}$ and $\boldsymbol{\Sigma}^{-1}$. For example Ansley [1978, 1979] gives a method and a program based on calculating the Cholesky decomposition of the dispersion matrix of an artificial MA model. The FORTRAN subroutines of Gardner et al. [1980] use Kalman filter recursions, and suggestions for improvement are given by Kohn and Ansley [1984]. Extensions to time series with missing observations are given by Jones [1980]. The fast algorithm of Mélard [1984] is basically a composite of several of the others; FORTRAN subroutines implementing the method are given by him.

None of the above methods enables calculation of the derivatives of the log-likelihood L, so that an optimization method which does not require derivatives is required for maximum-likelihood estimation (Section 13.5). This can be a least-squares method (Section 14.5) if L is represented as a sum of squares as in Ansley [1979] and Mélard [1984]. Since the function evaluations are time-consuming, maximum-likelihood calculations can be slow. The recursive methods of McGilchrist et al. [1981] do provide derivatives of L, but we have seen no comparisons of their method with the better-known algorithms.

Box and Jenkins [1976: Chapter 7] describe methods for computing an approximation to L and its derivatives for use in calculating approximate maximum-likelihood estimates. The FORTRAN subroutines of McCleod and Holanda Sales [1983] implement this method of calculating the value of L, but with extensions to cover seasonal models and the use of an approximation to the $|\boldsymbol{\Sigma}|^{-1/2}$ term in the likelihood introduced by McCleod [1977]. The approximate calculations of McCleod and Holanda Sales [1983] are generally more efficient computationally than exact likelihood calculations except when some parameters are close to the admissibility (i.e. stationarity and invertibility) boundaries. Their algorithm is designed to be used in conjunction with a least-squares algorithm that does not require derivatives to calculate approximate maximum-likelihood estimates.

More recently, Godolphin [1984] described an iterative method for calculating estimates which maximize an alternative approximation to L. His method, which uses the data through the sample autocorrelations, appears to be efficient and easy to program. Godolphin claimed that it often gives results which agree with maximum-likelihood estimates to several decimal places. The approach is asymptotically equivalent to maximum likelihood, and Godolphin questions the strength of the evidence in favor of maximum likelihood over the approximations in small samples. The method may have problems if the parameters are close to the admissibility boundaries.

As a method for calculating exact maximum-likelihood estimates in nonlinear models, the iterated two-stage approach may be too slow to be practical. Of course this situation could be improved by using an efficient approximate method of estimating the time-series parameters, such as that of Godolphin [1984], until the final few iterations. However, several specialist methods for exact maximum-likelihood estimation have been suggested for linear models with ARMA errors, and we now consider these briefly.

Consider the linear model

$$\mathbf{y} = \mathbf{X}\boldsymbol{\beta} + \boldsymbol{\varepsilon} \tag{6.126}$$

with ARMA errors and $\mathscr{D}[\boldsymbol{\varepsilon}] = \sigma_a^2 \mathbf{V}_a$, where $\mathbf{V}_a = \mathbf{V}_a(\boldsymbol{\phi}, \boldsymbol{\xi})$. Under normality assumptions, the log-likelihood function is

$$L(\boldsymbol{\beta}, \boldsymbol{\phi}, \boldsymbol{\xi}, \sigma_a^2) = \text{const} - \frac{n}{2}\log\sigma_a^2 - \tfrac{1}{2}\log|\mathbf{V}_a| - \frac{1}{2\sigma_a^2}(\mathbf{y} - \mathbf{X}\boldsymbol{\beta})'\mathbf{V}_a^{-1}(\mathbf{y} - \mathbf{X}\boldsymbol{\beta}). \tag{6.127}$$

For fixed $\boldsymbol{\phi}$ and $\boldsymbol{\xi}$, the maximum-likelihood estimator (MLE) of $\boldsymbol{\beta}$ is the generalized least-squares estimator

$$\tilde{\boldsymbol{\beta}} = (\mathbf{X}'\mathbf{V}_a^{-1}\mathbf{X})^{-1}\mathbf{X}'\mathbf{V}_a^{-1}\mathbf{y},$$

and the MLE of σ_a^2 is

$$\tilde{\sigma}_a^2 = \frac{1}{n}(\mathbf{y} - \mathbf{X}\tilde{\boldsymbol{\beta}})'\mathbf{V}_a^{-1}(\mathbf{y} - \mathbf{X}\tilde{\boldsymbol{\beta}}).$$

Then inserting $\tilde{\boldsymbol{\beta}}$ and $\tilde{\sigma}_a^2$ into (6.127), we obtain the concentrated negative log-likelihood for $\boldsymbol{\phi}$ and $\boldsymbol{\xi}$ as

$$-M(\boldsymbol{\phi}, \boldsymbol{\xi}) = \text{const} + \frac{n}{2}\log \tilde{\sigma}_a^2 + \tfrac{1}{2}\log|\mathbf{V}_a|. \tag{6.128}$$

The linear-model techniques of Harvey and Phillips [1979] and Kohn and Ansley [1985] are clever methods of calculating (6.128) without forming and inverting $\mathbf{V}_a$. Both papers express (6.128) as a sum of squares of the form $\sum_i r_i(\boldsymbol{\phi}, \boldsymbol{\xi})^2$ which can be minimized using a least-squares algorithm that does not require derivatives. Harvey and Phillips [1979] employ Kalman filter recursions. The method of Kohn and Ansley [1985] is apparently more efficient and has the attractive feature that it combines an ARMA maximum-likelihood algorithm for time series such as that of Ansley [1979] or Mélard [1984], with an ordinary least-squares regression subroutine. An alternative method due to McGilchrist et al. [1981] also involves recursive calculations. Both the function values and derivatives of the concentrated likelihood are calculated, and the likelihood is maximized by a slight variant of Fisher's scoring technique.

The exploitation of the linear nature of the model is central to all of the three methods above—in particular the closed-form expression for $\tilde{\sigma}_a^2$ and clever ways of computing it. For nonlinear models $\mathbf{y} = \mathbf{f}(\boldsymbol{\theta}) + \boldsymbol{\varepsilon}$ and fixed $(\boldsymbol{\phi}, \boldsymbol{\xi})$, $\tilde{\boldsymbol{\theta}}$ has to be obtained iteratively before $\tilde{\sigma}_a^2$ and the concentrated likelihood can be evaluated. The linear ideas cannot therefore be simply adapted to give an efficient technique for nonlinear models. Nevertheless, the standard linearization technique which follows can always be used. At the kth iteration, the model function is linearized about the current approximation $\boldsymbol{\theta}^{(k)}$ to $\hat{\boldsymbol{\theta}}$ to obtain the linear model

$$y_i = f(\mathbf{x}_i; \boldsymbol{\theta}^{(k)}) + \frac{\partial f(\mathbf{x}_i; \boldsymbol{\theta}^{(k)})}{\partial \boldsymbol{\theta}'}(\boldsymbol{\theta} - \boldsymbol{\theta}^{(k)}) + \varepsilon_i,$$

or equivalently

$$\mathbf{y}^{(k)} = \mathbf{F}_{\bullet}(\boldsymbol{\theta}^{(k)})\boldsymbol{\theta} + \boldsymbol{\varepsilon}, \tag{6.129}$$

where

$$y_i^{(k)} = y_i - f(\mathbf{x}_i; \boldsymbol{\theta}^{(k)}) + \frac{\partial f(\mathbf{x}_i; \boldsymbol{\theta}^{(k)})}{\partial \boldsymbol{\theta}'}\boldsymbol{\theta}^{(k)}.$$

The next iterate $\boldsymbol{\theta}^{(k+1)}$ is obtained by applying the linear-model algorithm to (6.129). If starting values are sufficiently close to the maximum-likelihood estimators so that the method converges, the method will also produce the correct asymptotic variance–covariance matrices for $\hat{\boldsymbol{\theta}}$, $\hat{\boldsymbol{\phi}}$, $\hat{\boldsymbol{\xi}}$, and $\hat{\sigma}_a^2$.

6.7 FITTING AND DIAGNOSIS OF ERROR PROCESSES

6.7.1 Choosing an Error Process

In practice, the nature of the serially correlated error process is unknown. The methods that have been suggested for modeling the process are, in general, ad hoc and involve the fitting of stationary time-series models to the ordinary least-squares (OLS) residuals. There are some partial theoretical justifications for this. For instance, Hannan [1970: Chapter VII] showed that for linear models, with some regularity conditions on the design matrix, the estimated autocorrelations from the OLS residuals converge to the true autocorrelations as $n \to \infty$.

As in linear modeling, serially correlated errors are often first detected from plots of the OLS residuals (e.g. $\hat{\varepsilon}_i$ versus $\hat{\varepsilon}_{i-1}$), or by formal tests such as the Durbin–Watson test. The latter test was derived for linear regression models to test the hypothesis that $\phi = 0$ in the AR(1) error model $\varepsilon_i = \phi\varepsilon_{i-1} + a_i$. The test statistic is

$$\begin{aligned} D &= \frac{\sum_{i=2}^{n}(\hat{\varepsilon}_i - \hat{\varepsilon}_{i-1})^2}{\sum_{i=1}^{n}\hat{\varepsilon}_i^2} \\ &= \frac{\sum_{i=2}^{n}\hat{\varepsilon}_i^2 + \sum_{i=2}^{n}\hat{\varepsilon}_{i-1}^2 - 2\sum_{i=2}^{n}\hat{\varepsilon}_i\hat{\varepsilon}_{i-1}}{\sum_{i=1}^{n}\hat{\varepsilon}_i^2} \\ &\approx 2(1 - \hat{\rho}_1), \end{aligned} \tag{6.130}$$

where $\hat{\rho}_1$ is the usual lag-1 sample autocorrelation (6.27). As $\hat{\rho}_1$ takes values between -1 and $+1$, the test statistic takes values approximately 0 to 2 for positive correlation and 2 to 4 for negative correlation. The exact limits for a significant test statistic D depend on the design matrix $\mathbf{X}$. However, Durbin and Watson [1951] calculated bounds d_L and d_U on the lower limit for D, and these bounds only depend on the number of regressors. To test $H_0: \phi = 0$ versus $H_1: \phi > 0$ at significance level α, the hypothesis is "accepted" if $D > d_U$ and rejected if $D < d_L$, and the test is inconclusive if $d_L \le D \le d_U$. To test $H_1: \phi < 0$ we replace D by $4 - D$. To test versus $H_1: \phi \ne 0$ at the 2α level, we "accept" H_0 if D and $4 - D$ are greater than d_U, reject H_0 if D or $4 - D$ is less than d_L, and do neither otherwise. Tables of d_L and d_U are given by Durbin and Watson [1951] and reproduced in Abraham and Ledolter [1983: Table D, p. 433]. Durbin and Watson [1971] described methods that used the data to calculate the exact limits numerically. These can be employed if the test is inconclusive. An approximate procedure which is sometimes used is to treat the inconclusive region as part of the rejection region. The applicability of the Durbin–Watson test procedure to nonlinear regression models does not seem to have been established rigorously although, as Amemiya [1983: 355] noted, the asymptotic equivalence between the nonlinear model $f(\mathbf{x}; \boldsymbol{\theta})$ and its linear approximation

$$f(\mathbf{x}; \boldsymbol{\theta}^*) + \frac{\partial f(\mathbf{x}; \theta^*)}{\partial \boldsymbol{\theta}'}(\boldsymbol{\theta} - \boldsymbol{\theta}^*)$$

about $\boldsymbol{\theta}^*$, the true value of $\boldsymbol{\theta}$, suggests that the test will be approximately valid. Although the Durbin–Watson test was derived to detect AR(1) departures from an uncorrelated error sequence, King [1983] noted that it has good power properties against MA(1) departures and is, in fact, approximately a locally best invariant test in this situation. Moreover, he stated that this type of behavior is common in tests designed to detect AR(1) disturbances. He also proposed a test for linear models that can be more powerful than the Durbin–Watson test for detecting AR(1) departures from zero correlation, and a test procedure for distinguishing between an AR(1) and an MA(1) error structure. However, we will focus upon more general methods. Recently Robinson [1985] has developed score tests for uncorrelated errors versus AR(q_1) alternatives which can be used with missing observations in an equally spaced time series.

The construction of ARMA(q_1, q_2) models for observations from a stationary time series is described in most books on time-series analysis that deal with the time domain (Box–Jenkins) method—for example, Box and Jenkins [1976] and Abraham and Ledolter [1983]. Only a brief overview is given here, bearing in mind that the methods are applied to an estimated time series $\{\hat{\varepsilon}_i\}$. We begin by introducing the partial autocorrelation functions.

The partial autocorrelation between ε_i and ε_{i-k}, denoted by ϕ_{kk}, of an autoregressive process can be thought of as the regression coefficient of ε_{i-k} in the representation

$$\varepsilon_i = \phi_{k1}\varepsilon_{i-1} + \phi_{k2}\varepsilon_{i-2} + \cdots + \phi_{kk}\varepsilon_{i-k} + a_i. \tag{6.131}$$

Sample estimates can be obtained by fitting autoregressive processes of orders 1, 2, 3,... by the methods described earlier and picking out the estimates $\hat{\phi}_{11}$, $\hat{\phi}_{22}, \dots, \hat{\phi}_{kk}$ of the last coefficient fitted at each stage. A recursive formula is also available, namely:

(i) $\hat{\phi}_{11} = \hat{\rho}_1$.
(ii) For $k = 2, 3, \dots$

$$\hat{\phi}_{kk} = \frac{\hat{\rho}_k - \sum_{j=1}^{k-1} \hat{\phi}_{k-1,j}\hat{\rho}_{k-j}}{1 - \sum_{j=1}^{k-1} \hat{\phi}_{k-1,j}\hat{\rho}_j}, \tag{a}$$

$$\hat{\phi}_{kj} = \hat{\phi}_{k-1,j} - \hat{\phi}_{kk}\hat{\phi}_{k-1,k-j} \qquad (j = 1, 2, \dots, k-1), \tag{b}$$

where

$$\hat{\rho}_k = \frac{\sum_{i=k+1}^{n} \hat{\varepsilon}_i \hat{\varepsilon}_{i-k}}{\sum_{i=1}^{n} \hat{\varepsilon}_i^2} \tag{6.132}$$

is the usual sample autocorrelation. This second method is rather sensitive to roundoff errors in computation and is unreliable when the parameters are close to the nonstationarity boundaries. Most computer packages use double precision when computing the estimates.

For an AR(q_1) process, $\rho_k \neq 0$ for all k, but for $k \geq q_1 + 1$, $\phi_{kk} = 0$ and the $\hat{\phi}_{kk}$ are approximately i.i.d. $N(0, n^{-1})$. However, for an MA(q_1) process the reverse situation applies. We have $\phi_{kk} \neq 0$ for all k, but for $k \geq q_2 + 1$, $\rho_k = 0$ and the $\hat{\rho}_k$ are approximately $N(0, v_k)$, where

$$v_k = \frac{1}{n}\left(1 + 2\sum_{s=1}^{q_2} \rho_s^2\right). \tag{6.133}$$

We can estimate v_k by replacing ρ_s by $\hat{\rho}_s$. If all the ε_t's are uncorrelated, then $\text{var}[\hat{\rho}_k] \approx n^{-1}$. These results are useful for providing a rough check on whether the ϕ_{kk} or ρ_k are approximately zero beyond some particular lag.

To choose the orders q_1 and q_2 of an ARMA(q_1, q_2) process to be tentatively entertained, both Box and Jenkins [1976: Chapter 6] and Abraham and Ledolter [1983: Section 5.5] recommend plotting the sample autocorrelation function or SACF ($\hat{\rho}_k$ versus k) and the sample partial autocorrelation function or SPACF ($\hat{\phi}_{kk}$ versus k). The orders are then obtained by matching the patterns in these plots to the patterns expected in theoretical ACF and PACF plots as described in Table 6.3. The patterns can be obscured by the random error in the sample estimates and the (sometimes strong) autocorrelation between them. Occasional moderately large estimated autocorrelations can occur after the theoretical function has damped out. At this stage it may be desirable to overfit the model slightly so that q_1 and/or q_2 are bigger than appears necessary. The model could

Table 6.3 Properties of the ACF and the PACF for Various ARMA Models[a]

Model	ACF	PACF
AR(1)	Exponential or oscillatory decay	$\phi_{kk} = 0$ for $k > 1$
AR(2)	Exponential or sine wave decay	$\phi_{kk} = 0$ for $k > 2$
AR(q_1)	Exponential and/or sine-wave decay	$\phi_{kk} = 0$ for $k > q_1$
MA(1)	$\rho_k = 0$ for $k > 1$	Dominated by damped exponential
MA(2)	$\rho_k = 0$ for $k > 2$	Dominated by damped exponential or sine wave
MA(q_2)	$\rho_k = 0$ for $k > q_2$	Dominated by linear combination of damped exponentials and/or sine waves
ARMA(1, 1)	Tails off. Exponential decay from lag 1	Tails off. Dominated by exponential decay from lag 1
ARMA(q_1, q_2)	Tails off after $q_2 - q_1$ lags Exponential and/or sine wave decay after $q_2 - q_1$ lags	Tails off after $q_1 - q_2$ lags Dominated by damped exponentials and/or sine waves after $q_1 - q_2$ lags

[a]From Abraham and Ledolter [1983] with permission from John Wiley and Sons.

Table 6.4 Partial Autocorrelations of the Ordinary Least-Squares Residuals[a]

k	$\hat{\phi}_{kk}$
1	0.417
2	0.262
3	0.155
4	0.085
5	0.011

[a]Reproduced from C. A. Glasbey, "Nonlinear Regression with Autoregressive Time Series Errors". Biometrics 36: 135–140, 1980. With permission from The Biometric Society.

Table 6.5 Partial Autocorrelations for the Residuals When AR(3) is Fitted[a]

k	$\hat{\phi}_{kk}$
1	−0.015
2	−0.029
3	0.024
4	0.066
5	0.008

[a]Reproduced from C. A. Glasbey, "Nonlinear Regression with Autoregressive Time Series Errors". Biometrics 36: 135–140, 1980. With permission from The Biometric Society.

then be simplified, if necessary, after the formal estimation of the parameters. More recently a generalized PACF was introduced, but its use has been questioned (Davies and Petruccelli [1984]).

Example 6.2 **(*continued*)** Table 6.4, adapted from Glasbey [1980], gives the estimates of the partial autocorrelations for the ordinary least-squares residuals in Example 6.2 following (6.87). Using the approximate standard error of $n^{-1/2} = 0.09$, we see that only $\hat{\phi}_{11}$ and $\hat{\phi}_{22}$ differ from zero by more than two standard errors, although $\hat{\phi}_{33}$ is quite large. Glasbey [1980] then fitted an AR(3) error process to the data and calculated the partial autocorrelations for the new $\hat{a}$ residuals (Table 6.5). These autocorrelations are small, thus indicating a satisfactory model for the correlation structure. Also, with an AR(3) model, $\hat{\phi}_3 = 0.18$ with a standard error of 0.09 (see Table 6.2). This is just significant at the 5% level. Glasbey therefore retained the AR(3) process. ■

6.7.2 Checking the Error Process

a Overfitting

One simple way of checking the adequacy of the fitted model is to try fitting one more ARMA parameter, with perhaps an additional AR parameter first, followed by a further MA parameter. A likelihood-ratio test can then be carried out to formally test whether the additional parameter is zero.

b Use of Noise Residuals

Assuming the model to be true, the random variables $\{a_i\}$ (commonly called a "white noise" process) are uncorrelated, so that it is natural to check whether there is still serial correlation present in the $\hat{a}_i$'s. The methods we describe involve treating $\{\hat{\varepsilon}_i\}$ as though it were the actual time series $\{\varepsilon_i\}$. For an AR(q_1) process, we have from (6.52)

$$\hat{a}_i = \hat{\varepsilon}_i - \hat{\phi}_1 \hat{\varepsilon}_{i-1} - \cdots - \hat{\phi}_{q_1} \hat{\varepsilon}_{i-q_1}, \qquad i \geq q_1 + 1.$$

With an ARMA(q_1, q_2) process we can obtain the $\hat{a}_i$'s recursively by using (6.112), namely

$$\hat{a}_i = \hat{\varepsilon}_i - \sum_{r=1}^{q_1} \hat{\phi}_i \hat{\varepsilon}_{i-r} + \sum_{s=1}^{q_2} \hat{\phi}_s \hat{a}_{i-s},$$

and the starting values $\hat{a}_j = 0$ for $j \leq 0$. However, as the use of artificial starting values could make the initial $\hat{a}_i$'s unreliable, improvements can be made by forecasting these values using the backward process (Box and Jenkins [1976: Section 6.4]).

Let $\hat{\rho}_k(\hat{\mathbf{a}})$ be the sample autocorrelation of the $\hat{a}_i$ residuals being used. For ordinary time series, the usual variance estimate of $1/n$ for an autocorrelation can be substantially larger than the true variance of $\hat{\rho}_k(\hat{\mathbf{a}})$ for small k. Its use, therefore, can lead to the omission of important autocorrelations at small lags. Also, for small lags, the $\hat{\rho}_k(\hat{\mathbf{a}})$ can be strongly correlated. For higher values of k the variance estimate of $1/n$ works well. An improved estimate of the large-sample dispersion matrix of $[\hat{\rho}_1(\hat{\mathbf{a}}), \ldots, \hat{\rho}_k(\hat{\mathbf{a}})]'$ is given by Abraham and Ledolter [1983: Section 5.7.1] and McCleod [1978].

An overall test of model adequacy [i.e. $\rho_k(\hat{\mathbf{a}}) = 0$ for all k], usually called the Box–Pierce portmanteau test, uses the test statistic (with suitable K)

$$Q^* = n \sum_{k=1}^{K} \hat{\rho}_k^2(\hat{\mathbf{a}}),$$

which has a large sample chi-square distribution with $K - q_1 - q_2$ degrees of freedom under the null hypothesis. Pierce [1971b] proved that this result still holds when applied to a linear regression model with ARMA errors.

More recently Ljung and Box [1978] and Ansley and Newbold [1979] advocated a slight modification to Q^*, namely

$$Q = n(n+2) \sum_{k=1}^{K} \frac{\hat{\rho}_k^2(\hat{\mathbf{a}})}{n-k}, \tag{6.134}$$

which is asymptotically equivalent but more closely approximates the chi-square distribution in smaller samples. It apparently works satisfactorily for $K \geq 20$. Unfortunately the portmanteau test can have low power against important alternatives (see Newbold [1981]), and a slightly less general test can be obtained which has more power against ARMA alternatives. Godfrey [1979] derived efficient score (Lagrange multiplier) tests (Section 12.4) for detecting ARMA (q_1+r, q_2) or ARMA(q_1, q_2+r) departures from an ARMA(q_1, q_2) model. Godfrey's test statistics can be calculated as follows. Define $w_i = z_i = 0$ for $i \leq 0$, and recursively calculate

$$w_i = -\hat{\varepsilon}_i + \hat{\xi}_1 w_{i-1} + \cdots + \hat{\xi}_{q_2} w_{i-q_2}$$

and

$$z_i = \hat{a}_i + \hat{\xi}_1 z_{i-1} + \cdots + \hat{\xi}_{q_2} z_{i-q_2}$$

for $i = 1, 2, \ldots, n$. To test H_0 that the time series is ARMA(q_1, q_2) against the alternative H_A that it is ARMA(q_1+r, q_2), we fit the regression equation

$$\begin{aligned} \hat{a}_i = {} & \alpha_1 w_{i-1} + \alpha_2 w_{i-2} + \cdots + \alpha_{q_1+r} w_{i-q_1-r} \\ & + \beta_1 z_{i-1} + \cdots + \beta_{q_2} z_{i-q_2} + \text{error}. \end{aligned} \tag{6.135}$$

The Lagrange-multiplier statistic is n times the coefficient of determination R^2 (c.f. Seber [1977: 111]), and it is asymptotically χ_r^2 when H_0 is true. To test H_0 against the alternative H_B that the series is ARMA(q_1, q_2+r), the regression equation

$$\hat{a}_i = \alpha_1 w_{i-1} + \cdots + \alpha_{q_1} w_{i-q_1} + \beta_1 z_{i-1} + \cdots + \beta_{q_2+r} z_{i-q_2-r} + \text{error}$$

is used instead of (6.135). Godfrey found that the asymptotic theory provided better approximations to the finite-sample distributions of his statistics than for the portmanteau test in the examples he simulated. He also found the Lagrange-multiplier test has substantially greater power than the portmanteau test when H_0 was false, even with some ARMA series that differed from the specification used in deriving the test [e.g. H_0:ARMA(1, 0), H_B:ARMA(2, 0), actual series ARMA(2, 1)]. Poskitt and Tremayne [1980] proved that the Lagrange-multiplier tests for ARMA(q_1+r, q_2), ARMA(q_1, q_2+r), and ARMA(q_1+u, q_2+v) with $r = \max(u, v)$ were asymptotically equivalent, thus showing a wider applicability of Godfrey's test. Hosking [1980] gave a unified treatment of a range of Lagrange-multiplier tests for departures from specified ARMA models. He also included the portmanteau test which he derived as a Lagrange-multiplier test against a specific alternative (which is not ARMA).

In conclusion it should be noted that the above theory refers to observed time series. For nonlinear models we use $\{\hat{\varepsilon}_i\}$ and assume that it is sufficiently close to the unknown series $\{\varepsilon_i\}$. Clearly the above comments will still hold asymptotically for nonlinear models, though larger samples will be needed for finite-sample validity.

CHAPTER 7

Growth Models

7.1 INTRODUCTION

In this and the following chapters we make a fairly detailed study of several classes of model that, historically, have provided a good deal of motivation for the statistical development of nonlinear regression. This chapter deals primarily with sigmoidal growth models, though there are sections on the closely related agricultural yield-density models. The following chapter, Chapter 8, discusses compartmental and sums-of-exponentials models, while Chapter 9 considers change-of-phase models.

Models for growth data and compartmental models have developed over a long period of time. As time progressed the main emphasis shifted from the development of families of curves to describe trend or average behavior, to "objective" ways of fitting these models, such as by nonlinear least squares. Recently, there has been more emphasis on the problems involved in using asymptotic results to make inferences from finite samples and on finding more realistic ways to model the stochastic behavior of the data, as the assumptions required for least squares are often untenable. By concentrating on only a few areas of modeling in some detail, we hope to give some insight into the process of statistical model building.

Frequently a dichotomy is made between empirical and mechanistic models for modeling trends in data. An empirical model is one described by a (possibly parametric) family of functions that is sufficiently flexible so that a member of family fits the data well. A mechanistic model, on the other hand, is described by a family of functions that is deduced from the mathematics of the mechanism producing the data. In reality we find that the distinctions between the two types of model are blurred. Models which were originally thought of as mechanistic are often based upon such oversimplified assumptions that they are little more than empirical and yet may be given undue credence because of their mechanistic origins. The changepoint models of Chapter 9 are unashamedly empirical.

An empirical growth curve is a scatterplot of some measure of the size of an object or individual against time x. It is generally assumed that, apart from some form of random fluctuation, the underlying growth follows a smooth curve. This curve, the theoretical growth curve, is usually assumed to belong to a known parametric family of curves $f(x;\boldsymbol{\theta})$ and the principal aim is to estimate the parameters $\boldsymbol{\theta}$.

The analysis of growth data is important in many fields of study. Biologists are interested in the description of growth and in trying to understand its underlying mechanisms. Chemists are interested in the formulation of the product of a chemical reaction over time. In agriculture there are obvious economic and management advantages in knowing how large things grow, how fast they grow, and how these factors respond to environmental conditions or treatments. Normal infant growth is of interest in medicine, as is the growth of tumors and the effect of treatments upon such growth. Social scientists are interested, for example, in the growth of populations, new political parties, the food supply, and energy demand. The same types of model occur when the explanatory variable x is no longer time but increasing intensity of some other factor, such as the growth of smog with increased solar radiation, weight gains with increased nutrients in the diet, increase of nutrient in the blood with nutrient in the diet, changes in crop yield with increased density of planting (Section 7.6), and dose–response curves in bioassay (Finney [1978]) or radioligand assay (Finney [1976]). As most of the mechanistic modeling has been done in a biological context, we shall use the same context in this chapter.

While single growth curves may be of interest in their own right, very often growth-curve data are collected to determine how growth responds to various treatments or other covariate information. It is then essential to reduce each individual growth curve to a small number of parameters so that changing patterns of growth can be understood in terms of changes in these parameters. Indeed, each growth curve may be summarized by its parameter estimates as a single low-dimensional multivariate observation. These observations may then by subject to an analysis of variance or to a regression analysis (Rodda [1981]).

There have been two main approaches to the analysis of growth-curve data, namely statistical and biological. The so-called "statistical" approach is purely empirical and involves fitting polynomial curves to the data using multivariate models. The theory of such models is described by Seber [1984: Section 9.7]. These polynomial methods can provide useful predictive information and may be the best approach if the growth information has been collected over a limited range of the growth cycle. However, their parameters have no physical interpretation, and they do not model subject-matter knowledge of the growth process, for example the fact that the size of an animal tends to stabilize after a certain age. Richards [1969] states that "the polynomial curves commonly adopted in statistical work are usually quite inappropriate for growth studies."

For biological growth the mechanistic approach is often termed "biological." This approach seeks a model with a biological basis and biologically interpretable parameters. We discuss several types of "biological" models. Unfortunately, as Sandland and McGilchrist [1979] state, "At the current state of knowledge about the vastly complex web of biology, biochemistry, nutrition and environment in which growth is embedded, such biological bases are at best a crude approximation and at worst a chimera." This statement is supported by the discussions of plant growth contained in Richards [1969] and Causton and Venus [1981]. The models which follow are better thought of as empirical models which have known nonlinear behavior built into them and which have physically meaningful parameters. An interesting account of the historical development of growth models is given by Sandland [1983].

7.2 EXPONENTIAL AND MONOMOLECULAR GROWTH CURVES

The simplest organisms begin to grow by the binary splitting of cells. With x denoting time and f the size, this leads to exponential growth of the form

$$\frac{df}{dx} = \kappa f, \qquad \text{or} \quad f(x) = e^{\kappa(x-\gamma)}.$$

Thus the growth rate is proportional to the current size f, and for $\kappa > 0$ growth is unlimited. Exponential growth for small x is a feature of many growth models. The *time-power* model

$$f(x) = \alpha x^{\beta}, \tag{7.1}$$

which does not increase as fast as the exponential, is also sometimes useful. However, both of these models lead to unlimited growth, whereas biological growth almost invariably stabilizes over time, so that the final size

$$\alpha = \lim_{x \to \infty} f(x) \tag{7.2}$$

exists and is finite. As a result $df/dx \to 0$ as $x \to \infty$.

Perhaps the simplest assumption leading to limited growth is the assumption that the growth rate is proportional to the size remaining, i.e.,

$$\frac{df}{dx} = \kappa(\alpha - f) \tag{7.3}$$

for some $\kappa > 0$. The general solution to (7.3) can be parametrized as

$$f(x) = \alpha - (\alpha - \beta)e^{-\kappa x}, \tag{7.4}$$

where $x > 0$ and $\kappa > 0$. If the curve is to describe growth (i.e. increase), we require $\alpha > \beta > 0$. With this parametrization α is the final size, $\beta\,[= f(0)]$ is the initial size, and κ acts as a scale parameter on x, thus governing the rate of growth. A more usual parametrization of (7.4) is

$$f(x) = \alpha - \beta e^{-\kappa x} \tag{7.5}$$

with β replacing $\alpha - \beta$. In line with Section 7.3 below, we can also write it in the form

$$f(x) = \alpha(1 - e^{-\kappa(x-\gamma)}). \tag{7.6}$$

The model is usually called the *monomolecular* growth model. Gallucci and Quinn [1979] call it the Von Bertalanffy model, but his name is more usually attached to another model (7.16) described below. The same curve arises in other applications. If we reparametrize (7.4) by replacing $\alpha - \beta$ by $-\beta$ and $e^{-\kappa}$ by $\rho (0 < \rho < 1)$, we obtain

$$f(x) = \alpha + \beta \rho^x, \qquad 0 < \rho < 1, \tag{7.7}$$

in which form it is known as the *asymptotic regression* model (Stevens [1951]). The relationship is also Newton's law of cooling for a body cooling over time. When used to model crop yield versus the rate of fertilizer, it has been known as Mitcherlich's law. In conclusion we recall that the monomolecular model was used in Chapter 3 as an example for discussing the effects of ill-conditioning, experimental design, and choice of parametrization on parameter estimation.

7.3 SIGMOIDAL GROWTH MODELS

7.3.1 General Description

For many types of growth data, the growth rate does not steadily decline, but rather increases to a maximum before steadily declining to zero. This is shown in the growth curve by an S-shaped, or *sigmoidal*, pattern (Fig. 7.1*a*). Such a model adds another recognizable feature to the curve, the position of the point of inflection being the time when the growth rate is greatest (Fig. 7.1*b*). In future sections we shall use w_M as in Fig. 7.1*b* for the maximum growth rate. The three models described in the next three sections achieve this sigmoidal behavior by modeling the current growth rate as the product of functions of the current size and remaining growth, namely

$$\frac{df}{dx} \propto g(f)[h(\alpha) - h(f)], \tag{7.8}$$

where g and h are increasing functions with $g(0) = h(0) = 0$. Most of the growth curves to follow are monomolecular for a simple transformation of either size f (e.g. the Richards curve) or of time x (e.g. the Weibull model).

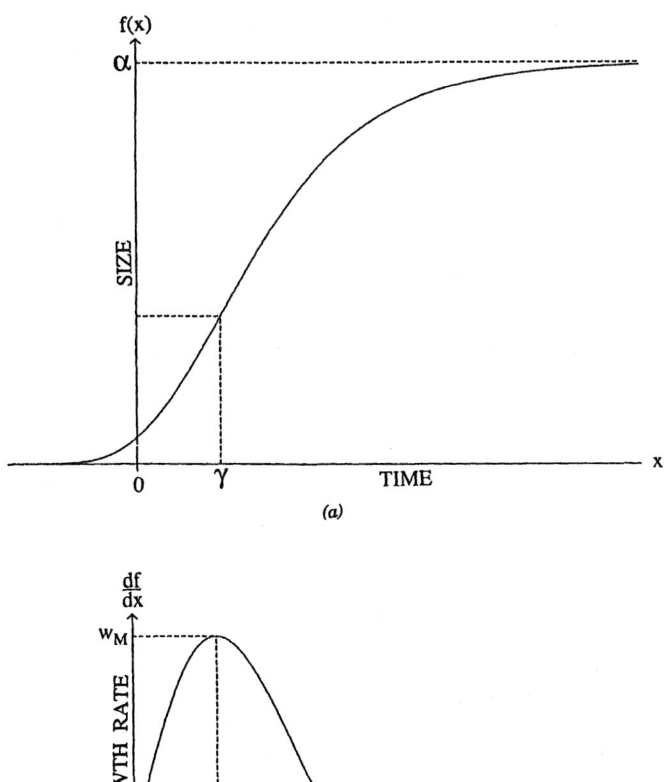

Figure 7.1 Sigmoidal growth model. (*a*) Growth curve with α = final size and γ = point of inflection. (*b*) Growth-rate curve with w_M = maximum growth rate.

7.3.2 Logistic (Autocatalytic) Model

The simplest form of (7.8) is with $g(f) = h(f) = f$, so that

$$\frac{df}{dx} = \frac{\kappa}{\alpha} f(\alpha - f), \tag{7.9}$$

where $\kappa > 0$ and $0 < f < \alpha$. We have used κ/α for the proportionality constant so that the parameters are separately interpretable later. Thus from (7.9) the relative growth rate $f^{-1}\,df/dx$ decreases linearly in f until f approaches α. Equation (7.9) has a general solution which can be written as

$$f(x) = \frac{\alpha}{1 + e^{-\kappa(x-\gamma)}}, \qquad -\infty < x < \infty, \tag{7.10}$$

the so-called *logistic model*. The curve has asymptotes $f = 0$ as $x \to -\infty$ and $f = \alpha$ as $x \to \infty$, which are, of course, never actually attained. This causes few difficulties in practice, because at the time at which growth begins to be monitored we have $f > 0$. From (7.9) it is easily seen that the growth rate is at a maximum when $f = \alpha/2$ and, from (7.10) with $f = \alpha/2$, this occurs when $x = \gamma$. Thus if $\gamma > 0$, the point of inflection for $f(x)$ is visible, as in Fig. 7.1. The maximum growth rate is $w_M = \kappa\alpha/4$, and the growth rate is now symmetrical about $x = \gamma$, leading to a symmetrical sigmoidal growth curve. Again κ acts as a scale parameter on x, thus influencing the growth rate. The initial size is $f(0) = \alpha/(1 + e^{\kappa\gamma})$, and Fletcher [1974] parametrizes $f(x)$ in terms of $f(0)$ and w_M to obtain

$$f(x) = \frac{\alpha}{1 + \dfrac{\alpha - f(0)}{f(0)} e^{-4w_M x/\alpha}}. \tag{7.11}$$

Probably the most common parametrization is

$$f(x) = \frac{\alpha}{1 + \beta e^{-\kappa x}}, \tag{7.12}$$

where $\beta = e^{\kappa\gamma}$ in (7.10). Note that if $f(x)$ takes the form (7.12), then $f^* = b - f^{-1}$ is monomolecular [c.f. (7.6)] for $b > 1/f(0)$. Here the constant b has been included to preserve the range restrictions we used for monomolecular parameters described after (7.4).

The logistic equation (7.9) was derived purely empirically by Verhulst in 1838 to describe growth in the size of a population or organ. However, the equation also describes the rate of consumption of a monomolecular chemical substance as it breaks down in an autocatalytic reaction, and there were early suggestions (Robertson [1908]) that the biochemical nature of the growth process was like this.

7.3.3 Gompertz Growth Curve

The Gompertz model is frequently used for population studies and animal growth in situations where growth is not symmetrical about the point of inflection. The growth rate is

$$\frac{df}{dx} = \kappa f(\log \alpha - \log f) \qquad (\kappa > 0, \quad \alpha > 0), \tag{7.13}$$

and the relative growth rate now declines with log(size). From (7.13) we obtain

$$f(x) = \alpha \exp\{-e^{-\kappa(x-\gamma)}\}. \tag{7.14}$$

The point of inflection for (7.14) is at time $x = \gamma$ with $f = \alpha/e$, at which point the maximum growth rate, $w_M = \kappa\alpha/e$, occurs. For the Gompertz curve $\log f$ is monomolecular. Also, the Gompertz function has the property that any power of f is Gompertz. This means that several measures of size such as weight, surface area, and length may all give rise to growth curves of the same shape. The growth curve depicted in Fig. 7.1 is, in fact, a Gompertz curve.

The Gompertz model, based upon a model given by Gompertz in 1825 for the hazard in a life table, was used as a growth model by Wright [1926]. Its initial formulation was largely empirical, but later Medawar [1940] derived it as a growth model for the heart of a chicken.

7.3.4 Von Bertalanffy Model

The model

$$\frac{df}{dx} = \eta f^{\delta} - \xi f \tag{7.15}$$

was proposed as a mechanistic model for animal growth by von Bertalanffy in 1938. The growth rate of an animal with weight f is the difference between the metabolic forces of *anabolism* and *catabolism.* Here anabolism is the synthesis of new material, and catabolism is the "continuous loss of building material as it takes place in any living organism" (von Bertalanffy [1957]). Catabolism is taken to be proportional to weight, and anabolism is taken to have an allometric relationship with weight (see Example 1.5 in Section 1.2). Von Bertalanffy [1957] assumed that anabolism is proportional to the metabolic rate. He found empirically that for a wide class of animals, the allometric power for the metabolic rate is $\frac{2}{3}$, giving

$$\frac{df}{dx} = \eta f^{2/3} - \xi f, \tag{7.16}$$

and it is this relationship that is usually termed the von Bertalanffy model. It is used extensively in fisheries research (see Seber [1986: p. 281] for some references). All the models of Sections 7.3.2 to 7.3.4 above are special cases of the Richards model to follow. We shall see that the von Bertalanffy model is similar in shape to the Gompertz model, but with its point of inflection at a proportion $(\frac{2}{3})^3 \approx 0.296$ of the final size α, compared with $1/e \approx 0.368$ for the Gompertz.

Nelder et al. [1960] also derived (7.15), but with different values of δ, from assumptions about plant growth being proportional to the intensity of light reaching the leaves. Von Bertalanffy found that, for a number of other animals, the metabolic rate was proportional to weight, thus leading to exponential growth. He allowed for values of δ in the range $\frac{2}{3} \le \delta \le 1$.

Unfortunately, the "biological basis" of these models has been taken too seriously and has led to the use of ill-fitting growth curves. For example, Sandland [1983] states: "Despite the crudity with which the metabolic processes above were modeled, equation (4) [our Equation (7.16)] has become ubiquitous in the fisheries growth literature and has acquired, in some circles, a status far beyond that accorded to purely empirical models." We recall also the quotation from Sandland and McGilchrist [1979] that we used in Section 7.1.

Vieira and Hoffman [1977] discussed the relative merits of the logistic and Gompertz curves. They argued that growth processes which are symmetrical about their point of inflection are rare, and that the Gompertz model may be better for asymmetrical curves. However, without a convincing mechanistic model, a point of inflection at $f = \alpha/e$ is as arbitrarily imposed as a point of inflection at $f = \alpha/2$. What is needed is a fourth parameter to allow for this uncertainty as to the position of the point of inflection, or equivalently, as to the degree of asymmetry in the growth curve.

7.3.5 Richards Curve

Richards [1959] doubted the theoretical validity and usefulness of von Bertalanffy's models as a description of the growth mechanism. However he noted that Equation (7.15) with δ treated as a free parameter provided a flexible family of curves with an arbitrarily placed point of inflection. It is easily seen that for the growth rate to be positive but the ultimate size limited, η and ξ in (7.15) have to be negative when $\delta > 1$. Imposing $df/dx = 0$ at $f = \alpha$, we can write (7.15) as

$$\frac{df}{dx} = \frac{\kappa}{1-\delta} f \left[\left(\frac{f}{\alpha} \right)^{\delta - 1} - 1 \right], \qquad \delta \neq 1, \tag{7.17}$$

where $\kappa = \eta(1-\delta)\alpha^{\delta-1}$. This model of Richards includes the monomolecular model ($\delta = 0$), the von Bertalanffy model ($\delta = \frac{2}{3}$), the logistic model ($\delta = 2$), and (by taking the limit as $\delta \to 1$) the Gompertz equation (7.13).

To solve (7.17) we employ a substitution

$$y = \begin{cases} \alpha^{1-\delta} - f^{1-\delta}, & \delta < 1, \\ f^{1-\delta} - \alpha^{1-\delta}, & \delta > 1. \end{cases}$$

In both cases (7.17) reduces to $dy/dx = -\kappa y$ which has solution $y = e^{-\kappa(x-c)}$.

Thus we obtain

$$f(x) = \begin{cases} [\alpha^{1-\delta} - e^{-\kappa(x-c)}]^{1/(1-\delta)}, & \delta < 1, \\ [\alpha^{1-\delta} + e^{-\kappa(x-c)}]^{1/(1-\delta)}, & \delta > 1. \end{cases}$$

We adopt the single equivalent parametrization (c.f. Jorgensen [1981])

$$f(x) = \alpha[1 + (\delta - 1)e^{-\kappa(x-\gamma)}]^{1/(1-\delta)}, \qquad \delta \neq 1, \tag{7.18}$$

which is valid for all $\delta \neq 1$. However, for $\delta < 1$ the further restriction $(1 - \delta)e^{\kappa\gamma} \leq 1$ is required to ensure that $0 \leq f \leq \alpha$. Using this parametrization (7.18), $f(x)$ and its derivatives of all orders tend to the corresponding derivatives of the Gompertz function (7.14) as $\delta \to 1$.

Differentiating (7.17),

$$\frac{d^2f}{dx^2} = \frac{\kappa}{1-\delta}\left[\delta\left(\frac{f}{\alpha}\right)^{\delta-1} - 1\right]\frac{df}{dx}, \tag{7.19}$$

which is zero at $f = \alpha/\delta^{1/(\delta-1)}$ when $\delta > 0$. Thus the parameter δ indirectly locates the point of inflection of the curve on the f-axis at a proportion $p = \delta^{1/(1-\delta)}$ of final size and thus controls the shape of the curve. A parameter such as p would be preferable, but there is no closed-form solution for δ in terms of p. Again the parameter γ locates the point of inflection on the x-axis, and now the maximum growth rate is

$$w_M = \kappa\alpha\delta^{\delta/(1-\delta)}.$$

As $\delta \to \infty$, the curve (7.18) tends to a truncated exponential. For $\delta \leq 0$ there is no point of inflection. Again κ acts as a scale factor on time x, so that for fixed δ it acts as a rate parameter. Unfortunately, with different-shaped curves there can be no simple (single-parameter) concept of comparative growth rates. However, Richards [1959] derived

$$\frac{1}{\alpha}\int_0^{\alpha} \frac{\kappa f}{1-\delta}\left[\left(\frac{f}{\alpha}\right)^{\delta-1} - 1\right] df = \frac{\alpha\kappa}{2(\delta+1)} \tag{7.20}$$

as an average growth rate. Thus, adjusting for final size, $\kappa/[2(\delta + 1)]$ ($= \lambda$, say) is a crude measure of "growth rate" for comparisons between growth curves with different shapes. Figure 7.2 shows the shapes of the growth curves for different values of δ, while Fig. 7.3 shows the corresponding growth rate curves $\dot{f} = df/dx$ as a function of f. The curves are standardized to have the same final size α and "growth rate" λ. Clearly, from the figure, the maximum growth rates for curves with the same value of λ and $0.4 \leq \delta \leq 4$ are very similar. This confirms the usefulness of the parameter λ for a wide range of curves.

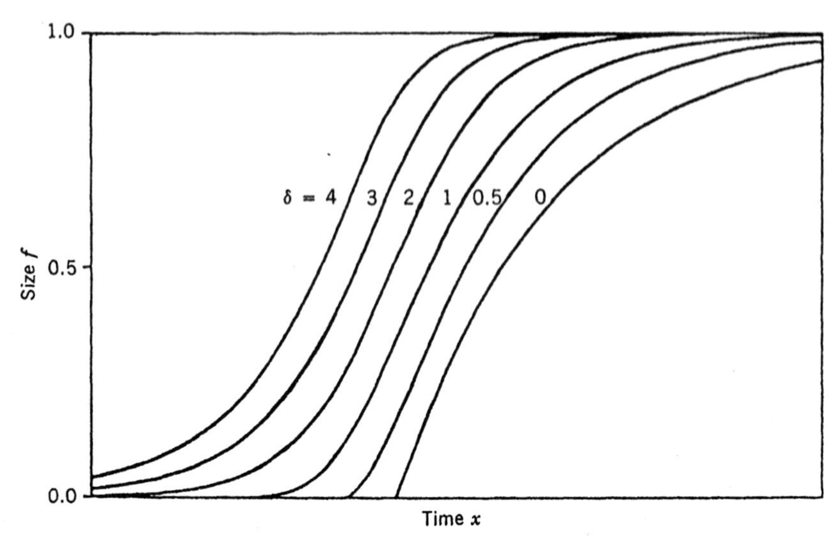

Figure 7.2 Richards family of growth curves for various δ. Here $\alpha = 1$ and λ is kept constant. The parameter γ is changed each time so that the curves do not sit on top of one another. From Richards [1959].

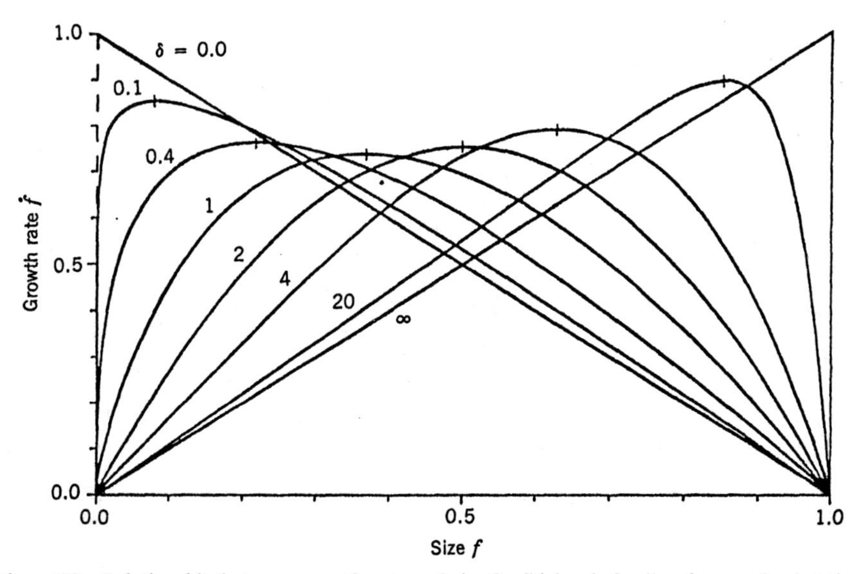

Figure 7.3 Relationship between growth rate and size for Richards family of curves having the δ values indicated. In all cases $\alpha = 1$ and λ is kept constant. The position of the maximum on each curve is shown. From Richards [1959].

We noted earlier that initial growth is often exponential, corresponding to the binary splitting of cells. This feature is present in the Richards family when $\delta > 1$, since from (7.17)

$$\lim_{f \to 0} \frac{1}{f} \frac{df}{dx} = \frac{\kappa}{\delta - 1},$$

which is constant, so that $f(x) \approx f(0) \exp\{\kappa x/(\delta - 1)\}$ when x is small. From (7.17) we can write df/dx as

$$\frac{df}{dx} = \frac{\kappa}{\delta - 1} f - \frac{\kappa}{\delta - 1} \frac{f^\delta}{\alpha^{\delta - 1}},$$

so that the second term damps the first (exponential) growth term as f increases until it overwhelms it and growth ceases. When $\delta < 1$, the initial relative growth rate is infinite as $f \to 0$.

For any member of the Richards family $f^{1-\delta}$ is of the monomolecular functional form (7.6). However, for $\delta < 1$ we need to use $b - f^{1-\delta}$, where $b > f(0)^{1-\delta}$, to fulfill the range restrictions given for monomolecular parameters. The Richards curves can also be parametrized in terms of $f_0 = f(x_0)$ for some x_0 as

$$f(x) = \{\alpha^{1-\delta} - (\alpha^{1-\delta} - f_0^{1-\delta}) e^{-\kappa(x - x_0)}\}^{1/(1-\delta)}. \tag{7.21}$$

7.3.6 Starting Values for Fitting Richards Models

The physical interpretability of many of the parameters means that crude initial estimates can often be obtained from a scatterplot of the growth data, perhaps with a freehand smooth curve added. For example, the final size α can often be guessed this way. The coordinates of the point of inflection, $x = \gamma$ and $f = \alpha\delta^{1/(1-\delta)}$, can then provide starting values for γ and δ. Finally, the intercept on the f-axis is

$$f(0) = \alpha[1 + (\delta - 1)e^{\kappa\gamma}]^{1/(1-\delta)},$$

which can be solved for κ.

If convergence cannot be obtained from the above starting values, more sophisticated refinements are available. Rearranging (7.18), we obtain a transformation $f^* = g(f)$, namely

$$f^* = -\log \frac{\left\{\left(\dfrac{f}{\alpha}\right)^{1-\delta} - 1\right\}}{\delta - 1} = \kappa x - \kappa\gamma \qquad (\delta \neq 1), \tag{7.22}$$

or, from (7.14),

$$f^* = -\log[\log(\alpha/f)] = \kappa x - \kappa\gamma \qquad (\delta = 1). \tag{7.23}$$

Thus for given α and δ, starting values for κ and γ can be obtained from a simple linear regression of $y^* = g(y)$ on x, where y is the observation taken on f. This can be improved upon, if necessary, by using a weighted regression with weights inversely proportional to the variance of y^*, using the result that

$$\text{var}[y^*] = \text{var}[g(y)] \approx \text{var}[y]\left[\frac{dg(y)}{dy}\right]^2_{y=E[y]}. \tag{7.24}$$

When the asymptote is not clear on the scatterplot, a range of several α-values or a grid of α and δ values can be used with κ and γ obtained by linear regression as above for each pair (α, δ). This gives us a set of possible starting values $(\alpha, \delta, \kappa, \gamma)$. The best of these can be chosen using the optimality criterion to be employed in fitting the model. Using least squares, for example, we can choose the 4-tuple $(\alpha, \delta, \kappa, \gamma)$ with the smallest value of

$$\sum_i [y_i - \alpha\{1 + (\delta - 1)e^{-\kappa(x_i - \gamma)}\}^{1/(1-\delta)}]^2. \tag{7.25}$$

Richards [1969] described a simple method for estimating α for the monomolecular model ($\delta = 0$) using three equally spaced time points x_1, x_2, and x_3, which should be widely separated. Let $\tilde{f}_i$ be the estimate of $f(x_i)$ obtained visually from a freehand curve through the scatterplot of y versus x. Then from (7.6) it is easily verified that

$$\frac{\alpha - \tilde{f}_2}{\alpha - \tilde{f}_1} = \frac{\alpha - \tilde{f}_3}{\alpha - \tilde{f}_2},$$

so that

$$\alpha = \frac{\tilde{f}_2^2 - \tilde{f}_1\tilde{f}_3}{2\tilde{f}_2 - \tilde{f}_1 - \tilde{f}_3}.$$

For any $\delta \neq 1$ the same formula applied to the $\tilde{f}_i^{1-\delta}$ gives an estimate of $\alpha^{1-\delta}$. For the Gompertz curve, the method applied to $\log \tilde{f}_i$ gives an estimate of $\log \alpha$.

Bad ill-conditioning and convergence problems have often been experienced in fitting the Richards model to data. Both problems can occur when insufficient of the curve is visible to provide a good initial estimate of the final size α (c.f. Sections 3.2 and 3.4.2). Data which give rise to only moderate ill-conditioning when logistic or Gompertz curves are fitted often give rise to pronounced ill-conditioning when the full four-parameter Richards model is fitted. An important reason for this is clear from Fig. 7.2. Curves with quite different δ-values look very similar, and we would anticipate difficulty in distinguishing between them with even a small amount of scatter. Indeterminancy in δ or the location of the point of inflection on the f-axis (at $x = \gamma$) leads to indeterminacy in γ. The parameters δ and κ are also tied together through the fairly stable growth-rate parameter $\lambda = \kappa/[2(\delta + 1)]$.

Ratkowsky [1983: Chapter 8] gives a fuller discussion of methods for finding starting values, including some more specialized techniques for the logistic and Gompertz models.

7.3.7 General Procedure for Sigmoidal Curves

An obvious way of describing a sigmoidal shape is to use the distribution function $F(x; \boldsymbol{\theta})$ of an absolutely continuous random variable with a unimodal distribution. Then

$$f(x) = \alpha F(\kappa(x - \gamma); \boldsymbol{\theta}) \tag{7.26}$$

gives us a sigmoidal growth curve for empirical use. The parametric family $F(x; \boldsymbol{\theta})$ can be chosen to give as much flexibility as desired. For the Richards family,

$$F(x; \delta) = [1 + (\delta - 1)e^{-x}]^{1/(1-\delta)}.$$

This is a slightly different parametrization of the asymmetric family of distributions described by Prentice [1976: Section 4] for binary-response data. It includes the logistic distribution ($\delta = 2$) and the extreme-value distribution ($\delta \to 1$). In Prentice [1976] this family is embedded in a two-parameter family of the $\log F$ distribution which gives control over both skewness and kurtosis. This two-parameter family also includes the normal, double-exponential, and exponential distributions.

In the parametrization (7.26) with given α and κ, different values of $f(0)$ are coped with by changing the time origin via γ, i.e. by shifting the standard curve horizontally. Another method is to shift the standard curve vertically using

$$f(x) = \beta + (\alpha - \beta)F(\kappa x; \boldsymbol{\theta}). \tag{7.27}$$

When this second method is applied to the logistic curve, the resulting curve is called the Riccati model because it is the solution of the Ricatti differential equation under certain restrictions on the range of the equation parameters. The Riccati equation is of the form

$$\frac{df}{dx} = af^2 + bf + c.$$

For details, see Levenbach and Reuter [1976]. One of the features these authors like about the Ricatti model is that it enables zero size to be attained at a finite time (with $\beta < 0$). This applies to any curve obtained by the second method above [c.f. (7.27)].

We can also combine (7.26) and (7.27) to obtain

$$f(x) = \beta + (\alpha - \beta)F(\kappa(x - \gamma); \boldsymbol{\theta}).$$

When this is applied to the logistic distribution function

$$F(z) = (1 + e^{-z})^{-1},$$

we get

$$f(x) = \beta + \frac{\alpha - \beta}{1 + \exp\{-\kappa(x - \gamma)\}}.$$

This is the four-parameter logistic curve widely used in radioligand assay (see Finney [1976], Ratkowsky and Reedy [1986]), where $x = \log(\text{dose})$.

7.3.8 Weibull Model

The Richards family can be obtained by assuming that a transformation of size, namely $f^{1-\delta}$, is monomolecular. The Weibull family can similarly be obtained by assuming that size is monomolecular for a transformation x^δ of x. The one-parameter Weibull distribution function is

$$F(x; \delta) = 1 - \exp(-x^\delta), \qquad x > 0. \tag{7.28}$$

This distribution function has a point of inflection at $x = [(\delta - 1)/\delta]^{1/\delta}$ and $F = 1 - \exp\{-(1 - \delta^{-1})\}$. In accordance with the previous section, a Weibull growth curve can then be constructed either using the first method (7.26), giving

$$f(x) = \alpha(1 - \exp\{-[\kappa(x - \gamma)]^\delta\}), \tag{7.29}$$

or using the second method (7.27) to obtain

$$f(x) = \alpha - (\alpha - \beta)\exp\{-(\kappa x)^\delta\}. \tag{7.30}$$

The two families of curves given by (7.29) and (7.30) are different. Equation (7.30) seems to be more commonly used. For example, it is this model that Ratkowsky [1983] calls the Weibull model. From (7.30) we have

$$\frac{df}{dx} = \delta\kappa(\alpha - f)\left(\log\frac{\alpha - f(0)}{\alpha - f}\right)^{1-\delta^{-1}}. \tag{7.31}$$

The general points made in the previous section about obtaining starting values for (7.30) are appropriate again here. From a scatterplot we would hope to be able to estimate α and also β [using $f(0) = \beta$]. Starting values for κ and δ can then be obtained by simple linear regression, since

$$\log\left(-\log\frac{\alpha-f}{\alpha-\beta}\right)=\delta\log\kappa+\delta\log x.$$

7.3.9 Generalized Logistic Model

This model, developed by Pearl and Reed [1925], incorporates skewed curves by making $f(x)$ logistic [c.f. (7.10)] for a transformation $g(x)$ of time x, namely

$$f(x)=\frac{\alpha}{1+\exp[\gamma'+\kappa g(x)]}. \tag{7.32}$$

The transformation they employed in practice was a cubic polynomial $g(x)=\beta_1 x + \beta_2 x^2+\beta_3 x^3$. This model has seldom been used. Only for some combinations of the β_i-values is the time scale monotonically increasing over the time range of interest. Also the parameters β_1, β_2, and β_3 are difficult to interpret.

Another possibility is to use a Box–Cox power transformation

$$g(x)=\frac{x^\lambda-1}{\lambda}.$$

This has been used in a generalized logistic quantal response curve by Miller et al. (1982). For this choice, $g(x)$ is a monotonically increasing function of x.

7.3.10 Fletcher Family

As we have previously noted, the Richards family can be thought of in terms of initial exponential growth which is increasingly damped as the size increases until all growth ceases. However, this interpretation does not apply to curves below the Gompertz ($\delta\le 1$). Fletcher [1974] defined a family of skewed growth curves which he called the "quadric law of damped exponential growth" with the aim of producing a family of curves which all have the damped exponential form. He also wanted to gain direct parametric control over the location of the point of inflection, η say, on the f-axis. The other desired parameters were the maximum rate of growth w_M and the final size α. This was achieved by taking a quadratic function in f, as used for the logistic of (7.9), and rotating the axes but keeping the curve fixed.

The general equation of a parabolic relationship between f and $\dot{f}=df/dx$ is

$$\dot{f}+af^2+bf+g\dot{f}^2+hf\dot{f}+c=0, \tag{7.33}$$

where $ag-h^2/4=0$ and

$$\begin{pmatrix} a & h & b\\ h & g & 1\\ b & 1 & c\end{pmatrix}\ne 0.$$

The parameters g and h are "rotation" terms ($g = h = 0$ for logistic growth). Fletcher [1974] found that to preserve positive growth, in terms of the parameters η, w_M, and α, we must have

$$a = \frac{4w_M}{\alpha(4\eta - \alpha)}, \qquad b = \frac{-4w_M}{4\eta - \alpha},$$
$$g = \frac{(\alpha - 2\eta)^2}{w_M \alpha(4\eta - \alpha)}, \qquad h = \frac{4(\alpha - 2\eta)}{\alpha(4\eta - \alpha)}, \tag{7.34}$$

and that under these conditions (7.33) has solution

$$f = \frac{g^2 - (4\eta - \alpha)^2}{8(\alpha - 2\eta)}.$$

Here g satisfies

$$(g - u_1)^{u_1}(g - u_2)^{-u_2} = \beta e^{4w_M x}, \tag{7.35}$$

where $u_1 = 4\eta - \alpha$ and $u_2 = 3\alpha - 4\eta$. The parameter β locates the curve on the x-axis, but in a complex way involving the other parameters. Fletcher chooses β so that f passes through the initial-condition point (x_0, f_0). The family is only defined for $\alpha/4 < \eta < 3\alpha/4$. Figure 7.4 shows three representations of the curve for $\alpha = 10$, $\eta = 7$, $w_M = 2$, and $(x_0, f_0) = (0, 0.1)$.

Fletcher's curves still have the interpretation of damped exponential growth throughout their range, $0.25 < \eta/\alpha < 0.75$, in contrast to the Richards model, which only has this property for part of the range, namely $\eta/\alpha > e^{-1} \approx 0.37$. However, Fletcher has lost direct parametric control of the location of the curve on the x-axis and, more importantly, there are no longer explicit formulae for $f(x)$ and its derivatives. This makes model fitting slower and more difficult, since $f(x)$ and its derivatives must be obtained by the numerical solution of nonlinear equations.

7.3.11 Morgan–Mercer–Flodin (MMF) Family

In the context of growth response to nutritional intake, Morgan et al. [1975] developed a family of sigmoidal growth curves based upon the hyperbolic function

$$f(x) = \frac{\beta\gamma + \alpha x^\delta}{\gamma + x^\delta},$$

which we shall write in the form

$$f(x) = \alpha - \frac{\alpha - \beta}{1 + (\kappa x)^\delta}. \tag{7.36}$$

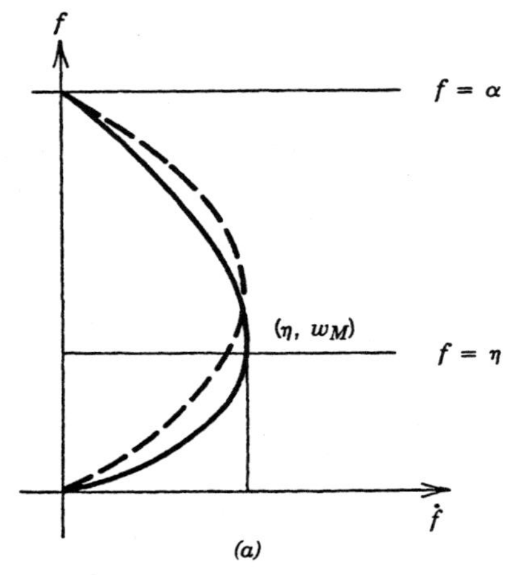

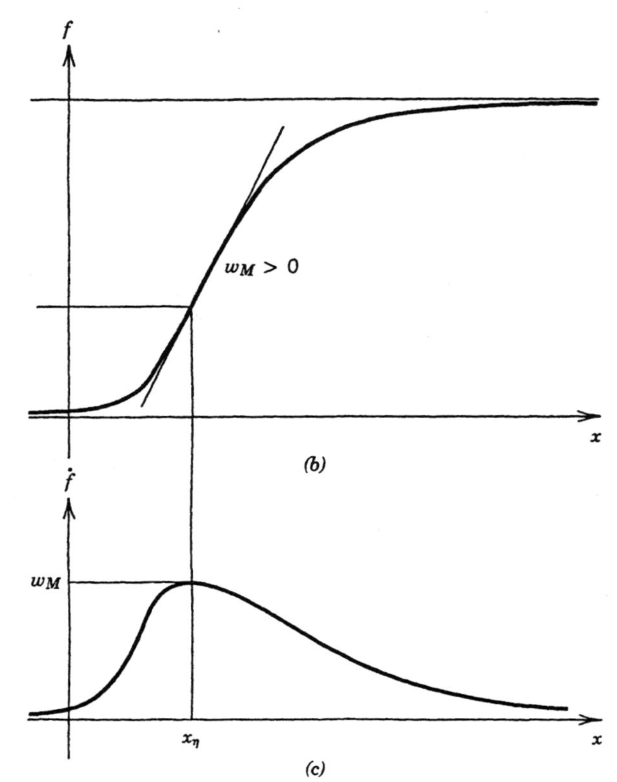

Figure 7.4 Example of a member of the Fletcher family. (*a*) Parabolic relationship (7.33) between size f and growth rate $\dot{f}$. (*b*) Growth curve. (*c*) Growth rate versus time x. Reproduced from R. I. Fletcher, "The Quadric Law of Damped Exponential Growth". Biometrics 30: 111–124, 1974. With permission from The Biometric Society.

With this parametrization, α is the final size, $f(0) = \beta$, δ determines the shape of the sigmoid, and κ is a scale parameter. In terms of Section 7.3.7, this family is equivalent to taking the distribution function

$$F(x;\delta) = \frac{x^\delta}{1+x^\delta}, \qquad 0 \le x < \infty, \tag{7.37}$$

and using the method of (7.27). This distribution has a point of inflection at $x = [(\delta-1)/(\delta+1)]^{1/\delta}$ and $F = (\delta-1)/(2\delta)$ for $\delta \ge 1$. There is no point of inflection for $\delta < 1$. All the MMF curves are "sublogistic" in the sense that the point of inflection of $F(x)$ is always below $\frac{1}{2}$. Now

$$\begin{aligned}\frac{dF}{dx} &= \delta F^{(\delta-1)/\delta}(1-F)^{(\delta+1)/\delta}\\ &\propto F^{1-1/\delta}(1-F)^{1+1/\delta},\end{aligned}$$

which may be compared with $F(1-F)$ for the logistic distribution. Thus $F(x)$ tends to the logistic in shape as $\delta \to \infty$. Starting values for the MMF curve (7.36) can be obtained as in Section 7.3.6 with α and β obtained from a scatter plot, and δ and κ obtained using the linearization

$$\log\frac{f-\beta}{\alpha-f} = \delta\log\kappa + \delta\log x$$

7.4 FITTING GROWTH MODELS: DETERMINISTIC APPROACH

Traditionally, growth curves with a trend relationship $y_i \approx f(x_i;\boldsymbol{\theta})$ have been fitted by least squares, possibly after transforming the relationship to stabilize the variance (Section 2.8). This is appropriate under the usual nonlinear regression model

$$y_i = f(x_i;\boldsymbol{\theta}) + \varepsilon_i \qquad (i = 1,2,\ldots,n), \tag{7.38}$$

where the ε_i are i.i.d with mean zero and variance σ^2. Simply adding a random or measurement error to the trend relationship is often called the "deterministic" approach. Such a model seems reasonable with cross-sectional data in which a single size measurement is taken on each individual or experimental unit and the "ages" of the individuals cover a range of x-values.

The use of (7.38) for growth data has been investigated at length by Ratkowsky [1983]. In Chapter 4 of his book he applied the Gompertz, logistic, Richards, Morgan–Mercer–Flodin (MMF), and Weibull models to four different sets of agricultural data using both the model (7.38) and the same model after a log transformation. Only the Richards model showed significant intrinsic curvature. This model had a significant intrinsic curvature for all four data sets given under both the untransformed and the logged model. The data sets used by Ratkowsky were too small to discriminate between the untransformed and logged models. The other four-parameter curves—the MMF and Weibull models—had a further advantage over the Richards curve in that their parameter estimates were more stable on changing the transformation. Thus Ratkowsky recommended the MMF and Weibull models in preference to the Richards model for vegetative growth data when three-parameter curves such as the logistic proved inadequate. Ratkowsky also searched for parametrizations of these models with low parameter-effects curvature. However, the parametrizations described in this chapter were not among those that he tried. Good parametrizations are sometimes data-dependent. Any confidence intervals for the new parameters need to be "detransformed" to give intervals for the parameters of interest.

Similar explorations of the monomolecular model are described in Chapter 5 of Ratkowsky [1983]. The model was applied to seven small data sets including growth curves, data on crop yield versus fertilizer application rate, and data on leaf production versus light irradiance. The intrinsic curvature was acceptably low in only three of the data sets. The parametrization

$$f(x) = \alpha^{-1} - \beta\gamma^{x} \tag{7.39}$$

had the consistently lowest parameter-effects curvative among the six parametrizations that he tried. Recently, Ratkowsky and Reedy (1986) made a similar study in which the four-parameter logistic model described at the end of Section 7.3.7 was applied to five sets of radioligand data. Several parametrizations were acceptable for all five data sets.

Before leaving the model (7.38), we note that there is a specialized literature on obtaining direct (i.e. noniterative) estimators for the parameters of the asymptotic or monomolecular model (7.39) and extensions of that model such as

$$f(x) = \alpha + \delta x + \beta x^{\gamma}$$

(see Shah and Khatri [1965, 1973], Brooks [1974]). Similar estimators for fitting

$$y_i = \alpha + \sum_{j=1}^{k} \beta_j \rho_j^i + \varepsilon_i$$

have been given (Agha [1971], Della Corte et al. [1974]).

We noted earlier that the standard least-squares approach is fairly natural for cross-sectional data. However, growth data are often collected longitudinally. Size measurements are taken on the same individual as it ages. In such cases, unless the time spacing between x_i and x_{i+1} is sufficiently large, it is usual to see long runs of residuals with the same sign, thus making the assumption of uncorrelated errors untenable. Glasbey [1979] noted that for animal growth, factors such as measurement error and how full the animal's stomach is at the time of weighing may often be approximately uncorrelated. However, contributions from uncontrollable environmental factors, illness, and variation of the nutritional quality of the diet would be expected to have a lingering effect. These problems can be partially accommodated in the deterministic approach by allowing for an autocorrelated error structure for the errors in (7.38) and using the methods of Chapter 6. Examples of this approach using growth data are also given there. We further noted in Chapter 6 that the use of ordinary least squares can lead to inefficient estimates and grossly underestimated standard errors if the error structure is autocorrelated.

Sometimes a mixture of cross-sectional and longitudinal data types are used. For example, Vieira and Hoffman [1977] estimated an average growth curve from monthly weight measurements on ten Holstein cows for the first 56 months of life. Here plausible correlation structures are even more complex. Berkey [1982] adopts an empirical Bayes approach for mixed data types.

7.5 STOCHASTIC GROWTH-CURVE ANALYSIS

7.5.1 Introduction

For longitudinal growth models, explaining scatter by simply tacking a random error onto the deterministic growth equation seems particularly unsatisfactory. Where variation is caused by an unpredictably varying environment, we would expect setbacks caused by such factors as a period of bad nutrition, a bout of sickness, and so on to affect growth patterns for some time into the future, and not just perturb the current measurement. We can incorporate such lingering effects by allowing autocorrelated errors, as in Section 7.4. However, even this is rather clumsy.

An attractive alternative approach comes from thinking of growth as an incremental process in which the current environmental conditions affect the current increment in size. This leads to the consideration of models of the form

$$\dot{y} = \frac{dy}{dx} = h(x, y; \theta) + \text{error}. \tag{7.40}$$

We shall find that even when the errors in such models are independent, the y-values are correlated. This is intuitively obvious, as (see below) for $x_2 > x_1$, $y(x_2)$ is $y(x_1)$ plus further increments. Sandland and McGilchrist [1979] state that the y-process is also nonstationary. The discussion in this section will be largely based on the work of Sandland and McGilchrist [1979], White and Brisbin [1980], and Garcia [1983]. These authors found that, in practice, the errors in relative growth rate are closer to having constant variance than the errors in the rates themselves. Thus we also consider models of the form

$$\frac{1}{y}\frac{dy}{dx}\left(=\frac{d\log y}{dx}\right) = h(x, y; \theta) + \text{error}. \tag{7.41}$$

White and Brisbin [1980] and Sandland and McGilchrist [1979] found that for their data sets, the usual determinstic models required complicated autocorrelation structures for the errors. In contrast, growth-rate or relative-growth-rate models like (7.40) and (7.41) with independent errors or a simple autocorrelation structure usually fit well.

A natural simplification of (7.41) is

$$\frac{\dot{y}}{y} = h(y) + \varepsilon(x). \tag{7.42}$$

The difficulty with this equation is that, in practice, we measure size and not the rate of change of size. The use of approximations to (7.42) has a long history. Suppose we have measurements (x_i, y_i), $i = 1, 2, \ldots, n$, on (7.42) with $x_1 < x_2 < \cdots < x_n$. People have tended to use least squares to fit models of the form

$$R_i = h(y_i; \boldsymbol{\theta}) + \varepsilon_i, \tag{7.43}$$

where R_i is some empirical estimate of the relative growth rate at time x_i. For the logistic curve, for example, the relative growth rate is of the form $\beta - \gamma y$ [c.f. (7.9)]. As early as 1927, Hotelling estimated the relative rate $\dot{y}_i/y_i$ by

$$R_i = \frac{y_{i+1} - y_i}{y_i(x_{i+1} - x_i)} = \frac{\Delta y_i}{y_i \Delta x_i}, \tag{7.44}$$

say. He sought to estimate β and γ by minimizing

$$\sum_{i=1}^{n} \left(\frac{\Delta y_i}{y_i \Delta x_i} - (\beta - \gamma y_i) \right)^2,$$

or more particularly its limiting form as $\Delta x_i \to 0$, namely

$$\int \left(\frac{1}{y} \frac{dy}{dx} - (\beta - \gamma y) \right)^2 dx$$

using the calculus of variations (see Sandland [1983]).

A more commonly used estimate of relative growth rate is

$$R_i = \frac{\log y_{i+1} - \log y_i}{x_{i+1} - x_i} = \frac{\displaystyle\int_{x_i}^{x_{i+1}} \frac{d \log y}{dx} dx}{\displaystyle\int_{x_i}^{x_{i+1}} dx}, \tag{7.45}$$

the average value of $\dot{y}/y$ over the interval $[x_i, x_{i+1})$, as noted by R. A. Fisher in 1920. More recently Levenbach and Reuter [1976] advocated estimating the parameters of longitudinal growth curves using (7.43). For a number of models (exponential, linear, modified exponential, logistic, and Ricatti growth) they derived estimates R_i based upon (x_i, y_i) and one or more neighboring points. These estimates have the property that if there is no random error, $R_i = \dot{y}_i/y_i$. They dealt only with the case of equally spaced x_i's, and many of the estimates can only be used if the y_i-sequence is monotonic.

Clearly autocorrelated errors are easily incorporated in the empirical approach using (7.43). For the remainder of this section, our treatment of rate modeling will be based on the interpretation of (7.40) and (7.41) as stochastic differential equations (c.f. Appendix C). Models in which the growth rate or relative rate are treated as functions of size y instead of time are discussed later in Section 7.5.3. Sandland and McGilchrist [1979] noted that Equation (7.42) is generally not tractable. For this reason, and also because of their lack of faith in the soundness of the biological basis of the standard growth models, they approximated the relative rate by a function of time x rather than size. Often they used a polynomial function of time. Their approach is described below.

7.5.2 Rates as Functions of Time

Sandland and McGilchrist [1979] considered models described by the stochastic differential equation (see Appendix C)

$$\frac{1}{y}\frac{dy}{dx} = \frac{d \log y}{dx} = h(x; \boldsymbol{\theta}) + \sigma(x; \boldsymbol{\eta})\varepsilon(x), \tag{7.46}$$

where $\varepsilon(x)$ is a stationary Gaussian random process with $\text{var}[\varepsilon(x)] = 1$, and $\sigma(x; \boldsymbol{\eta})$ is included to allow the variance of the errors to change over time. The error process $\varepsilon(x)$ is either delta-correlated (i.e. $\text{cov}[\varepsilon(x_i), \varepsilon(x_j)] = 0$ for $x_i \neq x_j$) or has a stationary autocorrelation function $\text{corr}[\varepsilon(u), \varepsilon(v)] = \rho(|u - v|)$. Alternatively, one could use

$$\frac{dy}{dx} = h(x; \boldsymbol{\theta}) + \sigma(x; \boldsymbol{\eta})\varepsilon(x). \tag{7.47}$$

For ease of notation we use (7.47). The relative-growth-rate model (7.46), which is probably more useful in practice, fits into the framework of (7.47) if we log the sizes (i.e. use $\log y$ instead of y) and take $h(x; \boldsymbol{\theta})$ as the theoretical *relative* growth rate written as a function of time x.

From Appendix C1 the solution to (7.47) is a Gaussian process

$$y(x) = y(x_0) + \int_{x_0}^{x} h(u; \boldsymbol{\theta})\, du + \int_{x_0}^{x} \sigma(u, \boldsymbol{\eta})\varepsilon(u)\, du \tag{7.48}$$

for which

$$E[y(x) | y_0] = y_0 + \int_{x_0}^{x} h(u; \boldsymbol{\theta})\, du, \tag{7.49}$$

where $y_0 = y(x_0)$. (The differences in the mean of the solution process under different stochastic integral representations discussed by Sandland and McGilchrist [1979] are resolved in Appendix C.) We discuss other aspects of the solution process under two headings: uncorrelated and autocorrelated errors.

a Uncorrelated Error Process

If $\varepsilon(x)$ is "white noise" (technically, the "derivative" of a Wiener process), then $y(x)$ has independent increments. In particular, setting $z_i = y_{i+1} - y_i$ [where $y_i = y(x_i)$] for $i = 1, 2, \ldots, n-1$, the z_i's are mutually independent. From (7.48) with x_{i+1} and x_i instead of x and x_0, and $E[\varepsilon(x)] = 0$, we have

$$E[z_i] = \int_{x_i}^{x_{i+1}} h(u; \boldsymbol{\theta})\, du. \tag{7.50}$$

Since $\operatorname{var}[\varepsilon(x)] = 1$, we also have

$$\begin{aligned} \operatorname{var}[z_i] &= \int_{x_i}^{x_{i+1}} \sigma^2(u; \boldsymbol{\eta})\, du \\ &= (x_{i+1} - x_i)\sigma^2 \qquad \text{if} \quad \sigma(u; \boldsymbol{\eta}) \equiv \sigma. \end{aligned} \tag{7.51}$$

Hence, from (7.48),

$$z_i = \int_{x_i}^{x_{i+1}} h(u; \boldsymbol{\theta})\, du + \varepsilon_i \qquad (i = 1, 2, \ldots, n-1), \tag{7.52}$$

where the ε_i are independently distributed [since $\varepsilon(x)$ has independent increments], and (7.52) can be fitted by weighted least squares using weights from (7.51).

It is also useful to have the variance–covariance matrix of the original responses y_i. Conditional on $y_0 = y(x_0)$, the size at some chosen time origin, we have from Appendix C1

$$\operatorname{var}[y_i | y_0] = \operatorname{var}[y_i - y_0 | y_0] = \int_{x_0}^{x_i} \sigma^2(u; \boldsymbol{\eta})\, du = \sigma_i^2, \tag{7.53}$$

say. Since

$$y_r - y_0 = z_0 + z_1 + \cdots + z_{r-1},$$

it follows that for $x_i \leq x_j$,

$$\begin{aligned} \operatorname{cov}[y_i, y_j | y_0] &= \operatorname{cov}[y_i - y_0, y_j - y_0 | y_0] \\ &= \operatorname{var}[y_i - y_0 | y_0] \\ &= \operatorname{var}[y_i | y_0]. \end{aligned} \tag{7.54}$$

The error structure considered in (7.47) models variation in the size of the growth increment due to stochastic variations in the environment. Garcia [1983] considered an additional source of error, namely errors in the measurement of y_i. He assumes we measure y_i^* given by

$$y_i^* = y_i + \tau_i,$$

where $\tau_1, \tau_2, \ldots, \tau_n$ are i.i.d. $N(0, \sigma_\tau^2)$ independently of the y_i. Thus $E[y_i^*] = E[y_i]$ and, for $\mathbf{y}^* = (y_1^*, y_2^*, \ldots, y_n^*)'$,

$$\mathscr{D}[\mathbf{y}^*] = \mathscr{D}[\mathbf{y}] + \sigma_\tau^2 \mathbf{I}_n.$$

The model can then be fitted using maximum likelihood or generalized least squares.

b Autocorrelated Error Processes

When $\varepsilon(x)$ has autocorrelation function $\text{corr}[\varepsilon(u), \varepsilon(v)] = \rho(|u-v|)$, we have from Appendix C1 [Equations (9), (11) and (12) with $x_i \leq x_j$],

$$\text{var}[y_i - y_j] = \int_{x_i}^{x_j} \int_{x_i}^{x_j} \sigma(u; \boldsymbol{\eta}) \sigma(v; \boldsymbol{\eta}) \rho(|u-v|)\, du\, dv. \tag{7.55}$$

Also, for any two nonoverlapping time intervals $[x_1, x_2)$ and $[x_3, x_4)$,

$$\text{cov}[y_2 - y_1, y_4 - y_3] = \int_{x_1}^{x_2} \int_{x_3}^{x_4} \sigma(u; \boldsymbol{\eta}) \sigma(v; \boldsymbol{\eta}) \rho(|u-v|)\, du\, dv. \tag{7.56}$$

Consider the special case in which $\sigma(u; \boldsymbol{\eta}) \equiv \sigma$ and $\rho(|u-v|) = e^{-\lambda|u-v|}$ (or $\rho^{|u-v|}$ for $0 < \rho < 1$). This is the continuous analogue of a positively autocorrelated AR(1) process (see Section 6.2). Evaluating (7.56) when $x_1 < x_2 < x_3 < x_4$, we have

$$\begin{aligned} \text{cov}[y_2 - y_1, y_4 - y_3] &= \int_{x_1}^{x_2} \left(\int_{x_3}^{x_4} \sigma^2 e^{-\lambda(u-v)}\, du \right) dv \\ &= \frac{\sigma^2}{\lambda^2} (e^{\lambda x_2} - e^{\lambda x_1})(e^{-\lambda x_3} - e^{-\lambda x_4}). \end{aligned} \tag{7.57}$$

From (7.55) and symmetry considerations,

$$\begin{aligned} \text{var}[y_2 - y_1] &= 2 \int_{x_1}^{x_2} \int_{x_1}^{v} \sigma^2 e^{-\lambda(v-u)}\, du\, dv \\ &= \frac{2\sigma^2}{\lambda^2} [\lambda(x_2 - x_1) + e^{\lambda(x_1 - x_2)} - 1] \\ &= \sigma_*^2, \end{aligned} \tag{7.58}$$

say. In the case of time intervals of equal length δ and distance $r\delta$ apart, i.e. $x_2 - x_1 = x_4 - x_3 = \delta$ and $x_3 - x_1 = r\delta$, then from (7.57) and (7.58) we have

$$\operatorname{cov}[y_2 - y_1, y_4 - y_3] = \sigma^2(e^{\lambda\delta} - 1)^2 e^{-(r+1)\lambda\delta}/\lambda^2$$

and

$$\operatorname{var}[y_2 - y_1] = \operatorname{var}[y_4 - y_3] = 2\sigma^2(\lambda\delta + e^{-\lambda\delta} - 1)/\lambda^2.$$

We now look at the correlation between $y_2 - y_1$ and $y_4 - y_3$ as δ becomes small. To maintain the spacing between x_1 and x_3 as the interval lengths shrink, we let $\delta \to 0$ and $r \to \infty$ in such a way that $\delta r = c$ for some positive constant c. Either using L'Hospital's rule several times or expanding the exponentials we find that, as $\delta \to 0$,

$$\operatorname{corr}(y_2 - y_1, y_4 - y_3) \to e^{-\lambda c} = e^{-\lambda(x_3 - x_1)}. \tag{7.59}$$

Suppose $x_{i+1} - x_i = \delta$ $(i = 1, 2, \ldots, n-1)$, and let $z_i = y_{i+1} - y_i$ as before. Then provided δ is not too large, we would expect from (7.59) that $\operatorname{corr}[z_i, z_j] \approx e^{-\lambda\delta|i-j|}$, which is of the form $\rho^{|i-j|}$. Hence we would expect the error structure of the z_i to be approximately AR(1).

Sandland and Gilchrist [1979] used the above analysis to motivate their methods for fitting stochastic growth models to longitudinal data collected at equally spaced time intervals. They worked with logged data, so that [c.f. (7.45)]

$$R_i = \frac{z_i}{x_{i+1} - x_i} = \frac{\log y_{i+1} - \log y_i}{x_{i+1} - x_i} \tag{7.60}$$

is the average relative growth rate, and they empirically fitted models of the form

$$R_i = \frac{1}{x_{i+1} - x_i} \int_{x_i}^{x_{i+1}} h(u; \boldsymbol{\theta})\, du + \varepsilon_i \tag{7.61}$$

where the ε_i's are possibly autocorrelated [e.g. ARMA(q_1, q_2)] errors. The trend relationship for R_i may be transformed to remove heteroscedasticity before fitting a model with additive autocorrelated errors. As

$$\int_{x_i}^{x_{i+1}} h(u; \boldsymbol{\theta})\, du \approx h(x_i; \boldsymbol{\theta})(x_{i+1} - x_i) \quad \text{implies} \quad R_i \approx h(x_i; \boldsymbol{\theta}),$$

they advocated plotting R_i versus x_i to help determine the form of $h(x; \boldsymbol{\theta})$ to use. They constructed models in which $h(x; \boldsymbol{\theta})$ is either polynomial in x, or consists of polynomial segments if the plots reveal a "change of phase" in the growth rate (see Chapter 9). The following example describes one of their analyses.

Table 7.1 Growth of the Son of the Count de Montbeillard[a]

Age (yr, mth [day])	Height (cm)	Age (yr, mth [day])	Height (cm)
0	51.4	9, 0	137.0
0, 6	65.0	9, 7[12]	140.1
1, 0	73.1	10, 0	141.9
1, 6	81.2	11, 6	141.6
2, 0	90.0	12, 0	149.9
2, 6	92.8	12, 8	154.1
3, 0	98.8	13, 0	155.3
3, 6	100.4	13, 6	158.6
4, 0	105.2	14,0	162.9
4, 7	109.5	14, 6[10]	169.2
5, 0	111.7	15, 0[2]	175.0
5, 7	115.5	15, 6[8]	177.5
6, 0	117.8	16, 3[8]	181.4
6, 6[19]	122.9	16, 6[6]	183.3
7, 0	124.3	17, 0[2]	184.6
7, 3	127.0	17, 1[9]	185.4
7, 6	128.9	17, 5[5]	186.5
8, 0	130.8	17, 7[4]	186.8
8, 6	134.3		

[a]Originally from Scammon [1927]. Reproduced from R. L. Sandland and C. A. McGilchrist, "Stochastic Growth Curve Analysis". Biometrics 35: 255–271. With permission from The Biometric Society.

Example 7.1 The data in Table 7.1 are a subset of the famous set of measurements taken on the height of the son of the Count de Montbeillard between 1759 and 1777. Only the first ten years of data were used in this analysis, and the observation at age 7 years 3 months was omitted to preserve approximately equal spacing of the x's. A plot of R_i versus age is given in Fig. 7.5. Initially Sandland and McGilchrist [1979] fitted cubic and quartic polynomials to the average relative growth rates R_i, but these gave unsatisfactory patterns of residuals even when an AR(1) error process was used. Inspection of Fig. 7.5 (and a cusum plot) suggested a log transform for the R_i's [c.f. (7.60)]. The reasons for this included the apparent heteroscedasticity in the R_i-versus-age plot, and the initial steep falloff in this plot during the first year of life, which they thought might be better fitted by an exponential function. With the log transformation and AR(1) errors, they stated that "The model fitted reasonably well from quadratic upwards". There was a fairly substantial gain in moving from a quadratic to a cubic model. The usual chi-square statistic ($= -2 \log$ likelihood) for model fitting dropped from 17.37 to 7.07, and the residual pattern looked better. We note that Sandland and

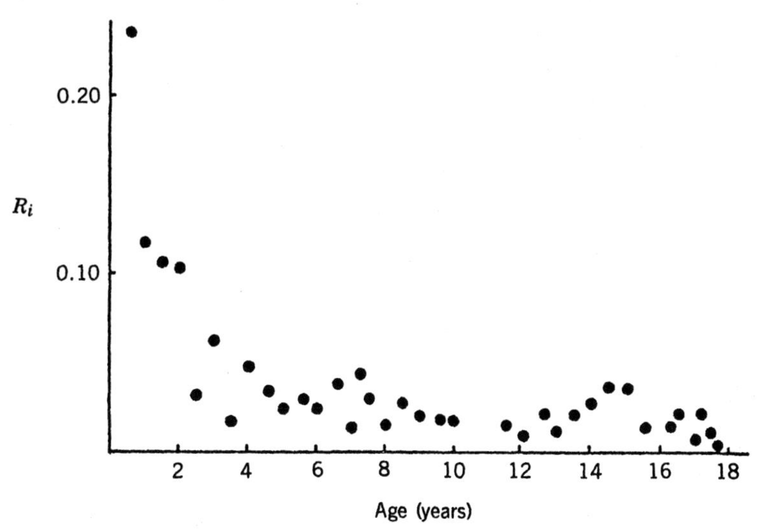

Figure 7.5 Plot of R_i versus x_i for data in Table 7.1. Reproduced from R. L. Sandland and C. A. McGilchrist, "Stochastic Growth Curve Analysis". Biometrics 35: 255–271, 1979. With permission from The Biometric Society.

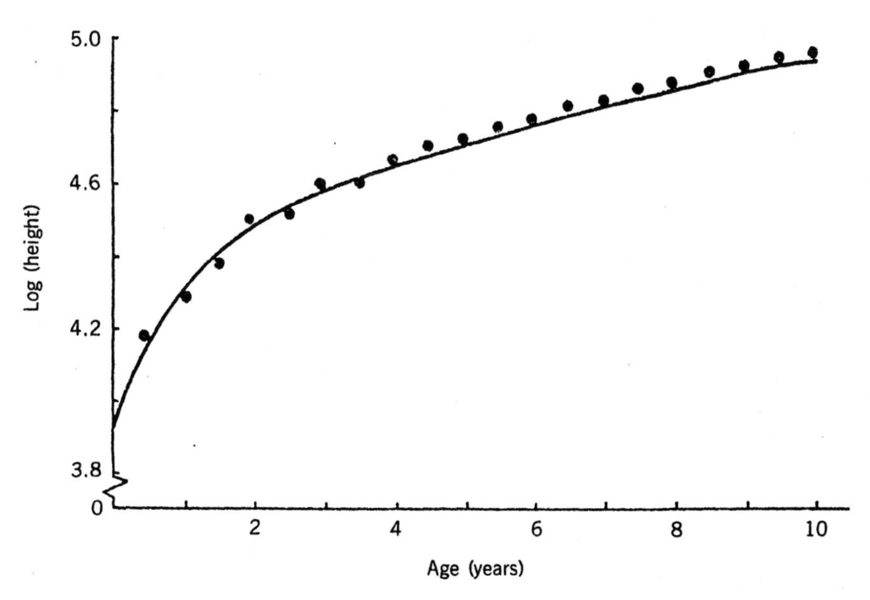

Figure 7.6 Log height data from Table 7.1 with fitted growth curve superimposed. Reproduced from R. L. Sandland and C. A. McGilchrist, "Stochastic Growth Curve Analysis". Biometrics 35: 255–271, 1979. With permission from The Biometric Society.

McGilchrist [1979] do not use the residuals that we have defined in Chapter 6. Instead they use "recursive residuals" that are a byproduct of their method of fitting. Finally, Fig. 7.6 is their plot "of expected values from the fitted model correcting for the bias induced by the transformation." ■

There is a difficulty in obtaining predicted or expected values for y in rate models that Sandland and McGilchrist do not mention. If we have a model for dy/dx, where y may be log(size), and wish to integrate this to find y, an unknown integration constant is introduced. More technically, from (7.49),

$$E[y(x)|y_0] = y_0 + \int_{x_0}^{x} h(u;\boldsymbol{\theta})\,du. \tag{7.62}$$

By adjusting y_0, the expected-value curve can be made to pass through any one of the data points. In the paper referenced in Section 7.5.1, Levenbach and Reuter [1976] used rate models to forecast time series. They described three possible approaches. One, which they call a last-point fit, sets y_0 $[=y(x_0)]$ so that the curve passes through the last data point (x_n, y_n). Another is to use a one-parameter least-squares fit to the data to determine the integration constant or, equivalently, y_0. A third approach is to use a least-squares fit to find an integration constant for each year of data and then use some sort of average of these constants across years. They stated that in their time-series forecasting work, this third approach has been found to work quite well. These ad hoc least-squares methods are less attractive when we recall that the y-sequence is autocorrelated. The "last-point fit" method is well justified in the case of an uncorrelated error process for the rates, because the forecast is then based upon the conditional distribution of the future y given $\mathbf{y}=(y_1, y_2, \ldots, y_n)'$, as it is easily shown that the conditional distribution of y given $\mathbf{y}$ is normal with mean

$$y_n + \int_{x_n}^{x} h(u;\boldsymbol{\theta})\,du \tag{7.63}$$

and variance

$$\int_{x_n}^{x} \sigma^2(u;\boldsymbol{\eta})\,du. \tag{7.64}$$

Yet another method for obtaining a forecast would be to take the unconditional expectation of y at x, i.e. to take the expectation of (7.49) with respect to y_0. However, we shall see that the distribution of y_0 typically contains parameters which are not contained in the rate model.

There is nothing to prevent the fitting of the usual nonlinear growth models discussed previously using the method of Sandland and McGilchrist [1979]. For notational simplicity we will express our models in terms of $z_i = y_{i+1} - y_i$ rather than converting to R_i. Thus we fit

$$z_i = \int_{x_i}^{x_{i+1}} h(u; \boldsymbol{\theta})\, du + \varepsilon_i \tag{7.65}$$

for $i = 1, 2, \ldots, n-1$. If y is size and the growth curve has equation $f(x; \boldsymbol{\theta})$, then $h(x; \boldsymbol{\theta}) = df(x; \boldsymbol{\theta})/dx$ and (7.65) becomes

$$z_i = f(x_{i+1}, \boldsymbol{\theta}) - f(x_i; \boldsymbol{\theta}) + \varepsilon_i. \tag{7.66}$$

However, we would usually be more interested in relative growth rates so that y is log(size). Thus we require $h(x; \boldsymbol{\theta}) = \dot{f}(x; \boldsymbol{\theta})/f(x; \boldsymbol{\theta}) = d \log f(x; \boldsymbol{\theta})/dx$. We illustrate the approach using the Richards model (7.18). For this model

$$f(x; \boldsymbol{\theta}) = \alpha[1 + (\delta - 1)e^{-\kappa(x-\gamma)}]^{1/(1-\delta)}$$

and

$$\int_{x_i}^{x_{i+1}} h(u; \boldsymbol{\theta})\, du = \log f(x_{i+1}; \boldsymbol{\theta}) - \log f(x_i; \boldsymbol{\theta})$$

$$= \frac{1}{1-\delta}\left[\log \frac{1 + (\delta - 1)e^{-\kappa(x_{i+1}-\gamma)}}{1 + (\delta - 1)e^{-\kappa(x_i-\gamma)}}\right], \tag{7.67}$$

which is independent of α. Thus there is no information in $z_1, z_2, \ldots, z_{n-1}$ about the limiting size α, which is often a parameter of primary importance. If $f(x_0; \boldsymbol{\theta})$ is known for some x_0, we could express α in terms of the other parameter estimates. Otherwise all our information is contained in $y_1, z_1, z_2, \ldots, z_{n-1}$. Taking (x_1, y_1) as our origin (x_0, y_0) and adopting the independent-error model, one approach would be to set

$$z_0 = y_1 = f(x_1; \boldsymbol{\theta}) + \varepsilon_0 \tag{7.68}$$

and use $z_0, z_1, \ldots, z_{n-1}$. However, for estimation purposes, care needs to be taken in considering how ε_0 relates to $\varepsilon_1, \varepsilon_2, \ldots, \varepsilon_{n-1}$. For example, an assumption like $\text{var}[\varepsilon_0] = \text{var}[\varepsilon_i]$ for $i \geq 1$, which is implicit in a least-squares analysis, may be unreasonable.

7.5.3 Rates as Functions of Size

In Section 7.5.1 we discussed the desirability of being able to work with models of the form

$$\frac{1}{y}\frac{dy}{dx} = h(y) + \varepsilon(x), \tag{7.69}$$

where $\varepsilon(x)$ is some error stochastic process. In fact, models of the form

$$\frac{dy}{dx} = h(y, \varepsilon) \tag{7.70}$$

are hardly ever tractable analytically. We describe two methods that have been used to circumvent this difficulty.

a A Tractable Differential Equation

One of the few stochastic differential equations that are tractable is the linear equation (Appendix C). Garcia [1983] exploited this to construct a tractable stochastic differential equation for fitting the Richards model. Rearranging (7.18),

$$f(x)^{1-\delta} = \alpha^{1-\delta}[1 + (\delta - 1)e^{-\kappa(x-\gamma)}] \tag{7.71}$$

and

$$\begin{aligned} \frac{df^{1-\delta}}{dx} &= -\alpha^{1-\delta}\kappa(\delta - 1)e^{-\kappa(x-\gamma)} \\ &= \kappa(\alpha^{1-\delta} - f^{1-\delta}). \end{aligned} \tag{7.72}$$

This defines a linear differential equation for $f^{1-\delta}$. Garcia [1983] used

$$\frac{dy^{1-\delta}}{dx} = \kappa(\alpha^{1-\delta} - y^{1-\delta}) + \sigma(x)\varepsilon(x) \tag{7.73}$$

subject to $y(x_0) = y_0$, where $\varepsilon(x)$ is "white noise". Writing $v = y^{1-\delta}$, (7.73) becomes

$$\frac{dv}{dx} = \kappa(\alpha^{1-\delta} - v) + \sigma(x)\varepsilon(x),$$

which can be solved using Appendix C3 with $\beta(x) \equiv \beta(= -\kappa)$ and $\xi(x) \equiv \xi(= \kappa\alpha^{1-\delta})$. If $y_0 = y(x_0)$, it follows from Equations (22) and (23) of Appendix C3 that

$$E[v(x)|v_0] = \alpha^{1-\delta} - (\alpha^{1-\delta} - v_0)e^{-\kappa(x-x_0)} \tag{7.74}$$

and

$$\text{cov}[v(x_i), v(x_j)|v_0] = \int_{x_0}^{\min(x_i,x_j)} e^{-\kappa(x_i+x_j-2u)}\sigma^2(u)\,du, \tag{7.75}$$

where $v_0 = y_0^{1-\delta}$. Garcia further allowed for measurement error in determining y and assumed that the error acted additively on $y^{1-\delta}$. Let $\tilde{y}$ be the measured value of y. Then the final model with observations $(x_i, \tilde{y}_i)$, $i = 1, 2, \ldots, n$, with $x_1 <$

$x_2 < \cdots < x_n$ is

$$\tilde{y}_i^{1-\delta} = \alpha^{1-\delta} - (\alpha^{1-\delta} - y_0^{1-\delta})e^{-\kappa(x_i - x_0)} + \xi_i + \tau_i, \tag{7.76}$$

where $\boldsymbol{\xi} = (\xi_1, \xi_2, \ldots, \xi_n)'$ is multivariate normal with mean vector $\mathbf{0}$ and dispersion matrix $\mathscr{D}[\boldsymbol{\xi}]$ with (i,j)th element given by (7.75). The measurement errors $\tau_1, \tau_2, \ldots, \tau_n$ are assumed to be i.i.d. $N(0, \sigma_\tau^2)$. If δ were known and the terms τ_i were ignored, the model parameters could be estimated by generalized least squares using $\mathscr{D}[\boldsymbol{\xi}]$. However, usually both δ and σ_τ^2 have to be estimated. Garcia [1983] used maximum-likelihood estimation. To find the joint density function of $\tilde{y}_1, \tilde{y}_2, \ldots, \tilde{y}_n$, Garcia found it convenient to transform to $z_1, z_2, \ldots, z_n$, where

$$z_i = \alpha^{1-\delta} - \tilde{y}_i^{1-\delta} - e^{-\kappa(x_i - x_{i-1})}(\alpha^{1-\delta} - \tilde{y}_{i-1}^{1-\delta}) \tag{7.77}$$

and $\tilde{y}_0 = y_0$. He was able to show that $\mathbf{z} = (z_1, z_2, \ldots, z_n)'$ is multivariate normal with $E[z_i] = 0$ and

$$\operatorname{cov}[z_i, z_j] = \begin{cases} \displaystyle\int_{x_0}^{x_1} e^{-2\kappa(x_1 - u)}\sigma^2(u)\,du + \sigma_\tau^2, & i = j = 1, \\ \displaystyle\int_{x_{i-1}}^{x_i} e^{-2\kappa(x_i - u)}\sigma^2(u)\,du & \\ \quad + (1 + e^{-2\kappa(x_i - x_{i-1})})\sigma_\tau^2, & i = j \neq 1, \\ -e^{-\kappa|x_i - x_j|}\sigma_\tau^2, & |i - j| = 1, \\ 0 & \text{otherwise.} \end{cases} \tag{7.78}$$

The likelihood function is thus the joint density

$$f(\tilde{y}_1, \tilde{y}_2, \ldots, \tilde{y}_n) = (2\pi)^{-n/2}|\mathbf{C}|^{-1/2}\exp(-\tfrac{1}{2}\mathbf{z}'\mathbf{C}^{-1}\mathbf{z})|J|, \tag{7.79}$$

where the variance–covariance matrix $\mathbf{C}$ has elements $c_{ij} = \operatorname{cov}[z_i, z_j]$ given by (7.78). Also

$$\begin{aligned} |J| &= \left|\det\left[\left(\frac{\partial z_i}{\partial \tilde{y}_j}\right)\right]\right| \\ &= |1 - \delta|^n \left(\prod_{i=1}^n \tilde{y}_i\right)^{-\delta}, \end{aligned} \tag{7.80}$$

where J is the Jacobian of the transformation. A very brief description of a computer program which performs the maximum-likelihood estimation is given by Garcia [1983], and details are available from him. The program uses a modified Newton algorithm to maximize the log-likelihood, but exploits the special structure of (7.79) to make the computation efficient. We note that in

Garcia's model, where the rate is a function of size rather than time, we are now able to estimate α [c.f. (7.67) and the discussion].

Garcia's method shares a difficulty with the use of Box–Cox transformations in the linear model (see Section 2.8.4). One transformation has to fulfill several purposes simultaneously. Here δ is the parameter value that characterizes the point of inflection of the growth curve as a proportion of final size. The same value must linearize the stochastic differential equation (7.73) and define the transformation scale on which measurement errors are additive [see (7.76)]. Garcia [1983] reported difficulties with the separate estimation of the components of variance from measurement error and the stochastically varying environment. However, in his experience, the estimates of the parameters of interest seemed to be fairly insensitive to changes in variance. The reader is referred to his paper for a further discussion of these issues.

b Approximating the Error Process

Rather than search for a tractable form of stochastic differential equation, White and Brisbin [1980] made a discrete approximation of the error process. We now describe their methods. They considered the additive and proportional error models

$$\frac{dy}{dx} = h(y;\boldsymbol{\theta}) + \varepsilon(x) \qquad \text{(additive error)} \tag{7.81}$$

and

$$\frac{dy}{dx} = h(y;\boldsymbol{\theta}) + y\varepsilon(x) \qquad \text{(proportional error)} \tag{7.82}$$

respectively. For size data y_i collected at time x_i $(i = 0, 1, \ldots, n)$, White and Brisbin approximate a Gaussian error structure by assuming that, for $i = 1, 2, \ldots, n$,

$$\varepsilon(x) = \varepsilon_i, \qquad x_{i-1} \le x < x_i, \tag{7.83}$$

where the ε_i are uncorrelated with mean zero and variance σ^2. They use nonlinear least-squares estimation, so that $\hat{\boldsymbol{\theta}}$ is found by minimizing $\sum_i \varepsilon_i^2$. The calculation of ε_i for given $\boldsymbol{\theta}$ involves finding ε_i such that the respective solutions of

$$\frac{dy}{dx} = h(y) + \varepsilon_i \tag{7.84}$$

and

$$\frac{dy}{dx} = h(y) + y\varepsilon_i \tag{7.85}$$

subject to $y(x_{i-1}) = y_{i-1}$ satisfy $y(x_i) = y_i$. In general, the differential equations must be solved numerically. A zero-finding algorithm is required to solve the following for ε_i:

$$y_i = \int_{x_{i-1}}^{x_i} [h(y) + \varepsilon_i]\, dx$$

and

$$y_i = \int_{x_{i-1}}^{x_i} [h(y) + y\varepsilon_i]\, dx.$$

In the special case of a logistic growth model, White and Brisbin [1980] give an analytic solution to (7.84). They also give an analytic solution in the more useful case of the Richards model under proportional errors (7.85). Before deriving their solution, we note that they call the combination of the two models above, together with the above method of nonlinear least-squares estimation, ADE (for additive discrete error) and PDE (proportional discrete error) respectively.

We can write the Richards model (7.17) as

$$\begin{aligned} \frac{dy}{dx} &= \frac{\kappa}{1-\delta} y \left[\left(\frac{y}{\alpha} \right)^{\delta - 1} - 1 \right] \qquad (\delta \neq 1) \\ &= \frac{\kappa}{1-\delta} \alpha^{1-\delta} y^{\delta} - \frac{\kappa}{1-\delta} y. \end{aligned} \tag{7.86}$$

When constrained to pass through (x_{i-1}, y_{i-1}), this has solution, from (7.21),

$$y = \{\alpha^{1-\delta} - (\alpha^{1-\delta} - y_{i-1}^{1-\delta}) e^{-\kappa(x - x_{i-1})}\}^{1/(1-\delta)}. \tag{7.87}$$

Let us write (7.86) in the form

$$\frac{dy}{dx} = a y^{\delta} - b y, \tag{7.88}$$

so that

$$\frac{a}{b} = \alpha^{1-\delta}, \qquad \kappa = (1-\delta) b. \tag{7.89}$$

The solution to (7.88) that passes through (x_{i-1}, y_{i-1}) is thus

$$y = \left[\frac{a}{b} - \left(\frac{a}{b} - y_{i-1}^{1-\delta} \right) e^{-(1-\delta) b (x - x_{i-1})} \right]^{1/(1-\delta)} \tag{7.90}$$

Now under the PDE model, (7.86) becomes

$$\begin{aligned} \frac{dy}{dx} &= \frac{\kappa}{1-\delta} \alpha^{1-\delta} y^{\delta} - \frac{\kappa}{1-\delta} y + \varepsilon y \\ &= a y^{\delta} - (b - \varepsilon) y. \end{aligned} \tag{7.91}$$

Thus comparing (7.91) with (7.88), we see that (7.90) becomes

$$y(x) = \left[\frac{a}{b-\varepsilon} - \left(\frac{a}{b-\varepsilon} - y_{i-1}^{1-\delta}\right)e^{-(1-\delta)(b-\varepsilon)(x-x_{i-1})}\right]^{1/(1-\delta)}. \tag{7.92}$$

For fixed $\boldsymbol{\theta} = (\alpha, \kappa, \delta)'$, we now solve

$$y_i = \left[\frac{a}{b-\varepsilon_i} - \left(\frac{a}{b-\varepsilon_i} - y_{i-1}^{1-\delta}\right)e^{-(1-\delta)(b-\varepsilon_i)(x_i-x_{i-1})}\right]^{1/(1-\delta)} \tag{7.93}$$

for ε_i and choose $\hat{\boldsymbol{\theta}}$ to minimize $\sum_i \varepsilon_i^2$ using a least-squares algorithm that does not require derivatives.

The ADE and PDE methods are clearly computationally intensive, particularly when there is no analytic solution for the differential equations. Instead of solving the differential equations, White and Brisbin [1980], for comparative purposes, use the empirical estimates of the derivatives

$$\frac{dy(x_i)}{dx} \approx \frac{y_{i+1} - y_i}{x_{i+1} - x_i},$$

$$\frac{1}{y}\frac{dy(x_i)}{dx} = \frac{d \log y}{dx} \approx \frac{\log y_{i+1} - \log y_i}{x_{i+1} - x_i},$$

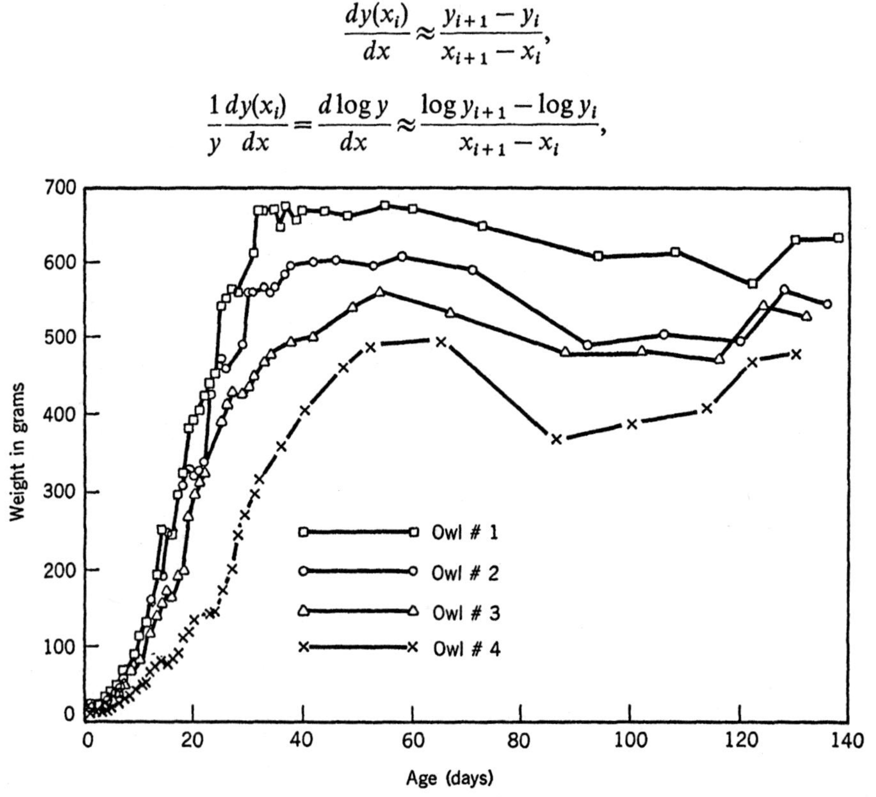

Figure 7.7 Growth data of four sibling barn owls. Owls are numbered sequentially according to the order of hatch. Although hatching occurred asynchronously, all birds are represented here synchronously on a common age axis for comparison purposes. From White and Brisbin [1980].

and fit the models

ADD:
$$\frac{y_{i+1} - y_i}{x_{i+1} - x_i} = h(y_i) + \varepsilon_i, \tag{7.94}$$

PDD:
$$\frac{\log y_{i+1} - \log y_i}{x_{i+1} - x_i} = \frac{h(y_i)}{y_i} + \varepsilon_i \tag{7.95}$$

Table 7.2 Parameter Estimates ($\pm$ Standard Deviations) and P-values for tests of Autocorrelation of Least-Squares Residuals for the Barn Owl Data in Fig. 7.7 using Five Different Models[a]

	Owl 1	Owl 2	Owl 3	Owl 4
		Sampling-Error Model		
α	652.6($\pm$7.6)	569.4($\pm$9.3)	528.2($\pm$6.3)	444.5($\pm$9.3)
b^c	−0.170($\pm$0.030)	−0.175($\pm$0.046)	−0.220($\pm$0.037)	−0.089($\pm$0.011)
δ	2.28($\pm$0.39)	2.16($\pm$0.50)	1.68($\pm$0.29)	3.86($\pm$1.30)
$\Pr^b$	0.0001	0.0001	0.010	0.031
		Additive-Discrete-Error (ADE) Model		
α	683.0($\pm$41.2)	613.8($\pm$53.0)	529.8($\pm$27.9)	480.1($\pm$30.1)
b	−0.354($\pm$0.535)	−0.617($\pm$1.96)	−0.224($\pm$0.177)	−0.144($\pm$0.076)
δ	1.39($\pm$0.73)	1.19($\pm$0.67)	1.66($\pm$0.73)	1.91($\pm$0.77)
Pr	0.35	0.39	0.32	0.20
		Additive-Discrete-Derivative (ADD) Model		
α	669.9($\pm$33.3)	597.6($\pm$44.3)	527.2($\pm$29.8)	468.2($\pm$24.9)
b	−0.391($\pm$0.536)	−0.946($\pm$4.22)	−0.338($\pm$0.416)	−0.151($\pm$0.073)
δ	1.38($\pm$0.63)	1.13($\pm$0.64)	1.39($\pm$0.60)	1.90($\pm$0.69)
Pr	0.37	0.40	0.36	0.07
		Proportional-Discrete-Error (PDE) Model		
α	702.4($\pm$104.4)	589.2($\pm$70.6)	518.4($\pm$64.4)	514.5($\pm$175.2)
b	−0.395($\pm$0.302)	−0.255($\pm$0.085)	−0.174($\pm$0.045)	−0.219($\pm$0.179)
δ	1.30($\pm$0.34)	1.63($\pm$0.44)	2.14($\pm$0.86)	1.40($\pm$0.56)
Pr	0.97	0.51	0.35	0.02
		Proportional-Discrete-Derivative (PDD) Model		
α	655.0($\pm$61.3)	579.3($\pm$66.8)	516.9($\pm$81.3)	466.3($\pm$110.8)
b	−0.205($\pm$0.045)	−0.264($\pm$0.859)	−0.196($\pm$0.069)	−0.123($\pm$0.039)
δ	1.91($\pm$0.51)	1.59($\pm$0.39)	1.78($\pm$0.65)	2.12($\pm$1.13)
Pr	0.96	0.52	0.35	0.01

[a] From White and Brisbin [1980].
[b] P-value from test for autocorrelation, as determined by the adequacy-of-fit test (Pierce [1971b]).
[c] The parameter $b = \kappa/(1 - \delta)$.

$(i = 0, 1, \ldots, n-1)$ by least squares. The reader will note that (7.95) is simply (7.43) with R_i estimated using (7.45).

Example 7.2 White and Brisbin [1980] fitted Richards curves using all four models ADE, ADD, PDE, and PDD to the growth data for four sibling brown owls shown in Fig. 7.7. They also fitted the "deterministic" Richards curve [c.f. (7.18) and (7.38)] directly to $y_0, y_1, \ldots, y_n$ by ordinary least squares and called the deterministic model the "sampling-error model." Using the residuals from the least-squares fit of this model, they tested for autocorrelation and found significant autocorrelations for all four birds (Table 7.2). When using the models ADE, ADD, PDE, and PDD there were no significant autocorrelations in the residuals for owls 1, 2, and 3. However, for owl 4 the autocorrelations were significant for the PDE and PDD models, and close to being significant for the ADD model. Parameter estimates are given for all the models in Table 7.2. There are some fairly large differences in parameter estimates between the sampling-error model and ADE. However, PDE and PDD are generally in reasonable agreement, and ADE and ADD show particularly good agreement. In view of the computational expense of the ADE method, White and Brisbin recommended ADD for future work. However, further work is required, even for modeling the owl data, as the residual plots for ADE and ADD show considerable heterogeneity of variance. The PDE and PDD residual plots are not given, but White and Brisbin [1980] say that they display an even larger heterogeneity of variance. ■

7.6 YIELD–DENSITY CURVES

7.6.1 Preliminaries

Several models which are closely related to growth curves are found in agriculture, where they are used for quantifying the relationship between the density of crop planting and crop yield. Although some of them can be derived from (usually oversimplified) mechanistic considerations, these models are again best thought of simply as families of curves which have been found empirically to fit data well.

Because of competition between plants for resources, the yield per plant tends to decrease with increasing density of planting. However, the overall crop yield may increase. The agronomist tends to be interested in yield–density curves for prediction, for finding the density of planting giving rise to maximum yield, and for the comparison of relationships under different conditions. Let w denote the yield per unit area, and x the density of planting. Scatterplots of w versus x tend to conform qualitatively to one of the two types of relationship depicated in Fig. 7.8. A monotonically increasing curve which appears to be leveling off to an asymptote is termed *asymptotic*, while a curve that rises and then falls again is said to be *parabolic*. A particular type of crop tends to give rise consistently to one form of curve or the other, although different parts of the same plant may give rise to different forms of curve. Willey and Heath [1969] presented an example showing an asymptotic curve for the yield of parsnips. However, when the parsnips were subjected to size grading, the resulting curves were parabolic. If observations are not taken at sufficiently large x-values it may be impossible to distinguish between the two forms. Because of factors like these, it is desirable to have families of curves capable of representing both shapes.

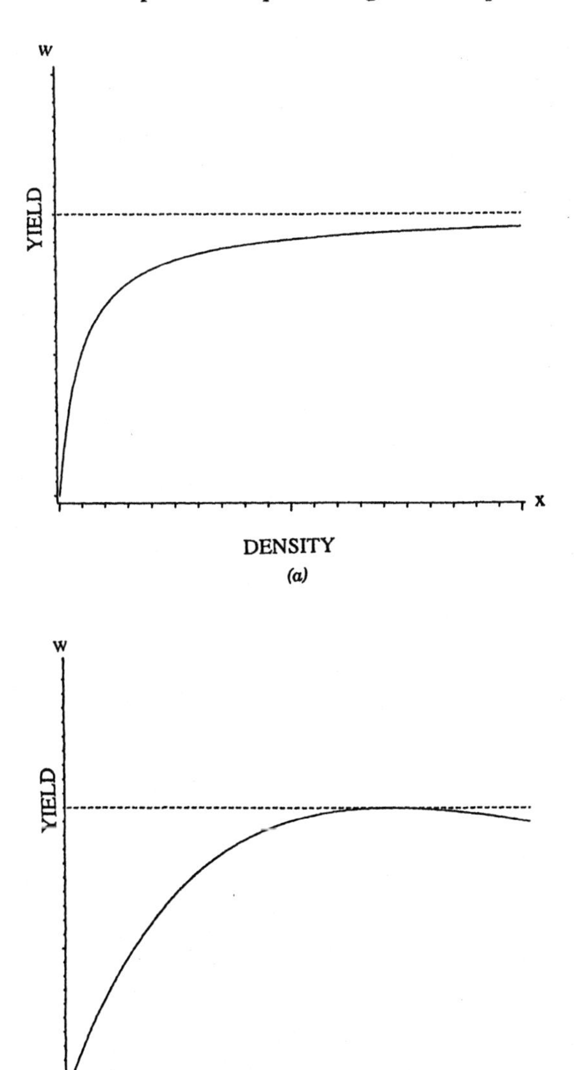

Figure 7.8 Yield–density curves: (*a*) asymptotic, (*b*) parabolic.

Let $y = w/x$, representing the average yield per plant if all plants survived. We will consider the following models for y:

Shinozaki and Kira [1956]:

$$f(x) = (\alpha + \beta x)^{-1}, \tag{7.96}$$

Holliday [1960a]:

$$f(x) = (\alpha + \beta x + \gamma x^2)^{-1}, \tag{7.97}$$

Bleasdale and Nelder [1960]:

$$f(x) = (\alpha + \beta x^{\phi})^{-1/\theta}, \tag{7.98}$$

Bleasdale's simplified equation:

$$f(x) = (\alpha + \beta x)^{-1/\theta}, \tag{7.99}$$

Farazdaghi and Harris [1968]:

$$f(x) = (\alpha + \beta x^{\phi})^{-1}. \tag{7.100}$$

The most commonly used asymptotic curve is (7.96), which is embedded in all the remaining families; for example, that of Holliday when $\gamma = 0$, and that of Bleasdale and Nelder when $\theta = \phi = 1$. The last two models are special cases of Bleasdale and Nelder ($\phi = 1$ and $\theta = 1$, respectively).

Before discussing the above models in more detail, we briefly describe one further stage in modeling this process. Crop yield is not only affected by the density of planting, or equivalently by the area available to each plant. The shape of that area also affects yield, a square tending to be more favorable than an elongated rectangle. A model was proposed by Berry [1967] which essentially extends Bleasdale's simplified equation to take shape into account, namely

$$f(d_1, d_2) = \left[\alpha + \beta\left(\frac{1}{d_1} + \frac{1}{d_2}\right) + \frac{\gamma}{d_1 d_2}\right]^{-1/\theta},$$

where d_1 is the spacing between plants within a row and d_2 is the distance between rows.

One further feature we should note before proceeding is the following. For all of our models, when x is small $f(x)$ decreases as x increases. However, in reality, if measurements are taken at densities low enough so that no competition between plants is felt, this decrease will not occur. Consequently the models may not fit well at low planting densities. For a more detailed discussion of planting densities and crop yields, the reader is referred to Willey and Heath [1969], Mead [1979], and Holliday [1960a, b].

7.6.2 Bleasdale–Nelder Model

The family of curves of Bleasdale and Nelder [1960] can be derived by assuming that with time t, plant growth follows a Richards growth curve (7.18), namely

$$y(t) = \alpha[1 + (\delta - 1)e^{-\kappa(t-\gamma)}]^{1/(1-\delta)}, \tag{7.101}$$

and that the (allometric) relationship between final size α and density x is of the form

$$\alpha x^{\rho} = c, \tag{7.102}$$

where c is a constant. Then, applying (7.21), at any time point t and density x,

$$\begin{aligned} y(t)^{1-\delta} &= y(0)^{1-\delta}e^{-\kappa t} + \alpha^{1-\delta}(1 - e^{-\kappa t}) \\ &= y(0)^{1-\delta}e^{-\kappa t} + c^{1-\delta}x^{(\delta-1)\rho}(1 - e^{-\kappa t}), \end{aligned}$$

which, for different plants measured at the same time t, is of the form

$$y^{-\theta} = \alpha + \beta x^{\phi}$$

with $\theta = \delta - 1$ and $\phi = (\delta - 1)\rho$.

Case (i): $\theta = \phi$. For this particular case the curve is asymptotic. The only curve of this form in common use is $\theta = \phi = 1$. When $\theta = \phi$, $\rho = 1$ in (7.102) in which case this equation is called "the law of constant final yield." Then for fixed θ, both α and β have physical interpretations. As $x \to 0$, $y \to \alpha^{-1/\theta}$, so that this quantity represents the average yield per plant in the given environment if no competition is present. As $x \to \infty$, $w = xy \to \beta^{-1/\theta} = w_{\max}$ and the latter represents the maximum yield per unit area. For an asymptotic relationship, the maximum yield cannot be attained, so that interest centers on the planting density x at which a proportion p of maximum yield is reached. This is given by

$$x^{\theta} = \frac{\alpha}{\beta}\frac{p^{\theta}}{1 - p^{\theta}}. \tag{7.103}$$

Case (ii): $\theta \neq \phi$. In this case the only values of interest are $\theta < \phi$, as these are the only values which produce the characteristic parabolic shape. As $x \to 0$, $y \to \alpha^{-1/\theta}$. However, the other parameters do not have a simple interpretation. The maximum value of w occurs when

$$x_{\max} = \left(\frac{\alpha\theta}{\beta(\phi - \theta)}\right)^{1/\phi} \tag{7.104}$$

and is equal to

$$w_{\max} = x_{\max} f(x_{\max}) = x_{\max}\left(\frac{\phi - \theta}{\alpha\phi}\right)^{1/\theta}. \tag{7.105}$$

The pivotal role of α/β in these equations suggests a parameterization

$$y^{-\theta} = \alpha(1 + \beta x^{\phi}). \tag{7.106}$$

For a more detailed description of this family of curves, see Mead [1970].

7.6.3 Holliday Model

Under the Shinozaki–Kiro model (7.96), a scatterplot of y^{-1} versus x should have a linear trend. Holliday [1960] added a quadratic term [c.f. (7.97)] to explain curvature in such a plot. The curve is asymptotic for $\gamma = 0$, where it reduces to (7.96). For $\gamma > 0$ the maximum yield w_{max} occurs at

$$x_{max} = (\alpha/\gamma)^{1/2},$$

where

$$w_{max} = \frac{x_{max}}{2\alpha + \beta x_{max}}.$$

Nelder [1966] has generalized the Holliday model to the class of inverse polynomials.

7.6.4 Choice of Model

With yield–density models there is no inherent difficulty in using the usual least-squares approach (after transforming to stabilize the variance: see Section 2.8) as there is with longitudinal growth data. Following Nelder [1963], there appears to be a consensus that the log transformation generally stabilizes the variance.

The Bleasdale–Nelder model (7.98) is seldom used in its complete form. Willey and Heath [1969] and Mead [1970] both state that the value of the ratio θ/ϕ is most important in determining the form of the yield–density curve. For a given value of θ/ϕ the individual parameters θ and ϕ tend to be badly determined; hence the common use of either Bleadale's simplified equation ($\phi = 1$) or the Farazdaghi–Harris model ($\theta = 1$). P. Gillis and D. A. Ratkowsky have, between them, carried out fairly extensive comparisons of yield–density models (see Gillis and Ratkowsky [1978], Ratkowsky [1983]) using simulations and fitting the models to large numbers of published data sets. Broadly they conclude that in terms of fit there is usually little difference between the models, but there are great differences in the intrinsic curvature. The Holliday model performed consistently well, but there were often intrinsic-curvature problems with Bleasdale's simplified equation. In the smaller study described by Ratkowsky [1983: Chapter 3], the Farazadaghi–Harris model (7.100) also had consistently greater intrinsic curvature than the Holliday model. It had pronounced parameter-effects curvature as well. However, no attempts to find a better parametrization were reported.

Mead [1970, 1979], in contrast, advocated the Bleasdale model for unspecified "biological advantages" and also because he found that in many practical situations in which one is interested in comparing curves, a single value of θ can be used for the whole set of curves. This simplifies comparisons. Generally, when θ is treated as fixed, nonlinearity problems tend to disappear.

7.6.5 Starting Values

Starting values for the Holliday model $E[y] = (\alpha + \beta x + \gamma x^2)^{-1}$ can be obtained by regressing y^{-1} on x and x^2. However, better estimates can be obtained by weighted least squares. Suppose the log transformation is the appropriate one to stabilize the variance, so that we have

$$\log y_i = \log f(x_i; \boldsymbol{\theta}) + \varepsilon_i, \tag{7.107}$$

where the ε_i are i.i.d with mean 0 and variance σ^2. To a first-order approximation $\text{var}[y] \approx \text{var}[\log y] \; \{E[y]\}^2 = \sigma^2 \{E[y]\}^2$. Thus for the Holliday model the weighted least-squares estimators minimize

$$\begin{aligned} \sum_i \frac{(y_i - E[y_i])^2}{\text{var}[y_i]} &\approx \frac{1}{\sigma^2} \sum_i \left(\frac{y_i}{E[y_i]} - 1 \right)^2 \qquad (7.108) \\ &= \frac{1}{\sigma^2} \sum_i (1 - \alpha y_i - \beta x_i y_i - \gamma x_i^2 y_i)^2. \end{aligned}$$

These estimates can be readily obtained by a multiple regression of a vector of 1's on y, xy, and $x^2 y$ without a constant term.

With the Bleasdale model, similar methods apply for fixed θ and ϕ. Here α and β can be simply estimated by regressing $y^{-\theta}$ on x^{ϕ}. However, it is virtually as easy to use weighted least squares again. If (7.98) is appropriate and $\text{var}[\log y] = \sigma^2$, then $\text{var}[\log y_i^{\theta}] = \theta^2 \, \text{var}[\log y_i] = \theta^2 \sigma^2$. Hence

$$E[y_i^{\theta}] \approx (\alpha + \beta x_i^{\phi})^{-1} \quad \text{and} \quad \text{var}[y_i^{\theta}] \approx \sigma^2 \theta^2 [E(y_i^{\theta})]^2.$$

From (7.108) the weighted least-squares criterion for y_i^{θ} now reduces to minimizing

$$\sum_{i=1}^{n} (1 - \alpha y_i^{\theta} - \beta x_i^{\phi} y_i^{\theta})^2,$$

which can be performed by regressing a vector of 1's on y^{θ} and $x^{\phi} y^{\theta}$ without a constant term. This method can be combined with a grid search on θ and ϕ if necessary.

CHAPTER 8

Compartmental Models

8.1 INTRODUCTION

The models of the previous chapter were, to a large extent, simply parametric families of curves which have been found to fit data from various subject areas very well. Any connections between the family of curves and underlying physical processes were often tenuous. In this chapter we discuss a class of models in which links with underlying physical processes can be much stronger.

Jacquez [1972] defined a compartmental system as "a system which is made up of a finite number of macroscopic subsystems, called compartments or pools, each of which is homogeneous and well mixed, and the compartments interact by exchanging materials. There may be inputs from the environment into one or more of the compartments, and there may be outputs (excretion) from one or more compartments into the environment." We shall depict compartments by boxes, and the exchange of material from one compartment to another by a directed arrow as in Fig. 8.1 (the constants labeling the arrows are defined in the

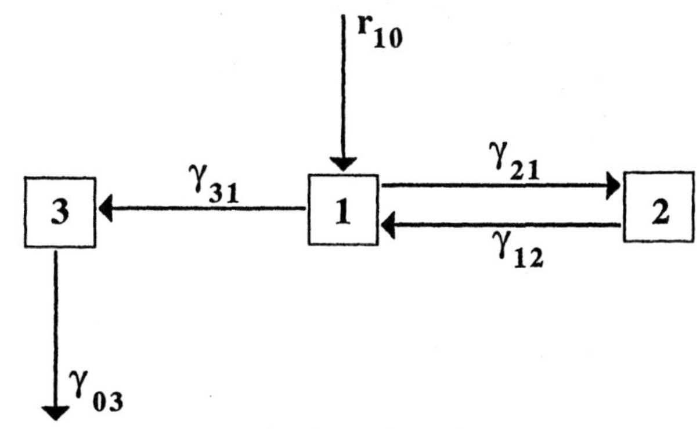

Figure 8.1 Three-compartment model, showing exchanges between compartments and to and from the environment.

next section). This diagram shows a system of three compartments in which there is a flow of material from the environment into compartment 1, an exchange between compartments 1 and 2, a flow of material from compartment 1 to compartment 3, and an excretion to the environment from compartment 3. A system with no interchange with the environment is said to be *closed*: otherwise the system is *open*. The compartments need not be physically separated. For example, in chemistry the compartments may refer to quantities of different substances within the same container, and the "exchange of material" to the way in which certain of the components are reacting to form others and vice versa. Many chemical reactions, e.g. $A \leftrightarrows B + C$, proceed in both directions but at different rates.

Our attention will be mainly limited to *linear* compartmental models, in which the rate of transfer of material from compartment j to any other compartment k is proportional to the current amount of material in compartment j. These models lead to linear systems of differential equations. However, many of the general considerations carry over to nonlinear compartmental modeling as well. We follow the historical development of the subject in that we begin with a deterministic treatment and introduce stochastic considerations later, leading finally to some stochastic process models.

The development and use of compartmental models began in the 1940s with the use of tracer experiments in research into the physiology of animals and humans. Measurements were made on the subsequent transport of the tracer through other parts of the body. The early tracers were dyes; with the introduction of radioactive tracers in the late 1940s, the subject began to expand more rapidly. Important early works include the books by Sheppard [1962] and Rescigno and Segre [1966] (Italian version, 1961). However, modern usage is much wider. Compartmental models are now an important tool in many branches of medicine, such as physiology (Jacquez [1972], Lambrecht and Rescigno [1983]) and pharmacokinetics (the study of how drugs circulate through the body over time—Wagner [1975]), and in many other branches of biology (Atkins [1969]), including the study of ecosystems in ecology (O'Neill [1979], Matis et al. [1979]). They are also useful in chemistry and biochemistry, in specialized areas of agriculture such as fertilizer response, in the study of population movements, and of the spread of epidemics. Kalbfleisch and Lawless [1985] use a (Markov process) compartmental model in which healthy individuals pass through a series of healthy and diseased states until they exit from the system via death. Matis and Hartley [1971] reference models in sociology and public health, and an anatomical model. Some idea of this diversity can be seen from the following examples.

Example 8.1 Batschelet et al. [1979] used the model depicted in Fig. 8.2 to describe the movement of lead in the human body. The input to the blood compartment is the intake of lead from the air, water, and diet. Excretion from the blood compartment was via urine, and excretion from the tissue compartment was via hair, nails, and perspiration. Measurements on the system were made

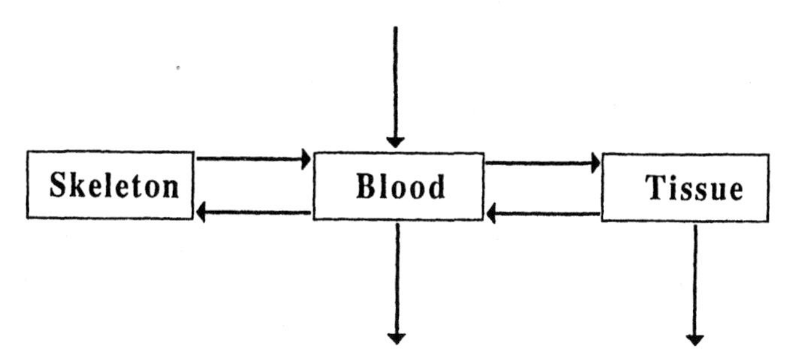

Figure 8.2 Movement of lead in the human body described by a three-compartment model.

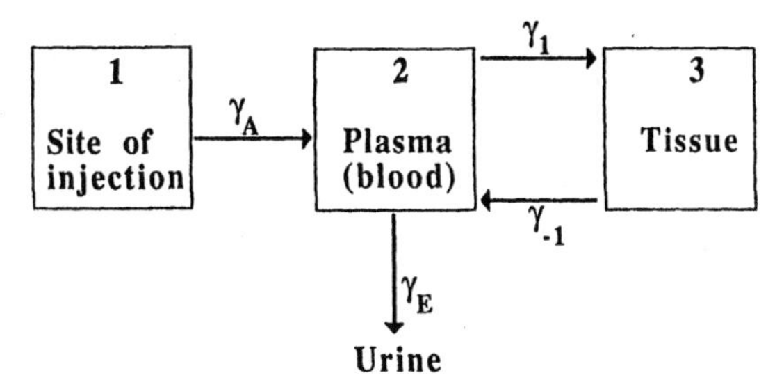

Figure 8.3 Progress of an injected drug.

possible by adding a small amount of a stable lead isotope, e.g. ^{204}Pb, to the diet as a tracer. ■

Example 8.2 Metzler [1971] discussed a model for the kinetics of drug movement when the drug is injected at an intramuscular site or given by oral dose at time $x = 0$, and then allowed to disperse through the body. Noting the comments of Haynes [1971], we use the representation of Allen [1983] for this model, which is given in Fig. 8.3. Measurements were taken on the plasma compartment over time via blood samples. ■

Example 8.3 Yet another variant of the three-compartment model is described by White [1982]. The model, depicted in Fig. 8.4, was developed in a study of how bees pick up radioactivity. The aim was to use honey bees to monitor radioactive contaminants in the environment. The bees were put in a sealed cage, and their water source was contaminated with radioactive tritium. ■

The previously referenced books contain many more examples. For instance, Anderson [1983] includes among many other models a nine-compartment model for the food chain in the Aleutian Islands.

Compartmental models typically involve "rate" constants like λ in the growth model $dN_x/dx = \lambda N_x$, where the growth rate is proportional to the population

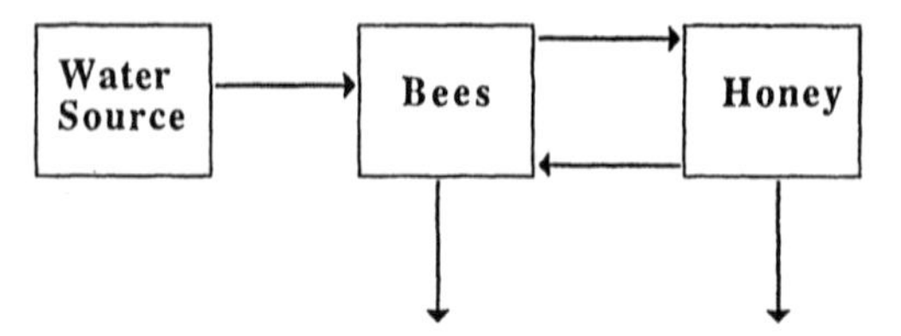

Figure 8.4 Transfer of radioactive tritium.

size N_x at time x. Thus

$$\lambda = \frac{dN_x/dx}{N_x} \tag{8.1}$$

or

$$dN_x = \lambda N_x \, dx. \tag{8.2}$$

This model has solution $N_x = N_0 e^{\lambda x}$, and compartmental models typically involve linear combinations of exponential terms.

8.2 DETERMINISTIC LINEAR MODELS

8.2.1 Linear and Nonlinear Models

Let $f_j(x)$ denote the amount of material in compartment j at time x, and let r_{j0} be the input rate to compartment j, that is, the rate at which material passes from the environment to compartment j. We define the fractional transfer coefficient γ_{jk} as the rate at which material is passing from compartment k into compartment j divided by the amount of material in k [c.f. (8.1)]: $j = 0$ corresponds to the environment. Note that the subscript convention for γ_{jk} is the opposite to that used in Markov chains. The parameters γ_{jk} and r_{10} are shown in Fig. 8.1. Heuristically, in the infinitesimal time interval $[x, x + dx)$ an amount of material $\gamma_{jk} f_k \, dx$ passes from compartment k into j. Also an amount $r_{j0} \, dx$ passes into compartment j from the environment. Thus in Fig. 8.1 the change in the quantity of material in compartment 1 is

$$df_1 = \gamma_{12} f_2 \, dx + r_{10} \, dx - \gamma_{31} f_1 \, dx - \gamma_{21} f_1 \, dx.$$

The rate of change is given by

$$\dot{f}_1 = \frac{df_1}{dx} = \gamma_{12} f_2 + r_{10} - (\gamma_{31} + \gamma_{21}) f_1.$$

Thus the whole system of Fig. 8.1 is described by the set of differential equations

$$\begin{aligned} \dot{f}_1 &= -(\gamma_{31}+\gamma_{21})f_1+\gamma_{12}f_2+r_{10}, \\ \dot{f}_2 &= \gamma_{21}f_1-\gamma_{12}f_2, \\ \dot{f}_3 &= \gamma_{31}f_1-\gamma_{03}f_3. \end{aligned} \tag{8.3}$$

In general the fractional rates γ_{jk} may depend on the time x, the state $\mathbf{f}$ of the system, and some unknown parameters $\boldsymbol{\theta}$. Furthermore, we could have m such compartments, with material being transferred between all compartments and between the environment and any compartment. Such a general model can be described by the system of differential equations

$$\dot{f}_j(x)=\sum_{\substack{k=1\\k\neq j}}^{m}\gamma_{jk}(\mathbf{f},x,\boldsymbol{\theta})f_k(x)-\sum_{\substack{k=0\\k\neq j}}^{m}\gamma_{kj}(\mathbf{f},x,\boldsymbol{\theta})f_j(x)+r_{j0}(x) \qquad (j=1,2,\ldots,m), \tag{8.4}$$

called the *mass-balance* equations. For any particular system, for example those in Figs. 8.1 to 8.4, certain of the γ_{jk} and r_{j0} are known to be zero.

In this chapter we focus almost entirely on linear systems, in which $\gamma_{jk}(\mathbf{f},x)\equiv\gamma_{jk}$, a constant. Frequently we shall also assume $r_{j0}(x)\equiv r_{j0}$ is constant. Thus (8.4) becomes

$$\begin{aligned} \dot{f}_j &= \sum_{\substack{k=1\\k\neq j}}^{m}\gamma_{jk}f_k-\sum_{\substack{k=0\\k\neq j}}^{m}\gamma_{kj}f_j+r_{j0} \\ &= \sum_{k=1}^{m}\alpha_{jk}f_k+r_{j0}, \end{aligned} \tag{8.5}$$

where

$$\alpha_{jk}=\gamma_{jk}, \qquad j\neq k,$$

and

$$\alpha_{jj}=-\sum_{\substack{k=0\\k\neq j}}^{m}\gamma_{kj}=-\sum_{\substack{k=0\\k\neq j}}^{m}\alpha_{kj}. \tag{8.6}$$

Writing (8.5) in matrix form gives us

$$\dot{\mathbf{f}}=\mathbf{A}\mathbf{f}+\mathbf{r}(x), \tag{8.7}$$

where $\mathbf{A}=[(\alpha_{jk})]$ is $m\times m$. In general the nonzero γ_{jk}'s will form part of the vector $\boldsymbol{\theta}$ of unknown parameters. The initial state of the system $\mathbf{f}(0)$ may be known or may contain unknown parameters. Experimental data used to fit a compartmental model typically consist of measurements on $f_j(x)$ taken for one or more compartments at a sequence of discrete time points. In many, if not most, experimental situations it is impossible to take measurements from all of the compartments. Indeed, at times the only measurement available is $\sum_j f_j(x)$, the total material in the system. The inability to take measurements in all the compartments can cause identifiability problems with the parameters $\boldsymbol{\theta}$. Identifiability in compartmental models is discussed in Section 8.4, and identifiability in general is discussed in Section 3.4. The purpose of a compartmental analysis is usually to estimate $\boldsymbol{\theta}$ and/or test the fit of a postulated model to experimental data, or to choose between competing models.

The matrix **A** in (8.7) satisfies the following conditions: (a) every off-diagonal element is nonnegative, (b) every diagonal element is nonpositive, and (c) the column sums are all nonpositive. Condition (c) holds because the kth column sum is $-\gamma_{0k}$, by (8.6). Any square matrix **A** satisfying (a)–(c) is called a *compartmental matrix*. Particular properties for the solution to (8.7) can be identified when **A** is a compartmental matrix. Some of these will be summarized in Section 8.3.3.

We note that linear systems are useful in their own right for describing transfers by diffusion, chemical reactions which follow first-order kinetics, and radioactive decay. They have a large number of applications in various fields, including pharmacology and ecosystems modeling. However, the most important reason for their popularity is that they describe the kinetics of tracer experiments.

8.2.2 Tracer Exchange in Steady-State Systems

Physical and biological systems are often in a so-called *steady state*, even if just for a brief period of time. Material is being exchanged between compartments, but the total amount of material, $f_j(x)$, in any particular compartment is constant, i.e., $\dot{f}_j(x) = 0$ for all x. Because the system is static, measurements of $f_j(x)$ reveal nothing of its kinetics. A change in environmental conditions or in input rates may cause the system to drift to a new steady state. Jacquez [1972: Section 5.2] discussed the effects of perturbing steady-state systems. Tracer experiments enable experimenters to investigate the kinetics of a steady-state system.

In a tracer experiment a small amount of labeled material is added to one or more compartments. This may be done as an effectively instantaneous injection (a *bolus* dose), as a series of bolus doses, or at a steady rate. The amounts of labeled material in accessible compartments are then measured at a series of discrete time points. We shall show that the tracer kinetics are approximately linear even when the underlying system is nonlinear. In biological systems tracers may be dyes, stable isotopes of elements which are identifiable by a different mass, or radioactive isotopes detectable by an instrument such as a Geiger counter. If stable or radioactive isotopes are used, they will be present in only very small quantities. Atkins [1969] stated the following as the properties of a perfect tracer:

(i) The system is unable to distinguish between the material and its tracer.

(ii) The amount of tracer is sufficiently small that it does not disturb the steady state.

(iii) The tracer, when added, is not in equilibrium, so that the tracer itself is not in a steady state.

(iv) If the tracer is an isotope, there should be no exchange of the isotope between labeled compounds and the other compounds.

Implicit in property (i) and in the original definition of a compartment as "well mixed" is that a bolus injection of tracer is instantaneously diffused throughout the compartment. If the diffusion through the compartment is not very fast compared with exchanges between compartments, the situation can often be approximated by replacing this compartment by two or more compartments. These and other practical considerations relating to the use of tracers are discussed in most books on compartmental models in various subject areas, e.g. Atkins [1969], Jacquez [1972], and Lambrecht and Rescigno [1983].

In this section we shall use $\mathbf{v}$ to represent the "mother" material in each compartment of the system and reserve $\mathbf{f}$ just for the amounts of tracer. The general compartmental system of differential equations (8.5) can be rewritten

$$\begin{aligned} \dot{v}_j(x) &= \sum_{k=1}^{m} \alpha_{jk}(\mathbf{v}, x) v_k(x) + r_{j0}(x) \\ &= N_j(\mathbf{v}, x) + r_{j0}(x), \end{aligned} \tag{8.8}$$

say, where

$$N_j(\mathbf{v}, x) = \sum_{k=1}^{m} \alpha_{jk}(\mathbf{v}, x) v_k, \tag{8.9}$$

and the α_{jk} are given by (8.6). Suppose that the input rate r_{j0} is constant. Then as the system is assumed to be in a steady state, we have $\dot{v}_j = 0$, $N_j(\mathbf{v}, x) = N_j(\mathbf{v})$, and, from (8.8),

$$0 = N_j(\mathbf{v}) + r_{j0}. \tag{8.10}$$

Suppose now that a small amount of tracer $f_j(0)$ is placed in compartment j at time 0, and that the tracer is added at rate $b_j(x)$. Corresponding to (8.8) we have the following equations for the combined tracer and mother material:

$$\dot{v}_j + \dot{f}_j = N_j(\mathbf{v} + \mathbf{f}, x) + r_{j0} + b_j(x). \tag{8.11}$$

We can expand N_j as a first-order Taylor series about $\mathbf{v}$ to obtain

$$\dot{v}_j + \dot{f}_j = N_j(\mathbf{v}) + \sum_k \frac{\partial N_j(\mathbf{v})}{\partial v_k} f_k + r_{j0} + b_j(x) + O(\|\mathbf{f}\|^2).$$

Using (8.10),

$$\dot{f}_j = \sum_k \frac{\partial N_j(\mathbf{v})}{\partial v_k} f_k + b_j(x) + O(\|\mathbf{f}\|^2). \tag{8.12}$$

From (8.10), $N_j(\mathbf{v})$ is independent of time x. Therefore $\alpha_{jk}^* = \partial N_j / \partial v_k$ is independent of both $\mathbf{f}$ and x. Hence from (8.12)

$$\dot{f}_j \approx \sum_{k=1}^{m} \alpha_{jk}^* f_k + b_j(x) \qquad (j = 1, 2, \ldots, m),$$

or

$$\dot{\mathbf{f}} \approx \mathbf{A}^* \mathbf{f} + \mathbf{b}(x), \tag{8.13}$$

and these are the equations of a linear system. If the original system was linear, i.e. $\alpha_{jk}(\mathbf{v}, x) \equiv \alpha_{jk}$, then from (8.8) $\alpha_{jk}^* = \partial N_j / \partial v_k = \alpha_{jk}$ and (8.13) holds exactly. As this is an equation in terms of the tracer $\mathbf{f}$ only, we have the intuitively obvious result that the tracer is exchanged according to the same rates as mother material. This means that in considering the transfer of tracer we can ignore the mother material entirely provided the latter is in a steady state. Unfortunately, real tracers are not exchanged at exactly the same rate as mother substance. Jacquez [1972: Section 5.3] notes that "radioactive isotopes differ in mass from their 'normal' isotopes and both rates of diffusion and the rates of chemical reactions are affected by mass." This effect is known as *isotope fractionation*. Frequently such effects are negligible. For further discussion the reader is referred both to Jacquez [1972] and Atkins [1969].

The tracer kinetics of (8.13) can be used to investigate the linearity of a system. If it is possible to manipulate the inputs r_{j0} to the system or change the environment in such a way that other steady states are achieved, then we can test for the constancy of the α_{jk}^*'s across experiments. In a nonlinear system the α_{jk}^*'s provide information about the fractional transfer rates at the steady-state values. Using (8.9),

$$\alpha_{jk}^* = \frac{\partial N_j(\mathbf{v})}{\partial v_k} = \alpha_{jk}(\mathbf{v}) + \sum_r \frac{\partial \alpha_{jr}(\mathbf{v})}{\partial v_k} v_r. \tag{8.14}$$

The practical usefulness of these nonlinear equations will depend on the complexity of the $\alpha_{jk}(\mathbf{v})$.

We have noted that tracer experiments lead to ordinary linear compartmental models of the form

$$\dot{\mathbf{f}}(x) = \mathbf{A}\mathbf{f}(x) + \mathbf{b}(x), \qquad \mathbf{f}(0) = \boldsymbol{\xi}, \tag{8.15}$$

where $\mathbf{A}$ is a compartmental matrix and $\mathbf{f}(x)$ is the amount of tracer present in the compartments at time x. Frequently only the concentration of tracer in a sample, the *specific abundance*, or (for radioactive tracers) the *specific activity*, can be measured. Apart from sampling and measurement errors, the concentration of the tracer is $f_j(x)/\text{vol}_j$, where vol_j is the volume of the jth compartment. The specific abundance is defined as $f_j(x)/v_j$, the ratio of the amount of tracer to the amount of mother substance, or equivalently the ratio of the concentrations of the tracer and the mother substance. Anderson [1983] also defines specific activity for radioactive tracers as $f_j(x)/v_j$. Atkins [1969] defines specific activity as proportional to this ratio, where the constant of proportionality allows for conversion from radioactive units to mass units and is the same for all compartments. As the constant cancels out of the kinetic equations, we shall follow Anderson [1983]. Jacquez [1972] allows for different conversion rates in different compartments, but we shall ignore this complication.

Whether our measurements are concentrations, specific abundances, or specific activities, we have measurements of the form $c_j = f_j(x)/v_j$, where only the interpretation of v_j alters. Thus from (8.13), with the star omitted from the notation for convenience, we can write

$$\dot{c}_j = \sum_k \alpha_{jk} \frac{v_k}{v_j} c_k + \frac{b_j(x)}{v_j},$$

or

$$\dot{\mathbf{c}} = \mathbf{V}^{-1}\mathbf{A}\mathbf{V}\mathbf{c} + \mathbf{V}^{-1}\mathbf{b}(x), \tag{8.16}$$

where $\mathbf{V} = \text{diag}(v_1, v_2, \ldots, v_m)$. This is again a linear system. The matrix $\mathbf{C} = \mathbf{V}^{-1}\mathbf{A}\mathbf{V}$ need not be compartmental, but it is similar to $\mathbf{A}$ and thus has the same eigenvalues: these are discussed in Section 8.3 below. When concentrations or specific-abundance measurements are taken, any unknown v_j's are included as part of $\boldsymbol{\theta}$. Thus $\boldsymbol{\theta}$ consists of the unknown fractional transfer rates γ_{jk}, any unknown components of the initial conditions, and any unknown steady-state amounts v_j of mother substance (or compartment volumes).

8.2.3 Tracers in Nonsteady-State Linear Systems

Even for linear systems, tracer experiments can result in systems of differential equations which do not have the simple linear form with constant coefficients. Consider the linear system

$$\dot{v}_j(x) = \sum_k \alpha_{jk} v_k(x) + r_j(x). \tag{8.17}$$

If we have added tracer so that at time x we have a total amount $f_j(x) + v_j(x)$ in the jth compartment, then

$$\dot{v}_j + \dot{f}_j = \sum_k \alpha_{jk}(v_k + f_k) + r_j(x) + b_j(x) \tag{8.18}$$

and, subtracting (8.17),

$$\dot{f}_j = \sum_k \alpha_{jk} f_k + b_j(x)$$

as before. However, if we can only measure concentrations, specific abundances, or specific activities, we have $c_j(x) = f_j(x)/v_j(x)$, or $f_j = c_j v_j$. Thus (8.18) becomes

$$\dot{v}_j + \dot{c}_j v_j + c_j \dot{v}_j = \sum_k \alpha_{jk}(c_k v_k + v_k) + r_j(x) + b_j(x).$$

Using (8.17) to remove the expression for $\dot{v}_j$ and to replace $\dot{v}_j$ in $c_j \dot{v}_j$, we obtain

$$\dot{c}_j = \sum_k \alpha_{jk} \frac{v_k}{v_j}(c_k - c_j) - \frac{c_j r_j}{v_j} + \frac{b_j}{v_j}, \qquad j = 1, 2, \ldots, m,$$

or

$$\dot{c}_j = -\left(\frac{r_j}{v_j} + \sum_{\substack{k \\ k \neq j}} \alpha_{jk} \frac{v_k}{v_j}\right) c_j + \sum_{\substack{k \\ k \neq j}} \alpha_{jk} \frac{v_k}{v_j} c_k + \frac{b_j}{v_j}$$

$$= \sum_k \beta_{jk} c_k + b_j v_j^{-1}, \tag{8.19}$$

say. Unlike (8.17) with constant coefficients, Equation (8.19) has coefficients β_{jk} which are now functions of x through the $v_j(x)$.

Example 8.4 Consider a single compartment system with

$$\dot{v} = \alpha v + r.$$

This has solution

$$v(x) = \left[v(0) + \frac{r}{\alpha} \right] e^{\alpha x} - \frac{r}{\alpha}.$$

Setting $c = f/v$, (8.19) becomes

$$\dot{c}(x) = -\frac{r}{v(x)} c(x) + \frac{b(x)}{v(x)}. \qquad \blacksquare$$

8.3 SOLVING THE LINEAR DIFFERENTIAL EQUATIONS

8.3.1 The Solution and its Computation

Our compartmental models have produced linear systems of differential equations of the form

$$\dot{\mathbf{f}}(x) = \mathbf{A}\mathbf{f}(x) + \mathbf{b}(x) \tag{8.20}$$

subject to

$$\mathbf{f}(0) = \boldsymbol{\xi}. \tag{8.21}$$

Here $\mathbf{b}(x)$ can refer either to an input vector $\mathbf{r}$ of material as in the compartmental model of (8.7) or to the input of tracer as in the tracer experiment of (8.13). Such equations are discussed in many books on differential equations and dynamical systems: see, for example, Hirsch and Smale [1974]. They involve the use of the matrix exponential $e^{\mathbf{B}}$ for a square matrix $\mathbf{B}$. This is defined as

$$e^{\mathbf{B}} = \mathbf{I} + \mathbf{B} + \frac{\mathbf{B}^2}{2!} + \frac{\mathbf{B}^3}{3!} + \cdots, \tag{8.22}$$

and this series converges for any $\mathbf{B}$. If $\mathbf{A}$ is any square matrix, the general solution of the homogeneous equation $\dot{\mathbf{f}} = \mathbf{A}\mathbf{f}$ is given by $\mathbf{f} = e^{\mathbf{A}x}\mathbf{c}$ for any $\mathbf{c}$. A particular solution to (8.20) is given by

$$\int_0^x e^{\mathbf{A}(x-u)}\mathbf{b}(u)\,du.$$

That these are solutions is verifiable directly by differentiation and using the expansion (8.22) in the case of $e^{\mathbf{A}x}\mathbf{c}$. Thus the complete solution to (8.20) which satisfies the initial conditions (8.21) is

$$\mathbf{f}(x) = e^{\mathbf{A}x}\boldsymbol{\xi} + \int_0^x e^{\mathbf{A}(x-u)}\mathbf{b}(u)\,du. \tag{8.23}$$

There are a number of methods for finding the form of $e^{\mathbf{A}x}$. We note that it is virtually never useful to sum the series (8.22) (Moler and Van Loan [1978]). Once $e^{\mathbf{A}x}$ and $e^{\mathbf{A}[x-u]}$ have been formed, the integration in (8.23) can be performed componentwise. A special case of interest is $\mathbf{b}(x) \equiv \mathbf{b}$, a constant, with $\mathbf{A}$ nonsingular. Then it can be shown that (8.23) reduces to

$$\mathbf{f} = e^{\mathbf{A}x}(\boldsymbol{\xi} + \mathbf{A}^{-1}\mathbf{b}) - \mathbf{A}^{-1}\mathbf{b}. \tag{8.24}$$

When the $m \times m$ matrix $\mathbf{A}$ has m linearly independent eigenvectors, it is possible to form the spectral decomposition

$$\mathbf{A} = \mathbf{S}\boldsymbol{\Lambda}\mathbf{S}^{-1}, \tag{8.25}$$

where $\boldsymbol{\Lambda} = \text{diag}(\lambda_1, \lambda_2, \ldots, \lambda_m)$ consists of the eigenvalues of $\mathbf{A}$, and the ith column of $\mathbf{S}$ is a right eigenvector of $\mathbf{A}$ corresponding to λ_i (see Hunter [1983: Chapter 4] or Rao [1973: Chapter 1] for details). In particular, this is possible if all the eigenvalues of $\mathbf{A}$ are distinct, as this implies that the eigenvectors are all linearly independent. When there are multiple eigenvalues, a necessary and sufficient condition for the existence of the spectral decomposition is that the dimension of the subspace spanned by the eigenvectors of each eigenvalue must equal the multiplicity of the eigenvalue. Since $\mathbf{A}^n = \mathbf{S}\mathbf{\Lambda}^n\mathbf{S}^{-1}$, we can apply (8.22) and obtain

$$e^{\mathbf{A}x} = \mathbf{S}e^{\mathbf{\Lambda}x}\mathbf{S}^{-1}, \tag{8.26}$$

where $e^{\mathbf{\Lambda}x} = \text{diag}(e^{\lambda_1 x}, e^{\lambda_2 x}, \ldots, e^{\lambda_m x})$. Writing the columns of $\mathbf{S}$ as $\mathbf{s}_1, \mathbf{s}_2, \ldots, \mathbf{s}_m$, and the rows of $\mathbf{S}^{-1}$ as $\mathbf{s}^{(1)\prime}, \mathbf{s}^{(2)\prime}, \ldots, \mathbf{s}^{(m)\prime}$, we have

$$\mathbf{A} = \sum_{j=1}^{m} \lambda_j \mathbf{s}_j \mathbf{s}^{(j)\prime} \tag{8.27}$$

and

$$e^{\mathbf{A}x} = \sum_{j=1}^{m} e^{\lambda_j x} \mathbf{s}_j \mathbf{s}^{(j)\prime}. \tag{8.28}$$

The solution to the homogeneous equations $\dot{\mathbf{f}} = \mathbf{A}\mathbf{f}$ and $\mathbf{f}(0) = \boldsymbol{\xi}$ can then be written

$$\begin{aligned}\mathbf{f} &= e^{\mathbf{A}x}\boldsymbol{\xi} \\ &= \sum_{j=1}^{m} \boldsymbol{\beta}_j e^{\lambda_j x},\end{aligned} \tag{8.29}$$

where $\boldsymbol{\beta}_j = (\mathbf{s}^{(j)\prime}\boldsymbol{\xi})\mathbf{s}_j$: the so-called *sum-of-exponentials model.* The nonhomogeneous equation with constant inputs has solution (8.24), which, using $\mathbf{A}^{-1} = \mathbf{S}\mathbf{\Lambda}^{-1}\mathbf{S}^{-1}$, can be written as

$$\mathbf{f} = \sum_{j=1}^{m} \boldsymbol{\beta}_j^* e^{\lambda_j x} - \mathbf{A}^{-1}\mathbf{b}, \tag{8.30}$$

where $\boldsymbol{\beta}_j^* = \{\mathbf{s}^{(j)\prime}(\boldsymbol{\xi} + \mathbf{b}/\lambda_j)\}\mathbf{s}_j$ for $j = 1, 2, \ldots, m$. The eigenvalues λ_j and eigenvectors $\mathbf{s}_j$ of $\mathbf{A}$ may be complex rather than real. Almost invariably, though, data from actual compartmental models reveal a nonoscillatory exponential decay form associated with the real and nonpositive eigenvalues λ_j. In Section 8.3.3. we shall discuss the eigenstructure of $\mathbf{A}$ and the nature of the solution when $\mathbf{A}$ is compartmental rather than a general matrix.

The spectral decomposition of (8.25) can be used to find a theoretical solution to the linear system of differential equations in the manner described by Hunter [1983: Chapter 4]. Since $\mathbf{S}^{-1}\mathbf{A} = \mathbf{\Lambda}\mathbf{S}^{-1}$, the rows of $\mathbf{S}^{-1}$ are left eigenvectors of $\mathbf{A}$. Eigenvalues and eigenvectors can be obtained by solving the equations $(\mathbf{A} - \lambda\mathbf{I}_m)\mathbf{u} = \mathbf{0}$ and $\mathbf{v}'(\mathbf{A} - \lambda\mathbf{I}_m) = \mathbf{0}'$ directly. However, it is often easier to find the eigenvalues as the roots of the characteristic equation

$$|\mathbf{A} - \lambda\mathbf{I}_m| = 0. \tag{8.31}$$

For any square $m \times m$ matrix $\mathbf{B}$ it is known that

$$\mathbf{B}(\text{adj}\,\mathbf{B}) = (\text{adj}\,\mathbf{B})\mathbf{B} = |\mathbf{B}|\mathbf{I}_m,$$

where adj $\mathbf{B}$ is the transposed matrix of cofactors. Furthermore, for any eigenvalue λ_j, $|\mathbf{A} - \lambda_j \mathbf{I}_m| = 0$, so that

$$(\mathbf{A} - \lambda_j \mathbf{I}_m)\,\text{adj}\,(\mathbf{A} - \lambda_j \mathbf{I}_m) = [\text{adj}\,(\mathbf{A} - \lambda_j \mathbf{I}_m)](\mathbf{A} - \lambda_j \mathbf{I}_m) = \mathbf{0}. \tag{8.32}$$

Thus *any* nonzero column of adj $(\mathbf{A} - \lambda_j \mathbf{I}_m)$ is a right eigenvector of $\mathbf{A}$ corresponding to λ_j, and *any* row is a left eigenvector. This gives us a method of obtaining eigenvectors. We shall consider just the case when the eigenvalues are all different. Then we need only one nonzero row (or column) of the adjoint matrix for each eigenvalue, and we can take $\mathbf{S}$ as a set of right eigenvectors with $\mathbf{s}_j$ corresponding to λ_j. If $\mathbf{C}$ is a set of left eigenvectors with row $\mathbf{c}^{(j)\prime}$ corresponding to λ_j, then we have

$$\mathbf{s}^{(j)} = \mathbf{c}^{(j)}/\mathbf{c}^{(j)\prime}\mathbf{s}_j \tag{8.33}$$

(as $\mathbf{S}^{-1}\mathbf{S} = \mathbf{I}_m$).

Example 8.5 Consider Example 8.2 in Section 8.1 with an initial quantity R of drug in the injection compartment 1 and no drug elsewhere. The kinetic equations are

$$\begin{aligned}
\dot{f}_1 &= -\gamma_A f_1, \\
\dot{f}_2 &= \gamma_A f_1 + \gamma_{-1} f_3 - (\gamma_1 + \gamma_E) f_2, \\
\dot{f}_3 &= \gamma_1 f_2 - \gamma_{-1} f_3
\end{aligned}$$

with initial conditions $\boldsymbol{\xi} = R(1,0,0)' = R\mathbf{e}_1$ and parameters

$$\boldsymbol{\theta} = (R, \gamma_A, \gamma_1, \gamma_{-1}, \gamma_E)'.$$

Hence

$$\mathbf{A} = \begin{bmatrix} -\gamma_A & 0 & 0 \\ \gamma_A & -(\gamma_1 + \gamma_E) & \gamma_{-1} \\ 0 & \gamma_1 & -\gamma_{-1} \end{bmatrix},$$

$$\mathbf{A} - \lambda \mathbf{I}_3 = \begin{bmatrix} -\gamma_A - \lambda & 0 & 0 \\ \gamma_A & -\gamma_1 - \gamma_E - \lambda & \gamma_{-1} \\ 0 & \gamma_1 & -\gamma_{-1} - \lambda \end{bmatrix},$$

and the characteristic polynomial is

$$|\mathbf{A}-\lambda\mathbf{I}_3| = -(\gamma_A+\lambda)\{\lambda^2+(\gamma_1+\gamma_{-1}+\gamma_E)\lambda+\gamma_{-1}\gamma_E\}=0$$

with eigenvalues

$$-\gamma_A \quad \text{and} \quad -\tfrac{1}{2}\{\gamma_1+\gamma_{-1}+\gamma_E \pm [(\gamma_1+\gamma_{-1}+\gamma_E)^2-4\gamma_{-1}\gamma_E]^{1/2}\}.$$

We write the final two eigenvalues as a and b, where $a+b=-(\gamma_1+\gamma_{-1}+\gamma_E)$ and $ab=\gamma_{-1}\gamma_E$. Then

$$\operatorname{adj}(\mathbf{A}-\lambda\mathbf{I}_3) = \begin{bmatrix} (\gamma_1+\gamma_E+\lambda)(\gamma_{-1}+\lambda)-\gamma_1\gamma_{-1} & 0 & 0 \\ \gamma_A(\gamma_{-1}+\lambda) & (\gamma_A+\lambda)(\gamma_{-1}+\lambda) & \gamma_{-1}(\gamma_A+\lambda) \\ \gamma_A\gamma_1 & \gamma_1(\gamma_A+\lambda) & (\gamma_A+\lambda)(\gamma_1+\gamma_E+\lambda) \end{bmatrix},$$

from which we read off right (column) eigenvectors corresponding to $-\gamma_A$, a, and b respectively, giving (after some slight manipulation and canceling common column factors)

$$\mathbf{S} = \begin{bmatrix} (\gamma_A+a)(\gamma_A+b) & 0 & 0 \\ \gamma_A(\gamma_{-1}-\gamma_A) & \gamma_{-1}+a & \gamma_{-1}+b \\ \gamma_A\gamma_1 & \gamma_1 & \gamma_1 \end{bmatrix}.$$

The left (row) eigenvectors give us (using rows 1 and 3)

$$\mathbf{C} = \begin{bmatrix} 1 & 0 & 0 \\ \gamma_A\gamma_1 & \gamma_1(\gamma_A+a) & (\gamma_A+a)(\gamma_1+\gamma_E+a) \\ \gamma_A\gamma_1 & \gamma_1(\gamma_A+b) & (\gamma_A+b)(\gamma_1+\gamma_E+b) \end{bmatrix}.$$

Thus, normalizing the rows of $\mathbf{C}$ using (8.33), we obtain

$$\mathbf{S}^{-1} = \begin{bmatrix} \dfrac{1}{(\gamma_A+a)(\gamma_A+b)} & 0 & 0 \\ \dfrac{\gamma_A}{(\gamma_A+a)(a-b)} & \dfrac{1}{a-b} & \dfrac{\gamma_1+\gamma_E+a}{\gamma_1(a-b)} \\ \dfrac{\gamma_A}{(\gamma_A+b)(b-a)} & \dfrac{1}{b-a} & \dfrac{\gamma_1+\gamma_E+b}{\gamma_1(b-a)} \end{bmatrix}.$$

Since $\boldsymbol{\xi} = R\mathbf{e}_1$, it follows from (8.29) that

$$\mathbf{f} = R\sum_{j=1}^{3} e^{\lambda_j x}\mathbf{s}_j\mathbf{s}^{(j)\prime}\mathbf{e}_1 = R(w_1, w_2, w_3)',$$

say, where

$$\begin{aligned}
w_1 &= e^{-\gamma_A x},\\
w_2 &= \gamma_A\left(\frac{\gamma_{-1}-\gamma_A}{(\gamma_A+a)(\gamma_A+b)}e^{-\gamma_A x} + \frac{\gamma_{-1}+a}{(\gamma_A+a)(a-b)}e^{ax} + \frac{\gamma_{-1}+b}{(\gamma_A+b)(b-a)}e^{bx}\right),\\
w_3 &= \gamma_A\gamma_1\left(\frac{1}{(\gamma_A+a)(\gamma_A+b)}e^{-\gamma_A x} + \frac{1}{(\gamma_A+a)(a-b)}e^{ax} + \frac{1}{(\gamma_A+b)(b-a)}e^{bx}\right).
\end{aligned} \tag{8.34}$$

We note that $f_1 = R\exp(-\gamma_A x)$ corresponds to simple exponential decay, as we would expect from the physical situation. The expression $f_2 = Rw_2$ differs from Allen [1983] and Metzler [1971] only in that we use a and b for the eigenvalues where they use $-\alpha$ and $-\beta$. We also note that methods for solving linear systems based upon Laplace transforms or the differential-operator method (Atkins [1969]) are often quicker and simpler than the spectral-decomposition method for small problems. ∎

Let us now consider the solution (8.23) for a general input function $\mathbf{b}(x)$. Using the spectral decomposition, we have

$$\int_0^x e^{\mathbf{A}(x-u)}\mathbf{b}(u)\,du = \mathbf{S}\int_0^x e^{\mathbf{\Lambda}(x-u)}\mathbf{S}^{-1}\mathbf{b}(u)\,du,$$

where

$$\left[\int_0^x e^{\mathbf{\Lambda}(x-u)}\mathbf{S}^{-1}\mathbf{b}(u)\,du\right]_j = \sum_k s^{jk}\int_0^x e^{\lambda_j(x-u)}b_k(u)\,du, \tag{8.35}$$

and $\mathbf{S}^{-1} = [(s^{jk})]$. Thus the solution (8.23) can be expressed in terms of fairly simple scalar integrals of the form (8.35). This fact is useful for both the analytic and the numerical computation of $\mathbf{f}(x)$.

The spectral decomposition provides a mechanism for the numerical computation of $e^{\mathbf{A}x}$, where standard eigenvalue–eigenvector routines such as those in EISPACK (Smith et al. [1976a]) can be used to perform the decomposition. Theoretically, the method breaks down when $\mathbf{A}$ does not have a complete set of linearly independent eigenvectors. Moler and Van Loan [1978] note that, in practice, accuracy problems occur if several eigenvectors are nearly linearly dependent so that the matrix $\mathbf{S}$ is ill conditioned. (Note that this problem does not occur when $\mathbf{A}$ is symmetric, as we can then always choose an orthogonal $\mathbf{S}$.)

Moler and Van Loan [1978] reviewed 19 algorithms for computing $e^{\mathbf{A}}$ and $e^{\mathbf{A}x}$.

Those that are suited for calculating $e^{\mathbf{A}x}$ for a large number of x-values appear to be based upon decompositions similar to (8.25), namely $\mathbf{A} = \mathbf{SBS}^{-1}$, so that $e^{\mathbf{A}x} = \mathbf{S}e^{\mathbf{B}x}\mathbf{S}^{-1}$, where, ideally, $e^{\mathbf{B}x}$ is easy to compute and $\mathbf{S}$ is well conditioned. As with the spectral decomposition, other methods like the use of Schur complements can be inaccurate if some of the eigenvalues of $\mathbf{A}$ are very close together. We should note at this point that this poses less of a problem to the compartmental modeler than to the numerical analyst. If two or more eigenvalues are close together, the modeler will often decide that the model is too complex and collapse the model to one with fewer compartments, and thus fewer eigenvalues. However, as we shall see in Section 8.3.2, features of compartmental systems called traps can introduce repeated zero eigenvalues. Also, hypotheses that specify the equality of some of the fractional transfer coefficients γ_{jk} can also induce repeated eigenvalues.

A decomposition which is sometimes suggested when there are r repeated eigenvalues is the Jordan canonical form (Kalbfleisch et al. [1983], Yang [1985]). Here $\mathbf{B}$ is block-diagonal, namely $\mathbf{B} = \text{diag}(\mathbf{J}_1, \mathbf{J}_2, \ldots, \mathbf{J}_r)$, where $\mathbf{J}_k$ corresponds to the kth distinct eigenvalue λ_k and is given by

$$\mathbf{J}_k = \begin{bmatrix} \lambda_k & 1 & 0 & \cdot & \cdot & \cdot & 0 \\ 0 & \lambda_k & 1 & 0 & \cdot & \cdot & 0 \\ \cdot & & \cdot & \cdot & & & \cdot \\ \cdot & & & \cdot & \cdot & & \cdot \\ \cdot & & & & \cdot & \cdot & \cdot \\ 0 & \cdot & \cdot & \cdot & 0 & \lambda_k & 1 \\ 0 & 0 & \cdot & \cdot & \cdot & 0 & \lambda_k \end{bmatrix}.$$

The dimension d_k of $\mathbf{J}_k$ is equal to the multiplicity of λ_k. Then $e^{\mathbf{B}x} = \text{diag}(e^{\mathbf{J}_1 x}, \ldots, e^{\mathbf{J}_r x})$, where

$$e^{\mathbf{J}_k x} = e^{\lambda_k x} \begin{bmatrix} 1 & x & x^2/2! & \cdots & x^{d_k - 1}/(d_k - 1)! \\ 0 & 1 & x & \cdots & x^{d_k - 2}/(d_k - 2)! \\ \vdots & \vdots & \vdots & & \vdots \\ 0 & 0 & 0 & \cdots & 1 \end{bmatrix}. \tag{8.36}$$

This method may be useful for obtaining theoretical results, but it is an extremely unstable and inaccurate method for numerical computations (Moler and Van Loan [1978]).

A stable decomposition algorithm has been developed by Bavely and Stewart [1979] in which eigenvalues are grouped into clusters of "nearly equal" eigenvalues. The matrix $\mathbf{B}$ is again block-diagonal, each block being an upper triangular matrix with dimension equal to the number of eigenvalues in the corresponding cluster.

8.3.2 Some Compartmental Structures

Another type of diagram used in the compartmental literature is the so-called connectivity diagram. Formally, this is a directed graph in which the nodes are the compartments and the directed edges are the nonzero flows connecting the compartments. For example, the connectivity diagram for Fig. 8.4 in Section 8.1 is given by Fig. 8.5. We shall use these to describe some structure types that have been isolated in the literature because specific results can be proved about their behavior. A detailed description of connectivity diagrams and their uses is given by Rescigno et al. [1983].

A k-compartment *catenary* system is a system as shown in Fig. 8.6. A k-compartment *mammillary* system is described by Fig. 8.7. The central compartment, labeled 1, is sometimes called the *mother compartment* and the remainder are called *daughters.*

A subset T of compartments is called a *trap* if, once material has arrived at T, it can never get out again. For example, in Fig. 8.8 $T_1 = \{4\}$ and $T_2 = \{5, 6, 7\}$ are traps, but $\{1\}$ is not, as material can escape to the environment. We can define a trap as a subset T of the compartment set $S = \{1, 2, \ldots, m\}$ for which $\gamma_{kj} = 0$ for all j in T and k not in T. A system that does not excrete to the environment is itself a trap. It is easily seen that T is a trap if and only if, by permuting the compartment

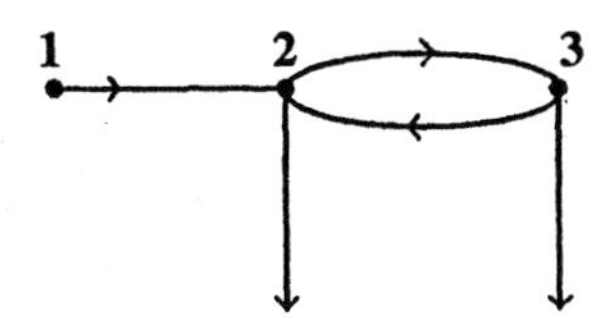

Figure 8.5 Connectivity diagram corresponding to Fig. 8.4.

Figure 8.6 A k-compartment catenary system.

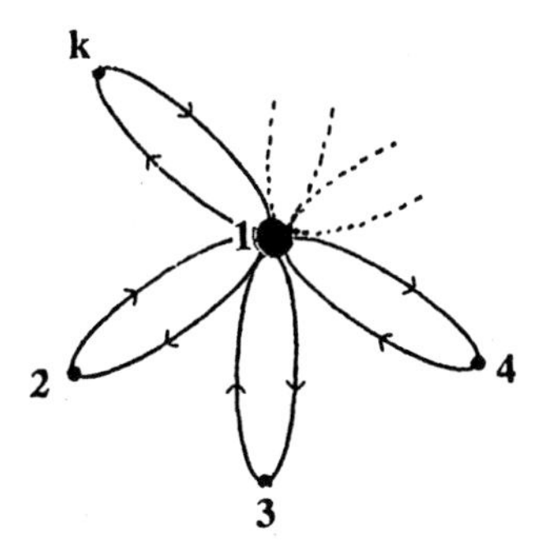

Figure 8.7 A k-compartment mammillary system.

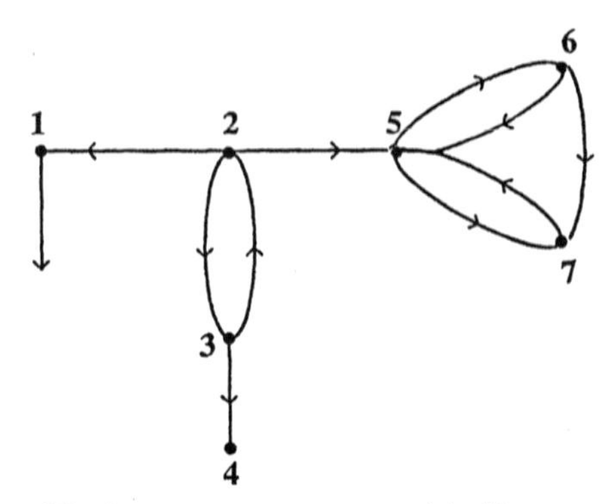

Figure 8.8 A seven-compartment model with two traps.

labels so that T consists of the last r components, $\mathbf{A}$ can be written in the form

$$\mathbf{A} = \begin{pmatrix} \mathbf{B} & \mathbf{0} \\ \mathbf{C} & \mathbf{D} \end{pmatrix} \begin{matrix} \}m-r \\ \}r \end{matrix}, \qquad \text{(columns: } m-r,\ r\text{)} \tag{8.37}$$

Here $\mathbf{D}$ is the compartmental matrix (see Section 8.2.1) for the compartmental subsystem consisting only of the compartments in T. If there is more than one trap, say $T_1, T_2, \ldots, T_p$, we can permute the compartment labels so that $S = S' \cup T_1 \cup \cdots \cup T_p$ and $\mathbf{A}$ has the form

$$\mathbf{A} = \begin{bmatrix} \mathbf{B} & \mathbf{0} & \mathbf{0} & \cdots & \mathbf{0} \\ \mathbf{C}_1 & \mathbf{D}_1 & \mathbf{0} & \cdots & \mathbf{0} \\ \mathbf{C}_2 & \mathbf{0} & \mathbf{D}_2 & \cdots & \mathbf{0} \\ \vdots & \vdots & \vdots & & \vdots \\ \mathbf{C}_p & \mathbf{0} & \mathbf{0} & \cdots & \mathbf{D}_p \end{bmatrix}. \tag{8.38}$$

If material can reach any compartment from any other compartment, then the system is called *strongly connected*. Its matrix cannot then be written in the form (8.37). Using terminology similar to that used in Markov chains, we call such a matrix *irreducible*.

8.3.3 Properties of Compartmental Systems

In this subsection we list some properties of compartmental systems. Proofs or references to proofs can be found in Anderson [1983: Section 12].

Let $\mathbf{A}$ be a compartmental matrix (see Section 8.2.1). Then the following apply:

1. The matrix $\mathbf{A}$ has no purely imaginary eigenvalues, and the real part (Re) of any eigenvalue is nonpositive.

2. (a) The eigenvalue, $\lambda_{(1)}$ say, of smallest magnitude is real and nonpositive.
   (b) This eigenvalue has a nonnegative eigenvector $\mathbf{v}$ (i.e. $v_i \geq 0$, all i).
   (c) For any other eigenvalue λ_j, $\text{Re}\,\lambda_j \leq \lambda_{(1)}$.
3. (Strongly connected systems.) If $\mathbf{A}$ is irreducible, result 2 can be strengthened to the following:
   (a) $\lambda_{(1)}$ has a strictly positive eigenvector $\mathbf{v}$ (i.e. $v_i > 0$, all i).
   (b) $\lambda_{(1)}$ is a simple (nonrepeated) eigenvalue.
   (c) $\text{Re}\,\lambda_j < \lambda_{(1)}$.

Thus when the spectral decomposition exists, all the exponential terms in (8.29) or (8.30) decay towards zero as $x \to \infty$ except possibly the term corresponding to $\lambda_{(1)}$. If $\lambda_{(1)}$ is not zero, this term is the one that decays most slowly.

We now revert to the general case where $\mathbf{A}$ is no longer restricted to being irreducible. The following bounds can be placed on $\lambda_{(1)}$.

4. Let $r_{\min}$ and $r_{\max}$ be the respective minimum and maximum row sums of $\mathbf{A}$. Recalling from (8.6) that

$$\alpha_{0j} = -\sum_{k=1}^{m} \alpha_{kj},$$

   then it can be shown that:
   (a) $r_{\min} \leq \lambda_{(1)} \leq \min(0, r_{\max})$.
   (b) $|\lambda_{(1)}| \leq \min_{1 \leq j \leq m} |\alpha_{jj}|$.
   (c) $\min_{1 \leq j \leq m} \alpha_{0j} \leq |\lambda_{(1)}| \leq \max_{1 \leq j \leq m} \alpha_{0j}$.

Anderson [1983] notes that for a system with a single exit in compartment j, bound (c) reduces to $0 \leq |\lambda_1| \leq \alpha_{0j}$. Numerical problems arise if the excretion rate α_{0j} is slow compared with other transfer rates. In this case $\mathbf{A}$ may appear singular.

We noted earlier that data from compartmental experiments almost always showed the nonoscillatory "exponential decay" associated with real eigenvalues in (8.29) or (8.30). A matrix $\mathbf{A}$ is said to be *symmetrizable* if there exists a positive definite $m \times m$ matrix $\mathbf{W}$ such that $\mathbf{WA}$ is symmetric. It can be shown (Anderson [1983]) that: (a) if $\mathbf{A}$ is symmetrizable, it has real eigenvalues; (b) for a catenary system as in Fig. 8.6, $\mathbf{A}$ is symmetrizable provided $\alpha_{j,j-1} \neq 0$ for $j = 2, 3, \ldots, m$; (c) for a mammillary system, $\mathbf{A}$ is symmetrizable; (d) if the longest circuit that can be taken from a compartment via other compartments and back again is of length 2, then $\mathbf{A}$ has real eigenvalues. It is also of interest to know when the eigenvalues are distinct.

5. (a) A mammillary system has distinct eigenvalues provided $\alpha_{1j} \neq \alpha_{1k}$ for all $j \neq k \neq 1$, and $\alpha_{1j}\alpha_{j1} > 0$ for all $j = 1, 2, \ldots, m$.
   (b) A catenary compartmental matrix has distinct eigenvalues when $\alpha_{j,j+1} \neq 0$, $\alpha_{j+1,j} \neq 0$ for $j = 1, 2, \ldots, m-1$.

In Section 8.3.1 we noted that traps introduce zero eigenvalues. Fife [1972] proved the following:

6. The system has (or is) a trap if and only if zero is an eigenvalue of **A**.

We note that a closed system has no exits and is therefore itself a trap. Hence a closed system has at least one zero eigenvalue, and (8.29) is thus of the form

$$\beta_1 + \sum_{j=2}^{m} \beta_j e^{\lambda_j x}.$$

Using the partition of **A** in (8.38), it is readily shown that:

7. Each trap induces a zero eigenvalue.

This justifies our earlier statement (Section 8.3.1) that traps can induce zero as a repeated eigenvalue of **A**.

We can rewrite result 6 as the following:

8. **A** is invertible if and only if the system contains no traps.

In particular, **A** is invertible for any strongly connected open system.

9. If **A** is invertible and $\mathbf{b}(x) \equiv \mathbf{b}$ a constant, then for any initial state $\mathbf{f}(0)$, $\mathbf{f}(x) \to -\mathbf{A}^{-1}\mathbf{b} = \mathbf{f}_e$, say, as $x \to \infty$.

This result is clear from (8.30), result 2, and the fact that zero is not an eigenvalue. If $\mathbf{f}(x) \to$ const as $x \to \infty$, then the system is said to be asymptotically stable. If we choose $\mathbf{f}(0) = \mathbf{f}_e$, then from (8.24) with $\boldsymbol{\xi} = -\mathbf{A}^{-1}\mathbf{b}$ we get $\mathbf{f}(x) \equiv \mathbf{f}_e$. Thus $\dot{\mathbf{f}}(x) = \mathbf{0}$, and the system is in a steady state.

*8.4 IDENTIFIABILITY IN DETERMINISTIC LINEAR MODELS

In Section 5.13 we discussed optimal experimental designs which involve choosing the values of the explanatory x-variables to give maximum precision for parameter estimates. In experimental design with a compartmental model this appears, at first sight, to relate to the choice of time points $x_1, x_2, \ldots, x_n$, say, at which to sample the system. However, there are other choices involved. There may be a choice of compartments to sample, and in tracer experiments there may be a choice of where tracer inputs are to be made, how they are made, and in what quantities.

It is seldom possible to sample all compartments. The number of compartments to which tracers can be added may also be severely limited. Thus it may not be possible to separately identify the fractional transfer coefficients (Section 8.2.1) or other relevant parameters even with complete "error-free" information from the deterministic system. This lack of identifiability carries over to the stochastic models we describe subsequently. If the lack of identifiability makes it impossible to estimate parameters or test hypotheses of particular interest, it would be desirable to know this fact before carrying out the experiment. Consequently, a large literature has grown up on *a priori identifiability* in deterministic differential-equation models, particularly in the linear and compartmental case. Approximately half of Anderson [1983] is devoted to this topic. There are eight papers, each partly review articles, in the Richard Bellman Memorial Volume of *Mathematical Biosciences*, Vol. 77, 1985. The literature can also be approached via Cobelli [1981]. We shall confine our attention to a few easily applied practical methods for linear compartmental models. Of course, identifiability of relevant parameters is only a first step. When measurements are taken with error, the identifiable model may still be plagued with the extreme ill-conditioning problems discussed in Section 3.4, which make the precise estimation of parameters practically impossible.

In keeping with the literature on identifiability, let us rewrite the linear system (8.20) as

$$\begin{aligned} \dot{\mathbf{f}} &= \mathbf{A}(\boldsymbol{\theta})\mathbf{f} + \mathbf{B}(\boldsymbol{\theta})\mathbf{u}, \\ \mathbf{f}(0) &= \mathbf{0}, \\ \boldsymbol{\eta} &= \mathbf{C}(\boldsymbol{\theta})\mathbf{f}, \end{aligned} \tag{8.39}$$

where $\boldsymbol{\eta}$ is defined below. Nonzero initial amounts (called $\boldsymbol{\xi}$ in our previous discussion) are coped with by treating them as part of the input process, i.e. as a bolus dose at time $x = 0$ in $\mathbf{u}$. The above representation (8.39) explicitly shows that $\mathbf{A}$ depends on the $p \times 1$ vector of unknown parameters $\boldsymbol{\theta}$. The inputs $\mathbf{b} = \mathbf{B}(\boldsymbol{\theta})\mathbf{u}$ are represented in terms of $q \le m$ known inputs $\mathbf{u} = (u_1, u_2, \ldots, u_q)'$ and a matrix $\mathbf{B}(\boldsymbol{\theta})$ which allots inputs to compartments. Usually $\mathbf{B}$ consists of 0's and 1's. For example, with two inputs going to the first and third compartments respectively of a three-compartment system we have

$$\mathbf{B} = \begin{bmatrix} 1 & 0 \\ 0 & 0 \\ 0 & 1 \end{bmatrix}.$$

However, $\mathbf{B}$ may contain unknown parameters, as when an injection dose is modeled as going into two compartments in unknown proportions p and $1 - p$ respectively.

Finally, it is possible to measure $\boldsymbol{\eta}$ of dimension $d \le m$. Typically $\boldsymbol{\eta}$ is simply a subset of $\mathbf{f}$, so that $\mathbf{C}$ is then a constant $d \times m$ matrix of 0's and 1's. However, it too may involve unknown parameters, as in the two-compartment model of Jennrich and Bright [1976], where

$$\eta(x) = \theta_4[f_1(x) + f_2(x)].$$

The a priori identifiability problem is whether $\boldsymbol{\theta}$ is uniquely determined by the function $\boldsymbol{\eta}(x)$. If it is, the system is called globally or structurally identifiable. In contrast, the system is said to be locally identifiable if for almost any solution $\boldsymbol{\theta}^*$ of (8.39) there is a neighborhood of $\boldsymbol{\theta}^*$ in which there are no other solutions. This local condition is usually sufficient in practice (Jacquez and Greif [1985]).

From (8.23) with $\boldsymbol{\xi} = \mathbf{0}$,

$$\boldsymbol{\eta}(x) = \int_0^x \mathbf{C}e^{\mathbf{A}(x-r)}\mathbf{B}\mathbf{u}(r)\,dr, \tag{8.40}$$

which has Laplace transform (or transfer function)

$$\begin{aligned}\mathbf{L}_{\boldsymbol{\eta}}(s) &= \mathbf{C}(s\mathbf{I}_m - \mathbf{A})^{-1}\mathbf{B}\mathbf{L}_{\mathbf{u}}(s) \\ &= \hat{\boldsymbol{\Phi}}(s)\mathbf{L}_{\mathbf{u}}(s),\end{aligned} \tag{8.41}$$

where $\mathbf{L}_{\mathbf{u}}(s)$ is the Laplace transform of $\mathbf{u}$. As $\mathbf{u}$ is known $\mathbf{L}_{\mathbf{u}}(s)$ is known. Jacquez and Greif [1985] call $\hat{\boldsymbol{\Phi}}(s)$ the transfer function of the experiment. In order to identify $\boldsymbol{\theta}$ it is necessary and sufficient to be able to identify $\boldsymbol{\theta}$ from $\hat{\boldsymbol{\Phi}}(s)$, i.e., different values of $\boldsymbol{\theta}$ cannot give the same function. Many useful approaches have been based upon $\hat{\boldsymbol{\Phi}}(s)$, but before developing this further, we look at a simple example.

Example 8.6 Consider the two-compartment model in Fig. 8.9 with input to and measurement of compartment 1 only. This example was considered by Bellman and Amström [1970]. The kinetic equations are (c.f. Section 8.2.1)

$$\begin{aligned}\dot{f}_1 &= -(\gamma_{01} + \gamma_{21})f_1 + \gamma_{12}f_2 + u, \\ \dot{f}_2 &= \gamma_{21}f_1 - (\gamma_{02} + \gamma_{12})f_2,\end{aligned}$$

with $\eta = f_1$ and

$$\mathbf{b} = \begin{pmatrix} 1 \\ 0 \end{pmatrix} u.$$

Then $\mathbf{C} = \mathbf{B}' = (1, 0)$, $\alpha_{11} = -\gamma_{21} - \gamma_{01}$, $\alpha_{22} = -\gamma_{12} - \gamma_{02}$,

$$\mathbf{A} = \begin{pmatrix} \alpha_{11} & \gamma_{12} \\ \gamma_{21} & \alpha_{22} \end{pmatrix},$$

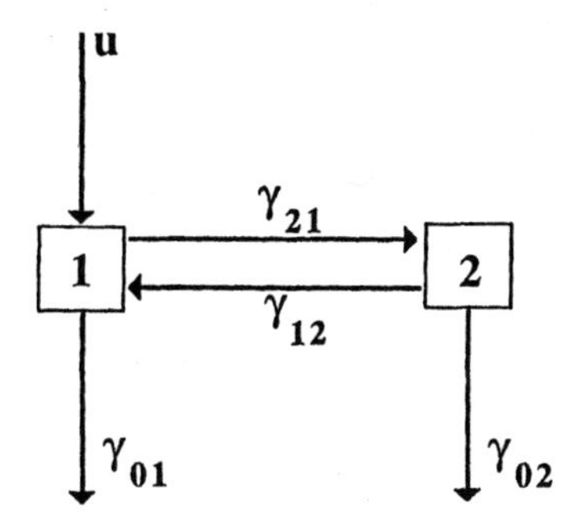

Figure 8.9 Two-compartment model with measurement on compartment 1.

and

$$\hat{\boldsymbol{\Phi}}(s) = \mathbf{C}(s\mathbf{I}_2 - \mathbf{A})^{-1}\mathbf{B}$$

$$= \frac{s - \alpha_{22}}{(s - \alpha_{11})(s - \alpha_{22}) - \gamma_{12}\gamma_{21}}$$

$$= \frac{s - c_1}{s^2 + c_2 s + c_3},$$

with no common factors for $\gamma_{12}\gamma_{21} \neq 0$. This gives us three equations in four unknowns, namely

$$c_1 = \alpha_{22} = -(\gamma_{12} + \gamma_{02}),$$

$$c_2 = -\alpha_{11} - \alpha_{22} = \gamma_{21} + \gamma_{01} + \gamma_{12} + \gamma_{02},$$

$$c_3 = \alpha_{11}\alpha_{22} - \gamma_{12}\gamma_{21} = (\gamma_{21}\gamma_{02} + \gamma_{01}\gamma_{12} + \gamma_{01}\gamma_{02}),$$

so there is no unique solution. The system is neither globally nor locally identifiable, as a continuum of different values of the γ_{jk} can give rise to the same $\hat{\boldsymbol{\Phi}}(s)$. However, as Anderson [1983] notes, if we could clamp off the exit from compartment 2 so that $\gamma_{02} = 0$, we would have

$$c_1 = -\gamma_{12},$$

$$c_2 = \gamma_{21} + \gamma_{12} + \gamma_{01},$$

$$c_3 = \gamma_{01}\gamma_{12}.$$

It is easily seen that this system has a unique solution, so that the system is globally identifiable. ■

The inverse Laplace transform of $\hat{\boldsymbol{\Phi}}(s) = \mathbf{C}(s\mathbf{I}_m - \mathbf{A})^{-1}\mathbf{B}$ is

$$\mathbf{R}(x) = \mathbf{C}e^{\mathbf{A}x}\mathbf{B}.$$

Thus identifying $\boldsymbol{\theta}$ from $\hat{\boldsymbol{\Phi}}(s)$ is equivalent to identifying $\boldsymbol{\theta}$ from $\mathbf{R}(x)$. As the number of compartments increases, direct use of $\hat{\boldsymbol{\Phi}}(s)$ or $\mathbf{C}e^{\mathbf{A}x}\mathbf{B}$ for the identifiability problem quickly involves the solution of very complicated nonlinear equations with tedious error prone derivations. Direct approaches like the above and other approaches in the literature can be facilitated by the use of symbolic algebra programs (Raksanyi et al. [1985]).

Some approaches to identifiability are based on the so-called *Markov parameters*

$$\mathbf{M}_k = \mathbf{C}\mathbf{A}^k\mathbf{B}, \qquad k = 0, 1, \dots. \tag{8.42}$$

Now by (8.22)

$$\mathbf{R}(x) = \mathbf{C}e^{\mathbf{A}x}\mathbf{B} = \sum_{j=0}^{\infty} \frac{\mathbf{M}_j x^j}{j!} \tag{8.43}$$

and thus

$$\mathbf{M}_k = \left[\frac{d^k\mathbf{R}(x)}{dx^k}\right]_{x=0}. \tag{8.44}$$

Suppose the characteristic polynomial of $\mathbf{A}$ is written

$$g(\lambda) = |\lambda\mathbf{I}_m - \mathbf{A}| = \sum_{j=0}^{m} d_j\lambda^j. \tag{8.45}$$

Then for integer $r > 0$,

$$d_m\mathbf{A}^{m+r} + d_{m-1}\mathbf{A}^{m+r-1} + \cdots + d_0\mathbf{A}^r = \mathbf{A}^r g(\mathbf{A}) = \mathbf{0}, \tag{8.46}$$

as $g(\mathbf{A}) = \mathbf{0}$ by the Cayley–Hamilton theorem (Appendix A2.1). Thus (8.46) provides a recursive formula for evaluating $\mathbf{A}^{m+r}$ in terms of lower powers of $\mathbf{A}$, so that, by induction, $\mathbf{A}^{m+r}$ can be written in terms of $\mathbf{I}_m$, $\mathbf{A}, \dots, \mathbf{A}^m$. Thus $\mathbf{M}_{m+r} = \mathbf{C}\mathbf{A}^{m+r}\mathbf{B}$ can be written in terms of $\mathbf{M}_0, \mathbf{M}_1, \dots, \mathbf{M}_m$.

We now consider the case where $\mathbf{B}(\boldsymbol{\theta}) \equiv \mathbf{B}$ and $\mathbf{C}(\boldsymbol{\theta}) \equiv \mathbf{C}$ are both known constant matrices. A necessary and sufficient condition for local identifiability based upon the Markov parameters is given by (8.50) below. Another approach is based upon the expansion of the transfer function $\hat{\boldsymbol{\Phi}}(s)$ as a ratio of two polynomials in s as we did in Example 8.6. It can be shown that (Anderson [1983: Section 17])

$$\hat{\boldsymbol{\Phi}}(s) = \frac{\mathbf{E}_{m-1} + s\mathbf{E}_{m-2} + \cdots + s^{m-1}\mathbf{E}_0}{|s\mathbf{I}_m - \mathbf{A}|}, \tag{8.47}$$

where $|s\mathbf{I}_m - \mathbf{A}| = a_m + a_{m-1}s + \cdots + a_1 s^{m-1} + s^m$, $\mathbf{E}_0 = \mathbf{M}_0 = \mathbf{C}\mathbf{B}$, $\mathbf{E}_j = [\mathbf{M}_j + a_1\mathbf{M}_{j-1} + \cdots + a_j\mathbf{M}_0]$ (a $d \times q$ matrix), and a_j is $(-1)^j$ times the sum of all the $j \times j$ principal minors of $\mathbf{A}$. Analogously to Example 8.6 (with $\mathbf{E}_0 = \mathbf{M}_0$ known, as

B and **C** are known), we can pick out the coefficients of s^k $(k = 1, 2, \ldots, m-1)$ in the numerator and denominator of (8.47), thus giving a maximum of $(m-1)dq + m$ nonlinear equations in p unknowns. The situation is complicated, since there may be cancellation between polynomials in the numerator and denominator. However, this formulation has been useful in constructing upper bounds on the number of independent equations which $\hat{\mathbf{\Phi}}(s)$ provides and thus finding necessary conditions for identifiability.

We now present a number of necessary conditions for global or structural identifiability for use when **B** and **C** are constant and **A** is a compartmental matrix. These conditions can be readily verified using the connectivity diagram. The two simplest necessary conditions for global identifiability are (Anderson [1983: Section 17]):

1. Each compartment must be reachable from at least one compartment receiving input. Otherwise we can never obtain information about transfers from a compartment which receives no input.
2. Any compartment possessing at least one path leaving it, whether to another compartment or an excretion, must be able to reach a sampled compartment.

Anderson [1983: Section 18] makes use of ν, the total number of independent equations in (8.47) available to be solved for $\boldsymbol{\theta}$. If $p = \dim(\boldsymbol{\theta}) > \nu$, then the ν nonlinear equations will not have a unique solution in any neighborhood. Anderson gives upper bounds N on ν. If $p > N$ the model is deemed unidentifiable.

CASE 1: Each input goes to just one compartment, and $\boldsymbol{\eta}$ is a subset of **f**. Let $I_1, I_2, \ldots, I_q$ denote the input compartments, and $O_1, O_2, \ldots, O_d$ denote the output compartments. Let $p(O_i, I_j)$ denote the minimum number of edges (steps) needed to traverse the connectivity diagram from I_j to O_i. Then it can be shown that

$$\nu \leq N = \sum_{i=1}^{d} \sum_{j=1}^{q} [m - p(O_i, I_j)] + m - t, \tag{8.48}$$

where t is the number of traps in the system. Here $[m - p(O_i, I_j)] = 0$ if O_i cannot be reached from I_j, and if $I_j = O_i$ then $p(O_i, I_j)$ is redefined to be 1.

The bound can be tightened if we consider the diagrams connected with subsystems, S_{ij} say, of compartments that both are reachable from I_j and can reach O_i. If the number of compartments in S_{ij} is $n(O_i, I_j)$, Anderson shows that

$$\nu \leq N = \sum_{i=1}^{d} \sum_{j=1}^{q} [n(O_i, I_j) - p(O_i, I_j)] + m - t. \tag{8.49}$$

CASE 2: An input may go to several compartments and a sampled output may come from several compartments. This case can be obtained from case 1 by replacing $p(O_i, I_j)$ in (8.48) and (8.49) by $p(\eta_i, u_j)$ taken from the shortest path from a compartment receiving part of u_j to a compartment contributing to η_i. Analogously we redefine $n(\eta_i, u_j)$.

Anderson [1983] tightens the bounds still further in the special case where

$$\eta = \sum_{j=1}^{m} f_j,$$

a so-called "washout" function.

Another useful condition which can be verified from the connectivity diagram is given by Eisenfeld [1985]. His Proposition 2.1 is stated as follows: Let α_{jk} and α_{rs} be transfer coefficients such that (a) every sequence from an input compartment to an output compartment which contains either one of the edges $k \to j$ or $s \to r$ contains the other, and (b) compartments j and s are excretory. Then α_{jk}, α_{rs}, α_{0k}, α_{0s} are identifiable parameters.

There are other methods, which we will not discuss, based upon the ideas of "controllability," "structural observability," and "exhaustive modeling" (Anderson [1983], Chapman and Godfrey [1985]), or "similarity transforms" (Jacquez and Greif [1985]); see also Vajda [1984]. We conclude this section with some verifiable criteria for local identifiability.

It may appear from (8.43) and the following discussion that $\mathbf{R}(x)$ is uniquely determined by the Markov parameters $\mathbf{M}_0, \mathbf{M}_1, \ldots, \mathbf{M}_m$. However, the coefficients d_r of the characteristic equation and in (8.46) are functions of the elements in $\mathbf{A}$ and thus of $\boldsymbol{\theta}$. In a 1979 paper Eisenfeld (see Eisenfeld [1986]) showed that at most the first $2m$ Markov matrix parameters are independent. Cobelli [1981] gave the following necessary and sufficient condition for local identifiability of $\boldsymbol{\theta}$:

$$\text{rank} \begin{bmatrix} \dfrac{\partial \mathbf{M}_0}{\partial \theta_1} & \cdots & \dfrac{\partial \mathbf{M}_0}{\partial \theta_p} \\ \dfrac{\partial \mathbf{M}_1}{\partial \theta_1} & \cdots & \dfrac{\partial \mathbf{M}_1}{\partial \theta_p} \\ \vdots & & \vdots \\ \dfrac{\partial \mathbf{M}_{2m-1}}{\partial \theta_1} & \cdots & \dfrac{\partial \mathbf{M}_{2m-1}}{\partial \theta_p} \end{bmatrix} = p. \tag{8.50}$$

He also gave a similar condition derived from the transfer function $\hat{\boldsymbol{\Phi}}(s)$ (c.f. Cobelli et al. [1979]). Similar criteria are discussed by Delforge [1984] and Eisenfeld [1986]. Anderson [1983: Section 22] also discussed this problem.

8.5 LINEAR COMPARTMENTAL MODELS WITH RANDOM ERROR

8.5.1 Introduction

The first step towards modeling the stochastic features of data being analyzed using compartmental models is to allow the measurement $f_j(x)$ on compartment j at time x to be made with error. Thus our ith measurement on f_j at time x_{ij} is

$$y_{ij} = f_j(x_{ij};\boldsymbol{\theta}) + \varepsilon_{ij}, \qquad i = 1, 2, \ldots, n_j, \quad j = 1, 2, \ldots, m, \tag{8.51}$$

where the ε_{ij} are uncorrelated with $E[\varepsilon_{ij}] = 0$ and $\text{var}[\varepsilon_{ij}] = \sigma^2$. This model justifies the long-established practice in compartmental analysis of estimating $\boldsymbol{\theta}$ by least squares. In line with Section 2.8.1, the relationship $y_{ij} \approx f_j(x_{ij};\boldsymbol{\theta})$ may have to be transformed before constructing a model such as (8.51). Methods for selecting a transformation are described in Section 2.8.

From (8.29) and (8.30), $f_j(x_{ij};\boldsymbol{\theta})$ can often be written as

$$f_j(x_{ij};\boldsymbol{\theta}) = \beta_{0j} + \sum_{r=1}^{m} \beta_{rj} e^{\lambda_r x_{ij}}, \tag{8.52}$$

where the β_{0j} are only needed if there are constant-rate inputs to the system and the jth element of $\mathbf{A}^{-1}\mathbf{b}$ is nonzero. In what follows we shall ignore the β_{0j}. If there is no exit from the system, then at least one eigenvalue, λ_1 say, is zero (Section 8.3.3, result 6). We shall call the set of β_{rj} and λ_j parameters the *exponential parameters*, and they are functions of the system parameters $\boldsymbol{\theta}$. The exponential parameters can be estimated by least squares, that is, by minimizing

$$\sum_j \sum_{i=1}^{n_j} \left(y_{ij} - \sum_{r=1}^{m} \beta_{rj} e^{\lambda_r x_{ij}} \right)^2, \tag{8.53}$$

where j ranges over the measured compartments. In compartmental modeling it has been common to estimate the exponential parameters by least squares, or even by crude methods such as exponential peeling described in Section 8.5.6 below, and then to solve for $\boldsymbol{\theta}$. Indeed, Anderson [1983: Section 23B] adopts this approach (see also Matis et al. [1983: 12]).

Example 8.7 We recall Example 8.5 and Equation (8.34). In Metzler [1971] measurements are taken from compartment 2, so that

$$f_2(x) = R\gamma_A \left(\frac{\gamma_{-1} - \gamma_A}{(\gamma_A + a)(\gamma_A + b)} e^{-\gamma_A x} + \frac{\gamma_{-1} + a}{(\gamma_A + a)(a - b)} e^{ax} + \frac{\gamma_{-1} + b}{(\gamma_A + b)(b - a)} e^{bx} \right) \tag{8.54}$$

$$= \beta_{12} e^{\lambda_1 x} + \beta_{22} e^{\lambda_2 x} + \beta_{32} e^{\lambda_3 x}. \tag{8.55}$$

Given estimates $\hat{\beta}_{12}, \hat{\beta}_{22}, \hat{\beta}_{32}, \hat{\lambda}_1, \hat{\lambda}_2$, and $\hat{\lambda}_3$ from the data, we could then solve for $\boldsymbol{\theta} = (R, \gamma_A, \gamma_1, \gamma_{-1}, \gamma_E)'$, where $\gamma_1 + \gamma_{-1} + \gamma_E = -(a+b)$ and $\gamma_{-1}\gamma_E = ab$, to obtain an estimate $\hat{\boldsymbol{\theta}}$.

However, a problem with this approach is immediately obvious. In our example there are more exponential parameters (six) than system parameters (five). If we also measured compartment 3, there would be nine exponential parameters and only five system parameters. It is thus extremely unlikely that the resulting nine equations in five unknowns would be consistent. Even if we have the same number of system parameters as exponential parameters and are able to solve for $\hat{\boldsymbol{\theta}}$, there is still the often difficult problem of finding its asymptotic dispersion matrix. We will therefore concentrate on methods that use $\boldsymbol{\theta}$ directly. ■

Let us write the model (8.51) in vector form, namely $\mathbf{y}^{(j)} = (y_{1j}, y_{2j}, \ldots, y_{n_j,j})'$, $\mathbf{f}^{(j)}(\boldsymbol{\theta}) = \{f_j(x_{1j}, \boldsymbol{\theta}), \ldots, f_j(x_{n_j,j}; \boldsymbol{\theta})\}'$, $\mathbf{y} = (\mathbf{y}^{(1)\prime}, \mathbf{y}^{(2)\prime}, \ldots, \mathbf{y}^{(m)\prime})'$, and $\mathbf{f}(\boldsymbol{\theta}) = \{\mathbf{f}^{(1)}(\boldsymbol{\theta})', \ldots, \mathbf{f}^{(m)}(\boldsymbol{\theta})'\}'$. If only $d < m$ compartments are measured, $\mathbf{y}$ and $\mathbf{f}$ are made up in the same way, but using only the $\mathbf{y}^{(j)}$ and $\mathbf{f}^{(j)}$ from measured compartments. To avoid confusion between $\mathbf{f}(\boldsymbol{\theta})$ as just defined and $\mathbf{f}$ defined by

$$\mathbf{f}'(x) = (f_1(x), f_2(x), \ldots, f_m(x))' = \mathbf{f}(x; \boldsymbol{\theta})'$$

as used prior to Section 8.5, we shall use $\mathbf{f}(\boldsymbol{\theta})$ and $\mathbf{f}(x)$ [or $\mathbf{f}(x;\boldsymbol{\theta})$] to distinguish between the two forms.

We can now write the model (8.51) as

$$\mathbf{y} = \mathbf{f}(\boldsymbol{\theta}) + \boldsymbol{\varepsilon}, \tag{8.56}$$

where $\mathscr{E}[\boldsymbol{\varepsilon}] = \mathbf{0}$, $\mathscr{D}[\boldsymbol{\varepsilon}] = \sigma^2 \mathbf{I}_n$, and n is the total number of observations ($= \sum_j n_j$ if all compartments are measured). The Gauss–Newton updating step for minimizing $[\mathbf{y} - \mathbf{f}(\boldsymbol{\theta})]'[\mathbf{y} - \mathbf{f}(\boldsymbol{\theta})]$ is [c.f. (2.29)]

$$\boldsymbol{\theta}^{(a+1)} = \boldsymbol{\theta}^{(a)} + [\mathbf{F}_{\cdot}'(\boldsymbol{\theta}^{(a)})\mathbf{F}_{\cdot}(\boldsymbol{\theta}^{(a)})]^{-1}\mathbf{F}_{\cdot}'(\boldsymbol{\theta}^{(a)})[\mathbf{y} - \mathbf{f}(\boldsymbol{\theta}^{(a)})],$$

and an estimate of the asymptotic variance–covariance matrix of $\hat{\boldsymbol{\theta}}$ is

$$\hat{\mathscr{D}}[\hat{\boldsymbol{\theta}}] = \hat{\sigma}^2[\mathbf{F}_{\cdot}'(\hat{\boldsymbol{\theta}})\mathbf{F}_{\cdot}(\hat{\boldsymbol{\theta}})]^{-1},$$

where

$$\mathbf{F}_{\cdot}(\boldsymbol{\theta}) = \frac{\partial \mathbf{f}(\boldsymbol{\theta})}{\partial \boldsymbol{\theta}'}. \tag{8.57}$$

Most of the other least-squares algorithms described in Chapter 14, for example the modified Gauss–Newton and the Levenburg–Marquardt methods, also make use of (8.57). We shall therefore devote much of our discussion, in particular Sections 8.5.2 and 8.5.3, to the computation of (8.57).

An added complication that we have introduced [c.f. (8.39)] is that instead of measuring samples from the compartments themselves, observations may be taken at time x on

$$\boldsymbol{\eta}(x;\boldsymbol{\theta}) = \mathbf{C}(\boldsymbol{\theta})\mathbf{f}(x;\boldsymbol{\theta}),$$

where $\mathbf{C}(\boldsymbol{\theta})$ is a $d \times n$ matrix ($d \le m$). Thus in place of (8.56) we have

$$\mathbf{y} = \boldsymbol{\eta}(\boldsymbol{\theta}) + \boldsymbol{\varepsilon}, \tag{8.58}$$

where $\boldsymbol{\eta}(\boldsymbol{\theta})$ is defined analogously to $\mathbf{f}(\boldsymbol{\theta})$ and we require $\partial\boldsymbol{\eta}(\boldsymbol{\theta})/\partial\boldsymbol{\theta}'$ in place of $\partial\mathbf{f}(\boldsymbol{\theta})/\partial\boldsymbol{\theta}'$. This adds no real difficulties since, by the chain rule,

$$\frac{\partial\eta_j(x;\boldsymbol{\theta})}{\partial\theta_r} = \sum_k \left(c_{jk}(\boldsymbol{\theta}) \frac{\partial f_k(x;\boldsymbol{\theta})}{\partial\theta_r} + \frac{\partial c_{jk}(\boldsymbol{\theta})}{\partial\theta_r} f_k(x;\boldsymbol{\theta}) \right).$$

We therefore concentrate on (8.56) and $\partial\mathbf{f}(\boldsymbol{\theta})/\partial\boldsymbol{\theta}'$.

8.5.2 Use of the Chain Rule

Let the vector $\boldsymbol{\delta}$ contain the exponential parameters $\{\lambda_j, \beta_{rj}\}$. When the differential equations (8.20) have been solved to obtain $\mathbf{f}(x;\boldsymbol{\theta})$ in an explicit analytical form as in Example 8.5 [see (8.34)], there are no theoretical difficulties in deriving $\partial\mathbf{f}(\boldsymbol{\theta})/\partial\boldsymbol{\theta}'$ by differentiation. However, the derivatives $\partial f_j(x_{ij};\boldsymbol{\theta})/\partial\delta_r$ with respect to the exponential parameters are much easier to obtain [c.f. (8.29) and (8.34)]. Using the chain rule (Appendix A10.3) gives us

$$\frac{\partial\mathbf{f}(\boldsymbol{\theta})}{\partial\boldsymbol{\theta}'} = \frac{\partial\mathbf{f}}{\partial\boldsymbol{\delta}'}\frac{\partial\boldsymbol{\delta}}{\partial\boldsymbol{\theta}'}, \tag{8.59}$$

or

$$\mathbf{F}_{\bullet}(\boldsymbol{\theta}) = \mathbf{G}_{\bullet}(\boldsymbol{\delta})\frac{\partial\boldsymbol{\delta}}{\partial\boldsymbol{\theta}'},$$

say. The calculation of $\mathbf{F}_{\bullet}(\boldsymbol{\theta})$ by calculating $\mathbf{G}_{\bullet}(\boldsymbol{\delta})$ followed by computer multiplication by $\partial\boldsymbol{\delta}/\partial\boldsymbol{\theta}'$ may be less open to human error. The situation would be further improved if we could eliminate analytic differentiation altogether, as in Section 8.5.3 below.

In some compartmental modeling, particularly in chemistry, the fractional transfer rates γ_{jk} are themselves functions of the experimental conditions such as temperature and pressure, as, for example, in the Arrhenius model (McLean [1985]). These functions have unknown parameters $\boldsymbol{\phi}$, so that $\boldsymbol{\theta} = \boldsymbol{\theta}(\boldsymbol{\phi})$, where $\dim(\boldsymbol{\phi}) < \dim(\boldsymbol{\theta})$, and $\boldsymbol{\phi}$ contains the parameters of interest. If we have a method of computing $\mathbf{F}_{\bullet}(\boldsymbol{\theta})$, we can then convert this to the required $\partial\mathbf{f}/\partial\boldsymbol{\phi}'$ using the same method, namely

$$\frac{\partial \mathbf{f}}{\partial \boldsymbol{\phi}'} = \mathbf{F}_{\bullet}(\boldsymbol{\theta}) \frac{\partial \boldsymbol{\theta}}{\partial \boldsymbol{\phi}'}.$$

In compartmental modeling, rate constants are positive. Hence Bates and Watts [1985a] and Bates et al. [1986] use the parameters $\phi_i = \log \theta_i$ so that $\theta_i = \exp(\phi_i)$ is forced to be positive. This can have the disadvantage that if a path between two compartments is not necessary for modeling the data (i.e. $\theta_i = 0$ for some i), or if $\hat{\theta}_i$ is negative, the minimization algorithm tends to make $\phi_i^{(a)}$ more and more negative as the iteration proceeds until it stalls. Bates and Watts [1985a: Response to Discussion] found this situation easy to recognize and found the transformation to be generally useful. They pointed out problems with another positivity transformation $\theta_i = \phi_i^2$, which can take the value of $\phi_i = 0$.

8.5.3 Computer-Generated Exact Derivatives

a Method of Bates et al.

We now present a method given by Bates et al. [1986] which generalizes Jennrich and Bright [1976] and Kalbfleisch et al. [1983]. This method enables us to compute $\mathbf{F}_{\bullet}(\boldsymbol{\theta}) = \partial \mathbf{f}(\boldsymbol{\theta})/\partial \boldsymbol{\theta}'$ given only $\mathbf{A} = [(\alpha_{rs}\{\boldsymbol{\theta}\})]$, the initial state $\boldsymbol{\xi} = \boldsymbol{\xi}(\boldsymbol{\theta})$ of the system, and the derivatives of these functions with respect to $\boldsymbol{\theta}$. Such functions are typically trivial to compute. Usually $\boldsymbol{\theta}$ will include the nonzero α_{rs}, so that these derivatives are either zero or one. Because the model can be expressed simply in this way, the method is well suited to computer package implementation, and a number of difficult and error-prone processes are avoided. For example, it is no longer necessary to solve the linear system (8.20) explicitly. Also it is no longer necessary to correctly differentiate complicated functions and to program the results. Both steps frequently involve errors and waste a lot of time.

Recall that the linear system (8.20) is

$$\dot{\mathbf{f}}(x) = \mathbf{A}\mathbf{f}(\mathbf{x}) + \mathbf{b}(x) \tag{8.60}$$

with solution (8.23), namely

$$\mathbf{f}(x) = e^{\mathbf{A}x}\boldsymbol{\xi} + \int_0^x e^{\mathbf{A}(x-u)}\mathbf{b}(u)\,du. \tag{8.61}$$

This can be written as

$$\mathbf{f}(x) = e^{\mathbf{A}x}\boldsymbol{\xi} + e^{\mathbf{A}x} * \mathbf{b}(x), \tag{8.62}$$

where $*$ denotes the convolution operator. (We note that Bates et al. [1986] include $\boldsymbol{\xi}$ in $\mathbf{b}(x)$ using a Dirac delta function at $x = 0$: our notation differs from that used in the above papers.) When $\mathbf{b}(x) \equiv \mathbf{b}$ $[= \mathbf{b}(\boldsymbol{\theta})]$, a constant input rate, then (8.61) reduces to [c.f. (8.24)]

$$\mathbf{f}(x) = e^{\mathbf{A}x}\boldsymbol{\xi} + (e^{\mathbf{A}x} - \mathbf{I}_m)\mathbf{A}^{-1}\mathbf{b}, \tag{8.63}$$

which has the familiar form $e^{\mathbf{A}x}\boldsymbol{\xi}$ if there are no inputs ($\mathbf{b} = \mathbf{0}$).

We now differentiate (8.60) with respect to θ_k to obtain

$$\frac{\partial \dot{\mathbf{f}}(x;\boldsymbol{\theta})}{\partial \theta_k} = \mathbf{A}\frac{\partial \mathbf{f}(x;\boldsymbol{\theta})}{\partial \theta_k} + [\mathbf{A}^{(k)}\mathbf{f}(x;\boldsymbol{\theta}) + \mathbf{b}^{(k)}(x;\boldsymbol{\theta})], \tag{8.64}$$

where $\mathbf{A}^{(k)} = \partial\mathbf{A}/\partial\theta_k$ and $\mathbf{b}^{(k)} = \partial\mathbf{b}/\partial\theta_k$. We can also add $\boldsymbol{\xi}^{(k)} = \partial\boldsymbol{\xi}/\partial\theta_k$. Then (8.64) is a linear system of the form (8.60) with $\partial\mathbf{f}/\partial\theta_k$ playing the role of $\mathbf{f}$, and with initial condition

$$\left[\frac{\partial \mathbf{f}(x)}{\partial \theta_k}\right]_{x=0} = \boldsymbol{\xi}^{(k)}.$$

Thus it has solution

$$\frac{\partial \mathbf{f}(x)}{\partial \theta_k} = e^{\mathbf{A}x}\boldsymbol{\xi}^{(k)} + e^{\mathbf{A}x} * [\mathbf{A}^{(k)}\mathbf{f}(x) + \mathbf{b}^{(k)}(x)], \tag{8.65}$$

where we suppress the dependence upon $\boldsymbol{\theta}$ for brevity. In the case of constant inputs we have $\mathbf{b}^{(k)}(x) \equiv \mathbf{b}^{(k)}$ and, by (8.63), (8.65) becomes

$$\begin{aligned}\frac{\partial \mathbf{f}(x)}{\partial \theta_k} &= e^{\mathbf{A}x}\boldsymbol{\xi}^{(k)} + e^{\mathbf{A}x} * [\mathbf{A}^{(k)}\mathbf{f}(x)] + (e^{\mathbf{A}x} - \mathbf{I})\mathbf{A}^{-1}\mathbf{b}^{(k)} \\ &= e^{\mathbf{A}x}\boldsymbol{\xi}^{(k)} + e^{\mathbf{A}x} * [\mathbf{A}^{(k)}\{e^{\mathbf{A}x}\boldsymbol{\xi} + (e^{\mathbf{A}x} - \mathbf{I})\mathbf{A}^{-1}\mathbf{b}\}] + (e^{\mathbf{A}x} - \mathbf{I})\mathbf{A}^{-1}\mathbf{b}^{(k)} \\ &= e^{\mathbf{A}x}\boldsymbol{\xi}^{(k)} + e^{\mathbf{A}x} * [\mathbf{A}^{(k)}e^{\mathbf{A}x}](\boldsymbol{\xi} + \mathbf{A}^{-1}\mathbf{b}) + (e^{\mathbf{A}x} - \mathbf{I})\mathbf{A}^{-1}(\mathbf{b}^{(k)} - \mathbf{A}^{(k)}\mathbf{A}^{-1}\mathbf{b}),\end{aligned} \tag{8.66}$$

since

$$e^{\mathbf{A}x} * [\mathbf{A}^{(k)}\mathbf{A}^{-1}\mathbf{b}] = (e^{\mathbf{A}x} - \mathbf{I})\mathbf{A}^{-1}\mathbf{A}^{(k)}\mathbf{A}^{-1}\mathbf{b}.$$

Often $\boldsymbol{\xi}$ and $\mathbf{b}$ will be known constants, so that $\boldsymbol{\xi}^{(k)} = \mathbf{0}$ and $\mathbf{b}^{(k)} = \mathbf{0}$ with resulting simplifications. For zero inputs ($\mathbf{b} = \mathbf{0}$), (8.66) becomes

$$\frac{\partial \mathbf{f}(x)}{\partial \theta_k} = e^{\mathbf{A}x}\boldsymbol{\xi}^{(k)} + e^{\mathbf{A}x} * [\mathbf{A}^{(k)}e^{\mathbf{A}x}]\boldsymbol{\xi}. \tag{8.67}$$

Bates et al. [1986] showed that similar expressions can be obtained for second derivatives if desired. For (8.66) and (8.67) to be useful we need to be able to compute these expressions, particularly the convolutions. This is done using the spectral decomposition $\mathbf{A} = \mathbf{S}\boldsymbol{\Lambda}\mathbf{S}^{-1}$ of (8.25) again as follows.

We showed in (8.35) that a convolution of the form

$$e^{\mathbf{A}x} * \mathbf{b}(x) = \int_0^x e^{\mathbf{A}(x-u)}\mathbf{b}(u)\,du$$

can be evaluated from scalar convolutions using the spectral decomposition. We can use this method to find $\mathbf{f}(x)$ of (8.62) and then apply it to (8.65) to get $\partial\mathbf{f}(x)/\partial\theta_k$. Useful explicit results can be obtained in the special case of constant input rates $[\mathbf{b}(x) \equiv \mathbf{b}]$. Using the spectral decomposition,

$$\begin{aligned} e^{\mathbf{A}x} * [\mathbf{A}^{(k)} e^{\mathbf{A}x}] &= \mathbf{S}e^{\Lambda x}\mathbf{S}^{-1} * [\mathbf{A}^{(k)}\mathbf{S}e^{\Lambda x}\mathbf{S}^{-1}] \\ &= \mathbf{S}\{e^{\Lambda x} * [(\mathbf{S}^{-1}\mathbf{A}^{(k)}\mathbf{S})e^{\Lambda x}]\}\mathbf{S}^{-1} \\ &= \mathbf{S}\{e^{\Lambda x} * [\mathbf{G}^{(k)}e^{\Lambda x}]\}\mathbf{S}^{-1} \\ &= \mathbf{S}\mathbf{B}^{(k)}\mathbf{S}^{-1}. \end{aligned} \tag{8.68}$$

Here $\mathbf{G}^{(k)} = \mathbf{S}^{-1}\mathbf{A}^{(k)}\mathbf{S}$ with (i, j)th element $g_{ij}^{(k)} = [\mathbf{S}^{-1}\mathbf{A}^{(k)}\mathbf{S}]_{ij}$, and

$$\mathbf{B}^{(k)} = e^{\Lambda x} * [\mathbf{G}^{(k)} e^{\Lambda x}] = \int_0^x e^{\Lambda(x-u)}\mathbf{G}^{(k)}e^{\Lambda u}\,du$$

with (i, j)th element

$$\begin{aligned} b_{ij}^{(k)} &= g_{ij}^{(k)} \int_0^x e^{\lambda_i(x-u)} e^{\lambda_j u}\,du \\ &= g_{ij}^{(k)} \lambda_{ij}(x), \end{aligned} \tag{8.69}$$

where

$$\lambda_{ij}(x) = \begin{cases} \dfrac{e^{\lambda_i x} - e^{\lambda_j x}}{\lambda_i - \lambda_j}, & \lambda_i \neq \lambda_j, \\ x e^{\lambda_i x}, & \lambda_i = \lambda_j. \end{cases} \tag{8.70}$$

Bates et al. [1986] state that "in practice, the condition $\lambda_i = \lambda_j$ is determined by comparing $|(\lambda_i - \lambda_j)x|$ to machine precision." This is done to prevent the subtraction $\exp(\lambda_i x) - \exp(\lambda_j x)$ from losing too many significant figures.

A disadvantage with the above procedure is that it uses the spectral decomposition, which is relatively unstable for nonsymmetric matrices and need not exist (see Section 8.3.1). However, Bates et al. [1986] state that it is possible to compute $e^{\mathbf{A}x}$ and $e^{\mathbf{A}x} * (\mathbf{A}^{(k)}e^{\mathbf{A}x})$ using the stable decomposition of Bavely and Stewart [1979]. The equivalent decomposition of $\mathbf{B}^{(k)}$ is much more complicated.

Example 8.8 The Brunhilda data of Jennrich and Bright [1976] are given in Table 8.1. It consists of measurements on the radioactivity of blood samples

Table 8.1 Radioactive Counts of Blood Samples Taken at Specified Times after a Bolus Injection of Radioactive Sulfate[a]

Time (min)	Count	Time (min)	Count	Time (min)	Count
2	151117	25	70593	90	53915
4	113601	30	67041	110	50938
6	97652	40	64313	130	48717
8	90935	50	61554	150	45996
10	84820	60	59940	160	44968
15	76891	70	57698	170	43602
20	73342	80	56440	180	42668

[a]From Jennrich and Bright [1976].

taken from a baboon called Brunhilda at a number of specified times after a bolus injection of radioactive sulfate. Jennrich and Bright analyzed these data using the three-compartment catenary model depicted in Fig. 8.10. The measurements were treated as coming from compartment 1, and the bolus injection was taken as going into compartment 1. Bates et al. [1986] reanalyzed these data with a free parameter θ_6 representing the initial radioactivity in compartment 1 immediately following the injection. The linear system defining the model is

$$\begin{aligned}
\dot{f}_1(x) &= -(\theta_1+\theta_2)f_1 + \theta_3 f_2, \\
\dot{f}_2(x) &= \theta_2 f_1 - (\theta_3+\theta_4)f_2 + \theta_5 f_3, \\
\dot{f}_3(x) &= \theta_4 f_2 - \theta_5 f_3,
\end{aligned} \tag{8.71}$$

subject to $\boldsymbol{\xi} = (\theta_6, 0, 0)'$. Since there was no further input of tracer, the simple version (8.67) of $\partial \mathbf{f}(x)/\partial\theta_k$ applies. The quantities required for applying the above procedure are

$$\boldsymbol{\xi}^{(1)} = \boldsymbol{\xi}^{(2)} = \cdots = \boldsymbol{\xi}^{(5)} = \mathbf{0}, \qquad \boldsymbol{\xi}^{(6)} = (1,0,0)',$$

$$\mathbf{A} = \begin{bmatrix} -(\theta_1+\theta_2) & \theta_3 & 0 \\ \theta_2 & -(\theta_3+\theta_4) & \theta_5 \\ 0 & \theta_4 & -\theta_5 \end{bmatrix},$$

$$\mathbf{A}^{(1)} = \begin{bmatrix} -1 & 0 & 0 \\ 0 & 0 & 0 \\ 0 & 0 & 0 \end{bmatrix}, \qquad \mathbf{A}^{(2)} = \begin{bmatrix} -1 & 0 & 0 \\ 1 & 0 & 0 \\ 0 & 0 & 0 \end{bmatrix}, \dots, \qquad \mathbf{A}^{(6)} = \mathbf{0}.$$

■

Suppose that we measure $\eta(x) = \mathbf{c}'(\boldsymbol{\theta})\mathbf{f}(x)$ at various time points x. We consider

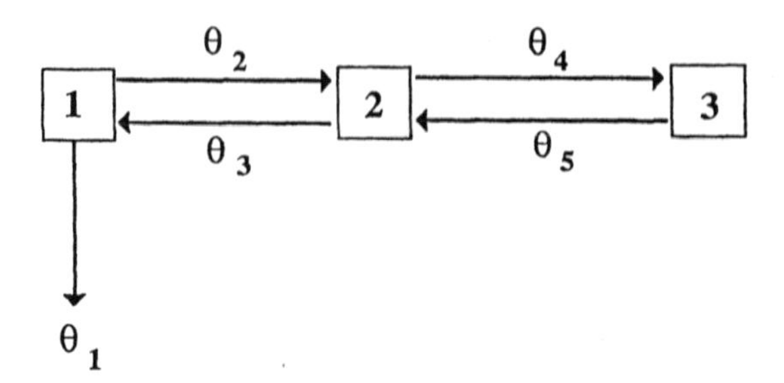

Figure 8.10 Three-compartment model for Brunhilda blood data.

here only the case of no input after time $x=0$, so that $\mathbf{f}(x)=e^{\mathbf{A}x}\boldsymbol{\xi}$. Using the chain rule,

$$\frac{\partial \eta(x)}{\partial \theta_k}=\mathbf{c}^{(k)\prime}\mathbf{f}(x)+\mathbf{c}'\frac{\partial \mathbf{f}(x)}{\partial \theta_k}, \tag{8.72}$$

where $\mathbf{c}^{(k)}=\partial\mathbf{c}(\boldsymbol{\theta})/\partial\theta_k$. We can evaluate (8.72) using (8.26) (8.67) and (8.68). However, the approach of Jennrich and Bright [1976] described below involves only one matrix of a similar form to $\mathbf{SBS}^{-1}$ for computing $\partial\eta/\partial\theta_k$ $(k=1,2,\ldots,p)$, rather than the p matrices (8.68) required by the above approach. It may, therefore, be possible to develop faster algorithms using Jennrich and Bright's approach.

b Method of Jennrich and Bright

Using the chain rule with η a function of $\mathbf{c}$, $\boldsymbol{\xi}$, and $\mathbf{A}=[(\alpha_{uv})]$, we have

$$\begin{aligned}\frac{\partial\eta}{\partial\theta_k}&=\frac{\partial\eta}{\partial\mathbf{c}'}\frac{\partial\mathbf{c}}{\partial\theta_k}+\frac{\partial\eta}{\partial\boldsymbol{\xi}'}\frac{\partial\boldsymbol{\xi}}{\partial\theta_k}+\sum_u\sum_v\frac{\partial\eta}{\partial\alpha_{uv}}\frac{\partial\alpha_{uv}}{\partial\theta_k}\\&=\mathbf{f}'(x)\mathbf{c}^{(k)}+\mathbf{c}'e^{\mathbf{A}x}\boldsymbol{\xi}^{(k)}+\sum_u\sum_v\frac{\partial\eta}{\partial\alpha_{uv}}\frac{\partial\alpha_{uv}}{\partial\theta_k}.\end{aligned} \tag{8.73}$$

We now show that the matrix $\partial\eta/\partial\mathbf{A}$ with (u,v)th element $\partial\eta/\partial\alpha_{uv}$ is given by

$$\frac{\partial\eta}{\partial\mathbf{A}}=(\mathbf{SMS}^{-1})', \tag{8.74}$$

where $\mathbf{M}$ has (u,v)th element

$$m_{uv}=\tilde{\xi}_u\lambda_{uv}\tilde{c}_v$$

with $\tilde{\boldsymbol{\xi}}=\mathbf{S}^{-1}\boldsymbol{\xi}$, $\tilde{\mathbf{c}}=\mathbf{S}'\mathbf{c}$, and λ_{uv} defined by (8.70). Firstly, since $\mathbf{f}(x)=e^{\mathbf{A}x}\boldsymbol{\xi}$, we have

$$\frac{\partial\eta}{\partial\alpha_{uv}}=\mathbf{c}'\frac{\partial e^{\mathbf{A}x}}{\partial\alpha_{uv}}\boldsymbol{\xi}.$$

Then we find $\partial e^{\mathbf{A}x}/\partial\alpha_{uv}$, essentially by using $\theta_k = \alpha_{uv}$ and just the second term in (8.67), namely

$$\begin{aligned}\frac{\partial\eta}{\partial\alpha_{uv}} &= \mathbf{c}'e^{\mathbf{A}x} * [\mathbf{A}^{(u,v)}e^{\mathbf{A}x}]\boldsymbol{\xi} \\ &= \mathbf{c}'\mathbf{S}\mathbf{B}^{(u,v)}\mathbf{S}^{-1}\boldsymbol{\xi} \qquad [\text{by (8.68)}] \\ &= \tilde{\mathbf{c}}'\mathbf{B}^{(u,v)}\tilde{\boldsymbol{\xi}}, \end{aligned} \tag{8.75}$$

where the superscripts (u,v) denote differentiation with respect to α_{uv}. Also $\mathbf{A}^{(u,v)}$ has all zero elements apart from unity in the (u,v)th position. Hence $\mathbf{S}^{-1}\mathbf{A}^{(u,v)}\mathbf{S}$ has (i,j)th element $g_{ij}^{(u,v)} = s^{iu}s_{vj}$ where $s^{iu} = (\mathbf{S}^{-1})_{iu}$, and from (8.69) and (8.75)

$$\frac{\partial\eta}{\partial\alpha_{uv}} = \sum_i\sum_j \tilde{c}_i s^{iu} s_{vj}\lambda_{ij}\tilde{\xi}_j.$$

This is the (u,v)th element of $(\mathbf{S}\mathbf{M}\mathbf{S}^{-1})'$.

We can readily generalize the method of Jennrich and Bright [1976] to $\boldsymbol{\eta}(x) = \mathbf{C}(\boldsymbol{\theta})\mathbf{f}(x)$ for a $q \times m$ matrix $\mathbf{C}(\boldsymbol{\theta})$ by applying (8.73) and (8.74) to each element of $\boldsymbol{\eta}$. Thus

$$\frac{\partial\eta_j}{\partial\mathbf{A}} = (\mathbf{S}\mathbf{M}^{(j)}\mathbf{S}^{-1})',$$

where $m_{uv}^{(j)} = \tilde{\xi}_u\lambda_{uv}\tilde{c}_v^{(j)}$, $\tilde{\mathbf{c}}^{(j)} = \mathbf{S}'\mathbf{c}^{(j)}$, and $\mathbf{c}^{(j)\prime}$ is the jth row of $\mathbf{C}(\boldsymbol{\theta})$. This requires q matrices of the form $\mathbf{S}\mathbf{M}^{(j)}\mathbf{S}^{-1}$.

c General Algorithm

The above methods of Bates et al. and Jennrich and Bright may be summarized as follows:

(i) Use the current value of $\boldsymbol{\theta}$ to compute $\mathbf{A}(\boldsymbol{\theta})$, $\boldsymbol{\xi}(\boldsymbol{\theta})$ [and $\mathbf{c}(\boldsymbol{\theta})$ if necessary].

(ii) Compute the spectral decomposition of $\mathbf{A}$, i.e., find $\boldsymbol{\Lambda}$ and $\mathbf{S}$.

(iii) For each time point x:

- **a.** Compute $\mathbf{f}(x)$ using $\boldsymbol{\Lambda}$ and $\mathbf{S}$ (c.f. Section 8.3.1).
- **b.** Compute $\partial\mathbf{f}(x)/\partial\theta_k$ (or $\partial\boldsymbol{\eta}/\partial\theta_k$) for $k = 1, 2, \ldots, p$.
- **c.** Use the results of substeps *a* and *b* in $\mathbf{f}(\theta)$ and $\mathbf{F}_{\bullet}(\theta)$ of Equations (8.56) and (8.57).

(iv) Update $\boldsymbol{\theta}$ using an appropriate algorithm such as Gauss–Newton.

For step (ii), Bates et al. [1986] recommend using standard eigenvalue–eigenvector routines such as those in EISPACK (Section 15.1.1, Smith et al. [1976a]), but note that such routines usually return a decomposition even when the spectral decomposition does not exist. The only warning of this, or of the instability of the computation, is an S-matrix with a large condition number. Advice on efficient implementation is given in Jennrich and Bright [1976].

Sometimes in kinetic modeling the measured time x does not correspond to effective time in the system. The linear equations (8.60) apply more realistically to $x = \max(x, \tau)$ for some unknown $\tau > 0$. The interval $[0, \tau]$ is referred to as "dead time" or "lag time." Bates and Watts [1985b] give simple adjustments to the above analysis to accommodate dead time.

8.5.4 The Use of Constraints

Working with the sums-of-exponentials model (8.52) is attractive because differentiation of $f_j(x_{ij}; \boldsymbol{\beta}, \boldsymbol{\lambda})$ is simple. In contrast, finding $f_j(x; \boldsymbol{\theta})$ explicitly and differentiating with respect to the $\boldsymbol{\theta}$-parameters can be a tedious, error-prone process. This was an important motivation for the development of the methods of the previous section. Allen [1983] uses an approach which can often be implemented using available software. He advocates using a constrained nonlinear least-squares program and imposing the relationship between $(\boldsymbol{\beta}, \boldsymbol{\lambda})$ and $\boldsymbol{\theta}$ as constraints. If hypotheses are to be tested which impose further constraints on $\boldsymbol{\theta}$, then the hypothesis constraints are simply added to the constraint list. This approach also allows the estimation of implicit functions of the parameters. The ideas used are completely general and can apply to any nonlinear model.

Example 8.9 We now return to the three-compartment model explored in Examples 8.2, 8.5, and 8.7. Measurements of drug concentrations taken on compartment 2 follow the model [c.f. (8.54)]

$$f_2(x) = R\gamma_A\left(\frac{\gamma_{-1} + a}{(\gamma_A + a)(a - b)}e^{ax} + \frac{\gamma_{-1} + b}{(\gamma_A + b)(b - a)}e^{bx} + \frac{\gamma_{-1} - \gamma_A}{(\gamma_A + a)(\gamma_A + b)}e^{-\gamma_A x}\right), \tag{8.76}$$

where

$$a + b = -(\gamma_1 + \gamma_{-1} + \gamma_E), \qquad ab = \gamma_{-1}\gamma_E. \tag{8.77}$$

We can solve (8.77) for a and b and substitute these values into (8.76) to obtain $f_2(x; \boldsymbol{\theta})$ in terms of just the system parameters $\boldsymbol{\theta} = (R, \gamma_1, \gamma_{-1}, \gamma_E, \gamma_A)'$. An algorithm such as Gauss–Newton then requires the derivatives $\partial \mathbf{f}/\partial \boldsymbol{\theta}'$. However, differentiating $f_2(x; \boldsymbol{\theta})$ with respect to the elements of $\boldsymbol{\theta}$ is clearly more difficult than differentiating $f_2(x)$ in (8.76) with respect to $\boldsymbol{\delta} = (a, b, \boldsymbol{\theta}')'$. Allen [1983] advocates fitting the latter model, written as

$$y = f(x; \boldsymbol{\delta}) + \varepsilon, \tag{8.78}$$

subject to the constraints

$$ab - \gamma_{-1}\gamma_E = 0, \tag{8.79a}$$

$$a + b + \gamma_1 + \gamma_{-1} + \gamma_E = 0 \tag{8.79b}$$

using a constrained nonlinear (generalized) least-squares program. This procedure is clearly less efficient in terms of computer time than fitting $y = f(x;\boldsymbol{\theta}) + \varepsilon$ directly. However, it is a less error-prone procedure and is therefore more efficient in terms of programming time—a factor which is assuming much more importance as computing costs fall.

The constrained method also allows us to estimate functions of the parameters and is particularly useful for those that can only be expressed implicitly. For example, by differentiating (8.76) we see that ρ, the time at which maximum concentration is achieved, is the solution of

$$0 = a\frac{\gamma_{-1}+a}{(\gamma_A+a)(a-b)}e^{a\rho} + b\frac{\gamma_{-1}+b}{(\gamma_A+b)(b-a)}e^{b\rho} - \frac{(\gamma_{-1}-\gamma_A)\gamma_A}{(\gamma_A+a)(\gamma_A+b)}e^{-\gamma_A\rho}. \tag{8.80}$$

The estimation of ρ is achieved by simply adding (8.80) as a further constraint and augmenting $\boldsymbol{\delta}$ to $(a, b, \boldsymbol{\theta}', \rho)'$.

In line with our earlier comments, we can take the restriction process a step further and write $f_2(x)$ in terms of its exponential parameters [c.f. (8.55)]

$$f_2(x) = \beta_1 e^{ax} + \beta_2 e^{bx} + \beta_3 e^{-\gamma_A x}$$

subject to the constraints

$$\beta_1 = R\gamma_A\frac{\gamma_{-1}+a}{(\gamma_A+a)(a-b)},$$

$$\beta_2 = R\gamma_A\frac{\gamma_{-1}+b}{(\gamma_A+b)(b-a)}, \tag{8.81}$$

and

$$\beta_3 = \frac{R\gamma_A(\gamma_{-1}-\gamma_A)}{(\gamma_A+a)(\gamma_A+b)},$$

together with (8.79). The above constraints can be simplified to

$$\begin{aligned} 0 &= \beta_1 + \beta_2 + \beta_3, \\ 0 &= \beta_1(\gamma_A+a)(\gamma_{-1}+b) + \beta_2(\gamma_{-1}+a)(\gamma_A+b), \\ 0 &= \beta_1(\gamma_A+a)(b-a) + R\gamma_A(\gamma_{-1}+a). \end{aligned} \tag{8.82}$$

If the estimation of ρ is also required, we can rewrite (8.80) as

$$0 = \beta_1 a e^{a\rho} + \beta_2 b e^{b\rho} - \beta_3\gamma_A e^{-\gamma_A\rho}, \tag{8.83}$$

and $\boldsymbol{\delta}$ becomes $(a, b, \beta_1, \beta_2, \beta_3, \boldsymbol{\theta}', \rho)'$. ■

In general, suppose we want to fit a model of the form

$$\mathbf{y} = \mathbf{f}(\boldsymbol{\theta}) + \boldsymbol{\varepsilon} \tag{8.84}$$

in which $\mathbf{f}(\boldsymbol{\theta})$ can be rewritten in terms of the parameters $\boldsymbol{\beta}$ [$= \boldsymbol{\beta}(\boldsymbol{\theta})$]. Suppose also that the derivatives $\partial f_j/\partial \theta_k$ are complicated, whereas the derivatives $\partial f_j/\partial \beta_k$ are of simple form. We can write $\boldsymbol{\delta} = (\boldsymbol{\beta}', \boldsymbol{\theta}')'$ (where redundancies in $\boldsymbol{\beta}$ and $\boldsymbol{\theta}$ are omitted) and express the equations linking $\boldsymbol{\beta}$ and $\boldsymbol{\theta}$ in terms of the restriction

$$\mathbf{h}(\boldsymbol{\delta}) = \mathbf{0}. \tag{8.85}$$

Writing $\mathbf{f}(\boldsymbol{\theta}) = \mathbf{g}(\boldsymbol{\delta})$, the method is to fit (8.84) by (generalized) least squares by solving the constrained problem

$$\text{Minimize} \quad \{\mathbf{y} - \mathbf{g}(\boldsymbol{\delta})\}'\boldsymbol{\Sigma}^{-1}\{\mathbf{y} - \mathbf{g}(\boldsymbol{\theta})\}$$
$$\text{subject to} \quad \mathbf{h}(\boldsymbol{\delta}) = \mathbf{0} \tag{8.86}$$

using a constrained least-squares program, with $\boldsymbol{\Sigma} = \mathbf{I}_n$ for ordinary least squares. Other functions of the parameters which are of interest, say $\boldsymbol{\rho}$, may be estimated by augmenting $\boldsymbol{\delta}$ to include $\boldsymbol{\rho}$, and augmenting $\mathbf{h}(\boldsymbol{\delta}) = \mathbf{0}$ to include the restrictions describing $\boldsymbol{\rho}$.

To use the constraints method for inference we still need the asymptotic distribution of $\hat{\boldsymbol{\delta}}$. Allen [1983] quotes a 1981 University of Kentucky thesis in which A. A. Lu proved the following: Consider the model $\mathbf{y} = \mathbf{g}(\boldsymbol{\delta}) + \boldsymbol{\varepsilon}$, where $\mathscr{E}[\boldsymbol{\varepsilon}] = \mathbf{0}$, $\mathscr{D}[\boldsymbol{\varepsilon}] = \boldsymbol{\Sigma}$ and $\boldsymbol{\delta}$ is subject to $\mathbf{h}(\boldsymbol{\delta}) = \mathbf{0}$. Then under essentially the same conditions on $\mathbf{g}(\boldsymbol{\delta})$ and $\boldsymbol{\varepsilon}$ as used by Jennrich [1969] (see Section 12.2), and conditions on $\mathbf{h}(\boldsymbol{\delta})$ that allow the use of the implicit-function theorem, $\hat{\boldsymbol{\delta}}$ [the solution to (8.86)] is asymptotically normal with mean $\boldsymbol{\delta}$. Writing $\mathbf{G}_{\bullet}(\boldsymbol{\delta}) = \partial \mathbf{g}(\boldsymbol{\delta})/\partial \boldsymbol{\delta}'$ and $\mathbf{H}_{\bullet}(\boldsymbol{\delta}) = \partial \mathbf{h}(\boldsymbol{\delta})/\partial \boldsymbol{\delta}'$, the asymptotic variance–covariance matrix is given by constrained linear model theory (c.f. Gerig and Gallant [1975]) with $\hat{\mathbf{G}} = \mathbf{G}_{\bullet}(\hat{\boldsymbol{\delta}})$ in place of the design matrix $\mathbf{X}$, and $\hat{\mathbf{H}} = \mathbf{H}_{\bullet}(\hat{\boldsymbol{\delta}})$ in place of the linear restriction matrix. Thus, asymptotically,

$$\mathscr{D}[\hat{\boldsymbol{\delta}}] = \sigma^2(\hat{\mathbf{G}}\hat{\mathbf{Q}})^+\boldsymbol{\Sigma}(\hat{\mathbf{G}}\hat{\mathbf{Q}})^{+\prime}, \tag{8.87}$$

where $\hat{\mathbf{Q}} = \mathbf{I} - \hat{\mathbf{H}}^+\hat{\mathbf{H}}$ and the superscript "+" denotes the (unique) Moore–Penrose generalized inverse of a matrix (Appendix A5). The number of degrees of freedom associated with $\{\mathbf{y} - \mathbf{g}(\hat{\boldsymbol{\delta}})\}'\boldsymbol{\Sigma}^{-1}\{\mathbf{y} - \mathbf{g}(\hat{\boldsymbol{\delta}})\}$, which reduces to σ^2 times the residual sum of squares when $\boldsymbol{\Sigma} = \sigma^2\mathbf{I}_n$, is $n - \text{rank}\,[\hat{\mathbf{G}}\hat{\mathbf{Q}}]$. The matrix (8.87) can be calculated after finding $\hat{\boldsymbol{\delta}}$ using a constrained least-squares program.

8.5.5 Fitting Compartmental Models without Using Derivatives

The nonlinear model (8.56) can be fitted in a straightforward fashion using a least-squares algorithm that requires only function values $\mathbf{f}(x; \boldsymbol{\theta})$ and not derivatives with respect to $\boldsymbol{\theta}$. If $\mathbf{f}(x;\boldsymbol{\theta})$ has not been obtained analytically, we can compute $\mathbf{f}(x;\boldsymbol{\theta}^{(a)})$ for all time points x using the spectral decomposition method or, preferably, a more stable decomposition such as that of Bavely and Stewart [1979]. The method of Berman et al. [1962] and the approach Berman and Weiss used in their popular SAAM programs (Berman and Weiss [1978]) are of interest. Rather than use decomposition methods to calculate $e^{\mathbf{A}x}\boldsymbol{\xi}$, the original system of differential equations is solved numerically (using a fourth-order Runge–Kutta method) so that the program can be used for nonlinear as well as linear systems of differential equations.

We have previously [c.f. (8.58)] discussed the situation where mixtures of outputs from compartments, namely

$$\boldsymbol{\eta}(x;\boldsymbol{\theta}) = \mathbf{C}(\boldsymbol{\theta})\mathbf{f}(x;\boldsymbol{\theta}) \tag{8.88}$$

are measured at various time points x. Here the $d \times m$ matrix $\mathbf{C}(\boldsymbol{\theta})$ typically contains parameters of interest. Berman et al. [1962] treat a special case of this. Expanding (8.88), we obtain

$$\eta_t(x;\boldsymbol{\theta}) = \sum_{j=1}^{m} c_{tj}(\boldsymbol{\theta}) f_j(x;\boldsymbol{\theta}), \qquad t = 1, 2, \ldots, d.$$

Suppose that the unknown parameters in $\mathbf{C}(\boldsymbol{\theta})$ are the c_{tj}'s themselves and that these do not appear in $\mathbf{f}(x;\boldsymbol{\theta})$. We thus write $\boldsymbol{\theta} = (\mathbf{c}', \boldsymbol{\phi}')'$ and $\mathbf{f} = \mathbf{f}(x;\boldsymbol{\phi})$. Also, measurements y_{it} are taken on $\eta_t(x;\boldsymbol{\theta})$ at times x_i for $i = 1, 2, \ldots, q$, so that

$$y_{it} = \sum_{j=1}^{m} c_{tj} f_j(x_i;\boldsymbol{\phi}) + \varepsilon_{it}. \tag{8.89}$$

Berman et al. [1962] treat (8.89) as a partially linear least squares problem (Section 5.10.2) with the current estimates $\mathbf{c}^{(a)}$ and $\boldsymbol{\phi}^{(a)}$ being updated as follows. We first fix $\mathbf{c} = \mathbf{c}^{(a)}$ in (8.89) and then obtain $\boldsymbol{\phi}^{(a+1)}$ from (8.89) by nonlinear least squares. Substituting $\boldsymbol{\phi}^{(a+1)}$ into $f_j(x_i;\boldsymbol{\phi})$ gives us (8.89) in the form of the linear model

$$\mathbf{y} = \mathbf{X}\mathbf{c} + \boldsymbol{\varepsilon}. \tag{8.90}$$

We then find $\mathbf{c}^{(a+1)}$ using (8.90) and linear least squares. This process is then iterated until convergence.

If some of the coefficients c_{tj} are known, then we arrange (8.89) in the form

$$\begin{aligned} z_{it} &= y_{it} - \sum_j{}^{+} c_{tj} f_j(x_i;\boldsymbol{\phi}) \\ &= \sum_j{}^{*} c_{tj} f_j(x_i;\boldsymbol{\phi}) + \varepsilon_{it}, \end{aligned}$$

or

$$\mathbf{z} = \mathbf{X}_1 \mathbf{c}_1 + \boldsymbol{\varepsilon}, \tag{8.91}$$

where $\sum^+$ and $\sum^*$ represent summation with respect to the known and unknown c_{ij} respectively. We now proceed with the iterations as before, except that given $\boldsymbol{\phi}^{(a+1)}$, we use (8.91) instead of (8.90) to compute $\mathbf{c}_1^{(a+1)}$.

8.5.6 Obtaining Initial Parameter Estimates

a All Compartments Observed with Zero or Linear Inputs

We recall that for a linear compartmental model with zero inputs, the deterministic system of linear differential equations is $\dot{\mathbf{f}}(x) = \mathbf{A}\mathbf{f}(x)$, or [c.f. (8.4)]

$$\begin{aligned} \dot{f}_j(x) &= \sum_k \alpha_{jk} f_k(x) \\ &= \sum_{\substack{k=1 \\ k \neq j}}^{m} \gamma_{jk} f_k(x) - \left(\sum_{\substack{k=0 \\ k \neq j}}^{m} \gamma_{kj} \right) f_j(x). \end{aligned} \tag{8.92}$$

In many modeling situations $\boldsymbol{\theta}$ consists of just the rate constants, the γ_{jk}'s. We can thus rewrite (8.92) in the form

$$\dot{f}_j(x) = \mathbf{z}_j'\boldsymbol{\theta}, \tag{8.93}$$

where $\mathbf{z}_j$ is a function of $\mathbf{f}(x)$. If estimates $\dot{f}_j^*(x)$ of $\dot{f}_j(x)$ and $f_j^*(x)$ of $f_j(x)$ can be formed at a number of time points x, then we can put all these estimates at different times and for different compartments into single vectors $\dot{\mathbf{f}}^*$ and $\mathbf{f}^*$ satisfying

$$\dot{\mathbf{f}}^* \approx \mathbf{Z}^*\boldsymbol{\theta}, \tag{8.94}$$

where $\mathbf{Z}^*$ is a function of $\mathbf{f}^*$. Here the rows of $\mathbf{Z}^*$ are the estimated $\mathbf{z}_j$'s of (8.93) evaluated at the different time points. An initial estimate or starting value for $\boldsymbol{\theta}$ can then be obtained by regressing $\dot{\mathbf{f}}^*$ on $\mathbf{Z}^*$.

Suppose observations are taken on all of the compartments, with compartment j being observed at times $x_{1j}, x_{2j}, \ldots, x_{n_j,j}$. Bates and Watts [1985a] use the midpoint times $\bar{x}_{ij} = (x_{ij} + x_{i+1,j})/2$ for $i = 1, 2, \ldots, n_j - 1$ and estimate

$$\dot{f}_j^*(\bar{x}_{ij}) = \frac{y_{i+1,j} - y_{ij}}{x_{i+1,j} - x_{ij}} \tag{8.95}$$

and

$$f_j^*(\bar{x}_{ij}) = \tfrac{1}{2}(y_{ij} + y_{i+1,j}).$$

Thus we can set $\dot{\mathbf{f}}_j^* = (\dot{f}_j^*(\bar{x}_{1j}), \ldots, \dot{f}_j^*(\bar{x}_{n_j-1,j}))'$ and form $\dot{\mathbf{f}}^* = (\dot{\mathbf{f}}_1^{*\prime}, \dot{\mathbf{f}}_2^{*\prime}, \ldots, \dot{\mathbf{f}}_m^{*\prime})'$. The row of $\mathbf{Z}^*$ corresponding to $\dot{f}_j^*(\bar{x}_{ij})$ is $\mathbf{z}_j'$ evaluated at $\mathbf{f}^*(\bar{x}_{ij})$.

The rate approximations above are based upon first differences and are fairly crude. McLean [1985] classified the technique as a "differential" technique for analyzing rate data as described in standard chemical-reaction engineering texts. He noted that the reliability of such approximations is highly dependent on the spacing of the x's and the scatter in the data. Bates and Watts [1985a] suggested that it may be possible to improve the derivative estimates by smoothing the data $(x_{1j}, y_{1j}), (x_{2j}, y_{2j}), \ldots, (x_{n_j,j}, y_{n_j,j})$ for each compartment. The estimates $\dot{f}_j^*(x)$ and $f_j^*(x)$ may then be obtained from the smoothed relationship. Bates and Watts suggested smoothing the response curve for each compartment using smoothing splines (Section 9.5.2). In a similar context Jacquez [1972: 110–111] suggested fitting a single exponential curve to successive groups of three or five points and using this to smooth the central term. Alternatively, lower-order polynomials can be used instead. The fitted curve can also provide the rate estimate at the central value.

In conclusion we note that (8.93) can be generalized to allow for inputs into the system which are linear in $\boldsymbol{\theta}$. The only change is in $\mathbf{z}_j$. Totally known inputs should be subtracted from $\dot{\mathbf{f}}^*$, analogously with (8.91).

b Some Compartments Unobserved

Bates and Watts [1985a] stated that when only a subset of the compartments are observed it may still be possible to use the approximate-rates method above provided other information is exploited, and they cite mass-balance relationships. Components from unobserved compartments in $\dot{\mathbf{f}}^*$ [c.f. (8.94)] should clearly be dropped. However, since any derivative may involve any of the other compartments, we still have the problem of estimating $f_j(x)$ terms from unobserved compartments as required by $\mathbf{Z}^*$ in (8.94).

McLean [1985] showed that the above technique of linear regression using rate estimates can still sometimes be exploited in nonlinear systems. Otherwise we are forced back to general methods for obtaining starting values as discussed in Section 15.2.1. If the coefficients γ_{jk} are functions of the parameters $\boldsymbol{\theta}$ of interest, then starting values γ_{jk}^* of the γ_{jk}'s obtained using the above method may still be useful if we can solve the equations $\gamma_{jk}^* = \gamma_{jk}(\boldsymbol{\theta}^*)$ for $\boldsymbol{\theta}^*$ exactly or approximately (e.g. a least-squares solution).

c Exponential Peeling

Starting values for sums-of-exponentials models

$$E[y] = \sum_{j=1}^{m} \beta_j e^{\lambda_j x} \qquad (\lambda_j \le 0 \quad \text{for all } j)$$

have traditionally been obtained using exponential curve peeling. Suppose the λ_j's are negative and $|\lambda_1| < |\lambda_2| < \cdots < |\lambda_m|$. For large x the term $\beta_1 e^{\lambda_1 x}$ will dominate, so that $E[y] \approx \beta_1 e^{\lambda_1 x}$. We can plot $\log y$ versus x. If the plot is linear there is only one exponential. Otherwise we look for a roughly linear portion at the end of the curve, as illustrated in Fig. 8.11a. A straight line is fitted to the points there, and estimates of β_1 and λ_1 are obtained. The points used for

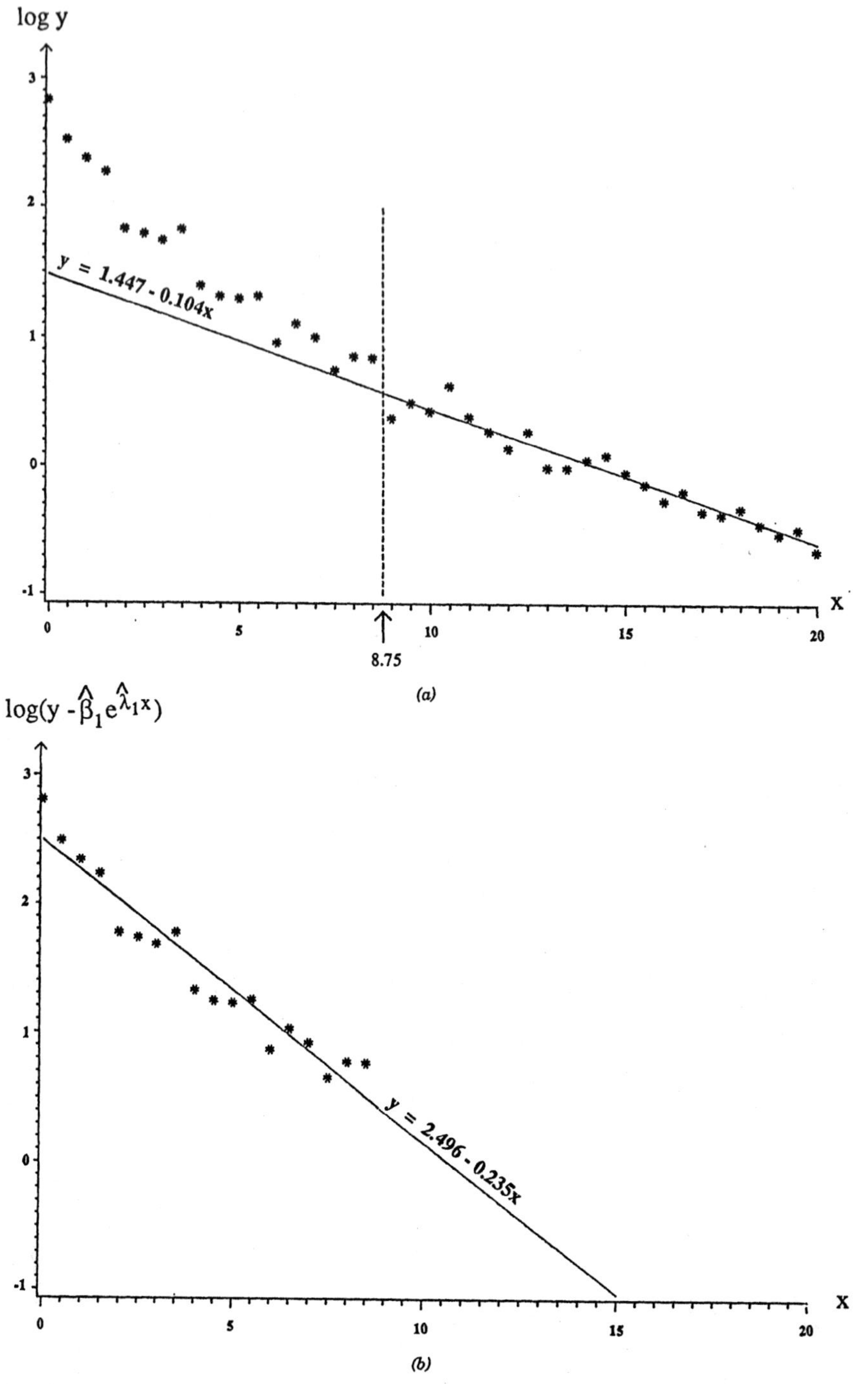

Figure 8.11 Exponential peeling. (*a*) $\log y$ versus x. Solid line is least-squares regression line for points with $x > 8.75$. (*b*) $\log\{\hat{y} - \hat{\beta}_1 \exp(\hat{\lambda}_1 x)\}$ versus x for points with $x < 8.75$, where $\hat{\beta}_1 = e^{1.447}$, $\hat{\lambda}_1 = -0.104$. Line is fitted least-squares regression line. Final estimates $\hat{\beta}_1 = e^{1.447}, \hat{\lambda}_1 = -0.104$, $\hat{\beta}_2 = e^{2.496}, \hat{\lambda}_2 = -0.235$.

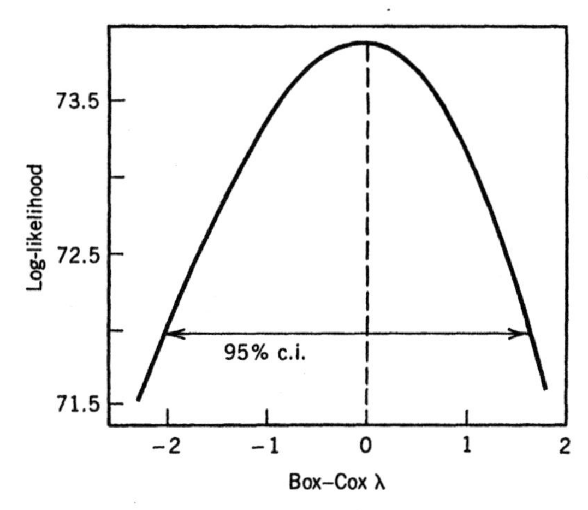

Figure 8.12 Concentrated log-likelihood function $M(\lambda)$ for Brunhilda data, showing approximate 95% confidence interval for λ. From Bates et al. [1986].

estimating β_1 and λ_1 are now discarded, and we plot $\log(y - \hat{\beta}_1 e^{\hat{\lambda}_1 x})$ versus x for the remaining points. For large x, $\beta_2 e^{\lambda_2 x}$ should now dominate, and the roughly linear portion at the end of the curve corresponds to $\log \beta_2 + \lambda_2 x$. The points on the linear portion are used to estimate β_2 and λ_2 and are then discarded. We then plot $\log(y - \hat{\beta}_1 e^{\hat{\lambda}_1 x} - \hat{\beta}_2 e^{\hat{\lambda}_2 x})$ versus x, and the process continues. The first linear plot we see tells us that this is the last pair of exponential parameters that can be fitted. In Fig. 8.11 only two exponentials were fitted. Exponential curve peeling is subjective and unreliable, but it can provide starting values for a least-squares analysis. For a more detailed description see Jacquez [1972: 103].

We further note that the sums-of-exponentials model is a partially linear model (Sections 5.10.2, 14.7), so that given starting values for $\boldsymbol{\lambda}$, starting values for $\boldsymbol{\beta}$ can be obtained by linear least squares.

Table 8.2 Parameter Estimates for Brunhilda Data[a]

	Logged Analysis		OLS
Parameter	Est. θ_i	95% Confidence Interval	Est. θ_i
θ_1	0.00941	(0.0085, 0.0104)	0.00972
θ_2	0.2848	(0.2324, 0.3491)	0.3011
θ_3	0.1923	(0.1642, 0.2253)	0.2022
θ_4	0.0342	(0.0244, 0.0481)	0.0384
θ_5	0.0627	(0.0525, 0.0749)	0.0667
θ_6[b]	2.434	(2.228 , 2.659)	2.489

[a]From Bates et al. [1986] with permission of North-Holland Physics Publishing Co., Amsterdam.
[b]Scaled by 10^{-5}.

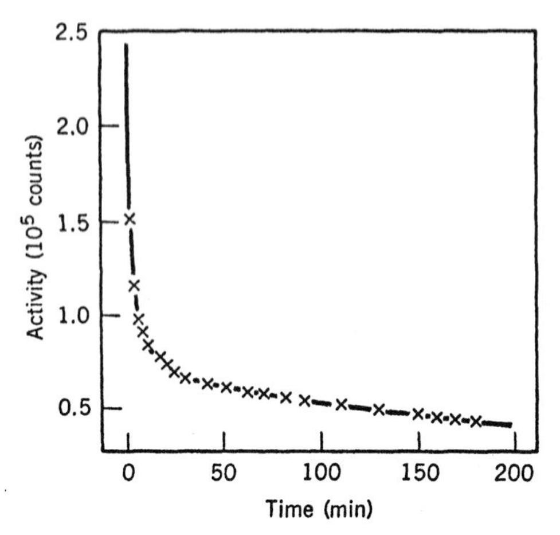

Figure 8.13 Brunhilda data with estimated curve superimposed (logged analysis). From Bates et al. [1986].

8.5.7 Brunhilda Example Revisited

We consider Example 8.8 [c.f. (8.71)] once again and describe the analysis of the data in Table 8.1 by Bates et al. [1986]. The linear system defining the model is given by (8.71). Bates et al. allowed for heterogeneity by using a Box–Cox transformation of both sides (Section 2.8.1), namely

$$y_i^{(\lambda)} = f_1(x_i; \boldsymbol{\theta})^{(\lambda)} + \varepsilon_i, \tag{8.96}$$

where the ε_i are assumed to be i.i.d. $N(0, \sigma^2)$. The concentrated log-likelihood, $M(\lambda) = \log L\{\lambda, \tilde{\theta}(\lambda), \tilde{\sigma}^2(\lambda)\}$, is plotted in Fig. 8.12. From this the maximum-likelihood estimate of λ is approximately -0.1, with an approximate 95% confidence interval for λ of $[-2, 1.75]$. These limits are very wide. Bates et al. selected the log transformation ($\lambda = 0$) and fitted

$$\log y_i = \log f_1(x_i; \boldsymbol{\theta}) + \varepsilon_i$$

by least squares. The data with the fitted curve superimposed are shown in Fig. 8.13. The parameter estimates together with a large-sample confidence interval for each parameter are given in Table 8.2. The last column of Table 8.2 gives parameter estimates from an ordinary least-squares analysis. These are close to the estimates obtained using the log transformation. This reflects the width of the confidence interval for λ and the fact that the log transformation is only gently nonlinear over the range of the data (c.f. Bates et al. [1986, Fig. 4b]). Bates et al. [1986] analyzed a second data set in which λ is much more precisely estimated (approximate confidence limits of -0.1 to 0.35). This time a logged analysis and ordinary least squares gave very different results. Also the log transformation was severely nonlinear over the range of the data.

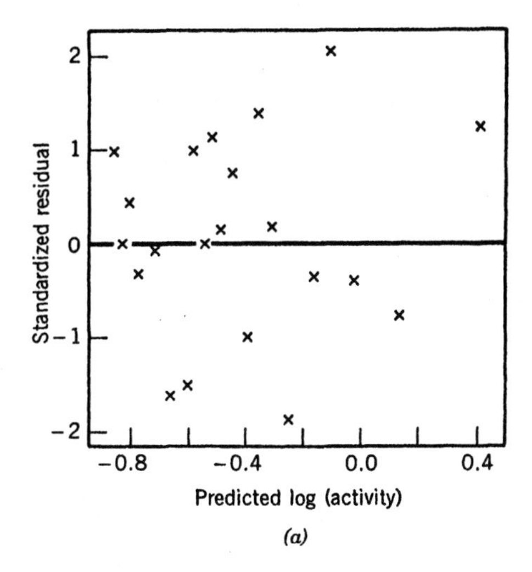

(a)

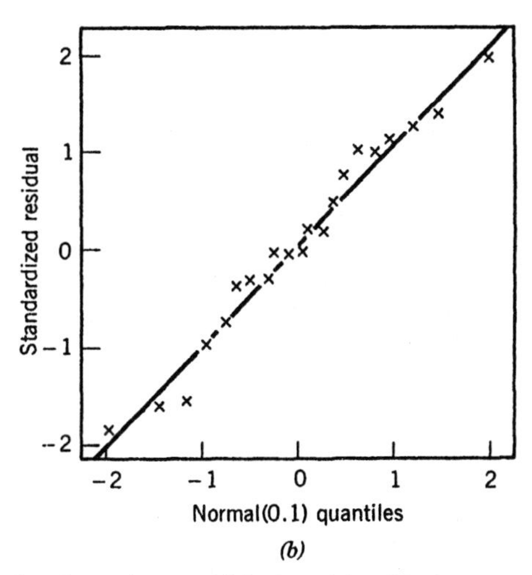

(b)

Figure 8.14 Residual plots from the Brunhilda data: (*a*) residuals versus fitted values; (*b*) normal probability plot. From Bates et al. [1986].

Residual plots for the logged analysis of the Brunhilda data are given in Fig. 8.14. They show no disturbing patterns. In contrast, Bates et al. reported fitting a two-compartment model to these data and obtained noticeable patterns in the residuals. A formal F-test confirmed the need for the three-compartment model.

Rather than treat θ_6 as a free parameter, Jennrich and Bright [1976] used an estimated value of $\theta_6 = 2 \times 10^5$. This value is outside the confidence interval given in Table 8.2. Moreover, Bates et al. [1986] state that the residuals from this fit were poorly behaved. Jennrich and Bright [1976] originally obtained starting values $\theta_1, \theta_2, \ldots, \theta_5$ using a limited knowledge of the system. They obtained θ_1 as minus the slope of the last part of the plot of log y versus time, and θ_2 as minus the slope of the first part. All other rates were set equal to θ_2. They also used weighted least squares with weights proportional to y_i^{-2}, which, to a first order approximation, is equivalent to using ordinary least squares with the log transformation [c.f. (2.146), $\lambda_2 = 2$]. They found that the starting values were not crucial and that a variety of values led to the same minimum of the error sum of squares.

8.5.8 Miscellaneous Topics

An impression may have arisen from the previous discussions that in compartmental analysis, subject-matter considerations suggest a compartmental model, which is then fitted to data to obtain parameter estimates. However, the route between subject-matter theory and the formulation of a suggested model is seldom this direct. Commonly the data themselves are used to determine how many compartments are required to fit them, as discussed below. Subject-matter considerations may suggest what these compartments could be thought of as representing. This in turn suggests possibly competing compartmental models, depending upon what pathways between compartments and to the environment are postulated as being open. It is therefore necessary to use the data to distinguish between possibly non-nested competing models and perhaps to design experiments that will efficiently distinguish between competing models.

Determining the number of compartments to use is done by fitting sums-of-exponentials models [c.f. (8.52)] to the data and determining how many exponential terms can be fitted. This problem is closely analogous to the variable selection problem in regression. Jacquez [1972: 109] suggested adding exponential terms until the reduction in the residual sum of squares with an additional term becomes small. Another suggestion has been to add exponential terms until the residuals first appear to have a reasonably random scatter. Smith et al. [1976b] developed a transform method which "produces a spectrum whose peaks correspond in amplitude and position to the various exponential components present in the data." However, this method works well only when σ, the standard deviation of the errors, is very small compared with $E[y]$. Values of σ of 1% and 0.1% of the maximum value of $E[y]$ were used, and even at these values spectral peaks can become very indistinct.

For the remainder of this section we deal briefly with identifiability problems and experimental design. In Section 8.4 we discussed the question of whether the system parameters $\boldsymbol{\theta}$ were identifiable from the entire function $\mathbf{f}(x)$. There is another general approach which looks at whether the parameters are locally identifiable from the function values $\mathbf{f}(x_1), \mathbf{f}(x_2), \ldots, \mathbf{f}(x_n)$ for a potential experiment. This approach, which allows the investigation of ill-conditioning, has already been described in Section 3.4. Several of the examples of experimental designs given by Endrenyi [1981b] involve simple compartmental models. Cobelli and Thomaseth [1985] reference further work on experimental design for compartmental models. This work deals with the optimal times $x_1, x_2, \ldots, x_n$ at which the systems should be observed, and specializes the ideas in Section 5.13. However, they also extend the ideas of optimal design to include the optimal choice of the inputs (e.g. tracer inputs) for a compartmental system.

8.6 MORE COMPLICATED ERROR STRUCTURES

In Section 8.5.1 we described models which essentially allowed for measurement errors in the monitoring of a deterministic process. These random errors were assumed to be uncorrelated. However, if we wish to allow for part of the stochastic variability to be due to fluctuations in the flow of the material itself, then it would seem appropriate to allow for correlated errors. For example, if more material than expected leaves compartment j along one path, we might expect less material to leave on another path. This would suggest that we should allow for correlations between measurements taken on different compartments at the same time x.

In this section we still imagine that the deterministic model describes the expected amount of material present in any compartment at any time x, so that we still consider models of the form

$$\mathbf{y} = \mathbf{f}(\boldsymbol{\theta}) + \boldsymbol{\varepsilon}, \tag{8.97}$$

where $\mathscr{E}[\boldsymbol{\varepsilon}] = \mathbf{0}$. However, we now allow for certain types of correlation between elements of $\boldsymbol{\varepsilon}$. Thus we assume that $\mathscr{D}[\boldsymbol{\varepsilon}] = \mathbf{V}$, where $\mathbf{V}$ is a positive definite matrix which may or may not depend on unknown parameters. For known $\mathbf{V}$, the Gauss–Newton updating step is

$$\boldsymbol{\theta}^{(a+1)} = \boldsymbol{\theta}^{(a)} + [\mathbf{F}_{.}'(\boldsymbol{\theta}^{(a)})\mathbf{V}^{-1}\mathbf{F}_{.}(\boldsymbol{\theta}^{(a)})]^{-1}\mathbf{F}_{.}'(\boldsymbol{\theta}^{(a)})\mathbf{V}^{-1}[\mathbf{y} - \mathbf{f}(\boldsymbol{\theta}^{(a)})] \tag{8.98}$$

and, asymptotically,

$$\mathscr{D}[\hat{\boldsymbol{\theta}}] = [\mathbf{F}_{.}'(\hat{\boldsymbol{\theta}})\mathbf{V}^{-1}\mathbf{F}_{.}(\hat{\boldsymbol{\theta}})]^{-1}. \tag{8.99}$$

Thus the work that we have done in Section 8.5 on computing $\mathbf{F}_{.}(\boldsymbol{\theta})$ applies also to this section. The computation of (8.98) and (8.99) using a Cholesky decomposition for $\mathbf{V}$ was described in Section 2.1.4.

Before continuing we recall the notation of Section 8.5.1. However, we simplify the situation there somewhat by assuming that all of the monitored compartments are observed at the same times $x_1, x_2, \ldots, x_n$. Thus y_{ij} is the measurement on compartment j at time x_i, $\mathbf{y}^{(j)} = (y_{1j}, y_{2j}, \ldots, y_{nj})'$ and $\mathbf{y} = (\mathbf{y}^{(1)\prime}, \mathbf{y}^{(2)\prime}, \ldots, \mathbf{y}^{(m)\prime})$. Similarly $\mathbf{f}^{(j)}(\boldsymbol{\theta}) = (f_j(x_1;\boldsymbol{\theta}), f_j(x_2;\boldsymbol{\theta}), \ldots, f_j(x_n;\boldsymbol{\theta}))'$, $\mathbf{f}(\boldsymbol{\theta}) = (\mathbf{f}^{(1)\prime}(\boldsymbol{\theta}), \mathbf{f}^{(2)\prime}(\boldsymbol{\theta}), \ldots, \mathbf{f}^{(m)\prime}(\boldsymbol{\theta}))'$, and $\mathbf{y} = \mathbf{f}(\boldsymbol{\theta}) + \boldsymbol{\varepsilon}$. Here $f_j(x;\boldsymbol{\theta})$, the expected amount of material in compartment j at time x, is defined by the system of linear differential equations which specifies the compartmental model. If $r < m$ compartments are observed, only the $\mathbf{y}^{(j)}$ and $\mathbf{f}^{(j)}(\boldsymbol{\theta})$ for the observed compartments are included in $\mathbf{y}$ and $\mathbf{f}(\boldsymbol{\theta})$.

It was suggested above that we should allow for correlation between measurements taken on different compartments at the same time x. Beauchamp and Cornell [1966] do this by assuming that

$$\operatorname{cov}(\varepsilon_{rj}, \varepsilon_{sk}) = \begin{cases} 0, & r \neq s, \\ \sigma_{jk}, & r = s. \end{cases}$$

This also assumes that measurements taken at different times are uncorrelated. Since, in general, the covariances σ_{jk} are unknown, Beauchamp and Cornell estimate them using the strongly consistent estimators

$$\tilde{\sigma}_{jk} = \frac{1}{n} \sum_{i=1}^{n} \tilde{\varepsilon}_{ij} \tilde{\varepsilon}_{ik}.$$

Here $\tilde{\varepsilon}_{ij}$ is the residual

$$\tilde{\varepsilon}_{ij} = y_{ij} - f_j(x_i; \hat{\boldsymbol{\theta}}^{(j)}), \tag{8.100}$$

where $\boldsymbol{\theta}^{(j)}$ is the subset of elements of $\boldsymbol{\theta}$ appearing in $f_j(x; \boldsymbol{\theta})$ and $\hat{\boldsymbol{\theta}}^{(j)}$ is the ordinary least-squares estimate of $\boldsymbol{\theta}^{(j)}$ obtained only from the observations $\mathbf{y}^{(j)}$ on the jth compartment. Let $\tilde{\mathbf{V}}$ be obtained from $\mathbf{V}$ $(= \mathscr{D}[\boldsymbol{\varepsilon}])$ by replacing σ_{jk} by $\tilde{\sigma}_{jk}$. As the latter estimators are strongly consistent as $n \to \infty$, the generalized least-squares estimator obtained using $\tilde{\mathbf{V}}$ has the same asymptotic distribution as that obtained using $\mathbf{V}$. The work of Beauchamp and Cornell [1966] is described in detail and in a more general setting in Section 11.2, where we use multivariate notation.

Example 11.2 in Section 11.2 describes their analysis of the data in Table 11.1. The passage of radioactive tracer was investigated using a two-compartment model. Tracer was injected into one compartment, and the proportion of tracer in each compartment was then monitored at various times after the injection. The model

$$y_{ij} = f_j(x_i; \boldsymbol{\theta}) + \varepsilon_{ij} \qquad (i = 1, 2, \ldots, n, \quad j = 1, 2)$$

was fitted, where $\boldsymbol{\theta} = (\theta_1, \theta_2, \theta_3)'$,

$$f_1(x; \boldsymbol{\theta}) = \theta_1 e^{-\theta_2 x} + (1 - \theta_1) e^{-\theta_3 x},$$
$$f_2(x; \boldsymbol{\theta}) = 1 - (\theta_1 + \theta_4) e^{-\theta_2 x} + (\theta_1 + \theta_4 - 1) e^{-\theta_3 x},$$

and

$$\theta_4 = \frac{(\theta_3 - \theta_2)\theta_1(1 - \theta_1)}{(\theta_3 - \theta_2)\theta_1 + \theta_2}.$$

As the measurements were taken over time, we would suspect serial correlation between the errors as well as correlations between compartments. The fitted curves of Beauchamp and Cornell are shown superimposed upon the data in Figs. 11.1 and 11.2. They do indeed show a lack of fit that appears consistent with serial correlations. This brings us back to a dilemma of empirical model building. Do we improve the fit by increasing the complexity of the compartmental model, or do we explain the lack of fit by using some form of serial-correlation structure?

In situations where we only have measurements on a single compartment, or on a single linear combination of compartments, serial correlation can be modeled using the methods of Chapter 6. To apply this approach to measurements on more than one compartment, some sort of joint time-series modeling of the errors in different compartments would be required. However, instead of an empirical approach along these lines, more truly stochastic models have been developed which can provide a credible correlation structure.

*8.7 MARKOV-PROCESS MODELS

8.7.1 Background Theory

We begin consideration of stochastic modeling by describing Markov-process models. These closely model the idealized compartmental system defined in Section 8.1 in terms of homogeneous well-mixed compartments exchanging material. We shall find that the mean behavior of the stochastic model is described by the deterministic models we have already developed.

Suppose that the material is composed of particles and that the stochastic nature of material transfer is caused by the independent random movement of particles according to a continuous-time Markov process (see Cox and Miller [1965: Chapter 4]). We shall find it convenient to treat the two processes "excretion to the environment" and "input from the environment" as separate processes. For the first process the environment is regarded as a "sink" for particles excreted from the compartments and is labeled compartment 0. However, particles arriving in the system from the outside are considered as a second independent process and not as coming from compartment 0.

Let $X(t)$ be the compartment occupied by any particular particle in the system at time t. A Markov process can be specified in terms of the $(m+1)\times(m+1)$ transition-probability matrix $\tilde{\mathbf{P}}(s,t)$ with (j,k)th element

$$p_{jk}(s,t) = \mathrm{pr}[X(t) = j \mid X(s) = k] \tag{8.101}$$

for $j = 0, 1, \ldots, m$ and $k = 0, 1, \ldots, m$. We note that $p_{j0}(s,t) = 0$ for $t > s$ and $j \neq 0$, and we define $p_{00}(s,t) = 1$. The subscript convention for compartmental models is used here for consistency within the chapter, and not that of the stochastic-process literature, which is the reverse. Thus $p_{jk}(s,t)$ is the probability that a particle resident in compartment k at time s is in compartment j at time t. The particles move independently according to (8.101). Under these assumptions the transfer probabilities do not depend on either the time the particle has been in the compartment, or the number of other particles in the compartments, or the previous history of the process.

Suppose we begin with $n_j(0)$ particles in compartment j, and let

$$\tilde{\mathbf{n}}(0) = (n_0(0), n_1(0), \ldots, n_m(0))' = (n_0(0), \mathbf{n}'(0))',$$

with $n_0(0) = 0$ (as the sink is empty at the start). We similarly define $\tilde{\mathbf{n}}(t) = (n_0(t), \mathbf{n}'(t))'$, which gives the number of particles in each compartment at time t. Let $\mathbf{z}(t)$ be the vector whose jth element is the number of particles in the jth compartment at time t that have come into the system from the environment after $t = 0$. In keeping with our separation of the input and output processes for the environment, we define $\mathbf{z}(0) = \mathbf{0}$. Let $\mathbf{u}_r(t)$ be the vector whose jth element is the number of particles in compartment j at time t (including $j = 0$) that originated from compartment r at time 0. These $\mathbf{u}_r(t)$ are independent. Also let

$$\mathbf{w}(t) = \sum_{r=1}^{m} \mathbf{u}_r(t). \tag{8.102}$$

This $(m+1) \times 1$ vector records the numbers of particles in each compartment, including the sink, that were originally present in the system at $t = 0$. Then clearly

$$\tilde{\mathbf{n}}(t) = \mathbf{w}(t) + \mathbf{z}(t) \tag{8.103}$$

with $\mathbf{z}(t)$ independent of $\mathbf{u}_1(t), \mathbf{u}_2(t), \ldots, \mathbf{u}_m(t)$ [and thus of $\mathbf{w}(t)$]. Also

$$\mathbf{u}_r(t) \sim \text{Multinomial}[n_r(0); p_{0r}(0,t), p_{1r}(0,t), \ldots, p_{mr}(0,t)]. \tag{8.104}$$

a No Environmental Input

For ease of exposition we first of all restrict ourselves to the case of zero input from the environment, i.e. $\mathbf{z}(t) = \mathbf{0}$. Then

$$\tilde{\mathbf{n}}(t) = \sum_{r=1}^{m} \mathbf{u}_r(t),$$

so that, from (8.104),

$$E[n_j(t)] = \sum_{r=1}^{m} n_r(0) p_{jr}(0,t) \qquad (j = 0, 1, \ldots, m), \tag{8.105}$$

or

$$\mathscr{E}[\tilde{\mathbf{n}}(t)] = \tilde{\mathbf{P}}(0,t)\tilde{\mathbf{n}}(0), \tag{8.106}$$

say. The Markov process can also be expressed in terms of the intensity matrix (or instantaneous transfer-rate matrix), defined as $\tilde{\mathbf{A}}(t)$ with (j,k)th elements

$(j = 0, 1, \ldots, m, k = 0, 1, \ldots, m)$

$$\alpha_{jk}(t) = \gamma_{jk}(t) = \lim_{\delta t \downarrow 0} \frac{p_{jk}(t, t + \delta t)}{\delta t} \qquad (j \neq k)$$

and

$$\alpha_{kk}(t) = -\sum_{\substack{j=0 \\ j \neq k}}^{m} \gamma_{jk}(t) = -\lim_{\delta t \downarrow 0} \frac{1 - p_{kk}(t, t + \delta t)}{\delta t}$$

as

$$\sum_{j=0}^{m} p_{jk}(t, t + \delta t) = 1.$$

Thus

$$\alpha_{00}(t) = 0, \qquad \alpha_{j0}(t) = 0 \quad (j = 1, 2, \ldots, m, \ t > 0). \tag{8.107}$$

We are now in a position to show that the deterministic compartmental model describes the mean behavior of the Markov-process model. Using the same sum-of-multinomials argument as above in (8.105), we have, for $j = 1, 2, \ldots, m$,

$$E[n_j(t + \delta t) | \tilde{\mathbf{n}}(t)] = \sum_{r=1}^{m} n_r(t) p_{jr}(t, t + \delta t)$$

so that, unconditionally,

$$E[n_j(t + \delta t)] = \sum_{r=1}^{m} E[n_r(t)] p_{jr}(t, t + \delta t). \tag{8.108}$$

In a similar fashion we have

$$\begin{aligned} E[n_0(t + \delta t)] &= E[n_0(t)] + \sum_{r=1}^{m} E[n_r(t)] p_{0r}(t, t + \delta t) \\ &= \sum_{r=0}^{m} E[n_r(t)] p_{0r}(t, t + \delta t), \end{aligned}$$

since $p_{00}(t, t + \delta t) = 1$. Thus for $j = 0, 1, \ldots, m$,

$$\begin{aligned} E[n_j(t + \delta t)] - E[n_j(t)] = {} & \{p_{jj}(t, t + \delta t) - 1\} E[n_j(t)] \\ & + \sum_{\substack{r=1 \\ r \neq j}}^{m} E[n_r(t)] p_{jr}(t, t + \delta t). \end{aligned}$$

Dividing by δt and taking limits, the definitions following (8.106) give us

$$\frac{dE[n_j(t)]}{dt} = \sum_{r=1}^{m} \alpha_{jr}(t) E[n_r(t)] \qquad (j = 0, 1, \ldots, m),$$

or

$$\frac{d}{dt}\mathscr{E}[\tilde{\mathbf{n}}(t)] = \tilde{\mathbf{A}}(t)\mathscr{E}[\tilde{\mathbf{n}}(t)]. \tag{8.109}$$

For (8.109) we have the constraint $\mathscr{E}[\tilde{\mathbf{n}}(0)] = \tilde{\mathbf{n}}(0) = \boldsymbol{\xi}$.

We shall now compare (8.109) with the deterministic model (8.7) for the no-input case $[\mathbf{r}(x) = \mathbf{0}]$. To do this we remove the effect of the sink from (8.109). Using (8.107), we see that $\tilde{\mathbf{A}}(t)$ has the form

$$\tilde{\mathbf{A}}(t) = \begin{pmatrix} 0 & \mathbf{u}' \\ \mathbf{0} & \mathbf{A}(t) \end{pmatrix}, \tag{8.110}$$

where $\mathbf{A}(t)$ corresponds to the matrix $\mathbf{A}$ of (8.7), since it deals only with transfers between the m actual compartments. Thus, omitting the first element (relating to the sink) from $\tilde{\mathbf{n}}(t)$, we see that (8.109) reduces to

$$\frac{d}{dt}\mathscr{E}[\mathbf{n}(t)] = \mathbf{A}(t)\mathscr{E}[\mathbf{n}(t)]$$

so that the deterministic model (8.7) describes the behavior of $\mathscr{E}[\mathbf{n}(t)]$. (We note that this is not true for stochastic compartmental models in general: c.f. Section 8.8.)

In what follows we shall restrict ourselves to mainly time-homogeneous Markov-process models which are the natural analogues of linear deterministic compartmental models. For these models

$$\alpha_{jk}(t) \equiv \alpha_{jk}, \qquad \tilde{\mathbf{A}}(t) \equiv \tilde{\mathbf{A}}, \quad \text{and} \quad \mathbf{P}(s,t) \equiv \mathbf{P}(t-s).$$

Then, from standard Markov-process theory (see Cox and Miller [1965: Section 4.5]), $\tilde{\mathbf{P}}(t)$ is the solution of the forward Kolmogorov equations

$$\dot{\tilde{\mathbf{P}}}(t) = \tilde{\mathbf{P}}(t)\tilde{\mathbf{A}} \qquad \text{subject to } \tilde{\mathbf{P}}(0) = \mathbf{I}_{m+1},$$

namely

$$\begin{aligned} \tilde{\mathbf{P}}(t) &= e^{\tilde{\mathbf{A}}t} \\ &= \begin{pmatrix} 0 & \mathbf{u}'e^{\mathbf{A}t} \\ \mathbf{0} & e^{\mathbf{A}t} \end{pmatrix} \qquad \text{[by (8.110)]} \\ &= \begin{pmatrix} 0 & \mathbf{u}'e^{\mathbf{A}t} \\ \mathbf{0} & \mathbf{P}(t) \end{pmatrix}, \end{aligned} \tag{8.111}$$

say. Also, from (8.106),

$$\mathscr{E}[\tilde{\mathbf{n}}(t)] = e^{\tilde{\mathbf{A}}'t}\tilde{\mathbf{n}}(0) \tag{8.112}$$

so that

$$\mathscr{E}[\mathbf{n}(t)] = e^{\mathbf{A}t}\mathbf{n}(0). \tag{8.113}$$

If a spectral decomposition of **A** is available, then from (8.29) we have

$$\mathscr{E}[\mathbf{n}(t)] = \sum_{r=1}^{m} \boldsymbol{\beta}_r e^{\lambda_r t}. \tag{8.114}$$

Typically $\mathbf{A} = \mathbf{A}(\boldsymbol{\theta})$ and some or all of the initial counts $\mathbf{n}(0)$ may be unknown and thus contribute to $\boldsymbol{\theta}$. We consider data of the form $n_j(t)$ $(j = 1, 2, \ldots, m)$ observed at times $t_1, t_2, \ldots, t_n$ to produce $y_{ij} = n_j(t_i)$. We form $\mathbf{y}^{(j)} = (y_{1j}, y_{2j}, \ldots, y_{nj})'$ and $\mathbf{y} = (\mathbf{y}^{(1)\prime}, \mathbf{y}^{(2)\prime}, \ldots, \mathbf{y}^{(m)\prime})'$ in the usual way, with the $\mathbf{y}^{(j)}$ from unobserved compartments omitted. The vector $\mathbf{f}(\boldsymbol{\theta})$ of expected responses is obtained by stacking $E[n_j(t_i)]$ analogously, and $\boldsymbol{\varepsilon}$ is formed from $\varepsilon_{ij} = n_j(t_i) - E[n_j(t_i)]$, thus giving us the model (8.97). Note also that we do not use the count of particles that have gone to the environment. This can be recovered from the original counts and the counts in the other compartments. Its use would introduce an exact linear dependence in the data.

Referring to the model (8.97), we have that $\mathscr{D}[\boldsymbol{\varepsilon}] = \mathbf{V}(\boldsymbol{\theta})$, which incorporates correlations across compartments and over time. This matrix has elements of the form $\operatorname{cov}[n_j(t_a), n_k(t_b)]$ where we may have $j = k$ and/or $t_a = t_b$. From the multinomial distributions (8.104) and $\tilde{\mathbf{n}}(t) = \sum_r \mathbf{u}_r(t)$ we have, for a time-homogeneous process,

$$\operatorname{var}[n_j(t)] = \sum_{r=1}^{m} n_r(0) p_{jr}(t)[1 - p_{jr}(t)] \tag{8.115}$$

and

$$\operatorname{cov}[n_j(t), n_k(t)] = -\sum_{r=1}^{m} n_r(0) p_{jr}(t) p_{kr}(t). \tag{8.116}$$

It remains to find $\operatorname{cov}[n_j(t_a), n_k(t_b)]$ for $t_a \neq t_b$. We shall write $\mu_j(t) = E[n_j(t)]$ and use $(t, t+\tau)$ rather than (t_a, t_b). The case when $t_a > t_b$ can be obtained by symmetry.

Given $\tilde{\mathbf{n}}(t)$, $\tilde{\mathbf{n}}(t+\tau)$ is the sum of independent multinomials $\mathbf{u}_k(t+\tau)$, with

$$\mathbf{u}_k(t+\tau) \mid \tilde{\mathbf{n}}(t) \sim \text{Multinomial}[n_k(t); p_{0k}(\tau), \ldots, p_{mk}(\tau)], \tag{8.117}$$

where $\mathbf{u}_k(t+\tau)$ denotes the number of particles in each compartment originating from the $n_k(t)$ particles in compartment k at time t. Hence, from (8.108) with τ instead of δt, we have the unconditional expectations

$$E[n_k(t+\tau)] = \sum_{r=1}^{m} \mu_r(t) p_{kr}(\tau).$$

Using the same conditional approach,

$$E[n_j(t)n_k(t+\tau)|\tilde{\mathbf{n}}(t)] = \sum_{r=1}^{m} n_j(t)n_r(t)p_{kr}(\tau),$$

so that

$$\begin{aligned} E[n_j(t)n_k(t+\tau)] &= E[n_j^2(t)]p_{kj}(\tau) + \sum_{r\neq j} E[n_j(t)n_r(t)]p_{kr}(\tau) \\ &= p_{kj}(\tau)\{\text{var}[n_j(t)] + \mu_j^2(t)\} \\ &\quad + \sum_{r\neq j}\{\text{cov}[n_j(t)n_r(t)] + \mu_j(t)\mu_r(t)\}p_{kr}(\tau). \end{aligned}$$

Hence for all $j = 1, 2, \ldots, m$ and $k = 1, 2, \ldots, m$,

$$\begin{aligned} \text{cov}[n_j(t), n_k(t+\tau)] &= E[n_j(t)n_k(t+\tau)] - \mu_j(t)\sum_r \mu_r(t)p_{kr}(\tau) \\ &= p_{kj}(\tau)\,\text{var}[n_j(t)] + \sum_{\substack{r=1\\ r\neq j}}^{m} \text{cov}[n_j(t), n_r(t)]\, p_{kr}(\tau), \end{aligned} \qquad (8.118)$$

which can be expressed in terms of the $n_r(0)$ and $p_{jr}(t)$ $(r, j = 1, 2, \ldots, m)$ using (8.115) and (8.116). Thus, summarizing, we have from (8.113)

$$\mathscr{E}[\mathbf{n}(t_i)] = e^{\mathbf{A}t_i}\mathbf{n}(0).$$

and the terms of $\text{cov}[n_j(t_a), n_k(t_b)]$ $(\tau = t_b - t_a)$ are functions of $\mathbf{n}(0)$ and [c.f. (8.111)]

$$\mathbf{P}(\tau) = e^{\mathbf{A}\tau}.$$

Hence $\mathscr{D}[\boldsymbol{\varepsilon}] = \mathbf{V}(\boldsymbol{\theta})$ can be expressed in terms of the elements of $\mathbf{A}$ and $\mathbf{n}(0)$.

A similar result to (8.118) was given by El-Asfouri et al. [1979], though they incorrectly omitted the second term. Our result (8.118) can be shown to agree with the equations (8) of Kodell and Matis [1976] in the two-compartment case that they discussed. Matis and Hartley [1971] give the covariance structure for data of the form $y_i = \sum_j n_j(t_i)$. This model is generalized by Faddy [1976] to the time-dependent case.

The above derivations apply equally to the time-dependent Markov process if we replace $p_{jk}(t)$ by $p_{jk}(0, t)$ and $p_{jk}(\tau)$ by $p_{jk}(t, t+\tau)$. The additional difficulties in the time-dependent case come in the computation of $P(s, t)$ and its derivatives with respect to $\boldsymbol{\theta}$.

b Input from the Environment

Until now we have avoided discussion of the distribution of $\mathbf{z}(t)$, the number of particles in each compartment at time t which have entered the system from the environment after $t = 0$. We shall assume that the input stream consists of independent Poisson processes to some or all of the compartments with the input process to compartment j having mean $b_j(t)$. Furthermore, we assume that once in the system, the immigrant particles move independently according to the same process as the original particles. Using simple probability arguments, Faddy [1977] showed that $z_0(t), z_1(t), \ldots, z_m(t)$ are independent Poisson processes with

$$E[z_j(t)] = \text{var}[z_j(t)] \tag{8.119}$$

$$= \sum_{k=1}^{m} \int_0^t b_k(u) p_{jk}(u, t)\, du \tag{8.120}$$

and

$$\text{cov}[z_j(t), z_k(t)] = 0. \tag{8.121}$$

Setting $b_0(t) \equiv 0$ and

$$\tilde{\mathbf{b}}(t) = (b_0(t), \mathbf{b}'(t))' = (b_0(t), b_1(t), \ldots, b_m(t))',$$

we have from (8.120) that

$$\mathscr{E}[\mathbf{z}(t)] = \int_0^t \tilde{\mathbf{P}}(u, t) \tilde{\mathbf{b}}(u)\, du. \tag{8.122}$$

For the *time-homogeneous* Markov process this reduces [using (8.111)] to

$$\mathscr{E}[\mathbf{z}(t)] = \int_0^t e^{\tilde{\mathbf{A}}(t-u)} \tilde{\mathbf{b}}(u)\, du, \tag{8.123}$$

the contribution of the input process to the deterministic model. Thus from $\tilde{\mathbf{n}}(t) = \mathbf{w}(t) + \mathbf{z}(t)$, and applying (8.112) [with $\tilde{\mathbf{n}}$ replaced by $\mathbf{w}$],

$$\mathscr{E}[\tilde{\mathbf{n}}(t)] = e^{\tilde{\mathbf{A}}t} \tilde{\mathbf{n}}(0) + \int_0^t e^{\tilde{\mathbf{A}}(t-u)} \tilde{\mathbf{b}}(u)\, du, \tag{8.124}$$

which may be compared with (8.23). To make the comparison more direct we can remove the entry for compartment 0 in (8.124), using the same argument as that following (8.109) and using $b_0(u) \equiv 0$. Thus

$$\mathscr{E}[\mathbf{n}(t)] = e^{\mathbf{A}t} \mathbf{n}(0) + \int_0^t e^{\mathbf{A}(t-u)} \mathbf{b}(u)\, du. \tag{8.125}$$

In addition to the variance and covariance expressions (8.119) and (8.121), we also need $\text{cov}[z_j(t), z_k(t+\tau)]$. We use the fact that the particles $\mathbf{z}(t)$ are dispersed in the usual multinomial way. Let $\mathbf{v}_r(t;t+\tau)$ denote the number of particles in each compartment at time $t+\tau$ originating from the $z_r(t)$ particles in compartment r at time t. Similarly let $\mathbf{z}(t;t+\tau)$ denote the numbers of particles in each compartment at time $t+\tau$ that have entered the system subsequently to time t. Then

$$\mathbf{z}(t+\tau)=\sum_{r=1}^{m}\mathbf{v}_r(t;t+\tau)+\mathbf{z}(t;t+\tau),$$

where, conditionally on $\mathbf{z}(t)$, all the variables on the right-hand side are statistically independent with

$$\mathbf{v}_j(t;t+\tau)\,|\,\mathbf{z}(t)\sim \text{Multinomial}\,[z_j(t);\,p_{0j}(\tau),p_{1j}(\tau),\ldots,p_{mj}(\tau)].$$

Hence

$$E[z_k(t+\tau)\,|\,\mathbf{z}(t)]=\sum_{r=1}^{m}z_r(t)p_{kr}(\tau)+E[z_k(t;t+\tau)\,|\,\mathbf{z}(t)],$$

so that

$$E[z_k(t+\tau)]=\sum_{r=1}E[z_r(t)]p_{kr}(\tau)+E[z_k(t;t+\tau)].$$

Using a similar argument,

$$E[z_j(t)z_k(t+\tau)]=\sum_{r=1}^{m}E[z_j(t)z_r(t)]p_{kr}(\tau)+E[z_j(t)]E[z_k(t;t+\tau)].$$

Thus

$$\begin{aligned}\text{cov}\,[z_j(t),z_k(t+\tau)]&=E[z_j(t)z_k(t+\tau)]-E[z_j(t)]E[z_k(t+\tau)]\\&=\sum_{r=1}^{m}\text{cov}\,[z_j(t),z_r(t)]p_{kr}(\tau)\\&=p_{kj}(\tau)\,\text{var}\,[z_j(t)],\end{aligned}\tag{8.126}$$

since the $z_j(t)$ $(j=1,2,\ldots,m)$ are independent Poisson processes and $z_0(t)\equiv 0$. Finally, from $\tilde{\mathbf{n}}(t)=\mathbf{w}(t)+\mathbf{z}(t)$ and the fact that $\mathbf{w}(t)$ and $\mathbf{z}(t)$ are independent, we have

$$\text{cov}\,[n_j(t_a),n_k(t_b)]=\text{cov}\,[w_j(t_a),w_k(t_b)]+\text{cov}\,[z_j(t_a),z_k(t_b)].$$

Thus from (8.115) and (8.116) [with $n_j(t)$ replaced by $w_j(t)$], and from (8.119) and (8.121), we have for $j,k=1,2,\ldots,m$

$$\text{var}\,[n_j(t)]=\sum_{r=1}^{m}n_r(0)p_{jr}(t)[1-p_{jr}(t)]+E[z_j(t)]\tag{8.127}$$

with

$$\operatorname{cov}[n_j(t), n_k(t)] = -\sum_{r=1}^{m} n_r(0) p_{jr}(t) p_{kr}(t). \tag{8.128}$$

Finally, from (8.118) [with $n_j(t)$ replaced by $w_j(t)$], (8.126) and (8.127),

$$\operatorname{cov}[n_j(t), n_k(t+\tau)] = p_{kj}(\tau) \operatorname{var}[n_j(t)] + \sum_{\substack{r=1 \\ r \neq j}}^{m} \operatorname{cov}[n_j(t), n_r(t)] p_{kr}(\tau), \tag{8.129}$$

which can be expressed in terms of $E[z_j(t)]$, $\mathbf{n}(0)$, and $\mathbf{P}(\tau)$ using (8.127) and (8.128).

As in the no-input case, our data take the form $y_{ij} = n_j(t_i)$ $(j = 1, 2, \ldots, m)$, the number of counts in the jth compartment at time t_i. These data are arranged in a single vector $\mathbf{y}$, and $\boldsymbol{\varepsilon} = \mathbf{y} - \mathscr{E}[\mathbf{y}]$. The elements of $\mathscr{E}[\mathbf{y}]$ are obtained from (8.125), namely

$$\mathscr{E}[\mathbf{n}(t_i)] = e^{\mathbf{A}t_i}\mathbf{n}(0) + \int_0^t e^{\mathbf{A}(t_i - u)}\mathbf{b}(u)\, du.$$

These are functions of $\mathbf{n}(0)$, the elements of $\mathbf{A}$, and the parameters of the input process $\mathbf{b}(t)$. We note that (8.111) still holds for a nonzero input, as it represents the movements of particles between compartments, so that $P(t) = e^{\mathbf{A}t}$. Hence the terms $\operatorname{cov}[n_j(t_a), n_k(t_b)]$ $(\tau = t_b - t_a)$ making up $\mathbf{V}(\boldsymbol{\theta}) = \mathscr{D}[\boldsymbol{\varepsilon}]$ are functions of $\mathbf{n}(0)$, $P(\tau) = \mathrm{e}^{\mathbf{A}\tau}$, and the Poisson rates in $\mathscr{E}[\mathbf{z}(t)]$.

We find that the moments given above are valid for both time-dependent and time-homogeneous Markov processes. If the internal process is time-dependent, the notational changes from $p_{jk}(t)$ to $p_{jk}(0, t)$ and from $p_{jk}(\tau)$ to $p_{jk}(t, t+\tau)$ are again required. We note also that the formulation above involving $\mathbf{z}$ makes it easy to allow a different Markov process for the transfer of immigrant particles from that used for the original particles.

Instead of a continuous input, another possibility is to have an input of batches at fixed times. In this case each batch is treated like a separate experiment and contributes moments of the same form as (8.105), (8.115), (8.116), and (8.118) [but with $n_j(t)$ replaced by $w_j(t)$]. We need only change $\mathbf{n}(0)$ to the number in the batch added and replace t in $p_{jk}(t)$ by the length of time the batch has been in the system.

8.7.2 Computational Methods

a Unconditional Generalized Least Squares

As in Chapter 6 and Section 2.8.8, we can take several approaches to estimating $\boldsymbol{\theta}$ for the model $\mathbf{y} = \mathbf{f}(\boldsymbol{\theta}) + \boldsymbol{\varepsilon}$ where $\mathscr{D}[\boldsymbol{\varepsilon}] = \mathbf{V}(\boldsymbol{\theta})$.

1. *Two-Stage Estimation.* Take a consistent estimator of $\boldsymbol{\theta}$ such as the ordinary least-squares estimator $\hat{\boldsymbol{\theta}}_{\text{OLS}}$, and choose $\boldsymbol{\theta}$ to minimize

$$[\mathbf{y}-\mathbf{f}(\boldsymbol{\theta})]'\mathbf{V}^{-1}(\hat{\boldsymbol{\theta}}_{\text{OLS}})[\mathbf{y}-\mathbf{f}(\boldsymbol{\theta})]. \tag{8.130}$$

2. *Iterated Two-Stage Estimation.* Use approach 1 iteratively, i.e., for $a = 1, 2, \ldots$ obtain $\boldsymbol{\theta}^{(a+1)}$ by minimizing

$$[\mathbf{y}-\mathbf{f}(\boldsymbol{\theta})]'\mathbf{V}^{-1}(\boldsymbol{\theta}^{(a)})[\mathbf{y}-\mathbf{f}(\boldsymbol{\theta})] \tag{8.131}$$

   with respect to $\boldsymbol{\theta}$, and iterate to convergence.

3. Minimize $[\mathbf{y}-\mathbf{f}(\boldsymbol{\theta})]'\mathbf{V}^{-1}(\boldsymbol{\theta})[\mathbf{y}-\mathbf{f}(\boldsymbol{\theta})]$ with respect to $\boldsymbol{\theta}$ directly.

Approach 2 was used by Matis and Hartley [1971], though their data were of the form $y_i = \sum_j n_j(t_i)$ rather than counts for individual components. Approach 3 will seldom be computationally feasible, though it was used by Kodell and Matis [1976] for the two-compartment model, in which they obtained analytic expressions for $\mathbf{V}^{-1}(\boldsymbol{\theta})$.

b Conditional Generalized Least Squares

Suppose we again observe $\mathbf{n}(t)$ at times $t_1 < t_2 < \cdots < t_n$ and we assume that there is no external input ($\mathbf{z}(t) \equiv \mathbf{0}$), i.e., the total number of counts over all the compartments including the sink is constant. Let $\mathbf{y}_i = (n_1(t_i), n_2(t_i), \ldots, n_m(t_i))' = \mathbf{n}(t_i)$, and define $v_i = t_i - t_{i-1}$ $(i = 1, 2, \ldots, m)$ with $t_0 = 0$. We recall that

$$\begin{aligned}\tilde{\mathbf{n}}(t_i) &= [n_0(t_i), \mathbf{n}'(t_i)]' \\ &= \sum_{r=1}^{m} \mathbf{u}_{i,r},\end{aligned} \tag{8.132}$$

say, where conditional on $\mathbf{y}_{i-1}$ [or equivalently $\tilde{\mathbf{n}}(t_{i-1})$, as the number of counts to the environment (sink) can be obtained by subtraction], the $\mathbf{u}_{i,r}$ $(r = 1, 2, \ldots, m)$ are independent, and

$$\mathbf{u}_{i,r}|\mathbf{y}_{i-1} \sim \text{Multinomial}[y_{i-1,r}; p_{0r}(v_i), p_{1r}(v_i), \ldots, p_{mr}(v_i)]. \tag{8.133}$$

Thus the conditional probability distribution $\text{pr}(\tilde{\mathbf{n}}(t_i)|\mathbf{y}_{i-1})$ is an m-fold convolution of multinomials. Also $\text{pr}(\mathbf{y}_i|\mathbf{y}_{i-1})$ is the same except that the (dependent) counts going to the environment at each stage are omitted in the description of the distribution. Theoretically we could then build up the joint distribution of $\mathbf{y}_1, \mathbf{y}_2, \ldots, \mathbf{y}_n$ as

$$\text{pr}[\mathbf{y}_1, \mathbf{y}_2, \ldots, \mathbf{y}_n|\tilde{\mathbf{n}}(0)] = \prod_{i=1}^{n} \text{pr}(\mathbf{y}_i|\mathbf{y}_{i-1})$$

and use this distribution for full maximum-likelihood estimation. Kalbfleisch et al. [1983] note that the convolutions make this approach computationally intractible and develop approximations.

Comparing (8.133) and (8.104), and noting that $\mathbf{y}_{i-1} = \mathbf{n}(t_{i-1})$, we see that we can replace $\mathbf{n}(0)$ by $\mathbf{y}_{i-1}$ in (8.105), (8.115), and (8.116) to give

$$\mathbf{f}_i(\boldsymbol{\theta}) = \mathscr{E}[\mathbf{y}_i|\mathbf{y}_{i-1}] = \mathbf{P}(v_i)\mathbf{y}_{i-1} \tag{8.134}$$

and

$$\begin{aligned}\boldsymbol{\Sigma}_i(\boldsymbol{\theta}) &= \mathscr{D}[\mathbf{y}_i|\mathbf{y}_{i-1}] \\ &= \operatorname{diag}\{\mathbf{P}(v_i)\mathbf{y}_{i-1}\} - \mathbf{P}(v_i)\operatorname{diag}\{\mathbf{y}_{i-1}\}\mathbf{P}'(v_i),\end{aligned} \tag{8.135}$$

where $\mathbf{P}(v_i)$ is the matrix of elements $p_{jk}(v_i)$ $(j = 1, 2, \ldots, m, k = 1, 2, \ldots, m)$. If there is no excretion to the environment, $\mathbf{y}_i$ is singular given $\mathbf{y}_{i-1}$ as $\mathbf{1}'_m\mathbf{y}_i = \mathbf{1}'_m\mathbf{y}_{i-1}$. We must then drop the elements of $\mathbf{y}_i$ and $\mathbf{f}_i(\boldsymbol{\theta})$ corresponding to one of the compartments, and drop the corresponding row and column of $\boldsymbol{\Sigma}_i(\boldsymbol{\theta})$ to obtain a nonsingular distribution for estimation.

It is possible to modify $\mathbf{f}_i(\boldsymbol{\theta})$ and $\boldsymbol{\Sigma}_i(\boldsymbol{\theta})$ to allow for input processes by using theory that is very similar to that of the previous section. There we incorporated inputs using the unconditional moments. Kalbfleisch et al. [1983] discussed incorporating input (or immigration) processes in their Section 5.

The conditional least-squares estimator of Kalbfleisch et al. [1983] minimizes

$$S_1(\boldsymbol{\theta}) = \sum_{i=1}^{n} [\mathbf{y}_i - \mathbf{f}_i(\boldsymbol{\theta})]'[\mathbf{y}_i - \mathbf{f}_i(\boldsymbol{\theta})]. \tag{8.136}$$

Conditional generalized least squares estimators are obtained from

$$S_2(\boldsymbol{\theta}) = \sum_{i=1}^{n} [\mathbf{y}_i - \mathbf{f}_i(\boldsymbol{\theta})]'\boldsymbol{\Sigma}_i^{-1}(\boldsymbol{\theta})[\mathbf{y}_i - \mathbf{f}_i(\boldsymbol{\theta})]. \tag{8.137}$$

Kalbfleisch et al. also take a normal approximation to the multinomials in (8.133) and thus to the conditional distribution of $\mathbf{y}_i$ given $\mathbf{y}_{i-1}$. If the approximating multivariate normal distribution of $\mathbf{y}_1, \mathbf{y}_2, \ldots, \mathbf{y}_n$ is obtained via (8.134) and (8.135), the resulting approximate -2(log-likelihood function) is

$$S_3(\boldsymbol{\theta}) = \sum_{i=1}^{n} \log\det\boldsymbol{\Sigma}_i(\boldsymbol{\theta}) + S_2(\boldsymbol{\theta}). \tag{8.138}$$

Kalbfleisch et al. consider four estimators $\hat{\boldsymbol{\theta}}_k$ $(k = 1, 2, 3, 4)$. Here $\hat{\boldsymbol{\theta}}_1$ minimizes $S_1(\boldsymbol{\theta})$ of (8.136), $\hat{\boldsymbol{\theta}}_3$ minimizes $S_3(\boldsymbol{\theta})$ of (8.138), and $\hat{\boldsymbol{\theta}}_4$ minimizes $S_2(\boldsymbol{\theta})$ of (8.137): the authors prefer $\hat{\boldsymbol{\theta}}_3$ to $\hat{\boldsymbol{\theta}}_4$. Their $\hat{\boldsymbol{\theta}}_2$ is an iterated two-stage estimator, and is obtained iteratively by choosing $\boldsymbol{\theta}^{(a+1)}$ to minimize

$$S_2^*(\boldsymbol{\theta}) = \sum_{i=1}^{n} [\mathbf{y}_i - \mathbf{f}_i(\boldsymbol{\theta})]'\boldsymbol{\Sigma}_i^{-1}(\boldsymbol{\theta}^{(a)})[\mathbf{y}_i - \mathbf{f}_i(\boldsymbol{\theta})]. \tag{8.139}$$

All the above estimates can obviously be obtained using minimization programs that do not use derivatives. However, for $\hat{\boldsymbol{\theta}}_1$ and $\hat{\boldsymbol{\theta}}_2$, the authors stress Gauss–Newton-based methods. If we write $\mathbf{y} = (\mathbf{y}_1', \mathbf{y}_2', \ldots, \mathbf{y}_n')'$, $\mathbf{f}(\boldsymbol{\theta}) = (\mathbf{f}_1'(\boldsymbol{\theta}), \mathbf{f}_2'(\boldsymbol{\theta}), \ldots, \mathbf{f}_n'(\boldsymbol{\theta}))'$, $\mathbf{F}_{\bullet i}^{(a)} = \partial \mathbf{f}_i(\boldsymbol{\theta}^{(a)})/\partial \boldsymbol{\theta}'$, and $\mathbf{F}_{\bullet}^{(a)} = (\mathbf{F}_{\bullet 1}^{(a)\prime}, \mathbf{F}_{\bullet 2}^{(a)\prime}, \ldots, \mathbf{F}_{\bullet n}^{(a)\prime})'$, then

$$S_2^*(\boldsymbol{\theta}) = [\mathbf{y} - \mathbf{f}(\boldsymbol{\theta})]' \boldsymbol{\Sigma}^{(a)-1} [\mathbf{y} - \mathbf{f}(\boldsymbol{\theta})], \tag{8.140}$$

where $\boldsymbol{\Sigma}^{(a)-1}$ is a block diagonal with blocks $\boldsymbol{\Sigma}_i^{-1}(\boldsymbol{\theta}^{(a)})$. Minimizing (8.140) is then a generalized least-squares problem, and the usual Gauss–Newton step is (Section 2.1.4)

$$\begin{aligned} \boldsymbol{\theta}^{(a+1)} &= \boldsymbol{\theta}^{(a)} + [\mathbf{F}_{\bullet}^{(a)\prime} \boldsymbol{\Sigma}^{(a)-1} \mathbf{F}_{\bullet}^{(a)}]^{-1} \mathbf{F}_{\bullet}^{(a)\prime} \boldsymbol{\Sigma}^{(a)-1} [\mathbf{y} - \mathbf{f}(\boldsymbol{\theta}^{(a)})] \\ &= \boldsymbol{\theta}^{(a)} + \left(\sum_{i=1}^{n} \mathbf{F}_{\bullet i}^{(a)\prime} \boldsymbol{\Sigma}_i^{-1}(\boldsymbol{\theta}^{(a)}) \mathbf{F}_{\bullet i}^{(a)} \right)^{-1} \\ &\quad \times \left(\sum_{i=1}^{n} \mathbf{F}_{\bullet i}^{(a)\prime} \boldsymbol{\Sigma}_i^{-1}(\boldsymbol{\theta}^{(a)}) [\mathbf{y}_i - \mathbf{f}_i(\boldsymbol{\theta}^{(a)})] \right). \end{aligned} \tag{8.141}$$

The standard estimate of the asymptotic dispersion matrix for $\hat{\boldsymbol{\theta}}_2$ is given by

$$\hat{\mathscr{D}}[\hat{\boldsymbol{\theta}}_2] = \left(\sum_{i=1}^{n} \mathbf{F}_{\bullet i}'(\hat{\boldsymbol{\theta}}_2) \boldsymbol{\Sigma}_i^{-1}(\hat{\boldsymbol{\theta}}_2) \mathbf{F}_{\bullet i}(\hat{\boldsymbol{\theta}}_2) \right)^{-1}. \tag{8.142}$$

We note that the Gauss–Newton method for $\hat{\boldsymbol{\theta}}_1$ is identical to (8.141) but with $\boldsymbol{\Sigma}^{(a)}$ replaced by $\mathbf{I}_n$. However, the asymptotic dispersion estimate then becomes

$$\begin{aligned} \hat{\mathscr{D}}[\hat{\boldsymbol{\theta}}_1] &= \left(\sum_{i=1}^{n} \mathbf{F}_{i\bullet}'(\hat{\boldsymbol{\theta}}_1) \mathbf{F}_{i\bullet}(\hat{\boldsymbol{\theta}}_1) \right)^{-1} \left(\sum_{i=1}^{n} \mathbf{F}_{i\bullet}'(\hat{\boldsymbol{\theta}}_1) \boldsymbol{\Sigma}_i(\hat{\boldsymbol{\theta}}_1) \mathbf{F}_{i\bullet}(\hat{\boldsymbol{\theta}}_1) \right) \\ &\quad \times \left(\sum_{i=1}^{n} \mathbf{F}_{i\bullet}'(\hat{\boldsymbol{\theta}}_1) \mathbf{F}_{i\bullet}(\hat{\boldsymbol{\theta}}_1) \right)^{-1}. \end{aligned} \tag{8.143}$$

Kalbfleisch et al. [1983] used (8.142) to provide asymptotic dispersion estimates for $\hat{\boldsymbol{\theta}}_3$ and $\hat{\boldsymbol{\theta}}_4$ also. Computational savings are possible if the times t_i are equally spaced (see Kalbfleisch et al. [1983: Section 4]).

As the $\mathbf{y}_i$'s are not independent, further work is needed to justify the use of $\hat{\mathscr{D}}[\hat{\boldsymbol{\theta}}_1]$ and $\hat{\mathscr{D}}[\hat{\boldsymbol{\theta}}_2]$. The authors identify two asymptotic limits as being of interest. If N, the number of particles in the system, is large, whereas n, the number of time points, is moderate, the limit as $N \to \infty$ is of interest. With this limit the authors state they can show that $\hat{\boldsymbol{\theta}}_2$ and $\hat{\boldsymbol{\theta}}_4$ are consistent and asymptotically normal. They also state that if the system is ergodic and a moderate number of individuals are followed over a large number of time points, the limit as $n \to \infty$ becomes relevant. Work on the consistency of univariate conditional least-squares estimators is referenced, as it may shed some light on this limit.

For a homogeneous Markov process we have from (8.113), and (8.134)

$$\begin{aligned}\mathbf{f}_i(\boldsymbol{\theta}) &= \mathbf{P}(v_i)\mathbf{y}_{i-1}\\ &= e^{\mathbf{A}(\boldsymbol{\theta})v_i}\mathbf{y}_{i-1},\end{aligned} \tag{8.144}$$

where the dependence on $\boldsymbol{\theta}$ is shown explicitly. As $\mathbf{f}_i(\boldsymbol{\theta})$ is of the form discussed in Section 8.5.3, the methods of that section can be used to compute $\mathbf{f}_i(\boldsymbol{\theta})$ and its derivatives $\mathbf{F}_{i\cdot}(\boldsymbol{\theta}) = \partial\mathbf{f}_i(\boldsymbol{\theta})/\partial\boldsymbol{\theta}'$ without having to find an analytic expression for $\mathbf{P}(v_i)$.

There are some disadvantages in the conditional approach. For instance, it can only be used in situations where all the compartments can be observed, and observed at the same times. However, the unconditional methods work for any subset of the compartments, and the formulae in the previous section can still be used to calculate the covariance structure of the observations $n_j(t_{ij})$, even when each compartment is observed at a different set of times.

The main advantage of the conditional approach is computational. From (8.131), and the first equation of (8.141) with $\mathbf{V}(\boldsymbol{\theta}^{(a)})$ instead of $\boldsymbol{\Sigma}^{(a)}$, we see that we need to find the inverse of the $nm \times nm$ matrix $\mathbf{V}(\boldsymbol{\theta}^{(a)})$. However, the conditional method only requires the inverses of n $m \times m$ block matrices, $\boldsymbol{\Sigma}_i(\boldsymbol{\theta}^{(a)})$ making up $\boldsymbol{\Sigma}^{(a)}$ in (8.141). Of course, we do not actually invert any matrices, but rather use the Cholesky decomposition approach of Section 2.1.4. For example, in (8.141) we find the Cholesky decomposition $\boldsymbol{\Sigma}_i(\boldsymbol{\theta}^{(a)}) = \mathbf{U}_i'\mathbf{U}_i$ for each i, then solve $\mathbf{U}_i'\mathbf{z}_i = \mathbf{y}_i$, $\mathbf{U}_i'\mathbf{g}_i(\boldsymbol{\theta}) = \mathbf{f}_i(\boldsymbol{\theta})$, and $\mathbf{U}_i'\mathbf{G}_{i\cdot}(\boldsymbol{\theta}) = \mathbf{F}_{i\cdot}(\boldsymbol{\theta})$ for $\mathbf{z}_i$, $\mathbf{g}_i(\boldsymbol{\theta})$, and $\mathbf{G}_{i\cdot}(\boldsymbol{\theta})$ [$= \partial\mathbf{g}_i(\boldsymbol{\theta})/\partial\boldsymbol{\theta}'$] respectively. Then (8.139) becomes

$$\begin{aligned}S_2^*(\boldsymbol{\theta}) &= \sum_{i=1}^{n} [\mathbf{z}_i - \mathbf{g}_i(\boldsymbol{\theta})]'[\mathbf{z}_i - \mathbf{g}_i(\boldsymbol{\theta})]\\ &= [\mathbf{z} - \mathbf{g}(\boldsymbol{\theta})]'[\mathbf{z} - \mathbf{g}(\boldsymbol{\theta})],\end{aligned} \tag{8.145}$$

where $\mathbf{z} = (\mathbf{z}_1', \mathbf{z}_2', \ldots, \mathbf{z}_n')'$ and $\mathbf{g}(\boldsymbol{\theta}) = (\mathbf{g}_1'(\boldsymbol{\theta}), \mathbf{g}_2'(\boldsymbol{\theta}), \ldots, \mathbf{g}_n'(\boldsymbol{\theta}))'$. Applying the Gauss–Newton method for ordinary least squares to (8.145) also requires

$$\mathbf{G}_{\cdot}(\boldsymbol{\theta}) = \frac{\partial g(\boldsymbol{\theta})}{\partial\boldsymbol{\theta}'} = [\mathbf{G}_{1\cdot}'(\boldsymbol{\theta}), \mathbf{G}_{2\cdot}'(\boldsymbol{\theta}), \ldots, \mathbf{G}_{n\cdot}'(\boldsymbol{\theta})]'.$$

As usual, the transformed problem results in the correct asymptotic dispersion matrix for $\hat{\boldsymbol{\theta}}$.

Example 8.10 Kodell and Matis [1976] simulated data from the so-called two-compartment open model depicted in Fig. 8.15. The data are given in Table 8.3. For this model

$$\boldsymbol{\theta} = (\gamma_{12}, \gamma_{21}, \gamma_{01}, \gamma_{02})'$$

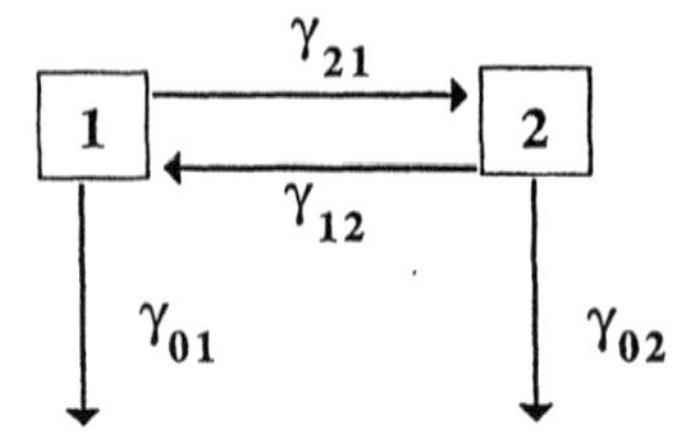

Figure 8.15

Table 8.3 Simulated Data[a] from the Two-Compartment Model in Fig. 8.15

t_i	$n_1(t_i)$	$n_2(t_i)$	t_i	$n_1(t_i)$	$n_2(t_i)$
0	1000	0	2.75	124	98
0.25	772	103	3.00	106	79
0.50	606	169	3.25	78	80
0.75	477	191	3.50	58	76
1.00	386	198	3.75	45	70
1.25	317	181	4.00	42	58
1.50	278	162	4.25	36	49
1.75	217	156	4.50	35	39
2.00	183	152	4.75	26	27
2.25	159	126	5.00	21	24
2.50	142	107			

[a]Reproduced from R. L. Kodell and J. H. Matis, "Estimating the Rate Constants in a Two-compartment Stochastic Model". Biometrics 32: 377–390, 1976. With permission from The Biometric Society.

and

$$\tilde{\mathbf{A}}(\boldsymbol{\theta}) = \begin{bmatrix} 0 & \gamma_{01} & \gamma_{02} \\ 0 & -\gamma_{01}-\gamma_{21} & \gamma_{12} \\ 0 & \gamma_{21} & -\gamma_{02}-\gamma_{12} \end{bmatrix},$$

where the first row and column refer to transitions to and from the environment. For this model the explicit form for $\tilde{\mathbf{P}}(t) = e^{\tilde{\mathbf{A}}t}$ is readily obtainable (though this not necessary for the computational procedures we have outlined). The eigenvalues of $\tilde{\mathbf{A}}$ are $\lambda_0 = 0$ and

$$\lambda_1, \lambda_2 = -\tfrac{1}{2}[\gamma_{01} + \gamma_{21} + \gamma_{02} + \gamma_{12} \pm \{(\gamma_{01} + \gamma_{21} - \gamma_{02} - \gamma_{12})^2 + 4\gamma_{12}\gamma_{21}\}^{1/2}].$$

Now

$$\tilde{\mathbf{P}}(t) = \begin{bmatrix} 1 & p_{01}(t) & p_{02}(t) \\ 0 & p_{11}(t) & p_{12}(t) \\ 0 & p_{21}(t) & p_{22}(t) \end{bmatrix},$$

Table 8.4 Parameter Estimates for Example 8.10[a,b]

Procedure	γ_{01}	γ_{21}	γ_{02}	γ_{12}
True Values	0.50	0.50	0.75	0.25
K. M.[a] (Method 3)	0.484	0.550	0.727	0.397
K. M.[a] (OLS)	0.502	0.565	0.691	0.482
$\hat{\theta}_1$[b]	0.506	0.568	0.683	0.498
$\hat{\theta}_2$	0.492	0.555	0.731	0.409
$\hat{\theta}_3$	0.495	0.561	0.730	0.425
$\hat{\theta}_4$	0.501	0.589	0.732	0.485

[a]Reproduced from R. L. Kodell and J. H. Matis, "Estimating the Rate Constants in a Two-compartment Stochastic Model". Biometrics 32: 377–390, 1976. With permission from The Biometric Society.
[b]Reproduced from J. D. Kalbfleisch, J. F. Lawless and W. M. Vollmer, "Estimation in Markov Models from Aggregate Data". Biometrics 39: 907–919, 1983. With permission from The Biometric Society. Definitions of $\hat{\theta}_k$ $(k = 1, 2, 3, 4)$ follow Equation (8.138).

and using the method of Example 8.5 following Equation (8.33), we find that

$$p_{11}(t) = \frac{1}{\lambda_2 - \lambda_1}[(\gamma_{01} + \gamma_{21} + \lambda_2)e^{\lambda_1 t} - (\gamma_{01} + \gamma_{21} + \lambda_1)e^{\lambda_2 t}]$$

and

$$p_{21}(t) = \frac{\gamma_{21}}{\lambda_2 - \lambda_1}(e^{\lambda_2 t} - e^{\lambda_1 t}).$$

The terms $p_{22}(t)$ and $p_{21}(t)$ are obtained by symmetry, and $p_{01}(t)$ and $p_{02}(t)$ follow from the fact that the column sums of $\tilde{\mathbf{P}}$ are unity.

The data in Table 8.3 were analyzed by both Kodell and Matis [1976] and Kalbfleisch et al. [1983]. Kodell and Matis used the unconditional method 3 following Equation (8.131). They also produced estimates using ordinary least squares (OLS). Table 8.4 gives the true parameter values together with estimates from each method. All the estimators are fairly similar. Kalbfleisch et al. give estimated standard errors for $\hat{\theta}_2$. ■

8.7.3 Discussion

From (8.114) and (8.124) we see that time-homogeneous Markov processes lead to models which are linear combinations of decaying exponentials ($e^{\lambda x}, \lambda < 0$). Purdue [1979] references a series of papers by M. F. Wise and also work by A. H. Marcus in which curves involving terms like $x^{-\alpha}$ or $x^{-\alpha}e^{-\beta x}$ fit some of the data much better (see also Wise [1979]). This type of behavior can be accommodated by generalizing Markov processes to semi-Markov processes. This is done as follows.

A Markov process can be defined in terms of the particles having exponentially distributed residency times within compartments. At the end of its residence time the particle is transferred to one of the other compartments (or excreted) with probabilities according to a Markov-chain transition matrix. A semi-Markov process is obtained by allowing for general distributions for residence times instead of being limited to exponential distributions. Purdue [1979] discusses these processes in addition to time-dependent Markov processes and gives references to earlier work (see also Marcus [1979]). Matis et al. [1983] discuss the use of gamma residence times. As particles still move independently in a semi-Markov process, the previous analyses of Sections 8.7.1 and 8.7.2, and particular the moments in terms of $p_{jk}(t)$, still apply. However, it becomes much more difficult to obtain or compute the $p_{jk}(t)$. More recent work on semi-Markov compartmental models has been done by Agrafiotis [1982], Mehata and Selvam [1981], and Parthasarathy and Mayilswami [1982, 1983].

In virtually all the examples we have seen, the material being transferred between compartments is composed of particles. Application of the theory above depends upon being able to count the particules in measured compartments. However, in situations where there are huge numbers of particles this cannot be done. There are no theoretical problems if the measurements taken, for example radioactive counts, are proportional to the number of particles, i.e., we observe $\mathbf{y}(t) = \phi^2 \mathbf{n}(t)$. As $\mathscr{E}[\mathbf{y}(t)] = \phi^2 \mathscr{E}[\mathbf{n}(t)]$ and $\mathscr{D}[\mathbf{y}(t)] = \phi^4 \mathscr{D}[\mathbf{n}(t)]$ our earlier theory applies with, at worst, the introduction of an additional parameter ϕ.

Unfortunately, for the above Markov-process and semi-Markov-process models, the coefficient of variation

$$\mathrm{CV}_j = \frac{\{\mathrm{var}[n_j(t)]\}^{1/2}}{E[n_j(t)]}$$

tends to zero as the number of particles involved becomes large (Matis and Wehrly [1979b]). Suppose there are no inputs to the Markov-process system, so that $n_j(t) = w_j(t)$. Setting $n_j(0) = \rho_j N$, it follows from (8.105) and (8.115) that $\mathrm{CV}_j \to 0$ at rate $N^{-1/2}$. However, in many biological systems involving huge numbers of particles the variability or scatter is still appreciable. This indicates that the component of the observed variability arising just from the independent movement of particles is often negligible compared with other stochastic variability in the system. Rescigno and Matis [1981] noted that the coefficient of variation due to particle movement in such a system can still be appreciable in compartments with small numbers of particles, and that in pharmacokinetic and cancer studies these "small' compartments are often of primary importance.

Particle-based models have also been very useful in population dynamics. However, they do not always explain adequately the variability in many applications, so that other stochastic approaches have been explored. We sketch some of these in the following section.

We note, in conclusion, the real-data example of Matis and Hartley [1971], which concerned the progress of plastic beads through the gastrointestinal tract of a sheep. The collected data were of the form (y_i, t_i), where $y_i = \sum_j n_j(t_i)$. In their analysis the authors allowed other sources of error in $\mathscr{D}[\mathbf{y}]$ in addition to that due to the Markov process, namely

$$\mathscr{D}[\mathbf{y}] = \mathbf{\Sigma}_{MP} + \mathbf{\Sigma}_{EP} + \mathbf{\Sigma}_{M}.$$

Here $\mathbf{\Sigma}_{MP}$ is the Markov process covariance matrix; $\mathbf{\Sigma}_{EP}$ comes from "end-period error"; and $\mathbf{\Sigma}_{M}$ was due to "some unfortunate mastication of the beads by the sheep." In a similar vein one could write (with MS referring to measurement)

$$\mathscr{D}[\mathbf{y}] = \mathbf{\Sigma}_{MP} + \mathbf{\Sigma}_{MS}.$$

Unfortunately, Matis and Hartley [1971] give no indication of how they estimated these other components of variance.

8.8 FURTHER STOCHASTIC APPROACHES

In the models of the previous section the stochastic nature of the system is due to the random movement of individual particles. Matis and Wehrly [1979a, b] call this P1 stochasticity. In addition they discuss allowing for the transfer rate coefficients to vary from particle to particle. They call this P2 stochasticity.

There has been another strand to stochastic compartmental modeling, developed by various researchers, which allows for random variation between experiments of the same type, e.g. experiments on different subjects or a repeated experiment on one subject. Matis and Wehrly [1979a, b] call the situation in which the initial quantities of material in the compartments is random R1 stochasticity, while R2 stochasticity allows the rate coefficients to vary randomly from experiment to experiment. However, for any particular experiment a single realization of the initial quantities is obtained and a single realization of the set of rate constants. These features (R1 and R2) therefore add no within-experiment variation. Matis and Wehrly [1979a, b] give references to the literature in R1/R2 modeling, most of which has centered on finding the unconditional distribution of $\mathbf{y}$. They use the standard expectation formulae to give the first and second moments of the unconditional distribution of $\mathbf{y}$ for different combinations of P1, P2, R1, and R2 stochasticity for the one-compartment model. An interesting feature of the results is that for models with P2 or R2 stochasticity, the behavior of the mean is not described by the deterministic model. Indeed, there are models in which the mean for the stochastic model always exceeds values obtained from the deterministic model evaluated at the mean transfer rates. Matis et al. [1983: Section 5] discuss the building of models that allow for clustering of particles.

Yet another approach to stochastic compartmental modeling has been through stochastic differential equations, described, for example, by White and Clark [1979] and Tiwari [1979]. These equations allow for random quantities of material transferred over time, perhaps caused by randomly varying environmental conditions. Tiwari [1979] gives a fairly general framework for stochastic differential-equation models and discusses an approach to estimation involving information theory (the principle of maximum entropy).

White and Clark [1979] discuss two models. For the one-compartment case the first is a discrete approximation to

$$\frac{dy(x)}{dx} = ky(x) + \varepsilon(x),$$

like that discussed in Section 7.5.3b. For observations $y_1, y_2, \ldots, y_n$ collected at times $x_1 < x_2 < \cdots < x_n$, White and Clark assume that $\varepsilon(x) = a_i$ for $x_{i-1} < x \leq x_i$ and that the a_i are i.i.d. $N(0, \sigma_a^2)$. They then use maximum-likelihood estimation. This is extended naturally to the multicompartment case as

$$\frac{d\mathbf{y}(x)}{dx} = \mathbf{K}\mathbf{y}(x) + \boldsymbol{\varepsilon}(x)$$

with the $\mathbf{a}_i$ i.i.d. multivariate normal $N_m(\mathbf{0}, \boldsymbol{\Sigma}_a)$. If we do not make the discrete approximation and allow $\boldsymbol{\varepsilon}(x)$ to be a Wiener process, we have a special case of the growth model of Garcia [1983]. The solution process is given by Arnold [1974: Section 8.5] and is simply a vector version of Appendix C3, Equation (19).

White and Clark's second model for the single compartment is

$$\frac{dy(x)}{dx} = k(x)y(x),$$

where $k(x)$ is randomly distributed over time x about a mean level k. They use the approximation $k(x) = k_i$ for $x_{i-1} < x \leq x_i$, where the k_i are i.i.d. $N(k, \sigma_k^2)$.

CHAPTER 9

Multiphase and Spline Regressions

9.1 INTRODUCTION

Consider a regression relationship between y and x in which the regression function $E[y|x] = f(x;\boldsymbol{\theta})$ is obtained by piecing together different curves over different intervals, i.e.

$$f(x;\boldsymbol{\theta},\boldsymbol{\alpha}) = \begin{cases} f_1(x;\boldsymbol{\theta}_1), & x \le \alpha_1; \\ f_2(x;\boldsymbol{\theta}_2), & \alpha_1 < x \le \alpha_2; \\ \vdots & \vdots \\ f_D(x;\boldsymbol{\theta}_D), & \alpha_{D-1} < x. \end{cases} \tag{9.1}$$

Here $f(x;\boldsymbol{\theta})$ is defined by different functions on different intervals, and typically the end points of the intervals are unknown and must be estimated. The D submodels will be referred to as phase models or regimes, and the end points as changepoints or joinpoints. The problem is also called piecewise regression. The models are intended for situations in which (a) the number of regimes is small, (b) the behavior of $E[y|x]$ within a regime can be well described by a simple parametric function such as a straight line, and (c) there are fairly abrupt changes between regimes. Such behavior is demonstrated in the examples to follow.

Models of the form (9.1) also arise in spline regression. Here the emphasis is on modeling underlying smooth regression relationships rather than describing abrupt or structural changes in the regression. In spline regression the individual phase models $f_d(x;\boldsymbol{\theta}_d)$ $(d = 1,2,\dots,D)$ are polynomials, and more stringent conditions are imposed on $f(x;\boldsymbol{\theta},\boldsymbol{\alpha})$ at the changepoints α_d—for example, continuous second derivatives. The changepoints are called knots in the spline literature. In the early parts of this chapter we confine our attention to change-of-phase models, postponing the motivation of spline models until Section 9.5.

Example 9.1 Figure 9.1 plots pollen concentration at various depths of a lake sediment core. Concentration measurements were taken at depth units $x = 1$,

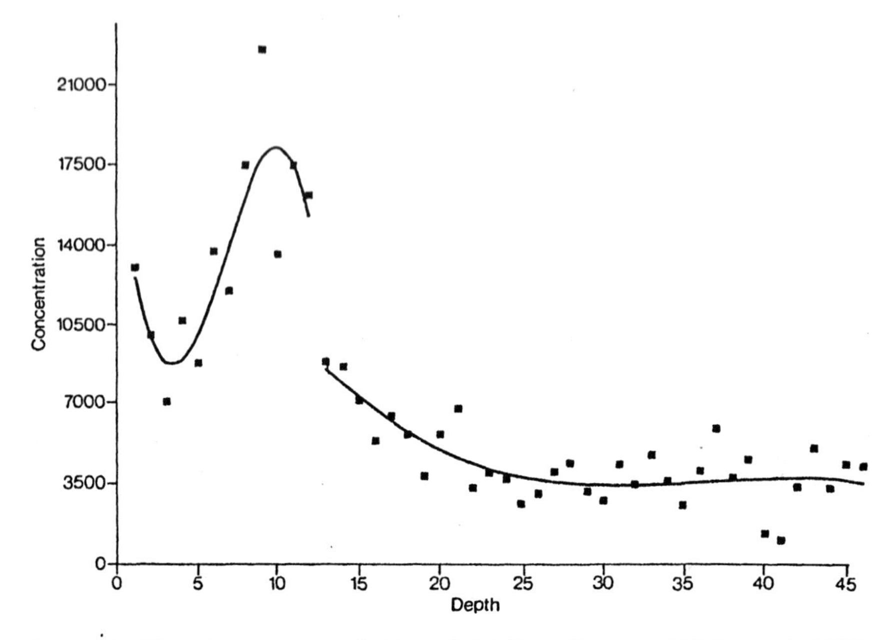

Figure 9.1 The pollen-count example, Example 9.1. From Esterby and El-Shaarawi [1981].

$2,\ldots,46$. There appears to be an abrupt change in the concentration–depth relationship between $x = 12$ and $x = 13$. Esterby and El-Shaarawi [1981] model this relationship as

$$y_i = f_1(x_i; \boldsymbol{\theta}_1) + \varepsilon_i, \qquad i = 1, 2, \ldots, m,$$

$$y_i = f_2(x_i; \boldsymbol{\theta}_2) + \varepsilon_i, \qquad i = m+1, m+2, \ldots, n.$$

The observation number m after which the regime changes is unknown and must be estimated from the data. They take f_1 and f_2 as polynomials in x and make no assumption about the continuity of the relationship. From the plot it is fairly clear that $m = 12$. They wisely do not make the continuity assumption that $f_1(\alpha; \boldsymbol{\theta}_1) = f_2(\alpha; \boldsymbol{\theta}_2)$ for some α satisfying $x_m \leq \alpha < x_{m+1}$. There is an apparent discontinuity in the trend between the first 12 points and the remaining points. This may represent a fairly abrupt shift in pollen concentrations because of the abrupt change in the levels of airborne pollen. If the relationship between x and y is continuous, or even smooth, there are insufficient data to model the transition in trend between the first 12 and remaining points. If a continuity assumption were to be imposed, either this would end up imposing an unwarrantedly severe constraint on the $f_1(x; \boldsymbol{\theta}_1)$ and $f_2(x; \boldsymbol{\theta}_2)$ leading to a bad fit, or we would have to use more complicated functional forms for f_1 and f_2 than would otherwise be necessary. In Fig. 9.1 the fitted curves f_1 and f_2 are both cubic. ■

Although it is not our main emphasis, in Section 9.2 below we discuss change-of-phase models in which no continuity assumption is made.

Example 9.2 Figure 9.2 from Lerman [1980] plots $\log_e$(primordia number) versus the number of days since wheat was sown. The number of primordia is a measure of the development of the shoot apex. The pattern in the plot is clearly well summarized by a pair of straight lines intersecting near $x = 48$ days. Lerman [1980] notes that "there are good biological grounds for believing that the transition occurs at an identifiable stage in the plant's development, namely at the end of spikelet initiation." We might simplistically write the model

$$E[y|x] = \begin{cases} \beta_{10} + \beta_{11}x, & x \le \alpha, \\ \beta_{20} + \beta_{21}x, & x > \alpha. \end{cases} \tag{9.2}$$

However, to impose continuity at the joinpoint, thus avoiding a situation as shown in Fig. 9.3, we must further impose the constraint

$$\beta_{10} + \beta_{11}\alpha = \beta_{20} + \beta_{21}\alpha. \tag{9.3}$$

■

Example 9.3 Figure 9.4, adapted from Smith and Cook [1980], plots the reciprocal of serum-creatinine (corrected for body weight) versus time following a

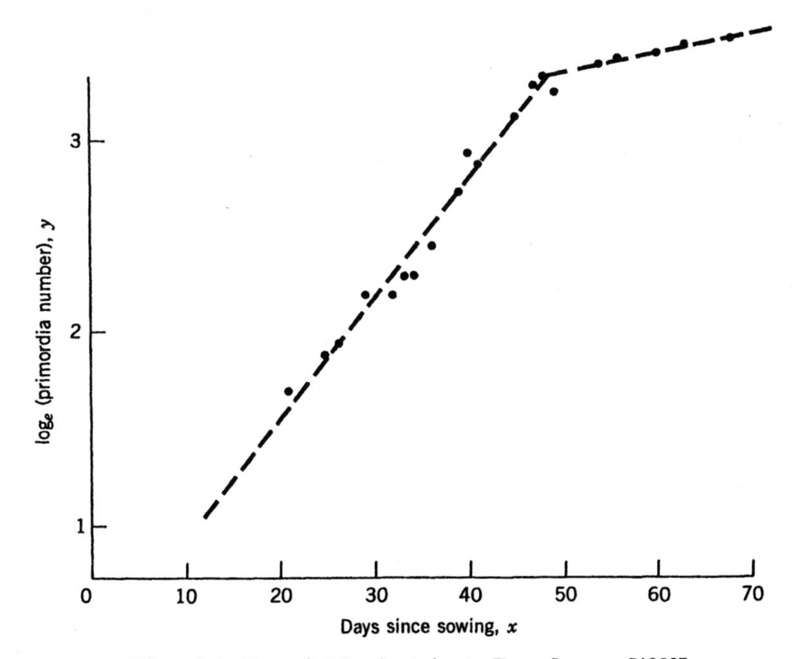

Figure 9.2 Example 9.2, wheat shoots. From Lerman [1980].

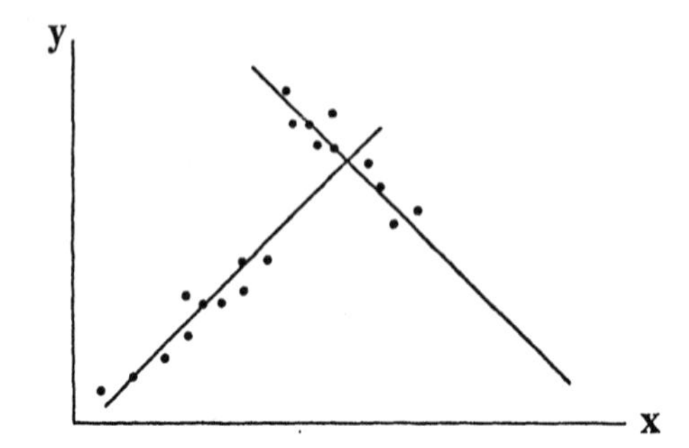

Figure 9.3 Two linear regimes which are not continuous.

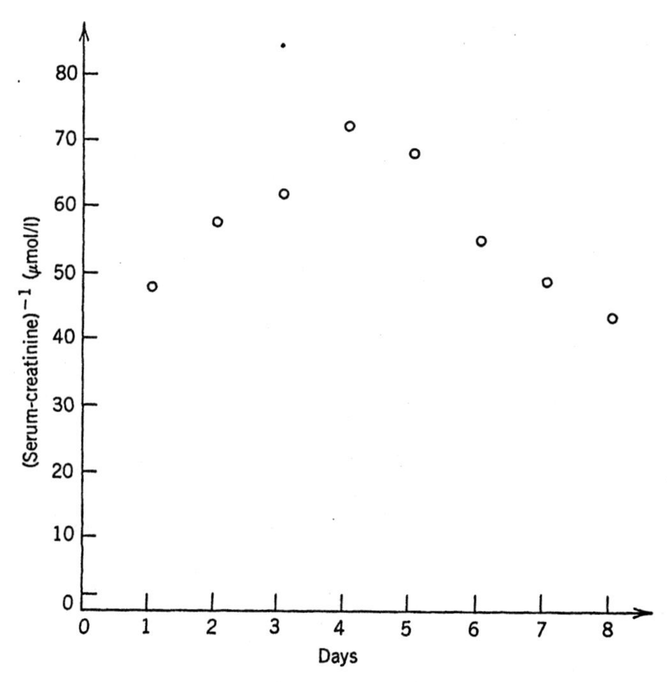

Figure 9.4 Plot of (serum-creatinine)$^{-1}$ versus time (arbitrary origin). From Smith and Cook [1980: patient A only].

renal transplant. This shows behavior similar to the previous example. The transition from a positively to a negatively sloped line is interpreted as the time when rejection of the kidney takes place. ■

An abrupt change, i.e. a change which appears continuous at the join but has discontinuous first derivatives, tends to be visible in a scatterplot. Higher-order continuity at the joins tends to arise in spline modeling (Section 9.5). In

Section 9.3 we discuss inference from models which have various orders of continuity at the changepoints.

In their discussion of Example 9.3, Smith and Cook [1980] note that "it might be argued that rejection is not an instantaneous process, so that the sharp intersection of the two lines is a fiction...." They go on to state that, in their case, any transition period is usually very short compared with the interval between measurements, so that the intersecting lines provide a reasonable approximation. However, in Example 9.4 to follow, there are more data points in the region of change, and we see what appears to be two linear regimes with a smooth transition.

Example 9.4 Watts and Bacon [1974: Example 2] give a data set from a study of how fast sediment settles in a tank of fluid. The sediment was thoroughly mixed through the liquid by agitation. At various times t after agitation ceased, the experimenter measured the depth of clear liquid at the top of the tank. The data are plotted in Fig. 9.5. ∎

Example 9.5 Figure 9.6, adapted from Prunty [1983], plots yield of corn in bushels per case versus quantity of fertilizer applied. The data were obtained in a series of experiments in Ontario, Oregon. ∎

The pattern in Fig. 9.6 is again two lines with a smooth transition, but the transition is not so well characterized by the data points. In Section 9.4 we discuss families of curves which summarize patterns consisting of two or more line segments with smooth transitions between regimes. A feature of these curves is a

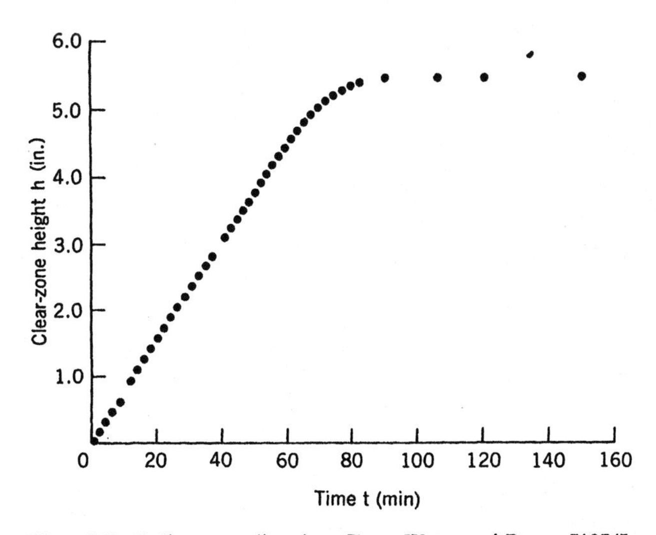

Figure 9.5 Sediment settling data. From Watts and Bacon [1974].

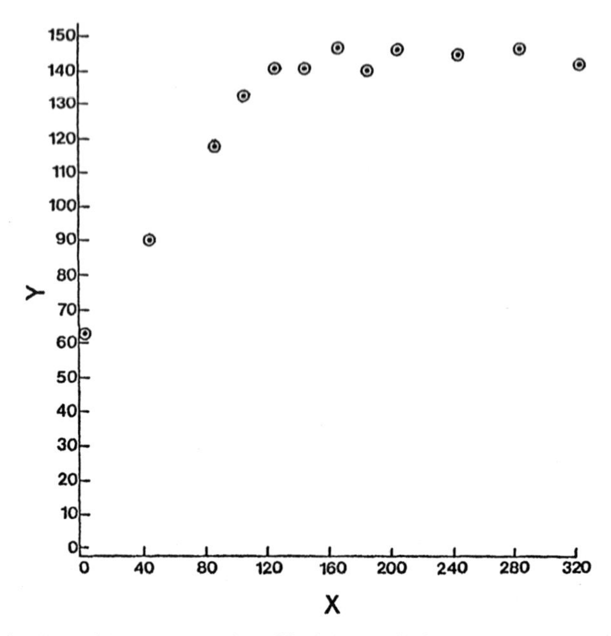

Figure 9.6 Yield of corn (y) versus quantity of fertilizer applied (x). From Prunty [1983: data points only].

parameter γ which controls the sharpness of the transition. Also, as $\gamma \to 0$ the smooth curve tends to intersecting linear regimes.

Finally, Section 9.5 discusses the building of regression models using splines.

9.2 NONCONTINUOUS CHANGE OF PHASE FOR LINEAR REGIMES

9.2.1 Two Linear Regimes

Consider fitting the model

$$E[y|x] = \begin{cases} \beta_{10} + \beta_{11}x, & x \leq \alpha, \\ \beta_{20} + \beta_{21}x, & x > \alpha, \end{cases}$$

to data in which $x_1 < x_2 < \cdots < x_n$. If $E[y|x]$ is not constrained to be continuous, we cannot hope to estimate the changepoint α. At best we might estimate the sequence number, $i = \kappa$, say, in the x_i-sequence after which the transition takes place. The model then becomes

$$y_i = \begin{cases} \beta_{10} + \beta_{11}x_i + \varepsilon_{1i}, & i = 1, 2, \ldots, \kappa, \\ \beta_{20} + \beta_{21}x_i + \varepsilon_{2i}, & i = \kappa + 1, \ldots, n. \end{cases} \tag{9.4}$$

If κ is known, (9.4) is an ordinary linear regression problem. We treat the case in which κ is unknown. Because the condition $x_1 < x_2 < \cdots < x_n$ is of little value in developing procedures for inference, most of the literature is written in terms of an arbitrary, rather than a monotonic, x-sequence. Only the subscript order is important. The model (9.4) was first studied by Quandt [1958].

Generalizing (9.4) to multiple linear regression, we consider the following model. The sequence of data points $(\mathbf{x}_1, y_1)$, $(\mathbf{x}_2, y_2), \ldots, (\mathbf{x}_n, y_n)$ is expressed according to some natural ordering. This may be the time order of data collection or according to the value of some other explanatory variable. It may be reasonable to assume, or suspect, that after some unknown sequence number $i = \kappa$ there has been a structural change in the relationship between y and $\mathbf{x}$. Thus

$$\begin{aligned} y_i &= \mathbf{x}_i'\boldsymbol{\beta}_1 + \varepsilon_{1i}, \qquad i = 1, 2, \ldots, \kappa, \\ y_i &= \mathbf{x}_i'\boldsymbol{\beta}_2 + \varepsilon_{2i}, \qquad i = \kappa + 1, \ldots, n, \end{aligned} \tag{9.5}$$

where $\boldsymbol{\beta}_i$ is $p \times 1$ $(i = 1, 2)$. Moreover, we assume independent errors ε_{di} with $\varepsilon_{di} \sim N(0, \sigma_d^2)$. Usually we make the further assumption $\sigma_1^2 = \sigma_2^2$, so that there is a change in the regression but not in the residual variability. For $\boldsymbol{\beta}_1$ and $\boldsymbol{\beta}_2$ to be estimable (i.e., the number of observations on each submodel is at least p), we restrict κ to the range $\kappa = p, p+1, \ldots, n-p$.

There are three important problems to address: (a) testing whether there is indeed a change of phase, or whether a single regression can describe the data; (b) making inferences about the changepoint κ; and (c) making inferences about regression parameters $\boldsymbol{\beta}_1$, $\boldsymbol{\beta}_2$, σ_1^2, and σ_2^2. Most of the attention in the literature has been given to problems (a) and (b), particularly in the special case of a change in location, i.e. $E[y_i] = \mu_1$ for $i \le \kappa$, $E[y_i] = \mu_2$ for $i > \kappa$.

For fixed κ, (9.5) can be written in matrix form as independent regressions

$$\begin{aligned} \mathbf{y}_1^{(\kappa)} &= \mathbf{X}_1^{(\kappa)}\boldsymbol{\beta}_1 + \boldsymbol{\varepsilon}_1^{(\kappa)}, \\ \mathbf{y}_2^{(\kappa)} &= \mathbf{X}_2^{(\kappa)}\boldsymbol{\beta}_2 + \boldsymbol{\varepsilon}_2^{(\kappa)}, \end{aligned} \tag{9.6}$$

where $\mathbf{y}_1^{(\kappa)}$ is $\kappa \times 1$ and $\mathbf{y}_2^{(\kappa)}$ is $(n-\kappa) \times 1$. The likelihood function is

$$\begin{aligned} \ell(\boldsymbol{\beta}_1, \boldsymbol{\beta}_2, \sigma_1^2, \sigma_2^2, \kappa) &= (2\pi)^{-n/2}(\sigma_1^2)^{-\kappa/2}(\sigma_2^2)^{-(n-\kappa)/2} \\ &\quad \times \prod_{d=1}^{2} \exp\left\{ -\frac{1}{2\sigma_d^2}(\mathbf{y}_d^{(\kappa)} - \mathbf{X}_d^{(\kappa)}\boldsymbol{\beta}_d)'(\mathbf{y}_d^{(\kappa)} - \mathbf{X}_d^{(\kappa)}\boldsymbol{\beta}_d) \right\}. \end{aligned} \tag{9.7}$$

Since for fixed κ we have two independent linear models, the maximum-likelihood estimators are

$$\hat{\boldsymbol{\beta}}_d^{(\kappa)} = (\mathbf{X}_d^{(\kappa)\prime}\mathbf{X}_d^{(\kappa)})^{-1}\mathbf{X}_d^{(\kappa)\prime}\mathbf{y}_d^{(\kappa)} \qquad (d = 1, 2), \tag{9.8}$$

$$\hat{\sigma}_1^2(\kappa) = \frac{\hat{S}_1^{(\kappa)}}{\kappa}, \tag{9.9a}$$

$$\hat{\sigma}_2^2(\kappa) = \frac{\hat{S}_2^{(\kappa)}}{n-\kappa}, \tag{9.9b}$$

where

$$\hat{S}_d^{(\kappa)} = (\mathbf{y}_d^{(\kappa)} - \mathbf{X}_d^{(\kappa)}\hat{\boldsymbol{\beta}}_d^{(\kappa)})'(\mathbf{y}_d^{(\kappa)} - \mathbf{X}_d^{(\kappa)}\hat{\boldsymbol{\beta}}_d^{(\kappa)}) \tag{9.10}$$

is the residual sum of squares for the dth model ($d = 1, 2$). If $\sigma_1^2 = \sigma_2^2 = \sigma^2$ then

$$\hat{\sigma}^{2(\kappa)} = \frac{\hat{S}_1^{(\kappa)} + \hat{S}_2^{(\kappa)}}{n}. \tag{9.11}$$

Substituting these values in $\log \ell$, we can concentrate the log-likelihood function (see Section 2.2.3) with respect to κ to obtain

$$M(\kappa) = -\frac{n}{2} - \frac{n}{2}\log 2\pi - \frac{\kappa}{2}\log\frac{\hat{S}_1^{(\kappa)}}{\kappa} - \frac{n-\kappa}{2}\log\frac{\hat{S}_2^{(\kappa)}}{n-\kappa} \qquad (\sigma_1^2 \neq \sigma_2^2), \tag{9.12}$$

$$M(\kappa) = -\frac{n}{2} - \frac{n}{2}\log 2\pi - \frac{n}{2}\log\frac{\hat{S}_1^{(\kappa)} + \hat{S}_2^{(\kappa)}}{n} \qquad (\sigma_1^2 = \sigma_2^2 = \sigma^2). \tag{9.13}$$

The maximum-likelihood estimate, $\hat{\kappa}$, of κ is obtained by maximizing $M(\kappa)$ over $\kappa = p, p+1, \ldots, n-p$. In particular, under the assumption $\sigma_1^2 = \sigma_2^2$, $(= \sigma^2$, say), $\hat{\kappa}$ minimizes $\hat{S}_1^{(\kappa)} + \hat{S}_2^{(\kappa)}$.

9.2.2 Testing for a Two-Phase Linear Regression

Consider testing for a single regression, i.e.

$$H_0: \mathbf{y} = \mathbf{X}\boldsymbol{\beta} + \boldsymbol{\varepsilon}, \qquad \boldsymbol{\varepsilon} \sim N_n(\mathbf{0}, \sigma^2\mathbf{I}_n). \tag{9.14}$$

versus the two-phase alternative (9.6). The maximized log-likelihood for the model (9.14) [c.f. Equation (2.47)] is

$$-\frac{n}{2} - \frac{n}{2}\log 2\pi - \frac{n}{2}\log\frac{\tilde{S}}{n}, \tag{9.15}$$

where $\tilde{S}$ is the residual sum of squares from (9.14). Thus the -2(log-likelihood ratio) for the test is, using (9.12) and (9.15).

$$L_1 = \left[n\log\frac{\tilde{S}}{n} - \kappa\log\frac{\hat{S}_1^{(\kappa)}}{\kappa} - (n-\kappa)\log\frac{\hat{S}_2^{(\kappa)}}{(n-\kappa)} \right]_{\kappa=\hat{\kappa}}. \tag{9.16}$$

As Worsley [1983] notes, this tests for a change in variance as well as for a change in regression. For the remainder of this discussion we follow the literature and further assume $\sigma_1^2 = \sigma_2^2 = \sigma^2$. Then, from (9.13) and (9.15), the -2(log-likelihood ratio) is

$$
\begin{aligned}
L_2 &= n \log\left[\frac{\tilde{S}}{\hat{S}_1^{(\kappa)} + \hat{S}_2^{(\kappa)}}\right]_{\kappa=\hat{\kappa}} \\
&= n \max_{p \le \kappa \le n-p} \log \frac{\tilde{S}}{\hat{S}_1^{(\kappa)} + \hat{S}_2^{(\kappa)}}. \qquad (9.17)
\end{aligned}
$$

Under standard asymptotic theory (see LR of Section 5.9.2) L_2 would be asymptotically χ_{p+1}^2. However, standard theory does not apply here, because κ takes only discrete values and (9.14) is also true if the change is outside the range of the data. The likelihood ratio does not have a limiting distribution, but tends to infinity as n increases (Hawkins [1980], Quandt [1960]).

For known κ the usual F-test statistic for $H_0: \boldsymbol{\beta}_1 = \boldsymbol{\beta}_2$ in the model (9.6) is, with $\sigma_1^2 = \sigma_2^2$,

$$
\begin{aligned}
F_\kappa &= \frac{[\tilde{S} - (\hat{S}_1^{(\kappa)} + \hat{S}_2^{(\kappa)})]/p}{(\hat{S}_1^{(\kappa)} + \hat{S}_2^{(\kappa)})/(n-2p)} \\
&= \left(\frac{\tilde{S}}{\hat{S}_1^{(\kappa)} + \hat{S}_2^{(\kappa)}} - 1\right)\frac{n-2p}{p}, \qquad (9.18)
\end{aligned}
$$

which has an $F_{p,n-2p}$ distribution when H_0 is true. An intuitively appealing procedure is to base a test upon

$$
F_{\max} = \max_{p \le \kappa \le n-p} F_\kappa. \qquad (9.19)
$$

Moreover, (9.19) is clearly equivalent to the likelihood-ratio procedure of (9.17). Beckman and Cook [1979] and Worsley [1983] form approximations to the null distribution of $F_{\max}$. Now $F_{\max}$ is the maximum of $n - 2p + 1$ correlated F-statistics F_κ which will be very highly correlated for adjacent values of κ. It can be shown that if $\boldsymbol{\beta}_1 = \boldsymbol{\beta}_2$ ($= \boldsymbol{\beta}$, say), the F-statistic F_κ for testing $H_0: \boldsymbol{\beta}_1 = \boldsymbol{\beta}_2$ in (9.6) does not depend upon the parameters $\boldsymbol{\beta}$ and σ^2, although it clearly depends upon the changepoint κ used. Thus the null distribution of $F_{\max}$ is independent of $\boldsymbol{\beta}$ and σ^2 and only depends upon the design matrix $\mathbf{X}$ of explanatory variables. For any particular data set we can therefore simulate the distribution of $F_{\max}$ using arbitrary $\boldsymbol{\beta}$ and σ^2 values (e.g. $\boldsymbol{\beta} = \mathbf{0}$, $\sigma^2 = 1$). The hypothesis of a single regression can be rejected at the $100\alpha\%$ level if $F_{\max}$ is in the upper $100\alpha\%$ of simulated values.

Beckman and Cook [1979] simulated a simple linear regression ($p = 2$) with four configurations of x-sequence on the interval $0 \le x \le 1$ to estimate the 90% percentile of the $F_{\max}$-distribution in each case. Configurations with larger variances tended to give larger $F_{\max}$-values. For sample sizes over 10 the estimated percentiles were based upon 10,000 simulations. For smaller sample sizes 100,000 simulations were used. Since $F_{\max}$ is the maximum of $n-2p+1$ F-statistics, the usual Bonferroni modification is to compare $F_{\max}$ with $F_{p,n-2p}^{\alpha/(n-2p+1)}$ or, equivalently, multiply the P-value by $n-2p+1$ (see below). Thus the Bonferroni approximation to the 90th percentile of the distribution of $F_{\max}$ is $F_{p,n-2p}^{0.1/(n-2p+1)}$. Beckman and Cook also report these Bonferroni upper bounds. However, the Bonferroni bounds are usually unduly conservative. For $10 \le n \le 27$ the Bonferroni 90th percentile is usually about one higher than the highest 90th percentile for the four configurations. It should also be noted that the F-statistic does not tend to a limiting distribution as n increases, as Beckman and Cook [1979] suggest. In fact it tends to infinity (Hawkins [1980]).

Worsley [1983] employs an improved Bonferroni inequality. Let A_κ be the event that $F_\kappa > u$. The standard Bonferroni inequality is

$$\text{pr}\left[\bigcup_\kappa A_\kappa\right] \le \sum_\kappa \text{pr}[A_\kappa] \tag{9.20}$$

thus leading to

$$\text{pr}[F_{\max} > u] = \text{pr}\left[\bigcup_{\kappa=p}^{n-p} A_\kappa\right] \le (n-2p+1)\,\text{pr}[F_{p,n-2p} > u].$$

However, Worsley [1982] gave an improved inequality

$$\text{pr}\left[\bigcup_{\kappa=p}^{n-p} A_\kappa\right] \le \sum_{\kappa=p}^{n-p} \text{pr}[A_\kappa] - \sum_{\kappa=p}^{n-p-1} \text{pr}[A_\kappa \cap A_{\kappa+1}]. \tag{9.21}$$

Worsley [1983] found it convenient to work with

$$V_\kappa = \frac{\tilde{S} - (\hat{S}_1^{(\kappa)} + \hat{S}_2^{(\kappa)})}{\tilde{S}} \quad \text{and} \quad V_{\max} = \max_{p \le \kappa \le n-p} V_\kappa \tag{9.22}$$

in preference to F_κ and $F_{\max}$. For fixed κ, V_κ has a beta distribution, Beta(a,b), with parameters $a = \frac{1}{2}p$ and $b = \frac{1}{2}(n-2p)$. Worsley gave an improved Bonferroni upper bound for $\text{pr}[V_{\max} > v]$ derived from (9.21). He also gave a simpler approximation to this expression, namely

$$\text{pr}[V_{\max} > v] = 1 - G_{p,\nu}(v) + \frac{2}{\pi(n-p)} g_{p+1,\nu+1}(v)\left[\sum_{\kappa=p}^{n-p-1} \tau_\kappa - \frac{1}{6}\left\{\frac{(\nu-1)v}{(2p+2)(1-v)} - 1\right\}\sum_{\kappa=p}^{n-p-1} \tau_\kappa^3\right], \tag{9.23}$$

where $G_{p,\nu}(\cdot)$ and $g_{p,\nu}(\cdot)$ are, respectively, the distribution function and density function of a Beta$(\frac{1}{2}p, \frac{1}{2}\nu)$ random variable, $\nu = n - 2p$,

$$\tau_\kappa^2 = \mathbf{x}'_{\kappa+1}(\mathbf{X}_2^{(\kappa)\prime}\mathbf{X}_2^{(\kappa)})^{-1}\mathbf{X}'\mathbf{X}(\mathbf{X}_1^{(\kappa+1)\prime}\mathbf{X}_1^{(\kappa+1)})^{-1}\mathbf{x}_{\kappa+1}, \tag{9.24}$$

and $\mathbf{x}_{\kappa+1}$ is the $(\kappa+1)$th row of the complete design matrix $\mathbf{X}$. Worsley [1983] checked the accuracy of the approximation (9.23) in two situations and found a vast improvement over the simple Bonferroni bounds. The new bounds were recommended as giving accurate conservative tests for significance levels $\alpha \le 0.1$ and sample sizes $10 \le n \le 50$.

Farley et al. [1975] compared three competitive test methods for testing for a two-phase regression. The first, "Chow's test," simply takes $\kappa = n/2$ and uses the usual F-test for $\boldsymbol{\beta}_1 = \boldsymbol{\beta}_2$. The second was the likelihood-ratio test as above (i.e. based on L_2, $F_{\max}$, or $V_{\max}$). The third method was that of Farley and Hinich [1970], though Farley et al. [1975] gave a simpler interpretation, which we now present. The equations (9.6) can be rewritten as

$$\begin{aligned} \mathbf{y}_1^{(\kappa)} &= \mathbf{X}_1^{(\kappa)}\boldsymbol{\beta} + \boldsymbol{\varepsilon}_1^{(\kappa)}, \\ \mathbf{y}_2^{(\kappa)} &= \mathbf{X}_2^{(\kappa)}(\boldsymbol{\beta} + \boldsymbol{\delta}) + \boldsymbol{\varepsilon}_2^{(\kappa)}, \end{aligned} \tag{9.25}$$

and the test for $H_0: \boldsymbol{\beta}_1 = \boldsymbol{\beta}_2$ is replaced by a test of $H_0: \boldsymbol{\delta} = \mathbf{0}$. Now from (9.25)

$$E[y_i] = \sum_j x_{ij}\beta_j + \sum_j z_{ij}^{(\kappa)}\delta_j, \tag{9.26}$$

where

$$z_{ij}^{(\kappa)} = \begin{cases} 0, & i \le \kappa, \\ x_{ij} & \text{otherwise.} \end{cases} \tag{9.27}$$

We could write (9.26) in matrix form as

$$\mathbf{y} = \mathbf{X}\boldsymbol{\beta} + \mathbf{Z}^{(\kappa)}\boldsymbol{\delta} + \boldsymbol{\varepsilon}. \tag{9.28}$$

Thus for known κ we could use (9.28) and test for a change by testing $H_0: \boldsymbol{\delta} = \mathbf{0}$. However, κ is unknown. If we think of every value of κ as being equally likely a priori, then $\text{pr}[\kappa = i] = 1/n$ (or $\text{pr}[\kappa = i] = 1/(n - 2p + 1)$ if $p \le \kappa \le n - p$) and

$$E[y_i] = \sum_j x_{ij}\beta_j + \sum_j E[z_{ij}^{(\kappa)}]\delta_j,$$

where

$$E[z_{ij}^{(\kappa)}] = \sum_{r=1}^{n} z_{ij}^{(r)}\frac{1}{n} = \frac{i-1}{n}x_{ij}.$$

The Farley–Hinich test uses the linear model

$$\mathbf{y} = \mathbf{X}\boldsymbol{\beta} + \mathbf{Z}\boldsymbol{\delta} + \boldsymbol{\varepsilon}, \tag{9.29}$$

where $z_{ij} = (i-1)x_{ij}/n$. The test statistic is the usual F-test statistic for testing $H_0: \boldsymbol{\delta} = \mathbf{0}$, which has an $F_{p,n-2p}$ distribution when H_0 is true. This F-test is thus much easier to use than the likelihood-ratio test. The Farley–Hinich test is also the likelihood-ratio test for a linear (in i) shift in the parameter values. However, the Farley–Hinich procedure experiences difficulty when there may be a change both in the intercept and in the coefficient of a trended x-variable. As an extreme case, suppose $x_{i1} = 1$ for all i (corresponding to the intercept) and $x_{ij} = i$. Then $z_{i1} = (i-1)/n$, so that the jth column of $\mathbf{X}$ and the first column of $\mathbf{Z}$ are linearly dependent. Thus β_j and δ_1 are not separately estimable.

Farley et al. [1975] performed rather limited simulations to investigate the comparative powers of the Chow, likelihood-ratio, and Farley–Hinich tests using a model with one untrended and one trended continuous explanatory (x) variable and a sample size of 60. They introduced a change in the coefficient of one explanatory variable at a time. The change was of the order of six standard errors of the estimated coefficient, and on different runs the change was introduced at a different point $\kappa = i$ in the sequence. The Chow and the Farley–Hinich tests were modified so that they were only trying to detect changes in coefficient for the explanatory variables and not in the intercept. From comments in their paper (*p.* 310) it does not seem that the likelihood-ratio test was similarly modified. Several patterns emerged. As might be expected, the Chow test was most powerful if the change occurred near the middle of the data sequence, but its power dropped off much more quickly than for the other methods as κ moved away from that central region. The Farley–Hinich test was less powerful than the Chow test for changes at central i-values, with a maximum power loss of 14%. The likelihood-ratio test was less powerful than the Farley–Hinich test for a range of central i-values, but more powerful toward the extremes of the data sequence. Farley et al. [1975] included a cautionary note (*p.* 315) that only substantial shifts in regression parameters can be detected with reasonable power.

A special case of the two-phase problem is one in which a change in intercept only is conjectured. Worsley [1983] specializes his improved Bonferroni approximation to the likelihood-ratio test to this case. Quantiles for the likelihood-ratio test were calculated from simulations by Maronna and Yohai [1978] in the case of a simple linear regression. These authors also investigated the asymptotic behavior of the likelihood-ratio statistic. The Farley–Hinich procedure is specialized by using only the first column of $\mathbf{Z}$ in (9.29), which corresponds to the initial column of ones in $\mathbf{X}$ for the intercept term. However, as we have already noted, there may be difficulties if some columns in $\mathbf{X}$ are strongly trended, and the method cannot be used if there is an x-variable which is linear in i $(i = 1, 2, \ldots, n)$.

Sequential control-chart approaches to detecting general (rather than simply two-phase) changes in multiple-regression parameters, when data are collected sequentially over time, are discussed by Brown et al. [1975]. These methods are based upon the residuals for a one-step-ahead prediction, i.e. $\hat{\varepsilon}_{(i)} = y_i - \mathbf{x}_i'\hat{\boldsymbol{\beta}}^{(i)}$, where $\hat{\boldsymbol{\beta}}^{(i)}$ is the least-squares estimator of $\boldsymbol{\beta}$ from $(\mathbf{x}_1, y_1), \ldots, (\mathbf{x}_{i-1}, y_{i-1})$. Brown et al. call these residuals recursive residuals and advocate the use of cusum plots of the standardized residuals and their squares. Cusum methods for detecting changes in the intercept only are discussed by Schweder [1976]. A nonparametric approach to testing for a change of phase in regression is given by Sen [1980, 1983].

For a discussion of the very special problem of testing for a change of mean, (i.e. $E[y_i] = \mu_1$ for $i \leq \kappa$, $E[y_i] = \mu_2$ for $i > \kappa$), see James et al. [1987], Siegmund [1986: Section 3.4], Worsley [1979], Hawkins [1977], and the survey of Zacks [1983]. For nonparametric approaches to testing for changes in location the reader is referred to Wolfe and Schechtman [1984], Schechtman and Wolfe [1985], Schechtman [1983], Sen [1983], and Pettitt [1979]. A Bayesian approach to detecting a change is given by Broemeling and Chin Choy [1981].

9.2.3 Parameter Inference

The asymptotic distribution of the maximum-likelihood estimator $\hat{\kappa}$ of the changepoint has not been worked out at the time of writing. Hinkley [1970] obtained the asymptotic distribution of $\hat{\kappa}$ for the special case of a change in normal mean, using the theory of random walks. In particular, $\hat{\kappa}$ is not consistent, as the numbers of observations both preceding and following y_κ tend to infinity. As Hinkley states, this is to be expected, because observations distant from the changepoint give negligible information about κ. Hinkley [1970] advocated inverting the likelihood-ratio test of $H_0: \kappa = \kappa_0$ to obtain a confidence set for κ rather than using the asymptotic distribution of $\hat{\kappa}$, because the use of $\hat{\kappa}$ involves a loss of information.

The -2(log-likelihood ratio) statistic for testing $H_0: \kappa = \kappa_0$ in the model (9.6) is, from (9.13) and (5.92),

$$
\begin{aligned}
2[M(\hat{\kappa}) - M(\kappa_0)] &= n \log \frac{\hat{S}_1^{(\kappa_0)} + \hat{S}_2^{(\kappa_0)}}{\hat{S}_1^{(\hat{\kappa})} + \hat{S}_2^{(\hat{\kappa})}} \\
&= n \max_{p \leq \kappa \leq n-p} \log \frac{\hat{S}_1^{(\kappa_0)} + \hat{S}_2^{(\kappa_0)}}{\hat{S}_1^{(\kappa)} + \hat{S}_2^{(\kappa)}}.
\end{aligned}
\tag{9.30}
$$

The asymptotic distribution of the likelihood-ratio statistic under H_0 was discussed by Hinkley [1970] for the change-of-means problem together with a minor adjustment needed for the theory to apply to a change of intercept in a simple linear regression. Another approach, based upon cusums, was used by Hinkley [1971a] (see also Hinkley [1972]). Siegmund [1986: Section 3.5] compared confidence sets for κ obtained by inverting the likelihood-ratio test with other approaches for the change-in-means problem (c.f. Cobb [1978] and Hinkley and Schechtman [1987]).

For the regression model, work on inference about the changepoint κ has been primarily in the Bayesian framework, apart from Esterby and El-Shaarawi [1981], who derived a marginal likelihood function for κ in the sense of Kalbfleisch and Sprott [1970]. Esterby and El-Shaarawi advocated direct use of the likelihood function to obtain a range of plausible values for κ. We now describe some of the Bayesian work.

Ferreira [1975] gave marginal posterior distributions for all of the parameters of the two-phase simple linear regression model including the changepoint. He put independent uniform priors on the regression coefficients and $\log\sigma^2$. An arbitrary independent discrete prior was placed upon κ, including the special case of a discrete uniform prior. Holbert and Broemeling [1977] also analyzed this situation. Chin Choy and Broemeling [1980] gave posterior distributions for $\boldsymbol{\beta}_1, \boldsymbol{\beta}_2, \sigma^2$, and κ in (9.6). They placed a multivariate normal prior upon $\boldsymbol{\beta} = (\boldsymbol{\beta}_1', \boldsymbol{\beta}_2')'$ and a dependent inverse gamma prior upon σ^2. An independent discrete uniform prior was placed upon κ. Posterior means and covariances are also obtained for the parameters (Broemeling and Chin Choy [1981]) and Moen et al. [1985] generalized this work to multivariate regressions. The priors above were chosen largely for mathematical convenience. Considerations for choosing priors for changepoint problems were discussed by Booth and Smith [1982], and Pettitt [1981] combined nonparametric and Bayesian ideas to obtain a posterior distribution for the changepoint. Hsu [1982] explored robust Bayesian inference in the two-phase regression model using the (parametric) family of exponential power distributions for the regression error ε_i. This family includes both long- and short-tailed alternatives to the normal distribution. Guttman and Menzefricke [1982] applied Bayesian decision theory to point estimation of the changepoint. Earlier Bayesian work is referenced by the authors above.

As we have seen, posterior distributions for the regression parameters $\boldsymbol{\beta}_1, \boldsymbol{\beta}_2$, and σ^2 have been given for several sets of prior assumptions. Sampling-theory inference about these parameters appears to be unexplored. The maximum-likelihood estimators, $\hat{\boldsymbol{\beta}}_j^{(\hat{k})}$ from (9.8) and $\hat{\sigma}^2(\hat{k}) = (\hat{S}_1^{(\hat{k})} + \hat{S}_2^{(\hat{k})})/n$, are available, but the usual large-sample theory does not apply. Nevertheless, it seems likely that ad hoc solutions such as separately applying ordinary linear regression theory to the data points before and after the range of plausible values for κ should give reasonable answers.

9.2.4 Further Extensions

Efficient computation of maximum-likelihood estimates for multiphase regression is difficult, particularly when the algorithm must also determine the number of phases to be fitted. Algorithms have been proposed by Hawkins [1976] and by Ertel and Fowlkes [1976], who also reference earlier work.

A problem which is slightly related to two-phase regression involves so-called "switching regressions" (Quandt and Ramsey [1978]). Here, for $i = 1, 2, \ldots, n$,

$$y_i = \begin{cases} \boldsymbol{\beta}_1'\mathbf{x}_i + \varepsilon_{1i} & \text{with probability } \lambda, \\ \boldsymbol{\beta}_2'\mathbf{x}_i + \varepsilon_{2i} & \text{with probability } 1-\lambda, \end{cases} \tag{9.31}$$

where $\varepsilon_{1i} \sim N(0, \sigma_1^2)$, $\varepsilon_{2i} \sim N(0, \sigma_2^2)$, and all errors are independent.

9.3 CONTINUOUS CASE

9.3.1 Models and Inference

Let us return to our original change-of-phase model (9.1) with its single explanatory variable x. We assume

$$y_i = f(x_i; \boldsymbol{\theta}, \boldsymbol{\alpha}) + \varepsilon_i \qquad (i = 1, 2, \ldots, n), \tag{9.32}$$

where $E[\varepsilon_i] = 0$, $\text{var}[\varepsilon_i] = \sigma^2$, and the ε_i's are independent. We employ a specialized version of (9.1) in (9.32), namely

$$f(x; \boldsymbol{\theta}, \boldsymbol{\alpha}) = \begin{cases} \boldsymbol{\theta}_1' \mathbf{g}_1(x), & a \le x \le \alpha_1, \\ \boldsymbol{\theta}_2' \mathbf{g}_2(x), & \alpha_1 < x \le \alpha_2, \\ \vdots & \vdots \\ \boldsymbol{\theta}_D' \mathbf{g}_D(x), & \alpha_{D-1} < x \le b, \end{cases} \tag{9.33}$$

where $\boldsymbol{\theta}_d$ has dimension p_d $(d = 1, 2, \ldots, D)$ and $\boldsymbol{\theta}' = (\boldsymbol{\theta}_1', \boldsymbol{\theta}_2', \ldots, \boldsymbol{\theta}_D')$. Moreover, we impose continuity on $f(x; \boldsymbol{\theta}, \boldsymbol{\alpha})$ at the changepoints α_d, so that

$$\boldsymbol{\theta}_d' \mathbf{g}_d(\alpha_d) = \boldsymbol{\theta}_{d+1}' \mathbf{g}_{d+1}(\alpha_d) \tag{9.34}$$

for $d = 1, 2, \ldots, D - 1$. To allow for the possible imposition of greater degrees of smoothness, let m_d be the order of the first derivative of $f(x; \boldsymbol{\theta}, \boldsymbol{\alpha})$ with respect to x which is discontinuous at $x = \alpha_d$. In other words, $\partial^s f / \partial x^s$ is continuous at α_d for $s = 0, 1, \ldots, m_{d-1}$ but discontinuous for $s = m_d$.

An important special case of (9.34) is joining two or more line segments [c.f. the model (9.2)], in which case $\mathbf{g}_d'(x) = (1, x)$ and $m_d = 1$ for all d. However, (9.33) also allows for polynomial segments, and other functions can be useful. For example, Lerman [1980] fits a segment of the form $g(x) = x/(x - 1)$ as part of an agricultural change-of-phase model. He also considers segments of the form $g(x) = \cos x$.

For *fixed* $\boldsymbol{\alpha}$, the model (9.33) is a linear regression model subject to linear constraints. Thus the exact methods of inference for partial linear models discussed in Section 5.10 can be applied. One of these methods is specialized in Section 9.3.4 to test for no change of phase. We now discuss asymptotic results.

If the errors ε_i are normally distributed, maximum-likelihood estimation for model (9.32) is the same as least-squares estimation (Section 2.2.1). The error sum of squares is

$$S(\boldsymbol{\theta}, \boldsymbol{\alpha}) = \sum_{d=1}^{D} \sum_{x_i \in I_d} [y_i - \boldsymbol{\theta}_d' \mathbf{g}_d(x_i)]^2, \tag{9.35}$$

where

$$I_d = \{x_i : \alpha_{d-1} < x_i \le \alpha_d\}, \tag{9.36}$$

$\alpha_0 \equiv a$ and $\alpha_D \equiv b$. The usual asymptotic theory of Section 2.1 does not apply, because $S(\boldsymbol{\theta}, \boldsymbol{\alpha})$ may not even have continuous first derivatives with respect to the α_d's, and this makes the derivation of asymptotic results very difficult. Hinkley [1969, 1971b] derived asymptotic results for the special case of two line segments and normally distributed errors.

Feder [1975a, b] gave a rigorous treatment for the full model (9.33). The only further requirement on the error distribution, in addition to the ε_i being i.i.d. with mean zero and variance σ^2, was $E[|\varepsilon_i|^{2+\delta}] < \infty$ for some $\delta > 0$. His most important condition is that no two adjacent phase models are identical at the true values $\boldsymbol{\theta}^*$ and $\boldsymbol{\alpha}^*$, respectively, of $\boldsymbol{\theta}$ and $\boldsymbol{\alpha}$. In particular, this means that the number of phases D must be known. Furthermore, the asymptotic sequence in which $(x_1, x_2, \ldots, x_n)$ is embedded must be such that as $n \to \infty$ the number of observations falling into each interval I_d of (9.36) also tends to infinity, and the pattern of x_i's must be such that each curve is identifiable. Under the above conditions Feder [1975a] proved the consistency of the least-squares estimators $\hat{\boldsymbol{\theta}}$ and $\hat{\boldsymbol{\alpha}}$ of $\boldsymbol{\theta}^*$ and $\boldsymbol{\alpha}^*$, and also the consistency of

$$\hat{\sigma}^2 = \frac{1}{n} S(\hat{\boldsymbol{\theta}}, \hat{\boldsymbol{\alpha}}). \tag{9.37}$$

Another consistent estimator is the usual nonlinear estimate

$$s^2 = \frac{S(\hat{\boldsymbol{\theta}}, \hat{\boldsymbol{\alpha}})}{n - P}, \tag{9.38}$$

where P is the number of "free parameters". Now in the model (9.33) we have $p + D - 1$ parameters, $D - 1$ continuity constraints, and m_T smoothing constraints, where

$$p = \sum_{d=1}^{D} p_d, \qquad m_T = \sum_{d=1}^{D-1} (m_d - 1),$$

and the m_T "smoothing" constraints are given by

$$\boldsymbol{\theta}_d' \frac{d^s \mathbf{g}_d(\alpha_d)}{dx^s} = \boldsymbol{\theta}_{d+1}' \frac{d^s g_{d+1}(\alpha_d)}{dx^s} \qquad (s = 1, 2, \ldots, m_d - 1, \quad d = 1, 2, \ldots, D - 1).$$

Thus $P = p - m_T$.

The asymptotic distributions of $\hat{\boldsymbol{\theta}}$ and $\hat{\boldsymbol{\alpha}}$ depend upon whether the m_d's are odd or even. If all the m_d's are odd, then asymptotically (see Feder [1975a: Theorem 4.13])

$$\sqrt{n}(\hat{\boldsymbol{\theta}} - \boldsymbol{\theta}^*) \sim N_p(\mathbf{0}, \sigma^2 \mathbf{G}^{-1}), \tag{9.39}$$

where

$$\mathbf{G} = \int_a^b \frac{\partial f(x;\boldsymbol{\theta}^*,\boldsymbol{\alpha}^*)}{\partial \boldsymbol{\theta}} \frac{\partial f(x;\boldsymbol{\theta}^*,\boldsymbol{\alpha}^*)}{\partial \boldsymbol{\theta}'} dH(x), \tag{9.40}$$

[see (9.33) for f] and $H(x)$ is the limiting distribution function for the asymptotic behavior of the x-sequence. If any m_d is even, then $\hat{\boldsymbol{\theta}}_d$ and $\hat{\boldsymbol{\theta}}_{d+1}$ are not asymptotically normal in general (Feder [1975*a*: 76]). However, (9.39) covers the important special case of intersecting lines ($m_d = 1$ for all d) and the case where second-order continuity is imposed on polynomial segments ($m_d = 3$ for all d). This latter case will be important in Section 9.5. However, (9.39) does not cover first-order continuity ($m_d = 2$), which we discuss in more detail in Section 9.3.3. We now consider the asymptotic distribution of $\hat{\boldsymbol{\alpha}}$.

If all m_d's are odd, then asymptotically (Feder [1975a: Theorem 4.17])

$$\sqrt{n}\begin{bmatrix} (\hat{\alpha}_1 - \alpha_1^*)^{m_1} \\ (\hat{\alpha}_2 - \alpha_2^*)^{m_2} \\ \vdots \\ (\hat{\alpha}_{D-1} - \alpha_{D-1}^*)^{m_{D-1}} \end{bmatrix} \sim N_{D-1}(\mathbf{0}, \sigma^2 \mathbf{A}_* \mathbf{G}^{-1} \mathbf{A}_*'), \tag{9.41}$$

where $\mathbf{A}_* = \mathbf{A}(\boldsymbol{\theta}^*, \boldsymbol{\alpha}^*)$ is a $(D-1) \times p$ matrix which we now describe. Let us write the rth row of $\mathbf{A} = \mathbf{A}(\boldsymbol{\theta}, \boldsymbol{\alpha})$ as $(\mathbf{a}_{r1}', \mathbf{a}_{r2}', \ldots, \mathbf{a}_{rD}')$, where $\mathbf{a}_{rd}$ is a vector of dimension p_d given by

$$\mathbf{a}_{rd} = \begin{cases} c_r^{-1}\mathbf{g}_r(\alpha_r), & d = r, \\ -c_r^{-1}\mathbf{g}_{r+1}(\alpha_r), & d = r+1, \\ \mathbf{0} & \text{otherwise,} \end{cases} \tag{9.42}$$

Here

$$c_r = \frac{1}{m_r!}[\boldsymbol{\theta}_{r+1}'\mathbf{g}_{r+1}^{(m_r)}(\alpha_r) - \boldsymbol{\theta}_r'\mathbf{g}_r^{(m_r)}(\alpha_r)]$$

and $\mathbf{g}_r^{(m)}(x) = d^m\mathbf{g}_r(x)/dx^m$. The results (9.39) and (9.41) require that the individual phase functions $\boldsymbol{\theta}_r^{*\prime}\mathbf{g}_r(x)$ have m_r continuous derivatives at α_r and m_{r-1} continuous derivatives at α_{r-1}. This is no real restriction, as most functions we wish to use for our individual phase models, such as polynomials, are infinitely differentiable on all of $[a, b]$.

For sample data $(x_1, y_1), \ldots, (x_n, y_n)$, $\mathbf{G}$ can be estimated by

$$\hat{\mathbf{G}} = \frac{1}{n}\sum_{i=1}^{n} \frac{\partial f(x_i;\hat{\boldsymbol{\theta}},\hat{\boldsymbol{\alpha}})}{\partial \boldsymbol{\theta}} \frac{\partial f(x_i;\hat{\boldsymbol{\theta}},\hat{\boldsymbol{\alpha}})}{\partial \boldsymbol{\theta}'}. \tag{9.43}$$

Using (9.33), it is easy to show that $\hat{\mathbf{G}}$ is a diagonal block matrix with D blocks, one for each $\boldsymbol{\theta}_d$. The dth block is the $p_d \times p_d$ matrix

$$\hat{\mathbf{G}}_d = \frac{1}{n} \sum_{i:x_i \in \hat{I}_d} \mathbf{g}_d(x_i)\mathbf{g}_d'(x_i), \tag{9.44}$$

where

$$\hat{I}_d = \{x_i : \hat{\alpha}_{d-1} < x_i \le \hat{\alpha}_d\}.$$

Thus, asymptotically, if all the m_d's are odd, we can treat the $\hat{\boldsymbol{\theta}}_d$'s as independent multivariate normals with

$$\hat{\boldsymbol{\theta}}_d \sim N_{p_d}(\boldsymbol{\theta}_d, \hat{\sigma}^2(\mathbf{X}_d'\mathbf{X}_d)^{-1}), \tag{9.45}$$

where

$$\mathbf{X}_d'\mathbf{X}_d = \sum_{x_i \in \hat{I}_d} \mathbf{g}_d(x_i)\mathbf{g}_d(x_i)'.$$

The inverse of $\mathbf{X}_d'\mathbf{X}_d$ can be obtained from a separate linear regression of y on $\mathbf{g}_d(x)$ for all the data points with $\hat{\alpha}_{d-1} < x_i \le \hat{\alpha}_d$. However, following Lerman [1980], it is reasonable to omit any data points lying within computational accuracy of $\hat{\alpha}_{d-1}$ or $\hat{\alpha}_d$ in computing $\mathbf{X}_d'\mathbf{X}_d$.

If each $m_d = 1$, $\hat{\boldsymbol{\alpha}}$ is asymptotically normal with mean $\boldsymbol{\alpha}^*$ and variance–covariance matrix estimated by $\sigma^2\hat{\mathbf{A}}(n\hat{\mathbf{G}})^{-1}\hat{\mathbf{A}}'$, where $\hat{\mathbf{A}} = \mathbf{A}(\hat{\boldsymbol{\theta}}, \hat{\boldsymbol{\alpha}})$.

Example 9.6 Consider the case of two intersecting lines

$$y_i = \begin{cases} \beta_{10} + \beta_{11}x_i + \varepsilon_i, & x_i \le \alpha, \\ \beta_{20} + \beta_{21}x_i + \varepsilon_i, & x_i > \alpha, \end{cases} \tag{9.46}$$

with a continuity constraint $\beta_{10} + \beta_{11}\alpha = \beta_{20} + \beta_{21}\alpha$. Let us write the x_i's below $\hat{\alpha}$ as $x_{11}, x_{12}, \ldots, x_{1n_1}$ and those after $\hat{\alpha}$ as $x_{21}, x_{22}, \ldots, x_{2n_2}$. Then from (9.45) and standard linear regression theory (e.g. Seber [1977: 179]) we have, asymptotically, for $d = 1, 2$,

$$\begin{aligned} \hat{\beta}_{d1} &\sim N\left(\beta_{d1}^*, \frac{\sigma^2}{\sum_j (x_{dj} - \bar{x}_{d\cdot})^2}\right), \\ \hat{\beta}_{d0} &\sim N\left(\beta_{d0}^*, \frac{\sigma^2 \sum_j x_{dj}^2}{n_d \sum_j (x_{dj} - \bar{x}_{d\cdot})^2}\right), \end{aligned} \tag{9.47}$$

and

$$\operatorname{cov}[\hat{\beta}_{d0}, \hat{\beta}_{d1}] = -\frac{\sigma^2 \bar{x}_{d\cdot}}{\sum_j (x_{dj} - \bar{x}_{d\cdot})^2}. \tag{9.48}$$

Also, from the linear constraints we have

$$\hat{\alpha} = \frac{\hat{\beta}_{10} - \hat{\beta}_{20}}{\hat{\beta}_{21} - \hat{\beta}_{11}}. \tag{9.49}$$

As $\mathbf{g}_d(x) = (1, x)'$ for $d = 1, 2$, we find that

$$\mathbf{A} = \frac{1}{\beta_{21} - \beta_{11}}(1, \alpha, -1, -\alpha). \tag{9.50}$$

Hence from (9.41) with each $m_d = 1$ we have that $\hat{\alpha}$ is asymptotically normal with mean α^* and variance

$$\begin{aligned}
\text{var}[\hat{\alpha}] &= \sigma^2 \mathbf{A} \begin{pmatrix} \mathbf{X}_1'\mathbf{X}_1 & \mathbf{0} \\ \mathbf{0} & \mathbf{X}_2'\mathbf{X}_2 \end{pmatrix}^{-1} \mathbf{A}' \\
&= (\beta_{21} - \beta_{11})^{-2} \{ (\text{var}[\hat{\beta}_{10}] + \text{var}[\hat{\beta}_{20}]) \\
&\quad + 2\alpha(\text{cov}[\hat{\beta}_{10}, \hat{\beta}_{11}] + \text{cov}[\hat{\beta}_{20}, \hat{\beta}_{21}]) \\
&\quad + \alpha^2(\text{var}[\hat{\beta}_{11}] + \text{var}[\hat{\beta}_{21}]) \}.
\end{aligned} \tag{9.51}$$

This expression can also be obtained directly by applying the so-called δ-method (i.e. a Taylor expansion) to $\hat{\alpha}$ of (9.49) and recalling that $\hat{\beta}_{10}$ and $\hat{\beta}_{11}$ are asymptotically independent of $\hat{\beta}_{20}$ and $\hat{\beta}_{21}$. ■

In his empirical experience with the two-line-segment problem, Hinkley [1969, 1971b] found that the asymptotic normal distributions for the $\hat{\theta}_r$'s were good approximations in moderate-sized samples but that the asymptotic normal approximation for $\hat{\alpha}$ tended to be poor. Both Hinkley [1969] and Feder [1975a] suggested refinements to the asymptotic distribution of $\hat{\alpha}$ which give better small-sample properties. However, Hinkley recommends use of the likelihood-ratio statistic for inferences about α or joint inferences about $(\boldsymbol{\theta}, \alpha)$.

In the discussion following (9.38) we saw that the dimension of the parameter space was $P = p - m_T$. Suppose, now, that we have a hypothesis H_0 that restricts the dimension of the parameter space to $P - q$, so that q independent restrictions are imposed. Under normality assumptions the -2(log-likelihood ratio) statistic for testing H_0 is, using (2.62) and (5.92),

$$\text{LR} = n \log \frac{S(\hat{\boldsymbol{\theta}}, \hat{\boldsymbol{\alpha}})}{S(\tilde{\boldsymbol{\theta}}, \tilde{\boldsymbol{\alpha}})}, \tag{9.52}$$

where $(\tilde{\boldsymbol{\theta}}, \tilde{\boldsymbol{\alpha}})$ are the maximum-likelihood (least-squares) estimators of $(\boldsymbol{\theta}, \boldsymbol{\alpha})$ subject to the constraints imposed by H_0. Then Feder [1975b: Theorem 3.2] proved that, under certain conditions,

$$\text{LR} \sim \chi_q^2$$

asymptotically when H_0 is true. The error distribution need not be normal, but $E(|\varepsilon_i|^{2+\delta})$ must be finite for some $\delta > 0$. The m_d's, defined following (9.34), must all be odd. An important additional condition is that the parameters of the model must remain identified under H_0. Feder [1975b: 90] noted that the distribution of LR might be better approximated in finite samples by a multiple of the asymptotically equivalent $F_{q,n-p}$ distribution. Thus (c.f. Section 5.3) we might use

$$F_0 = \frac{S(\tilde{\boldsymbol{\theta}}, \tilde{\boldsymbol{\alpha}}) - S(\hat{\boldsymbol{\theta}}, \hat{\boldsymbol{\alpha}})}{qs^2} \tag{9.53}$$

as a test statistic with an approximate $F_{q,n-p}$ null distribution.

We can use (9.53) to make inferences about $\boldsymbol{\alpha}$. Consider a two-phase regression model with $(m-1)$th-order continuity. To test $H_0: \alpha = \alpha_0$ (i.e. $q = 1$) we require $S(\tilde{\boldsymbol{\theta}}, \tilde{\boldsymbol{\alpha}})$ [$= S(\tilde{\boldsymbol{\theta}}(\alpha_0), \alpha_0)$, say]. For fixed α_0 we have two linear regressions

$$\begin{aligned} \mathbf{y}_1 &= \mathbf{X}_1\boldsymbol{\theta}_1 + \boldsymbol{\varepsilon}_1, \\ \mathbf{y}_2 &= \mathbf{X}_2\boldsymbol{\theta}_2 + \boldsymbol{\varepsilon}_2, \end{aligned} \tag{9.54}$$

where $\mathbf{y}_1$ consists of the κ_0 observations, say, with $x_i \le \alpha_0$; $\mathbf{y}_2$ consists of $n - \kappa_0$ observations; $\mathbf{X}_1$ has rows $\mathbf{g}_1'(x_i)$ [$x_i \le \alpha_0$]; and $\mathbf{X}_2$ has rows $\mathbf{g}_2'(x_i)$ [$x_i > \alpha_0$]. Let us write (9.54) as

$$\mathbf{y} = \begin{pmatrix} \mathbf{X}_1 & \mathbf{0} \\ \mathbf{0} & \mathbf{X}_2 \end{pmatrix} \boldsymbol{\theta} + \boldsymbol{\varepsilon}. \tag{9.55}$$

Then $S(\tilde{\boldsymbol{\theta}}(\alpha_0), \alpha_0)$ is the residual sum of squares for (9.55) subject to the linear constraints $\boldsymbol{\theta}_1'\mathbf{g}_1^{(s)}(\alpha_0) = \boldsymbol{\theta}_2'\mathbf{g}_2^{(s)}(\alpha_0)$ [$s = 0, 1, \ldots, m-1$], or $\mathbf{B}\boldsymbol{\theta} = \mathbf{0}$, where

$$\mathbf{B} = \begin{bmatrix} \mathbf{g}_1'(\alpha_0) & -\mathbf{g}_2'(\alpha_0) \\ \vdots & \vdots \\ \mathbf{g}_1^{(m-1)\prime}(\alpha_0) & -\mathbf{g}_2^{(m-1)\prime}(\alpha_0) \end{bmatrix}, \tag{9.56}$$

$\mathbf{g}_d^{(s)}(x) = d^s\mathbf{g}_d(x)/dx^s$, and $\mathbf{g}_d^{(0)}(x) = \mathbf{g}_d(x)$. From Appendix D2.4 we can treat $\mathbf{B}\boldsymbol{\theta} = \mathbf{0}$ as a "hypothesis" constraint and obtain the relationship

$$S(\tilde{\boldsymbol{\theta}}(\alpha_0), \alpha_0) = S(\breve{\boldsymbol{\theta}}(\alpha_0), \alpha_0) + \breve{\boldsymbol{\theta}}'\mathbf{B}'[\mathbf{B}(\mathbf{X}'\mathbf{X})^{-1}\mathbf{B}']^{-1}\mathbf{B}\breve{\boldsymbol{\theta}}, \tag{9.57}$$

where $\breve{\boldsymbol{\theta}}(\alpha_0)$ and $S(\breve{\boldsymbol{\theta}}(\alpha_0), \alpha_0)$ are, respectively, the "unconstrained" least-squares estimator of $\boldsymbol{\theta}$ and the corresponding residual sum of squares obtained by fitting

(9.55) without the constraints. We can now test $\alpha = \alpha_0$ using (9.57) and (9.53) with $q = 1$, namely

$$F(\alpha_0) = \frac{S(\tilde{\boldsymbol{\theta}}(\alpha_0), \alpha_0) - S(\hat{\boldsymbol{\theta}}, \hat{\alpha})}{s^2},$$

where $s^2 = S(\hat{\boldsymbol{\theta}}, \hat{\alpha})/(n - P)$ and $P = p - m + 1 = p_1 + p_2 - m + 1$.

Following Hinkley [1971b], we can obtain a $100(1-\delta)\%$ confidence interval for α by inverting the F-test for H_0, namely

$$\{\alpha: F(\alpha) \leq F^{\delta}_{1,n-P}\}. \tag{9.58}$$

Unfortunately, $S(\tilde{\theta}(\alpha), \alpha)$ [and hence $F(\alpha)$] is not necessarily monotonic in α for α on either side of $\hat{\alpha}$, so that (9.58) may not be a single interval. Unless the likelihood function for α is distinctly bimodal, Hinkley advocates using the interval (α_L, α_U), where α_L and α_U are the smallest and largest values of α satisfying (9.58).

Example 9.6 (continued) For the two-line-segment model (9.46), $m = 1, \boldsymbol{\theta} = (\beta_{10}, \beta_{11}, \beta_{20}, \beta_{21})'$, and $\mathbf{B} = (1, \alpha_0, -1, -\alpha_0)$. Let

$$R(\alpha_0) = S(\tilde{\boldsymbol{\theta}}(\alpha_0), \alpha_0) - S(\breve{\boldsymbol{\theta}}(\alpha_0), \alpha_0)$$

and

$$c_d = \left(n_d \sum_{i=1}^{n_d} (x_{di} - \bar{x}_{d\cdot})^2 \right)^{-1}.$$

Then, from (9.57),

$$R(\alpha_0) = \frac{[\breve{\beta}_{10} - \breve{\beta}_{20} + \alpha_0(\breve{\beta}_{11} - \breve{\beta}_{21})]^2}{\sum_{d=1}^{2} c_d(\sum_i x_{di}^2 - 2\alpha_0 n_d \bar{x}_{d\cdot} + \alpha_0^2 n_d)}, \tag{9.59}$$

and the condition $F(\alpha) < F^{\delta}_{1,n-4}$ is equivalent to

$$\begin{aligned} R(\alpha) &< S(\hat{\boldsymbol{\theta}}, \hat{\alpha}) - S(\breve{\boldsymbol{\theta}}(\alpha), \alpha) + s^2 F^{\delta}_{1,n-4} \\ &= \breve{w}, \end{aligned} \tag{9.60}$$

say. Then a boundary point of (9.58) satisfies

$$\begin{aligned} \alpha^2[(\breve{\beta}_{11} - \breve{\beta}_{21})^2 - \breve{w}(c_1 n_1 + c_2 n_2)] + 2\alpha[(\breve{\beta}_{10} - \breve{\beta}_{20})(\breve{\beta}_{11} - \breve{\beta}_{21}) \\ + \breve{w}(n_1 \bar{x}_{1\cdot} c_1 + n_2 \bar{x}_{2\cdot} c_2)] + (\breve{\beta}_{10} - \breve{\beta}_{20})^2 - \breve{w} \sum_d \sum_i c_d x_{di}^2 = 0. \end{aligned} \tag{9.61}$$

However, the $\breve{\beta}_{rs}$ and $\breve{w}$ depend on where the split of the form (9.54) occurs and therefore depend on α. Hinkley's [1971b] procedure for finding α_L is to begin with a split at x_2 and then move on to x_3, x_4, etc. until a root of (9.61) is located in $[x_i, x_{i+1})$. If no such root exists for $i \le \hat{\kappa}$, where $x_{\hat{\kappa}}$ is the largest x_i not exceeding $\hat{\alpha}$, then we technically would set $\alpha_L = -\infty$. In a similar fashion α_U is obtained by working down from x_{n-2}. If no such root exists for $i \ge \hat{\kappa} + 1$, then $\alpha_U = +\infty$.

At this point it is convenient to summarize the results obtained thus far for Example 9.6. In the next section we shall discuss methods of computing $\hat{\boldsymbol{\theta}} = (\hat{\beta}_{10}, \hat{\beta}_{11}, \hat{\beta}_{20}, \hat{\beta}_{21})'$ and $\hat{\alpha}$. Once $\hat{\alpha}$ is available, we can find variance estimates as follows. Using $\hat{\alpha}$, we can split the data in two and fit a separate (unconstrained) regression line for each set of data. The standard variances for these estimates are then asymptotically the same as the var$[\hat{\beta}_{rs}]$, and var$[\hat{\alpha}]$ follows from (9.51). Confidence intervals for α are then obtained from (9.61). To test hypotheses about the elements of $\boldsymbol{\theta}$, e.g. $H_0: \beta_{11} = 0$, we can use either (9.52) or (9.53), which requires minimizing $S(\boldsymbol{\theta}, \alpha)$ subject to H_0. Another possibility is via (9.47), though, as suggested by experience with other nonlinear models, methods such as (9.53) based on the likelihood ratio are probably more reliable. ■

The restriction that the parameters remain identified under H_0, which was used in Feder's [1975b] proof that the likelihood-ratio statistic is asymptotically χ^2_q, rules out some very important hypotheses. For example, it rules out the hypothesis $H_0: \boldsymbol{\theta}_r' \mathbf{g}_r(x) = \boldsymbol{\theta}_{r+1}' \mathbf{g}_{r+1}(x)$ of no change of model from the rth segment to the next. Under such a hypothesis α_r, the intersection of the two regimes, becomes unidentifiable. In particular, Feder's result does not cover testing for no change in a two-phase linear regression model (9.46). Indeed, Feder [1975b] investigated the special case of this model in which the first line is $E[y] = 0$ (i.e. $\beta_{10} = \beta_{11} = 0$ and $q = 2$). He found the asymptotic null distribution of the likelihood-ratio statistic (9.52) for testing no change "to be that of the maximum of a number of correlated χ^2_1 and χ^2_2 random variables and to vary with the configuration of the observation points of the independent variable" rather than being asymptotically χ^2_2. Naive use of the χ^2_2 distribution was found to be conservative.

For the general two-line problem (9.46) there are four independent parameters for $E[y]$ (five parameters $\boldsymbol{\theta}$ and α, less one for the constraint), which is reduced to two (the common intercept and slope) under the no-change hypothesis H_0. Thus $q = 2$ once again. However, Hinkley [1971b] reported strong empirical evidence that the null distribution of (9.52) was asymptotically χ^2_3. This suggests the use of $F_0 = [S_1 - S(\hat{\boldsymbol{\theta}}, \hat{\alpha})]/3s^2$ [c.f. (9.53)], where S_1 is the residual sum of squares from fitting a single line, as an $F_{3,n-4}$-test for H_0. This is in strong contrast to the noncontinuous case, where the F-statistic does not have a limiting distribution [see the comments following (9.17)]. We consider "exact" tests for no change of phase in Section 9.3.4.

Before concluding this section we reference some related work. Seber [1977: Section 7.6] and Sprent [1961] consider the model (9.46) when it is known that α lies between x_d and x_{d+1}. Smith and Cook [1980] give a Bayesian analysis of (9.46) along with a method for handling the continuity constraint. A posterior probability for the validity of the constraint is also introduced. Gbur and Dahm [1985] discuss (9.46) for the case when the x_i's are measured with error. A multivariate version of (9.46) is considered by Heath and Anderson [1983]. Finally, Shaban [1980] gives an annotated bibliography on two-phase regression and the changepoint problem.

9.3.2 Computation

There have been a variety of computational techniques suggested for change-of-phase models. Suppose the individual phase models are straight lines. In Section 9.4 we shall discuss algorithms [e.g. following (9.117)] in which the change-of-phase model $f(x; \boldsymbol{\theta}, \boldsymbol{\alpha})$ of (9.33) is approximated by a smooth function $h(x; \boldsymbol{\theta}, \boldsymbol{\alpha}, \gamma)$ which tends to $f(x; \boldsymbol{\theta}, \boldsymbol{\alpha})$ as $\gamma \to 0$. The function h is chosen so that it has continuous derivatives when γ is fixed. Thus for fixed γ, h can be fitted using an ordinary nonlinear least-squares program. The least-squares estimates $\hat{\boldsymbol{\theta}}$ and $\hat{\boldsymbol{\alpha}}$ for f are obtained by progressively reducing γ. We now discuss more direct techniques.

We wish to find $\hat{\boldsymbol{\theta}}$ and $\hat{\boldsymbol{\alpha}}$ to minimize $S(\boldsymbol{\theta}, \boldsymbol{\alpha})$ given by (9.35) subject to the continuity constraints (9.34), which are linear in $\boldsymbol{\theta}$ at the phase boundaries. Because (9.33) is a partially linear model (being linear when $\boldsymbol{\alpha}$ is known), we have a separable least-squares problem (Section 14.7) which can be solved by minimizing in two stages. Firstly we fix $\boldsymbol{\alpha}$, thus giving a partition of the x_i's of the form (9.36) and the regression model

$$\begin{bmatrix} \mathbf{y}_1 \\ \mathbf{y}_2 \\ \vdots \\ \mathbf{y}_D \end{bmatrix} = \begin{bmatrix} \mathbf{X}_1 & \mathbf{0} & \cdots & \mathbf{0} \\ \mathbf{0} & \mathbf{X}_2 & \cdots & \mathbf{0} \\ \vdots & \vdots & & \vdots \\ \mathbf{0} & \mathbf{0} & \cdots & \mathbf{X}_D \end{bmatrix} \begin{bmatrix} \boldsymbol{\theta}_1 \\ \boldsymbol{\theta}_2 \\ \vdots \\ \boldsymbol{\theta}_D \end{bmatrix} + \boldsymbol{\varepsilon}, \tag{9.62}$$

where the rows of $\mathbf{X}_d$ are $\mathbf{g}_d'(x_i)$ for $x_i \in I_d$ of (9.36). Secondly, we find $\tilde{\boldsymbol{\theta}}(\boldsymbol{\alpha})$, the constrained least-squares estimate of $\boldsymbol{\theta}$, which minimizes $S(\boldsymbol{\theta}(\boldsymbol{\alpha}), \boldsymbol{\alpha})$ for (9.62) subject to the linear constraints (9.34). This can be done as in (9.57), where the case $D = 2$ is considered. Finally, we find $\hat{\boldsymbol{\alpha}}$ to minimize $S(\tilde{\boldsymbol{\theta}}(\boldsymbol{\alpha}), \boldsymbol{\alpha})$ and thus obtain $\tilde{\boldsymbol{\theta}}(\hat{\boldsymbol{\alpha}}) = \hat{\boldsymbol{\theta}}$. Lerman [1980] advocates the use of a grid search (Section 13.2.3) to do this. In special cases such as segmented polynomials with smooth joins, $S(\tilde{\boldsymbol{\theta}}(\boldsymbol{\alpha}), \boldsymbol{\alpha})$ can be used much more efficiently (Section 9.3.3).

Hudson [1966] derived an algorithm for two-phase regression in which the model function, but not its derivatives, is continuous at the changepoint. The algorithm is not appropriate when we have a continuous first derivative at the changepoint. A method for polynomial models with continuous first derivatives is described in Section 9.3.3. Our description below of Hudson's method for finding $\hat{\alpha}$ to minimize $S(\tilde{\boldsymbol{\theta}}(\alpha), \alpha)$ and calculating this minimum value draws heavily on the paragraphs including Equations (9.54) to (9.57).

Suppose that the data are split so that we have κ observations for the first model and $n-\kappa$ for the second. We now allow for the situation where the split is not necessarily in the interval containing the join at $x=\alpha$, i.e., κ is no longer constrained to satisfy $x_\kappa \le \alpha < x_{\kappa+1}$. In order for the p_d-dimensional vector $\boldsymbol{\theta}_d$ to be identifiable ($d=1,2$) we must have $p_1 \le \kappa \le n-p_2$. Let $\check{\boldsymbol{\theta}}^{(\kappa)} = (\check{\boldsymbol{\theta}}_1^{(\kappa)\prime}, \check{\boldsymbol{\theta}}_2^{(\kappa)\prime})'$, where $\check{\boldsymbol{\theta}}_d^{(\kappa)}$ is the unconstrained least-squares estimate of $\boldsymbol{\theta}_d$, and let $S_1(\check{\boldsymbol{\theta}}^{(\kappa)})$ be the corresponding residual sum of squares after fitting the two segments. Suppose that the two unconstrained fitted models intersect at $\check{\alpha}^{(\kappa)}$, so that

$$\check{\boldsymbol{\theta}}_1^{(\kappa)\prime}\mathbf{g}_1(\check{\alpha}^{(\kappa)}) = \check{\boldsymbol{\theta}}_2^{(\kappa)\prime}\mathbf{g}_2(\check{\alpha}^{(\kappa)}). \tag{9.63}$$

For given α and κ we further define $\tilde{\boldsymbol{\theta}}^{(\kappa)}(\alpha)$ to be the constrained least-squares estimate of $\boldsymbol{\theta}$ (i.e., the joinpoint is constrained to be α) with corresponding residual sum of squares $S(\tilde{\boldsymbol{\theta}}^{(\kappa)}(\alpha), \alpha)$. For each κ let $\tilde{\alpha}^{(\kappa)}$ be the value of α which minimizes $S(\tilde{\boldsymbol{\theta}}^{(\kappa)}(\alpha), \alpha)$, and denote $\tilde{\boldsymbol{\theta}}^{(\kappa)} = \tilde{\boldsymbol{\theta}}^{(\kappa)}(\tilde{\alpha}^{(\kappa)})$.

Having set up the notation, let us assume for the moment that κ is correctly known. Then Hudson [1966] showed that if $\check{\alpha}^{(\kappa)} \in (x_\kappa, x_{\kappa+1})$ and (9.63) does not hold for first derivatives of the $\mathbf{g}_d$, then $\tilde{\alpha}^{(\kappa)} = \check{\alpha}^{(\kappa)}$ and $\tilde{\boldsymbol{\theta}}^{(\kappa)} = \check{\boldsymbol{\theta}}^{(\kappa)}$. Moreover, as the constraint is obeyed at the unconstrained minimum, the constrained residual sum of squares is the same as the unconstrained residual sum of squares, i.e.

$$S(\tilde{\boldsymbol{\theta}}^{(\kappa)}, \tilde{\alpha}^{(\kappa)}) = S(\check{\boldsymbol{\theta}}^{(\kappa)}, \check{\alpha}^{(\kappa)}) \quad [= S_1(\check{\boldsymbol{\theta}}^{(\kappa)})].$$

Following Hudson, we call a minimum located in $(x_\kappa, x_{\kappa+1})$ a type-I minimum. If there is no solution to (9.63) in $(x_\kappa, x_{\kappa+1})$, then Hudson showed that, for known κ, $S(\tilde{\theta}^{(\kappa)}(\alpha), \alpha)$ is minimized for $\alpha \in [x_\kappa, x_{\kappa+1}]$ at $\alpha = x_\kappa$ or $x_{\kappa+1}$. However, in this situation we must perform the constrained minimization explicitly, i.e. compute $S(\tilde{\theta}^{(\kappa)}(\alpha), \alpha)$ using (9.57) for each α. We shall call an endpoint minimum of this type a type-II minimum.

In practice κ is unknown and must be found using a systematic search. As $\hat{\alpha}$ solves

$$\min_{\kappa=p_1,\ldots,n-p_2} \quad \min_{x_\kappa \le \alpha < x_{\kappa+1}} S(\tilde{\boldsymbol{\theta}}^{(\kappa)}(\alpha), \alpha), \tag{9.64}$$

we need to search through all the type-I and type-II minima to see which one gives the smallest value of $S(\boldsymbol{\theta}, \alpha)$. We thus have the following algorithm of Hudson:

1. Find the lowest type-I minimum.
Set $S(\hat{\boldsymbol{\theta}}, \hat{\alpha}) = \infty$.
For $\kappa = p_1, p_1+1, \ldots, n-p_2$:

a. Compute the unconstrained least squares estimate $\check{\boldsymbol{\theta}}^{(\kappa)}$ and find $S_1(\check{\boldsymbol{\theta}}^{(\kappa)})$.

b. Solve (9.63) for $\check{\alpha}^{(\kappa)}$.

c. Suppose $\check{\alpha}^{(\kappa)} \in (x_\kappa, x_{\kappa+1})$ [type-I minimum], otherwise go to step d. Then if $S_1(\check{\boldsymbol{\theta}}^{(\kappa)}) < S(\hat{\boldsymbol{\theta}}, \hat{\alpha})$ set:

(i) $\hat{\boldsymbol{\theta}} = \breve{\boldsymbol{\theta}}^{(\kappa)}$,

(ii) $\hat{\alpha} = \breve{\alpha}^{(\kappa)}$, and

(iii) $S(\hat{\boldsymbol{\theta}}, \hat{\alpha}) = S_1(\breve{\boldsymbol{\theta}}^{(\kappa)})$.

d. If $\breve{\alpha}^{(\kappa)} \notin (x_\kappa, x_{\kappa+1})$ [type-II minimum], then store $S_1(\breve{\boldsymbol{\theta}}^{(\kappa)})$ for future reference.

Now that we have the lowest type-I minimum, we search through the intervals that do not have a type-I minimum for a lower type-II minimum. We need to look at the interval $[x_\kappa, x_{\kappa+1}]$ only if the unconstrained residual sum of squares $S_1(\breve{\boldsymbol{\theta}}^{(\kappa)})$ is lower than the current value of $S(\hat{\boldsymbol{\theta}}, \hat{\alpha})$, as the constrained sum of squares is larger than the unconstrained. Thus we proceed as follows:

2. Find the lowest type-II minimum.
For $\kappa = p_1, p_1 + 1, \ldots, n - p_2$: If there is no type-I minimum and $S(\breve{\boldsymbol{\theta}}^{(\kappa)})$ is less than the current value of $S(\hat{\boldsymbol{\theta}}, \hat{\alpha})$, then:

a. Compute $S(\tilde{\boldsymbol{\theta}}^{(\kappa)}(x_\kappa), x_\kappa)$ and $S(\tilde{\boldsymbol{\theta}}^{(\kappa)}(x_{\kappa+1}), x_{\kappa+1})$, and choose the smaller (say, the former).

b. If $S(\tilde{\boldsymbol{\theta}}^{(\kappa)}(x_\kappa), x_\kappa) < S(\hat{\boldsymbol{\theta}}, \hat{\alpha})$, then set

(i) $\hat{\alpha} = x_\kappa$,

(ii) $S(\hat{\boldsymbol{\theta}}, \hat{\alpha}) = S(\tilde{\boldsymbol{\theta}}^{(\kappa)}(x_\kappa), x_\kappa)$.

This algorithm has been refined for the case of two straight lines by Hinkley [1969, 1971b]. It has been extended in a straightforward manner to fitting three phases by Williams [1970]. A difficulty with the algorithm, identified by Lerman [1980], occurs when (9.63) cannot be solved analytically. Lerman went on the propose the grid-search method described at the beginning of this subsection. Equation (9.63) is readily solved for lines and quadratics in x. If the phase models are cubic (respectively, quartic) polynomials, solution of (9.63) involves finding the roots of a cubic (respectively, quartic). Explicit expressions for these roots can be obtained from Van der Waerden [1953: Section 58].

In discussing type-I maxima we ruled out the case where the slopes of the two segments are equal at $\hat{\alpha}^{(\kappa)}$. Such a situation would occur "by accident" and would therefore have a low probability. It is certainly ruled out when the segments are straight lines or constants, as the derivative constraint would imply that the segments are identical. However, for general models, such a contingency (called a type-III join) is discussed by Hudson, who suggests a procedure for finding a type-III minimum.

9.3.3 Segmented Polynomials

Hartley's [1961] proof of the convergence of the modified Gauss–Newton algorithm (also called Hartley's method: see Section 14.2.1) requires continuous second derivatives of f with respect to all of the parameters. However, Gallant and Fuller [1973] state a theorem proved by Gallant [1971] which proves convergence of Hartley's method from an interior starting point to a stationary point of the sum of squares. The proof requires only continuous first derivatives of f with respect to its parameters. Conditions imposed include: (a) the minimum does not lie on the boundary of the parameter space, and (b) the first-derivative matrix $\mathbf{F}_{\bullet}(\boldsymbol{\theta}, \boldsymbol{\alpha})$ of $\mathbf{f}$ with respect to the parameters retains full rank throughout the iterations. In this section we take advantage of these results to determine a subclass of segmented polynomial models which can be fitted using a standard nonlinear least-squares program and for which the standard asymptotic inference results of Chapter 5 can be used.

A segmented polynomial regression model is a special case of the model (9.33) with continuity constraints of various orders in which the individual phase models are polynomials in x, i.e.

$$\boldsymbol{\theta}_d' \mathbf{g}_d(x) = \sum_{j=0}^{p_d - 1} \theta_{dj} x^j, \tag{9.65}$$

or equivalently $\mathbf{g}_d(x) = (1, x, x^2, \ldots, x^{p_d - 1})'$. Again we assume $m_d - 1$ continuous derivatives at the joinpoints α_d, $d = 1, \ldots, D-1$, i.e.

$$\left[\frac{d^s}{dx^s} \sum_{j=0}^{p_d - 1} \theta_{dj} x^j \right]_{x = \alpha_d} = \left[\frac{d^s}{dx^s} \sum_{j=0}^{p_{d+1} - 1} \theta_{d+1,j} x^j \right]_{x = \alpha_d}, \qquad s = 0, 1, \ldots, m_d - 1. \tag{9.66}$$

In certain cases segmented polynomial models can be written as a linear combination of terms $1, x, x^2, \ldots, x^v$ $[v = \max_d (p_d - 1)]$ and terms of the form $(x - \alpha_d)_+^r$ $(r = 1, 2, \ldots, v)$ in such a way that all the continuity constraints are taken care of implicitly. Here

$$z_+^r = \begin{cases} z^r, & z \geq 0, \\ 0, & z < 0. \end{cases} \tag{9.67}$$

For example, if we can write our model as

$$f(x) = \sum_j \phi_j x^j + \sum_d \sum_r \phi_{dr} (x - \alpha_d)_+^r \tag{9.68}$$

with no value of r appearing which is less than 2, then f has continuous derivatives with respect to all of its parameters (and x) and can be fitted using a standard nonlinear least-squares program. We qualify these remarks later in Section 9.3.3b.

Instead of using terms $(x - \alpha_d)_+^r$, it is sometimes more convenient to work with terms of the form $(\alpha_d - x)_+^r$. Both these terms have continuous derivatives with respect to either α_d or x of orders $0, 1, \ldots, r - 1$.

A spline function of degree q (see Section 9.5) is a segmented polynomial model in which every polynomial phase model has degree q and there are $q - 1$

continuous derivatives of x at the joins. It is well known (e.g. de Boor [1978], Smith [1979]) that a q-spline can be written as [c.f. (9.130)]

$$f(x)=\sum_{j=0}^{q}\phi_j x^j+\sum_{d=1}^{D-1}\xi_d(x-\alpha_d)_+^q. \tag{9.69}$$

Thus q-splines for $q \geq 2$ fit into the above framework. In the important special case of cubic splines ($q=3$), $f(x)$ has continuous second derivatives with respect to all of its parameters. We now look at some simple models which were explored by Gallant and Fuller [1973].

Consider grafting two quadratic segments ($D=2, p_1=p_2=3$) to preserve continuous first derivatives of f. Using (9.65) and imposing continuity constraints (9.66) on f and its derivative, we obtain

$$\theta_{10}+\theta_{11}\alpha_1+\theta_{12}\alpha_1^2=\theta_{20}+\theta_{21}\alpha_1+\theta_{22}\alpha_1^2 \tag{9.70}$$

and

$$\theta_{11}+2\theta_{12}\alpha_1=\theta_{21}+2\theta_{22}\alpha_1. \tag{9.71}$$

Using (9.70) and (9.71) to express θ_{10} and θ_{11} in terms of the other parameters, it is readily established that

$$f_1(x)=f_2(x)+(\theta_{12}-\theta_{22})(\alpha_1-x)^2. \tag{9.72}$$

Thus we can express $f(x)$ in reparametrized form as

$$f(x)=\phi_1+\phi_2 x+\phi_3 x^2+\phi_4(\alpha_1-x)_+^2, \tag{9.73}$$

where $\phi_1=\theta_{20}$, $\phi_2=\theta_{21}$, $\phi_3=\theta_{22}$, and $\phi_4=\theta_{12}-\theta_{22}$. The initial polynomial term thus expresses the second quadratic. If we replace $(\alpha_1-x)_+^2$ by $(x-\alpha_1)_+^2$ in (9.73), the polynomial term is the first quadratic. Two other models are reparametrized by Gallant and Fuller [1973]. The quadratic–linear model ($D=2, p_1=3, p_2=2$) subject to first-order continuity is

$$f(x)=\phi_1+\phi_2 x+\phi_3(\alpha_1-x)_+^2, \tag{9.74}$$

where $\phi_1=\theta_{20}$, $\phi_2=\theta_{21}$, and $\phi_3=\theta_{12}$. The $\phi_1+\phi_2 x$ part is thus the second (linear) phase. This time we cannot write down the initial quadratic first and use terms of the form $(x-\alpha_1)_+^r$. To describe a quadratic going to a line we would have to use

$$f(x)=\phi_1+\phi_2 x+\phi_3 x^2+\phi_4(x-\alpha_1)_+^2$$

with a constraint $\phi_4=-\phi_3$, and we do not want to use constraints. For the quadratic–quadratic–linear model ($D=3, p_1=p_2=3, p_3=2$) with first-order continuity, we again have to work from the right-hand segment and express $f(x)$ as

$$f(x)=\phi_1+\phi_2 x+\phi_3(\alpha_2-x)_+^2+\phi_4(\alpha_1-x)_+^2, \tag{9.75}$$

where $\phi_1=\theta_{30}$, $\phi_2=\theta_{31}$, $\phi_3=\theta_{22}$, and $\phi_4=\theta_{12}-\theta_{22}$. Similarly, we would write a linear–quadratic–quadratic model working from the left-hand phase as

$$f(x)=\phi_1+\phi_2 x+\phi_3(x-\alpha_1)_+^2+\phi_4(x-\alpha_2)_+^2. \tag{9.76}$$

As Gallant and Fuller [1973] note, the functions $1,x,x^2,\dots,x^v,(\alpha_d-x)_+^r$ [$v=\max_d(p_d-1)$, $d=1,2,\dots,D-1$, $r=2,3,\dots,v$] form a basis for grafted polynomials with first-order continuity. If we wish to require higher-order continuity at α_d, e.g. m_d-1 continuous derivatives, only terms $(\alpha_d-x)_+^r$ with $r\geq m_d$ are included in the basis. However, it is not generally possible to use the basis without imposing constraints on the parameters ϕ_j unless the orders of the polynomial phase models are either nonincreasing ($p_1\geq p_2\geq\cdots\geq p_D$) or nondecreasing ($p_1\leq p_2\leq\cdots\leq p_D$). If they are nondecreasing, we work from the left-hand side using terms of the form $(x-\alpha_d)_+^r$. If they are nonincreasing, we work from the right-hand side using terms of the form $(\alpha_d-x)_+^r$.

The parameters ϕ_j are sometimes simple linear functions of the parameters θ_{dj} expressing changes in polynomial coefficients as we pass from phase to phase, but they may also be nonlinear functions of the parameters. Consider a cubic–cubic model with continuous first derivatives. We might write such a model as

$$f(x)=\phi_0+\phi_1 x+\phi_2 x^2+\phi_3 x^3+\phi_4(x-\alpha)_+^2+\phi_5(x-\alpha)_+^3, \tag{9.77}$$

where the polynomial represents the first cubic. It is clear that $\phi_5=\theta_{23}-\theta_{13}$, the difference in the cubic terms, but the quadratic term of the second model is $\theta_{22}=\phi_2+\phi_4-3\phi_5\alpha$. Thus $\phi_4=\theta_{22}-\theta_{12}+3(\theta_{23}-\theta_{13})\alpha$, a nonlinear function of $\boldsymbol{\theta}$ and α.

Models that can be written in the form (9.68) with all the terms $(x-\alpha_d)_+^r$ [or $(\alpha_d-x)_+^r$] having $r\geq 2$ will now be discussed under two headings: (a) inference and (b) computation. Recall that models constructed in this way have continuous derivatives with respect to x and all of their parameters.

a Inference

The consistency of the least-squares estimators of $\boldsymbol{\phi}$ and $\boldsymbol{\alpha}$ follow from the results of Feder [1975a] quoted in Section 9.3.1. However, the standard asymptotic results for the least-squares estimators $\hat{\boldsymbol{\phi}}$ and $\hat{\boldsymbol{\alpha}}$ do not follow from his general treatment in cases where m_d is even (e.g. $m_d=2$ with continuous first derivatives only). However, Eubank [1984] states that Gallant [1974] has proved that for continuously differentiable models of the form (9.68), and under certain regularity conditions, the least-squares estimators $(\hat{\boldsymbol{\phi}},\hat{\boldsymbol{\alpha}})$ are asymptotically normal with dispersion matrix estimated by

$$\hat{\sigma}^2[\mathbf{F}_.'(\hat{\boldsymbol{\phi}}, \hat{\boldsymbol{\alpha}})\mathbf{F}_.(\hat{\boldsymbol{\phi}}, \hat{\boldsymbol{\alpha}})]^{-1} \tag{9.78}$$

in the usual way. Gallant [1971] considered the quadratic–quadratic case in detail and gave conditions upon the x_i-sequence under which these results hold.

Example 9.7 The data given in Table 9.1 and plotted in Fig. 9.7 consist of 72 measurements on the age in months (x) and the weight/height ratio (y) of preschool boys. From the plot an obvious initial choice for $f(x; \boldsymbol{\phi}, \boldsymbol{\alpha})$ is the quadratic–linear model (9.74) with join α_1, say. In addition, Gallant and Fuller [1973] fitted the quadratic–quadratic–linear model (9.75) with joins α_1 and α_2, by least squares. For (9.74), their modified Gauss–Newton algorithm converged in 5 iterations. The authors do not detail how they arrived at their starting values, but for fixed α_1 (9.74) is an ordinary linear multiple regression model. Therefore one need only specify a starting value for α_1 obtained by eye (Gallant and Fuller used $\alpha_1 = 12$), and then $\phi_1^{(1)}$, $\phi_2^{(1)}$, and $\phi_3^{(1)}$ are obtained by linear regression. Similarly,

Table 9.1 Weight/Height (W/H) versus Age for Preschool Boys[a]

W/H	Age	W/H	Age	W/H	Age
0.46	0.5	0.88	24.5	0.92	48.5
0.47	1.5	0.81	25.5	0.96	49.5
0.56	2.5	0.83	26.5	0.92	50.5
0.61	3.5	0.82	27.5	0.91	51.5
0.61	4.5	0.82	28.5	0.95	52.5
0.67	5.5	0.86	29.5	0.93	53.5
0.68	6.5	0.82	30.5	0.93	54.5
0.78	7.5	0.85	31.5	0.98	55.5
0.69	8.5	0.88	32.5	0.95	56.5
0.74	9.5	0.86	33.5	0.97	57.5
0.77	10.5	0.91	34.5	0.97	58.5
0.78	11.5	0.87	35.5	0.96	59.5
0.75	12.5	0.87	36.5	0.97	60.5
0.80	13.5	0.87	37.5	0.94	61.5
0.78	14.5	0.85	38.5	0.96	62.5
0.82	15.5	0.90	39.5	1.03	63.5
0.77	16.5	0.87	40.5	0.99	64.5
0.80	17.5	0.91	41.5	1.01	65.5
0.81	18.5	0.90	42.5	0.99	66.5
0.78	19.5	0.93	43.5	0.99	67.5
0.87	20.5	0.89	44.5	0.97	68.5
0.80	21.5	0.89	45.5	1.01	69.5
0.83	22.5	0.92	46.5	0.99	70.5
0.81	23.5	0.89	47.5	1.04	71.5

[a]From Eppright et al. [1972] and reproduced by Gallant [1977b].

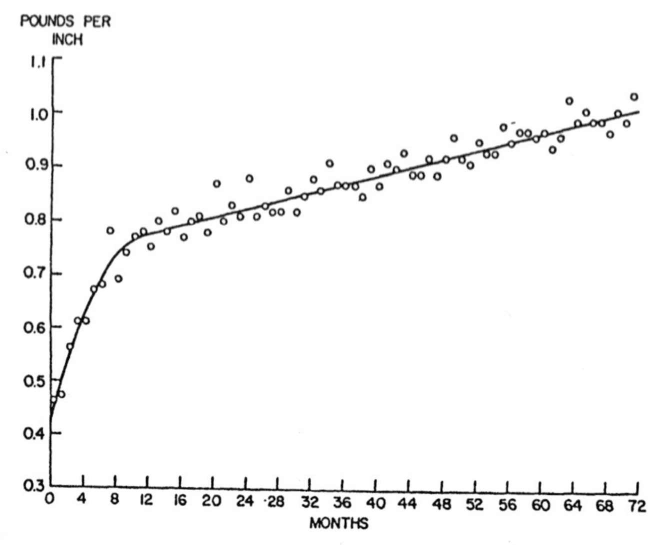

Figure 9.7 Weight–height ratio versus age for preschool boys with fitted quadratic–linear model. From Gallant [1977b].

to fit (9.75) the two joinpoints are obtained by eye (Gallant and Fuller used $\alpha_1^{(1)} = 4$ and $\alpha_2^{(1)} = 12$), and the remaining parameters are obtained by linear regression. It is clear by inspection of the plot that α_1 and α_2 are not clearly defined, as α_1 was in (9.74). It is not surprising, therefore, that the optimization problem is much more difficult. Gallant and Fuller's program took 64 iterations to fit the quadratic–quadratic–linear model. All of the iterates are displayed in their paper. The least-squares estimates and residual sums of squares for both models are given in Table 9.2. We have already discussed the joinpoints. The parameters ϕ_1 and ϕ_2 are respectively the intercept and slope of the linear segment, and these values are very similar for the two models. There is very little difference in model fit in terms of the residual sum of squares, and Gallant and

Table 9.2 Models Fitted to the Data of Table 9.1[a]

	Quadratic–Linear Model	Quadratic–Quadratic–Linear Model
$\hat{\phi}_1$	0.7292	0.7314
$\hat{\phi}_2$	0.003969	0.003927
$\hat{\phi}_3$	−0.002197	−0.0007999
$\hat{\alpha}_2$	—	14.7977
$\hat{\phi}_4$	—	−0.002071
$\hat{\alpha}_1$	11.8313	8.3188
RSS[b]	0.03790	0.03754

[a]From Gallant and Fuller [1973].
[b]Residual sum of squares.

Fuller go on to test (9.74) against (9.75). We could regard this as a partial test of fit of the quadratic–linear model. Thus we test

$$H_0: \phi_4 = 0 \quad (\text{i.e. } \theta_{12} = \theta_{22}) \quad \text{and} \quad \alpha_1 = \alpha_2.$$

From Section 9.3.1 the standard asymptotic F-test results do not hold. Gallant and Fuller appeal to the analogy with linear model theory and suggest the F-test statistic

$$F_0 = \frac{[S(\tilde{\boldsymbol{\phi}}, \tilde{\boldsymbol{\alpha}}) - S(\hat{\boldsymbol{\phi}}, \hat{\boldsymbol{\alpha}})]/2}{S(\hat{\boldsymbol{\phi}}, \hat{\boldsymbol{\alpha}})/(n-6)}, \tag{9.79}$$

where $S(\hat{\boldsymbol{\phi}}, \hat{\boldsymbol{\alpha}})$ is the residual sum of squares under (9.75), and $S(\tilde{\boldsymbol{\phi}}, \tilde{\boldsymbol{\alpha}})$ is the residual sum of squares for the reduced model under the hypothesis, i.e. (9.74). The postulated approximate null distribution is $F_{2,n-6}$ $(n = 72)$. Here $F_0 = 0.31$, which provides no evidence at all against H_0. ■

We note that experimental design for segmented polynomial models has been discussed by Park [1978].

b Computation

Jupp [1978] studied the special case of the model (9.68) where one fits cubic splines by least squares. The changepoints (knots) are subject to an ordering constraint $a < \alpha_1 < \alpha_2 < \cdots < \alpha_{D-1} < b$. The parameter space for $\boldsymbol{\alpha}$ is thus $\mathscr{A}$, say, the set of $\boldsymbol{\alpha}$-values satisfying the constraints. Jupp found that the sum-of-squares surface typically had many local minima and that algorithms tend to become trapped near the boundary of $\mathscr{A}$ and often appear to converge there, quite far from the stationary point on that boundary. Jupp [1978] used a separable nonlinear least-squares algorithm of the type described in Section 14.7.2. He found performance to be vastly improved if $\boldsymbol{\alpha}$ was replaced by the transformed form $\mathbf{w}$, where

$$w_d = \log_e \frac{\alpha_{d+1} - \alpha_d}{\alpha_d - \alpha_{d-1}}, \qquad d = 1, \ldots, D-1, \tag{9.80}$$

where $\alpha_0 = a$, $\alpha_D = b$. In the spline literature, the basis $(x - \alpha_d)_+^r$ for the space of spline functions is known to lead to ill-conditioned design matrices for fitting the model (9.69), which is an ordinary (i.e. unconstrained) linear regression for fixed $\boldsymbol{\alpha}$. A more stable parametrization is in terms of so-called B-splines (see Jupp [1978] for details).

9.3.4 Exact Tests for No Change of Phase in Polynomials

In this section we consider just polynomial phase models. Initially we discuss testing for no change in the two-phase model using the method discussed following Equation (5.28). A two-phase polynomial model with polynomials of degrees $p_d - 1$ $(d = 1, 2)$ and a single continuity constraint at α can be written (when $p_1 \geq p_2$) as

$$f(x; \boldsymbol{\theta}, \alpha) = \sum_{j=0}^{p_2-1} \theta_{2j} x^j + \sum_{j=1}^{p_1-1} \phi_j (\alpha - x)_+^j, \tag{9.81}$$

where the parameters ϕ_j are functions of $\boldsymbol{\theta}_1$, $\boldsymbol{\theta}_2$, and α. Recall that higher-order continuity is obtained by dropping the initial few $(\alpha - x)_+^j$ terms. To modify (9.81) if $p_1 < p_2$, see the discussion following (9.76).

We wish to test the fit of the single-phase model

$$f(x; \boldsymbol{\theta}) = \sum_{j=0}^{p_2-1} \theta_{2j} x^j = \mathbf{x}'\boldsymbol{\theta}_2. \tag{9.82}$$

If α were known, we could fit the model

$$\mathbf{y} = \mathbf{X}\boldsymbol{\theta}_2 + \mathbf{Z}\boldsymbol{\delta} + \boldsymbol{\varepsilon}, \tag{9.83}$$

where the ith row of $\mathbf{X}$ is $\mathbf{x}_i'$, $\boldsymbol{\delta} = (\phi_1, \ldots, \phi_{p_1-1})$, and $z_{ij} = (\alpha - x_i)_+^j$. The usual linear-regression F-test of $H_0: \boldsymbol{\delta} = \mathbf{0}$ would then be applicable. However, α is unknown. For any choice of $\mathbf{Z}$, the F-test of $\boldsymbol{\delta} = \mathbf{0}$ gives an "exact" test of fit for model (9.82) in the sense that it has the correct significance level. The important consideration is to choose $\mathbf{Z}$ so that the test has good power against the change-of-phase alternative (9.81). Gallant [1977b] suggests choosing several values $a_1, a_2, \ldots, a_t$ covering a range of plausible values of α. One could then choose the rows of $\mathbf{Z}$ to be $[(a_1 - x_i)_+, \ldots, (a_t - x_i)_+, \ldots, (a_1 - x_i)_+^{p_1-1}, \ldots, (a_t - x_i)_+^{p_1-1}]$. As this may result in a high-dimensional $\boldsymbol{\delta}$, Gallant uses $\mathbf{Z}^*$ formed by taking the first few principal components of the columns of $\mathbf{Z}$. Gallant [1977b] gives an application of this test for no change.

The idea also applies to testing the fit of the nonlinear model

$$y = f(x; \boldsymbol{\theta}) + \varepsilon \tag{9.84}$$

given that

$$y = f(x; \boldsymbol{\theta}) + \sum_{j=1}^{p_1-1} \phi_j (\alpha - x)_+^j + \varepsilon. \tag{9.85}$$

In this case we form

$$y_i = f(x_i; \boldsymbol{\theta}) + \mathbf{z}_i'\boldsymbol{\delta} + \varepsilon_i, \tag{9.86}$$

where $\mathbf{z}_i'$ is the ith row of $\mathbf{Z}$, and use the asymptotic likelihood-ratio test or F-test for $H_0: \boldsymbol{\delta} = \mathbf{0}$. The approach is justified provided the usual asymptotic theory applies to the model (9.84).

Example 9.7 (continued) Gallant [1977b] applied this method to Example 9.7 of Section 9.3.3 to test the fit of the quadratic–linear model (9.74)

$$f(x;\boldsymbol{\theta}) = \phi_1 + \phi_2 x + \phi_3(\alpha_1 - x)_+^2$$

versus the quadratic–quadratic–linear model (9.75) [with α_1, α_2 interchanged]

$$f(x;\boldsymbol{\theta}) = \phi_1 + \phi_2 x + \phi_3(\alpha_1 - x)_+^2 + \phi_4(\alpha_2 - x)_+^2.$$

Gallant [1977b] took $\alpha_2 = 4, 8, 12$ as plausible values of α_2 (see Fig. 9.7). He used a single z-variable which was the first principle component of $(4-x)_+^2$, $(8-x)_+^2$, and $(12-x)_+^2$, namely

$$z = [2.08(4-x)_+^2 + 14.07(8-x)_+^2 + 39.9(12-x)_+^2] \times 10^{-4},$$

and fitted

$$y_i = \phi_1 + \phi_2 x_i + \phi_3(\alpha - x_i)_+^2 + \delta z_i + \varepsilon_i, \tag{9.87}$$

$i = 1, 2, \ldots, 72$, by nonlinear least squares. He then performed a likelihood-ratio test of $H_0: \delta = 0$ using the F-statistic (5.23), namely

$$\tilde{F}_2 = \frac{S(\tilde{\boldsymbol{\phi}}(0), \tilde{\alpha}(0), 0) - S(\hat{\boldsymbol{\phi}}, \hat{\alpha}, \hat{\delta})}{S(\hat{\boldsymbol{\phi}}, \hat{\alpha}, \hat{\delta})/(n-5)} = \frac{0.03790 - 0.03769}{0.03769/(72-5)} = 0.37,$$

which we compare with $F_{1,67}$. This F-statistic is fairly similar to the previous one [following (9.79)] and provides no evidence against the quadratic–linear model. ■

9.4 SMOOTH TRANSITIONS BETWEEN LINEAR REGIMES

9.4.1 The sgn Formulation

Let us reconsider the simple continuous two-phase regression problem

$$E[y|x] = \begin{cases} \beta_{10} + \beta_{11}x, & x \le \alpha, \\ \beta_{20} + \beta_{21}x, & x > \alpha, \end{cases} \tag{9.88}$$

where $\beta_{10} + \beta_{11}\alpha = \beta_{20} + \beta_{21}\alpha$. Bacon and Watts [1971] note that this parametrization is not a sensitive one for detecting changes in slope and prefer the reparametrization

$$E[y|x] = \theta_0 + \theta_1(x - \alpha) + \theta_2|x - \alpha| \tag{9.89}$$

$$= \theta_0 + \theta_1(x - \alpha) + \theta_2(x - \alpha)\,\text{sgn}(x - \alpha), \tag{9.90}$$

where

$$\operatorname{sgn}(z) = \begin{cases} -1, & z < 0, \\ 0, & z = 0, \\ +1, & z > 0. \end{cases} \tag{9.91}$$

Under this parametrization $\theta_0 = E[y|\alpha]$, i.e., the changepoint is (α, θ_0). Also $\beta_{11} = \theta_1 - \theta_2$ and $\beta_{21} = \theta_1 + \theta_2$, so that $\theta_1 = (\beta_{11} + \beta_{21})/2$, the average slope, while $\theta_2 = (\beta_{21} - \beta_{11})/2$, or half the difference in slopes.

In Examples 9.4 and 9.5 of Section 9.1 we saw smooth transitions between approximately linear regimes. The transitions ranged from gradual to fairly abrupt. It is clearly desirable to have a family of curves providing a smooth transition between linear regimes—a family which gives parametric control of the "sharpness" of the transition. Bacon and Watts [1971] approached the problem by replacing $\operatorname{sgn}(\cdot)$ in (9.90) by a member of a family of approximating smooth functions $\operatorname{trn}(\cdot)$ which obeyed certain conditions, to obtain

$$E[y|x] = \theta_0 + \theta_1(x - \alpha) + \theta_2(x - \alpha)\operatorname{trn}\left(\frac{x - \alpha}{\gamma}\right). \tag{9.92}$$

We use the modified and more concise version of these conditions due to Griffiths and Miller [1973]:

$$\text{(i)} \quad \lim_{z \to \pm\infty} [z \operatorname{trn}(z) - |z|] = 0, \tag{9.93}$$

$$\text{(ii)} \quad \lim_{\gamma \to 0} \operatorname{trn}(z/\gamma) = \operatorname{sgn}(z), \tag{9.94}$$

$$\text{(iii)} \quad \operatorname{trn}(0) = \operatorname{sgn}(0) = 0. \tag{9.95}$$

It is clear from (9.89) and (9.90) that to maintain asymptotic linearity as we move away from the changepoint, $z \operatorname{trn}(z)$ should approach $z \operatorname{sgn}(z)$ $(= |z|)$ arbitrarily closely as $z \to \pm\infty$, as ensured by the condition (9.93). The condition (9.94) ensures that the piecewise linear model (9.90) is a limiting form of the smooth transition model. Finally, the condition (9.95) ensures that the smooth model passes through (α, θ_0), the intersection between the two lines (which are the asymptotes to the curve).

One family that obeys the conditions (9.93) to (9.95) is

$$\operatorname{trn}(z) = \tanh z = \frac{e^z - e^{-z}}{e^z + e^{-z}}. \tag{9.96}$$

Figure 9.8 plots $z \operatorname{trn}(z/\gamma) = z \tanh(z/\gamma)$ versus z for various values of γ, showing the convergence to $z \operatorname{sgn}(z) = |z|$ as $\gamma \to 0$. Bacon and Watts [1971] reported that, in their experience, the fitted models were insensitive to the choice of transition function among the functions they tried that obeyed the conditions (9.93)–(9.95). These authors noted that for (9.92), with trn replaced by tanh, the radius of curvature for the curve at the changepoint $x = \alpha$ is (c.f. Appendix B2)

$$R = \frac{\gamma}{2|\theta_2|}(1 + |\theta_1|^2)^{3/2}, \tag{9.97}$$

thus giving a more explicit interpretation to γ.

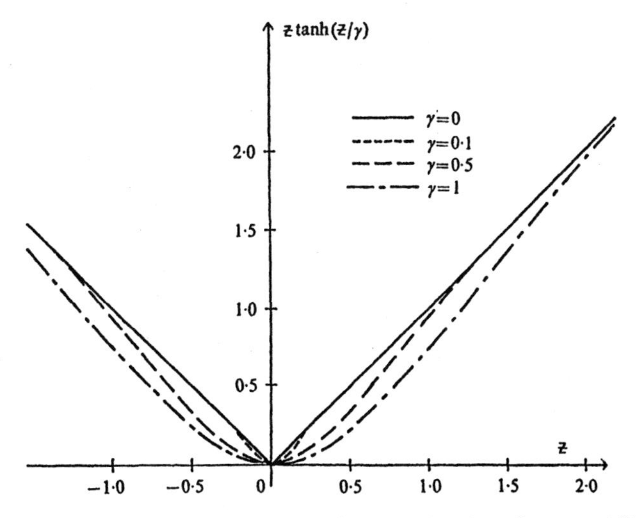

Figure 9.8 Plots of $z\tanh(z/\gamma)$ versus z for several values of γ. From Bacon and Watts [1971].

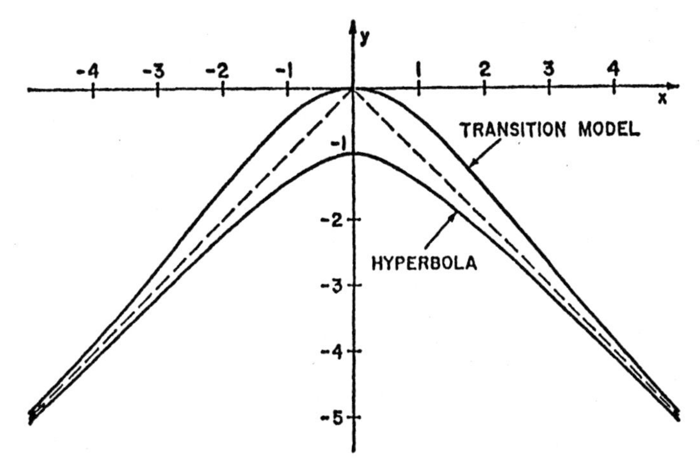

Figure 9.9 Comparison of the behavior of the tanh transition model (9.92), (9.96) with that of the hyperbolic model (9.99). The parameters are $\alpha = 0$; $\theta_0 = 0$; $\gamma = 2$ (tanh), $\gamma = 1$ (hyperbolic), and $\gamma = 0$ (asymptotes). The asymptotes (9.88), shown dashed, have zero intercepts and slopes ± 1. From Watts and Bacon [1974].

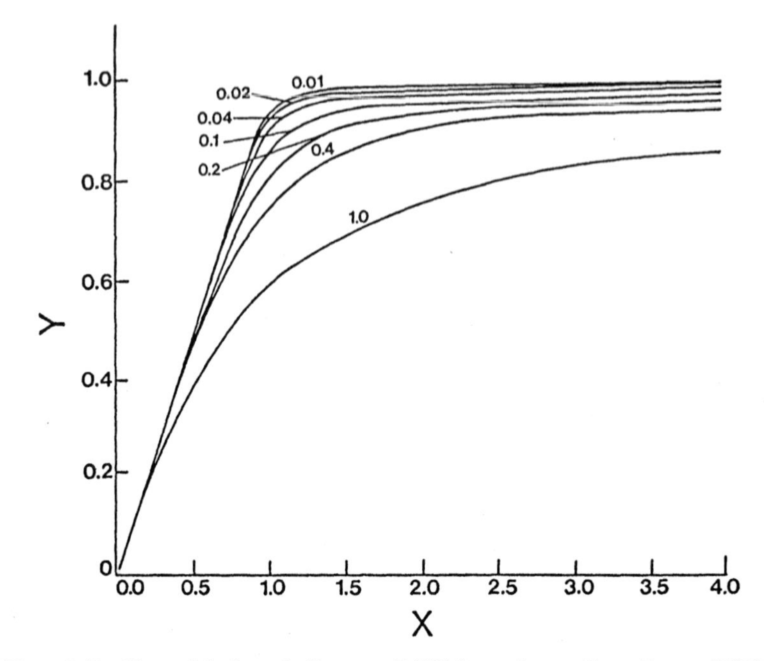

Figure 9.10 Plots of the hyperbolic curve (9.100) for various γ. From Prunty [1983].

In Figs. 9.8 and 9.9 the transition model bulges on the outside of the corner created by the intersection of the two lines. It is compelled to do so by the condition (9.95), which forces the (smooth) transition model through the intersection of the asymptotes (9.90). However, for data plotted in Figs. 9.5 and 9.6, the tendency is to cut the corner analogously to the curve labeled "hyperbola" in Fig. 9.9. For this reason Griffiths and Miller [1973] decided to drop the condition (9.95). They proposed replacing sgn(z) by

$$\text{hyp}(z;\gamma)=\frac{1}{z}\sqrt{z^2+\gamma}, \tag{9.98}$$

which satisfies only the first two conditions (9.93) and (9.94). This leads to the model

$$E[y|x]=\theta_0+\theta_1(x-\alpha)+\theta_2\sqrt{(x-\alpha)^2+\gamma}. \tag{9.99}$$

This is the equation of a "bent," rather than rectangular, hyperbola. Like the tanh model, it also has asymptotes (9.89) as $\gamma\to 0$. If we set $\theta_0=\theta_1=0$, $\theta_2=1$, $\alpha=0$, and $\gamma=1$, we obtain the simple hyperbola $(E[y])^2-x^2=1$. The model (9.99) was proposed independently by Watts and Bacon [1974], though they replaced γ in (9.99) by $\gamma^2/4$. This was done so that the radius of curvature of (9.99) at $x=\alpha$ would again be given by (9.97). For simplicity we will retain the parametrization in (9.99). The behavior of the hyperbolic model is shown in Fig. 9.10. The plotted curves are of

$$y = f(x) = 1 + \frac{1}{1 + \sqrt{1-\gamma}}[(x-\alpha) - \sqrt{(x-\alpha)^2 + \gamma}], \tag{9.100}$$

where $\alpha = \sqrt{1-\gamma}$, for various values of γ. The parametrization was chosen to keep $f(0) = 0, \dot{f}(0) = 1$, and $f(x) \to 1, \dot{f}(x) \to 0$ as $x \to \infty$, where $\dot{f}$ is the derivative. Because of the constraints, the joinpoint $\alpha = \sqrt{1-\gamma}$ drifts. We note that as $\gamma \to 0$, $f(x)$ tends to the pair of straight lines $y = x$ and $y = 1$.

Prior to Equation (9.97) we stated that the transition model was insensitive to choice of transition function. However, Griffiths and Miller [1973] explain that the transition model and the hyperbolic model can give rather different results, particularly for the parameter measuring the difference in slopes.

Prunty [1983] generalized (9.99) to

$$E[y|x] = \theta_0 + \theta_1(x-\alpha) + \theta_2\{|x-\alpha|^\delta + \gamma\}^{1/\delta}. \tag{9.101}$$

He notes that as $\delta \to \infty$,

$$1 + x - (x^\delta + 1)^{1/\delta} \to \min(1, x)$$

for $x \geq 0$. This is the same limit as for (9.100) as $\gamma \to 0$, which can be seen by setting $\gamma = 0$, $\alpha = 1$ and noting that $\sqrt{(x-1)^2} = 1 - x$ for $x < 1$ in (9.100). Hence the curves for (9.101) show a very similar pattern to those in Fig. 9.10. However, the formulation (9.101) is not differentiable, and we do not pursue this model further, although a very similar idea is used in the next subsection.

The ideas we have presented generalize readily to more than two line segments. The D-segment model

$$E[y|x] = \begin{cases} \beta_{10} + \beta_{11}x, & x \leq \alpha_1, \\ \beta_{20} + \beta_{21}x, & \alpha_1 \leq x \leq \alpha_2, \\ \vdots & \vdots \\ \beta_{D0} + \beta_{D1}x, & \alpha_{D-1} \leq x, \end{cases} \tag{9.102}$$

can be written

$$E[y|x] = \theta_0 + \theta_1 x + \sum_{j=2}^{D} \theta_j(x - \alpha_{j-1})\,\text{sgn}(x - \alpha_{j-1}), \tag{9.103}$$

$$= \theta_0 + \theta_1 x + \sum_{j=2}^{D} \theta_j|x - \alpha_{j-1}|. \tag{9.104}$$

Here

$$\beta_{11} = \theta_1 - \sum_{j=2}^{D} \theta_j \quad \text{and} \quad \beta_{d1} = \sum_{j=1}^{d} \theta_j - \sum_{j=d+1}^{D} \theta_j \quad (d = 2, 3, \ldots, D-1),$$

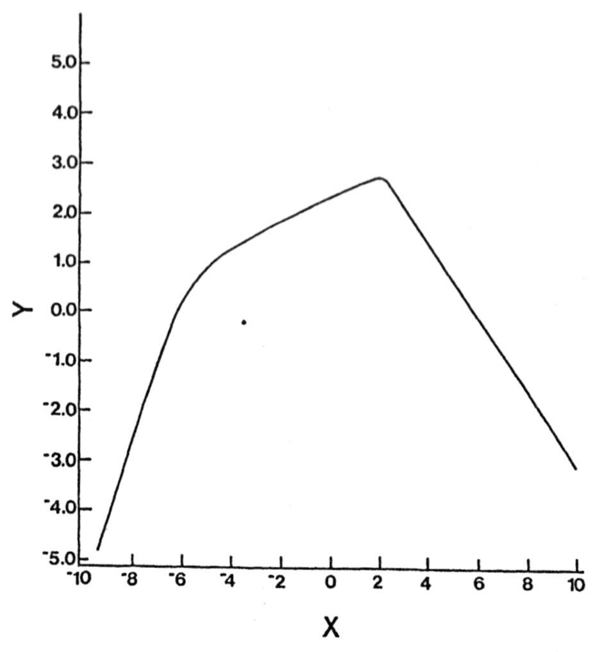

Figure 9.11 Plot of $y = 8 + 0.5x - 0.75[(x+6)^2 + 1]^{1/2} - 0.5[(x-2)^2 + 0.1]^{1/2}$. From Prunty [1983].

so that $\theta_d = (\beta_{d1} - \beta_{d-1,1})/2$. The line segments then intersect at the point $(\alpha_d, \theta_0 + \theta_1 \alpha_d + \sum_j \theta_j |\alpha_d - \alpha_{j-1}|)$. The hyperbolic approximation to (9.103) is

$$E[y|x] = \theta_0 + \theta_1 x + \sum_{j=2}^{D} \theta_j \sqrt{(x - \alpha_{j-1})^2 + \gamma_{j-1}}, \tag{9.105}$$

where a different degree of curvature γ_{j-1} can be estimated for each transition (see Fig. 9.11).

Following Griffiths and Miller [1975] (see also Bunke and Schulze [1985]) we can link up D nonlinear segments of the form $f_j(x; \boldsymbol{\theta}_j)$ in a similar way. Such submodels may be written in combined form as

$$E[y|x] = \frac{1}{2}\Bigg(f_1(x; \boldsymbol{\theta}_1) + f_D(x; \boldsymbol{\theta}_D) + \sum_{j=2}^{D} [f_j(x; \boldsymbol{\theta}_j) - f_{j-1}(x; \boldsymbol{\theta}_{j-1})] \operatorname{sgn}(x - \alpha_{j-1}) \Bigg). \tag{9.106}$$

Smooth transitions are accomplished by replacing $\operatorname{sgn}(x - \alpha_{j-1})$ by, for example, $\operatorname{hyp}((x - \alpha_{j-1}); \gamma)$ given by (9.98) or $\tanh[(x - \alpha_{j-1})/\gamma]$.

We note in passing that Bunke and Schulze [1985] smooth sgn(z) using

$$\operatorname{trn}(z) = \begin{cases} -1, & z \leq -\gamma \\ \frac{15}{8}z - \frac{10}{8}z^3 + \frac{3}{8}z^5, & -\gamma < z < \gamma, \\ 1, & z \geq \gamma. \end{cases}$$

This method is similar in spirit to the method explored in detail in Section 9.4.2a. However, it has the property that trn(0) = 0, thus forcing the smooth model to bulge outside the corner as the model (9.96) does.

9.4.2 The max–min Formulation

From Fig. 9.12 it is clear that an alternative way of writing the continuous two-phase regression problem (9.88) is either

$$E[y|x] = \max(\beta_{10} + \beta_{11}x, \beta_{20} + \beta_{21}x), \tag{9.107}$$

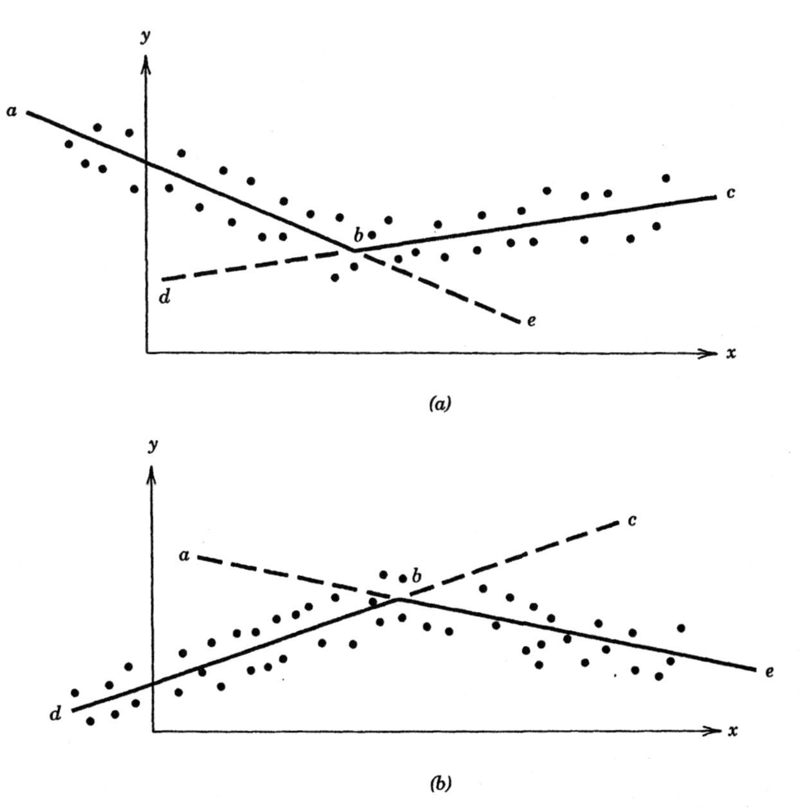

Figure 9.12 The two forms of the continuous two-phase regression model. From Tishler and Zang [1981b].

or

$$E[y|x] = \min(\beta_{10} + \beta_{11}x, \beta_{20} + \beta_{21}x). \tag{9.108}$$

Continuity of $E[y|x]$ is implicit in this formulation and the location of the changepoint, α, need not be represented. Since $\min(b_1, b_2) = -\max(-b_1, -b_2)$, methods need only be developed to cope with (9.107). Tishler and Zang [1981a, b] smooth the max formulation (9.107) rather than the sgn formulation (9.90). One advantage is the immediate generalization to changes of regime with a multivariate explanatory variable $\mathbf{x}$, for example

$$E[y|\mathbf{x}] = \max(\boldsymbol{\beta}_1'\mathbf{x}, \boldsymbol{\beta}_2'\mathbf{x}), \tag{9.109}$$

where $\mathbf{x}$ has dimension k, say. When $k = 3$, i.e. we have two intersecting planes, we find that the analogue of a changepoint is a change line of intersection. Again, the continuous two-plane problem reduces to using the maximum or the minimum of $(\boldsymbol{\beta}_1'\mathbf{x}, \boldsymbol{\beta}_2'\mathbf{x})$. Obviously (9.109) can be further generalized to

$$E[y|x] = \max(\boldsymbol{\beta}_1'\mathbf{x}, \boldsymbol{\beta}_2'\mathbf{x}, \ldots, \boldsymbol{\beta}_D'\mathbf{x}), \tag{9.110}$$

a formulation which can be useful for econometric models (see Tishler and Zang [1981a]). However, in Section 9.4.2c below we shall use (9.109) to modify the sgn formulation so that it, too, will handle multivariate $\mathbf{x}$-variables.

Having motivated the change in formulation, we return to the problem of smoothing (9.107). Tishler and Zang [1981a, b] use two different approaches. We first describe the approach given in their [1981a] paper.

a Smoothing max (0, *z*)

Let

$$u(z) = \max(0, z), \qquad z \in \mathbb{R} \tag{9.111}$$
$$(= z_+).$$

We may rewrite (9.109) as

$$E[y|x] = \boldsymbol{\beta}_1'\mathbf{x} + u(\boldsymbol{\beta}_2'\mathbf{x} - \boldsymbol{\beta}_1'\mathbf{x}). \tag{9.112}$$

Any lack of differentiability of (9.112) is due to a lack of differentiability of $u(z)$, which is continuously differentiable (of all orders) except for the kink at $z = 0$. Tishler and Zang [1981a] replace $u(\cdot)$ in (9.112) by an approximation in which the kink is smoothed out as in Fig. 9.13. They use

$$u_\gamma(z) = \begin{cases} 0, & z \le -\gamma, \\ (z+\gamma)^2/4\gamma, & -\gamma \le z \le \gamma, \\ z, & z \ge \gamma, \end{cases} \tag{9.113}$$

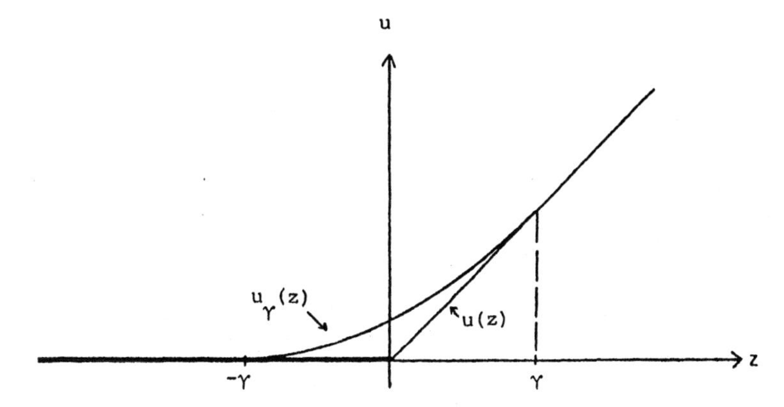

Figure 9.13 Plots of $u(z)$ and $u_\gamma(z)$ of (9.113). From Tishler and Zang [1981a].

for some positive γ. Clearly $\lim_{\gamma \to 0} u_\gamma(z) = u(z)$ and $u_\gamma(z)$ has continuous first, but not second, derivatives. Our two-phase regression model now becomes

$$E[y|x] = \boldsymbol{\beta}_1'\mathbf{x} + u_\gamma(\boldsymbol{\beta}_2'\mathbf{x} - \boldsymbol{\beta}_1'\mathbf{x}). \tag{9.114}$$

We note that (9.114) is *identical* to the two-phase linear model for all $\mathbf{x}$ except for $\mathbf{x}$ satisfying $|\boldsymbol{\beta}_2'\mathbf{x} - \boldsymbol{\beta}_1'\mathbf{x}| < \gamma$. This is in contrast to the tanh and hyperbolic models of the previous subsection, for which the two lines were merely asymptotes. Higher orders of continuity for $u_\gamma(z)$ can be obtained by using higher-order polynomials in (9.113). For example (Zang [1980])

$$u_\gamma(z) = \frac{-z^4}{16\gamma^3} + \frac{3z^2}{8\gamma} + \frac{z}{2} + \frac{3\gamma}{16}, \qquad -\gamma \le z \le \gamma, \tag{9.115}$$

has continuous second derivatives. Zang [1980] also gives expressions which will enable $u_\gamma(z)$ to have continuous third or fourth derivatives. However, Tishler and Zang [1981a] only use (9.113). As with the models of Section 9.4.1, (9.114) can be viewed as a model for two linear regimes with a smooth transition. We can treat γ as a free parameter of the model which is to be estimated. This parameter governs the sharpness of the transition between linear regimes, and as $\gamma \to 0$ the model tends to intersecting straight lines. We note that the changepoint, or more generally the hyperplane at which the change of regimes takes place, is no longer a parameter that must be estimated. However, once $\boldsymbol{\beta}_1$ and $\boldsymbol{\beta}_2$ have been estimated, the changepoint (hyperplane) can be estimated as the intersection of $\hat{\boldsymbol{\beta}}_1'\mathbf{x}$ and $\hat{\boldsymbol{\beta}}_2'\mathbf{x}$.

The more general model for D linear segments, given by (9.110), can be rewritten

$$E(y|x] = \boldsymbol{\beta}_1'\mathbf{x} + u[\boldsymbol{\beta}_2'\mathbf{x} - \boldsymbol{\beta}_1'\mathbf{x} + u[\cdots u[\boldsymbol{\beta}_{D-1}'\mathbf{x} - \boldsymbol{\beta}_{D-2}'\mathbf{x} + u(\boldsymbol{\beta}_D'x - \boldsymbol{\beta}_{D-1}'\mathbf{x})]\cdots]]. \tag{9.116}$$

Each nested $u(\cdot)$ can be replaced by a smooth $u_{\gamma_j}(\cdot)$, where different γ_j parameters allow varying degrees of abruptness of transition from regime to regime. Each γ_j is to be estimated from the data.

However, Tishler and Zang [1981a] did not present the above method primarily as a smooth-transition model. Their main objective was to approximate two-phase or multiphase linear regression models so that the error sum of squares

$$S(\boldsymbol{\beta}) = \sum_{i=1}^{n} \left(y_i - \max_{1 \le j \le D} (\boldsymbol{\beta}_j' \mathbf{x}_i) \right)^2 \tag{9.117}$$

could be minimized using standard iterative methods for smooth objective functions, e.g. quasi-Newton methods (Section 13.4.2). Let $S_\gamma(\boldsymbol{\beta}) = \sum_i \{y_i - E[y_i|\mathbf{x}_i]\}^2$, where $E[y|\mathbf{x}]$ is given by (9.116) with $u(\cdot)$ replaced by $u_\gamma(\cdot)$. Their minimization strategy is broadly as follows:

1. Select a starting value $\gamma^{(1)}$ for γ (e.g. $0.5\bar{y}$).
2. For $a = 1, 2, \ldots$:
   (a) Obtain $\boldsymbol{\beta}^{(a)}$ to minimize $S_\gamma(\boldsymbol{\beta})$ with $\gamma = \gamma^{(a)}$, using $\boldsymbol{\beta}^{(a-1)}$ as a starting value.
   (b) Choose $\gamma^{(a+1)} < \gamma^{(a)}$.

Tishler and Zang used a quasi-Newton algorithm for step (a), although a specialist least-squares algorithm (see Chapter 14) such as a modified Gauss–Newton algorithm could also be used. The procedure stops when it is decided that, at $\gamma = \gamma^{(a)}$, $S(\boldsymbol{\beta}^{(a)}) = S_\gamma(\boldsymbol{\beta}^{(a)})$. In the case of only two segments [see (9.112)] we can see that this must occur except in the unlikely event of the fitted lines intersecting exactly at a data point, as (9.112) and (9.114) are identical except for $|\boldsymbol{\beta}_1'\mathbf{x} - \boldsymbol{\beta}_2'\mathbf{x}| < \gamma$. For small enough γ this region will contain no data points. However, it is important that γ not be allowed to become too small, as $S_\gamma(\boldsymbol{\beta}) \to S(\boldsymbol{\beta})$ when $\gamma \to 0$ and does not have continuous derivatives. Methods for minimizing smooth functions will then experience difficulties. For algorithmic details, such as the method for reducing γ at each step, the reader is referred to Tishler and Zang [1981a] and Zang [1980].

b Limiting Form for max$\{z_i\}$

Let us recall the D-linear-segment problem (9.110), namely $E[y|x] = \max(\boldsymbol{\beta}_1'\mathbf{x}, \ldots, \boldsymbol{\beta}_D'\mathbf{x})$, and its specialization (9.109) when $D = 2$. The approach of Tishler and Zang [1981b] is based upon the well-known result (e.g. Hardy et al. [1952: 15]), that for positive real numbers $z_1, z_2, \ldots, z_D$,

$$\lim_{\delta \to \infty} \left(\sum_{j=1}^{D} z_j^\delta \right)^{1/\delta} = \max_{1 \le j \le D} z_j. \tag{9.118}$$

Because this result requires positive z_i, Tishler and Zang [1981b] modify the two-phase regression model from

$$y = \max(\boldsymbol{\beta}_1'\mathbf{x}, \boldsymbol{\beta}_2'\mathbf{x}) + \varepsilon \tag{9.119}$$

to

$$B + y = \max(B + \boldsymbol{\beta}_1'\mathbf{x}, B + \boldsymbol{\beta}_2'\mathbf{x}) + \varepsilon, \tag{9.120}$$

for which a smooth-transition approximation is

$$B + y = [(B + \boldsymbol{\beta}_1'\mathbf{x})^{\delta} + (B + \boldsymbol{\beta}_2'\mathbf{x})^{\delta}]^{1/\delta} + \varepsilon. \tag{9.121}$$

From (9.118) with $D = 2$ we see that (9.120) is a limiting case of (9.121) as $\delta \to \infty$. Tishler and Zang do not treat B as a free parameter, but simply as a number which is used in an ad hoc way to keep everything positive. It is important that B is not too large, otherwise the error-sum-of-squares surface becomes very flat and is thus difficult to minimize. Tishler and Zang [1981b] suggest, as a rule of thumb, that a value of B slightly larger than $\max|y_i|$ be used. Additionally, to avoid overflow errors in the computation of $E[B + y_i|\mathbf{x}_i]$ while fitting (9.121) when δ is large, Tishler and Zang [1981b] compute this value as

$$E[B + y_i|\mathbf{x}_i] = c_i\left[\left(\frac{B + \boldsymbol{\beta}_1'\mathbf{x}_i}{c_i}\right)^{\delta} + \left(\frac{B + \boldsymbol{\beta}_2'\mathbf{x}_i}{c_i}\right)^{\delta}\right]^{1/\delta}, \tag{9.122}$$

where $c_i = \max(|B + \boldsymbol{\beta}_1'\mathbf{x}_i|, |B + \boldsymbol{\beta}_2'\mathbf{x}_i|)$.

The parameter δ in (9.121), like γ in the models (9.92), (9.99), and (9.114), determines the abruptness of the transition between regimes. However, the two-intersecting-line model is approached as $\delta \to \infty$, in contrast to $\gamma \to 0$. Analogously with the earlier models, δ could be estimated from the data by least squares, i.e. $(\boldsymbol{\beta}_1, \boldsymbol{\beta}_2, \delta)$ chosen to minimize

$$S(\boldsymbol{\beta}, \delta) = \sum_{i=1}^{n} [B + y_i - \{(B + \boldsymbol{\beta}_1'\mathbf{x}_i)^{\delta} + (B + \boldsymbol{\beta}_2'\mathbf{x}_i)^{\delta}\}^{1/\delta}]^2. \tag{9.123}$$

In contrast, Tishler and Zang [1981b] again view (9.120) largely as a computational device for fitting (9.119) within an iterative process, such as the following:

1. Begin with starting values $\boldsymbol{\beta}^{(1)}$ and $\delta^{(1)}$ (moderately large and positive).
2. For $a = 1, 2, \ldots$:
   (i) Starting from $\boldsymbol{\beta}^{(a)}$, minimize the function $S_{\delta}^{(a)}(\boldsymbol{\beta}) = S(\boldsymbol{\beta}, \delta^{(a)})$ with respect to $\boldsymbol{\beta}$ to obtain $\boldsymbol{\beta}^{(a+1)}$.
   (ii) Update $\delta^{(a+1)} = \delta^{(a)} + 20$, say.

The minimization step uses a method for smooth functions such as the quasi-

Newton method. Iterations cease, together with the increasing of $\delta^{(a)}$, when $\| \boldsymbol{\beta}^{(a)} - \boldsymbol{\beta}^{(a-1)} \|$ is sufficiently small. Extensions of the model and algorithm details are discussed in Tishler and Zang [1981b]. These authors state (p. 119) that large odd values of δ create regions in which $S_\delta(\boldsymbol{\beta})$ is very steep, a factor which can cause numerical problems.

We reiterate that, although Tishler and Zang do not really present (9.121) as a model for data with a smooth transition between linear regimes such as in Examples 9.4 and 9.5 in Section 9.1, with δ as a parameter, it is an obvious candidate for that. In such circumstances a smaller value of δ may provide the best fit to the data.

c *Extending* sgn *to a Vector of Regressors*

In this section we have expressed the changepoint model in the form $E[y|\mathbf{x}] = \max(\boldsymbol{\beta}_1'\mathbf{x}, \boldsymbol{\beta}_2'\mathbf{x})$. Following (9.119) we tried a smooth approximation to the maximum function resulting in (9.121). However,

$$\max(\boldsymbol{\beta}_1'\mathbf{x}, \boldsymbol{\beta}_2'\mathbf{x}) = \tfrac{1}{2}(\boldsymbol{\beta}_1 + \boldsymbol{\beta}_2)'\mathbf{x} + \tfrac{1}{2}(\boldsymbol{\beta}_1 - \boldsymbol{\beta}_2)'\mathbf{x}\,\mathrm{sgn}[(\boldsymbol{\beta}_1 - \boldsymbol{\beta}_2)'\mathbf{x}], \tag{9.124}$$

so that we can again approximate $\mathrm{sgn}(z)$ using a smooth transition function such as $\tanh(z/\gamma)$ or $\mathrm{hyp}(z) = (z^2 + \gamma)^{1/2}/z$ as in (9.98). Using the hyperbolic approximation, (9.124) becomes

$$E[y|x] = \tfrac{1}{2}(\boldsymbol{\beta}_1 + \boldsymbol{\beta}_2)'\mathbf{x} + \tfrac{1}{2}[\{(\boldsymbol{\beta}_1 - \boldsymbol{\beta}_2)'\mathbf{x}\}^2 + \gamma]^{1/2}. \tag{9.125}$$

We have thus obtained a model which behaves qualitatively the same way as (9.121). However, it does not have the clumsy feature embodied in the unknown constant B which had to be added to (9.121) to preserve positivity. Moreover, in the special case of two lines, with $\mathbf{x} = (1, x)'$, it is straightforward to show that where (9.89) and (9.90) define the maximum of two lines (rather than the minimum), (9.124) is simply a reparametrization of (9.90). Equation (9.125) is the equivalent reparametrization of (9.99). Thus in vector form, (9.125) is a generalization of the usual hyperbolic model (9.99).

9.4.3 Examples

All four smooth-transition two-regime models, namely the tanh version of (9.92) and the models (9.99), (9.114), and (9.121), have been applied to the two data sets given by Bacon and Watts [1971]. The data sets, which come originally from the Queen's University Ph.D. thesis of R. A. Cook, concern how the stagnant-surface-layer height of a controlled flow of water behaves when flowing down an "inclined channel using different surfactants." For both data sets x is the logarithm of the flow rate in g/cm-sec and y is the logarithm of the height of the stagnant band in centimeters. Both sets of data are given in Table 9.3. Data set 1 is plotted in Fig. 9.14.

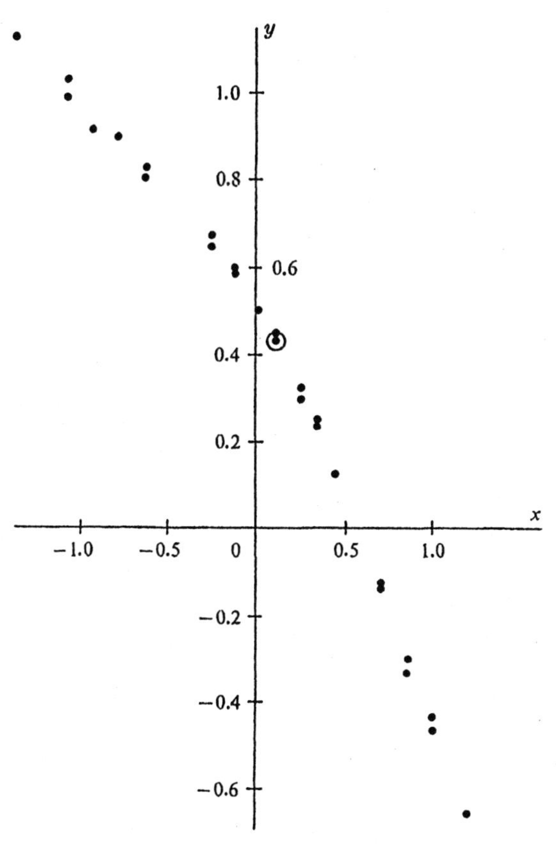

Figure 9.14 Stagnant-band-height data set 1 of Table 9.3. Plot of $y = \log_e$(band height in cm) versus $x = \log_e$(flow rate in g/cm-sec). From Bacon and Watts [1971].

The models above have all been fitted by least squares, and the least-squares estimates and the residual sum of squares are presented in Table 9.4. Also results have been converted to the θ-parametrization of (9.89) and (9.92). For models (9.114) and (9.121), the joinpoints were obtained as the intersection of the two lines, which are asymptotes in the case of (9.121). For all the models except (9.121), the parameter estimates tabulated are least-squares estimates. The model (9.121) was fitted by Tishler and Zang [1981b] using the strategy described following (9.123), so it is really a piecewise linear fit and thus its residual sum of squares should be approximately 0.009. However, for these data there is strong evidence that the model should have a smooth transition. If we use the hyperbolic model and test $H_0: \gamma = 0$ using the F-test [see Equation (5.23)], we find

$$F = \frac{(0.00938 - 0.00515)/1}{0.00515/24} = 19.7,$$

Table 9.3 Stagnant-Band-Height Data[a]

Data Set 1

x	y	x	y	x	y	x	y
0.11	0.44	0.11	0.43	0.34	0.25	−1.08	0.99
−0.80	0.90	0.11	0.43	1.19	−0.65	−1.08	1.03
0.01	0.51	−0.63	0.81	0.59	−0.01	0.44	0.13
−0.25	0.65	−0.63	0.83	0.85	−0.30	0.34	0.24
−0.25	0.67	−1.39	1.12	0.85	−0.33	0.25	0.30
−0.12	0.60	−1.39	1.12	0.99	−0.46	—	—
−0.12	0.59	0.70	−0.13	0.99	−0.43	—	—
−0.94	0.92	0.70	−0.14	0.25	0.33	—	—

Data Set 2

x	y	x	y	x	y
−0.40	0.61	0.29	0.31	−0.58	0.67
−0.40	0.69	0.29	0.31	−1.39	1.06
−0.40	0.64	−1.51	1.15	−1.39	1.10
−0.80	0.83	−1.51	1.13	0.33	0.17
−0.80	0.80	0.01	0.44	0.46	0.10
−1.08	0.93	0.01	0.44	0.46	0.10
−1.08	0.95	−0.40	0.55	1.19	−0.73
−0.12	0.52	−0.40	0.60	1.19	−0.73
−0.12	0.54	−0.80	0.87	1.03	−0.56
0.11	0.42	−0.80	0.89	1.03	−0.60
0.11	0.39	−0.80	0.89	0.85	−0.36
0.01	0.49	−1.08	0.99	0.85	−0.36
0.01	0.46	−1.08	0.99	0.64	−0.12
−0.25	0.55	0.11	0.34	0.44	0.14
−0.25	0.58	0.11	0.40	0.80	−0.33
0.21	0.34	−0.58	0.73	1.19	−0.80

[a]From Bacon and Watts [1971] with permission of the Biometrika Trustees. $x = \log$ (flow rate in g/cm-sec), $y = \log$ (band height in cm).

which is greater than $F_{1,24}^{0.01} = 7.8$. The models (9.92) and (9.114) clearly give even higher F-statistics because of their smaller residual sums of squares. There is little to choose between the smooth models in terms of goodness of fit.

Tishler and Zang [1981b] use the parameter estimates given in Table 9.4 and determine the mean squared error for fitting a piecewise linear model using these parameter estimates. Their method of fitting (9.121) gives a much lower mean squared error, namely 33×10^{-5} compared with 133×10^{-5} for (9.92) and 263 $\times 10^{-5}$ for (9.99). This is scarcely surprising, since their method is designed to minimize this mean squared error. As also noted by Griffiths [1983], this method of comparison is clearly unfair to (9.92) and (9.99).

Table 9.4 Parameter Estimates and Residual Sum of Squares for Least-Squares Fit of Five Models with Two Regimes

	Smooth-Transition Models				
	(9.92)[a] tanh	(9.99)[a] Hyperbolic	(9.114)[b] u_γ	(9.121)[c]	(9.90)[a] Piecewise Linear
			Data Set 1		
$\hat{\theta}_0$	0.484	0.586	0.546	0.531	0.529
$\hat{\theta}_1$	−0.728	−0.735	−0.731	−0.720	−0.719
$\hat{\theta}_2$	−0.287	−0.359	−0.333	−0.300	−0.297
$\hat{\alpha}$	0.047	0.063	0.056	0.037	0.036
$\hat{\gamma}$	0.622	0.096	0.285	—	—
δ	—	—	—	200	—
B	—	—	—	2.0	—
RSS[d]	0.00490	0.00515	0.00482	—	0.00938
			Data Set 2		
$\hat{\theta}_0$	—	—	0.333	0.335	0.333
$\hat{\theta}_1$	—	—	−0.816	−0.817	−0.816
$\hat{\theta}_2$	—	—	−0.348	−0.350	−0.348
$\hat{\alpha}$	—	—	0.251	0.251	0.252
$\hat{\gamma}$	—	—	0.01	—	—
δ	—	—	—	200	—
B	—	—	—	2.0	—
RSS[d]	—	—	0.0520	0.0520	0.0520

[a]From Griffiths and Miller [1973].
[b]Calculated from the results of Tishler and Zang [1981a].
[c]From Tishler and Zang [1981b].
[d]Residual sum of squares.

The smooth-transition models give very similar results except for the parameter $\hat{\theta}_2$ measuring the average difference in slopes. The difference is caused by the fact that the tanh model (9.92) bulges outside the corner, while the hyperbolic models (9.99) and (9.114) cut the corner, as is explained by Griffiths and Miller [1973]. The smooth-transition models fit considerably better than the piecewise linear model. For data set 2, every model gives essentially the same answers. Griffiths and Miller [1973] reported only the piecewise linear fit. For the models (9.92) and (9.99), the parameter γ was so small in each case that they were essentially piecewise linear anyway.

Before concluding, we note that Watts and Bacon [1974] analyze the data in Fig. 9.5 on p. 437 using the hyperbolic model. This is a particularly interesting analysis because they are forced to use an AR(1) error structure (c.f. Section 6.2).

9.4.4 Discussion

We have seen no discussion about asymptotic inference for the smooth-transition models of Section 9.4. However, all are smooth nonlinear regression models in which $E[y|x] = f(x; \boldsymbol{\theta})$ has continuous second derivatives with respect to all of its parameters. This is even true of the models of Section 9.4.2a provided we use (9.115) to define $u_\gamma(z)$. We rewrite this function as $u_z(\gamma)$ to stress that we wish to consider it as a function of γ for fixed z. Thus

$$\begin{aligned}
&\text{if} \quad z \geq 0 \quad \text{then} \quad u_z(\gamma) = \begin{cases} z, & \gamma \leq z, \\ A(\gamma, z), & \gamma \geq z, \end{cases} \\
&\text{if} \quad z < 0 \quad \text{then} \quad u_z(\gamma) = \begin{cases} 0, & \gamma \leq -z, \\ A(\gamma, z), & \gamma \geq -z, \end{cases}
\end{aligned} \tag{9.126}$$

where $A(\gamma, z)$ is $u_\gamma(z)$ of (9.115). Differentiating this function with respect to γ, it is easily seen that $u_z(\gamma)$ has continuous second derivatives. The situation is a little complicated with the model (9.121) because of the positivity constraints required in raising arguments to the power δ. Bacon and Watts [1971] develop Bayesian methods of inference about the parameter γ and the changepoint α.

On the problem of inferring whether there really has been a change of regime, the smooth models are of no particular help, because we do not have a full set of identifiable parameters if there is no change of regime.

There appears to be little to choose between the models. It may often be possible, by looking at a scatterplot, to choose between the tanh model (9.92), which bulges outside the corner, and the other models, which cut the corner. Moreover, as Griffiths and Miller [1973] note, the tanh and hyperbolic models can give quite different estimates of the difference-in-slopes parameter even when both fit the data well, and this fact seems disturbing. However, it seems disturbing only because of a too literal interpretation of these models as two line segments with a smooth transition. If we want a very literal interpretation like this, we should clearly use a model like (9.114). We would, however, recommend the extended "hyperbolic" model (9.125) as replacement for (9.121). The model (9.125) is a reparametrization of the usual hyperbolic model (9.99) which allows it to be extended to a change of regression planes rather than regression lines.

Even when the underlying model has a smooth transition, the two-phase linear regression model ($\gamma = 0$) can be expected to fit well whenever there are very few data points in the vicinity of the changepoint. A reparametrization of the form $\gamma \to e^\phi$ (or $\gamma \to \phi^2$) should be used to prevent the minimization algorithm stepping to impossible nonpositive values. An indication that $\hat{\gamma}$ $(= e^{\hat{\phi}})$ should be zero will be signaled by $\phi^{(a)} \to -\infty$. In any situation where there are not many data points estabilishing the nature of the transition, the parameter γ (or δ) can be expected to be estimated with poor precision. When γ is used as a free parameter, the sum-of-squares surface tends to be badly conditioned, leading to problems with minimization. Following Miller [1979: 46], it is advisable initially to hold γ fixed and first minimize with respect to the other parameters. We note further that the models (9.90), (9.92), and (9.99) give rise to separable least-squares problems (see Section 14.7), since for fixed (α, γ) these are linear regression models in the other parameters. This fact is useful both for finding starting values and for the possibility of using specialist minimization software. From a guessed changepoint $\alpha^{(1)}$, starting values for the linear-regression parameters for the two line segments of (9.90) can be obtained by ordinary linear least-squares regression. These values can then be used as starting points for fitting (9.114), (9.121), or (9.125).

9.5 SPLINE REGRESSION

9.5.1 Fixed and Variable Knots

A spline function (q-spline) is a piecewise or segmented polynomial of degree q with $q-1$ continuous derivatives at the changepoints. In the spline literature the changepoints are called knots. We consider only functions of a single variable x. Thus a spline function is a special case of the polynomial phase models discussed in Section 9.3. For example, continuous-line-segment models are linear splines. Cubic splines, or cubic polynomials with continuous second derivatives at the joins, are commonly used in approximation theory.

In our previous discussion of multiphase models we were primarily concerned with models to deal with trend relationships between y and x characterized by abrupt changes between well-behaved (e.g. linear or low-order polynomial) regimes. In this section the emphasis shifts to using such models to approximate smooth regression functions. Our discussion draws heavily upon the review paper of Eubank [1984].

Suppose $y = f(x) + \varepsilon$, but $f(x)$ is completely unknown. When faced with a well-behaved curved trend in a scatterplot, most satisticians would fit a low-order polynomial in x and go on to make inferences which are conditional upon the truth of the polynomial model. This is done largely because the family of polynomials is a family of simple curves which is flexible enough to approximate a large variety of shapes. More technically, one could appeal to the Stone–Weierstrass theorem (see Royden [1968: 174]), from which it follows that any continuous function on $[a, b]$ can be approximated arbitrarily closely by a polynomial. In practical terms, a better approximation can be obtained by increasing the order of the polynomial. The cost incurred for doing this is the introduction of additional parameters and some oscillation between data points.

Spline functions can be viewed as another way of improving a polynomial approximation. Suppose we are prepared to assume that $f(x)$ has $q+1$ continuous derivatives on $[a, b]$. Then by Taylor's theorem,

$$f(x) = \sum_{j=0}^{q} \phi_j x^j + r(x), \tag{9.127}$$

where

$$\sum_{j=0}^{q} \phi_j x^j = \sum_{j=0}^{q} \frac{1}{j!} \frac{d^j f(a)}{dx^j} (x-a)^j,$$

and the remainder is given by

$$\begin{aligned} r(x) &= \frac{1}{q!} \int_a^x \frac{d^{q+1} f(t)}{dt^{q+1}} (x-t)^q \, dt \\ &= \frac{1}{q!} \int_a^b \frac{d^{q+1} f(t)}{dt^{q+1}} (x-t)_+^q \, dt, \end{aligned} \tag{9.128}$$

where $z_+^q = z^q$ if $z \geq 0$ and 0 otherwise. Use of an order-q polynomial model ignores the remainder term. Instead of improving the polynomial approximation by increasing the order of the polynomial, let us approximate the remainder term. To do this we consider a partition $a \equiv \alpha_0 < \alpha_1 < \cdots < \alpha_{D-1} < \alpha_D \equiv b$ and approximate the $(q+1)$th derivative of f by a step function to obtain

$$r(x) \approx \frac{1}{q!} \sum_{d=1}^{D-1} \xi_d (x - \alpha_d)_+^q. \tag{9.129}$$

The final term of the summation has been omitted as $(x-b)_+^q = 0$ for $x \leq b$. The approximation to $f(x)$ resulting from (9.129) is of the form

$$f(x) \approx \sum_{j=0}^{q} \phi_j x^j + \sum_{d=1}^{D-1} \xi_d (x - \alpha_d)_+^q. \tag{9.130}$$

It should come as no surprise after Section 9.3.3 that the family of functions (9.130) is precisely the family of q-splines with knots at $\alpha_1, \ldots, \alpha_{D-1}$. De Boor [1978] gives a proof of this equivalence. Thus we have two parametrizations of q-splines. The first is of the form (9.33) with the phase models all being polynomials of order q and having $m_d = q$ at every change. The second is (9.130). The spline approximation can clearly be improved either by increasing the number of knots, or by allowing the data to determine the position of the knots, thus improving the approximation of the integral in (9.128). The advantage of using splines over increasing the order of the polynomial, in some situations, is that the oscillatory behavior of high-order polynomials can be avoided with splines.

Example 9.8 Figure 9.15a plots a property of titanium as it changes with temperature. Superimposed are several polynomials and a cubic spline with five knots fitted by least squares. The oscillatory nature of the polynomials limits their usefulness as approximating functions. However, the comparision is a little unfair. The seventh-order polynomial has 8 free parameters. The knots for the spline have been placed optimally, so that the spline model has 14 free parameters

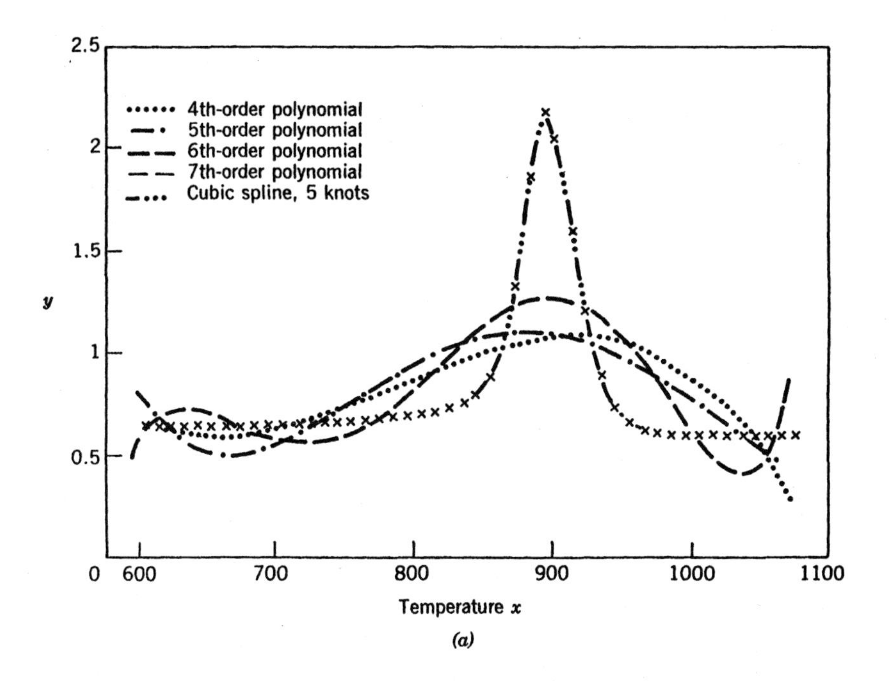

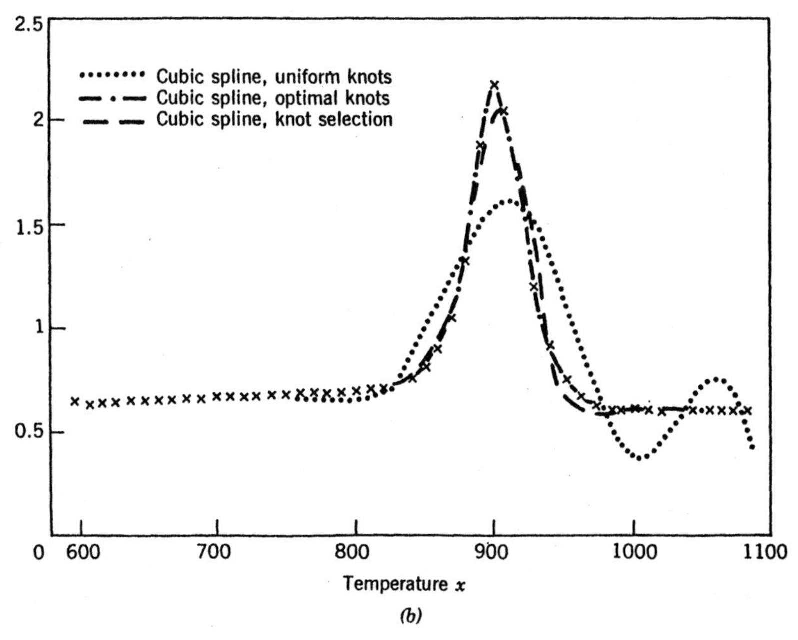

Figure 9.15 Titanium heat data (crosses): (*a*) least-squares fit by polynomials and a cubic spline; (*b*) least-squares fit by cubic splines with uniform and free knots. From Eubank [1984].

including the knots (or changepoints). Figure 9.15*b* illustrates the improvement to be obtained by letting the α_D's be free parameters. The dotted line corresponds to equally spaced α_d's. ■

In the same spirit in which polynomial regression models are used, inferences can be made which are conditional upon the truth of the spline model (9.130). A primary purpose for fitting spline models is prediction within the range of the data. Extrapolation is particularly dangerous in the unstructured situations in which spline modeling is usually considered.

Following the literature, we consider the use of (9.130): (a) when the knots are treated as fixed constants and (b) when the knots are treated as free parameters.

a *Fixed Knots*

When the positions of the knots α_d are treated as fixed, the model (9.130) is clearly a linear regression model. Placement of the knots is discussed by Wold [1974]. Equation (9.130) is an attractive parametrization of the spline model. The model is easily fitted using an ordinary regression program. Testing whether a knot can be removed and the same polynomial equation used to explain two adjacent segments can be tested by testing $H_0: \xi_d = 0$, which is one of the *t*-test statistics always printed by a regression program. However, this parametrization can lead to ill-conditioned design matrices. The parametrization in terms of the model (9.33), namely a linear model subject to linear constraints, has the advantage of parameters which are easily interpretable in terms of the individual phase models. However, a more sophisticated program is required to perform restricted least squares, and if the number of knots is large, a large number of unnecessary parameters are introduced. A useful alternative parametrization which leads to easy evaluation and well-conditioned design matrices is via so-called *B*-splines (see Eubank [1984: 441 ff.], Wold [1974], Jupp [1978]).

Fixed-spline regression models can clearly be checked using the standard methods of regression diagnostics. The problems of particular interest with splines are: (a) checking for the omission of knots (the recursive residuals of Brown et al. [1975] can be useful here) and (b) checking whether the spline model is too smooth. Figure 9.16 shows an example in which a smoothing spline model smooths out an obvious sharp corner. A particular version of the linear-regression variable selection problem involves the adding and subtracting of knots. For further details and an introduction to the literature the reader is referred to Eubank [1984], Wegman and Wright [1983], and Smith [1979].

b *Variable Knots*

Spline regression models with variable (or free) knots are a special case of the models of Section 9.3, in particular those described in Section 9.3.1 and Section 9.3.3. One problem that we did not mention there is the "estimation" of $D-1$, the number of knots to be used in (9.130). Ertel and Fowlkes [1976] use a C_p-like statistic which requires an estimate of σ^2 to be obtained by overfitting. Other

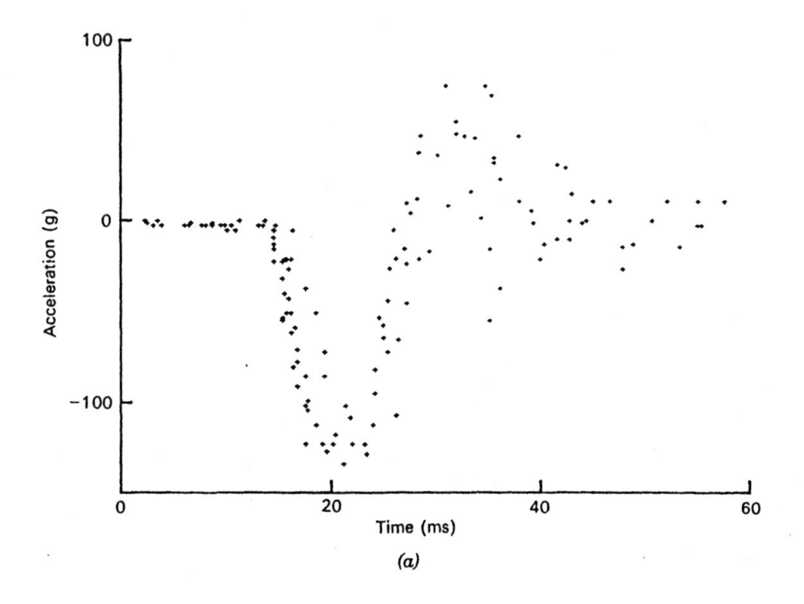

(a)

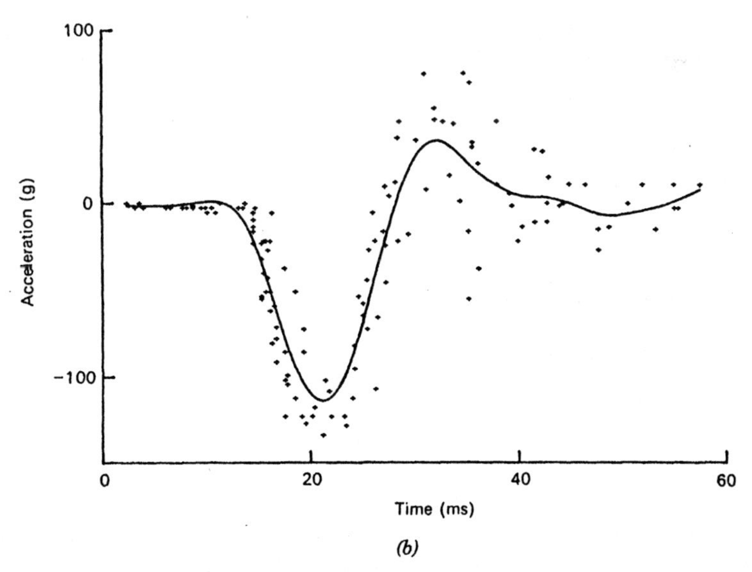

(b)

Figure 9.16 Motorcycle-helment impact data: (*a*) raw data; (*b*) with a cubic smoothing spline superimposed. From Silverman [1985].

approaches include Eubank [1984] (generalized cross validation) and Brannigan [1981] (Akaike's information criterion). Agarwal and Studden [1978, 1980] give an analysis in which an asymptotic expression for the integrated mean squared error is minimized with respect to the allocation of the x's, the placing of the knots, and the number of knots. Eubank [1984: 473] quotes some asymptotic results that suggest that variable-knot spline functions are more parsimonious than the smoothing-spline models to follow.

9.5.2 Smoothing Splines

Let us assume we have data of the form (x_i, y_i), $i = 1, \ldots, n$, with $a \leq x_1 < \cdots < x_n \leq b$, generated by a regression model

$$y_i = f(x_i) + \varepsilon_i \tag{9.131}$$

with $E[\varepsilon_i] = 0$, $\text{var}[\varepsilon_i] = \sigma^2$, $i = 1, 2, \ldots, n$. We wish to estimate $f(x)$, but the only structure we wish to impose upon the model comes from our belief that the regression function has a certain degree of smoothness. This is a nonparametric regression problem and, as such, is outside the scope of the book. However, because of its relationship with the other spline models of this chapter, we give a brief introduction to smoothing-spline approaches to the nonparametric regression problem.

Silverman [1985] states that we would typically be interested in the estimation of f and interesting features of f such as its maximum value. In applications to growth data, the maximum value of the derivative of f representing the maximum growth rate may be of interest. It would clearly be desirable to be able to obtain interval estimates for the underlying function, and even more so to obtain prediction intervals for y at values of x within the range of the data, as the model itself is not particularly meaningful. Much early work concentrated simply on the estimation of a smooth curve as a "data summary."

A smooth curve can be drawn that goes through all of the data points. However, we wish to filter out very local variability, due to random error, from grosser trends. One approach is to choose as $\hat{f}$ the function f with m derivatives which minimizes

$$S(f) = \sum_{i=1}^{n} [y_i - f(x_i)]^2 + \lambda \int_a^b \left(\frac{df^m(x)}{dx^m} \right)^2 dx. \tag{9.132}$$

The first term assures fidelity to the data in a least-squares sense. The second term is a "roughness penalty" imposed so that, for example, the curve $\hat{f}$ does not have too many bends. The parameter λ adjusts the relative weighting given to the error sum of squares and the roughness penalty. It thus controls the degree of smoothing. The roughness penalty is proportional to the average size of the squared mth derivative over the region of interest. The parameter m controls the "order" of smoothing. The most commonly used value is $m = 2$, which penalizes

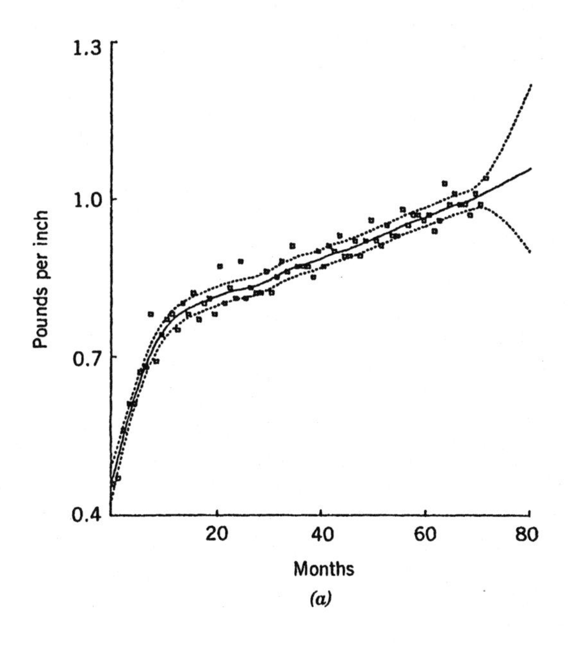

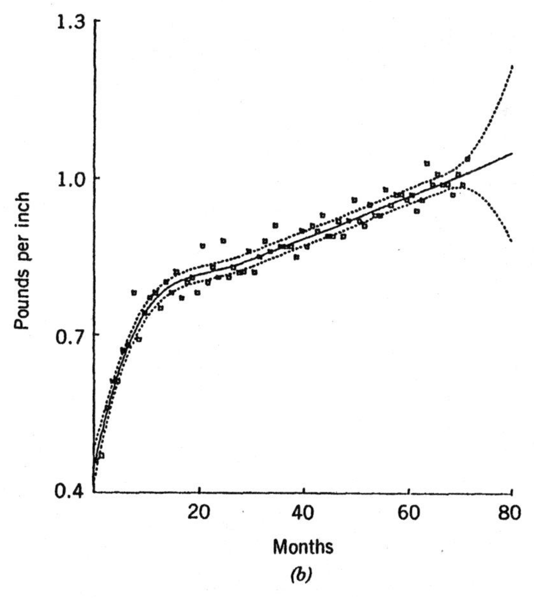

Figure 9.17 Smoothing splines with 95% confidence limits fitted to data on weight/height versus age: (*a*) $m = 2$ (cubic); (*b*) $m = 3$. (A square indicates observations; a solid line indicates the fitted spline; broken lines indicate the 95% confidence limits.) From Wecker and Ansley [1983].

any bending of the function. When $m = 3$, nonzero second derivatives attract no penalty, but *changes* in the second derivative, or changes in curvature, are penalized. Perhaps contrary to immediate intuition, choosing $m = 3$ often results in smoother curves than $m = 2$ because the minimum is found within a smoother class (see Fig. 9.17 above). The value $m = 2$ is the smallest giving visual smoothness. If the data cannot be fitted exactly by a polynomial of degree less than m, the unique minimizer of (9.132) is a spline of order $q = 2m - 1$ with possible knots at each x_i (Schoenberg [1964]). We call such a spline a smoothing spline. Silverman [1984] showed that, asymptotically, minimizing (9.132) behaves like a variable-bandwidth kernel smoothing method, thus establishing relationships with other forms of nonparametric regression (c.f. the discussion of Silverman [1985], Friedman [1984], Cleveland [1979], Velleman [1980], Cressie [1986] and his discussion of Silverman [1985], Wegman and Wright [1983: Section 5], and Eubank [1988].

An important ingredient of (9.132) is the specification of the smoothing parameter λ. The cross-validatory choice minimizes

$$CV(\lambda) = \frac{1}{n} \sum_{i=1}^{n} [y_i - \hat{f}_\lambda^{(-i)}(x_i)]^2, \tag{9.133}$$

where $\hat{f}_\lambda^{(-i)}$ is the estimator of f obtained using the value λ and omitting (x_i, y_i) from the data set. Thus $\hat{f}_\lambda^{(-i)}(x_i)$ is the predicted value of y at x_i obtained using all the data except (x_i, y_i). A popular choice of λ minimizes a slight variant of (9.133). This method, called generalized cross validation, was introduced by Craven and Wahba [1979], who showed that it minimized the average squared error at the design points. The IMSL package (see Section 15.1) has a cubic-spline smoothing routine in which λ is obtained by generalized cross validation. An approximation which involves much less computation is given by Silverman [1985]. Cross-validatory methods are particularly in tune with viewing a smoothing spline not as estimating a model, but as producing a useful predictive equation. Figure 9.16,

Table 9.5 Maximum-Likelihood Estimates[a] of Parameters for the Model (9.134) Applied to the Data of Table 9.1. in Section 9.3.3a

	Model	
Parameter	m = 2	m = 3
$\hat{\phi}_0$	0.4556	0.4468
$\hat{\phi}_1$	0.0383	0.0466
$\hat{\phi}_2$		−0.0017
$\hat{\lambda}$	0.0186	2.1087×10^{-4}
$\hat{\sigma}$	0.0250	0.0249

[a]From Wecker and Ansley [1983].

taken from Silverman [1985], plots accelerometer readings on motorcycle helmets versus time for simulated motorcycle crashes. Figure 9.16*a* just plots the data, while Fig. 9.16*b* has a cubic smoothing spline superimposed.

Inference about f from (9.131) and (9.132) as they stand is not possible, because (9.132) is a data-smoothing criterion; it does not define a statistical model. Wahba [1978, 1983] imposes the prior distribution of f, given $\boldsymbol{\phi}$, determined by

$$f(x|\boldsymbol{\phi}) = \sum_{j=0}^{m-1} \phi_j \frac{x^j}{j!} + \frac{\sigma}{\sqrt{\lambda}} \int_0^x \frac{(x-u)^{m-1}}{(m-1)!} dW(u), \qquad (9.134)$$

where $W(u)$ is a Wiener process (Appendix C1). Furthermore, a diffuse prior was placed upon $\boldsymbol{\phi}$, i.e. $\boldsymbol{\phi} \approx N_n(\mathbf{0}, \xi \mathbf{I}_m)$ with $\xi \to \infty$. Under these conditions, Wahba [1978] found that for fixed λ the posterior mean of $f(x)$ is the smoothing spline which minimizes (9.132). An interval estimate can be obtained using the posterior variance (with λ set equal to its generalized cross-validatory estimate). A small-scale simulation bore out the efficacy of this approach. However, the reader should refer to Wahba's discussion of Silverman [1985] for some qualifications.

Wecker and Ansley [1983] interpret (9.134) as a stochastic model for $f(x)$. The realization of $f(x)$ is measured with error at each x_i-value. Under the assumption of normal errors ε_i, $(\phi, \lambda, \sigma^2)$ are estimated by maximum likelihood using Kalman filtering methods, which lead to great reductions in computation. Parameter estimates are given in Table 9.5. This method again leads to point and interval estimates for f. Wecker and Ansley [1983] have applied their techniques to the data of Example 9.7 in Section 9.3.3. The results are plotted in Fig. 9.17. Model 2 uses $m = 2$, and model 3 uses $m = 3$.

Another Bayesian approach, but with a finite-dimensional prior, has been suggested by Silverman [1985]. However, his prior is concentrated on the space of spline functions with knots at the observed x_i's. Thus the underlying model function is constrained to lie in the space of solution functions to (9.132). For comparative discussions of this and the earlier proposals see Silverman [1985] and the discussants, and Eubank [1984]. Although we have written this section in terms of distinct x_i-values (i.e. unreplicated data), tied x_i's can be handled (Silverman [1985], Wecker and Ansley [1983]). Silverman [1985] has generalized (9.132) to a weighted least-squares formulation; Silverman [1985] and Eubank [1984] discuss model diagnostics and robust parallels of (9.132).

Many examples to which smoothing splines have been applied use time as the explanatory variable. Thus, there is always a strong possibility of serially correlated errors. Several ideas for extensions in this direction are given in the discussion to Silverman [1985]. A new algorithm for spline smoothing has been given by Kohn and Ansley [1987]. For further discussion of all these issues and references to the literature, the reader is referred to Wegman and Wright [1983], Eubank [1984], and Silverman [1985]. In a new book, Eubank [1988] gives detailed attention to many of the issues raised in this section.

CHAPTER 10*

Errors-in-Variables Models

10.1 INTRODUCTION

In Section 1.4 we discussed the effects that random x-variables (regressors) have on nonlinear models. Two types of model were considered under the headings of functional relationships and structural relationships respectively. In each case an underlying relationship is assumed to exist either among mathematical or random variables, but the variables are measured with error. Such models are usually referred to as errors-in-variables models. A third variation was also alluded to in Example 1.8 [see (1.26)], in which the response could not be expressed explicitly in terms of the regressors. This type of model is called an implicit model, and it will be discussed for the case of random regressors.

A major problem with the above models is that in addition to the main parameter vector $\boldsymbol{\theta}$ of interest, there are also nuisance (incidental) parameters relating to the distribution of the random regressors, for example the mean values of the regressors. This creates complications with any asymptotic theory, because the number of parameters tends to infinity as n, the number of observations, tends to infinity. For example Neyman and Scott [1948] (see also Moran [1971]) showed that when the responses y_i do not come from identical populations, the maximum-likelihood estimates need not be consistent. Furthermore, even when they are consistent, their asymptotic variance–covariance matrix is generally not given by the inverse of the expected information matrix [c.f. (2.48)] when incidental parameters are present. The incidental parameters can also cause identifiability problems. One way of overcoming these problems is to use replication, and this extension is considered for the above models.

Associated with implicit models there is a problem which can arise when modeling, for example, chemical reactions in which some of the measurements are unobservable. This can be dealt with, theoretically, by using the implicit equations to eliminate the unobserved variables, though this can lead to extremely complex equations which are difficult to use. Alternatively, the unobserved variables can be kept in to aid interpretation and maintain simplicity, and these are estimated along with the parameters of interest. An algorithm for doing this is discussed in detail in Section 10.6.

The chapter concludes with a brief discussion of a fourth type of model alluded to in Section 1.5, which is usually referred to, in the linear case, as Berkson's model or the controlled errors-in-variables model. Here the experimenter endeavors to set the x-variables at certain targeted values but is unable to achieve them exactly because of errors in the system.

10.2 FUNCTIONAL RELATIONSHIPS: WITHOUT REPLICATION

Suppose that the mathematical variables μ and $\boldsymbol{\xi} = (\xi^{(1)}, \xi^{(2)}, \ldots, \xi^{(k)})'$ satisfy a functional relationship

$$\mu = f(\boldsymbol{\xi}; \boldsymbol{\theta}), \tag{10.1}$$

where f is single-valued with finite continuous first and second derivatives with respect to $\boldsymbol{\xi}$ and $\boldsymbol{\theta}$ in a closed bounded set. (This notation of using superscripts in the $\xi^{(j)}$ is to facilitate the extension of the theory to the replicated case later on.) The parameters $\boldsymbol{\theta}$ are often called the *structural* parameters, and $\boldsymbol{\xi}$ the *incidental* parameters. It is assumed that μ and $\boldsymbol{\xi}$ are measured with error, so that the random variables y and $x^{(1)}, x^{(2)}, \ldots, x^{(k)}$ are observed instead, where

$$y = \mu + \varepsilon \quad \text{and} \quad x^{(j)} = \xi^{(j)} + \delta^{(j)} \quad (j = 1, 2, \ldots, k). \tag{10.2}$$

Suppose that n sets of observations are made, so that y and $x^{(j)}$ now become y_i and $x_i^{(j)}$ $(i = 1, 2, \ldots, n)$. Equations (10.2) and (10.1) now become

$$y_i = \mu_i + \varepsilon_i, \qquad x_i^{(j)} = \xi_i^{(j)} + \delta_i^{(j)}$$

and

$$\mu_i = f(\boldsymbol{\xi}_i; \boldsymbol{\theta}), \qquad \text{where} \quad \boldsymbol{\xi}_i = (\xi_i^{(1)}, \xi_i^{(2)}, \ldots, \xi_i^{(k)})'.$$

The first of these three equations has the vector representation $\mathbf{y} = \boldsymbol{\mu} + \boldsymbol{\varepsilon}$. However, the second equation, for the x's, can be expressed two ways, depending on whether the elements of the matrix $[(x_i^{(j)})]$ are listed by column or row:

$$\begin{aligned}[(x_i^{(j)})] &= (\mathbf{x}^{(1)}, \mathbf{x}^{(2)}, \ldots, \mathbf{x}^{(k)}) \quad \text{(column representation)},\\ &= \begin{bmatrix} \mathbf{x}_1' \\ \mathbf{x}_2' \\ \vdots \\ \mathbf{x}_n' \end{bmatrix} \quad \text{(row representation)},\end{aligned}$$

where $\mathbf{x}^{(j)} = \boldsymbol{\xi}^{(j)} + \boldsymbol{\delta}^{(j)}$ for $j = 1, 2, \ldots, k$. Using the column listing, as in Dolby and

Freeman [1975], we define

$$\begin{aligned} \mathbf{x} &= (\mathbf{x}^{(1)\prime}, \mathbf{x}^{(2)\prime}, \ldots, \mathbf{x}^{(k)\prime})', \\ \boldsymbol{\xi} &= (\boldsymbol{\xi}^{(1)\prime}, \boldsymbol{\xi}^{(2)\prime}, \ldots, \boldsymbol{\xi}^{(k)\prime})', \end{aligned} \tag{10.3}$$

and the augmented error vector

$$\mathbf{e} = (\boldsymbol{\delta}^{(1)\prime}, \boldsymbol{\delta}^{(2)\prime}, \ldots, \boldsymbol{\delta}^{(k)\prime}, \boldsymbol{\varepsilon}')' \tag{10.4}$$

$$= (\boldsymbol{\delta}', \boldsymbol{\varepsilon}')', \tag{10.5}$$

with $m = n(k+1)$ components. This leads to the model

$$\mathbf{y} = \boldsymbol{\mu} + \boldsymbol{\varepsilon}, \qquad \mathbf{x} = \boldsymbol{\xi} + \boldsymbol{\delta}, \quad \text{and} \quad \boldsymbol{\mu} = \mathbf{f}(\boldsymbol{\xi}; \boldsymbol{\theta}). \tag{10.6}$$

It is further assumed that $\mathbf{e} \sim N_m(\mathbf{0}, \boldsymbol{\Sigma})$, where $\boldsymbol{\Sigma} = \sigma^2 \mathbf{V}$, $\mathbf{V}$ is a known $m \times m$ matrix, and σ^2 is an unknown constant. In addition to the p structural parameters $\boldsymbol{\theta}$ we now have nk incidental parameters $\boldsymbol{\xi}^{(j)}$ $(j = 1, 2, \ldots, k)$ to be estimated. Without replication, we find from related work on linear models that we have to assume $\mathbf{V}$ is known to avoid inconsistencies and irregularities in the maximum-likelihood equations (Lindley [1947], Dolby and Lipton [1972: 122], Dolby [1976b: 43]). If $\mathbf{V}$ is diagonal, then this is equivalent to assuming that we know the ratios of the variances, a somewhat unrealistic constraint often used for estimating linear functional relationships.

To find the maximum-likelihood estimates of $\boldsymbol{\theta}, \boldsymbol{\xi}$, and σ^2 we begin with the log-likelihood function

$$L(\boldsymbol{\theta}, \boldsymbol{\xi}, \sigma^2) = \text{const} - \frac{n}{2} \log \sigma^2 - \frac{\sigma^2}{2} \mathbf{e}' \mathbf{V} \mathbf{e}. \tag{10.7}$$

For the case $k = 1$, Dolby [1972, 1976a] derived the maximum-likelihood equations and gave an iterative Fisher "scoring" method for their solution. Egerton and Laycock [1979] discussed the straightforward extension to general k, but with a row ordering applied to $[(x_i^{(j)})]$ instead of the column ordering used above. Since $\mathbf{x}_i = \boldsymbol{\xi}_i + \boldsymbol{\delta}_i$, they defined $\mathbf{x} = (\mathbf{x}_1', \mathbf{x}_2', \ldots, \mathbf{x}_n')'$, with a similar ordering applied to $\boldsymbol{\delta}$ [c.f. (10.5)]. However, since we discuss the replicated case in Section 10.3 using the approach of Dolby and Freeman [1975], we shall use their notation as described by (10.3) to (10.6). We first define the following matrices:

$$\mathbf{V}^{-1} = \begin{pmatrix} \mathbf{F} & \mathbf{G} \\ \mathbf{G}' & \mathbf{H} \end{pmatrix}^{-1} = \begin{pmatrix} \mathbf{A} & \mathbf{B} \\ \mathbf{B}' & \mathbf{C} \end{pmatrix} \begin{matrix} \}\, nk \\ \}\, n \end{matrix}, \tag{10.8}$$

$$\mathbf{N} = \left[\left(\frac{\partial \mu_i}{\partial \theta_r} \right) \right] \qquad (i = 1, 2, \ldots, n, \quad r = 1, 2, \ldots, p), \tag{10.9}$$

$$\mathbf{M} = \begin{pmatrix} \mathbf{M}_1 \\ \mathbf{M}_2 \\ \vdots \\ \mathbf{M}_n \end{pmatrix}, \tag{10.10}$$

$$\mathbf{M}_j = \text{diag}\left(\frac{\partial \mu_1}{\partial \xi_1^{(j)}}, \frac{\partial \mu_2}{\partial \xi_2^{(j)}}, \dots, \frac{\partial \mu_n}{\partial \xi_n^{(j)}} \right), \qquad j = 1, 2, \dots, k, \tag{10.11}$$

$$\mathbf{Q} = \mathbf{A} + \mathbf{MB}' + \mathbf{BM}' + \mathbf{MCM}', \tag{10.12}$$

where **A**, **B**, and **C** can be obtained from **V** using Appendix A3.1. Then, from (10.5),

$$\mathbf{e}'\mathbf{V}^{-1}\mathbf{e} = (\mathbf{x} - \boldsymbol{\xi})'\mathbf{A}(\mathbf{x} - \boldsymbol{\xi}) + 2(\mathbf{x} - \boldsymbol{\xi})'\mathbf{B}[\mathbf{y} - \mathbf{f}(\boldsymbol{\xi}; \boldsymbol{\theta})] \\ + [\mathbf{y} - \mathbf{f}(\boldsymbol{\xi}; \boldsymbol{\theta})]'\mathbf{C}[\mathbf{y} - \mathbf{f}(\boldsymbol{\xi}; \boldsymbol{\theta})]. \tag{10.13}$$

The partial derivatives of $L(\boldsymbol{\theta}, \boldsymbol{\xi}, \sigma^2)$ in (10.7) with respect to $\boldsymbol{\theta}$, $\boldsymbol{\xi}$, and σ^2 yield the following likelihood equations respectively (c.f. Appendix A10.5):

$$\mathbf{N}'\{\mathbf{B}'(\mathbf{x} - \hat{\boldsymbol{\xi}}) + \mathbf{C}(\mathbf{y} - \hat{\boldsymbol{\mu}})\} = \mathbf{0}, \tag{10.14}$$

$$(\mathbf{A} + \mathbf{MB}')(\mathbf{x} - \hat{\boldsymbol{\xi}}) + (\mathbf{B} + \mathbf{MC})(\mathbf{y} - \hat{\boldsymbol{\mu}}) = \mathbf{0}, \tag{10.15}$$

and

$$\hat{\sigma}^2 = \frac{1}{n(k+1)} \hat{\mathbf{e}}'\mathbf{V}^{-1}\hat{\mathbf{e}}, \tag{10.16}$$

where $\hat{\mu}_i = f(\hat{\boldsymbol{\xi}}_i; \hat{\boldsymbol{\theta}})$ and $\hat{\mathbf{e}}' = (\mathbf{x}' - \hat{\boldsymbol{\xi}}', \mathbf{y}' - \hat{\boldsymbol{\mu}}')$. Using the method of scoring (Appendix A14), the following iterative scheme for solving (10.14) and (10.15) can be used (Dolby [1976a]):

$$\begin{pmatrix} \boldsymbol{\theta}_{(a+1)} \\ \boldsymbol{\xi}_{(a+1)} \end{pmatrix} = \begin{pmatrix} \boldsymbol{\theta}_{(a)} \\ \boldsymbol{\xi}_{(a)} \end{pmatrix} + \mathbf{W}_{(a)}^{-1} \begin{bmatrix} \mathbf{N}'\mathbf{B}' & \mathbf{N}'\mathbf{C} \\ \mathbf{A} + \mathbf{MB}' & \mathbf{B} + \mathbf{MC} \end{bmatrix}_{(a)} \begin{bmatrix} \mathbf{x} - \boldsymbol{\xi}_{(a)} \\ \mathbf{y} - \boldsymbol{\mu}_{(a)} \end{bmatrix}. \tag{10.17}$$

Here $\mathbf{W}_{(a)}$, the (expected) information matrix $\mathbf{W}$ for $\boldsymbol{\theta}$ and $\boldsymbol{\xi}$ evaluated at the ath approximations $\boldsymbol{\theta}_{(a)}$ and $\boldsymbol{\xi}_{(a)}$ of the maximum-likelihood estimates, is given by [c.f. (10.29) below]

$$\mathbf{W} = \begin{bmatrix} \mathbf{N}'\mathbf{CN} & \mathbf{N}'(\mathbf{B}' + \mathbf{CM}') \\ (\mathbf{B} + \mathbf{MC})\mathbf{N} & \mathbf{Q} \end{bmatrix}. \tag{10.18}$$

Also Dolby [1972] assumed (incorrectly, as we shall see below) that, asymptotically,

$$\mathscr{D}\begin{pmatrix} \hat{\boldsymbol{\theta}} \\ \hat{\boldsymbol{\xi}} \end{pmatrix} = \sigma^2 \mathbf{W}^{-1}, \tag{10.19}$$

and [c.f. (10.31) with $R = 1$]

$$\begin{aligned} \mathscr{D}[\hat{\boldsymbol{\theta}}] &= \sigma^2[\mathbf{N}'\{\mathbf{C} - (\mathbf{B}' + \mathbf{C}\mathbf{M}')\mathbf{Q}^{-1}(\mathbf{B} + \mathbf{M}\mathbf{C})\}\mathbf{N}]^{-1} \\ &= \sigma^2[\mathbf{N}'\boldsymbol{\psi}\mathbf{N}]^{-1}, \end{aligned} \tag{10.20}$$

say. Although (10.19) and (10.20) are generally invalid for the above model the same equations hold (apart from division by the number of replicates) for the replicated model discussed in the next section. That is why (10.19) is valid for the limiting process $\sigma^2 \to 0$ (Johansen [1983; 1743]).

Unfortunately there are some problems associated with the above analysis. To begin with, the maximum-likelihood estimator $\hat{\sigma}^2$ of (10.16) is inconsistent. This result is well known for the linear case (Kendall and Stuart [1973: 403]), and seems to be a typical problem where there are incidental parameters to be estimated (Moran [1971: 245], Johansen [1983: 166]). However, by adapting the argument in Kendall and Stuart [1973], Egerton and Laycock [1979] proposed using the estimator

$$\tilde{\sigma}^2 = \frac{1}{n-p} \hat{\mathbf{e}}'\mathbf{V}^{-1}\hat{\mathbf{e}}, \tag{10.21}$$

which is effectively our usual s^2 but with $\boldsymbol{\xi}$ estimated by $\hat{\boldsymbol{\xi}}$. The estimator $\tilde{\sigma}^2$ will be a consistent estimator if the other parameter estimates are consistent.

A second problem is that, even with an adjustable step length, the above iterative scheme may lead to convergence onto the wrong estimates of the parameters. Egerton and Laycock [1979] suggest that, without recourse to further calculations, the obvious starting values $(\boldsymbol{\xi}_0, \boldsymbol{\theta}_0)$ to choose are $\boldsymbol{\xi}_0 = \mathbf{x}$ along with $\boldsymbol{\theta}_0 = \hat{\boldsymbol{\theta}}(\boldsymbol{\xi}_0)$, when this is simple to evaluate. However, they give the polynomial example $\mu = \theta\xi(1 - \xi^2)$ where this starting choice does not lead to convergence to the maximum-likelihood estimates. Fortunately, with polynomials of this form, namely

$$f(\boldsymbol{\xi}; \boldsymbol{\theta}) = \theta_1 P_1(\boldsymbol{\xi}) + \theta_2 P_2(\boldsymbol{\xi}) + \cdots + \theta_p P_p(\boldsymbol{\xi}), \tag{10.22}$$

where $P_j(\boldsymbol{\xi})$ is a polynomial in $\boldsymbol{\xi}$, they show that a better procedure is available. Starting with $\boldsymbol{\xi} = \boldsymbol{\xi}_0$, it is seen that the likelihood equations for $\hat{\boldsymbol{\theta}}(\boldsymbol{\xi}_0)$ are the ordinary weighted least-squares equations, the model (10.22) now being linear in $\boldsymbol{\theta}$. A new estimate of $\boldsymbol{\xi}$ is then found by solving a sequence of polynomial equations using a standard package, and the details are given by Egerton and Laycock [1979]. For the general nonlinear model, however, a more efficient and stable algorithm for finding $\hat{\boldsymbol{\theta}}$ and the $\hat{\xi}_i$ is given by Schwetlick and Tiller [1985]. Although this algorithm, described in Section 11.6, was developed for multiresponse models, it can be applied to the above by simply setting $d = 1$, d being the dimension of the response vector. The basis of the algorithm is the Gauss–Newton method, but with a Marquardt modification for just the $\boldsymbol{\theta}$ part of the iteration.

A third difficulty arises over the use of maximum-likelihood estimators in the presence of incidental parameters. We noted at the beginning of this chapter that even if the maximum-likelihood estimators are consistent, their asymptotic variance–covariance matrix is not generally given by the inverse of the expected information matrix. Patefield [1977, 1978] showed this to be the case for a straight-line functional relationship and gave an expression for the correct asymptotic variance–covariance matrix of the slope and intercept estimates for both uncorrelated and correlated errors. He also showed that a consistent estimator of this matrix is not obtained by replacing parameters by their maximum-likelihood estimators. The same problems will certainly arise with nonlinear models, so that, as already noted above, (10.19) and (10.20) are no longer true and will usually lead to underestimates of the var$[\hat{\theta}_r]$. However, the above problems tend to disappear when the model is sufficiently replicated. This is because in the usual asymptotic maximum-likelihood theory it is the number of replicates that goes to infinity while the total number of unknown parameters remains fixed.

Dolby [1972] also introduced a generalized least-squares technique similar in spirit to the one proposed by Sprent [1969]. Under normality, Dolby showed that his technique is approximately equivalent to maximizing the so-called relative likelihood function of $\boldsymbol{\theta}$ (Kalbfleisch and Sprott [1970]), the latter being proportional to the concentrated likelihood function (see Section 2.2.3). If $z_i = y_i - f(x_i; \boldsymbol{\theta})$ and $\boldsymbol{\psi}^*$ is $\boldsymbol{\psi}$ of (10.20) with $\mathbf{M}$ evaluated at $\boldsymbol{\xi} = \mathbf{x}$, then his procedure is to minimize $\mathbf{z}'\boldsymbol{\psi}^*\mathbf{z}$ with respect to $\boldsymbol{\theta}$. Dolby also gave two other methods for estimating $\mathbf{M}$. He finally showed that, for straight-line functional relationships, his generalized least-squares method, the method of maximum relative likelihood, and maximum likelihood are all exactly equivalent under normality, thus extending similar results obtained by Sprent [1969]. The properties of Sprent's generalized least-squares estimates for the straight line are discussed further by Mak [1983].

10.3 FUNCTIONAL RELATIONSHIPS: WITH REPLICATION

The model considered above was $\mathbf{y} = \boldsymbol{\mu} + \boldsymbol{\varepsilon}$ and $\mathbf{x} = \boldsymbol{\xi} + \boldsymbol{\delta}$, where $\mu_i = f(\xi_i; \boldsymbol{\theta})$, and $\mathbf{x} = (\mathbf{x}^{(1)\prime}, \mathbf{x}^{(2)\prime}, \ldots, \mathbf{x}^{(k)\prime})'$ represents n observations on each of k regressors. It was assumed that $\mathbf{e} = (\boldsymbol{\delta}', \boldsymbol{\varepsilon}')' \sim N_m(\mathbf{0}, \boldsymbol{\Sigma})$, where $m = n(k+1)$, $\boldsymbol{\Sigma} = \sigma^2\mathbf{V}$, and $\mathbf{V}$ is known. The assumption of known $\mathbf{V}$ is often unrealistic, but it was required for a satisfactory solution of the maximum-likelihood equations (c.f. Dolby and Lipton [1972: Equation (3)]) and the attainment of a genuine maximum.

However, no assumption about $\boldsymbol{\Sigma}$ is needed if the model is replicated R times $(R \geq m)$, namely

$$\mathbf{y}_{(r)} = \boldsymbol{\mu} + \boldsymbol{\varepsilon}_{(r)}, \quad \mathbf{x}_{(r)} = \boldsymbol{\xi} + \boldsymbol{\delta}_{(r)} \qquad (r = 1, 2, \ldots, R),$$

where $\mathbf{x}_{(r)} = (\mathbf{x}_r^{(1)\prime}, \mathbf{x}_r^{(2)\prime}, \ldots, \mathbf{x}_r^{(k)\prime})'$. If $\mathbf{e}_r = (\boldsymbol{\delta}'_{(r)}, \boldsymbol{\varepsilon}'_{(r)})'$, then the log-likelihood function is given by

$$L(\boldsymbol{\theta}, \boldsymbol{\xi}, \boldsymbol{\Sigma}^{-1}) = \text{const} + \tfrac{1}{2}R \log|\boldsymbol{\Sigma}^{-1}| - \frac{1}{2}\sum_{r=1}^{R} \mathbf{e}_r'\boldsymbol{\Sigma}^{-1}\mathbf{e}_r \tag{10.23}$$

$$= \text{const} + \tfrac{1}{2}R \log|\boldsymbol{\Sigma}^{-1}| - \frac{1}{2}\text{tr}\left[\boldsymbol{\Sigma}^{-1} \sum_{r=1}^{R} \mathbf{e}_r\mathbf{e}_r'\right]. \tag{10.24}$$

Corresponding to the δ and ε parts of $\mathbf{e}_r$, the matrices $\boldsymbol{\Sigma}$ and $\boldsymbol{\Sigma}^{-1}$ can be partitioned in a conformable manner as in (10.8). For convenience we use the same notation as before, but with $\mathbf{V}$ replaced by $\boldsymbol{\Sigma}$. Differentiating (10.24) with respect to $\boldsymbol{\theta}$, $\boldsymbol{\xi}$, and $\boldsymbol{\Sigma}^{-1}$ leads to the maximum-likelihood equations [c.f. (10.9) to (10.12) for definitions of matrices]

$$\hat{\mathbf{N}}'[\hat{\mathbf{B}}'(\bar{\mathbf{x}} - \hat{\boldsymbol{\xi}}) + \hat{\mathbf{C}}(\bar{\mathbf{y}} - \hat{\boldsymbol{\mu}})] = \mathbf{0}, \tag{10.25}$$

$$(\hat{\mathbf{A}} + \hat{\mathbf{M}}\hat{\mathbf{B}}')(\bar{\mathbf{x}} - \hat{\boldsymbol{\xi}}) + (\hat{\mathbf{B}} + \hat{\mathbf{M}}\hat{\mathbf{C}})(\bar{\mathbf{y}} - \hat{\boldsymbol{\mu}}) = \mathbf{0}, \tag{10.26}$$

and

$$\hat{\boldsymbol{\Sigma}} = \frac{1}{R}\sum_{r=1}^{R} \begin{pmatrix} \mathbf{x}_r - \hat{\boldsymbol{\xi}} \\ \mathbf{y}_r - \hat{\boldsymbol{\mu}} \end{pmatrix} \begin{pmatrix} \mathbf{x}_r - \hat{\boldsymbol{\xi}} \\ \mathbf{y}_r - \hat{\boldsymbol{\mu}} \end{pmatrix}' \tag{10.27}$$

$$= \begin{pmatrix} \hat{\mathbf{A}} & \hat{\mathbf{B}} \\ \hat{\mathbf{B}}' & \hat{\mathbf{C}} \end{pmatrix}^{-1},$$

where $\bar{\mathbf{x}} = \sum_r \mathbf{x}_{(r)}/R$, and on. To use the scoring method for solving the above equations (see Appendix A14) we require the (expected) information matrix

$$\mathbf{I}(\boldsymbol{\phi}) = -\mathscr{E}\left[\frac{\partial^2 L}{\partial \boldsymbol{\phi}\, \partial \boldsymbol{\phi}'}\right]$$

$$= \mathscr{E}\left[\left(\frac{\partial L}{\partial \boldsymbol{\phi}}\right)\left(\frac{\partial L}{\partial \boldsymbol{\phi}}\right)'\right] \tag{10.28}$$

for $\boldsymbol{\phi} = [\boldsymbol{\theta}', \boldsymbol{\xi}', (\text{vech}\, \boldsymbol{\Sigma})']'$, where vech $\boldsymbol{\Sigma}$ is a column vector of the distinct elements of $\boldsymbol{\Sigma}$, defined in Appendix A13.2. Differentiating (10.23) with respect to $\boldsymbol{\theta}$, we have

$$\frac{\partial L}{\partial \boldsymbol{\theta}} = -\sum_{r=1}^{R} (\mathbf{0}, \mathbf{N}')\boldsymbol{\Sigma}^{-1}\mathbf{e}_r$$

[which leads to (10.25)]. Hence, from $\mathscr{E}[\sum_r \mathbf{e}_r \mathbf{e}_r'] = R\boldsymbol{\Sigma}$ we have

$$\mathscr{E}\left[\left(\frac{\partial L}{\partial \boldsymbol{\theta}}\right)\left(\frac{\partial L}{\partial \boldsymbol{\theta}}\right)'\right] = \mathscr{E}\left[\sum_{r=1}^{R}\sum_{s=1}^{R} (\mathbf{0}, \mathbf{N}')\boldsymbol{\Sigma}^{-1}\mathbf{e}_r\mathbf{e}_s'\boldsymbol{\Sigma}^{-1}(\mathbf{0}, \mathbf{N}')'\right]$$
$$= R(\mathbf{0}, \mathbf{N}')\boldsymbol{\Sigma}^{-1}(\mathbf{0}, \mathbf{N}')'$$
$$= R\mathbf{N}'\mathbf{C}\mathbf{N}.$$

We similarly find that

$$\mathscr{E}\left[\left(\frac{\partial L}{\partial \boldsymbol{\theta}}\right)\left(\frac{\partial L}{\partial \boldsymbol{\xi}}\right)'\right] = R(\mathbf{0}, \mathbf{N}')\boldsymbol{\Sigma}^{-1}(\mathbf{I}, \mathbf{M})'$$
$$= R\mathbf{N}'(\mathbf{B}' + \mathbf{C}\mathbf{M}') \qquad [\text{by (10.8) with } \mathbf{V} = \boldsymbol{\Sigma}]$$

and

$$\mathscr{E}\left[\left(\frac{\partial L}{\partial \boldsymbol{\xi}}\right)\left(\frac{\partial L}{\partial \boldsymbol{\xi}}\right)'\right] = R(\mathbf{I}, \mathbf{M})\boldsymbol{\Sigma}^{-1}(\mathbf{I}, \mathbf{M})'$$
$$= R\mathbf{Q} \qquad [\text{by (10.12)}].$$

Turning our attention to $\boldsymbol{\Sigma}$ we find that $E[\partial^2 L/\partial\sigma_{ij}\,\partial v] = 0$, where v is any element of $\boldsymbol{\theta}$ or $\boldsymbol{\xi}$. For example

$$\frac{\partial^2 \mathbf{e}_r'\boldsymbol{\Sigma}^{-1}\mathbf{e}_r}{\partial\sigma_{ij}\partial\boldsymbol{\theta}} = \frac{\partial}{\partial\sigma_{ij}}\left[(\mathbf{0}, \mathbf{N}')\boldsymbol{\Sigma}^{-1}\mathbf{e}_r\right]$$
$$= (\mathbf{0}, \mathbf{N}')\frac{\partial\boldsymbol{\Sigma}^{-1}}{\partial\sigma_{ij}}\mathbf{e}_r,$$

which has zero expectation, as $\mathscr{E}[\mathbf{e}_r] = \mathbf{0}$. Combining the above results, the (expected) information matrix (10.28) takes the form

$$\mathbf{I}(\boldsymbol{\phi}) = R\begin{pmatrix} \mathbf{N}'\mathbf{C}\mathbf{N} & \mathbf{N}'(\mathbf{B}' + \mathbf{C}\mathbf{M}') & \mathbf{0} \\ (\mathbf{B} + \mathbf{M}\mathbf{C})\mathbf{N} & \mathbf{Q} & \mathbf{0} \\ \mathbf{0} & \mathbf{0} & \boldsymbol{\Delta} \end{pmatrix} \tag{10.29}$$

$$= R\begin{pmatrix} \mathbf{W} & \mathbf{0} \\ \mathbf{0} & \boldsymbol{\Delta} \end{pmatrix}, \tag{10.30}$$

say. For large R, the asymptotic variance–covariance matrix of the maximum-likelihood estimates $\hat{\boldsymbol{\phi}}$ is given by

$$\mathbf{I}^{-1}(\boldsymbol{\phi}) = \frac{1}{R}\begin{pmatrix} \mathbf{W}^{-1} & \mathbf{0} \\ \mathbf{0} & \boldsymbol{\Delta}^{-1} \end{pmatrix},$$

where

$$\mathbf{W}^{-1} = \begin{pmatrix} \mathbf{W}_{11} & \mathbf{W}_{12} \\ \mathbf{W}_{21} & \mathbf{W}_{22} \end{pmatrix}^{-1} = \begin{pmatrix} \mathbf{T}_{11} & \mathbf{T}_{12} \\ \mathbf{T}_{21} & \mathbf{T}_{22} \end{pmatrix}.$$

Hence using Appendix A3.1(2),

$$\begin{aligned} \mathscr{D}[\hat{\boldsymbol{\theta}}] &\approx \frac{1}{R}\mathbf{T}_{11} \\ &= \frac{1}{R}(\mathbf{W}_{11} - \mathbf{W}_{12}\mathbf{W}_{22}^{-1}\mathbf{W}_{21})^{-1} \\ &= \frac{1}{R}[\mathbf{N}'\{\mathbf{C} - (\mathbf{B}' + \mathbf{C}\mathbf{M}')\mathbf{Q}^{-1}(\mathbf{B} + \mathbf{M}\mathbf{C})\}\mathbf{N}]^{-1} \\ &= \frac{1}{R}[\mathbf{N}'\boldsymbol{\psi}\mathbf{N}]^{-1}, \end{aligned} \tag{10.31}$$

say, where $\boldsymbol{\psi}$ is expressed in terms of the block matrices making up $\boldsymbol{\Sigma}^{-1}$ [c.f. (10.8)]. We now show that $\boldsymbol{\psi}$ can be expressed in terms of $\boldsymbol{\Sigma}$, namely

$$\begin{aligned} \boldsymbol{\psi} &= \boldsymbol{\Theta}^{-1} \\ &= (\mathbf{H} - \mathbf{M}'\mathbf{G} - \mathbf{G}'\mathbf{M} + \mathbf{M}'\mathbf{F}\mathbf{M})^{-1}, \end{aligned} \tag{10.32}$$

where

$$\boldsymbol{\Sigma} = \begin{pmatrix} \mathbf{F} & \mathbf{G} \\ \mathbf{G}' & \mathbf{H} \end{pmatrix}.$$

To prove (10.32) we observe that

$$\begin{pmatrix} \mathbf{I}_{kn} & \mathbf{M} \\ \mathbf{0} & \mathbf{I}_n \end{pmatrix} \boldsymbol{\Sigma}^{-1} \begin{pmatrix} \mathbf{I}_{kn} & \mathbf{0} \\ \mathbf{M}' & \mathbf{I}_n \end{pmatrix} = \begin{pmatrix} \mathbf{Q} & \mathbf{B} + \mathbf{M}\mathbf{C} \\ \mathbf{B}' + \mathbf{C}\mathbf{M}' & \mathbf{C} \end{pmatrix} \tag{10.33}$$

and

$$\begin{pmatrix} \mathbf{I}_{kn} & \mathbf{0} \\ -\mathbf{M}' & \mathbf{I}_n \end{pmatrix} \boldsymbol{\Sigma} \begin{pmatrix} \mathbf{I}_{kn} & -\mathbf{M} \\ \mathbf{0} & \mathbf{I}_n \end{pmatrix} = \begin{pmatrix} \mathbf{F} & \mathbf{G} - \mathbf{F}\mathbf{M} \\ \mathbf{G}' - \mathbf{M}'\mathbf{F} & \boldsymbol{\Theta}^{-1} \end{pmatrix}. \tag{10.34}$$

Since the left-hand sides of (10.33) and (10.34) are inverses of each other, then so are the right-hand sides. Equation (10.32) now follows from the inverse of (10.33) using Appendix A3.1(1).

With slight changes in notation, the above theory is based on Dolby and Freeman [1975] who proposed an iterative method based on scoring, namely if

$\mathbf{v}' = (\boldsymbol{\theta}', \boldsymbol{\xi}')$ then from the left-hand sides of (10.25) and (10.26) we have

$$\begin{aligned}\delta \mathbf{v}_{(a)} &= \mathbf{v}_{(a+1)} - \mathbf{v}_{(a)} \\ &= \frac{1}{R} \mathbf{W}_{(a)}^{-1} \begin{pmatrix} \mathbf{N}'\mathbf{B}' & \mathbf{N}'\mathbf{C} \\ \mathbf{A} + \mathbf{M}\mathbf{B}' & \mathbf{B} + \mathbf{M}\mathbf{C} \end{pmatrix}_{(a)} \begin{pmatrix} \bar{\mathbf{x}} - \boldsymbol{\xi}_{(a)} \\ \bar{\mathbf{y}} - \boldsymbol{\mu}_{(a)} \end{pmatrix} \\ &= \frac{1}{R} \mathbf{W}_{(a)}^{-1} \mathbf{u}_{(a)}, \end{aligned} \qquad (10.35)$$

say. The steps are as follows:

1. Use $\bar{\mathbf{x}}$ $(= \boldsymbol{\xi}_{(0)})$ as an initial estimate of $\boldsymbol{\xi}$.
2. Choose an initial approximation $\boldsymbol{\theta}_{(0)}$ of $\boldsymbol{\theta}$.
3. Using these initial estimates ($\mathbf{v}_{(0)}$, say), calculate the elements of $\boldsymbol{\mu}_{(0)}$ from $\mu_i = f(\boldsymbol{\xi}_i; \boldsymbol{\theta})$.
4. "Invert" the estimate of $\boldsymbol{\Sigma}$ from (10.27) using $\boldsymbol{\xi}_{(0)}$ and $\boldsymbol{\mu}_{(0)}$, thus giving estimates of **A**, **B**, and **C**, the block matrices of $\boldsymbol{\Sigma}^{-1}$. (In practice one would use a Cholesky decomposition method).
5. Obtain an estimate of **M** at $\mathbf{v}_{(0)}$ [c.f. (10.11)], and calculate **u** of (10.35).
6. Compute **W** of (10.30), and solve $\mathbf{W}\mathbf{k} = R^{-1}\mathbf{u}$ to get $\mathbf{k} = \delta\mathbf{v}_{(0)}$. Add $\delta\mathbf{v}_{(0)}$ to the initial estimate to obtain the next approximation $\mathbf{v}_{(1)}$ $(= \mathbf{v}_{(0)} + \delta\mathbf{v}_{(0)})$. Repeat from step 3.

The above method can also be used for the special case of the linear model

$$\mu = \theta_0 + \theta_1 \xi_1 + \cdots + \theta_{p-1} \xi_{p-1} = (1, \boldsymbol{\xi}')\boldsymbol{\theta},$$

with $k = p - 1$. Dolby and Freeman [1975] showed that the maximum-likelihood method is equivalent to a generalized least-squares procedure, thus linking up the least-squares methods of Brown [1957] and Sprent [1966] with the maximum-likelihood methods of Barnett [1970], Villegas [1961], and Dolby and Lipton [1972]. The latter paper is a special case of the above general theory ($k = 2$), but with a different ordering for the elements of **e**, namely [c.f. (10.4)]

$$\mathbf{e} = (\delta_1^{(1)}, \varepsilon_1, \delta_2^{(1)}, \varepsilon_2, \ldots, \delta_n^{(1)}, \varepsilon_n)'.$$

When $\boldsymbol{\Sigma}$ is completely unknown, the restriction $R \geq n(k+1)$ may lead to a large number of replications. However, the number of replicates may be reduced if $\boldsymbol{\Sigma}$ is partially known, the simplest case being when $\boldsymbol{\Sigma}$ is diagonal. The most common situation is when the errors associated with a given experimental unit are correlated but are independent of those from other units. Recalling that $\delta_i^{(j)}$ is the ith element of $\boldsymbol{\delta}^{(j)}$, this means that the vectors $(\delta_i^{(1)}, \delta_i^{(2)}, \ldots, \delta_i^{(k)}, \varepsilon_i)'$ are mutually independent for $i = 1, 2, \ldots, n$ and have dispersion matrices $\boldsymbol{\Sigma}_i^*$, say. Now if

$\Sigma_i^* = \Sigma^* \; (i = 1, 2, \ldots, n)$, then (Appendix A13.1)

$$\Sigma = \Sigma^* \otimes \mathbf{I}_n.$$

In this case Dolby and Freeman [1975] showed that the maximum-likelihood estimate of Σ is $\hat{\Sigma}^* \otimes \mathbf{I}_n$, where

$$\hat{\Sigma}^* = \frac{1}{nR} \sum_{r=1}^{R} \sum_{i=1}^{n} \mathbf{w}_{ir} \mathbf{w}_{ir}',$$

$$\mathbf{w}_{ir} = (x_{ir}^{(1)} - \hat{\xi}_i^{(1)}, x_{ir}^{(2)} - \hat{\xi}_i^{(2)}, \ldots, x_{ir}^{(k)} - \hat{\xi}_i^{(k)}, y_i - \hat{\mu}_i)',$$

and $x_{ir}^{(j)}$ is the ith element of $\mathbf{x}_r^{(j)}$.

Finally we note there is a general formulation of the nonlinear problem given by Wolter and Fuller [1982]. The main application of this formulation appears to be the replicated model, and estimates are obtained with useful properties. However, Σ is assumed known.

10.4 IMPLICIT FUNCTIONAL RELATIONSHIPS: WITHOUT REPLICATION

In Example 1.8 of Section 1.2 we saw that the response y could not be expressed explicitly as a function of x, so that the relationship between y and x was described by an implicit function. Usually the relationship is between the true values of the variables, but the variables are measured with error. A very general formulation of this model is given by Britt and Luecke [1973], and the following discussion is based on their paper, with some changes in notation. Suppose the $m \times 1$ vector of expected ("true") values $\boldsymbol{\zeta}$ of all the variables satisfies the implicit functional relationship

$$\mathbf{g}(\boldsymbol{\zeta}; \boldsymbol{\theta}) = \mathbf{0}, \tag{10.36}$$

where $\mathbf{g}$ is a $q \times 1$ vector of functions, $\boldsymbol{\theta}$ is our usual $p \times 1$ vector of unknown parameters, and $p < q \leq m$. The measurements on $\boldsymbol{\zeta}$ contain random experimental errors, so that we actually observe

$$\mathbf{z} = \boldsymbol{\zeta} + \boldsymbol{\varepsilon}. \tag{10.37}$$

Here $\mathbf{z}$ includes both x and y variables, as the distinction between response and regressor variables is unnecessary here. It is further assumed that $\boldsymbol{\varepsilon} \sim N_m(\mathbf{0}, \sigma^2 \mathbf{V}_0)$, where $\mathbf{V}_0$ is a *known* positive definite matrix and σ^2 is an unknown constant. The matrix $\mathbf{V}_0$ needs to be known for much the same reasons as given for the explicit model of Section 10.2. In fact, by taking $g(\boldsymbol{\zeta}; \boldsymbol{\theta}) = \mu - f(\boldsymbol{\xi}; \boldsymbol{\theta})$ in (10.1) and $\mathbf{z} = (y, \mathbf{x}')'$, the explicit model is clearly a special case of the above implicit model.

Example 10.1 A common special case of (10.37) is when we have a single nonlinear relationship $g(\xi, \eta; \theta) = 0$ between the expected values ξ and η of the random variables x and y, respectively. For n pairs of observations we have

$$\begin{aligned} g_1(\boldsymbol{\zeta}; \boldsymbol{\theta}) &= g(\xi_1, \eta_1; \boldsymbol{\theta}) = 0, \\ g_2(\boldsymbol{\zeta}; \boldsymbol{\theta}) &= g(\xi_2, \eta_2; \boldsymbol{\theta}) = 0, \\ &\vdots \\ g_n(\boldsymbol{\zeta}; \boldsymbol{\theta}) &= g(\xi_n, \eta_n; \boldsymbol{\theta}) = 0, \end{aligned} \tag{10.38}$$

i.e. $\mathbf{g}(\boldsymbol{\zeta}; \boldsymbol{\theta}) = \mathbf{0}$, where $\boldsymbol{\zeta} = (\xi_1, \eta_1, \xi_2, \eta_2, \ldots, \xi_n, \eta_n)'$. Also $\mathbf{z}' = (x_1, y_1, x_2, y_2, \ldots, x_n, y_n)$, $m = 2n$, and $q = n$. Usually the pairs (x_i, y_i) are mutually independent, so that $\mathbf{z} \sim N_{2n}(\boldsymbol{\zeta}, \sigma^2 \mathbf{V}_0)$, where

$$\mathbf{V}_0 = \text{diag}(\mathbf{V}, \mathbf{V}, \ldots, \mathbf{V}) = \mathbf{I}_n \otimes \mathbf{V},$$

and $\mathbf{V}$ is a *known* 2×2 positive definite matrix. In some cases x_i and y_i are also independent, so that $\mathbf{V}$ is diagonal.

The above special case (10.38) can be extended by allowing more than one function of g and more than two variables ξ and η. For example we could replace ξ_i by $\boldsymbol{\xi}_i$ and x_i by $\mathbf{x}_i$. Alternatively, we could have the double response model

$$\begin{aligned} g_1(\boldsymbol{\zeta}; \boldsymbol{\theta}) &= g_{(1)}(\xi_1, \eta_1; \boldsymbol{\theta}), \\ g_2(\boldsymbol{\zeta}; \boldsymbol{\theta}) &= g_{(2)}(\xi_1, \nu_1; \boldsymbol{\theta}), \\ g_3(\boldsymbol{\zeta}; \boldsymbol{\theta}) &= g_{(1)}(\xi_2, \eta_2; \boldsymbol{\theta}), \\ g_4(\boldsymbol{\zeta}; \boldsymbol{\theta}) &= g_{(2)}(\xi_2, \nu_2; \boldsymbol{\theta}), \\ &\vdots \end{aligned} \tag{10.39}$$

where for every ξ there are two responses η and ν, each satisfying its own implicit equation: y_i is then replaced by a bivariate vector $\mathbf{y}_i$ in $\mathbf{z}$.

The model (10.38) can be more simply formulated by defining $\mathbf{z}_i = (x_i, y_i)'$, $\boldsymbol{\zeta}_i = (\xi_i, \eta_i)'$ and writing

$$\mathbf{z}_i = \boldsymbol{\zeta}_i + \boldsymbol{\varepsilon}_i \qquad (i = 1, 2, \ldots, n). \tag{10.40}$$

The equations (10.38) now take the form $g_i(\boldsymbol{\zeta}; \boldsymbol{\theta}) = g(\boldsymbol{\zeta}_i; \boldsymbol{\theta}) = 0$, $i = 1, 2, \ldots, n$. ■

Returning to the general formulation (10.36), we now require some assumptions to ensure that $\mathbf{g}$ is suitably behaved. We assume the following in a suitably large neighborhood of the true values of $\boldsymbol{\zeta}$ and $\boldsymbol{\theta}$:

1. $\mathbf{g}$ is a continuous function from $\mathbb{R}^{m+p}$ to $\mathbb{R}^q$.

2. The partial derivative of each component of $\mathbf{g}(\boldsymbol{\zeta};\boldsymbol{\theta})$ with respect to each argument in both $\boldsymbol{\zeta}$ and $\boldsymbol{\theta}$ exists and is continuous.
3. The second partial derivative of each component of $\mathbf{g}$ with respect to each pair of arguments exists and is bounded.
4. Let

$$\mathbf{S} = \left[\left(\frac{\partial g_j(\boldsymbol{\zeta};\boldsymbol{\theta})}{\partial \zeta_k}\right)\right] \quad \text{and} \quad \mathbf{T} = \left[\left(\frac{\partial g_j(\boldsymbol{\zeta};\boldsymbol{\theta})}{\partial \theta_k}\right)\right]. \tag{10.41}$$

Then the $q \times m$ matrix $\mathbf{S}$ has rank q, and the $q \times p$ matrix $\mathbf{T}$ has rank p. The matrices $\mathbf{S}$ and $\mathbf{T}$ are sometimes referred to as the Jacobians of $\mathbf{g}$ with respect to $\boldsymbol{\zeta}$ and $\boldsymbol{\theta}$, respectively.

Armed with the above assumptions, we shall consider the problem of finding the maximum-likelihood estimators of the unknown parameters. The density function of $\mathbf{z}$ in (10.37) is given by

$$p(\mathbf{z}|\boldsymbol{\zeta},\boldsymbol{\theta};\sigma^2) = (2\pi\sigma^2)^{-m/2}|\mathbf{V}_0|^{-1/2}\exp\left\{-\frac{1}{2\sigma^2}(\mathbf{z}-\boldsymbol{\zeta})'\mathbf{V}_0^{-1}(\mathbf{z}-\boldsymbol{\zeta})\right\}, \tag{10.42}$$

where $\mathbf{g}(\boldsymbol{\zeta};\boldsymbol{\theta}) = \mathbf{0}$. To maximize the log-likelihood $\log p(\cdot)$ subject to the restrictions (or equivalently minimize $-\log p$, as in Britt and Luecke [1973]), we equate to zero the derivatives of the function

$$\tfrac{1}{2}m\log\sigma^2 + \frac{1}{2\sigma^2}(\mathbf{z}-\boldsymbol{\zeta})'\mathbf{V}_0^{-1}(\mathbf{z}-\boldsymbol{\zeta}) + \boldsymbol{\lambda}'\mathbf{g}(\boldsymbol{\zeta};\boldsymbol{\theta}) \tag{10.43}$$

with respect to σ^2, $\boldsymbol{\zeta}, \boldsymbol{\theta}$, and the Lagrange multiplier $\boldsymbol{\lambda}$. These equations are (Appendix A10.4)

$$\frac{m}{2\sigma^2} - \frac{1}{2\sigma^4}(\mathbf{z}-\boldsymbol{\zeta})'\mathbf{V}_0^{-1}(\mathbf{z}-\boldsymbol{\zeta}) = 0, \tag{10.44}$$

$$-\frac{1}{\sigma^2}\mathbf{V}_0^{-1}(\mathbf{z}-\boldsymbol{\zeta}) + \mathbf{S}'\boldsymbol{\lambda} = \mathbf{0}, \tag{10.45}$$

$$\mathbf{T}'\boldsymbol{\lambda} = \mathbf{0}, \tag{10.46}$$

$$\mathbf{g}(\boldsymbol{\zeta};\boldsymbol{\theta}) = \mathbf{0}. \tag{10.47}$$

If $\hat{\sigma}^2, \hat{\boldsymbol{\zeta}}, \hat{\boldsymbol{\theta}}$, and $\hat{\boldsymbol{\lambda}}$ are the solutions of the above equations, then (10.44) leads to $\hat{\sigma}^2 = m^{-1}(\mathbf{z}-\hat{\boldsymbol{\zeta}})'\mathbf{V}_0^{-1}(\mathbf{z}-\hat{\boldsymbol{\zeta}})$. However, this estimator is inconsistent, and a more appropriate estimator is [c.f. (10.21)]

$$\tilde{\sigma}^2 = \frac{1}{q-p}(\mathbf{z}-\hat{\boldsymbol{\zeta}})'\mathbf{V}_0^{-1}(\mathbf{z}-\hat{\boldsymbol{\zeta}}). \tag{10.48}$$

Here $q-p$ is obtained from $m-(m+p-q)$, where $m+p-q$ is the number of free parameters [q are lost because of the constraints (10.47)]. To solve (10.45) to (10.47) we can use a linearization technique. We first expand $\mathbf{g}(\hat{\boldsymbol{\zeta}};\hat{\boldsymbol{\theta}})$ [$=\mathbf{0}$, by (10.47)] about the ath approximation $(\boldsymbol{\zeta}^{(a)},\boldsymbol{\theta}^{(a)})$ of $(\boldsymbol{\zeta},\boldsymbol{\theta})$, giving

$$\mathbf{g}(\boldsymbol{\zeta}^{(a)};\boldsymbol{\theta}^{(a)})+\mathbf{S}_a(\hat{\boldsymbol{\zeta}}-\boldsymbol{\zeta}^{(a)})+\mathbf{T}_a(\hat{\boldsymbol{\theta}}-\boldsymbol{\theta}^{(a)})\approx\mathbf{0}, \qquad (10.49)$$

where $\mathbf{S}_a$ is $\mathbf{S}$ evaluated at the approximation, and so on. The corresponding approximations for (10.45) and (10.46) are

$$(\mathbf{z}-\boldsymbol{\zeta}^{(a)}+\boldsymbol{\zeta}^{(a)}-\hat{\boldsymbol{\zeta}})-\mathbf{V}_0\mathbf{S}_a'\hat{\boldsymbol{\lambda}}\hat{\sigma}^2\approx\mathbf{0}, \qquad (10.50)$$

$$\mathbf{T}_a'\boldsymbol{\lambda}\approx\mathbf{0}. \qquad (10.51)$$

Multiplying (10.50) by $\mathbf{S}_a$ and substituting in (10.49) leads to

$$\mathbf{g}(\boldsymbol{\zeta}^{(a)};\boldsymbol{\theta}^{(a)})+\mathbf{S}_a(\mathbf{z}-\boldsymbol{\zeta}^{(a)})+\mathbf{T}_a(\hat{\boldsymbol{\theta}}-\boldsymbol{\theta}^{(a)})-\mathbf{S}_a\mathbf{V}_0\mathbf{S}_a'\hat{\boldsymbol{\lambda}}\hat{\sigma}^2\approx\mathbf{0}. \qquad (10.52)$$

Since $\mathbf{V}_0$ is positive definite and the rows of $\mathbf{S}_a$ are linearly independent (assumption 4), $\mathbf{A}_a=\mathbf{S}_a\mathbf{V}_0\mathbf{S}_a'$ is also positive definite (Appendix A4.5), and therefore nonsingular. Premultiplying (10.52) by $\mathbf{T}_a'\mathbf{A}_a^{-1}$, using (10.51), and premultiplying by $\mathbf{B}_a^{-1}=(\mathbf{T}_a'\mathbf{A}_a^{-1}\mathbf{T}_a)^{-1}$ gives us, approximately,

$$\begin{aligned}\delta\boldsymbol{\theta}^{(a)}&=\hat{\boldsymbol{\theta}}-\boldsymbol{\theta}^{(a)}\\&=-\mathbf{B}_a^{-1}\mathbf{T}_a'\mathbf{A}_a^{-1}[\mathbf{g}(\boldsymbol{\zeta}^{(a)};\boldsymbol{\theta}^{(a)})+\mathbf{S}_a(\mathbf{z}-\boldsymbol{\zeta}^{(a)})].\end{aligned} \qquad (10.53)$$

From (10.52) we solve for $\hat{\boldsymbol{\lambda}}\hat{\sigma}^2$ and substitute in (10.50) to get, approximately,

$$\begin{aligned}\delta\boldsymbol{\zeta}^{(a)}&=\hat{\boldsymbol{\zeta}}-\boldsymbol{\zeta}^{(a)}\\&=\mathbf{z}-\boldsymbol{\zeta}^{(a)}-\mathbf{V}_0\mathbf{S}_a'\mathbf{A}_a^{-1}[\mathbf{g}(\boldsymbol{\zeta}^{(a)};\boldsymbol{\theta}^{(a)})+\mathbf{S}_a(\mathbf{z}-\boldsymbol{\zeta}^{(a)})+\mathbf{T}_a\delta\boldsymbol{\theta}^{(a)}].\end{aligned} \qquad (10.54)$$

The new iterations are given by

$$\boldsymbol{\theta}^{(a+1)}=\boldsymbol{\theta}^{(a)}+\delta\boldsymbol{\theta}^{(a)} \quad \text{and} \quad \boldsymbol{\zeta}^{(a+1)}=\boldsymbol{\zeta}^{(a)}+\delta\boldsymbol{\zeta}^{(a)},$$

and the process continues until $\|\boldsymbol{\zeta}^{(a+1)}-\boldsymbol{\zeta}^{(a)}\|$ and $\|\boldsymbol{\theta}^{(a+1)}-\boldsymbol{\theta}^{(a)}\|$ are sufficiently small: $\hat{\sigma}^2$ is given by (10.48) evaluated at the final value of $\boldsymbol{\zeta}$. The steps (10.53) and (10.54) were also obtained independently by Bard [1974: 158].

The above procedure is equivalent to minimizing a linearized version of (10.43) where $\mathbf{g}$ is replaced by

$$\mathbf{g}(\boldsymbol{\zeta}^{(a)};\boldsymbol{\theta}^{(a)})+\mathbf{S}_a(\boldsymbol{\zeta}-\boldsymbol{\zeta}^{(a)})+\mathbf{T}_a(\boldsymbol{\theta}-\boldsymbol{\theta}^{(a)}).$$

A starting value for the algorithm may be obtained using an approximate version of the above algorithm due to Deming [1943]. He expanded $\mathbf{g}$ about $(\mathbf{z}, \boldsymbol{\theta}^{(a)})$ rather than $(\boldsymbol{\zeta}^{(a)}, \boldsymbol{\theta}^{(a)})$ and obtained

$$\delta\boldsymbol{\theta}^{(a)} = -[\mathbf{T}_a'(\mathbf{S}_z\mathbf{V}_0\mathbf{S}_z')^{-1}\mathbf{T}_a]^{-1}\mathbf{T}_a'(\mathbf{S}_z\mathbf{V}_0\mathbf{S}_z')^{-1}\mathbf{g}(\mathbf{z};\boldsymbol{\theta}^{(a)}) \tag{10.55}$$

and

$$\delta\boldsymbol{\zeta}^{(a)} = \mathbf{z} - \boldsymbol{\zeta}^{(a)} - \mathbf{V}_0\mathbf{S}_z'(\mathbf{S}_z\mathbf{V}_0\mathbf{S}_z')^{-1}[\mathbf{g}(\mathbf{z};\boldsymbol{\theta}^{(a)}) + \mathbf{T}_a\,\delta\boldsymbol{\theta}^{(a)}]. \tag{10.56}$$

Here $\mathbf{S}_z$ represents $\mathbf{S}$ evaluated at $(\mathbf{z}, \boldsymbol{\theta}^{(a)})$. We note that $\boldsymbol{\zeta}^{(a)}$ does not occur in $\delta\boldsymbol{\theta}^{(a)}$, so that the procedure is much quicker. The only reason for computing $\delta\boldsymbol{\zeta}^{(a)}$ is to help determine when convergence has been obtained. The method does not give a true minimum, though it sometimes gives an answer which is close to the true minimum. However, it has the same structure as the full algorithm, and it can therefore be incorporated into the same computer program to find initial starting values for the full algorithm. We note that (10.55) and (10.56) follow from (10.53) and (10.54) on setting $\boldsymbol{\zeta}^{(a)} = \mathbf{z}$.

We now turn our attention to confidence intervals. Since $(\boldsymbol{\zeta}, \boldsymbol{\theta})$ will be close to $(\boldsymbol{\zeta}^*, \boldsymbol{\theta}^*)$, the true values of the parameters, we have the Taylor expansion

$$\begin{aligned}\mathbf{0} &= \mathbf{g}(\boldsymbol{\zeta}^*;\boldsymbol{\theta}^*)\\ &\approx \mathbf{g}(\hat{\boldsymbol{\zeta}};\hat{\boldsymbol{\theta}}) + \hat{\mathbf{S}}(\boldsymbol{\zeta}^* - \hat{\boldsymbol{\zeta}}) + \hat{\mathbf{T}}(\boldsymbol{\theta}^* - \hat{\boldsymbol{\theta}})\\ &= \hat{\mathbf{S}}(\boldsymbol{\zeta}^* - \hat{\boldsymbol{\zeta}}) + \hat{\mathbf{T}}(\boldsymbol{\theta}^* - \hat{\boldsymbol{\theta}}),\end{aligned}$$

where $\hat{\mathbf{S}}$ represents $\mathbf{S}$ evaluated at the maximum-likelihood estimates. Combining the above equation with the maximum-likelihood equations, Britt and Luecke [1973] showed, using similar algebra to the above, that (with $\mathbf{A} = \mathbf{S}\mathbf{V}_0\mathbf{S}'$ and $\mathbf{B} = \mathbf{T}'\mathbf{A}^{-1}\mathbf{T}$)

$$\begin{aligned}\hat{\boldsymbol{\theta}} - \boldsymbol{\theta}^* &\approx -\hat{\mathbf{B}}^{-1}\hat{\mathbf{T}}'\hat{\mathbf{A}}^{-1}\hat{\mathbf{S}}(\mathbf{z} - \boldsymbol{\zeta}^*)\\ &= -\hat{\mathbf{K}}(\mathbf{z} - \boldsymbol{\zeta}^*),\end{aligned} \tag{10.57}$$

say. They then stated that if $\hat{\mathbf{K}}$ is approximately constant and equal to $\mathbf{K}$, say, in a suitable neighborhood of $(\boldsymbol{\zeta}^*, \boldsymbol{\theta}^*)$, then

$$\begin{aligned}\mathscr{D}[\hat{\boldsymbol{\theta}}] &\approx \mathscr{D}[\mathbf{K}\mathbf{z}]\\ &= \sigma^2\mathbf{K}\mathbf{V}_0\mathbf{K}'\\ &= \sigma^2\mathbf{B}^{-1}\mathbf{T}'\mathbf{A}^{-1}\mathbf{S}\mathbf{V}_0\mathbf{S}'\mathbf{A}^{-1}\mathbf{T}\mathbf{B}^{-1}\\ &= \sigma^2\mathbf{B}^{-1}\mathbf{B}\mathbf{B}^{-1}\\ &= \sigma^2\mathbf{T}'(\mathbf{S}\mathbf{V}_0\mathbf{S}')^{-1}\mathbf{T}.\end{aligned} \tag{10.58}$$

However, we now show that (10.58) is generally invalid, so that an expression for $\mathscr{D}[\hat{\boldsymbol{\theta}}]$ is still needed.

The problem with (10.58) is highlighted by first noting that an explicit relationship $\boldsymbol{\mu} = \mathbf{f}(\boldsymbol{\xi}; \boldsymbol{\theta})$ can be written in the implicit form

$$\mathbf{g}(\boldsymbol{\zeta}; \boldsymbol{\theta}) = \boldsymbol{\mu} - \mathbf{f}(\boldsymbol{\xi}; \boldsymbol{\theta}) = \mathbf{0}. \tag{10.59}$$

It then transpires that the linearization procedure used above and the scoring method of Section 10.2 are equivalent (Dolby [1976a]). This extends the result of Appendix A15 proved by Ratkowsky and Dolby [1975] for ordinary, as opposed to errors-in-variables, nonlinear models. Also, equating $\mathbf{T} = -\mathbf{N}$ and $\mathbf{S} = (-\mathbf{M}, \mathbf{I})$ for the model (10.59), Dolby [1976a] showed that $\mathbf{S}\mathbf{V}_0\mathbf{S}'$ of (10.58) (called $\boldsymbol{\Theta}$ in his notation) is the same as $\boldsymbol{\psi}^{-1}$ of (10.20): thus (10.58) and (10.20) are the same. However, it is known that (10.20) is incorrect for linear models as $n \to \infty$, so that (10.58) is also likely to be inappropriate for any implicit model in addition to the special case of (10.59). This suggests that the argument following (10.57) is suspect.

Unfortunately, the equivalence between the two iterative procedures indicates that some of the problems encountered in Section 10.2 will be compounded with implicit functions. For instance, the above linearization procedure may not converge. If it does converge, then assumptions 1 to 3 above on $\mathbf{g}(\boldsymbol{\zeta}; \boldsymbol{\theta})$ will guarantee that the result is a stationary point of the likelihood function. However, the stationary point $(\hat{\boldsymbol{\zeta}}, \hat{\boldsymbol{\theta}})$ may not be a maximum.

It should be noted that for many of the applications of the above theory, like (10.38) and (10.39), the matrices $\mathbf{S}$ and $\mathbf{T}$ of (10.41) take a simple form. For example $\mathbf{S}$ is diagonal for (10.38).

When σ^2 is known, we simply omit Equation (10.44) from the analysis so that the estimates of $\boldsymbol{\zeta}$ and $\boldsymbol{\theta}$ remain the same.

Example 10.2 (Britt and Luecke [1973]). The following model, describing the compressibility of helium kept at a constant temperature of 273.15 K, was fitted to the $m = 12$ measurements ($\mathbf{z}$) in Table 10.1:

$$g_i(\boldsymbol{\zeta}; \boldsymbol{\theta}) = \zeta_{i+1}\zeta_i \sum_{r=1}^{2} (\theta_r \zeta_{i+1}^{r-1} - \theta_3 \zeta_i^{r-1}) + \zeta_i - \theta_3 \zeta_{i+1} = 0$$

$$(i = 1, 2, \ldots, q: \quad q = 11).$$

Table 10.1 Gas Pressures for Helium at 273.15 K measured by the Burnett Method[a]

Expt.	Pressure (atm)	Expt.	Pressure (atm)	Expt.	Pressure (atm)
1	683.599	5	89.0521	9	14.2908
2	391.213	6	55.9640	10	9.1072
3	233.607	7	35.3851	11	5.8095
4	143.091	8	22.4593	12	3.7083

[a] From Blancett et al. [1970] with permission of North-Holland Physics Publishing Co., Amsterdam.

Table 10.2 Estimated Parameters[a] for Helium Data of Table 10.1

	Estimate	
Parameter	Approximate Algorithm	General Algorithm
θ_1 (cm^3/g-mole)	11.9517521	11.9517622
θ_2 (cm^3/g-mole)	113.9623863	113.9619475
θ_3 = volume ratio	1.5648810	1.5648810

[a]From Britt and Luecke [1973].

Here $\boldsymbol{\theta} = (\theta_1, \theta_2, \theta_3)'$, where θ_1 and θ_2 are called virial coefficients and θ_3 is a volume ratio. The experimental data were obtained by the Burnett method in which the pressure of a contained gas at constant temperature is measured before and after expansion into a larger volume. Successive gas expansions are continued until the pressure is below a practical operating level. The matrix $\mathbf{V}_0(=\mathscr{D}[\mathbf{z}]/\sigma^2)$ was assumed to be diagonal.

Starting values of $\theta_1 = 0, \theta_2 = 0$, and $\theta_3 = 1.5$, the latter being estimated from the physical dimensions of the Burnett-cell volumes, were used for the general algorithm. However, the algorithm failed to converge because the estimates of the pressures ζ_i were totally unreasonable after the first iteration, so that the subsequent linearization (10.49) about $(\boldsymbol{\zeta}^{(a)}, \boldsymbol{\theta}^{(a)})$ was not valid. The Deming approximate algorithm was not subject to this problem, since the linearization is about $(\mathbf{z}, \boldsymbol{\theta}^{(a)})$, and convergence was obtained in four interations. Using the Deming estimates as starting values, the general algorithm was applied, and it converged in two iterations, bringing only slight changes in the estimates (see Table 10.2). ■

The above model and associated iterative procedures are very general and encompass a wide range of models. A simpler format, however, for the more common special case (10.40) is given below in Section 10.5.1. The model is derived for R_i replicates of each $\mathbf{z}_i$: we simply set each $R_i = 1$. Another algorithm, giving rise to the same estimates but using a Bayesian approach, is described in Section 10.5.2. Once again we set $R_i = 1$.

An interesting special case arises when some of the variables (labeled $\mathbf{x}$, say) are measured without error. For a single implicit relationship we then have

$$h(\boldsymbol{\eta}_i, \mathbf{x}_i; \boldsymbol{\theta}) = 0, \tag{10.60}$$

where $\mathbf{x}_i$ is measured without error, and

$$\mathbf{y}_i = \boldsymbol{\eta}_i + \boldsymbol{\varepsilon}_i \qquad (i = 1, 2, \ldots, n).$$

To fit this into the general framework we set

$$g_i(\boldsymbol{\zeta};\boldsymbol{\theta}) = h(\boldsymbol{\eta}_i, \mathbf{x}_i;\boldsymbol{\theta}) \qquad (i = 1, 2, \ldots, n)$$

and

$$\mathbf{z} = \boldsymbol{\zeta} + \boldsymbol{\varepsilon}, \tag{10.61}$$

where $\boldsymbol{\zeta} = (\boldsymbol{\eta}'_1, \boldsymbol{\eta}'_2, \ldots, \boldsymbol{\eta}'_n)'$ and $\mathbf{z} = (\mathbf{y}'_1, \mathbf{y}'_2, \ldots, \mathbf{y}'_n)'$. We can also go a step further in generalization and replace h in (10.60) by a vector function $\mathbf{h}$. This particular model was considered by Bard [1974: 24, 68, 154–159] under the title of "exact structural model," and he developed a general algorithm which can be shown to reduce to the one given above by Britt and Luecke [1973].

Sometimes the equations $\mathbf{g}(\boldsymbol{\zeta};\boldsymbol{\theta}) = \mathbf{0}$ are such that part of $\mathbf{z}$ is not accessible to the experimenter. If the model is well defined, then it will be possible, in theory at least, to eliminate the corresponding part of $\boldsymbol{\zeta}$ from the equations and reduce the problem to one which can be analyzed using the above methods. However, if the equations are complicated, it may be more convenient to retain the original equations and estimate the inaccessible part of $\boldsymbol{\zeta}$ as an intermediate step. Methods for doing this are described by Sachs [1976], and we discuss his approach in Section 10.6 below. This idea is akin to estimating missing observations in linear models.

10.5 IMPLICIT FUNCTIONAL RELATIONSHIPS: WITH REPLICATION

In the previous section we considered a very general implicit model without replication. Such a formulation, however, is not so appropriate when considering replication, and we now follow the notation of (10.40). Let

$$\mathbf{z}_{ir} = \boldsymbol{\zeta}_i + \boldsymbol{\varepsilon}_{ir} \qquad (i = 1, 2, \ldots, n, \quad r = 1, 2, \ldots, R_i), \tag{10.62}$$

where the $\boldsymbol{\varepsilon}_{ir}$ are i.i.d. $N_d(\mathbf{0}, \boldsymbol{\Sigma})$. Also, suppose

$$\mathbf{g}(\boldsymbol{\zeta}_i;\boldsymbol{\theta}) = \mathbf{0} \qquad (i = 1, 2, \ldots, n), \tag{10.63}$$

$\mathbf{g}$ being a $q_0 \times 1$ vector of functions ($q_0 \leq d$). [In the notation of (10.36), $q = nq_0$ and $m = nd$.] We shall consider both maximum-likelihood and Bayesian methods of estimation below.

10.5.1 Maximum-Likelihood Estimation

We now adapt the method of Britt and Luecke [1973] of Section 10.4 to fit in with the new notation in (10.62) and (10.63). As the mathematics for the replicated case is very similar to that of the unreplicated case already considered, we shall only sketch the details. We begin by assuming $\boldsymbol{\Sigma} = \sigma^2 \mathbf{V}_1$, where $\mathbf{V}_1$ is known. Then (10.43) becomes

$$\tfrac{1}{2} N d \log \sigma^2 + \frac{1}{2\sigma^2} \sum_{i=1}^{n} \sum_{r=1}^{R_i} (\mathbf{z}_{ir} - \boldsymbol{\zeta}_i)' \mathbf{V}_1^{-1} (\mathbf{z}_{ir} - \boldsymbol{\zeta}_i) + \sum_{i=1}^{n} \boldsymbol{\lambda}_i' \mathbf{g}(\boldsymbol{\zeta}_i; \boldsymbol{\theta}), \qquad (10.64)$$

where $N = \sum_i R_i$, the total number of observations. We define the following matrices by their (j, k)th elements:

$$\mathbf{S}_i = \left[\left(\frac{\partial g_j(\boldsymbol{\zeta}_i; \boldsymbol{\theta})}{\partial (\boldsymbol{\zeta}_i)_k} \right) \right] \qquad (10.65)$$

and

$$\mathbf{T}_i = \left[\left(\frac{\partial g_j(\boldsymbol{\zeta}_i; \boldsymbol{\theta})}{\partial \theta_k} \right) \right]. \qquad (10.66)$$

The maximum-likelihood equations are now [c.f. (10.44) to (10.47)]

$$\frac{Nd}{2\sigma^2} - \frac{1}{2\sigma^4} \sum_i \sum_r (\mathbf{z}_{ir} - \boldsymbol{\zeta}_i)' \mathbf{V}_1^{-1} (\mathbf{z}_{ir} - \boldsymbol{\zeta}_i) = 0,$$

$$-\frac{1}{\sigma^2} \mathbf{V}_1^{-1} \sum_r (\mathbf{z}_{ir} - \boldsymbol{\zeta}_i) + \mathbf{S}_i' \boldsymbol{\lambda}_i = \mathbf{0} \qquad (i = 1, 2, \ldots, n), \qquad (10.67)$$

$$\sum_i \mathbf{T}_i' \boldsymbol{\lambda}_i = \mathbf{0}, \qquad (10.68)$$

and (10.63). Expanding $\mathbf{g}(\hat{\boldsymbol{\zeta}}_i; \hat{\boldsymbol{\theta}})$ about the ath approximation $(\boldsymbol{\zeta}_i^{(a)}, \boldsymbol{\theta}^{(a)})$ of the maximum-likelihood estimates $(\hat{\boldsymbol{\zeta}}_i, \hat{\boldsymbol{\theta}})$ gives

$$\mathbf{g}(\boldsymbol{\zeta}_i^{(a)}; \boldsymbol{\theta}^{(a)}) + \mathbf{S}_{ia}(\hat{\boldsymbol{\zeta}}_i - \boldsymbol{\zeta}_i^{(a)}) + \mathbf{T}_{ia}(\hat{\boldsymbol{\theta}} - \boldsymbol{\theta}^{(a)}) \approx \mathbf{0},$$

where $\mathbf{S}_{ia}$ is $\mathbf{S}_i$ evaluated at the approximation, etc. Using the same arguments that produced (10.50) to (10.54), and defining

$$\mathbf{A}_{ia} = \mathbf{S}_{ia} \mathbf{V}_1 \mathbf{S}_{ia}', \qquad \mathbf{B}_{ia} = \mathbf{T}_{ia}' \mathbf{A}_{ia}^{-1} \mathbf{T}_{ia}, \qquad \mathbf{C}_a = \sum_{i=1}^{n} R_i \mathbf{B}_{ia},$$

we obtain

$$\begin{aligned} \delta\boldsymbol{\theta}^{(a)} &= \hat{\boldsymbol{\theta}} - \boldsymbol{\theta}^{(a)} \\ &= -\mathbf{C}_a^{-1} \sum_{i=1}^{n} R_i \mathbf{T}_{ia}' \mathbf{A}_{ia}^{-1} [\mathbf{g}(\boldsymbol{\zeta}_i^{(a)}; \boldsymbol{\theta}^{(a)}) + \mathbf{S}_{ia}(\bar{\mathbf{z}}_i - \boldsymbol{\zeta}_i^{(a)})] \end{aligned} \qquad (10.69)$$

and

$$\begin{aligned} \delta\boldsymbol{\zeta}_i^{(a)} &= \hat{\boldsymbol{\zeta}}_i - \boldsymbol{\zeta}_i^{(a)} \\ &= \bar{\mathbf{z}}_i - \boldsymbol{\zeta}_i^{(a)} - \mathbf{V}_1 \mathbf{S}_{ia}' \mathbf{A}_{ia}^{-1} [\mathbf{g}(\boldsymbol{\zeta}_i^{(a)}; \boldsymbol{\theta}^{(a)}) + \mathbf{S}_{ia}(\bar{\mathbf{z}}_i - \boldsymbol{\zeta}_i^{(a)}) + \mathbf{T}_{ia} \delta\boldsymbol{\theta}^{(a)}], \end{aligned} \qquad (10.70)$$

where $\bar{\mathbf{z}}_i = \sum_r \mathbf{z}_{ir}/R_i$. The new iterations are given by

$$\boldsymbol{\theta}^{(a+1)} = \boldsymbol{\theta}^{(a)} + \delta\boldsymbol{\theta}^{(a)} \quad \text{and} \quad \boldsymbol{\zeta}_i^{(a+1)} = \boldsymbol{\zeta}_i^{(a)} + \delta\boldsymbol{\zeta}_i^{(a)} \quad (i = 1, 2, \ldots, n).$$

The analogous expression to (10.57) is

$$\begin{aligned}\hat{\boldsymbol{\theta}} - \boldsymbol{\theta}^* &\approx -\hat{\mathbf{C}}^{-1} \sum_{i=1}^{n} R_i \hat{\mathbf{T}}_i' \hat{\mathbf{A}}_i^{-1} \hat{\mathbf{S}}_i (\bar{\mathbf{z}}_i - \boldsymbol{\zeta}_i^*) \\ &= -\hat{\mathbf{C}}^{-1} \sum_{i=1}^{n} R_i \hat{\mathbf{T}}_i' (\hat{\mathbf{S}}_i \mathbf{V}_1 \hat{\mathbf{S}}_i')^{-1} \hat{\mathbf{S}}_i (\bar{\mathbf{z}}_i - \boldsymbol{\zeta}_i^*),\end{aligned} \tag{10.71}$$

so that

$$\mathscr{D}[\hat{\boldsymbol{\theta}}] \approx \sigma^2 \mathbf{C}^{-1} \mathbf{C} \mathbf{C}^{-1} = \sigma^2 \mathbf{C}^{-1}, \tag{10.72}$$

where $\mathbf{C} = \sum_i R_i \mathbf{T}_i' (\mathbf{S}_i \mathbf{V}_1 \mathbf{S}_i')^{-1} \mathbf{T}_i$ is evaluated at the true values of the parameters. If we let each $R_i \to \infty$ in such a way that R_i/N is constant, we can apply the central limit theorem to each $\bar{\mathbf{z}}_i$, and hence to $\hat{\boldsymbol{\theta}}$, so that

$$\hat{\boldsymbol{\theta}} - \boldsymbol{\theta}^* \sim N_p(\mathbf{0}, \sigma^2 \mathbf{C}^{-1}) \qquad \text{asymptotically.} \tag{10.73}$$

Also, the maximum-likelihood estimate of σ^2, which is again inconsistent, can be adjusted to give the consistent estimator

$$\tilde{\sigma}^2 = \frac{\sum_i \sum_r (\mathbf{z}_{ir} - \hat{\boldsymbol{\zeta}}_i)' \mathbf{V}_1^{-1} (\mathbf{z}_{ir} - \hat{\boldsymbol{\zeta}}_i)}{Nd - (dn + p - nq_0)}. \tag{10.74}$$

The Deming approximate algorithm follows from (10.69) and (10.70) on setting $\boldsymbol{\zeta}_i^{(a)} = \bar{\mathbf{z}}_i$.

If $\boldsymbol{\Sigma}$ is unknown, it can be estimated by the unbiased estimate

$$\tilde{\boldsymbol{\Sigma}} = \frac{1}{N-n} \sum_{i=1}^{n} \sum_{r=1}^{R_i} (\mathbf{x}_{ir} - \bar{\mathbf{x}}_i)(\mathbf{x}_{ir} - \bar{\mathbf{x}}_i)', \tag{10.75}$$

which does not depend on the validity of the restrictions (10.63). Since σ^2 can be incorporated with $\boldsymbol{\lambda}_i$ in (10.67), the iterative procedure does not depend on σ^2. We can therefore set $\sigma^2 = 1$ and put $\mathbf{V}_1 = \tilde{\boldsymbol{\Sigma}}$ in (10.69) and (10.70).

In conclusion we note that, apart from (10.72) and (10.73), the above theory holds when there are no replicates and each R_i is unity. This special case duplicates the theory of the previous section, but in a more convenient format.

10.5.2 Bayesian Estimation

Assuming σ^2 is known (i.e., we can set $\sigma^2 = 1$), we now reanalyze (10.62) using a Bayesian method proposed by Reilly and Patino-Leal [1981]. The first step is to obtain an expression for the posterior density of $\boldsymbol{\theta}$ given the data, and this is done iteratively. If $\tilde{\boldsymbol{\zeta}}_i$ is an approximation for $\boldsymbol{\zeta}_i$, then we have a Taylor expansion for (10.63), namely

$$\begin{aligned} \mathbf{0} &= \mathbf{g}(\boldsymbol{\zeta}_i; \boldsymbol{\theta}) \\ &\approx \mathbf{g}(\tilde{\boldsymbol{\zeta}}_i; \boldsymbol{\theta}) + \tilde{\mathbf{S}}_i(\boldsymbol{\zeta}_i - \tilde{\boldsymbol{\zeta}}_i), \end{aligned} \tag{10.76}$$

or

$$\tilde{\mathbf{S}}_i \boldsymbol{\zeta}_i \approx \tilde{\mathbf{S}}_i \tilde{\boldsymbol{\zeta}}_i - \mathbf{g}(\tilde{\boldsymbol{\zeta}}_i; \boldsymbol{\theta}) = \tilde{\mathbf{b}}_i, \tag{10.77}$$

say, where [c.f. (10.65)]

$$\tilde{\mathbf{S}}_i = \left[\left(\frac{\partial g_j(\boldsymbol{\zeta}_i; \theta)}{\partial (\boldsymbol{\zeta}_i)_k} \right) \right]_{\boldsymbol{\zeta}_i = \tilde{\boldsymbol{\zeta}}_i}.$$

The constraints on $\boldsymbol{\zeta}_i$ are now approximately linear, being of the form $\tilde{\mathbf{S}}_i \boldsymbol{\zeta}_i \approx \tilde{\mathbf{b}}_i$. However, these constraints can be more conveniently expressed in the "freedom equation" form (c.f. Appendix E)

$$\boldsymbol{\zeta}_i = \mathbf{W}_i \boldsymbol{\alpha}_i + \tilde{\mathbf{S}}_i'(\tilde{\mathbf{S}}_i \tilde{\mathbf{S}}_i')^{-1} \tilde{\mathbf{b}}_i,$$

where $\mathbf{W}_i$ is a $d \times (d - q_0)$ matrix of rank $d - q_0$ satisfying $\tilde{\mathbf{S}}_i \mathbf{W}_i = \mathbf{0}$ (q_0 being the dimension of $\mathbf{g}$) and $\boldsymbol{\alpha}_i$ is unconstrained. It can then be shown that the likelihood function is approximately proportional to

$$\begin{aligned} &\exp\left\{ -\frac{1}{2} \sum_{i=1}^{n} R_i (\tilde{\mathbf{S}}_i \bar{\mathbf{z}}_i - \tilde{\mathbf{b}}_i)' (\tilde{\mathbf{S}}_i \mathbf{V}_1 \tilde{\mathbf{S}}_i')^{-1} (\tilde{\mathbf{S}}_i \bar{\mathbf{z}}_i - \tilde{\mathbf{b}}_i) \right\} \\ &\qquad \times \exp\left\{ -\frac{1}{2} \sum_{i=1}^{n} R_i (\boldsymbol{\alpha}_i - \hat{\boldsymbol{\alpha}}_i)' (\mathbf{W}_i' \mathbf{V}_1^{-1} \mathbf{W}_i)^{-1} (\boldsymbol{\alpha}_i - \hat{\boldsymbol{\alpha}}_i) \right\}, \end{aligned}$$

where $\hat{\boldsymbol{\alpha}}_i = (\mathbf{W}_i' \mathbf{V}_1^{-1} \mathbf{W}_i)^{-1} \mathbf{W}_i' \mathbf{V}_1^{-1} \mathbf{y}_i$, $\mathbf{y}_i = \bar{\mathbf{z}}_i - \tilde{\mathbf{S}}_i' (\tilde{\mathbf{S}}_i \tilde{\mathbf{S}}_i')^{-1} \tilde{\mathbf{b}}_i$, and $\bar{\mathbf{z}}_i = \sum_i \mathbf{z}_{ir} / R_i$.

The authors then use Jeffrey's prior [c.f. (2.100)] for each $\boldsymbol{\alpha}_i$, namely

$$\begin{aligned} p(\boldsymbol{\alpha}_i) &\propto \left| \mathscr{E}\left(-\frac{\partial^2 L(\boldsymbol{\alpha}_i)}{\partial \boldsymbol{\alpha}_i \partial \boldsymbol{\alpha}_i'} \right) \right|^{1/2} \\ &\propto |\mathbf{W}_i' \mathbf{V}_1^{-1} \mathbf{W}_i|^{1/2}, \end{aligned}$$

where $L(\boldsymbol{\alpha}_i)$ is the log-likelihood function expressed as a function of just $\boldsymbol{\alpha}_i$. Using an independent uniform prior for $\boldsymbol{\theta}$, the joint posterior density of $\boldsymbol{\theta}$ and the $\boldsymbol{\alpha}_i$ is obtained by multiplying the likelihood function by $\prod_i p(\boldsymbol{\alpha}_i)$. We then find that the $\boldsymbol{\alpha}_i$ can be integrated out using the properties of the multivariate normal distribution. The matrices $\mathbf{W}_i$ conveniently disappear, and we end up with the following approximation for the posterior density of $\boldsymbol{\theta}$:

$$p(\boldsymbol{\theta}|\text{data}) \approx k \exp\left\{ -\frac{1}{2} \sum_{i=1}^{n} R_i (\tilde{\mathbf{S}}_i \bar{\mathbf{z}}_i - \tilde{\mathbf{b}}_i)' (\tilde{\mathbf{S}}_i \mathbf{V}_1 \tilde{\mathbf{S}}_i')^{-1} (\tilde{\mathbf{S}}_i \bar{\mathbf{z}}_i - \tilde{\mathbf{b}}_i) \right\}. \quad (10.78)$$

From (10.77)

$$\tilde{\mathbf{S}}_i \bar{\mathbf{z}}_i - \tilde{\mathbf{b}}_i = \tilde{\mathbf{S}}_i \bar{\mathbf{z}}_i - \tilde{\mathbf{S}}_i \tilde{\boldsymbol{\zeta}}_i + \mathbf{g}(\tilde{\boldsymbol{\zeta}}_i; \boldsymbol{\theta}), \quad (10.79)$$

and an approximation for (10.78) is obtained by setting $\tilde{\boldsymbol{\zeta}}_i = \bar{\mathbf{z}}_i$, namely

$$p(\boldsymbol{\theta}|\text{data}) \approx k \exp\left\{ -\frac{1}{2} \sum_{i=1}^{n} R_i \mathbf{g}(\bar{\mathbf{z}}_i; \boldsymbol{\theta})' (\bar{\mathbf{S}}_i \mathbf{V}_1 \bar{\mathbf{S}}_i')^{-1} \mathbf{g}(\bar{\mathbf{z}}_i; \boldsymbol{\theta}) \right\}, \quad (10.80)$$

where $\bar{\mathbf{S}}_i$ is $\mathbf{S}_i$ evaluated at $\boldsymbol{\zeta}_i = \bar{\mathbf{z}}_i$.

Reilly and Patino-Leal [1981] also give an iterative method for finding the exact posterior density for any value of $\boldsymbol{\theta}$. They replace $\tilde{\boldsymbol{\zeta}}_i$ by a sequence of iterations $\boldsymbol{\zeta}_i^{(a)}$ which converges to a root of (10.63), so that the linear constraint (10.77) approaches (10.63). Their sequence is

$$\boldsymbol{\zeta}_i^{(a+1)} = \bar{\mathbf{z}}_i - \mathbf{V}_1 \mathbf{S}_i^{(a)\prime} (\mathbf{S}_i^{(a)} \mathbf{V}_1 \mathbf{S}_i^{(a)\prime})^{-1} [\mathbf{g}(\boldsymbol{\zeta}_i^{(a)}; \boldsymbol{\theta}) + \mathbf{S}_i^{(a)} (\bar{\mathbf{z}}_i - \boldsymbol{\zeta}_i^{(a)})], \quad (10.81)$$

where $\mathbf{S}_i^{(a)}$ is $\mathbf{S}_i$ evaluated at $\boldsymbol{\zeta}_i^{(a)}$, with a starting value of $\boldsymbol{\zeta}_i^{(1)} = \bar{\mathbf{z}}_i$. If the process converges so that $\boldsymbol{\zeta}_i^{(a)} \to \hat{\boldsymbol{\zeta}}_i$, say, then $\mathbf{S}_i^{(a)} \to \mathbf{S}(\hat{\boldsymbol{\zeta}}_i)$ [$= \hat{\mathbf{S}}_i$, say] and (10.81) becomes

$$\bar{\mathbf{z}}_i - \hat{\boldsymbol{\zeta}}_i = \mathbf{V}_1 \hat{\mathbf{S}}_i' (\hat{\mathbf{S}}_i \mathbf{V}_1 \hat{\mathbf{S}}_i')^{-1} [\mathbf{g}(\hat{\boldsymbol{\zeta}}_i; \boldsymbol{\theta}) + \hat{\mathbf{S}}_i (\bar{\mathbf{z}}_i - \hat{\boldsymbol{\zeta}}_i)].$$

Premultiplying this equation by $\hat{\mathbf{S}}_i$ leads to $\mathbf{g}(\hat{\boldsymbol{\zeta}}_i; \boldsymbol{\theta}) = \mathbf{0}$, and we have checked that the limit $\hat{\boldsymbol{\zeta}}_i$ satisfies the constraints. The exact posterior density is obtained by taking the limit of (10.78) as $\tilde{\boldsymbol{\zeta}}_i \to \hat{\boldsymbol{\zeta}}_i$. Using (10.79) and $\mathbf{g}(\hat{\boldsymbol{\zeta}}_i; \boldsymbol{\theta}) = \mathbf{0}$, we thus get

$$p(\boldsymbol{\theta}|\text{data}) = k \exp\left\{ -\frac{1}{2} \sum_i R_i [\hat{\mathbf{S}}_i (\bar{\mathbf{z}}_i - \hat{\boldsymbol{\zeta}}_i)]' (\hat{\mathbf{S}}_i \mathbf{V}_1 \hat{\mathbf{S}}_1')^{-1} [\hat{\mathbf{S}}_i (\bar{\mathbf{z}}_i - \hat{\boldsymbol{\zeta}}_i)] \right\}. \quad (10.82)$$

Having established the posterior density of $\boldsymbol{\theta}$, we can use the mode $\hat{\boldsymbol{\theta}}$, say, of (10.82) as an estimate of $\boldsymbol{\theta}$. Again iteration is needed to find this mode, so that using (10.81), the overall algorithm requires a double iteration, one for each of $\boldsymbol{\theta}$ and $\boldsymbol{\zeta}_i$. For every iterated value of $\boldsymbol{\theta}$ it is necessary to iterate (10.81) to convergence for each i.

When the dimension of $\mathbf{g}$ is greater than one, we can avoid finding the matrix inverse in (10.81) and (10.82) by using the usual Cholesky factorization method. Letting $\mathbf{S}_i^{(a)} \mathbf{V}_1 \mathbf{S}_i^{(a)\prime} = \mathbf{U}_i' \mathbf{U}_i$ be the Cholesky decomposition, where $\mathbf{U}_i$ is upper triangular, we solve the lower triangular systems of equations

$$\begin{aligned} \mathbf{U}_i' \mathbf{W}_i &= \mathbf{S}_i^{(a)} \mathbf{V}_1, \\ \mathbf{U}_i' \mathbf{w}_i &= \mathbf{g}(\boldsymbol{\zeta}_i^{(a)}; \boldsymbol{\theta}) + \mathbf{S}_i^{(a)} (\bar{\mathbf{z}}_i - \boldsymbol{\zeta}_i^{(a)}), \end{aligned} \quad (10.83)$$

for the $\mathbf{W}_i$ and $\mathbf{w}_i$ $(i = 1, 2, \ldots, n)$. Equation (10.81) now becomes

$$\begin{aligned}\boldsymbol{\zeta}_i^{(a+1)} &= \bar{\mathbf{z}}_i - \mathbf{W}_i'\mathbf{U}_i(\mathbf{U}_i'\mathbf{U}_i)^{-1}\mathbf{U}_i'\mathbf{w}_i \\ &= \bar{\mathbf{z}}_i - \mathbf{W}_i'\mathbf{w}_i,\end{aligned}$$

since $\mathbf{U}_i$ is nonsingular. Let $\hat{\mathbf{U}}_i$, $\hat{\mathbf{W}}_i$, and $\hat{\mathbf{w}}_i$ denote the converged values of $\mathbf{U}_i$, $\mathbf{W}_i$, and $\mathbf{w}_i$, respectively. Then, using $\mathbf{g}(\hat{\boldsymbol{\zeta}}_i; \boldsymbol{\theta}) = \mathbf{0}$, we have from (10.83) that $\hat{\mathbf{S}}_i(\bar{\mathbf{z}}_i - \hat{\boldsymbol{\zeta}}_i) = \hat{\mathbf{U}}_i'\hat{\mathbf{w}}_i$, and substituting in (10.82) leads to

$$p(\boldsymbol{\theta}\,|\,\text{data}) = k \exp\left\{ -\frac{1}{2}\sum_i R_i \hat{\mathbf{w}}_i'\hat{\mathbf{w}}_i \right\}. \qquad (10.84)$$

The authors used the stability of $\mathbf{w}_i'\mathbf{w}_i$ as their criterion of convergence. Although recognizing that starting with $\boldsymbol{\zeta}_i^{(1)} = \bar{\mathbf{z}}_i$ cannot guarantee convergence, they found that it is almost always effective and that convergence is rapid. Usually no more than ten iterations were required, and often no more than three or four for relative tolerances as small as one part in 10^{10} on $\mathbf{w}_i'\mathbf{w}_i$.

An interesting feature of the above Bayesian method is that for the particular choice of the prior distributions, it leads to the same estimates as the maximum-likelihood estimates of Section 10.5.1. That is why the same notation has been used for $\hat{\boldsymbol{\theta}}$ and $\hat{\boldsymbol{\zeta}}_i$ in both cases. How do the two methods compare? The Britt–Luecke algorithm for the maximum-likelihood estimates appears to be more straightforward in that there are just two iteration formulae (10.69) and (10.70). A major drawback is the computation of $\mathbf{C}_a$ and its inverse in (10.69), where

$$\mathbf{C}_a = \sum_i R_i \mathbf{T}_{ia}'(\mathbf{S}_{ia}\mathbf{V}_1\mathbf{S}_{ia}')^{-1}\mathbf{T}_{ia}.$$

However, given a good algorithm for the iterates of $\boldsymbol{\theta}$ leading to the maximization of (10.78) as $\tilde{\boldsymbol{\zeta}}_i \to \hat{\boldsymbol{\zeta}}_i$, the method of Reilly and Patino-Leal is efficient and stable.

The Bayesian method has other advantages over the maximum-likelihood method. The latter relies on an asymptotic theory for $\mathscr{D}[\hat{\boldsymbol{\theta}}]$ which does not apply in the unreplicated case. However, the Bayesian method leads to a posterior density which can be studied using, for example, the methods of Section 2.7.3. The posterior density is valid for the unreplicated case also.

Example 10.3 (Reilly and Patino-Leal [1981]). A radiographic technique is available for investigating prosthetic hip performance. To evaluate the relative success of major surgical repair via implants, an analysis of the performance of hip protheses in motion was undertaken. X-ray images of the prostheses were made under conditions such that the beam of X-rays struck the photographic plate an oblique angle, distorting the image of the spherical ball to an ellipse. To estimate the distortion introduced, an ellipse of the form

$$g(\boldsymbol{\zeta}_i; \boldsymbol{\theta}) = \begin{pmatrix} \zeta_{i1} - \theta_1 \\ \zeta_{i2} - \theta_2 \end{pmatrix}' \begin{pmatrix} \theta_3 & \theta_4 \\ \theta_4 & \theta_5 \end{pmatrix} \begin{pmatrix} \zeta_{i1} - \theta_1 \\ \zeta_{i2} - \theta_2 \end{pmatrix} - 1 = 0 \qquad (10.85)$$

Table 10.3 Digitized X-Ray Image Data[a]

i	z_{i1}	z_{i2}	i	z_{i1}	z_{i2}
1	0.5	−0.12	11	1.34	−5.97
2	1.2	−0.60	12	0.9	−6.32
3	1.6	−1.00	13	−0.28	−6.44
4	1.86	−1.4	14	−0.78	−6.44
5	2.12	−2.54	15	−1.36	−6.41
6	2.36	−3.36	16	−1.90	−6.25
7	2.44	−4.0	17	−2.50	−5.88
8	2.36	−4.75	18	−2.88	−5.50
9	2.06	−5.25	19	−3.18	−5.24
10	1.74	−5.64	20	−3.44	−4.86

[a]From Reilly and Patino-Leal [1981].

was fitted to data, an example (unreplicated) set of which is shown in Table 10.3. The data are the measured z_1, z_2 coordinates of points covering about half of the circumference of the elliptical image of the ball. They may be expressed in the (unreplicated) form

$$\mathbf{z}_i = \begin{pmatrix} z_{i1} \\ z_{i2} \end{pmatrix} = \begin{pmatrix} \zeta_{i1} \\ \zeta_{i2} \end{pmatrix} + \begin{pmatrix} \varepsilon_{i1} \\ \varepsilon_{i2} \end{pmatrix} \qquad (i = 1, 2, \dots, 20),$$

as in (10.40). It was known from past experience that all the ε_{ij} are independently distributed with common variance equal to 0.0001, i.e. $\mathbf{V}_1 = 0.0001\mathbf{I}_2$.

Point estimates of the θ_r $(r = 1, 2, 3, 4)$ were obtained using both the approximate and the exact posterior density functions. In each case the starting values, given in Table 10.4, were obtained by guessing θ_1 and θ_2 by eye and then substituting three of the measured sets of coordinates into (10.85) and solving for θ_3, θ_4, and θ_5. For the approximate method, the negative logarithm of (10.80) was minimized using a variable-metric (quasi-Newton) method, and convergence was

Table 10.4 Point Estimates for Ellipse Parameters[a]

		Final Value	
Parameter	Starting Value	Approximate Method	Exact Method
θ_1	−0.57	−1.00788	−0.99938
θ_2	−3.4	−2.92395	−2.93105
θ_3	0.1	0.08745	0.08757
θ_4	5.7×10^{-4}	0.01646	0.01623
θ_5	0.082	0.07962	0.07975

[a]From Reilly and Patino-Leal [1981].

reached in 26 iterations. The "exact" approach minimized the negative logarithm of (10.82) using a variable-metric method for the $\boldsymbol{\theta}$-iterations along with (10.81). Convergence required 24 iterations on $\boldsymbol{\theta}$. The two sets of estimates are given in Table 10.4. ■

In concluding this section we now prove a lemma which may be skipped by those readers not interested in the details.

Lemma The maximum-likelihood and Bayesian methods lead to the same estimates.

Proof: Considering first the Bayesian method, we expand $\mathbf{g}(\boldsymbol{\zeta}_i^{(a)};\boldsymbol{\theta})$ about the ath iterate $\boldsymbol{\theta}^{(a)}$, namely

$$\mathbf{g}(\boldsymbol{\zeta}_i^{(a)};\boldsymbol{\theta}) \approx \mathbf{g}(\boldsymbol{\zeta}_i^{(a)};\boldsymbol{\theta}^{(a)}) + \mathbf{T}_{ia}(\boldsymbol{\theta}-\boldsymbol{\theta}^{(a)}), \tag{10.86}$$

where $\mathbf{T}_{ia}$ is $\mathbf{T}_i$ of (10.66) evaluated at $(\boldsymbol{\zeta}_i^{(a)}, \boldsymbol{\theta}_i^{(a)})$. Using (10.86) in (10.81), and replacing $\boldsymbol{\theta}$ by $\boldsymbol{\theta}^{(a)}$ in $\mathbf{S}_i^{(a)}$ to get $\mathbf{S}_{ia}$, we find that (10.81) becomes (10.70).

We now replace $\tilde{\boldsymbol{\zeta}}_i$ by $\boldsymbol{\zeta}_i^{(a)}$ in (10.77) and (10.78) to get the approximation

$$p(\boldsymbol{\theta}|\text{data}) \approx k \exp\left\{ -\frac{1}{2}\sum_{i=1}^{n} R_i[\mathbf{S}_i^{(a)}(\bar{\mathbf{z}}_i-\boldsymbol{\zeta}_i^{(a)}) + \mathbf{g}(\boldsymbol{\zeta}_i^{(a)};\boldsymbol{\theta})]' \right.$$
$$\left. \times (\mathbf{S}_i^{(a)}\mathbf{V}_1\mathbf{S}_i^{(a)\prime})^{-1}[\mathbf{S}_i^{(a)}(\bar{\mathbf{z}}_i-\boldsymbol{\zeta}_i^{(a)}) + \mathbf{g}(\boldsymbol{\zeta}_i^{(a)};\boldsymbol{\theta})] \right\}. \tag{10.87}$$

Using (10.86) once again and replacing $\mathbf{S}_i^{(a)}\,[=\mathbf{S}_i(\boldsymbol{\zeta}_i^{(a)};\boldsymbol{\theta})]$ by $\mathbf{S}_{ia}\,[=\mathbf{S}_i(\boldsymbol{\zeta}_i^{(a)};\boldsymbol{\theta}^{(a)})]$ in (10.87) gives us

$$p(\boldsymbol{\theta}|\text{data}) \approx k \exp\left\{ -\frac{1}{2}\sum_{i=1}^{n} R_i[\mathbf{K}_{ia} + \mathbf{T}_{ia}(\boldsymbol{\theta}-\boldsymbol{\theta}^{(a)})]' \right.$$
$$\left. \times (\mathbf{S}_{ia}\mathbf{V}_1\mathbf{S}_{ia}')^{-1}[\mathbf{K}_{ia} + \mathbf{T}_{ia}(\boldsymbol{\theta}-\boldsymbol{\theta}^{(a)})] \right\}, \tag{10.88}$$

where $\mathbf{K}_{ia} = \mathbf{S}_{ia}(\bar{\mathbf{z}}_i - \boldsymbol{\zeta}_i^{(a)}) + \mathbf{g}(\boldsymbol{\zeta}_i^{(a)}, \boldsymbol{\theta}^{(a)})$. The negative logarithm of (10.88) is a quadratic in $\boldsymbol{\theta}$ which is minimized when $\boldsymbol{\theta}-\boldsymbol{\theta}^{(a)}$ is equal to (c.f. Appendix A10.4)

$$\delta\boldsymbol{\theta}^{(a)} = -\left[\sum_i R_i \mathbf{T}_{ia}'(\mathbf{S}_{ia}\mathbf{V}_1\mathbf{S}_{ia}')^{-1}\mathbf{T}_{ia}\right]^{-1} \sum_i R_i \mathbf{T}_{ia}'(\mathbf{S}_{ia}\mathbf{V}_1\mathbf{S}_{ia}')^{-1}\mathbf{K}_{ia}.$$

This equation is the same as (10.69). Hence the Bayesian estimators are obtained by iterating on two equations which are the same as the two equations (10.69) and (10.70) used for finding the maximum-likelihood estimates. We also note that the Bayes approximate method is the same as the Deming approximate algorithm, as they are both obtained by setting $\boldsymbol{\zeta}_i^{(a)} = \bar{\mathbf{z}}_i$. ■

10.6 IMPLICIT RELATIONSHIPS WITH SOME UNOBSERVABLE RESPONSES

10.6.1 Introduction

Sachs [1976] considered the implicit model $\mathbf{h} = \mathbf{h}(\boldsymbol{\eta}, \mathbf{x}; \boldsymbol{\theta}) = \mathbf{0}$ for a titration experiment, where

$$\begin{aligned} h_1 &= \frac{\theta_3 k_4}{k_4 + x} - \eta_2 - \eta_3 - \eta_4 = 0, \\ h_2 &= \left(\frac{10^{-\eta_1}}{\beta} + \eta_3 + 2\eta_4 - \frac{10^{-(k_5 - \eta_1)}}{\beta} \right) \left(\frac{k_4 + x}{k_6 x} \right) - 1 = 0, \\ h_3 &= \frac{\eta_1 10^{-\eta_1}}{\beta \eta_3 \theta_1} - 1 = 0, \\ h_4 &= \frac{\eta_3 10^{-\eta_2}}{\beta^3 \eta_4 \theta_2} - 1 = 0. \end{aligned} \tag{10.89}$$

Here $\boldsymbol{\eta} = (\eta_1, \eta_2, \eta_3, \eta_4)'$, $\boldsymbol{\theta} = (\theta_1, \theta_1, \theta_3)'$, x is measured with negligible error, and β is a complicated function of $\boldsymbol{\eta}$, x, and θ, being the solution of the two equations

$$\alpha = \frac{1}{2} \left(\frac{2k_7 k_4}{k_4 + x} + \frac{k_6 x}{k_4 + x} + \frac{10^{-(k_5 - \eta_1)}}{\beta} + \frac{10^{-\eta_1}}{\beta} + \eta_3 + \eta_4 \right)$$

and

$$\beta = 10^{-0.519[\sqrt{\alpha}/(1 + \sqrt{\alpha}) - 0.2\alpha]}.$$

The quantities k_i $(i = 4, 5, 6, 7)$ are known, and only η_1 is measured (with error), so that η_2, η_3, and η_4 are inaccessible. It would be theoretically possible to use the four equations $\mathbf{h} = \mathbf{0}$ to eliminate η_2, η_3, and η_4 to get a single constraint $h(\eta_{i1}, \mathbf{x}_i; \boldsymbol{\theta}) = 0$ as in (10.60), and use the Britt–Luecke algorithm to estimate $\boldsymbol{\theta}$. However, the resulting equation would be very complex, and it would be difficult to derive analytically the partial derivatives needed for the algorithms of Sections 10.4 and 10.5. Instead, it seems preferable to develop a method which utilizes the original equations. Such methods are described by Sachs [1976], and the following discussion is based on his paper.

10.6.2 Least-Squares Estimation

Suppose that we have the general implicit-function model

$$\mathbf{h}(\boldsymbol{\eta}_i, \mathbf{x}_i; \boldsymbol{\theta}) = \mathbf{0} \qquad (i = 1, 2, \ldots, n), \tag{10.90}$$

where $\mathbf{h}$ is $q \times 1$, $\mathbf{x}_i$ is the ith measurement on a vector of v variables measured *without* error, and $\boldsymbol{\eta}_i = (\eta_{i1}, \eta_{i2}, \ldots, \eta_{iq})'$ is the unknown mean of the vector of q

response variables $\mathbf{y}_i$. However, only the tth elements of the $\mathbf{y}_i$ are observed, so that

$$y_{it} = \eta_{it} + \varepsilon_{it} \qquad (i = 1, 2, \ldots, n).$$

Let $\mathbf{u}_t = (y_{1t}, y_{2t}, \ldots, y_{nt})'$, $\mathbf{v}_t = (\eta_{1t}, \eta_{2t}, \ldots, \eta_{nt})'$, and $\mathbf{e}_t = (\varepsilon_{1t}, \varepsilon_{2t}, \ldots, \varepsilon_{nt})' = \mathbf{u}_t - \mathbf{v}_t$. It is assumed that $\mathscr{E}[\mathbf{e}_t] = \mathbf{0}$ and $\mathscr{D}[\mathbf{e}_t] = \sigma^2 \mathbf{W}$, where $\mathbf{W} = \operatorname{diag}(w_1, w_2, \ldots, w_n)$ is known and σ^2 is unknown. The least-squares estimate $\hat{\boldsymbol{\theta}}$ (and maximum-likelihood estimate under normality) is obtained by minimizing

$$Q = \mathbf{e}_t' \mathbf{W}^{-1} \mathbf{e}_t \tag{10.91}$$

subject to (10.90). In the course of finding $\hat{\boldsymbol{\theta}}$, we also obtain estimates $\hat{\mathbf{v}}_t$ and $\hat{\mathbf{e}}_t$. An estimate of σ^2 is then given by

$$\hat{\sigma}_e^2 = \frac{1}{n-p} \hat{\mathbf{e}}_t' \mathbf{W}^{-1} \hat{\mathbf{e}}_t.$$

Methods for finding $\hat{\boldsymbol{\theta}}$ are given by Sachs [1976], and we now outline his general procedure.

Assuming that it is possible to express each $\boldsymbol{\eta}_i$ uniquely as a function of $\boldsymbol{\theta}$, say $\boldsymbol{\eta}_i = \mathbf{f}(\mathbf{x}_i; \boldsymbol{\theta})$, we can define the $n \times p$ matrix

$$\begin{aligned} \mathbf{N}_{\boldsymbol{\theta}} &= \frac{\partial \mathbf{v}_t}{\partial \boldsymbol{\theta}'} \\ &= \left[\left(\frac{\partial \eta_{it}}{\partial \theta_j} \right) \right] \\ &= \left[\left(\frac{\partial f_t(\mathbf{x}_i; \boldsymbol{\theta})}{\partial \theta_j} \right) \right] \end{aligned} \tag{10.92}$$

as the Jacobian of $\mathbf{v}_t$ with respect to $\boldsymbol{\theta}$. (In the algorithm that follows we do not need to know f, only that it exists.) The least-squares estimate $\hat{\boldsymbol{\theta}}$ of $\boldsymbol{\theta}$ is then given by the following normal equations:

$$\begin{aligned} 0 &= \frac{\partial}{\partial \theta_j} \left(\sum_{i=1}^{n} w_i^{-1} (y_{it} - \eta_{it})^2 \right) \Bigg|_{\hat{\boldsymbol{\theta}}} \\ &= -2 \sum_{i=1}^{n} \left(\frac{\partial \eta_{it}}{\partial \theta_j} w_i^{-1} (y_{it} - \eta_{it}) \right) \Bigg|_{\hat{\boldsymbol{\theta}}} \qquad (j = 1, 2, \ldots, p), \end{aligned}$$

or

$$\hat{\mathbf{N}}_{\boldsymbol{\theta}}' \mathbf{W}^{-1} \hat{\mathbf{e}}_t = \mathbf{0}. \tag{10.93}$$

If $\boldsymbol{\theta}^{(a)}$ is an approximation for $\hat{\boldsymbol{\theta}}$, then we have the Taylor expansion

$$\begin{aligned}\hat{\mathbf{e}}_t &= \mathbf{u}_t - \hat{\mathbf{v}}_t \\ &\approx \mathbf{u}_t - \mathbf{v}_t^{(a)} - \mathbf{N}_\theta^{(a)}(\hat{\boldsymbol{\theta}} - \boldsymbol{\theta}^{(a)}) \\ &= \mathbf{e}_t^{(a)} - \mathbf{N}_\theta^{(a)}(\hat{\boldsymbol{\theta}} - \boldsymbol{\theta}^{(a)}),\end{aligned} \tag{10.94}$$

say. Substituting this approximation for $\hat{\mathbf{e}}_t$ into (10.93), and using the approximation $\mathbf{N}_\theta^{(a)} \approx \hat{\mathbf{N}}_\theta$, we have

$$\mathbf{N}_\theta^{(a)\prime}\mathbf{W}^{-1}[\mathbf{e}_t^{(a)} - \mathbf{N}_\theta^{(a)}(\hat{\boldsymbol{\theta}} - \boldsymbol{\theta}^{(a)})] \approx \mathbf{0},$$

or

$$\hat{\boldsymbol{\theta}} - \boldsymbol{\theta}^{(a)} \approx (\mathbf{N}_\theta^{(a)\prime}\mathbf{W}^{-1}\mathbf{N}_\theta^{(a)})^{-1}\mathbf{N}_\theta^{(a)\prime}\mathbf{W}^{-1}\mathbf{e}_t^{(a)}. \tag{10.95}$$

Hence an iterative (Gauss–Newton) procedure for finding $\hat{\boldsymbol{\theta}}$ is given by

$$\boldsymbol{\theta}^{(a+1)} = \boldsymbol{\theta}^{(a)} + (\mathbf{N}_\theta^{(a)\prime}\mathbf{W}^{-1}\mathbf{N}_\theta^{(a)})^{-1}\mathbf{N}_\theta^{(a)\prime}\mathbf{W}^{-1}(\mathbf{u}_t - \mathbf{v}_t^{(a)}). \tag{10.96}$$

To carry out this iteration we need a method of finding $\mathbf{v}_t^{(a)}$ and $\mathbf{N}_\theta^{(a)}$. Such a method is described below.

Given $\boldsymbol{\theta}^{(a)}$, there is a unique $\boldsymbol{\eta}_i^{(a)}$ $(i = 1, 2, \ldots, n)$ such that $\boldsymbol{\eta}_i^{(a)} = \mathbf{f}(\mathbf{x}_i; \boldsymbol{\theta}^{(a)})$. This can be found by solving

$$\mathbf{h}(\boldsymbol{\eta}_i^{(a)}, \mathbf{x}_i; \boldsymbol{\theta}^{(a)}) = \mathbf{0} \tag{10.97}$$

iteratively for $\boldsymbol{\eta}_i^{(a)}$. We then select from the $\boldsymbol{\eta}_i^{(a)}$ the elements

$$\mathbf{v}_t^{(a)} = (\eta_{1t}^{(a)}, \eta_{2t}^{(a)}, \ldots, \eta_{nt}^{(a)})'.$$

Using a Taylor expansion in the neighborhood of $\boldsymbol{\eta}_i^{(a,b)}$, an approximation for $\boldsymbol{\eta}_i^{(a)}$, we have

$$\begin{aligned}\mathbf{0} &= \mathbf{h}(\boldsymbol{\eta}_i^{(a)}, \mathbf{x}_i; \boldsymbol{\theta}^{(a)}) \\ &\approx \mathbf{h}(\boldsymbol{\eta}_i^{(a,b)}, \mathbf{x}_i; \boldsymbol{\theta}^{(a)}) + \mathbf{H}_{\eta_i}^{(a,b)}(\boldsymbol{\eta}_i^{(a)} - \boldsymbol{\eta}_i^{(a,b)}),\end{aligned}$$

or

$$\boldsymbol{\eta}_i^{(a)} - \boldsymbol{\eta}_i^{(a,b)} \approx -[\mathbf{H}_{\eta_i}^{(a,b)}]^{-1}\mathbf{h}(\boldsymbol{\eta}_i^{(a,b)}, \mathbf{x}_i; \boldsymbol{\theta}^{(a)}),$$

where $\mathbf{H}_{\eta_i}^{(a,b)}$ is $\partial\mathbf{h}/\partial\boldsymbol{\eta}'$ evaluated at $(\boldsymbol{\eta}_i^{(a,b)}, \boldsymbol{\theta}^{(a)})$. This suggests the Newton iterative procedure

$$\boldsymbol{\eta}_i^{(a,b+1)} = \boldsymbol{\eta}_i^{(a,b)} - [\mathbf{H}_{\eta_i}^{(a,b)}]^{-1}\mathbf{h}(\boldsymbol{\eta}_i^{(a,b)}, \mathbf{x}_i; \boldsymbol{\theta}^{(a)}) \tag{10.98}$$

for finding $\boldsymbol{\eta}_i^{(a)}$.

To find $\mathbf{N}_\theta^{(a)}$ [c.f. (10.92)] we note that

$$\begin{aligned}\mathbf{0} &= \mathbf{h}(\boldsymbol{\eta}, \mathbf{x}; \boldsymbol{\theta}) \\ &= \mathbf{h}[\mathbf{f}(\mathbf{x}; \boldsymbol{\theta}), \mathbf{x}; \boldsymbol{\theta}].\end{aligned}$$

Applying the chain rule of partial differentiation gives us

$$\sum_{j=1}^{q} \frac{\partial h_i}{\partial \eta_j} \frac{\partial f_j}{\partial \theta_r} + \frac{\partial h_i}{\partial \theta_r} = 0 \qquad (i = 1, 2, \ldots, q, \quad r = 1, 2, \ldots, p),$$

or

$$\mathbf{H}_\eta \mathbf{F}_\theta + \mathbf{H}_\theta = \mathbf{0},$$

so that $\mathbf{F}_\theta = -\mathbf{H}_\eta^{-1}\mathbf{H}_\theta$. If the above Jacobians are evaluated at $(\boldsymbol{\eta}_i^{(a)}, \mathbf{x}_i, \boldsymbol{\theta}^{(a)})$, we write

$$\mathbf{F}_{\theta i}^{(a)} = -[\mathbf{H}_{\eta_i}^{(a)}]^{-1}\mathbf{H}_{\theta i}^{(a)}, \tag{10.99}$$

where

$$\mathbf{F}_{\theta i} = \begin{bmatrix} \dfrac{\partial f_1(\mathbf{x}_i; \boldsymbol{\theta})}{\partial \theta_1} & \dfrac{\partial f_1(\mathbf{x}_i; \boldsymbol{\theta})}{\partial \theta_2} & \cdots & \dfrac{\partial f_1(\mathbf{x}_i; \boldsymbol{\theta})}{\partial \theta_p} \\ \dfrac{\partial f_2(\mathbf{x}_i; \boldsymbol{\theta})}{\partial \theta_1} & \dfrac{\partial f_2(\mathbf{x}_i; \boldsymbol{\theta})}{\partial \theta_2} & \cdots & \dfrac{\partial f_2(\mathbf{x}_i; \boldsymbol{\theta})}{\partial \theta_p} \\ \vdots & \vdots & & \vdots \\ \dfrac{\partial f_q(\mathbf{x}_i; \boldsymbol{\theta})}{\partial \theta_1} & \dfrac{\partial f_q(\mathbf{x}_i; \boldsymbol{\theta})}{\partial \theta_2} & \cdots & \dfrac{\partial f_q(\mathbf{x}_i; \boldsymbol{\theta})}{\partial \theta_p} \end{bmatrix}.$$

Comparing the above with $\mathbf{N}_\theta$ of (10.9), we see that the ith row of $\mathbf{N}_\theta^{(a)}$ is the same as the tth row of $\mathbf{F}_{\theta i}^{(a)}$. Hence $\mathbf{N}_\theta^{(a)}$ can be computed using (10.99).

10.6.3 The Algorithm

The three stages of the above method can now be put together. Given $\boldsymbol{\theta}^{(a)}$ and $\boldsymbol{\eta}_i^{(a,0)}$ $(i = 1, 2, \ldots, n)$, we proceed as follows. For $i = 1, 2, \ldots, n$ carry out steps 1 to 4 below:

1. For $b = 0, 1, \ldots$ until $\| \boldsymbol{\eta}_i^{(a,b+1)} - \boldsymbol{\eta}_i^{(a,b)} \|$ is sufficiently small, compute

$$\boldsymbol{\eta}_i^{(a,b+1)} = \boldsymbol{\eta}_i^{(a,b)} - [\mathbf{H}_{\eta_i}^{(a,b)}]^{-1}\mathbf{h}(\boldsymbol{\eta}_i^{(a,b)}, \mathbf{x}_i; \boldsymbol{\theta}^{(a)}).$$

   Take $\boldsymbol{\eta}_i^{(a)}$ and $\mathbf{H}_{\eta_i}^{(a)}$ as $\boldsymbol{\eta}_i^{(a,b+1)}$ and $\mathbf{H}_{\eta_i}^{(a,b)}$, respectively, from the final iteration.
2. Compute $\varepsilon_{it}^{(a)} = u_{it} - \eta_{it}^{(a)}$, the ith element of $\mathbf{e}_t^{(a)}$.
3. Compute $\mathbf{N}_\theta^{(a)}|_i = -[\mathbf{H}_{\eta_i}^{(a)}]^{-1}\mathbf{H}_{\theta i}^{(a)}|_t$, where$|_u$ denotes the uth row.
4. Set $\boldsymbol{\eta}_i^{(a+1,0)} = \boldsymbol{\eta}_i^{(a)}$.

We can now obtain a new estimate, $\boldsymbol{\theta}^{(a+1)}$, given by

$$\boldsymbol{\theta}^{(a+1)} = \boldsymbol{\theta}^{(a)} + [\mathbf{N}_\theta^{(a)\prime}\mathbf{W}^{-1}\mathbf{N}_\theta^{(a)}]^{-1}\mathbf{N}_\theta^{(a)\prime}\mathbf{W}^{-1}\mathbf{e}_t^{(a)}.$$

The whole of the above cycle is then repeated until $\boldsymbol{\theta}^{(a)}$ converges, to $\hat{\boldsymbol{\theta}}$ say. To start the process we require suitable initial approximations $\boldsymbol{\theta}^{(0)}$ and $\boldsymbol{\eta}_i^{(0,0)}$ $(i = 1, 2, \ldots, n)$. The algorithm can be applied with a variable or fixed number of iterations for the Newton procedure in step 1.

In order for the various stages of the above algorithm to be valid, at least the following assumptions are required:

a. Over a suitably large range of values Ω of $(\boldsymbol{\eta}, \mathbf{x}, \boldsymbol{\theta})$:

(i) $\mathbf{h}(\boldsymbol{\eta}, \mathbf{x}; \boldsymbol{\theta}) = \mathbf{0}$.

(ii) The Jacobian of $\mathbf{h}$ with respect to $(\boldsymbol{\eta}, \mathbf{x}, \boldsymbol{\theta})$ exists and is continuous.

(iii) The $q \times q$ matrix $\mathbf{H}_\eta$ is nonsingular. Then the implicit-function theorem (e.g. Cooper and Steinberg [1970: 108]) guarantees the existence of a set of q functions $\mathbf{f}$ which are unique, continuous, and differentiable in Ω such that $\boldsymbol{\eta} = \mathbf{f}(\mathbf{x}; \boldsymbol{\theta})$.

(iv) $\mathbf{N}_\theta$ has rank p.

b. If Q is defined by (10.91), then the Hessian $[(\partial^2 Q/\partial\theta_r\, \partial\theta_s)]$ is positive definite in a neighborhood of $\hat{\boldsymbol{\theta}}$. This assumption implies that if the process converges to $\hat{\boldsymbol{\theta}}$, then $\hat{\boldsymbol{\theta}}$ is at least a local minimum.

This completes the list of assumptions.

To find confidence intervals for the elements of $\boldsymbol{\theta}$ we require an expression for $\mathscr{D}[\hat{\boldsymbol{\theta}}]$. For large n, $\hat{\boldsymbol{\theta}}$ will be close to the true value of $\boldsymbol{\theta}$ and we have the Taylor expansion

$$\begin{aligned}\mathbf{e}_t &= \mathbf{u}_t - \mathbf{v}_t \\ &\approx \mathbf{u}_t - \hat{\mathbf{v}}_t - \hat{\mathbf{N}}_\theta(\boldsymbol{\theta} - \hat{\boldsymbol{\theta}}). \\ &= \hat{\mathbf{e}}_t - \hat{\mathbf{N}}_\theta(\boldsymbol{\theta} - \hat{\boldsymbol{\theta}}).\end{aligned}$$

Substituting for $\hat{\mathbf{e}}_t$ in (10.93) gives us

$$\hat{\mathbf{N}}'_\theta\mathbf{W}^{-1}[\mathbf{e}_t + \hat{\mathbf{N}}_\theta(\boldsymbol{\theta} - \hat{\boldsymbol{\theta}})] \approx \mathbf{0},$$

or

$$\begin{aligned}\hat{\boldsymbol{\theta}} - \boldsymbol{\theta} &\approx (\hat{\mathbf{N}}'_\theta\mathbf{W}^{-1}\hat{\mathbf{N}}_\theta)^{-1}\hat{\mathbf{N}}'_\theta\mathbf{W}^{-1}\mathbf{e}_t \\ &= \hat{\mathbf{A}}_\theta\mathbf{e}_t,\end{aligned}$$

say. Sachs [1976] effectively assumes that $\hat{\mathbf{A}}_\theta \approx \mathbf{A}_\theta$, so that

$$\begin{aligned}\mathscr{D}[\hat{\boldsymbol{\theta}}] &\approx \mathbf{A}_\theta\mathscr{D}[\mathbf{e}_t]\mathbf{A}'_\theta \\ &= \sigma^2\mathbf{A}_\theta\mathbf{W}\mathbf{A}'_\theta \\ &= \sigma^2(\mathbf{N}'_\theta\mathbf{W}^{-1}\mathbf{N}_\theta)^{-1}. \qquad (10.100)\end{aligned}$$

Table 10.5 Potentiometric Titration of *N, N*-Dimethylaminoethylamine[a]

i	u_{i1}	$\hat{\varepsilon}_{i1}$	x_i	$10^3\hat{\eta}_{i2}$	$10^3\hat{\eta}_{i3}$	$10^3\hat{\eta}_{i4}$
1	10.002	0.0153	1.695	31.636	8.050	1.701×10^{-3}
2	9.820	0.0229	2.402	27.575	10.856	3.550×10^{-3}
3	9.739	0.0219	2.743	25.693	12.161	4.781×10^{-3}
4	9.643	0.0249	3.195	23.276	13.838	6.834×10^{-3}
5	9.559	0.0248	3.599	21.190	15.286	9.161×10^{-3}
6	9.457	0.0246	4.104	18.678	17.030	12.901×10^{-3}
7	9.339	0.0211	4.678	15.944	18.927	18.669×10^{-3}
8	9.276	0.0213	4.990	14.511	19.921	22.724×10^{-3}
9	9.145	0.0162	5.589	11.858	21.757	33.172×10^{-3}
10	8.994	0.0032	6.190	9.323	23.504	49.247×10^{-3}
11	8.830	−0.0017	6.790	6.915	25.148	76.017×10^{-3}
12	8.731	−0.0081	7.090	5.759	25.927	97.027×10^{-3}
13	8.660	−0.0086	7.294	4.993	26.437	1.164×10^{-1}
14	8.497	−0.0234	7.660	3.665	27.295	1.690×10^{-1}
15	8.230	−0.0542	8.097	2.194	28.150	3.004×10^{-1}
16	7.991	−0.0798	8.386	1.357	28.478	4.971×10^{-1}
17	7.735	−0.0814	8.671	75.282×10^{-2}	28.387	8.912×10^{-1}
18	7.430	−0.0566	9.070	34.022×10^{-2}	27.434	18.449×10^{-1}
19	7.194	−0.0264	9.531	17.294×10^{-2}	25.772	32.109×10^{-1}
20	7.026	−0.0101	9.980	10.532×10^{-2}	24.023	45.915×10^{-1}
21	6.889	−0.0078	10.410	7.098×10^{-2}	22.338	59.041×10^{-1}
22	6.777	−0.0019	10.840	5.003×10^{-2}	20.676	71.925×10^{-1}
23	6.675	0.0028	11.280	3.594×10^{-2}	19.009	84.810×10^{-1}
24	6.578	0.0055	11.730	2.603×10^{-2}	17.344	97.659×10^{-1}
25	6.474	0.0055	12.230	1.835×10^{-2}	15.542	11.154
26	6.378	0.0085	12.725	1.296×10^{-2}	13.808	12.489
27	6.280	0.0104	13.229	90.100×10^{-4}	12.094	13.809
28	6.184	0.0145	13.726	61.768×10^{-4}	10.451	15.072
29	6.078	0.0137	14.225	41.009×10^{-4}	8.849	16.305
30	5.959	0.0091	14.727	25.911×10^{-4}	7.282	17.509
31	5.837	0.0137	15.221	15.361×10^{-4}	5.783	18.661
32	5.686	0.0147	15.720	8.062×10^{-4}	4.311	19.792
33	5.489	0.0158	16.220	3.407×10^{-4}	2.877	20.894
34	5.249	0.0161	16.634	1.171×10^{-4}	1.721	21.782
35	5.045	0.0128	16.859	47.424×10^{-6}	1.107	22.253
36	4.852	0.0064	17.000	20.326×10^{-6}	72.908×10^{-2}	22.542
37	4.656	0.0080	17.101	8.251×10^{-6}	46.650×10^{-2}	22.741
38	4.552	0.0068	17.140	5.153×10^{-6}	36.925×10^{-2}	22.814
39	4.453	0.0024	17.170	3.341×10^{-6}	29.764×10^{-2}	22.867
40	4.220	0.0017	17.228	1.151×10^{-6}	17.499×10^{-2}	22.953

[a]From Sachs [1976].

Table 10.6 Parameter Estimates and Their Estimated Standard Deviations for the Data of Table 10.5[a]

Parameter	Estimate	Standard Deviation
θ_1	5.2592×10^{-10}	0.0848×10^{-10}
θ_2	1.0665×10^{-6}	0.0169×10^{-6}
θ_3	0.043051	0.000011
		$\hat{\sigma}_e = 0.027$

[a] From Sachs [1976]. Known constants: $k_4 = 20.000$, $k_5 = 13.685$, $k_6 = 0.09975$, $k_7 = 0.09658$.

In concluding the above theoretical development a word of caution is appropriate. First, we do not know if the algorithm will converge or not. Conditions for convergence have yet to be established. Sachs [1976] notes that other algorithms can be used instead of the Gauss–Newton and Newton stages. He also discusses extensions of the theory to the multivariate case when more than one response is accessible. Second, the problem of finding the correct asymptotic variance–covariance matrix for $\hat{\boldsymbol{\theta}}$ in previous unreplicated models [see the comments following (10.58)] suggests that (10.100) is suspect.

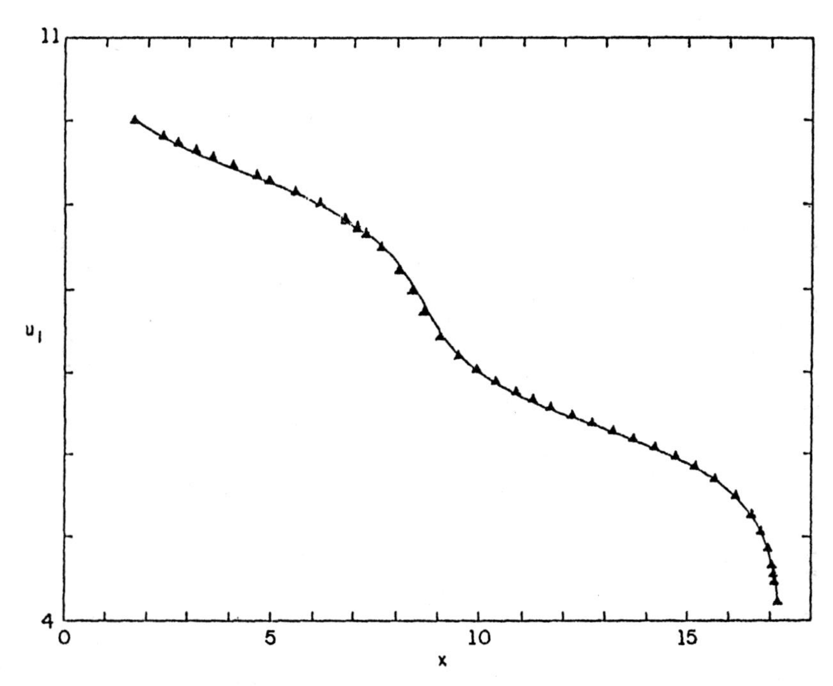

Figure 10.1 The (u_{i1}, x_i) data are plotted as filled triangles. The solid line is the titration curve computed with the parameter values of Table 10.6. From Sachs [1976].

Example 10.4 We now return to the model of Sachs [1976] introduced at the beginning of Section 10.6.1 [see (10.89)]. In terms of the general theory, $t = 1$, and 40 observation pairs (x_i, u_{i1}) are listed in Table 10.5. The known constants were $k_4 = 20.00$, $k_5 = 13.685$, $k_6 = 0.09975$, and $k_7 = 0.09658$. Reasonable initial estimates were available (Table 10.6), and, using just one iteration in the Newton step, convergence was achieved in 10 iterations. The final estimates are given in Table 10.6. The parameter β was also calculated iteratively throughout the course of the computation.

By eliminating η_2, η_3, and η_4 from (10.89), it is possible to express η_1 in terms of x, namely $\eta_1 = f(x; \boldsymbol{\theta}, \beta)$. To evaluate how well the model fits the data, the data pairs (x_i, u_{i1}) were plotted as filled triangles in Fig. 10.1. The fitted model $\hat{\eta}_1 = f(x; \hat{\boldsymbol{\theta}}, \hat{\beta})$ was plotted as the solid curve, and we see that this curve fits the data well. ∎

10.7 STRUCTURAL AND ULTRASTRUCTURAL MODELS

In Section 10.2 we investigated the model in which n observations were made on a functional relationship, namely

$$\mu_i = f(\boldsymbol{\xi}_i; \boldsymbol{\theta}) \qquad (i = 1, 2, \ldots, n) \tag{10.101}$$

with $y_i = \mu_i + \varepsilon_i$ and $\mathbf{x}_i = \boldsymbol{\xi}_i + \boldsymbol{\delta}_i$. Suppose that μ_i and $\boldsymbol{\xi}_i$ are regarded as random variables rather than mathematical variables, so that (10.101) becomes

$$v_i = f(\mathbf{u}_i; \boldsymbol{\theta}), \tag{10.102}$$

where $\mathbf{u}_i$ has some specified distribution with mean $\boldsymbol{\alpha}$. This model (10.102), in which the $\mathbf{u}_i$ have a common mean (usually the $\mathbf{u}_i$ are a random sample), is called a *structural* relationship. Dolby [1976b] proposed a generalization of this, which he called an ultrastructural relationship, in which the $\mathbf{u}_i$ have different means $\boldsymbol{\alpha}_i$. This generalization represents a synthesis of the functional and structural models. It is a functional relationship when $\mathscr{D}[\mathbf{u}_i] = \mathbf{0}$, i.e. $\mathbf{u}_i \equiv \boldsymbol{\alpha}_i$ ($= \boldsymbol{\xi}_i$ in Section 10.2), and a structural relationship when $\boldsymbol{\alpha}_i = \boldsymbol{\alpha}$.

The case when f is linear with a single regressor variable, namely

$$v_i = \theta_1 + \theta_2 u_i, \tag{10.103}$$

has received considerable attention in the literature (e.g. Moran [1971], Dolby [1976b], Seber [1977: 210], Bickel and Ritov [1987]). One helpful feature of the linear model is that the joint unconditional distribution of (x_i, y_i) can be found and a likelihood function constructed. For example, if $y_i = v_i + \varepsilon_i$, $x_i = u_i + \delta_i$, $\varepsilon_i \sim N(0, \sigma_\varepsilon^2)$, $\delta_i \sim N(0, \sigma_\delta^2)$, $u_i \sim N(\alpha_i, \sigma_u^2)$, and the last three random variables are mutually independent, then (x_i, y_i) have a bivariate normal distribution with mean ϕ_i and variance–covariance matrix Σ, where

$$\phi_i = \begin{pmatrix} \alpha_i \\ \theta_1 + \theta_2\alpha_i \end{pmatrix} \quad \text{and} \quad \Sigma = \begin{pmatrix} \sigma_\delta^2 + \sigma_u^2 & \theta_2\sigma_u^2 \\ \theta_2\sigma_u^2 & \sigma_\varepsilon^2 + \theta_2^2\sigma_u^2 \end{pmatrix}.$$

Setting $\alpha_i = \alpha$ leads to a linear structural model, and setting $\sigma_u^2 = 0$ leads to a linear functional model. However, this linear ultrastructural model and its special cases have problems associated with the solution of the maximum-likelihood equations unless certain variance ratios are known. When $\sigma_u^2 = 0$, we find from the theory of the straight-line functional model that we must know $\sigma_\varepsilon^2/\sigma_\delta^2$; otherwise the parameters are not identifiable.

When $\sigma_\varepsilon^2/\sigma_\delta^2$ and $\sigma_u^2/\sigma_\delta^2$ are both known, Dolby [1976b] gives the maximum-likelihood equations and estimates for the ultrastructural model. However, he incorrectly used the information matrix to find the asymptotic variance–covariance matrix of the parameter estimates. Patefield [1978] pointed out that the maximum-likelihood estimates can be inconsistent in some cases, and gave modifications. He also noted that the information matrix is not appropriate for variance estimation and gave the correct asymptotic variance–covariance matrix. When $\sigma_\varepsilon^2/\sigma_\delta^2$ is known and $\sigma_u^2/\sigma_\delta^2$ is unknown, Dolby [1976b] showed that the maximum-likelihood equations have no solution. However, Gleser [1985] proved that maximum-likelihood estimates still exist, but they occur at the part of the boundary of the parameter space defined by requiring $\sigma_u^2 = 0$. Hence the maximum-likelihood estimates are identical to those of the corresponding parameters in the linear functional model with known ratio $\sigma_\varepsilon^2/\sigma_\delta^2$. Multivariate versions of these linear models are given by Morton [1981], Chan and Mak [1983], and Amemiya and Fuller [1984].

The problem of having to know the variance ratios can be avoided by using replication. One such ultrastructural model is considered by Cox [1976] and Dolby [1976b]. It takes the form

$$v_{ir} = \theta_1 + \theta_2 u_{ir} \qquad (r = 1, 2, \ldots, R), \tag{10.104}$$

together with $y_{ir} = v_{ir} + \varepsilon_{ir}$, $x_{ir} = u_{ir} + \delta_{ir}$, $\varepsilon_{ir} \sim N(0, \sigma_\varepsilon^2)$, $\delta_{ir} \sim N(0, \sigma_\delta^2)$, and $u_{ir} \sim N(\alpha_i, \sigma_u^2)$, where the ε_{ir}, δ_{ir}, and u_{ir} are mutually independent. However, when $\alpha_i = \alpha$, this model does not reduce to the usual replicated structural model and the parameters become unidentifiable. Maximum-likelihood estimation of the usual replicated model (with v_{ir} and u_{ir} replaced by v_i and u_i) is considered in detail by Chan and Mak [1979].

When we generalize (10.104) to the nonlinear case

$$v_{ir} = g(u_{ir}; \boldsymbol{\theta}),$$

we run into difficulties, as the distribution of v_{ir} can no longer be readily expressed in terms of the distribution of u_{ir}. The likelihood function therefore becomes complicated, and approximate methods have to be sought (Brown [1978]).

Although we have not examined nonlinear models in this section, it is hoped that the above discussion of the simplest linear case will highlight some of the difficulties involved in developing a nonlinear theory.

10.8 CONTROLLED VARIABLES

We now consider a model which lies, in some sense, between the functional and structural models. Let

$$y = f(\mathbf{x}; \boldsymbol{\theta}) + \varepsilon, \tag{10.105}$$

where $\mathbf{x}$ is a vector of k random regressor variables, $E[\varepsilon] = 0$, and $\mathrm{var}[\varepsilon] = \sigma^2$. However, $\mathbf{x}$, although set on a known target value of $\mathbf{x}_0$, is unknown because of an unknown error $\boldsymbol{\delta} = (\delta_1, \delta_2, \ldots, \delta_n)'$ given by

$$\mathbf{x} = \mathbf{x}_0 + \boldsymbol{\delta}.$$

Here we assume that $\boldsymbol{\delta}$ is independent of ε, $\mathscr{E}[\boldsymbol{\delta}] = \mathbf{0}$, and $\mathscr{D}[\boldsymbol{\delta}] = \lambda^2 \boldsymbol{\Delta}(\mathbf{x}_0)$, where $\boldsymbol{\Delta}$ is a known positive definite matrix function of $\mathbf{x}_0$ (and possibly $\boldsymbol{\theta}$), and λ^2 is a *known* constant. We note that if $\boldsymbol{\Delta} \neq \mathbf{I}_k$ then λ is not unique. Our model now becomes

$$y = f(\mathbf{x}_0 + \boldsymbol{\delta}; \boldsymbol{\theta}) + \varepsilon,$$

a nonlinear generalization of Berkson's [1950] model (c.f. Section 1.5). This model may be compared with the functional-relationship model

$$y = f(\boldsymbol{\xi}; \boldsymbol{\theta}) + \varepsilon$$

and the structural-relationship model

$$y = f(\mathbf{x} - \boldsymbol{\delta}; \boldsymbol{\theta}) + \varepsilon,$$

$\mathbf{x}$ being known. To analyze (10.105) we follow the treatment of Fedorov [1972: Section 1.6; 1974], who used a slightly more general format in his 1974 paper. To begin with, we assume that for $r, s, t = 1, 2, \ldots, k$,

$$E[\delta_r \delta_s \delta_t] = \lambda^3 c_{rst}, \tag{10.106}$$

where c_{rst} is bounded. We then find approximations for the mean and variance of $f(\mathbf{x}; \boldsymbol{\theta})$ by using a Taylor expansion about $\mathbf{x}_0$, namely

$$f(\mathbf{x}_0 + \boldsymbol{\delta}; \boldsymbol{\theta}) = f(\mathbf{x}_0; \boldsymbol{\theta}) + \left\{ \frac{\partial f(\mathbf{x}; \boldsymbol{\theta})}{\partial \mathbf{x}} \right\}_{\mathbf{x}_0}' \boldsymbol{\delta} + \tfrac{1}{2} \boldsymbol{\delta}' \left\{ \frac{\partial^2 f(\mathbf{x}; \boldsymbol{\theta})}{\partial \mathbf{x}\, \partial \mathbf{x}'} \right\}_{\mathbf{x}_0} \boldsymbol{\delta} + \cdots \tag{10.107}$$

Now $\mathscr{E}[\boldsymbol{\delta}] = \mathbf{0}$, so that $E[v] = 0$ and, using $\text{tr}[\mathbf{AB}] = \text{tr}[\mathbf{BA}]$ for conformable matrices,

$$\begin{aligned}
\text{var}[v] &= E[v^2] \\
&= E[\text{tr}(vv)] \\
&= E\left[\text{tr}\left(\boldsymbol{\delta}\boldsymbol{\delta}'\left\{\frac{\partial f(\mathbf{x};\boldsymbol{\theta})}{\partial \mathbf{x}}\frac{\partial f(\mathbf{x};\boldsymbol{\theta})}{\partial \mathbf{x}'}\right\}_{\mathbf{x}_0}\right)\right] \\
&= \text{tr}\left(\mathscr{E}[\boldsymbol{\delta}\boldsymbol{\delta}']\left\{\frac{\partial f}{\partial \mathbf{x}}\frac{\partial f}{\partial \mathbf{x}'}\right\}_{\mathbf{x}_0}\right) \\
&= \lambda^2 \,\text{tr}\left[\boldsymbol{\Delta}(\mathbf{x}_0)\left\{\frac{\partial f}{\partial \mathbf{x}}\frac{\partial f}{\partial \mathbf{x}'}\right\}_{\mathbf{x}_0}\right].
\end{aligned} \tag{10.108}$$

Using similar algebra,

$$E[w] = \lambda^2 \,\text{tr}\left[\boldsymbol{\Delta}(\mathbf{x}_0)\left\{\frac{\partial^2 f}{\partial \mathbf{x}\,\partial \mathbf{x}'}\right\}_{\mathbf{x}_0}\right],$$

$$E[vw] = O(\lambda^3) \quad \text{and} \quad \text{var}[w] = O(\lambda^4).$$

Hence

$$E[f(\mathbf{x};\boldsymbol{\theta})] = f(\mathbf{x}_0;\boldsymbol{\theta}) + \tfrac{1}{2}E[w] + O(\lambda^3)$$

and

$$\begin{aligned}
\text{var}[f(\mathbf{x};\boldsymbol{\theta})] &= \text{var}[v + \tfrac{1}{2}w] + O(\lambda^3) \\
&= \text{var}[v] + O(\lambda^3).
\end{aligned}$$

Finally, we have from (10.105)

$$\begin{aligned}
a(\mathbf{x}_0;\boldsymbol{\theta}) &= E[y] \\
&= E[f(\mathbf{x};\boldsymbol{\theta})] \\
&= f(\mathbf{x}_0;\boldsymbol{\theta}) + \tfrac{1}{2}\lambda^2 \,\text{tr}\left[\boldsymbol{\Delta}(\mathbf{x}_0)\left\{\frac{\partial^2 f}{\partial \mathbf{x}\,\partial \mathbf{x}'}\right\}_{\mathbf{x}_0}\right] + O(\lambda^3),
\end{aligned} \tag{10.109}$$

and, since $\mathbf{x}$ is independent of ε,

$$\begin{aligned}
b(\mathbf{x}_0;\boldsymbol{\theta}) &= \text{var}[y] \\
&= \text{var}[\varepsilon + f(\mathbf{x};\boldsymbol{\theta})] \\
&= \sigma^2 + \text{var}[f(\mathbf{x};\boldsymbol{\theta})] \\
&= \sigma^2 + \lambda^2 \,\text{tr}\left(\boldsymbol{\Delta}(\mathbf{x}_0)\left\{\frac{\partial f}{\partial \mathbf{x}}\frac{\partial f}{\partial \mathbf{x}'}\right\}_{\mathbf{x}_0}\right) + O(\lambda^3).
\end{aligned} \tag{10.110}$$

Suppose that we have n independent pairs $(\mathbf{x}_i, y_i)$, with $\mathbf{x}_i = \mathbf{x}_{0i} + \boldsymbol{\delta}_i$ $(i = 1, 2, \ldots, n)$, from the model (10.105), and σ^2 is also known. Then Fedorov proposed the following weighted least-squares iterative procedure. If $\boldsymbol{\theta}^{(a)}$ is the ath iteration, then $\boldsymbol{\theta}^{(a+1)}$ is obtained by minimizing

$$\sum_{i=1}^{n} [b(\mathbf{x}_{0i}; \boldsymbol{\theta}^{(a)})]^{-1}[y_i - a(\mathbf{x}_{0i}; \boldsymbol{\theta})]^2 = \sum_i [b_i(\boldsymbol{\theta}^{(a)})]^{-1}[y_i - a_i(\boldsymbol{\theta})]^2$$

$$= [\mathbf{y} - \mathbf{a}(\boldsymbol{\theta})]' \mathbf{B}^{-1}(\boldsymbol{\theta}^{(a)})[\mathbf{y} - \mathbf{a}(\boldsymbol{\theta})],$$

say, with respect to $\boldsymbol{\theta}$, using the approximations (10.109) and (10.110). Here $\mathbf{a}(\boldsymbol{\theta})$ has elements $a_i(\boldsymbol{\theta})$, and $\mathbf{B}(\boldsymbol{\theta})$ is the diagonal matrix with elements $b_i(\boldsymbol{\theta})$. Given that the approximations are valid, this procedure is an example of the iteratively reweighted least-squares method described in Sections 2.2.2 and 2.8.8. The theory of Section 2.8.8 applies here. In particular, if $\boldsymbol{\theta}^{(a)}$ converges to $\tilde{\boldsymbol{\theta}}$, then, under fairly general conditions, $\sqrt{n}(\tilde{\boldsymbol{\theta}} - \boldsymbol{\theta}^*)$ is asymptotically $N_p(\mathbf{0}, \boldsymbol{\Omega})$, $\boldsymbol{\theta}^*$ being the true value of $\boldsymbol{\theta}$. The matrix $\boldsymbol{\Omega}$ is consistently estimated by

$$\tilde{\boldsymbol{\Omega}} = \frac{1}{n}[\mathbf{A}_{\bullet}'(\tilde{\boldsymbol{\theta}}) \mathbf{B}^{-1}(\tilde{\boldsymbol{\theta}}) \mathbf{A}_{\bullet}(\tilde{\boldsymbol{\theta}})]^{-1},$$

where $\mathbf{A}_{\bullet}(\boldsymbol{\theta}) = [(\partial a_i(\boldsymbol{\theta})/\partial \theta_j)]$. However, the usefulness of the above theory is severely restricted by the requirement of knowing both σ^2 and λ^2.

If we can ignore w in (10.107), then the model (10.105) becomes, for n observations,

$$\begin{aligned} y_i &= f(\mathbf{x}_{0i}; \boldsymbol{\theta}) + v_i + \varepsilon_i \\ &= f(\mathbf{x}_{0i}; \boldsymbol{\theta}) + \varepsilon_i^*, \end{aligned} \tag{10.111}$$

which is now a fixed-regressor model with $E[\varepsilon_i^*] = 0$ and

$$\operatorname{var}[\varepsilon_i^*] = \sigma^2 + \lambda^2 \operatorname{tr}\left[\boldsymbol{\Delta}(\mathbf{x}_{0i}) \left\{ \frac{\partial f}{\partial \mathbf{x}} \frac{\partial f}{\partial \mathbf{x}'} \right\}_{\mathbf{x}_{0i}} \right]. \tag{10.112}$$

Although the ε_i^* will be mutually independent, the variance of ε_i^* will vary with i, and this heterogeneity will need to be taken into account. However, if one suspects that the errors in $\mathbf{x}$ are small compared with ε, then it may be appropriate to fit (10.111) by ordinary least squares and then carry out the usual residual plots to detect heterogeneity.

CHAPTER 11

Multiresponse Nonlinear Models

11.1 GENERAL MODEL

In the previous chapters we considered the single-response model

$$y_i = f(\mathbf{x}_i; \boldsymbol{\theta}) + \varepsilon_i \qquad i = 1, 2, \ldots, n.$$

A natural extension of this is to replace the single response y_i by a vector of responses $\mathbf{y}_i$. The same regressor variables $\mathbf{x}_i$ will usually apply to each element y_{ij} of $\mathbf{y}_i$, but the functional form of $E[y_{ij}]$ will generally vary with j. We therefore have the multiresponse nonlinear model

$$y_{ij} = f_j(\mathbf{x}_i; \boldsymbol{\theta}) + \varepsilon_{ij} \qquad (i = 1, 2, \ldots, n, \quad j = 1, 2, \ldots, d), \tag{11.1}$$

which represents d nonlinear models each observed at n values of $\mathbf{x}$. This model can also be expressed in vector form

$$\mathbf{y}_i = \mathbf{f}(\mathbf{x}_i; \boldsymbol{\theta}) + \boldsymbol{\varepsilon}_i, \tag{11.2}$$

where the $\boldsymbol{\varepsilon}_i = (\varepsilon_{i1}, \varepsilon_{i2}, \ldots, \varepsilon_{id})'$ are usually assumed to be i.i.d. $N_d(\mathbf{0}, \boldsymbol{\Sigma})$. The formulation (11.1) also includes the compartmental models of Chapter 8, where we have measurements y_{ij} $(j = 1, 2, \ldots, d)$ taken on d of m compartments at n timepoints $x_1, x_2, \ldots, x_n$. We now have two explanatory regressor variables, time and the compartment measured, so we can write $E[y_{ij}] = f(x_i, j; \boldsymbol{\theta})$. This is a special case of (11.1) with $f_j(x_i; \boldsymbol{\theta}) = f(x_i, j; \boldsymbol{\theta})$.

If the model (11.2) were linear, it would take the form

$$\mathbf{y}_i = \mathbf{B}'\mathbf{x}_i + \boldsymbol{\varepsilon}_i \qquad (i = 1, 2, \ldots, n), \tag{11.3}$$

which can be expressed as

$$\begin{bmatrix} \mathbf{y}_1' \\ \mathbf{y}_2' \\ \vdots \\ \mathbf{y}_n' \end{bmatrix} = \begin{bmatrix} \mathbf{x}_1' \\ \mathbf{x}_2' \\ \vdots \\ \mathbf{x}_n' \end{bmatrix} \mathbf{B} + \begin{bmatrix} \boldsymbol{\varepsilon}_1' \\ \boldsymbol{\varepsilon}_2' \\ \vdots \\ \boldsymbol{\varepsilon}_n' \end{bmatrix},$$

or, in matrix form,

$$\mathbf{Y} = \mathbf{XB} + \mathbf{U}. \tag{11.4}$$

We can also express (11.1) in much the same way, namely

$$\mathbf{Y} = \mathbf{G}(\boldsymbol{\theta}) + \mathbf{U}, \tag{11.5}$$

where $\mathbf{G}(\boldsymbol{\theta})_{ij} = f_j(\mathbf{x}_i; \boldsymbol{\theta})$. However, it is also convenient to express this model in a "rolled-out" form in which the responses are listed in a single vector $\mathbf{y}$ $(= \operatorname{vec} \mathbf{Y})$, model by model. For the jth model we have

$$\begin{bmatrix} y_{1j} \\ y_{2j} \\ \vdots \\ y_{nj} \end{bmatrix} = \begin{bmatrix} f_j(\mathbf{x}_1; \boldsymbol{\theta}) \\ f_j(\mathbf{x}_2; \boldsymbol{\theta}) \\ \vdots \\ f_j(\mathbf{x}_n; \boldsymbol{\theta}) \end{bmatrix} + \begin{bmatrix} \varepsilon_{1j} \\ \varepsilon_{2j} \\ \vdots \\ \varepsilon_{nj} \end{bmatrix},$$

or

$$\mathbf{y}^{(j)} = \mathbf{f}^{(j)}(\boldsymbol{\theta}) + \boldsymbol{\varepsilon}^{(j)} \qquad (j = 1, 2, \ldots, d), \tag{11.6}$$

say, and the normality assumptions now take the form $\boldsymbol{\varepsilon}^{(j)} \sim N_n(\mathbf{0}, \sigma_{jj}\mathbf{I}_n)$ and $\mathscr{E}[\boldsymbol{\varepsilon}^{(j)}\boldsymbol{\varepsilon}^{(k)\prime}] = \sigma_{jk}\mathbf{I}_n$, where $[(\sigma_{jk})] = \boldsymbol{\Sigma}$. Then using

$$\mathbf{y} = \operatorname{vec} \mathbf{Y} = (\mathbf{y}^{(1)\prime}, \mathbf{y}^{(2)\prime}, \ldots, \mathbf{y}^{(d)\prime})',$$

we have

$$\mathbf{y} = \mathbf{f}(\boldsymbol{\theta}) + \boldsymbol{\varepsilon}, \tag{11.7}$$

say, where $\boldsymbol{\varepsilon} \sim N_{nd}(\mathbf{0}, \boldsymbol{\Omega})$ and

$$\begin{aligned} \boldsymbol{\Omega} &= \boldsymbol{\Sigma} \otimes \mathbf{I}_n \\ &= \begin{bmatrix} \sigma_{11}\mathbf{I}_n & \sigma_{12}\mathbf{I}_n & \cdots & \sigma_{1d}\mathbf{I}_n \\ \sigma_{21}\mathbf{I}_n & \sigma_{22}\mathbf{I}_n & \cdots & \sigma_{2d}\mathbf{I}_n \\ \vdots & \vdots & & \vdots \\ \sigma_{d1}\mathbf{I}_n & \sigma_{d2}\mathbf{I}_n & \cdots & \sigma_{dd}\mathbf{I}_n \end{bmatrix}. \end{aligned} \tag{11.8}$$

The above system actually allows for a more general formulation. For instance, the x-observations, $\boldsymbol{\theta}$, and the number of observations could vary from model to model, so that instead of (11.1) we have

$$y_{ij} = f_j(\mathbf{x}_{ij}; \boldsymbol{\theta}_j) + \varepsilon_{ij} \qquad (i = 1, 2, \ldots, n_j, \quad j = 1, 2, \ldots, d). \tag{11.9}$$

Example 11.1 An example of (11.9) is the following pair of equations investigated by Gallant [1975c]:

$$f_1(\mathbf{x}_{i1}; \boldsymbol{\theta}_1) = \theta_{11} + \theta_{12} x_{i11} + \theta_{13} \exp(\theta_{14} x_{i12})$$

and

$$f_2(x_{i2}; \boldsymbol{\theta}_2) = \theta_{21} + \theta_{22} \exp(\theta_{23} x_{i2}).$$

By setting $g_j(\mathbf{x}_{ij}; \boldsymbol{\theta}) = f_j(\mathbf{x}_{ij}; \boldsymbol{\theta}_j)$, where

$$\boldsymbol{\theta} = (\boldsymbol{\theta}'_1, \boldsymbol{\theta}'_2, \ldots, \boldsymbol{\theta}'_d)', \tag{11.10}$$

we can still express (11.9) in the form of (11.7). If the $\boldsymbol{\theta}_j$'s are not all different, then $\boldsymbol{\theta}$ is the vector of all the distinct elements making up the $\boldsymbol{\theta}_j$'s. With this more general formulation (11.2) becomes

$$\mathbf{y}_i = \mathbf{f}_i(\boldsymbol{\theta}) + \boldsymbol{\varepsilon}_i. \tag{11.11}$$

The model (11.6) also includes this generalization. ■

11.2 GENERALIZED LEAST SQUARES

We now drop the assumption that the $\boldsymbol{\varepsilon}_i$ are multivariate normal and assume they are i.i.d. with mean $\mathbf{0}$ and variance–covariance matrix $\boldsymbol{\Sigma}$. If $\boldsymbol{\Sigma}$ were known, an estimate of $\boldsymbol{\theta}$ could be obtained by minimizing

$$\begin{aligned} T(\boldsymbol{\theta}) &= \sum_{i=1}^{n} [\mathbf{y}_i - \mathbf{f}_i(\boldsymbol{\theta})]' \boldsymbol{\Sigma}^{-1} [\mathbf{y}_i - \mathbf{f}_i(\boldsymbol{\theta})] \\ &= [\mathbf{y} - \mathbf{f}(\boldsymbol{\theta})]' \boldsymbol{\Omega}^{-1} [\mathbf{y} - \mathbf{f}(\boldsymbol{\theta})] \end{aligned} \tag{11.12}$$

with respect to $\boldsymbol{\theta}$, when $\boldsymbol{\Omega}^{-1} = \boldsymbol{\Sigma}^{-1} \otimes \mathbf{I}_n$, to get a generalized least-squares estimate of $\boldsymbol{\theta}$. However, as $\boldsymbol{\Sigma}$ is unknown, Beauchamp and Cornell [1966] suggested using the method of Zellner [1962] whereby $\boldsymbol{\Sigma}$ is replaced by a consistent estimator as follows. Let $\overset{+}{\boldsymbol{\theta}}{}^{(j)}$ be the least-squares estimate of $\boldsymbol{\theta}$ obtained by fitting the jth single response model (11.6). Then $\boldsymbol{\varepsilon}^{(j)}$ is estimated by

$$\mathbf{e}_j = \mathbf{y}^{(j)} - \mathbf{f}^{(j)}(\overset{+}{\boldsymbol{\theta}}{}^{(j)}), \tag{11.13}$$

and σ_{rs} can be estimated by

$$\breve{\sigma}_{rs} = \frac{1}{n}\mathbf{e}_r'\mathbf{e}_s \qquad (r, s = 1, 2, \ldots, d). \tag{11.14}$$

We now minimize (11.12) using $\breve{\boldsymbol{\Sigma}} = [(\breve{\sigma}_{rs})]$ as an estimate of $\boldsymbol{\Sigma}$ to obtain an estimate, $\breve{\boldsymbol{\theta}}$ say, of $\boldsymbol{\theta}$. Define

$$\breve{\mathbf{W}} = \frac{1}{n}\mathbf{F}_{\bullet}'(\breve{\boldsymbol{\theta}})[\breve{\boldsymbol{\Sigma}}^{-1} \otimes \mathbf{I}_n]\mathbf{F}_{\bullet}(\breve{\boldsymbol{\theta}}),$$

where

$$\mathbf{F}_{\bullet}(\boldsymbol{\theta}) = \frac{\partial \mathbf{f}(\boldsymbol{\theta})}{\partial \boldsymbol{\theta}'} \tag{11.15}$$

$$= \left[\left(\frac{\partial f_r(\boldsymbol{\theta})}{\partial \theta_s}\right)\right]$$

$$= \begin{bmatrix} \mathbf{F}_{\bullet 1}(\boldsymbol{\theta}) \\ \mathbf{F}_{\bullet 2}(\boldsymbol{\theta}) \\ \vdots \\ \mathbf{F}_{\bullet d}(\boldsymbol{\theta}) \end{bmatrix} \tag{11.16}$$

and $\mathbf{F}_{\bullet j}(\boldsymbol{\theta}) = [(\partial f_u^{(j)}/\partial \theta_v)] = [\partial f_j(\mathbf{x}_u; \boldsymbol{\theta})/\partial \theta_v)]$, the usual matrix of derivatives for the jth model. Then, under appropriate regularity conditions (see Section 12.5), $\sqrt{n}(\breve{\boldsymbol{\theta}} - \boldsymbol{\theta})$ is asymptotically normal with mean zero and variance–covariance matrix for which $\breve{\mathbf{W}}^{-1}$ is a strongly consistent estimator.

For convenience the steps are summarized as follows:

1. Minimize $\|\mathbf{y}^{(j)} - \mathbf{f}^{(j)}(\boldsymbol{\theta})\|^2$ with respect to $\boldsymbol{\theta}$ to obtain $\overset{+}{\boldsymbol{\theta}}{}^{(j)}$, for each $j = 1, 2, \ldots, d$.
2. Calculate $\mathbf{e}_j$ from (11.13), and form $\breve{\boldsymbol{\Sigma}} = [(\breve{\sigma}_{rs})]$ using (11.14).
3. Form the model (11.7), and minimize

$$\breve{T}(\boldsymbol{\theta}) = [\mathbf{y} - \mathbf{f}(\boldsymbol{\theta})]'(\breve{\boldsymbol{\Sigma}}^{-1} \otimes \mathbf{I}_n)[\mathbf{y} - \mathbf{f}(\boldsymbol{\theta})] \tag{11.17}$$

with respect to $\boldsymbol{\theta}$ to get $\breve{\boldsymbol{\theta}}$.

The minimization of $\breve{T}(\boldsymbol{\theta})$ can be carried out in various ways. For example, we can use the usual Taylor expansion of $\mathbf{f}(\boldsymbol{\theta})$ for $\boldsymbol{\theta}$ close to an approximation $\boldsymbol{\theta}^{(a)}$,

namely

$$\begin{aligned}\mathbf{f}(\boldsymbol{\theta}) &\approx \mathbf{f}(\boldsymbol{\theta}^{(a)}) + \mathbf{F}_{\bullet}(\boldsymbol{\theta}^{(a)})(\boldsymbol{\theta} - \boldsymbol{\theta}^{(a)}) \\ &= \mathbf{f}^{(a)} + \mathbf{F}_{\bullet}^{(a)}(\boldsymbol{\theta} - \boldsymbol{\theta}^{(a)}),\end{aligned}$$

say, where $\mathbf{F}_{\bullet}(\boldsymbol{\theta})$ is given by (11.16). Substituting into $\check{T}(\boldsymbol{\theta})$ gives

$$\check{T}(\boldsymbol{\theta}) \approx [\mathbf{y} - \mathbf{f}^{(a)} - \mathbf{F}_{\bullet}^{(a)}(\boldsymbol{\theta} - \boldsymbol{\theta}^{(a)})]'\check{\boldsymbol{\Omega}}^{-1}[\mathbf{y} - \mathbf{f}^{(a)} - \mathbf{F}_{\bullet}^{(a)}(\boldsymbol{\theta} - \boldsymbol{\theta}^{(a)})],$$

which can be readily minimized with respect to $\boldsymbol{\theta}$. The next approximation for $\check{\boldsymbol{\theta}}$ is given by $\boldsymbol{\theta}^{(a+1)} = \boldsymbol{\theta}^{(a)} + \boldsymbol{\delta}^{(a)}$, where

$$\boldsymbol{\delta}^{(a)} = [\mathbf{F}_{\bullet}^{(a)\prime}\check{\boldsymbol{\Omega}}^{-1}\mathbf{F}_{\bullet}^{(a)}]^{-1}\mathbf{F}_{\bullet}^{(a)\prime}\check{\boldsymbol{\Omega}}^{-1}[\mathbf{y} - \mathbf{f}^{(a)}]. \tag{11.18}$$

Beauchamp and Cornell [1966] also considered the case where the jth model is observed at only n_j of the n values of $\mathbf{x}$. We can still use the model (11.6), but some of the elements of each $\mathbf{y}^{(j)}$ may be missing, so that $\mathbf{y}^{(j)}$ now has dimension n_j. Then σ_{rr} is estimated by

$$\dot{\sigma}_{rr} = \frac{1}{n_r}\mathbf{e}_r'\mathbf{e}_r,$$

as in (11.14), but we have to make an adjustment for estimating σ_{rs} $(r \neq s)$. If the rth and sth models have n_{rs} values of $\mathbf{x}$ in common, then we can use these values to find least-squares estimates, $\dot{\boldsymbol{\theta}}_{r(s)}$ and $\dot{\boldsymbol{\theta}}_{s(r)}$ respectively of $\boldsymbol{\theta}$, and compute the corresponding residuals $\mathbf{e}_{r(s)}$ and $\mathbf{e}_{s(r)}$. Then σ_{rs} is estimated by

$$\dot{\sigma}_{rs} = \frac{1}{n_{rs}}\mathbf{e}_{r(s)}'\mathbf{e}_{s(r)} \qquad (r \neq s),$$

and $\boldsymbol{\Sigma}$ is estimated by $\dot{\boldsymbol{\Sigma}} = [(\dot{\sigma}_{rs})]$.

Example 11.2 Beauchamp and Cornell [1966] applied their theory to the model

$$y_{ij} = f_j(x_i; \boldsymbol{\theta}) + \varepsilon_{ij} \qquad (i = 1, 2, \ldots, n, \quad j = 1, 2),$$

where $\boldsymbol{\theta}' = (\theta_1, \theta_2, \theta_3)$,

$$\begin{aligned}f_1(x_i; \boldsymbol{\theta}) &= \theta_1 \exp(-\theta_2 x_i) + (1 - \theta_1)\exp(-\theta_3 x_i) \\ f_2(x_i; \boldsymbol{\theta}) &= 1 - (\theta_1 + \theta_4)\exp(-\theta_2 x_i) + (\theta_1 + \theta_4 - 1)\exp(-\theta_3 x_i).\end{aligned}$$

This model, from Galambos and Cornell [1962], is based on a compartmental analysis in humans involving radioactive sulfate as a tracer. In Table 11.1 the observation y_{ij} refers to the proportion of injected radioactive tracer in compartment j of the body; x_i refers to the time in hours. Here $d=2$, $n_1=8$, $n_2=9$, $n=9$, and $n_{12}=8$.

An initial estimate $\boldsymbol{\theta}^{(0)}=(0.381, 0.021, 0.197)'$ was found graphically. Using the data y_{i1} $(i=2,3,\ldots,9)$ from the first model only, the least-squares estimate $\overset{+}{\boldsymbol{\theta}}{}^{(1)}=(0.555, 0.031, 0.171)'$ was found using Hartley's modified Gauss–Newton procedure (c.f. Section 14.2.1 with $\boldsymbol{\theta}^{(0)}$ as a starting value). Similarly, using the data y_{i2} $(i=1,2,\ldots,9)$ from the second model, the estimate $\overset{+}{\boldsymbol{\theta}}{}^{(2)}=(0.061, 0.007, 0.093)'$ was found. Since y_{11}, the first element of $\mathbf{y}^{(1)}$, was not observed, the latter calculation was repeated but with y_{12}, the first element of $\mathbf{y}^{(2)}$, deleted to give $\dot{\boldsymbol{\theta}}_{2(1)}=(0.060, 0.007, 0.093)'$. We note that $\dot{\boldsymbol{\theta}}_{1(2)}=\overset{+}{\boldsymbol{\theta}}{}^{(1)}$.

Table 11.1 Data on Proportions y_{i1} and y_{i2} of Radioactive Tracer in Two Different Compartments at Various Times x_i[a]

i	x_i (hours)	y_{i1}	y_{i2}
1	0.33		.03
2	2	.84	.10
3	3	.79	.14
4	5	.64	.21
5	8	.55	.30
6	12	.44	.40
7	24	.27	.54
8	48	.12	.66
9	72	.06	.71

[a]From Beauchamp and Cornell [1966].

Table 11.2 Estimated Error Vectors[a]

i (1)	$\mathbf{e}_1$ (2)	$\mathbf{e}_2$ (3)	$\mathbf{e}_{1(2)}$ (4)	$\mathbf{e}_{2(1)}$ (5)
1		.0128		
2	.0026	.0034	.0026	.0036
3	.0184	.0011	.0184	.0014
4	−.0237	−.0034	−.0237	−.0030
5	.0049	−.0038	.0049	−.0034
6	.0020	.0059	.0020	.0061
7	.0014	−.0048	.0014	.0050
8	−.0030	.0044	−.0030	.0042
9	.0022	−.0020	.0020	−.0018

[a]From Beauchamp and Cornell [1966].

From these estimates the residuals were calculated (see Table 11.2) giving

$$\dot{\sigma}_{11} = \sum(2)^2/8 = 0.1189 \times 10^{-3},$$
$$\dot{\sigma}_{22} = \sum(3)^2/9 = 0.3179 \times 10^{-4},$$
$$\dot{\sigma}_{12} = \sum(4)(5)/8 = 0.9753 \times 10^{-5},$$

where $\sum(a)(b)$ represents the sum of the products of corresponding elements in columns (a) and (b) of Table 11.2. The estimate $\dot{\boldsymbol{\Sigma}}$ was then used in the iterative process (11.18), using $\boldsymbol{\theta}^{(0)}$ as a starting value. Eleven iterations were carried out to give $\breve{\boldsymbol{\theta}} = (0.06751, 0.00706, 0.08393)'$ correct to four decimal places. Graphs showing the original data and the fitted equation for each response variable are given in Figs. 11.1 and 11.2. ■

In the case of Example 11.1 above, where $\boldsymbol{\theta}_j$ varies with the model, we find that $\mathbf{F}_{\bullet}(\boldsymbol{\theta})$ of (11.16) now takes the form

$$\mathbf{F}_{\bullet}(\boldsymbol{\theta}) = \text{diag}[\mathbf{F}_{\bullet 1}(\boldsymbol{\theta}_1), \mathbf{F}_{\bullet 2}(\boldsymbol{\theta}_2), \ldots, \mathbf{F}_{\bullet d}(\boldsymbol{\theta}_d)], \tag{11.19}$$

and steps 1–3 of the above algorithm still apply. As noted by Gallant [1975c], the generalized least-squares problem associated with (11.17) can be reduced to an (unweighted) least-squares problem using the standard reduction of Section 2.1.4 based on the Cholesky decomposition $\dot{\boldsymbol{\Sigma}} = \mathbf{U}'\mathbf{U}$, where $\mathbf{U}$ is upper triangular.

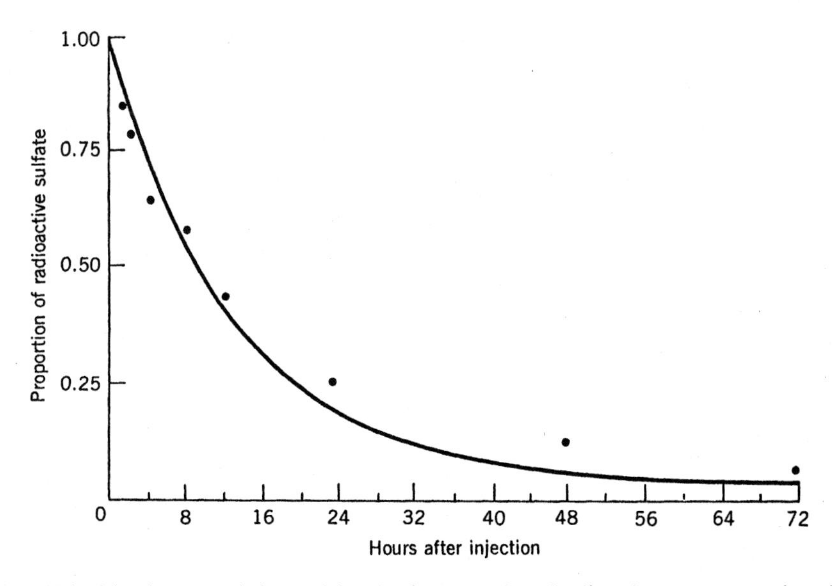

Figure 11.1 Fitted curve and observed data for the proportion of radioactive tracer versus time after injection for the first compartment. From Beauchamp and Cornell [1966].

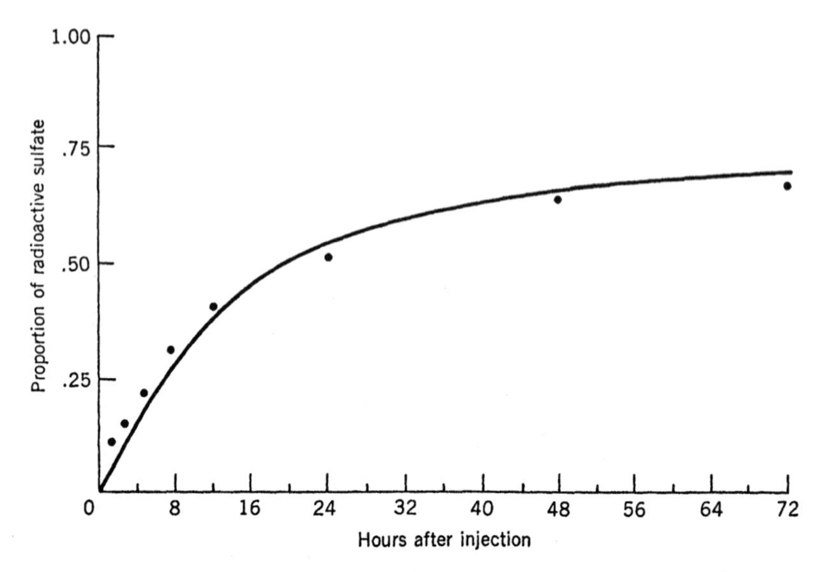

Figure 11.2 Fitted curve and observed data for the proportion of radioactive tracer versus time after injection for the second compartment. From Beauchamp and Cornell [1966].

Defining $\mathbf{R} = (\mathbf{U}')^{-1}$, $\mathbf{z}_i = \mathbf{R}\mathbf{y}_i$, and $\mathbf{k}_i(\boldsymbol{\theta}) = \mathbf{R}\mathbf{f}_i(\boldsymbol{\theta})$, we have, from (11.12),

$$\begin{aligned} \check{T}(\boldsymbol{\theta}) &= \sum_{i=1}^{n} [\mathbf{y}_i - \mathbf{f}_i(\boldsymbol{\theta})]'\check{\boldsymbol{\Sigma}}^{-1}[\mathbf{y}_i - \mathbf{f}_i(\boldsymbol{\theta})] \\ &= \sum_{i=1}^{n} [\mathbf{z}_i - \mathbf{k}_i(\boldsymbol{\theta})]'[\mathbf{z}_i - \mathbf{k}_i(\boldsymbol{\theta})] \\ &= [\mathbf{z} - \mathbf{k}(\boldsymbol{\theta})]'[\mathbf{z} - \mathbf{k}(\boldsymbol{\theta})], \end{aligned} \tag{11.20}$$

say, where $\mathbf{z}' = (\mathbf{z}_1', \mathbf{z}_2', \ldots, \mathbf{z}_n')$ and so on. Gallant [1975c: 40] also noted that $\check{\boldsymbol{\theta}}$, the minimum of (11.20), is asymptotically more efficient than the model-by-model estimate $(\overset{+}{\boldsymbol{\theta}}^{(1)\prime}, \overset{+}{\boldsymbol{\theta}}^{(2)\prime}, \ldots, \overset{+}{\boldsymbol{\theta}}^{(d)\prime})'$ [obtained by minimizing $\| \mathbf{y}^{(j)} - \mathbf{f}^{(j)}(\boldsymbol{\theta}_j) \|^2$ for $j = 1, 2, \ldots, d$] in the sense that the difference between their asymptotic variance–covariance matrices is positive definite.

11.3 MAXIMUM-LIKELIHOOD INFERENCE

11.3.1 Estimation

In the previous section we did not need to assume multivariate normality of the $\boldsymbol{\varepsilon}_i$. If we bring back this assumption, then the joint distribution of the "rolled-out" vector $\mathbf{y}$ ($= \operatorname{vec} \mathbf{Y}$) of responses is

$$p(\mathbf{y} \mid \boldsymbol{\theta}, \boldsymbol{\Sigma}) = (2\pi)^{-nd/2} |\boldsymbol{\Sigma}|^{-n/2} \exp\{-\tfrac{1}{2} Q(\boldsymbol{\theta}, \boldsymbol{\Sigma})\},$$

where

$$Q(\boldsymbol{\theta}, \boldsymbol{\Sigma}) = \sum_{i=1}^{n} [\mathbf{y}_i - \mathbf{f}_i(\boldsymbol{\theta})]'\boldsymbol{\Sigma}^{-1}[\mathbf{y}_i - \mathbf{f}_i(\boldsymbol{\theta})]$$
$$= \operatorname{tr}\{\boldsymbol{\Sigma}^{-1}\mathbf{V}(\boldsymbol{\theta})\}, \tag{11.21}$$

and

$$\mathbf{V}(\boldsymbol{\theta}) = \sum_{i=1}^{n} [\mathbf{y}_i - \mathbf{f}_i(\boldsymbol{\theta})][\mathbf{y}_i - \mathbf{f}_i(\boldsymbol{\theta})]'$$
$$= [\mathbf{Y} - \mathbf{G}(\boldsymbol{\theta})]'[\mathbf{Y} - \mathbf{G}(\boldsymbol{\theta})]. \tag{11.22}$$

Here $\mathbf{Y} = [(y_{ij})]$ and $\mathbf{G}(\boldsymbol{\theta}) = [(f_j\{\mathbf{x}_i; \boldsymbol{\theta}\})]$. Apart from a constant, the log-likelihood function is given by

$$L(\boldsymbol{\theta}, \boldsymbol{\Sigma}) = -\frac{n}{2}\log|\boldsymbol{\Sigma}| - \tfrac{1}{2}\operatorname{tr}\{\boldsymbol{\Sigma}^{-1}\mathbf{V}(\boldsymbol{\theta})\}, \tag{11.23}$$

so that maximizing the likelihood is equivalent to minimizing

$$L_1(\boldsymbol{\theta}, \boldsymbol{\Sigma}) = \log|\boldsymbol{\Sigma}| + \operatorname{tr}\{\boldsymbol{\Sigma}^{-1}\mathbf{V}(\boldsymbol{\theta})/n\}. \tag{11.24}$$

This minimizing can be done in two steps. We first fix $\boldsymbol{\theta}$ and then minimize $L_1(\boldsymbol{\theta}, \boldsymbol{\Sigma})$ with respect to $\boldsymbol{\Sigma}$. The resulting value of $\boldsymbol{\Sigma}$, a function of $\boldsymbol{\theta}$, is then substituted back into $L_1(\boldsymbol{\theta}, \boldsymbol{\Sigma})$, and the resulting function of $\boldsymbol{\theta}$ is then minimized with respect to $\boldsymbol{\theta}$. This technique of first eliminating $\boldsymbol{\Sigma}$ to obtain a function of $\boldsymbol{\theta}$ is called concentrating the likelihood, and it is considered in detail in Section 2.2.3. From Appendix A7.1 we see that $L_1(\boldsymbol{\theta}, \boldsymbol{\Sigma})$ is minimized for fixed $\boldsymbol{\theta}$ when $\boldsymbol{\Sigma} = \mathbf{V}(\boldsymbol{\theta})/n$, and the concentrated log-likelihood is $M(\boldsymbol{\theta}) = -(n/2)M_1(\boldsymbol{\theta})$, where

$$M_1(\boldsymbol{\theta}) = L_1\{\boldsymbol{\theta}, \mathbf{V}(\boldsymbol{\theta})/n\}$$
$$= \log|\mathbf{V}(\boldsymbol{\theta})/n| + d. \tag{11.25}$$

Hence the maximum-likelihood estimate $\hat{\boldsymbol{\theta}}$ of $\boldsymbol{\theta}$ is obtained by minimizing $|V(\boldsymbol{\theta})|$: a method for doing this, due to Bates and Watts [1987], is described in Section 11.4.2. Also, from Section 2.2.3, the maximum-likelihood estimate of $\boldsymbol{\Sigma}$ is $\hat{\boldsymbol{\Sigma}} = \mathbf{V}(\hat{\boldsymbol{\theta}})/n$. Since $\hat{\boldsymbol{\theta}}$ minimizes $L_1(\boldsymbol{\theta}, \boldsymbol{\Sigma})$ when $\boldsymbol{\Sigma} = \hat{\boldsymbol{\Sigma}}$, it follows that $L_1(\boldsymbol{\theta}, \hat{\boldsymbol{\Sigma}})$ is minimized by minimizing

$$\operatorname{tr}\{\hat{\boldsymbol{\Sigma}}^{-1}\mathbf{V}(\boldsymbol{\theta})\} = \sum_{i=1}^{n} [\mathbf{y}_i - \mathbf{f}_i(\boldsymbol{\theta})]'\hat{\boldsymbol{\Sigma}}^{-1}[\mathbf{y}_i - \mathbf{f}_i(\boldsymbol{\theta})].$$

Hence from (11.12) we see that $\hat{\boldsymbol{\theta}}$ can also regarded as a generalized least-squares estimate.

Under fairly general regularity conditions (c.f. Section 12.5) it can be shown that, asymptotically, $\sqrt{n}(\hat{\boldsymbol{\theta}} - \boldsymbol{\theta}) \sim N(\mathbf{0}, \mathbf{A}^{-1}(\boldsymbol{\theta}))$, where

$$\mathbf{A}(\boldsymbol{\theta}) = \lim_{n \to \infty} \left\{ -\frac{1}{n} \mathscr{E}\left[\frac{\partial^2 M(\boldsymbol{\theta})}{\partial \boldsymbol{\theta}\, \partial \boldsymbol{\theta}'} \right] \right\}.$$

A method of estimating $\mathbf{A}^{-1}(\boldsymbol{\theta})$ is also described at the end of Section 12.5. These asymptotic results can be used for hypothesis testing and constructing confidence regions and intervals, along the lines described in Chapter 5.

At this point it is appropriate to discuss dimensions. We note that $|\mathbf{V}(\boldsymbol{\theta})| \geq 0$, as the $d \times d$ matrix $\mathbf{V}(\boldsymbol{\theta})$ is positive semidefinite. In order for the maximum-likelihood method to work, we must have $\mathbf{V}(\boldsymbol{\theta})$ nonsingular. From (11.22), rank $\mathbf{V}(\boldsymbol{\theta}) = \text{rank}\{\mathbf{Y} - \mathbf{G}(\boldsymbol{\theta})\} \leq \min(n, d)$, so that we require $n \geq d$. We also need $n > p$; otherwise we could fit one response perfectly with the corresponding column of $\mathbf{Y} - \mathbf{G}(\boldsymbol{\theta})$ being $\mathbf{0}$. Then $|\mathbf{V}(\boldsymbol{\theta})| = 0$ no matter how well the other responses have been fitted.

In conclusion we stress that it is important to check the estimated model to see if it fits the data adequately. This can be done by looking at appropriate plots of the residuals $\mathbf{y}^{(j)} - \mathbf{f}^{(j)}(\hat{\boldsymbol{\theta}})$ for each of the d single-response models. Provided intrinsic curvature is not a problem, standard residual plots for linear regression models can be used (e.g. Seber [1977: Section 6.6]). Some linear multivariate techniques (c.f. Seber [1984: 408]) may also be appropriate.

11.3.2 Hypothesis Testing

The general theory for univariate models in Chapter 5 extends naturally to multivariate models. To test a hypothesis $H_0: \mathbf{a}(\boldsymbol{\theta}) = \mathbf{0}$ of q constraints, we can use the likelihood-ratio statistic based on concentrated likelihoods [c.f. (5.92)], namely

$$\begin{aligned} \text{LR} &= 2[M(\hat{\boldsymbol{\theta}}) - M(\tilde{\boldsymbol{\theta}})] \\ &= n[M_1(\tilde{\boldsymbol{\theta}}) - M_1(\hat{\boldsymbol{\theta}})] \\ &= n[\log|\mathbf{V}(\tilde{\boldsymbol{\theta}})| - \log|\mathbf{V}(\hat{\boldsymbol{\theta}})|] \qquad \text{[by (11.25)]}. \end{aligned}$$

Here $\tilde{\boldsymbol{\theta}}$ and $\hat{\boldsymbol{\theta}}$ are the restricted (by H_0) and unrestricted maximum-likelihood estimates of $\boldsymbol{\theta}$, obtained by minimizing $|\mathbf{V}(\boldsymbol{\theta})|$ using, for example, the method of Bates and Watts (Section 11.4.2). When H_0 is true, LR is approximately χ_q^2. The estimate $\tilde{\boldsymbol{\theta}}$ can be obtained by expressing H_0 in the form of freedom equations as in Section 5.9.3. However, if this is not possible analytically, then the method could be computationally expensive.

Previously, in the absence of a good algorithm for minimizing a determinant, several approximate methods were used. One possibility is to use a standard likelihood-ratio statistic, but with $\boldsymbol{\Sigma}$ "pre-estimated" by a consistent estimator $\hat{\boldsymbol{\Sigma}}_c$.

This leads to [c.f. (11.23)]

$$\begin{aligned}\mathrm{LR}_c &= 2[L(\hat{\boldsymbol{\theta}}_c, \hat{\boldsymbol{\Sigma}}_c) - L(\tilde{\boldsymbol{\theta}}_c, \hat{\boldsymbol{\Sigma}}_c)] \\ &= Q[\tilde{\boldsymbol{\theta}}_c, \hat{\boldsymbol{\Sigma}}_c] - Q[\hat{\boldsymbol{\theta}}_c, \hat{\boldsymbol{\Sigma}}_c],\end{aligned}$$

which is approximately χ_q^2 when H_0 is true. Here $\tilde{\boldsymbol{\theta}}_c$ and $\hat{\boldsymbol{\theta}}_c$ are the restricted and unrestricted values of $\boldsymbol{\theta}$, respectively, which maximize $L(\boldsymbol{\theta}, \hat{\boldsymbol{\Sigma}}_c)$, that is, which minimize $Q(\boldsymbol{\theta}, \hat{\boldsymbol{\Sigma}}_c)$ [c.f. (11.21) and (11.23)]. This minimization is discussed in Section 11.2 above. Various choices of $\hat{\boldsymbol{\Sigma}}_c$ are possible, for example $\check{\boldsymbol{\Sigma}}$ of (11.14). Another possibility is to minimize $Q(\boldsymbol{\theta}, \mathbf{I}_d)$ with respect to $\boldsymbol{\theta}$ and then use the residuals as in (11.14). Gallant [1987: 326] proposed an F-approximation for LR_c.

In conclusion we note that confidence intervals can be constructed by "inverting" test statistics as in Section 5.4.

11.4 BAYESIAN INFERENCE

11.4.1 Estimates from Posterior Densities

In Section 2.7.2 we derived the posterior distribution of $\boldsymbol{\theta}$ for a single-response normal nonlinear model. The arguments there extend naturally to multiresponse models. Following Jeffreys [1961], Box and Tiao [1973: 428] used independent noninformative priors (see Section 2.7.1) for $\boldsymbol{\theta}$ and $\boldsymbol{\Sigma}$, namely

$$\begin{aligned}p(\boldsymbol{\theta}, \boldsymbol{\Sigma}) &= p(\boldsymbol{\theta})p(\boldsymbol{\Sigma}) \\ &\propto |\boldsymbol{\Sigma}|^{-(d+1)/2}. \end{aligned} \tag{11.26}$$

Assuming multivariate normality of the data, they showed that the posterior distribution of $\boldsymbol{\theta}$ for the model (11.1) is given by

$$p(\boldsymbol{\theta}|\mathbf{y}) \propto |\mathbf{V}(\boldsymbol{\theta})|^{-n/2}, \tag{11.27}$$

where $\mathbf{V}(\boldsymbol{\theta})$ is given by (11.22). An estimate of $\boldsymbol{\theta}$ can then be obtained by, for example, maximizing (11.27), that is, by minimizing $|\mathbf{V}(\boldsymbol{\theta})|$. Therefore, with the prior (11.26) we find that the posterior mode of (11.27) is the same as the maximum-likelihood estimate. However, it is important to consider the whole posterior distribution and not just the mode, as we see from the following example.

Example 11.3 Box and Draper [1965] considered the following model:

$$y_{ij} = f_j(x_i; \theta) + \varepsilon_{ij} \qquad (i = 1, 2, \ldots, 10, \quad j = 1, 2), \tag{11.28}$$

with $\theta = \log \phi$, $f_1(x; \theta) = e^{-\phi x}$, $f_2(x; \theta) = 1 - e^{-\phi x}$, and the pairs $(\varepsilon_{i1}, \varepsilon_{i2})'$ i.i.d.

Table 11.3 Data Simulated from the Model (11.28)[a]

	Data Set 1		Data Set 2	
x_i	y_{i1}	y_{i2}	y_{i1}	y_{i2}
$\frac{1}{2}$	.907	.127	.907	.142
$\frac{1}{2}$	.915	.064	.915	.079
1	.801	.134	.801	.160
1	.825	.200	.825	.188
2	.649	.274	.649	.315
2	.675	.375	.675	.416
4	.446	.570	.446	.624
4	.468	.535	.468	.589
8	.233	.792	.233	.838
8	.187	.803	.187	.849

[a]From Box and Draper [1965] with permission of the Biometrika Trustees.

$N_2(\mathbf{0}, \boldsymbol{\Sigma})$. The data are given in Table 11.3, data set 1, and the following discussion is based on their paper.

Using (11.27), we have that

$$\begin{aligned} p(\theta|\mathbf{y}) &= p(\theta|\mathbf{y}^{(1)}, \mathbf{y}^{(2)}) \\ &\propto |(v_{rs})|^{-5}, \end{aligned}$$

where $v_{rs} = \sum_i [y_{ir} - f_r(x_i; \theta)][y_{is} - f_s(x_i; \theta)]$. If only a single response vector $\mathbf{y}^{(1)}$ or $\mathbf{y}^{(2)}$ is available, then $p(\theta|\mathbf{y}^{(j)}) \propto v_{jj}^{-5}$, but if both $\mathbf{y}^{(1)}$ and $\mathbf{y}^{(2)}$ are available, then

$$p(\theta|\mathbf{y}) \propto (v_{11}v_{22} - v_{12}^2)^{-5}.$$

Box and Draper simulated the model (11.28) with $\sigma_{11} = 0.0004$, $\sigma_{22} = 0.0016$, and $\sigma_{12} = 0.0004$. When the posterior distributions from $\mathbf{y}^{(1)}$ and $\mathbf{y}^{(2)}$ taken separately are plotted (see Fig. 11.3), they present consistent information about θ. As expected, the precision of an estimate (e.g. the mode) from the first response, which has smaller variance, is greater than that from the second. Also, although the two responses are correlated to some extent ($\rho = \sigma_{12}/\sqrt{\sigma_{11}\sigma_{22}} = 0.5$), the two responses together provide a posterior distribution which is sharper than either of the two from individual responses.

The authors considered a second example, data set 2 in Table 11.3, in which $\mathbf{y}^{(1)}$ is the same as for set 1 but $\mathbf{y}^{(2)}$ is generated using a different value of θ. The three corresponding posterior distributions are again presented (Fig. 11.4), and we see that they give conflicting information. The posterior density for $\mathbf{y}^{(1)}$ alone gives evidence about θ conflicting with that obtained from $\mathbf{y}^{(2)}$ alone. Furthermore, $\mathbf{y}^{(1)}$ and $\mathbf{y}^{(2)}$ together provide less precise information about θ than $\mathbf{y}^{(1)}$

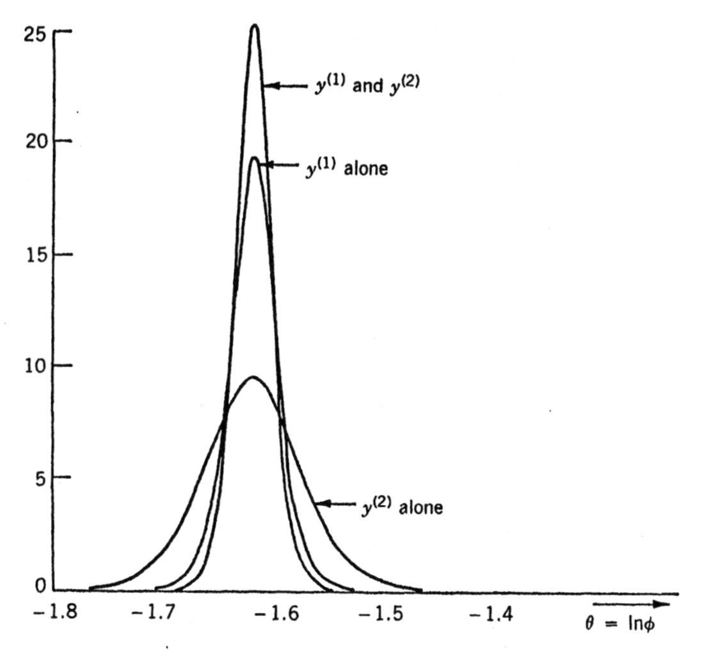

Figure 11.3 Posterior distributions for data set 1. From Box and Draper [1965] with permission of the Biometrika Trustees.

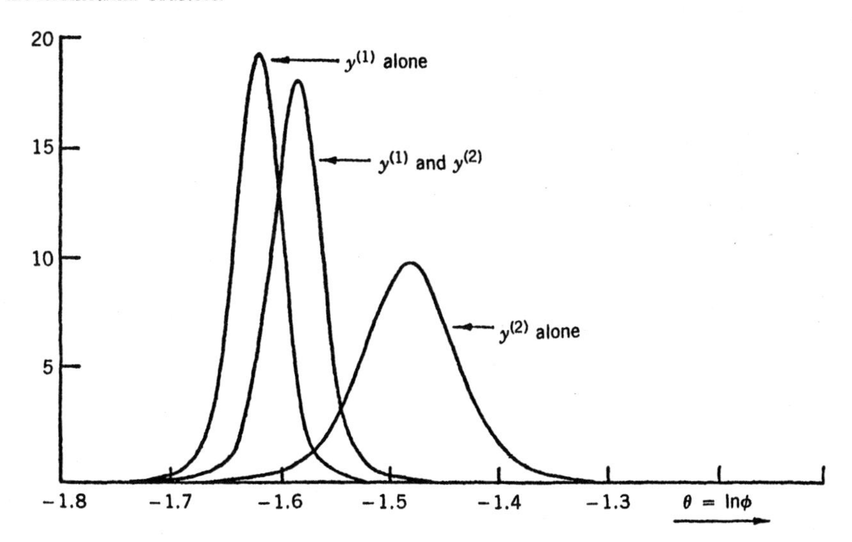

Figure 11.4 Posterior distributions for data set 2. From Box and Draper [1965] with permission of the Biometrika Trustees.

alone. This conflicting evidence is due to a lack of fit of the model: in this case a single value of θ is not adequate for both responses. Box and Draper [1965] concluded that, since the validity of a postulated model is always open to question, the experimenter should:

(i) check the individual fit of each response by analyzing residuals, and
(ii) consider the consistency of the information from the different responses by comparing posterior densities.

Such care should be exercised even more with multiparameter models; Box and Draper give an interesting example with three response variables and two parameters. ■

11.4.2 H.P.D. Regions

We now turn our attention to the problem of finding h.p.d. regions for $\boldsymbol{\theta}$. The following discussion is based on Bates and Watts [1985a]. Since $\hat{\boldsymbol{\theta}}$ is found by minimizing $d(\boldsymbol{\theta}) = |\mathbf{V}(\boldsymbol{\theta})|$ with respect to $\boldsymbol{\theta}$, we have $\partial d(\boldsymbol{\theta})/\partial\boldsymbol{\theta} = \mathbf{0}$ at $\boldsymbol{\theta} = \hat{\boldsymbol{\theta}}$. Hence for $\boldsymbol{\theta}$ close to $\hat{\boldsymbol{\theta}}$ we have the Taylor expansion [c.f. (1.9) and (1.11)]

$$\begin{aligned} d(\boldsymbol{\theta}) &\approx d(\hat{\boldsymbol{\theta}}) + \left[\frac{\partial d(\boldsymbol{\theta})}{\partial \boldsymbol{\theta}}\right]'_{\hat{\boldsymbol{\theta}}}(\boldsymbol{\theta}-\hat{\boldsymbol{\theta}}) + \tfrac{1}{2}(\boldsymbol{\theta}-\hat{\boldsymbol{\theta}})'\left[\frac{\partial^2 d(\boldsymbol{\theta})}{\partial\boldsymbol{\theta}\,\partial\boldsymbol{\theta}'}\right]_{\hat{\boldsymbol{\theta}}}(\boldsymbol{\theta}-\hat{\boldsymbol{\theta}}) \\ &= d(\hat{\boldsymbol{\theta}}) + \tfrac{1}{2}(\boldsymbol{\theta}-\hat{\boldsymbol{\theta}})'\hat{\mathbf{H}}(\boldsymbol{\theta}-\hat{\boldsymbol{\theta}}), \end{aligned} \tag{11.29}$$

where $\hat{\mathbf{H}} = \mathbf{H}(\hat{\boldsymbol{\theta}})$ and $\mathbf{H}(\boldsymbol{\theta})$ is the Hessian of $d(\boldsymbol{\theta})$. Using the above in (11.27) and setting $\nu = n - p$ gives us

$$\begin{aligned} p(\boldsymbol{\theta}|\mathbf{y}) &\propto [d(\boldsymbol{\theta})]^{-n/2} \\ &\approx [d(\hat{\boldsymbol{\theta}})]^{-n/2}\left(1 + \frac{(\boldsymbol{\theta}-\hat{\boldsymbol{\theta}})'\hat{\mathbf{H}}(\boldsymbol{\theta}-\hat{\boldsymbol{\theta}})}{2d(\hat{\boldsymbol{\theta}})}\right)^{-n/2} \\ &\propto [1 + \nu^{-1}(\boldsymbol{\theta}-\hat{\boldsymbol{\theta}})'\{2d(\hat{\boldsymbol{\theta}})\hat{\mathbf{H}}^{-1}\nu^{-1}\}^{-1}(\boldsymbol{\theta}-\hat{\boldsymbol{\theta}})]^{-(\nu+p)/2}. \end{aligned} \tag{11.30}$$

From Appendix A9 we see that (11.30) represents a p-dimensional t-distribution, $t_p(\nu, \hat{\boldsymbol{\theta}}, 2d(\hat{\boldsymbol{\theta}})\hat{\mathbf{H}}^{-1}\nu^{-1})$, and a $100(1-\alpha)\%$ h.p.d. region for $\boldsymbol{\theta}$ is given approximately by

$$\{\boldsymbol{\theta} : (\boldsymbol{\theta}-\hat{\boldsymbol{\theta}})'\hat{\mathbf{H}}(\boldsymbol{\theta}-\hat{\boldsymbol{\theta}}) \le 2pd(\hat{\boldsymbol{\theta}})\nu^{-1}F^{\alpha}_{p,\nu}\}. \tag{11.31}$$

Bates and Watts [1985a] noted that for uniresponse models, regions of the form (11.31) are exact for linear models and approximate for nonlinear models, whereas in the multiresponse case, (11.31) is always approximate and its validity requires further investigation. If $\mathbf{C} = 2d(\hat{\boldsymbol{\theta}})(\nu\hat{\mathbf{H}})^{-1}$, then we have from Appendix A9 that

$$\frac{\theta_r - \hat{\theta}_r}{\sqrt{c_{rr}}} \sim t_\nu, \tag{11.32}$$

from which we can construct a marginal h.p.d. interval for θ_r. Bates and Watts gave an approximation for $\mathbf{H} = [(h_{rs})]$, which we now briefly describe.

Referring to (11.5), let $\mathbf{Z} = \mathbf{Y} - \mathbf{G}(\boldsymbol{\theta})$, so that $d(\boldsymbol{\theta}) = |\mathbf{Z}'\mathbf{Z}|$. Define

$$\mathbf{Z}_r = \frac{\partial \mathbf{Z}}{\partial \theta_r} = \left[\left(\frac{\partial z_{ij}}{\partial \theta_r}\right)\right] \quad \text{and} \quad k(\boldsymbol{\theta}) = \tfrac{1}{2}\log d(\boldsymbol{\theta}).$$

Then Bates and Watts [1985a] proved that

$$\begin{aligned} k_r(\boldsymbol{\theta}) &= \frac{\partial k(\boldsymbol{\theta})}{\partial \theta_r} \\ &= \operatorname{tr}[\mathbf{Z}^+\mathbf{Z}_r], \end{aligned}$$

and

$$\begin{aligned} k_{rs}(\boldsymbol{\theta}) &= \frac{\partial^2 k(\boldsymbol{\theta})}{\partial \theta_r \partial \theta_s} \\ &= -\operatorname{tr}[\mathbf{Z}^+\mathbf{Z}_r\mathbf{Z}^+\mathbf{Z}_s] + \operatorname{tr}[\mathbf{Z}^+(\mathbf{Z}^+)'\mathbf{Z}_r'(\mathbf{I}_n - \mathbf{Z}\mathbf{Z}^+)\mathbf{Z}_s] + \operatorname{tr}[\mathbf{Z}^+\mathbf{Z}_{rs}], \quad (11.33) \end{aligned}$$

where $\mathbf{Z}_{rs} = \partial\mathbf{Z}/\partial\theta_r\partial\theta_s$ and $\mathbf{Z}^+$ is the Moore–Penrose generalized inverse (Appendix A5) of $\mathbf{Z}$. Hence the gradient $\mathbf{g}(\boldsymbol{\theta})$ and Hessian $\mathbf{H}(\boldsymbol{\theta})$ of $d(\boldsymbol{\theta})$ are given by

$$g_r(\boldsymbol{\theta}) = \frac{\partial d(\boldsymbol{\theta})}{\partial \theta_r} = 2d(\boldsymbol{\theta})k_r$$

and

$$h_{rs}(\boldsymbol{\theta}) = \frac{\partial^2 d(\boldsymbol{\theta})}{\partial \theta_r \partial \theta_s} = 2d(\boldsymbol{\theta})[k_{rs} + 2k_r k_s].$$

Bates and Watts use an approximation for $\mathbf{H}$ obtained by setting $\mathbf{Z}_{rs} = \mathbf{0}$ in (11.33).

The estimate $\hat{\boldsymbol{\theta}}$ can now be found using a standard Newton algorithm: from (2.36) we have

$$\boldsymbol{\delta}^{(a)} = \boldsymbol{\theta}^{(a+1)} - \boldsymbol{\theta}^{(a)} = -(\mathbf{H}^{-1}\mathbf{g})_{\boldsymbol{\theta}^{(a)}}.$$

Bates and Watts [1987] give the following compact scheme for computing k_r and the approximation to k_{rs} in terms of a matrix $\mathbf{J}_r$ which is found as follows. Let

$$\mathbf{Z} = \mathbf{Q}\mathbf{R} = \mathbf{Q}\begin{pmatrix} \mathbf{R}_{11} \\ \mathbf{0} \end{pmatrix}$$

be a QR decomposition of $\mathbf{Z}$ (c.f. Appendix A8.3), where $\mathbf{Q}$ is an $n \times n$ orthogonal matrix and $\mathbf{R}_{11}$ is a $d \times d$ upper triangular matrix. Then

$$\mathbf{J}_r = \mathbf{Q}'\mathbf{Z}_r\mathbf{R}_{11}^{-1} \qquad (r = 1, 2, \ldots, p),$$

is obtained by solving the triangular system $\mathbf{J}_r\mathbf{R}_{11} = \mathbf{Q}'\mathbf{Z}_r$, from which we can calculate

$$k_r(\boldsymbol{\theta}) = \operatorname{tr}[\mathbf{J}_r]$$

and

$$k_{rs}(\boldsymbol{\theta}) = -\sum_{u=1}^{d}\sum_{v=1}^{d}\{\mathbf{J}_r\}_{uv}\{\mathbf{J}_s\}_{vu} + \sum_{u=d+1}^{n}\sum_{v=1}^{d}\{\mathbf{J}_r\}_{uv}\{\mathbf{J}_s\}_{uv}.$$

If $\mathbf{H}^{(a)} = \mathbf{H}(\boldsymbol{\theta}^{(a)})$ is positive definite, then we can use the Cholesky decomposition $\mathbf{H}^{(a)} = \mathbf{U}'\mathbf{U}$, where $\mathbf{U}$ is upper triangular. Bates and Watts suggest the standard technique of finding $\boldsymbol{\delta}^{(a)}$ by solving

$$\mathbf{U}'\mathbf{U}\boldsymbol{\delta}^{(a)} = -\mathbf{g}(\boldsymbol{\theta}^{(a)})$$

for $\mathbf{U}\boldsymbol{\delta}^{(a)}$, and then solving for $\boldsymbol{\delta}^{(a)}$. They also propose stopping the iterations when

$$\frac{\nu\|\mathbf{U}\boldsymbol{\delta}^{(a)}\|^2}{2pd(\hat{\boldsymbol{\theta}})} < \tau^2,$$

where τ, the tolerance, is set at 0.001. For further details and subroutines see Bates and Watts [1984, 1987].

An important application of multiresponse methods is in the kinetic modeling of complex chemical reactions. Although in some situations only one of the responses may be of any real interest, the use of other responses can improve the precision of estimation and shed light on the modeling process (Ziegel and Gorman [1980]). Frequently a kinetic network can be modeled by a linear system of simultaneous differential equations of the form $\dot{\mathbf{s}} = \mathbf{A}\mathbf{s}$, and Bates and Watts [1985a, b] show how to apply the above methods to such systems. This work is discussed in Chapter 8.

11.4.3 Missing Observations

If some of the observations in $\mathbf{y} = \operatorname{vec}\mathbf{Y}$ are missing, then, with $\mathbf{y}$ suitably rearranged, we can write $\mathbf{y}' = (\mathbf{y}_0', \mathbf{y}_m')$ where $\mathbf{y}_0$ are the observations actually observed and $\mathbf{y}_m$ are the missing observations. Using a noninformative prior for $\mathbf{y}_m$ [namely, $p(\mathbf{y}_m)$ a constant], M. J. Box et al. [1970] showed that under the assumptions which led to (11.27),

$$p(\boldsymbol{\theta}, \mathbf{y}_m) \propto |\mathbf{V}(\boldsymbol{\theta}, \mathbf{y}_m)|^{-n/2}, \tag{11.34}$$

which is the same as (11.27) but with $\mathbf{V}$ now regarded as a function of both $\boldsymbol{\theta}$ and $\mathbf{y}_m$. Estimates of $\boldsymbol{\theta}$ and $\mathbf{y}_m$ can be found by minimizing $|\mathbf{V}|$ with respect to $\boldsymbol{\theta}$ and $\mathbf{y}_m$. However, as the estimate of $\mathbf{y}_m$ will generally not be of interest in itself, we can integrate out $\mathbf{y}_m$ to obtain the marginal posterior density function of $\boldsymbol{\theta}$. This will usually require numerical integration, though the approximate methods of Section 2.7.2 are applicable here.

The above method is a convenient one when there are only a few missing values. However, with several missing values, the total number of parameters to be estimated may be very large. An alternative procedure is to use the posterior density $p(\boldsymbol{\theta}, \boldsymbol{\Sigma} | \mathbf{y}_0)$ in which only the observed responses appear. This density is given by

$$p(\boldsymbol{\theta}, \boldsymbol{\Sigma} | \mathbf{y}_0) \propto p(\mathbf{y}_0 | \boldsymbol{\theta}, \boldsymbol{\Sigma}) p(\boldsymbol{\theta}, \boldsymbol{\Sigma}), \tag{11.35}$$

and we can once again use the prior (11.26). The likelihood function for the observed data, $p(\mathbf{y}_0 | \boldsymbol{\theta}, \boldsymbol{\Sigma})$, is readily obtained from the "full" likelihood $p(\mathbf{y} | \boldsymbol{\theta}, \boldsymbol{\Sigma})$, as any subset of a multivariate normal is also multivariate normal. The above approach is advocated by Stewart and Sørensen [1981], and the reader is referred to their paper for details and examples.

*11.5 LINEAR DEPENDENCIES

11.5.1 Dependencies in Expected Values

We now consider a rather specialized problem, that of handling possible linear dependencies in the model. Our discussion is based on G. E. P. Box et al. [1973] and McLean et al. [1979], and relates to stoichiometry, the branch of chemistry concerned with the proportions of substances in compounds and reactions. In chemical reactions it is not uncommon for linear relationships to exist among the expected ("true") values of the responses because of the requirement of material and energy balances or steady-state conditions. Using the model (11.1), we might have the constraint

$$\begin{aligned} a_0 &= E[y_{i1}] + E[y_{i2}] + \cdots + E[y_{id}] \\ &= \mathbf{1}_d' \mathscr{E}[\mathbf{y}_i]. \end{aligned} \tag{11.36}$$

If there are m such linear constraints, they can be written as

$$\mathbf{A} \mathscr{E}[\mathbf{y}_i] = \mathbf{a}_0 \qquad (i = 1, 2, \ldots, n), \tag{11.37}$$

where $\mathbf{A}$ is a known $m \times d$ matrix of rank m and $\mathbf{a}_0$ is a known $m \times 1$ vector. For example, in a given chemical process it may follow that to maintain the carbon balance, a certain linear relationship must exist among the amounts of the substituents in all the experimental runs. A second relationship may exist to maintain the nitrogen balance, and so on. Unfortunately the experimenter is not usually given the full details of the process, but must learn about them by interrogating the data. Careful experimental planning is also important.

11.5.2 Dependencies in the Data

Linear dependencies may exist in the data as well as the expected values, for a number of reasons. First, the experimenter, in realizing that formulae like (11.36) do not hold for the actual observations because of experimental error, may force his or her observations to fit the constraint by normalizing the data. For example, if the original observations are z_{ij} and if $y_{ij} = a_0 z_{ij}/\sum_r z_{ir}$, then

$$y_{i1} + y_{i2} + \cdots + y_{id} = a_0. \tag{11.38}$$

In the case of m constraints like (11.37), the experimenter may measure the d variables and then force m_1 of the constraints to hold for the data by normalizing or otherwise adjusting the data. A second possibility is that the investigator may measure only $d-1$ of the response variables, say the first $d-1$, and then calculate the dth (y_{id}) using a relationship like (11.38). In the case of m_1 constraints the experimenter may measure only $d - m_1$ responses and then use the constraints to calculate the remaining m_1 variables.

Suppose, then, that the first m_1 relationships of (11.37) are used for normalizing or for calculating m_1 responses. Then

$$\mathbf{A}_1 \mathbf{y}_i = \mathbf{a}_{01} \qquad (i = 1, 2, \ldots, n), \tag{11.39}$$

where $\mathbf{A}' = (\mathbf{A}_1', \mathbf{A}_2')$, $\mathbf{a}_0' = (\mathbf{a}_{01}', \mathbf{a}_{02}')$, and $\mathbf{A}_1$ is $m_1 \times d$. Combining (11.39) with (11.37) leads to

$$\mathbf{A}_1(\mathbf{y}_i - \mathscr{E}[\mathbf{y}_i]) = \mathbf{0}, \tag{11.40}$$

so that from (11.22)

$$\mathbf{A}_1 \mathbf{V}(\boldsymbol{\theta}) = \mathbf{0} \tag{11.41}$$

and $\mathbf{V}(\boldsymbol{\theta})$ has m_1 zero eigenvalues. McLean et al. [1979] mentioned two other possible scenarios which are somewhat similar. First, the standardization may be such that it varies with the response vectors, so that in contrast to (11.39) we have

$$\mathbf{A}_1 \mathbf{y}_i = \mathbf{b}_i. \tag{11.42}$$

Taking expected values, we have $\mathbf{A}_1 \mathscr{E}[\mathbf{y}_i] = \mathbf{b}_i$, so that both (11.40) and (11.41) still hold. Second, (11.39) may hold, but in the process of standardizing the responses we may find that $\mathscr{E}[\mathbf{y}_i] \neq \mathbf{f}_i(\boldsymbol{\theta})$, contrary to (11.11), so that

$$\mathbf{A}_1[\mathbf{y}_i - \mathbf{f}_i(\boldsymbol{\theta})] \neq \mathbf{0} \tag{11.43}$$

and $\mathbf{V}(\boldsymbol{\theta})$ is nonsingular.

Box et al. [1973] stress that normalizing or adjusting the data in the manner described above should be avoided where possible. It is preferable for the experimenter to try and measure each response independently rather than forcing responses to satisfy theoretical relationships that he or she believes to be true. Furthermore, independent information on each of the m relationships will provide better estimates of the parameters as well as allowing the model to be checked. Unfortunately, in some situations, the analytical procedures or equipment used necessarily makes use of such relationships. This is the case, for example, with chemical-composition data obtained from gas chromatography, where it is only possible to calculate relative percentages.

When it is impossible to determine all the responses independently or to avoid normalized data, careful note should be made of which are actually measured, which are obtained by calculation, and which of the expectation relationships (11.37) are used in obtaining the data. Ignoring such dependencies can lead to interpretation problems and to a poor fit of the model to the data.

Although a careful study of the system should, in principle, reveal all the dependencies in the expected values, Box et al. [1973] have found that dependencies are frequently overlooked and must be looked for empirically. They proposed using the following eigenvalue analysis for the situation described by (11.40).

11.5.3 Eigenvalue Analysis

As $\mathbf{a}_0$ in (11.37) is usually unknown, we can eliminate it by working with the variables $d_{ij} = y_{ij} - \bar{y}_{\cdot j}$, $\mathbf{d}_i = \mathbf{y}_i - \bar{\mathbf{y}}$. Define the $n \times d$ matrix $\mathbf{D} = (\mathbf{d}_1, \mathbf{d}_2, \ldots, \mathbf{d}_n)'$ (the transpose of the D-matrix used by Box et al.); then $\mathbf{D}'\mathbf{D}$ is positive semidefinite. Let $0 \leq \lambda_1 \leq \lambda_2 \leq \cdots \leq \lambda_d$ be the ordered eigenvalues of $\mathbf{D}'\mathbf{D}$, with corresponding orthonormal eigenvectors $\mathbf{z}_k$. Hence

$$\mathbf{D}'\mathbf{D}\mathbf{z}_k = \lambda_k \mathbf{z}_k \qquad (k = 1, 2, \ldots, d),$$

and premultiplying by $\mathbf{z}_k'$ gives us

$$\|\mathbf{D}\mathbf{z}_k\|^2 = \lambda_k. \tag{11.44}$$

Now if there are m_1 exact linear relationships in the data, namely $\mathbf{A}_1\mathbf{y}_i = \mathbf{a}_{01}$ $(i = 1, 2, \ldots, n)$, then $\mathbf{A}_1\mathbf{d}_i = \mathbf{0}$ and

$$\mathbf{A}_1\mathbf{D}' = \mathbf{A}_1(\mathbf{d}_1, \mathbf{d}_2, \ldots, \mathbf{d}_n) = \mathbf{0}. \tag{11.45}$$

Since $\mathbf{A}_1\mathbf{D}'\mathbf{D} = \mathbf{0}$, the first m_1 eigenvalues of $\mathbf{D}'\mathbf{D}$ will be zero. Then, from (11.44), $\mathbf{D}\mathbf{z}_k = \mathbf{0}$ $(k = 1, 2, \ldots, m_1)$, so that

$$\mathbf{Z}_1\mathbf{D}' = \mathbf{0}, \tag{11.46}$$

where $\mathbf{Z}_1 = (\mathbf{z}_1, \mathbf{z}_2, \ldots, \mathbf{z}_{m_1})'$. Comparing (11.45) and (11.46), we have

$$\mathbf{A}_1 = \mathbf{K}_1\mathbf{Z}_1 \tag{11.47}$$

for some nonsingular $d \times d$ matrix $\mathbf{K}_1$, as the rows of $\mathbf{A}_1$ and the rows of $\mathbf{Z}_1$ form a basis for the same vector space.

In addition to the m_1 exact linear relationships in the data, there may be a further $m_2 = m - m_1$ linear relationships in the expected values, namely $\mathbf{A}_2 \mathscr{E}[\mathbf{d}_i] = \mathbf{0}$, or

$$\mathbf{A}_2 \mathbf{d}_i = \mathbf{u}_i \qquad (i = 1, 2, \ldots, n), \tag{11.48}$$

where $\mathscr{E}[\mathbf{u}_i] = \mathbf{0}$. Unless, however, $\mathbf{A}_2$ is given by the theory of the experiment, the existence of an $\mathbf{A}_2$ can only be inferred from the data. Suppose that each row of $\mathbf{A}_2$ is split into two components, one in $\mathscr{R}[\mathbf{A}_1'] \,(= \mathscr{R}[\mathbf{Z}_1'])$ and one perpendicular to $\mathscr{R}[\mathbf{A}_1']$. This split gives us the matrix decomposition

$$\mathbf{A}_2 = \mathbf{A}_{2(1)} + \mathbf{A}_{2(1)}^{\perp},$$

where $\mathbf{A}_{2(1)} \mathbf{d}_i = \mathbf{0}$ [by (11.45)]. Subtracting this last equation from (11.48), we have $\mathbf{A}_{2(1)}^{\perp} \mathbf{d}_i = \mathbf{u}_i$, i.e.

$$\mathbf{A}_{2(1)}^{\perp} \mathbf{D}' = (\mathbf{u}_1, \mathbf{u}_2, \ldots, \mathbf{u}_n) = \mathbf{U}', \tag{11.49}$$

say. Subtracting off the expected value of the above expression gives us

$$\mathbf{A}_{2(1)}^{\perp} (\mathbf{D}' - \mathscr{E}[\mathbf{D}']) = \mathbf{U}',$$

so that the $\mathbf{u}_i$ are linear combinations of the $\mathbf{d}_i - \mathscr{E}[\mathbf{d}_i]$, and therefore of the "error" vectors $\boldsymbol{\varepsilon}_i$. Hence, to the order of magnitude of the errors, (11.49) implies $\mathbf{A}_{2(1)}^{\perp} \mathbf{D}' \approx \mathbf{0}$. As in (11.46), this will produce a corresponding set of constraints $\mathbf{Z}_2 \mathbf{D}' \approx \mathbf{0}$, where $\mathbf{Z}_2 = (\mathbf{z}_{m_1+1}, \mathbf{z}_{m_1+2}, \ldots, \mathbf{z}_{m_1+m_2})'$. Since $\| \mathbf{z}_k' \mathbf{D}' \|^2 = \lambda_k$ [by (11.44)], the existence of m_2 $(= m - m_1)$ such constraints will be heralded by the corresponding eigenvalues $\lambda_{m_1+1}, \lambda_{m_1+2}, \ldots, \lambda_{m_1+m_2}$ being sufficiently small. Also, corresponding to (11.47) we have the approximate relationship

$$\mathbf{A}_{2(1)}^{\perp} \approx \mathbf{K}_2 \mathbf{Z}_2,$$

where $\mathbf{K}_2$ is nonsingular. Although $\mathbf{Z}_2$ comes from the random matrix $\mathbf{D}'\mathbf{D}$, it can be treated as being approximately constant because of the above equation. Hence from $\mathscr{E}[\mathbf{A}_{2(1)}^{\perp} \mathbf{d}_i] = \mathbf{0}$ we have $\mathbf{K}_2 \mathbf{Z}_2 \mathscr{E}[\mathbf{d}_i] \approx \mathbf{0}$, or

$$\mathbf{Z}_2 \mathscr{E}[\mathbf{d}_i] \approx \mathbf{0}. \tag{11.50}$$

In order to determine m_2 we need to look at the positive eigenvalues λ_{m_1+1}, $\lambda_{m_1+2}, \ldots$ and draw a cutting line when they get too large. Box et al. [1973: Appendix] showed that if (11.50) holds, then

$$\begin{aligned} E[\lambda_k] &= E[\mathbf{z}_k' \mathbf{D}'\mathbf{D} \mathbf{z}_k] \\ &\approx (n-1) \mathbf{z}_k' \boldsymbol{\Sigma} \mathbf{z}_k, \qquad k = m_1 + 1, \ldots, m_1 + m_2. \end{aligned}$$

Here $\boldsymbol{\Sigma}$, the variance–covariance matrix of the $\boldsymbol{\varepsilon}_i$, must be estimated from previous experiments or replicates. However, only an order-of-magnitude value is required, so that a fairly rough method of estimating $\boldsymbol{\Sigma}$ will be sufficient. For example, if high correlations among the responses are not expected, then we can approximate $\boldsymbol{\Sigma}$ by $\mathbf{I}_d\bar{\sigma}^2$, where $\bar{\sigma}^2 = \sum_i \sigma_{ii}/d$, and we have

$$\begin{aligned} E[\lambda_k] &\approx (n-1)\mathbf{z}_k'\mathbf{I}_d\mathbf{z}_k\bar{\sigma}^2 \\ &= (n-1)\bar{\sigma}^2. \end{aligned} \tag{11.51}$$

One suggested estimate of $\bar{\sigma}^2$ is RSS/$(nd-p)$, where RSS is the minimum value of $\sum\sum[y_{ij} - f_j(\mathbf{x}_i;\boldsymbol{\theta})]^2$.

11.5.4 Estimation Procedures

Returning now to the original problem of estimating $\boldsymbol{\theta}$, we saw earlier in this chapter that the maximum-likelihood estimate (or posterior mode) $\hat{\boldsymbol{\theta}}$ is the value of $\boldsymbol{\theta}$ which minimizes $|\mathbf{V}_d(\boldsymbol{\theta})|$ of (11.27). The subscript d, representing the number of response variables, is now added for notational convenience. When there are m_1 linear constraints among the responses, there will generally be problems in using the above determinant criterion because, for example, of (11.41). To avoid these, the set of responses should be reduced to just the $d-m_1$ responses that have been independently determined (or, more generally, any $d-m_1$ independent linear combinations of them). We then minimize the corresponding determinant $|\mathbf{V}_{d-m_1}(\boldsymbol{\theta})|$. Any relationship, however, among the expected values of the responses will not directly affect the fitting of the model and therefore will not be detected at this stage. We see, therefore, that a preliminary eigenvalue–eigenvector analysis is a useful tool for (i) uncovering exact relationships in the data which must be allowed for in the subsequent analysis, and (ii) suggesting linear relationships among the expected values which may throw some light on the adequacy of the model being considered. Clearly any relationships found or suggested by the analysis should be satisfactorily explained before any further analysis of the data.

The above methodology of Box et al. [1973] refers specifically to the case described by (11.39). Here, both $\mathbf{D}'\mathbf{D}$ and $\mathbf{V}(\boldsymbol{\theta})$ have m_1 zero eigenvalues, so that the eigenvalues of $\mathbf{D}'\mathbf{D}$ will tell us about the singularities of $\mathbf{V}(\boldsymbol{\theta})$. However, for the situation described by (11.42), we still have (11.40) and (11.41), but now

$$\mathbf{A}_1\mathbf{d}_i = \mathbf{A}_1(\mathbf{y}_i - \bar{\mathbf{y}}) = \mathbf{b}_i - \bar{\mathbf{b}}.$$

This will only be zero for all i if $\mathbf{b}_1 = \mathbf{b}_2 = \cdots = \mathbf{b}_n$. If the $\mathbf{b}_i$'s are not all equal, then we have the situation where $\mathbf{D}'\mathbf{D}$ is nonsingular, but $\mathbf{V}(\boldsymbol{\theta})$ has m_1 zero eigenvalues. On the other hand, we could have the situation described by (11.43), in which $\mathbf{V}(\boldsymbol{\theta})$ is nonsingular but $\mathbf{D}'\mathbf{D}$ has m_1 zero eigenvalues. However, as McLean et al. [1979] point out, the latter result may be acceptable or even preferable if the linear relationships in the data have resulted from normalizing or adjusting procedures, and these procedures are not desired in the model. In this case we would not reduce the number of responses, but simply go ahead with minimizing $|\mathbf{V}_d(\boldsymbol{\theta})|$. The elimination of linear dependencies would result in an unnecessary loss of precision in the parameter estimates.

It is clear from the above discussion that the eigenvalues of $\mathbf{D}'\mathbf{D}$ will shed light on possible problems with minimizing $|\mathbf{V}(\boldsymbol{\theta})|$. However, $\mathbf{D}'\mathbf{D}$ may be singular and $\mathbf{V}(\boldsymbol{\theta})$ nonsingular, and vice versa, so that it is important to look at the eigenvalues of $\mathbf{V}(\boldsymbol{\theta})$ as well. A scheme for looking at both sets of eigenvalues is described by McLean et al. [1979]. If $\mathbf{V}(\boldsymbol{\theta})$ is singular, there is the problem of how to remove the singularities. Box et al. [1973] state that, apart from roundoff error, it makes no difference which response variables involved in the underlying linear relationships are deleted as long as the remaining response variables are linearly independent. They noted that when the data do not satisfy a linear relationship exactly, the location and precision of the parameter estimates will depend to some extent on how the singularity is removed. In such cases McLean et al. [1979] suggest that if the removal of a singularity could be achieved by deleting one of two regressor variables whose variances are substantially different, then intuitively the precision of the estimated parameters and the fits of the individual response models would be affected by which response variable is removed. They then give an example which supports this view.

Example 11.4 Box et al. [1973] gave an interesting example of a complex chemical reaction in which the proposed reaction model is pictured in Fig. 11.5. The following detailed discussion is based closely on their article. The chemical reaction is the isomerization of α-pinene (y_1) to dipentene (y_2) and allo-ocimen (y_3), the latter in turn yielding α- and β-pyrone (y_4) and a dimer (y_5). The unknown parameters θ_r ($r = 1, 2, \ldots, 5$) are the rate constants of the reactions. This process was studied by Fuguitt and Hawkins [1947], who reported the concentrations of the reactant and the four products (Table 11.4). If the chemical reaction orders are known, then mathematical models can be derived which give

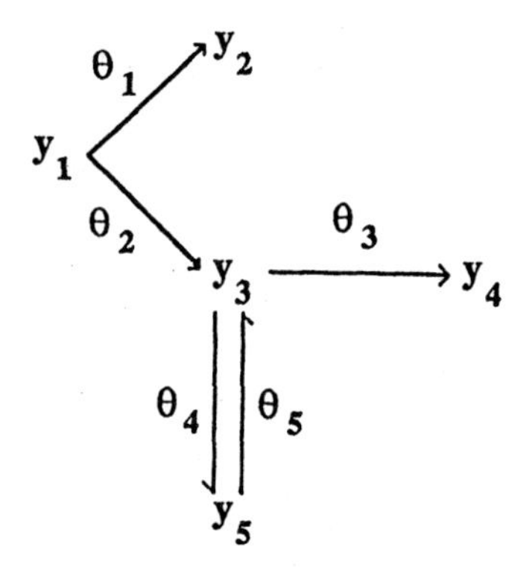

Figure 11.5 Thermal isomerization of α-pinene (y_1).

Table 11.4 Data on Percentage Concentration versus Time for the Isomerization of α-Pinene at 189.5°C[a]

Time x_i (min)	y_{i1} α-Pinene	y_{i2} Dipentene	y_{i3} allo-Ocimene	y_{i4} Pyronene	y_{i5} Dimer
1230	88.35	7.3	2.3	0.4	1.75
3060	76.4	15.6	4.5	0.7	2.8
4920	65.1	23.1	5.3	1.1	5.8
7800	50.4	32.9	6.0	1.5	9.3
10680	37.5	42.7	6.0	1.9	12.0
15030	25.9	49.1	5.9	2.2	17.0
22620	14.0	57.4	5.1	2.6	21.0
36420	4.5	63.1	3.8	2.9	25.7

[a]From G.E.P. Box et al. [1973].

the concentrations of various substances as a function of time. In particular Hunter and McGregor [1967] assumed first-order kinetics and derived the following equations for the five responses:

$$E[y_{i1}] = y_{10}\exp(-\phi t_i),$$

$$E[y_{i2}] = \frac{\theta_1 y_{10}}{\phi}[1 - \exp(-\phi t_i)],$$

$$E[y_{i3}] = C_1\exp(-\phi t_i) + C_2\exp(\beta t_i) + C_3\exp(\gamma t_i),$$

$$E[y_{i4}] = \theta_3\left\{\frac{C_1}{\phi}[1 - \exp(-\phi t_i)] + \frac{C_2}{\beta}[\exp(\beta t_i) - 1] + \frac{C_3}{\gamma}[\exp(\gamma t_i) - 1]\right\},$$

$$E[y_{i5}] = \theta_4\left\{\frac{C_1}{(\theta_5 - \phi)}\exp(-\phi t_i) + \frac{C_2}{(\theta_5 + \beta)}\exp(\beta t_i) + \frac{C_3}{(\theta_5 + \gamma)}\exp(\gamma t_i)\right\}.$$

Here y_{10} is the value of y_1 at $t = 0, \alpha = \theta_3 + \theta_4 + \theta_5, \beta = \frac{1}{2}[-\alpha + (\alpha^2 - 4\theta_3\theta_5)^{1/2}]$, $\gamma = \frac{1}{2}[-\alpha - (\alpha^2 - 4\theta_3\theta_5)^{1/2}]$, $\phi = \theta_1 + \theta_2$, $C_1 = \theta_2 y_{10}(\theta_5 - \phi)/[(\phi + \beta)(\phi + \gamma)]$, $C_2 = \theta_2 y_{10}(\theta_5 + \beta)/[(\phi + \beta)(\beta - \gamma)]$, and $C_3 = \theta_2 y_{10}(\theta_5 + \gamma)/[(\phi + \gamma)(\gamma - \beta)]$. An estimate of $\boldsymbol{\theta} = (\theta_1, \theta_2, \ldots, \theta_5)'$ can be obtained by minimizing the determinant criterion $|\mathbf{V}_5(\boldsymbol{\theta})|$. However, such an estimate leads to a model which provides an unsatisfactory fit to the data (Fig. 11.6). This example shows how an analysis of multiresponse data can lead to useless estimates if linear dependencies are ignored.

It transpires that, because of experimental difficulties, y_4 was not measured independently but rather was assumed to constitute 3% of y_1. This leads to the exact linear relationship

$$0.03y_{i1} + y_{i4} = 3. \tag{11.52}$$

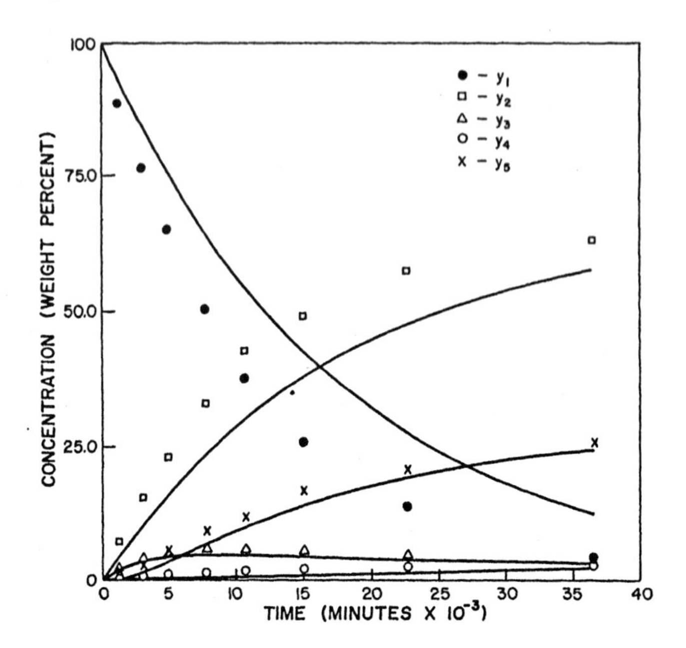

Figure 11.6 Data fit with parameter values which, ignoring dependencies, minimize $|\mathbf{V}_5|$. From Box et al. [1973].

Further algebraic manipulations of the original first-order differential equations produced two relationships among the expected responses, namely

$$E[y_{i1}] + E[y_{i2}] + E[y_{i3}] + E[y_{i4}] + E[y_{i5}] = a_1 \tag{11.53}$$

and

$$E[y_{i1}] + \left(1 + \frac{\theta_2}{\theta_1}\right) E[y_{i2}] = a_2, \tag{11.54}$$

where a_1 and a_2 are constants. Equation (11.53) simply expresses an overall mass balance for the system, while (11.54) results from the fact that the isomerizations of α-pinene (y_1) are assumed irreversible. The experimenters in this study also chose to report their data in "normalized" weight form, thus perhaps using (11.53) to force the following relationship among the observed responses:

$$y_{i1} + y_{i2} + y_{i3} + y_{i4} + y_{i5} = 100. \tag{11.55}$$

There are therefore three linear relationships in the system, (11.52) and (11.55) for the data, and (11.54) for expected values. The first two are exact apart from rounding error. In fact it was the presence of the rounding error which led to $|\mathbf{V}_5(\boldsymbol{\theta})|$ not being exactly zero, and this made it technically possible to obtain a spurious minimum for $|\mathbf{V}_5(\boldsymbol{\theta})|$.

Table 11.5 Eigenvalues and Eigenvectors of D′D[a]

		Eigenvalues		
λ_1	λ_2	λ_3	λ_4.	λ_5
0.0013	0.0168	1.21	25.0	9660.
		Eigenvectors		
$\mathbf{z}_1$	$\mathbf{z}_2$	$\mathbf{z}_3$	$\mathbf{z}_4$	$\mathbf{z}_5$
−.169	.476	−.296	.057	.809
−.211	.490	−.611	−.224	−.540
−.161	.435	.640	−.612	−.013
.931	.364	−.010	.004	−.024
−.185	.459	.360	.756	−.231

[a]From Box et al. [1973].

Having found the linear relationships theoretically, we now see whether an eigenvalue–eigenvector analysis will show them up. The eigenvalues λ_k and the corresponding eigenvectors $\mathbf{z}_k$ of $\mathbf{D}'\mathbf{D}$ are listed in Table 11.5. The mean variance, $\bar{\sigma}^2$, is estimated by

$$\frac{\text{RSS}}{nd-p}=\frac{19.87}{40-5}=0.6.$$

To check for exact linear relationships among the responses, we look for zero eigenvalues. Although there are none which are exactly zero, λ_1 and λ_2 are both very small and much smaller than [c.f. (11.51)]

$$(n-1)\bar{\sigma}^2 \approx 4.2, \tag{11.56}$$

which we would expect from a linear expectation relationship. We therefore suspect that $\mathbf{z}_1$ and $\mathbf{z}_2$ provide linear relationships among the responses that are exact except for rounding error. To obtain an estimate of $E[\lambda_k]$ when there is only rounding error present, we can assume that the rounding error has a uniform distribution with range -0.5 to $+0.5$ of the last digit reported. The variance of the distribution, σ_{re}^2 say, is then given by the range squared divided by 12. In this example all responses were rounded to the same number of significant figures, so that

$$E[\lambda_k] \approx (n-1)\mathbf{z}_k'\mathbf{I}\mathbf{z}_k\sigma_{\text{re}}^2 = (n-1)\sigma_{\text{re}}^2. \tag{11.57}$$

The data in Table 11.4 were reported to the nearest 0.1%, so that the range is from

-0.05 to $+0.05$, or 0.10. Hence

$$E[\lambda_k] \approx 7(0.10)^2/12 = 0.006.$$

Both λ_1 and λ_2 are of this order of magnitude, thus helping to confirm that $\mathbf{z}_1$ and $\mathbf{z}_2$ represents exact dependencies among the responses. Since we know the matrix $\mathbf{A}_1$ from theoretical considerations, namely

$$\mathbf{A}_1 = \begin{pmatrix} \mathbf{a}_1' \\ \mathbf{a}_2' \end{pmatrix} = \begin{pmatrix} 0.03 & 0 & 0 & 1 & 0 \\ 1 & 1 & 1 & 1 & 1 \end{pmatrix}, \tag{11.58}$$

we can also check to see if the plane determined by the two rows of $\mathbf{A}_1$ is the same as the plane determined by $\mathbf{z}_1$ and $\mathbf{z}_2$. The respective cosines of the angles made by each of the vectors $\mathbf{a}_1$ and $\mathbf{a}_2$ with the latter plane are 0.9999 and 0.9993. The four vectors may therefore be regarded as coplanar.

Box et al. [1973] noted that if the true underlying relationships are not known, then it may be possible to use the λ_k to provide some idea as to what they might be. For instance, if the smallest of the "zero" eigenvalues is much smaller than the others, then its corresponding eigenvector may correspond fairly closely to the most exact of the linear relationships (apart from a constant multiplier as $\|\mathbf{z}_1\| = 1$). However, this probably would not be the case if the "zero" eigenvalues were of nearly equal magnitude. In the above example $\lambda_1 = 0.0013$ is much smaller than $\lambda_2 = 0.0168$, $\mathbf{z}_1 = (-0.169, -0.211, -0.161, 0.931, -0.185)'$ has some resemblance to $\mathbf{a}_1$ of (11.58), and $\mathbf{z}_2 = (0.476, 0.490, 0.435, 0.364, 0.459)'$ is similar to that component of $\mathbf{a}_2$ which is orthogonal to $\mathbf{z}_1$, namely (0.465, 0.468, 0.464, 0.363, 0.466)'.

Since λ_3 $(= 1.21)$ is of similar order of magnitude to (11.56), it would seem that $\mathbf{z}_3$ represents a linear relationship among the expected values. We can check this as follows. From (11.54) we have the vector

$$\left(1, 1 + \frac{\theta_2}{\theta_1}, 0, 0, 0\right), \tag{11.59}$$

which can be evaluated using the current estimates of θ_1 and θ_2. Taking the component of (11.59) which is perpendicular to both $\mathbf{z}_1$ and $\mathbf{z}_2$ (i.e. perpendicular to $\mathscr{R}[\mathbf{A}_1']$) and normalizing, we have

$$\mathbf{a}_{2(1)}^{\perp} = (-0.308, -0.665, 0.482, 0.008, 0.482)'.$$

This is very similar to

$$z_3 = (-0.296, -0.611, 0.640, -0.010, 0.360)'.$$

The confirmation of an expectation relationship such as (11.54) provides a useful check on the hypothesized model structure. For example, if either of the isomerization reactions had been reversible, then (11.54) would not have been true.

Having established the existence of two linear relationships among the responses, we now consider the problem of estimating $\boldsymbol{\theta}$. The first step is to remove these two "singularities" by dropping two of the responses so as to leave a mathematically independent subset of the responses. Often the structure of the problem and of the dependencies may suggest a natural way of dropping responses. In the above example the first relationship (11.52) indicates that y_1 and y_4 are natural candidates, especially y_4, as it is obtained from y_1. The second relationship (11.55) contains all five responses, so that one of the additional four responses can be dropped. Another approach to the problem is to use three linearly independent combinations of all five responses, where the linear combinations adequately define the three-dimensional subspace orthogonal to the two-dimensional "singular plane," $\mathscr{R}[\mathbf{A}_1']$.

Suppose that the theoretical relationships are unknown, so that we must rely on the information provided by $\mathbf{z}_1$ and $\mathbf{z}_2$. If $\mathbf{Z}=(\mathbf{z}_1,\mathbf{z}_2,\ldots,\mathbf{z}_5)'$, we can make the rotation $\mathbf{Z}\mathbf{y}_i$ to produce new coordinates $g_{ik}=\mathbf{z}_k'\mathbf{y}_i$ $(k=1,2,\ldots,5)$. The first two coordinates are (approximately) constant because of the linear relationships, and the last three $(k=3,\ 4,\ 5)$ can be used to give three independent linear combinations. Also the new variables have zero sample correlations. To see this, define $\mathbf{u}=\mathbf{y}_i$ with probability n^{-1} $(i=1,2,\ldots,n)$. Then $\mathscr{E}[\mathbf{u}]=\bar{\mathbf{y}}$, $\mathscr{D}[\mathbf{u}]=n^{-1}\sum_i(\mathbf{y}_i-\bar{\mathbf{y}})(\mathbf{y}_i-\bar{\mathbf{y}})'=n^{-1}\mathbf{D}'\mathbf{D}$, and

$$\begin{aligned}\mathscr{C}[\mathbf{z}_r'\mathbf{u},\mathbf{z}_s'\mathbf{u}]&=\mathbf{z}_r'\mathscr{D}[\mathbf{u}]\mathbf{z}_s \qquad \text{[by (1.2)]}\\&=n^{-1}\mathbf{z}_r'\mathbf{D}'\mathbf{D}\mathbf{z}_s\\&=n^{-1}\mathbf{z}_r'\lambda_s\mathbf{z}_s\\&=n^{-1}\lambda_s\delta_{rs}.\end{aligned}\tag{11.60}$$

We now estimate $\boldsymbol{\theta}$ by minimizing the determinant of

$$\mathbf{V}_3(\boldsymbol{\theta})=\sum_{i=1}^{n}(\mathbf{g}_i-\mathscr{E}[\mathbf{g}_i])(\mathbf{g}_i-\mathscr{E}[\mathbf{g}_i])',\tag{11.61}$$

Table 11.6 Estimates of Rate Constants[a]

	Rate Constants ($10^{-5}\,\text{min}^{-1}$)				
Minimization of	$\hat{\theta}_1$	$\hat{\theta}_2$	$\hat{\theta}_3$	$\hat{\theta}_4$	$\hat{\theta}_5$
RSS	5.93	2.96	2.05	27.5	4.00
$\lvert\mathbf{V}_3\rvert$:					
Empirical eigenvectors	5.95	2.84	0.43	31.3	5.74
Theoretical eigenvectors	5.95	2.85	0.50	31.5	5.89

[a]From Box et al. [1973].

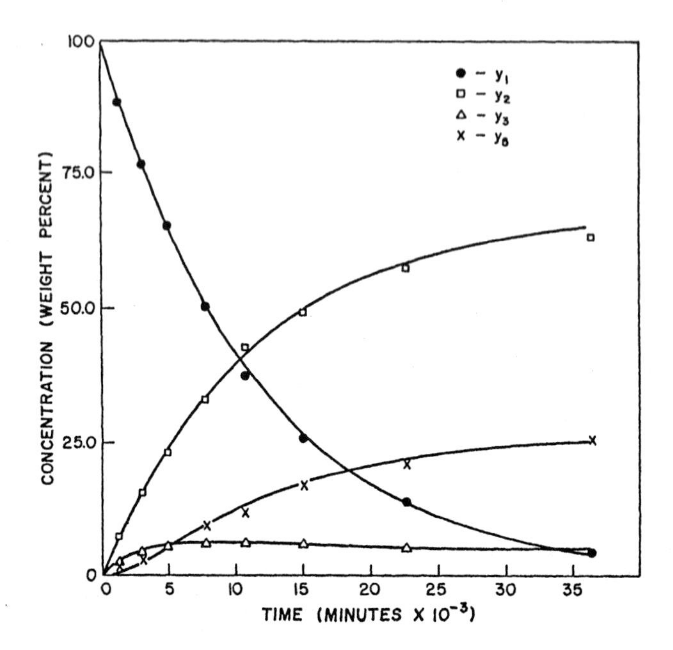

Figure 11.7 Data fit with parameter values which, allowing for three linear relationships, minimize $|\mathbf{V}_5|$. From Box et al. [1973].

where $\mathbf{g}_i' = (g_{i3}, g_{i4}, g_{i5})$-and $E[g_{ik}] = \mathbf{z}_k'\mathbf{f}_i(\boldsymbol{\theta})$. The estimates are given in row 2 of Table 11.6. With the two dependencies removed, we now see from Fig. 11.7 that the model fits the data very well, in contrast to Fig. 11.6.

When the theoretical relationships have been uncovered it is better, however, to use three independent linear combinations of the five responses which are orthogonal to the *true* singularity plane, $\mathscr{R}[\mathbf{A}_1']$, than the approximate one given by $\mathbf{z}_1$ and $\mathbf{z}_2$ (although these will differ only slightly, as the roundoff error is relatively small). For this purpose Box et al. [1973] used as the three linear combinations the vector components $\mathbf{a}_3, \mathbf{a}_4$, and $\mathbf{a}_5$ of $\mathbf{z}_3, \mathbf{z}_4$, and $\mathbf{z}_5$ which are orthogonal to $\mathbf{a}_1$ and $\mathbf{a}_2$. We again minimize $|\mathbf{V}_3(\boldsymbol{\theta})|$ with $g_{ik} = \mathbf{a}_k'\mathbf{f}_i(\boldsymbol{\theta})$ $(k = 3, 4, 5)$, and the estimates of the θ_r are in row 3 in Table 11.6. As expected, these are close to the values in row 2. The authors noted that the estimate of θ_3 is imprecise, as is reflected by an extremely flat surface of $|\mathbf{V}_3(\boldsymbol{\theta})|$ in the θ_3-direction. This was to be expected, since most of the information on the θ_3 rate constant is contained in the singularity plane and, in particular, in y_4.

In practice, if any dependencies which are detected empirically can be justified theoretically, then the theoretical relationships will be used as above. Otherwise the empirical relationships are used. ■

*11.6 FUNCTIONAL RELATIONSHIPS

So far in this chapter we have assumed that the vector of k explanatory variables $\mathbf{x}$ is measured without error, or at least has a variance–covariance matrix which is negligible compared with $\boldsymbol{\Sigma}$, the variance–covariance matrix of the $d \times 1$ response vector $\mathbf{y}$. However, in many situations this is not the case and an error or random fluctuation in $\mathbf{x}$ must be allowed for in the analysis. One such model which incorporates these so-called errors in variables is the functional-relationship model, in which we have mathematical variables $\boldsymbol{\mu}$ and $\boldsymbol{\xi}$ satisfying a vector functional relationship

$$\boldsymbol{\mu} = \mathbf{f}(\boldsymbol{\xi}; \boldsymbol{\theta}). \tag{11.62}$$

Both $\boldsymbol{\mu}$ and $\boldsymbol{\xi}$ are measured with error, so that what are observed are $\mathbf{y}$ and $\mathbf{x}$ respectively. If we have n observations on this model, then, in contrast to (11.2), we have

$$\mathbf{y}_i = \mathbf{f}(\boldsymbol{\xi}_i; \boldsymbol{\theta}) + \boldsymbol{\varepsilon}_i \tag{11.63}$$

and

$$\mathbf{x}_i = \boldsymbol{\xi}_i + \boldsymbol{\delta}_i \qquad (i = 1, 2, \ldots, n). \tag{11.64}$$

Setting $\mathbf{e}_i = (\boldsymbol{\varepsilon}_i' \ \boldsymbol{\delta}_i')'$, we assume that the $\mathbf{e}_i$ are independently distributed with mean zero and positive definite variance–covariance matrix $\mathbf{V}_i$.

In order to estimate $\boldsymbol{\theta}$ we now have to estimate the nuisance parameters $\boldsymbol{\xi}_i$ $(i = 1, 2, \ldots, n)$ as well, and this cannot be done without some knowledge of the $\mathbf{V}_i$. Assuming, therefore, that the $\mathbf{V}_i$ are known, or at least estimates of the $\mathbf{V}_i$ are available, we can find estimates of the unknown parameters using the method of generalized least squares. This consists of minimizing

$$\sum_{i=1}^{n} \mathbf{e}_i' \mathbf{V}_i^{-1} \mathbf{e}_i = \sum_{i=1}^{n} \begin{pmatrix} \mathbf{y}_i - \mathbf{f}(\boldsymbol{\xi}_i; \boldsymbol{\theta}) \\ \mathbf{x}_i - \boldsymbol{\xi}_i \end{pmatrix}' \mathbf{V}_i^{-1} \begin{pmatrix} \mathbf{y}_i - \mathbf{f}(\boldsymbol{\xi}_i; \boldsymbol{\theta}) \\ \mathbf{x}_i - \boldsymbol{\xi}_i \end{pmatrix} \tag{11.65}$$

with respect to $\boldsymbol{\theta}$ and the $\boldsymbol{\xi}_i$. If the $\mathbf{e}_i$ are also normally distributed then generalized least squares is equivalent to maximum-likelihood estimation. Schwetlick and Tiller [1985], in discussing the above model, gave a modified Gauss–Newton-type method for carrying out the minimization of (11.65), called "a partially regularized Marquardt-like method." This modification was introduced to overcome stability and convergence problems that can occur with the Gauss–Newton procedure. Their method is now discussed in detail.

The first step consists of transforming the model so that $\mathbf{V}_i$ becomes $\mathbf{I}_{d+k}$, and then rolling out the transformed responses into a single $nd \times 1$ vector, as in (11.7). Let $\mathbf{W}_i = \mathbf{V}_i^{-1}$, and let $\mathbf{W}_i = \mathbf{R}_i'\mathbf{R}_i$ be the Cholesky decomposition

(Appendix A8.2) of $\mathbf{W}_i$, where

$$\mathbf{R}_i = \begin{pmatrix} \mathbf{R}_{i3} & \mathbf{R}_{i2} \\ \mathbf{0} & \mathbf{R}_{i1} \end{pmatrix} \begin{matrix} \}d \\ \}k \end{matrix} \qquad (i = 1, 2, \ldots, n), \tag{11.66}$$

$\mathbf{R}_{i3}$ is a $d \times d$ upper triangular nonsingular matrix (corresponding to $\mathbf{y}_i$) and $\mathbf{R}_i$ is a $k \times k$ upper triangular nonsingular matrix (corresponding to $\mathbf{x}_i$). Then

$$\begin{aligned} \sum_i \mathbf{e}_i' \mathbf{V}_i^{-1} \mathbf{e}_i &= \sum_i \mathbf{e}_i' \mathbf{W}_i \mathbf{e}_i \\ &= \sum_i (\mathbf{R}_i \mathbf{e}_i)'(\mathbf{R}_i \mathbf{e}_i) \\ &= \sum_i (\| \mathbf{R}_{i1} \boldsymbol{\delta}_i \|^2 + \| \mathbf{R}_{i2} \boldsymbol{\delta}_i + \mathbf{R}_{i3} \boldsymbol{\varepsilon}_i \|^2). \end{aligned}$$

If $\boldsymbol{\xi} = (\boldsymbol{\xi}_1', \boldsymbol{\xi}_2', \ldots, \boldsymbol{\xi}_n')'$ and $\boldsymbol{\zeta} = (\boldsymbol{\xi}', \boldsymbol{\theta}')'$, we have

$$\sum_i \mathbf{e}_i' \mathbf{V}_i^{-1} \mathbf{e}_i = \mathbf{g}'(\boldsymbol{\zeta}) \mathbf{g}(\boldsymbol{\zeta}),$$

where

$$\begin{aligned} \mathbf{g}(\boldsymbol{\zeta}) &= \left[\begin{array}{c} \mathbf{R}_{11}(\boldsymbol{\xi}_1 - \mathbf{x}_1) \\ \mathbf{R}_{21}(\boldsymbol{\xi}_2 - \mathbf{x}_2) \\ \vdots \\ \mathbf{R}_{n1}(\boldsymbol{\xi}_n - \mathbf{x}_n) \\ \hline \mathbf{R}_{12}(\boldsymbol{\xi}_1 - \mathbf{x}_1) + \mathbf{R}_{13}[\mathbf{f}(\boldsymbol{\xi}_1; \boldsymbol{\theta}) - \mathbf{y}_1] \\ \mathbf{R}_{22}(\boldsymbol{\xi}_2 - \mathbf{x}_2) + \mathbf{R}_{23}[\mathbf{f}(\boldsymbol{\xi}_2; \boldsymbol{\theta}) - \mathbf{y}_2] \\ \vdots \\ \mathbf{R}_{n2}(\boldsymbol{\xi}_n - \mathbf{x}_n) + \mathbf{R}_{n3}[\mathbf{f}(\boldsymbol{\xi}_n; \boldsymbol{\theta}) - \mathbf{y}_n] \end{array} \right] \\ &= \begin{pmatrix} \mathbf{g}_1(\boldsymbol{\zeta}) \\ \mathbf{g}_2(\boldsymbol{\zeta}) \end{pmatrix}, \end{aligned} \tag{11.67}$$

say. If $\mathbf{J}(\boldsymbol{\zeta}^{(a)})$ is the $n(k+d) \times (nk+p)$ Jacobian, $\partial \mathbf{g}/\partial \boldsymbol{\zeta}'$ [see (1.10)], evaluated at the ath iteration, then the Gauss–Newton method uses the linear approximation

$$\mathbf{g}(\boldsymbol{\zeta}) \approx \mathbf{g}(\boldsymbol{\zeta}^{(a)}) + \mathbf{J}(\boldsymbol{\zeta}^{(a)})(\boldsymbol{\zeta} - \boldsymbol{\zeta}^{(a)}) \tag{11.68}$$

when minimizing $\mathbf{g}'\mathbf{g}$. The next iteration is given by

$$\boldsymbol{\zeta}^{(a+1)} = \boldsymbol{\zeta}^{(a)} - \Delta \boldsymbol{\zeta}^{(a)},$$

where, suppressing the argument $\zeta^{(a)}$ of the functions for notational convenience,

$$\Delta\zeta = (\mathbf{J}'\mathbf{J})^{-1}\mathbf{J}'\mathbf{g}. \tag{11.69}$$

This is the solution of the linear least-squares problem

$$\underset{\Delta\zeta}{\text{Minimize}} \, \|\mathbf{g} - \mathbf{J}\Delta\zeta\|^2. \tag{11.70}$$

We note that $\mathbf{J}$ takes the form

$$\mathbf{J} = (\mathbf{J}_1, \mathbf{J}_2) \tag{11.71}$$

$$= \begin{pmatrix} \mathbf{J}_{11}, \mathbf{0} \\ \mathbf{J}_{21}, \mathbf{J}_{22} \end{pmatrix} \tag{11.72}$$

$$= \begin{bmatrix} \dfrac{\partial \mathbf{g}_1}{\partial \boldsymbol{\xi}}, \dfrac{\partial \mathbf{g}_1}{\partial \boldsymbol{\theta}} \\ \dfrac{\partial \mathbf{g}_2}{\partial \boldsymbol{\xi}}, \dfrac{\partial \mathbf{g}_2}{\partial \boldsymbol{\theta}} \end{bmatrix}$$

$$= \left[\begin{array}{cc|c} \nabla & 0 & 0 \\ & \ddots & \vdots \\ 0 & \nabla & 0 \\ \hline \square & 0 & \square \\ & \ddots & \vdots \\ 0 & \square & \square \end{array}\right]. \tag{11.73}$$

Here $\mathbf{J}_{11}$ is a block-diagonal matrix with blocks $\mathbf{R}_{i1}(=\mathbf{J}_{11,i}$ say) which are upper triangular, $\mathbf{J}_{21}$ is a block-diagonal matrix with rectangular blocks

$$\mathbf{J}_{21,i} = \mathbf{R}_{i2} + \mathbf{R}_{i3}\frac{\partial \mathbf{f}(\boldsymbol{\xi}_i; \boldsymbol{\theta})}{\partial \boldsymbol{\xi}_i'}, \tag{11.74}$$

and $\mathbf{J}_{22}$ has vertical blocks

$$\mathbf{J}_{22,i} = \mathbf{R}_{i3}\frac{\partial \mathbf{f}(\boldsymbol{\xi}_i; \boldsymbol{\theta})}{\partial \boldsymbol{\theta}'}.$$

We now proceed to solve (11.70) using appropriate transformations to simplify the computations. Ignoring the zeros in $\mathbf{J}_1$ [see, (11.73)], we can apply a QR decomposition (Appendix A8.3) using, for example, Householder reflections (c.f. Lawson and Hanson [1974] or Dongarra et al. [1979]) to each triangle–rectangle vertical pair, namely

$$\begin{bmatrix} \triangledown \\ \square \end{bmatrix} = \begin{pmatrix} \mathbf{J}_{11,i} \\ \mathbf{J}_{21,i} \end{pmatrix} = \mathbf{Q}_i \begin{pmatrix} \mathbf{K}_{11,i} \\ \mathbf{0} \end{pmatrix},$$

or

$$\mathbf{Q}_i' \begin{pmatrix} \mathbf{J}_{11,i} \\ \mathbf{J}_{21,i} \end{pmatrix} = \begin{pmatrix} \mathbf{K}_{11,i} \\ \mathbf{0} \end{pmatrix}. \tag{11.75}$$

Here $\mathbf{Q}_i'$ is orthogonal, being a product of, say, Householder reflections, and $\mathbf{K}_{11,i}$ is upper triangular. By inserting diagonal ones and off-diagonal zeros into each $\mathbf{Q}_i'$, we obtain an augmented orthogonal matrix $\mathbf{Q}'_{i(\text{AUG})}$ which now operates on all the rows of the ith vertical block in (11.73) and not just on the triangle–rectangle pair. For example,

$$\mathbf{Q}_1' = \begin{bmatrix} \square & \square \\ \square & \square \end{bmatrix} \qquad \text{operating on} \qquad \begin{bmatrix} \triangledown \\ \square \end{bmatrix}$$

becomes

$$\mathbf{Q}'_{1(\text{AUG})} = \begin{bmatrix} \square & \mathrm{O} & \square & \mathrm{O} \\ \mathrm{O} & \mathbf{I} & \mathrm{O} & \mathrm{O} \\ \square & \mathrm{O} & \square & \mathrm{O} \\ \mathrm{O} & \mathrm{O} & \mathrm{O} & \mathbf{I} \end{bmatrix} \qquad \text{operating on} \begin{bmatrix} \triangledown \\ \mathrm{O} \\ \square \\ \mathrm{O} \end{bmatrix}$$

These augmented matrices can be multiplied together to give a single orthogonal matrix $\mathbf{Q}'$ such that

$$\mathbf{Q}'\mathbf{J} = \mathbf{Q}' \begin{pmatrix} \mathbf{J}_{11} & \mathbf{0} \\ \mathbf{J}_{21} & \mathbf{J}_{22} \end{pmatrix} = \begin{pmatrix} \mathbf{K}_{11} & \mathbf{K}_{12} \\ \mathbf{0} & \mathbf{K}_{22} \end{pmatrix} = \mathbf{K}, \tag{11.76a}$$

say, where $\mathbf{K}$ takes the form

$$\mathbf{K} = \begin{bmatrix} \triangledown & & \mathrm{O} & \square \\ & \ddots & & \vdots \\ \mathrm{O} & & \triangledown & \square \\ & & & \square \\ & \mathrm{O} & & \vdots \\ & & & \square \end{bmatrix}$$

Letting

$$\mathbf{h} = \mathbf{Q}'\mathbf{g} = \mathbf{Q}'\begin{pmatrix} \mathbf{g}_1 \\ \mathbf{g}_2 \end{pmatrix} = \begin{pmatrix} \mathbf{h}_1 \\ \mathbf{h}_2 \end{pmatrix},$$

we find that [c.f. (11.70)]

$$\begin{aligned} \|\mathbf{g} - \mathbf{J}\,\Delta\boldsymbol{\zeta}\|^2 &= \|\mathbf{Q}'(\mathbf{g} - \mathbf{J}\,\Delta\boldsymbol{\zeta})\|^2 \\ &= \|\mathbf{h} - \mathbf{K}\,\Delta\boldsymbol{\zeta}\|^2 \\ &= \|\mathbf{h}_1 - \mathbf{K}_{11}\,\Delta\boldsymbol{\xi} - \mathbf{K}_{12}\,\Delta\boldsymbol{\theta}\|^2 + \|\mathbf{h}_2 - \mathbf{K}_{22}\,\Delta\boldsymbol{\theta}\|^2. \end{aligned} \quad (11.76b)$$

To minimize (11.76b) we first calculate $\Delta\boldsymbol{\theta}$ as the solution of the linear least-squares problem

$$\underset{\Delta\boldsymbol{\theta}}{\text{Minimize}}\,\|\mathbf{h}_2 - \mathbf{K}_{22}\,\Delta\boldsymbol{\theta}\|^2. \quad (11.77)$$

Then, using this value of $\Delta\boldsymbol{\theta}$, we choose $\Delta\boldsymbol{\xi}$ to make the first term of (11.76b) zero, i.e., we solve the upper triangular system

$$\mathbf{K}_{11}\Delta\boldsymbol{\xi} = \mathbf{h}_1 - \mathbf{K}_{12}\,\Delta\boldsymbol{\theta} \quad (11.78)$$

for $\Delta\boldsymbol{\xi}$. This system decouples into n upper triangular systems of the form

$$\mathbf{K}_{11,i}\Delta\boldsymbol{\xi}_i = \mathbf{h}_{1,i} - \mathbf{K}_{12,i}\,\Delta\boldsymbol{\theta},$$

where the index i refers to the corresponding blocks in $\Delta\boldsymbol{\xi}, \mathbf{h}_1$, and $\mathbf{K}_{12}$, respectively. These are solved rapidly and accurately using back substitution. We note that once $\mathbf{J}$ has been reduced to $\mathbf{K}$, the orthogonal matrix $\mathbf{Q}$ does not need to be recovered, as it does not appear in (11.76b).

If there is an (almost) rank deficiency in $\mathbf{J}$ at any stage of the iteration process or the linear approximation (11.68) is unsatisfactory because of strong nonlinearity in the model, then the Gauss–Newton method will experience convergence problems. However, it is well known that these problems can be lessened using a modification due to Marquardt [1963] (c.f. Section 14.2.2). We now replace (11.69) by

$$\Delta\boldsymbol{\zeta}(\lambda) = (\lambda^2\mathbf{I}_c + \mathbf{J}'\mathbf{J})^{-1}\mathbf{J}'\mathbf{g}, \qquad c = nk + p,$$

that is, (11.70) becomes

$$\underset{\Delta\boldsymbol{\zeta}}{\text{Minimize}}\left\|\begin{pmatrix} \mathbf{g} \\ \mathbf{0} \end{pmatrix} - \begin{pmatrix} \mathbf{J} \\ \lambda\mathbf{I}_c \end{pmatrix}\Delta\boldsymbol{\zeta}\right\|^2$$

using a suitable λ. Unfortunately, as $\mathbf{J}$ is large, this will be prohibitively expensive. However, the $\boldsymbol{\xi}$ part of the iteration is not so critical, as we have good initial estimates $\mathbf{x}_i$ of $\boldsymbol{\xi}_i$ $(i = 1, 2, \ldots, n)$. Also, the $\boldsymbol{\xi}$ part of $\mathbf{J}$ contains the constant nonsingular matrix $\mathbf{J}_{11} = \operatorname{diag}(\mathbf{R}_{11}, \mathbf{R}_{21}, \ldots, \mathbf{R}_{n1})$, so that it is only necessary to apply the Marquardt modification to the $\boldsymbol{\theta}$ part of $\mathbf{J}$. Schwetlick and Tiller [1985] show that such a modification is to find $\mathbf{K}$ as in (11.76a), but (11.77) and (11.78) now become

$$\underset{\Delta\boldsymbol{\theta}}{\text{Minimize}} \left\| \begin{pmatrix} \mathbf{h}_2 \\ \mathbf{0} \end{pmatrix} - \begin{pmatrix} \mathbf{K}_{22} \\ \lambda \mathbf{I}_p \end{pmatrix} \Delta\boldsymbol{\theta} \right\|^2 \tag{11.79}$$

and

$$\mathbf{K}_{11}\, \Delta\boldsymbol{\xi} = (1 + \lambda^2)^{-1} \mathbf{h}_1 - \mathbf{K}_{12}\, \Delta\boldsymbol{\theta}. \tag{11.80}$$

Further details of the above algorithm, including starting values, choice of λ, etc., are given in their paper.

CHAPTER 12*

Asymptotic Theory

12.1 INTRODUCTION

Throughout this book we have made frequent use of large-sample Taylor expansions for linearizing nonlinear theory. We have also assumed that various estimators such as least-squares and maximum-likelihood estimators are asymptotically normal. The validity of such results requires certain regularity assumptions to hold, and in this chapter we collect together some of this underlying theory. As most of the theory is very technical, we give just an overview of the main principles involved, without worrying about proofs and technical details.

Least-squares estimation is considered first, and we rely heavily on Jennrich [1969], Malinvaud [1970a, b], and Wu [1981], and the helpful review papers of Gallant [1975a] and Amemiya [1983]. After a brief interlude with maximum-likelihood estimation, nonlinear hypothesis tests are discussed. The likelihood-ratio, Wald, and Lagrange-multiplier tests are shown to be asymptotically equivalent. Finally, multivariate estimation is considered.

12.2 LEAST-SQUARES ESTIMATION

12.2.1 Existence of Least-Squares Estimate

Consider our usual nonlinear model

$$y_i = f(\mathbf{x}_i; \boldsymbol{\theta}^*) + \varepsilon_i \qquad (i = 1, 2, \ldots, n), \tag{12.1}$$

where $\boldsymbol{\theta}^*$ is the true value of the p-dimensional vector parameter $\boldsymbol{\theta}$ and $\mathbf{x}_i$ is a $k \times 1$ vector in $\mathbb{R}^k$. We shall assume the following:

A(1). The ε_i are i.i.d. with mean zero and variance σ^2 ($\sigma^2 > 0$).
A(2). For each i, $f_i(\boldsymbol{\theta}) = f(\mathbf{x}_i; \boldsymbol{\theta})$ is a continuous function of $\boldsymbol{\theta}$ for $\boldsymbol{\theta} \in \Theta$.
A(3). Θ is a closed, bounded (i.e. compact) subset of $\mathbb{R}^p$.

Any value of $\boldsymbol{\theta}$ which minimizes

$$S_n(\boldsymbol{\theta}) = \sum_{i=1}^{n} [y_i - f_i(\boldsymbol{\theta})]^2 \tag{12.2}$$

subject to $\boldsymbol{\theta} \in \Theta$ is called a least-squares estimator and is denoted by $\hat{\boldsymbol{\theta}}_n$. In studying the properties of $\hat{\boldsymbol{\theta}}_n$, the boundedness of Θ in assumption A(3) is not a serious restriction, as most parameters are bounded by the physical constraints of the system being modeled. For example, if θ_1 is the mean of a Poisson distribution, it can theoretically lie in the range $(0, \infty)$. But in reality we know it will be an interior point of $[0, c]$, where c is a very large number that can be specified in a given situation. As noted by Wu [1981], we can go a step further and assume that Θ is a set with just a finite number of elements. In actual computation we can only search for the minimum of $S_n(\boldsymbol{\theta})$ over a finite set, say to the eighth decimal place.

Given assumptions A(1) to A(3), Jennrich [1969] showed that a least-squares estimator exists, that is, there exists a measurable function $\hat{\boldsymbol{\theta}}_n = \hat{\boldsymbol{\theta}}(y_1, y_2, \ldots, y_n)$ of the observations which minimizes $S_n(\boldsymbol{\theta})$. We now ask under what conditions $\hat{\boldsymbol{\theta}}_n$ is (i) a weakly consistent estimator of $\boldsymbol{\theta}^*$ (i.e., plim $\hat{\boldsymbol{\theta}}_n = \boldsymbol{\theta}^*$, or given $\delta > 0$, $\lim \text{pr}[\|\hat{\boldsymbol{\theta}}_n - \boldsymbol{\theta}^*\| < \delta] = 1$), and (ii) a strongly consistent estimator ($\hat{\boldsymbol{\theta}}_n \to \boldsymbol{\theta}^*$ a.s., that is, $\text{pr}[\hat{\boldsymbol{\theta}}_n \to \boldsymbol{\theta}^*] = 1$). Strong consistency implies weak consistency, but not vice versa.

12.2.2. Consistency

Amemiya [1983] gives a helpful sketch of the necessary ingredients for the consistency of $\hat{\boldsymbol{\theta}}_n$. He indicates that the essential step is to prove that $\boldsymbol{\theta}^*$ uniquely minimizes plim $n^{-1}S_n(\boldsymbol{\theta})$. In that case, if n is sufficiently large so that $n^{-1}S_n(\boldsymbol{\theta})$ is close to plim $n^{-1}S_n(\boldsymbol{\theta})$, then $\hat{\boldsymbol{\theta}}_n$, which minimizes the former, will be close to $\boldsymbol{\theta}^*$, which minimizes the latter. This gives us weak consistency. Now

$$\begin{aligned}
n^{-1}S_n(\boldsymbol{\theta}) &= n^{-1}\sum_i [y_i - f_i(\boldsymbol{\theta})]^2 \\
&= n^{-1}\sum_i [y_i - f_i(\boldsymbol{\theta}^*) + f_i(\boldsymbol{\theta}^*) - f_i(\boldsymbol{\theta})]^2 \\
&= n^{-1}\sum [\varepsilon_i + f_i(\boldsymbol{\theta}^*) - f_i(\boldsymbol{\theta})]^2 \\
&= n^{-1}\sum_i \varepsilon_i^2 + 2n^{-1}\sum_i \varepsilon_i[f_i(\boldsymbol{\theta}^*) - f_i(\boldsymbol{\theta})] \\
&\quad + n^{-1}\sum_i [f_i(\boldsymbol{\theta}^*) - f_i(\boldsymbol{\theta})]^2 \\
&= C_1 + C_2 + C_3,
\end{aligned} \tag{12.3}$$

say. By a law of large numbers, plim $C_1 = \sigma^2$ and, subject to certain conditions on $n^{-1}\sum_i f_i(\boldsymbol{\theta}^*) f_i(\boldsymbol{\theta})$ [see assumption A(4a) below], plim $C_2 = 0$. Hence plim $n^{-1}S_n(\boldsymbol{\theta})$ will have a unique minimum at $\boldsymbol{\theta}^*$ if lim C_3 also has a unique minimum at $\boldsymbol{\theta}^*$ [see assumption A(4b) below]. Weak consistency would then be proved. The strong consistency of $\hat{\boldsymbol{\theta}}_n$ is obtained by using "almost sure" limits rather than limits in probability in the above.

With these ideas in mind, we define

$$B_n(\boldsymbol{\theta}, \boldsymbol{\theta}_1) = \sum_{i=1}^{n} f_i(\boldsymbol{\theta}) f_i(\boldsymbol{\theta}_1), \tag{12.4}$$

$$D_n(\boldsymbol{\theta}, \boldsymbol{\theta}_1) = \sum_{i=1}^{n} [f_i(\boldsymbol{\theta}) - f_i(\boldsymbol{\theta}_1)]^2, \tag{12.5}$$

and add the following assumption:

A(4). (a) $n^{-1} B_n(\boldsymbol{\theta}, \boldsymbol{\theta}_1)$ converges uniformly for all $\boldsymbol{\theta}$, $\boldsymbol{\theta}_1$ in Θ to a function $B(\boldsymbol{\theta}, \boldsymbol{\theta}_1)$ [which is continuous if (A2) and (A3) hold]. This implies, by expanding (12.5), that $n^{-1} D_n(\boldsymbol{\theta}, \boldsymbol{\theta}_1)$ converges uniformly to $D(\boldsymbol{\theta}, \boldsymbol{\theta}_1) = B(\boldsymbol{\theta}, \boldsymbol{\theta}) + B(\boldsymbol{\theta}_1, \boldsymbol{\theta}_1) - 2B(\boldsymbol{\theta}, \boldsymbol{\theta}_1)$.

(b) It is now further assumed that $D(\boldsymbol{\theta}, \boldsymbol{\theta}^*) = 0$ if and only if $\boldsymbol{\theta} = \boldsymbol{\theta}^*$ [i.e., $D(\boldsymbol{\theta}, \boldsymbol{\theta}^*)$ is "positive definite"].

We shall also be interested in the two following related assumptions:

A(4′). $D_n(\boldsymbol{\theta}, \boldsymbol{\theta}_1) \to \infty$ as $n \to \infty$ for all $\boldsymbol{\theta}, \boldsymbol{\theta}_1$ in Θ such that $\boldsymbol{\theta} \neq \boldsymbol{\theta}_1$.

A(5′). ε_i is a continuous random variable with range $(-\infty, \infty)$.

Given assumptions A(1) to A(3), Jennrich [1969] (see also Johansen [1984]) proved that A(4) is sufficient for $\hat{\boldsymbol{\theta}}_n$ and $\hat{\sigma}_n^2 = S_n(\hat{\boldsymbol{\theta}}_n)/n$ to be strongly consistent estimators of $\boldsymbol{\theta}^*$ and σ^2 respectively. However, if we tighten up A(3) so that Θ is now a set with a finite number of elements, then (Wu [1981]) A(4′) is sufficient for $\hat{\boldsymbol{\theta}}_n$ to be strongly consistent. If we add assumption A(5′) as well, then A(4′) is also necessary, as strong consistency now implies A(4′). This means that if we can find a pair $\boldsymbol{\theta}$ and $\boldsymbol{\theta}_1$ in finite Θ such that $\lim D_n(\boldsymbol{\theta}, \boldsymbol{\theta}_1) < \infty$, then $\hat{\boldsymbol{\theta}}_n$ is not consistent. For example, Malinvaud [1970a] considered the decay model

$$y_i = e^{-\theta i} + \varepsilon_i \qquad (i = 1, 2, \ldots, n), \tag{12.6}$$

which assumes that $\theta \in \Theta$, a finite subset of $(0, 2\pi)$, and $\varepsilon_i \in (-\infty, \infty)$. Since

$$\begin{aligned} \lim D_n(\theta, \theta_1) &= \sum_{i=1}^{\infty} (e^{-\theta i} - e^{-\theta_1 i})^2 \\ &= \frac{e^{-2\theta}}{1 - e^{-2\theta}} + \frac{e^{-2\theta_1}}{1 - e^{-2\theta_1}} - \frac{2e^{-(\theta + \theta_1)}}{1 - e^{-(\theta + \theta_1)}} < \infty \end{aligned}$$

for any θ $(\neq \theta_1)$ in $(0, 2\pi)$, $\hat{\theta}_n$ is not consistent. A further example is provided by

Jennrich [1969], namely

$$y_i = \frac{\theta}{i} + \varepsilon_i \qquad (i = 1, 2, \ldots, n), \tag{12.7}$$

where $\theta \in \Theta$, a finite subset of $(0, \infty)$, and $\varepsilon_i \in (-\infty, \infty)$. Then since

$$D_n(\theta, \theta_1) = \sum_{i=1}^{n} \frac{(\theta - \theta_1)^2}{i^2}$$

converges for any $\theta \neq \theta_1$ in $(0, \infty)$, $\hat{\theta}_n$ is not consistent. It is perhaps not surprising that these two examples cause problems, in that for large enough i, $E[y_i]$ becomes much smaller than ε_i and in the limit becomes indistinguishable from zero.

It should be noted that if, contrary to A(5′), ε_i is bounded, then strong consistency need not imply A(4′). For example, suppose that ε_i is uniformly distributed on $[-1, 1]$. Adapting an example of Wu [1981: 505], let $\Theta = \{0, 1\}$ and $f(\mathbf{x}_i; \theta) = x_{i1} + \theta x_{i2}$, where $x_{r2} > 2$ for some r, $x_{i1} > 0$ for all i, and $\sum_i x_{i2}^2$ converges. Given $n \geq r$, we can identify the true value θ^* correctly by observing y_r, which will be less than $x_{r_1} + 1$ when $\theta^* = 0$ and greater than $x_{r_1} + 1$ when $\theta^* = 1$. Then $S_n(\theta)$ of (12.2) can only be evaluated at one of the two possible values of θ, namely the true value of θ, so that $\hat{\theta}_n = \theta^*$. Thus $\hat{\theta}_n$ is strongly consistent, but $D_n(\theta, \theta_1) = \sum_i x_{i2}^2$ converges as $n \to \infty$.

Wu [1981] proves a more general result about consistency which states that, given a weaker version of A(1) to A(3) (including Θ being *any* subset of $\mathbb{R}^p$ and not just one with a finite number of elements), together with assumption A(5′) and the assumption that the distribution of ε_i has so-called "finite Fisher information", then $D_n(\boldsymbol{\theta}, \boldsymbol{\theta}_1) \to \infty$ is a necessary condition for the existence of *any* weakly consistent estimator. Therefore, if his assumptions are satisfied and $\lim D_n(\boldsymbol{\theta}, \boldsymbol{\theta}_1) < \infty$, then it is not possible to find a weakly consistent estimator by least squares, or by any other method. Hence if ε_i satisfies this finite-information property for the examples (12.6) and (12.7), then a weakly consistent estimator does not exist in either case.

In practice, however, the distribution of ε_i is unknown and discrete (as continuous variables are measured to the nearest unit), and the practical range of ε_i is finite. This means that the normal distribution does not exactly describe real measurements. But it can, of course, be a useful approximation to reality if the probability in the tails of the distribution lying outside the true range is negligible. For the same reason, if the observations suggest that the distribution of ε_i is unimodal and "suitably" shaped, it will be reasonable to define ε_i to have the range $(-\infty, \infty)$ and thus satisfy A(5′). An obvious exception to this is when ε_i has a known distribution like the uniform distribution on $[-a, a]$ $(a > 0)$. A typical practical constraint is that $y_i > 0$, especially when y_i is some physical measurement, so that $\varepsilon_i > -f(\mathbf{x}_i; \boldsymbol{\theta}^*)$.

Summing up, we see that assumptions A(1) to A(4) are sufficient for strong consistency. Assumptions A(1) to A(3) can generally be justified on practical grounds, but A(4) can cause problems. In models like (12.6) and (12.7) where $x_i = i$, the limit in A(4) can be investigated directly. However, in most cases we simply have n observations and it is difficult to see how to build up an appropriate infinite sequence $\{x_i\}$. It would be helpful if A(4) could be split into two easily verified assumptions, one concerning the form of f and the other relating to the sequence $\{x_i\}$. Various possibilities have been proposed. Perhaps the simplest, given by Jennrich [1969: 635], is to assume that $f(\mathbf{x};\boldsymbol{\theta})$ is bounded and continuous on $\mathscr{X} \times \Theta$, where $\mathscr{X} \subset \mathbb{R}^k$, and that $\mathbf{x}_1, \mathbf{x}_2, \ldots, \mathbf{x}_n$ is a random sample from some distribution with distribution function $H(\mathbf{x})$. Then, for $\mathbf{x} \in \mathscr{X}$, we can define the empirical distribution function $H_n(\mathbf{x}) = c/n$, where c is the number of $\mathbf{x}_i$'s less than or equal to $\mathbf{x}$ ($\mathbf{a}$ less than or equal to $\mathbf{b}$ means $a_i \le b_i$ for all i). Since $H_n(\mathbf{x}) \to H(\mathbf{x})$ with probability one, it follows from the Helly–Bray theorem that, with probability one,

$$\begin{aligned} n^{-1} D_n(\boldsymbol{\theta}, \boldsymbol{\theta}_1) &= \int [f(\mathbf{x};\boldsymbol{\theta}) - f(\mathbf{x};\boldsymbol{\theta}_1)]^2 \, dH_n(\mathbf{x}) \\ &\to \int [f(\mathbf{x};\boldsymbol{\theta}) - f(\mathbf{x};\boldsymbol{\theta}_1)]^2 \, dH(x) \\ &= D(\boldsymbol{\theta}, \boldsymbol{\theta}_1) \end{aligned} \tag{12.8}$$

uniformly for all $\boldsymbol{\theta}$, $\boldsymbol{\theta}_1$ in Θ. Hence A(4a) holds. If we also add the assumption that $f(\mathbf{x};\boldsymbol{\theta}) = f(\mathbf{x};\boldsymbol{\theta}^*)$ with probability one implies $\boldsymbol{\theta} = \boldsymbol{\theta}^*$, then using the integral representation in (12.8) we see that assumption A(4b) also holds.

There are a number of variations on the above replacement for A(4a): Jennrich [1969] gives two. For example, the $\mathbf{x}_i$ need not be a random sample, but $H_n(\mathbf{x})$ converges to a distribution function. Alternatively, if the $\mathbf{x}_i$ are a random sample, then it could be assumed that f is uniformly bounded in $\boldsymbol{\theta}$ by a function which is square-integrable with respect to H. Malinvaud [1970a] generalizes the first idea of Jennrich by introducing the concept of weak convergence of measure and replaces the assumption that f is bounded by the assumption that $\mathscr{X}$ is compact (since a continuous function on a compact set is bounded). Gallant [1977a, 1987] generalizes the second idea of Jennrich by considering Cesaro summability. In discussing these generalizations Amemiya [1983] concludes that the best procedure is to leave A(4) as it is and try to verify it directly. However, there still remains the problem of defining the sequence $\{\mathbf{x}_i\}$.

One other approach, suggested by Gallant [1975a], is to assume that the $\mathbf{x}_i$ are replicates of a fixed set of points. For example, given a fixed set of points $\mathbf{x}_1^*, \mathbf{x}_2^*, \ldots, \mathbf{x}_T^*$, we could choose $\mathbf{x}_1 = \mathbf{x}_1^*, \ldots, \mathbf{x}_T = \mathbf{x}_T^*$, $\mathbf{x}_{T+1} = \mathbf{x}_1^*$, $\mathbf{x}_{T+2}^* = \mathbf{x}_2$, etc. Disproportionate allocation can be included by allowing some of the $\mathbf{x}_i^*$ to be equal. Then $H_n(\mathbf{x}) \to H(\mathbf{x})$, where $H(\mathbf{x}) = c^*/T$, c^* being the number of $\mathbf{x}_i^*$ less than or equal to $\mathbf{x}$.

As a summary, we can now list an alternative set of assumptions for the strong consistency of $\hat{\boldsymbol{\theta}}_n$ and $\hat{\sigma}_n^2$. These assumptions are a slightly stronger version of those given by Malinvaud [1970a, Theorem 3; 1970b, Chapter 9] (see also Gallant [1975a]):

B(1). The ε_i are i.i.d. with mean zero and variance σ^2 ($\sigma^2 > 0$).
B(2). The function $f(\mathbf{x};\boldsymbol{\theta})$ is continuous on $\mathscr{X} \times \Theta$.
B(3). $\mathscr{X}$ and Θ are closed, bounded (compact) subsets of $\mathbb{R}^k$ and $\mathbb{R}^p$ respectively.
B(4). The observations $\mathbf{x}_i$ are such that $H_n(\mathbf{x}) \to H(\mathbf{x})$ with probability one, where $H_n(\mathbf{x})$ is the empirical distribution function and $H(\mathbf{x})$ is a distribution function.
B(5). If $f(\mathbf{x};\boldsymbol{\theta}) = f(\mathbf{x};\boldsymbol{\theta}^*)$ with probability one, then $\boldsymbol{\theta} = \boldsymbol{\theta}^*$.

If $\mathbf{x}$ is not a random variable, then the statement "with probability one" is omitted in B(4) and B(5).

Other sets of assumptions can be used, and the reader is referred to Crowder [1986] for a general discussion of consistency.

Example 12.1 Gallant [1975b, 1987] simulated the following one-way analysis-of-covariance model:

$$E[y_{ij}] = \mu_i + \gamma e^{\beta z_{ij}} \qquad (i = 1, 2, \quad j = 1, 2, \ldots, J), \tag{12.9}$$

consisting of two treatments and a single concomitant variable, age ($= z$), which affects the response exponentially. This model can be recast in the form

$$E[y_i] = f(\mathbf{x}_i;\boldsymbol{\theta}), \qquad i = 1, 2, \ldots, n,$$

where

$$f(\mathbf{x};\boldsymbol{\theta}) = \theta_1 x_1 + \theta_2 x_2 + \theta_4 e^{\theta_3 x_3}. \tag{12.10}$$

Here $n = 2J$, $\boldsymbol{\theta} = (\theta_1, \theta_2, \theta_3, \theta_4)' = (\mu_1 - \mu_2, \mu_2, \beta, \gamma)'$, $x_3 = z$, $\mathbf{x} = (x_1, x_2, x_3)'$, and (x_1, x_2) are indicator variables taking the values (1, 1) for the first treatment and (0, 1) for the second. The observations on (x_1, x_2) therefore represent J replicates at each of two points, while x_3 can be regarded as a random observation from an age distribution determined by the random allocation of experimental units to the two treatments from a population of units. Although $\mathbf{x}$ is now a combination of replicate data and random observations, it is clear that $H_n(\mathbf{x}) \to H(\mathbf{x})$ for suitably defined $H(\mathbf{x})$: thus B(4) is satisfied. Under general conditions, the remaining assumptions of B(1) to B(5) will also hold, so that $\hat{\boldsymbol{\theta}}_n$ will be strongly consistent. ■

12.2.3 Asymptotic Normality

Assuming at least the weak consistency of the least-squares estimator, we now find conditions under which $\hat{\boldsymbol{\theta}}_n$ is asymptotically normal. This can be achieved by carrying out an expansion of $\partial S_n(\boldsymbol{\theta})/\partial\boldsymbol{\theta}$ or a linear expansion of $f_i(\boldsymbol{\theta})$ as in Section 2.1.2. The latter approach is used by Jennrich [1969]. However, we shall use the former expansion and follow Amemiya [1983].

Using the mean-value theorem and assuming that appropriate continuous derivatives exist, we have (with high probability for large n)

$$\begin{aligned}\mathbf{0} &= \frac{\partial S_n(\hat{\boldsymbol{\theta}}_n)}{\partial\boldsymbol{\theta}} \\ &= \frac{\partial S_n(\boldsymbol{\theta}^*)}{\partial\boldsymbol{\theta}} + \frac{\partial^2 S_n(\tilde{\boldsymbol{\theta}}_n)}{\partial\boldsymbol{\theta}\,\partial\boldsymbol{\theta}'}(\hat{\boldsymbol{\theta}} - \boldsymbol{\theta}^*), \qquad (12.11)\end{aligned}$$

where $\partial S_n(\boldsymbol{\theta}^*)/\partial\boldsymbol{\theta}$ represents $\partial S_n(\boldsymbol{\theta})/\partial\boldsymbol{\theta}$ evaluated at $\boldsymbol{\theta} = \boldsymbol{\theta}^*$, etc., and $\tilde{\boldsymbol{\theta}}_n$ lies between $\hat{\boldsymbol{\theta}}_n$ and $\boldsymbol{\theta}^*$. Then, from (12.11),

$$\sqrt{n}(\hat{\boldsymbol{\theta}}_n - \boldsymbol{\theta}^*) = -\left[\frac{1}{n}\frac{\partial^2 S_n(\tilde{\boldsymbol{\theta}}_n)}{\partial\boldsymbol{\theta}\,\partial\boldsymbol{\theta}'}\right]^{-1} \frac{1}{\sqrt{n}}\frac{\partial S_n(\boldsymbol{\theta}^*)}{\partial\boldsymbol{\theta}}, \qquad (12.12)$$

and we need to prove that (i) $n^{-1/2}\partial S_n(\boldsymbol{\theta}^*)/\partial\boldsymbol{\theta}$ is asymptotically normal, and (ii) $n^{-1}\partial^2 S_n(\tilde{\boldsymbol{\theta}}_n)/\partial\boldsymbol{\theta}\,\partial\boldsymbol{\theta}'$ converges in probability to a nonsingular matrix. Dealing with (i) first, we have

$$\frac{\partial S_n(\boldsymbol{\theta})}{\partial\boldsymbol{\theta}} = -2\sum_{i=1}^{n}[y_i - f_i(\boldsymbol{\theta})]\frac{\partial f_i(\boldsymbol{\theta})}{\partial\boldsymbol{\theta}}, \qquad (12.13)$$

so that

$$\begin{aligned}\frac{1}{\sqrt{n}}\frac{\partial S_n(\boldsymbol{\theta}^*)}{\partial\boldsymbol{\theta}} &= \frac{-2}{\sqrt{n}}\sum_{i=1}^{n}\varepsilon_i\frac{\partial f_i(\boldsymbol{\theta}^*)}{\partial\boldsymbol{\theta}} \\ &= \frac{-2}{\sqrt{n}}\mathbf{F}'_{\cdot}(\boldsymbol{\theta}^*)\boldsymbol{\varepsilon}, \qquad (12.14)\end{aligned}$$

where $\mathbf{F}_{\cdot}(\boldsymbol{\theta}) = [(\partial f_i(\boldsymbol{\theta})/\partial\theta_r)]$ Since (12.14) represents a weighted average of the i.i.d. ε_i, we can apply an appropriate central limit theorem and obtain

$$\frac{1}{\sqrt{n}}\frac{\partial S_n(\boldsymbol{\theta}^*)}{\partial\boldsymbol{\theta}} \sim N_p(\mathbf{0}, 4\sigma^2\boldsymbol{\Omega}) \quad \text{asymptotically}, \qquad (12.15)$$

provided that

$$\begin{aligned}\boldsymbol{\Omega} &= \lim\frac{1}{n}\sum_{i=1}^{n}\left\{\frac{\partial f_i(\boldsymbol{\theta}^*)}{\partial\boldsymbol{\theta}}\frac{\partial f_i(\boldsymbol{\theta}^*)}{\partial\boldsymbol{\theta}'}\right\} \\ &= \lim\frac{1}{n}\mathbf{F}'_{\cdot}(\boldsymbol{\theta}^*)\mathbf{F}_{\cdot}(\boldsymbol{\theta}^*) \qquad (12.16)\end{aligned}$$

exists and is nonsingular.

To consider (ii) we differentiate (12.13) to get

$$\begin{aligned}
\frac{1}{n}\frac{\partial^2 S_n(\tilde{\boldsymbol{\theta}}_n)}{\partial\boldsymbol{\theta}\,\partial\boldsymbol{\theta}'} &= \frac{2}{n}\sum_i \frac{\partial f_i(\tilde{\boldsymbol{\theta}}_n)}{\partial\boldsymbol{\theta}}\frac{\partial f_i(\tilde{\boldsymbol{\theta}}_n)}{\partial\boldsymbol{\theta}'} \\
&\quad -\frac{2}{n}\sum_i [f_i(\boldsymbol{\theta}^*) - f_i(\tilde{\boldsymbol{\theta}}_n)]\frac{\partial^2 f_i(\tilde{\boldsymbol{\theta}}_n)}{\partial\boldsymbol{\theta}\,\partial\boldsymbol{\theta}'} - \frac{2}{n}\sum_i \varepsilon_i \frac{\partial^2 f_i(\tilde{\boldsymbol{\theta}}_n)}{\partial\boldsymbol{\theta}\,\partial\boldsymbol{\theta}'} \\
&= \mathbf{G}_1 - \mathbf{G}_2 - \mathbf{G}_3, \qquad (12.17)
\end{aligned}$$

say. If $\hat{\boldsymbol{\theta}}_n$ is strongly consistent, then $\tilde{\boldsymbol{\theta}}_n \to \boldsymbol{\theta}^*$ with probability one. Hence, with probability approaching one as $n\to\infty$, $\tilde{\boldsymbol{\theta}}_n$ will be in Θ^*, an open neighborhood. Now, if $h_n(\boldsymbol{\theta})$ is any sequence of continuous functions and $h_n(\boldsymbol{\theta})$ converges almost surely to a certain nonstochastic function $h(\boldsymbol{\theta})$ uniformly in $\boldsymbol{\theta}$, then

$$\operatorname{plim} h_n(\tilde{\boldsymbol{\theta}}_n) = h(\operatorname{plim}\tilde{\boldsymbol{\theta}}_n) = h(\boldsymbol{\theta}^*).$$

Therefore to prove (ii), we apply the above result to each element of the matrix equation (12.17) and show that each of the three terms on the right-hand side of (12.17) converges almost surely to a nonstochastic function uniformly in $\boldsymbol{\theta}$. Sufficient conditions for this are

$$\frac{1}{n}\sum_{i=1}^{n}\frac{\partial f_i(\boldsymbol{\theta})}{\partial\theta_r}\frac{\partial f_i(\boldsymbol{\theta})}{\partial\theta_s} \quad\text{and}\quad \frac{1}{n}\sum_{i=1}^{n}\left[\frac{\partial^2 f_i(\boldsymbol{\theta})}{\partial\theta_r\,\partial\theta_s}\right]^2 \text{ converge uniformly}$$

for $\boldsymbol{\theta}\in\Theta^*$ and $r,s=1,2,\ldots,p$. These conditions imply that plim $\mathbf{G}_a = \mathbf{0}$ $(a=2,3)$, and, using (12.16), we see that (12.17) leads to

$$\operatorname{plim}\frac{1}{n}\frac{\partial^2 S_n(\tilde{\boldsymbol{\theta}}_n)}{\partial\boldsymbol{\theta}\,\partial\boldsymbol{\theta}'} = 2\boldsymbol{\Omega}. \qquad (12.18)$$

Hence from (12.12), (12.14), and (12.18) we see that

$$\sqrt{n}(\hat{\boldsymbol{\theta}} - \boldsymbol{\theta}^*) \approx -(2\boldsymbol{\Omega})^{-1}\frac{1}{\sqrt{n}}\frac{\partial S_n(\boldsymbol{\theta}^*)}{\partial\boldsymbol{\theta}} \qquad (12.19)$$

$$\to N_p(\mathbf{0}, \sigma^2\boldsymbol{\Omega}^{-1}) \qquad [\text{by } (12.15)]. \qquad (12.20)$$

Thus (i) and (ii) imply (12.20). Also, from (12.14) and (12.16), (12.19) leads to

$$\hat{\boldsymbol{\theta}}_n - \boldsymbol{\theta}^* \approx [\mathbf{F}_\cdot'(\boldsymbol{\theta}^*)\mathbf{F}_\cdot(\boldsymbol{\theta}^*)]^{-1}\mathbf{F}_\cdot'(\boldsymbol{\theta}^*)\boldsymbol{\varepsilon}, \qquad (12.21)$$

which is obtained by a different route in Section 2.1.2.

We now summarize the above discussion by listing the additional assumptions needed for (12.20) to hold.

A(5). $\boldsymbol{\theta}^*$ is an interior point of Θ. (Technically this means that $\boldsymbol{\theta}^*$ does not lie on the boundary but belongs in an open subset of Θ.) Let Θ^* be an open neighborhood of $\boldsymbol{\theta}^*$ in Θ.

A(6). The first and second derivatives, $\partial f_i(\boldsymbol{\theta})/\partial\theta_r$ and $\partial^2 f_i(\boldsymbol{\theta})/\partial\theta_r\,\partial\theta_s$ $(r,s=1,2,\ldots,p)$, exist and are continuous for all $\boldsymbol{\theta}\in\Theta^*$.

A(7). $(1/n)\sum_{i=1}^{n}(\partial f_i(\boldsymbol{\theta})/\partial\boldsymbol{\theta})(\partial f_i(\boldsymbol{\theta})/\partial\boldsymbol{\theta}')$ $[=n^{-1}\mathbf{F}_{\bullet}'(\boldsymbol{\theta})\mathbf{F}_{\bullet}(\boldsymbol{\theta})]$ converges to some matrix $\boldsymbol{\Omega}(\boldsymbol{\theta})$ uniformly in $\boldsymbol{\theta}$ for $\boldsymbol{\theta}\in\Theta^*$.

A(8). $(1/n)\sum_{i=1}^{n}[\partial^2 f_i(\boldsymbol{\theta})/\partial\theta_r\,\partial\theta_s]^2$ converges uniformly in $\boldsymbol{\theta}$ for $\boldsymbol{\theta}\in\Theta^*$ $(r,s=1,2,\ldots,p)$.

A(9). $\boldsymbol{\Omega}=\boldsymbol{\Omega}(\boldsymbol{\theta}^*)$ is nonsingular.

If assumptions A(1) to A(9) hold, then

$$\sqrt{n}(\hat{\boldsymbol{\theta}}_n-\boldsymbol{\theta}^*)\sim N_p(\mathbf{0},\sigma^2\boldsymbol{\Omega}^{-1})\qquad\text{asymptotically.}\tag{12.22}$$

Also $n^{-1}\mathbf{F}_{\bullet}(\hat{\boldsymbol{\theta}}_n)'\mathbf{F}_{\bullet}(\hat{\boldsymbol{\theta}}_n)$ is a strongly consistent estimator of $\boldsymbol{\Omega}$. This follows from A(7) and the strong consistency of $\hat{\boldsymbol{\theta}}_n$ [by A(1) to A(4)]. Finally, using (12.16) we have, for large n,

$$\hat{\boldsymbol{\theta}}_n-\boldsymbol{\theta}^*\sim N_p(\mathbf{0},\sigma^2[\mathbf{F}_{\bullet}'(\boldsymbol{\theta}^*)\mathbf{F}_{\bullet}(\boldsymbol{\theta}^*)]^{-1})\qquad\text{approximately.}\tag{12.23}$$

Jennrich [1969] (see also Johansen [1984: Chapter 5]) proved asymptotic normality using a similar set of assumptions: A(7) and A(8) are replaced by the stronger assumption:

A(8′). $n^{-1}\sum_i g_i(\boldsymbol{\theta})h_i(\boldsymbol{\theta}_1)$ converges uniformly for $\boldsymbol{\theta},\boldsymbol{\theta}_1$ in Θ, where g_i and h_i are each any one of $f_i(\boldsymbol{\theta})$, $\partial f_i(\boldsymbol{\theta})/\partial\theta_r$, and $\partial^2 f_i(\boldsymbol{\theta})/\partial\theta_r\,\partial\theta_s$ $(r,s=1,2,\ldots,p)$.

(The limit is known as the tail product.)

He also noted that although $\hat{\boldsymbol{\theta}}_n$ is asymptotically normal under the above assumptions, it is not generally asymptotically efficient. However, asymptotic efficiency is obtained if the ε_i are normally distributed.

If, instead, we use the approach basically due to Malinvaud [1970b] which led to assumptions B(1) to B(4) in Section 12.2.2, then we come up with the following additional assumptions:

B(6). $\boldsymbol{\theta}^*$ is an interior point of Θ.

B(7). The functions $\partial f(\mathbf{x};\boldsymbol{\theta})/\partial\theta_r$ and $\partial^2 f(\mathbf{x};\boldsymbol{\theta})/\partial\theta_r\,\partial\theta_s$ $(r,\ s=1,2,\ldots,p)$ are continuous on $\mathscr{X}\times\Theta$.

B(8). The matrix $\boldsymbol{\Omega}=[(\omega_{rs}(\boldsymbol{\theta}^*))]$, where

$$\omega_{rs}(\boldsymbol{\theta}^*)=\int\frac{\partial f(\mathbf{x};\boldsymbol{\theta}^*)}{\partial\theta_r}\frac{\partial f(\mathbf{x};\boldsymbol{\theta}^*)}{\partial\theta_s}\,dH(\mathbf{x}),\tag{12.24}$$

is nonsingular.

Gallant [1975a: see References 5 and 6 there] showed that assumptions B(1) to B(8) are sufficient for (12.20) to hold. A further set of assumptions is given by Gallant [1987: Chapter 4]. Another approach is to fix n and let $\sigma^2 \to 0$ (Johansen [1983: 174]; see also Section 12.2.6).

Some of the above results are reviewed by Bunke et al. [1977]. The work of Jennrich and Malinvaud has been extended by Bunke and Schmidt [1980] and Zwanzig [1980], respectively, to the cases of weighted least squares and nonidentical distributions of the ε_i. Their approach is via a partially linear approximating function (see also Bunke [1980] and Chanda [1976]).

Example 12.2 Consider the model $f(x;\boldsymbol{\theta}) = \theta_1 \exp(\theta_2 x)$, and suppose that assumptions B(1) to B(7) hold. To prove B(8) we note that

$$\begin{aligned}\mathbf{a}'\boldsymbol{\Omega}\mathbf{a} &= \sum_r \sum_s a_r a_s \omega_{rs}(\boldsymbol{\theta}^*) \\ &= \int \left(a_1 \frac{\partial f(x;\boldsymbol{\theta}^*)}{\partial \theta_1} + a_2 \frac{\partial f(x;\boldsymbol{\theta}^*)}{\partial \theta_2} \right)^2 dH(x) \\ &= \int (a_1 + a_2 \theta_1^* x)^2 \exp(2\theta_2^* x)\, dH(x).\end{aligned}$$

Then the above expression is zero if and only if $a_1 + a_2\theta_1^* x = 0$ with probability one, that is, if $a_1 = a_2 = 0$. Hence $\boldsymbol{\Omega}$ is positive definite, and therefore nonsingular, so that B(8) is true. ■

12.2.4 Effects of Misspecification

So far we have assumed that $E[y_i] = f(\mathbf{x}_i;\boldsymbol{\theta})$ is the correct model, and we obtained $\hat{\boldsymbol{\theta}}_n$ by minimizing

$$S_n(\boldsymbol{\theta}) = \sum_{i=1}^{n} [y_i - f(\mathbf{x}_i;\boldsymbol{\theta})]^2.$$

However, in many experimental situations, e.g. growth-curve analysis, the true model is unknown and f is simply selected on empirical grounds from a range of possible models. Suppose, then, that the true model is actually

$$y_i = g(\mathbf{x}_i;\boldsymbol{\theta}) + \varepsilon_i \qquad (i = 1, 2, \ldots, n), \tag{12.25}$$

where $\mathbf{x}_i \in \mathscr{X}$ and $\boldsymbol{\theta} \in \Theta$. Also define

$$q(\mathbf{x};\boldsymbol{\theta}) = [g(\mathbf{x};\boldsymbol{\theta}^*) - f(\mathbf{x};\boldsymbol{\theta})]^2,$$

where $\boldsymbol{\theta}^*$ is, once again, the true value of $\boldsymbol{\theta}$. We now consider the effects of this misspecification on the asymptotic properties of $\hat{\boldsymbol{\theta}}_n$ and $\hat{\sigma}_n^2 = n^{-1} S_n(\hat{\boldsymbol{\theta}}_n)$, where $\hat{\boldsymbol{\theta}}_n$ minimizes $S_n(\boldsymbol{\theta})$.

Following White [1981], we first make some assumptions which are similar to B(1) to B(5) of Section 12.2.2:

C(1). The ε_i are i.i.d. with mean zero and variance σ^2 ($\sigma^2 > 0$).

C(2). $g(\mathbf{x};\boldsymbol{\theta})$ and $f(\mathbf{x};\boldsymbol{\theta})$ are continuous on $\mathscr{X} \times \Theta$.

C(3). $\mathscr{X}$ and Θ are closed, bounded (compact) subsets of $\mathbb{R}^k$ and $\mathbb{R}^p$ respectively.

C(4). The observations $\mathbf{x}_1, \mathbf{x}_2, \ldots, \mathbf{x}_n$ are realizations of a random sample from a distribution with distribution function $H(\mathbf{x})$.

C(5). $q(\mathbf{x},\boldsymbol{\theta}) \le m(\mathbf{x})$ in $\mathscr{X} \times \Theta$, where $m(\mathbf{x})$ is integrable with respect to H, i.e.,

$$\int m(\mathbf{x})\, dH(\mathbf{x}) < \infty,$$

C(6). The prediction mean square error

$$T(\boldsymbol{\theta}) = \int q(\mathbf{x},\boldsymbol{\theta})\, dH(\mathbf{x})$$

has a unique minimum at $\boldsymbol{\theta} = \boldsymbol{\theta}^\dagger$.

Then given assumptions C(1) to C(6), White [1981] showed that $\hat{\boldsymbol{\theta}}_n$ and $\hat{\sigma}_n^2$ are strongly consistent estimates of $\boldsymbol{\theta}^\dagger$ and $T(\boldsymbol{\theta}^\dagger)$. Before giving further assumptions to ensure asymptotic normality, we define

$$q_r(\boldsymbol{\theta}) = \frac{\partial q(\mathbf{x},\boldsymbol{\theta})}{\partial \theta_r}$$

$$q_r^{(i)}(\boldsymbol{\theta}) = \frac{\partial q(\mathbf{x}_i,\boldsymbol{\theta})}{\partial \theta_r},$$

$$q_{rs}(\boldsymbol{\theta}) = \frac{\partial^2 q(\mathbf{x},\boldsymbol{\theta})}{\partial \theta_r \partial \theta_s},$$

$$q_{rs}^{(i)}(\boldsymbol{\theta}) = \frac{\partial^2 q(\mathbf{x}_i,\boldsymbol{\theta})}{\partial \theta_r \partial \theta_s}.$$

Similar equations are defined for $g(\mathbf{x};\boldsymbol{\theta})$. We also define the matrices $\mathbf{A}_n(\boldsymbol{\theta})$ and $\mathbf{B}_n(\boldsymbol{\theta})$ with (r,s)th elements given respectively by

$$a_{nrs}(\boldsymbol{\theta}) = \frac{1}{n}\sum_{i=1}^{n} q_{rs}^{(i)}(\boldsymbol{\theta})$$

and

$$b_{nrs}(\boldsymbol{\theta}) = \frac{1}{n}\sum_{i=1}^{n} q_r^{(i)}(\boldsymbol{\theta})q_s^{(i)}(\boldsymbol{\theta}).$$

When they exist, define $\mathbf{A}(\boldsymbol{\theta})$ and $\mathbf{B}(\boldsymbol{\theta})$ with elements

$$a_{rs}(\boldsymbol{\theta}) = E[q_{rs}^{(i)}(\boldsymbol{\theta})]$$

and

$$b_{rs}(\boldsymbol{\theta}) = E[q_r^{(i)}(\boldsymbol{\theta})q_s^{(i)}(\boldsymbol{\theta})].$$

We now add the following assumptions:

C(7). Each $q_r(\boldsymbol{\theta})$ and $q_{rs}(\boldsymbol{\theta})$ is continuous on $\mathscr{X} \times \Theta$.
C(8). For all $\mathbf{x}$ and $\boldsymbol{\theta}$ in $\mathscr{X} \times \Theta$, $|q_r(\boldsymbol{\theta})q_s(\boldsymbol{\theta})|$ and $|q_{rs}(\boldsymbol{\theta})|$ are dominated by integrable functions of $\mathbf{x}$ (r, $s = 1, 2, \ldots, p$).
C(9). $\boldsymbol{\theta}^\dagger$ is an interior point of Θ.
C(10). $\mathbf{A}(\boldsymbol{\theta}^\dagger)$ and $\mathbf{B}(\boldsymbol{\theta}^\dagger)$ are nonsingular.

Then given assumptions C(1) to C(10), it can be shown that, asymptotically,

$$\sqrt{n}(\hat{\boldsymbol{\theta}}_n - \boldsymbol{\theta}^\dagger) \sim N_p\{\mathbf{0}, \mathbf{C}(\boldsymbol{\theta}^\dagger)\},$$

where $\mathbf{C}(\boldsymbol{\theta}) = \mathbf{A}^{-1}(\boldsymbol{\theta})\mathbf{B}(\boldsymbol{\theta})\mathbf{A}^{-1}(\boldsymbol{\theta})$. Also $\hat{\mathbf{C}}_n = \hat{\mathbf{A}}_n^{-1}\hat{\mathbf{B}}_n\hat{\mathbf{A}}_n^{-1}$ is a strongly consistent estimator of $\mathbf{C}(\boldsymbol{\theta}^\dagger)$, where

$$(\hat{\mathbf{A}}_n)_{rs} = \frac{2}{n}\sum_i [g_r^{(i)}(\hat{\boldsymbol{\theta}}_n)g_s^{(i)}(\hat{\boldsymbol{\theta}}_n) + g_{rs}^{(i)}(\hat{\boldsymbol{\theta}}_n)\hat{\varepsilon}_i],$$

$$(\hat{\mathbf{B}}_n)_{rs} = \frac{4}{n}\sum_i [\hat{\varepsilon}_i^2 g_r^{(i)}(\hat{\boldsymbol{\theta}}_n)g_s^{(i)}(\hat{\boldsymbol{\theta}}_n)],$$

and

$$\hat{\varepsilon}_i = y_i - g(\mathbf{x}_i; \hat{\boldsymbol{\theta}}_n).$$

White [1981] actually proved more general results using weaker assumptions. For example, $q(\mathbf{x}, \boldsymbol{\theta})$ can be a more general function. However, the above assumptions are similar in spirit to B(1) to B(8). White [1981] also gave two tests of the hypothesis that the chosen model is true up to an independent error. Further extensions of the theory and general misspecification tests are given by H. White [1982].

12.2.5 Some Extensions

White [1980] extended the above theory to the case where the regressor variables $\mathbf{x}_i$ are random and independent of ε_i, the $\{\mathbf{x}_i, \varepsilon_i\}$ are independent but not identically distributed, and estimation is by weighted least squares. The essential details of the asymptotic theory are summarized briefly by Amemiya [1983: 358], who also considers autocorrelated errors. For example, suppose that the $\mathbf{x}_i$ are nonstochastic, but the errors follow a general stationary process [c.f. (6.7)]

$$\varepsilon_i = \sum_{r=0}^{\infty} \psi_r a_{i-r} \qquad (i = 1, 2, \ldots, n),$$

where the $\{a_r\}$ are i.i.d. with $E[a_r] = 0$ and $\text{var}[a_r] = \sigma_a^2$, the ψ's satisfy the stationarity condition $\sum_r |\psi_r| < \infty$, and the spectral density $g(\omega)$ of the $\{\varepsilon_i\}$ is continuous. Given that $\mathscr{D}[\boldsymbol{\varepsilon}] = \boldsymbol{\Sigma}$, Amemiya [1983] proves that the latter assumptions, together with A(2) to A(9) above, are sufficient for (12.21) to still hold. Hence

$$\sqrt{n}(\hat{\boldsymbol{\theta}}_n - \boldsymbol{\theta}^*) \to N_p(\mathbf{0}, \sigma^2 \lim n^{-1} (\mathbf{F}_{\cdot}'\mathbf{F}_{\cdot})^{-1} \mathbf{F}_{\cdot}'\boldsymbol{\Sigma}\mathbf{F}_{\cdot}(\mathbf{F}_{\cdot}'\mathbf{F}_{\cdot})^{-1}),$$

where $\mathbf{F}_{\cdot} = \mathbf{F}_{\cdot}(\boldsymbol{\theta}^*)$. However, as noted in Chapter 6, the ordinary least-squares estimate $\hat{\boldsymbol{\theta}}_n$ can be extremely inefficient. Further generalizations of least-squares theory are given by Burguete et al. [1982] and White and Domowitz [1984].

12.2.6 Asymptotics with Vanishingly Small Errors

Up till now we have developed our theory in terms of the limiting process $n \to \infty$. A different type of asymptotics, however, is possible if we fix n and let $\sigma^2 \to 0$. This approach corresponds to repeating the whole experiment R times and letting $R \to \infty$. It assumes that each ε_i is small and expansions are carried out in terms of powers of the ε_i. Such a method was pioneered by Beale [1960], and his work was refined by Johansen [1983, 1984]. For completeness we briefly summarize the main results of "small-error" asymptotics and relate Johansen's work to that of Chapters 4 and 5 of this book.

Using the notation of Section 4.6, we obtained from (4.173) the following expression for $\hat{\boldsymbol{\theta}}$ as $\sigma^2 \to 0$:

$$\begin{aligned}\hat{\boldsymbol{\theta}} - \boldsymbol{\theta}^* &= \mathbf{K}\hat{\boldsymbol{\gamma}} \\ &\approx \mathbf{K}(\mathbf{z} + \mathbf{B}_{\mathbf{w}}\mathbf{z} - \tfrac{1}{2}\mathbf{z}' A_{\cdot\cdot}^T \mathbf{z}),\end{aligned}$$

where $\mathbf{z} = \mathbf{Q}_p'\boldsymbol{\varepsilon}$ and $\mathbf{w} = \mathbf{Q}_{n-p}'\boldsymbol{\varepsilon}$. The first term above is linear and the other two terms are quadratic in the ε_i. It can be shown that the above approximation is the one given by Johansen [1983: Lemma A.1]. We then used this approximation to derive the approximate bias and variance–covariance matrix of $\hat{\boldsymbol{\theta}}$ [Equations (4.174) and (4.176)]. Also, from (4.157), we have

$$\begin{aligned}\mathbf{f}(\hat{\boldsymbol{\theta}}) - \mathbf{f}(\boldsymbol{\theta}) &= \boldsymbol{\varepsilon} - \hat{\boldsymbol{\varepsilon}} \\ &\approx \mathbf{Q}_p\mathbf{z} + \mathbf{Q}_p\mathbf{B}_{\mathbf{w}}\mathbf{z} + \tfrac{1}{2}\mathbf{Q}_{n-p}\mathbf{Q}_{n-p}'(\mathbf{z}' G_{\cdot\cdot}\mathbf{z}),\end{aligned}$$

which is also given by Johansen. The linear term is $\mathbf{Q}_p\mathbf{z} = \mathbf{Q}_p\mathbf{Q}_p'\boldsymbol{\varepsilon} = \mathbf{P}_{\mathbf{F}}\boldsymbol{\varepsilon}$, where $\mathbf{P}_{\mathbf{F}} = \mathbf{F}_{\cdot}(\mathbf{F}_{\cdot}'\mathbf{F}_{\cdot})^{-1}\mathbf{F}_{\cdot}'$ is the projection matrix projecting onto the range of $\mathbf{F}_{\cdot}$. Thus we have the first-order approximations (to order σ)

$$\sigma^{-1}[\mathbf{f}(\hat{\boldsymbol{\theta}}) - \mathbf{f}(\boldsymbol{\theta})] = \frac{\mathbf{P}_F \boldsymbol{\varepsilon}}{\sigma} + \mathbf{O}(\sigma)$$

and

$$\begin{aligned}\sigma^{-1}\hat{\boldsymbol{\varepsilon}} &= \frac{\boldsymbol{\varepsilon} - \mathbf{Q}_p \mathbf{z}}{\sigma} + \mathbf{O}(\sigma) \\ &= \frac{(\mathbf{I}_n - \mathbf{P}_F)\boldsymbol{\varepsilon}}{\sigma} + \mathbf{O}(\sigma),\end{aligned}$$

where $\boldsymbol{\varepsilon}/\sigma \sim N_n(\mathbf{0}, \mathbf{I}_n)$. These results form Johansen's [1983] Theorem 5.3 and are essentially the same as those given for the case of $n \to \infty$ [c.f. (2.16) and (2.17)].

By taking higher-order terms in the ε_i, Johansen used characteristic functions to prove that, to order σ^3, $S(\hat{\boldsymbol{\theta}})/\sigma^2 = \hat{\boldsymbol{\varepsilon}}'\hat{\boldsymbol{\varepsilon}}/\sigma^2$ and $[S(\boldsymbol{\theta}) - S(\hat{\boldsymbol{\theta}})]/\sigma^2 = (\boldsymbol{\varepsilon}'\boldsymbol{\varepsilon} - \hat{\boldsymbol{\varepsilon}}'\hat{\boldsymbol{\varepsilon}})/\sigma^2$ are independently distributed as $[1 + \sigma^2\gamma_2/(n-p)]\chi^2_{n-p}$ and $(1 - \sigma^2\gamma_2/p)\chi^2_p$, respectively, where γ_2 is given by (5.27). This work has been extended by Hamilton and Wiens [1987], and some of the extensions are discussed in Section 5.3. The limiting process $\sigma^2 \to 0$ is also referred to in Section 2.8.3 and in the paragraph following Equation (10.20).

12.3 MAXIMUM-LIKELIHOOD ESTIMATION

In Section 2.2 we discussed the method of maximum-likelihood estimation for the nonlinear model $y_i = f(\mathbf{x}_i; \boldsymbol{\theta}) + \varepsilon_i$ when the errors ε_i are normal or nonnormal. We found that with normal errors, the maximum-likelihood estimate $\hat{\boldsymbol{\theta}}$ is the same as the least-squares estimate and the theory of Section 12.1 applies. When the errors are not normally distributed but the likelihood is a known function of $\mathscr{E}[\mathbf{y}]$, methods like IRLS of Section 2.2.2 can be used for finding $\hat{\boldsymbol{\theta}}$. Unfortunately, readily verifiable conditions for the uniqueness, consistency, asymptotic normality, etc. are not easily available from the literature for the general nonnormal case. The reason for this is that the y_i are not identically distributed, having different expected values. The general case of the y_i being independently but not identically distributed is considered by Bradley and Gart [1962] and Hoadley [1971], but the assumptions postulated are complex and difficult to verify. Questions of existence and uniqueness in some special cases are considered by Wedderburn [1976], Pratt [1981], and Burridge [1981]. A helpful indication of the assumptions required is given by Cox and Hinkley [1974].

12.4 HYPOTHESIS TESTING

Suppose that the response vector of observations $\mathbf{y}$ has probability density $p(\mathbf{y}|\boldsymbol{\theta})$ and log-likelihood function $L(\boldsymbol{\theta}) = \log p(\mathbf{y}|\boldsymbol{\theta})$. In Section 5.9.1 we considered the problem of testing $H_0:\mathbf{a}(\boldsymbol{\theta}) = [a_1(\boldsymbol{\theta}), a_2(\boldsymbol{\theta}), \ldots, a_q(\boldsymbol{\theta})]' = \mathbf{0}$ determined by q independent constraints. Three test statistics for H_0 were defined, namely (5.92), (5.93), and (5.94), with $L(\boldsymbol{\theta})$ replaced by the concentrated log-likelihood function $M(\boldsymbol{\theta})$ (to get rid of the nuisance parameter σ^2). In this section we shall retain $L(\boldsymbol{\theta})$ to indicate the full generality of what follows. The three statistics are then

$$\text{(Likelihood ratio) LR} = 2[L(\hat{\boldsymbol{\theta}}) - L(\tilde{\boldsymbol{\theta}})], \tag{12.26}$$

$$\text{(Wald's test) W} = \mathbf{a}'(\hat{\boldsymbol{\theta}})(\mathbf{A}\mathbf{D}^{-1}\mathbf{A}')_{\hat{\boldsymbol{\theta}}}^{-1}\mathbf{a}(\hat{\boldsymbol{\theta}}), \tag{12.27}$$

$$\text{(Lagrange multiplier) LM} = \left(\frac{\partial L}{\partial \boldsymbol{\theta}'}\mathbf{D}^{-1}\frac{\partial L}{\partial \boldsymbol{\theta}}\right)_{\tilde{\boldsymbol{\theta}}}, \tag{12.28}$$

where

$$\mathbf{D} = -\frac{\partial^2 L(\boldsymbol{\theta})}{\partial \boldsymbol{\theta}\, \partial \boldsymbol{\theta}'},$$

and

$$\mathbf{A} = \left[\left(\frac{\partial a_i(\boldsymbol{\theta})}{\partial \theta_j}\right)\right]$$

is a $q \times p$ matrix of rank q (by the independence of the constraints). The estimates $\hat{\boldsymbol{\theta}}$ and $\tilde{\boldsymbol{\theta}}$ represent the respective unrestricted and restricted (by H_0) maximum-likelihood estimates of $\boldsymbol{\theta}$. These are also the least squares estimates under the assumption of normality. (We have dropped the dependence on n used in Section 12.1 for notational convenience; e.g. $\hat{\boldsymbol{\theta}}_n$ becomes $\hat{\boldsymbol{\theta}}$.) Following the method of Seber [1980: Chapter 11], we now demonstrate that the three statistics are asymptotically equivalent and asymptotically distributed as χ_q^2 when H_0 is true. This is done by showing that the model and hypothesis are asymptotically equivalent to a linear normal model and a linear hypothesis.

The three statistics defined above are asymptotically equal to three statistics which are identical for the linear model. Seber [1980] gives a rigorous derivation for the case when the y_i are independently and identically distributed. The extension to the case when the y_i have different means, as in the nonlinear regression model, will follow in a similar fashion. For rigor one should use $\sqrt{n}\hat{\boldsymbol{\theta}}$, $\sqrt{n}\tilde{\boldsymbol{\theta}}$, $\sqrt{n}\mathbf{a}(\hat{\boldsymbol{\theta}})$, $n^{-1}\mathbf{D}$, and $n^{-1/2}\partial L/\partial\boldsymbol{\theta}$ in the asymptotics, as in Seber [1980]. However, to reduce technical detail and simplify the discussion, we shall generally avoid the various multiplying factors involving n.

We shall assume that appropriate regularity conditions hold, so that $\hat{\boldsymbol{\theta}}$ and $\tilde{\boldsymbol{\theta}}$ are asymptotically normal. Also, $\mathbf{D}$ is assumed to be positive definite in a neighborhood of $\boldsymbol{\theta}^*$, the true value of $\boldsymbol{\theta}$ (see Milliken and DeBruin [1978] for a discussion of the assumptions needed). Now as $n \to \infty$, $\hat{\boldsymbol{\theta}}$ and $\tilde{\boldsymbol{\theta}}$ will be consistent estimators of $\boldsymbol{\theta}^*$ and some vector $\boldsymbol{\theta}^\dagger$, respectively, where $\mathbf{a}(\boldsymbol{\theta}^\dagger) = \mathbf{0}$. If H_0 is true, then $\boldsymbol{\theta}^\dagger = \boldsymbol{\theta}^*$. It is convenient to consider a limiting sequence of alternative hypotheses, since for a *fixed* alternative the power of a test will tend to unity as $n \to \infty$. Suppose then that $\boldsymbol{\theta}^* \to \boldsymbol{\theta}^\dagger$ in such a way that $\sqrt{n}(\boldsymbol{\theta}^* - \boldsymbol{\theta}^\dagger) \to \boldsymbol{\delta}$, where $\boldsymbol{\delta}$ is fixed. Then, for large enough n, the vectors $\hat{\boldsymbol{\theta}}, \tilde{\boldsymbol{\theta}}, \boldsymbol{\theta}^*$, and $\boldsymbol{\theta}^\dagger$ will all be close to each other in a neighborhood, $\Theta^\dagger$ say, of $\boldsymbol{\theta}^\dagger$. Let $\hat{\mathbf{D}}$, $\tilde{\mathbf{D}}$, $\mathbf{D}_*$, and $\mathbf{D}_\dagger$ represent the positive definite matrix $\mathbf{D}$ evaluated at each of the four values of $\boldsymbol{\theta}$: similar definitions apply to $\mathbf{A}$. Using a Taylor expansion, and defining $\partial L(\hat{\boldsymbol{\theta}})/\partial\boldsymbol{\theta}$ to be $\partial L(\boldsymbol{\theta})/\partial\boldsymbol{\theta}$ evaluated at $\hat{\boldsymbol{\theta}}$, we have

$$\mathbf{0} = \frac{\partial L(\hat{\boldsymbol{\theta}})}{\partial\boldsymbol{\theta}} \approx \frac{\partial L(\boldsymbol{\theta}^\dagger)}{\partial\boldsymbol{\theta}} - \mathbf{D}_\dagger(\hat{\boldsymbol{\theta}} - \boldsymbol{\theta}^\dagger),$$

or

$$\hat{\boldsymbol{\theta}} - \boldsymbol{\theta}^\dagger \approx \mathbf{D}_\dagger^{-1}\frac{\partial L(\boldsymbol{\theta}^\dagger)}{\partial\boldsymbol{\theta}}. \tag{12.29}$$

With an appropriate version of the multivariate central limit theorem, we find that $\mathbf{v} = \hat{\boldsymbol{\theta}} - \boldsymbol{\theta}^\dagger \sim N_p(\mathbf{0}, \mathbf{D}_\dagger^{-1})$ approximately. Thus

$$\hat{\boldsymbol{\theta}} - \boldsymbol{\theta}^* \approx \boldsymbol{\theta}^\dagger - \boldsymbol{\theta}^* + \mathbf{v}, \qquad \mathbf{v} \sim N_p(\mathbf{0}, \mathbf{D}_\dagger^{-1}). \tag{12.30}$$

Turning our attention to $\tilde{\boldsymbol{\theta}}$, we need to differentiate

$$L(\boldsymbol{\theta}) + \boldsymbol{\lambda}'\mathbf{a}(\boldsymbol{\theta})$$

with respect to $\boldsymbol{\theta}$, where $\boldsymbol{\lambda}$ is a vector of Lagrange multipliers. Thus $\tilde{\boldsymbol{\theta}}$ satisfies

$$\frac{\partial L(\tilde{\boldsymbol{\theta}})}{\partial\boldsymbol{\theta}} + \tilde{\mathbf{A}}'\tilde{\boldsymbol{\lambda}} = \mathbf{0} \tag{12.31}$$

and

$$\mathbf{a}(\tilde{\boldsymbol{\theta}}) = \mathbf{0}. \tag{12.32}$$

Since $\tilde{\boldsymbol{\theta}}$ is close to $\hat{\boldsymbol{\theta}}$ (in $\Theta^\dagger$) and $\hat{\mathbf{D}}$ is close to $\mathbf{D}_\dagger$, we have the Taylor expansions

$$\frac{\partial L(\tilde{\boldsymbol{\theta}})}{\partial\boldsymbol{\theta}} \approx \frac{\partial L(\hat{\boldsymbol{\theta}})}{\partial\boldsymbol{\theta}} - \hat{\mathbf{D}}(\tilde{\boldsymbol{\theta}} - \hat{\boldsymbol{\theta}}) \tag{12.33}$$

$$\approx -\mathbf{D}_\dagger(\tilde{\boldsymbol{\theta}} - \hat{\boldsymbol{\theta}})$$

$$= -\mathbf{D}_\dagger(\tilde{\boldsymbol{\theta}} - \boldsymbol{\theta}^\dagger) + \mathbf{D}_\dagger(\hat{\boldsymbol{\theta}} - \boldsymbol{\theta}^\dagger). \tag{12.34}$$

Also, since $\mathbf{a}(\boldsymbol{\theta}^\dagger) = \mathbf{0}$,

$$\mathbf{0} = \mathbf{a}(\tilde{\boldsymbol{\theta}}) - \mathbf{a}(\boldsymbol{\theta}^\dagger)$$

$$\approx \mathbf{A}_\dagger(\tilde{\boldsymbol{\theta}} - \boldsymbol{\theta}^\dagger). \tag{12.35}$$

Substituting (12.34) into (12.31) and approximating $\tilde{\mathbf{A}}$ by $\mathbf{A}_\dagger$ gives us the equations

$$\mathbf{D}_\dagger(\tilde{\boldsymbol{\theta}} - \boldsymbol{\theta}^\dagger) + \mathbf{D}_\dagger(\hat{\boldsymbol{\theta}} - \boldsymbol{\theta}^\dagger) + \mathbf{A}'_\dagger\tilde{\boldsymbol{\lambda}} \approx \mathbf{0} \tag{12.36}$$

and

$$\mathbf{A}_\dagger(\tilde{\boldsymbol{\theta}} - \boldsymbol{\theta}^\dagger) \approx \mathbf{0}. \tag{12.37}$$

Since $\mathbf{D}_\dagger$, and therefore $\mathbf{D}_\dagger^{-1}$, is positive definite, we have (Appendix A4.3) $\mathbf{D}_\dagger^{-1} = \mathbf{V}_\dagger\mathbf{V}'_\dagger$, where $\mathbf{V}_\dagger$ is nonsingular. Let

$$\begin{aligned} \boldsymbol{\varepsilon} &= \mathbf{V}_\dagger^{-1}\mathbf{v} \sim N(\mathbf{0}, \mathbf{I}_p), \\ \mathbf{z} &= \mathbf{V}_\dagger^{-1}(\hat{\boldsymbol{\theta}} - \boldsymbol{\theta}^\dagger), \\ \boldsymbol{\beta} &= \mathbf{V}_\dagger^{-1}(\boldsymbol{\theta}^* - \boldsymbol{\theta}^\dagger), \\ \tilde{\boldsymbol{\beta}} &= \mathbf{V}_\dagger^{-1}(\tilde{\boldsymbol{\theta}} - \boldsymbol{\theta}^\dagger). \end{aligned} \tag{12.38}$$

Then multiplying (12.30) by $\mathbf{V}_\dagger^{-1}$ we have, asymptotically,

$$\mathbf{z} = \boldsymbol{\beta} + \boldsymbol{\varepsilon}, \qquad \boldsymbol{\varepsilon} \sim N_p(\mathbf{0}, \mathbf{I}_p). \tag{12.39}$$

Multiplying (12.36) by $\mathbf{V}'_\dagger$ gives us the asymptotic equations

$$-\tilde{\boldsymbol{\beta}} + \mathbf{z} + (\mathbf{A}_\dagger\mathbf{V}_\dagger)'\tilde{\boldsymbol{\lambda}} = \mathbf{0} \tag{12.40}$$

and

$$\mathbf{A}_\dagger\mathbf{V}_\dagger\tilde{\boldsymbol{\beta}} = \mathbf{0},$$

where $\mathbf{A}_\dagger\mathbf{V}_\dagger$ is a $q \times p$ matrix of rank q, as $\mathbf{V}_\dagger$ is nonsingular. But these are precisely the restricted least-squares equations for testing the linear hypothesis $\mathbf{A}_\dagger\mathbf{V}_\dagger\boldsymbol{\beta} = \mathbf{0}$ for the linear model (12.39). This leads us to the following lemma.

Lemma Let $\mathbf{z} = \boldsymbol{\beta} + \boldsymbol{\varepsilon}$, where $\boldsymbol{\varepsilon} \sim N_p(\mathbf{0}, \mathbf{I}_p)$, and consider $H_0\colon \mathbf{C}\boldsymbol{\beta} = \mathbf{0}$, where $\mathbf{C}$ is $q \times p$ of rank q. Let $\tilde{\boldsymbol{\lambda}}$ and $\tilde{\boldsymbol{\beta}}$ be the restricted least-squares solutions of

$$\mathbf{z} - \tilde{\boldsymbol{\beta}} + \mathbf{C}'\tilde{\boldsymbol{\lambda}} = \mathbf{0} \tag{12.41}$$

and

$$\mathbf{C}\tilde{\boldsymbol{\beta}} = \mathbf{0}. \tag{12.42}$$

Then H_0 can be tested using

$$\begin{aligned} G &= (\mathbf{z} - \tilde{\boldsymbol{\beta}})'(\mathbf{z} - \tilde{\boldsymbol{\beta}}) & (12.43) \\ &= \mathbf{z}'\mathbf{C}'(\mathbf{C}\mathbf{C}')^{-1}\mathbf{C}\mathbf{z} & (12.44) \\ &= \tilde{\boldsymbol{\lambda}}'\mathbf{C}\mathbf{C}'\tilde{\boldsymbol{\lambda}}, & (12.45) \end{aligned}$$

where $G \sim \chi_q^2(\Delta)$, the noncentral chi-square distribution with noncentrality parameter $\Delta = \boldsymbol{\beta}'\mathbf{C}'(\mathbf{C}\mathbf{C}')^{-1}\mathbf{C}\boldsymbol{\beta}$.

Proof: To find the restricted least-squares estimate $\tilde{\boldsymbol{\beta}}$, we differentiate $\|\mathbf{z} - \boldsymbol{\beta}\|^2 + \boldsymbol{\lambda}'\mathbf{C}\boldsymbol{\beta}$. Since $\boldsymbol{\lambda}'\mathbf{C}\boldsymbol{\beta} = \boldsymbol{\beta}'\mathbf{C}'\boldsymbol{\lambda}$, we can use Appendix A10.4 and obtain (12.41). Multiplying this equation by $\mathbf{C}$ and using (12.42), we obtain $\tilde{\boldsymbol{\lambda}} = -(\mathbf{C}\mathbf{C}')^{-1}\mathbf{C}\mathbf{z}$. Substituting in (12.41) gives us $\tilde{\boldsymbol{\beta}} = (\mathbf{I}_p - \mathbf{P})\mathbf{z}$, where $\mathbf{P} = \mathbf{C}'(\mathbf{C}\mathbf{C}')^{-1}\mathbf{C}$ is symmetric and idempotent of rank q (by Appendix A11.4). Hence

$$\begin{aligned} G &= \mathbf{z}'\mathbf{P}^2\mathbf{z} \\ &= \mathbf{z}'\mathbf{P}\mathbf{z} \\ &= \mathbf{z}'\mathbf{C}'(\mathbf{C}\mathbf{C}')^{-1}\mathbf{C}\mathbf{z} \\ &= \tilde{\boldsymbol{\lambda}}'\mathbf{C}\mathbf{C}'\tilde{\boldsymbol{\lambda}}. \end{aligned}$$

By Appendix A12.1 we have $E[G] = \operatorname{tr}\mathbf{P} + \Delta = q + \Delta$, where $\Delta = \boldsymbol{\beta}'\mathbf{P}\boldsymbol{\beta}$. Hence, using Appendix A11.5, we finally have $G \sim \chi_q^2(\Delta)$. ∎

To prove the asymptotic equality of our three large-sample tests we now apply the above lemma with $\mathbf{C} = \mathbf{A}_\dagger\mathbf{V}_\dagger$. Firstly, using a quadratic Taylor expansion and $\partial L(\hat{\boldsymbol{\theta}})/\partial\boldsymbol{\theta} = \mathbf{0}$, we have, from (12.26), with $\hat{\mathbf{D}}$ approximated by $\mathbf{D}_\dagger$,

$$\begin{aligned} -\tfrac{1}{2}\text{LR} &= L(\tilde{\boldsymbol{\theta}}) - L(\hat{\boldsymbol{\theta}}) \\ &\approx \frac{\partial L(\hat{\boldsymbol{\theta}})}{\partial\boldsymbol{\theta}'}(\tilde{\boldsymbol{\theta}} - \hat{\boldsymbol{\theta}}) - \tfrac{1}{2}(\tilde{\boldsymbol{\theta}} - \hat{\boldsymbol{\theta}})'\hat{\mathbf{D}}(\tilde{\boldsymbol{\theta}} - \hat{\boldsymbol{\theta}}) \\ &\approx -\tfrac{1}{2}(\tilde{\boldsymbol{\theta}} - \hat{\boldsymbol{\theta}})'(\mathbf{V}_\dagger')^{-1}\mathbf{V}_\dagger^{-1}(\tilde{\boldsymbol{\theta}} - \hat{\boldsymbol{\theta}}) \\ &= -\tfrac{1}{2}(\mathbf{z} - \tilde{\boldsymbol{\beta}})'(\mathbf{z} - \tilde{\boldsymbol{\beta}}) \qquad \text{[by (12.38)]} \\ &= -\tfrac{1}{2}G, \end{aligned}$$

say. We now turn our attention to W. Since $\mathbf{a}(\boldsymbol{\theta}^\dagger) = \mathbf{0}$,

$$\begin{aligned} \mathbf{a}(\hat{\boldsymbol{\theta}}) &\approx \mathbf{A}_\dagger(\hat{\boldsymbol{\theta}} - \boldsymbol{\theta}^\dagger) \\ &= \mathbf{A}_\dagger\mathbf{V}_\dagger\mathbf{V}_\dagger^{-1}(\hat{\boldsymbol{\theta}} - \boldsymbol{\theta}^\dagger) \\ &= \mathbf{C}\mathbf{z}. \end{aligned}$$

Approximating $\hat{\mathbf{A}}$ and $\hat{\mathbf{D}}$ by $\mathbf{A}_\dagger$ and $\mathbf{D}_\dagger$ respectively,

$$\begin{aligned} \hat{\mathbf{A}}\hat{\mathbf{D}}^{-1}\hat{\mathbf{A}}' &\approx \mathbf{A}_\dagger\mathbf{D}_\dagger^{-1}\mathbf{A}_\dagger' \\ &= \mathbf{A}_\dagger\mathbf{V}_\dagger\mathbf{V}_\dagger'\mathbf{A}_\dagger' \\ &= \mathbf{C}\mathbf{C}', \end{aligned} \tag{12.46}$$

so that from (12.27),

$$\mathbf{W} \approx \mathbf{z}'\mathbf{C}'(\mathbf{C}\mathbf{C}')^{-1}\mathbf{C}\mathbf{z} = G.$$

Finally, using (12.28) and (12.31) leads to

$$\begin{aligned} \mathrm{LM} &= \tilde{\boldsymbol{\lambda}}'\tilde{\mathbf{A}}\tilde{\mathbf{D}}^{-1}\tilde{\mathbf{A}}'\tilde{\boldsymbol{\lambda}} \\ &\approx \tilde{\boldsymbol{\lambda}}'\mathbf{A}_\dagger\mathbf{D}_\dagger^{-1}\mathbf{A}_\dagger'\tilde{\boldsymbol{\lambda}} \\ &= \tilde{\boldsymbol{\lambda}}'\mathbf{C}\mathbf{C}'\tilde{\boldsymbol{\lambda}} \qquad [\text{by } (12.46)] \\ &= G. \end{aligned} \tag{12.47}$$

Thus by the lemma, LR, W, and LM are approximately equivalent to G, which is $\chi_q^2(\Delta)$. Here [c.f. (12.38) and (12.46)]

$$\begin{aligned} \Delta &= \boldsymbol{\beta}'\mathbf{C}'(\mathbf{C}\mathbf{C}')^{-1}\mathbf{C}\boldsymbol{\beta} \\ &\approx (\boldsymbol{\theta}^* - \boldsymbol{\theta}^\dagger)'(\mathbf{V}_\dagger^{-1})'\mathbf{V}_\dagger'\mathbf{A}_\dagger'[\mathbf{A}_\dagger\mathbf{D}_\dagger^{-1}\mathbf{A}_\dagger']^{-1}\mathbf{A}_\dagger\mathbf{V}_\dagger\mathbf{V}_\dagger^{-1}(\boldsymbol{\theta}^* - \boldsymbol{\theta}^\dagger) \\ &\approx \boldsymbol{\delta}'\mathbf{A}_\dagger'[\mathbf{A}_\dagger(n^{-1}\mathbf{D}_\dagger)^{-1}\mathbf{A}_\dagger']^{-1}\mathbf{A}_\dagger\boldsymbol{\delta}, \end{aligned}$$

where $\sqrt{n}(\boldsymbol{\theta}^* - \boldsymbol{\theta}^\dagger) \to \boldsymbol{\delta}$. When H_0 is true, $\boldsymbol{\theta}^* = \boldsymbol{\theta}^\dagger$, $\boldsymbol{\delta} = \mathbf{0}$, and $\Delta = 0$. We note in passing that, by (12.47), LM is a function of the Lagrange multiplier $\tilde{\boldsymbol{\lambda}}$, whence it gets its name.

The above method can also be used when H_0 has the freedom-equation specification $\boldsymbol{\theta} = \boldsymbol{\theta}(\boldsymbol{\alpha})$, where $\boldsymbol{\alpha}$ is $(p-q) \times 1$. In this case $\tilde{\boldsymbol{\theta}} = \boldsymbol{\theta}(\tilde{\boldsymbol{\alpha}})$, where $\tilde{\boldsymbol{\alpha}}$ maximizes $L[\boldsymbol{\theta}(\boldsymbol{\alpha})]$, and we assume $\tilde{\boldsymbol{\theta}} \to \boldsymbol{\theta}(\boldsymbol{\alpha}^\dagger) = \boldsymbol{\theta}^\dagger$. If the $p \times (p-q)$ matrix $\mathbf{T} = [(\partial\theta_i/\partial\alpha_j)]_{\alpha^\dagger}$ has rank $p-q$, then, under appropriate regularity conditions, the corresponding asymptotic linear model and hypothesis take the form (Seber [1964, 1967])

$$\mathbf{z} = \boldsymbol{\beta} + \boldsymbol{\varepsilon}, \qquad \boldsymbol{\varepsilon} \sim N_p(\mathbf{0}, \mathbf{I}_p)$$

and

$$H_0 : \boldsymbol{\beta} = \mathbf{V}_\dagger^{-1}\mathbf{T}_\dagger\boldsymbol{\gamma}.$$

The same linearization can be extended to allow for identifiability constraints on the parameters and to handle nested hypotheses (c.f. Seber [1967]). For further details relating to the underlying assumptions in the normal case see Johansen [1984: Section 5.3].

12.5 MULTIVARIATE ESTIMATION

The multivariate nonlinear model takes the form (Chapter 11)

$$\mathbf{y}_i = \mathbf{f}_i(\boldsymbol{\theta}) + \boldsymbol{\varepsilon}_i, \qquad i = 1, 2, \ldots, n,$$

where the $\boldsymbol{\varepsilon}_i$ are i.i.d. with mean $\mathbf{0}$ and variance–covariance matrix $\boldsymbol{\Sigma}$. Also $\mathbf{f}_i(\boldsymbol{\theta}) = \mathbf{f}(\mathbf{x}_i; \boldsymbol{\theta})$ [c.f. (11.1) and (11.2)] or has the more general formulation given by (11.9). Two methods of estimation for this model have been widely used, maximum-likelihood estimation and generalized least-squares estimation. If the $\boldsymbol{\varepsilon}_i$ are assumed to be multivariate normal, then the maximum-likelihood estimate $\hat{\boldsymbol{\theta}}_n$ is the value of $\boldsymbol{\theta}$ which maximizes the joint distribution of the $\mathbf{y}_i$. From Section 11.3 this is obtained by minimizing $|\mathbf{V}(\boldsymbol{\theta})|$, where

$$\mathbf{V}(\boldsymbol{\theta}) = \sum_{i=1}^{n} [\mathbf{y}_i - \mathbf{f}_i(\boldsymbol{\theta})][\mathbf{y}_i - \mathbf{f}_i(\boldsymbol{\theta})]'. \tag{12.48}$$

The maximum-likelihood estimator of $\boldsymbol{\Sigma}$ is then

$$\hat{\boldsymbol{\Sigma}}_n = \frac{1}{n} \mathbf{V}(\hat{\boldsymbol{\theta}}_n). \tag{12.49}$$

When normality is not assumed, we follow the econometric literature and call $\hat{\boldsymbol{\theta}}_n$ the quasi-maximum-likelihood estimator (QMLE).

To describe the generalized least-squares method we begin with a matrix $\mathbf{S}_n$ which is positive definite, with probability one if random. Then a generalized least-squares (GLS) estimator, also called the minimum-distance estimator or MDE in the econometric literature, is the value $\tilde{\boldsymbol{\theta}}_n(\mathbf{S}_n)$ of $\boldsymbol{\theta}$ which minimizes

$$\sum_{i=1}^{n} [\mathbf{y}_i - \mathbf{f}_i(\boldsymbol{\theta})]' \mathbf{S}_n [\mathbf{y}_i - \mathbf{f}_i(\boldsymbol{\theta})].$$

We now discuss the asymptotic properties of $\tilde{\boldsymbol{\theta}}_n(\mathbf{S}_n)$ and $\hat{\boldsymbol{\theta}}_n$ as estimators of the true value $\boldsymbol{\theta}^*$.

If $\mathbf{S}_n = \mathbf{S}$ for all n, then the following conditions, which are multivariate analogues of assumptions A(1) to A(4) in Section 12.2.1 above, are sufficient for the strong consistency of $\tilde{\boldsymbol{\theta}}_n(\mathbf{S})$:

a(1). The $\boldsymbol{\varepsilon}_i$ are i.i.d. with mean $\mathbf{0}$ and positive definite dispersion matrix $\boldsymbol{\Sigma}$.

a(2). For each i, the elements of $\mathbf{f}_i(\boldsymbol{\theta})$ are continuous functions of $\boldsymbol{\theta}$ for $\boldsymbol{\theta} \in \Theta$, a subset of $\mathbb{R}^p$.

a(3). Θ is a closed, bounded (compact) subset of $\mathbb{R}^p$.

a(4). $\lim n^{-1} \sum_{i=1}^{n} \mathbf{f}_i(\boldsymbol{\theta}) \mathbf{f}_i(\boldsymbol{\theta}_1)'$ exists, and the convergence is uniform for all $\boldsymbol{\theta}, \boldsymbol{\theta}_1$ in Θ;

$$\lim n^{-1} \sum_{i=1}^{n} [\mathbf{f}_i(\boldsymbol{\theta}) - \mathbf{f}_i(\boldsymbol{\theta}^*)][\mathbf{f}_i(\boldsymbol{\theta}) - \mathbf{f}_i(\boldsymbol{\theta}^*)]'$$

is positive definite for all $\boldsymbol{\theta}$ ($\neq \boldsymbol{\theta}^*$) in Θ.

If a(1) to a(4) hold, then Phillips [1976] showed that $\tilde{\boldsymbol{\theta}}_n(\mathbf{S})$ and the residual matrix

$$\tilde{\boldsymbol{\Sigma}}_n(\mathbf{S}) = \frac{1}{n} \sum_{i=1}^{n} [\mathbf{y}_i - \mathbf{f}_i(\tilde{\boldsymbol{\theta}}_n(\mathbf{S}))][\mathbf{y}_i - \mathbf{f}_i(\tilde{\boldsymbol{\theta}}_n(\mathbf{S}))]'$$

are strongly consistent estimators of $\boldsymbol{\theta}^*$ and $\boldsymbol{\Sigma}^*$ respectively, where $\boldsymbol{\Sigma}^*$ is the true value of $\boldsymbol{\Sigma}$.

For the more general situation in which $\mathbf{S}_n$ depends on n, we need to add just one further assumption for $\tilde{\boldsymbol{\theta}}_n(\mathbf{S}_n)$ and $\tilde{\boldsymbol{\Sigma}}_n(\mathbf{S}_n)$ to be strongly consistent, namely:

I. $\mathbf{S}_n \to \mathbf{S}$ with probability one, where $\mathbf{S}$ is positive definite.

Malinvaud [1970b] proved the same result, but for weak consistency, using a slightly different set of assumptions from a(1) to a(4), namely the multivariate analogues of B(1) to B(5) in Section 12.2.3.

In practice we could choose $\mathbf{S}_n$ to be the inverse of the residual matrix and use the following iteration process, called the iterated GLS procedure (Barnett [1976]) or the Zellner iteration (Amemiya [1983]), after Zellner [1962]:

$$\boldsymbol{\theta}_n^{(a+1)} = \tilde{\boldsymbol{\theta}}(\boldsymbol{\Sigma}_n^{(a)-1}), \tag{12.50}$$

where

$$\boldsymbol{\Sigma}_n^{(a)} = n^{-1} \sum_{i=1}^{n} [\mathbf{y}_i - \mathbf{f}_i(\boldsymbol{\theta}_n^{(a)})][\mathbf{y}_i - \mathbf{f}_i(\boldsymbol{\theta}_n^{(a)})]'. \tag{12.51}$$

Here the starting matrix $\boldsymbol{\Sigma}_n^{(1)}$ is chosen to be an arbitrary positive definite matrix independent of n, for example $\mathbf{I}_d$, and $\boldsymbol{\theta}_n^{(a+1)}$ is obtained by minimizing

$$\sum_{i=1}^{n} [\mathbf{y}_i - \mathbf{f}_i(\boldsymbol{\theta})]'(\boldsymbol{\Sigma}_n^{(a)})^{-1}[\mathbf{y} - \mathbf{f}_i(\boldsymbol{\theta})]$$

with respect to $\boldsymbol{\theta}$. Two questions come to mind. Firstly, does the iteration process converge, and secondly, is the limit of $\tilde{\boldsymbol{\theta}}_n^{(a)}$ as $a \to \infty$ independent of the starting value $\tilde{\boldsymbol{\Sigma}}_n^{(1)}$ (given that the process converges)? Phillips [1976] showed that under certain conditions, and for large enough n, the answer to both questions is yes. He also proved that as $a \to \infty$, $\boldsymbol{\theta}^{(a)}$ converges to the QMLE $\hat{\boldsymbol{\theta}}_n$. These sufficient conditions are assumptions a(1) to a(4) above together with matrix analogues a(5) to a(9) of A(5), A(6), A(8′), and A(9). For example A(8′) is replaced by a(8′), in which $n^{-1}\sum g_i(\boldsymbol{\theta})h_i(\boldsymbol{\theta}_1)$ becomes $n^{-1}\sum \mathbf{g}_i(\boldsymbol{\theta})\mathbf{h}_i(\boldsymbol{\theta}_1)'$. Also, A(9) now takes the form:

a(9). For any positive definite $p \times p$ matrix $\mathbf{R}$, the matrix $\mathbf{M}_n(\boldsymbol{\theta}, \mathbf{R})$ with (r, s)th element

$$n^{-1} \sum_{i=1}^{n} \left(\frac{\partial \mathbf{f}_i(\boldsymbol{\theta})}{\partial \theta_r} \right)' \mathbf{R} \left(\frac{\partial f_i(\boldsymbol{\theta})}{\partial \theta_s} \right) \tag{12.52}$$

has a positive definite limit as $n \to \infty$.

Consider now just the ath iteration. We find that given a(1) to a(9), $\boldsymbol{\theta}_n^{(a+1)}$ and $\boldsymbol{\Sigma}_n^{(a)}$ are strongly consistent and $\boldsymbol{\theta}_n^{(a+1)}$ is asymptotically normal. To see this we set $\boldsymbol{\Sigma}_n^{(1)} = \mathbf{I}_d$ and use similar arguments to those given in Section 12.2.3 to show that $\boldsymbol{\theta}_n^{(1)}$ is asymptotically normal. Then $\boldsymbol{\Sigma}_n^{(1)}$ is strongly consistent, so that assumption I holds and $\boldsymbol{\theta}_n^{(2)}$ is asymptotically normal. This argument is repeated until we reach $\boldsymbol{\theta}_n^{(a+1)}$. Unfortunately, we cannot assume that consistency and asymptotic normality automatically apply to the limit $\hat{\boldsymbol{\theta}}_n$ of $\boldsymbol{\theta}_n^{(a)}$. Although this route to the QMLE $\hat{\boldsymbol{\theta}}_n$ gives us a computational method for finding $\hat{\boldsymbol{\theta}}_n$, it does not lead to an asymptotic theory. Even if we express $\hat{\boldsymbol{\theta}}_n$ directly as an MDE, namely $\tilde{\boldsymbol{\theta}}_n(\hat{\boldsymbol{\Sigma}}_n^{-1})$ [see (12.50)], its strong consistency does not follow just from a(1) to a(4): assumption I is also needed with this approach. Unfortunately, we cannot immediately assert that $\hat{\boldsymbol{\Sigma}}_n$ converges with probability one to a positive definite matrix, the requirement of assumption I. It is a question of which you prove first, the consistency of $\hat{\boldsymbol{\theta}}_n$ or the consistency of $\hat{\boldsymbol{\Sigma}}_n$. Interestingly enough, however, assumption I is not required, and Phillips [1976] was able to show *directly* that a(1) to a(4) are sufficient for the strong consistency of $\hat{\boldsymbol{\theta}}_n$. It then follows that $\hat{\boldsymbol{\Sigma}}_n$ is a strongly consistent estimator of $\boldsymbol{\Sigma}^*$.

Barnett [1976] makes similar comments about the relative merits of the MDE and the maximum-likelihood estimate, the QMLE under normality. He suggests that the emphasis should shift from minimum-distance (generalized least-squares) estimation to maximum-likelihood estimation, as good algorithms for the latter are available. In making the comparison he notes that, apart from some general lemas on consistency, Malinvaud's [1970b] results apply solely to the MDE and are not appropriate for maximum-likelihood estimation. Assuming multivariate normality of the $\boldsymbol{\varepsilon}_i$, Barnett then proceeds to give a general theory for maximum-likelihood estimation based on the m-dimensional parameter vector $\boldsymbol{\phi} = (\boldsymbol{\theta}', \boldsymbol{\sigma}')'$, where $m = p + \frac{1}{2}d(d+1)$ and $\boldsymbol{\sigma} = \text{vech}\,\boldsymbol{\Sigma}$, the vector of distinct elements of $\boldsymbol{\Sigma}$ (in the upper triangle; c.f. Appendix A13.2). [In his notation $\boldsymbol{\theta} = (\boldsymbol{\gamma}', \boldsymbol{\sigma}')'$.] We now briefly summarize some of his results.

Barnett [1976] begins by effectively assuming some version of either a(1) to a(4) or Malinvaud's corresponding assumptions, so that the maximum-likelihood estimator $\hat{\boldsymbol{\phi}}_n$ of $\boldsymbol{\phi}$ is weakly consistent. He then adds the following from Malinvaud [1970b]:

c(1). For any positive definite $p \times p$ matrix $\mathbf{R}$, the matrix $\mathbf{M}_n(\boldsymbol{\theta}^*, \mathbf{R})$ of (12.52) is nonsingular and converges to a nonsingular matrix $\mathbf{M}(\boldsymbol{\theta}^*, \mathbf{R})$ as $n \to \infty$.

c(2). The first three derivatives of each element of $\mathbf{f}_i(\boldsymbol{\theta})$ are bounded uniformly in i and $\boldsymbol{\theta}$, $\boldsymbol{\theta} \in \Theta$.

If $l(\boldsymbol{\phi})$ is the likelihood function, being proportional to the joint distribution of the $\mathbf{y}_i$, and

$$\mathbf{J}_n(\boldsymbol{\phi}) = -\underset{\boldsymbol{\phi}}{\mathscr{E}}\left(\frac{\partial^2 \log l(\boldsymbol{\phi})}{\partial \boldsymbol{\phi}\, \partial \boldsymbol{\phi}'}\right) \tag{12.53}$$

is the (expected) information matrix, then $n^{-1}\mathbf{J}_n(\boldsymbol{\phi})$ converges to a nonsingular matrix, $\mathbf{J}(\boldsymbol{\phi})$ say, as $n \to \infty$. Barnett shows that if $\hat{\boldsymbol{\phi}}_n$ is a consistent solution of the maximum-likelihood equations, then $\sqrt{n}(\hat{\boldsymbol{\phi}}_n - \boldsymbol{\phi}^*)$ is asymptotically $N_m(\mathbf{0}, \mathbf{J}^{-1}(\boldsymbol{\phi}^*))$. Since $\mathbf{J}_n(\boldsymbol{\phi})$ and $\mathbf{J}(\boldsymbol{\phi})$ are known to be block-diagonal matrices with the blocks corresponding to $\boldsymbol{\theta}$ and $\boldsymbol{\sigma}$ respectively, $\hat{\boldsymbol{\theta}}_n$ and $\hat{\boldsymbol{\sigma}}_n$ are asymptotically independent. Hence it follows that $\sqrt{n}(\hat{\boldsymbol{\theta}}_n - \hat{\boldsymbol{\theta}}^*)$ is asymptotically $N_p(\mathbf{0}, \mathbf{A}^{-1}(\boldsymbol{\theta}^*))$, where

$$\mathbf{A}(\boldsymbol{\theta}) = \lim n^{-1} \underset{\boldsymbol{\theta}}{\mathscr{E}}\left[-\frac{\partial^2 \log l(\boldsymbol{\phi})}{\partial \boldsymbol{\theta}\, \partial \boldsymbol{\theta}'} \right]. \tag{12.54}$$

The subscript $\boldsymbol{\theta}$ is added to emphasize that the expected value is calculated using $\boldsymbol{\theta}$ and not $\boldsymbol{\theta}^*$ in the density function. A similar result holds for $\hat{\boldsymbol{\sigma}}_n = \operatorname{vech} \hat{\boldsymbol{\Sigma}}_n$.

Barnett also proves that a weakly consistent estimator of $\mathbf{J}^{-1}(\boldsymbol{\phi}^*)$ is $n\hat{\mathbf{H}}^{-1}$, where

$$\hat{\mathbf{H}} = \left[-\frac{\partial^2 \log l(\boldsymbol{\phi})}{\partial \boldsymbol{\phi}\, \partial \boldsymbol{\phi}'} \right]_{\hat{\boldsymbol{\phi}}_n}. \tag{12.55}$$

Unfortunately, $\hat{\mathbf{H}}$ is not block-diagonal, and some method of computing the diagonal blocks of $\hat{\mathbf{H}}^{-1}$ is needed. Barnett shows that a concentrated-likelihood method can be used for finding these blocks directly. For example, to find the estimator of $\mathbf{A}^{-1}(\boldsymbol{\theta})$ we use the concentrated log-likelihood function for $\boldsymbol{\theta}$, namely $M(\boldsymbol{\theta})$ [$= -(n/2)M_1(\boldsymbol{\theta})$; c.f. (11.25)]. Then the upper diagonal block $(\hat{\mathbf{H}}^{-1})_{11}$, say, of $\hat{\mathbf{H}}^{-1}$ corresponding to $\boldsymbol{\theta}$ is given by

$$(\hat{\mathbf{H}}^{-1})_{11} = \left[-\frac{\partial^2 M(\boldsymbol{\theta})}{\partial \boldsymbol{\theta}\, \partial \boldsymbol{\theta}'} \right]_{\hat{\boldsymbol{\theta}}_n}.$$

The above multivariate problem has also been studied extensively by Gallant [1987: Chapter 5], who gives a different set of assumptions.

CHAPTER 13

Unconstrained Optimization

13.1 INTRODUCTION

Estimation of the parameters of a nonlinear model typically involves the optimization (maximization or minimization) of a function. We may require the parameter values that minimize a weighted or unweighted sum of squares, maximize a likelihood function, or maximize a posterior density function. For nonlinear regression models with additive errors of the form

$$y_i = f(\mathbf{x}_i; \boldsymbol{\theta}) + \varepsilon_i, \qquad i = 1, 2, \ldots, n,$$

a typical problem is to find the value of $\boldsymbol{\theta}$ which minimizes a function $h(\boldsymbol{\theta})$ having one of the following forms:

$$h(\boldsymbol{\theta}) = \sum_{i=1}^{n} [y_i - f(\mathbf{x}_i; \boldsymbol{\theta})]^2 = \sum_{i=1}^{n} r_i(\boldsymbol{\theta})^2, \tag{13.1}$$

$$h(\boldsymbol{\theta}) = \sum_{i=1}^{n} |y_i - f(\mathbf{x}_i; \boldsymbol{\theta})| = \sum_{i=1}^{n} |r_i(\boldsymbol{\theta})|, \tag{13.2}$$

$$h(\boldsymbol{\theta}) = \sum_{i=1}^{n} \rho(r_i(\boldsymbol{\theta})) \qquad \text{(robust loss function)}. \tag{13.3}$$

In (13.1) $\boldsymbol{\theta}$ is chosen to be the least-squares estimator, while in (13.3) we minimize a loss function $\rho(\cdot)$, usually one of the so-called robust loss functions. Bayesian methods also lead to optimization problems where the goal might be to minimize the Bayes risk or else find the mode of the posterior distribution of $\boldsymbol{\theta}$.

The optimization problems arising in estimation for nonlinear models can seldom be solved analytically. Almost invariably the optimal value of $\boldsymbol{\theta}$ must be located by iterative techniques using a computer. Fortunately, good computer programs for optimization are widely available in a number of the major numerical subroutine libraries, in some of the statistical packages, and as published programs in journals and technical reports (see Chapter 15). Consequently statisticians need not, and in general probably should not, write their own software. However, an understanding of how the major algorithms work is very valuable when one wishes to choose an algorithm for a particular problem. We need to be aware of the limitations of our chosen algorithms and have some idea of what might go wrong. In this section we describe the major techniques for the unconstrained optimization of a general function—unconstrained in the sense that there are no restrictions placed on $\boldsymbol{\theta}$. Particular techniques for specialized functions of the form (13.1) to (13.3) above are discussed in Chapter 14. In both chapters attention is restricted to the important concepts, and major results are given without proof. Readers interested in a more comprehensive coverage or in the details of the algorithms will need to consult the general literature. Good introductory references are the books by Nash and Walker-Smith [1987], Scales [1985], Dennis and Schnabel [1983], Fletcher [1980], and Gill et al. [1981], and Chapters 10 and 11 of Kennedy and Gentle [1980]. All contain algorithms laid out step by step to some extent, though their descriptions are more detailed in Dennis and Schnabel [1983], Scales [1985], and Kennedy and Gentle [1980]. Some older books and review articles such as Nash [1979], Brodlie [1977], Dennis [1977], Powell [1976], Lill [1976], and Murray [1972a] are still valuable. Constrained optimization problems in which $h(\boldsymbol{\theta})$ is optimized subject to $\boldsymbol{\theta}$ belonging to some defined subregion of Euclidean space occur less frequently in statistical work and are not described here. For an introduction to constrained optimization, also known as "mathematical programming," see Scales [1985], Fletcher [1981], and Gill et al. [1981]. A method for optimization subject to linear equality constraints is described in Appendix E.

The reader who is not interested in technical details can skip to the summary of Section 13.7 and the decision tree in Fig. 13.5.

13.2 TERMINOLOGY

13.2.1 Local and Global Minimization

This chapter discusses the minimization of $h(\boldsymbol{\theta})$, a real-valued function of p real variables $\boldsymbol{\theta} = (\theta_1, \theta_2, \ldots, \theta_p)'$. A separate discussion of maximization is unnecessary, as $\hat{\boldsymbol{\theta}}$ maximizes $h(\boldsymbol{\theta})$ if and only if it minimizes $-h(\boldsymbol{\theta})$. Initially we restrict attention to unconstrained minimization, where we wish to minimize $h(\boldsymbol{\theta})$ over $\mathbb{R}^p$. Most estimation methods require a global minimum of $h(\boldsymbol{\theta})$, namely a point $\hat{\boldsymbol{\theta}}$ such that $h(\boldsymbol{\theta}) \geq h(\hat{\boldsymbol{\theta}})$ for all $\boldsymbol{\theta}$ in $\mathbb{R}^p$. Unfortunately, global minimization is possible only for very restrictive classes of functions such as convex functions (Dixon and Szegö [1975, 1978]). In most cases, the best we can hope for is a numerical algorithm that will converge to a local (relative) minimum.

Intuitively, $\hat{\boldsymbol{\theta}}$ is a local minimum if $h(\boldsymbol{\theta})$ cannot be decreased in a small neighborhood of $\hat{\boldsymbol{\theta}}$. There are two types of local minimum. If there exists $\delta > 0$

such that for $\boldsymbol{\theta} \neq \hat{\boldsymbol{\theta}}$,

$$\|\hat{\boldsymbol{\theta}} - \boldsymbol{\theta}\| < \delta \quad \Rightarrow \quad h(\boldsymbol{\theta}) > h(\hat{\boldsymbol{\theta}}),$$

where $\|\cdot\|$ is some vector norm, then $\hat{\boldsymbol{\theta}}$ is said to be a *strong local minimum*. A *weak local minimum* occurs if

$$\|\hat{\boldsymbol{\theta}} - \boldsymbol{\theta}\| < \delta \quad \Rightarrow \quad h(\boldsymbol{\theta}) \geq h(\hat{\boldsymbol{\theta}})$$

and the equality holds for at least one $\boldsymbol{\theta} \neq \hat{\boldsymbol{\theta}}$.

Except for the direct search methods, all the minimization algorithms for "general" functions that we shall discuss are derived under the assumption that h has continuous second derivatives (and is therefore continuous with continuous first derivatives). We use the following notation. The *gradient vector* of h at $\boldsymbol{\theta}$ is the vector of first derivatives of h,

$$\mathbf{g}(\boldsymbol{\theta}) = \frac{\partial h(\boldsymbol{\theta})}{\partial \boldsymbol{\theta}} = \left(\frac{\partial h(\boldsymbol{\theta})}{\partial \theta_1}, \frac{\partial h(\boldsymbol{\theta})}{\partial \theta_2}, \ldots, \frac{\partial h(\boldsymbol{\theta})}{\partial \theta_p} \right)', \tag{13.4}$$

while the *Hessian matrix* at $\boldsymbol{\theta}$ is the $p \times p$ matrix of second derivatives

$$\mathbf{H}(\boldsymbol{\theta}) = \frac{\partial^2 h(\boldsymbol{\theta})}{\partial \boldsymbol{\theta}\, \partial \boldsymbol{\theta}'} = \left[\left(\frac{\partial^2 h(\boldsymbol{\theta})}{\partial \theta_i\, \partial \theta_j} \right) \right]. \tag{13.5}$$

When a sequence of values $\boldsymbol{\theta}^{(1)}, \boldsymbol{\theta}^{(2)}, \ldots$ is under discussion, we use the notation $h^{(a)} = h(\boldsymbol{\theta}^{(a)})$, $\mathbf{g}^{(a)} = \mathbf{g}(\boldsymbol{\theta}^{(a)})$, and $\mathbf{H}^{(a)} = \mathbf{H}(\boldsymbol{\theta}^{(a)})$ together with $\hat{h} = h(\hat{\boldsymbol{\theta}})$, $\hat{\mathbf{g}} = \mathbf{g}(\hat{\boldsymbol{\theta}})$, and $\hat{\mathbf{H}} = \mathbf{H}(\hat{\boldsymbol{\theta}})$.

For a function with continuous second derivatives, the conditions

N1. $\hat{\mathbf{g}} = \mathbf{g}(\hat{\boldsymbol{\theta}}) = \mathbf{0}$,
N2. $\hat{\mathbf{H}} = \mathbf{H}(\hat{\boldsymbol{\theta}})$ is positive semidefinite (Appendix A4)

are *necessary* for $\hat{\boldsymbol{\theta}}$ to be a local minimum of $h(\boldsymbol{\theta})$, while the conditions

S1. $\hat{\mathbf{g}} = \mathbf{0}$,
S2. $\hat{\mathbf{H}}$ is positive definite

are *sufficient* for $\hat{\boldsymbol{\theta}}$ to be a strong local minimum of $h(\boldsymbol{\theta})$. The corresponding conditions for a local maximum are obtained by replacing "positive (semi)definite" by "negative (semi)definite" above. From the example $h(\theta) = \theta^4$ we see that the sufficient conditions are not also necessary. If $\hat{\mathbf{g}} = \mathbf{0}$, then $\hat{\boldsymbol{\theta}}$ is called a stationary point of $h(\boldsymbol{\theta})$. A stationary point that is neither a local minimum nor a local maximum is called a saddle point (see Fig. 13.1). Algorithms that approximate the Hessian or do not use it at all seek out a stationary point.

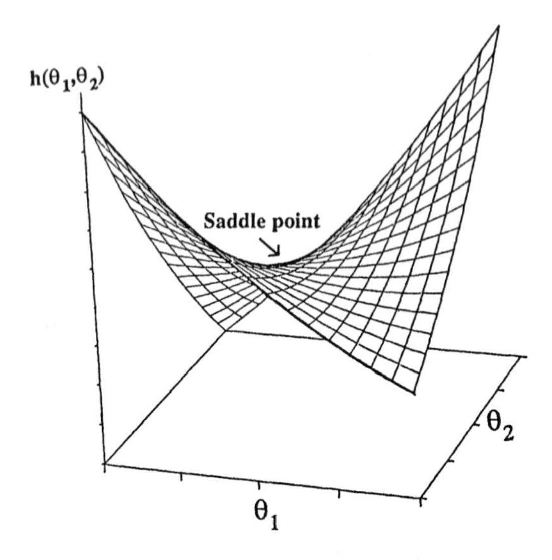

Figure 13.1 Saddle point.

Having converged to $\hat{\boldsymbol{\theta}}$, the algorithm may incorporate a random search in the vicinity of $\hat{\boldsymbol{\theta}}$ to try and ensure that $\hat{\boldsymbol{\theta}}$ is a local minimum and not just a saddle point.

13.2.2 Quadratic Functions

Perhaps the simplest form of function that can possess a minimum is the quadratic function

$$q(\boldsymbol{\theta}) = a + \mathbf{b}'\boldsymbol{\theta} + \tfrac{1}{2}\boldsymbol{\theta}'\mathbf{G}\boldsymbol{\theta}, \tag{13.6}$$

where $a, \mathbf{b}$, and $\mathbf{G}$ are independent of $\boldsymbol{\theta}$. Differentiating, we see that $q(\boldsymbol{\theta})$ has a gradient vector

$$\mathbf{g}(\boldsymbol{\theta}) = \frac{\partial q(\boldsymbol{\theta})}{\partial \boldsymbol{\theta}} = \mathbf{b} + \mathbf{G}\boldsymbol{\theta} \tag{13.7}$$

and Hessian

$$\mathbf{H}(\boldsymbol{\theta}) = \frac{\partial^2 q(\boldsymbol{\theta})}{\partial \boldsymbol{\theta}\, \partial \boldsymbol{\theta}'} = \mathbf{G}. \tag{13.8}$$

Since any stationary point $\hat{\boldsymbol{\theta}}$ must satisfy $\mathbf{g}(\hat{\boldsymbol{\theta}}) = \mathbf{0}$, $q(\boldsymbol{\theta})$ possesses a stationary point if $-\mathbf{b}$ is contained in the space spanned by the columns of $\mathbf{G}$. In particular, for nonsingular $\mathbf{G}$,

$$\hat{\boldsymbol{\theta}} = -\mathbf{G}^{-1}\mathbf{b}. \tag{13.9}$$

Suppose then that a stationary point $\hat{\boldsymbol{\theta}}$ exists for $q(\boldsymbol{\theta})$, and let $\lambda_1, \lambda_2, \ldots, \lambda_p$ be the eigenvalues of the symmetric matrix $\mathbf{G}$ with $\mathbf{u}_1, \mathbf{u}_2, \ldots, \mathbf{u}_p$ a corresponding orthonormal set of eigenvectors (Appendix A2.2). Then

$$\begin{aligned} q(\hat{\boldsymbol{\theta}} + \alpha \mathbf{u}_i) &= q(\hat{\boldsymbol{\theta}}) + \alpha \mathbf{u}_i'(\mathbf{G}\hat{\boldsymbol{\theta}} + \mathbf{b}) + \tfrac{1}{2}\alpha^2 \mathbf{u}_i'\mathbf{G}\mathbf{u}_i \\ &= q(\hat{\boldsymbol{\theta}}) + \tfrac{1}{2}\alpha^2 \lambda_i, \end{aligned} \tag{13.10}$$

since $\mathbf{G}\hat{\boldsymbol{\theta}} + \mathbf{b} = \mathbf{0}$ and $\mathbf{G}\mathbf{u}_i = \lambda_i \mathbf{u}_i$. Hence the change in the value of q when we move away from $\hat{\boldsymbol{\theta}}$ in the direction $\pm \mathbf{u}_i$ depends on the sign of λ_i. If $\lambda_i > 0$, q increases; if $\lambda_i < 0$, q decreases; and if $\lambda_i = 0$, q is constant along this line. The speed with which q changes is directly proportional to the size of λ_i, so that if λ_i is small, q changes very slowly along the line through $\hat{\boldsymbol{\theta}}$ in the direction $\pm \mathbf{u}_i$. When all the eigenvalues of $\mathbf{G}$ are positive, corresponding to $\mathbf{G}$ being positive definite, $\hat{\boldsymbol{\theta}}$ is the unique global minimum of $q(\boldsymbol{\theta})$. By substituting $\hat{\boldsymbol{\theta}}$ from (13.9) we can verify that

$$q(\boldsymbol{\theta}) = \tfrac{1}{2}(\boldsymbol{\theta} - \hat{\boldsymbol{\theta}})'\mathbf{G}(\boldsymbol{\theta} - \hat{\boldsymbol{\theta}}) + a - \tfrac{1}{2}\mathbf{b}'\mathbf{G}^{-1}\mathbf{b}. \tag{13.11}$$

Thus the contours of $q(\boldsymbol{\theta})$ are ellipsoids centered at $\hat{\boldsymbol{\theta}}$. Also the principal axes of these ellipsoids lie along the directions of the eigenvectors of $\mathbf{G}$, with the length of

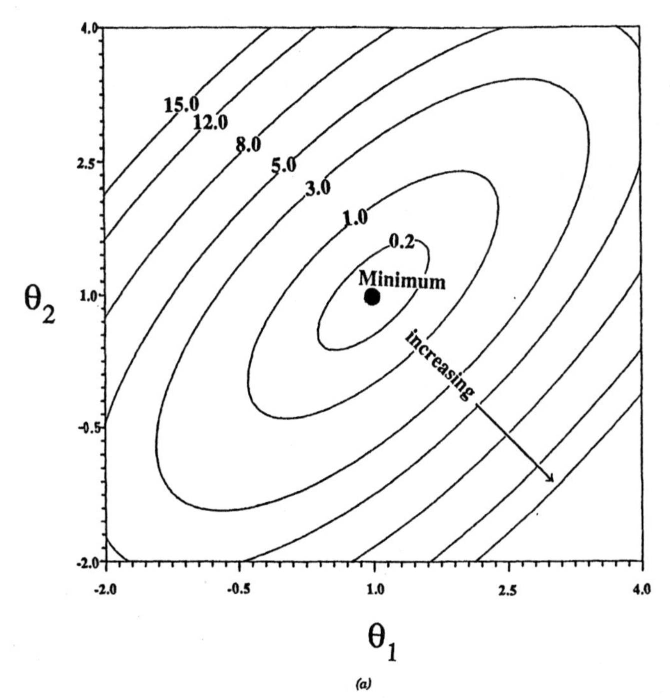

(a)

Figure 13.2 Contours for a quadratic: (*a*) positive eigenvalues; (*b*) negative eigenvalues; (*c*) positive and negative eigenvalues. The functions used are (*a*) $h(\boldsymbol{\theta}) = (\theta_1 - 1)^2 - 1.4(\theta_1 - 1)(\theta_2 - 1) + (\theta_2 - 1)^2$, (*b*) $h(\boldsymbol{\theta}) = 20 - (\theta_1 - 1)^2 + 1.4(\theta_1 - 1)(\theta_2 - 1) - (\theta_2 - 1)^2$, (*c*) $h(\boldsymbol{\theta}) = 100 + (\theta_1 - 1)^2 - 0.6(\theta_1 - 1)(\theta_2 - 1) + (\theta_2 - 1)^2$.

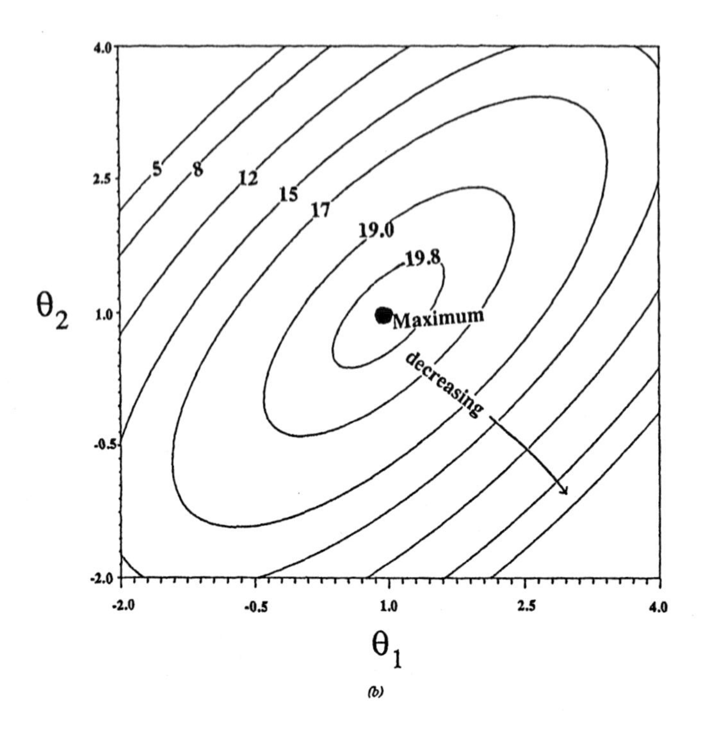

(b)

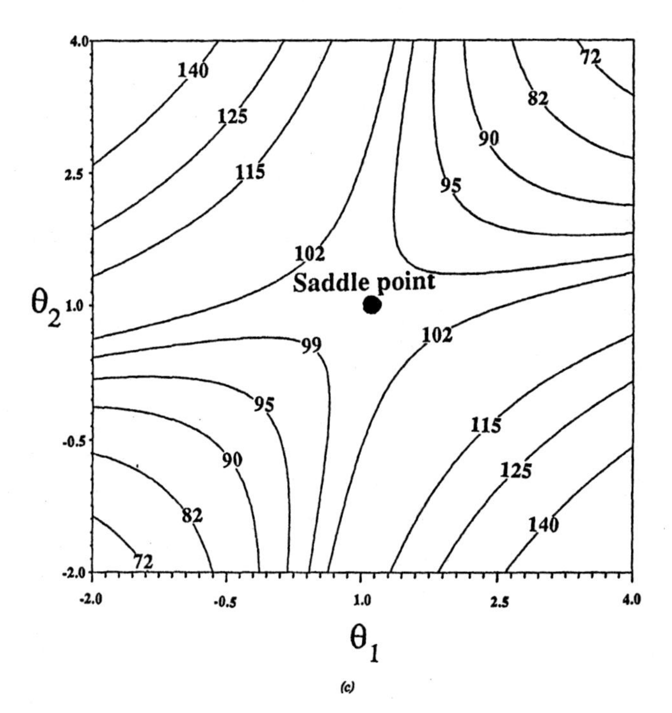

(c)

Figure 13.2 (*Continued*)

the axis along $\mathbf{u}_i$ being proportional to $\lambda_i^{-1/2}$ (Appendix A6). Analogous results for a maximum hold if all the eigenvalues are negative. If $\mathbf{G}$ has eigenvalues of both signs ($\mathbf{G}$ is indefinite), then $\hat{\boldsymbol{\theta}}$ is a saddle point and $q(\boldsymbol{\theta})$ is unbounded above and below. Figure 13.2 depicts contours for three different quadratic functions illustrating the situation when $p = 2$. These ideas are used to predict the local behavior of a general function in a small neighborhood of a stationary point.

Quadratic functions are of particular importance in nonlinear optimization for three main reasons. Firstly, any general minimization algorithm should work well on such a simple function. Secondly, as a consequence of Taylor's theorem, a general function can be well approximated by a quadratic function in a neighborhood of the minimum. Thirdly, the best minimization algorithms use quadratic approximations to the general function $h(\boldsymbol{\theta})$ as part of the iterative process.

13.2.3 Iterative Algorithms

Two noniterative algorithms for finding $\hat{\boldsymbol{\theta}}$ to minimize $h(\boldsymbol{\theta})$ are the grid search and the random search. The *grid search* consists simply of constructing a p-dimensional rectangular block of points to cover the region where $\hat{\boldsymbol{\theta}}$ is known to be, evaluating $h(\boldsymbol{\theta})$ at each of the points, and taking the smallest value as $h(\hat{\boldsymbol{\theta}})$. This can give $\hat{\boldsymbol{\theta}}$ to the precision of one "cell". Clearly the grid must be fine enough so that $h(\boldsymbol{\theta})$ is "fairly constant" within each cell. Halving the cell edge length leads to 2^p times as many function evaluations. A *random search* consists of randomly generating a sequence of points lying within a specified region, evaluating $h(\boldsymbol{\theta})$ at each point, and choosing the minimum. Both methods are extremely costly, since they make a large number of unnecessary function evaluations and there is no scope for learning from past values.

The algorithms which are of practical value are iterative. An initial guess $\boldsymbol{\theta}^{(1)}$ is provided, from which the algorithm generates a sequence of points $\boldsymbol{\theta}^{(2)}, \boldsymbol{\theta}^{(3)}, \ldots$ which are intended to converge to a local minimum $\hat{\boldsymbol{\theta}}$. Practically useful algorithms ensure that $h(\boldsymbol{\theta})$ is reduced at each iteration so that $h(\boldsymbol{\theta}^{(a+1)}) < h(\boldsymbol{\theta}^{(a)})$. A method which always satisfies this condition is called a *descent method*. The process is then terminated after a finite number of steps, using a stopping rule to decide when it is close enough to the minimum (Section 15.2.4).

For brevity we shall not concern ourselves very much with mathematical proofs of convergence. Most proofs assume conditions which are difficult to verify in practical applications and ignore, for example, the effects of roundoff error in computations. There are a number of direct search algorithms which have been found to work extremely well in practice but for which no mathematical convergence proofs have yet been produced.

a Convergence Rates

Although an algorithm may produce a convergent sequence $\{\boldsymbol{\theta}^{(a)}\}$, it may still be of little value if it converges too slowly. The concept of speed of convergence has been approached mathematically via the *asymptotic convergence rate*. This is the largest integer r such that

$$0 \leq \lim_{a \to \infty} \frac{\|\boldsymbol{\theta}^{(a+1)} - \hat{\boldsymbol{\theta}}\|}{\|\boldsymbol{\theta}^{(a)} - \hat{\boldsymbol{\theta}}\|^r} < \infty, \tag{13.12}$$

where $\|\cdot\|$ is any standard norm, usually the Euclidean norm, which we shall denote by $\|\cdot\|_2$. If the rate r is 1 and the limit above is strictly less than 1, the sequence is said to have a *linear* rate of convergence. If the limit is zero when $r = 1$, the algorithm is said to exhibit *superlinear* convergence. The best that is achieved in practice is *quadratic* convergence, when $r = 2$. In this case, after $\boldsymbol{\theta}^{(a)}$ gets close enough to $\hat{\boldsymbol{\theta}}$, the number of correct figures in $\boldsymbol{\theta}^{(a)}$ approximately doubles at each iteration. Unfortunately the term "quadratically convergent" is sometimes also applied to an algorithm which converges in a small finite number of steps to the minimum of a quadratic function, though this concept is perhaps better known as quadratic termination.

b Descent Methods

The iterative process can usually be thought of in terms of a point moving in $\mathbb{R}^p$ from $\boldsymbol{\theta}^{(1)}$ to $\boldsymbol{\theta}^{(2)}$ to $\boldsymbol{\theta}^{(3)}$ etc. along straight-line steps. The ath iteration consists of the computation of the ath step $\boldsymbol{\delta}^{(a)}$, whereupon $\boldsymbol{\theta}^{(a+1)}$ becomes available as

$$\boldsymbol{\theta}^{(a+1)} = \boldsymbol{\theta}^{(a)} + \boldsymbol{\delta}^{(a)}. \tag{13.13}$$

The ath step $\boldsymbol{\delta}^{(a)}$ is usually computed in two stages: first computing the direction for the step and then deciding how far to proceed in that direction. Thus

$$\boldsymbol{\delta}^{(a)} = \rho^{(a)}\mathbf{d}^{(a)}, \tag{13.14}$$

where the vector $\mathbf{d}^{(a)}$ is called the *direction* of the step and $\rho^{(a)}$ is the *step length.* Usually $\rho^{(a)}$ is chosen to approximately minimize $h(\boldsymbol{\theta})$ along the line $\boldsymbol{\theta}^{(a)} + \rho\mathbf{d}^{(a)}$, a process known as a *line search.* If $\rho^{(a)}$ is the exact minimum at each iteration, the algorithm is said to employ *exact line searches*; otherwise it employs *approximate* (*partial*) *line searches.* When exact line searches are not employed, it is important to ensure that $h(\boldsymbol{\theta})$ is sufficiently reduced at each iteration so that there is convergence to a local minimum (Gill et al. [1981: 100]). Otherwise the process might appear to converge simply by taking progressively smaller steps. As the object is to minimize $h(\boldsymbol{\theta})$, it is natural to move from $\boldsymbol{\theta}^{(a)}$ in a direction that, at least initially, leads downhill. A *descent direction* $\mathbf{d}$ is one for which

$$\mathbf{g}^{(a)\prime}\mathbf{d} = \frac{\partial}{\partial \rho} h(\boldsymbol{\theta}^{(a)} + \rho\mathbf{d})|_{\rho=0} < 0. \tag{13.15}$$

Using a Taylor expansion, we have

$$h(\boldsymbol{\theta}^{(a)} + \rho\mathbf{d}) \approx h(\boldsymbol{\theta}^{(a)}) + \rho\mathbf{g}^{(a)\prime}\mathbf{d} + O(\rho^2), \tag{13.16}$$

so that when (13.15) holds, a small enough step ρ in direction $\mathbf{d}$ must decrease $h(\boldsymbol{\theta})$. As previously noted, the line search should ensure that the reduction obtained is not too small.

The following important theorem suggests how descent directions might be calculated.

Theorem 13.1 A direction $\mathbf{d}$ is a descent direction at $\boldsymbol{\theta}$ if and only if there exists a positive definite matrix $\mathbf{R}$ such that

$$\mathbf{d} = -\mathbf{R}\mathbf{g}.$$

Proof: When $\mathbf{d} = -\mathbf{R}\mathbf{g}$, $\mathbf{g}'\mathbf{d} = -\mathbf{g}'\mathbf{R}\mathbf{g} < 0$ and (13.15) is satisfied. ■

Using the above theorem, (13.13) and (13.14) may be written as

$$\boldsymbol{\theta}^{(a+1)} = \boldsymbol{\theta}^{(a)} - \rho^{(a)}\mathbf{R}^{(a)}\mathbf{g}^{(a)}. \tag{13.17}$$

One might use the direction for which the downhill gradient at $\boldsymbol{\theta}^{(a)}$ is steepest, i.e. for which the decrease in h is a maximum for unit $\mathbf{d}$. From (13.16) we wish to maximize $(\mathbf{g}^{(a)\prime}\mathbf{d})^2/\mathbf{d}'\mathbf{d}$, and this leads to (Appendix A7.2) $\mathbf{d} \propto -\mathbf{g}^{(a)}$ [corresponding to $\mathbf{R} = \mathbf{I}_p$ in (13.17)]. The choice of $\mathbf{d} = -\mathbf{g}^{(a)}$ coupled with an exact line search at each iteration is called the *method of steepest descent*. However, the direction of steepest descent depends entirely upon the scaling of the variables $\boldsymbol{\theta}$. Figure 13.3 shows the behavior of the steepest-descent algorithm with exact line searches on the quadratic function $h(\theta_1, \theta_2) = \theta_1^2 + 40\theta_2^2$. This pattern is known as *hemstitching*. However, if we replace θ_2 by $\phi_2 = \sqrt{40}\theta_2$, then $f(\theta_1, \phi_2) = \theta_1^2 + \phi_2^2$. The contours of this function are circles, and the steepest-descent direction finds the minimum of $(0, 0)$ in a single step from any starting point.

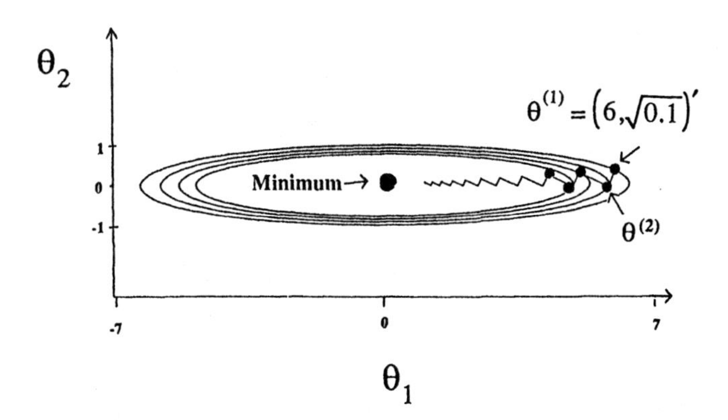

Figure 13.3 Progress of the steepest-descent algorithm minimizing $h(\theta_1, \theta_2) = \theta_1^2 + 40\theta_2^2$ from $\boldsymbol{\theta}^{(1)\prime} = (6, \sqrt{0.1})$. Contours of h are drawn at the levels $h(\boldsymbol{\theta}^{(1)})$, $h(\boldsymbol{\theta}^{(2)})$, $h(\boldsymbol{\theta}^{(3)})$, and $h(\boldsymbol{\theta}^{(4)})$.

For a quadratic function of the form (13.6) with $\mathbf{G}$ positive definite, $\mathbf{H}^{(a)} = \mathbf{H} = \mathbf{G}$, $\mathbf{g}^{(a)} = \mathbf{b} + \mathbf{G}\boldsymbol{\theta}^{(a)}$, and, from (13.9),

$$\begin{aligned}\hat{\boldsymbol{\theta}} &= -\mathbf{G}^{-1}\mathbf{b} \\ &= -\mathbf{G}^{-1}(\mathbf{g}^{(a)} - \mathbf{G}\boldsymbol{\theta}^{(a)}),\end{aligned}$$

or

$$\hat{\boldsymbol{\theta}} = \boldsymbol{\theta}^{(a)} - \mathbf{H}^{-1}\mathbf{g}^{(a)}. \tag{13.18}$$

For a general function $h(\boldsymbol{\theta})$, the step

$$\boldsymbol{\delta}^{(a)} = -\mathbf{H}^{(a)-1}\mathbf{g}^{(a)} \tag{13.19}$$

is called the *Newton step*, and the direction of the step, the *Newton direction*. This direction is one which, for a quadratic function, takes into account the "relative scaling" of the parameters and leads straight to the minimum. We shall note in Section 13.3 that the algorithm $\boldsymbol{\theta}^{(a+1)} = \boldsymbol{\theta}^{(a)} + \boldsymbol{\delta}^{(a)}$ with $\boldsymbol{\delta}^{(a)}$ given by (13.19) for general $h(\boldsymbol{\theta})$ is called *Newton's method*. When applied to a twice continuously differentiable function, Newton's method can be shown to be *locally convergent*, i.e., it converges provided $\boldsymbol{\theta}^{(1)}$ is sufficiently close to $\boldsymbol{\theta}$. The rate of convergence is quadratic (Dennis and Schnabel [1983: Theorem 5.2.1]).

The fastest and most robust algorithms for the minimization of a general smooth function employ the Newton direction $\mathbf{d}^{(a)} = -\mathbf{H}^{(a)-1}\mathbf{g}^{(a)}$, or an approximation to it, as the basic direction of search. In contrast to the direction of steepest descent, the Newton direction is scale-invariant. To see this, suppose we transform from parameters $\boldsymbol{\theta}$ to $\boldsymbol{\phi} = \mathbf{A}\boldsymbol{\theta}$, where $\mathbf{A}$ is a nonsingular $p \times p$ matrix, and let $w(\boldsymbol{\phi}) = h(\mathbf{A}^{-1}\boldsymbol{\phi}) = h(\boldsymbol{\theta})$. In terms of the $\boldsymbol{\phi}$ parameters, with $\boldsymbol{\phi}^{(a)} = \mathbf{A}\boldsymbol{\theta}^{(a)}$,

$$\begin{aligned}\mathbf{d}_{\phi}^{(a)} &= -\left[\left(\frac{\partial^2 w(\boldsymbol{\phi})}{\partial\boldsymbol{\phi}\,\partial\boldsymbol{\phi}'}\right)^{-1}\frac{\partial w(\boldsymbol{\phi})}{\partial\boldsymbol{\phi}}\right]_{\boldsymbol{\phi}^{(a)}} \\ &= -\left[\left\{(\mathbf{A}^{-1})'\mathbf{H}(\boldsymbol{\theta})\mathbf{A}^{-1}\right\}^{-1}(\mathbf{A}^{-1})'\frac{\partial h(\boldsymbol{\theta})}{\partial\boldsymbol{\theta}}\right]_{\boldsymbol{\theta}^{(a)}} \qquad \text{(by Appendix A10.4)} \\ &= -\mathbf{A}[\mathbf{H}^{-1}(\boldsymbol{\theta})\mathbf{g}(\boldsymbol{\theta})]_{\boldsymbol{\theta}^{(a)}},\end{aligned}$$

so that

$$\mathbf{A}^{-1}\mathbf{d}_{\phi}^{(a)} = -\mathbf{H}^{(a)-1}\mathbf{g}^{(a)} = \mathbf{d}_{\theta}^{(a)}.$$

Hence the search direction in the $\boldsymbol{\theta}$ parameter space is the same whether we calculate it directly or transform the Newton direction obtained under the $\boldsymbol{\phi}$ parametrization. However, this is a theoretical result applying to exact arithmetic. It is important, in practice, to watch out for the relative scaling of the parameters, as bad scaling leads to ill-conditioned Hessians, and the Newton direction may not be computed accurately enough; e.g., it may not even be a descent direction. Gill et al. [1981: Sections 7.5.1, 8.7] discuss the difficulties caused by bad scaling in general optimization problems and suggest remedies.

Fletcher [1980: Theorem 2.4.1] proved a result that gives some guiding principles for the construction of minimization algorithms using partial (approximate) line searches. Let $\mathbf{g}(\boldsymbol{\theta})$ be uniformly continuous on $\{\boldsymbol{\theta}: h(\boldsymbol{\theta}) < h(\boldsymbol{\theta})^{(1)}\}$. Suppose the following two conditions hold:

(i) for some η, $0 < \eta < \frac{1}{2}$, we have

$$\eta |\mathbf{g}^{(a)\prime}\boldsymbol{\delta}^{(a)}| \le h^{(a)} - h^{(a+1)} \le (1-\eta)|\mathbf{g}^{(a)\prime}\boldsymbol{\delta}^{(a)}|; \tag{13.20}$$

(ii) if $\xi^{(a)}$ is the angle between $\boldsymbol{\delta}^{(a)}$ and $-\mathbf{g}^{(a)}$, i.e.,

$$\cos \xi^{(a)} = \frac{-\mathbf{g}^{(a)\prime}\boldsymbol{\delta}^{(a)}}{\|\mathbf{g}^{(a)}\|_2 \|\boldsymbol{\delta}^{(a)}\|_2},$$

where $\|\cdot\|_2$ is the Euclidean norm, then $\xi^{(a)} \le \pi/2 - \mu$ for some $\mu > 0$ and all a.

Under these conditions we find that either $\mathbf{g}^{(a)} = \mathbf{0}$ for some $\mathbf{a}$, $\mathbf{g}^{(a)} \to \mathbf{0}$ (convergence to a stationary point), or $h^{(a)} \to -\infty$. Condition (ii) implies that the search direction must be bounded away from an angle of 90° with the steepest-descent direction $-\mathbf{g}^{(a)}$ or, equivalently, be bounded away from becoming tangential to the contour at $\boldsymbol{\theta}^{(a)}$. This can be achieved by choosing $\mathbf{d}^{(a)} = -\mathbf{R}^{(a)}\mathbf{g}^{(a)}$, where the eigenvalues of the positive definite matrix $\mathbf{R}^{(a)}$ are bounded away from 0 and ∞. The left-hand side of condition (i) ensures a "sufficient decrease" in $h(\boldsymbol{\theta})$, while the right-hand side prevents the step lengths from becoming too large.

In conclusion we note that the result above is a global convergence result. Whether or not the algorithm converges does not depend on the location of the starting value $\boldsymbol{\theta}^{(1)}$.

c Line Searches

The basic problem is to find the size $\rho^{(a)}$ of the step to be taken along the direction $\mathbf{d}^{(a)}$ by choosing $\rho^{(a)}$ $(= \hat{\rho}$ say) to "minimize" the one-dimensional function

$$h(\rho) = h(\boldsymbol{\theta}^{(a)} + \rho \mathbf{d}^{(a)}). \tag{13.21}$$

Clearly an inefficient line-search method will slow down the entire minimization algorithm, so that efficiency is important. The extent of minimization required is also important. Many algorithms only require that h be sufficiently decreased and not that an accurate approximation to the minimum be obtained.

Although accurate line searches usually lead to a decrease in the number of iterations required by the minimization algorithm, the increased work involved in minimizing $h(\rho)$ may be inefficient. Therefore, in what follows it should be understood that as soon as a value of ρ is found that satisfies the conditions for a sufficient decrease in h, e.g. (13.20), the search will stop. The best line-search methods, sometimes called safeguarded polynomial interpolation methods, combine two distinct elements, *sectioning* or *bracketing*, and *polynomial interpolation.*

A sectioning algorithm will begin with an interval (ρ_1, ρ_2) known to contain, or bracket, the minimum $\hat{\rho}$, and steadily decrease the length of the bracketing interval until the accuracy required for the estimate of $\hat{\rho}$ is obtained. When the assumptions governing this algorithm are correct, the method tends to be reliable but fairly slow. If first derivatives are available so that $\dot{h}(\rho)$ $[= dh/d\rho]$ can be computed from

$$\dot{h}(\rho) = \mathbf{g}(\boldsymbol{\theta}^{(a)} + \rho \mathbf{d}^{(a)})' \mathbf{d}^{(a)}, \tag{13.22}$$

the sectioning can be based upon the well-known method of bisection for solving $\dot{h}(\rho) = 0$. The length of the bracketing interval is halved at each iteration. When only function values are to be used, sectioning methods again begin with two points (ρ_1, ρ_2) bracketing $\hat{\rho}$ but assume that $h(\rho)$ is unimodal on this interval. By placing two points c and d inside the interval i.e. $\rho_1 < c < d < \rho_2$, (ρ_1, d) must bracket $\hat{\rho}$ if $h(c) < h(d)$ and (c, ρ_2) must bracket $\hat{\rho}$ if $h(c) > h(d)$, thus giving a new bracketing interval of reduced length. The *golden-section* search locates c and d at proportions $1 - \tau$ and τ respectively along the interval (ρ_1, ρ_2), where $\tau = 2/(1 + \sqrt{5}) \approx 0.62$. This choice ensures that one of the intermediate points for the current interval is in the correct position for the next interval, so that only one new function evaluation is required at each iteration (Gill et al. [1981: p. 90]). The length of the bracketing interval is reduced by approximately one-third at each iteration.

Polynomial interpolation methods tend to be much faster than sectioning (function-comparison) methods when they work. However, they are not as reliable. Several, usually two or three, approximations to the minimum are maintained at each iteration. A quadratic or cubic approximation to $h(\rho)$ is formed that agrees with $h(\rho)$ at these points, and with the derivatives at one or more of them if derivatives are being used. Then the minimum of the approximating polynomial is taken as a new approximation to $\hat{\rho}$, and one of the old points is discarded. Algorithms using only function evaluations usually use quadratic interpolation, whereas if derivatives are employed, either quadratic or cubic interpolation may be used. The choice of interpolation formula can depend upon the relative expense of computing $h(\rho)$ and $\dot{h}(\rho)$.

Safeguarded polynomial interpolation methods try to combine the speed of polynomial interpolation with the safety of sectioning. There are many ways of combining these algorithms, and there is no general agreement on the optimal way to do it. One can ensure that the new point produced by the interpolation method does not fall outside the bracketing interval, and then the new point can be used to update the bracket. If this does not sufficiently reduce the length of the bracket, a sectioning step can be taken instead. Care must also be taken so that successive approximations to $\hat{\rho}$ do not become too close together, or rounding errors in the computation of differences between function values can cause problems. More detailed comments and references are available from Fletcher [1980] and Gill et al. [1981]. In his discussion of this topic, Scales [1985: Chapter 2] includes detailed step-by-step algorithms.

13.3 SECOND-DERIVATIVE (MODIFIED NEWTON) METHODS

13.3.1 Step-Length Methods

Modified Newton methods make use of both the gradient vector and the Hessian matrix. However, if one or both of these are unavailable, the same methods with finite-difference approximations used for the derivatives can still perform well (see Sections 13.4.1 and 13.5.1). It is generally accepted by specialists in optimization that when second derivatives are available, a good modified Newton algorithm is the most robust and reliable method for minimizing a general smooth function. Usually it will take fewer iterations to reach minimum. This is contrary to the perceptions of many statisticians whose ideas have been influenced by accounts of the failings of the classical Newton (or Newton–Raphson method). The step-length philosophy behind the modifications discussed in this section consists of modifying the Hessian, if necessary, to find a descending search direction along which a step is taken. The length of the step is determined by a line search.

The classical Newton method for a general function is based on taking a second-order truncated Taylor series approximation

$$\begin{aligned} h(\boldsymbol{\theta}^{(a)} + \boldsymbol{\delta}) &\approx h(\boldsymbol{\theta}^{(a)}) + \mathbf{g}^{(a)\prime}\boldsymbol{\delta} + \tfrac{1}{2}\boldsymbol{\delta}'\mathbf{H}^{(a)}\boldsymbol{\delta} \\ &= q_h(\boldsymbol{\delta}), \end{aligned} \tag{13.23}$$

say, and choosing the ath step to minimize $q_h(\boldsymbol{\delta})$. By (13.9) the minimum occurs at

$$\boldsymbol{\delta}^{(a)} = -\mathbf{H}^{(a)-1}\mathbf{g}^{(a)} \tag{13.24}$$

when $\mathbf{H}^{(a)}$ is positive definite. This is the Newton step once again, and the resulting method is called the *Newton method*. Statisticians tend to call it the Newton–Raphson method, particularly when it is used to optimize the log-likelihood function. As we have previously noted, if the starting value $\boldsymbol{\theta}^{(1)}$ is sufficiently close to a local minimum $\hat{\boldsymbol{\theta}}$, the algorithm will converge at a quadratic rate. Unmodified, however, the algorithm is not a good general-purpose minimization algorithm, for the following two reasons. Firstly, even if $\mathbf{H}^{(a)}$ is positive definite, so that the step is taken in a descent direction, the quadratic approximation may not be adequate at a distance $\|\boldsymbol{\delta}^{(a)}\|$ from $\boldsymbol{\theta}^{(a)}$. Thus the Newton step may be too long and may even increase $h(\boldsymbol{\theta})$. This can be remedied by simply including an adjustable step length $\rho^{(a)}$ in the algorithm rather than using the direct Newton step with $\rho^{(a)} = 1$. A common strategy is to try $\rho^{(a)} = 1$ and then "backtrack" or progressively reduce the step until an acceptable step has been formed [e.g. acceptable in the sense of [13.20)].

A second reason for failure with the classical Newton method is that although the Hessian is positive definite at a local minimum in most applications, it need not be positive definite at each iteration. Hence the method does not ensure descent directions. If negative definite Hessians are encountered, the method may converge to a local maximum. Indeed, $\mathbf{H}^{(a)}$ may be indefinite (eigenvalues of both signs) or even singular.

A modified Newton method performs the *a*th iteration as

$$\boldsymbol{\theta}^{(a+1)} = \boldsymbol{\theta}^{(a)} - \rho^{(a)}\bar{\mathbf{H}}^{(a)-1}\mathbf{g}^{(a)}, \tag{13.25}$$

where $\bar{\mathbf{H}}^{(a)}$ is a positive definite matrix which is $\mathbf{H}^{(a)}$ itself or is formed by modifying $\mathbf{H}^{(a)}$ in some way. To prevent the algorithm crisscrossing a ridge in a hemstitching pattern (as steepest descent did in Fig. 13.3) and thus making slow progress, it is desirable that $\bar{\mathbf{H}}^{(a)}$ make use of the information contained in $\mathbf{H}^{(a)}$ on the relative scaling of the parameters. By using the unmodified Hessian whenever it is positive definite, the modified algorithm reverts to finding Newton directions as the minimum is approached, thus taking advantage of the Newton method's quadratic convergence.

a Directional Discrimination

An early approach to modifying the Hessian, which illustrates a number of the issues involved, is called *directional discrimination* by Bard [1974]. An example is the method of Greenstadt [1967]. Directional discrimination methods involve using a spectral decomposition of the Hessian, namely (Appendix A2.2)

$$\mathbf{H} = \mathbf{T}\boldsymbol{\Lambda}\mathbf{T}', \tag{13.26}$$

where $\boldsymbol{\Lambda} = \mathrm{diag}(\lambda_1, \lambda_2, \ldots, \lambda_p)$, the λ_i are the eigenvalues, and $\mathbf{T}$ is an orthogonal matrix with ith column $\mathbf{t}_i$ an eigenvector corresponding to λ_i. We can also write $\mathbf{H} = \sum_i \lambda_i \mathbf{t}_i \mathbf{t}_i'$. Now the quadratic approximation to $h(\boldsymbol{\theta} + \boldsymbol{\delta})$ about $\boldsymbol{\theta}$ is

$$q_h(\boldsymbol{\delta}) = h(\boldsymbol{\theta}) + \mathbf{g}'\boldsymbol{\delta} + \tfrac{1}{2}\boldsymbol{\delta}'\mathbf{H}\boldsymbol{\delta}. \tag{13.27}$$

However, if we parametrize using $\boldsymbol{\phi} = \mathbf{T}'\boldsymbol{\delta}$, we obtain

$$q_h(\boldsymbol{\phi}) = h(\boldsymbol{\theta}) + \sum_{i=1}^{p} [(\mathbf{g}'\mathbf{t}_i)\phi_i + \tfrac{1}{2}\lambda_i\phi_i^2],$$

which can be minimized by minimizing with respect to each variable ϕ_i

separately. The Newton step is

$$\begin{aligned}\boldsymbol{\delta}_{\text{Newton}} &= -\mathbf{H}^{-1}\mathbf{g} \\ &= -\mathbf{T}\boldsymbol{\Lambda}^{-1}\mathbf{T}'\mathbf{g} \\ &= -\sum_{i=1}^{p} \lambda_i^{-1}(\mathbf{g}'\mathbf{t}_i)\mathbf{t}_i,\end{aligned}$$

and this minimizes $q_h(\boldsymbol{\delta})$ for positive definite $\mathbf{H}$. For any $\boldsymbol{\phi}$,

$$\boldsymbol{\delta} = \mathbf{T}\boldsymbol{\phi} = \sum_{i=1}^{p} \phi_i \mathbf{t}_i,$$

so that the component of $\boldsymbol{\delta}$ along the direction $\mathbf{t}_i$ has magnitude $|\phi_i|$. For general $\mathbf{H}$, $\mathbf{t}_i$ is a direction of positive curvature if $\lambda_i > 0$ (see Fig. 13.4*a*) and $q_h(\boldsymbol{\phi})$ is minimized by taking $\phi_i = -\lambda_i^{-1}(\mathbf{g}'\mathbf{t}_i)$. If $\lambda_i < 0$, $\mathbf{t}_i$ is said to be a direction of negative curvature, which is defined as $\mathbf{t}_i'\mathbf{H}\mathbf{t}_i\,(=\mathbf{t}_i'\lambda_i\mathbf{t}_i = \lambda_i) < 0$; see Fig. 13.4*b*. The Newton step simply solves $\partial q_h(\boldsymbol{\phi})/\partial\phi_i = 0$. However, for $\lambda_i < 0$, q_h is a negative quadratic in ϕ_i, so that ϕ_i satisfying $\partial q_h/\partial\phi_i = 0$ maximizes $q_h(\phi_i)$. In other words, instead of trying to minimize h, the component of $\boldsymbol{\delta}_{\text{Newton}}$ along the direction $\mathbf{t}_i$ attempts to find the maximum of h at the top of the ridge (Fig. 13.4*b*). Greenstadt's suggestion was simply to reverse the direction of this component and take $\phi_i = -|\lambda_i|^{-1}\mathbf{g}'\mathbf{t}_i$. The length of the component has not been changed, since even though $q_h(\boldsymbol{\delta})$ can be reduced indefinitely along $\mathbf{t}_i$, too long a step could be dangerous. The quadratic approximation is unlikely to be close to $h(\boldsymbol{\theta} + \boldsymbol{\delta})$ as we get too far from $\boldsymbol{\theta}$. Very small eigenvalues make the corresponding component of the step very long, so that similar considerations apply. Finally, zero eigenvalues have to be coped with. For example, Greenstadt [1967] proposed using $\bar{\mathbf{H}} = \mathbf{TBT}'$, where $\mathbf{B}$ is diagonal with its ith diagonal element being the maximum of $|\lambda_i|$ and η, for some small $\eta > 0$. Bard [1974] refers to several papers with similar approaches.

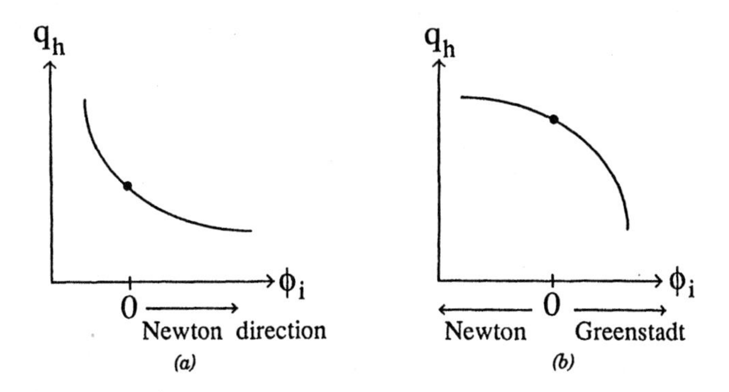

Figure 13.4 Direction of Newton step: (*a*) positive curvature ($\lambda_i > 0$); (*b*) negative curvature ($\lambda_i < 0$).

Unfortunately, the overheads involved in computing the spectral decomposition make such algorithms inefficient. More efficient implementations of these methods compute the Newton direction using an $\mathbf{L}_1\mathbf{D}\mathbf{L}_1'$ Cholesky factorization of the Hessian, where $\mathbf{L}_1$ is lower triangular (Appendix A8.2). The spectral decomposition is used only if a negative pivot is encountered during the factorization. However, these methods are still not competitive with some of the methods to follow.

b Adding to the Hessian

Another way of modifying the Hessian involves adding a positive definite matrix $\mathbf{A}$ to $\mathbf{H}$, namely

$$\bar{\mathbf{H}} = \mathbf{H} + \mathbf{A}, \tag{13.28}$$

where $\mathbf{A}$ is large enough for $\bar{\mathbf{H}}$ also to be positive definite. This idea was used by Levenberg [1944] and Marquardt [1963] in least-squares problems. The first application to the Newton method was by Goldfeld et al. [1966], who replaced $\mathbf{H}^{(a)}$ by

$$\bar{\mathbf{H}}^{(a)} = \mathbf{H}^{(a)} + \nu^{(a)}\mathbf{I}_p, \tag{13.29}$$

where $\nu^{(a)} > 0$. Depending on the size of $\nu^{(a)}$, the method behaves more like the Newton method ($\nu^{(a)} \to 0$) or steepest descent ($\nu^{(a)} \to \infty$). It is desirable that $\nu^{(a)}$ become zero as the minimum is approached to take advantage of the fast convergence properties of the Newton method.

To choose $\nu^{(a)}$, Goldfeld et al. [1966] compute an eigenvalue decomposition of $\mathbf{H}$ and choose $\nu^{(a)}$ a little larger than the magnitude of the most negative eigenvalue. Modern methods use a matrix factorization such as an $\mathbf{L}_1\mathbf{D}\mathbf{L}_1'$ Cholesky decomposition (Appendix A8.2), which is useful for solving the linear equations $\bar{\mathbf{H}}\boldsymbol{\delta} = -\mathbf{g}$ and which reveals, during computation, whether or not a matrix is positive definite. However, this remark needs qualification. Because of roundoff errors in computation, it is virtually impossible to know whether or not a badly conditioned matrix with one or more very small eigenvalues is positive definite. Consequently, ill-conditioned matrices are also modified. Since it is desirable for $\nu^{(a)}$ to be kept as small as possible, several trial values may be needed on any particular iteration.

The most commonly used modified Newton step-length method is that of Murray [1972b], developed in several papers with P. E. Gill. This algorithm is implemented in the NAG library (see Section 15.1.2) and is based upon a modified Cholesky decomposition of $\mathbf{H}$. Gill and Murray modify a standard algorithm for Cholesky factorization to produce an $\mathbf{L}_1\mathbf{D}\mathbf{L}_1'$ decomposition of $\mathbf{H}$ (if $\mathbf{H}$ is "sufficiently" positive definite), or of $\bar{\mathbf{H}} = \mathbf{H} + \mathbf{E}$, where $\mathbf{E}$ is diagonal with positive diagonal elements (otherwise). The factorization is computed in a single pass and thus avoids the additional overheads of needing several trial $\mathbf{E}$-matrices.

It is also arranged so that each element in $\mathbf{L}_1$ is bounded by a positive number β chosen to reconcile three criteria: (i) the procedure should be numerically stable, (ii) it should not modify sufficiently positive definite matrices, and (iii) when modification is necessary, it should produce a correction matrix $\mathbf{E}$ which is as small as possible. Fletcher [1980: 37] argues that Gill and Murray's approach does not adequately handle situations in which $\boldsymbol{\theta}^{(a)}$ is close to a saddle point (i.e. $\mathbf{g}^{(a)}$ small, $\mathbf{H}^{(a)}$ indefinite) and in his Chapter 5 produces numerical examples where their method makes slow progress in the vicinity of a saddle point. However, Gill et al. [1981: 107] overcome this by searching along a direction of negative curvature in the vicinity of a saddle point. Such a direction can be computed easily from the modified Cholesky decomposition.

A direction of negative curvature is a descent direction $\mathbf{d}$ in which $\mathbf{d}'\mathbf{H}\mathbf{d} < 0$, so that the downhill gradient at $\boldsymbol{\theta}^{(a)}$ in the direction $\mathbf{d}$ is becoming steeper. Such a direction is clearly an attractive search direction. Scales [1985: Sections 3.3.4, 3.3.5] gives fairly detailed algorithms for the Gill–Murray approaches. Fletcher and Freeman [1977], however, found that it is not satisfactory to use directions of negative curvative on successive iterations. For references to other algorithms for computing directions of negative curvature see Gill et al. [1981: Section 4.4]. They describe or reference work in which other stable algorithms for factorizing a symmetric indefinite matrix have been used as the basis for a modified Newton algorithm.

13.3.2 Restricted Step Methods

Restricted step methods, also known as *trust-region methods*, approach the modification of the Newton step from a very different viewpoint from that of step-length methods. They are closer in spirit to the work of Levenberg [1944], Marquardt [1963], and Goldfeld et al. [1966]. The quadratic approximation $q_h^{(a)}(\boldsymbol{\delta})$ to $h(\boldsymbol{\theta}^{(a)} + \boldsymbol{\delta})$ [c.f. (13.27)] is regarded as being adequate for practical purposes only in a neighborhood of $\boldsymbol{\delta} = \mathbf{0}$ called a *trust region*. Therefore, instead of minimizing $q_h^{(a)}(\boldsymbol{\delta})$ globally as the Newton method does, a restricted step method chooses the ath step $\boldsymbol{\delta}^{(a)}$ to minimize $q_h^{(a)}(\boldsymbol{\delta})$ over the trust region. This can be done even when $\mathbf{H}^{(a)}$ is indefinite or singular. In most practical applications a trust region of the form

$$\{\boldsymbol{\delta} : \|\boldsymbol{\delta}\|_2 \le k^{(a)}\}, \tag{13.30}$$

for some $k^{(a)} > 0$, is used. The trust region is expanded, maintained, or contracted for the $(a+1)$th iteration depending on how well $q_h^{(a)}(\boldsymbol{\delta}^{(a)})$ approximates $h(\boldsymbol{\theta}^{(a)} + \boldsymbol{\delta}^{(a)})$. A typical modern implementation of a restricted step method employs the following sequence of steps (adapted from Fletcher [1980: 78]) during the ath iteration:

1. Find $\boldsymbol{\delta}^{(a)}$ to minimize $q_h^{(a)}(\boldsymbol{\delta})$ subject to $\|\boldsymbol{\delta}\|_2 \le k^{(a)}$.

2. Compute

$$r_a = \frac{h(\theta^{(a)}) - h(\theta^{(a)} + \delta^{(a)})}{h(\theta^{(a)}) - q_h^{(a)}(\delta^{(a)})}.$$

3. If $r_a \leq 0$, reduce $k^{(a)}$ and return to step 1; otherwise
   a. if r_a is small (e.g. $r_a \leq 0.25$), choose some $k^{(a+1)} < k^{(a)}$;
   b. if r_a is moderate (e.g. $0.25 \leq r_a \leq 0.75$), retain $k^{(a+1)} = k^{(a)}$;
   c. if r_a is sufficiently large (e.g. $r_a > 0.75$), choose some $k^{(a+1)} > k^{(a)}$, thus enlarging the trust region.
4. $\theta^{(a+1)} = \theta^{(a)} + \delta^{(a)}$.

A number of rules of varying levels of sophistication are employed for expanding or contracting the trust region. Eventually, as a local minimum with positive definite Hessian is approached, the Newton step is always taken so that quadratic convergence is retained.

Step 1 also needs further comment. If $\mathbf{H}^{(a)}$ is positive definite, then the unconstrained minimum of $q_h^{(a)}(\delta)$, which gives the Newton step, is also the constrained minimum if $\| -\mathbf{H}^{(a)-1}\mathbf{g}^{(a)} \|_2 \leq k^{(a)}$. Otherwise, if there is no local minimum of $q_h^{(a)}$ in $\| \delta \|_2 \leq k^{(a)}$, the minimizer $\delta^{(a)}$ is found on the boundary of $\| \delta \|_2 = k^{(a)}$ regardless of $\mathbf{H}^{(a)}$. It can be shown by introducing a Lagrange multiplier (Fletcher [1980: Section 5.2]) that, in almost every situation, the minimizer is of the form

$$\delta(\nu) = -(\mathbf{H}^{(a)} + \nu \mathbf{I}_n)^{-1}\mathbf{g}^{(a)} \tag{13.31}$$

for some $\nu > \max(0, -\lambda_{(n)})$, where $\lambda_{(n)}$ is the smallest eigenvalue of $\mathbf{H}^{(a)}$. As ν varies from $-\lambda_{(n)}$ to ∞, $\| \delta(\nu) \|_2$ decreases monotonically from ∞ to zero. Therefore $\delta^{(a)}$ is obtained as $\delta^{(a)} = \delta(\nu^{(a)})$, where $\nu^{(a)}$ (approximately) solves $\| \delta(\nu) \| = k^{(a)}$. A detailed and efficient algorithm for solving this nonlinear equation is given by Dennis and Schnabel [1983]. The exceptional case occurs when $\mathbf{g}^{(a)}$ is orthogonal to the subspace generated by the eigenvectors of $\mathbf{H}^{(a)}$ corresponding to $\lambda_{(n)}$, and $k^{(a)}$ exceeds a certain threshold. A suggestion for dealing with this situation is given by Fletcher [1980: 85], who goes on to state that this aspect of the problem has not been adequately researched. Early restricted step algorithms tended to control $\nu^{(a)}$ itself rather than the radius of the trust region $k^{(a)}$.

Largely because of their association with the classical Newton method, the modified Newton methods have often been thought to perform well only close to a local minimum. Consequently, "hybrid" algorithms have been written which use steepest descent until the iterates approach the minimum. However, we find that well-implemented modified Newton methods are the most reliable methods for minimizing a general function possessing continuous second derivatives. Examples exist where steepest descent fails to make progress at a point far from the minimum whereas a modified Newton method converges quite satisfactorily (see the Rosenbrock function in Gill et al. [1981]). Modified Newton methods tend to have fewer iterations than their competitors and have very fast local convergence. Also, by using the curvature information, they are able to detect and move away from saddle points. However, the calculation of second derivations may be expensive, making each iteration very time-consuming. With complicated functions, both the differentiation and the programming of the resulting formulae require considerable effort from the user, with the increased risk of making mistakes. For the remainder of this chapter we consider methods that use only function and gradient evaluations.

13.4 FIRST-DERIVATIVE METHODS

13.4.1 Discrete Newton Methods

Modified Newton methods can be successfully implemented using finite differences of the gradient to approximate the Hessian matrix. The resulting methods are called *discrete Newton methods.* An increment b_i is taken along the ith unit vector $\mathbf{e}_i$ (with ith element unity and the remaining elements zero), and the ith column of $\mathbf{H}^{(a)}$ is estimated by

$$\frac{1}{b_i}[\mathbf{g}(\boldsymbol{\theta}^{(a)} + b_i\mathbf{e}_i) - \mathbf{g}(\boldsymbol{\theta}^{(a)})]. \tag{13.32}$$

To approximate the derivative of the ith column, especially as $\mathbf{g}(\boldsymbol{\theta}^{(a)}) \to \mathbf{0}$, a small value of b_i is necessary. However, if b_i is too small, too many significant figures will be lost in computing the difference in (13.32) using finite arithmetic. Fortunately, a reasonable range of b_i-values give good results in practice. Details of computing the finite differences and strategies for choosing b_i are given by Gill et al. [1981: Section 8.6]; see also Dennis and Schnabel [1983: Algorithm A5.6.1, p. 320].

Discrete Newton methods retain the advantages of modified Newton methods to a large extent. However, they require p additional gradient vector evaluations to approximate the Hessian, and still require a modification of the Hessian when it is not positive definite, together with the solution of linear equations to obtain $\boldsymbol{\delta}^{(a)}$. Therefore, unless a known pattern of sparseness in the Hessian can be exploited, discrete Newton methods tend to be inefficient compared with the following quasi-Newton methods when p is larger than about 10.

13.4.2 Quasi-Newton Methods

Rather than calculate an approximation to the Hessian at the point $\boldsymbol{\theta}^{(a)}$, quasi-Newton methods, otherwise known as *variable-metric* or *secant* methods, build up an approximation to the Hessian gradually as the iterations proceed. A current approximation $\mathbf{B}^{(a)}$ of $\mathbf{H}^{(a)}$ is maintained and used to define the direction $\mathbf{d}^{(a)}$ of the *a*th step by solving

$$\mathbf{B}^{(a)}\mathbf{d}^{(a)} = -\mathbf{g}^{(a)}. \tag{13.33}$$

The actual step $\boldsymbol{\delta}^{(a)} = \rho^{(a)}\mathbf{d}^{(a)}$ is obtained after a line search along $\mathbf{d}^{(a)}$ to find $\rho^{(a)}$.

A Taylor expansion of $\mathbf{g}(\boldsymbol{\theta}^{(a)} + \boldsymbol{\delta}^{(a)})$ about $\boldsymbol{\theta}^{(a)}$ yields

$$\mathbf{g}(\boldsymbol{\theta}^{(a)} + \boldsymbol{\delta}^{(a)}) \approx \mathbf{g}(\boldsymbol{\theta}^{(a)}) + \mathbf{H}^{(a)}\boldsymbol{\delta}^{(a)},$$

so that

$$\begin{aligned}\mathbf{H}^{(a)}\boldsymbol{\delta}^{(a)} &\approx \mathbf{g}(\boldsymbol{\theta}^{(a)} + \boldsymbol{\delta}^{(a)}) - \mathbf{g}(\boldsymbol{\theta}^{(a)}) \\ &= \mathbf{g}^{(a+1)} - \mathbf{g}^{(a)},\end{aligned} \tag{13.34}$$

where strict equality holds in (13.34) if $h(\boldsymbol{\theta})$ is quadratic [by (13.7)]. Ideally we would like to update the approximations so that (13.34) is satisfied with $\mathbf{H}^{(a)}$ replaced by $\mathbf{B}^{(a)}$. However, this is not possible, since $\mathbf{B}^{(a)}$ is used to determine $\boldsymbol{\delta}^{(a)}$. Instead $\mathbf{B}^{(a+1)}$ is chosen to satisfy

$$\mathbf{B}^{(a+1)}\boldsymbol{\delta}^{(a)} = \mathbf{g}^{(a+1)} - \mathbf{g}^{(a)}, \tag{13.35}$$

which is called the *quasi-Newton condition.* Since an approximation to the Hessian is being built up, it is desirable to preserve the good properties of $\mathbf{B}^{(a)}$. This is done by forming

$$\mathbf{B}^{(a+1)} = \mathbf{B}^{(a)} + \mathbf{E}^{(a)}, \tag{13.36}$$

where $\mathbf{E}^{(a)}$ is a "small" update matrix, usually of rank 1 or 2, such that (13.35) holds. Iteration usually begins with $\mathbf{B}^{(1)} = \mathbf{I}_p$, so that the first step is the steepest-descent step $\mathbf{d}^{(1)} = -\mathbf{g}^{(1)}$.

The simplest systematic correction $\mathbf{E}^{(a)}$ is a rank-one update of the form $\mathbf{E}^{(a)} = \beta\mathbf{r}\mathbf{r}'$, where, to satisfy (13.35),

$$\mathbf{E}^{(a)} = \frac{(\boldsymbol{\gamma}^{(a)} - \mathbf{B}^{(a)}\boldsymbol{\delta}^{(a)})(\boldsymbol{\gamma}^{(a)} - \mathbf{B}^{(a)}\boldsymbol{\delta}^{(a)})'}{(\boldsymbol{\gamma}^{(a)} - \mathbf{B}^{(a)}\boldsymbol{\delta}^{(a)})'\boldsymbol{\delta}^{(a)}} \tag{13.37}$$

with

$$\boldsymbol{\gamma}^{(a)} = \mathbf{g}^{(a+1)} - \mathbf{g}^{(a)}. \tag{13.38}$$

For a quadratic function with nonsingular Hessian $\mathbf{H}$, the exact Hessian is built up after at most p steps provided $\mathbf{B}^{(a)}$ does not become singular at any stage. The rank-one update was briefly popular in the later 1960's. It gave the minimum of a quadratic in at most $p+1$ steps without the necessity for exact line searches, which are required to ensure this result for existing rank-two methods. However, $\mathbf{B}^{(a)}$ may lose positive definiteness and stop generating descent directions. As a consequence of this, and because of the development of rank-two methods without exact line searches, the rank-one update has been virtually abandoned.

Most of the better-known quasi-Newton algorithms use updates belonging to the Broyden single-parameter rank-2 family (Broyden [1967]), for which

$$\mathbf{E}^{(a)} = -\frac{\mathbf{B}^{(a)}\boldsymbol{\delta}^{(a)}\boldsymbol{\delta}^{(a)\prime}\mathbf{B}^{(a)}}{\boldsymbol{\delta}^{(a)\prime}\mathbf{B}^{(a)}\boldsymbol{\delta}^{(a)}} + \frac{\boldsymbol{\gamma}^{(a)}\boldsymbol{\gamma}^{(a)\prime}}{\boldsymbol{\gamma}^{(a)\prime}\boldsymbol{\delta}^{(a)}} + \beta^{(a)}\mathbf{r}^{(a)}\mathbf{r}^{(a)\prime}, \tag{13.39}$$

where

$$\mathbf{r}^{(a)} = (\boldsymbol{\delta}^{(a)\prime}\mathbf{B}^{(a)}\boldsymbol{\delta}^{(a)})^{1/2}\left(\frac{\boldsymbol{\gamma}^{(a)}}{\boldsymbol{\gamma}^{(a)\prime}\boldsymbol{\delta}^{(a)}} - \frac{\mathbf{B}^{(a)}\boldsymbol{\delta}^{(a)}}{\boldsymbol{\delta}^{(a)\prime}\mathbf{B}^{(a)}\boldsymbol{\delta}^{(a)}}\right), \tag{13.40}$$

$\beta^{(a)}$ is an arbitrary real parameter, and $\boldsymbol{\gamma}^{(a)}$ is given by (13.38). The first quasi-Newton algorithm, namely the Davidon–Fletcher–Powell (DFP) algorithm due to Davidon [1959] and improved by Fletcher and Powell [1963], uses $\beta^{(a)} = 1$. If exact line searches are used, $\mathbf{B}^{(a)}$ remains positive definite as iterations proceed provided

(i) $\boldsymbol{\delta}^{(a)\prime}\boldsymbol{\gamma}^{(a)} > 0$ and

(ii) $$\beta^{(a)} > \frac{-(\boldsymbol{\delta}^{(a)\prime}\boldsymbol{\gamma}^{(a)})^2}{(\boldsymbol{\delta}^{(a)\prime}\mathbf{B}^{(a)}\boldsymbol{\delta}^{(a)})(\boldsymbol{\gamma}^{(a)\prime}\mathbf{B}^{(a)-1}\boldsymbol{\gamma}^{(a)}) - (\boldsymbol{\delta}^{(a)\prime}\boldsymbol{\gamma}^{(a)})^2}$$

at each iteration (Brodlie [1977: 243]). Brodlie further states that condition (i) is always satisfied in the case of an exact line search. Furthermore, using exact arithmetic, quadratic termination is achieved, since a quadratic function is minimized in at most p iterations with $\mathbf{B}^{(p+1)}$ being equal to the Hessian.

For a general nonlinear function $h(\boldsymbol{\theta})$, Dixon [1972] proved that when exact line searches are used, all nondegenerate members of the Broyden family generate the same search directions, and therefore the same sequence of steps, irrespective of the choice of $\beta^{(a)}$. However, some updates tend to achieve better reductions than others with a "natural" step $\rho^{(a)} = 1$. This makes line searches easier. Some choices of $\beta^{(a)}$ lead to updates which are more stable numerically in finite arithmetic. Another complicating factor is that for general functions, quasi-Newton methods tend to be more efficient with less stringent line-search criteria, even though the quadratic termination property above is thereby lost. The update formula generally considered to be best uses $\beta^{(a)} = 0$. This is called the BFGS formula, as it was introduced independently by Broyden [1970], Fletcher [1970], Goldfarb [1970], and Shanno [1970]. An initial value of $\mathbf{B}^{(1)} = \mathbf{I}_p$ is usually used, so that the first step is a steepest-descent step.

A rank-two modification of a positive definite matrix causes a rank-two modification of its inverse. Let $\mathbf{C}^{(a)} = \mathbf{B}^{(a)-1}$. Then, for any member of the Broyden family, the update of $\mathbf{C}^{(a)}$ corresponding to the update of $\mathbf{B}^{(a)}$ is obtained from (13.39) and (13.40) as follows. We replace $\mathbf{B}^{(a)}$ by $\mathbf{C}^{(a)}$, interchange $\boldsymbol{\delta}^{(a)}$ and $\boldsymbol{\gamma}^{(a)}$, and replace $\beta^{(a)}$ by

$$\mu^{(a)} = \frac{\beta^{(a)} - 1}{\beta^{(a)} - 1 - \beta^{(a)} b^{(a)}}, \tag{13.41}$$

where

$$b^{(a)} = \frac{(\boldsymbol{\gamma}^{(a)\prime} \mathbf{C}^{(a)} \boldsymbol{\gamma}^{(a)})(\boldsymbol{\delta}^{(a)\prime} \mathbf{B}^{(a)} \boldsymbol{\delta}^{(a)})}{(\boldsymbol{\gamma}^{(a)\prime} \boldsymbol{\delta}^{(a)})^2} \tag{13.42}$$

(Fletcher [1980: 62]). In particular, the DFP formula corresponds to $\mu^{(a)} = 0$, and the BGFS to $\mu^{(a)} = 1$. The above symmetry relating $\mathbf{B}^{(a)}$ and $\mathbf{C}^{(a)}$ is not surprising, as the quasi-Newton condition (13.35) may be written either as

$$\mathbf{B}^{(a+1)} \boldsymbol{\delta}^{(a)} = -\boldsymbol{\gamma}^{(a)} \tag{13.43}$$

or as

$$\mathbf{C}^{(a+1)} \boldsymbol{\gamma}^{(a)} = -\boldsymbol{\delta}^{(a)}. \tag{13.44}$$

In earlier work the approximation $\mathbf{C}^{(a)}$ to $\mathbf{H}^{(a)-1}$ was maintained rather than an approximation to $\mathbf{H}^{(a)}$, since the search direction (13.33) is then obtained by matrix–vector multiplication, namely

$$\mathbf{d}^{(a)} = -\mathbf{C}^{(a)} \mathbf{g}^{(a)}, \tag{13.45}$$

instead of by the more expensive operation of solving a set of linear equations. (We note that most books use the symbol $\mathbf{H}^{(a)}$ to represent $\mathbf{C}^{(a)}$, the approximation to the inverse Hessian.)

When sufficiently accurate, though not necessarily exact, line searches are employed with DFP or BFGS updates, the positive definiteness of each $\mathbf{B}^{(a)}$ is preserved. However, this is a theoretical result assuming exact arithmetic. In practice, because of roundoff error, the computed matrix may not be positive definite, so that the resultant search direction is not a descent direction. Gill and Murray [1972] introduced the idea of maintaining and updating $\mathbf{LDL}'$ Cholesky factors of $\mathbf{B}^{(a)}$. Gill and Murray [1977] review algorithms for updating factors when a rank-one change is made and a rank-two change is just the sum of two rank-one changes. Once the factorization has been updated, (13.33) can be solved by one forward solution and one back substitution (Gill et al. [1981: Section 2.2.4.4]), taking a comparable number of operations to the matrix multiplication in (13.45). The advantages of this approach are increased numerical stability and the ease of preserving the positive definiteness of $\mathbf{B}^{(a+1)}$ by ensuring that the diagonal elements of $\mathbf{D}$ are positive. This method is described in detail by Scales [1985: Section 3.5]. On most problems, earlier programs which do not use factorization will cause no difficulties, and indeed, Grandinetti [1978] found on practical advantage in using factorizations even on some ill-conditioned problems.

We have seen how quasi-Newton methods attempt to build up an approximation to the Hessian over a number of iterations ($p+1$ for a quadratic function). Clearly this approach cannot work in regions where the curvature, and hence the Hessian, is changing rapidly. Although quasi-Newton methods produce descent directions, progress through such a region is likely to be considerably slower than for a modified Newton method. Similar problems can arise in regions where the Hessian is indefinite. However, quasi-Newton methods are generally fast and reliable. Superlinear convergence has been established under suitable conditions on the function for the DFP and BFGS algorithms with exact line searches, and for the BFGS algorithm with less stringent line searches (see Gill et al. [1981] for references). Checks of whether $\hat{\boldsymbol{\theta}}$ is indeed a minimum can be made by performing a random search in the vicinity of $\hat{\boldsymbol{\theta}}$ or by estimating the Hessian at the solution by finite differences. Extraction of the approximate Hessian at the solution, for use in statistical estimation, can be difficult using some of the commercial subroutine libraries. It may be necessary for the user to calculate a finite-difference approximation to $\mathbf{H}(\hat{\boldsymbol{\theta}})$. Algorithms A5.6.1 and A5.6.2 of Dennis and Schnabel [1983] perform these calculations using gradient and function values respectively.

Before continuing, we note that Dennis and Schnabel [1983] use the term "quasi-Newton" in a more general sense than we do here. Our meaning agrees with earlier literature. Dennis and Schnabel broaden the definition of quasi-Newton to include the modified Newton methods of Section 13.3, and call variable-metric methods "secant" methods. These authors also give a trust-region (restricted-step) secant method.

*13.4.3 Conjugate-Gradient Methods

For a very large problem, use of a quasi-Newton algorithm may be impractical or expensive because of the necessity of storing an approximation to the Hessian matrix. The conjugate-gradient methods described here are much less costly in this regard and typically only require storage of three or four p-dimensional vectors. They have been used successfully on problems with as many as 4000 parameters. However, they tend to be both less efficient and less robust than quasi-Newton methods and are therefore recommended only for large problems, $p \geq 100$ say. From a statistical viewpoint one should be suspicious of any estimation procedure which uses large numbers of parameters and be doubtful of the applicability of asymptotic theory without vast quantities of data. However, large unconstrained optimization problems often arise as subproblems in solving constrained optimization problems.

The methods we now consider are based upon the concept of conjugate directions. We say that vectors $\mathbf{r}_1$ and $\mathbf{r}_2$ are conjugate with respect to a positive definite matrix $\mathbf{G}$ if $\mathbf{r}_1'\mathbf{G}\mathbf{r}_2 = 0$. Any set of vectors $\mathbf{r}_1, \mathbf{r}_2, \ldots, \mathbf{r}_k$ for $k \leq p$ which are mutually conjugate with respect to $\mathbf{G}$ are linearly independent, since left-multiplying $\sum_i c_i \mathbf{r}_i = \mathbf{0}$ by $\mathbf{r}_j'\mathbf{G}$ implies that $c_j \mathbf{r}_j'\mathbf{G}\mathbf{r}_j = 0$ or $c_j = 0$.

Consider the quadratic function

$$q(\boldsymbol{\theta}) = a + \mathbf{b}'\boldsymbol{\theta} + \tfrac{1}{2}\boldsymbol{\theta}'\mathbf{G}\boldsymbol{\theta},$$

where $\mathbf{G}$ is positive definite. It can be shown (e.g. Scales [1985: Section 3.4.2]) that q is minimized by minimizing with respect to each of a set of p mutually conjugate directions $\mathbf{r}_1, \mathbf{r}_2, \ldots, \mathbf{r}_p$ using exact line searches. The search directions can be used in any order. In their conjugate-gradient algorithm, Fletcher and Reeves [1964] used the following directions for the $\mathbf{r}_i$'s:

$$\mathbf{r}^{(1)} = -\mathbf{g}^{(1)}, \tag{13.46}$$

$$\mathbf{r}^{(a+1)} = -\mathbf{g}^{(a+1)} + \beta^{(a)}\mathbf{r}^{(a)} \tag{13.47}$$

for $a = 2, 3, \ldots, p$, where

$$\beta^{(a)} = \frac{\mathbf{g}^{(a+1)\prime}\mathbf{g}^{(a+1)}}{\mathbf{g}^{(a)\prime}\mathbf{g}^{(a)}}. \tag{13.48}$$

For $q(\boldsymbol{\theta})$, a step of length $\rho^{(a)} = -\mathbf{g}^{(a)\prime}\mathbf{r}^{(a)}/\mathbf{r}^{(a)\prime}\mathbf{G}\mathbf{r}^{(a)}$ minimizes q along $\mathbf{r}^{(a)}$. Fletcher and Reeves chose these directions to try and associate conjugacy properties with the steepest-descent method in an attempt to improve steepest descent's reliability and efficiency. In addition to conjugacy, Fletcher [1980: Theorem 4.1.1] proved that this set of directions has further properties, namely (i) they have orthogonal gradients, i.e. $\mathbf{g}^{(a)\prime}\mathbf{g}^{(b)} = 0$ for $a \neq b$; (ii) they are descent directions of search; and (iii) q is minimized in $k \leq p$ steps, where k is the largest integer for which $\mathbf{g}^{(a)} \neq \mathbf{0}$.

A method for minimizing a general nonlinear function is thus obtained as

$$\boldsymbol{\theta}^{(a+1)} = \boldsymbol{\theta}^{(a)} + \rho^{(a)}\mathbf{r}^{(a)}, \tag{13.49}$$

where $\mathbf{r}^{(a)}$ is defined as in (13.46) to (13.48), for $1 \leq a \leq p$, and $\rho^{(a)}$ is obtained using an accurate line search. The method no longer terminates in p steps, but is iterative. Traditionally $\mathbf{r}^{(a)}$ is reset to the steepest-descent direction $-\mathbf{g}^{(a)}$ after every p iterations. This strategy is motivated by the idea that when iterates enter the neighborhood of the solution in which $h(\boldsymbol{\theta})$ is closely approximated by a quadratic, then the reset method can be expected to converge rapidly. However, with continued application of (13.47) this may not occur.

Another formula for calculating $\beta^{(a)}$ is the Polak–Ribiere formula (Polak [1971])

$$\beta^{(a)} = \frac{(\mathbf{g}^{(a+1)} - \mathbf{g}^{(a)})'\mathbf{g}^{(a+1)}}{\mathbf{g}^{(a)\prime}\mathbf{g}^{(a)}}, \tag{13.50}$$

which is identical to the Fletcher–Reeves formula (13.48) for a quadratic function but not in general. There are indications (Powell [1977]) that the Polak–Ribiere formula should be used when p is very large. Scales [1985: Section 3.4.4] gives a conjugate-gradient algorithm with resets in which either (13.48) or (13.50) can be used.

The conjugate-gradient method as described above can be shown to be p-step superlinearly convergent, i.e.

$$\lim_{a\to\infty} \frac{\|\boldsymbol{\theta}^{(ap+p)} - \hat{\boldsymbol{\theta}}\|}{\|\boldsymbol{\theta}^{(ap)} - \hat{\boldsymbol{\theta}}\|} = 0,$$

on a wide class of functions when implemented with restarts in exact arithmetic and with exact line searches. However, this means little if p is large, and in practice conjugate-gradient algorithms tend to exhibit only a linear convergence rate (Gill et al. [1981: Section 4.8.3]).

Because of the difficulties in minimizing $h(\boldsymbol{\theta})$ along $\mathbf{r}^{(a)}$ to a high level of accuracy, conjugate-gradient algorithms are often implemented with inexact linear searches (Scales [1985: Section 3.4.9]), but care then has to be taken to ensure each $\mathbf{r}^{(a)}$ is a descent direction. It is possible to restart with $\mathbf{r}^{(a)} = -\mathbf{g}^{(a)}$ each time the descent criterion is violated, but if this occurs too frequently the algorithm can be slowed drastically. Often it is found that little reduction in $h(\boldsymbol{\theta})$ tends to be made at the restarting iteration in comparison with other iterations. Powell [1977] gives an efficient algorithm using restarts other than steepest descent at the expense of increased storage requirements.

One research idea is *preconditioning*, in which an attempt is made to rescale the problem so that the distinct eigenvalues of the Hessian tend to cluster close to unity. If possible, this greatly speeds the algorithm. Another idea follows from the fact that the "approximate inverse Hessian" formed by a single update of the identity matrix using the BFGS formula [c.f. $\mathbf{C}^{(a)}$ preceeding (13.41)] results in the conjugate-gradient direction. *Limited-memory quasi-Newton* methods attempt to store as many previous update vectors as possible and use them directly rather than storing an approximate Hessian. For further discussion and references see Fletcher [1980] and Gill et al. [1981].

13.5 METHODS WITHOUT DERIVATIVES

It is at present generally agreed that, unless storage problems make them impractical, the fastest and most reliable methods for minimizing a smooth function without the use of derivatives are quasi-Newton methods implemented with finite-difference estimates of the derivatives. Nevertheless, we shall briefly discuss several other methods that are widely available as well.

13.5.1 Nonderivative Quasi-Newton Methods

A natural and inexpensive estimate of $g_i(\boldsymbol{\theta}) = \partial h(\boldsymbol{\theta})/\partial \theta_i$ is given by the forward-difference formula

$$\hat{g}_i(\boldsymbol{\theta}, \xi_i) = \frac{h(\boldsymbol{\theta} + \xi_i \mathbf{e}_i) - h(\boldsymbol{\theta})}{\xi_i} \tag{13.51}$$

for some small $\xi_i > 0$. There are two major sources of error in the use of (13.51) as an estimate of $g_i(\boldsymbol{\theta})$. Firstly, there is the truncation error caused by truncating the Taylor series expansion of $h(\boldsymbol{\theta} + \xi_i \mathbf{e}_i)$ after the linear term, an error of order ξ_i. Secondly, there is the cancellation error caused by computing the subtraction in (13.51) in finite arithmetic, an error of order $1/\xi_i$. Consequently ξ_i is chosen to minimize an estimate of the sum of the two errors. Now an estimate of $\partial^2 h(\boldsymbol{\theta})/\partial\theta_i^2$ is required to estimate the truncation error. Stewart [1967], for example, re-estimates the $\{\xi_i\}$ at each iteration using the (i, i)th element of the current approximation to the Hessian as an estimate of $\partial^2 h(\boldsymbol{\theta})/\partial\theta_i^2$. Gill and Murray [1972] use a finite-difference approximation to the Hessian at the first iteration and find that it is usually adequate to use the set of intervals obtained for the first iteration thereafter. Both methods produce estimates of $g_i(\boldsymbol{\theta})$ which are adequate for practical purposes provided $g_i(\boldsymbol{\theta})$ is not too small. However, $g_i(\boldsymbol{\theta}) \to \theta$ as $\boldsymbol{\theta} \to \hat{\boldsymbol{\theta}}$, and the approximate derivatives become inadequate when the difference in function values becomes small relative to ξ_i. Large cancellation errors then begin to invalidate the result. At this point a central-difference approximation to $g_i(\boldsymbol{\theta})$ may be used, namely

$$\tilde{g}_i(\boldsymbol{\theta}, \xi_i) = \frac{h(\boldsymbol{\theta} + \xi_i \mathbf{e}_i) - h(\boldsymbol{\theta} - \xi_i \mathbf{e}_i)}{2\xi_i}. \tag{13.52}$$

The truncation error in this approximation is of order ξ_i^2, and it can therefore be used with larger values of ξ_i to avoid the cancellation errors. Forward differences as in (13.51) are used initially, because $\mathbf{g}(\boldsymbol{\theta})$ is estimated with half the number of function evaluations that are necessary with central differences.

The use of finite-difference approximations to the gradient necessitates changes in strategy in an implementation of a quasi-Newton algorithm. Gradient evaluations become relatively expensive, and therefore the line search should not require further gradient evaluations. Additionally, more accurate line searches should be employed, since this typically reduces the number of iterations required and hence the number of gradient evaluations. The resulting algorithms are slightly more prone to failure in ill-conditioned problems or problems where the parameters are badly scaled than their exact-derivative counterparts. However, they are generally reliable and have the fast convergence property of exact derivative quasi-Newton methods. Gill et al. [1981: Sections 4.6, 8.6] contains an excellent detailed discussion of the practical considerations.

*13.5.2 Direction-Set (Conjugate-Direction) Methods

Conjugate directions for a quadratic function (13.6) can be obtained without the use of derivatives. We recall from Section 13.4.3 that once such a set of directions has been obtained, a quadratic function can be minimized by minimizing along each of the p directions in any order.One mechanism for obtaining conjugate directions is through the parallel-subspace property, which is as follows (Fletcher [1980: Theorem 4.2.1]).

Theorem 13.2 Let Θ_1 and Θ_2 be two linear subspaces generated by linearly independent direction vectors $\mathbf{d}^{(1)}, \mathbf{d}^{(2)}, \ldots, \mathbf{d}^{(k)}$ for some $k < p$ about two points $\mathbf{v}^{(1)}$ and $\mathbf{v}^{(2)}$. Thus, for $u = 1, 2$,

$$\Theta_u = \left\{ \boldsymbol{\theta} : \boldsymbol{\theta} = \mathbf{v}^{(u)} + \sum_{j=1}^{k} \alpha_j \mathbf{d}^{(j)}, \text{ each } \alpha_j \text{ in } \mathbb{R} \right\}.$$

If $\tilde{\boldsymbol{\theta}}^{(u)}$ minimizes

$$q(\boldsymbol{\theta}) = a + \mathbf{b}'\boldsymbol{\theta} + \tfrac{1}{2}\boldsymbol{\theta}'\mathbf{G}\boldsymbol{\theta}$$

over Θ_u $(u = 1, 2)$, then $\tilde{\boldsymbol{\theta}}^{(2)} - \tilde{\boldsymbol{\theta}}^{(1)}$ is conjugate to each of the $\mathbf{d}^{(j)}$ with respect to $\mathbf{G}$, the Hessian of $q(\boldsymbol{\theta})$.

Proof: Since $\tilde{\boldsymbol{\theta}}^{(u)}$ minimizes $q(\boldsymbol{\theta})$ for $\boldsymbol{\theta} \in \Theta_u$, it also minimizes $q(\boldsymbol{\theta})$ subject to $\boldsymbol{\theta} = \tilde{\boldsymbol{\theta}}^{(u)} + \alpha_j \mathbf{d}^{(j)}$ for each j. Furthermore,

$$\frac{\partial}{\partial \alpha_j} q(\boldsymbol{\theta}^{(u)} + \alpha_j \mathbf{d}^{(j)}) = 0 \qquad \text{at} \quad \alpha_j = 0 \quad (\text{i.e. } \boldsymbol{\theta} = \tilde{\boldsymbol{\theta}}^{(u)})$$

implies $\mathbf{g}'(\tilde{\boldsymbol{\theta}}^{(u)})\mathbf{d}^{(j)} = 0$ for each j, where $\mathbf{g}(\boldsymbol{\theta}) = \partial q / \partial \boldsymbol{\theta} = \mathbf{b} + \mathbf{G}\boldsymbol{\theta}$. Thus we have

$$(\tilde{\boldsymbol{\theta}}^{(2)} - \tilde{\boldsymbol{\theta}}^{(1)})'\mathbf{G}\mathbf{d}^{(j)} = [\mathbf{g}'(\tilde{\boldsymbol{\theta}}^{(2)}) - \mathbf{g}'(\tilde{\boldsymbol{\theta}}^{(1)})]\mathbf{d}^{(j)} = 0. \qquad \blacksquare$$

We note that for $\tilde{\boldsymbol{\theta}}^{(2)} - \tilde{\boldsymbol{\theta}}^{(1)}$ to be nonzero, $\mathbf{v}^{(2)} - \mathbf{v}^{(1)}$ cannot be in the space spanned by $\mathbf{d}^{(1)}, \mathbf{d}^{(2)}, \ldots, \mathbf{d}^{(k)}$.

The above theorem forms the basis for Powell's [1964] basic algorithm which not only constructs p mutually conjugate directions for a quadratic function in p steps, but also forms the basis for a general-purpose minimization algorithm. To start the algorithm, p independent directions $\mathbf{r}^{(1)}, \mathbf{r}^{(2)}, \ldots, \mathbf{r}^{(p)}$ are required, and $\boldsymbol{\theta}^{(1)}$ is chosen to minimize h from an initial value $\boldsymbol{\theta}^{(0)}$ along $\mathbf{r}^{(p)}$. The algorithm then proceeds as follows (paraphrasing Fletcher [1980: 72]). For $a = 1, 2, \ldots$ perform the following:

1. Find $\tilde{\boldsymbol{\theta}} = \boldsymbol{\theta}^{(a)} + \sum_{j=1}^{p} \alpha_j \mathbf{r}^{(j)}$ by minimizing $h(\boldsymbol{\theta})$ from $\boldsymbol{\theta}^{(a)}$ along the search directions $\mathbf{r}^{(1)}, \mathbf{r}^{(2)}, \ldots, \mathbf{r}^{(p)}$ sequentially; i.e., starting from $\boldsymbol{\theta}^{(a)}$, minimize along $\mathbf{r}^{(1)}$, then from this minimum minimize along $\mathbf{r}^{(2)}$, etc. (Note that if h is quadratic, then for $a < p$, the directions $\mathbf{r}^{(p-a+1)}, \mathbf{r}^{(p-a+2)}, \ldots, \mathbf{r}^{(p)}$ are mutually conjugate, as shown below.)

2. Relabel $\mathbf{r}^{(j+1)}$ as $\mathbf{r}^{(j)}$ for $j = 1, 2, \ldots, p-1$, and set $\mathbf{r}^{(p)} = \tilde{\boldsymbol{\theta}} - \boldsymbol{\theta}^{(a)}$ as the new conjugate direction.
3. Obtain $\boldsymbol{\theta}^{(a+1)}$ by minimizing h from $\tilde{\boldsymbol{\theta}}$ along $\mathbf{r}^{(p)}$.

It is asserted in the algorithm that at the beginning of the ath iteration ($a < p$), the a directions $\mathbf{r}^{(p-a+1)}, \mathbf{r}^{(p-a+2)}, \ldots, \mathbf{r}^{(p)}$ are mutually conjugate if $h(\boldsymbol{\theta})$ is quadratic. We show inductively that this is true. It is trivially true for $a = 1$, as we have just one vector $\mathbf{r}^{(p)}$, and we assume it is true for general a. Now $\boldsymbol{\theta}^{(a)}$ is obtained from $\boldsymbol{\theta}^{(a-1)}$ by minimizing from $\boldsymbol{\theta}^{(a-1)}$ along the directions $\mathbf{r}^{(1)}_{\text{old}}$, $\mathbf{r}^{(1)}, \ldots, \mathbf{r}^{(p)}$ sequentially. Here $\mathbf{r}^{(1)}_{\text{old}}$ is the first direction, now discarded, from the previous iteration, and $\mathbf{r}^{(1)}$ is $\mathbf{r}^{(2)}_{\text{old}}$ relabeled, etc. Thus $\boldsymbol{\theta}^{(a)}$ is obtained by minimizing h from $\mathbf{v}^{(1)}$ along the directions $\mathbf{r}^{(1)}, \mathbf{r}^{(2)}, \ldots, \mathbf{r}^{(p)}$ sequentially, where $\mathbf{v}^{(1)}$ is found by minimizing from $\boldsymbol{\theta}^{(a-1)}$ along the direction $\mathbf{r}^{(1)}_{\text{old}}$. Also, taking $\mathbf{v}^{(2)} = \boldsymbol{\theta}^{(a)}$, $\tilde{\boldsymbol{\theta}}$ minimizes h from $\mathbf{v}^{(2)}$ along the directions $\mathbf{r}^{(1)}, \ldots, \mathbf{r}^{(p)}$. Thus from Theorem 13.2 above, $\tilde{\boldsymbol{\theta}} - \boldsymbol{\theta}^{(a)}$ is conjugate to $\mathbf{r}^{(1)}, \ldots, \mathbf{r}^{(p)}$, so that $\{\mathbf{r}^{(p-a+1)}, \ldots, \mathbf{r}^{(p)}, \tilde{\boldsymbol{\theta}} - \boldsymbol{\theta}^{(a)}\}$—or, relabeling, $\{\mathbf{r}^{(p-a)}, \mathbf{r}^{(p-a+1)}, \ldots, \mathbf{r}^{(p)}\}$—are mutually conjugate. Hence the assertion is true for $a+1$ iterations.

Unfortunately, the direction set can become linearly dependent, so that some directions in $\mathbb{R}^p$ are no longer searched. Therefore, rather than dropping the oldest direction, Powell [1964] discards a direction chosen so that the remaining p direction vectors are, in a sense to be defined, most strongly conjugate.

Let $\mathbf{R}$ be any $p \times p$ matrix whose columns $\mathbf{r}^{(1)}, \mathbf{r}^{(2)}, \ldots, \mathbf{r}^{(p)}$ are scaled so that $\mathbf{r}^{(i)\prime}\mathbf{G}\mathbf{r}^{(i)} = 1$ for a positive definite matrix $\mathbf{G}$. Theorem 4.2.2 of Fletcher [1980] states that the determinant of $\mathbf{R}$ is maximized for all such matrices when $\mathbf{r}^{(1)}, \ldots, \mathbf{r}^{(p)}$ are mutually conjugate with respect to $\mathbf{G}$. For a quadratic function, the quantities $\mathbf{r}^{(i)\prime}\mathbf{G}\mathbf{r}^{(i)}$ can be calculated during the line search along $\mathbf{r}^{(i)}$ without any knowledge of $\mathbf{G}$ being necessary. The vectors of the direction set can therefore be scaled so that $\mathbf{r}^{(i)\prime}\mathbf{G}\mathbf{r}^{(i)} = 1$. Powell discards the member of the direction set for which the determinant of the matrix formed by the remaining p directions, suitably scaled, is largest. Besides preserving the "most conjugate" direction set, this strategy preserves the linear independence of the direction set.

There have been a number of modifications of those basic ideas. For a quadratic function, after p steps of Powell's basic algorithm the matrix $\mathbf{R}$ formed as in the previous paragraph has the property $\mathbf{R}'\mathbf{G}\mathbf{R} = \mathbf{I}_p$, so that the Hessian can be estimated by $(\mathbf{R}\mathbf{R}')^{-1}$. Brent [1973] proceeds by finding the eigenvectors of $(\mathbf{R}\mathbf{R}')^{-1}$ and uses these to define a new initial set of directions for a subsequent cycle of p iterations, and so on. Other algorithms make use of another important result due to Powell [1972]. If $\bar{\mathbf{R}}$ is formed by making an orthogonal transformation of $\mathbf{r}^{(1)}, \ldots, \mathbf{r}^{(p)}$ and rescaling the resulting directions so that $\bar{\mathbf{r}}^{(i)\prime}\mathbf{G}\bar{\mathbf{r}}^{(i)} = 1$, then $|\bar{\mathbf{R}}| \geq |\mathbf{R}|$. Hence orthogonal transformations of the direction set can improve conjugacy. Nazareth [1977] abandons "exact" line searches along search directions, but uses rotations which produce direction sets that converge to conjugacy. Other algorithms which use orthogonal transformations to improve conjugacy have been given by Rhead [1974], Brodlie [1975], and Coope [1976]. These algorithms tend to be better than those of Powell [1964], but are still slower than quasi-Newton methods implemented with finite-difference approximations to derivatives.

13.5.3 Direct Search Methods

Direct search methods do not attempt to either approximate derivatives numerically or approximate $h(\boldsymbol{\theta})$ in the vicinity of $\boldsymbol{\theta}^{(a)}$ by a smooth function. Instead they rely only upon the comparison of function values. A good survey of such methods is given by Swann [1972]. In general, for smooth functions, no direct search methods appear to be competitive with well- implemented discrete quasi-Newton methods or conjugate-direction methods in terms of rate of convergence, efficiency, or reliability. Gill et al. [1981] recommend such methods only for nonsmooth functions. We shall describe the most successful of these algorithms, the "simplex" algorithm of Nelder and Mead [1965]. This algorithm has been popular with many statisticians, perhaps largely because it is simple and easy to program. It is not to be confused with the more famous simplex algorithm of linear programming and is sometimes called the *polytope algorithm* to make this distinction clear.

The polytope algorithm requires $p+1$ starting values $\boldsymbol{\theta}^{(1)}, \boldsymbol{\theta}^{(2)}, \dots, \boldsymbol{\theta}^{(p+1)}$ which are mutually equidistant in p-dimensional space, thus forming the corners of a regular polytope or simplex. For $p = 2$ the polytope is a triangle, and for $p = 3$ it is a tetrahedron. The polytope is moved through space toward the minimum by progressively shifting the corner with the highest function value. The regularity or equidistant property is not preserved as the polytope moves, since an adaptive polytope is more efficient. Therefore at each stage $(p+1)$ points $\boldsymbol{\theta}^{(1)}, \dots, \boldsymbol{\theta}^{(p+1)}$ are retained, together with their function values, and are ordered so that

$$h^{(1)} \leq h^{(2)} \leq \cdots \leq h^{(p+1)},$$

where $h^{(r)} = h(\boldsymbol{\theta}^{(r)})$. The worst point $\boldsymbol{\theta}^{(p+1)}$ is reflected through the centroid $\mathbf{c}$ of the remaining points, namely

$$\mathbf{c} = \frac{1}{p} \sum_{i=1}^{p} \boldsymbol{\theta}^{(i)}. \tag{13.53}$$

The new trial point is

$$\boldsymbol{\theta}_T = \mathbf{c} + \alpha(\mathbf{c} - \boldsymbol{\theta}^{(p+1)}), \tag{13.54}$$

where $\alpha > 0$, called the reflection coefficient, is a parameter of the algorithm. The action to be taken depends upon $h_T = h(\boldsymbol{\theta}_T)$ as follows:

1. If $h^{(1)} \leq h_T \leq h^{(p)}$, replace $\boldsymbol{\theta}^{(p+1)}$ by $\boldsymbol{\theta}_T$ and reorder.
2. If $h_T < h^{(1)}$, the search direction is very promising, so try an expanded step.

Compute

$$\boldsymbol{\theta}_E = \mathbf{c} + \beta(\boldsymbol{\theta}_T - \mathbf{c}), \tag{13.55}$$

where $\beta > 1$ is a parameter called the expansion coefficient. If $h(\boldsymbol{\theta}_E) < h_T$, then $\boldsymbol{\theta}_E$ replaces $\boldsymbol{\theta}^{(p+1)}$; otherwise $\boldsymbol{\theta}_T$ replaces $\boldsymbol{\theta}^{(p+1)}$.

3. If $h_T > h^{(p)}$, so that $\boldsymbol{\theta}_T$ is still the worst point, take a (smaller) contracted step, giving

$$\boldsymbol{\theta}_C = \begin{cases} \mathbf{c} + \gamma(\boldsymbol{\theta}^{(p+1)} - \mathbf{c}) & \text{if} \quad h_T \geq h^{(p+1)}, \\ \mathbf{c} + \gamma(\boldsymbol{\theta}_T - \mathbf{c}) & \text{if} \quad h_T < h^{(p+1)}, \end{cases} \tag{13.56}$$

where γ $(0 < \gamma < 1)$ is the contraction coefficient. Replace $\boldsymbol{\theta}^{(p+1)}$ by $\boldsymbol{\theta}_C$ if $h(\boldsymbol{\theta}_C) < \min\{h(\boldsymbol{\theta}_T), h(\boldsymbol{\theta}^{(p+1)})\}$; otherwise contract still further.

Occasionally the polytope can be replaced by a regular polytope to prevent it from becoming too distorted or from collapsing into a lower-dimensional subspace. A possible stopping rule is to accept $\boldsymbol{\theta}^{(1)}$ if p iterations have not produced a lower point. O'Neill [1971] gives a computer program for the algorithm; Parkinson and Hutchinson [1972] and Gill et al. [1981: 95] consider modifications. A more recent discussion is given by Olsson and Nelson [1975]. Spendley [1969] adapts the algorithm for least-squares problems.

13.6 METHODS FOR NONSMOOTH FUNCTIONS

Methods for smooth functions, particularly quasi-Newton methods, still tend to perform well on functions with a few isolated discontinuities provided these do not occur in a close neighborhood of the minimum. When the discontinuities in the gradient form a structured pattern, special methods can be developed, for example for L_1 (least absolute deviations) problems. Here $h(\boldsymbol{\theta})$ has the form

$$h(\boldsymbol{\theta}) = \sum_{i=1}^{n} |h_i(\boldsymbol{\theta})|. \tag{13.57}$$

The L_1 problem will be discussed in Section 14.6.2. Finally, the most generally useful methods for nonsmooth problems are the direct search methods such as the simplex (polytope) method of Section 13.5.3.

13.7 SUMMARY

Suppose $h(\boldsymbol{\theta})$, a smooth function of p variables (parameters), is to be minimized. Unless p is very large (say $p > 250$), the most effective algorithms for minimizing h

are iterative algorithms based upon the classical Newton method in which

$$\begin{aligned} \boldsymbol{\theta}^{(a+1)} &= \boldsymbol{\theta}^{(a)} + \boldsymbol{\delta}^{(a)}, \\ \boldsymbol{\delta}^{(a)} &= -\mathbf{H}(\boldsymbol{\theta}^{(a)})^{-1}\mathbf{g}(\boldsymbol{\theta}^{(a)}), \end{aligned} \tag{13.58}$$

where $\mathbf{g}(\boldsymbol{\theta}) = \partial h(\boldsymbol{\theta})/\partial\boldsymbol{\theta}$ is the gradient vector, and $\mathbf{H}(\boldsymbol{\theta}) = \partial^2 h(\boldsymbol{\theta})/\partial\boldsymbol{\theta}\,\partial\boldsymbol{\theta}'$ is called the Hessian matrix. In practice this algorithm is modified to improve stability and convergence. Line searches (Section 13.2.3c) may be incorporated using $\boldsymbol{\delta}^{(a)} = -\rho^{(a)}\mathbf{H}^{(a)-1}\mathbf{g}^{(a)}$, where $\rho^{(a)}$ (> 0) is chosen so that $\boldsymbol{\theta}^{(a+1)}$ approximately minimizes h from $\boldsymbol{\theta}^{(a)}$ along the direction $-\mathbf{H}^{(a)-1}\mathbf{g}^{(a)}$. The Hessian $\mathbf{H}^{(a)}$ is modified, if necessary, to form $\bar{\mathbf{H}}^{(a)}$ which is positive definite. This ensures that the direction $-\bar{\mathbf{H}}^{(a)-1}\mathbf{g}^{(a)}$ leads, initially, downhill. Alternatively, a restricted-step or trust-region strategy may be used to modify (13.58). Such algorithms are called

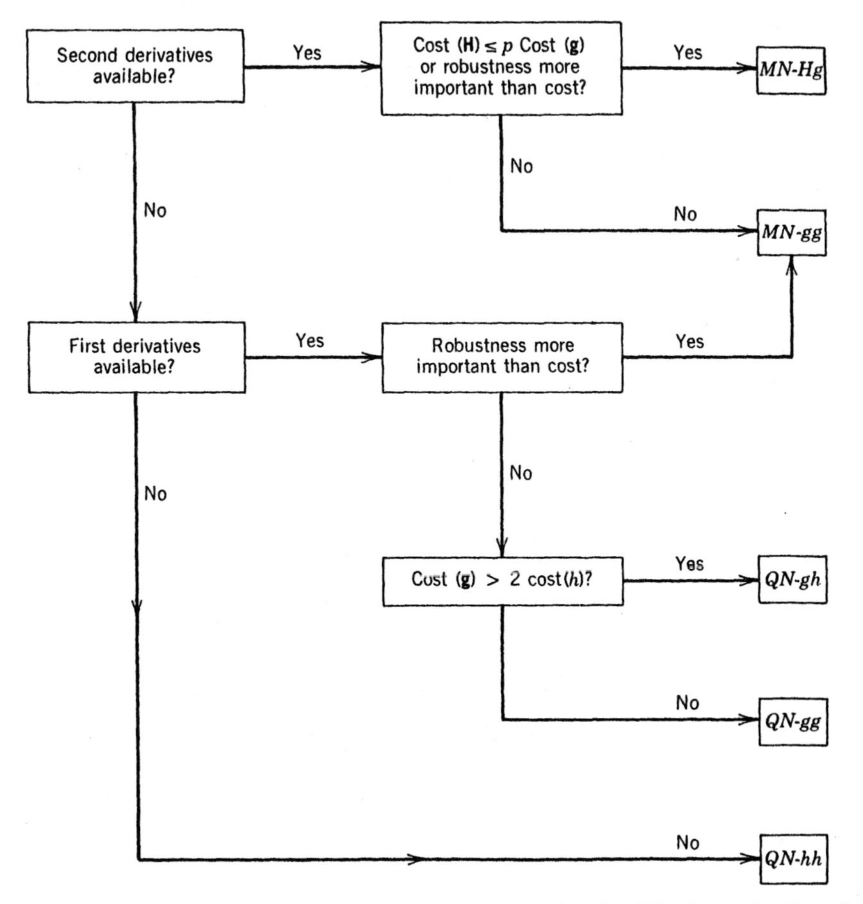

Figure 13.5 Decision tree for methods of minimizing a smooth function $h(\boldsymbol{\theta})$ where p, the dimension of $\boldsymbol{\theta}$, is not very large (e.g. $p \leq 250$). Adapted from Scales [1985].

modified Newton (MN) algorithms. When analytic expressions are used for both $\mathbf{H}^{(a)}$ and $\mathbf{g}^{(a)}$ (Section 13.3) we will label the method MN-*Hg*.

However, analytic expressions for $\mathbf{H}(\boldsymbol{\theta})$ may not be available or we may not wish to go to the trouble of obtaining them. In this case $\mathbf{H}(\boldsymbol{\theta}^{(a)})$ can be approximated in the MN algorithm by using finite differences in $\mathbf{g}(\boldsymbol{\theta})$ (Section 13.4.1). We shall label these methods MN-*gg*. Finally, $\mathbf{g}$ may not be available analytically but can be approximated using finite differences in $h(\boldsymbol{\theta})$ (Section 13.5.1). These are labeled MN-*hh*.

Quasi-Newton (QN) methods employ a different strategy to approximate $\mathbf{H}(\boldsymbol{\theta}^{(a)})$ when analytic expressions are unavailable (Section 13.4.2). They attempt to gradually build up an approximation to the Hessian as iterations proceed using changes in the gradient. Quasi-Newton methods can be implemented using analytic gradient expressions (QN-*gg*) or using finite differences in $h(\boldsymbol{\theta})$ to approximate $\mathbf{g}(\boldsymbol{\theta}^{(a)})$ (Section 13.5.1). The latter is denoted by QN-*hh*. Scales [1985] uses QN-*gh* to refer to the use of $\mathbf{g}(\boldsymbol{\theta})$ in the basic iterations and just $h(\boldsymbol{\theta})$ in line searches.

Quality subroutine libraries such as NAG include advice leading to the "best" algorithm for one's problem. Often this takes the form of a decision tree (e.g. Gill et al. [1981: Fig. 8*a*]). Our Figure 13.5 is a much less detailed decision tree and is adapted from Scales [1985: Fig. 3.2].

For problems involving very large numbers of parameters, conjugate-gradient methods (Section 13.4.3) are often recommended, largely because they require much less computer storage space. Finally, we note that in Section 13.6 we gave a brief discussion of methods for nonsmooth problems.

CHAPTER 14

Computational Methods for Nonlinear Least Squares

14.1 GAUSS–NEWTON ALGORITHM

The computational problem most often encountered in this book is the nonlinear least-squares problem: find $\boldsymbol{\theta}$ to minimize

$$\begin{aligned} S(\boldsymbol{\theta}) &= \sum_{i=1}^{n} r_i(\boldsymbol{\theta})^2 \qquad (14.1) \\ &= \sum_{i=1}^{n} [y_i - f_i(\boldsymbol{\theta})]^2 \\ &= \|\mathbf{y} - \mathbf{f}\|^2. \end{aligned}$$

Computational methods for adapting ordinary least-squares minimization algorithms to solve weighted or generalized least-squares problems have already been discussed (Section 2.1.4).

Although the least-squares problem (14.1) can be solved using any of the methods of the previous chapter, with $h(\boldsymbol{\theta}) = S(\boldsymbol{\theta})$, algorithms have been developed which exploit the special structure of (14.1) as a sum of squares. These algorithms are based upon a modification to the Newton method introduced by Gauss in 1809 and known as the Gauss–Newton algorithm (GN).

In Section 2.1.3 we derived the Gauss–Newton algorithm by taking a linear Taylor series approximation to $f_i(\boldsymbol{\theta})$ about $\boldsymbol{\theta}^{(a)}$ to obtain

$$f_i(\boldsymbol{\theta}) \approx f_i(\boldsymbol{\theta}^{(a)}) + \frac{\partial f_i(\boldsymbol{\theta}^{(a)})}{\partial \boldsymbol{\theta}'}(\boldsymbol{\theta} - \boldsymbol{\theta}^{(a)}). \qquad (14.2)$$

Using the approximation (14.2) in (14.1), the minimization problem was converted to a linear least-squares problem, namely [c.f. (2.28)]

$$\text{Minimize } \| \mathbf{r}^{(a)} - \mathbf{F}_{\bullet}^{(a)}(\boldsymbol{\theta} - \boldsymbol{\theta}^{(a)}) \|^2, \tag{14.3}$$

where $\mathbf{r}^{(a)} = \mathbf{y} - \mathbf{f}(\boldsymbol{\theta}^{(a)})$ and $\mathbf{F}_{\bullet}^{(a)}$ is $\partial \mathbf{f}/\partial \boldsymbol{\theta}'$ evaluated at $\boldsymbol{\theta}^{(a)}$. Now (14.3) has solution

$$\boldsymbol{\theta} - \boldsymbol{\theta}^{(a)} = (\mathbf{F}_{\bullet}^{(a)\prime}\mathbf{F}_{\bullet}^{(a)})^{-1}\mathbf{F}_{\bullet}^{(a)\prime}\mathbf{r}^{(a)}$$

leading to the Gauss–Newton algorithm

$$\boldsymbol{\theta}^{(a+1)} = \boldsymbol{\theta}^{(a)} + \boldsymbol{\delta}^{(a)},$$

where

$$\boldsymbol{\delta}^{(a)} = (\mathbf{F}_{\bullet}^{(a)\prime}\mathbf{F}_{\bullet}^{(a)})^{-1}\mathbf{F}_{\bullet}^{(a)\prime}\mathbf{r}^{(a)}. \tag{14.4}$$

Functions of the form (14.1) also occur outside the context of nonlinear least-squares regression, and numerical analysts prefer to discuss the problem in terms of (14.1) instead of $\| \mathbf{y} - \mathbf{f} \|^2$. In fact, most prefer to use $S(\boldsymbol{\theta})/2$. Expressed in terms of (14.1), the gradient and Hessian of $S(\boldsymbol{\theta})$ are, respectively,

$$\begin{aligned} g(\boldsymbol{\theta}) &= \frac{\partial S(\boldsymbol{\theta})}{\partial \boldsymbol{\theta}} \\ &= 2\sum_{i=1}^{n} r_i(\boldsymbol{\theta})\frac{\partial r_i(\boldsymbol{\theta})}{\partial \boldsymbol{\theta}}, \qquad (14.5) \\ &= 2\mathbf{J}'\mathbf{r}, \qquad (14.6) \end{aligned}$$

and

$$\begin{aligned} \mathbf{H}(\boldsymbol{\theta}) &= \frac{\partial S(\boldsymbol{\theta})}{\partial \boldsymbol{\theta}\,\partial \boldsymbol{\theta}'} \\ &= 2\sum_{i=1}^{n} \frac{\partial r_i(\boldsymbol{\theta})}{\partial \boldsymbol{\theta}} \frac{\partial r_i(\boldsymbol{\theta})}{\partial \boldsymbol{\theta}'} + 2\sum_{i=1}^{n} r_i(\boldsymbol{\theta})\frac{\partial^2 r_i(\boldsymbol{\theta})}{\partial \boldsymbol{\theta}\,\partial \boldsymbol{\theta}'} \\ &= 2(\mathbf{J}'\mathbf{J} + \mathbf{A}). \qquad (14.7) \end{aligned}$$

Here

$$\begin{aligned} \mathbf{J} &= \mathbf{J}(\boldsymbol{\theta}) \\ &= \left[\left(\frac{\partial r_i}{\partial \theta_j}\right)\right] \\ &= \frac{\partial \mathbf{r}}{\partial \boldsymbol{\theta}'} \\ &= -\mathbf{F}_{\bullet}(\boldsymbol{\theta}), \qquad (14.8) \end{aligned}$$

the $n \times p$ matrix with jth column $\partial \mathbf{r}/\partial \theta_j$, and

$$\mathbf{A} = \mathbf{A}(\boldsymbol{\theta}) = \sum_{i=1}^{n} r_i(\boldsymbol{\theta}) \frac{\partial^2 r_i(\boldsymbol{\theta})}{\partial \boldsymbol{\theta}\, \partial \boldsymbol{\theta}'}. \tag{14.9}$$

Thus the Newton step is [c.f. (2.36) and (13.19)]

$$\text{N:} \qquad \boldsymbol{\delta}^{(a)} = -\mathbf{H}^{(a)-1}\mathbf{g}^{(a)} = -(\mathbf{J}^{(a)\prime}\mathbf{J}^{(a)} + \mathbf{A}^{(a)})^{-1}\mathbf{J}^{(a)\prime}\mathbf{r}^{(a)}. \tag{14.10}$$

Since $\mathbf{J}^{(a)} = -\mathbf{F}_{\bullet}^{(a)}$, we have $\mathbf{F}_{\bullet}^{(a)\prime}\mathbf{F}_{\bullet}^{(a)} = \mathbf{J}^{(a)\prime}\mathbf{J}^{(a)}$ and (14.4) becomes

$$\text{GN:} \qquad \boldsymbol{\delta}^{(a)} = -(\mathbf{J}^{(a)\prime}\mathbf{J}^{(a)})^{-1}\mathbf{J}^{(a)\prime}\mathbf{r}^{(a)}. \tag{14.11}$$

Thus the Gauss–Newton algorithm is obtained from the Newton algorithm (N) by ignoring part of the Hessian, namely $\mathbf{A}(\boldsymbol{\theta})$ of (14.9). This dropping of $\mathbf{A}(\boldsymbol{\theta})$ from the Hessian matrix is responsible for both the strengths and the weaknesses of Gauss–Newton-based algorithms.

Although we have written the Gauss–Newton step in the form (14.11), in good implementations $\mathbf{J}'\mathbf{J}$ is never formed. Instead $\boldsymbol{\delta}^{(a)}$ is obtained by solving the linear least-squares problem (14.3), namely

$$\underset{\boldsymbol{\delta}}{\text{Minimize}} \, \| \mathbf{r}^{(a)} + \mathbf{J}^{(a)}\boldsymbol{\delta} \|^2 \tag{14.12}$$

using numerically stable methods based upon factorizations of $\mathbf{J}^{(a)}$. These factorizations may be orthogonal (QR) factorizations (Appendix A8.3) derived from Householder or Givens transformations, or the singular-value decomposition (Appendix A8.1). For descriptions of QR factorizations see Kennedy and Gentle [1980], Golub and Van Loan [1983: Chapter 6], Lawson and Hanson [1974], Maindonald [1984: Chapter 4], and the algorithms in Dennis and Schnabel [1983].

The principal strengths of Gauss–Newton-based algorithms come from their requiring only the first derivatives of the $r_i(\boldsymbol{\theta})$ stored in $\mathbf{J}$. No separate approximation to the Hessian need be stored. Thus each iteration is less expensive than its Newton counterpart in terms of both time and storage. In problems where $\hat{\mathbf{A}} = \mathbf{A}(\hat{\boldsymbol{\theta}})$ is small compared with $\hat{\mathbf{J}}'\hat{\mathbf{J}}$ in (14.7), Gauss–Newton-based algorithms often take a similar number of steps to a full modified Newton algorithm and are thus much more efficient. They are also faster than the quasi-Newton methods, which are the fastest first-derivative methods for general functions.

We can begin to understand the potential weaknesses of Gauss–Newton-based algorithms by looking at the local convergence properties of Gauss–Newton itself. Recall that we derived the Gauss–Newton algorithm from the Newton algorithm by assuming that $\mathbf{A}(\boldsymbol{\theta})$ was small compared with $\mathbf{J}(\boldsymbol{\theta})'\mathbf{J}(\boldsymbol{\theta})$ in the Hessian (14.7) and could therefore be ignored.

Example 14.1 Dennis [1977: Example 7.1] considered fitting the model

$$y_i = e^{\theta x_i} + \varepsilon_i$$

to the four points $(x_i, y_i) = (1, 2), (2, 4), (2, -11), (3, 8)$. We note that for $i = 1, 2, 4$, $y_i = 2^{x_i} = e^{(\log_e 2)x_i}$, while the third point is a gross outlier. Dennis found that $\hat{\theta} = \log_e 0.188$, $\hat{J}'\hat{J} = 4.58 \times 10^{-2}$ and $\hat{A} = 18 \times 10^{-2}$. Thus, rather than being small compared with $\hat{J}'\hat{J}$, $\hat{A}$ is the dominant part of the Hessian. Dennis further showed that, for the Gauss–Newton algorithm in a small enough neighborhood of $\hat{\theta}$,

$$\| \theta^{(a+1)} - \hat{\theta} \| \approx 4 \| \theta^{(a)} - \hat{\theta} \|.$$

Thus the error in $\theta^{(a)}$ actually increases by a factor of four at each iteration, and GN diverges. Dennis further stated that if the number of good points was increased from three, this bad behavior of GN would be intensified. Readers uninterested in technical details may omit the remainder of this section.

By comparing (14.10) and (14.11) we might expect GN to behave like the Newton algorithm N in any iteration in which $\mathbf{A}^{(a)}$ is small compared with $\mathbf{J}^{(a)\prime}\mathbf{J}^{(a)}$. We would expect ultimate Newton-like behavior if $\hat{\mathbf{A}} = \mathbf{A}(\hat{\boldsymbol{\theta}}) = \mathbf{0}$. Indeed, Dennis and Schnabel [1983: Corollary 10.2.2] proved local convergence of GN at an asymptotically quadratic rate [c.f. (13.12) with $r = 2$] when $\hat{\mathbf{A}} = \mathbf{0}$. From (14.9), $\hat{\mathbf{A}} = \mathbf{0}$ if (i) $\mathbf{r}(\hat{\boldsymbol{\theta}}) = \mathbf{0}$, corresponding to a perfect fit of the nonlinear regression model to the data, or if (ii) each $\partial^2 r_i(\boldsymbol{\theta})/\partial\boldsymbol{\theta}\,\partial\boldsymbol{\theta}' = \mathbf{0}$. The latter occurs if each $r_i(\boldsymbol{\theta})$ [or equivalently $f_i(\boldsymbol{\theta})$] is linear in $\boldsymbol{\theta}$ in the vicinity of $\hat{\boldsymbol{\theta}}$. Thus $\hat{\mathbf{A}}$ may be interpreted as a composite measure of the size of the residuals and the nonlinearity of the problem. Close-to-Newton behavior might be expected if either (i) or (ii) is approximately true, thus producing a value of $\hat{\mathbf{A}}$ which is small compared with $\hat{\mathbf{J}}'\hat{\mathbf{J}}$.

Let λ be the smallest eigenvalue of $\hat{\mathbf{J}}'\hat{\mathbf{J}}$, and let σ be a parameter that is interpreted as $\| \hat{\mathbf{A}} \|_2$ $(= \sqrt{\lambda_1}$, where λ_1 is the maximum eigenvalue of $\hat{\mathbf{A}}'\hat{\mathbf{A}})$ by Dennis and Schnabel [1983: 224]. Under fairly weak conditions these authors, in their Theorem 10.2.1, proved local convergence of GN if $\sigma/\lambda < 1$. Moreover, they proved that for $\boldsymbol{\theta}^{(1)}$ within a small enough neighborhood of $\hat{\boldsymbol{\theta}}$,

$$\| \boldsymbol{\theta}^{(a+1)} - \hat{\boldsymbol{\theta}} \| \leq \frac{1}{2}\left(\frac{\sigma}{\lambda} + 1\right) \| \boldsymbol{\theta}^{(a)} - \hat{\boldsymbol{\theta}} \|, \tag{14.13}$$

thus proving a linear convergence rate for GN. If $\sigma/\lambda > 1$, GN may not be even locally convergent. In fact, Osborne [1972: Theorem 2, Corollary 2] showed that when $\boldsymbol{\theta} = \hat{\boldsymbol{\theta}} + \alpha\mathbf{h}$ ($\|\mathbf{h}\| = 1$), GN is divergent, regardless of the starting point $\boldsymbol{\theta}^{(1)}$ (c.f. Example 14.1 above), if

$$\underset{\mathbf{h}}{\text{Minimum}} \left\| (\mathbf{J}'\mathbf{J})^{-1} \frac{d\mathbf{J}'}{d\alpha} \mathbf{r} \right\|_{\boldsymbol{\theta}} > 1 \tag{14.14}$$

for all directions $\mathbf{h}$ through $\hat{\boldsymbol{\theta}}$.

The considerations of the previous paragraph lead us to discuss Gauss–Newton-based algorithms under two headings: "small-residual problems" and "large-residual problems." In doing this we follow the literature, although by "small-residual problems" we really mean problems for which $\hat{\mathbf{A}}$ is small. As we have seen, $\hat{\mathbf{A}}$ is small if $\mathbf{r}(\hat{\boldsymbol{\theta}})$ is small or the functions $r_i(\hat{\boldsymbol{\theta}})$ are nearly linear in the vicinity of $\hat{\boldsymbol{\theta}}$. However, there can be other causes. For example, if the second derivatives of the $r_i(\boldsymbol{\theta})$ are approximately equal at $\hat{\boldsymbol{\theta}}$, then $\hat{\mathbf{A}}$ [given by (14.9) at $\boldsymbol{\theta}=\hat{\boldsymbol{\theta}}$] will tend to be small because of the cancellation of positive and negative residuals.

Gauss–Newton-based algorithms for small-residual problems (Section 14.2) are derived by taking the GN step as a search direction and modifying the basic GN algorithm in the same ways that we modified the classical N algorithm in Section 13.3. This gives such popular least-squares algorithms as Hartley's [1961] method and the Marquardt [1963] method. In Section 14.3 we discuss much more recent work on methods for large-residual problems. There, an attempt is made to more closely approximate the full Newton direction of search by approximating $\mathbf{A}^{(a)}$ in (14.10).

By applying (14.6), the GN step (14.11) can be written as

$$\boldsymbol{\delta}^{(a)} = -\tfrac{1}{2}(\mathbf{J}^{(a)\prime}\mathbf{J}^{(a)})^{-1}\mathbf{g}^{(a)}. \tag{14.15}$$

By Appendix A4.6, $\mathbf{J}^{(a)\prime}\mathbf{J}^{(a)}$ is positive semidefinite and thus has nonnegative eigenvalues. If it is nonsingular, it is also positive definite and, by Theorem 13.1 of Section 13.2.3b, the direction of the GN step (14.15) is a descent direction. However, $\mathbf{J}^{(a)}$ may not be of full column rank, in which case (14.11) and (14.15) are not defined. A GN step can still be obtained using the formulation (14.12), but the solution to (14.12) is no longer unique. However, in statistical work a singular $\mathbf{J}^{(a)\prime}\mathbf{J}^{(a)}$, and particularly a singular $\hat{\mathbf{J}}'\hat{\mathbf{J}}$, tends to signal a lack of identifiability of the parameters $\boldsymbol{\theta}$ in the nonlinear regression model.

If $\mathbf{J}^{(a)}$ is ill conditioned, there are numerical problems in solving (14.12). However, ill-conditioning of $\hat{\mathbf{J}}=\mathbf{J}(\hat{\boldsymbol{\theta}})$ leads to more fundamental problems. When $\hat{\mathbf{J}}'\hat{\mathbf{J}}$ has some (comparatively speaking) very small eigenvalues, it follows from (14.13) and the surrounding discussion that the GN algorithm may be *locally* divergent. Even if it is convergent, the ultimate reduction in $\|\hat{\boldsymbol{\theta}}^{(a)}-\hat{\boldsymbol{\theta}}\|$ at each step depends upon the ratio σ/λ, where λ is the smallest eigenvalue of $\hat{\mathbf{J}}'\hat{\mathbf{J}}$ [see (14.13)]. Thus if λ is small, the ultimate convergence of GN may be extremely slow.

14.2 METHODS FOR SMALL-RESIDUAL PROBLEMS

14.2.1 Hartley's Method

To make practical use of the Gauss–Newton algorithm, modifications similar to those applied to the Newton method in Section 13.3 must be made. The method, variously known as the modified Gauss–Newton method, the damped Gauss–

Newton method, or Hartley's method, introduced by Hartley [1961], merely uses a line search along the Gauss–Newton direction, i.e.

$$\boldsymbol{\delta}^{(a)} = \rho^{(a)}\mathbf{d}^{(a)}, \tag{14.16}$$

where $\mathbf{d}^{(a)}$ minimizes [c.f. (14.11), (14.12)]

$$\| \mathbf{r}^{(a)} + \mathbf{J}^{(a)}\mathbf{d} \|^2 \tag{14.17}$$

with respect to $\mathbf{d}$. Further work has led to improved implementations, for example the use of stable numerical methods that do not form $\mathbf{J}^{(a)\prime}\mathbf{J}^{(a)}$ to minimize (14.17), as discussed near (14.12). Dennis [1977] recommends solving (14.17) using the singular-value decomposition method, which, although more expensive than the Householder or Givens transformation methods, yields much more useful information in the case of singular or near-singular $\mathbf{J}^{(a)\prime}\mathbf{J}^{(a)}$. Of course when $\mathbf{J}^{(a)\prime}\mathbf{J}^{(a)}$ is singular, (14.17) no longer has a unique solution. One possibility, suggested by Ben-Israel, is to use the solution of (14.17) with minimum Euclidean norm. In this case the solution does not have a component in the null space of $\mathbf{J}'\mathbf{J}$ and therefore belongs to the range space of $\mathbf{J}'\mathbf{J}$. A detailed exposition of this idea follows equation (14.51). Care must be taken to avoid searching for a minimum only in a subspace of $\mathbb{R}^p$ over a number of iterations. Catering for a rank-deficient $\mathbf{J}^{(a)\prime}\mathbf{J}^{(a)}$ is a complicated problem in practice because of difficulties inherent in making decisions about the rank of a matrix in the presence of roundoff errors. In general, implementations of Hartley's method are not as robust as good implementations of the Levenberg–Marquardt technique to follow.

14.2.2 Levenberg–Marquardt Methods

Using ideas due to Levenberg [1944] and Marquardt [1963], Levenberg–Marquardt algorithms allow for singular or ill-conditioned matrices $\mathbf{J}^{(a)\prime}\mathbf{J}^{(a)}$ by modifying the Gauss–Newton step (14.11) to

$$\boldsymbol{\delta}^{(a)} = -(\mathbf{J}^{(a)\prime}\mathbf{J}^{(a)} + \eta^{(a)}\mathbf{D}^{(a)})^{-1}\mathbf{J}^{(a)\prime}\mathbf{r}^{(a)}, \tag{14.18}$$

where $\mathbf{D}^{(a)}$ is a diagonal matrix with positive diagonal elements. Often, for simplicity, $\mathbf{D}^{(a)} = \mathbf{I}_p$. A popular choice is to set the diagonal elements of $\mathbf{D}^{(a)}$ to be the same as those of $\mathbf{J}^{(a)\prime}\mathbf{J}^{(a)}$, so that the method is approximately invariant under rescaling of the $\boldsymbol{\theta}$. Again, good modern implementations do not compute $\mathbf{J}^{(a)\prime}\mathbf{J}^{(a)}$, but instead calculate $\boldsymbol{\delta}^{(a)}$ as the solution of the linear least-squares problem

$$\underset{\boldsymbol{\delta}}{\text{Minimize}} \left\| \begin{pmatrix} \mathbf{r}^{(a)} \\ \mathbf{0} \end{pmatrix} + \begin{pmatrix} \mathbf{J}^{(a)} \\ (\eta^{(a)}\mathbf{D}^{(a)})^{1/2} \end{pmatrix} \boldsymbol{\delta} \right\|^2 \tag{14.19}$$

using numerically stable methods. However, this issue is less critical with

Levenberg–Marquardt algorithms than it is with Hartley's method because $\mathbf{J}'\mathbf{J} + \eta\mathbf{D}$ is better conditioned than $\mathbf{J}'\mathbf{J}$.

When $\mathbf{D}^{(a)} = \mathbf{I}_p$, the Levenberg–Marquardt direction [i.e. the direction of $\boldsymbol{\delta}^{(a)}$ in (14.18)] interpolates between the Gauss–Newton direction ($\eta^{(a)} \to 0$) and the steepest-descent direction ($\eta^{(a)} \to \infty$). Also, as the direction tends to steepest descent, the step length $\|\boldsymbol{\delta}^{(a)}\|$ tends to zero. For $\eta^{(a)} > 0$, $\mathbf{J}^{(a)\prime}\mathbf{J}^{(a)} + \eta^{(a)}\mathbf{D}^{(a)}$ is positive definite, as $\mathbf{D}^{(a)}$ is positive definite. Thus, from Theorem 13.1 of Section 13.2.3b, $\boldsymbol{\delta}^{(a)}$ of (14.18) is a descent direction. As we have noted, as $\eta^{(a)} \to \infty$ the length $\boldsymbol{\delta}^{(a)}$ tends to zero. Thus by choosing $\eta^{(a)}$ large enough we can reduce $S(\boldsymbol{\theta}) = \sum_i r_i^2(\boldsymbol{\theta})$. However, if $\eta^{(a)}$ is too large for too many iterations, the algorithm will take too many small steps and thus make little progress.

Levenberg–Marquardt algorithms differ in how they choose and update $\eta^{(a)}$. Originally Levenberg [1944] chose $\eta^{(a)}$ to minimize $S(\boldsymbol{\theta})$ for $\boldsymbol{\theta}$ of the form $\boldsymbol{\theta}^{(a)} + \boldsymbol{\delta}^{(a)}$, where $\boldsymbol{\delta}^{(a)}$ is given by (14.18). This has been abandoned because each trial value of $\eta^{(a)}$ requires the (expensive) solution of another least-squares problem (14.19). Marquardt [1963] adopted a much cheaper strategy. Initially a small positive value was taken, e.g. $\eta^{(1)} = 0.01$. If, at the ath iteration, the step $\boldsymbol{\delta}^{(a)}$ of (14.18) reduces $S(\boldsymbol{\theta})$, he sets $\boldsymbol{\theta}^{(a+1)} = \boldsymbol{\theta}^{(a)} + \boldsymbol{\delta}^{(a)}$ and divides η by a factor, e.g. $\eta^{(a+1)} = \eta^{(a)}/10$, to push the algorithm closer to Gauss–Newton [and to bigger steps for the next iteration, by (14.18)]. If within the ath iteration the step $\boldsymbol{\delta}^{(a)}$ does not reduce $S(\boldsymbol{\theta})$, he progressively increases $\eta^{(a)}$ by a factor, e.g. $\eta^{(a)} \to 10\eta^{(a)}$, each time recomputing $\boldsymbol{\delta}^{(a)}$ until a reduction in $S(\boldsymbol{\theta})$ is achieved. In contrast to Levenberg's approach above, we see that (14.19) needs to be solved several times with different η within some of the iterations only.

Some of the more modern Levenberg–Marquardt algorithms use a full trust-region approach (c.f. Section 13.3.2) to update $\eta^{(a)}$ in place of the ad hoc method used by Marquardt [1963] (see Fletcher [1971], Moré [1977]). Nash [1977] incorporates a line search into a Levenberg–Marquardt algorithm. As a result of some theoretical analysis, Dennis [1977: 281] recommends the use of "a cursory line search" in such algorithms. Dennis and Schnabel [1983: 228] particularly recommend the implementation of Moré [1977], which appears in MINPACK (Section 15.1.2). There is general agreement that, in practice, Levenberg–Marquardt algorithms have been proved to be good general-purpose algorithms for least-squares problems. Generally they are robust and work well. However, on large-residual problems their linear convergence can be very slow and they may even fail. Readers uninterested in technical details may omit the remainder of this section.

Nazareth [1980: Section 2.3] highlights a reason why Levenberg–Marquardt algorithms may sometimes do poorly. If $\mathbf{J}^{(a)}$ has rank q ($< p$), so that $\mathbf{J}^{(a)\prime}\mathbf{J}^{(a)}$ is singular, then (14.17) does not have a unique solution. However, if the minimum-norm solution is used, it will belong to the range space of $\mathbf{J}^{(a)\prime}\mathbf{J}^{(a)}$, and the modified Gauss–Newton algorithm may get "bogged down" in a subregion of $\mathbb{R}^p$. We now show that a Levenberg–Marquardt search direction has the same property when $\mathbf{D}^{(a)} = \mathbf{I}_p$ in (14.18).

Suppose the parameters $\boldsymbol{\theta}$ are permuted so that the nonzero eigenvalues of $\mathbf{J}'\mathbf{J}$

are the first q eigenvalues. Then the range space of $\mathbf{J}'\mathbf{J}$ is of dimension q and the null space has dimension $p-q$ (by Appendix A1.2). We take a singular-value decomposition of $\mathbf{J}$, namely (Appendix A8.1)

$$\mathbf{J}=\mathbf{U}\begin{pmatrix}\mathbf{S}_1 & \mathbf{0}\\ \mathbf{0} & \mathbf{0}\end{pmatrix}\mathbf{T}', \tag{14.20}$$

where $\mathbf{U}$ and $\mathbf{T}$ are orthogonal $n\times n$ and $p\times p$ matrices respectively, and $\mathbf{S}_1$ is diagonal with positive elements $\sqrt{\lambda_1}, \sqrt{\lambda_2},\dots,\sqrt{\lambda_q}$ $(q<p)$. Partition $\mathbf{T}=(\mathbf{T}_1,\mathbf{T}_2)$, where $\mathbf{T}_1$ consists of the first q columns of $\mathbf{T}$. Then

$$\mathbf{J}'\mathbf{J}=(\mathbf{T}_1,\mathbf{T}_2)\begin{pmatrix}\mathbf{\Lambda}_1 & \mathbf{0}\\ \mathbf{0} & \mathbf{0}\end{pmatrix}\begin{pmatrix}\mathbf{T}_1'\\ \mathbf{T}_2'\end{pmatrix}=\mathbf{T}_1\mathbf{\Lambda}_1\mathbf{T}_1', \tag{14.21}$$

where $\mathbf{\Lambda}_1=\operatorname{diag}(\lambda_1,\lambda_2,\dots,\lambda_q)$ consists of the nonzero eigenvalues of $\mathbf{J}'\mathbf{J}$. This is the spectral decomposition of $\mathbf{J}'\mathbf{J}$. As $\mathbf{T}_1'\mathbf{T}_2=\mathbf{0}$, we have from (14.21) and (14.20) respectively

$$\mathbf{J}'\mathbf{J}\mathbf{T}_2=\mathbf{0} \tag{14.22}$$

and

$$\mathbf{J}\mathbf{T}_2=\mathbf{0}. \tag{14.23}$$

From (14.22), the $p-q$ orthogonal columns of $\mathbf{T}_2$ are all in the null space of $\mathbf{J}'\mathbf{J}$ and thus form a basis. The q columns of $\mathbf{T}_1$ then form a basis of the range space of $\mathbf{J}'\mathbf{J}$ (by Appendix A1.3). Any vector direction $\mathbf{d}$ can be written $\mathbf{d}=\mathbf{d}_1+\mathbf{d}_2$, where $\mathbf{d}_1$ and $\mathbf{d}_2$ lie in the range and null spaces respectively of $\mathbf{J}'\mathbf{J}$. Thus $\mathbf{d}_1=\mathbf{T}_1\mathbf{u}$ for some $\mathbf{u}$, and $\mathbf{d}_2=\mathbf{T}_2\mathbf{v}$ for some $(p-q)$-vector $\mathbf{v}$. From (14.18), when $\mathbf{D}=\mathbf{I}_p$ the Levenberg–Marquardt step $\mathbf{d}$ satisfies

$$\mathbf{J}'\mathbf{J}\mathbf{d}+\eta\mathbf{d}=-\mathbf{J}'\mathbf{r}. \tag{14.24}$$

Thus writing $\mathbf{d}=\mathbf{T}_1\mathbf{u}+\mathbf{T}_2\mathbf{v}$ and using (14.22), Equation (14.24) becomes

$$\mathbf{J}'\mathbf{J}\mathbf{T}_1\mathbf{u}+\eta\mathbf{T}_1\mathbf{u}+\eta\mathbf{T}_2\mathbf{v}=-\mathbf{J}'\mathbf{r}. \tag{14.25}$$

Multiplying by $\mathbf{T}_2'$, and using (14.23) and $\mathbf{T}_2'\mathbf{T}_1=\mathbf{0}$, we get

$$\mathbf{0}=\eta\mathbf{T}_2'\mathbf{T}_2\mathbf{v}=\eta\mathbf{v}. \tag{14.26}$$

Thus for $\eta>0$, we have $\mathbf{v}=\mathbf{0}$ and $\mathbf{d}$ has no component in the null space of $\mathbf{J}'\mathbf{J}$. Hence $\mathbf{d}$ belongs to the range space of $\mathbf{J}'\mathbf{J}$.

By contrast, consider the Newton step (14.10). For this step $\mathbf{d}$ satisfies

$$\mathbf{J}'\mathbf{J}\mathbf{d}+\mathbf{A}\mathbf{d}=-\mathbf{J}'\mathbf{r}. \tag{14.27}$$

If we write $\mathbf{d} = \mathbf{T}_1\mathbf{u} + \mathbf{T}_2\mathbf{v}$ as before and repeat the steps following (14.24), we obtain

$$\mathbf{T}_2'\mathbf{A}\mathbf{T}_2\mathbf{v} = -\mathbf{T}_2'\mathbf{A}\mathbf{T}_1\mathbf{u}. \tag{14.28}$$

This imposes no requirement that $\mathbf{v} = \mathbf{0}$, so that the Newton step need not be confined to the range space of $\mathbf{J}'\mathbf{J}$.

14.2.3 Powell's Hybrid Method

Before concluding our discussion on small-residual problems we should mention Powell's [1970] hybrid method, which is roughly competitive with, and qualitatively similar to, the Levenberg–Marquardt method. The algorithm is given by Scales [1985: Section 4.2.3]. At each iteration, the Gauss–Newton step $\boldsymbol{\delta}_{\text{GN}}^{(a)}$ is calculated and used if it falls within a progressively updated trust region $\|\boldsymbol{\delta}^{(a)}\| \leq \Delta^{(a)}$. Otherwise, let $\boldsymbol{\delta}_{\text{SD}}^{(a)} = -\rho^{(a)}\mathbf{g}^{(a)}$ be a step in the steepest-descent direction with $\rho^{(a)} = \frac{1}{2}\|\mathbf{J}^{(a)\prime}\mathbf{r}^{(a)}\|^2/\|\mathbf{J}^{(a)}\mathbf{J}^{(a)\prime}\mathbf{r}^{(a)}\|^2$. The step size $\rho^{(a)}$ is readily shown to minimize the quadratic approximation [c.f. (14.12)] of $S(\boldsymbol{\theta}^{(a)} + \boldsymbol{\delta})$ along the direction of steepest descent, i.e. with respect to ρ when $\boldsymbol{\delta} = -\rho\mathbf{g}$. If $\boldsymbol{\delta}_{\text{SD}}^{(a)}$ lies within the trust region, i.e. $\|\boldsymbol{\delta}_{\text{SD}}^{(a)}\| \leq \Delta^{(a)}$, then $\boldsymbol{\delta}^{(a)}$ is obtained by searching the line joining $\boldsymbol{\delta}_{\text{SD}}^{(a)}$ and $\boldsymbol{\delta}_{\text{GN}}^{(a)}$. This is the so-called *dog-leg* step. Finally, if $\|\boldsymbol{\delta}_{\text{SD}}^{(a)}\| > \Delta^{(a)}$ a steepest-descent step is taken to the edge of the trust region.

14.3 LARGE-RESIDUAL PROBLEMS

14.3.1 Preliminaries

The central idea underlying Gauss–Newton-based algorithms for minimizing $S(\boldsymbol{\theta}) = \mathbf{r}(\boldsymbol{\theta})'\mathbf{r}(\boldsymbol{\theta})$ is the approximation of the Hessian $2(\mathbf{J}'\mathbf{J} + \mathbf{A})$ of (14.7) by $2\mathbf{J}'\mathbf{J}$, which involves assuming that $\mathbf{A} = \mathbf{A}(\boldsymbol{\theta})$ given by (14.9) is small in comparison with $\mathbf{J}'\mathbf{J}$. In the previous section we found that the best Gauss–Newton-based algorithms were based upon the Levenberg–Marquardt method and noted that such algorithms are good general-purpose algorithms for least-squares problems. However, they sometimes run into difficulties when $\hat{\mathbf{A}} = \mathbf{A}(\hat{\boldsymbol{\theta}})$ is large, the so-called "large-residual" problem. Levenberg–Marquardt algorithms generally converge, but convergence can be unacceptably slow. Occasionally they will fail.

To build algorithms which are faster and more robust on problems with large $\hat{\mathbf{A}}$ it is necessary to incorporate information about $\mathbf{A}(\boldsymbol{\theta})$ into the approximation to the Hessian. Recall from (14.9) that

$$\mathbf{A}(\boldsymbol{\theta}) = \sum_{i=1}^{n} r_i(\boldsymbol{\theta}) \frac{\partial^2 r_i(\boldsymbol{\theta})}{\partial\boldsymbol{\theta}\,\partial\boldsymbol{\theta}'}.$$

This involves the calculation of n $p \times p$ matrices of second derivatives. Explicit calculation of $\mathbf{A}(\boldsymbol{\theta})$ from analytic expressions for the second derivatives leads us back to the full Newton method. Gill and Murray [1978] point out that there has been little attention given in the literature to nonlinear least-squares minimization in which second derivatives can be provided. Yet they note that in many statistical problems in which r_i takes the form

$$r_i = y_i - f(\mathbf{x}_i; \boldsymbol{\theta}),$$

the values of all of the r_i and their first and second derivatives can be evaluated using a *single* subroutine which accepts y_i, $\mathbf{x}_i$, and $\boldsymbol{\theta}$ as parameters. In many of the models described in this book, obtaining second derivatives of $f(\mathbf{x}_i; \boldsymbol{\theta})$ with respect to $\boldsymbol{\theta}$ is not particularly onerous. Gill and Murray [1978] give a hybrid modified Newton/Gauss–Newton algorithm which can be useful when $\hat{\mathbf{A}}$ is not small. We describe this algorithm in Section 14.3.3. When analytic second derivatives are not available, the approximation of the np^2 second derivatives in $\mathbf{A}(\boldsymbol{\theta})$ by finite differences is generally considered unacceptably time-consuming.

A common strategy (Chambers [1973]) has been to switch from a Levenberg–Marquardt algorithm to a quasi-Newton method for general functions if the former proves inadequate. McKeown [1975], for example, demonstrates a range of large-residual least-squares test problems in which each $r_i(\boldsymbol{\theta})$ is quadratic in $\boldsymbol{\theta}$ and for which quasi-Newton algorithms for general functions are more efficient than (specialized) Levenberg–Marquardt algorithms. Gauss–Newton-type information about the Hessian can be incorporated to a very limited extent by starting the quasi-Newton approximations to the Hessian with $\mathbf{B}^{(1)} = 2\mathbf{J}^{(1)\prime}\mathbf{J}^{(1)}$ rather than $\mathbf{B}^{(1)} = \mathbf{I}_p$. However, the quasi-Newton methods approximate all of the Hessian $\mathbf{H}(\boldsymbol{\theta}) = 2(\mathbf{J}'\mathbf{J} + \mathbf{A})$. This ignores the fact that part of the Hessian, namely $\mathbf{J}'\mathbf{J}$, is available analytically. The techniques below use quasi-Newton methods to approximate just the unknown part of the Hessian, namely $\mathbf{A}(\boldsymbol{\theta})$. To distinguish quasi-Newton methods for general functions from the methods which follow, we shall call the former *full* quasi-Newton methods.

The following two sections are rather technical. Readers interested only in the central ideas may skip to the summary section 14.3.4.

14.3.2 Quasi-Newton Approximation of A(θ)

Let $\tilde{\mathbf{A}}^{(a)}$ be the approximation to $\mathbf{A}$ at the ath iteration. Then the approximate Hessian is

$$\mathbf{B}^{(a)} = 2[\mathbf{J}^{(a)\prime}\mathbf{J}^{(a)} + \tilde{\mathbf{A}}^{(a)}]. \tag{14.29}$$

The step $\boldsymbol{\delta}^{(a)} = \boldsymbol{\theta}^{(a+1)} - \boldsymbol{\theta}^{(a)}$ is formed using one of the standard modifications of the Newton method described in Section 13.3. For example, a step-length method (Section 13.3.1) uses

$$\boldsymbol{\delta}^{(a)} = \rho^{(a)}\mathbf{d}^{(a)}, \qquad \mathbf{d}^{(a)} = -\bar{\mathbf{B}}^{(a)-1}\mathbf{g}^{(a)}. \tag{14.30}$$

Here $\bar{\mathbf{B}}^{(a)}$ is obtained by modifying $\mathbf{B}^{(a)}$ if necessary to make it sufficiently positive definite, and $\rho^{(a)}$, the step length, is obtained using a line search along the direction $\mathbf{d}^{(a)}$. More often in the literature, a trust-region (or Levenberg–Marquardt) method as described in Section 13.3.2 has been used. However, we still have to decide how to initialize and update the approximation $\tilde{\mathbf{A}}^{(a)}$.

Suppose we apply the quasi-Newton condition (13.35) to $\mathbf{B}^{(a+1)}$. This gives

$$\mathbf{B}^{(a+1)}\boldsymbol{\delta}^{(a)} = \boldsymbol{\gamma}^{(a)} = \mathbf{g}^{(a+1)} - \mathbf{g}^{(a)}, \tag{14.31}$$

namely

$$2\{\mathbf{J}^{(a+1)\prime}\mathbf{J}^{(a+1)} + \tilde{\mathbf{A}}^{(a+1)}\}\boldsymbol{\delta}^{(a)} = 2\{\mathbf{J}^{(a+1)\prime}\mathbf{r}^{(a+1)} - \mathbf{J}^{(a)\prime}\mathbf{r}^{(a)}\}.$$

Hence

$$\tilde{\mathbf{A}}^{(a+1)}\boldsymbol{\delta}^{(a)} = \tilde{\boldsymbol{\gamma}}^{(a)}, \tag{14.32}$$

where

$$\tilde{\boldsymbol{\gamma}}^{(a)} = \mathbf{J}^{(a+1)\prime}\mathbf{r}^{(a+1)} - \mathbf{J}^{(a)\prime}\mathbf{r}^{(a)} - \mathbf{J}^{(a+1)\prime}\mathbf{J}^{(a+1)}\boldsymbol{\delta}^{(a)}. \tag{14.33}$$

Equation (14.32) is identical in form to the quasi-Newton condition. However, different methods define $\tilde{\boldsymbol{\gamma}}^{(a)}$ in (14.32) slightly differently. The form (14.33) was used in Dennis's *Broyden–Dennis method* published in 1973. We originally obtained the quasi-Newton condition from [c.f. (13.34)]

$$\mathbf{H}^{(a)}\boldsymbol{\delta}^{(a)} \approx \mathbf{g}^{(a+1)} - \mathbf{g}^{(a)},$$

which gives us

$$\mathbf{A}^{(a)}\boldsymbol{\delta}^{(a)} \approx \mathbf{J}^{(a+1)\prime}\mathbf{r}^{(a+1)} - \mathbf{J}^{(a)\prime}\mathbf{r}^{(a)} - \mathbf{J}^{(a)\prime}\mathbf{J}^{(a)}\boldsymbol{\delta}^{(a)}. \tag{14.34}$$

However, $\mathbf{A}^{(a)}$ is required to obtain the step $\boldsymbol{\delta}^{(a)}$, so, arguing as we did following (13.34), we cannot impose the condition (14.34) upon $\tilde{\mathbf{A}}^{(a)}$. If, instead, we impose (14.34) on $\tilde{\mathbf{A}}^{(a+1)}$, we obtain $\tilde{\mathbf{A}}^{(a+1)}\boldsymbol{\delta}^{(a)} = \tilde{\boldsymbol{\gamma}}^{(a)}$ as before, but with

$$\tilde{\boldsymbol{\gamma}}^{(a)} = \mathbf{J}^{(a+1)\prime}\mathbf{r}^{(a+1)} - \mathbf{J}^{(a)\prime}\mathbf{r}^{(a)} - \mathbf{J}^{(a)\prime}\mathbf{J}^{(a)}\boldsymbol{\delta}^{(a)}. \tag{14.35}$$

This form for $\tilde{\boldsymbol{\gamma}}^{(a)}$ was used by Betts [1976].

Finally, in their method, Dennis et al. [1981a, b] note that

$$\begin{aligned}\mathbf{A}(\boldsymbol{\theta}) &= \sum_{i=1}^{n} r_i(\boldsymbol{\theta})\frac{\partial^2 r_i(\boldsymbol{\theta})}{\partial\boldsymbol{\theta}\,\partial\boldsymbol{\theta}'},\\ &= \sum_{i=1}^{n} r_i(\boldsymbol{\theta})\mathbf{H}_i,\end{aligned} \tag{14.36}$$

say, where $\mathbf{H}_i$ is the Hessian of $r_i(\boldsymbol{\theta})$. Suppose that $\tilde{\mathbf{H}}_i^{(a)}$ is an approximation to $\mathbf{H}_i$.

If we impose the quasi-Newton condition upon $\tilde{\mathbf{H}}_i^{(a)}$, we obtain

$$\tilde{\mathbf{H}}_i^{(a+1)}\boldsymbol{\delta}^{(a)} = \frac{\partial r_i(\boldsymbol{\theta}^{(a+1)})}{\partial \boldsymbol{\theta}} - \frac{\partial r_i(\boldsymbol{\theta}^{(a)})}{\partial \boldsymbol{\theta}}.$$

Defining

$$\tilde{\mathbf{A}}^{(a+1)} = \sum_{i=1}^{n} r_i^{(a+1)}\tilde{\mathbf{H}}_i^{(a+1)}, \tag{14.37}$$

then

$$\begin{aligned} \tilde{\mathbf{A}}^{(a+1)}\boldsymbol{\delta}^{(a)} &= \sum_{i=1}^{n} r_i^{(a+1)}[(\text{row } i \text{ of } \mathbf{J}^{(a+1)}) - (\text{row } i \text{ of } \mathbf{J}^{(a)})]' \\ &= (\mathbf{J}^{(a+1)} - \mathbf{J}^{(a)})'\mathbf{r}^{(a+1)}. \end{aligned} \tag{14.38}$$

Again we have $\tilde{\mathbf{A}}^{(a+1)}\boldsymbol{\delta}^{(a)} = \tilde{\boldsymbol{\gamma}}^{(a)}$, but now

$$\tilde{\boldsymbol{\gamma}}^{(a)} = (\mathbf{J}^{(a+1)} - \mathbf{J}^{(a)})'\mathbf{r}^{(a+1)}. \tag{14.39}$$

Setting $u(t) = r_i(\boldsymbol{\theta}^{(a)} + t\boldsymbol{\delta}^{(a)})$, we can use

$$u(1) = u(0) + \int_0^1 \dot{u}(t)\, dt,$$

to get

$$r_i^{(a+1)} = r_i^{(a)} + \int_0^1 \frac{\partial r_i(\boldsymbol{\theta}^{(a)} + t\boldsymbol{\delta}^{(a)})}{\partial \boldsymbol{\theta}'}\boldsymbol{\delta}^{(a)}\, dt.$$

Thus

$$\begin{aligned} \mathbf{r}^{(a+1)} &= \mathbf{r}^{(a)} + \left(\int_0^1 \mathbf{J}(\boldsymbol{\theta}^{(a)} + t\boldsymbol{\delta}^{(a)})\, dt\right)\boldsymbol{\delta}^{(a)} \\ &= \mathbf{r}^{(a)} + \bar{\mathbf{J}}^{(a)}\boldsymbol{\delta}^{(a)}, \end{aligned} \tag{14.40}$$

say. Using (14.40) to replace $\mathbf{r}^{(a+1)}$ in the term $\mathbf{J}^{(a)\prime}\mathbf{r}^{(a+1)}$ of (14.39), the latter equation becomes

$$\tilde{\boldsymbol{\gamma}}^{(a)} = \mathbf{J}^{(a+1)\prime}\mathbf{r}^{(a+1)} - \mathbf{J}^{(a)\prime}\mathbf{r}^{(a)} - \mathbf{J}^{(a)\prime}\bar{\mathbf{J}}^{(a)}\boldsymbol{\delta}^{(a)}.$$

This result, noted by Nazareth [1980], shows that (14.39) is much more similar to (14.33) and (14.35) than is first apparent.

To summarize, by comparing (14.32) with (14.31) we see that the latter is a quasi-Newton condition which we would like to impose upon our updated approximation to $\mathbf{A}(\boldsymbol{\theta})$. Three very similar candidates, (14.33), (14.35), and (14.39), have been proposed for $\tilde{\gamma}^{(a)}$. Dennis et al. [1981a: 351] stated that they tested all three and that (14.39) came out "as the slight but clear winner."

Proposed algorithms have differed in terms of the quasi-Newton update formulae imposed. Betts [1976] used the symmetric rank-one formula $\tilde{\mathbf{A}}^{(a+1)} = \tilde{\mathbf{A}}^{(a)} + \mathbf{E}^{(a)}$, where the update $\mathbf{E}^{(a)}$ is given by (13.37) with $\mathbf{B}^{(a)}$ and $\boldsymbol{\gamma}^{(a)}$ replaced by $\tilde{\mathbf{A}}^{(a)}$ and $\tilde{\boldsymbol{\gamma}}^{(a)}$ of (14.35), respectively. The Broyden–Dennis method employs the Powell–symmetric-Broyden (PSB) update

$$\mathbf{E}^{(a)} = \frac{(\tilde{\boldsymbol{\gamma}}^{(a)} - \tilde{\mathbf{A}}^{(a)}\boldsymbol{\delta}^{(a)})\boldsymbol{\delta}^{(a)\prime} + \boldsymbol{\delta}^{(a)}(\tilde{\boldsymbol{\gamma}}^{(a)} - \tilde{\mathbf{A}}^{(a)}\boldsymbol{\delta}^{(a)})'}{\boldsymbol{\delta}^{(a)\prime}\boldsymbol{\delta}^{(a)}} - \frac{(\tilde{\boldsymbol{\gamma}}^{(a)} - \tilde{\mathbf{A}}^{(a)}\boldsymbol{\delta}^{(a)})'\boldsymbol{\delta}^{(a)}}{(\boldsymbol{\delta}^{(a)\prime}\boldsymbol{\delta}^{(a)})^2}\boldsymbol{\delta}^{(a)}\boldsymbol{\delta}^{(a)\prime}. \tag{14.41}$$

However, Dennis et al. [1981a, b] use a slight variant of the Davidon–Fletcher–Powell (DFP) update, namely

$$\mathbf{E}^{(a)} = \frac{(\tilde{\boldsymbol{\gamma}}^{(a)} - \tilde{\mathbf{A}}^{(a)}\boldsymbol{\delta}^{(a)})\boldsymbol{\gamma}^{(a)\prime} + \boldsymbol{\gamma}^{(a)}(\tilde{\boldsymbol{\gamma}}^{(a)} - \tilde{\mathbf{A}}^{(a)}\boldsymbol{\delta}^{(a)})'}{\boldsymbol{\delta}^{(a)\prime}\boldsymbol{\gamma}^{(a)}} - \frac{\boldsymbol{\delta}^{(a)\prime}(\tilde{\boldsymbol{\gamma}}^{(a)} - \tilde{\mathbf{A}}^{(a)}\boldsymbol{\delta}^{(a)})}{(\boldsymbol{\delta}^{(a)\prime}\boldsymbol{\gamma}^{(a)})^2}\boldsymbol{\gamma}^{(a)}\boldsymbol{\gamma}^{(a)\prime}, \tag{14.42}$$

where $\boldsymbol{\gamma}^{(a)} = \mathbf{g}^{(a+1)} - \mathbf{g}^{(a)} = 2(\mathbf{J}^{(a+1)\prime}\mathbf{r}^{(a+1)} - \mathbf{J}^{(a)\prime}\mathbf{r}^{(a)})$. We note that the multiplicative factor 2 in the formulae for $\boldsymbol{\gamma}^{(a)}$ can be omitted without affecting (14.42).

Quasi-Newton methods for approximating $\mathbf{A}(\boldsymbol{\theta})$ are not as securely based as the quasi-Newton methods for building up an approximation to the Hessian. The matrix $\mathbf{A}(\boldsymbol{\theta})$ need not be positive definite even at $\hat{\boldsymbol{\theta}}$, and may in fact be singular. However, $\mathbf{A}(\boldsymbol{\theta})$ is always symmetric, so it is natural to require the updates to be symmetric. Heuristically, if $\tilde{\mathbf{A}}^{(a)}$ is a reasonable approximation to $\mathbf{A}^{(a)}$ and if $\mathbf{A}(\boldsymbol{\theta})$ is not changing fast, then it is desirable that the update $\mathbf{E}^{(a)}$ to form $\tilde{\mathbf{A}}^{(a+1)} = \tilde{\mathbf{A}}^{(a)} + \mathbf{E}^{(a)}$ be "small." Let us denote by $\|\mathbf{C}\|_F = (\sum\sum c_{ij}^2)^{1/2}$ the Frobenius norm of a square matrix $\mathbf{C}$. If $\mathbf{R}$ is a symmetric, positive definite square matrix, then it has a symmetric positive definite square root $\mathbf{R}^{1/2}$. We define $\|\mathbf{C}\|_{F,\mathbf{R}} = \|\mathbf{R}^{-1/2}\mathbf{C}\mathbf{R}^{-1/2}\|_F$. Dennis et al. [1981a] showed that this weighted Frobenius norm is a natural analogue of the Frobenius norm for a matrix when the standard inner product $\|\boldsymbol{\theta}\| = (\boldsymbol{\theta}'\boldsymbol{\theta})^{1/2}$ is replaced by $\|\boldsymbol{\theta}\|_{\mathbf{R}} = (\boldsymbol{\theta}'\mathbf{R}\boldsymbol{\theta})^{1/2}$.

Theorem 3.1 of Dennis et al. [1981a] gives the following justification of the update (14.42). Consider the class Q of symmetric update matrices $\mathbf{E}$ such that

$$(\tilde{\mathbf{A}}^{(a)} + \mathbf{E})\boldsymbol{\delta}^{(a)} = \tilde{\boldsymbol{\gamma}}^{(a)}$$

(the quasi-Newton condition for $\tilde{\mathbf{A}}^{(a+1)}$). Then it transpires that $\mathbf{E} = \mathbf{E}^{(a)}$ of (14.42) has minimum weighted Frobenius norm $\|\mathbf{E}\|_{F,\mathbf{R}}$ for any weighting matrix $\mathbf{R}$ for which

$$\mathbf{R}\boldsymbol{\delta}^{(a)} = \boldsymbol{\gamma}^{(a)}. \tag{14.43}$$

We note that (14.43) is the quasi-Newton condition (14.31) satisfied by approximations to the Hessian. Referring to the above choice of norm, Dennis et al. [1981a] state, "Thus we hope the metric being used is not too different from that induced by the natural scaling of the problem." A starting value of $\mathbf{A}^{(1)} = \mathbf{0}$ is used by these authors and others because Gauss–Newton-based algorithms usually make good progress initially.

Dennis et al. [1981a: 351] observe that "It is well known by now that the update methods do not generate approximations that become arbitrarily accurate as the iteration proceeds." One of the problems is that on a "zero-residual" problem, $\mathbf{A}(\boldsymbol{\theta}) \to \mathbf{0}$ as $\boldsymbol{\theta} \to \hat{\boldsymbol{\theta}}$ and therefore $\tilde{\mathbf{A}}^{(a)}$ should tend to zero. As an extreme example suppose $\mathbf{r}^{(a+1)} = \mathbf{0}$. Then applying (14.32) using $\tilde{\boldsymbol{\gamma}}^{(a)}$ of (14.39),

$$\tilde{\mathbf{A}}^{(a+1)}\boldsymbol{\delta}^{(a)} = \tilde{\boldsymbol{\gamma}}^{(a)} = \mathbf{0}.$$

If we look at any of the update matrices $\mathbf{E}^{(a)}$, (14.42) for example, then $\mathbf{E}^{(a)}\mathbf{v} = \mathbf{0}$ for any vector $\mathbf{v}$ orthogonal to both $\boldsymbol{\delta}^{(a)}$ and $\boldsymbol{\gamma}^{(a)}$. Thus $\tilde{\mathbf{A}}^{(a+1)}$ is the same as $\tilde{\mathbf{A}}^{(a)}$ on the orthogonal complement of $\{\boldsymbol{\delta}^{(a)}, \boldsymbol{\gamma}^{(a)}\}$. Dennis et al. obtained a substantial improvement in performance by using so-called "sizing" of $\tilde{\mathbf{A}}^{(a)}$ before updating to obtain $\tilde{\mathbf{A}}^{(a+1)}$. They use

$$\tilde{\mathbf{A}}^{(a+1)} = \tau^{(a)}\tilde{\mathbf{A}}^{(a)} + \mathbf{E}^{(a)}, \tag{14.44}$$

where $\mathbf{E}^{(a)}$ is obtained from (14.42) but with $\tau^{(a)}\tilde{\mathbf{A}}^{(a)}$ replacing $\tilde{\mathbf{A}}^{(a)}$. Here $\tau^{(a)}$ is a deflating factor given by

$$\tau^{(a)} = \min\left(1, \frac{|\boldsymbol{\delta}^{(a)\prime}\tilde{\boldsymbol{\gamma}}^{(a)}|}{|\boldsymbol{\delta}^{(a)\prime}\tilde{\mathbf{A}}^{(a)}\boldsymbol{\delta}^{(a)}|}\right). \tag{14.45}$$

Nazareth [1980] tested a variety of methods for least-squares problems on a set of eleven test functions. He compared a Levenberg–Marquardt algorithm (LM) with Bett's method (B), the Broyden–Dennis method (BD), and an early (1977) version of the Dennis–Gay–Welsch method (DGW). To this list he added a full quasi-Newton method for general functions (DQN) and a hybrid algorithm of his own (H) in which the Hessian is approximated by

$$\tilde{\mathbf{H}}^{(a)} = 2\phi^{(a)}\mathbf{J}^{(a)\prime}\mathbf{J}^{(a)} + (1 - \phi^{(a)})\mathbf{B}^{(a)}, \tag{14.46}$$

a weighted average of the Gauss–Newton and quasi-Newton approximations. Taking $\phi^{(1)} = 1$, the initial step of H is (modified) Gauss–Newton. If $\phi^{(a)} \to 0$ as a increases, then H tends to DQN and inherits its faster (superlinear) ultimate rate of convergence. The strategy for choosing $\phi^{(a)}$ is based upon a comparison of the actual reduction in $S(\boldsymbol{\theta}) = \mathbf{r}(\boldsymbol{\theta})'\mathbf{r}(\boldsymbol{\theta})$ at the previous iteration with the reduction predicted by the quadratic approximation employing $\tilde{\mathbf{H}}^{(a)}$.

Recall that a Levenberg-Marquardt (LM) algorithm for least squares can be thought of as the Gauss–Newton method implemented with a trust-region (restricted-step) strategy. Nazareth coded all the methods above with a trust-region strategy, so that the differences between methods were confined to differences between approximations to the Hessian. He measured the numbers of function evaluations and the numbers of Jacobian evaluations as measures of efficiency. Six zero-residual problems and five "large-residual" problems were used. Before summarizing his conclusions, we note that these measures do not take account of the fact that methods H, B, BD, and DGW in particular require much more extensive "housekeeping" per iteration than LM.

Several lessons emerge from Nazareth's experiments. The specialized large-residual least-squares algorithms B, BD, and DGW were clear winners on the large-residual problems, with DGW being the best of the three. On the small-residual problems LM tended to be most efficient except for one problem, in which LM took twice as many function evaluations as DQN. On this occasion the hybrid version H did better than both its constituents. We detect three broad themes in Nazareth's results. A Gauss–Newton approximation to the Hessian appears to be best for zero-residual problems. The DGW augmented Gauss–Newton approximation appears to be best for large residual problems. Some form of hybridization of the approximate Hessians can produce a general-purpose algorithm which performs better overall than its components.

In their NL2SOL code, Dennis et al. [1981b] use a different approach to hybridization. The quadratic approximation to $S(\boldsymbol{\theta})$ used by the Newton method is

$$q_s^{(a)}(\boldsymbol{\theta}) = S(\boldsymbol{\theta}^{(a)}) + \mathbf{g}^{(a)\prime}(\boldsymbol{\theta} - \boldsymbol{\theta}^{(a)}) + \tfrac{1}{2}(\boldsymbol{\theta} - \boldsymbol{\theta}^{(a)})'\mathbf{H}^{(a)}(\boldsymbol{\theta} - \boldsymbol{\theta}^{(a)}) \tag{14.47}$$

[c.f. (2.33), (13.23)]. Consider the two approximations to $\mathbf{H}^{(a)}$, namely $\mathbf{H}_{\text{GN}}^{(a)} = 2\mathbf{J}^{(a)\prime}\mathbf{J}^{(a)}$ (Gauss–Newton) and $\mathbf{H}_{\text{DGW}}^{(a)} = 2(\mathbf{J}^{(a)\prime}\mathbf{J}^{(a)} + \tilde{\mathbf{A}}^{(a)})$. When substituted into (14.47), either form gives a quadratic approximation to $S(\boldsymbol{\theta})$. Rather than average the approximate Hessians as Nazareth [1980] does, DGW uses one or the other, depending upon which quadratic approximation more accurately predicted the change in $S(\boldsymbol{\theta})$ achieved at the previous iteration. We recall from Section 13.3.2 that the updating of the trust-region radius also depends upon how well the quadratic approximation predicts the change in $S(\boldsymbol{\theta})$. The interaction of these two factors is quite complicated, and the reader is referred to Dennis et al. [1981a] for further discussion.

14.3.3 The Gill–Murray Method

The corrected Gauss–Newton method of Gill and Murray [1978] is another augmented Gauss–Newton method but is qualitatively different from the Betts, Broyden–Dennis, and Dennis–Gay–Welsch methods. It is derived using a singular-value decomposition of $\mathbf{J}^{(a)}$, namely (Appendix A8.1)

$$\mathbf{J}^{(a)} = \mathbf{U}^{(a)} \begin{pmatrix} \mathbf{S}^{(a)} \\ \mathbf{0} \end{pmatrix} \mathbf{T}^{(a)\prime}, \tag{14.48}$$

in which $\mathbf{U}^{(a)}$ and $\mathbf{T}^{(a)}$ are orthogonal $n \times n$ and $p \times p$ matrices respectively, and $\mathbf{S}^{(a)} = \mathrm{diag}(s_1, \ldots, s_p)$ with $s_1 \geq s_2 \geq \cdots \geq s_p$. The Newton step (14.10) is defined by

$$(\mathbf{J}^{(a)\prime}\mathbf{J}^{(a)} + \mathbf{A}^{(a)})\boldsymbol{\delta}_{\mathrm{N}}^{(a)} = -\mathbf{J}^{(a)\prime}\mathbf{r}^{(a)}. \tag{14.49}$$

Substituting (14.48) in (14.49), we have

$$(\mathbf{T}^{(a)}\mathbf{S}^{(a)2}\mathbf{T}^{(a)\prime} + \mathbf{A}^{(a)})\boldsymbol{\delta}_{\mathrm{N}}^{(a)} = -\mathbf{T}^{(a)}(\mathbf{S}^{(a)}, \mathbf{0})\mathbf{U}^{(a)\prime}\mathbf{r}^{(a)},$$

or

$$(\mathbf{S}^{(a)2} + \mathbf{T}^{(a)\prime}\mathbf{A}^{(a)}\mathbf{T}^{(a)})\mathbf{T}^{(a)\prime}\boldsymbol{\delta}_{\mathrm{N}} = -\mathbf{S}^{(a)}\bar{\mathbf{r}}^{(a)}, \tag{14.50}$$

where $\bar{\mathbf{r}}^{(a)}$ denotes the first p components of $\mathbf{U}^{(a)\prime}\mathbf{r}^{(a)}$.

The Gauss–Newton step is obtained by assuming $\mathbf{A}^{(a)} = \mathbf{0}$. Thus

$$\mathbf{S}^{(a)2}\mathbf{T}^{(a)\prime}\boldsymbol{\delta}_{\mathrm{GN}}^{(a)} = -\mathbf{S}^{(a)}\bar{\mathbf{r}}^{(a)}. \tag{14.51}$$

If $\mathbf{J}^{(a)\prime}\mathbf{J}^{(a)}$, and thus $\mathbf{S}^{(a)}$, are of full rank p, (14.51) becomes

$$\boldsymbol{\delta}_{\mathrm{GN}}^{(a)} = -\mathbf{T}^{(a)}\mathbf{S}^{(a)-1}\bar{\mathbf{r}}^{(a)}. \tag{14.52}$$

However, if $\mathrm{rank}(\mathbf{J}^{(a)\prime}\mathbf{J}^{(a)}) = q < p$, then, theoretically, the last $p - q$ singular values are all zero, i.e. $s_{q+1} = s_{q+2} = \cdots = s_p = 0$. Let us partition

$$\mathbf{S}^{(a)} = \begin{pmatrix} \mathbf{S}_1 & \mathbf{0} \\ \mathbf{0} & \mathbf{S}_2 \end{pmatrix}, \qquad \bar{\mathbf{r}}^{(a)} = \begin{pmatrix} \bar{\mathbf{r}}_1 \\ \bar{\mathbf{r}}_2 \end{pmatrix}, \quad \text{and} \quad \mathbf{T}^{(a)\prime}\boldsymbol{\delta}_{\mathrm{GN}}^{(a)} = \begin{pmatrix} \mathbf{t}_1 \\ \mathbf{t}_2 \end{pmatrix}, \tag{14.53}$$

where $\mathbf{S}_1$ is $q \times q$ and $\bar{\mathbf{r}}_1$ and $\mathbf{t}_1$ are $q \times 1$. With $\mathbf{S}_2 = \mathbf{0}$, (14.51) imposes no restrictions upon $\mathbf{t}_2$, but requires $\mathbf{S}_1^2\mathbf{t}_1 = -\mathbf{S}_1\bar{\mathbf{r}}_1$ or $\mathbf{t}_1 = -\mathbf{S}_1^{-1}\bar{\mathbf{r}}_1$. As $\mathbf{T}^{(a)}$ is orthogonal, $\|\mathbf{T}^{(a)}\boldsymbol{\delta}\| = \|\boldsymbol{\delta}\|$ and the solution to (14.51) with minimum Euclidean norm is given by taking $\mathbf{t}_2 = \mathbf{0}$. Thus

$$\boldsymbol{\delta}_{\mathrm{GN}}^{(a)} = -\mathbf{T}^{(a)} \begin{pmatrix} \mathbf{S}_1^{-1}\bar{\mathbf{r}}_1 \\ \mathbf{0} \end{pmatrix}. \tag{14.54}$$

This solution is in a subspace of $\mathbb{R}^p$, the subspace spanned by the first q columns of $\mathbf{T}^{(a)}$. From the relationship between the singular-value decomposition of $\mathbf{J}^{(a)}$ and the spectral decomposition of $\mathbf{J}^{(a)\prime}\mathbf{J}^{(a)}$ (c.f. Appendices A8.1 and A2.2), the first q columns of $\mathbf{T}^{(a)}$ consist of eigenvectors corresponding to nonzero eigenvalues of

$\mathbf{J}^{(a)\prime}\mathbf{J}^{(a)}$. Thus the subspace above is the range space of $\mathbf{J}^{(a)\prime}\mathbf{J}^{(a)}$ [c.f. the discussion following (14.21)].

The result that $\mathbf{S}_2 = \mathbf{0}$ when $\operatorname{rank}(\mathbf{J}^{(a)\prime}\mathbf{J}^{(a)}) = q < p$ holds only if computations are done exactly. When performed in finite (floating-point) arithmetic, roundoff errors will ensure nonzero values for $s_{q+1}, \ldots, s_p$. However, they will typically be small. Thus any algorithm using the above method must use some cutoff point at which it is decided that smaller singular values should be regarded as zero. The number of "zero" singular values therefore gives an estimate of $(p-q)$; then (14.54) can be used. A cutoff value can be decided upon more sensibly if something is known about the relative scaling of the variables in the problem. Still, badly ill-conditioned full-rank matrices will be found to be rank-deficient also.

In the "corrected Gauss–Newton" method of Gill and Murray [1978] the above analysis is modified. The value q above is no longer the rank of $\mathbf{J}^{(a)}$, but a parameter of the method called the *grade*. Although q is not the rank of $\mathbf{J}^{(a)}$, the method is partly motivated by the idea that the most important deficiency of the Gauss–Newton direction as an approximation of the Newton direction is that it contains no approximation to the component of the latter in the null space of $\mathbf{J}^{(a)\prime}\mathbf{J}^{(a)}$.

Let q be the "grade" of $\mathbf{J}^{(a)}$: its choice will be discussed later. Let us retain the partitioning of $\mathbf{S}^{(a)}$ and $\bar{\mathbf{r}}^{(a)}$ in (14.53) and similarly write $\mathbf{T}^{(a)} = (\mathbf{T}_1, \mathbf{T}_2)$ where $\mathbf{T}_1$ consists of the first q columns of $\mathbf{T}^{(a)}$. As the columns of $\mathbf{T}^{(a)}$ form a basis for $\mathbb{R}^p$, we may write

$$\begin{aligned}\boldsymbol{\delta}_{\mathrm{N}} &= \mathbf{T}^{(a)}\mathbf{c} = (\mathbf{T}_1, \mathbf{T}_2)\mathbf{c}\\ &= \mathbf{T}_1\mathbf{c}_1 + \mathbf{T}_2\mathbf{c}_2,\end{aligned} \tag{14.55}$$

where $\mathbf{c}_1$ is a q-vector and $\mathbf{c}_2$ a $(p-q)$-vector. Using partitions and $\mathbf{T}_1'\mathbf{T}_2 = \mathbf{0}$, we can rewrite (14.50) showing the first q and remaining $p-q$ equations separately, namely

$$(\mathbf{S}_1^2 + \mathbf{T}_1'\mathbf{A}^{(a)}\mathbf{T}_1)\mathbf{c}_1 + \mathbf{T}_1'\mathbf{A}^{(a)}\mathbf{T}_2\mathbf{c}_2 = -\mathbf{S}_1\bar{\mathbf{r}}_1, \tag{14.56}$$

$$\mathbf{T}_2'\mathbf{A}^{(a)}\mathbf{T}_1\mathbf{c}_1 + (\mathbf{S}_2^2 + \mathbf{T}_2'\mathbf{A}^{(a)}\mathbf{T}_2)\mathbf{c}_2 = -\mathbf{S}_2\bar{\mathbf{r}}_2. \tag{14.57}$$

Gill and Murray assume that terms involving $\mathbf{A}^{(a)}$ are small compared with $\mathbf{S}_1^2$, which contains the largest q eigenvalues of $\mathbf{J}^{(a)\prime}\mathbf{J}^{(a)}$. We cannot make such an assumption about $\mathbf{S}_2^2$, however, as this contains the smaller eigenvalues of $\mathbf{J}^{(a)\prime}\mathbf{J}^{(a)}$, of which some may be computed estimates of zero eigenvalues. Therefore we approximate (14.56) by $\mathbf{S}_1^2\mathbf{c}_1 = -\mathbf{S}_1\bar{\mathbf{r}}_1$ to obtain

$$\tilde{\mathbf{c}}_1 = -\mathbf{S}_1^{-1}\bar{\mathbf{r}}_1. \tag{14.58}$$

This estimate $\tilde{\mathbf{c}}_1$ is substituted into (14.57), giving

$$(\mathbf{S}_2^2 + \mathbf{T}_2'\mathbf{A}^{(a)}\mathbf{T}_2)\tilde{\mathbf{c}}_2 = -\mathbf{S}_2\bar{\mathbf{r}}_2 - \mathbf{T}_2'\mathbf{A}^{(a)}\mathbf{T}_1\tilde{\mathbf{c}}_1, \tag{14.59}$$

which is then solved for $\tilde{\mathbf{c}}_2$. Finally, the "corrected" Gauss–Newton step is

$$\boldsymbol{\delta}_{\mathrm{C}}^{(a)} = \mathbf{T}_1\tilde{\mathbf{c}}_1 + \mathbf{T}_2\tilde{\mathbf{c}}_2. \tag{14.60}$$

Thus the step is made up of the Gauss–Newton step in the space spanned by the eigenvectors in $\mathbf{T}_1$ (corresponding to the larger eigenvalues of $\mathbf{J}^{(a)\prime}\mathbf{J}^{(a)}$) and an approximate Newton step in the space spanned by the columns of $\mathbf{T}_2$ (corresponding to the smaller eigenvalues of $\mathbf{J}^{(a)\prime}\mathbf{J}^{(a)}$). Of course if $q = p$, then $\mathbf{S}_1 = \mathbf{S}$ and $\boldsymbol{\delta}_{\mathrm{C}}^{(a)} = \boldsymbol{\delta}_{\mathrm{GN}}^{(a)}$. Gill and Murray present three approaches to solving (14.59), corresponding to: (a) the use of explicit second derivatives for $\mathbf{A}^{(a)}$, (b) approximating $\mathbf{T}_2'\mathbf{A}^{(a)}$ using finite differences, and (c) approximating $\mathbf{A}^{(a)}$ by a quasi-Newton method.

a Explicit Second Derivatives

When explicit second derivatives are available, Gill and Murray's method has several advantages over a full modified Newton method. For example, one continues to take the cheaper Gauss–Newton steps while good progress is being made. There are also advantages in terms of increased numerical stability resulting from the separation of the larger singular values of $\mathbf{J}^{(a)}$ into $\mathbf{S}_1$ and the smaller singular values into $\mathbf{S}_2$ (see Gill and Murray [1978] for details).

b Finite-Difference Approximation

Using the method of Section 13.4.1, p finite differences in the Jacobian matrix $\mathbf{J}$ would be required to approximate $\mathbf{A}^{(a)}$, one along each coordinate direction of $\boldsymbol{\theta}$. However, $\mathbf{A}^{(a)}$ enters (14.59) only in the form $\mathbf{T}_2'\mathbf{A}^{(a)}$. This enables Gill and Murray [1978] to reduce the number of finite differences required to $p - q$, which is typically very small. Instead of taking differences along each coordinate direction, i.e.

$$\frac{1}{\tau}[\mathbf{J}(\boldsymbol{\theta}^{(a)} + \tau\mathbf{e}_s) - \mathbf{J}(\boldsymbol{\theta}^{(a)})],$$

where $\mathbf{e}_s$ has 1 for the sth component and zeros elsewhere, Gill and Murray [1978] take differences along the directions defined by the columns of $\mathbf{T}_2$. We now derive their results. Recall that $\mathbf{T}^{(a)} = (\mathbf{T}_1, \mathbf{T}_2) = [(t_{rs})]$, and denote the sth column of $\mathbf{T}^{(a)}$ by $\mathbf{t}_s$ $(= \mathbf{T}^{(a)}\mathbf{e}_s)$. Let $\boldsymbol{\xi}$ satisfy $\boldsymbol{\theta} = \mathbf{T}^{(a)}\boldsymbol{\xi}$, and let h be an arbitrary continuously differentiable function of $\boldsymbol{\theta}$. Then for τ small and $k(\boldsymbol{\xi}) = h(\mathbf{T}^{(a)}\boldsymbol{\xi})$ we have

$$\begin{aligned}\frac{h(\boldsymbol{\theta} + \tau\mathbf{t}_s) - h(\boldsymbol{\theta})}{\tau} &= \frac{k(\boldsymbol{\xi} + \tau\mathbf{e}_s) - k(\boldsymbol{\xi})}{\tau} \\ &\approx \frac{\partial k(\boldsymbol{\xi})}{\partial \xi_s}\end{aligned}$$

$$
\begin{aligned}
&= \sum_r \frac{\partial h}{\partial \theta_r} \frac{\partial \theta_r}{\partial \xi_s} \\
&= \sum_r \frac{\partial h}{\partial \theta_r} t_{rs} \\
&= \mathbf{t}_s' \frac{\partial h(\boldsymbol{\theta})}{\partial \boldsymbol{\theta}}.
\end{aligned} \tag{14.61}
$$

Applying (14.61) to each element of $\mathbf{J} = [(\partial r_i/\partial \theta_j)]$, we get

$$
\frac{1}{\tau}\left(\frac{\partial r_i(\boldsymbol{\theta}^{(a)} + \tau \mathbf{t}_s)}{\partial \theta_j} - \frac{\partial r_i(\boldsymbol{\theta}^{(a)})}{\partial \theta_j} \right) \approx \mathbf{t}_s' \text{ (jth column of } \mathbf{H}_i^{(a)}),
$$

where $\mathbf{H}_i = \partial^2 r_i / \partial \boldsymbol{\theta}\, \partial \boldsymbol{\theta}'$. Hence

$$
\frac{1}{\tau}[\mathbf{J}(\boldsymbol{\theta}^{(a)} + \tau \mathbf{t}_s) - \mathbf{J}(\boldsymbol{\theta}^{(a)})] \approx \begin{bmatrix} \mathbf{t}_s'\mathbf{H}_1^{(a)} \\ \mathbf{t}_s'\mathbf{H}_2^{(a)} \\ \vdots \\ \mathbf{t}_s'\mathbf{H}_n^{(a)} \end{bmatrix}, \tag{14.62}
$$

and premultiplying (14.62) by $\mathbf{r}^{(a)\prime}$ gives us

$$
\begin{aligned}
\frac{1}{\tau}\mathbf{r}^{(a)\prime}\{\mathbf{J}(\boldsymbol{\theta}^{(a)} + \tau \mathbf{t}_s) - \mathbf{J}(\boldsymbol{\theta}^{(a)})\} &\approx \sum_{i=1}^{n} \mathbf{t}_s'(r_i^{(a)}\mathbf{H}_i^{(a)}) \\
&= \mathbf{t}_s'\mathbf{A}^{(a)}.
\end{aligned} \tag{14.63}
$$

The rows of $\mathbf{T}_2'\mathbf{A}^{(a)}$ are $\mathbf{t}_s'\mathbf{A}^{(a)}$ $(s = q+1, q+2, \ldots, p)$, thus proving our earlier assertion that $\mathbf{T}_2'\mathbf{A}^{(a)}$ can be obtained using only $p - q$ finite differences in $\mathbf{J}$.

c Quasi-Newton Approximation

In a third algorithm, Gill and Murray [1978] maintain a quasi-Newton approximation of $\mathbf{A}^{(a)}$ as in Section 14.3.2. They use the form (14.33) of $\tilde{\boldsymbol{\gamma}}^{(a)}$ in defining the quasi-Newton condition (14.32) together with a modification of the BFGS update formula (Section 13.4.2)

$$
\mathbf{E}^{(a)} = -\frac{1}{\boldsymbol{\delta}^{(a)\prime}\mathbf{W}^{(a)}\boldsymbol{\delta}^{(a)}}\mathbf{W}^{(a)}\boldsymbol{\delta}^{(a)}\boldsymbol{\delta}^{(a)\prime}\mathbf{W}^{(a)} + \frac{\boldsymbol{\gamma}^{(a)}\boldsymbol{\gamma}^{(a)\prime}}{\boldsymbol{\delta}^{(a)\prime}\boldsymbol{\gamma}^{(a)}}, \tag{14.64}
$$

where

$$
\mathbf{W}^{(a)} = \mathbf{J}^{(a+1)\prime}\mathbf{J}^{(a+1)} + \mathbf{A}^{(a)}
$$

and

$$
\boldsymbol{\gamma}^{(a)} = (\mathbf{J}^{(a+1)\prime}\mathbf{r}^{(a+1)} - \mathbf{J}^{(a)\prime}\mathbf{r}^{(a)}).
$$

In the context of their basic algorithm, Gill and Murray [1978] found the update (14.64) to be the best of several they tried, including Betts' update.

We recall that the "corrected" Gauss–Newton step $\boldsymbol{\delta}_C^{(a)}$ of (14.60) is the sum of two components: a Gauss–Newton step in the q-dimensional subspace spanned by eigenvectors corresponding to the largest q eigenvalues of $\mathbf{J}^{(a)\prime}\mathbf{J}^{(a)}$, and a Newton step in the orthogonal complement of this subspace, namely the subspace spanned by eigenvectors corresponding to the smallest $p-q$ eigenvalues. The way q is chosen for any particular iteration is clearly very important in implementing the corrected method. However, the choice of q is not as critical, as it is in the Gauss–Newton method, where it estimates the rank of $\mathbf{J}^{(a)}$ and determines a subspace in which the step must be taken. The "corrected" step $\boldsymbol{\delta}_C^{(a)}$ can be interpreted as interpolating between the Gauss–Newton step $(q=p)$ and the Newton step $(q=0)$.

Gill and Murray begin iterations with $q=p$ and maintain this value as long as adequate progress is being made by the Gauss–Newton method. The value of q is reduced by at least one at every iteration in which the proportional reduction in $S(\boldsymbol{\theta})$ is less than 1%. The value of q is maintained when the reduction is between 1% and 10% and increased by at least one if the reduction achieved is better than 10%. Adjustments of "at least one" are made as the authors try to keep clusters of singular values of similar size together, either in $\mathbf{S}_1$ or $\mathbf{S}_2$, and thus may move several singular values at the same time. As q decreases, the subspace in which Newton's method is used is enlarged.

Gill and Murray implement their method as a step-length method. The step $\boldsymbol{\delta}_C^{(a)}$ of (14.60) becomes a direction of step, the length being determined by a line search. Equation (14.59) is solved for $\tilde{\mathbf{c}}_2$ using a modified Cholesky decomposition which may modify $\mathbf{S}_2^2+\mathbf{T}_2'\mathbf{A}^{(a)}\mathbf{T}_2$ to ensure that it is positive definite (c.f. Section 13.3.1b). A further test

$$-\frac{\mathbf{g}^{(a)\prime}\boldsymbol{\delta}_C^{(a)}}{\|\mathbf{g}^{(a)}\|\;\|\boldsymbol{\delta}_C^{(a)}\|}<\phi,$$

for some small ϕ, is applied to ensure that $\boldsymbol{\delta}_C^{(a)}$ is a descent direction. Failing this, q is set to zero and a full modified Newton step is taken. Readers are referred to Gill and Murray [1978] for further discussion of implementation details, including a method for moving off saddle points using directions of negative curvature.

Gill and Murray [1978] tested their three algorithms (second derivatives, finite differences, and quasi-Newton) on 23 standard least-squares test problems, of which about half were "zero-residual" problems. Almost all of Nazareth's [1980] 11 problems were contained in their list. They also tested a version of Betts' algorithm (see Section 14.3.2). As measures of performance they used the number of iterations to convergence and the number of function evaluations; Jacobian evaluations were not counted. They found that second derivatives performed better than finite differences, which performed better than the two quasi-Newton algorithms. The latter were roughly comparable. For most problems, the differences in performance were fairly small. In the experience of Gill and Murray on large practical problems, the finite-difference and quasi-Newton algorithms usually require similar numbers of iterations and function evaluations, but occasionally the finite-difference algorithm will successfully find a solution to a problem for which the quasi-Newton algorithms fail to make progress.

14.3.4 Summary

Most specialized algorithms for least-squares problems, like the popular Levenberg–Marquardt algorithms, are based upon the Gauss–Newton modification to the Newton algorithm. Thus the Hessian matrix $2(\mathbf{J}'\mathbf{J} + \mathbf{A})$ of (14.7) is approximated by $2\mathbf{J}'\mathbf{J}$. This, in essence, is equivalent to assuming that $\mathbf{A}(\boldsymbol{\theta})$ is negligible in comparison with $\mathbf{J}'\mathbf{J}$. However, in many nonlinear least-squares problems $\mathbf{A}$ is not negligible, and for these problems even well-implemented Levenberg–Marquardt algorithms may converge painfully slowly or even fail. In such situations most analysts then revert to using a minimization algorithm for general functions, usually a quasi-Newton algorithm. However, Gill and Murray [1978] note that the calculation of second derivatives and the use of a full modified Newton algorithm is often a real possibility and should be considered.

Two alternative general methods have been proposed for improving on the Gauss–Newton approximation to the Hessian and thus forging more robust specialized least-squares algorithms. The first method (Section 14.3.2) is to maintain a quasi-Newton approximation $\tilde{\mathbf{A}}^{(a)}$ to just $\mathbf{A}^{(a)}$. The best algorithm of this type appears to be NL2SOL, written by Dennis et al. [1981a, b]. It is an algorithm which approximates the Hessian by $2\mathbf{J}^{(a)\prime}\mathbf{J}^{(a)}$ or $2(\mathbf{J}^{(a)\prime}\mathbf{J}^{(a)} + \tilde{\mathbf{A}}^{(a)})$, depending upon which quadratic approximation (14.47) gives the best prediction of the reduction in sum of squares achieved at the previous iteration. The second method (Section 14.3.3) is due to Gill and Murray [1978]. It involves a dynamic partitioning of the eigenvalues of $\mathbf{J}^{(a)\prime}\mathbf{J}^{(a)}$ into a set of dominant eigenvalues and the complementary set of lesser eigenvalues. The search direction used by the method is composed of a Gauss–Newton step in the space spanned by the eigenvectors corresponding to the dominant set of eigenvalues and an (approximate) Newton step in the space spanned by eigenvectors corresponding to the lesser eigenvalues. Initially all the eigenvalues belong to the dominant set, so that Gauss–Newton search directions are used. If at any stage poor progress is being made, more and more eigenvalues are transferred to the lesser set, so that the algorithm behaves more like a full modified Newton algorithm.

We have seen no comparative performance figures for NL2SOL and the Gill–Murray method (GM). In particular, Nazareth [1980] did not include Gill and Murray's method in his comparisons. He notes, though, that $\mathbf{A}^{(a)}$ can be significant when $\mathbf{J}^{(a)\prime}\mathbf{J}^{(a)}$ is not close to being singular, and that in this case GM makes no Newton-type correction to the Gauss–Newton direction, whereas the methods of Section 14.3.2 (e.g. NL2SOL) do. However, this results from an oversimplification of GM. It ignores the fact that the composition of the classes of dominant and lesser eigenvalues (or singular values) depends on the reductions being made in the sum of squares, and not on the condition number of $\mathbf{J}^{(a)\prime}\mathbf{J}^{(a)}$.

Nazareth's remark also ignores the fact that the quasi-Newton approximations $\tilde{\mathbf{A}}^{(a)}$ to $\mathbf{A}^{(a)}$ are not good element-by-element approximations. Thus any choice between NL2SOL and GM should be made on empirical grounds.

Algorithms such as NL2SOL and the GM method are motivated by a desire to gain improved performance over Gauss–Newton-based algorithms on larger-residual problems without sacrificing the efficiency of Gauss–Newton-based algorithms on zero- or very small-residual problems. Both are hybrid algorithms which revert to Gauss–Newton search directions while these directions are working well. They both, therefore, achieve the above objective. There are costs involved, however, in terms of increased storage and longer, more complicated computer code. With all the features described by Dennis et al. [1981a, b], NL2SOL occupies 2360 lines of FORTRAN code excluding comments. In integrated statistical packages such as SAS, program size is an important consideration that has to be traded off against other factors such as reliability. This may be the reason that PROC NLIN in SAS, and most other statistical packages, still use simple Gauss–Newton-based algorithms.

14.4 STOPPING RULES

14.4.1 Convergence Criteria

Stopping rules, or convergence criteria, are discussed in general in Section 15.2.4. There we recommend the concurrent use of several criteria. In this section we look at some special criteria for least-squares problems, and their statistical interpretation.

Least-squares minimization software is often used to solve nonlinear equations; e.g., $\mathbf{r}(\boldsymbol{\theta}) = \mathbf{0}$ is solved by minimizing $S(\boldsymbol{\theta}) = \mathbf{r}(\boldsymbol{\theta})'\mathbf{r}(\boldsymbol{\theta})$. Thus some criterion of the form

$$S(\boldsymbol{\theta}^{(a+1)}) < \tau_A \quad \text{or} \quad S(\boldsymbol{\theta}^{(a+1)}) < \tau_A S(\boldsymbol{\theta}^{(1)}) \tag{14.65}$$

is required for these cases. Dennis et al. [1981a] only use additional criteria if

$$S(\boldsymbol{\theta}^{(a)}) - S(\boldsymbol{\theta}^{(a+1)}) \leq 2[S(\boldsymbol{\theta}^{(a)}) - q_S^{(a)}(\boldsymbol{\theta}^{(a+1)})], \tag{14.66}$$

where $q_S^{(a)}(\boldsymbol{\theta})$ is the quadratic approximation (14.47) to $S(\boldsymbol{\theta})$ at $\boldsymbol{\theta}^{(a)}$. They argue that the other tests in use rely heavily upon the quadratic approximation and that this approximation is bad if (14.66) fails to hold.

Two convergence criteria often used are to stop at the ath iteration if

$$0 \leq \frac{S(\boldsymbol{\theta}^{(a)}) - S(\boldsymbol{\theta}^{(a+1)})}{S(\boldsymbol{\theta}^{(a)})} < \tau_S \tag{14.67}$$

or

$$\max_{1 \le j \le p} \frac{|\theta_j^{(a)} - \theta_j^{(a+1)}|}{|\theta_j^{(a)}|} < \tau_\theta, \tag{14.68}$$

where τ_S and τ_θ are small positive numbers. Similar, but slightly better, criteria are described in Section 15.2.4. In discussing such "relative-change" rules, Bates and Watts [1981b] note that many rules are just termination criteria rather than convergence criteria, since they do not necessarily indicate whether or not a local minimum has been reached. For example, Himmelblau [1972] recommended that both (14.67) and (14.68) be used, as compliance with one criterion need not necessarily imply compliance with the other. Bates and Watts [1981b], however, emphasize that even compliance with both criteria does not guarantee convergence. These relative-change criteria only indicate how the algorithm is progressing. Compliance with (14.67) could simply imply that $S(\boldsymbol{\theta})$ is very flat near $\boldsymbol{\theta}^{(a)}$, while (14.68) could be satisfied because the algorithm is taking small steps. Both types of behavior can occur at the same time, e.g. the progress of a steepest-descent algorithm along a flat ridge. For these reasons it is difficult to decide upon τ_S and τ_θ in (14.67) and (14.68).

14.4.2 Relative Offset

As Dennis [1977] noted, at the minimum [c.f. (14.6), (14.8)]

$$\begin{aligned} \mathbf{0} &= \mathbf{g}(\hat{\boldsymbol{\theta}}) \\ &= 2\mathbf{J}(\hat{\boldsymbol{\theta}})'\mathbf{r}(\hat{\boldsymbol{\theta}}) \\ &= -2\mathbf{F}_.(\hat{\boldsymbol{\theta}})'[\mathbf{y} - \mathbf{f}(\hat{\boldsymbol{\theta}})], \end{aligned}$$

so that the residual vector $\hat{\mathbf{r}} = \mathbf{r}(\hat{\boldsymbol{\theta}})$ is orthogonal to the columns of $\hat{\mathbf{J}}$ (or $\hat{\mathbf{F}}_.$). He therefore proposed a rule which declared convergence if the cosine of the angle between the residual vector and the jth column of $\mathbf{J}^{(a)}$ is smaller in absolute value than some τ for $j = 1, 2, \ldots, p$. Here the cosine of the angle between two vectors $\mathbf{u}$ and $\mathbf{v}$ is

$$\cos(\mathbf{u}, \mathbf{v}) = \frac{\mathbf{u}'\mathbf{v}}{\|\mathbf{u}\| \; \|\mathbf{v}\|}. \tag{14.69}$$

As $\hat{\mathbf{r}}$ is orthogonal to the columns of $\hat{\mathbf{F}}_.$, the orthogonal projection, $\hat{\mathbf{r}}_T$, of $\hat{\mathbf{r}}$ onto the range (column) space of $\hat{\mathbf{F}}_.$ is given by (Appendix A11.4)

$$\begin{aligned} \hat{\mathbf{r}}_T &= \hat{\mathbf{F}}_.(\hat{\mathbf{F}}_.'\hat{\mathbf{F}}_.)^{-1}\hat{\mathbf{F}}_.'\hat{\mathbf{r}} \\ &= \mathbf{0}. \end{aligned} \tag{14.70}$$

Let $\mathbf{r}_T^{(a)}$ be defined by (14.70) with $\hat{\boldsymbol{\theta}}$ replaced by $\boldsymbol{\theta}^{(a)}$. A criterion similar in spirit to the criterion of Dennis [1977] is to declare convergence if $\mathbf{r}_T^{(a)}$ is "small." We note that $\mathbf{r}_T^{(a)}$ is the orthogonal projection of the residual vector $\mathbf{r}^{(a)} = \mathbf{y} - \mathbf{f}^{(a)}$ onto the tangent plane at $\boldsymbol{\theta}^{(a)}$. Bates and Watts [1981b] declare convergence if

$$P^{(a)} = \frac{\| \mathbf{r}_T^{(a)} \|}{\rho} < \tau_P, \tag{14.71}$$

where

$$\rho = \left(\frac{pS(\hat{\boldsymbol{\theta}})}{n-p} \right)^{1/2}. \tag{14.72}$$

They call $P^{(a)}$ the *relative offset* at $\boldsymbol{\theta}^{(a)}$.

Bates and Watts motivate the use of the scale factor ρ in the denominator by referring to the confidence region based upon the tangent-plane approximation for the solution locus in the neighborhood of $\hat{\boldsymbol{\theta}}$. From (4.11) and (4.10), the confidence region is given by

$$\left\{ \boldsymbol{\theta} : \| \mathbf{f}(\boldsymbol{\theta}) - \mathbf{f}(\hat{\boldsymbol{\theta}}) \|^2 \leq \frac{pS(\hat{\boldsymbol{\theta}}) F^{\alpha}_{p,n-p}}{n-p} \right\}, \tag{14.73}$$

or

$$\{ \boldsymbol{\theta} : \| \boldsymbol{\mu} - \hat{\boldsymbol{\mu}} \|^2 \leq \rho^2 F^{\alpha}_{p,n-p} \}, \tag{14.74}$$

where

$$\mathbf{f}(\boldsymbol{\theta}) - \mathbf{f}(\hat{\boldsymbol{\theta}}) \approx \mathbf{F}_{\bullet}(\hat{\boldsymbol{\theta}})(\boldsymbol{\theta} - \hat{\boldsymbol{\theta}}), \tag{14.75}$$

the right-hand side of the above equation being a vector in the range of $\mathbf{F}_{\bullet}(\hat{\boldsymbol{\theta}})$, the tangent plane at $\hat{\boldsymbol{\theta}}$. The confidence region (14.74), expressed in terms of $\boldsymbol{\mu} - \hat{\boldsymbol{\mu}}$, gives rise to a disk on the tangent plane with radius proportional to ρ. Suppose, then, that convergence is declared at a value of $\boldsymbol{\theta}^{(a)}$ close to $\hat{\boldsymbol{\theta}}$. This has two effects on the resulting disk.

The first is that we use $\mathbf{f}(\hat{\boldsymbol{\theta}}) = \mathbf{f}(\boldsymbol{\theta}^{(a)})$ which means that the center is shifted by $\mathbf{d} = \mathbf{f}(\hat{\boldsymbol{\theta}}) - \mathbf{f}(\boldsymbol{\theta}^{(a)})$. Then, to the order of the approximation (14.75),

$$\mathbf{d} = \mathbf{F}_{\bullet}(\boldsymbol{\theta}^{(a)})(\hat{\boldsymbol{\theta}} - \boldsymbol{\theta}^{(a)}). \tag{14.76}$$

Now $[\mathbf{y} - \mathbf{f}(\hat{\boldsymbol{\theta}})]_T^{(a)}$, the projection of $\mathbf{y} - \mathbf{f}(\hat{\boldsymbol{\theta}})$ onto the tangent plane at $\boldsymbol{\theta}^{(a)}$, is approximately the same as its projection $\hat{\mathbf{r}}_T$ onto the tangent plane at $\hat{\boldsymbol{\theta}}$, namely zero [by (14.70)]. Also, (14.76) implies that $\mathbf{d}$ belongs to the range of $\mathbf{F}_{\bullet}^{(a)}$, the tangent plane at $\boldsymbol{\theta}^{(a)}$, so that

$$\begin{aligned} \mathbf{d} &= \mathbf{d}_T^{(a)} \\ &\approx [\mathbf{F}_{\bullet}^{(a)}(\hat{\boldsymbol{\theta}} - \boldsymbol{\theta}^{(a)})]_T^{(a)} \\ &\approx [\mathbf{y} - \mathbf{f}(\hat{\boldsymbol{\theta}}) + \mathbf{F}_{\bullet}^{(a)}(\hat{\boldsymbol{\theta}} - \boldsymbol{\theta}^{(a)})]_T^{(a)} \end{aligned}$$

$$\approx [\mathbf{y} - \mathbf{f}(\hat{\boldsymbol{\theta}}) + \mathbf{f}(\hat{\boldsymbol{\theta}}) - \mathbf{f}(\boldsymbol{\theta}^{(a)})]_T^{(a)}$$
$$= [\mathbf{y} - \mathbf{f}(\boldsymbol{\theta}^{(a)})]_T^{(a)}$$
$$= \mathbf{r}_T^{(a)}.$$

The second effect is that by (2.19) and (2.20)

$$\begin{aligned} \| \mathbf{f}(\boldsymbol{\theta}) - \mathbf{f}(\boldsymbol{\theta}^{(a)}) \|^2 &\approx S(\boldsymbol{\theta}) - S(\boldsymbol{\theta}^{(a)}) \\ &\geq S(\boldsymbol{\theta}) - S(\hat{\boldsymbol{\theta}}) \\ &\approx \| \mathbf{f}(\boldsymbol{\theta}) - \mathbf{f}(\hat{\boldsymbol{\theta}}) \|^2, \end{aligned}$$

so that the radius of (14.73) is increased slightly. Therefore, to summarize, the use of $\boldsymbol{\theta}^{(a)}$ instead of $\hat{\boldsymbol{\theta}}$ shifts the center of the disk by approximately $\mathbf{r}_T^{(a)}$ and increases the radius slightly. Comparing the length of the shift with the radius of the disk, and removing the factor $F_{p,n-p}^{\alpha}$ to avoid dependencies on the confidence level [i.e. using ρ of (14.72)], we arrive at the criterion (14.71). Bates and Watts [1981b] suggest setting $\tau_p = 0.001$; the disk will not be materially affected by a shift of the order of 0.001 times the radius.

In the case of replicated data we now have

$$y_{ij} = f_i(\boldsymbol{\theta}) + \varepsilon_{ij} \qquad (i = 1, 2, \ldots, n, \quad j = 1, 2, \ldots, J_i),$$

or

$$\mathbf{y} = \mathbf{f}(\boldsymbol{\theta}) + \boldsymbol{\varepsilon},$$

where $\mathbf{y} = (y_{11}, y_{12}, \ldots, y_{1J_1}, \ldots, y_{n1}, y_{n2}, \ldots, y_{nJ_n})'$ and $\mathbf{f}(\boldsymbol{\theta}) = \mathscr{E}[\mathbf{y}]$. Then

$$\begin{aligned} \| \mathbf{r}^{(a)} \|^2 &= \| \mathbf{y} - \mathbf{f}(\boldsymbol{\theta}^{(a)}) \|^2 \\ &= \sum_i \sum_j [y_{ij} - f_i(\boldsymbol{\theta}^{(a)})]^2 \\ &= \sum_i \sum_j (y_{ij} - \bar{y}_{i\cdot})^2 + \sum_i J_i [\bar{y}_{i\cdot} - f_i(\boldsymbol{\theta}^{(a)})]^2 \\ &= \mathbf{v}'\mathbf{v} + \mathbf{w}'\mathbf{w}, \end{aligned}$$

say, and

$$\mathbf{y} - \mathbf{f}(\boldsymbol{\theta}^{(a)}) = \mathbf{v} + \mathbf{w},$$

where $\mathbf{v}$ and $\mathbf{w}$ are orthogonal. In addition

$$\mathbf{v}'\mathbf{f}(\boldsymbol{\theta}) = \sum_i \sum_j (y_{ij} - \bar{y}_{i\cdot}) f_i(\boldsymbol{\theta}) = 0,$$

so that $\mathbf{v}$ is perpendicular to the solution locus $\mathbf{f}(\boldsymbol{\theta})$ and therefore to any tangent plane. Hence $\mathbf{v}$ inflates the length of $\mathbf{r}^{(a)}$ without contributing to $\mathbf{r}_T^{(a)}$. On this basis,

Bates and Watts [1981b] proposed modifying the denominator of $P^{(a)}$ in (14.71) and defining the relative offset for replicated observations as

$$P_{\text{rep}}^{(a)} = \frac{\| \mathbf{r}_T^{(a)} \|}{[p\{S(\boldsymbol{\theta}^{(a)}) - \mathbf{v}'\mathbf{v}\}/(N-p-\nu)]^{1/2}}, \tag{14.77}$$

where $N = \sum_i J_i$ and where $\nu = N - n$, the number of degrees of freedom of

$$\mathbf{v}'\mathbf{v} = \sum_i \sum_j (y_{ij} - \bar{y}_{i\cdot})^2.$$

14.4.3 Comparison of Criteria

The relative-offset criterion has some clear advantages over the various relative-change criteria. It provides an absolute measure of convergence in which the convergence decision is linked to the accuracy required in the resulting statistical inferences, in this case the radius of a confidence region. This prevents both premature termination and unproductive computation. By contrast, the tolerances τ_S and τ_θ in the relative-change criteria (14.67) and (14.68) are generally arbitrary. Bates and Watts [1981b] give two examples from the literature where the least-squares estimates need to be quoted to several more decimal places to obtain an acceptably low offset.

The ingredients for the relative offset, namely $\mathbf{F}_{\cdot}^{(a)}$ (or $\mathbf{J}^{(a)}$) and $\mathbf{r}^{(a)}$, are available when using most specialized least-squares programs, particularly those based upon the Gauss–Newton method. Suppose that instead of calculating $\mathbf{J}^{(a)\prime}\mathbf{J}^{(a)} = \mathbf{F}_{\cdot}^{(a)\prime}\mathbf{F}_{\cdot}^{(a)}$, the algorithm stores a QR factorization of $\mathbf{F}_{\cdot}^{(a)}$ $(= -\mathbf{J}^{(a)})$ of the form (Appendix A8.3)

$$\mathbf{F}_{\cdot}^{(a)} = \mathbf{Q}_p^{(a)}\mathbf{R}_{11}^{(a)},$$

where $\mathbf{R}_{11}^{(a)}$ is a $p \times p$ upper triangular matrix and $\mathbf{Q}_p^{(a)}$ is $n \times p$ with p orthonormal columns. Then from (14.70), with $\boldsymbol{\theta}^{(a)}$ replacing $\hat{\boldsymbol{\theta}}$,

$$\begin{aligned} \| \mathbf{r}_T^{(a)} \| &= \| \mathbf{Q}_p^{(a)}\mathbf{R}_{11}^{(a)}(\mathbf{R}_{11}^{(a)\prime}\mathbf{Q}_p^{(a)\prime}\mathbf{Q}_p^{(a)}\mathbf{R}_{11}^{(a)})^{-1}\mathbf{R}_{11}^{(a)\prime}\mathbf{Q}_p^{(a)\prime}\mathbf{r}^{(a)} \| \\ &= \| \mathbf{Q}_p^{(a)}\mathbf{Q}_p^{(a)\prime}\mathbf{r}^{(a)} \| \\ &= \| \mathbf{Q}_p^{(a)\prime}\mathbf{r}^{(a)} \|, \end{aligned} \tag{14.78}$$

thus providing for a calculation of (14.71).

The relative offset cannot be used in zero-residual problems, i.e. problems in which $\mathbf{r}(\hat{\boldsymbol{\theta}})'\mathbf{r}(\hat{\boldsymbol{\theta}})$ is theoretically zero. In this case the residual vector is due entirely to roundoff error at the minimum, so that its direction is virtually random. It is precisely to cope with such situations that an additional criterion such as (14.65) must also be imposed.

Another case where orthogonality cannot be achieved is where the solution locus is finite in extent but, with the current parametrization, $\mathbf{y}$ lies "off the edge" (Bates and Watts [1981b: 182]). These cases are often discovered when one parameter is forced to $\pm\infty$ or the derivative matrix becomes singular (e.g. Meyer and Roth [1972: Example 5], where the minimum occurs at $\theta_1 = \infty$). In such circumstances Bates and Watts note that the data analyst should realize that the model does not fit the data well and that alternative models should be used.

Dennis et al. [1981a] and Dennis and Schnabel [1983: 234] have a criterion which is similar to relative offset. As we have previously noted, at $\hat{\boldsymbol{\theta}}$ the residual vector $\hat{\mathbf{r}}$ is orthogonal to the columns of $\hat{\mathbf{F}}_.$. These columns span the tangent plane to $S(\boldsymbol{\theta})$ at $\boldsymbol{\theta} = \hat{\boldsymbol{\theta}}$. Thus $\hat{\mathbf{r}}_T$ of (14.70), the orthogonal projection of $\hat{\mathbf{r}}$ onto the tangent plane at $\hat{\boldsymbol{\theta}}$, is zero. Dennis et al. look at this from a different perspective. As $\boldsymbol{\theta}^{(a)} \to \hat{\boldsymbol{\theta}}$, the residual vector $\mathbf{r}^{(a)}$ tends to become orthogonal to the tangent plane at $\boldsymbol{\theta}^{(a)}$. In particular $\mathbf{r}^{(a)}$ and $\mathbf{r}_T^{(a)}$, its orthogonal projection onto the tangent plane at $\boldsymbol{\theta}^{(a)}$, tend to become orthogonal. Thus they terminate the iterative process when the angle between $\mathbf{r}^{(a)}$ and $\mathbf{r}_T^{(a)}$ gets close enough to 90°, i.e. when [c.f. (14.69)]

$$\cos(\mathbf{r}^{(a)}, \mathbf{r}_T^{(a)}) = \frac{\mathbf{r}^{(a)\prime}\mathbf{r}_T^{(a)}}{\|\mathbf{r}^{(a)}\| \, \|\mathbf{r}_T^{(a)}\|} < \tau_c. \tag{14.79}$$

However, we have reservations about this criterion. Firstly, τ_c is arbitrary and not related to any resulting statistical inferences. Secondly, we are unsure of its computational stability, since $\|\mathbf{r}_T^{(a)}\| \to 0$ as $\boldsymbol{\theta}^{(a)} \to \hat{\boldsymbol{\theta}}$.

An assumption has been made in the previous discussion that τ_θ in the relative-change criterion (14.68) is merely arbitrary. This is not strictly true. The tolerance τ_θ can indeed be chosen using statistical considerations. For example, we might wish to stop when $\delta_j^{(a)} = \theta_j^{(a+1)} - \theta_j^{(a)}$ becomes much smaller than the current approximation to the standard error (s.e.) of $\hat{\theta}_j$, e.g. stop if

$$\max_{1 \le j \le p} \frac{|\delta_j^{(a)}|}{\text{s.e.}(\hat{\theta}_j)} < \tau_\theta. \tag{14.80}$$

Dennis et al. [1981a: 357] quote a suggestion of J. W. Pratt. Let $\mathbf{V}^{(a)}$ be the current approximation to the asymptotic variance–covariance matrix of $\hat{\boldsymbol{\theta}}$ [e.g. $\sigma^{(a)2}\{\mathbf{F}_.^{(a)\prime}\mathbf{F}_.^{(a)}\}^{-1}$]. Pratt considers all possible linear combinations $\boldsymbol{\ell}'\boldsymbol{\delta}^{(a)}$ and suggests stopping if (Appendix A7.2)

$$\max_{\boldsymbol{\ell} \ne \mathbf{0}} \frac{(\boldsymbol{\ell}'\boldsymbol{\delta}^{(a)})^2}{\boldsymbol{\ell}'\mathbf{V}^{(a)}\boldsymbol{\ell}} = \boldsymbol{\delta}^{(a)\prime}\mathbf{V}^{(a)-1}\boldsymbol{\delta}^{(a)} < \tau_\theta. \tag{14.81}$$

These criteria are not sufficient by themselves, since $\boldsymbol{\delta}^{(a)}$ may be small simply because the algorithm has taken a short step, and not because $\boldsymbol{\theta}^{(a)}$ is close to $\hat{\boldsymbol{\theta}}$. However, when coupled with (14.66) to ensure a reasonable quadratic approximation, and when a natural rather than shortened Gauss–Newton or approximate Newton step has been taken, (14.80) and (14.81) become attractive criteria. They are also applicable outside the confines of the least-squares problem.

14.5 DERIVATIVE-FREE METHODS

The most obvious way of constructing a derivative-free least-squares algorithm is to take any of the previous first-derivative methods and use a forward-difference approximation to the first derivatives, i.e., the jth column of $\mathbf{J}$ is approximated by

$$(\mathbf{J})_j = \frac{\mathbf{r}(\boldsymbol{\theta} + \xi_j \mathbf{e}_j) - \mathbf{r}(\boldsymbol{\theta})}{\xi_j}. \tag{14.82}$$

Scales [1985: 125] gives a simple rule for choosing ξ_j based on earlier work by Brown and Dennis [1972]. The rule is

$$\xi_j = \min\{S(\boldsymbol{\theta}), \xi'_j\},$$

where

$$\xi'_j = \begin{cases} 0.01\varepsilon^{1/2} & \text{if} \quad \theta_j < 10\varepsilon^{1/2}, \\ 0.001|\theta_j| & \text{if} \quad \theta_j \geq 10\varepsilon^{1/2}, \end{cases}$$

and ε is the machine relative precision. However, Dennis et al. [1981b] use a much more elaborate strategy (see also Gill et al. [1981: Section 8.6]). Most of the major subroutine libraries discussed in Section 15.1.2 have finite-difference versions of their nonlinear least-squares routines available.

Finite-difference approximations to the np entries of $\mathbf{J}$ are costly unless evaluations of the residual functions $r_i(\boldsymbol{\theta})$ are cheap. However, we have seen several types of a problem where the evaluation of $f(x_i; \boldsymbol{\theta})$, and hence of $r_i(\boldsymbol{\theta}) = y_i - f(x_i; \boldsymbol{\theta})$, is very expensive. For example, for compartmental models described by nonlinear systems of differential equations, function values are formed by the numerical solution of those equations. Models with ARMA (q_1, q_2) autocorrelated error structures (Section 6.6) also result in least-squares problems where $r_i(\boldsymbol{\theta})$ is very expensive to compute. Consequently other methods have been developed that essentially try to build up derivative information as the iterations proceed.

As we have seen in Section 14.1, the Gauss–Newton algorithm for the model $\mathbf{y} = \mathbf{f}(\boldsymbol{\theta}) + \boldsymbol{\varepsilon}$ approximates $\mathbf{f}(\boldsymbol{\theta})$, for $\boldsymbol{\theta}$ close to $\boldsymbol{\theta}^{(a)}$, by the linear function

$$\mathbf{f}_L(\boldsymbol{\theta}) = \mathbf{f}(\boldsymbol{\theta}^{(a)}) + \mathbf{F}_.(\boldsymbol{\theta}^{(a)})(\boldsymbol{\theta} - \boldsymbol{\theta}^{(a)}). \tag{14.83}$$

This describes the tangent plane to the surface $\mathbf{f}(\boldsymbol{\theta})$ at $\boldsymbol{\theta}^{(a)}$. The next iterate, $\boldsymbol{\theta}^{(a+1)}$, is chosen to minimize $\| \mathbf{y} - \mathbf{f}_L(\boldsymbol{\theta}) \|^2$, i.e., $\boldsymbol{\theta}^{(a+1)}$ is the point on the approximating tangent hyperplane closest to $\mathbf{y}$. For consistency with the rest of the chapter we write this in terms of $\mathbf{r}(\boldsymbol{\theta}) = \mathbf{y} - \mathbf{f}(\boldsymbol{\theta})$, so that corresponding to (14.83), $\mathbf{r}(\boldsymbol{\theta})$ is approximated by

$$\mathbf{r}_L^{(a)}(\boldsymbol{\theta}) = \mathbf{r}(\boldsymbol{\theta}^{(a)}) + \mathbf{J}^{(a)}(\boldsymbol{\theta} - \boldsymbol{\theta}^{(a)}). \tag{14.84}$$

Now (14.84) defines the approximating tangent hyperplane to $\mathbf{r}(\boldsymbol{\theta})$ at $\boldsymbol{\theta} = \boldsymbol{\theta}^{(a)}$, and $\boldsymbol{\theta}^{(a+1)}$ is the point on this plane closest to the origin, i.e., $\boldsymbol{\theta}^{(a+1)}$ minimizes $\|\mathbf{r}_L^{(a)}(\boldsymbol{\theta})\|^2$.

In contrast, secant methods approximate $\mathbf{r}(\boldsymbol{\theta})$ by a secant hyperplane $\mathbf{r}_S^{(a)}(\boldsymbol{\theta})$ in $\mathbb{R}^p$ rather than the tangent hyperplane. One popular secant method is the DUD (Doesn't Use Derivatives) algorithm of Ralston and Jennrich [1978], which appears in several of the major statistical packages [e.g. in PROC NLIN in SAS, and the BMDP program BMDPAR]. DUD maintains a current set of $p+1$ approximations to $\hat{\boldsymbol{\theta}}$, which we will denote by $\boldsymbol{\theta}_1^{(a)}, \boldsymbol{\theta}_2^{(a)}, \ldots, \boldsymbol{\theta}_{p+1}^{(a)}$, at the ath iteration. This set is maintained so as not to become coplanar [i.e. lie in a $(p-1)$-dimensional hyperplane]. Now $\mathbf{r}_S^{(a)}(\boldsymbol{\theta})$ is chosen to agree with $\mathbf{r}(\boldsymbol{\theta})$ at $\boldsymbol{\theta}_1^{(a)}, \ldots, \boldsymbol{\theta}_{p+1}^{(a)}$. The order of the subscripts is from the oldest to the youngest, so that $\boldsymbol{\theta}_1^{(a)}$ has been in the "current set" for the largest number of past iterations. In the basic algorithm a Gauss–Newton-type step is taken: thus $\boldsymbol{\theta}^{(a+1)}$ minimizes $\|\mathbf{r}_S^{(a)}\|^2$. Then $\boldsymbol{\theta}^{(a+1)}$ replaces one of the points in the "current set", if possible the oldest. A line search can also be incorporated in the algorithm so that the $(a+1)$th step is now to $\boldsymbol{\theta}_{p+1}^{(a)} + \rho^{(a)}(\boldsymbol{\theta}^{(a+1)} - \boldsymbol{\theta}_{p+1}^{(a)})$, where $\rho^{(a)}$ is determined by the line search. Before discussing strategies for initiating and updating the "current set," we look at the computation of $\mathbf{r}_S^{(a)}(\boldsymbol{\theta})$ and $\boldsymbol{\theta}^{(a+1)}$.

As $\boldsymbol{\theta}_1^{(a)}, \ldots, \boldsymbol{\theta}_{p+1}^{(a)}$ are not coplanar, it can be shown that the vectors $\boldsymbol{\theta}_j^{(a)} - \boldsymbol{\theta}_{p+1}^{(a)}$ $(j = 1, 2, \ldots, p)$ are linearly independent and thus form a basis for $\mathbb{R}^p$. Any $\boldsymbol{\theta} \in \mathbb{R}^p$ can be written as

$$\begin{aligned}\boldsymbol{\theta} &= \boldsymbol{\theta}_{p+1}^{(a)} + \boldsymbol{\theta} - \boldsymbol{\theta}_{p+1}^{(a)} \\ &= \boldsymbol{\theta}_{p+1}^{(a)} + \sum_{j=1}^{p} \alpha_j(\boldsymbol{\theta}_j^{(a)} - \boldsymbol{\theta}_{p+1}^{(a)}) \\ &= \boldsymbol{\theta}_{p+1}^{(a)} + \mathbf{P}^{(a)}\boldsymbol{\alpha}, \end{aligned} \tag{14.85}$$

say, where the matrix $\mathbf{P}^{(a)}$ has jth column $\boldsymbol{\theta}_j^{(a)} - \boldsymbol{\theta}_{p+1}^{(a)}$ $(j = 1, 2, \ldots, p)$ and is nonsingular. Define

$$\mathbf{r}_S^{(a)}(\boldsymbol{\theta}) = \mathbf{r}(\boldsymbol{\theta}_{p+1}^{(a)}) + \mathbf{Q}^{(a)}\boldsymbol{\alpha}, \tag{14.86}$$

where the jth column of $\mathbf{Q}^{(a)}$ is $\mathbf{r}(\boldsymbol{\theta}_j^{(a)}) - \mathbf{r}(\boldsymbol{\theta}_{p+1}^{(a)})$ $(j = 1, 2, \ldots, p)$. It has the property that $\mathbf{r}_S^{(a)}(\boldsymbol{\theta}_j^{(a)}) = \mathbf{r}(\boldsymbol{\theta}_j^{(a)})$ (e.g., for $j \neq p+1$ take $\boldsymbol{\alpha} = \mathbf{e}_j$). Using (14.85), we can express (14.86) as the secant hyperplane

$$\mathbf{r}_S^{(a)}(\boldsymbol{\theta}) = \mathbf{r}(\boldsymbol{\theta}_{p+1}^{(a)}) + (\mathbf{Q}^{(a)}\mathbf{P}^{(a)-1})(\boldsymbol{\theta} - \boldsymbol{\theta}_{p+1}^{(a)}), \tag{14.87}$$

which passes through the points $\{\boldsymbol{\theta}_j^{(a)}, \mathbf{r}(\boldsymbol{\theta}_j^{(a)})\}, j = 1, 2, \ldots, p+1$. This representation emphasizes the similarity between the secant method and the Gauss–Newton method [c.f. (14.84)]. In the secant method, $\mathbf{Q}^{(a)}\mathbf{P}^{(a)-1}$ is used to approximate the Jacobian $\mathbf{J}^{(a)}$ of Gauss–Newton. Rather than solve the least-squares problem "minimize $\|\mathbf{r}_S^{(a)}(\boldsymbol{\theta})\|^2$" using QR factorizations or a singular-value decomposition, Ralston and Jennrich [1978] recover $\boldsymbol{\theta}^{(a+1)}$ by minimizing $\|\mathbf{r}_S^{(a)}(\boldsymbol{\theta})\|^2$ in two stages. Firstly, they find $\hat{\boldsymbol{\alpha}}$ to minimize $\|\mathbf{r}^{(a)}(\boldsymbol{\alpha})\|^2$ in (14.86), namely

$$\hat{\boldsymbol{\alpha}} = -(\mathbf{Q}^{(a)\prime}\mathbf{Q}^{(a)})^{-1}\mathbf{Q}^{(a)\prime}\mathbf{r}(\boldsymbol{\theta}_{p+1}^{(a)}). \tag{14.88}$$

To compute the matrix inversion they use Gauss–Jordan pivoting with pivots chosen using a stepwise regression method described by Jennrich and Sampson [1968]. Secondly, $\boldsymbol{\theta}^{(a+1)}$ is obtained by evaluating (14.85) at $\boldsymbol{\alpha} = \hat{\boldsymbol{\alpha}}$.

Ralston and Jennrich [1978] start the "current set" using a single user-supplied starting value $\boldsymbol{\theta}^{(1)}$, which becomes $\boldsymbol{\theta}_{p+1}^{(1)}$. Then $\boldsymbol{\theta}_1^{(1)}, \ldots, \boldsymbol{\theta}_p^{(1)}$ are obtained from $\boldsymbol{\theta}_{p+1}^{(1)}$ by displacing the ith element $\theta_{p+1,i}^{(1)}$ of $\boldsymbol{\theta}_{p+1}^{(1)}$ by a nonzero number h_i, e.g. $h_i = 0.1\theta_{p+1,i}^{(1)}$. The residual vector is evaluated at each $\boldsymbol{\theta}_j^{(1)}, j = 1, 2, \ldots, p+1$, and these vectors are relabeled so that $S(\boldsymbol{\theta}_1^{(1)}) \geq \cdots \geq S(\boldsymbol{\theta}_{p+1}^{(1)})$, where $S(\boldsymbol{\theta}) = \mathbf{r}(\boldsymbol{\theta})'\mathbf{r}(\boldsymbol{\theta})$. We next consider the strategy used for updating the "current set" in DUD.

Ralston and Jennrich [1978] state that, "Theoretically, if the current set of parameter vector differences spans the parameter space, the new set will span it also, if and only if the component of $\boldsymbol{\alpha}$ corresponding to the discarded parameter vector is nonzero." They discard the oldest vector $\boldsymbol{\theta}_1^{(a)}$ if $|\alpha_1| \geq 10^{-5}$. If $|\alpha_1| < 10^{-5}$, two members of the set are replaced as follows. Firstly $\boldsymbol{\theta}_i^{(a)}$, the oldest component for which $|\alpha_i| \geq 10^{-5}$, is discarded. Secondly, $\boldsymbol{\theta}_1^{(a)}$ is replaced by $(\boldsymbol{\theta}_1^{(a)} + \boldsymbol{\theta}^{(a+1)})/2$, where $\boldsymbol{\theta}^{(a+1)}$ minimizes the norm of (14.87). The latter is done to stop old vectors remaining in the set indefinitely.

Earlier we mentioned using line searches with DUD. Because DUD was primarily intended for problems in which evaluations of the residual vector, and hence the sum of squares, are expensive, line searches are used sparingly: for details see Ralston and Jennrich [1978: 8]. DUD can be thought of as an approximate modified Gauss–Newton algorithm. It therefore experiences the same types of difficulty with large residual problems. In their numerical-performance comparisons, Ralston and Jennrich compared the behavior of DUD with the published behavior of other least-squares algorithms. However, they did not standardize stopping rules and implementation details, so that their results have to be treated with caution. Still, DUD behaved very efficiently on some of the standard test problems.

Fletcher [1980: Section 6.3] described a class of algorithms generalizing the previous secant idea, which he called the *generalized secant method.* An approximation $\tilde{\mathbf{J}}^{(a)}$ to $\mathbf{J}$ is maintained using the change in residual vector $\mathbf{r}^{(a+1)} - \mathbf{r}^{(a)}$ to provide an approximation to the derivative along $\boldsymbol{\delta}^{(a)} = \boldsymbol{\theta}^{(a+1)} - \boldsymbol{\theta}^{(a)}$ as follows. Since, to a first-order approximation,

$$\mathbf{r}^{(a+1)} - \mathbf{r}^{(a)} = \mathbf{J}^{(a)}(\boldsymbol{\theta}^{(a+1)} - \boldsymbol{\theta}^{(a)}),$$

analogously to quasi-Newton updating of the Hessian [c.f. (13.35)], we require $\tilde{\mathbf{J}}^{(a+1)}$ to satisfy

$$\tilde{\mathbf{J}}^{(a+1)}\boldsymbol{\delta}^{(a)} = \mathbf{r}^{(a+1)} - \mathbf{r}^{(a)} = \boldsymbol{\gamma}^{(a)}, \tag{14.89}$$

say. For $a \leq p$ this is achieved using an update formula due to Barnes, namely

$$\tilde{\mathbf{J}}^{(a+1)} = \tilde{\mathbf{J}}^{(a)} + \mathbf{E}^{(a)}$$

where

$$\mathbf{E}^{(a)} = \frac{(\boldsymbol{\gamma}^{(a)} - \tilde{\mathbf{J}}^{(a)}\boldsymbol{\delta}^{(a)})\mathbf{w}^{(a)\prime}}{\mathbf{w}^{(a)\prime}\boldsymbol{\delta}^{(a)}} \tag{14.90}$$

and $\mathbf{w}^{(a)}$ is any vector orthogonal to $\boldsymbol{\delta}^{(1)},\dots,\boldsymbol{\delta}^{(a-1)}$. This update has the hereditary property of preserving past derivative information, as $\mathbf{E}^{(j)}\boldsymbol{\delta}^{(k)} = \mathbf{0}$ for $k<j$ and thus

$$\tilde{\mathbf{J}}^{(a+1)}\boldsymbol{\delta}^{(j)} = \mathbf{r}^{(j+1)} - \mathbf{r}^{(j)} \qquad (j \leq a). \tag{14.91}$$

[If $\mathbf{r}(\boldsymbol{\theta})$ is linear, so that $\mathbf{J}^{(a)} \equiv \mathbf{J}$, then $\mathbf{J}^{(p+1)} = \mathbf{J}$ and the next Gauss–Newton step solves the least-squares problem.] Since we can only retain derivative information along p linearly independent directions, for $a > p$ an old direction has to be dropped when the new direction $\boldsymbol{\delta}^{(a)}$ is added to the current set of p directions. This is achieved by choosing $\mathbf{w}^{(a)}$ in (14.90) orthogonal to the $p-1$ $\boldsymbol{\delta}$'s one wishes to retain. As with DUD, a strategy is required for discarding directions from the current set of $\boldsymbol{\delta}$'s in a way that prevents the set from coming close to being dependent. Fletcher [1980] suggests a startup strategy in which the first p steps are obtained from searches along the coordinate directions. The algorithm of Powell [1965] is almost in the class above. However, instead of obtaining his derivative information along $\boldsymbol{\delta}^{(a)}$ from the full step as in (14.90), he obtains it from the two lowest points found in the line search used to obtain $\boldsymbol{\theta}^{(a+1)}$. The experiments of Ralston and Jennrich [1978] suggest that DUD is superior to Powell's method in terms of numbers of function evaluations.

The method of Broyden [1965] is also similar to the generalized secant method. Broyden uses $\boldsymbol{\delta}^{(a)}$ in place of $\mathbf{w}^{(a)}$ in (14.90). By doing this the hereditary property (14.91) is lost, as is finite termination on a linear least-squares problem. Fletcher [1980: p. 109] notes that for linear least-squares problems Broyden's formula has the property that

$$\begin{aligned}(\tilde{\mathbf{J}}^{(a+1)} - \mathbf{J})' &= \left(\mathbf{I}_p - \frac{\boldsymbol{\delta}^{(a)}\boldsymbol{\delta}^{(a)\prime}}{\boldsymbol{\delta}^{(a)\prime}\boldsymbol{\delta}^{(a)}}\right)(\tilde{\mathbf{J}}^{(a)} - \mathbf{J})' \\ &= (\mathbf{I}_p - \mathbf{P})(\tilde{\mathbf{J}}^{(a)} - \mathbf{J})',\end{aligned} \tag{14.92}$$

where the symmetric idempotent matrix $\mathbf{P}$, and therefore $\mathbf{I}_p - \mathbf{P}$, is a projection matrix (Appendix A11.2). Since projecting a vector generally reduces its length, we can expect from (14.92) that the error in $\tilde{\mathbf{J}}$ is usually reduced. Fletcher also notes that the denominator of Broyden's update is $\boldsymbol{\delta}^{(a)\prime}\boldsymbol{\delta}^{(a)}$, which prevents the update from becoming arbitrarily large, as can happen with (14.90).

14.6 RELATED PROBLEMS

14.6.1 Robust Loss Functions

Rather than choose $\boldsymbol{\theta}$ to minimize the sum of squared residuals $\sum_i r_i(\boldsymbol{\theta})^2$, we can find a robust ($M$-) estimator (c.f. Section 2.6) to minimize an appropriate loss function. Such functions are chosen to downweight the effects upon the fit of observations with large residuals. Thus $\boldsymbol{\theta}$ is chosen to minimize a loss function of the form

$$h(\boldsymbol{\theta}) = \sum_{i=1}^{n} \rho\left(\frac{r_i(\boldsymbol{\theta})}{\sigma}\right), \tag{14.93}$$

where σ is some measure of dispersion. For least squares, $\rho_{\text{LS}}(t) = t^2$, $\dot{\rho}_{\text{LS}}(t) = 2t$, and $\ddot{\rho}_{\text{LS}}(t) = 2$. A well-known robust loss function is that of Huber [1973], which, for some constant $A > 0$, has

$$\rho_H(t) = \begin{cases} t^2, & |t| \leq A, \\ 2A|t| - A^2, & |t| \geq A. \end{cases} \tag{14.94}$$

Then

$$\dot{\rho}_H(t) = \begin{cases} 2t, & |t| \leq A, \\ 2A, & |t| > A \end{cases} \quad \text{and} \quad \ddot{\rho}_H(t) = \begin{cases} 2, & |t| \leq A, \\ 0, & |t| > A. \end{cases}$$

Another is that of Beaton and Tukey [1974], for which

$$\rho_{BT}(t) = \begin{cases} \frac{1}{3}A^2\{1 - [1 - (t/A)^2]^3\}, & |t| \leq A, \\ \frac{1}{3}A^2, & |t| \geq A, \end{cases}$$

$$\dot{\rho}_{BT}(t) = \begin{cases} 2t[1 - (t/A)^2]^2, & |t| \leq A, \\ 0, & |t| \geq A, \end{cases}$$

$$\ddot{\rho}_{BT}(t) = \begin{cases} 2[1 - (t/A)^2][1 - 5(t/A)^2], & |t| \leq A, \\ 0, & |t| \geq A. \end{cases} \tag{14.95}$$

The Beaton–Tukey loss function has continuous second derivatives, whereas the Huber loss function only has a continuous first derivative. Many others in common use have discontinuities in both derivatives at $|t| = A$ for some A. The point A is called a knot.

We shall deal only with methods for minimizing a function of the form (14.93) which can be used when some of the r_i's are nonlinear. More specialized methods are possible when each $r_i(\boldsymbol{\theta})$ is linear, for example $r_i(\boldsymbol{\theta}) = \mathbf{y} - \mathbf{x}_i'\boldsymbol{\theta}$. For a discussion of these methods see Kennedy and Gentle [1980: Section 11.3].

For linear residual functions, the sum of squares is convex in $\boldsymbol{\theta}$ and has only a single minimum provided the design matrix $\mathbf{X} = (\mathbf{x}_1', \ldots, \mathbf{x}_n')'$ is of full rank. This does not apply to robust loss functions for linear models. The function $h(\boldsymbol{\theta})$ of (14.93) may have several local minima in addition to the global minimum. This increase in the number of local minima can be expected to carry over to the nonlinear situation to some extent.

Another feature of the formulation (14.93) that requires discussion is the scale factor σ. In the usual least-squares problem, minimizing $\sum r_i(\boldsymbol{\theta})^2$ and minimizing $\sum [r_i(\boldsymbol{\theta})/\sigma]^2$ are equivalent. However, with robust loss functions a factor σ is used to scale the residuals, and this scaling affects the relative penalties (or "downweighting") accorded to each residual. In rare problems σ may be known. However, it is usually unknown and is estimated for the $(a+1)$th iteration by some robust measure of the spread of the residuals such as

$$s^{(a+1)} = \frac{1}{n} \sum_{i=1}^{n} |r_i(\boldsymbol{\theta}^{(a)})|. \tag{14.96}$$

For a discussion of robust regression with particular reference to linear models see Dutter [1977, 1978].

Let us now return to the problem of minimizing (14.93). The gradient vector and Hessian of $h(\boldsymbol{\theta})$ are given by

$$\mathbf{g}(\boldsymbol{\theta}) = \frac{1}{\sigma} \mathbf{J}(\boldsymbol{\theta})' \mathbf{v}(\boldsymbol{\theta}), \tag{14.97}$$

and

$$\mathbf{H}(\boldsymbol{\theta}) = \frac{1}{\sigma^2} [\mathbf{J}(\boldsymbol{\theta})' \mathbf{D}(\boldsymbol{\theta}) \mathbf{J}(\boldsymbol{\theta})] + \mathbf{A}(\boldsymbol{\theta}), \tag{14.98}$$

where

$$\mathbf{v}(\boldsymbol{\theta}) = \left[\dot{\rho}\left\{\frac{r_1(\boldsymbol{\theta})}{\sigma}\right\}, \ldots, \dot{\rho}\left\{\frac{r_n(\boldsymbol{\theta})}{\sigma}\right\} \right]', \tag{14.99}$$

$$\mathbf{D}(\boldsymbol{\theta}) = \operatorname{diag}\left[\ddot{\rho}\left\{\frac{r_1(\boldsymbol{\theta})}{\sigma}\right\}, \ldots, \ddot{\rho}\left\{\frac{r_n(\boldsymbol{\theta})}{\sigma}\right\} \right], \tag{14.100}$$

$$\mathbf{A}(\boldsymbol{\theta}) = \frac{1}{\sigma} \sum_{i=1}^{n} \dot{\rho}\left\{\frac{r_i(\boldsymbol{\theta})}{\sigma}\right\} \frac{\partial^2 r_i(\boldsymbol{\theta})}{\partial \boldsymbol{\theta}\, \partial \boldsymbol{\theta}'} \tag{14.101}$$

[c.f. (14.6), (14.7), and (14.9)]. One can, of course minimize $h(\boldsymbol{\theta})$ using general minimization algorithms. For example, Welsch and Becker [1975] obtained good results using a modified Newton algorithm. However, Dennis [1977] and Dennis and Welsch [1978] explore special-purpose algorithms that parallel the least-squares algorithms of Sections 14.1 to 14.3. For example, if we omit the term $\mathbf{A}(\boldsymbol{\theta})$ from the Hessian (14.98) in the Newton step [c.f. (14.10)], we obtain a Gauss–

Newton step

$$\boldsymbol{\delta}_{\mathrm{GN}}^{(a)} = -\sigma(\mathbf{J}^{(a)\prime}\mathbf{D}^{(a)}\mathbf{J}^{(a)})^{-1}\mathbf{J}^{(a)\prime}\mathbf{v}^{(a)}. \tag{14.102}$$

Dennis [1977: Theorem 9.2] considers the Levenberg–Marquardt modification of (14.102) and proves local convergence at a rate which is at least linear under conditions similar to those imposed on the Gauss–Newton algorithm in the discussion surrounding Equation (14.13). Again the important factor is the relative sizes of the minimum eigenvalue of $\mathbf{J}'\mathbf{D}\mathbf{J}$ and the value of $\|\mathbf{A}(\boldsymbol{\theta})\|$. The robust loss functions have the effect of reducing the size of the large residuals, and one might expect that this would increase the reliability of the Gauss–Newton method. However, the results are equivocal because at the same time [since generally $\ddot{\rho}(t) \leq 2$] the smallest eigenvalue of $\mathbf{J}'\mathbf{D}\mathbf{J}$ tends to be smaller than the smallest eigenvalue of $\mathbf{J}'\mathbf{J}$. There are other disadvantages, too. As it may happen that $\ddot{\rho}(t) < 0$, $\mathbf{J}'\mathbf{D}\mathbf{J}$ need not be positive (semi)definite and may not generate descent directions. It is also more likely to be singular.

Some of the drawbacks above can be eliminated by replacing $\ddot{\rho}(t)$ in the definition of $\mathbf{D}$ in (14.100) by $w(t) = \dot{\rho}(t)/t$. Since we generally have $\dot{\rho}(0) = 0$,

$$w(t) = \frac{\dot{\rho}(t)}{t} = \frac{\dot{\rho}(t) - \dot{\rho}(0)}{t - 0} \tag{14.103}$$

may then be interpreted as a secant approximation of $\ddot{\rho}(t)$. Dennis [1977] states that for all the specific choices of ρ that he has seen, $w(t) \geq \max\{0, \ddot{\rho}(t)\}$. Denote by $\tilde{D}$ the diagonal matrix obtained by making this change, that is, replacing $\ddot{\rho}(r_i/\sigma)$ by $\sigma\dot{\rho}(r_i/\sigma)/r_i$. Then (14.102) becomes

$$\boldsymbol{\delta}^{(a)} = -(\mathbf{J}^{(a)\prime}\tilde{\mathbf{D}}^{(a)}\mathbf{J}^{(a)})^{-1}\mathbf{J}^{(a)\prime}\tilde{\mathbf{D}}^{(a)}\mathbf{r}^{(a)}. \tag{14.104}$$

This is the solution of the weighted least squares problem

$$\underset{\boldsymbol{\delta}}{\text{Minimize}}\,(\mathbf{r}^{(a)} + \mathbf{J}^{(a)}\boldsymbol{\delta})'\tilde{\mathbf{D}}^{(a)}(\mathbf{r}^{(a)} + \mathbf{J}^{(a)}\boldsymbol{\delta}), \tag{14.105}$$

with positive diagonal weight matrix $\tilde{\mathbf{D}}^{(a)}$. Hence this modification of the Gauss–Newton method results in an iteratively reweighted least-squares algorithm. As $\tilde{\mathbf{D}}$ has positive elements, $\mathbf{J}'\tilde{\mathbf{D}}\mathbf{J}$ is positive semidefinite (Appendix A4.5). If $\mathbf{J}$ has full rank, then $\mathbf{J}'\tilde{\mathbf{D}}\mathbf{J}$ is positive definite and the method generates descent directions.

However, Dennis and Welsch [1978] state that Gauss–Newton-type algorithms for robust nonlinear regression "still lack reliability." This comment includes methods based on (14.102), (14.105), or the use of $\mathbf{J}'\mathbf{J}$ in place of $\mathbf{J}'\mathbf{D}\mathbf{J}$ in (14.102).

The approximate Newton method of Dennis et al. [1981a, b] (see Section 14.3.2) generalizes to the robust problem fairly directly. This method was based upon using quasi-Newton approximations to $\mathbf{A}(\boldsymbol{\theta})$ given by (14.9) [c.f. (14.101) in the robust case]. Using the argument following (14.36), the quasi-Newton condition becomes

$$\tilde{\mathbf{A}}^{(a+1)}\boldsymbol{\delta}^{(a)} = \frac{1}{\sigma}(\mathbf{J}^{(a+1)} - \mathbf{J}^{(a)})'\mathbf{v}^{(a+1)} = \tilde{\boldsymbol{\gamma}}^{(a)}, \tag{14.106}$$

say. In the update (14.42), $\boldsymbol{\gamma}^{(a)} = \mathbf{g}^{(a+1)} - \mathbf{g}^{(a)}$ with $\mathbf{g}$ given in (14.97). Dennis and Welsch [1978] give no discussion of how re-estimating the scale factor at each iteration affects the algorithm, nor how it might be modified to accommodate this. They state that their methods using this type of quasi-Newton algorithm "have been competitive with full Newton methods and more reliable than Gauss–Newton methods."

Dennis and Welsch note that stopping criteria based upon changes in $\boldsymbol{\theta}$ are not greatly affected by the choice of scale factor σ, but those based upon changes in gradient are [by (14.97)]. They note that at the minimum, $\mathbf{v}(\hat{\boldsymbol{\theta}})$ is orthogonal to the columns of $\hat{\mathbf{J}}$, and they use this to define a stopping rule (c.f. the orthogonality of $\hat{\mathbf{r}}$ and the columns of $\hat{\mathbf{J}}$ in Section 14.4). The criterion they propose is

$$\max_{1 \le j \le p} \cos(\mathbf{u}_j, \mathbf{v}) < \tau, \tag{14.107}$$

where $\mathbf{u}_j$ is the jth column of $\hat{\mathbf{J}}$ and $\cos(\mathbf{u}, \mathbf{v})$ is defined in (14.69). This is the direct analogue of the stopping criterion for least squares proposed by Dennis [1977] [see the paragraph preceding (14.69)]. In the least-squares case, Dennis has since abandoned this criterion in favor of (14.79). A direct analogue to (14.79) is

$$\frac{\mathbf{v}^{(a)\prime}\mathbf{v}_T^{(a)}}{\|\mathbf{v}^{(a)}\| \, \|\mathbf{v}_T^{(a)}\|} < \tau, \tag{14.108}$$

where $\mathbf{v}_T^{(a)} = \mathbf{J}^{(a)}(\mathbf{J}^{(a)\prime}\mathbf{J}^{(a)})^{-1}\mathbf{J}^{(a)\prime}\mathbf{v}^{(a)}$.

14.6.2 L_1-Minimization

Another important loss function which is sometimes used as a robust loss function is $\rho(t) = |t|$, so that

$$h(\boldsymbol{\theta}) = \sum_{i=1}^{n} |r_i(\boldsymbol{\theta})|, \tag{14.109}$$

the sum of absolute residuals. When the residual functions are linear, the L_1 problem can be transformed to a linear-programming problem, and this is a basis for many algorithms. For a survey in the linear-regression case see Kennedy and Gentle [1980: Chapter 11] and Narula and Wellington [1982]. We describe only the method proposed by Schlossmacher [1973] for the linear case, which applies equally well to nonlinear regression.

Schlossmacher's [1973] method uses iteratively reweighted least squares. At the ath iteration, $\boldsymbol{\theta}^{(a+1)}$ is chosen to minimize

$$h^{(a)}(\boldsymbol{\theta}) = \sum_{i=1}^{n} w_i^{(a)} r_i^2(\boldsymbol{\theta}), \tag{14.110}$$

where usually $w_i^{(a)} = |r_i(\boldsymbol{\theta}^{(a)})|^{-1}$. The process terminates when $\boldsymbol{\theta}^{(a+1)} \approx \boldsymbol{\theta}^{(a)}$. For linear problems, the least-absolute-deviations surface passes through one or more observations, and therefore zero residuals result. Also, during the iterations some residuals may be zero or very close to zero. For observations where $r_i(\boldsymbol{\theta}^{(a)}) \approx 0$, Schlossmacher sets $w_i^{(a)} = 0$. He notes that this is justified in that the effect of an almost zero residual on the total sum will be negligible. On the other hand, including $w_i^{(a)} = |r_i(\boldsymbol{\theta}^{(a)})|^{-1}$ leads to a very large weight, which can cause numerical instability. If the residual becomes important on a subsequent iteration, it is reinstated. Other approaches are to try a new starting point and/or a new starting set of weights. It is possible that none of these approaches will solve the problem, though Gill et al. [1981: 98] state that this is unlikely to occur in practice.

Other approaches to the nonlinear L_1 problem include Osborne and Watson [1971] (which involves the solution of a sequence of linear L_1 problems), El-Attar et al. [1979], and Murray and Overton [1980].

14.7 SEPARABLE LEAST-SQUARES PROBLEMS

14.7.1 Introduction

A separable least-squares problem is one in which we can partition $\boldsymbol{\theta} = (\boldsymbol{\beta}', \boldsymbol{\gamma}')'$ so that for fixed $\boldsymbol{\gamma}$, $S(\boldsymbol{\theta}) = S(\boldsymbol{\beta}, \boldsymbol{\gamma})$ is easily minimized with respect to $\boldsymbol{\beta}$. The most common problem of this type comes from partially linear models (Section 5.10.2)

$$\mathbf{y}(\boldsymbol{\gamma}) = \mathbf{X}(\boldsymbol{\gamma})\boldsymbol{\beta} + \boldsymbol{\varepsilon}, \tag{14.111}$$

in which the elements of $\mathbf{y}$ and the design matrix $\mathbf{X}$ may be functions of $\boldsymbol{\gamma}$. Here

$$\mathbf{r}(\boldsymbol{\beta}, \boldsymbol{\gamma}) = \mathbf{y}(\boldsymbol{\gamma}) - \mathbf{X}(\boldsymbol{\gamma})\boldsymbol{\beta} \tag{14.112}$$

and

$$S(\boldsymbol{\beta}, \boldsymbol{\gamma}) = \|\mathbf{r}(\boldsymbol{\beta}, \boldsymbol{\gamma})\|^2. \tag{14.113}$$

For $\boldsymbol{\gamma}$ fixed, the minimization of $S(\boldsymbol{\beta}, \boldsymbol{\gamma})$ with respect to $\boldsymbol{\beta}$ is a linear least-squares problem which can be solved using any of the numerically stable techniques mentioned in Section 14.1.

Let $\boldsymbol{\beta}$ and $\boldsymbol{\gamma}$ have dimensions p_1 and p_2 $(= p - p_1)$ respectively. Denote by $\hat{\boldsymbol{\theta}} = (\hat{\boldsymbol{\beta}}', \hat{\boldsymbol{\gamma}}')'$ the value of $\boldsymbol{\theta}$ which minimizes $S(\boldsymbol{\theta})$. Then suppose that for fixed $\boldsymbol{\gamma}$,

$\boldsymbol{\beta} = \mathbf{b}(\boldsymbol{\gamma})$ minimizes S, so that

$$\left.\frac{\partial S(\boldsymbol{\beta}, \boldsymbol{\gamma})}{\partial \boldsymbol{\beta}}\right|_{\boldsymbol{\beta}=\mathbf{b}(\boldsymbol{\gamma})} = \mathbf{0}. \tag{14.114}$$

Define the *concentrated sum of squares* as

$$M(\boldsymbol{\gamma}) = S(\mathbf{b}(\boldsymbol{\gamma}), \boldsymbol{\gamma}). \tag{14.115}$$

The idea underlying methods for separable functions is that we can obtain $\hat{\boldsymbol{\gamma}}$ by minimizing $M(\boldsymbol{\gamma})$ and then recover $\hat{\boldsymbol{\beta}}$, if necessary, as $\hat{\boldsymbol{\beta}} = \mathbf{b}(\hat{\boldsymbol{\gamma}})$. This can be justified for the case when (14.114) has a unique solution for each $\boldsymbol{\gamma}$ by using Theorem 2.2 of Section 2.2.3. The notation is different there because of a different emphasis, i.e., the parameters are $(\boldsymbol{\theta}', \boldsymbol{\tau}')'$ rather than $\boldsymbol{\theta} = (\boldsymbol{\beta}', \boldsymbol{\gamma}')'$. However, Theorem 2.2 is basically a general statement about a function and its derivatives, and the proof makes no use of the special properties of a log-likelihood function. It also extends to stationary values: a stationary value of $M(\boldsymbol{\gamma})$, as given by its derivative equated to zero, leads to a stationary value of $S(\boldsymbol{\theta})$.

The above argument does not deal with multiple solutions to (14.114). Golub and Pereyra [1973] give a rigorous treatment of the problem for a slight simplification of the partially linear model (14.113), namely $\mathbf{y}(\boldsymbol{\gamma}) \equiv \mathbf{y}$. To allow for $\mathbf{X}(\boldsymbol{\gamma})$ to be less than full rank, they take $\mathbf{b}(\boldsymbol{\gamma})$ to be the solution of the normal equations obtained using the Moore–Penrose generalized inverse (Appendix D1.5). In their Theorem 2.1 these authors prove that if the rank of $\mathbf{X}(\boldsymbol{\gamma})$ is constant over the parameter space, a stationary point (respectively, the global minimizer) of S can be obtained from a stationary point (respectively, the global minimizer) of M as outlined above.

It has been recognized for some time that the construction of (14.115) for the separable problem can be of value for partially linear models when using an algorithm which does not employ derivatives (Lawton and Sylvestre [1971]). Recognition that a least-squares problem is separable is useful for obtaining starting values even if a general least-squares algorithm is to be used. A starting value, $\boldsymbol{\gamma}^{(1)}$ say, need only be supplied for $\boldsymbol{\gamma}$. If required, a corresponding starting value for $\boldsymbol{\beta}$ is then $\boldsymbol{\beta}^{(1)} = \mathbf{b}(\boldsymbol{\gamma}^{(1)})$.

14.7.2 Gauss–Newton for the Concentrated Sum of Squares

In this subsection we consider just the partially linear model (14.112). Then, for fixed $\boldsymbol{\gamma}$, $\boldsymbol{\beta} = \mathbf{b}(\boldsymbol{\gamma})$ solves the linear least-squares problem (14.113). To allow for $\mathbf{X}(\boldsymbol{\gamma})$ possibly not having full rank we use Moore–Penrose generalized inverses (Appendix D1.5). A solution to the linear least-squares problem

$$\underset{\boldsymbol{\beta}}{\text{Minimize}} \, \| \mathbf{y}(\boldsymbol{\gamma}) - \mathbf{X}(\boldsymbol{\gamma})\boldsymbol{\beta} \|^2 \tag{14.116}$$

for fixed γ is

$$\mathbf{b}(\gamma) = \mathbf{X}^+(\gamma)\mathbf{y}(\gamma). \tag{14.117}$$

Writing $\mathbf{r}_M(\gamma) = \mathbf{r}(\mathbf{b}(\gamma), \gamma)$, we have

$$\begin{aligned} \mathbf{r}_M(\gamma) &= \mathbf{y}(\gamma) - \mathbf{X}(\gamma)\mathbf{X}^+(\gamma)\mathbf{y}(\gamma) \\ &= [\mathbf{I}_n - \mathbf{P}(\gamma)]\mathbf{y}(\gamma), \end{aligned} \tag{14.118}$$

where

$$\mathbf{P}(\gamma) = \mathbf{X}(\gamma)\mathbf{X}^+(\gamma) \tag{14.119}$$

is the projection matrix onto the range space of $\mathbf{X}(\gamma)$ (Appendix A11.4). Moreover,

$$\begin{aligned} M(\gamma) &= \|\mathbf{r}_M(\gamma)\|^2 \\ &= \|[\mathbf{I}_n - \mathbf{P}(\gamma)]\mathbf{y}(\gamma)\|^2 \\ &= \mathbf{y}(\gamma)'[\mathbf{I}_n - \mathbf{P}(\gamma)]\mathbf{y}(\gamma). \end{aligned} \tag{14.120}$$

To set up Gauss–Newton-based algorithms for $M(\gamma)$ we need only the $n \times p_2$ Jacobian matrix

$$\mathbf{J}_M(\gamma) = \frac{\partial \mathbf{r}_M(\gamma)}{\partial \gamma'}. \tag{14.121}$$

This Jacobian has jth column

$$[\mathbf{J}_M(\gamma)]_j = -\frac{\partial \mathbf{P}(\gamma)}{\partial \gamma_j}\mathbf{y}(\gamma) + [\mathbf{I}_n - \mathbf{P}(\gamma)]\frac{\partial \mathbf{y}(\gamma)}{\partial \gamma_j}, \tag{14.122}$$

where it is shown below that

$$\frac{\partial \mathbf{P}(\gamma)}{\partial \gamma_j} = [\mathbf{I}_n - \mathbf{P}(\gamma)]\frac{\partial \mathbf{X}(\gamma)}{\partial \gamma_j}\mathbf{X}^+(\gamma) + \left([\mathbf{I}_n - \mathbf{P}(\gamma)]\frac{\partial \mathbf{X}(\gamma)}{\partial \gamma_j}\mathbf{X}^+(\gamma)\right)'. \tag{14.123}$$

We now establish (14.123) using the clever proof of Golub and Pereyra [1973: Lemma 4.1]. For convenience we drop the explicit dependence of matrices on γ. Recall, also, that $\mathbf{P}$ projects onto the range space of $\mathbf{X}$, so that $\mathbf{PX} = \mathbf{X}$. Hence

$$\frac{\partial(\mathbf{PX})}{\partial \gamma_j} = \frac{\partial \mathbf{P}}{\partial \gamma_j}\mathbf{X} + \mathbf{P}\frac{\partial \mathbf{X}}{\partial \gamma_j} = \frac{\partial \mathbf{X}}{\partial \gamma_j}, \tag{14.124}$$

which leads to

$$\frac{\partial \mathbf{P}}{\partial \gamma_j}\mathbf{X} = (\mathbf{I}_n - \mathbf{P})\frac{\partial \mathbf{X}}{\partial \gamma_j}. \tag{14.125}$$

Also $\mathbf{P} = \mathbf{X}\mathbf{X}^+$, so that

$$\begin{aligned}\frac{\partial \mathbf{P}}{\partial \gamma_j}\mathbf{P} &= \left(\frac{\partial \mathbf{P}}{\partial \gamma_j}\mathbf{X}\right)\mathbf{X}^+ \\ &= (\mathbf{I}_n - \mathbf{P})\frac{\partial \mathbf{X}}{\partial \gamma_j}\mathbf{X}^+. \end{aligned} \tag{14.126}$$

However, as $\mathbf{P}$ is symmetric,

$$\left(\frac{\partial \mathbf{P}}{\partial \gamma_j}\mathbf{P}\right)' = \mathbf{P}\frac{\partial \mathbf{P}}{\partial \gamma_j}. \tag{14.127}$$

Hence using the fact that $\mathbf{P}$ is idempotent, together with (14.126) and (14.127), we have

$$\begin{aligned}\frac{\partial \mathbf{P}}{\partial \gamma_j} &= \frac{\partial \mathbf{P}^2}{\partial \gamma_j} \\ &= \frac{\partial \mathbf{P}}{\partial \gamma_j}\mathbf{P} + \mathbf{P}\frac{\partial \mathbf{P}}{\partial \gamma_j} \\ &= (\mathbf{I}_n - \mathbf{P})\frac{\partial \mathbf{X}}{\partial \gamma_j}\mathbf{X}^+ + \left((\mathbf{I}_n - \mathbf{P})\frac{\partial \mathbf{X}}{\partial \gamma_j}\mathbf{X}^+\right)', \end{aligned} \tag{14.128}$$

thus establishing (14.123). The proof given here also applies with other generalized inverses. It is not limited to the Moore–Penrose inverse.

Golub and Pereyra [1973] describe in detail how to implement a Levenberg–Marquardt modification of the Gauss–Newton algorithm using $\mathbf{J}_M(\boldsymbol{\gamma})$ in the case $\mathbf{y}(\boldsymbol{\gamma}) \equiv \mathbf{y}$. Refinements have been suggested by Krogh [1974] and Corradi and Stefanini [1978] which allow for general $\mathbf{y}(\boldsymbol{\gamma})$ (see also Ruhe and Wedin [1980]). Kaufman and Pereyra [1978] have developed separable methods for problems with equality constraints.

14.7.3 Intermediate Method

An obvious way of adapting the Gauss–Newton algorithm for the full problem "minimize $S(\boldsymbol{\theta})$" to the separable problem is as follows. The last p_2 elements of the Gauss–Newton step for the full problem, $\boldsymbol{\delta}_\gamma$ say, relate to $\boldsymbol{\gamma}$, so that we can use the updates

$$\boldsymbol{\gamma}^{(a+1)} = \boldsymbol{\gamma}^{(a)} + \boldsymbol{\delta}_\gamma^{(a)} \tag{14.129}$$

and

$$\boldsymbol{\beta}^{(a+1)} = \mathbf{b}(\boldsymbol{\gamma}^{(a+1)}). \tag{14.130}$$

Separability is therefore used just to obtain an improved update of $\boldsymbol{\beta}$.

Suppose we partition $\mathbf{J}(\boldsymbol{\theta}) = \mathbf{J}(\boldsymbol{\beta}, \boldsymbol{\gamma})$ as $(\mathbf{J}_1, \mathbf{J}_2)$, where $\mathbf{J}_i$ is $n \times p_i$. We may write the Gauss–Newton step

$$\boldsymbol{\delta} = (\mathbf{J}'\mathbf{J})^{-1}\mathbf{J}'\mathbf{r}$$

as

$$\begin{pmatrix} \boldsymbol{\delta}_\beta \\ \boldsymbol{\delta}_\gamma \end{pmatrix} = \begin{pmatrix} \mathbf{J}_1'\mathbf{J}_1 & \mathbf{J}_1'\mathbf{J}_2 \\ \mathbf{J}_2'\mathbf{J}_1 & \mathbf{J}_2'\mathbf{J}_2 \end{pmatrix}^{-1} \begin{pmatrix} \mathbf{J}_1' \\ \mathbf{J}_2' \end{pmatrix} \mathbf{r},$$

so that by Appendix A3.1(1),

$$\boldsymbol{\delta}_\gamma = [\mathbf{J}_2'\mathbf{J}_2 - \mathbf{J}_2'\mathbf{J}_1(\mathbf{J}_1'\mathbf{J}_1)^{-1}\mathbf{J}_1'\mathbf{J}_2]^{-1}\mathbf{J}_2'\mathbf{r}. \tag{14.131}$$

For the partially linear problem (14.113),

$$\mathbf{J}_1 = \frac{\partial \mathbf{r}}{\partial \boldsymbol{\beta}'} = -\frac{\partial \mathbf{f}}{\partial \boldsymbol{\beta}'} = -\mathbf{X}(\boldsymbol{\gamma}), \tag{14.132}$$

and if $(\mathbf{J}_2)_j$ is the jth column of $\mathbf{J}_2$,

$$(\mathbf{J}_2)_j = \frac{\partial \mathbf{r}}{\partial \gamma_j} = \frac{\partial \mathbf{y}(\boldsymbol{\gamma})}{\partial \gamma_j} - \frac{\partial \mathbf{X}(\boldsymbol{\gamma})}{\partial \gamma_j}\boldsymbol{\beta}. \tag{14.133}$$

This is the method proposed by Barham and Drane [1972]. The method of Kaufman [1975] is essentially a modern implementation of this idea (see Ruhe and Wedin [1980]). However, *QR* factorizations are used to compute the components of the Gauss–Newton step, and the $\mathbf{J}'\mathbf{J}$ matrices are never formed. Kaufman's method also incorporates a Levenberg–Marquardt modification. Corradi and Stefanini [1978] use both a step-length modification and a Levenberg–Marquardt modification.

Ruhe and Wedin [1980] reconcile the algorithm above and the algorithm of Golub and Pereyra [1973] in the previous section, namely Gauss–Newton on $M(\boldsymbol{\gamma})$, as follows. From (2.69)

$$\begin{aligned} \mathbf{g}_M(\boldsymbol{\gamma}) &= \frac{\partial M(\boldsymbol{\gamma})}{\partial \boldsymbol{\gamma}} \\ &= \left.\frac{\partial S(\boldsymbol{\beta}, \boldsymbol{\gamma})}{\partial \boldsymbol{\gamma}}\right|_{\boldsymbol{\beta} = \mathbf{b}(\boldsymbol{\gamma})} \\ &= 2\mathbf{J}_2'\mathbf{r}\big|_{\boldsymbol{\beta} = \mathbf{b}(\boldsymbol{\gamma})}. \end{aligned} \tag{14.134}$$

From (2.68), with appropriate notational changes,

$$\mathbf{H}_M(\boldsymbol{\gamma}) = \frac{\partial^2 M}{\partial \boldsymbol{\gamma}\, \partial \boldsymbol{\gamma}'} = (\mathbf{H}_{22} - \mathbf{H}_{21}\mathbf{H}_{11}^{-1}\mathbf{H}_{12})|_{\boldsymbol{\beta}=\mathbf{b}(\boldsymbol{\gamma})}, \tag{14.135}$$

where $\mathbf{H}_{22} = \partial^2 S/\partial\boldsymbol{\gamma}\,\partial\boldsymbol{\gamma}'$, $\mathbf{H}_{12} = \partial^2 S/\partial\boldsymbol{\beta}\,\partial\boldsymbol{\gamma}'$, etc. Then the Newton step for $M(\boldsymbol{\gamma})$ is

$$\begin{aligned}\boldsymbol{\delta}_\gamma &= -\mathbf{H}_M(\boldsymbol{\gamma})^{-1}\mathbf{g}_M(\boldsymbol{\gamma}) \\ &= -2[\{\mathbf{H}_{22} - \mathbf{H}_{21}\mathbf{H}_{11}^{-1}\mathbf{H}_{12}\}^{-1}\mathbf{J}_2'\mathbf{r}]_{\boldsymbol{\beta}=\mathbf{b}(\boldsymbol{\gamma})}. \end{aligned} \tag{14.136}$$

Now from (14.7) we have $\mathbf{H} = 2(\mathbf{J}'\mathbf{J} + \mathbf{A})$, where $\mathbf{A} = \sum r_i \partial^2 r_i/\partial\boldsymbol{\theta}\,\partial\boldsymbol{\theta}'$, so that $\mathbf{H}_{22} = 2(\mathbf{J}_2'\mathbf{J}_2 + \mathbf{A}_{22})$, $\mathbf{H}_{21} = 2(\mathbf{J}_2'\mathbf{J}_1 + \mathbf{A}_{21})$, etc. Clearly (14.131) corresponds to neglecting all of the A-terms in (14.136). Consider now only the partially linear model. From (14.132) and (14.133) we see that $\mathbf{A}_{11} = \mathbf{0}$ and that $\mathbf{A}_{21}$ involves only first derivatives. The closest we can get to $\mathbf{H}_M(\boldsymbol{\gamma})$ using only first derivatives is to neglect $\mathbf{A}_{22}$ and use in (14.136)

$$\begin{aligned}\tfrac{1}{2}\tilde{\mathbf{H}}_M &= \mathbf{J}_2'\mathbf{J}_2 - (\mathbf{J}_2'\mathbf{J}_1 + \mathbf{A}_{21})(\mathbf{J}_1'\mathbf{J}_1)^{-1}(\mathbf{J}_1'\mathbf{J}_2 + \mathbf{A}_{21}') \\ &= \mathbf{J}_2'\mathbf{J}_2 - \mathbf{J}_2'\mathbf{J}_1(\mathbf{J}_1'\mathbf{J}_1)^{-1}\mathbf{J}_1'\mathbf{J}_2 + \mathbf{A}_{21}(\mathbf{J}_1'\mathbf{J}_1)^{-1}\mathbf{A}_{21}' \\ &\quad + \mathbf{A}_{21}(\mathbf{J}_1'\mathbf{J}_1)^{-1}\mathbf{J}_1'\mathbf{J}_2 + \mathbf{J}_2'\mathbf{J}_1(\mathbf{J}_1'\mathbf{J}_1)^{-1}\mathbf{A}_{21}'. \end{aligned} \tag{14.137}$$

Ruhe and Wedin [1980] prove that using Gauss–Newton on $M(\boldsymbol{\gamma})$ (c.f. Golub and Pereyra [1973]) is equivalent to using the first three terms of (14.137), whereas (14.131) uses only the first two.

14.7.4 The NIPALS Method

The nonlinear interactive partial least-squares (NIPALS) technique, due to Wold and described by Wold and Lyttkens [1969] and Frisen [1979], alternately holds one of $\boldsymbol{\beta}$ and $\boldsymbol{\gamma}$ fixed and minimizes with respect to the other—i.e., for $a = 2, 3, \ldots,$

$$\text{find } \boldsymbol{\beta}^{(a+1)} \text{ to minimize } S(\boldsymbol{\beta}, \boldsymbol{\gamma}^{(a)}); \tag{14.138}$$

$$\text{find } \boldsymbol{\gamma}^{(a+1)} \text{ to minimize } S(\boldsymbol{\beta}^{(a+1)}, \boldsymbol{\gamma}). \tag{14.139}$$

This technique was designed for the subclass of separable problems in which *both* (14.138) and (14.139) are linear least-squares problems, i.e., neither subproblem requires an iterative solution. For example, the models

$$y = \gamma + \beta x + \gamma\beta x^2 + \varepsilon$$

and

$$y = (\boldsymbol{\gamma}'\mathbf{x}_1)(\boldsymbol{\beta}'\mathbf{x}_2) + \varepsilon$$

fall into this class.

Ruhe and Wedin [1980] investigated this method when (14.138) is a linear least-squares problem but (14.139) is nonlinear. They compared the convergence rate of NIPALS, both theoretically and empirically, with the methods of Sections 14.7.2 and 14.7.3, and found NIPALS to be considerably slower.

14.7.5 Discussion

Ruhe and Wedin [1980] investigated the (asymptotic) convergence rates of the separable algorithms, and found that the convergence rates of Gauss–Newton for the full problem and the methods of Sections 14.7.2 and 14.7.3 are of the same order. For example, if Gauss–Newton for the full problem is superlinearly convergent, then both the other algorithms are superlinearly convergent as well.

The empirical evidence from, for instance, Barham and Drane [1972], Golub and Pereyra [1973], Kaufman [1975], and Corradi and Stefanini [1978] shows that the separable algorithms are often, but not always, faster than the corresponding method for minimizing the original function. They may also succeed on difficult problems, such as sums of exponentials, where other methods fail.

We have noted that Golub and Pereyra's [1973] method, slightly improved by Krogh [1974], applied Gauss–Newton to the reduced problem $M(\gamma)$. Kaufman's [1975] algorithm, which is equivalent to using the γ-component from the Gauss–Newton step for the full problem, allows some reduction in computation. Her algorithm is largely an adaptation of Golub and Pereyra [1973] with some steps simplified. In the rather limited empirical comparisons carried out by Kaufman [1975] and Ruhe and Wedin [1980], both Kaufman's method and Golub and Pereyra's almost always took the same number of iterations, but Kaufman's iterations are cheaper (about a 20% reduction in computer time in the two examples of Kaufman [1975]). However, the search directions generated by Golub and Pereyra's method are closer to the Newton direction in the sense of using three terms of (14.137) instead of just two. It remains to be seen whether there is some improvement in reliability.

CHAPTER 15

Software Considerations

15.1 SELECTING SOFTWARE

There has long been a need for an extensive source of practical advice for researchers and data analysts who have practical problems to solve using optimization software. In particular, such people should be aware of the difficulties that can be encountered and how to circumvent then. This need has largely been fulfilled by Chapters 7 and 8 of Gill et al. [1981]. Chapter 7 of Dennis and Schnabel [1983] is also useful. In this chapter we raise most of the main issues. However, any reader who is having difficulties with a particular problem, or who wants a deeper understanding, should consult Gill et al. [1981].

15.1.1 Choosing a Method

The method chosen for minimizing a given function will depend on the software available to the data analyst. For example, the analyst might prefer a Gauss–Newton-based algorithm for a least-squares problem, but the quasi-Newton algorithm he or she already has will do the job. The additional efficiency of the Gauss–Newton method must be weighed against the expense of acquiring it, or the time required to type and test a published program. These costs will be too great unless many similar problems are expected to be encountered. This type of consideration should be born in mind during the following discussion.

Chambers [1973] suggested that the statistician should have access to, in decreasing order of importance:

1. a recent quasi-Newton method with provision for finite-difference approximations to first derivatives;
2. a direct-search method such as a polytope (simplex) method;
3. a special routine for nonlinear least squares; and
4. a modified Newton method for problems with inexpensive second derivatives.

This still seems to be good advice. There have been improvements made in each of these classes of algorithm over the intervening years, but no conceptually new ones have displaced them in importance. One advance which is worthy of mention, however, is in the development and availability of methods for problems with bounds or linear inequality constraints of the form $\mathbf{a}_i'\boldsymbol{\theta} \leq c_i$ upon the parameters. These programs can be used for unconstrained optimization by making the bounds very large, with little additional cost in terms of efficiency, though there is additional cost in terms of program size. Such programs can be much more efficient, particularly when the analyst has sufficient idea of where the minimum is, so that useful bounds can be placed upon the parameters.

Until now we have discussed minimal software requirements. At the other end of the scale, many if not most statisticians working in universities, government departments, and large industrial establishments now have access to high-quality numerical subroutine libraries such as NAG or IMSL. These libraries have a wide spectrum of optimization routines for different types of problem. Well-documented libraries give advice on particular routines for particular problems. For example, the NAG library uses decision trees to guide the user to a relevant subroutine. They also have two user interfaces to some types of routine, one for inexperienced users and another for experienced users which gives them more control over the parameters of the algorithm.

One general rule is to choose an algorithm that makes as much use of the structure of the particular problem as possible. Another is to choose a method that uses as much derivative information as you can reasonably supply. We saw in Chapter 13, as we progressed from second-derivative methods to derivative-free methods for minimizing smooth functions, that the later methods attempt to approximate the earlier methods and this leads to decreased reliability. The use of second derivatives, for example, can enable the algorithm to check whether the stationary point is a minimum, to steer around saddle points, and to check the condition number of the Hessian at the computed minimum. (In Section 15.3.3 we discuss further the problems associated with ill-conditioning.) Quasi-Newton algorithms do not always give good approximations to the Hessian, and in some library programs it is not easy for the user to gain access to the approximate Hessian for statistical purposes. These are all factors to be weighed against the personal work involved in deriving and programming second derivatives and the attendant risks of making mistakes. We recall the ordering given by Gill et al. [1981] for optimization methods in terms of reliability:

1. modified Newton using second derivatives;
2. modified Newton without second derivatives;
3. quasi-Newton;
4. discrete quasi-Newton;
5. conjugate gradient;
6. conjugate directions (direction set); and
7. polytope (simplex).

The reader is also referred to the decision tree of Fig. 13.5 in Section 13.7. Special methods also exist for specialized applications such as least-squares problems (Chapter 14) and large problems with sparse Hessians.

15.1.2 Some Sources of Software

To echo the advice of Chambers [1973], we do not recommend the writing of one's own optimization software as a casual project. It is an undertaking which requires a substantial investment because small implementation details can have a large effect on both efficiency and reliability.

All of the major statistical packages [e.g. SAS, BMDP, SPSS, and some less well-known packages such as NONLIN (*American Statistician*, 1986, p. 52)] contain programs for fitting nonlinear models by least squares, usually a variant of Marquardt's method (American Statistical Association [1982]). While the implementations are possibly not as sophisticated as that of Moré [1977], they have the advantage of easy use and easy data manipulation.

In this section we give other important sources of software for nonlinear optimization, namely the major subroutine libraries and the Association of Computing Machinery (ACM) algorithms. We divide the libraries into two categories: commercial and noncommercial. The commercial libraries provide a higher level of support. The noncommercial libraries are available for little more than handling changes. Some libraries may not be available for certain machines.

a Commercial Libraries

NAG: The Numerical Algorithms Group Ltd, NAG Central Office, Mayfield House, 265 Banbury Road, Oxford OX2 7DE, England.

IMSL: IMSL Incorporated, 2500 Park West One, 2500 City West Boulevard, Houston, Texas 77042-3020, U.S.A.

MINOS: Systems Optimization Laboratory, Operations Research Dept., Stanford University, Stanford, CA 94305, U.S.A.

b Noncommercial Libraries

Harwell: Computer Science and Systems Division, AERE, Harwell, Oxfordshire, England.

MINPACK: Argonne National Laboratory, 9700 South Cass Avenue, Argonne, Illinois 60439, U.S.A. [Complementary libraries: EISPAC (eigensystems), LINPAC (linear algebra)]

NPL: National Physical Laboratory, Teddington, TW11 0LW, England.

c ACM *Algorithms*

The ACM algorithms are described in the journal *ACM Transactions on Mathematical Software* (TOMS), which gives a brief description together with calling sequences etc. The algorithms are available on payment of handling charges from: ACM Algorithms Distribution Service, IMSL Inc., 2500 ParkWest One, 2500 CityWest Boulevard, Houston, Texas 77036, U.S.A. and are distributed on tape or floppy disk. Order forms are in each issue of TOMS. The ACM algorithms policy is described by Krogh [1986]. A feature of the ACM software is that it is intended to be easily transportable between machines. A list of algorithms March 1980–June 1986 is given in TOMS Vol. 12, 1986, p. 181, and for March 1975–December 1979 in TOMS Vol. 6, 1980, p. 134. In particular, we draw the reader's attention to algorithms 630, 611, 573, 566, and 500. Algorithm 611 of Gay [1983a] is an unconstrained-minimization package implementing modified Newton and quasi-Newton methods (including finite-difference approximations to derivatives) using a trust-region approach. Algorithm 573 is Dennis et al.'s [1981b] NL2SOL nonlinear least-squares program mentioned at the end of Section 14.3.2 (see also Gay [1983b]). Algorithm 500 is an earlier unconstrained-minimization package of Shanno and Phua [1976]. Part of that program was updated by Shanno and Phua [1980]. We note that Miller [1979: 42] found Shanno and Phua's convergence criterion could be disastrous when fitting mixtures of exponentials.

The software in major libraries, and even in TOMS to some extent, has the advantage over homegrown products of extensive testing, heavy usage, and updating as users have found problems. An excellent discussion of the testing of unconstrained-optimization software is given by Moré et al. [1981]. Their paper also contains a large number of test problems and describes ACM algorithm 566. This is a collection of subroutines for testing such software, with special attention paid to reliability and robustness. Hiebert [1981] tested the nonlinear least-squares software of some of the major libraries (e.g. NAG, MINPACK, NPL, and a version of NL2SOL), again with special regard to reliability and robustness. At the time of publication, some of the results were already out of date. However, Hiebert makes many valuable comments about both the routines and program testing. He notes, for example, that the standard test problems tend to be well scaled and have starting points close to the solution. To remedy this, three starting values were used: $\boldsymbol{\theta}^{(1)}$, $10\boldsymbol{\theta}^{(1)}$, and $100\boldsymbol{\theta}^{(1)}$, where $\boldsymbol{\theta}^{(1)}$ was the "usual" starting value. The parameters were also transformed to different orders of magnitude. These changes made the test problems much more difficult. Hoaglin et al. [1982] advocated and demonstrated the use of exploratory data-analysis techniques in analyzing the results of test data.

A very recent book, Nash and Walker-Smith [1987], comes with a floppy disk of optimization software that runs on an IBM PC.

15.2 USER-SUPPLIED CONSTANTS

Optimization programs invariably require the values of a number of algorithm parameters. Some may be built into the program while others have to be supplied by the user. However some may have the option of being supplied by the user who can thus taylor the algorithm to work efficiently on a particular type of problem.

Ideally, the users should be allowed a great deal of control if they want it. However, default values for these constants should be built in which are designed to work reasonably well on a wide spectrum of problems.

15.2.1 Starting Values

a Initial Parameter Value

The starting value $\boldsymbol{\theta}^{(1)}$, which is an initial guess at the minimum $\hat{\boldsymbol{\theta}}$, can sometimes be suggested by prior information, e.g. similar problems in the past. Sometimes there will be a starting value that tends to work well for a class of problems. An example is the fitting of logistic regression models (Section 1.6), where the starting value $\boldsymbol{\theta}^{(1)} = \mathbf{0}$ is almost always a good one. Indeed, the formulation of Fisher's scoring algorithm for generalized linear models as an iteratively reweighted least-squares method suggests a uniform starting mechanism for this whole class of models (see McCullagh and Nelder [1983]). However, it is very difficult to say anything about producing good starting values in general.

For the classical Newton or Gauss–Newton methods, a good starting value is crucial to obtaining convergence. However, with good implementations of modern methods which use line searches or trust regions and modify approximations to the Hessian to ensure descent directions, convergence can often be obtained from the first guess that comes to mind. Indeed, the optimization methods themselves tend to be far better and more efficient at seeking out a minimum than the various ad hoc procedures that are often suggested for finding starting values. Exceptions to this rule can occur when there are no safeguards to prevent the initial steps going into extreme regions of the $\boldsymbol{\theta}$-space from which recovery is impossible. For example, overflow or underflow (Section 15.3.2) halts execution or makes calculations meaningless. Difficulty may also occur with highly ill-conditioned problems when the curvature is changing rapidly in some directions but not others. In our experience, optimization methods used within some statistical packages (e.g. for maximum likelihood) are often very unsophisticated and not robust against the choice of starting value.

Methods that are sometimes suggested include a grid search or a random search over a defined rectangular region of the parameter space. If no sensible bounds can be suggested for a parameter θ_r, a transformed parameter can be used. For example, $\phi = e^{\theta}/(1 + e^{\theta})$ and $\phi = \arctan \theta$ both satisfy $0 < \phi < 1$. These procedures are expensive for high-dimensional problems; for instance, m levels along each of the p dimensions requires m^p function evaluations. They should perhaps be used only after several failures from guessed starting values.

The dimension of a computation can be reduced if the problem is separable, i.e. if we can partion $\boldsymbol{\theta} = (\boldsymbol{\beta}'\boldsymbol{\gamma}')'$ so that, for fixed $\boldsymbol{\gamma}$, $h(\boldsymbol{\theta}) = h(\boldsymbol{\beta}, \boldsymbol{\gamma})$ can be minimized exactly or cheaply with respect to $\boldsymbol{\beta}$ to obtain $\hat{\boldsymbol{\beta}} = \mathbf{b}(\boldsymbol{\gamma})$ [c.f. Section 14.7]. Then a starting value is required for just $\boldsymbol{\gamma}$, say $\boldsymbol{\gamma}^{(1)}$, which leads to $\boldsymbol{\theta}^{(1)} = [\mathbf{b}(\boldsymbol{\gamma}^{(1)})', \boldsymbol{\gamma}^{(1)\prime}]'$. This can result in a substantial reduction in the dimension required for a grid search. For example, in the growth model

$$y = \beta_1 + \beta_2 x + \beta_3 \gamma^x + \varepsilon,$$

we see that for fixed γ, $\boldsymbol{\beta}$ can be recovered by linear least squares. The dimension 4 of $\boldsymbol{\theta} = (\beta_1, \beta_2, \beta_3, \gamma)'$ is then reduced to the dimension 1 of γ.

Sometimes the model to be fitted is an extension or a simplification of a previously fitted model. Any relationships between the parameters of the two models should be exploited to obtain starting values for the new model.

In Chapter 7 we discussed starting values for fitting growth and yield–density models, and in Chapter 8 we considered starting values for compartmental models. The methods suggested there came from a combination of the methods discussed here and a knowledge of the form of the model. This dual approach is needed for other models as well.

b Initial Approximation to the Hessian

Quasi-Newton algorithms may offer a choice of an initial approximation to the Hessian. The usual default is $\mathbf{B}^{(1)} = \mathbf{I}_p$. However, in a least-squares problem, the Gauss–Newton approximation to the Hessian at $\boldsymbol{\theta}^{(1)}$, namely $\mathbf{J}^{(1)\prime}\mathbf{J}^{(1)}$ (c.f. Section 14.1), might be preferred. However, it is essential that $\mathbf{B}^{(1)}$ be positive definite.

15.2.2 Control Constants

a Step-Length Accuracy

Many programs allow the user to control the accuracy of the line searches [c.f. (13.21)]. Gill et al. [1981: 101] do this by requiring a sufficient reduction in the directional derivative along direction $\mathbf{d}^{(a)}$ at $\boldsymbol{\theta}^{(a+1)}$ from the value at $\boldsymbol{\theta}^{(a)}$, namely

$$|\mathbf{g}(\boldsymbol{\theta}^{(a)} + \rho \mathbf{d}^{(a)})'\mathbf{d}^{(a)}| \leq \tau |\mathbf{g}^{(a)\prime}\mathbf{d}^{(a)}|,$$

where $0 \leq \tau < 1$. A value of τ close to zero imposes a very accurate line search. Some programs may allow control over the method actually used for the line searches, e.g. whether first derivatives or only function values are used.

b Maximum Step Length

This could be the radius of the initial trust region or, in a step-length method, simply a device to prevent the first few steps of the algorithm from getting into extreme areas of the parameter space from which recovery is difficult. This is a more direct approach than that used by Miller [1979], who suggested fairly severe damping [$\eta = 0.5$ or 1 in (14.18)] during the initial stages of a Levenberg–Marquardt algorithm. The maximum step length can also provide a check for the divergence that occurs if $h(\boldsymbol{\theta})$ is unbounded below or asymptotically tends to a finite lower bound. Dennis and Schnabel [1983: 161] declare divergence if the maximum step length has to be used for five consecutive steps.

c Maximum Number of Function Evaluations

A constant is always sought from the user to prevent too much computational expense if little progress is being made. It may be a limit on the number of function evaluations (n_h) or the number of iterations (n_{It}). A limit on function evaluations is better in that it also counts the work being done in line searches. Gill et al. [1981: 295] suggest that n_h should have a lower bound of $5p$ for a quasi-Newton method using exact gradient evaluations.

15.2.3 Accuracy

a Precision of Function Values

A program may ask for an upper bound on the absolute error ε_A and /or the relative error ε_R in computations of $h(\boldsymbol{\theta})$. Gill et al. [1981: Section 8.5] list four areas in which a sophisticated program will use ε_A, namely (i) the specification of termination (stopping) criteria; (ii) determining the minimum separation between points to be used in line searches; (iii) the computation of the length of interval to be used in calculating finite-difference approximations to derivatives; and (iv) the rescaling of the parameters to lessen ill-conditioning (see Section 15.3.3). They note that use of a value for ε_A that is unrealistically small can cause unnecessary further iterations which make no real progress, thus indicating that the algorithm has failed. In contrast, too large a value may cause the algorithm to stop too soon.

Gill et al. [1981] make some very interesting observations. Using Taylor series expansions, they show in their Section 8.2.2 that if

$$|h(\boldsymbol{\theta}) - h(\hat{\boldsymbol{\theta}})| \approx \varepsilon_A,$$

then

$$\|\boldsymbol{\theta} - \hat{\boldsymbol{\theta}}\|^2 \approx \frac{2\varepsilon_A}{\mathbf{d}'\hat{\mathbf{H}}\mathbf{d}} \tag{15.1}$$

and

$$\|\mathbf{g}\|^2 \approx 2\varepsilon_A \frac{\|\hat{\mathbf{H}}\mathbf{g}\|^2}{\mathbf{d}'\hat{\mathbf{H}}\mathbf{d}}, \tag{15.2}$$

where $\mathbf{d} = (\boldsymbol{\theta} - \hat{\boldsymbol{\theta}})/\|\boldsymbol{\theta} - \hat{\boldsymbol{\theta}}\|$, a unit vector. Thus for well-conditioned problems

$$\begin{aligned} \|\boldsymbol{\theta} - \hat{\boldsymbol{\theta}}\| &= O(\sqrt{\varepsilon_A}), \\ \|\mathbf{g}\| &= O(\sqrt{\varepsilon_A}). \end{aligned} \tag{15.3}$$

They note that the relationship $\varepsilon_A = \varepsilon_R|h(\boldsymbol{\theta})|$ is useful only when ε_R is small compared with $|h(\boldsymbol{\theta})|$. Otherwise, in practice, ε_A is often much larger than this relationship would suggest.

In some problems, the accuracy of computing $h(\boldsymbol{\theta})$ depends largely on some particular intermediate calculation (the least accurate one) of roughly known accuracy. For situations where this is not true, Gill et al. [1981: Section 5.2.2] describe a number of methods for estimating ε_A, including the simple expedient of recalculating $h(\boldsymbol{\theta}^{(1)})$ using higher precision (more significant figures). As ε_A may change drastically over the path of the algorithm, these authors in their Section 8.5.3 advocate internally updating ε_A using

$$\varepsilon_A^{(a)} = (1 + |h(\boldsymbol{\theta}^{(a)})|) \max\left(\varepsilon_M, \frac{\varepsilon_A^{(1)}}{1 + |h(\boldsymbol{\theta}^{(1)})|} \right), \tag{15.4}$$

where ε_M (the machine relative precision) is essentially 10^{-m}, m being the number of significent digits used to represent θ_r by the computer. It may be necessary to recalculate ε_A near $\hat{\boldsymbol{\theta}}$. We should stress here that ε_A and ε_R refer only to errors in the *calculation* of $h(\boldsymbol{\theta})$ and should be small even when fitting models to data known only to two or three significant figures. Some people have the dangerous misconception (see Section 15.3.3) that because the data are known only to a small number of significant figures, little accuracy is required in computations. If the data are inaccurate, we shall probably wish to solve several precise optimization problems in which the data are perturbed slightly. We can then inspect the effect of the perturbations on the solution and any resulting statistical inferences (sensitivity analysis).

b Magnitudes

The program may ask for the typical magnitudes of $h(\boldsymbol{\theta})$ and each θ_r to be encountered during iterations. These magnitudes can be used in termination criteria and in rescaling the parameters (Section 15.3.3).

15.2.4 Termination Criteria (Stopping Rules)

When minimizing a smooth function, we should observe convergence in three sequences. As $a \to \infty$ we should have $\boldsymbol{\theta}^{(a)} \to \hat{\boldsymbol{\theta}}$, so that $\boldsymbol{\theta}^{(a)} - \boldsymbol{\theta}^{(a-1)} \to 0$, $h^{(a)} \to \hat{h}$ (which implies $h^{(a)} - h^{(a-1)} \to 0$), and finally $\mathbf{g}^{(a)} \to \mathbf{0}$. Let us consider the last sequence. Even if the exact value of $\hat{\boldsymbol{\theta}}$ were representable in the computer's floating-point number system, it would be unlikely that the computed value of $\hat{\mathbf{g}} = \mathbf{g}(\hat{\boldsymbol{\theta}})$ would be zero, because of roundoff errors in the computations. Also there would probably be values of $\boldsymbol{\theta} \neq \hat{\boldsymbol{\theta}}$ at which the *computed* value of the gradient was closer to zero than the computed $\hat{\mathbf{g}}$. The best we can do is to find a value of $\boldsymbol{\theta}$ for which the gradient is "small" and use that as an estimate of $\hat{\boldsymbol{\theta}}$.

Different programs use different convergence criteria, and very often these criteria are inadequate. It is clear (see Hiebert [1981]) that convergence criteria for all of the sequences involved should be satisfied to help prevent the algorithm from claiming to have found a solution when it has not done so. We base the following discussion on the criteria suggested by Gill et al. [1981: Section 8.2.3], Dennis et al. [1981a: Section 6], and Dennis and Schnabel [1983: Section 7.2].

Suppose we have an approximate Newton algorithm (e.g. modified Newton, quasi-Newton, Gauss–Newton). Then the step $\boldsymbol{\delta}^{(a)} = \boldsymbol{\theta}^{(a+1)} - \boldsymbol{\theta}^{(a)}$ is determined using a quadratic function $q^{(a)}(\boldsymbol{\theta})$ which is intended to approximate $h(\boldsymbol{\theta})$ in the vicinity of $\boldsymbol{\theta}^{(a)}$. Dennis et al. [1981a] make a rough check upon the adequacy of the quadratic approximation as a precursor to the more usual tests. They argue that the analysis underlying the other tests relies heavily on the quadratic approximation. The check performed is

$$0 < h(\boldsymbol{\theta}^{(a-1)}) - h(\boldsymbol{\theta}^{(a)}) \leq 2[h(\boldsymbol{\theta}^{(a-1)}) - q^{(a-1)}(\boldsymbol{\theta}^{(a)})], \tag{15.5}$$

i.e., the reduction in h achieved at the last iteration is no more than twice that predicted by the quadratic approximation. We now turn our attention to the convergence of the three sequences above, to answer the question "Do we stop after the $(a-1)$th iteration and accept $\boldsymbol{\theta}^{(a)}$?"

a Convergence of Function Estimates

Let τ_h be the relative accuracy required in the computed value of $h(\hat{\boldsymbol{\theta}})$ (e.g. $\tau_h = 10^{-6}$ if we require six significant figures in $\hat{h}$). Clearly we must have $\tau_h > \varepsilon_R$, the relative error in computing $h^{(a)}$. The basic test is for a relative change in h of no more than τ_h, i.e. $0 < h^{(a-1)} - h^{(a)} < \tau_h |h^{(a)}|$. However, this criterion is inadequate in cases where $h^{(a)} \to 0$. Gill et al. [1981: 307] state that, in their experience, a small value of h is usually the result of cancellation and that they therefore prefer to use similar measures of error when $h^{(a)} = 0$ to those used when $h^{(a)} = 1$. They therefore use

$$h^{(a-1)} - h^{(a)} \leq \tau_h(1 + |h^{(a)}|). \tag{15.6}$$

As Gill et al. note, there is an implicit scaling here. If we rescale h, e.g. $h \to 10^{-10}h$, Equation (15.6) may always be satisfied. In the spirit of Dennis and Schnabel [1983], (15.6) would be replaced by

$$h^{(a-1)} - h^{(a)} \leq \tau_h \max(|h^{(a)}|, \lambda_h), \tag{15.7}$$

where λ_h is a user-supplied constant giving a guess at the "usual" order of magnitude of h. The default value is $\lambda_h = 1$. Thus if $\hat{h} = 0$, as in some problems, then we should not stop until $h^{(a)} < \tau_h \lambda_h$. Thus if a criterion like (15.6) is used, h should be rescaled so that $\lambda_h = 1$. We note that the right-hand sides of (15.6) and (15.7) should be no smaller than ε_A, the absolute error in the computation of $h^{(a)}$. Instead of using a measure based upon $(h^{(a-1)} - h^{(a)})/|h^{(a)}|$ as (15.6) and (15.7) do, Dennis et al. [1981a] use a measure based on the predicted relative change in h at the next iteration, namely $[h^{(a)} - q^{(a)}(\boldsymbol{\theta}^{(a+1)})]/|h^{(a)}|$, where $\boldsymbol{\theta}^{(a+1)} - \boldsymbol{\theta}^{(a)} = -\mathbf{H}^{(a)-1}\mathbf{g}^{(a)}$. Thus the approximate Newton step, without line search etc., is computed, but not used if all the convergence criteria are met.

b Convergence of Gradient Estimates

We require $\mathbf{g}^{(a)} \to \mathbf{0}$. However, the absolute size of $\mathbf{g}^{(a)}$ depends both on the scaling of $h(\boldsymbol{\theta})$ and the scaling of $\boldsymbol{\theta}$. Now the ratio $\|\mathbf{g}^{(a)}\|/|h^{(a)}|$ is independent of the scaling of h. Using (15.3), we might impose a limit of the form $\sqrt{\tau_h}$ on this ratio. However, in the experience of Gill et al. [1981: 307] this is too stringent, and they use $\tau_h^{1/3}$ instead. Their termination criterion for $\mathbf{g}^{(a)}$ is thus

$$\|\mathbf{g}^{(a)}\| \leq \tau_h^{1/3}(1 + |h^{(a)}|), \tag{15.8}$$

where, as in (15.6), the "1 + " has been added to cope with those problems with $h^{(a)} \to 0$. Thus again [c.f. (15.7)] there is the implicit assumption that $\lambda_h \approx 1$. The norm these authors advocate for usual use is the infinity norm, i.e.

$$\max_j g_j^{(a)} \leq \tau_h^{1/3}(1 + |h^{(a)}|).$$

Under the transformation $\boldsymbol{\theta} \to 10^{-10}\boldsymbol{\theta}$, $h^{(a)}$ remains constant, but $\mathbf{g}^{(a)}$, and thus the ratio $\|\mathbf{g}^{(a)}\|/|h^{(a)}|$, becomes smaller by a factor of 10^{-10}. Hence, with a "badly" scaled $\boldsymbol{\theta}$ parametrization, (15.8) can be virtually automatically satisfied. For this reason Dennis and Schnabel [1983] impose a limit on

$$\begin{aligned} \text{rel grad}_j^{(a)} &= \frac{\text{relative change in } h}{\text{relative change in } \theta_j} \\ &= \lim_{\delta \to 0} \frac{[h(\boldsymbol{\theta}^{(a)} + \delta \mathbf{e}_j) - h(\boldsymbol{\theta}^{(a)})]/h(\boldsymbol{\theta}^{(a)})}{\delta/\theta_j^{(a)}} \\ &= \frac{g_j^{(a)}\theta_j^{(a)}}{h^{(a)}}. \end{aligned}$$

Their criterion is

$$\max_{1 \leq j \leq p} \left| \frac{g_j^{(a)} \max(|\theta_j^{(a)}|, \lambda_{\theta j})}{\max(|h^{(a)}|, \lambda_h)} \right| \leq \tau_g, \tag{15.9}$$

where $\lambda_{\theta j}$ is a user-supplied constant giving the "usual" order of magnitude of θ_j. The numerator and denominator of (15.9) have been modified to allow for both $\theta_j^{(a)} \to 0$ and $h^{(a)} \to 0$ [c.f. (15.7)].

The criterion (15.9) is invariant under the rescaling of h and the independent rescaling of individual θ_j's. If a criterion like (15.8) is used, the θ_j's should be scaled so that each θ_j typically has order of magnitude unity. Following (15.8), we could set $\tau_g = \tau_h^{1/3}$ in (15.9). To trap occurrences in which $\boldsymbol{\theta}^{(1)}$ is, by chance, close to $\hat{\boldsymbol{\theta}}$, Gill et al. [1981: 307] initially check if $\|\mathbf{g}^{(a)}\| < \varepsilon_A$.

c Convergence of Parameter Estimates

Gill et al. [1981] use

$$\|\boldsymbol{\theta}^{(a)} - \boldsymbol{\theta}^{(a-1)}\| < \sqrt{\tau_h}(1 + \|\boldsymbol{\theta}^{(a)}\|). \tag{15.10}$$

This is essentially a bound on relative changes in $\boldsymbol{\theta}$. The factor $\sqrt{\tau_h}$ is motivated largely by (15.3). If $\boldsymbol{\theta}^{(a)} \to \mathbf{0}$, the absolute bound $\|\boldsymbol{\theta}^{(a)} - \boldsymbol{\theta}^{(a-1)}\| < \sqrt{\tau_h}$ becomes operative. As above, there is an implicit assumption that the elements of $\boldsymbol{\theta}$ are of order of magnitude unity. Rather than expect users to rescale their parametrizations to achieve this, Dennis and Schnabel [1983] ask the user for a value $\lambda_{\theta j}$ which is the rough magnitude of θ_j in the problem and use

$$\max_{1 \le j \le p} \frac{|\theta_j^{(a)} - \theta_j^{(a-1)}|}{\max(|\theta_j^{(a)}|, \lambda_{\theta j})} < \tau_\theta. \tag{15.11}$$

Following (15.10), we might take $\tau_\theta = \sqrt{\tau_h}$.

In optimization problems arising in statistics, the approximate inverse Hessian is often an approximate variance–covariance matrix $\hat{\mathbf{V}}$ for $\hat{\boldsymbol{\theta}}$. Now if s.e.$(\hat{\theta}_j)$ denotes the standard error of $\hat{\theta}_j$, then (by Appendix A7.2)

$$\begin{aligned}\max_{\mathbf{b} \ne \mathbf{0}} \frac{|\mathbf{b}'(\hat{\boldsymbol{\theta}} - \boldsymbol{\theta}^{(a)})|}{\text{s.e.}(\mathbf{b}'\hat{\boldsymbol{\theta}})} &= \max_{\mathbf{b} \ne \mathbf{0}} \frac{|\mathbf{b}'(\hat{\boldsymbol{\theta}} - \boldsymbol{\theta}^{(a)})|}{(\mathbf{b}'\hat{\mathbf{V}}\mathbf{b})^{1/2}} \\ &= [(\hat{\boldsymbol{\theta}} - \boldsymbol{\theta}^{(a)})'\hat{\mathbf{V}}^{-1}(\hat{\boldsymbol{\theta}} - \boldsymbol{\theta}^{(a)})]^{1/2}.\end{aligned}$$

Thus if we require

$$[(\boldsymbol{\theta}^{(a)} - \boldsymbol{\theta}^{(a-1)})'\mathbf{V}^{(a)-1}(\boldsymbol{\theta}^{(a)} - \boldsymbol{\theta}^{(a-1)})]^{-1/2} < \tau_\theta, \tag{15.12}$$

the error in $\boldsymbol{\theta}^{(a)}$ as an estimate of $\hat{\boldsymbol{\theta}}$ will make very little difference to the confidence intervals for any linear combination of the θ_j's. We have previously explored this idea in the least-squares context following (14.80). The criterion (15.12) is independent of any rescaling of $\boldsymbol{\theta}$ of the form $\boldsymbol{\theta} \to \mathbf{A}\boldsymbol{\theta}$ for nonsingular $\mathbf{A}$ and has the advantage of a statistical interpretation. A corresponding criterion for $\mathbf{g}^{(a)}$ suggested by D. A. Belsley is

$$|\mathbf{g}^{(a)\prime}\mathbf{V}^{(a)}\mathbf{g}^{(a)}|^{1/2} < \tau_g, \tag{15.13}$$

which is related to score or Lagrange-multiplier tests.

d Discussion

Many programs use one rather than all of the checks described above. As noted in Section 14.4, small $\mathbf{g}$ and small relative changes in h may indicate that the algorithm is traversing a region in which the surface is fairly flat. Small relative changes in $\boldsymbol{\theta}$ may only indicate that, for some reason, the algorithm is taking short steps. Even imposing the conditions of Sections 15.2.4a to c above simultaneously may not be enough, because both types of behavior can occur simultaneously far from the minimum—for example, the progress of the steepest-descent algorithm through a fairly flat curving valley. The further check (15.5) on the adequacy of the quadratic approximation provides an additional safeguard.

If the problem has special structure, such as the least-squares problem in Section 14.4, better stopping rules can be formulated.

We stress that the termination criteria in common usage make assumptions about the scaling of h and $\boldsymbol{\theta}$. As the form of the criterion is not under user control, the problem should be rescaled to make these assumptions reasonable, for example so that the expected order of magnitude of h and each element of $\boldsymbol{\theta}$ is unity. Termination criteria that are too exacting, e.g. requiring more accuracy than the roundoff error in computations permit, will cause the program to report failure.

e *Checking for a Minimum*

A number of the algorithms we have described have a tendency to steer around saddle points or can be modified to do so. Nevertheless, if the Hessian (or a suitable finite-difference approximation) is available, Gill et al. [1981: 307] impose a further check on the solution, namely that $\mathbf{H}^{(a)}$ be "sufficiently positive definite." In the context of a modified Newton algorithm as described in Section 13.3.1, $\mathbf{H}^{(a)}$ is sufficiently positive definite if no modifications to the ordinary $\mathbf{LDL}'$ Cholesky factorization have to be made when computing the modified Cholesky decomposition. When second derivatives are not available (e.g. in quasi-Newton algorithms), some programs have the option of performing a local (e.g. random) search in the vicinity of the computed "minimum." The solution will be accepted if no better point is found.

15.3 SOME CAUSES AND MODES OF FAILURE

There are two main types of failure. In this section we are concerned with the first and most obvious type, that signaled by the program. A more insidious type of failure occurs when the program mistakenly thinks it has found a solution. We discuss some methods for checking a solution in Section 15.4. Some causes of failure, such as overstringent convergence criteria, have already been discussed in Sections 15.1 and 15.2.

15.3.1 Programming Error

Miller [1979: 41] described the incorporation of some "spying code" into a frequently used nonlinear least-squares algorithm designed to report on user problems. One of the most common causes of failure was incorrect differentiation or the incorrect programming of derivatives. The algorithm now incorporates an option for users to check their derivatives against finite-difference approximations, and Miller notes that this feature has by now saved many hours of computing time. Presumably there has been a similar saving in human time. Some of the major libraries such as NAG now also have this feature.

The onus is still entirely upon the user to correctly define and program the function $h(\boldsymbol{\theta})$. Besides carefully checking the program code, the user can check that the program gives the expected results at several points where the correct value of $h(\boldsymbol{\theta})$ is known. These points should not be so special that parts of the program code are not used. One cause of error is specifying a model containing parameters which are not identifiable. Whereas this problem may be diagnosed early with the classical Newton method because of singular Hessian matrices, the problem can be masked by methods which modify the Hessian.

Most optimization techniques assume that $h(\boldsymbol{\theta})$ is smooth, yet the computed version of $h(\boldsymbol{\theta})$ is necessarily discontinuous. This does not generally cause problems except in cases of bad scaling (Section 15.3.3) provided that the function and its derivatives are computed accurately. We noted earlier that in nonlinear least squares with $h(\boldsymbol{\theta}) = \sum_i [y_i - f(\mathbf{x}_i; \boldsymbol{\theta})]^2$ there is a temptation to think that because $\mathbf{y}$ and $\mathbf{x}$ are not measured accurately, $f(\mathbf{x}; \boldsymbol{\theta})$ need not be computed accurately. Inaccurate calculations induce large discontinuities in $h(\boldsymbol{\theta})$ which are particularly dangerous when finite-difference approximations to derivatives are being used. Miller [1979: Section 2] gives an indication of the staggering amount of accuracy that can be lost in least-squares calculations. Discontinuities in gradient calculations can lead to convergence to a spurious local minimum. The need to preserve smoothness has implications when the calculation involves iterative subproblems such as integrals, where fixed rather than adaptive quadrature should be used, or rational approximations. For further details see Gill et al. [1981, Section 7.3].

15.3.2 Overflow and Underflow

Exponential overflow occurs when a calculated number is too large in absolute value to be stored in the computer's floating-point number format. Computer systems commonly treat this fault either by halting execution, or by setting the answer to the largest representable number. Subsequent calculations are likely to be meaningless. Underflow occurs when a calculated nonzero number is too small to be represented. Again execution may halt or the result may be set to zero. This too may invalidate subsequent calculations or cause other fatal errors such as dividing by zero.

Under- and overflow problems are most likely to occur in calculating functions involving exponentials, and they commonly occur after a large initial step. One preventative measure is to decrease the maximum step size. The user should also look for ways of programming the calculation which are less susceptible to overflow. For example one can use

$$\theta_1^k e^{-\theta_2} = \exp\{(k \log \theta_1) - \theta_2\}.$$

15.3.3 Difficulties Caused by Bad Scaling (Parametrization)

The optimization literature refers to choosing the parametrization of $h(\boldsymbol{\theta})$ as "scaling the variables." In Section 15.2.4 we discussed how the scaling of h and $\boldsymbol{\theta}$

affects termination criteria. Changing the parametrization of $h(\boldsymbol{\theta})$ can affect the practical rate of convergence of the algorithm, and indeed, whether or not the program fails (see Example 3.3 of Section 3.2). Let $\boldsymbol{\theta} = \mathbf{A}\boldsymbol{\phi}$ for some nonsingular matrix $\mathbf{A}$. Then reparametrizing using $\boldsymbol{\phi}$, we obtain

$$h_2(\boldsymbol{\phi}) = h(\boldsymbol{\theta}) = h(\mathbf{A}\boldsymbol{\phi}),$$

$$\mathbf{g}_2(\boldsymbol{\phi}) = \mathbf{A}'\mathbf{g}(\boldsymbol{\theta}),$$

and

$$\mathbf{H}_2(\boldsymbol{\phi}) = \mathbf{A}'\mathbf{H}(\boldsymbol{\theta})\mathbf{A}.$$

Thus the relative magnitudes of the elements of the gradient vector have been changed, and so, as it turns out, has the condition number of the Hessian.

The Newton direction solves $\mathbf{H}^{(a)}\mathbf{d}^{(a)} = -\mathbf{g}^{(a)}$. If $\mathbf{H}^{(a)}$ is very ill conditioned, the Newton direction is almost orthogonal to the steepest-descent direction or, equivalently, almost tangential to the contour line. For an ill-conditioned matrix, the computed Newton direction will be particularly inaccurate. In extreme cases the computed direction may lead uphill, even though the actual direction is a descent direction. In less extreme cases, the angle between the computed direction and the contour line may be so small that little reduction in $h(\boldsymbol{\theta})$ is possible along this direction. Either way, failure results because a linear search fails to reduce $h(\boldsymbol{\theta})$ sufficiently. Similar effects occur in methods that approximate the Hessian. This is why well-implemented versions require a matrix defining a step direction to be "sufficiently positive definite." However, in a badly parametrized problem where $\mathbf{H}^{(a)}$ is badly ill-conditioned, $\mathbf{H}^{(a)}$ will often be substantially modified and this will slow convergence as $\boldsymbol{\theta}^{(a)}$ approaches $\hat{\boldsymbol{\theta}}$.

Ideally one would reparametrize so that the elements of $\mathbf{g}(\boldsymbol{\theta})$ were of similar magnitude and $\mathbf{H}(\boldsymbol{\theta})$ was well conditioned. These goals may not be achievable simultaneously. Bad scaling can cause some elements of a finite-difference approximation to $\mathbf{g}$ to be formed mainly from "noise," particularly if $\mathbf{g}$ is small. Gill et al. [1981: Section 8.7.1.3] discuss these issues and suggest some rescaling strategies to improve finite-difference approximations. In Chapter 3 we saw how nonlinear reparametrizations could improve conditioning. The only general rescaling strategy in widespread use is rescaling so that the elements of $\boldsymbol{\theta}$ have the same rough order of magnitude. Even this is difficult, as there may be large changes in the order of magnitude of a single parameter over the course of minimization. The algorithms of Dennis and Schnabel [1983] internally perform a transformation $\theta_j \rightarrow \theta_j/\lambda_{\theta j}$, where $\lambda_{\theta j}$ is a typical value of θ_j. If it is thought that the order of magnitude of θ_j may change substantially, these authors suggest (p. 278) a midway value. For some more sophisticated ideas see Gill et al. [1981: Section 8.7].

15.4 CHECKING THE SOLUTION

When presented with an apparent solution $\boldsymbol{\theta}^{(a)}$ for some a by an optimization program, the user can check that the following apply:

(i) $\|\mathbf{g}^{(a)}\|$ is much smaller than $\|\mathbf{g}^{(1)}\|$;
(ii) the final iterates show a fast rate of convergence to $\boldsymbol{\theta}^{(a)}$, and the function values display a similar rate of convergence to $h^{(a)}$;
(iii) the Hessian is well conditioned; and
(iv) (to check for a minimum) either the Hessian is positive definite or a local search cannot further reduce $h(\boldsymbol{\theta})$ by more than ε_A, the estimated error in the computation of $h(\boldsymbol{\theta})$.

If all of these apply, one can be fairly confident of the solution. A slowing down of the rate of convergence in condition (ii) is likely as the limiting accuracy of the computations is approached. Gill et al. [1981: Section 8.3] explore these ideas in much more detail. To investigate condition (ii) they look at quantities

$$\xi^{(b+1)} = h^{(b-1)} - h^{(b)} \qquad (b = 2, 3, \ldots, a).$$

They note that for superlinear convergence $\xi^{(b+1)} \approx \xi^{(b)r}$ for some $r > 1$, and for fast linear convergence $\xi^{(b+1)} \approx \xi^{(b)}/M$, where $M > 2$. If $h^{(a)}$ is large, they recommend looking at $\xi^{(b)}/|h^{(b)}|$. They further note that with Newton-type methods, slow convergence can result from errors in calculating the Hessian matrix.

If the user doubts the optimality of the solution, a number of approaches are possible. The parameters can be rescaled to improve conditioning. The same algorithm can be used with different starting points; for example, $\boldsymbol{\theta}^{(a)} + \alpha(\boldsymbol{\theta}^{(a)} - \boldsymbol{\theta}^{(1)})$ gives a point on the opposite side of $\boldsymbol{\theta}^{(a)}$ from the original starting point. The parameters of the line search (if applicable) can be changed. A different algorithm can be tried. In addition, Miller [1979: Section 2(f)] notes that with least-squares problems, it is often obvious from a comparison of observed and fitted values that a local minimum has been found. For a discussion of possible approaches to finding a global optimum see Dixon and Szegö [1975, 1978].

APPENDIX A

Vectors and Matrices

Below we list a number of matrix results used in this book. Most of the omitted proofs are given, for example, in Appendix A of Seber [1977, 1984]. The notation is given in Section 1.1.

A1 RANK

1. For any matrix $\mathbf{A}$, $\text{rank}[\mathbf{A}'\mathbf{A}] = \text{rank}[\mathbf{A}]$.
2. If a matrix $\mathbf{A}$ has n columns, then

$$\text{rank}[\mathbf{A}] + \text{nullity}[\mathbf{A}] = n,$$

where nullity$[\mathbf{A}]$ is the dimension of the null space

$$\mathcal{N}[\mathbf{A}] = \{\mathbf{x}:\mathbf{A}\mathbf{x} = \mathbf{0}\}.$$

3. If $\mathbf{A}$ is symmetric, $\mathcal{R}[\mathbf{A}] = \{\mathcal{N}[\mathbf{A}]\}^{\perp}$.

A2 EIGENVALUES

1. Let $\mathbf{A}$ be any $n \times n$ symmetric matrix with eigenvalues $\lambda_1, \lambda_2, \ldots, \lambda_n$, and let

$$g(\lambda) = |\lambda\mathbf{I}_n - \mathbf{A}| = \prod_{i=1}^{n} (\lambda - \lambda_i)$$

be the characteristic polynomial. Then:

(i) $\text{tr}\,\mathbf{A} = \sum_{i=1}^{n} \lambda_i$;

(ii) $|\mathbf{A}| = \prod_{i=1}^{n} \lambda_i$.

(iii) *Cayley-Hamilton theorem.* $g(\mathbf{A}) = \mathbf{0}$.

2. *Principal-Axis Theorem.* If $\mathbf{A}$ is an $n \times n$ symmetric matrix, then there exists an orthogonal matrix $\mathbf{T} = [\mathbf{t}_1, \mathbf{t}_2, \ldots, \mathbf{t}_n]$ such that $\mathbf{T}'\mathbf{A}\mathbf{T} = \mathbf{\Lambda}$, where $\mathbf{\Lambda} = \text{diag}(\lambda_1, \lambda_2, \ldots, \lambda_n)$. Here $\lambda_1, \lambda_2, \ldots, \lambda_n$ are the eigenvalues of $\mathbf{A}$, and $\mathbf{A}\mathbf{t}_i = \lambda_i \mathbf{t}_i$. The eigenvectors $\mathbf{t}_i$ form an orthonormal basis for $\mathbb{R}^n$. The factorization $\mathbf{A} = \mathbf{T}\mathbf{\Lambda}\mathbf{T}'$ is known as the spectral decomposition of $\mathbf{A}$.

A3 PATTERNED MATRICES

1. If All inverses exist, then

$$\begin{pmatrix} \mathbf{A}_{11} \mathbf{A}_{12} \\ \mathbf{A}_{21} \mathbf{A}_{22} \end{pmatrix}^{-1} = \begin{pmatrix} \mathbf{A}_{11}^{-1} + \mathbf{B}_{12}\mathbf{B}_{22}^{-1}\mathbf{B}_{21} & -\mathbf{B}_{12}\mathbf{B}_{22}^{-1} \\ -\mathbf{B}_{22}^{-1}\mathbf{B}_{21} & \mathbf{B}_{22}^{-1} \end{pmatrix} \tag{1}$$

$$= \begin{pmatrix} \mathbf{C}_{11}^{-1} & -\mathbf{C}_{11}^{-1}\mathbf{C}_{12} \\ -\mathbf{C}_{21}\mathbf{C}_{11}^{-1} & \mathbf{A}_{22}^{-1} + \mathbf{C}_{21}\mathbf{C}_{11}^{-1}\mathbf{C}_{12} \end{pmatrix}, \tag{2}$$

where $\mathbf{B}_{22} = \mathbf{A}_{22} - \mathbf{A}_{21}\mathbf{A}_{11}^{-1}\mathbf{A}_{12}$, $\mathbf{B}_{12} = \mathbf{A}_{11}^{-1}\mathbf{A}_{12}$, $\mathbf{B}_{21} = \mathbf{A}_{21}\mathbf{A}_{11}^{-1}$, $\mathbf{C}_{11} = \mathbf{A}_{11} - \mathbf{A}_{12}\mathbf{A}_{22}^{-1}\mathbf{A}_{21}$, $\mathbf{C}_{12} = \mathbf{A}_{12}\mathbf{A}_{22}^{-1}$, and $\mathbf{C}_{21} = \mathbf{A}_{22}^{-1}\mathbf{A}_{21}$.

2. $$\begin{vmatrix} \mathbf{A} & \mathbf{B} \\ \mathbf{C} & \mathbf{D} \end{vmatrix} = \begin{cases} |\mathbf{D}|\,|\mathbf{A} - \mathbf{B}\mathbf{D}^{-1}\mathbf{C}| & \text{if } \mathbf{D}^{-1} \text{ exists,} \\ |\mathbf{A}|\,|\mathbf{D} - \mathbf{C}\mathbf{A}^{-1}\mathbf{B}| & \text{if } \mathbf{A}^{-1} \text{ exists.} \end{cases}$$

3. $|\mathbf{A} + \mathbf{u}\mathbf{u}'| = |\mathbf{A}|(1 + \mathbf{u}'\mathbf{A}^{-1}\mathbf{u})$ ($\mathbf{A}$ nonsingular).

4. $(\mathbf{A} + \mathbf{u}\mathbf{v}')^{-1} = \mathbf{A}^{-1} - \dfrac{\mathbf{A}^{-1}\mathbf{u}\mathbf{v}'\mathbf{A}^{-1}}{1 + \mathbf{v}'\mathbf{A}^{-1}\mathbf{u}}$ ($\mathbf{A}$ nonsingular).

A4 POSITIVE DEFINITE AND SEMIDEFINITE MATRICES

An $n \times n$ symmetric matrix $\mathbf{A}$ is said to be positive semidefinite† (p.s.d.) if and only if $\mathbf{x}'\mathbf{A}\mathbf{x} \geq 0$ for all $\mathbf{x}$. If $\mathbf{x}'\mathbf{A}\mathbf{x} > 0$ for all $\mathbf{x}$, $\mathbf{x} \neq \mathbf{0}$, then $\mathbf{A}$ is said to be positive definite (p.d.).

1. A symmetric matrix $\mathbf{A}$ is p.s.d. if and only if its eigenvalues are nonnegative. It is p.d. if and only if its eigenvalues are all positive.
2. A p.s.d. matrix $\mathbf{A}$ is p.d. if and only if it is nonsingular.

†Some authors use the term *nonnegative definite.*

3. $\mathbf{A}$ is p.d. if and only if there exists a nonsingular $\mathbf{R}$ such that $\mathbf{A} = \mathbf{R}'\mathbf{R}$.
4. Given $\mathbf{A}$ is p.d., there exists a p.d. matrix $\mathbf{A}^{1/2}$ such that $(\mathbf{A}^{1/2})^2 = \mathbf{A}$.
5. If $\mathbf{A}$ is p.d. then so is $\mathbf{CAC}'$, when $\mathbf{C}$ has linearly independent rows.
6. $\mathbf{B}'\mathbf{B}$ is p.s.d. for any matrix $\mathbf{B}$. If $\mathbf{B}$ is $n \times p$ of rank p, then $\mathbf{B}'\mathbf{B}$ is p.d.

A5 GENERALIZED INVERSES

A generalized inverse of an $m \times n$ matrix $\mathbf{A}$ is defined to be any $n \times m$ matrix $\mathbf{A}^-$ that satisfies the condition (a) $\mathbf{AA}^-\mathbf{A} = \mathbf{A}$. Such a matrix always exists. If $\mathbf{A}^-$ also satisfies three more conditions, namely (b) $\mathbf{A}^-\mathbf{AA}^- = \mathbf{A}^-$, (c) $(\mathbf{AA}^-)' = \mathbf{AA}^-$, (d) $(\mathbf{A}^-\mathbf{A})' = \mathbf{A}^-\mathbf{A}$, then $\mathbf{A}^-$ is unique and it is called the Moore–Penrose inverse. We denote this inverse by $\mathbf{A}^+$. (See also Appendices A8.1, A11.4, and D1.5).

A6 ELLIPSOIDS

If $\mathbf{A}$ is an $m \times m$ positive definite matrix and $\delta > 0$, then $\mathbf{x}'\mathbf{Ax} = \delta^2$ represents an ellipsoid. This ellipsoid can be rotated so that its axes are parallel to the coordinate axes. The rotation is accomplished by an orthogonal matrix $\mathbf{T}$ where $\mathbf{T}'\mathbf{AT} = \mathrm{diag}(\lambda_1, \lambda_2, \ldots, \lambda_m)$ and the λ_i are the (positive) eigenvalues of $\mathbf{A}$. If we put $\mathbf{x} = \mathbf{Ty}$, the ellipsoid becomes

$$\mathbf{y}'\mathbf{T}'\mathbf{ATy} = \lambda_1 y_1^2 + \lambda_2 y_2^2 + \cdots + \lambda_m y_m^2 = \delta^2.$$

Setting all the y_i's equal to zero except y_r gives us $a_r = \delta/\sqrt{\lambda_r}$ as the length of the rth semimajor axis. The volume of the ellipsoid is then given by

$$\begin{aligned}
v &= \frac{\pi^{m/2}}{\Gamma(m/2+1)} a_1 a_2 \cdots a_m \\
&= \frac{\pi^{m/2}\delta^m}{\Gamma(m/2+1)(\prod_i \lambda_i)^{1/2}} \\
&= \frac{\pi^{m/2}\delta^m}{\Gamma(m/2+1)|\mathbf{A}|^{1/2}} \qquad \text{(by Appendix A2.1)} \\
&\propto |\mathbf{A}|^{-1/2}.
\end{aligned}$$

A7 OPTIMIZATION

1. Consider the matrix function f given by

$$f(\mathbf{\Sigma}) = [\log|\mathbf{\Sigma}| + \mathrm{tr}|\mathbf{\Sigma}^{-1}\mathbf{A}],$$

where $\mathbf{A}$ is positive definite. Then, subject to $\boldsymbol{\Sigma}$ being positive definite, $f(\boldsymbol{\Sigma})$ is minimized uniquely at $\boldsymbol{\Sigma} = \mathbf{A}$.

2. *Cauchy–Schwartz Inequality.* If $\mathbf{B}$ is positive definite, then for any $\mathbf{a}$

$$\sup_{\mathbf{x}:\mathbf{x} \neq \mathbf{0}} \frac{(\mathbf{a}'\mathbf{x})^2}{\mathbf{x}'\mathbf{B}\mathbf{x}} = \mathbf{a}'\mathbf{B}^{-1}\mathbf{a}.$$

The supremum occurs when $\mathbf{x} \propto \mathbf{B}^{-1}\mathbf{a}$.

3. If $\mathbf{A}$ is any symmetric matrix with minimum and maximum eigenvalues $\lambda_{\min}$ and $\lambda_{\max}$ respectively, then

$$\lambda_{\min}\mathbf{x}'\mathbf{x} \leq \mathbf{x}'\mathbf{A}\mathbf{x} \leq \lambda_{\max}\mathbf{x}'\mathbf{x}$$

for all $\mathbf{x}$, with the bounds being attained.

A8 MATRIX FACTORIZATIONS

1. *Singular-Value Decomposition* (SVD). Any $m \times n$ matrix $\mathbf{A}$ of rank r [$r \leq p = \min(m, n)$] can be expressed in the form

$$\mathbf{A} = \mathbf{M}\boldsymbol{\Delta}\mathbf{N}',$$

where $\mathbf{M}$ is an $m \times m$ orthogonal matrix, $\mathbf{N}$ is an $n \times n$ orthogonal matrix, and $\boldsymbol{\Delta} = [(\delta_{ij})]$ is an $m \times n$ matrix with $\delta_{11} \geq \delta_{12} \geq \cdots \geq \delta_{rr} > \delta_{r+1,r+1} = \delta_{r+2,r+2} = \cdots = \delta_{pp} = 0$ and all off-diagonal elements zero. The δ_{ii} are called the singular values of $\mathbf{A}$, and the positive δ_{ii} are the positive square roots of the nonzero eigenvalues of $\mathbf{A}'\mathbf{A}$ (or $\mathbf{A}\mathbf{A}'$). Also, the first r columns of $\mathbf{N}$ are the eigenvectors of $\mathbf{A}'\mathbf{A}$ corresponding to the positive δ_{ii}. If $\mathbf{A}$ is symmetric with ordered eigenvalues λ_i, then $\delta_{ii} = |\lambda_i|$. If $\mathbf{A}$ is positive definite, $\lambda_i = \delta_{ii}$. The Moore–Penrose generalized inverse of $\mathbf{A}$ is given by

$$\mathbf{A}^+ = \mathbf{N}\boldsymbol{\Delta}^+\mathbf{M}',$$

where $\boldsymbol{\Delta}^+$ is $\boldsymbol{\Delta}'$ with the first r diagonal elements δ_{ii} replaced by δ_{ii}^{-1}.

2. *Cholesky Decomposition.* If $\mathbf{A}$ is an $n \times n$ positive definite matrix, then there exists an $n \times n$ upper triangular matrix $\mathbf{U} = [(u_{ij})]$ such that

$$\mathbf{A} = \mathbf{U}'\mathbf{U}.$$

The matrix $\mathbf{U}$ is unique if its diagonal elements are all positive or all negative. Let $\mathbf{D}_1 = \operatorname{diag}(u_{11}, u_{22}, \ldots, u_{nn})$, and let

$$\mathbf{U}_1 = \mathbf{D}_1^{-1}\mathbf{U} = \begin{bmatrix} 1 & \tilde{u}_{12} & \tilde{u}_{13} & \cdots & \tilde{u}_{1n} \\ 0 & 1 & \tilde{u}_{23} & \cdots & \tilde{u}_{2n} \\ \vdots & \vdots & \vdots & & \vdots \\ 0 & 0 & \cdots & \cdots & 1 \end{bmatrix}.$$

Then

$$\mathbf{A} = \mathbf{U}_1'\mathbf{D}_1^2\mathbf{U}_1 = \mathbf{U}_1'\mathbf{D}\mathbf{U}_1,$$

where $\mathbf{D}$ is a diagonal matrix with positive elements. Some writers prefer to use a lower triangular matrix $\mathbf{L} = \mathbf{U}'$ (and $\mathbf{L}_1 = \mathbf{U}_1'$), so that $\mathbf{A} = \mathbf{L}\mathbf{L}' = \mathbf{L}_1\mathbf{D}\mathbf{L}_1'$. However, there is another factorization which is the "transpose" of the above, namely $\mathbf{A} = \tilde{\mathbf{L}}'\tilde{\mathbf{L}} = \tilde{\mathbf{U}}\tilde{\mathbf{U}}'$, where $\tilde{\mathbf{L}}$ $(= \tilde{\mathbf{U}}')$ is lower triangular with positive diagonal elements.

3. *QR decomposition.* Any $n \times p$ matrix of rank p can be expressed in the form

$$\begin{aligned}\mathbf{A} &= \mathbf{QR} \\ &= (\mathbf{Q}_p, \mathbf{Q}_{n-p})\begin{pmatrix}\mathbf{R}_{11} \\ \mathbf{0}\end{pmatrix} \\ &= \mathbf{Q}_p\mathbf{R}_{11},\end{aligned}$$

where $\mathbf{Q}$ is orthogonal, $\mathbf{Q}_p$ has p columns, and $\mathbf{R}_{11}$ is a $p \times p$ upper triangular matrix of rank p. Since

$$\mathbf{A}'\mathbf{A} = \mathbf{R}_{11}'\mathbf{Q}_p'\mathbf{Q}_p\mathbf{R}_{11} = \mathbf{R}_{11}'\mathbf{R}_{11}$$

is positive definite, $\mathbf{R}_{11}'\mathbf{R}_{11}$ is the Cholesky decomposition of $\mathbf{A}'\mathbf{A}$. If the diagonal elements of $\mathbf{R}_{11}$ are all positive or all negative, then $\mathbf{R}_{11}$ is unique. In this case $\mathbf{Q}_p = \mathbf{A}\mathbf{R}_{11}^{-1}$ is also unique and the decomposition $\mathbf{A} = \mathbf{Q}_p\mathbf{R}_{11}$ is unique. However, the matrix $\mathbf{Q}_{n-p}$ is not unique, as any permutation of its columns will still give $\mathbf{A} = \mathbf{QR}$.

A9 MULTIVARIATE *t*-DISTRIBUTION

An $m \times 1$ vector of random variables $\mathbf{y} = (y_1, y_2, \ldots, y_m)'$ is said to have a multivariate t-distribution if its probability density function is given by

$$p(\mathbf{y}) = \frac{\Gamma(\frac{1}{2}[\nu + m])}{(\pi\nu)^{m/2}\Gamma(\frac{1}{2}\nu)|\boldsymbol{\Sigma}|^{1/2}}[1 + \nu^{-1}(\mathbf{y} - \boldsymbol{\mu})'\boldsymbol{\Sigma}^{-1}(\mathbf{y} - \boldsymbol{\mu})]^{-(\nu+m)/2},$$

where $\boldsymbol{\Sigma}$ is an $m \times m$ positive definite matrix. We shall write $\mathbf{y} \sim t_m(\nu, \boldsymbol{\mu}, \boldsymbol{\Sigma})$. This distribution has the following properties:

(i) If $\boldsymbol{\Sigma} = [(\sigma_{ij})]$ then $(y_r - \mu_r)/\sqrt{\sigma_{rr}} \sim t_\nu$.

(ii) The distribution of $(\mathbf{y} - \boldsymbol{\mu})'\boldsymbol{\Sigma}^{-1}(\mathbf{y} - \boldsymbol{\mu})/m$ is $F_{m,\nu}$.

A10 VECTOR DIFFERENTIATION

Let $a = a(\boldsymbol{\alpha})$, $\mathbf{b} = \mathbf{b}(\boldsymbol{\alpha})$, and $\boldsymbol{\alpha} = \boldsymbol{\alpha}(\boldsymbol{\beta})$. Using the notation of Section 1.1, we have the following results:

1. $$\frac{\partial a}{\partial \beta_r} = \sum_j \frac{\partial a}{\partial \alpha_j} \frac{\partial \alpha_j}{\partial \beta_r}$$

   or, using (1.10),

   $$\frac{\partial a}{\partial \boldsymbol{\beta}} = \left(\frac{\partial \boldsymbol{\alpha}}{\partial \boldsymbol{\beta}'}\right)' \left(\frac{\partial a}{\partial \boldsymbol{\alpha}}\right).$$

   A more "natural" version of this result is

   $$\frac{\partial a}{\partial \boldsymbol{\beta}'} = \frac{\partial a}{\partial \boldsymbol{\alpha}'} \frac{\partial \boldsymbol{\alpha}}{\partial \boldsymbol{\beta}'}.$$

2. $$\frac{\partial^2 a}{\partial \beta_r \partial \beta_s} = \frac{\partial}{\partial \beta_r} \left\{ \sum_j \frac{\partial a}{\partial \alpha_j} \frac{\partial \alpha_j}{\partial \beta_s} \right\}$$

   $$= \sum_j \sum_k \frac{\partial^2 a}{\partial \alpha_k \partial \alpha_j} \frac{\partial \alpha_k}{\partial \beta_r} \frac{\partial \alpha_j}{\partial \beta_s} + \sum_j \frac{\partial a}{\partial \alpha_j} \frac{\partial^2 \alpha_j}{\partial \beta_r \partial \beta_s},$$

   so that, by (1.11),

   $$\frac{\partial^2 a}{\partial \boldsymbol{\beta} \partial \boldsymbol{\beta}'} = \left(\frac{\partial \boldsymbol{\alpha}}{\partial \boldsymbol{\beta}'}\right)' \left(\frac{\partial^2 a}{\partial \boldsymbol{\alpha} \partial \boldsymbol{\alpha}'}\right) \left(\frac{\partial \boldsymbol{\alpha}}{\partial \boldsymbol{\beta}'}\right) + \sum_j \frac{\partial a}{\partial \alpha_j} \frac{\partial^2 \alpha_j}{\partial \boldsymbol{\beta} \partial \boldsymbol{\beta}'}.$$

3. $$\frac{\partial \mathbf{b}}{\partial \boldsymbol{\beta}'} = \left[\left(\frac{\partial b_r}{\partial \beta_s}\right)\right]$$

   $$= \left[\left(\sum_j \frac{\partial b_r}{\partial \alpha_j} \frac{\partial \alpha_j}{\partial \beta_s}\right)\right]$$

   $$= \left(\frac{\partial \mathbf{b}}{\partial \boldsymbol{\alpha}'}\right) \left(\frac{\partial \boldsymbol{\alpha}}{\partial \boldsymbol{\beta}'}\right).$$

4. If $\boldsymbol{\alpha} = \mathbf{B}\boldsymbol{\beta}$, then:

   (i) $\dfrac{\partial \boldsymbol{\alpha}}{\partial \boldsymbol{\beta}'} = \mathbf{B}$, so that $\dfrac{\partial a}{\partial \boldsymbol{\beta}} = \mathbf{B}' \dfrac{\partial a}{\partial \boldsymbol{\alpha}}$ (by Appendix A10.1);

(ii) $\frac{\partial^2 a}{\partial \boldsymbol{\beta} \partial \boldsymbol{\beta}'} = \left(\frac{\partial \boldsymbol{\alpha}}{\partial \boldsymbol{\beta}'}\right)' \left(\frac{\partial^2 a}{\partial \boldsymbol{\alpha} \partial \boldsymbol{\alpha}'}\right) \frac{\partial \boldsymbol{\alpha}}{\partial \boldsymbol{\beta}'}$ (by Appendix A10.2)

$$= \mathbf{B}' \frac{\partial^2 a}{\partial \boldsymbol{\alpha} \partial \boldsymbol{\alpha}'} \mathbf{B}.$$

5. If $\mathbf{c}$ and $\mathbf{D}$ are, respectively, a vector and symmetric matrix of constants, then

(i) $\frac{\partial(\boldsymbol{\alpha}'\mathbf{c})}{\partial \boldsymbol{\alpha}} = \mathbf{c}$,

(ii) $\frac{\partial(\boldsymbol{\alpha}'\mathbf{D}\boldsymbol{\alpha})}{\partial \boldsymbol{\alpha}} = 2\mathbf{D}\boldsymbol{\alpha}$.

A11 PROJECTION MATRICES

1. Given Ω, a vector subspace of $\mathbb{R}^n$ (n-dimensional Euclidean space), every $n \times 1$ vector $\mathbf{y}$ can be expressed uniquely in the form

$$\mathbf{y} = \mathbf{u}_y + \mathbf{v}_y,$$

where $\mathbf{u}_y \in \Omega$ and $\mathbf{v}_y \in \Omega^\perp$, the orthogonal complement of Ω.

2. There is a unique matrix $\mathbf{P}_\Omega$ such that for every $\mathbf{y} \in \mathbb{R}^n$, $\mathbf{u}_y = \mathbf{P}_\Omega \mathbf{y}$. $\mathbf{P}_\Omega$ is called the orthogonal projection of $\mathbb{R}^n$ onto Ω. It has the following properties:

(i) $\Omega = \mathscr{R}[\mathbf{P}_\Omega]$.

(ii) $\mathbf{P}_\Omega$ is symmetric and idempotent, i.e. $\mathbf{P}_\Omega^2 = \mathbf{P}_\Omega$.

(iii) rank $\mathbf{P}_\Omega = \text{tr}\, \mathbf{P}_\Omega =$ dimension $[\Omega]$,

3. $\mathbf{v}_y = (\mathbf{I}_n - \mathbf{P}_\Omega)\mathbf{y}$ and $\mathbf{I}_n - \mathbf{P}_\Omega$ represents the orthogonal projection of $\mathbb{R}^n$ onto $\Omega^\perp$.

4. If $\Omega = \mathscr{R}[\mathbf{X}]$, where $\mathbf{X}$ is $n \times p$ of rank r, then

$$\mathbf{P}_\Omega = \mathbf{P}_\mathbf{X} = \mathbf{X}(\mathbf{X}'\mathbf{X})^- \mathbf{X}' = \mathbf{X}\mathbf{X}^+,$$

where $(\mathbf{X}'\mathbf{X})^-$ is a generalized inverse of $\mathbf{X}'\mathbf{X}$, and $\mathbf{X}^+$ is the Moore–Penrose generalized inverse of $\mathbf{X}$ (see Appendices A5, A8.1). If $r = p$, then $(\mathbf{X}'\mathbf{X})^-$ becomes $(\mathbf{X}'\mathbf{X})^{-1}$ and

$$\mathbf{P}_\mathbf{X} = \mathbf{X}(\mathbf{X}'\mathbf{X})^{-1}\mathbf{X}'.$$

In this case rank $\mathbf{P}_\mathbf{X} = \text{tr}\, \mathbf{P}_\mathbf{X} = p$ [by Appendix A11.2(iii)].

5. If $\mathbf{y} \sim N_n(\boldsymbol{\mu}, \sigma^2\mathbf{I}_n)$ and Ω is p-dimensional, then $Q = \mathbf{y}'\mathbf{P}_\Omega\mathbf{y}/\sigma^2$ has a noncentral chi-square distribution with p degrees of freedom and noncentrality parameter $\Delta = \boldsymbol{\mu}'\mathbf{P}_\Omega\boldsymbol{\mu}/\sigma^2$ [denoted by $\chi_p^2(\Delta)$]. We note from Appendix A12.1 that $E[Q] = p + \Delta$.
6. If $\boldsymbol{\varepsilon} \sim N_n(\mathbf{0}, \sigma^2\mathbf{I}_n)$ and Ω is p-dimensional, then $\boldsymbol{\varepsilon}'\mathbf{P}_\Omega\boldsymbol{\varepsilon}/\sigma^2$ and $\boldsymbol{\varepsilon}'(\mathbf{I}_n - \mathbf{P}_\Omega)\boldsymbol{\varepsilon}/\sigma^2$ are independently distributed as χ_p^2 and χ_{n-p}^2 respectively. Hence

$$\frac{n-p}{p}\frac{\boldsymbol{\varepsilon}'\mathbf{P}_\Omega\boldsymbol{\varepsilon}}{\boldsymbol{\varepsilon}'(\mathbf{I}_n - \mathbf{P}_\Omega)\boldsymbol{\varepsilon}} \sim F_{p,n-p}.$$

7. If $\omega \subset \Omega$ then

$$\mathbf{P}_\Omega - \mathbf{P}_\omega = \mathbf{P}_{\omega^\perp \cap \Omega}.$$

A12 QUADRATIC FORMS

1. If $E[\mathbf{y}] = \boldsymbol{\mu}$ and $D[\mathbf{y}] = \boldsymbol{\Sigma}$, then

$$E[\mathbf{y}'\mathbf{A}\mathbf{y}] = \operatorname{tr}[\mathbf{A}\boldsymbol{\Sigma}] + \boldsymbol{\mu}'\mathbf{A}\boldsymbol{\mu}.$$

When $\boldsymbol{\Sigma} = \sigma^2\mathbf{I}_n$, $\operatorname{tr}[\mathbf{A}\boldsymbol{\Sigma}] = \sigma^2 \operatorname{tr}[\mathbf{A}]$.
2. If $\boldsymbol{\varepsilon} \sim N_n(\mathbf{0}, \sigma^2\mathbf{I}_n)$ then

$$\operatorname{cov}[\boldsymbol{\varepsilon}'\mathbf{A}\boldsymbol{\varepsilon}, \boldsymbol{\varepsilon}'\mathbf{B}\boldsymbol{\varepsilon}] = 2\sigma^4 \operatorname{tr}[\mathbf{A}\mathbf{B}].$$

A13 MATRIX OPERATORS

1. Let $\mathbf{A}$ and $\mathbf{B}$ be $m \times m$ and $n \times n$ matrices, respectively; then

$$\mathbf{A} \otimes \mathbf{B} = \begin{bmatrix} a_{11}\mathbf{B} & a_{12}\mathbf{B} & \cdots & a_{1m}\mathbf{B} \\ a_{21}\mathbf{B} & a_{22}\mathbf{B} & \cdots & a_{2m}\mathbf{B} \\ \vdots & \vdots & & \vdots \\ a_{m1}\mathbf{B} & a_{m2}\mathbf{B} & \cdots & a_{mm}\mathbf{B} \end{bmatrix}$$

is called the Kronecker (direct or tensor) product of $\mathbf{A}$ and $\mathbf{B}$.
2. Let $\mathbf{A} = (\mathbf{a}_1, \mathbf{a}_2, \ldots, \mathbf{a}_n)$ be an $m \times n$ matrix with columns $\mathbf{a}_i$. Then the "rolled-out" vector $\mathbf{a} = (\mathbf{a}_1', \mathbf{a}_2', \ldots, \mathbf{a}_n')'$ is denoted by $\mathbf{a} = \operatorname{vec}\mathbf{A}$. If $\mathbf{A}$ is an $m \times m$ symmetric matrix, then the rolled-out vector of *distinct* elements, namely

$$\mathbf{b} = (a_{11}, a_{12}, a_{22}, a_{13}, a_{23}, a_{33}, \ldots, a_{mm})$$

is denoted by $\mathbf{b} = \operatorname{vech}\mathbf{A}$.

A14 METHOD OF SCORING

Consider a model with log-likelihood function $L(\boldsymbol{\gamma})$. Then Fisher's method of scoring for finding $\hat{\boldsymbol{\gamma}}$, the maximum-likelihood estimate of $\boldsymbol{\gamma}$, is given by the iterative process

$$\boldsymbol{\gamma}^{(a+1)} - \boldsymbol{\gamma}^{(a)} = -\left\{\mathscr{E}\left[\frac{\partial^2 L}{\partial \boldsymbol{\gamma}\, \partial \boldsymbol{\gamma}'}\right]\right\}^{-1}_{\boldsymbol{\gamma}^{(a)}} \left(\frac{\partial L}{\partial \boldsymbol{\gamma}}\right)_{\boldsymbol{\gamma}^{(a)}}. \tag{1}$$

This algorithm can be regarded as a Newton method for maximizing $L(\boldsymbol{\gamma})$, but with the Hessian replaced by its expected value. The expected Hessian is usually simpler and more likely to be positive definite because of the relationship

$$-\mathscr{E}\left[\frac{\partial^2 L}{\partial \boldsymbol{\gamma}\, \partial \boldsymbol{\gamma}'}\right] = \mathscr{E}\left[\frac{\partial L}{\partial \boldsymbol{\gamma}} \frac{\partial L}{\partial \boldsymbol{\gamma}'}\right],$$

which holds under fairly general conditions.

We can now apply this method to the nonlinear model

$$\mathbf{y} = \mathbf{f}(\boldsymbol{\theta}) + \boldsymbol{\varepsilon},$$

where $\boldsymbol{\varepsilon} \sim N_n(\mathbf{0}, \sigma^2 \mathbf{V})$, $\mathbf{V}$ is known, and $\boldsymbol{\gamma} = (\boldsymbol{\theta}', \sigma^2)'$. Here

$$L(\boldsymbol{\theta}, \sigma^2) = \text{const} - \frac{n}{2}\log \sigma^2 - \tfrac{1}{2}\log|\mathbf{V}| - \frac{1}{2\sigma^2}[\mathbf{y} - \mathbf{f}(\boldsymbol{\theta})]'\mathbf{V}^{-1}[\mathbf{y} - \mathbf{f}(\boldsymbol{\theta})],$$

and we find that

$$-\mathscr{E}\left[\frac{\partial^2 L}{\partial \boldsymbol{\gamma}\, \partial \boldsymbol{\gamma}'}\right] = \begin{bmatrix} -\mathscr{E}\left(\dfrac{\partial^2 L}{\partial \boldsymbol{\theta}\, \partial \boldsymbol{\theta}'}\right) & \mathbf{0} \\ \mathbf{0}' & \dfrac{n}{\sigma^4} \end{bmatrix}.$$

Also from A10.5, with $\mathbf{F}_{.} = \partial \mathbf{f}/\partial \boldsymbol{\theta}'$, we have

$$\frac{\partial L}{\partial \boldsymbol{\theta}} = \frac{1}{\sigma^2}\mathbf{F}_{.}'\mathbf{V}^{-1}[\mathbf{y} - \mathbf{f}(\boldsymbol{\theta})]$$

and

$$\begin{aligned} -\mathscr{E}\left[\frac{\partial^2 L}{\partial \boldsymbol{\theta}\, \partial \boldsymbol{\theta}'}\right] &= \mathscr{E}\left[\frac{\partial L}{\partial \boldsymbol{\theta}} \frac{\partial L}{\partial \boldsymbol{\theta}'}\right] \\ &= \frac{1}{\sigma^4}\mathscr{E}[\mathbf{F}_{.}'\mathbf{V}^{-1}\{\mathbf{y} - \mathbf{f}(\boldsymbol{\theta})\}\{\mathbf{y} - \mathbf{f}(\boldsymbol{\theta})\}'\mathbf{V}^{-1}\mathbf{F}_{.}] \\ &= \frac{1}{\sigma^2}\mathbf{F}_{.}'\mathbf{V}^{-1}\mathbf{F}_{.} \qquad \text{[by Equation (1.1)]}. \end{aligned}$$

Hence the first p elements of (1) lead to

$$\boldsymbol{\theta}^{(a+1)} - \boldsymbol{\theta}^{(a)} = (\mathbf{F}_{\cdot}'\mathbf{V}^{-1}\mathbf{F}_{\cdot})^{-1}\mathbf{F}_{\cdot}'\mathbf{V}^{-1}\{\mathbf{y} - \mathbf{f}(\boldsymbol{\theta})\}|_{\boldsymbol{\theta}^{(a)}}. \tag{2}$$

Now using a concentrated-likelihood approach [c.f. (2.62)], it is readily shown that maximizing $L(\boldsymbol{\theta}, \sigma^2)$ is equivalent to minimizing

$$S(\boldsymbol{\theta}) = \{\mathbf{y} - \mathbf{f}(\boldsymbol{\theta})\}'\mathbf{V}^{-1}\{\mathbf{y} - \mathbf{f}(\boldsymbol{\theta})\},$$

so that $\hat{\boldsymbol{\theta}}$ is also the generalized least-squares estimate. Using the Taylor expansion

$$\mathbf{f}(\boldsymbol{\theta}) \approx \mathbf{f}(\boldsymbol{\theta}^{(a)}) + \mathbf{F}_{\cdot}(\boldsymbol{\theta}^{(a)})(\boldsymbol{\theta} - \boldsymbol{\theta}^{(a)}),$$

we find that the Gauss–Newton method of minimizing $S(\boldsymbol{\theta})$ leads once again to (2) [c.f. Equation (2.37)]. This equivalence of the Gauss–Newton and the scoring algorithms was demonstrated by Ratkowsky and Dolby [1975].

APPENDIX B

Differential Geometry

B1 DERIVATIVES FOR CURVES

1. Let $\mathbf{e}_1, \mathbf{e}_2$, and $\mathbf{e}_3$ be an orthonormal basis in $\mathbb{R}^3$, and let

$$\mathbf{r}(b) = r_1(b)\mathbf{e}_1 + r_2(b)\mathbf{e}_2 + r_3(b)\mathbf{e}_3$$

represent a curve defined by the parameter b, $b \in \mathscr{B} \subset \mathbb{R}$. If

$$\dot{\mathbf{r}} = \frac{d\mathbf{r}}{db} = \frac{dr_1(b)}{db}\mathbf{e}_1 + \frac{dr_2(b)}{db}\mathbf{e}_2 + \frac{dr_3(b)}{db}\mathbf{e}_3$$

exists and is continuous for $b \in \mathscr{B}$, then $\mathbf{r}(b)$ is said to be a regular parametric representation. In this case the length of arc, s, measured from some fixed point on the curve can also be used as a parameter. Setting $\mathbf{r} = \mathbf{r}(s)$, we define

$$\mathbf{r}^1 = \frac{d\mathbf{r}}{ds} = \frac{dr_1(s)}{ds}\mathbf{e}_1 + \frac{dr_2(s)}{ds}\mathbf{e}_2 + \frac{dr_3(s)}{ds}\mathbf{e}_3,$$

so that the "dot" and "one" represent differentiation with respect to b and s respectively.

2. If $\|\mathbf{r}(b)\|$ is constant for all $b \in \mathscr{B}$, then, at any point on the curve, $\dot{\mathbf{r}}(b) \perp \mathbf{r}(b)$ and $\mathbf{r}^1(s) \perp \mathbf{r}(s)$.

Proof: Differentiating $\mathbf{r}'\mathbf{r} = \|\mathbf{r}\|^2 = c$ with respect to b gives us $\dot{\mathbf{r}}'\mathbf{r} + \mathbf{r}'\dot{\mathbf{r}} = 0$, or $\dot{\mathbf{r}}'\mathbf{r} = 0$, so that $\dot{\mathbf{r}} \perp \mathbf{r}$. Setting $b = s$, we have $\mathbf{r}^1 \perp \mathbf{r}$. [These results are also geometrically obvious. Let P, Q, and R be neighboring points as indicated in Fig. B1; R is included for later discussion. Then constant $\|\mathbf{r}\|$ implies that $OP = OQ$, so that $\delta\mathbf{r}$ becomes perpendicular to $\mathbf{r}$ as $Q \to P$.] ■

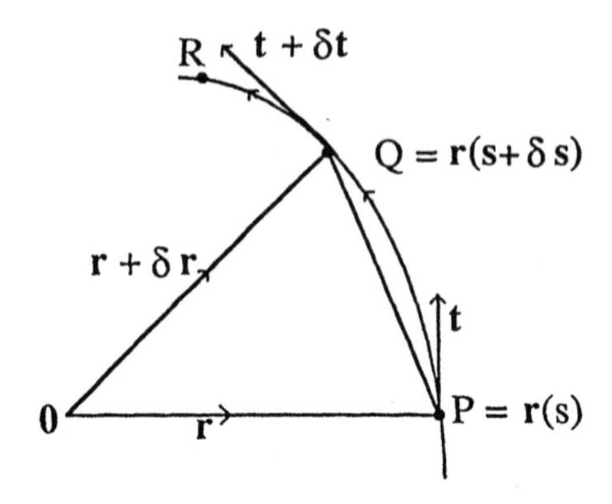

Figure B1.

3. Let P in Fig. B1 also have representation $\mathbf{r}(b)$. Then $\mathbf{r}^1(s) = \mathbf{t}(s)$ and $\dot{\mathbf{r}}(b) = \dot{s}\mathbf{t}[s(b)]$, where $\mathbf{t}$ is the unit tangent vector at P in the direction of increasing s. Also $\|\dot{\mathbf{r}}(b)\|^2 = \dot{s}^2$.

   Proof: Let $\delta s > 0$. Then

$$\left\|\frac{\delta \mathbf{r}}{\delta s}\right\| = \frac{\|\delta \mathbf{r}\|}{\delta s} = \frac{\text{chord } PQ}{\text{arc } PQ} \to 1,$$

so that $\|\mathbf{r}^1\| = 1$. Furthermore, the direction of $\delta \mathbf{r}$ is that of the chord PQ, so that, in the limit, $\mathbf{r}^1$ is a tangent at P pointing in the direction of increasing s, i.e., $\mathbf{r}^1 = \mathbf{t}(s)$ with $\|\mathbf{t}(s)\| = 1$. Finally

$$\dot{\mathbf{r}}(b) = \frac{d\mathbf{r}}{ds}\frac{ds}{db} = \mathbf{r}^1[s(b)]\dot{s} = \mathbf{t}[s(b)]\dot{s}.$$

■

B2 CURVATURE

1. The plane PQR in Fig. B1 as Q and R tend to P is called the *osculating plane* at P, and the plane perpendicular to $\mathbf{t}$ is called the *normal plane*. Then we have the following;
   (a) $\mathbf{r}^{11}(s) = \mathbf{t}^1(s) = \kappa(s)\mathbf{n}(s)$, where $\mathbf{r}^{11}$ is called the *curvature vector* and $|\kappa(s)|$, its length, is called the (absolute) *curvature* at P. The vector $\mathbf{n}(s)$, called the *unit principal normal*, is a unit vector in the normal plane parallel to $\mathbf{r}^{11}$ but with orientation chosen in relation to $\mathbf{t}(s)$ so that the pair $[\mathbf{t}(s), \mathbf{n}(s)]$ forms the usual orientation of $\mathbb{R}^2$ as in Fig. B2.
   (b) The osculating plane is the plane through $\mathbf{t}$ and $\mathbf{n}$.

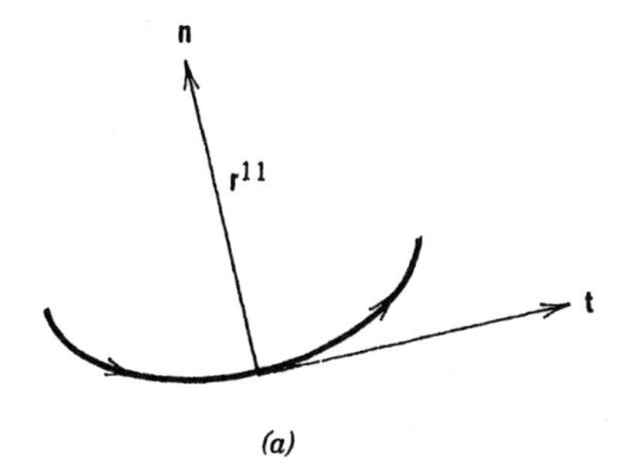

(a)

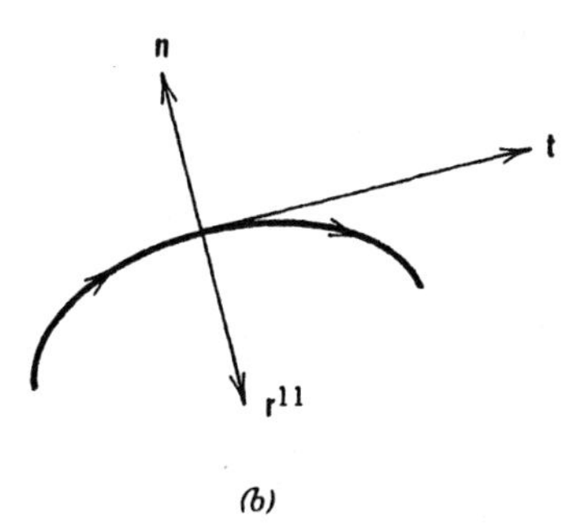

(b)

Figure B2.

Proof: (a): Since $\| \mathbf{t}(s) \| = 1$, we have $\mathbf{t}^1 \perp \mathbf{t}$ by Appendix B1.2, and $\mathbf{t}^1$ lies in the normal plane, i.e., $\mathbf{t}^1 \propto \mathbf{n}$ for some $\mathbf{n}$. (b): $\delta\mathbf{t}$ lies in the plane of $\mathbf{t}$ and $\mathbf{t} + \delta\mathbf{t}$. Also, the direction of QR tends to that of $\mathbf{t} + \delta\mathbf{t}$ as R tends to Q, and the direction of PQ tends to that of $\mathbf{t}$ as Q tends to P. Thus in the limit, both $\delta\mathbf{t}$ (which tends to the direction of $\mathbf{n}$) and $\mathbf{t}$ lie in the osculating plane. ■

2. $\rho(s) = 1/|\kappa(s)|$ is called the radius of curvature, and it can be shown that it is the limiting radius of the circle through P, Q, and R as Q and R tend to P. It is clear that the radius of curvature of a circle is the radius of the circle.

3. (a) $\dfrac{|\mathbf{n}'\ddot{\mathbf{r}}[s(b)]|}{\| \dot{\mathbf{r}}[s(b)] \|^2} = \dfrac{|\mathbf{n}'\ddot{\mathbf{r}}|}{\dot{s}^2} = |\kappa[s(b)]|$ and is invariant with respect to the parametric system used.

(b) $\dfrac{|\mathbf{t}'\ddot{\mathbf{r}}[s(b)]|}{\| \dot{\mathbf{r}}[s(b)] \|^2} = \dfrac{|\ddot{s}|}{\dot{s}^2}.$

Proof: Using Appendix B1.3,

$$\ddot{\mathbf{r}} = \frac{d\{\dot{s}\mathbf{t}[s(b)]\}}{db}$$
$$= \ddot{s}\mathbf{t} + \dot{s}\mathbf{t}^1\dot{s}$$
$$= \ddot{s}\mathbf{t} + \dot{s}^2\kappa\mathbf{n},$$

so that $|\mathbf{n}'\ddot{\mathbf{r}}| = \dot{s}^2|\kappa| = \|\dot{\mathbf{r}}\|^2|\kappa|$ (from Appendix B1.3). When $b = s$, we have $\|\mathbf{r}^1(s)\|^2 = \|\mathbf{t}(s)\|^2 = 1$ and $|\mathbf{n}'\mathbf{r}^{11}|/\|\mathbf{r}^1\|^2 = |\kappa|$ [by 1(a) above]. Thus $|\kappa|$ is a property of the curve at P and does not depend on the parametric system used. To prove (b) we note that $\mathbf{t}'\ddot{\mathbf{r}} = \ddot{s}$. This tangential component of curvature does depend on the parametric system, since when $s = b$, $\ddot{s} = 0$ and (b) is zero. ■

4. From the expression for $\ddot{\mathbf{r}}$ above we can write

$$\ddot{\mathbf{r}} = \ddot{\mathbf{r}}^T + \ddot{\mathbf{r}}^N,$$

where, from the proof of statement 3 above, $\ddot{\mathbf{r}}^T = \ddot{s}\mathbf{t}$ is the tangential component of the curvature vector $\ddot{\mathbf{r}}$, and $\ddot{\mathbf{r}}^N = \dot{s}^2\kappa\mathbf{n}$ is the normal component. We see that

$$\ddot{s} = |\mathbf{t}'\ddot{\mathbf{r}}| = \|\ddot{\mathbf{r}}^T\| = \|\mathbf{P}_t\ddot{\mathbf{r}}\|$$

and

$$\dot{s}^2|\kappa| = |\mathbf{n}'\ddot{\mathbf{r}}| = \|\ddot{\mathbf{r}}^N\| = \|(\mathbf{I}_3 - \mathbf{P}_t)\ddot{\mathbf{r}}\|,$$

where $\mathbf{P}_t = \mathbf{t}\mathbf{t}'$ is the projection matrix projecting $\mathbb{R}^3$ orthogonally onto the tangent vector $\mathbf{t}$ (Appendix A11.4).

B3 TANGENT PLANES

Consider a three-dimensional surface $\mathbf{r} = \mathbf{r}(a_1, a_2)$ determined by two parameters a_1 and a_2. For each point in the (a_1, a_2) plane there is a point $\mathbf{r}$ (called P, say) on the surface. If a curve with parametric representations $a_1 = a_1(b)$ and $a_2 = a_2(b)$ is traced out in the (a_1, a_2) plane, then a corresponding curve $\mathbf{r}[a_1(b), a_2(b)]$ will be traced out on the surface.

1. At the point P we have

$$\dot{\mathbf{r}} = \frac{d\mathbf{r}}{db}$$
$$= \frac{\partial \mathbf{r}}{\partial a_1}\dot{a}_1 + \frac{\partial \mathbf{r}}{\partial a_2}\dot{a}_2$$
$$= \mathbf{r}_1\dot{a}_1 + \mathbf{r}_2\dot{a}_2,$$

which is a linear combination of $\mathbf{r}_1$ and $\mathbf{r}_2$. Hence for all curves through P it follows from B1.3 that their tangents lie in a plane determined by $\mathbf{r}_1$ and $\mathbf{r}_2$, called the tangent plane.

2. The normal to the surface at P is defined to be a unit vector perpendicular to the tangent plane with direction given by the usual right-hand rule for vector products, i.e. $\mathbf{n} = \mathbf{r}_1 \times \mathbf{r}_2 / \| \mathbf{r}_1 \times \mathbf{r}_2 \|$. The intersection of any plane containing the normal with the surface is called a *normal section*. Moreover, the normal to the surface is also the normal to the curve of intersection, and we can apply Appendix B2.3(a). The curvature $|\kappa|$ is then called the *normal curvature* for this particular curve of intersection.
3. The above theory can be generalized to surfaces in $\mathbb{R}^n$. If $\mathbf{r} = \mathbf{r}(a_1, a_2, \ldots, a_{n-1})$ is an $(n-1)$-dimensional submanifold, then we have an $(n-1)$-dimensional tangent plane spanned by the vectors $\partial \mathbf{r}/\partial a_i$ $(i = 1, 2, \ldots, n-1)$, and also a unique normal to this tangent plane. However, if $\mathbf{r} = \mathbf{r}(a_1, a_2, \ldots, a_p)$, where $p < n-1$, then instead of a unique normal to the tangent plane we have an $(n-p)$-dimensional normal subspace. However, normal and tangential components of the curvature vector can still be defined by projecting onto the tangent plane and perpendicular to the tangent plane as in Appendix B2.4. Thus, given the curve $\mathbf{r}(b) = \mathbf{r}[a_1(b), a_2(b), \ldots, a_p(b)]$, we can follow Bates and Watts [1980] and define the two curvatures

$$K^T = \frac{\| \mathbf{P}_T \ddot{\mathbf{r}}(b) \|}{\| \dot{\mathbf{r}}(b) \|^2}$$

and

$$K^N = \frac{\| (\mathbf{I}_n - \mathbf{P}_T) \ddot{\mathbf{r}}(b) \|}{\| \dot{\mathbf{r}}(b) \|^2},$$

where $\mathbf{P}_T$ is the symmetric idempotent matrix representing the projection of $\mathbb{R}^n$ onto the tangent plane at $\mathbf{r}(b)$. Because of the lack of a unique normal to the tangent plane, the concept of radius of curvature does not carry over directly. Furthermore, the reciprocal of a curvature is not, technically, a radius of curvature in this case.

B4 MULTIPLICATION OF 3-DIMENSIONAL ARRAYS

Consider the $n \times p \times p$ array $W_{\cdot\cdot} = \{(\mathbf{w}_{rs})\}$ made up of a $p \times p$ array of n-dimensional vectors $\mathbf{w}_{rs} (r, s = 1, 2, \ldots, p)$. If w_{irs} is the ith element of $\mathbf{w}_{rs}$, then the matrix of ith elements $\mathbf{W}_{i\cdot\cdot} = [(w_{irs})]$ is called the ith face of $W_{\cdot\cdot}$. We now define two types of multiplication. Firstly, if $\mathbf{B}$ and $\mathbf{C}$ are $p \times p$ matrices, then

$$V_{\cdot\cdot} = \{(\mathbf{v}_{rs})\} = \mathbf{B} W_{\cdot\cdot} \mathbf{C}$$

denotes the array with ith face $\mathbf{V}_{i..} = \mathbf{B}\mathbf{W}_{i..}\mathbf{C}$, i.e.

$$\mathbf{v}_{rs} = \sum_{\alpha}\sum_{\beta} b_{r\alpha}\mathbf{w}_{\alpha\beta}c_{\beta s}.$$

Secondly, if $\mathbf{D}$ is a $q \times n$ matrix, then we define square-bracket multiplication by the equation

$$[\mathbf{D}][\mathscr{W}_{..}] = \{(\mathbf{D}\mathbf{w}_{rs})\},$$

where the right-hand side is a $q \times p \times p$ array. The following properties can be deduced.

1. $[\mathbf{D}][\mathbf{B}\mathscr{W}_{..}\mathbf{C}] = \mathbf{B}[\mathbf{D}][\mathscr{W}_{..}]\mathbf{C}$.

Proof:

$$\begin{aligned}[\mathbf{D}][\mathbf{B}\mathscr{W}_{..}\mathbf{C}] &= [\mathbf{D}]\left[\left\{\left(\sum_{\alpha}\sum_{\beta} b_{r\alpha}\mathbf{w}_{\alpha\beta}c_{\beta s}\right)\right\}\right] \\ &= \left\{\left(\sum_{\alpha}\sum_{\beta} b_{r\alpha}\mathbf{D}\mathbf{w}_{\alpha\beta}c_{\beta s}\right)\right\} \\ &= \mathbf{B}\{(\mathbf{D}\mathbf{w}_{rs})\}\mathbf{C} \\ &= \mathbf{B}[\mathbf{D}][\mathscr{W}_{..}]\mathbf{C}.\end{aligned}$$

■

2. $[\mathbf{D}\mathbf{B}][\mathscr{W}_{..}] = [\mathbf{D}][\{(\mathbf{B}\mathbf{w}_{rs})\}] = [\mathbf{D}][[\mathbf{B}][\mathscr{W}_{..}]]$.

Proof:

$$\begin{aligned}[\mathbf{D}\mathbf{B}][\mathscr{W}_{..}] &= \{(\mathbf{D}\mathbf{B}\mathbf{w}_{rs})\} \\ &= [\mathbf{D}][\{(\mathbf{B}\mathbf{w}_{rs})\}] \\ &= [\mathbf{D}][[\mathbf{B}][\mathscr{W}_{..}]].\end{aligned}$$

■

3. $\mathbf{a}'\mathscr{W}_{..}\mathbf{b} = \sum_u \sum_v a_u b_v \mathbf{w}_{uv}$.
4. $[\mathbf{d}'][\mathscr{W}_{..}]$ is a matrix with (r, s)th element $\sum_i d_i(\mathbf{w}_{rs})_i$.

B5 INVARIANCE OF INTRINSIC CURVATURE

The curvature K_δ^N of (4.16) is invariant under a bijective (one-to-one) transformation of the parameter space.

Proof: Let $\boldsymbol{\theta} = \mathbf{u}(\boldsymbol{\phi})$ where $\mathbf{u}$ is bijective with inverse $\mathbf{u}^{-1}$. Then $\boldsymbol{\phi} = \mathbf{u}^{-1}(\boldsymbol{\theta}) = \mathbf{v}(\boldsymbol{\theta})$, say. Let $\hat{\boldsymbol{\phi}} = \mathbf{v}(\hat{\boldsymbol{\theta}})$ and $f(\boldsymbol{\theta}) = f(\mathbf{u}(\boldsymbol{\phi})) = g(\boldsymbol{\phi})$, say. The quadratic Taylor expansion in the $\boldsymbol{\phi}$-system corresponding to (4.5) in the $\boldsymbol{\theta}$-system is

$$\begin{aligned} g(\boldsymbol{\phi}) - g(\hat{\boldsymbol{\phi}}) &\approx \hat{\mathbf{G}}_{.}(\boldsymbol{\phi} - \hat{\boldsymbol{\phi}}) + \tfrac{1}{2}(\boldsymbol{\phi} - \hat{\boldsymbol{\phi}})\hat{G}_{..}(\boldsymbol{\phi} - \hat{\boldsymbol{\phi}}) \\ &= \hat{\mathbf{G}}_{.}\boldsymbol{\Delta} + \tfrac{1}{2}\boldsymbol{\Delta}'\hat{G}_{..}\boldsymbol{\Delta}, \end{aligned}$$

say. Now

$$\begin{aligned} \hat{\mathbf{G}}_{.} &= \left[\left(\frac{\partial g_i}{\partial \phi_j}\right)\right]_{\boldsymbol{\phi}=\hat{\boldsymbol{\phi}}} \\ &= \left[\left(\sum_a \frac{\partial f_i}{\partial \theta_a}\frac{\partial \theta_a}{\partial \phi_j}\right)\right]_{\boldsymbol{\phi}=\hat{\boldsymbol{\phi}}} \\ &= \hat{\mathbf{F}}_{.}\hat{\boldsymbol{\Theta}}_{.}, \end{aligned}$$

where $\hat{\boldsymbol{\Theta}}_{.} = [(\partial \theta_i / \partial \phi_j)]_{\boldsymbol{\phi}=\hat{\boldsymbol{\phi}}}$ is nonsingular, as $\mathbf{u}$ is bijective. Hence the projection matrix onto $\mathscr{R}[\hat{\mathbf{G}}_{.}]$ is given by (Appendix A11.4)

$$\begin{aligned} \hat{\mathbf{P}}_G &= \hat{\mathbf{G}}_{.}(\hat{\mathbf{G}}_{.}'\hat{\mathbf{G}}_{.})^{-1}\hat{\mathbf{G}}_{.}' \\ &= \hat{\mathbf{F}}_{.}\hat{\boldsymbol{\Theta}}_{.}(\hat{\boldsymbol{\Theta}}_{.}'\hat{\mathbf{F}}_{.}'\hat{\mathbf{F}}_{.}\hat{\boldsymbol{\Theta}}_{.})^{-1}\hat{\boldsymbol{\Theta}}_{.}'\hat{\mathbf{F}}_{.}' \\ &= \hat{\mathbf{F}}_{.}(\hat{\mathbf{F}}_{.}'\hat{\mathbf{F}}_{.})^{-1}\hat{\mathbf{F}}_{.}' \\ &= \hat{\mathbf{P}}_F. \end{aligned} \tag{1}$$

Also

$$\begin{aligned} \mathbf{g}_{rs} &= \frac{\partial^2 \mathbf{g}}{\partial \phi_r \partial \phi_s} \\ &= \frac{\partial}{\partial \phi_r}\sum_b \frac{\partial \mathbf{f}}{\partial \theta_b}\frac{\partial \theta_b}{\partial \phi_s} \\ &= \sum_a \sum_b \frac{\partial^2 \mathbf{f}}{\partial \theta_a \partial \theta_b}\frac{\partial \theta_a}{\partial \phi_r}\frac{\partial \theta_b}{\partial \phi_s} + \sum_b \frac{\partial \mathbf{f}}{\partial \theta_b}\frac{\partial^2 \theta_b}{\partial \phi_r \partial \phi_s}. \end{aligned}$$

Hence

$$\begin{aligned} \hat{\mathbf{g}}_{rs} &= [\mathbf{g}_{rs}]_{\boldsymbol{\phi}=\hat{\boldsymbol{\phi}}} \\ &= \left[\sum_a \sum_b \mathbf{f}_{ab}\frac{\partial \theta_a}{\partial \phi_r}\frac{\partial \theta_b}{\partial \phi_s}\right]_{\boldsymbol{\phi}=\hat{\boldsymbol{\phi}}} + \hat{\mathbf{F}}_{.}\hat{\boldsymbol{\theta}}_{rs}, \end{aligned}$$

where $\hat{\boldsymbol{\theta}}_{rs} = [\partial^2 \boldsymbol{\theta} / \partial \phi_r \partial \phi_s]_{\boldsymbol{\phi}=\hat{\boldsymbol{\phi}}}$. Since $(\mathbf{I}_n - \hat{\mathbf{P}}_F)\hat{\mathbf{F}}_{.} = \mathbf{0}$,

$$\begin{aligned} \mathbf{g}_{rs}^N &= (\mathbf{I}_n - \hat{\mathbf{P}}_F)\hat{\mathbf{g}}_{rs} \qquad \text{[by (1) and Appendix A11.3]} \\ &= \left[\sum_a \sum_b \mathbf{f}_{ab}^N \frac{\partial \theta_a}{\partial \phi_r}\frac{\partial \theta_b}{\partial \phi_s}\right]_{\boldsymbol{\phi}=\hat{\boldsymbol{\phi}}}. \end{aligned}$$

From the ith element of the above vector equation we have

$$\hat{\mathbf{G}}_{i\cdot\cdot} = \hat{\mathbf{\Theta}}'_{\cdot}\hat{\mathbf{F}}^{N}_{i\cdot\cdot}\hat{\mathbf{\Theta}}_{\cdot},$$

or, symbolically (Appendix B4),

$$\hat{G}_{\cdot\cdot} = \hat{\mathbf{\Theta}}'_{\cdot}\hat{F}^{N}_{\cdot\cdot}\hat{\mathbf{\Theta}}_{\cdot}.$$

For small $\boldsymbol{\delta}$ we can write $\boldsymbol{\delta} = \hat{\mathbf{\Theta}}\boldsymbol{\Delta}$, so that

$$\frac{\|\boldsymbol{\Delta}'\hat{G}^{N}_{\cdot\cdot}\boldsymbol{\Delta}\|}{\|\hat{\mathbf{G}}_{\cdot}\boldsymbol{\Delta}\|^2} = \frac{\|\boldsymbol{\Delta}'\hat{\mathbf{\Theta}}'_{\cdot}\hat{F}^{N}_{\cdot\cdot}\hat{\mathbf{\Theta}}_{\cdot}\boldsymbol{\Delta}\|}{\|\hat{\mathbf{F}}_{\cdot}\hat{\mathbf{\Theta}}_{\cdot}\boldsymbol{\Delta}\|^2}$$

$$= \frac{\|\boldsymbol{\delta}'\hat{F}^{N}_{\cdot\cdot}\boldsymbol{\delta}\|}{\|\hat{\mathbf{F}}_{\cdot}\boldsymbol{\delta}\|^2},$$

and invariance is proved. ■

APPENDIX C

Stochastic Differential Equations

For an introduction to this subject see Arnold [1974].

C1 RATE EQUATIONS

Equations of the form

$$\dot{y} = \frac{dy}{dx} = f(x; \boldsymbol{\theta}) + \sigma(x; \boldsymbol{\theta})\varepsilon(x), \tag{1}$$

where $\varepsilon(x)$ is so-called "white noise," are more properly written as

$$dy = f(x; \boldsymbol{\theta})\, dx + \sigma(x; \boldsymbol{\theta})\, dw(x), \tag{2}$$

where $w(x)$ is a Wiener process. The latter process is a Gaussian process [i.e., $w(x)$ is normally distributed] with independent increments. Thus if $x_1 < x_2 < x_3 < x_4$, then $w(x_2) - w(x_1)$ is statistically independent of $w(x_4) - w(x_3)$. Moreover,

$$w(x_i) - w(x_j) \sim N(0, |x_i - x_j|). \tag{3}$$

Technically, a Wiener process is not differentiable, so that Equation (2) is another way of writing its formal solution

$$y(x) = y_0 + \int_{x_0}^{x} f(u; \boldsymbol{\theta})\, du + \int_{x_0}^{x} \sigma(u; \boldsymbol{\theta})\, dw(u), \tag{4}$$

where $y_0 = y(x_0)$. Conditionally on the value of y_0, we find that $y(x)$ is a Gaussian process with independent increments. Also

$$E[y(x)|y_0] = y_0 + \int_{x_0}^{x} f(u; \boldsymbol{\theta})\, du, \tag{5}$$

and noting from (3) that, effectively, $\text{var}[dw(x)] \equiv dx$ we have

$$\begin{aligned}\text{var}[y(x)|y_0] &= \text{var}[y(x) - y_0|y_0] \\ &= \int_{x_0}^{x} \sigma^2(u;\theta)\,du. \qquad (6)\end{aligned}$$

This reduces to $(x - x_0)\sigma^2$ when $\sigma^2(u;\theta) \equiv \sigma^2$. Furthermore, for $x_i < x_j$,

$$\text{var}[y(x_j) - y(x_i)] = \int_{x_i}^{x_j} \sigma^2(u;\theta)\,du. \qquad (7)$$

Instead of (2) we now consider the model

$$dy = f(x;\theta)\,dx + \sigma(x;\theta)\varepsilon(x)\,dx, \qquad (8)$$

where $\varepsilon(x)$ is an autocorrelated Gaussian process with mean zero, variance unity, and autocorrelation function $\text{corr}[\varepsilon(x_i), \varepsilon(x_j)] = \rho(|x_i - x_j|)$. Such processes are continuous, so that the solution and its moments can be obtained by ordinary integration. Thus

$$y(x) = y_0 + \int_{x_0}^{x} f(u;\theta)\,du + \int_{x_0}^{x} \sigma(u;\theta)\varepsilon(u)\,du \qquad (9)$$

with

$$E[y(x)|y_0] = y_0 + \int_{x_0}^{x} f(u;\theta)\,du \qquad (10)$$

since

$$E\left[\int_{x_0}^{x} \sigma(u;\theta)\varepsilon(u)\,du\right] = \int_{x_0}^{x} \sigma(u;\theta)E[\varepsilon(u)]\,du = 0.$$

Also

$$\begin{aligned}\text{var}[y(x)|y_0] &= \text{var}[y(x) - y_0|y_0] \\ &= \text{var}\left[\int_{x_0}^{x} \sigma(u;\theta)\varepsilon(u)\,du\right] \\ &= E\left[\int_{x_0}^{x}\int_{x_0}^{x} \sigma(u;\theta)\sigma(v;\theta)\varepsilon(u)\varepsilon(v)\,du\,dv\right] \\ &= \int_{x_0}^{x}\int_{x_0}^{x} \sigma(u;\theta)\sigma(v;\theta)\rho(|u - v|)\,du\,dv \qquad (11)\end{aligned}$$

and, for $x_1 < x_2 < x_3 < x_4$,

$$\operatorname{cov}[y(x_2)-y(x_1), y(x_4)-y(x_3)] = \int_{x_1}^{x_2}\int_{x_3}^{x_4} \sigma(u;\boldsymbol{\theta})\sigma(v;\boldsymbol{\theta})\rho(|u-v|)\,du\,dv. \quad (12)$$

If we now allow $\rho(|u-v|)$ to tend to the Dirac delta function, our process now tends to a so-called "delta-correlated" Gaussian process or "white noise," i.e., the solution for the autocorrelated error process (8) tends to the earlier solution process for (2). For example, (11) becomes

$$\operatorname{var}[y(x)|y_0] = \int_{x_0}^{x} \sigma^2(u;\boldsymbol{\theta})\,du,$$

which is Equation (6).

C2 PROPORTIONAL RATE EQUATIONS

We now consider equations of the form

$$\dot{y}/y = f(x;\boldsymbol{\theta}) + \sigma(x;\boldsymbol{\theta})\varepsilon(x). \quad (13)$$

If we want $\varepsilon(x)$ to represent "white noise," we can write Equation (13) in two possible ways, namely

$$d\log y(x) = f(x;\boldsymbol{\theta})\,dx + \sigma(x;\boldsymbol{\theta})\,dw(x) \quad (14)$$

or

$$dy(x) = yf(x;\boldsymbol{\theta})\,dx + \sigma(x;\boldsymbol{\theta})y\,dw(x). \quad (15)$$

Equation (14) is of the form (2) so that $\log y(x)$, rather than $y(x)$ itself, is a Gaussian process with independent increments and moments given by (5) and (6). From ordinary differential-equation theory we would expect (14) and (15) to have the same solution. However, the solution to (15) depends on the definition used for a stochastic integral because of the nature of $w(x)$. Ito's integral, for example, depends upon the limits of sums of the form [with $w_i = w(x_i)$] (Arnold [1974: 168])

$$\int g(x;w)\,dw(x) \approx \sum_i g(x_i, w_i)(w_{i+1}-w_i)$$

as the time intervals $(x_{i+1}-x_i)\to 0$. Using this integral, the solution to (14) differs from the solution to (15) only in the mean term. For Ito's integral, (15) gives

$$E[\log y(x)|y_0] = \log y_0 + \int_{x_0}^{x} [f(u;\boldsymbol{\theta}) - \tfrac{1}{2}\sigma^2(u;\boldsymbol{\theta})]\,du, \quad (16)$$

which differs from the corresponding mean of the solution to (14) by the subtraction of the term $\frac{1}{2}\sigma^2(u; \boldsymbol{\theta})$. In contrast, Stratonovich's stochastic calculus uses an integral which is the limit of sums of the form

$$\int g(x, w)\, dw(x) \approx \sum_i g\left(x_i, \frac{w_i + w_{i+1}}{2}\right)(w_{i+1} - w_i)$$

as $x_{i+1} - x_i \to 0$ (Arnold [1974: 168]). With this calculus (15) gives

$$E[\log y(x) | y_0] = \log y_0 + \int_{x_0}^{x} f(u; \boldsymbol{\theta})\, du,$$

as for (14). For a third approach we can consider

$$dy(x) = f(x; \boldsymbol{\theta}) y(x)\, dx + \sigma(x; \boldsymbol{\theta}) y(x) \varepsilon(x)\, dx,$$

where $\varepsilon(x)$ is an autocorrelated Gaussian process. In this case the above equation can be manipulated as an ordinary differential equation to give

$$\log y(x) = \log y_0 + \int_{x_0}^{x} f(u; \boldsymbol{\theta})\, du + \int_{x_0}^{x} \sigma(u; \boldsymbol{\theta}) \varepsilon(u)\, du. \tag{17}$$

This is the solution process, whether we use (14) or (15). The moments of $\int \sigma(u; \boldsymbol{\theta}) \varepsilon(u)\, du$ are given by (10) and (11). If we let the autocorrelation function of $\varepsilon(x)$ tend to the delta function, the solution tends to the Stratonovich solution.

In this book we take the view that the (uncorrelated) white-noise solution should be viewed as part of the continuum of solutions for autocorrelated error processes. Many apparently uncorrelated error processes probably have low levels of correlation, so that it is desirable to have very similar results with no autocorrelation and with very low levels of autocorrelation. Consequently we ignore the additional term in $E[\log y(x)]$ introduced by Ito's calculus.

C3 FIRST-ORDER EQUATIONS

Let $\varepsilon(x)$ again denote "white noise" or the "derivative" of a Wiener process. The equation

$$\frac{dy}{dx} = \beta(x) y + \xi(x) + \sigma(x) \varepsilon(x) \tag{18a}$$

subject to $y(x_0) = y_0$ is more properly written as

$$dy = [\beta(x) y + \xi(x)]\, dx + \sigma(x)\, dw(x). \tag{18b}$$

This equation has solution (Arnold [1974: Section 8.2])

$$y(x) = \Phi(x)\left(y_0 + \int_{x_0}^{x} \Phi^{-1}(u)\xi(u)\,du + \int_{x_0}^{x} \Phi^{-1}(u)\sigma(u)\,dw(u) \right), \tag{19}$$

where

$$\Phi(x) = \exp\left\{ \int_{x_0}^{x} \beta(u)\,du \right\}.$$

Conditionally upon y_0, $y(x)$ is a Gaussian process with independent increments. It follows from (19) that

$$E[y(x)|y_0] = \Phi(x)\left(y_0 + \int_{x_0}^{x} \Phi^{-1}(u)\xi(u)\,du \right) \tag{20}$$

and

$$\operatorname{cov}[y(x_1), y(x_2)|y_0] = \Phi(x_1)\Phi(x_2) \int_{x_0}^{\min(x_1, x_2)} [\Phi^{-1}(u)\sigma(u)]^2\,du. \tag{21}$$

Equations (20) and (21) readily generalize to the multivariate case of vector $\mathbf{y}(x)$, as in compartmental modeling for example. The corresponding equations are given by Arnold [1974].

For the homogeneous-coefficients case, i.e. $\beta(x) \equiv \beta$ and $\xi(x) \equiv \xi$, the above reduce to

$$E[y(x)|y_0] = y_0 e^{\beta(x - x_0)} + \frac{\xi}{\beta}(e^{\beta(x - x_0)} - 1) \tag{22}$$

and

$$\operatorname{cov}[y(x_1), y(x_2)] = \int_{x_0}^{\min(x_1, x_2)} e^{\beta(x_1 + x_2 - 2u)}\sigma^2(u)\,du. \tag{23}$$

APPENDIX D

Multiple Linear Regression

We summarize the following theory for convenient reference. Proofs are given in Seber [1977].

D1 ESTIMATION

1. Let $\mathbf{y} = \mathbf{X}\boldsymbol{\beta} + \boldsymbol{\varepsilon}$, where $\boldsymbol{\varepsilon} \sim N_n(\mathbf{0}, \sigma^2\mathbf{I}_n)$, $\mathbf{X}$ is $n \times p$ of rank p, and $\boldsymbol{\beta}$ plays the same role as $\boldsymbol{\theta}$ in nonlinear models. The least-squares estimate $\hat{\boldsymbol{\beta}}$, which minimizes $S(\boldsymbol{\beta}) = \|\mathbf{y} - \mathbf{X}\boldsymbol{\beta}\|^2$, is the solution of the normal equations $\mathbf{X}'\mathbf{X}\hat{\boldsymbol{\beta}} = \mathbf{X}'\mathbf{y}$, namely $\hat{\boldsymbol{\beta}} = (\mathbf{X}'\mathbf{X})^{-1}\mathbf{X}'\mathbf{y}$. Then

$$\hat{\boldsymbol{\beta}} \sim N_p(\boldsymbol{\beta}, \sigma^2(\mathbf{X}'\mathbf{X})^{-1}).$$

2. An unbiased estimate of σ^2 is $s^2 = S(\hat{\boldsymbol{\beta}})/(n-p)$. Since $\mathbf{X}\hat{\boldsymbol{\beta}} = \mathbf{X}(\mathbf{X}'\mathbf{X})^{-1}\mathbf{X}'\mathbf{y} = \mathbf{P}_\mathbf{X}\mathbf{y}$, where $\mathbf{P}_\mathbf{X}$ is a projection matrix (Appendix A11), we have

$$S(\hat{\boldsymbol{\beta}}) = \|\mathbf{y} - \mathbf{X}\hat{\boldsymbol{\beta}}\|^2 = \mathbf{y}'(\mathbf{I}_n - \mathbf{P}_\mathbf{X})\mathbf{y} = \mathbf{y}'\mathbf{R}_\mathbf{X}\mathbf{y},$$

say. Also $\hat{\boldsymbol{\beta}}$ is statistically independent of $S(\hat{\boldsymbol{\beta}})$, and $S(\hat{\boldsymbol{\beta}})/\sigma^2 \sim \chi^2_{n-p}$.

3. The residual vector for the linear model is

$$\hat{\boldsymbol{\varepsilon}} = \mathbf{y} - \mathbf{X}\hat{\boldsymbol{\beta}} = (\mathbf{I}_n - \mathbf{P}_\mathbf{X})\mathbf{y} = (\mathbf{I}_n - \mathbf{P}_\mathbf{X})\boldsymbol{\varepsilon},$$

as $\mathbf{P}_\mathbf{X}\mathbf{X} = \mathbf{X}$. From the normal equations we have

$$\mathbf{X}'\hat{\boldsymbol{\varepsilon}} = \mathbf{0}.$$

Also $\mathscr{E}[\hat{\boldsymbol{\varepsilon}}] = \mathbf{0}$ and $\mathscr{D}[\hat{\boldsymbol{\varepsilon}}] = \sigma^2(\mathbf{I}_n - \mathbf{P}_\mathbf{X})$. In Chapters 13 and 14 we use $\mathbf{r}$ instead of $\hat{\boldsymbol{\varepsilon}}$ for notational convenience.

4. From the normal equations it can be shown that

$$\begin{aligned} S(\boldsymbol{\beta}) &= \|\mathbf{y} - \mathbf{X}\hat{\boldsymbol{\beta}} + \mathbf{X}(\hat{\boldsymbol{\beta}} - \boldsymbol{\beta})\|^2 \\ &= S(\hat{\boldsymbol{\beta}}) + (\hat{\boldsymbol{\beta}} - \boldsymbol{\beta})'\mathbf{X}'\mathbf{X}(\hat{\boldsymbol{\beta}} - \boldsymbol{\beta}). \end{aligned}$$

5. When $\mathbf{X}$ does not have full rank p, the normal equations do not have a unique solution. However, the solution with minimum norm, i.e. that which minimizes $\|\hat{\boldsymbol{\beta}}\|$, is $\hat{\boldsymbol{\beta}} = \mathbf{X}^{+}\mathbf{y}$, where $\mathbf{X}^{+}$ is the Moore–Penrose generalized inverse of $\mathbf{X}$ (c.f. Appendix A5). For further details see Seber [1977: 76–78].

D2 INFERENCE

1. From Appendix D1.1 we have $(\hat{\boldsymbol{\beta}} - \boldsymbol{\beta})'\mathbf{X}'\mathbf{X}(\hat{\boldsymbol{\beta}} - \boldsymbol{\beta})/\sigma^2 \sim \chi_p^2$, and from D1.2 it follows that

$$\begin{aligned} F &= (\hat{\boldsymbol{\beta}} - \boldsymbol{\beta})'\mathbf{X}'\mathbf{X}(\hat{\boldsymbol{\beta}} - \boldsymbol{\beta})/ps^2 \\ &= \frac{(\hat{\boldsymbol{\beta}} - \boldsymbol{\beta})'\mathbf{X}'\mathbf{X}(\hat{\boldsymbol{\beta}} - \boldsymbol{\beta})}{S(\hat{\boldsymbol{\beta}})}\frac{(n-p)}{p} \\ &= \frac{(\hat{\boldsymbol{\beta}} - \boldsymbol{\beta})'\mathbf{X}'\mathbf{X}(\hat{\boldsymbol{\beta}} - \boldsymbol{\beta})}{\mathbf{y}'\mathbf{y} - \hat{\boldsymbol{\beta}}'\mathbf{X}'\mathbf{X}\hat{\boldsymbol{\beta}}}\frac{(n-p)}{p} \\ &\sim F_{p,n-p}. \end{aligned}$$

2. A $100(1-\alpha)\%$ confidence region for $\boldsymbol{\beta}$ is given by

$$\{\boldsymbol{\beta} : (\boldsymbol{\beta} - \hat{\boldsymbol{\beta}})'\mathbf{X}'\mathbf{X}(\boldsymbol{\beta} - \hat{\boldsymbol{\beta}}) \leq ps^2 F_{p,n-p}^{\alpha}\}.$$

3. To test $H_0 : \boldsymbol{\beta} = \mathbf{0}$ we use

$$\begin{aligned} F &= \hat{\boldsymbol{\beta}}'\mathbf{X}'\mathbf{X}\hat{\boldsymbol{\beta}}/ps^2 \\ &= \frac{\mathbf{y}'\mathbf{P}_{\mathbf{X}}\mathbf{y}}{\mathbf{y}'(\mathbf{I}_n - \mathbf{P}_{\mathbf{X}})\mathbf{y}}\frac{n-p}{p} \qquad \text{(by Appendix D1.2)} \end{aligned}$$

which is distributed as $F_{p,n-p}$ when H_0 is true.

4. To test $H_0 : \mathbf{A}\boldsymbol{\beta} = \mathbf{c}$, where $\mathbf{A}$ is $q \times p$ of rank q, we find $\tilde{\boldsymbol{\beta}}_H$ such that

$$S(\tilde{\boldsymbol{\beta}}_H) = \min_{H_0} \|\mathbf{y} - \mathbf{X}\boldsymbol{\beta}\|^2.$$

This is given by

$$\tilde{\boldsymbol{\beta}}_H = \hat{\boldsymbol{\beta}} + (\mathbf{X}'\mathbf{X})^{-1}\mathbf{A}'[\mathbf{A}(\mathbf{X}'\mathbf{X})^{-1}\mathbf{A}']^{-1}(\mathbf{c} - \mathbf{A}\hat{\boldsymbol{\beta}}).$$

Also

$$S(\tilde{\boldsymbol{\beta}}_H) - S(\hat{\boldsymbol{\beta}}) = (\mathbf{A}\hat{\boldsymbol{\beta}} - \mathbf{c})'[\mathbf{A}(\mathbf{X}'\mathbf{X})^{-1}\mathbf{A}'](\mathbf{A}\hat{\boldsymbol{\beta}} - \mathbf{c}),$$

and the test statistic for H_0 is

$$\begin{aligned} F &= \frac{S(\tilde{\boldsymbol\beta}_H) - S(\hat{\boldsymbol\beta})}{S(\hat{\boldsymbol\beta})} \frac{(n-p)}{q} \\ &= \frac{S(\tilde{\boldsymbol\beta}_H) - S(\hat{\boldsymbol\beta})}{qs^2} \\ &= \frac{(\mathbf{A}\hat{\boldsymbol\beta} - \mathbf{c})'[\mathbf{A}(\mathbf{X}'\mathbf{X})^{-1}\mathbf{A}']^{-1}(\mathbf{A}\hat{\boldsymbol\beta} - \mathbf{c})}{qs^2}, \end{aligned}$$

which is distributed as $F_{q,n-p}$ when H_0 is true. This test is also equivalent to the likelihood-ratio test.

D3 PARAMETER SUBSETS

1. Suppose

$$\mathbf{X}\boldsymbol\beta = (\mathbf{X}_1, \mathbf{X}_2)\begin{pmatrix} \boldsymbol\beta_1 \\ \boldsymbol\beta_2 \end{pmatrix} = \mathbf{X}_1\boldsymbol\beta_1 + \mathbf{X}_2\boldsymbol\beta_2,$$

where $\boldsymbol\beta_i$ is $p_i \times 1$ $(p_1 + p_2 = p)$. We wish to find the distribution of $\hat{\boldsymbol\beta}_2$. Now

$$\begin{aligned} \mathscr{D}\begin{pmatrix} \hat{\boldsymbol\beta}_1 \\ \hat{\boldsymbol\beta}_2 \end{pmatrix} &= \sigma^2(\mathbf{X}'\mathbf{X})^{-1} \\ &= \sigma^2\begin{pmatrix} \mathbf{X}_1'\mathbf{X}_1 & \mathbf{X}_1'\mathbf{X}_2 \\ \mathbf{X}_2'\mathbf{X}_1 & \mathbf{X}_2'\mathbf{X}_2 \end{pmatrix}^{-1}. \end{aligned}$$

Hence, from (1) in Appendix A3.1,

$$\begin{aligned} \mathscr{D}[\hat{\boldsymbol\beta}_2] &= \sigma^2\mathbf{B}_{22}^{-1} \\ &= \sigma^2[\mathbf{X}_2'\mathbf{X}_2 - \mathbf{X}_2'\mathbf{X}_1(\mathbf{X}_1'\mathbf{X}_1)^{-1}\mathbf{X}_1'\mathbf{X}_2]^{-1} \\ &= \sigma^2[\mathbf{X}_2'(\mathbf{I}_n - \mathbf{P}_{\mathbf{X}_1})\mathbf{X}_2]^{-1}, \end{aligned} \tag{1}$$

where $\mathbf{P}_{\mathbf{X}_1} = \mathbf{X}_1(\mathbf{X}_1'\mathbf{X}_1)^{-1}\mathbf{X}_1'$ represents the orthogonal projection onto $\mathscr{R}[\mathbf{X}_1]$ (by Appendix A11). Furthermore

$$\begin{aligned} \hat{\boldsymbol\beta} &= (\mathbf{X}'\mathbf{X})^{-1}\mathbf{X}'\mathbf{y} \\ &= \begin{pmatrix} \mathbf{X}_1'\mathbf{X}_1 & \mathbf{X}_1'\mathbf{X}_2 \\ \mathbf{X}_2'\mathbf{X}_1 & \mathbf{X}_2'\mathbf{X}_2 \end{pmatrix}^{-1}\begin{pmatrix} \mathbf{X}_1'\mathbf{y} \\ \mathbf{X}_2'\mathbf{y} \end{pmatrix}, \end{aligned}$$

so that from (1) in Appendix A3.1

$$\begin{aligned}\hat{\boldsymbol{\beta}}_2 &= -\mathbf{B}_{22}^{-1}\mathbf{B}_{21}\mathbf{X}_1'\mathbf{y} + \mathbf{B}_{22}^{-1}\mathbf{X}_2'\mathbf{y} \\ &= [\mathbf{X}_2'(\mathbf{I}_n - \mathbf{P}_{\mathbf{X}_1})\mathbf{X}_2]^{-1}\mathbf{X}_2'(\mathbf{I}_n - \mathbf{P}_{\mathbf{X}_1})\mathbf{y}. \qquad (2)\end{aligned}$$

We can, of course, obtain (1) directly from $\hat{\boldsymbol{\beta}}_2$. Thus

$$\hat{\boldsymbol{\beta}}_2 \sim N_{p_2}(\boldsymbol{\beta}_2, \sigma^2\mathbf{B}_{22}^{-1})$$

and

$$(\hat{\boldsymbol{\beta}}_2 - \boldsymbol{\beta}_2)'\mathbf{B}_{22}(\hat{\boldsymbol{\beta}}_2 - \boldsymbol{\beta}_2)/\sigma^2 \sim \chi^2_{p_2}, \qquad (3)$$

where

$$\mathbf{B}_{22} = \mathbf{X}_2'(\mathbf{I}_n - \mathbf{P}_{\mathbf{X}_1})\mathbf{X}_2.$$

2. To test $\mathbf{H}_0\colon \boldsymbol{\beta}_2 = \boldsymbol{\beta}_{02}$, we use Appendix D2.4 with $\mathbf{A} = (\mathbf{0}, \mathbf{I}_{p_2})$ and $\mathbf{c} = \boldsymbol{\beta}_{02}$ to get

$$F = (\hat{\boldsymbol{\beta}}_2 - \boldsymbol{\beta}_{02})'\mathbf{B}_{22}(\hat{\boldsymbol{\beta}}_2 - \boldsymbol{\beta}_{02})/p_2 s^2,$$

which is distributed as $F_{p_2, n-p}$ when H_0 is true.

3. A $100(1-\alpha)\%$ confidence region for $\boldsymbol{\beta}_2$ is given by

$$\{\boldsymbol{\beta}_2 : (\boldsymbol{\beta}_2 - \hat{\boldsymbol{\beta}}_2)'\mathbf{X}_2'[\mathbf{I}_n - \mathbf{P}_{\mathbf{X}_1}]\mathbf{X}_2(\boldsymbol{\beta}_2 - \hat{\boldsymbol{\beta}}_2) \le p_2 s^2 F^{\alpha}_{p_2, n-p}\}$$

or

$$\{\boldsymbol{\beta}_2 : \| (\mathbf{I}_n - \mathbf{P}_{\mathbf{X}_1})\mathbf{X}_2(\boldsymbol{\beta}_2 - \hat{\boldsymbol{\beta}}_2) \|^2 \le p_2 s^2 F^{\alpha}_{p_2, n-p}\}.$$

This confidence region can also be obtained by applying Appendix D1.1 to the linear model

$$\mathbf{y} = (\mathbf{I}_n - \mathbf{P}_{\mathbf{X}_1})\mathbf{X}_2\boldsymbol{\beta}_2 + \boldsymbol{\varepsilon} = \mathbf{Z}_2\boldsymbol{\beta}_2 + \boldsymbol{\varepsilon}.$$

This is verified by showing that $\hat{\boldsymbol{\beta}}_2$ of (2) is also given by

$$\hat{\boldsymbol{\beta}}_2 = (\mathbf{Z}_2'\mathbf{Z}_2)^{-1}\mathbf{Z}_2'\mathbf{y}.$$

4. Using Appendix D2.4, we can express the test statistic of D3.2 for testing $H_0\colon \boldsymbol{\beta}_2 = \mathbf{0}$ in the form

$$\begin{aligned}F &= \frac{S(\tilde{\boldsymbol{\beta}}_H) - S(\hat{\boldsymbol{\beta}})}{S(\hat{\boldsymbol{\beta}})} \frac{n-p}{p_2} \\ &= \frac{S(\tilde{\boldsymbol{\beta}}_H) - S(\hat{\boldsymbol{\beta}})}{S(\tilde{\boldsymbol{\beta}}_H) - [S(\tilde{\boldsymbol{\beta}}_H) - S(\hat{\boldsymbol{\beta}})]} \frac{(n-p)}{p_2},\end{aligned}$$

where

$$\begin{aligned} S(\tilde{\boldsymbol{\beta}}_H) &= \min_{\boldsymbol{\beta}_1} \| \mathbf{y} - \mathbf{X}_1 \boldsymbol{\beta}_1 \|^2 \\ &= \mathbf{y}'(\mathbf{I}_n - \mathbf{P}_{\mathbf{X}_1})\mathbf{y} \\ &= \mathbf{y}'\mathbf{R}\mathbf{y}, \end{aligned}$$

say, and (from Appendix D3.2)

$$S(\tilde{\boldsymbol{\beta}}_H) - S(\hat{\boldsymbol{\beta}}) = \hat{\boldsymbol{\beta}}_2' \mathbf{X}_2' \mathbf{R} \mathbf{X}_2 \hat{\boldsymbol{\beta}}_2.$$

Thus we have the following representation for F:

$$F = \frac{\hat{\boldsymbol{\beta}}_2 \mathbf{X}_2' \mathbf{R} \mathbf{X}_2 \hat{\boldsymbol{\beta}}_2}{\mathbf{y}'\mathbf{R}\mathbf{y} - \hat{\boldsymbol{\beta}}_2' \mathbf{X}_2' \mathbf{R} \mathbf{X}_2 \hat{\boldsymbol{\beta}}_2} \frac{n-p}{p_2}.$$

APPENDIX E

Minimization Subject to Linear Constraints

Suppose we wish to minimize $h(\boldsymbol{\theta})$ subject to the linear constraints $\mathbf{A}\boldsymbol{\theta} = \mathbf{c}$. Then it is readily established that

$$\{\boldsymbol{\theta} : \mathbf{A}\boldsymbol{\theta} = \mathbf{c}\} = \{\boldsymbol{\theta} : \boldsymbol{\theta} = \boldsymbol{\theta}_c + \mathbf{u}, \mathbf{u} \in \mathcal{N}[\mathbf{A}]\}, \tag{1}$$

where $\boldsymbol{\theta}_c$ is any particular solution of $\mathbf{A}\boldsymbol{\theta} = \mathbf{c}$. As we see later, we can always obtain a $p \times q$ matrix $\mathbf{S}$ and a $p \times (p - q)$ matrix $\mathbf{Z}$ such that $(\mathbf{S}, \mathbf{Z})$ is nonsingular, $\mathbf{AS} = \mathbf{I}_q$, and $\mathbf{AZ} = \mathbf{0}$. Now $\mathbf{A}(\mathbf{Sc}) = \mathbf{c}$ and, as $\mathbf{AZ} = \mathbf{0}$ and $\mathbf{Z}$ has rank $p - q$, the columns of $\mathbf{Z}$ span $\mathcal{N}[\mathbf{A}]$. Thus, using (1),

$$\{\boldsymbol{\theta} : \mathbf{A}\boldsymbol{\theta} = \mathbf{c}\} = \{\boldsymbol{\theta} : \boldsymbol{\theta} = \boldsymbol{\theta}(\boldsymbol{\phi}) = \mathbf{Sc} + \mathbf{Z}\boldsymbol{\phi}, \boldsymbol{\phi} \in \mathbb{R}^{p-q}\}.$$

By reparametrizing $h(\boldsymbol{\theta})$ is in terms of $\boldsymbol{\phi}$, the constrained problem is converted into an unconstrained optimization problem of minimizing $k(\boldsymbol{\phi})$, where

$$k(\boldsymbol{\phi}) = h(\mathbf{Sc} + \mathbf{Z}\boldsymbol{\phi}).$$

Now, by Appendix A10.4,

$$\frac{\partial k(\boldsymbol{\phi})}{\partial \boldsymbol{\phi}} = \left[\mathbf{Z}' \frac{\partial h(\boldsymbol{\theta})}{\partial \boldsymbol{\theta}} \right]_{\boldsymbol{\theta}(\boldsymbol{\phi})} = \mathbf{Z}'\mathbf{g}(\mathbf{Sc} + \mathbf{Z}\boldsymbol{\phi})$$

and

$$\frac{\partial^2 k(\boldsymbol{\phi})}{\partial \boldsymbol{\phi}\, \partial \boldsymbol{\phi}'} = \left[\mathbf{Z}' \frac{\partial^2 h}{\partial \boldsymbol{\theta}\, \partial \boldsymbol{\theta}'} \mathbf{Z} \right]_{\boldsymbol{\theta}(\boldsymbol{\phi})} = \mathbf{Z}'\mathbf{H}(\mathbf{Sc} + \mathbf{Z}\boldsymbol{\phi})\mathbf{Z},$$

where $\mathbf{g}$ and $\mathbf{H}$ are the gradient and Hessian of h. Let $\mathbf{g}_k$ and $\mathbf{H}_k$ be the gradient and Hessian of $k(\boldsymbol{\phi})$. Then the unconstrained problem of minimizing $k(\boldsymbol{\phi})$ can be solved by supplying as much of the following information as necessary to an ordinary program for unconstrained optimization:

For any $\boldsymbol{\phi}$ calculate $\boldsymbol{\theta} = \mathbf{Sc} + \mathbf{Z}\boldsymbol{\phi}$. Then

1. $k(\boldsymbol{\phi}) = h(\boldsymbol{\theta})$.
2. $\mathbf{g}_k(\boldsymbol{\phi}) = \mathbf{Z}'\mathbf{g}(\boldsymbol{\theta})$.
3. $\mathbf{H}_k(\boldsymbol{\phi}) = \mathbf{Z}'\mathbf{H}(\boldsymbol{\theta})\mathbf{Z}$.

If Newton's method (Section 13.3) is used for minimizing $k(\boldsymbol{\phi})$, then the iterations are given by

$$\begin{aligned}\boldsymbol{\phi}^{(a+1)} &= \boldsymbol{\phi}^{(a)} - [\mathbf{H}_k(\boldsymbol{\phi}^{(a)})]^{-1}\mathbf{g}_k(\boldsymbol{\phi}^{(a)}) \\ &= \boldsymbol{\phi}^{(a)} - [\mathbf{Z}'\mathbf{H}(\boldsymbol{\theta}^{(a)})\mathbf{Z}]^{-1}\mathbf{Z}'\mathbf{g}(\boldsymbol{\theta}^{(a)}),\end{aligned}$$

where

$$\boldsymbol{\theta}^{(a)} = \mathbf{Sc} + \mathbf{Z}\boldsymbol{\phi}^{(a)}.$$

For the Gauss–Newton algorithm (Section 14.1) we have

$$\boldsymbol{\phi}^{(a+1)} = \boldsymbol{\phi}^{(a)} - [\mathbf{Z}'\mathbf{J}'(\boldsymbol{\theta}^{(a)})\mathbf{J}(\boldsymbol{\theta}^{(a)})\mathbf{Z}]^{-1}\mathbf{Z}'\mathbf{J}'(\boldsymbol{\theta}^{(a)})[\mathbf{y} - \mathbf{f}(\boldsymbol{\theta}^{(a)})].$$

If $h(\boldsymbol{\theta})$ is a negative log-likelihood function, then the inverse of the Hessian or expected Hessian of $k(\boldsymbol{\phi})$ at the minimum gives the asymptotic variance–covariance matrix of the maximum-likelihood estimator $\hat{\boldsymbol{\phi}}$ of $\boldsymbol{\phi}$. Using

$$\mathscr{D}[\tilde{\boldsymbol{\theta}}] = \mathscr{D}[\mathbf{Sc} + \mathbf{Z}\hat{\boldsymbol{\phi}}] = \mathbf{Z}\mathscr{D}[\hat{\boldsymbol{\phi}}]\mathbf{Z}',$$

the estimated variance–covariance matrix of $\tilde{\boldsymbol{\theta}}$, the constrained maximum likelihood estimator of $\boldsymbol{\theta}$, can be recovered, e.g.

$$\hat{\mathscr{D}}[\tilde{\boldsymbol{\theta}}] = \mathbf{Z}[\mathbf{Z}'\mathbf{H}(\tilde{\boldsymbol{\theta}})\mathbf{Z}]^{-1}\mathbf{Z}'.$$

In the case of least-squares estimation we can set $h(\boldsymbol{\theta}) = S(\boldsymbol{\theta})/2\sigma^2 = \|\mathbf{y} - \mathbf{f}(\boldsymbol{\theta})\|^2/2\sigma^2$. Then $\mathscr{E}[\mathbf{H}(\boldsymbol{\theta})] = \mathbf{F}_{.}'\mathbf{F}_{.}'/\sigma^2$, where $\mathbf{F}_{.} = \partial\mathbf{f}/\partial\boldsymbol{\theta}'$, and we can use

$$\hat{\mathscr{D}}[\tilde{\boldsymbol{\theta}}] = s^2\mathbf{Z}\{\mathbf{Z}'\mathbf{F}_{.}'(\tilde{\boldsymbol{\theta}})\mathbf{F}_{.}(\tilde{\boldsymbol{\theta}})\mathbf{Z}\}^{-1}\mathbf{Z}'.$$

The main advantage of the above approach is that it can be used for testing whether certain submodels of an umbrella model $\mathscr{E}[\mathbf{y}] = \mathbf{f}(\boldsymbol{\theta})$, each specified by an appropriate $\mathbf{A}$ and $\mathbf{c}$, are true. Instead of having to provide a separate code for each submodel, the same piece of code can be used in all cases. All that is required is the specification of $\mathbf{A}$ and $\mathbf{c}$ from which $\mathbf{Z}$ is computed. If $\mathbf{c} = \mathbf{0}$ then $\mathbf{S}$ is unnecessary. For an unconstrained fit, $q = 0$ and we take $\mathbf{Z} = \mathbf{I}_p$.

If an unconstrained model or a model with fewer constraints has already been fitted, and has minimum $\breve{\boldsymbol{\theta}}$, a starting value of $\boldsymbol{\phi}$ can be obtained by "projecting" $\breve{\boldsymbol{\theta}}$ onto $\{\boldsymbol{\theta}:\mathbf{A}\boldsymbol{\theta}=\mathbf{c}\}$. The quadratic approximation (2.33) to $h(\boldsymbol{\theta})$ in the vicinity of $\breve{\boldsymbol{\theta}}$ is

$$q_h(\boldsymbol{\theta}) = q(\breve{\boldsymbol{\theta}}) + \mathbf{g}'(\breve{\boldsymbol{\theta}})(\boldsymbol{\theta}-\breve{\boldsymbol{\theta}}) + \tfrac{1}{2}(\boldsymbol{\theta}-\breve{\boldsymbol{\theta}})'\mathbf{H}(\breve{\boldsymbol{\theta}})(\boldsymbol{\theta}-\breve{\boldsymbol{\theta}}).$$

A natural choice of starting value for the constrained problem is the value which minimizes q_h subject to $\mathbf{A}\boldsymbol{\theta}=\mathbf{c}$, or equivalently, subject to $\boldsymbol{\theta}=\mathbf{Sc}+\mathbf{Z}\boldsymbol{\phi}$. In the $\boldsymbol{\phi}$-parametrization we have

$$q_h(\boldsymbol{\phi}) = q(\breve{\boldsymbol{\theta}}) + \mathbf{g}'(\breve{\boldsymbol{\theta}})(\mathbf{Sc}+\mathbf{Z}\boldsymbol{\phi}-\breve{\boldsymbol{\theta}}) + \tfrac{1}{2}(\mathbf{Sc}+\mathbf{Z}\boldsymbol{\phi}-\breve{\boldsymbol{\theta}})'\mathbf{H}(\breve{\boldsymbol{\theta}})(\mathbf{Sc}+\mathbf{Z}\boldsymbol{\phi}-\breve{\boldsymbol{\theta}}),$$

which is minimized with respect to $\boldsymbol{\phi}$ by taking

$$\boldsymbol{\phi} = [\mathbf{Z}'\mathbf{H}(\breve{\boldsymbol{\theta}})\mathbf{Z}]^{-1}\mathbf{Z}'[\mathbf{H}(\breve{\boldsymbol{\theta}})(\breve{\boldsymbol{\theta}}-\mathbf{Sc})-\mathbf{g}(\breve{\boldsymbol{\theta}})].$$

The corresponding value of $\boldsymbol{\theta}$ is

$$\boldsymbol{\theta} = \mathbf{Sc} + \mathbf{Z}[\mathbf{Z}'\mathbf{H}(\breve{\boldsymbol{\theta}})\mathbf{Z}]^{-1}\mathbf{Z}'[\mathbf{H}(\breve{\boldsymbol{\theta}})(\breve{\boldsymbol{\theta}}-\mathbf{Sc})-\mathbf{g}(\breve{\boldsymbol{\theta}})].$$

We note that $\mathbf{g}(\breve{\boldsymbol{\theta}})=\mathbf{0}$ if $\breve{\boldsymbol{\theta}}$ is an unconstrained minimum.

In conclusion we discuss the choice of $\mathbf{S}$ and $\mathbf{Z}$. Clearly any $(\mathbf{S},\mathbf{Z})$ of full rank with $\mathbf{AS}=\mathbf{I}_q$ and $\mathbf{AZ}=\mathbf{0}$ can be used. A particularly useful choice can be obtained from a *QR* factorization of the $p\times q$ matrix $(p\geq q)$ $\mathbf{A}'$ (see Appendix A8.3). Let

$$\mathbf{A}' = (\mathbf{Q}_1,\mathbf{Q}_2)\begin{pmatrix}\mathbf{R}\\ \mathbf{0}\end{pmatrix} = \mathbf{Q}_1\mathbf{R},$$

where $\mathbf{Q}=(\mathbf{Q}_1,\mathbf{Q}_2)$ is orthogonal, $\mathbf{Q}_1$ has q columns, $\mathbf{Q}_2$ has $p-q$ columns, and $\mathbf{R}$ is a $q\times q$ upper triangular matrix. Then we can use

$$\mathbf{S}=\mathbf{Q}_1(\mathbf{R}^{-1})' \quad \text{and} \quad \mathbf{Z}=\mathbf{Q}_2.$$

For further information about this and other choices of $(\mathbf{S},\mathbf{Z})$, the reader is referred to Fletcher [1981: Sections 10.1 and 11.1].

References

Abraham, B. and Ledolter, J. (1983). *Statistical Methods for Forecasting*. Wiley: New York.

Agarwal, G. G., and Studden, W. J. (1978). Asymptotic design and estimation using linear splines. *Commun. Stat. Simul. Comput. B*, *7*, 309–319.

Agarwal, G. G., and Studden, W. J. (1980). Asymptotic integrated mean square error using least squares and bias as minimizing splines. *Ann. Stat.*, *8*, 1307–1325.

Agha, M. (1971). A direct method for fitting linear combinations of exponentials. *Biometrics*, *27*, 399–413.

Agrafiotis, G. K. (1982). On the stochastic theory of compartments: a semi-Markov approach for the analysis of the *k*-compartmental systems. *Bull. Math. Biol.*, *44*, 809–817.

Aguirre-Torres, V., and Gallant, A. R. (1983). The null and non-null asymptotic distribution of the Cox test for multivariate nonlinear regression. Alternatives and a new distribution-free Cox test. *J. Econometrics*, *21*, 5–33.

Allen, D. M. (1983). Parameter estimation for nonlinear models with emphasis on compartmental models. *Biometrics*, *39*, 629–637.

Amari, S. (1982). Differential geometry of curved exponential families—curvatures and information loss. *Ann. Stat.*, *10*, 357–385.

Amari, S. (1985). *Differential-Geometrical Methods in Statistics*. Lecture Notes in Statistics No. 28, 290 pp. Springer-Verlag: Berlin.

Amemiya, T. (1983). Non-linear regression models. *In* Z. Griliches and M. D. Intriligator (Eds.), *Handbook of Econometrics*, Vol. I, pp. 333–389. North-Holland: Amsterdam.

Amemiya, T., and Powell, J. L. (1981). A comparison of the Box–Cox maximum likelihood estimator and the non-linear two-stage least squares estimator. *J. Econometrics*, *17*, 351–381.

Amemiya, Y., and Fuller, W. A. (1984). Estimation for the multivariate errors-in-variables model with estimated error covariance matrix. *Ann. Stat.*, *12*, 497–509.

American Statistical Assoc. (1982). *Software for Nonlinear models, Proceedings of the Statistical Computing Section*, pp. 1–20 (papers by D. M. Allen, J. Sall, and C. M. Metzler).

Anderson, D. H. (1983). *Compartmental Modelling and Tracer Kinetics*. Lecture Notes in Biomathematics; Springer-Verlag: New York.

Angell, I. O., and Godolphin, E. J. (1978). Implementation of the direct representation for the maximum likelihood estimator of a Gaussian moving average process. *J. Stat. Comput. Simul.*, *8*, 145–160.

Ansley, C. F. (1978). Subroutine ARMA: Exact likelihood for univariate ARMA processes. Technical Report, Graduate School of Business, University of Chicago.

Ansley, C. F. (1979). An algorithm for the exact likelihood of a mixed autoregressive–moving average process. *Biometrika*, *66*, 59–65.

Ansley, C. F., and Newbold, P. (1979). On the finite sample distribution of residual autocorrelations in autoregressive–moving average models. *Biometrika*, *66*, 547–553.

Arnold, L. (1974). *Stochastic Differential Equations: Theory and Applications*. Wiley: New York.

Atkins, G. L. (1969). *Multicompartmental Models for Biological Systems*. Methuen: London.

Atkinson, A. C. (1970). A method for discriminating between models. *J. R. Stat. Soc. B*, *32*, 323–344.

Atkinson, A. C. (1972). Planning experiments to detect inadequate regression models. *Biometrika*, *59*, 275–293.

Atkinson, A. C. (1981). A comparison of two criteria for the design of experiments for discriminating between models. *Technometrics*, *23*, 301–305.

Atkinson, A. C. (1982). Developments in the design of experiments. *Internat. Stat. Rev.*, *50*, 161–177.

Atkinson, A. C., and Cox, D. R. (1974). Planning experiments for discriminating between models. *J. R. Stat. Soc. B*, *36*, 321–348.

Atkinson, A. C., and Fedorov, V. V. (1975a). The design of experiments for discriminating between two rival models. *Biometrika*, *62*, 57–70.

Atkinson, A. C., and Fedorov, V. V. (1975b). Optimal design: Experiments for discriminating between several models. *Biometrika*, *62*, 289–303.

Atkinson, A. C., and Hunter, W. G. (1968). The design of experiments for parameter estimation. *Technometrics*, *10*, 271–289.

Bacon, D. W., and Watts, D. G. (1971). Estimating the transition between two intersecting straight lines. *Biometrika*, *58*, 525–534.

Bailey, B. J. R. (1977). Tables of the Bonferroni *t* statistic. *J. Am. Stat. Assoc.*, *72*, 469–478.

Baker, R. J., and Nelder, J. A. (1978). *The GLIM System, Release 3*. Numerical Algorithms Group: Oxford.

Bard, Y. (1974). *Nonlinear Parameter Estimation*. Academic Press: New York.

Barham, R. H., and Drane, W. (1972). An algorithm for least squares estimation of nonlinear parameters when some of the parameters are linear. *Technometrics*, *14*, 757–766.

Barnett, V. D. (1970). Fitting straight lines—the linear functional relationship. *Appl. Stat.*, *19*, 135–144.

Barnett, W. A. (1976). Maximum likelihood and iterated Aitken estimation of nonlinear systems of equations. *J. Am. Stat. Assoc.*, *71*, 354–360.

Bates, D. M., and Hunter, W. G. (1984). Nonlinear models. *In* S. Kotz and N. L. Johnson (Eds.), *Encyclopedia of Statistical Sciences*, Vol. 5. Wiley: New York.

Bates, D. M., and Watts, D. G. (1980). Relative curvature measures of nonlinearity (with Discussion). *J. R. Stat. Soc. B*, *42*, 1–25.

Bates, D. M., and Watts, D. G. (1981a). Parameter transformations for improved approximate confidence regions in nonlinear least squares. *Ann. Stat.*, *9*, 1152–1167.

Bates, D. M., and Watts, D. G. (1981b). A relative offset orthogonality convergence criterion for nonlinear least squares. *Technometrics*, *123*, 179–183.

Bates, D. M., and Watts, D. G. (1984). A multi-response Gauss–Newton algorithm. *Commun. Stat. Simul. Comput. B*, *13*, 705–715.

Bates, D. M., and Watts, D. G. (1985a). Multiresponse estimation with special application to linear systems of differential equations (with Discussion). *Technometrics*, *27*, 329–360.

Bates, D. M., and Watts, D. G. (1985b). Fitting first order kinetic models quickly and easily. *J. Research Nat. Bureau Standards*, *90*, 433–439.

Bates, D. M., and Watts, D. G. (1987). A generalized Gauss–Newton procedure for multi-response parameter estimation. *SIAM J. Sci. and Stat. Comput.*, *8*, 49–55.

Bates, D. M., Hamilton, D. C., and Watts, D. G. (1983). Calculation of intrinsic and parameter-effects curvatures for nonlinear regression models. *Commun. Stat. Simul. Comput. B*, *12*, 469–477.

Bates, D. M., Wolf, D. A., and Watts, D. G. (1986). Nonlinear least squares and first order kinetics. *In* D. M. Allen (Ed.), *Computer Science and Statistics: The Interface*, pp. 71–81. Elsevier: New York.

Batschelet, E., Brand, L. and Steiner, A. (1979). On the kinetics of lead in the human body. *J. Math. Biol.*, *8*, 15–23.

Bavely, C. A., and Stewart, G. W. (1979). An algorithm for computing reducing subspaces by block diagonalization. *SIAM J. Numer. Anal.*, *16*, 359–367.

Beal, S. L. (1982). Bayesian analysis of nonlinear models. *Biometrics*, *38*, 1089–1092.

Beale, E. M. L. (1960). Confidence regions in non-linear estimation (with Discussion). *J. R. Stat. Soc. B*, *22*, 41–88.

Beaton, A. E., and Tukey, J. W. (1974). The fitting of power series, meaning polynomials, illustrated on hand-spectroscopic data (with Discussion). *Technometrics*, *16*, 147–192.

Beauchamp, J. J., and Cornell, R. G. (1966). Simultaneous nonlinear estimation. *Technometrics*, *8*, 319–326.

Beckman, R. J., and Cook, R. D. (1979). Testing for two-phase regressions. *Technometrics*, *21*, 65–69.

Bellman, R., and Amström, K. J. (1970). On structural identifiability. *Math. Biosci.*, *7*, 329–339.

Berkey, C. S. (1982). Bayesian approach for a nonlinear growth model. *Biometrics*, *38*, 953–961.

Berkson, J. (1950). Are there two regressions? *J. Am. Stat. Assoc.*, *45*, 164–180.

Berman, M., and Weiss, M. F. (1978). *SAAM Manual.* U.S. Department of Health Education and Welfare Publication No. (NIH) 78–180.

Berman, M., Shahn, E., and Weiss, M. F. (1962). The routine fitting of kinetic data to models. *Biophys. J.*, *2*, 275–287.

Berndt, E. R., and Savin, N. E. (1977). Conflict among criteria for testing hypotheses in the multivariate linear regression model. *Econometrica*, *45*, 1263–1278.

Berry, G. (1967). A mathematical model relating plant yield with arrangement for regularly spaced crops. *Biometrics*, *23*, 505–515.

Betts, J. T. (1976). Solving the nonlinear least squares problem: Application of a general method. *J. Opt. Theor. Appl.*, *18*, 469–484.

Bickel, P. J., and Doksum, K. A. (1981). An analysis of transformations revisited. *J. Am. Stat. Assoc.*, *76*, 296–311.

Bickel, P. J., and Ritov, Y. (1987). Efficient estimation in the errors in variables model. *Ann. Stat.*, *15*, 513–540.

Bird, H. A., and Milliken, G. A. (1976). Estimable functions in the nonlinear model. *Commun. Stat. Theor. Methods A*, *6*, 999–1012.

Blancett, A. L., Hall, K. R., and Canfield, F. B. (1970). Isotherms for the He–Ar system at 50, 0 and −50°C up to 700 atm. *Physica*, *47*, 75–91.

Bleasdale, J. K. A., and Nelder, J. A. (1960). Plant population and crop yield. *Nature*, *188*, 342.

Booth, N. B., and Smith, A. F. M. (1982). A Bayesian approach to retrospective identification of change-points. *J. Econometrics*, *19*, 7–22.

Box, G. E. P., and Coutie, G. A. (1956). Application of digital computers in the exploration of functional relationships. *Proc. I.E.E.*, *103*, Part B, Suppl. 1, 100–107.

Box, G. E. P., and Cox, D. R. (1964). The analysis of transformations (with Discussion). *J. R. Stat. Soc. B*, *26*, 211–252.

Box, G. E. P., and Cox, D. R. (1982). An analysis of transformations revisited, rebutted. *J. Am. Stat. Assoc.*, *77*, 209–210.

Box, G. E. P., and Draper, N. R. (1965). The Bayesian estimation of common parameters from several responses. *Biometrika*, *52*, 355–365.

Box, G. E. P., and Hill, W. J. (1967). Discrimination among mechanistic models. *Technometrics*, *9*, 57–71.

Box, G. E. P., and Hill, W. J. (1974). Correcting inhomogeneity of variance with power transformation weighting. *Technometrics*, *16*, 385–389.

Box, G. E. P., and Hunter, W. G. (1965). Sequential design of experiments for nonlinear models. *I.B.M. Scientific Computing Symposium in Statistics*, pp. 113–137.

Box, G. E. P., and Jenkins, G. M. (1976). *Time Series Analysis: Forecasting and Control*, 2nd ed. Holden-Day: San Francisco.

Box, G. E. P., and Lucas, H. L. (1959). Design of experiments in non-linear situations. *Biometrika*, *47*, 77–99.

Box, G. E. P., and Tiao, G. C. (1973). *Bayesian Inference in Statistical Analysis*. Addison-Wesley: Reading, Mass.

Box, G. E. P., Hunter, W. G., MacGregor, J. F., and Erjavec, J. (1973). Some problems associated with the analysis of multiresponse data. *Technometrics*, *15*, 33–51.

Box, M. J. (1968). The occurrence of replications in optimal designs of experiments to estimate parameters in non-linear models. *J. R. Stat. Soc. B*, *30*, 290–302.

Box, M. J. (1969). Planning experiments to test the adequacy of non-linear models. *Appl. Stat.*, *18*, 241–248.

Box, M. J. (1970a). Some experiences with a nonlinear experimental design criterion. *Technometrics*, *12*, 569–589.

Box, M. J. (1970b). Improved parameter estimation. *Technometrics*, *12*, 219–229.

Box, M. J. (1971a). Bias in nonlinear estimation (with Discussion). *J. R. Stat. Soc. B*, *33*, 171–201.

Box, M. J. (1971b). An experimental design criterion for precise estimation of a subset of the parameters in a nonlinear model. *Biometrika, 58,* 149–153.

Box, M. J., and Draper, N. R. (1971). Factorial designs, the $|X'X|$ criterion, and some related matters. *Technometrics, 13,* 731–742.

Box, M. J., and Draper, N. R. (1972). Estimation and design criteria for multiresponse non-linear models with non-homogeneous variance. *Appl. Stat., 21,* 13–24.

Box, M. J., Draper, N. R., and Hunter, W. G. (1970). Missing values in multiresponse, nonlinear model fitting. *Technometrics, 12,* 613–620.

Bradley, E. L. (1973). The equivalence of maximum likelihood and weighted least squares estimates in the exponential family. *J. Am. Stat. Assoc., 68,* 199–200.

Bradley, R. A., and Gart, J. J. (1962). The asymptotic properties of ML estimators when sampling from associated populations. *Biometrika, 49,* 205–214.

Brannigan, M. (1981). An adaptive piecewise polynomial curve fitting procedure for data analysis. *Commun. Stat. Theor. Methods A, 10,* 1823–1848.

Breiman, L., and Meisel, W. S. (1976). General estimates of the intrinsic variability of data in nonlinear regression models. *J. Am. Stat. Assoc., 71,* 301–307.

Brent, R. P. (1973). *Algorithms for Minimization without Derivatives.* Prentice-Hall: Englewood Cliffs, N.J.

Breusch, T. S. (1979). Conflict among criteria for testing hypotheses: Extensions and comments. *Econometrica, 47,* 203–207.

Breusch, T. S., and Pagan, A. R. (1980). The Lagrange multiplier test and its applications to model specification in econometrics. *Rev. Econom. Stud., 47,* 239–253.

Britt, H. I., and Luecke, R. H. (1973). The estimation of parameters in nonlinear implicit models. *Technometrics, 15,* 233–247.

Brodlie, K. W. (1975). A new direction set method for unconstrained minimization without evaluating derivatives. *J. Inst. Math. Appl., 15,* 385–396.

Brodlie, K. W. (1977). Unconstrained minimization. *In* D. A. H. Jacobs (Ed.), *The State of the Art in Numerical Analysis,* pp. 229–268. Academic Press: London.

Broemeling, L. D., and Chin Choy, J. H. (1981). Detecting structural change in linear models. *Commun. Stat. Theor. Methods A, 10,* 2551–2561.

Brooks, M. A. (1974). Exponential regression with correlated observations. *Biometrika, 61,* 109–115.

Brown, G. H. (1978). Calibration with an ultra-structural model. *Appl. Stat., 27,* 47–51.

Brown, K. M., and Dennis, J. E. (1972). Derivative free analogues of the Levenberg–Marquardt and Gauss algorithms for nonlinear least squares approximation. *Numer. Math., 18,* 289–297.

Brown, R. L. (1957). The bivariate structural relation. *Biometrika, 44,* 84–96.

Brown, R. L., Durbin, J., and Evans, J. M. (1975). Techniques for testing the constancy of regression relationships over time (with Discussion). *J. R. Stat. Soc. B, 37,* 149–192.

Broyden, C. G. (1965). A class of methods for solving nonlinear simultaneous equations. *Math. Comp., 19,* 577–593.

Broyden, C. G. (1967). Quasi-Newton methods and their application to function minimization. *Math. Comp., 21,* 368–381.

Broyden, C. G. (1970). The convergence of a class of double-rank minimization algorithms. Parts I and II. *J. Inst. Math. Appl., 6,* 76–90 and 222–231.

Bunke, H. (1980). Parameter estimation in nonlinear regression. *In* P. R. Krishnaiah (Ed.), *Handbook of Statistics*, Vol. 1, pp. 593–615. North-Holland: Amsterdam.

Bunke, H., and Schmidt, W. H. (1980). Asymptotic results on nonlinear approximation of regression functions and weighted least squares. *Math. Operationsforsch. Stat., Ser. Stat.*, *11*, 3–22.

Bunke, H., and Schulze, U. (1985). Approximation of change points in regression models. *In* T. Pukkila and S. Puntanen (Eds.), *Proceedings of the First International Tampere Seminar on Linear Statistical Models and Their Applications*, pp. 161–171. Dept. of Math. Sci./Stat., University of Tampere, Finland.

Bunke, H., Henschke, K., Strüby, R., and Wisotzki, C. (1977). Parameter estimation in nonlinear regression models. *Math. Operationsforsch. Stat., Ser. Stat.*, *8*, 23–40.

Burguete, J. F., Gallant, A. R., and Souza, G. (1982). On unification of the asymptotic theory of nonlinear econometric models (with Discussion). *Econometric Rev.*, *1*, 151–190.

Burridge, J. (1981). A note on maximum likelihood estimation for regression models using grouped data. *J. R. Stat. Soc. B*, *43*, 41–45.

Byron, R. P., and Bera, A. K. (1983). Linearized estimation of nonlinear single equation functions. *Internat. Econom. Rev.*, *24*, 237–248.

Carr, N. L. (1960). Kinetics of catalytic isomerization of *n*-pentane. *Indust. Eng. Chem.*, *52*, 391–396.

Carroll, R. J. (1980). A robust method for testing transformations to achieve approximate normality. *J. R. Stat. Soc. B*, *42*, 71–78.

Carroll, R. J. (1982a). Two examples of transformations when there are possible outliers. *Appl. Stat.*, *31*, 149–152.

Carroll, R. J. (1982b). Test for regression parameters in power transformation models. *Scand. J. Stat.*, *9*, 217–222.

Carroll, R. J. (1982c). Prediction and power transformations when the choice of power is restricted to a finite set. *J. Am. Stat. Assoc.*, *77*, 908–915.

Carroll, R. J., and Ruppert, D. (1981a). On prediction and the power transformation family. *Biometrika*, *68*, 609–615.

Carroll, R. J., and Ruppert, D. (1981b). On robust tests for heteroscedasticity. *Ann. Stat.*, *9*, 206–210.

Carroll, R. J., and Ruppert, D. (1982a). Robust estimation in heteroscedastic linear models. *Ann. Stat.*, *10*, 429–441.

Carroll, R. J., and Ruppert, D. (1982b). A comparison between maximum likelihood and generalized least squares in a heteroscedastic linear model. *J. Am. Stat. Assoc.*, *77*, 878–882.

Carroll, R. J., and Ruppert, D. (1984). Power transformations when fitting theoretical models to data. *J. Am. Stat. Assoc.*, *79*, 321–328.

Carroll, R. J., and Ruppert, D. (1985). Transformations in regression: A robust analysis. *Technometrics*, *27*, 1–12.

Carroll, R. J., and Ruppert, D. (1987). Diagnostics and robust estimation when transforming the model and the response. *Technometrics*, *29*, 287–299.

Carroll, R. J., and Ruppert, D. (1988). *Transformation and Weighting in Regression*. Chapman and Hall: London.

Causton, D. R., and Venus, J. C. (1981). *The Biometry of Plant Growth.* Edward Arnold: London.

Chambers, J. M. (1973). Fitting nonlinear models: Numerical techniques. *Biometrika, 60,* 1–13.

Chan, L. K., and Mak, T. K. (1979). Maximum likelihood estimation of a linear structural relationship with replication. *J. R. Stat. Soc. B, 41,* 263–268.

Chan, N. N., and Mak, T. K. (1983). Estimation of multivariate linear functional relationships. *Biometrika, 70,* 263–267.

Chanda, K. C. (1976). Efficiency and robustness of least squares estimators. *Sankhyā B, 38,* 153–163.

Chapman, M. J., and Godfrey, K. R. (1985). Some extensions to the exhaustive-modelling approach to structural identifiability. *Math. Biosci.,* 77, 305–323.

Charnes, A., Frome, E. L., and Yu, P. L. (1976). The equivalence of generalized least squares and maximum likelihood estimates in the exponential family. *J. Am. Stat. Assoc., 71,* 169–171.

Chin Choy., J. H., and Broemeling, L. D. (1980). Some Bayesian inferences for a changing linear model. *Technometrics, 22,* 71–78.

Clarke, G. P. Y. (1980). Moments of the least squares estimators in a non-linear regression model. *J. R. Stat. Soc. B, 42,* 227–237.

Clarke, G. P. Y. (1987a). Approximate confidence limits for a parameter function in nonlinear regression. *J. Am. Stat. Assoc., 82,* 221–230.

Clarke, G. P. Y. (1987b). Marginal curvatures and their usefulness in the analysis of nonlinear regression models. *J. Am. Stat. Assoc., 84,* 844–850.

Cleveland, W. S. (1979). Robust locally weighted regression and smoothing scatterplots. *J. Am. Stat. Assoc., 74,* 829–836.

Cobb, G. W. (1978). The problem of the Nile: Conditional solution to a change point problem. *Biometrika, 65,* 243–251.

Cobelli, C. (1981). A priori identifiability analysis in pharmacokinetic experiment design. *In* L. Endrenyi (Ed.), *Kinetic Data Analysis: Design and Analysis of Enzyme and Pharmacokinetic Experiments,* pp. 181–208. Plenum Press: New York.

Cobelli, C., and Thomaseth, K. (1985). Optimal input design for identification of compartmental models. Theory and applications to a model of glucose kinetics. *Math. Biosci.,* 77, 267–286.

Cobelli, C., Lepschy, A., and Romanin Jacur, G. (1979). Identification experiments and identifiability criteria for compartmental systems. *In* J. H. Matis, B. C. Patten, and G. C. White (Eds.), *Compartmental Analysis of Ecosystem Models,* Statistical Ecology Series Vol. 10, pp. 99–123. International Co-operative Publishing House: Fairland, Md.

Cochran, W. G. (1973). Experiments for nonlinear functions. *J. Am. Stat. Assoc., 68,* 771–781.

Cochrane, D., and Orcutt, G. H. (1949). Application of least squares regression to relationships containing autocorrelated error terms. *J. Am. Stat. Assoc., 44,* 32–61.

Cole, T. J. (1975). Linear and proportional regression models in the prediction of ventilatory function. *J. R. Stat. Soc. A, 138,* 297–337.

Cook, R. D. (1986). Assessment of local influence. *J. R. Stat. Soc. B, 48,* 133–155.

Cook, R. D., and Goldberg, M. L. (1986). Curvatures for parameter subsets in nonlinear regression. *Ann. Stat.*, *14*, 1399–1418.

Cook, R. D., and Tsai, C.-L. (1985). Residuals in nonlinear regression. *Biometrika*, *72*, 23–29.

Cook, R. D., and Wang, P. C. (1983). Transformations and influential cases in regression. *Technometrics*, *25*, 337–343.

Cook, R. D., and Witmer, J. A. (1985). A note on parameter-effects curvature. *J. Am. Stat. Assoc.*, *80*, 872–878.

Cook, R. D., Tsai, C.-L., and Wei, B. C. (1986). Bias in nonlinear regression. *Biometrika*, *73*, 615–623.

Coope, I. D. (1976). Conjugate direction algorithms for unconstrained optimization. Ph.D. Thesis, Dept. of Mathematics, University of Leeds.

Cooper, L., and Steinberg, D. (1970). *Introduction to Methods of Optimization.* W. B. Saunders: Philadelphia.

Cornish-Bowden, A. J. (1976). *Principles of Enzyme Kinetics.* Butterworths: London.

Corradi, C., and Stefanini, L. (1978). Computational experience with algorithms for separable nonlinear least squares problems. *Calcolo*, *15*, 317–330.

Cox, C. (1984). Generalized linear models—the missing link. *Appl. Stat.*, *33*, 18–24.

Cox, D. R. (1961). Tests of separate families of hypotheses. *Proc. 4th Berkeley Symp.*, *1*, 105–123.

Cox, D. R. (1962). Further results on tests of separate families of hypotheses. *J. R. Stat. Soc. B*, *24*, 406–424.

Cox, D. R. (1970). *Analysis of Binary Data.* Chapman and Hall: London.

Cox, D. R. (1977). Nonlinear models, residuals and transformations. *Math. Operationsforsch. Stat., Ser. Stat.*, *8*, 3–22.

Cox, D. R., and Hinkley, D. V. (1974). *Theoretical Statistics.* Chapman and Hall: London.

Cox, D. R., and Miller, H. D. (1965). *The Theory of Stochastic Processes.* Methuen: London.

Cox, N. R. (1976). The linear structural relation for several groups of data. *Biometrika*, *63*, 231–237.

Craven, P., and Wahba, G. (1979). Smoothing noisy data with spline functions. *Numer. Math.*, *31*, 377–403.

Cressie, N. (1986). Kriging nonstationary data. *J. Am. Stat. Assoc.*, *81*, 625–634.

Crowder, M. (1986). On consistency and inconsistency of estimating equations. *Econometric Theory*, *2*, 305–330.

Crowder, M. (1987). On linear and quadratic estimating functions. *Biometrika*, *74*, 591–597.

Currie, D. J. (1982). Estimating Michaelis–Menten parameters: Bias, variance and experimental design. *Biometrics*, *38*, 907–919.

Dagenais, M. G., and Liem, T. C. (1981). Numerical approximation of marginals from "well behaved" joint posteriors. *J. Stat. Comput. Simul.*, *11*, 273–279.

Davidon, W. C. (1959). Variable metric method for minimization. A.E.C. Res. and Dev. Report ANL-5990. Argonne National Laboratory, Argonne, Illinois.

Davies, N., and Petruccelli, J. D. (1984). On the use of the general partial autocorrelation function for order determination in ARMA(p, q) processes. *J. Am. Stat. Assoc.*, *79*, 374–377.

Davis, A. W., and Veitch, L. G. (1978). Approximate confidence intervals for nonlinear functions, with applications to a cloud-seeding experiment. *Technometrics*, *20*, 227–230.

Davis, P. J., and Rabinowitz, P. (1967). *Numerical Integration*. Blaisdell: Waltham, Mass.

Dawid, A. P. (1975). Discussions on Efron's paper. *J. R. Stat. Soc. B*, *46*, 1231–1234.

De Boor, C. (1978). *A Practical Guide to Splines*. Springer-Verlag: New York.

Delforge, J. (1984). On local identifiability of linear systems. *Math. Biosci.*, *70*, 1–37.

Della Corte, M., Buricchi, L., and Romano, S. (1974). On the fitting of linear combinations of exponentials. *Biometrics*, *30*, 367–369.

Deming, W. E. (1943). *Statistical Adjustment of Data*. Wiley: New York.

Dennis, J. E., Jr. (1977). Non-linear least squares and equations. *In* D. A. H. Jacobs (Ed.), *The State of the Art in Numerical Analysis*, pp. 269–312. Academic Press: London.

Dennis, J. E., Jr., and Schnabel, R. B. (1983). *Numerical Methods for Unconstrained Optimization and Nonlinear Equations*. Prentice-Hall: Englewood Cliffs, N.J.

Dennis, J. E., Jr., and Welsch, R. E. (1978). Techniques for nonlinear least squares and robust regression. *Commun. Stat. Simul. Computat. B*, *7*, 345–359.

Dennis, J. E., Jr., Gay, D. M., and Welsch, R. E. (1981a). An adaptive nonlinear least-squares algorithm. *ACM Trans. Math. Software*, *7*, 348–368.

Dennis, J. E., Jr., Gay, D. M., and Welsch, R. E. (1981b). Algorithm 573. NL2SOL—An adaptive nonlinear least squares algorithm. *ACM Trans. Math. Software*, *7*, 369–383.

Dixon, L. C. W. (1972). Quasi-Newton family generate identical points. *Math. Prog.*, *2*, 383–387.

Dixon, L. C. W., and Szegö, G. P. (Eds.). (1975). *Towards Global Optimization*, Vol. 1. North Holland: Amsterdam.

Dixon, L. C. W., and Szegö, G. P. (Eds.). (1978). *Towards Global Optimization*, Vol. 2. North Holland: Amsterdam.

Doksum, K. A., and Wong, C. W. (1983). Statistical tests based on transformed data. *J. Am. Stat. Assoc.*, *78*, 411–417.

Dolby, G. R. (1972). Generalized least squares and maximum likelihood estimation of non-linear functional relationships. *J. R. Stat. Soc. B*, *34*, 393–400.

Dolby, G. R. (1976a). The connection between methods of estimation in implicit and explicit nonlinear models. *Appl. Stat.*, *25*, 157–162.

Dolby, G. R. (1976b). The ultrastructural relation: A synthesis of the functional and structural relations. *Biometrika*, *63*, 39–50.

Dolby, G. R., and Freeman, T. G. (1975). Functional relationships having many independent variables and errors with multivariate normal distribution. *J. Multivar. Anal.*, *5*, 466–479.

Dolby, G. R., and Lipton, S. (1972). Maximum likelihood estimation of the general nonlinear functional relationship with replicated observations and correlated errors. *Biometrika*, *59*, 121–129.

Donaldson, J. R., and Schnabel, R. B. (1987). Computational experience with confidence regions and confidence intervals for nonlinear least squares. *Technometrics*, *29*, 67–82.

Dongarra, J. J., Bunch, J. R., Moler, C. B., and Stewart, G. W. (1979). *LINPack Users' Guide*, Society for Industrial and Applied Mathematics.

Draper, N. R., and Hunter, W. G. (1966). Design of experiments for parameter estimation in multiresponse situations. *Biometrika, 53*, 525–533.

Draper, N. R., and Hunter, W. G. (1967a). The use of prior distributions in the design of experiments for parameter estimation in nonlinear situations. *Biometrika, 54*, 147–153.

Draper, N. R., and Hunter, W. G. (1967b). The use of prior distributions in the design of experiments for parameter estimation in nonlinear situations: Multiresponse case. *Biometrika, 54*, 662–665.

Draper, N. R., and Smith, H. (1981). *Applied Regression Analysis*, 2nd ed. Wiley: New York.

Duan, N. (1983). Smearing estimate: A nonparametric retransformation method. *J. Am. Stat. Assoc., 78*, 605–610.

Duncan, G. T. (1978). An empirical study of jackknife constructed confidence regions in nonlinear regression. *Technometrics, 20*, 123–129.

Durbin, J. (1960a). The fitting of time-series models. *Rev. Internat. Stat. Inst., 28*, 233–243.

Durbin, J. (1960b). Estimation of parameters in time-series regression models. *J. R. Stat. Soc. B, 22*, 139–153.

Durbin, J., and Watson, G. S. (1951). Testing for serial correlation in least squares regression II. *Biometrika, 38*, 159–178.

Durbin, J., and Watson, G. S. (1971). Testing for serial correlation in least squares regression III. *Biometrika, 58*, 1–19.

Dutter, R. (1977). Numerical solution of robust regression problems: computational aspects, a comparison. *J. Stat. Comput. Simul., 5*, 207–238.

Dutter, R. (1978). Robust regression: LINWDR and NLWDR. *In* L. C. A. Corsten and J. Hermans (Eds.), *COMPSTAT1978, Proceedings in Computational Statistics*, pp. 74–80. Physica-Verlag: Vienna.

Eaves, D. M. (1983). On Bayesian nonlinear regression with an enzyme example. *Biometrika, 70*, 373–379.

Efron, B. (1975). Defining the curvature of a statistical problem (with applications to second order efficiency) (with Discussion), *Ann. Stat., 3*, 1189–1242.

Egerton, M. F., and Laycock, P. J. (1979). Maximum likelihood estimation of multivariate non-linear functional relationships. *Math, Operationsforsch. Statist., Ser. Stat., 10*, 273–280.

Eisenfeld, J. (1985). Remarks on Bellman's structural identifiability. *Math. Biosci., 77*, 229–243.

Eisenfeld, J. (1986). A simple solution to the compartmental structural identifiability problem. *Math. Biosci., 79*, 209–220.

El-Asfouri, S., McInnis, B. C., and Kapadia, A. S. (1979). Stochastic compartmental modelling and parameter estimation with application to cancer treatment follow-up studies. *Bull. Math. Biol., 41*, 203–215.

El-Attar, R. A., Vidyasagar, M., and Dutta, S. R. K. (1979). An algorithm for L_1-norm minimization with applications to nonlinear L_1 approximation. *SIAM J. Numer. Anal., 16*, 70–86.

El-Shaarawi, A., and Shah, K. R. (1980). Interval estimation in non-linear models. *Sankhyā B*, *42*, 229–232.

Endrenyi, L. (Ed.). (1981a). *Kinetic Data Analysis: Design and Analysis of Enzyme and Pharmacokinetic Experiments.* Plenum Press: New York.

Endrenyi, L. (1981b). Design of experiments for estimating enzyme and pharmacokinetic parameters. *In* L. Endrenyi (Ed.), *Kinetic Data Analysis: Design and Analysis of Enzyme and Pharmacokinetic Experiments*, pp. 137–167. Plenum Press: New York.

Endrenyi, L., and Kwong, F. H. F. (1981). Tests for the behaviour of experimental errors. *In* L. Endrenyi (Ed.), *Kinetic Data Analysis: Design and Analysis of Enzyme and Pharmacokinetic Experiments*, pp. 89–103. Plenum Press: New York.

Engle, R. F. (1984). Wald, likelihood ratio and Lagrange multiplier tests in econometrics. *In* Z. Griliches and M. D. Intriligator (Eds.), *Handbook of Econometrics*, Vol. II, pp. 775–826. Elsevier Science Publishers.

Eppright, E. S., Fox, H. M., Fryer, B. A., Lamkin, G. H., Vivian, V. M., and Fuller, E. S. (1972). Nutrition of infants and preschool children in the North Central Region of the United States of America, *World Rev. Nutrition and Dietetics*, *14*, 269–332.

Ertel, J. E., and Fowlkes, E. B. (1976). Some algorithms for linear spline and piecewise multiple linear regression. *J. Am. Stat. Assoc.*, *71*, 640–648.

Esterby, S. R., and El-Shaarawi, A. (1981). Interference about the point of change in a regression model. *Appl. Stat.*, *30*, 277–285.

Eubank, R. L. (1984). Approximate regression models and splines. *Commun. Stat. Theory Methods A*, *13*, 433–484.

Eubank, R. L. (1988), *Spline Smoothing and Nonparametric Regression.* Marcel Dekker: New York.

Evans, G. B. A., and Savin, N. E. (1982). Conflict among the criteria revisited: The W, LR and LM tests. *Econometrica*, *50*, 737–748.

Faddy, M. J. (1976). A note on the general time-dependent stochastic compartmental model. *Biometrics*, *32*, 443–448.

Faddy, M. J. (1977). Stochastic compartmental models as approximations to more general stochastic systems with the general stochastic epidemic as an example. *Adv. Appl. Prob.*, *9*, 448–461.

Fahrmeir, L., and Kaufmann, H. (1985). Consistency and asymptotic normality of the maximum likelihood estimator in generalized linear models. *Ann. Stat.*, *13*, 342–368.

Farazdaghi, H., and Harris, P. M. (1968). Plant competition and crop yield. *Nature*, *217*, 289–290.

Farley, J. U., and Hinich, M. J. (1970). A test for a shifting slope coefficient in a linear model. *J. Am. Stat. Assoc.*, *65*, 1320–1329.

Farley, J. U., Hinich, M., and McGuire, T. W. (1975). Some comparisons of tests for a shift in the slopes of a multivariate linear time series model. *J. Econometrics*, *3*, 297–318.

Feder, P. I. (1975a). On asymptotic distribution theory in segmented regression problems—identified case. *Ann. Stat.*, *3*, 49–83.

Feder, P. I. (1975b). The log likelihood ratio in segmented regression. *Ann. Stat.*, *3*, 84–97.

Fedorov, V. V. (1972). *Theory of Optimal Experiments.* Academic Press: New York.

Fedorov, V. V. (1974). Regression problems with controllable variables subject to error. *Biometrika*, *61*, 49–56.

Fedorov, V. V. (1975). Optimal experimental designs for discriminating two rival regression models. *In* J. N. Srivastava (Ed.), *A Survey of Statistical Design and Linear Models*, pp. 155–164. North Holland: Amsterdam.

Ferreira, P. E. (1975). A Bayesian analysis of a switching regression model: Known number of regimes. *J. Am. Stat. Assoc.*, *70*, 370–376.

Fife, D. (1972). Which linear compartmental systems contain traps? *Math. Biosci.*, *14*, 311–315.

Finney, D. J. (1976). Radioligand assay. *Biometrics*, *32*, 721–740.

Finney, D. J. (1978). *Statistical Method in Biological Assay*, 3rd ed. Griffin: London.

Firth, D. (1987). On the efficiency of quasi-likelihood estimation. *Biometrika*, *74*, 233–245.

Fisher, G., and McAleer, M. (1979). On the interpretation of the Cox test in econometrics. *Econom. Lett.*, *4*, 145–150.

Fletcher, R. (1970), A new approach to variable metric algorithms. *Comput. J.*, *13*, 317–322.

Fletcher, R. (1971). A modified Marquardt subroutine for nonlinear least squares. Harwell Report AERE-R6799.

Fletcher, R. (1980). *Practical Methods of Optimization: Vol. 1. Unconstrained Optimization.* Wiley: New York.

Fletcher, R. (1981). *Practical Methods of Optimization: Vol. 2. Constrained Optimization.* Wiley: New York.

Fletcher, R., and Freeman, T. L. (1977). A modified Newton method for minimization. *J. Optim. Theory Appl.*, *23*, 357–372.

Fletcher, R., and Powell, M. J. D. (1963). A rapidly convergent descent method for minimization. *Comput. J.*, *6*, 163–168.

Fletcher, R., and Reeves, C. M. (1964). Function minimization by conjugate gradients. *Comput. J.*, *7*, 149–154.

Fletcher, R. I. (1974). The quadric law of damped exponential growth. *Biometrics*, *30*, 111–124.

Fox, T., Hinkley, D., and Larntz, K. (1980). Jackknifing in nonlinear regression. *Technometrics*, *22*, 29–33.

Frappell, B. D. (1973). Plant spacing of onions. *J. Hort. Sci.*, *48*, 19–28.

Friedman, J. H. (1984). A variable span smoother. LCS Technical Report No. 5, SLAC PUB-3477, Stanford Linear Accelerator Center, Stanford, Calif.

Frisen, M. (1979). A note on alternating estimation in non-linear regression. *J. Stat. Comput. Simul.*, *9*, 19–23.

Frydman, R. (1978). Nonlinear estimation problems. Ph.D. Dissertation, Dept. of Economics, Columbia Univ.

Frydman, R. (1980). A proof of the consistency of maximum likelihood estimators of nonlinear regression models with autocorrelated errors. *Econometrica*, *48*, 853–860.

Fuguitt, R. E., and Hawkins, J. E. (1947). Rate of thermal isomerization of α-pinene in the liquid phase. *J. Am. Chem. Soc.*, *69*, 319–322.

Galamabos, J. T., and Cornell, R. G. (1962). Mathematical models for the study of the metabolic pattern of sulfate. *J. Lab. and Clinical Med.*, *60*, 53–63.

Galbraith, R. F., and Galbraith, J. I. (1974). On the inverses of some patterned matrices arising in the theory of stationary time series. *J. Appl. Prob.*, *11*, 63–71.

Gallant, A. R. (1971). Statistical Inference for Non-linear Regression Models. Ph.D. Thesis, Iowa State Univ.

Gallant, A. R. (1974). The theory of nonlinear regression as it relates to segmented polynomial regressions with estimated join points. Institute of Statistics Mimeograph Series. No. 925, North Carolina State Univ.

Gallant, A. R. (1975a). The power of the likelihood ratio test of location in nonlinear regression models. *J. Am. Stat. Assoc.*, *70*, 198–203.

Gallant, A. R. (1975b). Nonlinear regression. *Am. Stat.*, *29*(2), 73–81.

Gallant, A. R. (1975c). Seemingly unrelated nonlinear regressions. *J. Econometrics*, *3*, 35–50.

Gallant, A. R. (1976). Confidence regions for the parameters of a nonlinear regression model. Institute of Statistics Mimeograph Series No. 1077, North Carolina State Univ., Raleigh, N. C.

Gallant, A. R. (1977a). Three-stage least-squares estimation for a system of simultaneous, nonlinear, implicit functions. *J. Econometrics*, *5*, 71–88.

Gallant, A. R. (1977b). Testing a nonlinear specification, a nonregular case. *J. Am. Stat. Assoc.*, *72*, 523–530.

Gallant, A. R. (1980). Explicit estimators of parametric functions in nonlinear regression. *J. Am. Stat. Assoc.*, *75*, 182–193.

Gallant, A. R. (1987). *Nonlinear Statistical Models.* Wiley: New York.

Gallant, A. R., and Fuller, W. A. (1973). Fitting segmented polynomial regression models whose join points have to be estimated. *J. Am. Stat. Assoc.*, *68*, 144–147.

Gallant, A. R., and Goebel, J. J. (1975). Nonlinear regression with autoregressive errors. Ser. No. 986. Inst. of Statistics, North Carolina State Univ., March 1975 (mimeographed).

Gallant, A. R., and Goebel, J. J. (1976). Nonlinear regression with autocorrelated errors. *J. Am. Stat. Assoc.*, *71*, 961–967.

Gallant, A. R., and Holly, A. (1980). Statistical inference in an implicit, nonlinear, simultaneous equation model in the context of maximum likelihood estimation. *Econometrica*, *48*, 697–720.

Gallucci, V. F. and Quinn, T. J. (1979). Reparameterizing, fitting and testing a simple growth model. *Trans. Am. Fish. Soc.*, *108*, 14–25.

Garcia, O. (1983). A stochastic differential equation model for the height growth of forest stands. *Biometrics*, *39*, 1059–1072.

Gardner, G., Harvey, A. C., and Phillips, G. D. A. (1980). Algorithm AS154. An algorithm for exact maximum likelihood estimation of autoregressive–moving average models by means of Kalman filtering. *Appl. Stat.*, *29*, 311–317.

Gay, D. M. (1983a). Algorithm 611. Subroutines for unconstrained minimization using a model/trust-region approach. *ACM Trans. Math. Software*, *9*, 503–524.

Gay, D. M. (1983b). Remark on Algorithm 573. *ACM Trans. Math. Software*, *9*, 139.

Gbur, E. E., and Dahm, P. F. (1985). Estimation of the linear–linear segmented regression model in the presence of measurement error. *Commun. Stat. Theory Methods A*, *14*, 809–826.

Gerig, T. M., and Gallant, A. R. (1975). Computing methods for linear models subject to linear parametric constraints. *J. Stat. Comput. Simul.*, *3*, 283–296.

Gill, P. E., and Murray, W. (1972). Quasi-Newton methods for unconstrained optimization. *J. Inst. Math. Appl.*, *9*, 91–108.

Gill, P. E., and Murray, W. (1977). Modification of matrix factorizations after a rank-one change. *In* D. A. H. Jacobs (Ed.), *The State of the Art of Numerical Analysis*, pp. 55–83. Academic Press: London.

Gill, P. E., and Murray, W. (1978). Algorithms for the solution of the nonlinear least-squares problem. *SIAM J. Numer. Anal.*, *15*, 977–992.

Gill, P. E., Murray, W., and Wright, M. H. (1981). *Practical Optimization*. Academic Press: New York.

Gillis, P. R., and Ratkowsky, D. A. (1978). The behaviour of estimators of the parameters of various yield–density relationships. *Biometrics*, *34*, 191–198.

Glasbey, C A. (1979). Correlated residuals in non-linear regression applied to growth data. *Appl. Stat.*, *28*, 251–259.

Glasbey, C. A. (1980). Nonlinear regression with autoregressive time series errors. *Biometrics*, *36*, 135–140.

Gleser, L. J. (1985). A note on G. R. Dolby's unreplicated ultrastructural model. *Biometrika*, *72*, 117–124.

Godambe, V. P. (1960). An optimum property of regular maximum likelihood estimation. *Ann. Math. Stat.*, *31*, 1208–1211.

Godambe, V. P., and Thompson, M. E. (1978). Some aspects of estimating equations. *J. Stat. Plan. Infer.*, *2*, 95–104.

Godambe, V. P., and Thompson, M. E. (1984). Robust estimation through estimating equations. *Biometrika*, *71*, 115–125.

Godfrey, L. G. (1979). Testing the adequacy of a time series model. *Biometrika*, *66*, 67–72.

Godolphin, E. J. (1984). A direct representation for the large-sample maximum likelihood estimator of a Gaussian autoregressive–moving average process. *Biometrika*, *71*, 281–289.

Godolphin, E. J., and de Gooijer, J. G. (1982). On the maximum likelihood estimation of the parameters of a Gaussian moving average process. *Biometrika*, *69*, 443–451.

Goldberg, M. L., Bates, D. M., and Watts, D. G. (1983). Simplified methods of assessing nonlinearity. *Am. Stat. Assoc. Proc. Bus. Econ. Statistics Section*, 67–74.

Goldfarb, D. (1970). A family of variable metric methods derived by variational means. *Math. Comp.*, *26*, 23–26.

Goldfeld, S. M., and Quandt, R. E. (1965). Some tests for homoscedasticity. *J. Am. Stat. Assoc.*, *60*, 539–547.

Goldfeld, S. M., and Quandt, R. E. (1972). *Nonlinear Methods in Econometrics*. North-Holland: Amsterdam.

Goldfeld, S. M., Quandt, R. E., and Trotter, H. F. (1966). Maximization by quadratic hill climbing. *Econometrica*, *34*, 541–551.

Golub, G. H., and Pereyra, V. (1973). The differentiation of pseudo-inverses and nonlinear least squares problems whose variables separate. *SIAM J. Numer. Anal.*, *10*, 413–432.

Golub, G. H., and Van Loan, C. F. (1983). *Matrix Computations*. North Oxford Academic: Oxford.

Gonin, R., and Money, A. H. (1985a). Nonlinear L_p-norm estimation: Part 1. On the choice of the exponent, p, where the errors are additive. *Commun. Stat. Theory Methods A*, *14*, 827–840.

Gonin, R., and Money, A. H. (1985b). Nonlinear L_p-norm estimation: Part 2. The asymptotic distribution of the exponent, *p*, a function of the sample kurtosis. *Commun. Stat. Theory Methods A*, *14*, 841–849.

Gourieroux, C., Holly, A., and Montfort, A. (1980). Kuhn–Tucker, likelihood ratio and Wald tests for nonlinear models, with inequality constraints on the parameters. Discussion paper 770. Dept. of Economics, Harvard Univ., Cambridge, Mass.

Gourieroux, C., Holly, A., and Montfort, A. (1982). Likelihood ratio test, Wald test and Kunh–Tucker test in linear models with inequality constraints on the regression parameters. *Econometrica*, *50*, 63–80.

Gourieroux, C., Montfort, A., and Trognon, A. (1983). Testing nested or non-nested hypotheses. *J. Econometrics*, *21*, 83–115.

Grandinetti, L. (1978). Factorization versus nonfactorization in quasi-Newtonian algorithms for differentiable optimization. Univ. della Calabria, Dipartimento di Sistemi, Report N5.

Grassia, A., and de Boer, E. S. (1980). Some methods of growth curve fitting. *Math. Scientist*, *5*, 91–103.

Green, P. J. (1984). Iteratively reweighted least squares for maximum likelihood estimation, and some robust and resistant alternatives (with Discussion). *J. R. Stat. Soc. B*, *46*, 149–192.

Greenstadt, J. (1967). On the relative efficiencies of gradient methods. *Math. Comp.*, *21*, 360–367.

Gregory, A. W., and Veall, M. R. (1985). Formulating Wald tests of nonlinear restrictions. *Econometrica*, *53*, 1465–1468.

Griffiths, D. (1983). Letter to the Editors on "Piecewise regression with smooth transition". *Appl. Stat.*, *32*, 89–90.

Griffiths, D. A., and Miller, A. J. (1973). Hyperbolic regression—a model based on two-phase piecewise linear regression with a smooth transition between regimes. *Commun. Stat.*, *2*, 561–569.

Griffiths, D., and Miller, A. (1975). Letter to the Editor. *Technometrics*, *17*, 281.

Griffiths, D., and Sandland, R. (1984). Fitting generalized allometric models to multivariate growth data. *Biometrics*, *40*, 139–150.

Guttman, I., and Meeter, D. A. (1965). On Beale's measures of non-linearity. *Technometrics*, *7*, 623–637.

Guttman, I., and Menzefricke, U. (1982). On the use of loss functions in the change-point problem. *Ann. Inst. Stat. Math.*, *34*, Part A, 319–326.

Halperin, M. (1963). Confidence interval estimation in non-linear regression. *J. R. Stat. Soc. B*, *25*, 330–333.

Halperin, M. (1964). Note on interval estimation in non-linear regression when responses are correlated. *J. R. Stat. Soc. B*, *26*, 267–269.

Hamilton, D. (1986). Confidence regions for parameter subsets in nonlinear regression. *Biometrika*, *73*, 57–64.

Hamilton, D. C., and Watts, D. G. (1985). A quadratic design criterion for precise estimation in nonlinear regression models. *Technometrics*, *27*, 241–250.

Hamilton, D., and Wiens, D. (1986). Correction factors for *F* ratios in nonlinear regression. Technical Report 86–1, Statistics Division, Dalhousie Univ. Halifax, N. S.

Hamilton, D., and Wiens, D. (1987). Correction factors for F ratios in nonlinear regression *Biometrika, 74*, 423–425.

Hamilton, D. C., Watts, D. G., and Bates, D. M. (1982). Accounting for intrinsic nonlinearity in nonlinear regression parameter inference regions. *Ann. Stat., 10*, 386–393.

Hannan, E. J. (1970). *Multiple Time Series*. Wiley: New York.

Hannan, E. J. (1971). Nonlinear time series regression. *J. Appl. Prob., 8*, 767–780.

Hansen, L. P. (1982). Large sample properties of generalized method of moments estimators. *Econometrica, 50*, 1029–1054.

Hansen, L. P., and Singleton, K. J. (1982). Generalized instrumental variables estimation of nonlinear rational expectations models. *Econometrica, 50*, 1269–1286.

Hardy, G. H., Littlewood, J. E., and Polya, G., (1952). *Inequalities*. Cambridge Univ. Press: Cambridge.

Hartley, H. O. (1961). The modified Gauss–Newton method for the fitting of nonlinear regression functions by least squares. *Technometrics, 3*, 269–280.

Hartley, H. O. (1964). Exact confidence regions for the parameters in nonlinear regression laws. *Biometrika, 51*, 347–353.

Harvey, A. C., and Phillips, G. D. A. (1979). Maximum likelihood estimation of regression models with autoregressive–moving average disturbances. *Biometrika, 66*, 49–58.

Hawkins, D. M. (1976). Point estimation of the parameters of piecewise regression models. *Appl. Stat., 25*, 51–57.

Hawkins, D. M. (1977). Testing a sequence of observations for a shift in location. *J. Am. Stat. Assoc., 72*, 180–186.

Hawkins, D. M. (1980). A note on continuous and discontinuous segmented regressions. *Technometrics, 22*, 443–444.

Haynes, J. D. (1971). Comment on article on pharmacokinetics. *J. Am. Stat. Assoc., 66*, 53–54.

Heath, A. B., and Anderson, J. A. (1983). Estimation in a multivariate two-phase regression. *Commun. Stat. Theory Methods A, 12*, 809–828.

Hernandez, F., and Johnson, R. A. (1980). The large-sample behaviour of transformations to normality. *J. Am. Stat. Assoc., 75*, 855–861.

Hiebert, K. L. (1981). An evaluation of mathematical software that solves nonlinear least squares problems. *ACM Trans. Math. Software, 7*, 1–16.

Hildreth, C. (1969). Asymptotic distribution of maximum likelihood estimators in a linear model with autoregressive disturbances. *Ann. Stat., 40*, 583–594.

Hill, P. D. H. (1978). A review of experimental design procedures for regression model discrimination. *Technometrics, 10*, 145–160.

Hill, P. D. H. (1980). D-optimal designs for partially nonlinear regression models. *Technometrics, 22*, 275–276.

Hill, W. J., and Hunter, W. G. (1974). Design of experiments for subsets of parameters. *Technometrics, 16*, 425–434.

Himmelblau, D. M. (1972). A uniform evaluation of unconstrained optimization techniques. *In* F. A. Lootsma (Ed.), *Numerical Methods for Nonlinear Optimization*, pp. 69–97. Academic Press: London.

Hinkley, D. V. (1969). Inference about the intersection in two-phase regression. *Biometrika, 56*, 495–504.

Hinkley, D. V. (1970). Inference about the change-point in a sequence of random variables. *Biometrika, 57,* 1–17.

Hinkley, D. V. (1971a). Inference about the change-point from cumulative sum tests. *Biometrika, 58,* 509–523.

Hinkley, D. V. (1971b). Inference in two-phase regression. *J. Am. Stat. Assoc., 66,* 736–743.

Hinkley, D. V. (1972). Time ordered classification. *Biometrika, 59,* 509–523.

Hinkley, D. V. (1975). On power transformations to symmetry. *Biometrika, 62,* 101–112.

Hinkley, D. V. (1978). Improving the jackknife with special reference to correlation estimation. *Biometrika, 65,* 13–21.

Hinkley, D. V., and Runger, G. (1984). The analysis of transformed data (with Discussion). *J. Am. Stat. Assoc., 79,* 302–320.

Hinkley, D. V., and Schechtman, E. (1987). Conditional bootstrap methods in the mean-shift model. *Biometrika, 74,* 85–93.

Hirsch, M. W., and Smale, S. (1974). *Differential Equations, Dynamical Systems, and Linear Algebra.* Academic Press: New York.

Hoadley, B. (1971). Asymptotic properties of maximum likelihood estimators for the independent not identically distributed case. *Ann. Math. Stat., 42,* 1977–1991.

Hoaglin, D. C., Klema, V. C., and Peters, S. C. (1982). Exploratory data analysis in a study of the performance of nonlinear optimization routines. *ACM Trans. Math. Software, 8,* 145–162.

Holbert, D., and Broemeling, L. (1977). Bayesian inferences related to shifting sequences and two-phase regression. *Commun. Stat. Theory Methods A, 6,* 265–275.

Holliday, R. (1960a). Plant population and crop yield: Part I. *Field Crop Abstracts, 13,* 159–167.

Holliday, R. (1960b). Plant population and crop yield: Part II. *Field Crop Abstracts, 13,* 247–254.

Hong, D., and L'Esperance, W. L. (1973). Effects of autocorrelated errors on various least squares estimators: A Monte Carlo study. *Commun. Stat., 6,* 507–523.

Hood, W. C., and Koopmans, T. C. (Eds.) (1953). *Studies in Econometric Method.* Wiley: New York.

Hosking, J. R. M. (1980). Lagrange-multiplier tests of time-series models. *J. R. Stat. Soc. B, 42,* 170–181.

Hougaard, P. (1985). The appropriateness of the asymptotic distribution in a nonlinear regression model in relation to curvature. *J. R. Stat. Soc. B, 47,* 103–114.

Hsu, D. A. (1982). Robust inferences for structural shift in regression models. *J. Econometrics, 19,* 89–107.

Huber, P. J. (1972). Robust statistics: A review. *Ann. Math. Stat., 43,* 1041–1067.

Huber, P. J. (1973). Robust regression: Asymptotics, conjectures, and Monte Carlo. *Ann. Stat., 1,* 799–821.

Huber, P. J. (1977). Robust methods of estimation of regression coefficients. *Math. Operationsforsch. Stat., Ser. Stat., 8,* 141–153.

Hudson, D. J. (1966). Fitting segmented curves whose join points have to be estimated. *J. Am. Stat. Assoc., 61,* 1097–1129.

Hunter, J. J. (1983). *Mathematical Techniques of Applied Probability. Volume 1. Discrete Time Models: Basic Theory.* Academic Press: New York.

Hunter, W. G., and McGregor, J. F. (1967). The estimation of common parameters from several responses: Some actual examples. Unpublished Report, Dept. of Statistics, Univ. of Wisconsin.

Hunter, W. G., Hill, W. J., and Henson, T. L. (1969). Designing experiments for precise estimation of all or some of the constants in a mechanistic model. *Canad. J. Chem. Eng.*, *47*, 76–80.

Jacquez, J. A. (1972). *Compartmental Analysis in Biology and Medicine.* Elsevier: New York.

Jacquez, J. A., and Grief, P. (1985). Numerical parameter identifiability and estimability: Integrating identifiability, estimability and optimal sampling design. *Math. Biosci.*, *77*, 201–227.

James, B., James, K. L., and Siegmund, D. (1987). Tests for a change-point. *Biometrika*, *74*, 71–83.

Jeffreys, H. (1961). *Theory of Probability*, 3rd ed. Clarendon: Oxford.

Jennrich, R. I. (1969). Asymptotic properties of nonlinear least squares estimators. *Ann. Math. Stat.*, *40*, 633–643.

Jennrich, R. I., and Bright, P. B. (1976). Fitting systems of linear differential equations using computer generated exact derivatives. *Technometrics*, *18*, 385–392.

Jennrich, R. I., and Moore, R. H. (1975). Maximum likelihood estimation by means of nonlinear least squares. *Am. Stat. Assoc., Proc. Statist Comput. Section*, 57–65.

Jennrich, R. I., and Sampson, P. F. (1968). Application of stepwise regression to nonlinear estimation. *Technometrics*, *10*, 63–72.

Jobson, J. D., and Fuller, W. A. (1980). Least squares estimation when the covariance matrix and parameter vector are functionally related. *J. Am. Stat. Assoc.*, *75*, 176–181.

Johansen, S. (1983). Some topics in regression. *Scand. J. Stat.*, *10*, 161–194.

Johansen, S. (1984), *Functional Relations, Random Coefficients, and Nonlinear Regression with Application to Kinetic Data.* Lecture Notes in Statistics, No. 22, 126 pp. Springer-Verlag: Berlin.

John, J. A., and Draper, N. R. (1980). An alternative family of transformations. *Appl. Stat.*, *29*, 190–197.

Jones, R. H. (1980). Maximum likelihood fitting of ARMA models to time series with missing observations. *Technometrics*, *22*, 389–397.

Jørgensen, B. (1983). Maximum likelihood estimation and large-sample inference for generalized linear and nonlinear regression models. *Biometrika*, *70*, 19–28.

Jørgensen, B. (1984). The delta algorithm and GLIM. *Internat. Stat. Rev.*, *52*, 283–300.

Jorgensen, M. A. (1981). Fitting animal growth curves. *New Zealand Statistician*, *16*, 5–15.

Jupp, D. L. B. (1978). Approximation to data by splines with free knots. *SIAM J. Numer. Anal.*, *15*, 328–343.

Just, R. E., and Pope, R. D. (1978). Stochastic specification of production functions and economic implications. *J. Econometrics*, *7*, 67–86.

Kalbfleisch, J. D., and Lawless, J. F. (1985). The analysis of panel data under a Markov assumption. *J. Am. Stat. Assoc.*, *80*, 863–871.

Kalbfleisch, J. D., and Sprott, D. A. (1970). Applications of likelihood methods to models involving large numbers of parameters (with Discussion). *J. R. Stat. Soc. B*, *32*, 175–208.

Kalbfleisch, J. D., Lawless, J. F., and Vallmer, W. M. (1983). Estimation in Markov models from aggregate data. *Biometrics, 39*, 907–919.

Kale, B. K. (1963). Some remarks on a method of maximum-likelihood estimation proposed by Richards. *J. R. Stat. Soc. B, 25*, 209–212.

Kass, R. E. (1984). Canonical parameterizations and zero parameter-effects curvature. *J. R. Stat. Soc. B, 46*, 86–92.

Katz, D., Azen, S. P., and Schumitzky, A. (1981). Bayesian approach to the analysis of nonlinear models: Implementation and evaluation. *Biometrics, 37*, 137–142.

Katz, D., Schumitzky, A., and Azen, S. P. (1982). Reduction of dimensionality in Bayesian nonlinear regression with a pharmacokinetic application. *Math. Biosci., 59*, 47–56.

Kaufman, L. (1975). A variable projection method for solving separable nonlinear least squares problems. *BIT, 15*, 49–57.

Kaufman, L., and Pereyra, V. (1978). A method for separable nonlinear least squares problems with separable nonlinear equality constraints. *SIAM J. Numer. Anal., 15*, 12–20.

Kendall, M. G., and Stuart, A. (1973). *The Advanced Theory of Statistics*, Vol. 2, 3rd ed. Griffin: London.

Kennedy, W. J., Jr., and Gentle, J. E. (1980). *Statistical Computing*. Marcel Dekker: New York.

Kent, J. T. (1986). The underlying structure of nonnested hypothesis tests. *Biometrika, 73*, 333–343.

Khorasani, F., and Milliken, G. A. (1982). Simultaneous confidence bands for nonlinear regression models. *Commun, Stat. Theory Methods A, 11*, 1241–1253.

Khuri, A. I. (1984). A note on *D*-optimal designs for partially nonlinear regression models. *Technometrics, 26*, 59–61.

King, L. J. (1969). *Statistical Analysis in Geography*. Prentice-Hall: Englewood Cliffs, N. J.

King, M. L. (1983). Testing for autoregressive against moving average errors in the linear regression model. *J. Econometrics, 21*, 35–51.

Kobayashi, M. (1985). Comparison of efficiencies of several estimators for linear regressions with autocorrelated errors. *J. Am. Stat. Assoc., 80*, 951–953.

Kodell, R. L., and Matis, J. H. (1976). Estimating the rate constants in a two-compartment stochastic model. *Biometrics, 32*, 377–390.

Kohn, R., and Ansley, C. F. (1984). A note on Kalman filtering for the seasonal moving average model. *Biometrika, 71*, 648–650.

Kohn, R., and Ansley, C. F. (1985). Efficient estimation and prediction in time series regression models. *Biometrika, 72*, 694–697.

Kohn, R., and Ansley, C. F. (1987). A new algorithm for spline smoothing based on smoothing a stochastic process. *SIAM J. Sci. Stat. Comput., 8*, 33–48.

Kramer, W. (1980). Finite sample efficiency of ordinary least squares in the linear regression model with autocorrelated errors. *J. Am. Stat. Assoc., 75*, 1005–1009.

Kramer, W. (1982). Note on estimating linear trend when residuals are autocorrelated. *Econometrics, 50*, 1065–1067.

Krogh, F. T. (1974). Efficient implementation of a variable projection algorithm for nonlinear least squares problems. *Comm. ACM, 17*, 167–169.

Krogh, F. T. (1986). ACM algorithms policy. *ACM Trans. Math. Software, 12*, 171–174.

Kuhry, B., and Marcus, L. F. (1977). Bivariate linear models in biometry. *Syst. Zool., 26*, 201–209.

Lahiri, K., and Egy, D. (1981). Joint estimation and testing for functional form and heteroskedasticity. *J. Econometrics, 15*, 299–307.

Lambrecht, R. M., and Rescigno, A. (Eds.) (1983). *Tracer Kinetics and Physiologic Modeling*. Lecture Notes in Biomathematics 48. Springer-Verlag: New York.

Laüter, E. (1974). Experimental designs in a class of models. *Math. Operationsforsch. Stat., Ser. Stat., 5*, 379–398.

Laüter, E. (1976). Optimal multipurpose designs for regression models. *Math. Operationsforsch. Stat., Ser. Stat., 7*, 51–68.

Lawrance, A. J. (1987). A note on the variance of the Box–Cox regression transformation estimate. *App. Stat., 36*, 221–223.

Lawson, C. L., and Hanson, R. J. (1974). *Solving Least Squares Problems*. Prentice-Hall: Englewood Cliffs, N.J.

Lawton, W. H., and Sylvestre, E. A. (1971). Estimation of linear parameters in nonlinear regression. *Technometrics, 13*, 461–467.

Lee, A. J., and Scott, A. J. (1986). Ultrasound in ante-natal diagnosis. *In* R. J. Brook, G. C. Arnold, T. H. Hassard, and R. M. Pringle (Eds.), *The Fascination of Statistics*, pp. 277–293. Marcel Dekker: New York.

Leech, D. (1975). Testing the error specification in nonlinear regression. *Econometrica, 43*, 719–725.

Lerman, P. M. (1980). Fitting segmented regression models by grid search. *Appl. Stat., 29*, 77–84.

Levenbach, H., and Reuter, B. E. (1976). Forecasting trending time series with relative growth-rate models. *Technometrics, 18*, 261–272.

Levenberg, K. (1944). A method for the solution of certain problems in least squares. *Quart. Appl. Math., 2*, 164–168.

Liang, K.-Y., and Zeger, S. L. (1986). Longitudinal data analysis using generalized linear models. *Biometrika, 73*, 13–22.

Lieberman, G. J., Miller, R. G., Jr., and Hamilton, M. A. (1967). Unlimited simultaneous discrimination intervals in regression. *Biometrika, 54*, 133–145.

Lill, S. A. (1976). A survey of methods for minimizing sums of squares of nonlinear functions. *In* L. C. W. Dixon (Ed.), *Optimization in Action*, pp. 1–26. Academic Press: London.

Lindley, D. V. (1947). Regression lines and the linear functional relationship. *Suppl. J. Roy. Stat. Soc., 9*, 218–244.

Linssen, H. N. (1975). Nonlinearity measures: A case study. *Statist. Neerland., 29*, 93–99.

Ljung, G. M., and Box, G. E. P. (1979). The likelihood function of stationary autoregressive–moving average models. *Biometrika, 66*, 265–270.

Lowry, R. K., and Morton, R. (1983). An asymmetry measure for estimators in nonlinear regression models. *Proc. 44th Session International Statistical Institute, Madrid, Contributed Papers 1983, Vol. 1*, 351–354.

Lwin, T., and Maritz, J. S. (1980). A note on the problem of statistical calibration. *Appl. Stat., 29*, 135–141.

Lwin, T., and Maritz, J. S. (1982). An analysis of the linear-calibration controversy from the perspective of compound estimation. *Technometrics, 24*, 235–241.

McCleod, A. I. (1977). Improved Box–Jenkins estimators. *Biometrika, 64*, 531–534.

McCleod, A. I. (1978). On the distribution of residual autocorrelations in Box–Jenkins models. *J. R. Stat. Soc. B, 40*, 296–302.

McCleod, A. I., and Holanda Sales, P. R. (1983). Algorithm AS 191. An algorithm for approximate likelihood calculation of ARMA and seasonal ARMA models. *Appl. Stat., 32*, 211–217.

McCleod, I. (1975). Derivation of the theoretical autocovariance function of autoregressive–moving average time series. *Appl. Stat., 24*, 255–256. (Correction: *Appl. Stat., 26*, p. 194).

McCullagh, P. (1983). Quasi-likelihood functions. *Ann. Stat., 11*, 59–67.

McCullagh, P., and Nelder, J. A. (1983). *Generalized Linear Models.* Chapman and Hall: London.

McGilchrist, C. A., Sandland, R. L., and Hill, L. J. (1981). Estimation in regression models with stationary dependent errors. *Commun. Stat. Theory Methods A, 10*, 2563–2580.

McKeown, J. J. (1975). Specialised versus general-purpose algorithms for minimizing functions that are sums of squared terms. *Math. Prog., 9*, 57–68.

MacKinnon, J. G., White, H., and Davidson, R. (1983). Tests for model specification in the presence of alternative hypotheses. *J. Econometrics, 21*, 53–70.

McLean, D. D. (1985). Discussion of Bates and Watts. *Technometrics, 27*, 340–348.

McLean, D. D., Pritchard, D. J., Bacon, D. W., and Downie, J. (1979). Singularities in multiresponse modelling. *Technometrics, 21*, 291–298.

Maddala, G. S. (1977). *Econometrics.* McGraw-Hill: New York.

Madsen, L. T. (1979). The geometry of statistical models—a generalization of curvature. Research Report 79–1, Statist. Res. Unit, Danish Medical Res. Council.

Maindonald, J. H. (1984). *Statistical Computation.* Wiley: New York.

Mak, T. K. (1983). On Sprent's generalized least-squares estimator. *J. R. Stat. Soc. B, 45*, 380–383.

Malinvaud, E. (1970a). The consistency of nonlinear regressions. *Ann. Math. Stat., 41*, 956–969.

Malinvaud, E. (1970b). *Statistical Methods of Econometrics*, translated by A. Silvey. North-Holland: Amsterdam.

Mannervik, B. (1981). Design and analysis of kinetic experiments for discrimination between rival models. *In* L. Endrenyi (Ed.), *Kinetic Data Analysis: Design and Analysis of Enzyme and Pharmacokinetic Experiments*, pp. 235–270. Plenum Press: New York.

Marcus, A. H. (1979). Semi-Markov compartmental models in ecology and environmental health. *In* J. H. Matis, B. C. Patten, and G. C. White (Eds.), *Compartmental Analysis of Ecosystem Models*, Statistical Ecology Series, Vol. 10, pp. 261–278. International Co-operative Publishing House: Fairland, Md.

Maritz, J. S. (1981). *Distribution-Free Statistical Methods.* Chapman and Hall: London.

Maronna, R., and Yohai, V. J. (1978). A bivariate test for the detection of a systematic change in mean. *J. Am. Stat. Assoc., 73*, 640–649.

Marquardt, D. W. (1963). An algorithm for least-squares estimation of nonlinear parameters. *SIAM J. Appl. Math.*, *11*, 431–441.

Matis, J. H., and Hartley, H. O. (1971). Stochastic compartmental analysis: Model and least squares estimation from time series data. *Biometrics.*, *27*, 77–102.

Matis, J. H., and Wehrly, T. E. (1979a). Stochastic models of compartmental systems. *Biometrics*, *35*, 199–220.

Matis, J. H., and Wehrly, T. E. (1979b). An approach to a compartmental model with multiple sources of stochasticity for modelling ecological systems. *In* J. H. Matis, B. C. Patten, and G. C. White, *Compartmental Analysis of Ecosystem Models*, Statistical Ecology Series, Vol. 10, pp. 195–222. International Co-operative Publishing House: Fairland, Md.

Matis, J. H., Patten, B. C., and White, G. C. (Eds.) (1979). *Compartmental Analysis of Ecosystem Models*, Statistical Ecology Series, Vol. 10. International Co-operative Publishing House: Fairland, Md.

Matis, J. H., Wehrly, T. E., and Gerald, K. B. (1983). The statistical analysis of pharmacokinetic data. *In* R. M. Lambrecht and A. Rescigno (Eds.), *Tracer Kinetics and Physiologic Modeling*, pp. 1–58. Lecture Notes in Biomathematics 48. Springer-Verlag: New York.

Mead, R. (1970). Plant density and crop yield. *Appl. Stat.*, *19*, 64–81.

Mead, R. (1979). Competition experiments. *Biometrics*, *35*, 41–54.

Medawar, P. B. (1940). Growth, growth energy, and ageing of the chicken's heart. *Proc. Roy. Soc. London*, *129*, 332–355.

Mehata, K. M., and Selvam, D. P. (1981). A stochastic model for the *n*-compartment irreversible system. *Bull. Math. Biol.*, *43*, 549–561.

Mélard, G. (1984). Algorithm AS197. A fast algorithm for the exact likelihood of autoregressive–moving average models. *Appl. Stat.*, *33*, 104–110.

Metzler, C. M. (1971). Usefulness of the two-compartment open model in pharmacokinetics. *J. Am. Stat. Assoc.*, *66*, 49–53.

Meyer, R. R., and Roth, P. M. (1972). Modified damped least squares: An algorithm for non-linear estimation. *J. Inst. Math. Appl.*, *9*, 218–233.

Miller, A. J. (1979). Problems in nonlinear least squares estimation. *In* D. McNeil (Ed.), *Interactive Statistics*, pp. 39–51. North-Holland: Amsterdam.

Miller, D. M. (1984). Reducing transformation bias in curve fitting. *Am. Stat.*, *38*, 124–126.

Miller, R. G., Jr. (1981). *Simultaneous Statistical Inference*, 2nd ed. McGraw-Hill: New York.

Miller, R. G., Jr., Halks-Miller, M., Egger, M. J., and Halpern, J. W. (1982). Growth kinetics of glioma cells. *J. Am. Stat. Assoc.*, *77*, 505–514.

Milliken, G. A., and De Bruin, R. L. (1978). A procedure to test hypotheses for nonlinear models. *Commun. Stat. Theory Methods A*, *7*, 65–79.

Milliken, G. A., De Bruin, R. L., and Smith, J. E. (1977). A simulation study of hypothesis testing procedures for the Michaelis–Menten enzyme kinetic model. Tech. Report No. 36, Dept. of Statistics, Kansas State Univ., Manhattan, Kansas 66506.

Mizon, G. E. (1977). Inferential procedures in nonlinear models: An application in a UK industrial cross section study of factor substitution and returns to scale. *Econometrica*, *45*, 1221–1242.

Moen, D. H., Salazar, D., and Broemeling, L. D. (1985). Structural changes in multivariate regression models. *Commun. Stat. Theory Methods A*, *14*, 1757–1768.

Moler, C., and Van Loan, C. (1978). Nineteen dubious ways to compute the exponential of a matrix. *SIAM Rev.*, *20*, 801–836.

Moolgavkar, S. H., Lustbader, E. D., and Venzon, D. J. (1984). A geometric approach to nonlinear regression diagnostics with application to matched case-control studies. *Ann. Stat.*, *12*, 816–826.

Moran, P. A. P. (1971). Estimating structural and functional relationships. *J. Multivariate Anal.*, *1*, 232–255.

Moré, J. J. (1977). The Levenberg–Marquardt algorithm: Implementation and theory. *In* G. A. Watson (Ed.), *Numerical Analysis, Lecture Notes in Mathematics 630*, pp. 105–116. Springer-Verlag: Berlin.

Moré, J. J., Garbow, B. S., and Hillstrom, K. E. (1981). Testing unconstrained optimization software. *ACM Trans. Math. Software*, *7*, 17–41.

Morgan, P. H., Mercer, L. P., and Flodin, N. W. (1975). General model for nutritional responses of higher organisms. *Proc. Nat. Acad, Sci. U.S.A.*, *72*, 4327–4331.

Morton, R. (1981). Estimating equations for an ultrastructural relationship. *Biometrika*, *68*, 735–737.

Morton, R. (1987a). A generalized linear model with nested strata of extra-Poisson variation. *Biometrika*, *74*, 247–257.

Morton, R. (1987b). Asymmetry of estimators in nonlinear regression. *Biometrika*, *74*, 679–685.

Murray, W. (Ed.) (1972a). *Numerical Methods for Unconstrained Optimization*. Academic Press: London.

Murray, W. (1972b). Second derivative methods. *In* W. Murray (Ed.), *Numerical Methods for Unconstrained Optimization*, pp. 57–71. Academic Press: London.

Murray, W., and Overton, M. L. (1980). A projected Lagrangian algorithm for nonlinear L_1 optimization. Report SOL 80-4, Dept. of Operations Research, Stanford Univ.

Narula, S. C., and Wellington, J. F. (1982). The minimum sum of absolute errors regression: A state of the art survey. *Internat. Stat. Rev.*, *50*, 317–326.

Nash, J. C. (1977). Minimizing a nonlinear sum of squares function on a small computer. *J. Inst. Math. Appl.*,*19*, 231–237.

Nash, J. C. (1979). *Compact Numerical Methods for Computers: Linear Algebra and Function Minimisation*. Adam Hilger: Bristol.

Nash, J. C., and Walker-Smith, M. (1987). *Nonlinear Parameter Estimation: An Integrated System in BASIC*. Marcel Dekker: New York.

Naylor, J. C., and Smith, A. F. M. (1982). Applications of a method for the efficient computation of posterior distributions. *Appl. Stat.*, *31*, 214–225.

Nazareth, L. (1977). A conjugate-direction algorithm without line searches. *J. Optim. Theory Appl.*, *23*, 373–388.

Nazareth, L. (1980), Some recent approaches to solving large residual nonlinear least squares problems. *SIAM Rev.*, *22*, 1–11.

Neill, J. W., and Johnson, D. E. (1984). Testing for lack of fit in regression—a review. *Commun. Stat. Theory Methods A*, *13*, 485–511.

Nelder, J. A. (1961). The fitting of a generalization of the logistic curve. *Biometrics, 17*, 89–110.

Nelder, J. A. (1963). Yield–density relations and Jarvis's lucerne data. *J. Agric. Sci., 61*, 427–429.

Nelder, J. A. (1966). Inverse polynomials, a useful group of multifactor response functions. *Biometrics, 22*, 128–141.

Nelder, J. A., and Mead, R. (1965). A simplex method for function minimization. *Comput. J., 7*, 308–313.

Nelder, J. A., and Pregibon, D. (1987). An extended quasi-likelihood function. *Biometrika, 74*, 221–232.

Nelder, J. A., and Wedderburn, R. W. M. (1972). Generalized linear models. *J. R. Stat. Soc. A, 135*, 370–384.

Nelder, J. A., Austin, R. B., Bleasdale, J. K. A., and Salter, P. J. (1960). An approach to the study of yearly and other variation in crop yields. *J. Horticultural Sci., 35*, 73–82.

Newbold, P. (1981). Some recent developments in time series analysis. *Internat. Stat. Rev., 49*, 53–66.

Neyman, J., and Scott, E. (1948). Consistent estimates based on partially consistent observations. *Econometrica, 16*, 1–32.

Oberhofer, W. (1982). The consistency of nonlinear regression minimizing the L_1-norm. *Ann. Stat., 10*, 316–319.

Olsson, D. M., and Nelson, L. S. (1975). The Nelder–Mead simplex procedure for function minimization. *Technometrics, 17*, 45–51.

Olsson, G. (1965). Distance and human interaction. A migration study. *Geografiska Annaler, 47*, 3–43.

O'Neill, R. (1971). Algorithm AS47: Function minimization using a simplex procedure. *Appl. Stat., 20*, 338–345.

O'Neill, R. V. (1979). A review of linear compartmental analysis in ecosystem science. *In* J.H. Matis, B. C. Patten, and G. C. White (Eds.), *Compartmental Analysis of Ecosystem Models*, Statistical Ecology Series, Vol. 10, pp. 3–27. International Co-operative Publishing House: Fairland, Md.

Osborne, M. R. (1972). Some aspects of nonlinear least squares calculations. *In* F. A. Lootsma (Ed.), *Numerical Methods for Nonlinear Optimization*, pp. 171–189. Academic Press: London.

Osborne, M. R., and Watson, G. A. (1971). An algorithm for discrete nonlinear L_1 approximation. *Comput. J., 14*, 184–188.

Pagan, A. R., and Nicholls, D. F. (1976). Exact maximum likelihood estimation of regression models with finite order moving average errors. *Rev. Econom. Stud., 43*, 383–387.

Park, R. E., and Mitchell, B. M. (1980). Estimating the autocorrelated error model with trended data. *J. Econometrics, 13*, 185–201.

Park, S. H. (1978). Experimental designs for fitting segmented polynomial regression models. *Technometrics, 20*, 151–154.

Parkinson, J. M., and Hutchinson, D. (1972). An investigation into the efficiency of variants of the simplex method. *In* F. A. Lootsma (Ed.), *Numerical Methods for Nonlinear Optimization*, pp. 115–135. Academic Press: London.

Parthasarathy, P. R., and Mayilswami, P. (1982). Stochastic compartmental models with birth, death and immigration of particles. *Commun. Stat. Theory Methods A*, *11*, 1625–1642.

Parthasarathy, P. R., and Mayilswami, P. (1983). Sojourn time distribution in a stochastic compartmental system. *Commun. Stat. Theory Methods A*, *12*, 1619–1635.

Patefield, W. M. (1977). On the information matrix in the linear functional relationship problem. *Appl. Stat.*, *26*, 69–70.

Patefield, W. M. (1978). The unreplicated ultrastructural relation: Large sample properties. *Biometrika*, *65*, 535–540.

Pearl, R. and Reed, L. J. (1925). Skew-growth curves. *Proc. Nat. Acad. Sci. U.S.A.* *11*, 16–22.

Pericchi, L. R. (1981). A Bayesian approach to transformations to normality. *Biometrika*, *68*, 35–43.

Pesaran, M. H. (1973). Exact maximum likelihood estimation of a regression equation with a first-order moving-average error. *Rev. Econom. Stud.*, *40*, 529–535.

Pettitt, A. N. (1979). A non-parametric approach to the change-point problem. *Appl. Stat.*, *28*, 126–135.

Pettitt, A. N. (1981). Posterior probabilities for a change-point using ranks. *Biometrika*, *68*, 443–450.

Phadke, M. S., and Kedem, G. (1978). Computation of the exact likelihood function of multivariate moving-average models. *Biometrika*, *65*, 511–519.

Phillips, P. C. B. (1976). The iterated minimum distance estimator and the quasi-maximum likelihood estimator. *Econometrica*, *44*, 449–460.

Pierce, D. A. (1971a). Least squares estimation in the regression model with autoregressive–moving average errors. *Biometrika*, *58*, 299–312.

Pierce, D. A. (1971b). Distribution of residual autocorrelations in the regression model with autoregressive–moving average errors. *J. R. Stat. Soc. B*, *33*, 140–146.

Pierce, D. A. (1972). Least squares estimation in dynamic-disturbance time series models. *Biometrika*, *59*, 73–78.

Pike, D. J. (1984). Discussion on a paper by Steinberg and Hunter. *Technometrics*, *26*, 105–109.

Pimentel-Gomes, F. (1953). The use of Mitscherlich's regression law in the analysis of experiments with fertilizers. *Biometrics*, *9*, 498–516.

Polak, E. (1971). *Computational Methods in Optimization: A Unified Approach.* Academic Press: New York.

Poskitt, D. S., and Tremayne, A. R. (1980). Testing the specification of a fitted autoregressive–moving average model. *Biometrika*, *67*, 359–363.

Powell, M. J. D. (1964). An efficient method for finding the minimum of a function of several variables without calculating derivatives. *Comput. J.*, *7*, 155–162.

Powell, M. J. D. (1965). A method for minimizing a sum of squares of nonlinear functions without calculating derivatives. *Comput. J.*, *7*, 303–307.

Powell, M. J. D. (1970). A hybrid method for nonlinear equations. *In* P. Rabinowitz (Ed.), *Numerical Methods for Nonlinear Algebraic Equations*, pp. 87–114. Gordon and Breach: London.

Powell, M. J. D. (1972). Unconstrained minimization algorithms without computation of derivatives. AERE Harwell Report TP483.

Powell, M. J. D. (1976). A view of unconstrained optimization. *In* L. C. W. Dixon (Ed.), *Optimization in Action*, pp. 117–152. Academic Press: London.

Powell, M. J. D. (1977). Restart procedures for the conjugate gradient method. *Math. Prog.*, *12*, 241–254.

Prais, S. J., and Winsten, C. B. (1954). Trend estimators and serial correlation. Unpublished Cowles Commission discussion paper: Stat. No. 383, Chicago.

Pratt, J. W. (1981). Concavity of the log likelihood. *J. Am. Stat. Assoc.*, *76*, 103–106.

Prentice, R. L. (1976). A generalization of the probit and logit methods for dose–response curves. *Biometrics*, *32*, 761–768.

Pritchard, D. J., and Bacon, D. W. (1977). Accounting for heteroscedasticity in experimental design. *Technometrics*, *19*, 109–115.

Pritchard, D. J., Downie, J., and Bacon, D. W. (1977). Further consideration of heteroscedasticity in fitting kinetic models. *Technometrics*, *19*, 227–236.

Prunty, L. (1983). Curve fitting with smooth functions that are piecewise-linear in the limit. *Biometrics*, *39*, 857–866.

Purdue, P. (1979). Stochastic compartmental models: A review of the mathematical theory with ecological application. *In* J. H. Matis, B. C. Patten, and G. C. White (Eds.), *Compartmental Analysis of Ecosystem Models*, Statistical Ecology Series, Vol. 10, pp. 223–260. International Co-operative Publishing House: Fairland, Md.

Quandt, R. E. (1958). The estimation of the parameters of a linear regression system obeying two separate regimes. *J. Am. Stat. Assoc.*, *53*, 873–880.

Quandt, R. E. (1960). Tests of the hypothesis that a linear regression system obeys two separate regimes. *J. Am. Stat. Assoc.*, *55*, 324–330.

Quandt, R. E. (1974). A comparison of methods for testing non-nested hypotheses. *Rev. of Econom. and Stat.*, *56*, 92–99.

Quandt, R. E., and Ramsey, J. B. (1978). Estimating mixtures of normal distributions and switching regressions (with Discussion). *J. Am. Stat. Assoc.*, *73*, 730–752.

Raksanyi, A., Lecourtier, Y., and Walter, E. (1985). Identifiability and distinguishability testing via computer algebra. *Math. Biosci.*, *77*, 245–266.

Ralston, M. L., and Jennrich, R. I. (1978). Dud, a derivative-free algorithm for nonlinear least squares. *Technometrics*, *20*, 7–14.

Rao, C. R. (1947). Large sample tests of statistical hypotheses concerning several parameters with applications to problems of estimation. *Proc. Camb. Phil. Soc.*, *44*, 50–57.

Rao, C. R. (1959). Some problems involving linear hypotheses in multivariate analysis. *Biometrika*, *46*, 49–58.

Rao, C. R. (1973). *Linear Statistical Inference and Its Applications*, 2nd ed. Wiley: New York.

Rao, P., and Griliches, Z. (1969). Small-sample properties of several two-stage regression methods in the context of auto-correlated errors. *J. Am. Stat. Assoc.*, *64*, 253–272.

Ratkowsky, D. A. (1983). *Nonlinear Regression Modeling*. Marcel Dekker: New York.

Ratkowsky, D. A. (1988). *Handbook of Nonlinear Regression Models*. Marcel Dekker: New York.

Ratkowsky, D. A., and Dolby, G. R. (1975). Taylor series linearization and scoring for parameters in nonlinear regression. *Appl. Stat., 24,* 109–111.

Ratkowsky, D. A., and Reedy, T. J. (1986). Choosing near-linear parameters in the four-parameter logistic model for radioligand and related assays. *Biometrics, 42,* 575–582.

Reeds, J. (1975). Discussions on Efron's paper. *Ann. Stat., 3,* 1234–1238.

Reich, J. G. (1974). Analysis of kinetic and binding measurements. I. Information content of kinetic data. *Studia Biophys., 42,* 165–180.

Reich, J. G. (1981). On parameter redundancy in curve fitting of kinetic data. *In* L. Endrenyi (Ed.), *Kinetic Data Analysis: Design and Analysis of Enzyme and Pharmacokinetic Experiments,* pp. 39–50. Plenum Press: New York.

Reich, J. G., and Zinke, I. (1974). Analysis of kinetic and binding measurements IV. Redundancy of model parameters. *Studia Biophys., 43,* 91–107.

Reilly, P. M. (1976). The numerical computation of posterior distributions in Bayesian statistical inference. *Appl. Stat., 25,* 201–209.

Reilly, P. M., and Patino-Leal, H. (1981). A Bayesian study of the error-in-variables model. *Technometrics, 23,* 221–231.

Rescigno, A., and Matis, J. H. (1981). On the relevance of stochastic compartmental models to pharmacokinetic systems (Letter to the Editor). *Bull. Math. Biol., 43,* 245–247.

Rescigno, A., and Segre, G. (1966). *Drug and Tracer Kinetics.* Blaisdell: Waltham, Mass.

Rescigno, A., Lambrecht, R. M., and Duncan, C. C. (1983). Mathematical methods in the formulation of pharmacokinetic models. *In* R. M. Lambrecht and A. Rescigno (Eds.), *Tracer Kinetics and Physiologic Modeling,* pp. 59–119. Lecture Notes in Biomathematics 48; Springer-Verlag: New York.

Rhead, D. G. (1974). On a new class of algorithms for function minimization without evaluating derivatives, Dept. of Mathematics Report, Univ. of Nottingham.

Richards, F. J. (1959). A flexible growth function for empirical use. *J. Exp. Botany, 10,* 290–300.

Richards, F. J. (1969). The quantitative analysis of growth. *In* F. C. Stewart (Ed.), *Plant Physiology, Volume VA: Analysis of Growth: Behaviour of Plants and their Organs,* Chapter 1, pp. 3–76. Academic Press: New York.

Richards, F. S. G. (1961). A method of maximum-likelihood estimation. *J. R. Stat. Soc. B, 23,* 469–476.

Robertson, T. B. (1908). On the normal rate of growth of an individual and its biochemical significance. *Roux' Arch. Entwicklungsmech. Organismen, 25,* 581–614.

Robinson, P. M. (1985). Testing for serial correlation in regression with missing observations. *J. R. Stat. Soc. B, 47,* 429–437.

Rodda, B. E. (1981). Analysis of sets of estimates from pharmacokinetic studies. *In* L. Endrenyi (Ed.), *Kinetic Data Analysis: Design and Analysis of Enzyme and Pharmacokinetic Experiments,* pp. 285–298. Plenum Press: New York.

Rogers, A. J. (1986). Modified Lagrange multiplier tests for problems with one-sided alternatives. *J. Econometrics, 31,* 341–361.

Ross, G. J. S. (1970). The efficient use of function minimization in non-linear maximum likelihood estimation. *Appl. Stat., 19,* 205–221.

Ross, G. J. S. (1980). Uses of non-linear transformation in non-linear optimisation problems. *In* M. M. Barritt and D. Wishart (Eds.), *COMPSTAT 1980, Proceedings in Computational Statistics*, pp. 382–388. Physica-Verlag: Vienna.

Ross, W. H. (1987). The geometry of case deletion and the assessment of influence in nonlinear regression. *Canad. J. Stat.*, *15*, 91–103.

Rothenberg, T. J. (1980). Comparing alternative asymptotically equivalent tests. Invited paper presented at World Congress of the Econometric Society, Aix-en-Provence.

Royden, H. L. (1968). *Real Analysis*, 2nd ed. MacMillan: New York.

Ruhe, A., and Wedin, P. A. (1980). Algorithms for separable nonlinear least squares problems. *SIAM Rev.*, *22*, 318–337.

Sachs, W. H. (1976). Implicit multifunctional nonlinear regression analysis. *Technometrics*, *18*, 161–173.

St. John, R. C., and Draper, N. R. (1975). *D*-optimality for regression designs: A review. *Technometrics*, *17*, 15–23.

Salzer, H. E., Zucker, R., and Capuano, R. (1952). Table of the zeroes and weight factors of the first twenty Hermite polynomials. *J. Res. Nat. Bur. Standards*, *48*, 111–116.

Sandland, R. L. (1983). Mathematics and the growth of organisms—some historical impressions. *Math. Scientist*, *8*, 11–30.

Sandland, R. L., and McGilchrist, C. A. (1979). Stochastic growth curve analysis. *Biometrics*, *35*, 255–271.

Savin, N. E. (1976). Conflict among testing procedures in a linear regression model with autoregressive disturbances. *Econometrica*, *44*, 1303–1315.

Savin, N. E. (1984). Multiple hypothesis testing. *In* Z. Griliches and M. D. Intriligator (Eds.), *Handbook of Econometrics*, Vol. II, pp. 827–879. Elsevier Science Publishers.

Scales, L. E. (1985). *Introduction to Nonlinear Optimization*. Macmillan: London.

Scammon, R. E. (1927). The first seriatim study of human growth. *Am. J. Phys. Anthropol.*, *10*, 329–336.

Schechtman, E. (1983). A conservative nonparametric distribution-free confidence bound for the shift in the change-point problem. *Commun. Stat. Theory Methods A*, *12*, 2455–2464.

Schechtman, E., and Wolfe, D. A. (1985). Multiple change-points problem—nonparametric procedures for estimation of the points of change. *Commun. Stat. Simul. Comput. B*, *14*, 615–631.

Scheffé, H. (1959). *The Analysis of Variance*. Wiley: New York.

Schlossmacher, E. J. (1973). An iterative technique for absolute deviations curve fitting. *J. Am. Stat. Assoc.*, *68*, 857–859.

Schoenberg, I. J. (1964). Spline functions and the problem of graduation. *Proc. Nat. Acad. Sci. U.S.A.*, *52*, 947–950.

Schweder, T. (1976). Some "optimal" methods to detect structural shift or outliers in regression. *J. Am. Stat. Assoc.*, *71*, 491–501.

Schwetlick, H., and Tiller, V. (1985). Numerical methods for estimating parameters in nonlinear models with errors in the variables. *Technometrics*, *27*, 17–24.

Scott, A. J., Rao, J. N. K., and Thomas, D. R. (1989). Weighted least squares and quasi-likelihood estimation for categorical data under singular models. Report Series, Dept. Math. and Stat., Auckland Univ., New Zealand

Seber, G. A. F. (1964). The linear hypothesis and large sample theory. *Ann. Math. Stat.*, *35*, 773–779.

Seber, G. A. F. (1967). Asymptotic linearisation of nonlinear hypotheses. *Sankhyā A*, *29*, 183–190.

Seber, G. A. F. (1977). *Linear Regression Analysis.* Wiley: New York.

Seber, G. A. F. (1980). *The Linear Hypothesis: A General Theory.* Griffin's Statistical Monographs No. 19, 2nd ed. Griffin: London.

Seber, G. A. F. (1984). *Multivariate Observations.* Wiley: New York.

Seber, G. A. F. (1986). A review of estimating animal abundance. *Biometrics*, *42*, 267–292.

Sen, P. K. (1980). Asymptotic theory of some tests for a possible change in the regression slope occurring at an unknown time point. *Z. Wahrscheinlichkeitstheorie verw. Gebiete*, *52*, 203–218.

Sen, P. K. (1983). On some recursive residual rank tests for change-points. *In* M. Haseeth Rizvi, S. Jagdish, and D. S. Rustagi (Eds.), *Recent Advances in Statistics: Papers in Honor of Herman Chernoff on His Sixtieth Birthday*, pp. 371–391. Academic Press: New York.

Serfling, R. J. (1980). *Approximation Theorems of Mathematical Statistics.* Wiley: New York.

Shaban, S. A. (1980). Change-point problem and two-phase regression: An annotated bibliography. *Internat. Stat. Rev.*, *48*, 83–93.

Shah, B. K., and Khatri, C. G. (1965). A method of fitting the regression curve $E(y) = \alpha + \delta x + \beta\rho^x$. *Technometrics*, *7*, 59–65.

Shah, B. K., and Khatri, C. G. (1973). An extension to a method of fitting regression curve $E(y) = \alpha + \delta x + \beta\rho^x$. *Commun. Stat.*, *1*, 365–370.

Shanno, D. F. (1970). Conditioning of quasi-Newton methods for function minimization. *Math. Comp.*, *24*, 647–657.

Shanno, D. F., and Phua, K. H. (1976). Algorithm 500. Minimization of unconstrained multivariate functions. *ACM Trans. Math. Software*, *2*, 87–94.

Shanno, D. F., and Phua, K. H. (1980). Remark on Algorithm 500. *ACM Trans. Math. Software*, *6*, 618–622.

Sheppard, C. W. (1962). *Basic Principles of the Tracer Method.* Wiley: New York.

Shinozaki, K., and Kira, T. (1956). Intraspecific competition among higher plants. VII. Logistic theory of the C–D effect. *J. Inst. Polytech. Osaka City Univ. D7*, 35–72.

Siegmund, D. (1986). Boundary crossing probabilities and statistical applications. *Ann. Stat.*, *14*, 361–404.

Silverman, B. W. (1984). Spline smoothings: The equivalent variable kernel method. *Ann. Stat.*, *12*, 898–916.

Silverman, B. W. (1985). Some aspects of the spline smoothing approach to non-parametric regression curve fitting (with Discussion). *J. R. Stat. Soc. B*, *47*, 1–52.

Silvey, S. D. (1959). The Lagrangian multiplier test. *Ann. Math. Stat.*, *30*, 389–407.

Silvey, S. D. (1980). *Optimal Design: An Introduction to the Theory for Parameter Estimation.* Chapman and Hall: London.

Simonoff, J. S., and Tsai, C.-L. (1986). Jackknife-based estimators and confidence regions in nonlinear regression. *Technometrics*, *28*, 103–112.

Smith, A. F. M., and Cook, D. G. (1980). Straight lines with a change-point: A Bayesian analysis of some renal transplant data. *Appl. Stat.*, *29*, 180–189.

Smith, A. F. M., Skene, A. M., Shaw, J. E. H., Naylor, J. C., and Dransfield, M. (1985). The implementation of the Bayesian paradigm. *Commun. Stat. Theory Methods A, 14*, 1079–1102.

Smith, B. T., Boyle, J. M., Dongarra, J. J., Garbow, B. S., Ikebe, Y., Klema, V. C., and Moler, C. B. (1976a). *Matrix Eigensystem Routines—EISPACK Guide*. Springer-Verlag: Berlin.

Smith, M. R., Cohn-Sfetcu, S., and Buckmaster, H. A. (1976b). Decomposition of multicomponent exponential decays by spectral analytic techniques. *Technometrics, 18*, 467–482.

Smith, P. L. (1979). Splines as a useful and convenient statistical tool. *Am. Stat., 33*, 57–62.

Smith, R. L., and Naylor, J. C. (1987). A comparison of maximum likelihood and Bayesian estimators for the three-parameter Weibull distribution. *Appl. Stat., 36*, 358–369.

Snedecor, G. W., and Cochran, W. G. (1967). *Statistical Methods*, 6th ed. Iowa State Univ. Press: Ames.

Spendley, W. (1969). Nonlinear least squares fitting using a modified simplex minimization method. *In* R. Fletcher (Ed.), *Optimization*, pp. 259–270. Academic Press: London and New York.

Spergeon, E. F. (1949). *Life Contingencies*. Cambridge Univ. Press.

Spitzer, J. J. (1979). Small-sample properties of nonlinear least squares and maximum likelihood estimators in the context of autocorrelated errors. *J. Am. Stat. Assoc., 74*, 41–47.

Sprent, P. (1961). Some hypotheses concerning two phase regression lines. *Biometrics, 17*, 634–645.

Sprent, P. (1969). *Models in Regression and Related Topics*. Methuen: London.

Steece, B. M. (1979). A Bayesian approach to model adequacy. *Commun. Stat. Theory Methods A, 14*, 1393–1402.

Stein, C. (1962). Confidence sets for the mean of the multivariate normal distribution (with Discussion). *J. R. Stat. Soc. B, 24*, 265–296.

Stein, C. (1965). Approximation of improper prior measures by prior probability measures. *In* J. Neyman and L. LeCam (Eds.), *Bernoulli, Bayes, Laplace Volume*, pp. 217–240. Springer-Verlag OHG: Berlin.

Steinberg, D. M., and Hunter, W. G. (1984). Experimental design: Review and comment. *Technometrics, 26*, 71–97.

Stevens, W. L. (1951). Asymptotic regression. *Biometrics, 7*, 247–267.

Stewart, G. W. (1967). A modification of Davidon's method to accept difference approximations of derivatives. *J. Assoc. Comput. Mach, 14*, 72–83.

Stewart, W. E., and Sørensen, J. P. (1981). Bayesian estimation of common parameters from multiresponse data with missing observations. *Technometrics, 23*, 131–141.

Stirling, W. D. (1984). Iteratively reweighted least squares for models with a linear part. *Appl. Stat., 33*, 7–17.

Stirling, W. D. (1985). Algorithm AS212: Fitting the exponential curve by least squares. *Appl. Stat., 34*, 183–192.

Stone, C. J. (1977). Consistent non-parametric regression. *Ann. Stat., 5*, 595–645.

Stone, M. (1976). Strong inconsistency from uniform priors. *J. Am. Stat. Assoc., 71*, 114–116.

Stroud. T. W. F. (1971). On obtaining large-sample tests from asymptotically normal estimators. *Ann. Math. Stat.*, *42*, 1412–1424.

Sundararaj, N. (1978). A method for confidence regions for nonlinear models. *Austral. J. Stat.*, *20*, 270–274.

Swann, W. H. (1972). Direct search methods. *In* W. Murray (Ed.), *Numerical Methods for Unconstrained Optimization*, pp. 13–28. Academic Press: London.

Sweeting, T. J. (1984). On the choice of prior distribution for the Box–Cox transformed linear model. *Biometrika*, *71*, 127–134.

Taylor, W. E. (1981). On the efficiency of the Cochrane–Orcutt estimator. *J. Econometrics*, *17*, 67–82.

Tiede, J. J., and Pagano, M. (1979). The application of robust calibration to radioimmunoassay. *Biometrics*, *35*, 567–574.

Tierney, L., and Kadane, J. B. (1986). Accurate approximations for posterior moments and marginal densities. *J. Am. Stat. Assoc.*, *81*, 82–86.

Tishler, A., and Zang, I. (1981a). A new maximum likelihood algorithm for piecewise regression. *J. Am. Stat. Assoc.*, *76*, 980–987.

Tishler, A., and Zang, I. (1981b). A maximum likelihood method for piecewise regression models with a continuous dependent variable. *Appl. Stat.*, *30*, 116–124.

Tiwari, J. L. (1979). A modelling approach based on stochastic differential equations, the principle of maximum entropy, and Bayesian inference for parameters. *In* J. H. Matis, B. C. Patten, and G. C. White (Eds.), *Compartmental Analysis of Ecosystem Models*, Statistical Ecology Series, Vol. 10, pp. 167–194. International Co-operative Publishing House: Fairland, Md.

Vajda, S. (1984). Analysis of unique structural identifiability via submodels. *Math. Biosci.*, *71*, 125–146.

Valiant, R. (1985). Nonlinear prediction theory and the estimation of proportions in a finite population. *J. Am. Stat. Assoc.*, *80*, 631–641.

Van der Waerden, B. L. (1953). *Modern Algebra.* Frederick Ungar: New York.

Velleman, P. F. (1980). Definition and comparison of robust nonlinear data smoothing algorithms. *J. Am. Stat. Assoc.*, *75*, 609–615.

Verbyla, A. P. (1985). A note on the inverse covariance matrix of the autoregressive process. *Austral. J. Stat.*, *27*, 221–224.

Vieira, S., and Hoffman, R. (1977). Comparison of the logistic and the Gompertz growth functions considering additive and multiplicative error terms. *Appl. Stat.*, *26*, 143–148.

Villegas, C. (1961). Maximum likelihood estimation of a linear functional relationship. *Ann. Math. Stat.*, *32*, 1048–1062.

Von Bertalanffy, L. (1957). Quantitative laws in metabolism and growth. *Quart. Rev. Biol.*, *32*, 217–231.

Wagner, J. G. (1975). *Fundamentals of Clinical Pharmacokinetics.* Drug Intelligence Publications: Hamilton, Illinois.

Wahba, G. (1978). Improper priors, spline smoothing and the problem of guarding against model errors in regression. *J. R. Stat. Soc. B*, *40*, 364–372.

Wahba, G. (1983). Bayesian "confidence intervals" for the cross-validated smoothing spline. *J. R. Stat. Soc. B*, *45*, 133–150.

Wald, A. (1943). Test of statistical hypotheses concerning several parameters when the number of observations is large. *Trans. Am. Math. Soc.*, *54*, 426–482.

Watts, D. G. (1981). An introduction to nonlinear least squares. *In* L. Endrenyi (Ed.), *Kinetic Data Analysis: Design and Analysis of Enzyme and Pharmacokinetic Experiments*, pp. 1–24. Plenum Press: New York.

Watts, D. G., and Bacon, D. W. (1974). Using an hyperbola as a transition model to fit two-regime straight-line data. *Technometrics*, *16*, 369–373.

Wecker, W. E., and Ansley, C. F. (1983). The signal extraction approach to nonlinear regression and spline smoothing. *J. Am. Stat. Assoc.*, *78*, 81–89.

Wedderburn, R. W. M. (1974). Quasi-likelihood functions, generalized linear models and the Gauss–Newton method. *Biometrika*, *61*, 439–447.

Wedderburn, R. W. M. (1976). On the existence and uniqueness of the maximum likelihood estimates for certain generalized linear models. *Biometrika*, *63*, 27–32.

Wegman, E. J., and Wright, I. W. (1983). Splines in statistics. *J. Am. Stat. Assoc.*, *78*, 351–365.

Welsch, R. E., and Becker, R. A. (1975). Robust nonlinear regression using the DOGLEG algorithm. *In* J. W. Frane (Ed.), *Proc. Comp. Sci. and Stat. 8th Annual Symp. Interface*, pp. 272–279. Health Sciences Computing Facility: Univ. of California at Los Angeles.

White, G. C. (1982). Nonlinear computer models for transuranic studies. Technical Report, *TRAN-STAT*, Battelle Memorial Inst., Pacific Northwest Lab., Richland, Wash.

White, G. C., and Brisbin, I. L., Jr. (1980). Estimation and comparison of parameters in stochastic growth models for barn owls. *Growth*, *44*, 97–111.

White, G. C., and Clark, G. M. (1979). Estimation of parameters for stochastic compartment models. *In* J. H. Matis, B. C. Patten, and G. C. White (Eds.), *Compartmental Analysis of Ecosystem Models*, Statistical Ecology Series, Vol. 10, pp. 131–144. International Co-operative Publishing House: Fairland, Md.

White, G. C., and Ratti, J. T. (1977). Estimation and testing of parameters in Richards' growth model for Western Grebes. *Growth*, *41*, 315–323.

White, H. (1980). Nonlinear regression on cross-section data. *Econometrica*, *48*, 721–746.

White, H. (1981). Consequences and detection of misspecified nonlinear regression models. *J. Am. Stat. Assoc.*, *76*, 419–433.

White, H. (1982). Maximum likelihood estimation of misspecified models. *Econometrica*, *50*, 1–25.

White, H., and Domowitz, I. (1984). Nonlinear regression with dependent observations. *Econometrica*, *52*, 143–161.

Willey, R. W., and Heath, S. B. (1969). The quantitative relationships between plant population and crop yield. *Adv. Agronomy*, *21*, 281–321.

Williams, D. A. (1970). Discrimination between regression models to determine the pattern of enzyme synthesis in synchronous cell cultures. *Biometrics*, *26*, 23–32.

Williams, E. J. (1962). Exact fiducial limits in non-linear estimation. *J. R. Stat. Soc. B*, *24*, 125–139.

Wise, M. F. (1979). The need for rethinking on both compartments and modelling. *In* J. H. Matis, B. C. Patten, and G. C. White (Eds.), *Compartmental Analysis of Ecosystem Models*, Statistical Ecology Series, Vol. 10, pp. 279–293, International Co-operative Publishing House: Fairland, Md.

Wold, H., and Lyttkens, E. (1969). Nonlinear iterative partial least squares (NIPALS) estimation procedures. *Bull. Internat. Stat. Inst.*, *43*, 29–51.

Wold, S. (1974). Spline functions in data analysis. *Technometrics*, *16*, 1–11.

Wolfe, D. A., and Schechtman, E. (1984). Nonparametric statistical procedures for the changepoint problem. *J. Stat. Plan. Inf.*, *9*, 389–396.

Wolter, K. M., and Fuller, W. A. (1982). Estimation of nonlinear errors-in-variables models. *Ann. Stat.*, *10*, 539–548.

Wood, J. T. (1974). An extension of the analysis of transformations of Box and Cox. *Appl. Stat.*, *23*, 278–283.

Worsley, K. J. (1979). On the likelihood ratio test for a shift in location of normal populations. *J. Am. Stat. Assoc.*, *74*, 365–367.

Worsley, K. J. (1982). An improved Bonferroni inequality and applications. *Biometrika*, *69*, 297–302.

Worsley, K. J. (1983). Testing for a two-phase multiple regression, *Technometrics*, *25*, 35–42.

Wright, D. E. (1986). A note on the construction of highest posterior density intervals. *Appl. Stat.*, *35*, 49–53.

Wright, S. (1926). Book review. *J. Am. Stat. Assoc.*, *21*, 493–497.

Wu, C. F. (1981). Asymptotic theory of nonlinear least squares estimation. *Ann. Stat.*, *9*, 501–513.

Yang, S. (1985). Discussion of Bates and Watts. *Technometrics*, *27*, 349–351.

Zacks, S. (1983). Survey of classical and Bayesian approaches to the change-point problem: Fixed sample and sequential procedures of testing and estimation. *In* M. Haseeth Rizvi, S. Jagdish, and D. S. Rustagi (Eds.), *Recent Advances in Statistics: Papers in Honor of Herman Chernoff on His Sixtieth Birthday*. Academic Press: New York.

Zang, I. (1980). A smoothing-out technique for min-max optimization. *Math. Prog.*, *19*, 61–77.

Zellner, A. (1962). An efficient method of estimating seemingly unrelated regressions and tests for aggregation bias. *J. Am. Stat. Assoc.*, *57*, 348–368.

Zellner, A. (1971). *An Introduction to Bayesian Inference in Econometrics*. Wiley: New York.

Ziegel, E. R., and Gorman, J. W. (1980). Kinetic modelling with multiresponse data. *Technometrics*, *22*, 139–151.

Zwanzig, S. (1980). The choice of approximative models in nonlinear regression. *Math. Operationsforsch. Stat., Ser. Stat*, *11*, 23–47.

Author Index

Abraham, B., 287, 289, 318–320, 322, 711
Agarwal, G.G., 486, 711
Agha, M., 343, 711
Agrafiotis, G.K., 430, 711
Aguirre-Torres, V., 236, 711
Allen, D.M., 369, 381, 402, 404, 711
Amari, S., 162, 711
Amemiya, T., 19, 20, 70, 196, 228, 230, 234–235, 274, 318, 563–564, 567, 569, 574–575, 583, 711
Amemiya, Y., 524, 711
Amström, K.J., 388, 713
Anderson, D.H., 369, 375, 384–385, 387, 389–393, 711
Anderson, J.A., 455, 726
Angell, I.O., 306, 712
Ansley, C.F., 314–315, 317, 322, 487–489, 712, 729, 742
Arnold, L., 432, 695, 697–699, 712
Atkins, G.L., 368, 372–375, 381, 712
Atkinson, A.C., 236, 250–251, 253, 255, 260, 712
Austin, R.B., 332, 734
Azen, S.P., 56–57, 729

Bacon, D.W., 80–85, 257, 437, 465–468, 476–480, 545–546, 549–550, 712, 731, 736, 742
Bailey, B.J.R., 192, 712
Baker, R.J., 15, 712
Bard, Y., 64, 504, 508, 600–601, 712
Barham, R.H., 658, 660, 712
Barnett, V.D., 500, 712
Barnett, W.A., 583–585, 712
Bates, D.M., 128, 131, 133–136, 138–139, 144, 146–150, 153–154, 156–161, 163, 177, 182, 215, 217–219, 222–226, 239, 250, 261, 396–399, 401–402, 406–407, 409–411, 537–538, 542–544, 641–645, 691, 712–713, 724, 726
Batschelet, E., 368, 713
Bavely, C.A., 382, 398, 405, 713
Beal, S.L., 713
Beale, E.M.L., 98, 128, 133, 157–159, 182, 575, 713
Beaton, A.E., 650, 713
Beauchamp, J.J., 414, 415, 531, 533–536, 713
Becker, R.A., 651, 742
Beckman, R.J., 441–442, 713
Bellman, R., 388, 713
Belsey, D.A., 671
Bera, A.K., 48–49, 716
Berkey, C.S., 344, 713
Berkson, J., 13, 525, 713
Berman, M., 405, 713
Berndt, E.R., 232, 713
Berry, G., 9, 362, 713
Betts, J.T., 629, 631, 714
Bickel, P.J., 75, 523, 714
Bird, H.A., 102–103, 714
Blancett, A.L., 506, 714
Bleasdale, J.K.A., 332, 362–363, 714, 734
Booth, N.B., 446, 714
Box, G.E.P., 53–54, 63, 69, 71–72, 75–76, 78–81, 86, 195, 250–251, 254, 257, 260, 264, 287, 290, 294–295, 306, 308, 311, 314, 316, 319, 320, 322, 539–542, 545–556, 714, 730
Box, M.J., 65, 182, 250, 253, 259, 260, 544, 714, 715
Boyle, J.M., 381, 401, 740

Bradley, E.L., 45, 46, 715
Bradley, R.A., 576, 715
Brand, L., 368, 713
Brannigan, M., 486, 715
Breiman, L., 715
Brent, R.P., 614, 715
Breusch, T.S., 228, 232, 236, 715
Bright, P.B., 387–388, 396, 398–402, 411, 728
Brisbin, I.L., Jr., 272, 310, 344–345, 356–360, 742
Britt, H.I., 501, 503, 505–508, 715
Brodlie, K.W., 588, 607, 614, 715
Broemeling, L.D., 445–446, 715, 717, 727, 733
Brooks, M.A., 343, 715
Brown, G.H., 524, 715
Brown, K.M., 646, 715
Brown, R.L., 444, 484, 500, 715
Broyden, C.G., 607, 649, 715
Buckmaster, H.A., 412, 740
Bunch, J.R., 560, 720
Bunke, H., 470–471, 572, 716
Burguete, J.F., 234, 575, 716
Buricchi, L., 343, 719
Burridge, J., 576, 716
Byron, R.P., 48–49, 716

Canfield, F.B., 506, 714
Capuano, R., 61, 738
Carr, N.L., 77–78, 716
Carroll, R.J., 69, 72–74, 76, 83, 85–86, 716
Causton, D.R., 327, 717
Chambers, J.M., 628, 661, 663, 717
Chan, L.K., 524, 717
Chan, N.N., 524, 717
Chanda, K.C., 572, 717
Chapman, M.J., 392, 717
Charnes, A., 46, 717
Chin Choy, J.H., 445–446, 715, 717
Clark, G.M., 431–432, 742
Clarke, G.P.Y., 165, 183–184, 222–223, 235, 260, 717
Cleveland, W.S., 488, 717
Cobb, G.W., 445, 717
Cobelli, C., 387, 392, 412, 717
Cochran, W.G., 189, 250, 717, 740
Cochrane, D., 282, 717
Cohn-Sfetcu, S., 412, 740
Cole, T.J., 9, 717
Cook, D.G., 435–437, 454, 739
Cook R.A., 476
Cook, R.D., 73–74, 136–137, 145, 165–166, 168–171, 173–174, 177, 179, 181–182, 214, 221, 441–442, 713, 717–718
Coope, I.D., 614, 718
Cooper, L., 520, 718
Cornell, R.G., 414–415, 531, 533–536, 713, 722
Cornish-Bowden, A.J., 119, 718
Corradi, C., 657–658, 660, 718
Coutie, G.A., 195, 714
Cox, C., 46, 718
Cox, D.R., 6, 14, 69, 71, 72, 75–76, 236, 260, 415, 418, 576, 712, 714, 718
Cox, N.R., 524, 718
Craven, P., 488, 718
Cressie, N., 488, 718
Crowder, M., 46, 568, 718
Currie, D.J., 86, 718

Dagenais, M.G., 63, 718
Dahm, P.F., 455, 723
Davidon, W.C., 607, 718
Davidson, R., 236, 731
Davies, N., 321, 718
Davis, A.W., 719
Davis, P.J., 61, 719
Dawid, A.P., 162, 719
De Boer, E.S., 272, 725
De Boor, C., 459, 482, 719
De Bruin, R.L., 230, 577, 732
De Gooijer, J.G., 306, 724
Delforge, J., 392, 719
Della Corte, M., 343, 719
Deming, W.E., 505, 719
Dennis, J.E., Jr., 588, 596, 604–605, 609, 621–622, 624–625, 629, 631–633, 639–642, 645–646, 651–653, 661, 664, 666, 668–671, 674, 715, 719
Dixon, L.C.W., 588, 607, 675, 719
Doksum, K.A., 75–76, 714, 719
Dolby, G.R., 492–494, 496, 499–501, 506, 523–524, 686, 719, 737
Domowitz, I., 575, 742
Donaldson, J.R., 191–192, 195, 221–222, 719
Dongarra, J.J., 381, 401, 560, 720, 740
Downie, J., 80–85, 545–546, 549–550, 731, 736
Drane, W., 658, 660, 712
Dransfield, M., 61–63, 245, 740
Draper, N.R., 31, 65, 71–72, 109, 250, 257, 259–260, 539–542, 544, 714–715, 720, 728, 738

Duan, N., 87, 720
Duncan, C.C., 383, 737
Duncan, G.T., 207, 209–210, 212, 720
Durbin, J., 47, 285, 318, 444, 484, 715, 720
Dutta, S.R.K., 654, 720
Dutter, R., 651, 720

Eaves, D.M., 54, 720
Efron, B., 159–161, 163, 720
Egerton, M.F., 493, 495, 720
Egger, M.J., 339, 732
Egy, D., 77, 730
Eisenfeld, J., 392, 720
El-Asfouri, S., 420, 720
El-Attar, R.A., 654, 720
El-Shaarawi, A., 242–243, 434, 445, 721
Endrenyi, L., 204, 251–253, 412, 721
Engle, R.F., 228, 232, 234–235, 721
Eppright, E.S., 461, 721
Erjavec, J., 545–546, 714
Ertel, J.E., 446, 484, 721
Esterby, S.R., 434–435, 721
Eubank, R.L., 460, 481, 483–484, 486, 488–489, 721
Evans, G.B.A., 232, 721
Evans, J.M., 444, 484, 715

Faddy, M.J., 420–421, 721
Fahrmeir, L., 46, 721
Farazdaghi, H., 362, 721
Farley, J.U., 443–444, 721
Feder, P.I., 448–449, 451–452, 454, 460, 721
Fedorov, V.V., 88–89, 257–260, 525, 527, 712, 721–722
Ferreira, P.E., 446, 722
Fife, D., 386, 722
Finney, D.J., 326, 338, 722
Firth, D., 46, 722
Fisher, G., 236, 722
Fisher, R.A., 345
Fletcher, R., 232, 588, 597, 599, 603–604, 607–608, 610–611, 613–614, 625, 648–649, 709, 722
Fletcher, R.I., 330, 339–341, 722
Flodin, N.W., 340, 733
Fowlkes, E.B., 446, 484, 721
Fox, H.M., 461, 721
Fox, T., 210, 212, 722
Frappell, B.D., 189, 722
Freeman, T.G., 492–493, 499–501, 719
Freeman, T.L., 603, 722
Friedman, J.H., 488, 722
Frisen, M., 659, 722
Frome, E.L., 46, 717
Frydman, R., 277–278, 722
Fryer, B.A., 461, 721
Fuguitt, R.E., 550, 722
Fuller, E.S., 461, 721
Fuller, W.A., 86, 457, 459–463, 501, 524, 711, 723, 728, 743

Galamabos, J.T., 533, 722
Galbraith, J.I., 314, 722
Galbraith, R.F., 314, 722
Gallant, A.R., 20, 24, 66–68, 192, 197, 199, 201, 203, 230, 232, 234–237, 272, 274, 280, 301–305, 404, 457, 459–465, 531, 535–536, 539, 563, 567–568, 572, 575, 585, 711, 716, 723
Gallucci, V.F., 328, 723
Garbow, B.S., 381, 401, 664, 733, 740
Garcia, O., 344, 348, 354–356, 432, 723
Gardner, G., 315, 723
Gart, J.J., 576, 715
Gay, D.M., 629, 631–633, 639–640, 645–646, 652, 664, 668–669, 719, 723
Gbur, E.E., 455, 723
Gentle, J.E., 588, 621, 650, 653, 729
Gerald, K.B., 393, 430–431, 732
Gerig, T.M., 404, 723
Gill, P.E., 232, 588, 594, 597–599, 602–603, 605, 608–609, 611–612, 615–616, 618, 628, 633–639, 646, 654, 661–662, 666–670, 672–675, 724
Gillis, P.R., 158, 182, 364, 724
Glasbey, C.A., 272, 278–279, 285, 297–300, 321, 343, 724
Gleser, L.J., 524, 724
Godambe, V.P., 47–48, 724
Godfrey, K.R., 392, 717
Godfrey, L.G., 323, 724
Godolphin, E.J., 306, 316, 712, 724
Goebel, J.J., 272, 280, 301–305, 723
Goldberg, M.L., 128, 154, 156–157, 159, 165–166, 168–171, 173, 718, 724
Goldfarb, D., 607, 724
Goldfeld, S.M., 17, 204, 602–603, 724
Golub, G.H., 621, 655–660, 724
Gonin, R., 50, 724–725
Gorman, J.W., 544, 743
Gourieroux, C., 234, 236, 725
Grandinetti, L., 608, 725
Grassia, A., 272, 725
Green, P.J., 34, 37, 725
Greenstadt, J., 600–601, 725
Gregory, A.W., 234, 725

Greif, P., 388, 392, 728
Griffiths, D.A., 9, 466, 468–470, 478–480, 725
Griliches, Z., 282–283, 736
Guttman, I., 158–159, 446, 725

Halks-Miller, M., 339, 732
Hall, K.R., 506, 714
Halperin, M., 238, 241–244, 725
Halpern, J.W., 339, 732
Hamilton, D.C., 150, 177, 199, 200, 202–203, 206, 215, 217–218, 227–228, 234, 239, 261–269, 576, 713, 725–726
Hamilton, M.A., 246, 730
Hannan, E.J., 274, 318, 726
Hansen, L.P., 47–48, 726
Hanson, R.J., 560, 621, 730
Hardy, G.H., 474, 726
Harris, P.M., 362, 721
Hartley, H.O., 236–237, 368, 420, 424, 430–431, 457, 623–624, 726, 732
Harvey, A.C., 315, 317, 723, 726
Hawkins, D.M., 441–442, 445–446, 726
Hawkins, J.E., 550, 722
Haynes, J.D., 369, 726
Heath, A.B., 455, 726
Heath, S.B., 361–362, 364, 742
Henschke, K., 572, 716
Henson, T.L., 253, 728
Hernandez, F., 70, 726
Hiebert, K.L., 664, 668, 726
Hildreth, C., 278, 726
Hill, L.J., 315, 317, 731
Hill, P.D.H., 255, 260, 726
Hill, W.J., 78–81, 86, 253–255, 260, 714, 726, 728
Hillstrom, K.E., 664, 733
Himmelblau, D.M., 641, 726
Hinich, M.J., 443–444, 721
Hinkley, D.V., 72, 76, 160, 210, 212, 445, 448, 451, 453–454, 457, 576, 718, 722, 726–727
Hirsch, M.W., 377, 727
Hoadley, B., 576, 727
Hoaglin, D.C., 664, 727
Hoffman, R., 332, 344, 741
Holanda Sales, P.R., 316, 731
Holbert, D., 446, 727
Holliday, R., 362, 364, 727
Holly, A., 234, 236, 723, 725
Hong, D., 282, 284, 727
Hood, W.C., 38, 727
Hosking, J.R.M., 323, 727
Hougaard, P., 134, 184, 727
Hsu, D.A., 446, 727
Huber, P.J., 50, 650, 727
Hudson, D.J., 455–457, 727
Hunter, J.J., 377–378, 727
Hunter, W.G., 250–251, 253–255, 257, 259, 544–556, 712, 714–715, 720, 726, 728, 740
Hutchinson, D., 616, 734

Ikebe, Y., 381, 401, 740

Jacquez, J.A., 367–368, 372–375, 388, 392, 407, 409, 412, 728
James, B., 445, 728
James, K.L., 445, 728
Jeffreys, H., 53–54, 539, 728
Jenkins, G.M., 287, 290, 294–295, 306, 308, 311, 314, 316, 319–320, 322, 714
Jennrich, R.I., 33, 46, 387–388, 396, 398–402, 404, 411, 563–567, 569, 571–572, 647–649, 728, 736
Jobson, J.D., 86, 728
Johansen, S., 201, 495, 565, 571–572, 575–576, 581, 728
John, J.A., 71–72, 728
Johnson, D.E., 204, 733
Johnson, R.A., 70, 726
Jones, R.H., 315, 728
Jørgensen, B., 37, 46, 728
Jorgensen, M.A., 333, 728
Jupp, D.L.B., 463, 484, 728
Just, R.E., 17, 728

Kadane, J.B., 62–63, 741
Kalbfleisch, J.D., 368, 382, 396, 424–426, 429, 445, 496, 728–729
Kale, B.K., 39, 40, 729
Kapadia, A.S., 420, 720
Kass, R.E., 162, 164, 729
Katz, D., 56, 57, 729
Kaufman, L., 657–658, 660, 729
Kaufmann, H., 46, 721
Kedem, G., 306, 735
Kendall, M.G., 495, 729
Kennedy, W.J., Jr., 588, 621, 650, 653, 729
Kent, J.T., 236, 729
Khatri, C.G., 10, 343, 739
Khorasani, F., 194, 729
Khuri, A.I., 255, 729
King, L.J., 6, 729
King, M.L., 319, 729
Kira, T., 362, 739
Klema, V.C., 381, 401, 664, 727, 740
Kobayashi, M., 285, 729

Kodell, R.L., 420, 424, 427–429, 729
Kohn, R., 315, 317, 489, 729
Koopmans, T.C., 38, 727
Kramer, W., 282–283, 729
Krogh, F.T., 657, 660, 664, 729, 730
Kuhry, B., 13, 730
Kwong, F.H.F., 204, 721

Lahiri, K., 77, 730
Lambrecht, R.M., 368, 373, 383, 730, 737
Lamkin, G.H., 461, 721
Larntz, K., 210, 212, 722
Laüter, E., 251, 730
Lawless, J.F., 368, 382, 396, 424–426, 429, 728, 729
Lawrance, A.J., 72, 730
Lawson, C.L., 560, 621, 730
Lawton, W.H., 655, 730
Laycock, P.J., 493, 495, 720
Lecourtier, Y., 390, 736
Ledolter, J., 287, 289, 318–320, 322, 711
Lee, A.J., 104, 106, 730
Leech, D., 69, 730
Lepschy, A., 392, 717
Lerman, P.M., 435, 447, 450, 455, 457, 730
L'Esperance, W.L., 282, 284, 727
Levenbach, H., 337, 345, 352, 730
Levenberg, K., 602–603, 624–625, 730
Liang, K.-Y., 48, 730
Lieberman, G.J., 246, 730
Liem, T.C., 63, 718
Lill, S.A., 588, 730
Lindley, D.V., 493, 730
Linssen, H.N., 158, 730
Lipton, S., 493, 496, 500, 719
Littlewood, J.E., 474, 726
Ljung, G.M., 322, 730
Lowry, R.K., 181, 187–190, 730
Lu, A.A., 404
Lucas, H.L., 250–251, 254, 264, 714
Luecke, R.H., 501, 503, 505–508, 715
Lustbader, E.D., 37, 179, 733
Lwin, T., 247–250, 730–731
Lyttkens, E., 659, 743

McAleer, M., 236, 722
McCleod, A.I., 309, 316, 322, 731
McCullagh, P., 14, 42–43, 46, 665, 731
McGilchrist, C.A., 315, 317, 326, 332, 344–347, 349–352, 731, 738
MacGregor, J.F., 545–556, 714, 728
McGuire, T.W., 443–444, 721
McInnes, B.C., 420, 720
McKeown, J.J., 628, 731
MacKinnon, J.G., 236, 731
McLean, D.D., 395, 406–407, 545–546, 549–550, 731
Maddala, G.S., 20, 57, 731
Madsen, L.T., 162, 731
Maindonald, J.H., 621, 731
Mak, T.K., 496, 524, 717, 731
Malinvaud, E., 563, 565, 567–568, 571–572, 583–584, 731
Mannervik, B., 260, 731
Marcus, A.H., 429–430, 731
Marcus, L.F., 13, 730
Maritz, J.S., 47, 247–250, 730–731
Maronna, R., 444, 731
Marquardt, D.W., 561, 602–603, 623–625, 732
Matis, J.H., 368, 393, 420, 424, 427–431, 729, 732, 737
Mayilswami, P., 430, 735
Mead, R., 362, 364, 615, 732, 734
Medawar, P.B., 331, 732
Meeter, D.A., 158–159, 725
Mehata, K.M., 430, 732
Meisel, W.S., 715
Mélard, G., 315, 317, 732
Menzefricke, U., 446, 725
Mercer, L.P., 340, 733
Metzler, C.M., 369, 381, 393, 711, 732
Meyer, R.R., 189, 224–225, 645, 732
Miller, A.J., 466, 468–470, 479–480, 664, 666, 672–673, 675, 725, 732
Miller, D.M., 87, 732
Miller, H.D., 415, 418, 718
Miller, R.G., Jr., 196, 246–247, 339, 730, 732
Milliken, G.A., 102–103, 194, 230, 577, 714, 729, 732
Mitchell, B.M., 279, 282–285, 301, 734
Mizon, G.E., 232, 235, 732
Moen, D.H., 446, 733
Moler, C.B., 377, 381–382, 401, 560, 720, 733, 740
Money, A.H., 50, 724–725
Montfort, A., 234, 236, 725
Moolgavkar, S.H., 37, 179, 733
Moore, R.H., 46, 728
Moran, P.A.P., 13, 491, 495, 523, 733
Moré, J.J., 625, 663–664, 733
Morgan, P.H., 340, 733
Morton, R., 46, 181, 187–190, 524, 730, 733
Murray, W., 232, 588, 594, 597–599, 602–603, 605, 608–609, 611–612, 615–616, 618, 628, 633–639, 646, 654, 661–662, 666–670, 672–675, 724, 733

Narula, S.C., 653, 733
Nash, J.C., 588, 625, 664, 733
Naylor, J.C., 61–63, 245, 733, 740
Nazareth, L., 614, 625, 630, 632–633, 638–639, 733
Neill, J.W., 204, 733
Nelder, J.A., 14–16, 46, 256, 332, 362–364, 615, 665, 712, 714, 731, 734
Nelson, L.S., 616, 734
Newbold, P., 322–323, 712, 734
Neyman, J., 491, 734
Nicholls, D.F., 306–307, 734

Oberhofer, W., 50, 734
Olsson, D.M., 616, 734
Olsson, G., 6, 734
O'Neill, R., 616, 734
O'Neill, R.V., 368, 734
Orcutt, G.H., 282, 717
Osborne, M.R., 622, 654, 734
Overton, M.L., 654, 733

Pagan, A.R., 228, 236, 306–307, 715, 734
Pagano, M., 50–52, 246, 741
Park, R.E., 279, 282–285, 301, 734
Park, S.H., 463, 734
Parkinson, J.M., 616, 734
Parthasarathy, P.R., 430, 735
Patefield, W.M., 496, 524, 735
Patino-Leal, H., 510, 512–514, 737
Patten, B.C., 368, 732
Pearl, R., 339, 735
Pereyra, V., 655–660, 724, 729
Pericchi, L.R., 72, 735
Pesaran, M.H., 306, 735
Peters, S.C., 664, 727
Petruccelli, J.D., 321, 718
Pettitt, A.N., 445–446, 735
Phadke, M.S., 306, 735
Phillips, G.D.A., 315, 317, 723, 726
Phillips, P.C.B., 583–584, 735
Phua, K.H., 664, 739
Pierce, D.A., 310–314, 322, 359, 735
Pike, D.J., 255–256, 735
Pimentel-Gomes, F., 189, 735
Polak, E., 610, 735
Polya, G., 474, 726
Pope, R.D., 17, 728
Poskitt, D.S., 323, 735
Powell, J.L., 70, 711
Powell, M.J.D., 588, 607, 611, 613–614, 627, 649, 722, 735, 736
Prais, S.J., 279, 282, 736
Pratt, J.W., 576, 645, 736
Pregibon, D., 46, 734
Prentice, R.L., 337, 736
Pritchard, D.J., 80–85, 257, 545–546, 549–550, 731, 736
Prunty, L., 437–438, 468–470, 736
Purdue, P., 429–430, 736

Quandt, R.E., 17, 204, 236, 439, 441, 446, 602–603, 724, 736
Quinn, T.J., 328, 723

Rabinowitz, P., 61, 719
Raksanyi, A., 390, 736
Ralston, M.L., 647–649, 736
Ramsey, J.B., 446, 736
Rao, C.R., 228, 244, 377, 736
Rao, J.N.K., 44, 738
Rao, P., 282–283, 736
Ratkowsky, D.A., 100, 136, 158, 181–182, 185–188, 222, 337–338, 342–343, 364, 506, 686, 724, 736, 737
Ratti, J.T., 310, 742
Reed, L.J., 339, 735
Reeds, J., 162, 737
Reedy, T.J., 222, 338, 343, 737
Reeves, C.M., 610, 722
Reich, J.G., 119–125, 737
Reilly, P.M., 57–60, 510, 512–514, 737
Resigno, A., 368, 373, 383, 430, 730, 737
Reuter, B.E., 337, 345, 352, 730
Rhead, D.G., 614, 737
Richards, F.J., 7, 326–327, 332–334, 336, 737
Richards, F.S.G., 39, 737
Ritov, Y., 523, 714
Robertson, T.B., 330, 737
Robinson, P.M., 319, 737
Rodda, B.E., 326, 737
Rogers, A.J., 235, 737
Romanin Jacur, G., 392, 717
Romano, S., 343, 719
Ross, G.J.S., 111, 117, 126, 222, 225, 737
Ross, W.H., 164, 179, 738
Roth, P.M., 189, 224–225, 645, 732
Rothenberg, T.J., 232, 738
Royden, H.L., 481, 738
Ruhe, A., 657–660, 738
Runger, G., 76, 727
Ruppert, D., 69, 72–74, 76, 83, 85–86, 716

Sachs, W.H., 508, 516–517, 520–523, 738
St. John, R.C., 250, 738
Salazar, D., 446, 733
Sall, J., 711

Salter, P.J., 332, 734
Salzer, H.E., 61, 738
Sampson, P.F., 648, 728
Sandland, R.L., 9, 315, 317, 326–327, 332, 344–347, 349–352, 725, 731, 738
Savin, N.E., 196, 232, 713, 721, 738
Scales, L.E., 588, 599, 603, 608, 610–611, 617–618, 627, 646, 738
Scammon, R.E., 350, 738
Schechtman, E., 445, 727, 738, 743
Scheffé, H., 107, 738
Schlossmacher, E.J., 653–654, 738
Schmidt, W.H., 572, 715
Schnabel, R.B., 191–192, 195, 221–222, 588, 596, 604–605, 609, 621–622, 625, 645, 661, 666, 668–671, 674, 719
Schoenberg, I.J., 488, 738
Schulze, U., 470–471, 716
Schumitzky, A., 56–57, 729
Schweder, T., 444, 738
Schwetlick, H., 496, 557, 562, 738
Scott, A.J., 44, 104, 106, 730, 738
Scott, E., 491, 734
Seber, G.A.F., 5, 12, 32, 50, 80, 102, 104, 108, 194, 196, 207, 244–246, 248, 323, 326, 331, 450, 454, 523, 538, 577, 581, 677, 701–702, 739
Segre, G., 368, 737
Selvam, D.P., 430, 732
Sen, P.K., 445, 739
Serfling, R.J., 739
Shaban, S.A., 455, 739
Shah, B.K., 10, 343, 739
Shah, K.R., 242–243, 721
Shahn, E., 405, 713
Shanno, D.F., 607, 664, 739
Shaw, J.E.H., 61–63, 245, 740
Sheppard, C.W., 368, 739
Shinozaki, K., 362, 739
Siegmund, D., 445, 728, 739
Silverman, B.W., 485–486, 488–489, 739
Silvey, S.D., 228, 251, 739
Simonoff, J.S., 51, 212–213, 739
Singleton, K.J., 47, 726
Skene, A.M., 61–63, 245, 740
Smale, S., 377, 727
Smith, A.F.M., 61–63, 245, 435–437, 446, 454, 714, 733, 739, 740
Smith, B.T., 381, 401, 740
Smith, H., 31, 109, 720
Smith, J.E., 230, 732
Smith, M.R., 412, 740
Smith, P.L., 459, 484, 740
Smith, R.L., 62, 740
Snedecor, G.W., 189, 740
Sørensen, J.P., 545, 740
Souza, G., 234, 575, 716
Spendley, W., 616, 740
Spergeon, E.F., 10, 740
Spitzer, J.J., 282–284, 740
Sprent, P., 454, 496, 500, 740
Sprott, D.A., 445, 496, 728
Steece, B.M., 65–66, 740
Stefanini, L., 657–658, 660, 718
Stein, C., 57, 740
Steinberg, D.M., 250, 520, 718, 740
Steiner, A., 368, 713
Stevens, W.L., 328, 740
Stewart, G.W., 382, 398, 405, 560, 612, 713, 720, 740
Stewart, W.E., 545, 740
Stirling, W.D., 37, 46, 103, 740
Stone, C.J., 248–250, 740
Stone, M., 57, 740
Stroud, T.W.F., 228, 741
Strüby, R., 572, 716
Stuart, A., 495, 729
Studden, W.J., 486, 711
Sundararaj, N., 238, 741
Swann, W.H., 615, 741
Sweeting, T.J., 72, 741
Sylvestre, E.A., 655, 730
Szegö, G.P., 588, 675, 719

Taylor, W.E., 284, 741
Thomas, D.R., 44, 738
Thomaseth, K., 412, 717
Thompson, M.E., 48, 724
Tiao, G.C., 53–54, 63, 539, 714
Tiede, J.J., 50–52, 246, 741
Tierney, L., 62–63, 740
Tiller, V., 496, 557, 562, 738
Tishler, A., 471–479, 741
Tiwari, J.L., 431, 741
Tremayne, A.R., 323, 735
Trognon, A., 234, 236, 725
Trotter, H.F., 602–603, 724
Tsai, C.-L., 51, 174, 177, 179, 181–182, 212–213, 221, 718, 739
Tukey, J.W., 650, 713

Vajda, S., 392, 741
Valiant, R., 5, 6, 741
Vallmer, W.M., 382, 396, 424–426, 429, 729
Van der Waerden, B.L., 457, 741
Van Loan, C.F., 377, 381–382, 621, 724
Veall, M.R., 234, 725
Veitch, L.G., 719

Velleman, P.F., 488, 741
Venus, J.C., 327, 717
Venzon, D.J., 37, 179, 733
Verbyla, A.P., 293, 741
Vidyasagar, M., 654, 720
Vieira, S., 332, 344, 741
Villegas, C., 500, 741
Vivian, V.M., 461, 721
Von Bertalanffy, L., 331–332, 741

Wagner, J.G., 368, 741
Wahba, G., 488–489, 718, 741
Wald, A., 228, 742
Walker-Smith, M., 588, 664, 733
Walter, E., 390, 736
Wang, P.C., 73–74, 718
Watson, G.A., 654, 734
Watson, G.S., 318, 720
Watts, D.G., 92–95, 128, 131, 133–136, 138–139, 144, 146–150, 153–154, 156–161, 163, 177, 182, 215, 217–219, 222–226, 239, 261–269, 396–399, 401–402, 406–407, 409–411, 437, 465–468, 476–480, 537–538, 542–544, 641–645, 691, 712–713, 724–726, 742
Wecker, W.E., 487–489, 742
Wedderburn, R.W.M., 43, 46, 576, 734, 742
Wedin, P.A., 657–660, 738
Wegman, E.J., 484, 488–489, 742
Wehrly, T.E., 393, 430–431, 732
Wei, B.C., 182, 221, 718
Weiss, M.F., 405, 713
Wellington, J.F., 653, 733
Welsch, R.E., 629, 631–633, 639–640, 645–646, 651–653, 664, 668–669, 719, 742
White, G.C., 272, 310, 344–345, 356–360, 368–369, 431–432, 732, 742
White, H., 236, 573–575, 731, 742
Wiens, D., 200, 203, 206, 234, 576, 725–726
Willey, R.W., 361–362, 364, 742
Williams, D.A., 457, 742
Williams, E.J., 239, 742
Winsten, C.B., 279, 282, 736
Wise, M.F., 429, 742
Wisotzki, C., 572, 716
Witmer, J.A., 136–137, 145, 718
Wold, H., 659, 743
Wold, S., 484, 743
Wolf, D.A, 396–399, 401, 409–411, 713
Wolfe, D.A., 445, 738, 743
Wolter, K.M., 501, 743
Wong, C.W., 76, 719
Wood, J.T., 76, 743
Worsley, K.J., 440–445, 743
Wright, D.E., 64, 743
Wright, I.W., 484, 488–489, 742
Wright, M.H., 232, 588, 594, 597–599, 603, 605, 608–609, 611–612, 615–616, 618, 646, 654, 661–662, 666–670, 672–675, 724
Wright, S., 331, 743
Wu, C.F., 563–566, 743

Yang, S., 382, 743
Yohai, V.J., 444, 731
Yu, P.L., 46, 717

Zacks, S., 445, 743
Zang, I., 471–479, 741, 743
Zeger, S.L., 48, 730
Zellner, A., 55, 531, 583, 743
Ziegel, E.R., 544, 743
Zinke, I., 122–123, 125, 737
Zucker, R., 61, 738
Zwanzig, S., 572, 743

Subject Index

Absolute residuals, 653
Acceleration vector, 155
Accuracy of computation, 667, 669, 675
ACF, *see* Autocorrelation function (ACF)
ACM algorithms, 663
Adding to the Hessian, *see* Hessian, modified
Additive errors, 5, 13, 16, 210, 356. *See also* Constant absolute error
Adequacy of model, 225. *See also* Replicated model
  test for, 204
Adjoint matrix, *see* Matrix
Allometry, 8, 13, 22, 363
α-connection, *see* Connection coefficients
Alternative models, 235, 260, 364
Analysis of variance, 102
Approximate nonidentifiability, 103
AR(1) errors, 18, 275
  change-of-phase model with, 479
  choosing between estimators for, 282
  Cochrane–Orcutt estimator, 282
  conditional least squares estimator, 281
  condition for stationarity, 275
  iterated two-stage estimation, 280
  maximum likelihood estimation, 277
  two-stage estimation, 279
  unequally spaced time intervals, 285
AR(2) errors, 286
  condition for stationarity, 288
Arcing, *see* Parameter-effects curvature
ARMA(q_1,q_2) errors, 307
  backforecasting, 311
  checking error process, 321
  choosing error process, 318
  conditional least squares estimation, 310, 314
  expensive residuals, 646
  invertibility, 312
  iterated two-stage estimation, 314
  linear-model techniques, 316–317
  maximum likelihood estimation, 314
  missing observations, 315
  model fitting, 321
  stationarity, 312
  two-stage estimation, 307, 314
AR(q_1) errors, 289
  approximate maximum likelihood estimation, 294
  asymptotic variances, 296
  computation, 304
  conditional least squares estimation, 300
  condition for stationarity, 290
  iterated two-stage estimation, 297
  three-stage estimator, 302
  two-stage estimation, 301
Arrays, three-dimensional, 691. *See also* Square-bracket multiplication; Symmetric storage mode (SSM)
  derivative, 129
  multiplication of, 130, 691
Arrhenius model, 395
Association of Computing Machinery (ACM) algorithms, 663
Asymmetry measure, 187
Asymptotic convergence rate, *see* Convergence
Asymptotic likelihood methods, 196. *See also* Lagrange multiplier (LM) test; Likelihood ratio test; Maximum likelihood estimation; Maximum likelihood estimator; Wald test
Asymptotic model, *see* Growth curve; Yield-density curve

Asymptotic normality:
AR(1) maximum likelihood estimators, 278
change-of-phase estimators, 448–450, 460
conditional estimators for AR(q_1,q_2) model, 312
implicit functional relationships with replication and, 510
iteratively reweighted least-squares estimator, 88–89
least-squares estimator, 24, 127, 568
maximum-likelihood estimator, 33, 576
maximum quasi-likelihood estimator, 44
multi-response maximum-likelihood estimator, 538, 585
replicated least-squares estimator, 205
segmented polynomial estimators, 460–461
solution of unbiased estimating equations, 47
stochastic compartmental model and, 426
a-th iteration, 91
Autocatalytic model, *see* Growth curve
Autocorrelated errors, 18, 271
checking error process, 321
choosing error process, 318
Gaussian process, 348, **696**, 698
growth curves with, 343-345, 348
linear modes with, 18, 282, 316
unequal time intervals, 285
Autocorrelation function (ACF), 273, 290, 346
sample (SACF), 279, 319, 320
Autocovariance function, 273
sample, 301
Autoregressive moving-average (ARMA) model, 273. *See also* ARMA(q_1,q_2) errors
Autoregressive (AR) process, *see* AR(1) errors; AR(q_1) errors
Average growth rate, *see* Growth curve
Average mean square error (AMSE), 66

Backforecasting, 311, 322
Backward-shift operator, 275, 287, 311
Badly conditioned sum of squares, 480
Banana-shaped contours, 115
Bayesian estimation, 52
Box–Cox transformation parameter, 72
Box–Hill method, 80
empirical, for inverse predictions, 247
implicit functional relationship, 510
multiresponse models, 539, 549
splines, 489
two-phase regression, 446, 480
Bayesian hypothesis testing, 245
model adequacy using replication, 65
Beale's measures of nonlinearity, 157
Beaton–Tukey loss function, 650
Bent hyperbola, 468
Berkson's model, 13, 525
Bett's method, 629, 632, 638
BFGS quasi-Newton update, 607–609, 637
Bisection method, 598
Bleasedale–Nelder model, *see* Yield–density curve
Bleasedale's simplified model, *see* Yield–density curve
BMDPAR, 647
BMDP (statistical package), 647, 663
Bolus dose, 372. *See also* Tracer
Bonferroni confidence intervals, 192, 194, 203
Bonferroni inequalities, 422
improved, 442
Box–Cox (power) transformation, 69, 71, 339
two-parameter, 72
Box–Cox procedure, *see* Transformation to linearity
Box-Hill method, 79
Box-Perce portmanteau test, 322
Bracketing, 598
Britt–Luecke algorithm, 501, 516
Broyden–Dennis method, 629, 632
Broyden derivative-free update, 649
Broyden single-parameter rank-2 family, 607
Brunhilda experiment, *see* Data
B-splines, *see* Splines

$\mathscr{C}$ (covariance operator), 2
Calibration, *see* Inverse prediction
Calibration model, 51
Cancellation error, 612
Catenary system, *see* Compartmental models
Cauchy–Schwartz inequality, 680
Causes of failure, *see* Program failure
Cayley–Hamilton theorem, 390, **678**
Central difference approximation, 612
Central limit theorem, 25, 569
Chain-rule of differentiation, 395, 400, 519
Change of mean, 445
Change of phase, test for, 439, 440–443, 454, 463
Change-of-phase models, 433. *See also* Linear regimes; Segmented polynomial; Smooth transition models; Splines; Two-line-segment model; Two-phase (multiple) linear regression
asymptotic theory, 448–450
Bayesian methods for, 446, 480

computation, 455, 463, 474–475, 480
continuity assumption, 434–435
continuous case, 447
hypothesis tests for parameters, 451
stochastic growth curve and, 349
testing for no change, 440, 454, 463
two-phase, 452
Changepoint, 433. *See also* Knot
inference for, 452
noncontinuous, 439–440
Changepoint observation number (κ), 438, 445
Characteristic equation:
for eigenvalues, 378, 390, 677
for time series, 290, 305
Checking the solution, 675
Chi-square distribution, 2. *See also* Likelihood ratio test
idempotent matrix and, 684
noncentral, 229, 580, **684**
quadratic form and, 684
Choice of design points, *see* Design points
Cholesky decomposition, 28, 36, 68, 89, 138, 262, 512, 535, 544, 557, **680**
LDL′, 602, 608, **681**
modified, 638, 672
using lower triangular matrix, 304, **681**
with negative diagonal elements, 148, 265
Chow's test, 443
Closed system, *see* Compartmental models
Cobb-Douglas family, 16
Cochrane-Orcutt estimator, *see* AR(1) errors
Coefficient of variation, 430
Commercial program libraries, 663
Compansion, *see* Parameter-effects curvature
Comparison of test statistics, 198–200, 232, 234
Compartmental matrix, 372, 384
eigenvalues, of, 385
Compartmental models, 367. *See also* Linear compartmental model; Sum-of-exponentials model
a priori identifiability, 387
catenary system for, 383, 385
closed system, 368
connectivity diagram for, 383
globally identifiable, 388–389, 391
locally identifiable, 388, 412
mammillary system, 383, 385
mixtures measured, 395, 405
mother compartment, 383
multiresponse, 413, 529, 533–534
number of compartments, 412
one-compartment, 376, 412
open system, 368
optimal design and, 254, 264, 412
properties of, 384
rate constant for, 369
some compartments unobserved, 386, 407
starting values, 406
steady state, 372
strongly connected, 384
structural identifiability, 391
three-compartment, 7, 66, 367, 369, 379, 393, 399, 402
trap, 383, 386
two-compartment, 387–389, 411, 427
Concentrated (log) likelihood, 37
autocorrelated linear regression and, 317
for Box–Cox transformation, 71, 410
for Box–Hill model, 79
for finding maximum likelihood estimator, 42
hypothesis testing and, 228-229
multiresponse model and, 537, 585
two linear regimes and, 440
Concentrated sum of squares, 655
Gauss–Newton method for, 655
Concentric ellipsoids, 27
Conditional (generalized) least squares, 281, 300, 310, 314, 424
Conditional regressor model, 11
Confidence band for model, 194, 246
Confidence ellipse, 106–109
Confidence interval, 191
affected by curvature, 192
inverse prediction, 245
jackknife, 207, 213
for linear function, 191
for nonlinear function, 192, 235
prediction, 193
simultaneous, 194
single parameter, 192, 203
Confidence region(s), *see also* Linearization region
approximate, 25, 97, 130
effects of curvature on, 98, 214, 227
exact, 97, 112, 116, 194, 237
Hartley's method for, 237
jackknife, 207, 213
liner regression and, 97, 106, 167, 702, 704
for nonlinear functions, 235
for parameter subsets, 202, 227
related to stopping rules, 642, 644–645, 671
single elements, 203
summary of, 220

Conjugate directions, 609, 613
without derivatives, *see* Direction-set (conjugate directions) method
Conjugate-gradient method, 609, 618, 662
convergence rate, 611
Fletcher–Reeves formula, 610
Polak–Ribiere formula, 610
Connection coefficients, 159
alpha, 162
exponential, 163
mixture, 163
Constant absolute error, 15
Constant proportional error, *see* Multiplicative model; Proportional error
Constant relative error, **15**, 251–252
Constrained linear model, 404
Constrained optimization, 153, 588, 662
linear equality constraints, 202, 707
Constraint equation specification, *see* Nonlinear hypothesis
Continuity assumption for multiphase models, 434
Contours, 92–97, 110, 112, 115–118, 254, 591–593, 595, 674
Control constants, 666
number of function evaluation, 667
step-length, 667
Controlled regressors, 13, 525. *See also* Berkson's model
Convergence, 91
asymptotic convergence rate, 593–594
conjugate-gradient method, 611
criteria, *see* Stopping rules
of function estimates, 669
global, 597
of gradient estimates, 670
linear rate of, 611, 675
Newton's method and, 596
of parameter estimates, 670
p-step superlinear, 611
quadratic, 594, 596
quasi-Newton method and, 609
superlinear, 594, 609, 675
Convolution operator, 396
Cook-Witmer example, 136, 144, 172
Corrected Gauss-Newton method, 633
step, 636, 638
Correlation matrix, 274
Corr (correlation) notation, 2
Cosine angle between vectors, 641, 645
Cross-sectional data, *see* Growth curve
Cubic splines, *see* Splines
Curvature(s), 98, 131. *See also* Intrinsic curvature; Parameter-effects curvature; Statistical curvature (of Efron); Subset curvature(s)
computation of, 150
experimental design and, 100
for growth models, 343
mean square, 158
optimal designs and, 261
radius of, 133–134, 467, 689
reduced formulae for, 138
replication and, 146
scale-free, 133
subset, 165, 168
summary of formulae, 145
three-dimensional, 688
when negligible, 135, 149, 170
Curved exponential family, *see* Exponential family
Cusum methods for intercept changes, 444–445

$\mathfrak{D}$ (variance-covariance operator), 2
Damped exponential growth, 340
Damped Gauss–Newton method, *see* Modified Gauss–Newton methods
Data:
for birthweights of monozygote twins, 105
for Brunhilda tracer experiment, 399, 410
for calibration model of Tiede and Pagano, 51
for catalytic isomerization of *n*-pentane, 78, 82
for compressibility of helium, 506
for digitized X-ray image, 514
for heights of Count de Montbeillard's son, 350
for isomerization of α-pinene, 551
for megacycles to failure, 59
for Michaelis–Menten model, 93, 148
for potentiometric titration, 521
for radioactive tracer and two-compartment model, 534
for stagnant band heights, 478
for two-compartment model, 428
for weights and heights of preschool boys, 461
Data dependencies, 546
Davidon–Fletcher–Powell (DFP) quasi-Newton algorithm, 607–609, 631
Dead time, 402
Delta-correlated Gaussian process, 346, 697
Deming's algorithm, 505, 510
Dennis–Gay–Welsch method, 629–633. *See also* NL2SOL
Dependent variable, *see* Response variable

Derivative arrays, 129
Derivative-free methods, 611, 646. *See also* Direct search methods
conjugate directions, 612
quasi-Newton, 611
Descent direction, 594, 599, 609, 638
Descent method, 593–594
Design matrix, 650, 654
Design points, 100, 125
choice of, 100, 102, 115, 118
Determinant, notation, 1
product of eigenvalues, 678
Deterministic compartmental models, 370
Detransformation, 86, 343
DFP, *see* Davidon–Fletcher–Powell (DFP) quasi-Newton algorithm
Diagonal matrix, *see* Matrix
Differential geometry, 128, 131, 687
Differentiation with respect to vectors, *see* Vector differentiation
Diffuse prior, *see* Prior distribution(s)
Directional discrimination, 600
Directiion of negative curvature, 601, **603**, 638
Direction-set (conjugate directions) method, 612, 662
slower than finite quasi-Newton, 615
Direct procedure, 80
Direct product, *see* Kronecker product
Direct search methods, 615, 661
Discrete Newton method, *see* Modified Newton methods
Discrete quasi-Newton method, 605, 618, 662
Discrimination, *see* Inverse prediction
Dispersion matrix, definition, 2
Dog-leg step, 627
D-optimality, *see* Optimal experimental design(s)
D_s-optimality, *see* Optimal experimental design(s)
Dose-response curve, 326
Dot notation for differentiation, 3, 22
DUD (Doesn't Use Derivatives) algorithm, 647–649
Durbin–Watson test, 318
Dynamic model, *see* Lagged model

$\mathscr{E}$ (expectation), 2
Effective noise level, 262
Effects of curvatures on residuals, *see* Residuals
Efficient-score statistic, 199
Eigenvalues and eigenvectors, 106, 108, 122, 218, 240, 263, 377, 547, 591, 600–602, 622–623, 626, 635, 639, **677–680**. *See also* Compartmental matrix; Spectral decomposition
maximum and minimum eigenvalues, 680
EISPACK library, 381, 400, **663**, 740
Ellipse, 106
Ellipsoid(s), 27, 130, 217, 591, **679**
approximating, 93, 130
concentric, 27
length of axes, 218, 679
volume of, 679
Empirical growth curve, 325
Empirical model, 325
Errors-in-variables model, 11, 491
Error sum of squares, 21
Estimable function, 102–103
Estimating equations, *see* Unbiased estimating equations
Euclidean norm, *see* Norm
Euclidean space ($\mathbb{R}^n$), 1
Exact confidence region, *see* Confidence region(s)
Exact line search, *see* Line search
Exact tests, 201, 236–244, 463
Excess variance, *see* Least squares estimator
Expectation surface, 98, 128
Expected information matrix, 33, 53, 498
inverse not asymptotic variance-covariance matrix, 496
Expected value of quadratic, *see* Quadratic form(s)
Expected value transformation, 222–223, 225
Experimental design, *see* Opimal experimental design(s)
Explanatory variables, 4
Exponential family, 14
curved, 161
one-parameter, 160
regular, 45
Exponential growth curve, *see* Growth curve
Exponential models, 5, 100, 103, 192, 201, 302, 568. *See also* Exponential-plus-linear trend model; Growth curve, monomolecular
negative, 8, 92, 94, 111, 233, 251, 327
sum of, *see* Sum-of-exponentials models
Exponential overflow, *see* Program failure
Exponential parameters, 393
Exponential peeling, 407
Exponential-plus-linear trend model, 298. *See also* Compartmental models

Face of array, **130**, 136, 224
Failure of algorithm, *see* Program failure
Fanning, *see* Parameter-effects curvature
Farazdaghi-Harris model, *see* Yield-density curve
Farley–Hinich test, 443
F-distribution, notation, 2
Final size, 327
Finite differences, 27, 345, 349, 358, 605, 609, 612, 636, 639, 646, 661, 667, 673
Finite population, 5
Fisher's scoring algorithm, 35–36, 317, 493–494, 497, **685**
  when same as Newton's method, 36, 506, 686
Fixed knots, *see* Spline regression
Fixed-regressor model, 10
Fletcher–Reeves formula, *see* Conjugate-gradient method
Forward Kolmogorov equations, 418
Forward substitution, 30
Fractional transfer coefficient, 370
Freedom equation specification, 511, 581. *See also* Nonliner hypothesis
Freezing point model, 9
Frobenius norm, *see* Norm
F-test:
  corrected for intrinsic curvature, 200
  exact, 236–244
  linear hypothesis, 197–202
  model adequacy using replication, 32
  nonlinear hypothesis, 231
Funtional relationships, 11. *See also* Implicit functional relationships
  expressed implicitly, 506
  multivariate, 557
  without replication, 492
  with replication, 496

Garcia's model, *see* Growth curve
Gaussian process, 346, 695
Gauss–Jordan pivoting, 648
Gauss–Newton method, 26, 29, 43, 49, 88, 94–95, 394, 413, 426, 518, 561, **619**, 634, 639, 660, 708. *See also* Modified Gauss–Newton methods
  compared with Newton method, 26, 621
  for concentrated sum of squares, 655
  convergence of, 622, 660
  nonderivative, 646
  for robust loss, 652
  separable problem and, 657, 660
  starting values and, 665
  tangent-plane approximation and, 646
Gauss–Newton step, 29, 621, 634, 638
  corrected, 636, 638
  minimum Euclidean norm, 624, 534
Generalized cross-validation, 488
Generalized inverse, *see* Matrix
Generalized least squares, **27**, 43, 88, 274, 294. *See also* Weighted least squares
  constrained, 404
  functional relationships and, 496
  for Markov-process models, 423
  multiresponse models and, 531
Generalized linear model, **14**, 17, 46, 48
  link function for, 46
Generalized logistic curve, *see* Growth curve
Generalized secant method, 648
Gill–Murray method for least squares problems, 633, 639
  explicit second derivatives, 636
  finite-difference approximation, 636
  quasi-Newton approximation, 637
GLIM (statistical package), 15, 712
Globally identifiable, *see* Compartmental models
Global minimum, *see* Minimum
GLS, *see* Generalized least squares
Golden section, 598
Gompertz curve, *see* Growth curve
Grade, 635
Gradient vector, 26, 589
  finite difference approximation, 612
  of least squares problem, 620
  of quadratic, 590
  of robust loss function, 651
Grid search, 455, 593, 665
Growth curve, *see also* Stochastic growth curve; Yield–density curve
  asymptotic, 328, 343
  autocatalytic, 329
  autocorrelated errors, 343–344
  average growth rate, 333
  biological and statistical approaches, 326
  cross-sectional data, 342–344
  curvature(s) and, 343
  deterministic approach, 342
  exponential, 327
  final size, 327
  Fletcher family, 339
  Garcia's model for, 354
  generalized logistic, 279, 339
  Gompertz, 330, 332, 336–337, 342
  growth rate, 333, 337
  initial size, 330
  logistic, 329, 332, 337–338, 342, 357
  longitudinal data, 343–344, 349

maximum rate of growth, 329–331, 333, 339
monomolecular, 94, 111, 251, **328**, 332, 336, 343
Morgan–Mercer–Flodin (MMF), 340, 342
Riccati, 337
Richards, 78, 271, 278, **332**, 335, 337, 342, 353–354, 360, 363
sigmoidal, 328, 337
splines fitted to, 486
starting values, 335
time-power, 327
von Bertalanffy, 328, 331–332
Weibull, 338, 342

Hartley's method, *see* Modified Gauss–Newton methods
Hartley's method of exact inference, 236
Harwell library, 663
Hat matrix, 211
Hemstitching, 595, 600
Hermite polynomials, 61
Hessian, 26, 589
expected, 685
finite difference approximation, 609, 612, 636
Gauss–Newton (GN) approximation, 621, 631, 633
ill-conditioned, 596, 602, 674
improving GN approximation, 627–639, 653
initial approximation for, 666
of least squares problem, 620
modified, 600–603
of quadratic, 590
quasi-Newton approximation, 605, 618, 628, 637, 653
of robust loss, 651
sufficiently positive definite, 603, 674
update, *see* Update-matrix
well-conditioned, 675
Heterogeneity, *see* Variance heterogeneity
Highest posterior density (h.p.d.):
interval, 6, 64, 542
region, 63, 542
Hill equation, 120
Holliday model, *see* Yield–density curve
Hotelling's T^2 distribution, 207
H.p.d., *see* Highest posterior density (h.p.d.)
Huber's loss function, 73, 650
Hudson's algorithm for two-phase regression, 455
Hyperbolic model, *see* Smooth transition models

Hypothesis testing, *see also* Linear hypothesis; Nonlinear hypothesis
Bayesian, 245
exact test, 201, 236
general theory, 576
multiple, 235
nested, 235, 581
separate (non-nested), 235

Idempotent matrix, *see* Matrix
Identifiability, 102
for compartmental model, 387–391, 412
problems, 102
structural relationships and, 13
Identity matrix, 1
Ill-conditioning, 95, 110, 126. *See also* Hessian; Scaling, bad
linear models and, 103, 106
nonlinear models and, 110, 201
Richards model and, 336
Implicit functional relationships, 12
normal equations for, 517
without replication, 501
with replication, 508
some unobservable responses, 516
some variables measured without error, 507
Implicit-function theorem, 40, 520
IMSL library, 488, 662, **663**
Incidental parameters, 492
Inconsistent estimation, 491, 493, 495, 503
Independent variables, *see* Explanatory variables
Influence methods, 74, 179
Influential observation(s), 73, 214
Information matrix, *see* Expected information matrix
Information metric, 162, 164
Initial estimates, *see* Starting values
Initial size, *see* Growth curve
Input rate, 370
Instantaneous transfer-rate matrix, *see* Intensity matrix
Instrumental variables, 47
Intensity matrix, 416
Intrinsically nonlinear, 7
Intrinsic curvature, 98, **131**, 133
effect on linearized confidence regions, 214
invariant under one-to-one transformation, 133, 136, 215, **692**
negligible, 165
Invariant under one-to-one parametrizations, 133, 136, 179, 195, 200, **692**

Inverse prediction, 245
  empirical Bayes interval for, 247
  multiple, 246
  single, 245
Inverted gamma distribution, **2**, 55–56
IRLS, *see* Iteratively reweighted least squares (IRLS)
Isotope fractionation, *see* Tracer
Iterated two-stage least squares, *see* Two-stage estimation
Iterative algorithms, 593
Iteratively reweighted least squares (IRLS), 37, 46, 86, 88, 652, 654
Ito's integral, 697

Jackknife, 51, 206. *See also* Confidence interval; Confidence region(s)
  linear, 210
  one-at-a-time, 207, 210
Jacobian, 261, 355, 503, 517
  quadratic approximation of, 261
  unit, 291
Jacobian of residual vector (J), 620
  finite difference approximation, 636, 646
  ill-conditioned, 623
  quasi-Newton approximation, 648–649
  secant approximation, 647
Jeffrey's prior, *see* Prior distribution(s)
John–Draper transformations, 72
Joinpoint, *see* Changepoint; Knot
Jordan canonical form, 382

Kalman filter, 315, 317, 489
Kernel (null space) of matrix, *see* Matrix
Knot, 433, 481–482, 484. *See aso* Changepoint
Kronecker product, 684
Kurtosis, test for, 186

Lack of fit test, 203. *See also* Replicated model
Lag, 273
Lagged model, 19
Lagrange multiplier, 152, 503, 509, 578, 604
Lagrange multiplier (LM) test, 198, 228–234, 323, 577, 671
Lag time, *see* Dead time
LAM, *see* Linear approximation method (LAM)
Laplace's method, 62
Large-residual problems, 623, **627**
  Levenberg–Marquardt algorithm and, 627
  quasi-Newton method and, 628
Last-point fit method, *see* Stochastic growth curve
Latent roots and vectors, *see* Eigenvalues and eigenvectors
Least absolute errors regression, *see* Norm, L_1-minimization
Least-squares computation, 25–26, 91–97, 587, **619**
Least-squares estimation, 21, 619
  with constraints, 402–404
  GLS, *see* Generalized least squares
  OLS (ordinary least squares), 28
    with serial correlation, 272
  replicated model, 31
Least-squares estimator, 21
  asymptotic normality of, 24, 568
  bias, 182
  bias test, 186
  consistency of, 21, 274, 564
  dispersion matrix of, 183
  effect of autocorrelated errors, 272, 299, 303
  excess variance test for, 186
  existence of, 563
  misspecification and, 572
  normal equations for, 22
  residuals from, 174
  same as maximum-likelihood estimator, 21
  simulted sampling distribution of, 184
  small error asymptotics, 575
  tests for normality of, 186
Least-squares Jacobian (**J**), *see* Jacobian of residual vector (**J**)
Length of vector, *see* Vector norm (length)
Levenberg–Marquardt methods, *see also* Trust-region methods
  general problems, 602–604
  least squares problems, **624**, 627, 629, 633, 639, 652, 657–658, 666
  problem with, 625
Levenberg–Marquardt step, 624, 626
Lifted line, 132
Likelihood ratio test:
  compared with Lagrange multiplier test, 231, 577
  compared with Wald test, 231, 577
  corrected for intrinsic curvature, 200
  general, 196
  linear hypothesis and, 197–198
  for multiresponse model, 538
  nonlinear hypothesis and, 228–234, 577
  not asymptotically chi-square, 441–442, 454
  for two-phase regression, 444–445, 451

Limited memory quasi-Newton method, *see* Quasi-Newton methods
Limiting zero variance, *see* Vanishingly small errors
Linear approximation, 23. *See also* Linear Taylor expansion
Linear approximation method (LAM), 48
Linear compartmental model, 368
computer-generated exact derivatives, 396
deterministic, 376
identifiability, 386
with random errors, 393
some compartments unobserved, 407
Linear dependencies, 545
Linear differential equation, 376
Linear equality constraints, 707
Linear functional relationships, *see* Functional relationships
Linear hypothesis, 197, 202
comparison of test statistics for, 199, 231
computations for, 202
for parameter subset, 197
Linearization region, 25, **194**, 214
intrinsic curvature and, 214
parameter-effects curvature and, 218
Linearized model, 78, 82
unweighted lease squares inappropriate, 82
Linearizing transformations, *see* Transformation to linearity
Linear jackknife, *see* Jackknife
Linear-plus-exponential trend model, *see* Exponential-plus-linear trend model
Linear rate of convergence, *see* Convergence
Linear regimes, *see also* Two-line-segment model
noncontinuous phase changes, 435, 438
with smooth transitions, 437, 465
Linear regression, 75, 97, 103–109, 231, 241, **701**. *See also* Two-line-segment model; Two-phase (multiple) linear regression
with autocorrelated errors, 282, 313, 316
Bayesian, 54, 56
definition, 5
Linear search, *see* Line search
Linear structural relationships, *see* Structural relationships
Linear Taylor expansion, 23, 25, 43, 47, 70, 193, 258, 317, 504, 518, 619
Line search, 597, 666–667, 675. *See also* Natural step
approximate (partial), 594, 597, 614
exact (accurate), 594, 606-607, 609–610, 612
Hartley's method and, 624
sufficiently accurate, 608
using finite differences, 612
Link function, *see* Generalized linear model
LINPACK library, 663
LM test, *see* Lagrange multiplier (LM) test
L_1-norm minimization, *see* Norm
Locally identifiable, *see* Compartmental models
Local minimum, *see* Minimum
Logistic curve, 14
Logistic growth model, *see* Growth curve
Log-likelihood ratio, 196
Log-normal distribution, 70
Log transformation, 10–11, 16–17, 72, 279, 365, 410
Longitudinal data, 343–344
Lower triangular system, 29
LR test, *see* Likelihood ratio test

Makeham's model, 10, 343
Mammillary system, *see* Compartmental models
MA(q_2) errors, 305
characteristic equation, 305
maximum likelihood estimation, 306
two-stage estimation, 306
Marginal posterior distribution, *see* Posterior distribution
Marginal prior distribution, *see* Prior distribution(s)
Markov parameters, 390
Markov process compartmental models, 415
computational methods for, 423
discussion, 429
input from environment, 420
no environmental input, 416
time-dependent, 420
time-homogeneous, 418, 421
Marquardt's method, *see* Modified Gauss-Newton methods
Mass-balance equation, 371
Mathematical programming, *see* Constrained optimization
Matrix:
adjoint, 379
compartmental, 372
determinant of, 1
diagonal, 1
doubly symmetric, 292
eigenvalues of, 677
eigenvectors of, 678
exponential, 377
generalized inverse, 1, 44, 243, **679**
hat, 211

Matrix (*Continued*)
idempotent, 22, 24
ill-conditioned, 104
irreducible, 384
Moore–Penrose generalized inverse, 404, 543, 655, 657, **679**, 680, 683, 703
nonnegative definite, 678
null space (kernel), 1, 626, 677
optimization, 679
orthogonal, 678
patterned, inverse of, 41, 197–198, 211, 499, **678**
permutation, 151, 171
positive definite, 678
positive semidefinite, 678
projection, 1, 131, 137, 175–181, 239, 641, 649, 656, **683**, 703
range (column) space, 1, 128, 626
rank, 677
square root, 679
symmetrizable, 385
trace, 1, 677
transpose, 1
triangular, 276
Maximin formulation, 251
Maximum, global, 153. *See also* Minimum
Maximum likelihood estimation, 32, 576
asymptotic normality, 278
autocorrelated normal errors, 277, 294, 306, 314
multiresponse model, 536, 549
nonnormal data, 34
normal errors, 32
Maximum likelihood estimator, 32
asymptotic variance-covariance matrix, 34, 42
from concentrated likelihood, 42
same as least squares, 21, **33**
same as maximum quasi-likelihood estimator, 46
Maximum quasi-likelihood estimation, 42, 69
asymptotic normality, 44
Maximum rate of growth, *see* Growth curve
Maximum step-length, 666
Max-min formulation, *see* Smooth transition models
Mean square curvature, *see* Curvature(s)
Measures of nonlinearity, *see* Asymmetry measure; Beale's measures of nonlinearity; Curvature(s); Least-squares estimator
Mechanistic model, 325
Median, 87
M-estimator, 50, 650
Michaelis–Menten model, 86, 92, 225, 230
Minimization subject to linear constraints, 709
Minimum:
checking for, 672, 675
global, 92, 588, 591, 675
local, 92, 153, 588, 641
Minimum-distance estimator, 582
Minimum sum of absolute residuals, *see* Norm, L_1-minimization
MINOS library, 663
MINPACK library, 625, **663**, 664
Missing observations in time series, 315
Misspecified models, 572
Mitcherlich's law, 328
Mixture connection, *see* Connection coefficients
Model A, *see* Power-transform-both-sides (PTBS) model
Model B, *see* Power-transformed weighted least squares (PTWLS) model
Modified Adair equation, 119
Modified Cholesky decomposition, *see* Cholesky decomposition
Modified Gauss-Newton methods, 577, **623**, 661. *See also* Gauss–Newton method; Large residual problems; Levenberg–Marquardt methods
derivative-free, 646
Hartley's method (damped method), 457, 534, **623**
Marquardt's method, 561, **624–625**
Modified Newton methods, 355, **599–605**, 609, 617–618, 651, 661–662, 664, 672. *See also* Fisher's scoring algorithm; Gauss–Newton method; Modified Gauss-Newton methods; Newton method; Quasi-Newton methods
discrete (second derivative approximation), **605**, 618, 662
for least-squares, 628
most reliable, 604
nonderivative, 618
restricted step methods, 603
for robust loss functions, 651, 653
step length methods, 599
Monod–Wyman–Changeux kinetic equation, 125
Monomolecular growth curve, *see* Growth curve
Moore–Penrose generalized inverse, *see* Matrix
Morgan–Mercer–Flodin curve, *see* Growth curve

Mother compartment, *see* Compartmental models
Mother material, 373
Moving average model, *see* MA(q_2) errors
Multiphse model, *see* Change-of-phase models
Multiple hypothesis testing, 235
Multipe linear regression, *see* Linear regression
Multiplicative model, 16, 210
Multiresponse nonlinear model, 529
  asymptotic theory for, 538
  Bayesian inference for, 539
  compartmental model, 413, 529, 533–534
  functional relationships and, 557
  generalized least squares for, 531
  highest posterior density region for, 539, 542
  hypothesis testing, 538
  linear dependencies, 545
  maximum likelihood estimation, 536, 582
  missing observations, 533, 544
Multivariate estimation, 581
Multivariate normal distribution, notation, 2
  approximation to posterior distribution, *see* Posterior density function
Multivariate partially linear model, 243
Multivariate *t*-distribution, *see* *t*-distribution

NAG library, 602, 618, 662, **663**, 664
Naming curvature, 161
Natural parameter of exponential family, 160
Natural step, 607, 645
Nested hypotheses, *see* Hypothesis testing
Newton direction, 596
Newton method, 26, 92, 518, 543, **599**, 617, 708. *See also* Modified Newton methods
  modifications for reliability, 600
  reasons for failure, 599
  starting values, 665
Newton-Raphson method, 27, 34, **599**
Newton's law of cooling, 328
Newton step, 27, 596, 601, 621, 626, 634, 638, 669
  compared with Gauss-Newton, 26, 621, 634
NIPALS, 659
NL2SOL, 633, 639, **664**
No change, test for, 454
Noise residuals, 322
Noncentral chi-square distribution, *see* Chi-square distribution
Noncommercial program libraries, 663
Nonconstant variance, *see* Variance heterogeneity
Noncontinuous change of phase, 438
Noninformative prior, *see* Prior distribution(s)
Nonlinear hypothesis, 228, 576
  comparison of test statistics for, 232, 234
  constraint equation specification, 228
  freedom equation specification, 232, 581
Nonlinearity, *see* Measures of nonlinearity
NONLIN (statistical package), 663
Nonnested hypotheses, *see* Hypothesis testing
Nonparametric regression, 486, 488
Nonsmooth functions, 616
Norm:
  Euclidean, 594, 634
  Frobenius, 631
  L_1-minimization, 50, 616, 653
  vector, 1
Normal equations:
  implicit functional relationships and, 517
  least squares, ordinary, 22, 50, 216, 234
  linear models, 701
  quasi-likelihood, 43, 88
  replicated data and, 31, 205
  subset, 227
NPL library, 663–664
Nuisance parameter(s), 39, 58, 491
Nullity, 677
Null space of matrix, *see* Matrix

OLS (ordinary least squares), *see* Least-squares estimation
$\mathbf{O}_p$, $\mathbf{o}_p$, 3, 24, 44
Open system, *see* Compartmental models
Optimal experimental design(s), 116, 250. *See also* Parameter redundancy
  allowing for curvature, 260
  compartmental model and, 254, 264, 412
  competing models and, 260
  design criteria for, 250
  D-optimality, 250
  D_s-optimality, 253
  multivariate models, 259
  for parameter subsets, 253
  prior estimates for, 255
  sequential, 257, 263
  S-optimal, 251
  T-optimal, 260
Optimization, 587
Optimization subject to constraints, *see* Constrained optimization
Orthogonal complement, 1, 179, 181

Orthogonal decomposition of vectors, 133, 683
Orthogonal matrix, *see* Matrix
Orthogonal offset, 156
Orthogonal projection, *see* Matrix, projection
Outlier(s), 51, 73
Overflow, *see* Program failure

PACF, *see* Partial autocorrelation function (PACF)
Parallel subspace property, 613
Parameter-effects array, 147–149, 170
  arcing, 147
  compansion, 147
  fanning, 148, 170
  torsion, 148, 170
Parameter-effects curvature, 98–99, **131**, 133, 218
  noninformative prior and, 55
  scaled, 133
  transformation to reduce, 222
  zero, 215
Parameter redundancy, 119
  relationship to experimental design, 125
Parameters, **4**
Parametric family, 5
Parametrization, *see* Reparametrization; Scaling (parametrization)
Partial autocorrelation function (PACF), 319
  sample, 319–320
Partially linear models, 240, 405, 409, 447, 455, 654. *See also* Separable least squares
Patterned (partitioned) matrix, 678
Permutation matrix, *see* Matrix
Pharmacokinetic models, 251, 368
Phase model, *see* Regime
Piecewise regression, 433
Planar assumption, *see* Tangent-plane approximation
Point of inflection, 328, 332–333, 335, 338, 342
Poisson distribution example, 46
Polak-Ribiere method, *see* Conjugate-gradient method
Polynomial, segmented, *see* Splines
Polynomial estimator, 66
Polynomial interpolation, 598
Polytope algorithm, 615, 661–662
Portmanteau test, 322
Positive definite matrix, *see* Matrix
Positive semidefinite matrix, *see* Matrix
Posterior density function, 52
  implicit functional relationships and, 512
  moments of, 57, 62
  normal approximation to, 64, 80, 84
Posterior distribution, 55
  marginal, 57, 84
Posterior mode, 56, 64
  direct procedure for, 80
  equivalent to maximum likelihood estimator, 56, 80, 539
  staged procedure for, 81
Posterior moments, *see* Posterior density function
Powell's hybrid algorithm, 627
Powell–symmetric-Broyden update, 631
Power-transform-both-sides (PTBS) model, 69, 410
  inference from, 74
  robust estimation, 73
Power-transformed weighted least squares (PTWLS) model, 70, 77
Preconditioning, 611
Prediction bias after detransforming, 86
Prediction interval, *see* Confidence interval
Principal-axis theorem, 678
Principal components, 108, 461
Principle of maximum entropy, 432
Prior distribution(s), 52
  based on Wiener process, 489
  conjugate, 57
  improper, 54, 57
  Jeffrey's rule, 53–54, 511
  marginal, 60
  multiparameter controversy, 57
  multivariate normal, 257, 259, 446
  noninformative, 53–55, 57, 539, 544
  reference, 53
  for two-phase linear regression, 446
  uniform, 58, 67, 80, 446
Prior estimates for optimal design, 255
PROC NLIN, 640, 647
Program failure, 672
  bad scaling, 673
  exponential overflow, 673
  programming error, 396, 403, 605, **672**
  underflow, 673
Programming error, *see* Program failure
Projected residuals, 69, 81, **179**
  computation of, 181
  properties, 179
Projection matrix, *see* Matrix
Proportional error, 356
Proportional rate equations, 346, 697
PTBS model, *see* Power-transform-both-sides (PTBS) model
PTWLS model, *see* Power-transformed weighted least squares (PTWLS) model

Pure error sum of squares, 31, 204
Pure error test of fit, *see* Replicated model

QR decomposition, 138, 150, 171, 174, 543, 559, 621, 644, **681**
Quadratic approximation, 26–27, 92, 600, 603, 633, 640, 669, 671. *See also* Quadratic Taylor expansion
Quadratic convergence, *see* Convergence
Quadratic form(s):
  chi-square distribution and, 684
  covariance of two, 684
  expected value, 684
Quadratic function, 590, 610, 613
  minimum of, 591
Quadratic interpolation, 203
Quadratic Taylor expansion, 26, 129, 175, 215, 525, 599
Quadratic termination, 594, 607
Quadric law of damped exponential growth, 339
Quasi-likelihood estimation, *see* Maximum quasi-likelihood estimation
Quasi-maximum-likelihood estimation (QMLE), 582
Quasi-Newton condition, 606, 608, 629–630
Quasi-Newton methods, 71, 474, **605,** 617–618, 628, 639, 661–662, 664, 666, 672
  convergence rate, 609, 612
  full, **628,** 632
  ill-conditioned Hessian approximation, 608
  includes modified Newton, 609
  limited memory, 611
  nonderivative, 611, 615, 617–618
  and nonsmooth function, 616
  for part of Hessian, **628,** 639–640, 653
Quasi-Newton update, *see* Update-matrix

ρ, *see* Standard radius
Radioligand assay, 326, 338, 343
Radius of curvature, 133–134, 467, 689
Random-regressor model, *see* Errors-in-variables model
Random search, **593,** 609, 665, 672
Range space of matrix, *see* Matrix
Rank, *see* Matrix
Rate of convergence, *see* Convergence
Rectangular grid for prior, 57
Recursive residuals, 444, 484
Reduced formulae for curvatures, *see* Curvature(s)
Redundant additional regressors, 238
Redundant directions, 122
Reference, prior, *see* Prior distribution(s)
Regime, 433
Regressor variables, *see* Explanatory variables
Relative change criterion, *see* Stopping rules
Relative curvature, 128
Relative growth rate, 7, 344–345
Relative offset, *see* Stopping rules
Relative scaling, *see* Scaling (parameterization)
Reliability of optimization methods, 662
Reparametrization, *see also* Expected value transformation; Scaling (parametrization); Stable parameters
  with constraints, 402
  of growth models, 328, 330, 333, 335
  to reduce curvature effects, 222, 342
  to remove approximate nonidentifiability, 103
Replicated model, 30, 65, 82
  Bayesian test for model adequacy, 65
  confidence intervals, 204–206
  curvature for, 146
  least squares estimator for, 31
  normal equations, 31, 205
  relative offset stopping rule, 643
  test for model adequacy, 30, 65, 82
  weighted, 82
Residual plot, 17, 84, 203, 411
  weighted, 81, 84
Residuals, 174. *See also* Projected residuals; Recursive residuals
  approximate moments of, 177
  for Brunhilda example, 411
  effects of curvature on, 178
  lack of fit tested using ordered, 204
  quadratic approximation for, 174
  weighted, 81
Response variable, 4
Restricted step methods, 603. *See also* Trust-region methods
Riçcati model, *see* Growth curve
Richards curve, *see* Growth curve
Robust estimation, 47, 50
  for linear models, 651
  for model A, 73
Robust loss function, 50, 587, 650
Roughness penalty, 486

Saddle point, 589, 592, 603, 638, 672
Sample autocorrelation function (SACF), *see* Autocorrelation function (ACF)
Sample partial autocorrelation function (SPACF), *see* Partial autocorrelation function (PACF)

SAS (statistical package), 640, 647, 663
Scale factor, *see* Standard radius
Scaling (parametrization), 596, 632, 669–670, 672. *See also* Reparametrization
  bad, 596, 612, 670, 673
  effect on algorithm, 95, 674
  relative, 596
  well scaled, 664
Scheffé confidence intervals, 107
Score test, *see* Lagrange multiplier (LM) test
Scoring method, *see* Fisher's scoring algorithm
Secant approximation of second derivatives, 154
Secant hyperplane, 647
Secant methods, 605, 609, 652
  for least squares, 647
Sectioning, 598
Segmented polynomial, 5, 447, 457. *See also* Splines
  computation, *see* Change-of-phase models
  cubic-cubic, 434, 460
  inference, 447, 460
  linear–linear, *see* Two-line-segment model
  linear–quadratic–quadratic, 460
  no change of phase, 463
  quadratic–linear, 459, 463
  quadratic–quadratic, 459, 461
  quadratic–quadratic–linear, 459, 462
Semi-Markov process, 429
Sensitivity analysis, 121–125, 668
Separable least squares, 455, 463, **654**, 665. *See also* Partially linear models
Separate hypotheses, *see* Hypothesis testing
Sequential design, *see* Optimal experimental design(s)
Serially correlated errors, *see* Autocorrelated errors
Sgn formulation, *see* Smooth transition models
Sgn with vectors, *see* Smooth transition models
Sigmoidal curve, *see* Growth curve
Simplex algorithm, 615. *See also* Polytope algorithm
Simultaneous confidence band, 194, 246
Simultaneous confidence intervals, 194
Simultaneous equations model, 19
Single compartment model, 376
Singular value decomposition (SVD), 621, 624, 626, 634, **680**
Sizing, 632
Skewness, 185
  test for, 186
Slice, *see* Face of array
Small-error asymptotics, *see* Vanishingly small errors
Small-residual problems, 623
Smearing estimate, 87
Smoothing splines, *see* Splines
Smooth transition models, 465
  Bayesian methods, 480
  hyperbolic model, 468, 476, 479–480
  max-min formulation, 471, 479–480
  sgn formulation, 465
  sgn with vectors, 476
  tanh model, 466, 479–480
  trn formulation, 466
Solution locus, *see* Expectation surface
S-optimal design, 251
Sparse Hessians, 663
Specific abundance, *see* Tracer
Specific activity, *see* Tracer
Spectral decomposition, 377, 398, 600, 626, 634, 640, **678**
Spline regression, 433, 481
  Bayesian methods, 489
  fixed knots, 484
  number of knots, 482, 484
  placing of knots, 482, 484
  variable knots, 484
Splines, 5, 458, 481–482
  B-, 463, 484
  cubic, 459, 463, 481
  linear, 481
  segmented polynomial, 5, 458
  smoothing, 486
SPSS (statistical package), 663
Square bracket multiplication, 130, 142, 146, 169, 195, 216, **692**
Square root of matrix, *see* Matrix
SSM, *see* Symmetric storage mode (SSM)
Stable parameters, 117
Staged procedure, 81
Standard radius, 130, 144, 217, 220, 221, 642
Starting values:
  general discussion, 665
  linear compartmental models, 406, 412
  Richards curves, 335
  sum-of-exponentials models, 407
  yield-density curves, 365
Stationarity assumption, 273, 288, 290, 312
Stationary point, 589
Stationary time series, 272, 275, 288, 290, 312
Statistical curvature (of Efron), 160
Steepest descent method, 595, 604, 641
  conjugate directions and, 610

failure of, 604
step, 606–607, 627
Step-length method(s), 594, 599, 628, 638, 658, 666. *See also* Natural step
Stochastic compartmental models, 415, 429, 431. *See also* Markov process compartmental models
Stochastic differential equation(s), 346, **695**
autocorrelated errors, 348
compartmental models, 431
uncorrelated errors, 347
Stochastic growth curve, 344
last-point fit method, 352
longitudinal data, 344, 349
rate as function of size, 353
rate as function of time, 346
Stochastic integral representation, 347
Stochastic process compartmental model, 415, 431
Stoichiometry, 545
Stopping rules, 667–668. *See also* Convergence
for least squares, 640
relative change criteria, 641, 644, 671
relative offset, 641
statistical inference and, 641–645, 671
Stratonovich's stochastic calculus, 698
Strongly connected, *see* Compartmental models
Structural parameters, 492
Structural relationships, 12, 523
identifiability of, 13
linear, 12
Subset confidence region, *see* Confidence region(s)
Subset curvature(s), 165
computation of, 170
reduced formulae for, 168
total, 170
Sum-of-exponentials models, 66, 119, 242, 254, 264, 378, 381, 393, 402, 412, 664. *See also* Compartmental models
partially linear, 409
using constraints, 402
Sum of squares, 21
ill-conditioned, 201, 480
invariant to one-to-one reparametrization, 195
Superlinear convergence, *see* Convergence
Switching regressions, 446
Symmetric matrix, 1
Symmetric storage mode (SSM), **150**, 155, 172
Symmetrizable matrix, *see* Matrix
Tail product, 571
Tangent plane, 98, 690
basis for, 139
Tangent-plane approximation, 130, 238
criteria for validity, 135–136
planar assumption, 134
uniform coordinate assumption, 135
Tanh model, *see* Smooth transition models
Target transformation curvature array, 223
Taylor expansion, *see* Linear Taylor expansion; Quadratic Taylor expansion
t-distribution:
multivariate, 56–57, 542, **681**
univariate, 2
Tensor product, *see* Kronecker product
Termination criteria, *see* Stopping rules
Three-dimensional arrays, *see* Arrays, three-dimensional
Three-stage estimator, 302
Time power model, 327. *See also* Allometry
Torsion, *see* Parameter-effects curvature
Trace, *see* Matrix
Tracer, 368
bolus dose of, 372, 399
isotope fractionation, 374
multivariate two-compartment model, 414, 534
nonsteady-state system, 375
perfect, 372–373
specific abundance, 374
specific activity, 374
steady-state system, 372
Transfer function, 388
Transformation to linearity, 6, **9**. *See also* Detransformation; Linearized model
Box-Cox procedure, 73, 75
controversy about, 75
extensions of, 76
prediction and transformation bias, 86
Transformed model, 70. *See also* Power-transform-both-sides (PTBS) model
inference from, 74
Transpose, *see* Matrix
Trap, *see* Compartmental models
Triangular matrix, *see* Matrix
Triangular system of equations, 29
Trn formulation, *see* Smooth transition models
Truncation error, 612
Trust-region methods, **603**, 609, 617, 625, 627, 629, 633, 664, 666. *See also* Levenberg-Marquardt methods
Gauss-Newton method using, 625, 633

Two-line-segment model, 435
asymptotic theory, 448
Bayesian inference, 446
computation, 457, 474–475, 480
continuous change, 450
estimate of changepoint, 440, 446, 450, 453
estimates of other parameters, 445, 450
hypothesis test for parameters, 453
noncontinuous change, 438
with smooth transition, 437, 465–469, 472–476, 480
test for, 440
Two-phase (multiple) linear regression, 452, 455–457, 472
with smooth transition, 472–476
test for single phase, 464
Two-stage estimation, 279, 301, 306–307, 314, 423
iterated, 280, 297, 424

Ultrastructural model, 523
Unbiased estimating equations, 47
Unequally spaced time intervals, *see* AR(1) errors
Uniform coordinate assumption, *see* Tangent-plane approximation
Uniform prior distribution, *see* Prior distribution(s)
Update-matrix, 606, 649
for least squares Jacobian, 648–649
for part of least squares Hessian, 631–632, 637
rank one, 606
rank two, 606–608
User-supplied constants, 664

Vanishingly small errors, 74, 200, 495, 575
Variable knots, *see* Spline regression
Variable-metric method, *see* Quasi-Newton methods
Variance–covariance matrix, *see* Dispersion matrix, definition
Variance heterogeneity, 15–17, 68, 84, 86, 257, 360
demonstrated, 84
Variance tending to zero, *see* Vanishingly small errors
Vec, 157, 530, 536, 684
Vech, 497, 684
Vector differentiation, 3, 682
Vector norm (length), 1
Von Bertalanffy model, *see* Growth curve

Wald test, 228–234, 577
linear hypothesis, 199
Washout function, 392
Weibull distribution, 58, 338
Weibull model, *see* Growth curve
Weighted least squares, 27. *See also* Generalized least squares; Power-transformed weighted least squares (PTWLS) model
controlled variables and, 527
for model B, 77
optimal design and, 250–251
replicated data, 31, 205
Richards curve and, 336
robust estimation and, 652
stochastic growth curve, 347
yield-density curve, 365
Weighted residuals, *see* Residuals
White noise, 347, 354, 695, 697
Wiener process, 347, 354, 489, **695**
W test, *see* Wald test

Yield–density curve, 9, **360**
asymptotic, 360
Bleasedale–Nelder, 362–363
Bleasedale's simplified, 77, 362, 365
choice of, 364
Farazdaghi–Harris, 362–363
Holliday, 17, 362, 364–365
inverse polynomial, 364
parabolic, 360
Shinozaki–Kira, 362
starting values, 365
Yield–fertilizer model, 8
Yule–Walker equations, 290, 301

Zellner iteration, 583
Zellner's method, 531
Zero residual problems (nonlinear equations), 622, 633, 638, 640

WILEY SERIES IN PROBABILITY AND STATISTICS
ESTABLISHED BY WALTER A. SHEWHART AND SAMUEL S. WILKS

The ***Wiley Series in Probability and Statistics*** is well established and authoritative. It covers many topics of current research interest in both pure and applied statistics and probability theory. Written by leading statisticians and institutions, the titles span both state-of-the-art developments in the field and classical methods.

Reflecting the wide range of current research in statistics, the series encompasses applied, methodological and theoretical statistics, ranging from applications and new techniques made possible by advances in computerized practice to rigorous treatment of theoretical approaches.

This series provides essential and invaluable reading for all statisticians, whether in academia, industry, government, or research.

ABRAHAM and LEDOLTER · Statistical Methods for Forecasting
AGRESTI · Analysis of Ordinal Categorical Data
AGRESTI · An Introduction to Categorical Data Analysis
AGRESTI · Categorical Data Analysis, *Second Edition*
ALTMAN, GILL, and McDONALD · Numerical Issues in Statistical Computing for the Social Scientist
ANDĚL · Mathematics of Chance
ANDERSON · An Introduction to Multivariate Statistical Analysis, *Third Edition*
*ANDERSON · The Statistical Analysis of Time Series
ANDERSON, AUQUIER, HAUCK, OAKES, VANDAELE, and WEISBERG · Statistical Methods for Comparative Studies
ANDERSON and LOYNES · The Teaching of Practical Statistics
ARMITAGE and DAVID (editors) · Advances in Biometry
ARNOLD, BALAKRISHNAN, and NAGARAJA · Records
*ARTHANARI and DODGE · Mathematical Programming in Statistics
*BAILEY · The Elements of Stochastic Processes with Applications to the Natural Sciences
BALAKRISHNAN and KOUTRAS · Runs and Scans with Applications
BARNETT · Comparative Statistical Inference, *Third Edition*
BARNETT and LEWIS · Outliers in Statistical Data, *Third Edition*
BARTOSZYNSKI and NIEWIADOMSKA-BUGAJ · Probability and Statistical Inference
BASILEVSKY · Statistical Factor Analysis and Related Methods: Theory and Applications
BASU and RIGDON · Statistical Methods for the Reliability of Repairable Systems
BATES and WATTS · Nonlinear Regression Analysis and Its Applications
BECHHOFER, SANTNER, and GOLDSMAN · Design and Analysis of Experiments for Statistical Selection, Screening, and Multiple Comparisons
BELSLEY · Conditioning Diagnostics: Collinearity and Weak Data in Regression
BELSLEY, KUH, and WELSCH · Regression Diagnostics: Identifying Influential Data and Sources of Collinearity

*Now available in a lower priced paperback edition in the Wiley Classics Library.

BENDAT and PIERSOL · Random Data: Analysis and Measurement Procedures, *Third Edition*
BERRY, CHALONER, and GEWEKE · Bayesian Analysis in Statistics and Econometrics: Essays in Honor of Arnold Zellner
BERNARDO and SMITH · Bayesian Theory
BHAT and MILLER · Elements of Applied Stochastic Processes, *Third Edition*
BHATTACHARYA and JOHNSON · Statistical Concepts and Methods
BHATTACHARYA and WAYMIRE · Stochastic Processes with Applications
BILLINGSLEY · Convergence of Probability Measures, *Second Edition*
BILLINGSLEY · Probability and Measure, *Third Edition*
BIRKES and DODGE · Alternative Methods of Regression
BLISCHKE AND MURTHY (editors) · Case Studies in Reliability and Maintenance
BLISCHKE AND MURTHY · Reliability: Modeling, Prediction, and Optimization
BLOOMFIELD · Fourier Analysis of Time Series: An Introduction, *Second Edition*
BOLLEN · Structural Equations with Latent Variables
BOROVKOV · Ergodicity and Stability of Stochastic Processes
BOULEAU · Numerical Methods for Stochastic Processes
BOX · Bayesian Inference in Statistical Analysis
BOX · R. A. Fisher, the Life of a Scientist
BOX and DRAPER · Empirical Model-Building and Response Surfaces
*BOX and DRAPER · Evolutionary Operation: A Statistical Method for Process Improvement
BOX, HUNTER, and HUNTER · Statistics for Experimenters: An Introduction to Design, Data Analysis, and Model Building
BOX and LUCEÑO · Statistical Control by Monitoring and Feedback Adjustment
BRANDIMARTE · Numerical Methods in Finance: A MATLAB-Based Introduction
BROWN and HOLLANDER · Statistics: A Biomedical Introduction
BRUNNER, DOMHOF, and LANGER · Nonparametric Analysis of Longitudinal Data in Factorial Experiments
BUCKLEW · Large Deviation Techniques in Decision, Simulation, and Estimation
CAIROLI and DALANG · Sequential Stochastic Optimization
CHAN · Time Series: Applications to Finance
CHATTERJEE and HADI · Sensitivity Analysis in Linear Regression
CHATTERJEE and PRICE · Regression Analysis by Example, *Third Edition*
CHERNICK · Bootstrap Methods: A Practitioner's Guide
CHERNICK and FRIIS · Introductory Biostatistics for the Health Sciences
CHILÈS and DELFINER · Geostatistics: Modeling Spatial Uncertainty
CHOW and LIU · Design and Analysis of Clinical Trials: Concepts and Methodologies
CLARKE and DISNEY · Probability and Random Processes: A First Course with Applications, *Second Edition*
*COCHRAN and COX · Experimental Designs, *Second Edition*
CONGDON · Bayesian Statistical Modelling
CONOVER · Practical Nonparametric Statistics, *Second Edition*
COOK · Regression Graphics
COOK and WEISBERG · Applied Regression Including Computing and Graphics
COOK and WEISBERG · An Introduction to Regression Graphics
CORNELL · Experiments with Mixtures, Designs, Models, and the Analysis of Mixture Data, *Third Edition*
COVER and THOMAS · Elements of Information Theory
COX · A Handbook of Introductory Statistical Methods
*COX · Planning of Experiments
CRESSIE · Statistics for Spatial Data, *Revised Edition*
CSÖRGŐ and HORVÁTH · Limit Theorems in Change Point Analysis

*Now available in a lower priced paperback edition in the Wiley Classics Library.

DANIEL · Applications of Statistics to Industrial Experimentation
DANIEL · Biostatistics: A Foundation for Analysis in the Health Sciences, *Sixth Edition*
*DANIEL · Fitting Equations to Data: Computer Analysis of Multifactor Data, *Second Edition*
DASU and JOHNSON · Exploratory Data Mining and Data Cleaning
DAVID and NAGARAJA · Order Statistics, *Third Edition*
*DEGROOT, FIENBERG, and KADANE · Statistics and the Law
DEL CASTILLO · Statistical Process Adjustment for Quality Control
DETTE and STUDDEN · The Theory of Canonical Moments with Applications in Statistics, Probability, and Analysis
DEY and MUKERJEE · Fractional Factorial Plans
DILLON and GOLDSTEIN · Multivariate Analysis: Methods and Applications
DODGE · Alternative Methods of Regression
*DODGE and ROMIG · Sampling Inspection Tables, *Second Edition*
*DOOB · Stochastic Processes
DOWDY and WEARDEN · Statistics for Research, *Second Edition*
DRAPER and SMITH · Applied Regression Analysis, *Third Edition*
DRYDEN and MARDIA · Statistical Shape Analysis
DUDEWICZ and MISHRA · Modern Mathematical Statistics
DUNN and CLARK · Applied Statistics: Analysis of Variance and Regression, *Second Edition*
DUNN and CLARK · Basic Statistics: A Primer for the Biomedical Sciences, *Third Edition*
DUPUIS and ELLIS · A Weak Convergence Approach to the Theory of Large Deviations
*ELANDT-JOHNSON and JOHNSON · Survival Models and Data Analysis
ENDERS · Applied Econometric Time Series
ETHIER and KURTZ · Markov Processes: Characterization and Convergence
EVANS, HASTINGS, and PEACOCK · Statistical Distributions, *Third Edition*
FELLER · An Introduction to Probability Theory and Its Applications, Volume I, *Third Edition,* Revised; Volume II, *Second Edition*
FISHER and VAN BELLE · Biostatistics: A Methodology for the Health Sciences
*FLEISS · The Design and Analysis of Clinical Experiments
FLEISS · Statistical Methods for Rates and Proportions, *Second Edition*
FLEMING and HARRINGTON · Counting Processes and Survival Analysis
FULLER · Introduction to Statistical Time Series, *Second Edition*
FULLER · Measurement Error Models
GALLANT · Nonlinear Statistical Models
GHOSH, MUKHOPADHYAY, and SEN · Sequential Estimation
GIFI · Nonlinear Multivariate Analysis
GLASSERMAN and YAO · Monotone Structure in Discrete-Event Systems
GNANADESIKAN · Methods for Statistical Data Analysis of Multivariate Observations, *Second Edition*
GOLDSTEIN and LEWIS · Assessment: Problems, Development, and Statistical Issues
GREENWOOD and NIKULIN · A Guide to Chi-Squared Testing
GROSS and HARRIS · Fundamentals of Queueing Theory, *Third Edition*
*HAHN and SHAPIRO · Statistical Models in Engineering
HAHN and MEEKER · Statistical Intervals: A Guide for Practitioners
HALD · A History of Probability and Statistics and their Applications Before 1750
HALD · A History of Mathematical Statistics from 1750 to 1930
HAMPEL · Robust Statistics: The Approach Based on Influence Functions
HANNAN and DEISTLER · The Statistical Theory of Linear Systems
HEIBERGER · Computation for the Analysis of Designed Experiments
HEDAYAT and SINHA · Design and Inference in Finite Population Sampling
HELLER · MACSYMA for Statisticians

*Now available in a lower priced paperback edition in the Wiley Classics Library.

HINKELMAN and KEMPTHORNE: · Design and Analysis of Experiments, Volume 1: Introduction to Experimental Design
HOAGLIN, MOSTELLER, and TUKEY · Exploratory Approach to Analysis of Variance
HOAGLIN, MOSTELLER, and TUKEY · Exploring Data Tables, Trends and Shapes
*HOAGLIN, MOSTELLER, and TUKEY · Understanding Robust and Exploratory Data Analysis
HOCHBERG and TAMHANE · Multiple Comparison Procedures
HOCKING · Methods and Applications of Linear Models: Regression and the Analysis of Variance, *Second Edition*
HOEL · Introduction to Mathematical Statistics, *Fifth Edition*
HOGG and KLUGMAN · Loss Distributions
HOLLANDER and WOLFE · Nonparametric Statistical Methods, *Second Edition*
HOSMER and LEMESHOW · Applied Logistic Regression, *Second Edition*
HOSMER and LEMESHOW · Applied Survival Analysis: Regression Modeling of Time to Event Data
HØYLAND and RAUSAND · System Reliability Theory: Models and Statistical Methods
HUBER · Robust Statistics
HUBERTY · Applied Discriminant Analysis
HUNT and KENNEDY · Financial Derivatives in Theory and Practice
HUSKOVA, BERAN, and DUPAC · Collected Works of Jaroslav Hajek—with Commentary
IMAN and CONOVER · A Modern Approach to Statistics
JACKSON · A User's Guide to Principle Components
JOHN · Statistical Methods in Engineering and Quality Assurance
JOHNSON · Multivariate Statistical Simulation
JOHNSON and BALAKRISHNAN · Advances in the Theory and Practice of Statistics: A Volume in Honor of Samuel Kotz
JUDGE, GRIFFITHS, HILL, LÜTKEPOHL, and LEE · The Theory and Practice of Econometrics, *Second Edition*
JOHNSON and KOTZ · Distributions in Statistics
JOHNSON and KOTZ (editors) · Leading Personalities in Statistical Sciences: From the Seventeenth Century to the Present
JOHNSON, KOTZ, and BALAKRISHNAN · Continuous Univariate Distributions, Volume 1, *Second Edition*
JOHNSON, KOTZ, and BALAKRISHNAN · Continuous Univariate Distributions, Volume 2, *Second Edition*
JOHNSON, KOTZ, and BALAKRISHNAN · Discrete Multivariate Distributions
JOHNSON, KOTZ, and KEMP · Univariate Discrete Distributions, *Second Edition*
JUREČKOVÁ and SEN · Robust Statistical Procedures: Aymptotics and Interrelations
JUREK and MASON · Operator-Limit Distributions in Probability Theory
KADANE · Bayesian Methods and Ethics in a Clinical Trial Design
KADANE AND SCHUM · A Probabilistic Analysis of the Sacco and Vanzetti Evidence
KALBFLEISCH and PRENTICE · The Statistical Analysis of Failure Time Data, *Second Edition*
KASS and VOS · Geometrical Foundations of Asymptotic Inference
KAUFMAN and ROUSSEEUW · Finding Groups in Data: An Introduction to Cluster Analysis
KEDEM and FOKIANOS · Regression Models for Time Series Analysis
KENDALL, BARDEN, CARNE, and LE · Shape and Shape Theory
KHURI · Advanced Calculus with Applications in Statistics, *Second Edition*
KHURI, MATHEW, and SINHA · Statistical Tests for Mixed Linear Models
KLEIBER and KOTZ · Statistical Size Distributions in Economics and Actuarial Sciences
KLUGMAN, PANJER, and WILLMOT · Loss Models: From Data to Decisions

*Now available in a lower priced paperback edition in the Wiley Classics Library.

KLUGMAN, PANJER, and WILLMOT · Solutions Manual to Accompany Loss Models: From Data to Decisions
KOTZ, BALAKRISHNAN, and JOHNSON · Continuous Multivariate Distributions, Volume 1, *Second Edition*
KOTZ and JOHNSON (editors) · Encyclopedia of Statistical Sciences: Volumes 1 to 9 with Index
KOTZ and JOHNSON (editors) · Encyclopedia of Statistical Sciences: Supplement Volume
KOTZ, READ, and BANKS (editors) · Encyclopedia of Statistical Sciences: Update Volume 1
KOTZ, READ, and BANKS (editors) · Encyclopedia of Statistical Sciences: Update Volume 2
KOVALENKO, KUZNETZOV, and PEGG · Mathematical Theory of Reliability of Time-Dependent Systems with Practical Applications
LACHIN · Biostatistical Methods: The Assessment of Relative Risks
LAD · Operational Subjective Statistical Methods: A Mathematical, Philosophical, and Historical Introduction
LAMPERTI · Probability: A Survey of the Mathematical Theory, *Second Edition*
LANGE, RYAN, BILLARD, BRILLINGER, CONQUEST, and GREENHOUSE · Case Studies in Biometry
LARSON · Introduction to Probability Theory and Statistical Inference, *Third Edition*
LAWLESS · Statistical Models and Methods for Lifetime Data, *Second Edition*
LAWSON · Statistical Methods in Spatial Epidemiology
LE · Applied Categorical Data Analysis
LE · Applied Survival Analysis
LEE and WANG · Statistical Methods for Survival Data Analysis, *Third Edition*
LePAGE and BILLARD · Exploring the Limits of Bootstrap
LEYLAND and GOLDSTEIN (editors) · Multilevel Modelling of Health Statistics
LIAO · Statistical Group Comparison
LINDVALL · Lectures on the Coupling Method
LINHART and ZUCCHINI · Model Selection
LITTLE and RUBIN · Statistical Analysis with Missing Data, *Second Edition*
LLOYD · The Statistical Analysis of Categorical Data
MAGNUS and NEUDECKER · Matrix Differential Calculus with Applications in Statistics and Econometrics, *Revised Edition*
MALLER and ZHOU · Survival Analysis with Long Term Survivors
MALLOWS · Design, Data, and Analysis by Some Friends of Cuthbert Daniel
MANN, SCHAFER, and SINGPURWALLA · Methods for Statistical Analysis of Reliability and Life Data
MANTON, WOODBURY, and TOLLEY · Statistical Applications Using Fuzzy Sets
MARDIA and JUPP · Directional Statistics
MASON, GUNST, and HESS · Statistical Design and Analysis of Experiments with Applications to Engineering and Science, *Second Edition*
McCULLOCH and SEARLE · Generalized, Linear, and Mixed Models
McFADDEN · Management of Data in Clinical Trials
McLACHLAN · Discriminant Analysis and Statistical Pattern Recognition
McLACHLAN and KRISHNAN · The EM Algorithm and Extensions
McLACHLAN and PEEL · Finite Mixture Models
McNEIL · Epidemiological Research Methods
MEEKER and ESCOBAR · Statistical Methods for Reliability Data
MEERSCHAERT and SCHEFFLER · Limit Distributions for Sums of Independent Random Vectors: Heavy Tails in Theory and Practice
*MILLER · Survival Analysis, *Second Edition*
MONTGOMERY, PECK, and VINING · Introduction to Linear Regression Analysis, *Third Edition*

*Now available in a lower priced paperback edition in the Wiley Classics Library.

MORGENTHALER and TUKEY · Configural Polysampling: A Route to Practical Robustness
MUIRHEAD · Aspects of Multivariate Statistical Theory
MURRAY · X-STAT 2.0 Statistical Experimentation, Design Data Analysis, and Nonlinear Optimization
MURTHY, XIE, and JIANG · Weibull Models
MYERS and MONTGOMERY · Response Surface Methodology: Process and Product Optimization Using Designed Experiments, *Second Edition*
MYERS, MONTGOMERY, and VINING · Generalized Linear Models. With Applications in Engineering and the Sciences
NELSON · Accelerated Testing, Statistical Models, Test Plans, and Data Analyses
NELSON · Applied Life Data Analysis
NEWMAN · Biostatistical Methods in Epidemiology
OCHI · Applied Probability and Stochastic Processes in Engineering and Physical Sciences
OKABE, BOOTS, SUGIHARA, and CHIU · Spatial Tesselations: Concepts and Applications of Voronoi Diagrams, *Second Edition*
OLIVER and SMITH · Influence Diagrams, Belief Nets and Decision Analysis
PALTA · Quantitative Methods in Population Health: Extensions of Ordinary Regressions
PANKRATZ · Forecasting with Dynamic Regression Models
PANKRATZ · Forecasting with Univariate Box-Jenkins Models: Concepts and Cases
*PARZEN · Modern Probability Theory and Its Applications
PEÑA, TIAO, and TSAY · A Course in Time Series Analysis
PIANTADOSI · Clinical Trials: A Methodologic Perspective
PORT · Theoretical Probability for Applications
POURAHMADI · Foundations of Time Series Analysis and Prediction Theory
PRESS · Bayesian Statistics: Principles, Models, and Applications
PRESS · Subjective and Objective Bayesian Statistics, *Second Edition*
PRESS and TANUR · The Subjectivity of Scientists and the Bayesian Approach
PUKELSHEIM · Optimal Experimental Design
PURI, VILAPLANA, and WERTZ · New Perspectives in Theoretical and Applied Statistics
PUTERMAN · Markov Decision Processes: Discrete Stochastic Dynamic Programming
*RAO · Linear Statistical Inference and Its Applications, *Second Edition*
RENCHER · Linear Models in Statistics
RENCHER · Methods of Multivariate Analysis, *Second Edition*
RENCHER · Multivariate Statistical Inference with Applications
RIPLEY · Spatial Statistics
RIPLEY · Stochastic Simulation
ROBINSON · Practical Strategies for Experimenting
ROHATGI and SALEH · An Introduction to Probability and Statistics, *Second Edition*
ROLSKI, SCHMIDLI, SCHMIDT, and TEUGELS · Stochastic Processes for Insurance and Finance
ROSENBERGER and LACHIN · Randomization in Clinical Trials: Theory and Practice
ROSS · Introduction to Probability and Statistics for Engineers and Scientists
ROUSSEEUW and LEROY · Robust Regression and Outlier Detection
RUBIN · Multiple Imputation for Nonresponse in Surveys
RUBINSTEIN · Simulation and the Monte Carlo Method
RUBINSTEIN and MELAMED · Modern Simulation and Modeling
RYAN · Modern Regression Methods
RYAN · Statistical Methods for Quality Improvement, *Second Edition*
SALTELLI, CHAN, and SCOTT (editors) · Sensitivity Analysis
*SCHEFFE · The Analysis of Variance
SCHIMEK · Smoothing and Regression: Approaches, Computation, and Application
SCHOTT · Matrix Analysis for Statistics

*Now available in a lower priced paperback edition in the Wiley Classics Library.

SCHUSS · Theory and Applications of Stochastic Differential Equations
SCOTT · Multivariate Density Estimation: Theory, Practice, and Visualization
*SEARLE · Linear Models
SEARLE · Linear Models for Unbalanced Data
SEARLE · Matrix Algebra Useful for Statistics
SEARLE, CASELLA, and McCULLOCH · Variance Components
SEARLE and WILLETT · Matrix Algebra for Applied Economics
SEBER and LEE · Linear Regression Analysis, *Second Edition*
SEBER · Multivariate Observations
SEBER and WILD · Nonlinear Regression
SENNOTT · Stochastic Dynamic Programming and the Control of Queueing Systems
*SERFLING · Approximation Theorems of Mathematical Statistics
SHAFER and VOVK · Probability and Finance: It's Only a Game!
SMALL and McLEISH · Hilbert Space Methods in Probability and Statistical Inference
SRIVASTAVA · Methods of Multivariate Statistics
STAPLETON · Linear Statistical Models
STAUDTE and SHEATHER · Robust Estimation and Testing
STOYAN, KENDALL, and MECKE · Stochastic Geometry and Its Applications, *Second Edition*
STOYAN and STOYAN · Fractals, Random Shapes and Point Fields: Methods of Geometrical Statistics
STYAN · The Collected Papers of T. W. Anderson: 1943–1985
SUTTON, ABRAMS, JONES, SHELDON, and SONG · Methods for Meta-Analysis in Medical Research
TANAKA · Time Series Analysis: Nonstationary and Noninvertible Distribution Theory
THOMPSON · Empirical Model Building
THOMPSON · Sampling, *Second Edition*
THOMPSON · Simulation: A Modeler's Approach
THOMPSON and SEBER · Adaptive Sampling
THOMPSON, WILLIAMS, and FINDLAY · Models for Investors in Real World Markets
TIAO, BISGAARD, HILL, PEÑA, and STIGLER (editors) · Box on Quality and Discovery: with Design, Control, and Robustness
TIERNEY · LISP-STAT: An Object-Oriented Environment for Statistical Computing and Dynamic Graphics
TSAY · Analysis of Financial Time Series
UPTON and FINGLETON · Spatial Data Analysis by Example, Volume II: Categorical and Directional Data
VAN BELLE · Statistical Rules of Thumb
VIDAKOVIC · Statistical Modeling by Wavelets
WEISBERG · Applied Linear Regression, *Second Edition*
WELSH · Aspects of Statistical Inference
WESTFALL and YOUNG · Resampling-Based Multiple Testing: Examples and Methods for p-Value Adjustment
WHITTAKER · Graphical Models in Applied Multivariate Statistics
WINKER · Optimization Heuristics in Economics: Applications of Threshold Accepting
WONNACOTT and WONNACOTT · Econometrics, *Second Edition*
WOODING · Planning Pharmaceutical Clinical Trials: Basic Statistical Principles
WOOLSON and CLARKE · Statistical Methods for the Analysis of Biomedical Data, *Second Edition*
WU and HAMADA · Experiments: Planning, Analysis, and Parameter Design Optimization
YANG · The Construction Theory of Denumerable Markov Processes
*ZELLNER · An Introduction to Bayesian Inference in Econometrics
ZHOU, OBUCHOWSKI, and McCLISH · Statistical Methods in Diagnostic Medicine

*Now available in a lower priced paperback edition in the Wiley Classics Library.